EVOLUTION

A BIOLOGICAL AND PALAEONTOLOGICAL APPROACH

Addison-Wesley Publishing Company

Wokingham, England • Reading, Massachusetts • Menlo Park, California
New York • Don Mills, Ontario • Amsterdam • Bonn • Sydney • Singapore
Tokyo • Madrid • San Juan • Milan • Paris • Mexico City • Seoul • Taipei

in association with

THE S365 COURSE TEAM

Course Team Chair and General Editor

Peter Skelton

Authors

Iain Gilmour
Marion Hall
Tim Halliday
Stephen Hurry
Dave Martill
Heather McLannahan
Caroline M. Pond
Irene Ridge
Peter Sheldon
Peter Skelton
Charles Turner

Course Manager

Alastair Ewing

Other Contributor

Pat Murphy

Editors

Catherine Baker
Gerry Bearman
Margaret Swithenby

Text Processing

Pam Berry

Print Production Controller

Harry Dodd

Design and Illustration

Steve Best
Lesley Passey
Ros Porter

Course Secretaries

Janet Dryden
Valerie Shadbolt

BBC

David Jackson

First edition 1993. Reprinted 1994.

This book is a component of the Open University course S365 *Evolution*. Details of this and other Open University courses are available from the Central Enquiry Service, The Open University, PO Box 200, Milton Keynes, MK7 6YZ. Tel. 0908 653231.

Edited, designed and typeset by The Open University. Printed in Great Britain by The Bath Press, Avon

British Library Cataloguing in Publication Data

Evolution
I. Skelton, P.W.

ISBN 0-201-54423-7

1.1

4970C/s365copubi1.1

PREFACE

This interdisciplinary introduction to evolution is the basis for the Open University's third level Science Course, S365 *Evolution*. It concentrates upon the principles of evolutionary science, rather than an account of the history of life. Accordingly, it emphasizes the basic information and intellectual and data-handling skills involved in generating and testing evolutionary hypotheses. Objectives are listed for each Chapter, and tested by self-assessment questions (SAQs) answered at the back. The text's four-part structure progresses up the evolutionary hierarchy, from the principles governing evolution within species populations, or *micro-evolution* (Part I), via the evolution of species (Part II), to evolution beyond the species level, or *macroevolution* (Part III), finishing with two case studies (Part IV).

Part I After a brief introduction, Chapter 2 discusses adaptation in organisms — a subtle concept which lies at the heart of the theory of evolution by natural selection. Chapter 3 introduces the sources of variation between individuals (including genetic mechanisms) and the means for studying them. Chapter 4 then explores the role of natural selection in fashioning adaptations from such variation. Chapter 5 considers how differing patterns of reproduction arise and their evolutionary consequences, and Chapter 6 discusses what constitute the units upon which natural selection may be said to act. An important form of natural selection arising from competition for mates ('*sexual selection*') is covered in Chapter 7. Chapter 8 discusses evolutionary ecology, showing how patterns of natural selection are themselves governed by the environments of organisms, especially the other species with which they interact.

Part II This Part explores how the principles of microevolution may be extended to account for the evolution of new species. Chapter 9 concentrates on biologically derived models of speciation, and briefly touches on the extinction of species. Chapter 10 investigates the fossil record of microevolution and what it reveals about the origin of species.

Part III This Part moves further up the scale of perception to patterns of change in form and diversity within whole groups of species. Chapter 11 introduces the methods used to reconstruct evolutionary relationships, and Chapter 12 surveys what kinds of information the fossil record can contribute. Chapter 13 explores the relationship between geography and patterns of evolution. Controls on the evolution of form are considered in Chapter 14, and Chapter 15 analyses patterns of change in form and diversity in higher taxa, including mass extinctions.

Part IV This presents case studies of particular episodes in evolutionary history, illustrating many of the principles discussed earlier. Chapter 16 investigates what can be inferred about the origins and early evolution of life, and Chapter 17 explores the influence of humans on the evolution of other species, as well as ourselves.

The production of a text so broad in scope requires much consultation. We have been greatly helped by the constructive criticisms of drafts provided by experts outside the Open University, appointed as Chapter assessors, as well as by the external Course assessor, Professor Arthur Cain F.R.S. Their careful scrutiny and advice is gratefully acknowledged. However, the Course Team bears responsibility for any errors which remain. Finally, this Book is dedicated to Catherine Baker, a member of the Course Team who died in the final stages of production.

CONTENTS

PART IV CASE STUDIES

Part I
Microevolution

CHAPTER 1
INTRODUCTION TO EVOLUTION

Prepared for the Course Team by Stephen Hurry

Study Comment *Section 1.1 examines the four basic premises used by Darwin in his formulation of the theory of evolution by natural selection, and how he used these to explain the origin of species. Section 1.2 outlines the status of evolutionary ideas in the period immediately before the publication of* The Origin of Species. *The lines of evidence Darwin used to establish the fact of evolution are given in Section 1.3.*

Objectives When you have finished studying this and all subsequent Chapters, you should be familiar with the meanings of all the terms printed in **bold type**. When you have completed this Chapter, you should also be able to do the following:

1.1 Define the term 'biological evolution'.

1.2 Describe how biological evolution differs from inorganic change.

1.3 Explain what Darwin meant by 'the struggle for existence', 'natural selection', and 'fitness' and how Darwin's concept of 'fitness' differs from its meaning today.

1.4 Summarize Darwin's arguments in favour of the theory of evolution by natural selection and his explanation of the origin of species.

1.1 WHAT IS EVOLUTION?

Evolution in a general sense means cumulative change. So we can consider the evolution of galaxies, languages, motor cars and many other categories. Biological evolution (sometimes referred to as organic evolution), however, means change in the characteristics of descendent populations of organisms. Theories of biological evolution set out to explain diversity amongst organisms, the origin and history of that diversity, and the natural processes by which it has developed and by which it is sustained.

Central to our current theory of biological evolution is the idea that species of organisms originate as modified descendants of other species. Most scientists studying evolution today also believe that all living organisms on Earth ultimately share a common ancestry in simple organisms that arose thousands of millions of years ago.

From simple beginnings, evolution has given rise to some extremely complicated organisms. But because there are many simple organisms living today, biological evolution clearly cannot be viewed only as a process in which new, more sophisticated organisms replace older, simpler ones. Nevertheless, the overall trend towards greater complexity in biological evolution is in marked contrast to the nature of changes in the non-living world, which seem to involve as much breaking down as building up of patterns of organization. This stems from a unique feature of living organisms, which is that for the duration of their lives, they retain their organized character through widely changing conditions. They do this by drawing energy from their environment (from light in the case of plants, or food in the case of animals) and using it to sustain life. That is, they form or process nutrients, building their own bodies and the biochemical substances (such as enzymes and hormones) which control their life processes, and also they reproduce themselves. So, although an organism is physically distinct from its environment, it is nevertheless an open system through which energy and material flow.

With regard to evolution, the crucial feature of living organisms is their capacity to reproduce; life's continuity is preserved by descent. While some forms of biological reproduction can produce almost exact copies, most organisms reproduce sexually, a process which generates variable offspring and so allows the descent of one generation from another with modification. In fact, Darwin used the phrase 'descent with modification' to define evolution. How this process leads to the origin of new species and how in turn the evolution of new species has given rise to the bewildering diversity and complexity of living organisms today are questions we attempt to answer in this Book.

The foundations of modern evolutionary thinking were laid by Charles Darwin (1809–1882) and Alfred Russel Wallace (1823–1913) (see Figures 1.1 and 1.2), with their joint communications on the theory of evolution by natural selection presented in 1858 and published subsequently in the *Journal of the Proceedings of the Linnean Society.*

In November 1859, Darwin published his book *On the Origin of Species by Means of Natural Selection, or The Preservation of Favoured Races in the Struggle for Life* in which he set out his ideas in full. It is this book which is now generally referred to as *The Origin of Species*. Darwin succinctly summarized the theory in his Introduction to *The Origin of Species*:

'As many more individuals of each species are born than can possibly survive; and as, consequently, there is a frequently recurring struggle for existence, it follows that any being, if it vary however slightly in any manner profitable to itself, under the complex and sometimes varying conditions of life, will have a better chance of surviving, and thus be *naturally selected*. From the strong principle of inheritance, any selected variety will tend to propagate its new and modified form.'

Figure 1.1 *Portrait of Charles Darwin.*

Figure 1.2 *Portrait of Alfred Wallace.*

So the argument has four basic premises:

1 More individuals are produced than can survive.

2 There is a *struggle for existence*, because of the disparity between the number of individuals produced in reproduction and the number that can survive.

3 Individuals show *variation*. No two individuals are exactly the same. Those with advantageous features have a greater chance of survival in this struggle (*natural selection*).

4 As selected varieties will tend to produce offspring similar to themselves (*the principle of inheritance*), these varieties will become more abundant in subsequent generations.

Selected varieties may thus increase in frequency within a population, altering its character with time. Darwin then went on to argue that such favoured varieties could eventually become separate species. The gist of his argument was that any varieties that were able to exploit new aspects of the environment would not be subject to the same limiting constraints as other members of the population (such a constraint might be a particular kind of food on which the others rely). Natural selection would therefore favour the increasing divergence of such varieties, and as these supplanted the intermediate forms they would eventually become distinct species. As new species arose, others would become extinct through competition. This, then, is the central argument of *The Origin of Species* reduced to its bare bones.

The importance of the theory lay in the fact that it was the first evolutionary model that clearly separated the existence of variation from the direction of evolutionary change. There had been several earlier evolutionary theories, but these tended to assume that variation somehow arose in response to an organism's needs or because of some innate tendency towards complexity. The early evolutionists advanced a number of highly speculative explanations for this, but the majority of their contemporaries were sceptical of the demands made by the theories. In contrast, variation, according to the Darwin–Wallace theory, was undirected, that is it was not necessarily produced in response to needs, nor as the result of some inherent tendency towards an end, nor in response to some external directing agency. The origin of variation was described though never satisfactorily explained by Darwin or Wallace, but it could at least be thought of as due to perfectly natural processes. The way in which natural selection could then mould the course of evolution was supported by the elegantly logical argument given above, and this could in principle be tested.

As well as suggesting what the mechanism of evolution might be, *The Origin of Species* also presented several kinds of evidence that evolution had occurred, some of which are presented in Section 1.3. Thus, it swayed the consensus of biologists and palaeontologists (i.e. people who study fossils) in favour of evolution, and modern ideas on evolutionary mechanisms stem from it. Evolutionists today, however, would express some parts of the argument differently, as you will find out later. In particular, and somewhat ironically, few would now accept Darwin's explanation for the origin of species in exactly the way that he phrased it, although there is as yet no universal agreement on how new species do arise.

Evolutionary studies are still replete with controversy, as this Book will demonstrate. That is why it is so important at this stage that you have a clear idea of precisely what was said in the original formulation of the theory of natural selection, and so have a solid foundation for understanding later debates.

So let us now return to exactly what Darwin meant in the different steps of his argument in *The Origin of Species*. Before doing this, however, an important point must be made. Darwin was not always consistent in his usage of terms and this Section attempts to clarify his general intentions in order to provide a background for subsequent Chapters.

1.1.1 THE STRUGGLE FOR EXISTENCE

The inference of a struggle for existence was based on a simple comparison between a population's reproductive potential and the limitations imposed by the environment (food supply, predation, disease, etc.). Darwin was greatly influenced by an essay on the regulation of populations by famine, war and pestilence written by the Reverend Thomas Malthus in 1789. Darwin wrote:

> 'Every being, which during its natural lifetime produces several eggs or seeds, must suffer destruction during some period of its life, and during some season or occasional year, otherwise, on the principle of geometrical increase, its numbers would quickly become so inordinately great that no country could support the product. Hence, as more individuals are produced than can possibly survive, there must in every case be a struggle for existence, either one individual with another of the same species, or with the individuals of distinct species, or with the physical conditions of life.'

By a geometrical increase is meant an increase by a fixed percentage and hence by an increasingly large number in each successive time interval.

▶ Give an example of an organism in which an increase in the survival rate of offspring has led to a geometrical increase in population size.

The human population explosion is probably the best example, though any population will expand in this way when limiting factors such as food supply or predation are relaxed, and will continue to do so until other limiting factors begin to operate.

Darwin stressed that the struggle for existence was to be taken:

'...in a large and metaphorical sense, including dependence of one being on another, and including (which is more important) not only the life of the individual, but success in leaving progeny. Two canine animals in a time of dearth, may be truly said to struggle with each other which shall get food and live. But a plant on the edge of a desert is said to struggle for life against the drought.... A plant which annually produces a thousand seeds, of which on an average only one comes to maturity, may be more truly said to struggle with the plants of the same and other kinds which already clothe the ground.'

1.1.2 Natural selection

Darwin often used the term 'natural selection' as if to imply preferential *survival* of individuals with profitable variations (as in the quotation in Section 1.1). It is obvious, however, that reproduction is necessary if natural selection is to have any evolutionary effect, and Darwin took this for granted. That is why, for example, he considered success in the struggle for existence to consist of individual survival *and* reproduction, as you will see in this extract:

'Can it, then, be thought improbable...that...variations useful...in some way to each being in the great and complex battle of life, should occur in the course of many successive generations? If such do occur, can we doubt...that individuals having any advantage, however slight, over others, would have the best chance of surviving and of procreating their kind?'

He made little direct reference, however, to the effects of relative differences of reproductive ability in those that do survive. Nowadays, *relative* **fecundity**, as this is called, is considered just as important an aspect of natural selection as *relative* **viability** or the relative chances of survival.

For many species, even fecundity, meaning the number of zygotes (fertilized eggs) to which an individual contributes, does not alone provide a very good measure for assessing an individual's reproductive success. This is because the chances of a zygote surviving may depend upon the care given to it by one or both parents, or where and when the eggs are laid. For a zygote to survive and develop to maturity it is necessary, in many species, that it be nourished and protected. In other words, an individual's reproductive success depends not only on its ability to produce zygotes but also on its abilities as a parent. Many biologists are therefore less interested in fecundity than in the number of offspring that survive to reach maturity. This measure of an individual's reproductive success is now termed its *fitness*. Fitness is thus a measure that incorporates components of both fecundity and survival, illustrating the important point that survival and reproductive success are variables that often cannot easily be separated and defined independently. This topic is discussed in greater detail in Chapter 4.

Another point is that it is not the *absolute* value of each individual's fecundity or viability that is important in natural selection, but their *relative* value—individuals will be favoured by selection only if they are *more* fecund, and/or if they and their offspring are *more* viable than others in the same population.

The only sense in which Darwin gave thorough consideration to the effects of relative fecundity was in relation to what he called sexual selection. By this he meant selection that arises when some individuals gain an advantage in terms of the number of successful matings they achieve over other individuals of the same sex. Most commonly, sexual selection arises in competition between males for females. Darwin envisaged two kinds of sexual selection: that which involves actual fighting between male animals; and that which involves competition between males to attract females. The results of these two kinds of selection are the elaboration, respectively, of male weapons (such as the antlers of deer) and of special structures (such as the peacock's tail) that are associated with male courtship displays.

Darwin treated sexual selection as a process separate from natural selection, largely on the grounds that the two processes could sometimes actually work in opposite directions. For example, sexual selection might favour the evolution of a very long tail that makes a male especially attractive to females, while natural selection would penalize the possessors of long tails if they were less adept at escaping predators as a result. We shall return to the question of whether sexual selection and natural selection are discrete processes in Chapter 7.

This brings us to the problem of how a variation might be 'profitable' to an individual. What Darwin meant here was any variation that helped to *adapt* an organism to a way of life. This concept will be discussed in detail in Chapter 2. You should note two things here, though, about Darwin's understanding of adaptation.

First, he discussed adaptations relating to reproduction as well as to individual welfare. For example, he referred to the mutual adaptation between many flowers and the insects that pollinate them. This again emphasizes his assumed inclusion of reproduction in the process of natural selection and indeed shows that the effects of differential fecundity were implicit in his arguments, even if not explicitly discussed.

Secondly, in later editions of *The Origin of Species*, Darwin borrowed Herbert Spencer's expression, 'the survival of the fittest', as a synonym for natural selection. To Darwin, 'the fittest' were simply the best adapted; these, he *predicted*, were more likely to survive and leave more offspring than the less fit. Although the expression redoubled emphasis on survival, successful reproduction was still an assumed consequence of survival. The word 'fitness' has therefore undergone an important change of meaning since Darwin's day. Nowadays, fitness is defined in terms of the *outcome* of the process of natural selection, as noted above, rather than as some quality of organisms likely to be favoured by selection.

1.1.3 Variation and the principle of inheritance

Just as reproduction is an essential component of evolution by natural selection, of equal importance for the process of evolution is the inheritance of features on which natural selection can act. Neither Darwin nor Wallace succeeded in clearing up the problem of how variations were inherited. That had to await the work of Mendel, Morgan, Weismann and others, pioneers in the experimental study of genetics. Darwinian theory was modified in the late 1930s and 1940s by the inclusion of ideas on genetics and systematics; this modified form is referred to as the neo-Darwinian view.

There is not space here to go into the details of Darwin's beliefs as to the sources of variation. This was a problem he studied for many years and his failure to find a satisfactory solution was probably one of the reasons for the long delay between his initial working out of the theory of natural selection, which he did in 1838, and his publication of it in 1859. Suffice it to say that he was very much a 'pluralist'; that is, he felt that heritable variations had many causes, including sustained environmental influences on the reproductive system, the effects of habitual use or disuse of characters, and developmental linkage with other evolving features. These were all conceived as providing 'fuel' for natural selection.

He thus felt that variation might to some extent be directed, so influencing the course of evolution. Some people believed that an attribute acquired during an individual's life, such as great physical strength or a particular skill, could be inherited by that individual's offspring. This notion, referred to as the inheritance of acquired characters, was an important part of the evolutionary theory of Jean Baptiste de Lamarck (1744–1829), published in 1809. The idea is now discredited.

One point upon which Darwin was adamant was that *small* variations were the key to evolution by natural selection. He felt that large variations would be so much at odds with the subtle adaptation of organisms to their environments that it would be very unlikely that they could ever be advantageous and therefore favoured by selection:

> 'It may be doubted whether sudden and considerable deviations of structure such as we occasionally see in our domestic productions, more especially with plants, are ever permanently propagated in a state of nature. Almost every part of every organic being is so beautifully related to its complex conditions of life that it seems as improbable that any part should have been suddenly produced perfect, as that a complex machine should have been invented by man in a perfect state.'

Darwin argued that evolution did not proceed through the sudden appearance of 'monstrosities' that showed major differences from other members of a species. Instead, he maintained that there was always scope for slight variations to prove profitable, either as refinements of existing adaptations or by allowing the exploitation of new and different resources.

▶ If small variations are indeed the key to evolution, what does this imply for rates of evolution?

On the basis of small variations, evolution would have to be a gradual, slow process.

Viewed in this way, evolution by natural selection could not involve the sudden appearance of forms very different from their ancestors; therefore we should not expect any sharp discontinuities in lines of descent. Darwin's views on evolutionary rates had a major influence on later evolutionary studies, as you will see in the next Section.

1.1.4 THE ORIGIN OF SPECIES

Many of Darwin's contemporaries regarded species as fixed and unchanging entities. For them, a species consisted of similar individuals with severe limitations on the possible variation within any one species. Accordingly, each species was separated from every other species by a sharp discontinuity and was unchanging in time. This way of regarding a species concentrates on the type and sees individual variation as unimportant. Darwin had to show that this view was incorrect. In 1856 he wrote, in a letter to his friend Joseph Hooker, a world-renowned botanist:

> 'I have just been comparing definitions of species.... It is really laughable to see what ideas are prominent in various naturalists' minds, when they speak of 'species'; in some, resemblance is everything and descent of little weight—in some, resemblance seems to go for nothing, and Creation the reigning idea—in some, descent is the key—in some, sterility an unfailing test, with others it is not worth a farthing. It all comes, I believe, from trying to define the undefinable.'

In *The Origin of Species*, he wrote:

> 'I look at the term species, as one arbitrarily given...to a set of individuals closely resembling each other...it does not essentially differ from the term variety, which is given to less distinct and more fluctuating forms.'

This contrasted with the position of his opponents and opened the way for the development of the idea that evolution was a process of gradual change. Darwin concluded that varieties within species were in the initial stages of becoming new species, and the only distinction between them lay in their degree of difference from related varieties or species.

The confusion over the definition of the term species that Darwin commented on to Hooker persisted long after the publication of *The Origin of Species*. It was not until the 1940s that the modern definition, with its emphasis on populations of individuals which are capable of breeding with each other freely but which are **reproductively isolated** from other populations, became widely accepted (see also Chapter 9).

Darwin himself did not define species as reproductively isolated populations. He acknowledged that hybrid crosses between accepted species are generally sterile (if not inviable) and that mongrels between varieties are usually vigorous and fertile, but he stressed that this distinction between species and varieties is far from reliable. Rather, he believed that, as varieties diverge to the extent of becoming separate species, they gradually become reproductively incompatible.

This nevertheless cleared the way for the modern definition, by showing how reproductive isolation might evolve; no longer could sterility between individuals of different species, or of hybrids, be cited as an argument for the fixity of species, as it had been by some of Darwin's opponents.

Darwin's concept of the gradual conversion of varieties into species led him to believe that new species were most likely to form in large, widespread populations, because these would contain the most variation on which natural selection could work. Although he admitted that small, geographically isolated populations could give rise to new species, he felt that these made a limited contribution to evolutionary history. As you will find later, many (though not all) evolutionists today take the contrary view and regard geographical isolation as an essential condition for the formation of most new species.

An historical point worth noting is that the gradual evolution that Darwin envisaged required vast amounts of time. Only in the half-century or so preceding the publication of *The Origin of Species* had the majority of geologists begun to accept that the Earth was very ancient. Even so, most of Darwin's contemporaries doubted that it was more than a few tens of millions of years old and therefore they had reservations about whether natural selection alone could account for the evolutionary changes that had occurred. Such reservations have largely disappeared in the light of modern estimates of the age of the Earth (about 4 500 million years).

1.2 BEFORE DARWIN

Charles Darwin was not the first person to put forward a theory of evolution as he himself acknowledges in *An Historical Sketch on the Progress of Opinion on the Origin of Species* (first published with the 3rd edition of *The Origin of Species* in 1861). However, evolution, meaning the change through descent of one kind of organism into another by means of natural causes, was not a popular idea in the early part of the 19th century. The origin of the various kinds of living creature was generally ascribed to their creation by a divine being either (as some believed) in a single original act of creation or (as others believed) in a series of divine creative episodes. Species were regarded as static and unchanging since their creation. Changes in a variety of branches of knowledge, a rapid growth in the facts available in biology and geology, plus a dawning realization that the Bible could not be literally true had nevertheless prepared the ground for the growth of evolutionary theories.

In cosmology, for instance, Immanuel Kant in 1755 presented a view that the whole cosmos was a continuously changing dynamic system. In geology, the belief that fossils were the remains of once living creatures was established by the end of the 17th century, replacing the older belief that fossils simply grew in the rocks in which they were found. Sequences of fossils made it possible to believe that the Earth itself had a history, and the idea that some fossils were of organisms now extinct became widely accepted by the close of the 18th century. These and other changes made it possible to ask questions about the origin of organisms.

In France in 1809, Lamarck published a theory of evolution in his *Philosophie Zoologique*. He did not propose a theory of the common descent of the various kinds of organism (i.e. descent from a common ancestor) but suggested that organisms were capable of slowly and gradually changing with time into new species. The mechanism of evolution set out by Lamarck involved an inborn trend in all species to move from being simple towards being complex and an innate capacity in organisms to react to the special conditions in their environments so that they were in harmony with them. Lamarck's ideas became widely known but they do not seem to have been very influential even in France, where his ideas were strongly attacked and eventually ignored altogether. However, Lamarck helped ease the way to the eventual acceptance of evolutionary ideas.

A second book which also increased people's receptiveness to evolutionary ideas was *Vestiges of the Natural History of Creation*, published anonymously in England in 1844. The author of *Vestiges* tried to establish the idea that the fauna of the world evolved through geological time and that the changes were slow and gradual. The arguments used in *Vestiges* were savagely attacked by contemporary scientists, philosophers and religious leaders and doubts were cast on the whole thesis. Darwin wrote of *Vestiges* in the *Historical Sketch*:

> 'The work, from its powerful and brilliant style, though displaying in the earlier editions little accurate knowledge and a great want of scientific caution, immediately had a very wide circulation. In my opinion it has done excellent service in this country in calling attention to the subject, in removing prejudice and in thus preparing the ground for the reception of analogous views.'

1.3 THE EVIDENCE FOR EVOLUTION

Darwin had two aims in view when he wrote *The Origin of Species*. One was to show that biological evolution was a fact, and the other that the process of biological evolution was the result of variation and natural selection. In order to be convincing, Darwin had to show that the known facts of natural history could be explained by evolution but not convincingly by any alternative explanations. He drew his evidence from the areas of **biogeography**, **morphology** and **embryology**, **taxonomy**, **systematics** and **palaeontology**.

1.3.1 BIOGEOGRAPHY

Biogeography (the study of the geographical distributions of animals and plants) had thrown up a number of problems for biologists. That the geographical distribution of animals and plants was not random was accepted by all biologists, but the factors responsible for the observed distributions were not obvious. How does it come about that, for example, regions with similar climatic and other physical conditions but geographically distinct from each other had dissimilar plants and animals? Darwin gave the examples of Australia, South Africa and Western South America which have similar climates but dissimilar flora and fauna. Another example was that although conditions in Australia, South America and Europe are nearly the same, there was not a single native species of mammal common to all three. It was also recognized that within a single geographical region there might be wide variations in environmental conditions and considerable variation amongst the animals and plants; nevertheless the animals in any one region shared similarities as did the plants. However, these similarities were not shared with the flora and fauna of other environmentally similar parts of the world.

If the first set of observations was explained by supposing the special creation of different species in the different geographical regions, then it was not clear why the different species within any one region should share similarities with others in the same region, rather than be similar to species living in similar environments in other geographical regions. Darwin pointed out that common descent with modification through variation and natural selection would explain both the dissimilarities between the flora and fauna of different regions and the similarities between different species in the same region.

Darwin also drew attention to two peculiarities of the fauna and flora of oceanic islands (islands of volcanic origin, remote from any other land-masses and which have never had land connections with any continental land-mass, e.g. the Azores, Hawaiian Islands and Galápagos Islands). One such peculiarity is the small number of animal and plant species on oceanic islands when compared with the number of species in areas of similar size on the mainland nearest to the islands. The second is that the proportion of **endemic species** on oceanic islands (i.e. species which occur there and nowhere else) is much higher than the proportion of endemics in areas of similar size on the nearest mainland.

The Galápagos Islands and their finches The Galápagos are a group of islands some 600 miles off the west coast of South America (Figure 1.3). Darwin visited four of the islands in 1835 and made collections of the animals and plants he found. When he returned to England in 1836, the collections were handed over to experts who named and classified the specimens. The birds in the collections were handed over to the ornithologist John Gould. On the basis of Gould's work, it was established that of the 26 species of land birds known to live on the islands, 21 species were endemic. Included in Darwin's collection were 31 individual specimens of finch-like birds, all of which belonged to endemic species. Since 1835, the birds of the Galápagos have been extensively collected and intensively studied with the result that 14 species of finch are now recognized; 13 are endemic to the Galápagos (the remaining one species is found only on Cocos Island which lies about 400 miles north-east of the Galápagos (Figure 1.3)). David Lack, the noted ornithologist, popularized the name Darwin's finches (Figure 1.4) for the whole group in his book entitled *Darwin's Finches,* published in 1947.

With regard to the fauna and flora of oceanic islands, Darwin posed a number of questions. These can be neatly illustrated by reference to Darwin's finches. While being distinct from other species, the Galápagos finches did resemble a group of finches found in South America. Darwin asked:

> 'Why should this be so? Why should the species which are supposed to have been created in the Galápagos Archipelago, and nowhere else, bear so plainly the stamp of affinity to those created in America?'

He also pointed out that (1) each island had a distinct assemblage of species but these species were related to each other more closely than they were to species from the nearest mainland; (2) the assemblages of species found on the different islands were different (Table 1.1); and (3) populations of the same species on different islands had some differences from each other (Figure 1.5).

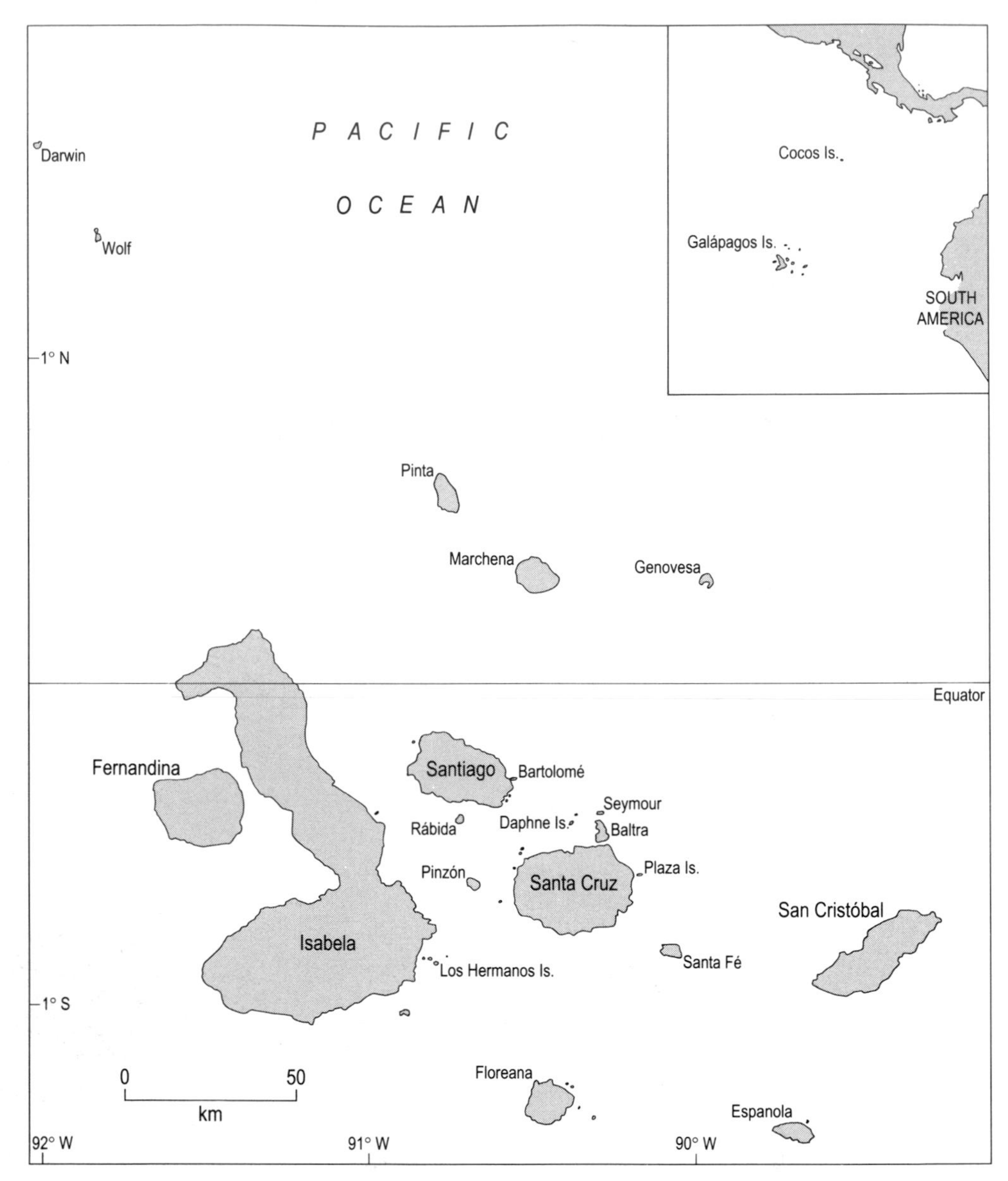

Figure 1.3 Map of the Galápagos Islands.

Figure 1.4 *Species of Darwin's finches on the Galápagos Islands.*

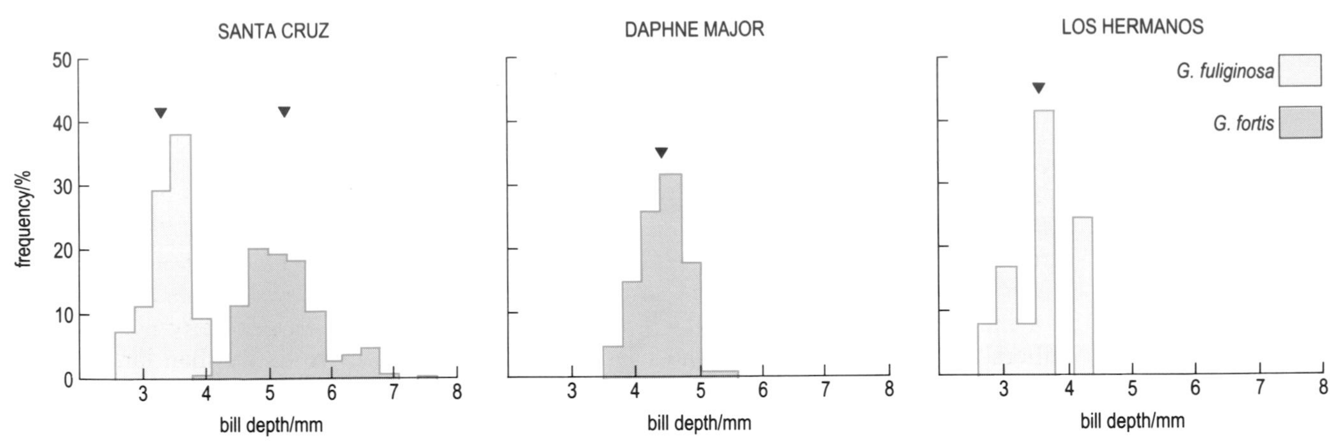

Figure 1.5 *Histograms of the beak sizes (beak depth) of two species of Darwin's finches on three Galápagos Islands. Average sizes are indicated by the triangles.*

Table 1.1 shows the distribution of the different species amongst 13 of the larger islands. In some cases, finches on different islands have evolved into subspecies. Only two islands listed in Table 1.1 have the same assemblage of species or subspecies.

Island	Geospiza magnirostris	Geospiza fortis	Geospiza fuliginosa	Geospiza difficilis	Geospiza scandens	Geospiza conirostris	Platyspiza crassirostris	Camarhynchus psittacula	Camarhynchus pauper	Camarhynchus parvulus	Cactospiza pallida	Cactospiza heliobates	Certhidea olivacea
Darwin	A	–	–	A	–	–	–	–	–	–	–	–	A
Wolf	A	–	–	A	–	–	–	–	–	–	–	–	A
Pinta	A	A	A	B	A	–	A	A	–	(A)	–	–	B
Marchena	A	A	A	–	B	–	A	A	–	–	–	–	B
Genovesa	A	–	–	B	–	A	–	–	–	–	–	–	C
Fernandina	A	A	A	C	–	–	A	B	–	A	A	A	D
Isabela	A	A	A	(C)	C	–	A	B	–	A	A	A	D
Santiago	A	A	A	C	D	–	A	C	–	A	B	–	D
Santa Cruz	A	A	A	(C)	C	–	A	C	–	A	B	–	D
Santa Fé	A	A	A	–	–	–	(A)	C	–	A	–	–	E
San Cristóbal	(A)	A	A	(C)	E	–	–	–	–	B	C	–	F
Floreana	(B)	A	A	(D)	C	–	A	C	A	A	–	–	G
Espanola	–	–	A	–	–	B	–	–	–	–	–	–	H

KEY
A–H = different subspecies
(A) = extinct population

Table 1.1 *Distribution of Darwin's finches amongst the Galápagos Islands.*

Figure 1.5 shows the variation in beak depths amongst two species of finch on three islands. Variation in beak depth (and length) has been shown to be important in relation to the diet of the finches. The explanation put forward by Darwin for this state of affairs involved the colonization of the islands by finches from the South American mainland and the subsequent evolution of differences between the birds on the different islands and between the island birds and the mainland birds. Finches are small birds and do not normally fly far, so colonization of the islands from the mainland could not have been frequent. Darwin contrasted the situation he found amongst the land birds with that amongst the 13 species of marine bird of the islands, only two of which were endemic. Marine birds generally are strong fliers and so could arrive at the islands more often and more easily than land birds.

The alternative view, that the differences between the birds of the islands and the mainland and between the birds of the various islands were the result of the special creation of each species on each island, seemed hardly plausible to Darwin; it could not explain the general resemblance between the finches on the various islands, and between them and the mainland finches. The situation could be better explained by the principle of descent with modification—the common descent of all the finches from colonists arriving from South America, and the common descent with modifications of the birds on the islands after the initial colonization.

1.3.2 Morphology and embryology

Morphology (the study of plant and animal form) involves the study of the structure of animals and plants and of the structure of parts of living things in relation to the functions they perform. The study of the similarities and differences in morphology between different organisms is called comparative anatomy.

One of the important conclusions reached by anatomists before the publication of *The Origin of Species* concerned the similarity of structure shown by members of the same group of living organism. It was realized, for instance, that in spite of wide differences in function and external appearance, all mammalian limbs display the same general pattern of internal structure. This is an example of **homology**. The mammal limb bones and the bones of the bird's wing shown in Figure 1.6 are homologues; they have the same skeletal elements in very different structures.

But having 'discovered' homology, the anatomists were unable to explain this phenomenon. Darwin pointed out that homology could be the result of the descent of all mammals from a common ancestor and the operation of natural selection on variations of the common pattern in the various environments in which the animals lived.

Darwin drew support for the principle of common descent with modification from embryology (the study of the early stages in the lives of organisms) in two ways. Between 1846 and 1854 he had made a study of the group of animals called the Cirripedia (Figure 1.7). This is the group of crustaceans (Arthropoda) to which the barnacles (Figure 1.7(b)), so common on rocky shores between the tide marks,

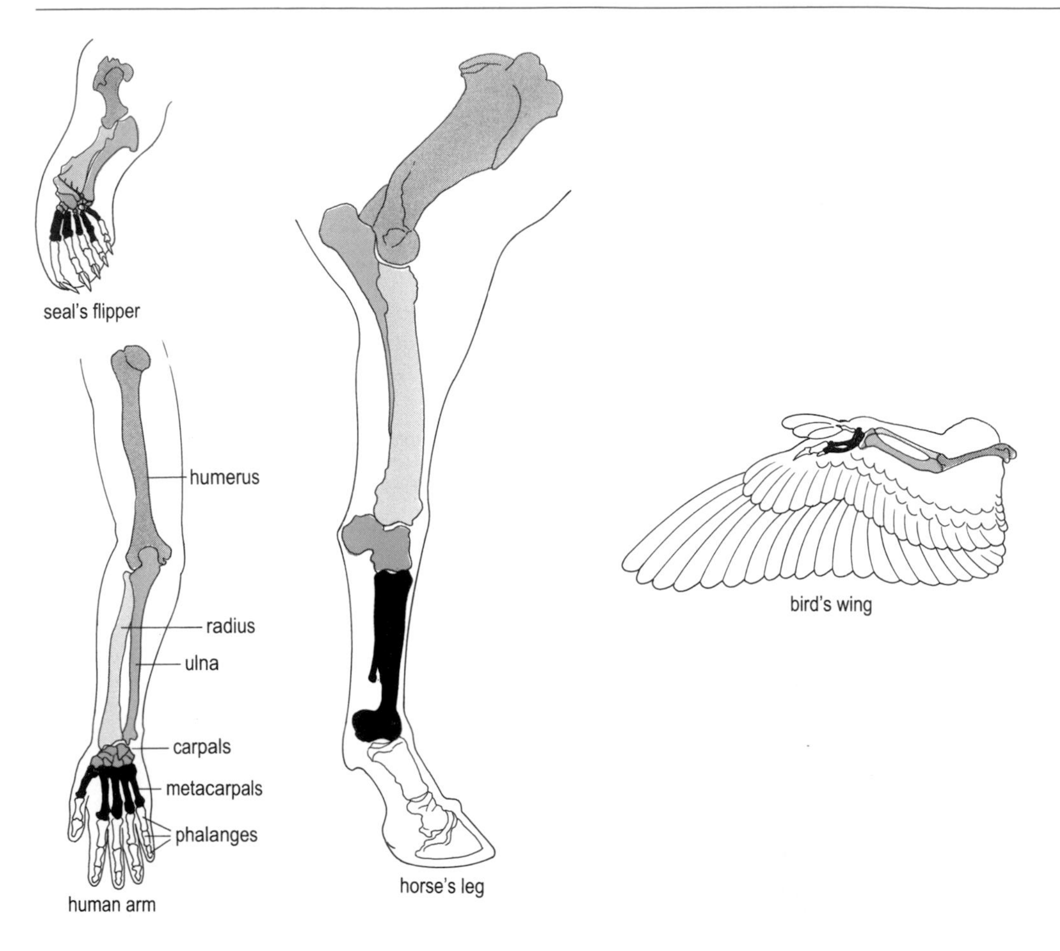

Figure 1.6 *Homologous limb bones in seal, human, horse and bird.*

belong. Another kind of barnacle (Figure 1.7(a)), the stalked barnacles, attach themselves to the hulls of ships and other large marine objects, while a third kind are parasites. Before 1839, it had been thought that, because many barnacles had shells around their bodies, they were some kind of mollusc. This is the group of animals (Mollusca) which includes the snails and bivalves. However, it was then discovered that the juvenile stages of barnacles were unlike any of the juvenile stages of the molluscs; yet the same juvenile stages were found in the life histories of all kinds of cirripedes. So that in spite of very wide differences in the appearance of adult barnacles, all of them shared a common appearance as juveniles.

Darwin gave another example of similarity in the early developmental stages of animals which were remarkably different in appearance as adults from amongst the vertebrates. He quoted a statement made by the great German zoologist Karl von Baer:

'The embryos of mammalia, of birds, lizards and snakes, probably also the chelonia [the turtles and tortoises], are in their earliest states exceedingly like one another, both as a whole and in the mode of the development of their parts; so much so that we can often

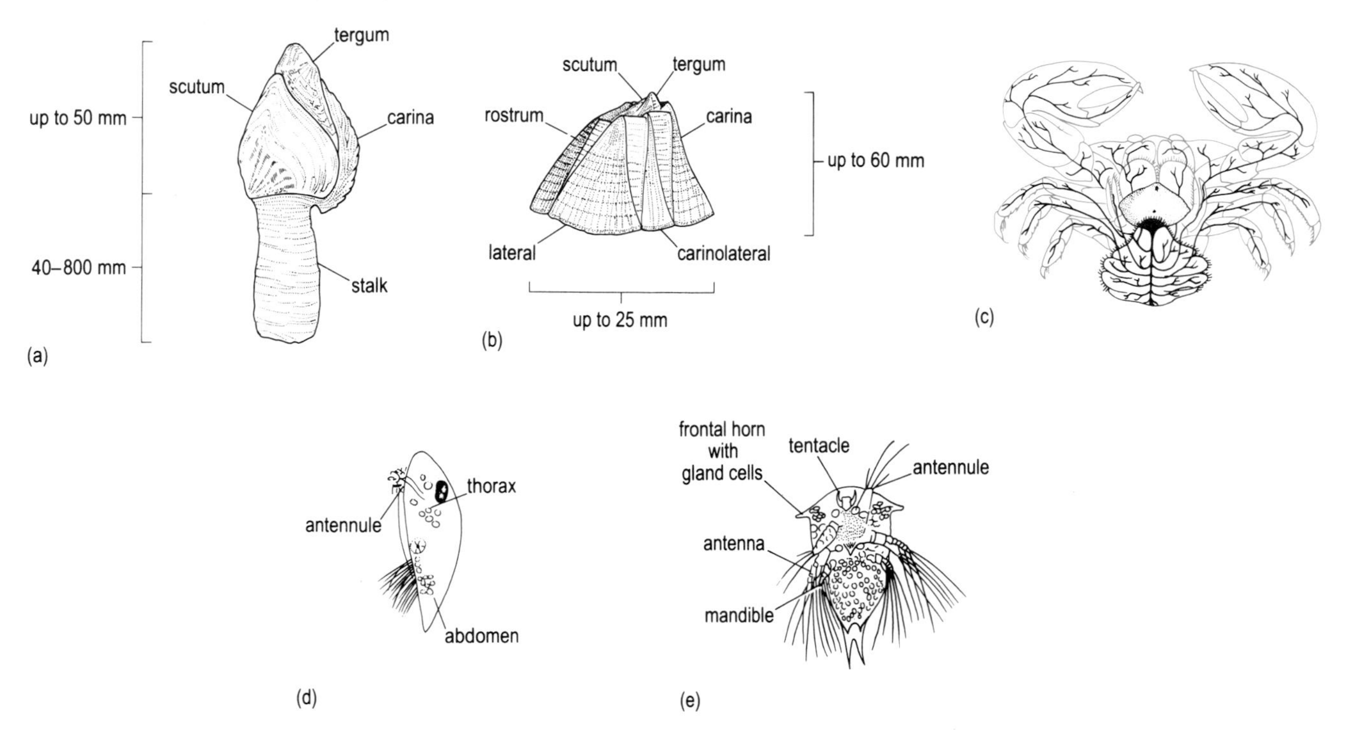

Figure 1.7 *Various kinds of cirripedes: (a) goose barnacle; (b) common barnacle; (c) parasitic form infesting a crab; (d) and (e) side and front view of a larva.*

distinguish the embryos only by their size. In my possession are two little embryos in spirit [preserved in alcohol], whose names I have omitted to attach and at present I am quite unable to to say to what class they belong. They may be lizards or small birds, or very young mammalia, so complete is the similarity in the mode of formation of the head and trunk in these animals.'

Figure 1.8 illustrates this point. The embryos of four different mammals are drawn to the same size and in three equivalent stages of development. Particularly in the first two stages, the embryos are very similar in appearance.

Darwin drew attention to the fact that, during the embryonic development of wings in bats and flippers in porpoises, the development of the bones follows a very similar sequence and yet the fully developed wing and flipper are very different in appearance and in detailed internal structure. Embryologists of Darwin's time had no explanation for this, but for Darwin the reason lay in the common descent of bats and porpoises.

He took pains to point out that similarities of structure amongst embryos of animals in the same class could not be due to a similarity in the conditions in which the embryos developed. Although it was possible to give this as a reason for the similarities between the juvenile stages of barnacles, all of which are free-living in the sea, it was not possible to do so with regard to the development of vertebrates. For example, in vertebrates the arteries in the throat region follow a similar pattern in their development: in young mammals developing in their mother's uterus; in young birds developing in shelled eggs; and in young frogs

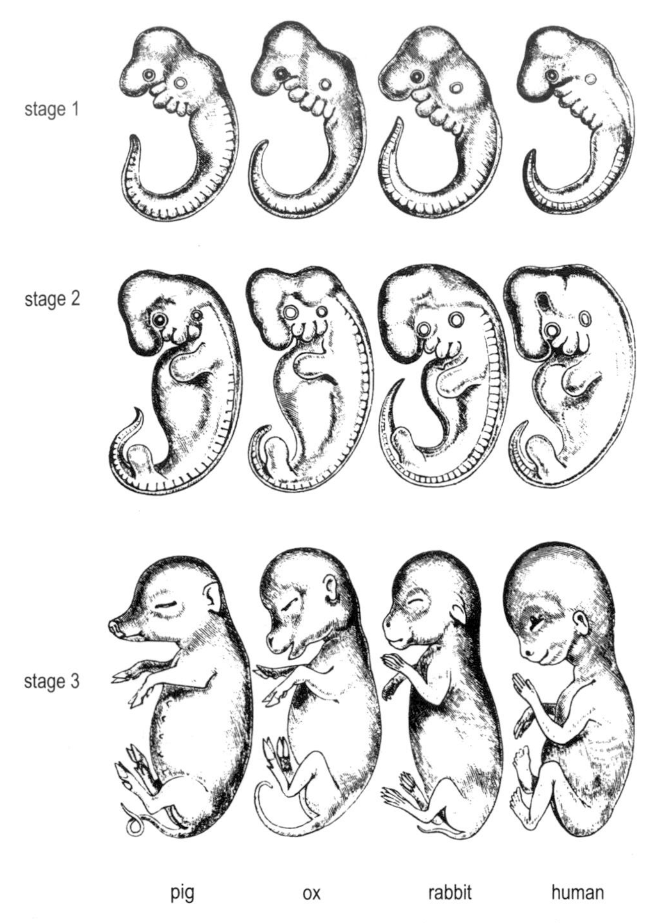

Figure 1.8 *Embryos of various mammals (not to scale).*

developing in spawn in a pond. But similarity in development under such widely different conditions could be explained if mammals, birds and frogs were all descended from a common ancestor. Indeed, it might be expected to occur.

1.3.3 TAXONOMY AND SYSTEMATICS

Taxonomy and systematics are the branches of biological sciences concerned with the description and naming of different species and the development of systems of classification of organisms, respectively.

Well before Darwin's time, taxonomists had recognized that 'natural affinities' could be detected between organisms, based on broad similarity and, in particular, on the shared possession of homologous features, and that these affinities could serve as the basis for classification. There was much debate, however, about which features best suited this purpose, and disagreement over the resulting classifications. Nevertheless, it was generally agreed that 'natural' systems were hierarchical and inclusive in character. Inclusive classifications involve grouping together similar objects to make larger units, and these in turn into larger units

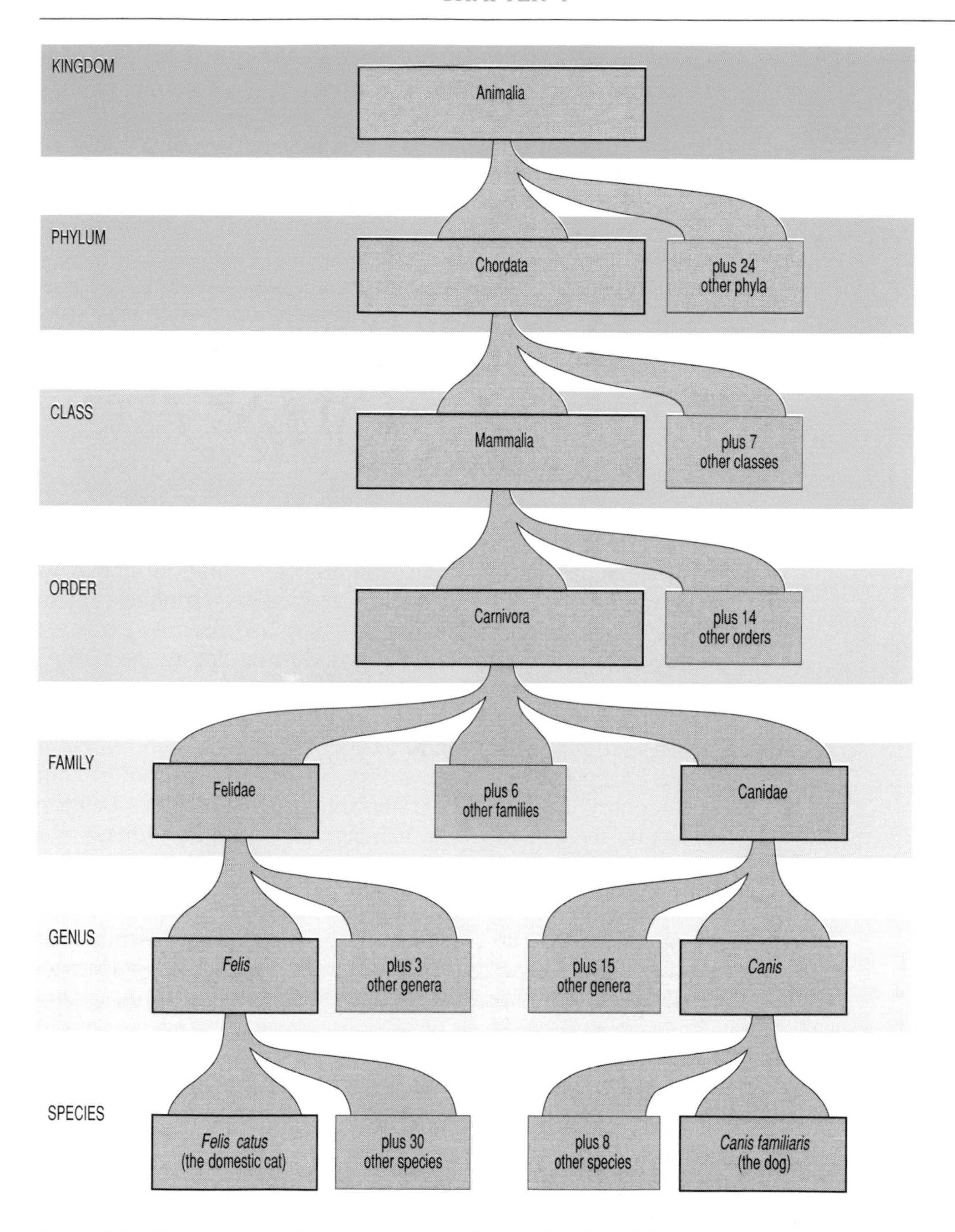

Figure 1.9 *The taxonomic hierarchy showing the classification of the domestic cat and dog.*

still. In biological terms, the smallest units are species, all members of which are similar. Groups of species which have some common features are placed in the same genus, groups of genera into the same family, groups of families into the same order, and so on. Each unit (except the smallest) contains within it subordinate units and each in turn (except the largest) is contained in a superior unit. This is illustrated by Figure 1.9 which shows the system applied to the classification of the domestic cat and dog.

Naturalists and philosophers were convinced that the taxonomic hierarchy was the reflection of some deeper order in nature. So a system of classification was required which reflected this deeper order, and this would be a Natural Classification. In Darwin's day, it was widely believed that a Natural Classification would show the pattern of creation. Darwin however proposed a different explanation:

'All the...difficulties in classification may be explained...on the view that the Natural System is founded on descent with modification;—that the characters which naturalists consider as showing the true affinity between any two or more species are those which have been inherited from a common parent, all true classification being genealogical;—that community of descent is the hidden bond which naturalists have been unconsciously seeking....'

1.3.4 Palaeontology

Darwin drew on known facts and agreed interpretations from palaeontology as evidence supporting the fact of biological evolution. It was well known that there were fossils of organisms such as ammonites, trilobites, mammoths and mastodons which had no living descendants—they had become extinct. Earlier in the 19th century, extinction had been supposed to be due to the overwhelming of the Earth by a series of catastrophes. By 1859, catastrophe as a means of explaining extinction had been abandoned, as had another earlier suggestion—that extinct forms were not in fact really extinct but persisted in remote and unexplored parts of the Earth. Darwin, however, as others had done before, attributed extinction to evolutionary change: the extinct organisms sometimes being replaced in the fossil record by their modified descendants or by allied forms.

Darwin also cited as evidence the fact that all known living and extinct species could be classified into existing classificatory groups or as intermediate forms sharing features with two or more existing groups of organism. This is of course what would be expected from the principle of common descent with modification. Since the recent re-evaluation and reconstruction of fossils from the Middle Cambrian Burgess Shale of Western Canada, however, we know that not all fossils can be easily accommodated in the way in which Darwin thought that they could be. Some of the Burgess Shale fossils are of organisms whose body plans are hard to relate to those of other organisms (see Chapter 14).

Similarly, common descent with modification would explain:

(a) why the observed differences between living and extinct forms became greater the older the extinct forms.

(b) why the fossils from any two consecutive strata are far more similar than are fossils from two that are farther apart.

(c) why the fauna living during any one great period of the Earth's history was intermediate in general character between the fauna of the immediately preceding (older) and the immediately succeeding (younger) period.

1.4 Conclusion

In *The Origin of Species*, Darwin intertwined arguments and evidence supporting the fact that biological evolution had taken place with arguments and evidence for natural selection as its mechanism. By about 1875, a relatively short period of time after the publication of *The Origin of Species* in 1859, the fact of evolution had been accepted by biologists, palaeontologists and others more or less worldwide. However, the part played in biological evolution by variation and natural selection did not command such universal agreement. As the rest of this Book will show, this is still a source of discussion and controversy today. What Darwin had succeeded in doing was to show that the principle of descent with modification was consistent with all the facts of natural history and that no other explanation was consistent. Part of the appeal and power of this principle lies in this very fact. As Theodosius Dobzhansky, one of the greatest biologists of recent times, wrote:

'Nothing in biology makes sense except in the light of evolution.'

Summary of Chapter 1

While not being first to propose a theory of evolution, Charles Darwin was, together with Wallace, the first to put forward a theory of evolution which relied upon natural causes rather than upon supernatural or divine causes. His theory set out to explain the origin of diversity amongst organisms, the origin and history of diversity and the natural processes by which it developed.

Darwin's argument has four basic premises:

1 More individuals are produced than can survive.

2 There is a *struggle for existence*, because of the disparity between the number of individuals produced in reproduction and the number that can survive.

3 Individuals show *variation*. No two individuals are exactly the same. Those with advantageous features have a greater chance of survival in this struggle (*natural selection*).

4 As selected varieties will tend to produce offspring similar to themselves (*the principle of inheritance*), these varieties will become more abundant in subsequent generations.

For Darwin there was no sharp distinction between a species and a variety. Indeed, he regarded varieties as incipient species. His views were in sharp contrast to the orthodox views that species were sharply distinct from each other and that they were unchanging and unchangeable.

Although evolutionists today express some parts of Darwin's arguments differently, his theory commands almost universal acceptance. In addition to providing an explanation for evolution, Darwin showed how evolution explained and brought order to the known facts in the fields of biogeography, comparative morphology and embryology, taxonomy and systematics and palaeontology. His success in this provided confirmation of the fact of evolution. Since publishing his theory in

1859, our knowledge of the causes and pathways of evolutionary change has increased enormously. In particular, our understanding of the sources of variation amongst organisms has grown with the development of the experimental study of genetics and inheritance.

FURTHER READING FOR CHAPTER 1

The key reference must be to *The Origin of Species* itself. There were several editions but the most widely used one is the 6th edition. This is generally available in the World's Classics series published by Oxford University Press.

Darwin, C. (1872) (6th edn) *On the Origin of Species by Means of Natural Selection, or The Preservation of Favoured Races in the Struggle for Life,* Murray, London.

If you are interested in the history of ideas about evolution and the background against which Darwin worked, the following are recommended:

Desmond, A. and Moore, J. (1991) *Darwin*, Michael Joseph, London, 808 pp.

Mayr, E. (1982) *The Growth of Biological Thought; Diversity, Evolution and Inheritance*, The Belknap Press of Harvard University Press.

Oldroyd, D. R. (1980) *Darwinian Impacts: An Introduction to the Darwinian Revolution*, The Open University Press.

Ruse, M. (1979) *The Darwinian Revolution; Science Red in Tooth and Claw*, University of Chicago Press.

For evolution in general, the following are recommended:

Dobzhansky, T., Ayala, F. J., Stebbins, G. L. and Valentine, J. W. (1977) *Evolution,* W. H. Freeman & Co.

Futuyma, D. J. (1986) (2nd edn) *Evolutionary Biology*, Sinauer Associates, Sunderland, Mass.

Grant, P. R. (1986) *Ecology and Evolution of Darwin's Finches*, Princeton University Press.

Lack, D. (1947) *Darwin's Finches*, Cambridge University Press, Cambridge.

SELF-ASSESSMENT QUESTIONS

SAQ 1.1 (*Objective 1.1*) In which, if any, of the statements (a)–(c) below is the term 'biological evolution' adequately defined?

(a) Biological evolution means changes in the appearance of organisms.

(b) Biological evolution means changes amongst organisms over several generations.

(c) Biological evolution means descent with modification, involving heritable features.

SAQ 1.2 (*Objectives 1.1 and 1.2*) Which, if any, of the statements (a)–(c) below is (are) correct?

(a) Biological variability of the sort upon which evolution depends arises independently of the requirements imposed on organisms by environmental change.

(b) Inorganic changes happen more rapidly than biological evolutionary changes.

(c) Evolutionary changes in biology are progressive and directional, because the most recently evolved organisms are better adapted and more nearly perfect than their ancestors.

SAQ 1.3 (*Objective 1.3*) Which, if any, of the statements (a)–(e) below is (are) correct?

(a) An organism's fitness can be measured by counting the number of offspring it leaves.

(b) The survivors of a struggle for existence are those that are the fittest.

(c) When two populations of an organism differ in fitness, competition between them will result in the survival of the fitter population.

(d) The concept of the struggle for existence in Darwin's theory includes the idea of relative reproductive success.

(e) Natural selection favours increasing diversity between individuals of a species; therefore those species which survive essentially unchanged for thousands of years must be exempt from natural selection.

SAQ 1.4 (*Objective 1.4*) Which, if any of the statements (a)–(g) below correctly represents Darwin's views on variation and species?

(a) Differences in the rate of evolution are largely due to differences in the sizes of variations: very large variations result in rapid evolutionary rates.

(b) Heritable variations may be due to any one of several different causes, some of which may be directed.

(c) The idea that characteristics acquired by parents in their lifetime and inherited by their offspring played an important part in evolution was promoted by both Lamarck and Darwin.

(d) An orthodox idea in Darwin's time was that the terms 'variety' and 'species' did not essentially differ in meaning.

(e) To succeed in persuading people that evolution had occurred, Darwin had to show that the terms 'species' and 'variety' differed very little in meaning.

(f) Darwin regarded each species as being reproductively isolated from every other species.

(g) Because the Earth was believed to be very old, Darwin had no difficulty persuading people that there had been sufficient time for evolution by natural selection to have produced all the known species of animals and plants.

CHAPTER 2

ADAPTATION

Prepared for the Course Team by Caroline Pond

Study Comment *Adaptation is a central concept in most aspects of modern biology: however, like many powerful ideas, that of adaptation has been misinterpreted and misused. This Chapter is an essay on the concept of adaptation and its many applications in the study of evolution. Most of the major points are illustrated with relevant examples. As adaptation is an almost universal property of organisms, the same principles could be illustrated by an enormous range of possible examples. Most of those used for this Chapter refer to mammals and other terrestrial vertebrates, many of which feature in museum and zoo exhibits, and in wildlife films. By studying these sources, you can verify for yourself many of the observations mentioned.*

Objectives When you have finished reading this Chapter, you should be able to do the following:

2.1 Describe the historical development and current usage of the terms 'adaptation' and 'function' in biology and palaeontology and explain the relationship of these terms to the neo-Darwinian concept of fitness.

2.2 Explain how the adaptive significance of a structure or habit is investigated.

2.3 Describe the kinds of evidence that can be used to identify adaptations in extinct species from fossil remains.

2.4 Describe how major anatomical adaptations involve physiological changes and thereby necessitate structural modifications to other organs, tissues and aspects of behaviour and life history.

2.5 Describe how the genetic basis of new adaptations may be investigated.

2.6 Give some examples of coevolved adaptations and describe some experiments that elucidate how the species involved are adapted to each other.

2.7 Describe some theories that have been proposed to explain the origin of novel structures in evolution.

2.8 Describe some adaptations to the climatic and biological conditions during the Quaternary Ice Age, and explain the palaeontological, experimental and comparative evidence that is used to investigate their evolutionary origins and physiological mechanisms.

2.1 Introduction

No organism can live everywhere, eat everything and breed anywhere, but all living species can, and all extinct species could at one time, flourish somewhere. Adaptation is the matching between an organism and its natural environment, and applies to all aspects of its biology: its anatomy, physiological and biochemical processes, behaviour, diet and life cycle. The idea is straightforward enough, but difficulties arise when we try to identify adaptations and to quantify their relative importance to the organism's survival and reproduction. The features most obvious to biologists may not be the organism's most important adaptations to its natural habitat. No two species ever live in the same place, at the same time or in exactly the same way, so adaptations, by definition, are never exactly similar. Each species has a different combination of characters, many, but not all, of which can be identified as adaptations to some feature of its habits or life cycle. Because each species is unique, it is very difficult to establish any generally applicable ways of investigating and identifying adaptations.

This Chapter illustrates the principles behind identifying adaptations that are, or have been, important in a species' natural habits, and for explaining their contribution to its evolution. The history of ideas often sheds light on the uses and abuses of a concept, so we begin by tracing briefly the origins of the idea of biological adaptation that was first developed by Aristotle more than 2000 years ago. Although most observant people appreciated that animals and plants were adapted to the environments in which they lived, the importance of the concept for evolution was largely overlooked until the early 19th century, but has now come to dominate much of biology.

2.1.1 The origin of the concept of adaptation

Until the end of the 17th century, biblical and other mythological theories about the origin of people, animals and plants, and of the age of the Earth, were almost universally accepted. The distribution and habits of animals and plants were studied mainly from the point of view of their role in economic activities such as hunting, agriculture and medicine. Some of the more prominent forms were regarded as having mystical powers or as exemplifying evil influences or moral virtues. Thus, ancient Egyptians worshipped cats, the 'sacred' ibis and certain snakes, and Australian aboriginals revere crocodiles. In Europe, toads, snakes and bats were associated with evil, lions with courage and storks and pelicans with fertility and maternity, to name but a few examples. However, such beliefs tended to obscure rather than extend the study of what we would now call natural history. Extinct organisms received even less attention: fossils were of little practical value, and very few people collected or studied them seriously. Most people believed that fossil species were a separate creation, not biologically related to, or directly comparable with, living forms.

By the middle of the 18th century, this picture was changing. There seemed to be far more different kinds of organisms than was previously thought. Increasing numbers of merchants and explorers returned from Africa, Asia and the Pacific with specimens and stories about extraordinary animals and plants. Deeper and more extensive mines, dug to extract coal or other minerals, turned up more fossils, some of them impressively large. Improved microscopes revealed more details of smaller and smaller micro-organisms. It was also suggested that there had been life on Earth for much longer than previously believed, and by the early 19th century most geologists had concluded that the Earth itself was much older than earlier estimates had suggested.

Although hunters, herbalists and loggers had amassed much information about economically important species—they needed such knowledge to find and harvest them successfully—collecting and observing wild plants and animals did not become a fashionable occupation for educated people until the mid-18th century. Two concepts proved particularly helpful in explaining, as well as documenting, the diversity of organisms: first, that there is a common basic plan for groups of organisms such as flowering plants, insects and vertebrates; and secondly that the many variations of that basic plan can be related to particular functions, e.g. gills and fins for aquatic animals, lungs and legs for terrestrial forms, etc. As biological knowledge grew, the concept was extended from very obvious examples ('teeth are for chewing') to relating specific features to particular habits ('the giraffe's neck is for reaching high branches' (Figure 2.7); 'this insect's wing pattern conceals it when it is resting on tree bark/leaves/soil, etc.').

Recognition of the intimate and elaborate relationship between organisms and their natural environment was one of the major advances towards understanding the mechanism of evolution and the nature and extent of biological diversity. But it was achieved gradually, and no scientist can be credited with creating and establishing the concept. Biologists such as Lamarck and Charles Darwin's grandfather, Erasmus Darwin, emphasized the role of inheritance in explaining the similarities between organisms, both living and extinct, and of the incorporation of gradual modifications in explaining their differences. They argued that these modifications were not random but arose in relation to the organism's diet and other habits: they were adaptations. There were then, and to some extent still are, contrasting opinions about exactly how an organism's structure comes to be adapted to its natural function. Although the major tenets of Lamarck's theories of the mechanism of evolution are now discredited, they established some ideas essential to our modern concepts of evolution and adaptation: the idea of the continuity of life, the fundamental similarity in both structure and habits between living and fossil species, and the importance of differences between species in understanding where and how they live, and their evolutionary history.

2.2 Adaptation and Function

Even when there is agreement on the *concept* of adaptation, identifying adaptations, that is, matching natural features to their normal functions, is not always straightforward. Hasty and uncritical assigning of functions to structures has sometimes brought the concept into disrepute. It is very difficult to establish universally applicable procedures for identifying structures as adaptations to particular functions, so we will have to 'explain' the concepts by using examples.

If living specimens are available for study, it is usually possible to establish the role of a structure, particularly if the entire life cycle of the organism can be

Figure 2.1 *The aye-aye* Daubentonia madagascariensis.

observed in the wild. The aye-aye *Daubentonia madagascariensis* (Figure 2.1) is a primitive primate that occurs in the rainforests of Madagascar. The forelimbs are very different from those of other primates: all the fingers are relatively long, but the third finger is also remarkably thin. Is this finger withered and useless, or is its unusual structure adaptive? Examination of a picture or skeleton alone tells us very little, but observations of the living animal in the wild reveal that they are lively, active animals that forage in trees at night, eating fruit and large arthropods that they locate by smell, and probably also by listening for faint sounds. When a wood-boring insect is located, the aye-aye bites through the bark and extracts the prey from holes and crevices using the third finger, which, in spite of being thin, is strong and agile. There is no doubt that the aye-aye's third finger is used mainly (but not necessarily exclusively) for feeding, and that its unusual features are an adaptation to a highly specialized method of catching prey.

Paws and claws Most biological structures cannot be so closely identified with particular habits or habitats; they have implications for several different aspects of the organism's biology, as the following example illustrates:

▶ Look carefully at the form and anatomical relations of the claws on the paws of cats and dogs. Then observe cats and dogs walking around, and when attacking and killing prey (or tease some friendly specimens into 'preying' upon an inanimate object such as an old shoe). Can you explain the form and anatomical position of the claws of cats (and other felids such as lions, tigers and leopards) and of dogs (and other canids such as wolves and jackals) in terms of their role in the animals' natural behaviour?

Cat claws, particularly those of the forelimbs, have sharp points, and are normally retracted into folds of furred skin. Consequently, cats can walk silently even on hard, shiny surfaces. When climbing on trees (or furniture) and when attacking 'prey', the claws are extended, forming sharp, curved hooks. Cats (and other felids) stalk their prey silently, then pounce, grabbing it with the forepaws as well as, or instead of, biting it. Dogs cannot retract their claws, which therefore knock audibly against hard surfaces and are usually blunt from wear. Their smaller, narrower paws are better suited to fast running. Dogs are less proficient at climbing trees and they chase their prey, grasping it only with the jaws. They may stand on their food while tearing it apart with the jaws, but they do not claw the prey as felids do.

Notice that at least three distinct features of hunting are directly related to retractible claws: silent walking, climbing trees and grasping prey with the forepaws. With a bit of imagination, one can suggest other ways in which such paws might benefit felids: they may be less susceptible to injury; less wear means that less rapid growth, and hence fewer nutrients, are necessary to maintain them; humans prefer soft, smooth paws, thereby promoting domestication and conservation of cats (see Chapter 17), etc. While we can dismiss the last of these suggestions as failing to explain the condition of paws in man-eating tigers, it is less easy to identify *the* principal adaptation of retractable claws among the other suggestions.

2.2.1 Comparative and experimental studies of adaptation

As the examples in Section 2.2 show, adaptation is clearly a difficult concept that can easily lead to speculation and unverifiable statements. Darwin's theory of evolution by natural selection brought adaptation to the forefront of evolutionary biology. The mechanism of natural selection is described in much greater detail in Chapter 4, and is outlined only very briefly here. A central tenet of his theory is that some *individuals* are fitter—a term used by Darwin to mean 'better adapted', or 'having qualities that promote survival and successful reproduction'. Fitter individuals survive and breed more successfully in their particular environment than other similar organisms. If the capacity for such fitness is inherited by the offspring, its frequency in the population will increase in successive generations by the action of natural selection (see Chapter 4). Over a long period of time, such adaptive changes accumulate and lead to the major structural differences that we see in the huge variety of organisms. Darwin's theory was not widely accepted during the fifty years after his death in 1882, but since the 1930s his ideas, particularly those on natural selection, have been re-examined and investigated experimentally. The revised version of Darwin's theories, sometimes called **neo-Darwinism**, tied the concept of adaptation to that of reproductive success, and R. A. Fisher's mathematical exposition of natural selection enabled biologists to quantify fitness. Fisher and other population geneticists have defined 'fitness' to describe the outcome of selection. Today, the most widely used way of describing adaptations quantitatively is to measure their capacity to promote viability (i.e. growth, foraging ability, longevity, etc.) and successful reproduction.

An obvious way to measure experimentally the contribution of an adaptive structure to fitness is to modify or inactivate it, and study the consequences for the organism's survival and reproduction. However, such procedures are impractical for most anatomical and physiological features which are adaptive in many different ways: tying one's arm behind one's back is a crude way to demonstrate that hands are useful.

Brood size and nestling survivorship Studies of organisms' behaviour and life history suffer from fewer such limitations. Birds feature prominently in such studies, not only because they are conspicuous and attractive, but also because individuals can be easily and reliably marked: leg rings can be attached in seconds, do not harm the wearer and are almost indestructible. No similarly satisfactory method of marking insects, fish or mammals has yet been devised.

One of the most extensively studied topics is the factors that determine brood size in birds and the relationship between brood size and chick survival. The number of eggs laid and nurtured varies between species: golden eagles normally raise one chick every other year, while small passerines such as great tits may lay more than a dozen eggs per clutch, and sometimes several clutches each breeding season. In many of the species that have been investigated, the hen lays another clutch if the original eggs are destroyed or experimentally removed, particularly if it is early in the incubation period. The process does not continue indefinitely, but some birds can be manipulated into laying four or five times more eggs than they would normally produce for a single clutch.

▶ Under what natural conditions could this capacity be adaptive?

Although most birds guard the nest and may try bravely to repel intruders, eggs are often subject to heavy predation: foxes, snakes, scavenging birds such as gulls and crows, and many other predators eat birds' eggs if they can find them. The capacity to replace lost eggs would maximize the number of chicks hatched per clutch. If clutch size is not normally limited by the capacity to form eggs, it must be an adaptation to some aspect of raising the chicks.

If all young have an equal chance of reaching adulthood and breeding successfully, then individuals that produce more offspring will make a greater genetic contribution to the next generation. The Darwinian theory of evolution by natural selection therefore predicts that parents produce as many offspring as they can successfully raise. However, as birds' eggs are relatively very large and full of energy-rich substances such as yolk (which is mostly fat), and most newly hatched chicks do not survive unless fed by the parents, parenthood severely taxes the adults' foraging capacity and energy reserves. Differences in brood size both within and between species can be related to a great variety of factors, such as latitude, season, nest type, food supply and the life expectancy of the parents. Figure 2.2(a) and (b) shows seasonal changes in the average brood size of three common British birds. Tits (Figure 2.2(a)) have a much shorter breeding season, and early breeders produce larger clutches, while blackbirds (Figure 2.2(b)) lay eggs over four months, with mean clutch size gradually increasing by up to 35%. These differences can be related to availability of the food that the parents bring to the nest. Tits bring mostly small caterpillars and aphids that are themselves feeding

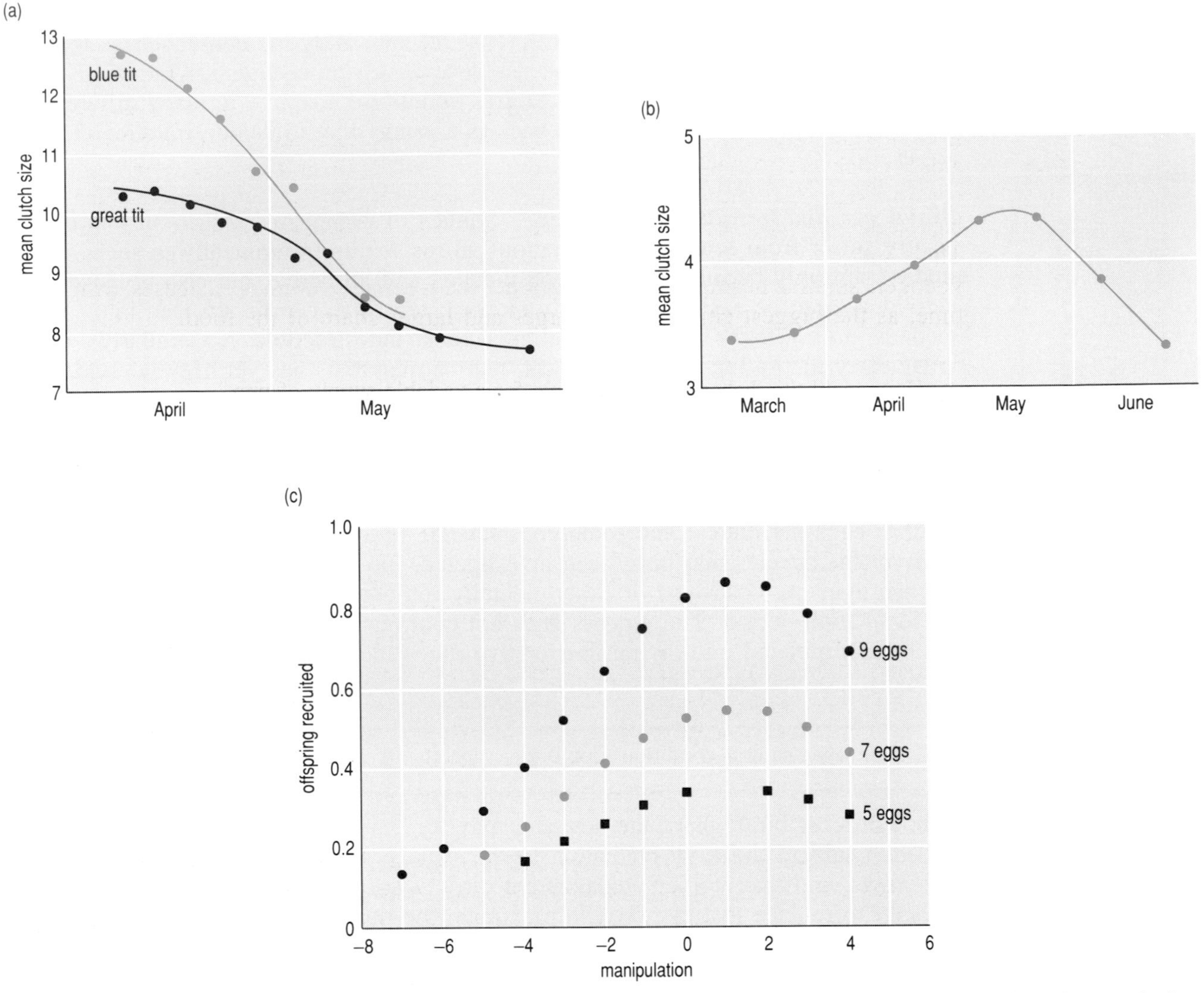

Figure 2.2 *The average number of eggs per clutch laid by: (a) two species of tit* Parus *spp.; (b) blackbird* Turdus merula *in southern Britain. (c) Effect of adding or removing eggs to wild great tit nests in which clutches of three different sizes had been laid, on the number of offspring that eventually reach adulthood. There are no data for adding one egg only.*

on new foliage. Such prey is more numerous but, probably more importantly, smaller and hence more suitable as food for nestlings earlier in the season. Blackbirds are larger than tits and feed their nestlings on worms, snails and large insects, all of which are accessible for several months, although they are more abundant at the end of the spring, enabling parents to raise larger broods later in the breeding season.

There are several other mechanisms by which fecundity is related to food supply. All owls are predators and many species, including those in Britain, feed mainly on small rodents such as voles and mice. The abundance of such prey varies greatly from year to year. Most birds, including tits and blackbirds (Figure 2.2), lay the whole clutch over a few days and incubation does not start until all the eggs are laid, so that all the chicks hatch more or less simultaneously. Owls lay

eggs at intervals over a period of several weeks. The total clutch size is very variable, with species such as the short-eared owl *Asio flammeus* producing up to 16 eggs per season, depending upon the food supply, the pair's experience as parents and probably other unknown factors. Each egg starts to develop as soon as it is laid and incubation takes about a month, but because laying is staggered, the eldest chick may have already hatched before the clutch is complete. All nestlings beg for food when hungry, but owls (and many other birds) preferentially feed the larger, more active ones. Bigger chicks may also take food from smaller ones. Therefore, the first chick to hatch is better nourished and grows faster than its younger, smaller nestmates. The difference between the siblings increases with time, as the biggest chick takes a larger and larger share of the food.

▶ How could this behaviour be adaptive for a variable supply of prey?

In years of abundance, the parents can find enough food for long enough to raise all the chicks to fledging. But when food is scarce, the parents feed only the larger, early-hatching chicks, which are able to grow to fledging. Those that hatch later usually die from starvation — unless a sudden change in the food supply enables the parents to find so much food that the larger chicks are satiated, leaving food for the smaller ones. The fact that in nearly all years a few, and in some years almost all, of the chicks starve to death on the nest may seem maladaptive. However, the whole reproductive strategy — variable clutch size, staggered laying, preferential feeding of larger chicks — maximizes the number of chicks that the parents can produce with an irregular, unpredictable food supply: parents are likely to raise at least one chick each season, and they can exploit food gluts very efficiently. In particularly good years, the parents may contribute as many as 14 fledglings to the next generation.

Similar observations have been made on at least 40 different species of birds. Many birds raise their young on insects, or, in the case of seabirds, small fish and invertebrates. The abundance and accessibility of these food sources depend greatly upon unpredictable factors such as weather and ocean currents. For example, swifts feed their young on small insects caught on the wing. Rain or strong winds keeps most small insects grounded, so foraging is slow. Swifts distribute food almost equally between the chicks, and when food is scarce all the nestlings in large clutches may be undernourished. In cool years, pairs that lay small clutches raise more offspring than those trying to raise larger broods, and are therefore 'fitter' in an evolutionary sense, but in warm years, fitness is maximized by laying more eggs. The variability of the environment means that no one strategy is always optimal, so no single habit becomes established in the population by natural selection.

The removal of one or two eggs from large clutches often goes unnoticed by the parents, and many species will incubate additional eggs and feed the hatchlings, particularly if both eggs and chicks resemble their own brood (see Section 2.3.2). Brood size can thus be experimentally increased or reduced (by adding or removing eggs), enabling biologists to find the 'optimum' clutch size for birds breeding under different environmental conditions. Great tits *Parus major* are common in gardens and woodland. They readily nest in or near human dwellings, often preferring artificial nest-boxes to natural holes in trees, and choosing the same nest site in successive seasons. These habits have enabled biologists to study natural variation in clutch size and the effects of manipulating clutch and brood size in great detail. A 20-year study of nearly a thousand nests in Wytham Wood near Oxford showed that clutch size varies from 5–13 eggs. Figure 2.2(c) shows the effect of experimentally adding or removing up to eight eggs from nests of three different initial clutch sizes on the numbers of chicks that reached adulthood and were observed to breed in the following year.

▶ Which of the parents of the clutches described in Figure 2.2(c) are maximizing the number of offspring produced?

All of them. In all three cases (and in the large number of others not shown on Figure 2.2(c)), broods close to the natural clutch size produced the greatest number of offspring, although, particularly for the smaller clutches, one chick more or less makes very little difference to the outcome. Mortality is always very high: as Figure 2.2(c) shows, only about one in ten of the eggs laid survives to become a mature adult. However, this experiment fails to explain the large differences in clutch size between breeding parents. Birds that laid the largest clutches raised more offspring than those that laid fewer eggs, regardless of how many eggs were removed or added. Also, observations on the breeding success of the same birds in successive years showed clearly that raising large broods had no effect on subsequent fecundity. The investigators concluded that some birds were more competent parents than others and that pairs laid clutches of a size appropriate to their ability to care for the young. How courting couples 'decide' what family size is best for them is unknown. This example illustrates the general point that adaptation, like selection, operates at the level of the individual (see Chapter 6). The optimum structure or habit for one organism is not necessarily identical to the optimum for another, even within the same species.

2.2.2 Interpretation of fossils

One of the most important applications of a thorough understanding of biological adaptation is the interpretation of fossils. This topic is covered in much greater detail in subsequent Chapters, so only the points relevant to the study of adaptation are discussed here. Most fossils are preserved skeletons or other hard parts, and provide a certain amount of information about the organism's structure. Trace fossils, such as footprints or burrows, provide information about organisms' behaviour, but it is often very difficult to establish by which species they were formed (see Chapters 10 and 12).

If there are living forms that closely resemble the fossil species or structure, the soft parts can be reconstructed from comparisons between **homologous structures**. Comparison of the internal anatomy of the paws of dogs (Figure 2.3(a)) and cats (Figure 2.3(b)) shows that in felids the claws are retracted upwards into a fold of skin over the last phalanges, the bones that bear the horny claws. The basic plan of the dog's paw shows that it is homologous with that of the felid, but the phalanges are much longer and stouter in cats than in dogs, and the joints between the last two phalanges have a larger bearing surface and can bend through a wider angle in felids than in canids.

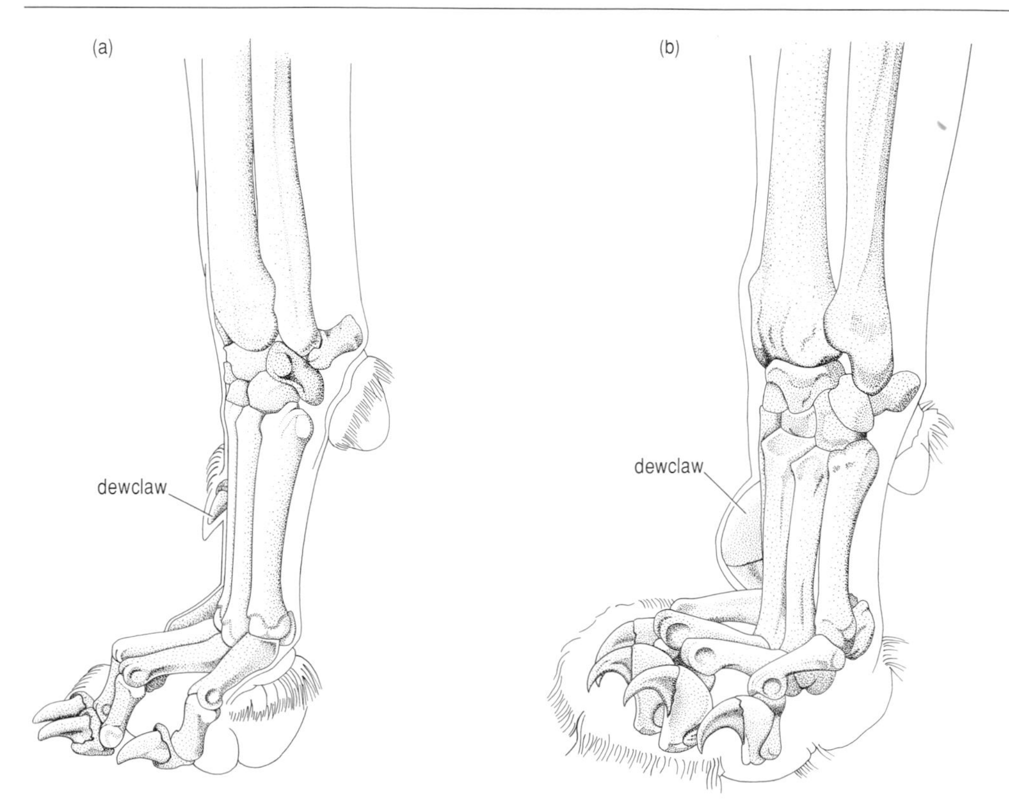

Figure 2.3 *The internal structure of the lower foreleg and paw of (a) canids (side view); and (b) felids. Note the vestigial dewclaws.*

▶ Why would properties of the bones and joints be more useful to palaeontologists than those of the soft tissues?

Features of the bones might be detectable in fossils but the form and dimensions of a fold of skin are unlikely to be preserved. Such features, and those of the skull and teeth, enable palaeontologists to assign fossils to the families Felidae or Canidae from the late Eocene (see Appendix 1) onwards. A relatively minor difference in the structure of the tip of the limb has proved to be central to major differences in how predation takes place and hence what kinds of animals are taken as prey. However, if we had been unable to make the detailed comparisons between living felids and canids (see Section 2.2), we might not have realized the adaptive significance of such anatomically minor, though functionally very important, differences in paw structure.

As well as comparison between homologous structures, studies of artefacts such as footprints and burrows, and analysis of the conditions of deposition of the sediments in which they occur (see Chapter 12), are often useful in interpreting adaptations in fossils. Functionally similar structures in obviously unrelated forms, e.g. fins on fish, whales and cuttlefish, are called **analogous** features: their similarities arise from common function, not common descent. Comparisons between such analogous features are often equally useful for establishing the origin and role of adaptations, particularly in extinct species that have no living descendants.

Figure 2.4 *Various ratite birds. (a) A reconstruction of the giant moa* Dinornis maximus *from New Zealand. Drawings of (b) African ostrich* Struthio camelus. *(c) Australian emu* Dromaius novaehollandiae. *(d) Kiwi* Apteryx owenii *(not to scale).*

Giant moas When 'key' adaptations such as paws and teeth are properly identified, they are powerful tools in reconstructing the structure and habits of extinct organisms from incomplete fossils. In 1839, Richard Owen, then Professor of Anatomy at the Royal College of Surgeons in London, published a reconstruction of a giant bird, which he named *Dinornis maximus*, based upon a single fragment of a limb bone found in New Zealand. The scientific community was very impressed when, three years later, an almost complete skeleton of *Dinornis* was unearthed, and was found to correspond remarkably precisely with Owen's reconstruction. This feat was possible because the internal structure of the bone was of a type found only in birds: there were cavities which in life contained air-filled sacs that assisted breathing as well as made the skeleton much lighter. The bone's size and relatively thick outer wall indicated that it came from a large, long-legged running bird. The only known birds with these characteristics are the ratites, a distinct group of flightless birds which includes the ostrich (Figure 2.4(b)), rhea, emu (Figure 2.4(c)), cassowary and kiwi (Figure 2.4(d)). Owen recognized the similarity and knew that ratites were among the few groups of terrestrial vertebrates to occur in New Zealand (kiwis) and Australia (emus and cassowaries), as well as in Africa (ostriches) and South America (rheas). Reconstructing *Dinornis* (Figure 2.4(a)) as a giant ratite was thus a plausible guess. *Dinornis* and several related flightless birds became extinct only a few hundred years ago, probably as a result of hunting, and were familiar to the

Maoris, the first humans to colonize New Zealand (see Chapter 17), who gave them the name moas.

However, when an entire group is extinct and there are no similar living forms, or where groups lack distinctive, unique features, reconstructing whole organisms from fossils becomes much more problematical. Dinosaurs were numerous and diverse throughout the Mesozoic, but they disappeared at the end of the Cretaceous, leaving no direct descendants except birds, which are anatomically and physiologically very different from their reptilian ancestors.

Dinosaurs Dinosaurs were a numerous and diverse group of reptiles, most of which were fairly large and which included the largest-known terrestrial animals. Although they were moderately abundant for more than 100 Ma, all of them have been extinct for the past 65 Ma, and some for much longer. Everything we know about them comes from bones, teeth, footprints and a few other fossil remains such as eggs. Our ideas about how they lived depend entirely upon making correct deductions from these fossils. As concepts and methods have advanced our ability to interpret adaptations, our image of dinosaur habits and habitats has changed.

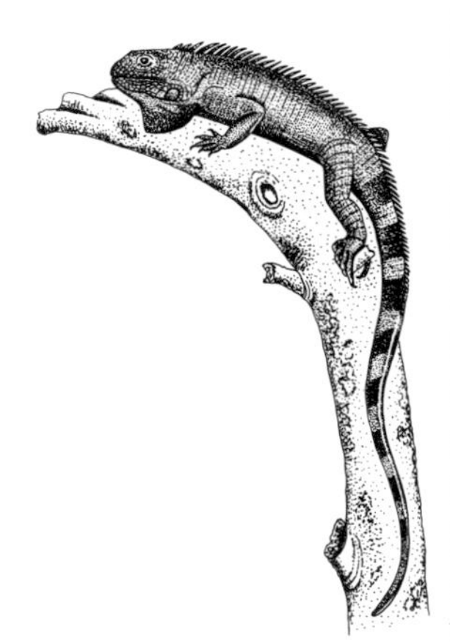

Iguanodon (Figure 2.5) was one of the first dinosaurs to be discovered (in Sussex, England, in the 1820s) and among the first to be identified correctly as a reptile, on the basis of the striking resemblance between its teeth and those of the large Central American plant-eating lizard *Iguana* (*Iguanodon* means 'iguana tooth'). It has since become one of the best-known dinosaur genera; fossilized fragments of literally hundreds of skeletons have been discovered in deposits of early Cretaceous age, including (in 1878) 39 adult specimens in a fissure filled with clay reached via a coal mine near Bernissart in Belgium. In 1851, the organizers of the Great Exhibition in London commissioned Richard Owen to advise the painter and sculptor Benjamin Waterhouse Hawkins on the building of life-sized models of several dinosaurs. The dominating influence in their reconstruction of *Iguanodon* (Figure 2.5(a)) was clearly the analogy with the living *Iguana*. The huge quadruped, almost 9 m long, has a sway back, drooping tail, scaly skin and even a prominent dewlap. Owen placed the dagger-like spine on the end of its nose, like a horn on a rhinoceros (see Section 2.3.3 and Figure 2.12). This concept was challenged in 1868 by Thomas Huxley, who proposed that *Iguanodon* was bipedal and, far from being heavy and sluggish, could run briskly. His main reasons were the resemblances, particularly of the pelvic girdle and hind limb, between these dinosaurs and birds. The reconstruction in Figure 2.5(b), based mainly upon the Belgian specimens and Huxley's interpretations, emphasizes the similarities between these dinosaurs and birds. *Iguanodon* has a relatively small head, a stiff, arched back and tail, massive hindquarters and the spine is now the thumb.

Several tracks of footprints have since been discovered, which confirm that most dinosaurs walked with the legs close together, directly underneath the body, more like a mammal than a modern lizard. Some were almost certainly made by dinosaurs walking bipedally, but those most likely to have been made by an *Iguanodon* show faint impressions of what may have been the forelimb, beside much deeper impressions of the hind foot. So perhaps there was more truth in Owen's reconstruction than Huxley and his successors have realized. As more

Figure 2.5 *Two reconstructions of* Iguanodon. *(a) According to Owen; (b) according to Huxley.*

information comes to light, and more sophisticated methods of reconstructing the size and arrangement of the muscles are developed, the interpretation of this and many other extinct species can be further refined by reference to features that resemble living forms.

Ammonites Very precise measurements of the physical properties of a structure often help to exclude or support ideas about its adaptive significance. Ammonites (see Plate 2.1) are cephalopod molluscs of the subclass Ammonoidea that first appeared at the beginning of the Jurassic and diversified into a great variety of forms and sizes before declining and becoming extinct at the end of the Cretaceous. The most frequently preserved component is the shell, which is nearly always coiled and typically disc-shaped. Most ammonite shells were 2 cm–0.5 m across, but a few species were more than 1 m in diameter. Many species were very abundant and some rock formations contain the remains of millions of specimens, together with those of marine reptiles such as ichthyosaurs and plesiosaurs, and fish, which became increasingly numerous and diverse during the Mesozoic. Their abundance and distinctive shape make ammonites among the most familiar fossils, but there are still many gaps in our understanding of how they lived, and why they disappeared after being so numerous and diverse for about 150 Ma.

Fortunately, there is a living genus of shelled cephalopod, the pearly *Nautilus* (Plate 14.1) which, although taxonomically only distantly related to ammonites, is similar in general form and in the shape of the chambered shell. Several species of *Nautilus* occur in the South Pacific and eastern Indian Ocean at depths of up to 650 m. Their soft parts are concentrated in the last chamber, and the head and tentacles protrude from the aperture. The soft tissues of fossil cephalopods were probably arranged in a similar way, and, as in *Nautilus*, the main function of the chambered shell was probably to adjust and maintain buoyancy. The animals probably spent much of the time stationary or drifting passively in the water, but comparison with *Nautilus* suggests they could also swim using jet propulsion created by rapid expulsion of water from the mantle cavity. Many ammonite shells had crinkles and lumps, called ribs and tubercles. The effect of shell size and shape and such external 'ornamentation' on the maximum swimming speed has been investigated experimentally. Fossilized ammonite shells (or life-sized models of them) were dragged through water at known speeds and the hydrodynamic drag that they generated was measured. Drag is similar to friction: it represents the energy of motion lost to the surrounding water. The pattern of flow of water over the shell could be seen by adding ink or a dye to the water.

At any one speed, laterally compressed shells (i.e. those in which the ratio of thickness to diameter was low) generate less drag than more globular or irregularly shaped shells. 'Ornaments' on the leading edge of the shell make the water flow become turbulent, and greatly increase the hydrodynamic drag, but those on the inside of each whorl, and near its centre, do not disrupt the smooth flow unless they are very large, or the shell is moving fast. In fact, elaborate 'sculpturing' and 'ornamentation' are more frequently found on the inner edges of the whorls, where they would have had least effect on swimming efficiency.

▶ Does this information explain the adaptive significance of shell ornamentation?

No. It only helps to explain the anatomical arrangement of shell ornamentation. Ornamentation must have another function, possibly strengthening the shell against implosion or providing protection from large predators. Such measurements can also be used to calculate the maximum swimming speeds of ammonites and a variety of living animals, including *Nautilus*, as shown on Figure 2.6.

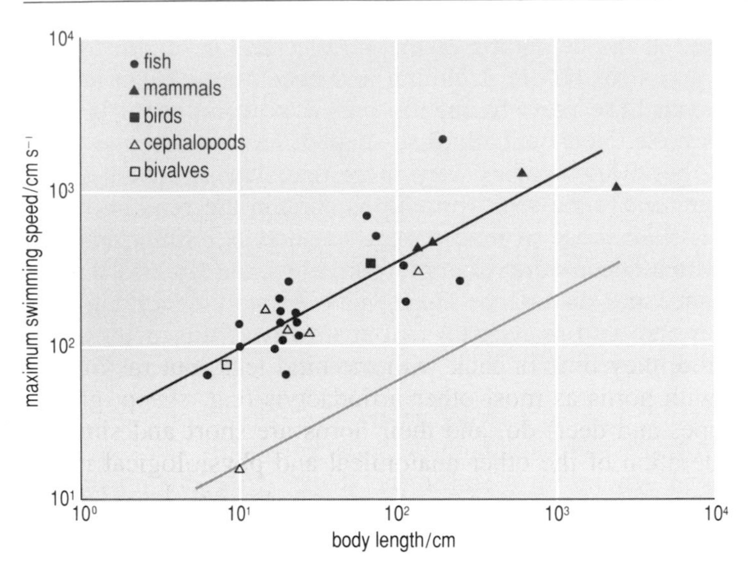

Figure 2.6 *The maximum swimming speed of various molluscs and vertebrates. The upper line is fitted to all the data except those from* Nautilus. *The lower line is the calculated speed of extinct ammonites. Note that both scales are logarithmic.*

▶ In terms of body lengths travelled per second, how does maximum swimming speed change with body size?

Figure 2.6 shows that animals of body length about 10 (10^1) cm swim at about 100 (10^2) cm sec^{-1}, which is equivalent to 10 body lengths per second. However, those about 10 m (10^3 cm) long travel only ten times faster, at about 1 body length per second.

Although larger animals have higher absolute swimming speeds than smaller ones, they swim more slowly in proportion to their body length. Maximum swimming speeds of all the modern cephalopods, such as squids and cuttlefish in which the shell is reduced and is internal, are comparable to those of fish and aquatic mammals and birds, but *Nautilus*, with its coiled, external shell, swims at only about one-tenth of the speed expected for its size. The green line in Figure 2.6 shows the calculated swimming speeds of extinct cephalopods; even the very largest probably could not swim faster than about 0.5 m sec^{-1}. Ammonites probably took only slow-moving prey, or obtained food by trapping rather than chasing other organisms.

More 'ornamentation' might strengthen the shell, but it would also increase drag, and so inevitably slow the maximum swimming speed. The basic body form of shelled cephalopods is apparently not adaptable to high swimming speeds in the same way as that of fish, ichthyosaurs and marine mammals. Although an impressive range of adaptations have evolved in a wide variety of organisms, basic body plans are not infinitely adaptable. The fish body plan has proved to be much more adaptable than that of ammonites, which declined as the bony fish diversified and increased in numbers (see Chapter 14).

2.2.3 INTEGRATING ADAPTATIONS

All external features are supported and maintained by internal physiological mechanisms. Large or complex structural adaptations may involve modifications of basic physiological processes such as blood circulation which in turn promote or exclude the evolution of other adaptive structures or behaviours. For example, the giraffe's long neck is clearly adaptive: giraffes can reach foliage on high branches that are inaccessible to shorter, non-climbing animals. With the head held high, they can probably also spot potential predators sooner. In spite of their size, giraffes are nervous animals that run away at the first sign of danger. When cornered, they bite, or kick with the hind legs, but they do not charge predators or fight with horns as most other artiodactyls (e.g. sheep, goats, bison, cattle, antelopes and deer) do, and their horns are short and simple (Figure 2.7). Careful consideration of the other anatomical and physiological modifications that the evolution of a very long neck entails shows that these habits are a direct consequence of the long neck. While feeding or walking, the giraffe's head is 3 m above its heart, so to achieve adequate blood circulation in the brain, the heart must generate a pressure of at least 222 mm of mercury (about 30 kilopascals). Biologists have measured the blood pressure of free-ranging giraffes by attaching pressure gauges linked with telemetric recorders to their necks. Values of up to 280 mm of mercury were recorded, more than twice the normal maximum for humans.

The animal must still be able to drink, and water is usually at ground level. Although the forelegs are normally splayed when drinking (Figure 2.7(a)), the head is up to 3 m below the level of the heart for several minutes. Without rapid, extensive adjustments to blood flow, the increased pressure caused by the head-down posture would be more than sufficient to rupture small blood vessels in the brain, causing internal bleeding.

Figure 2.7 *A giraffe* Giraffa camelopardalis*: (a) drinking; (b) galloping.*

▶ Vigorous exercise always involves increased blood pressure because the blood must perfuse small vessels in the muscles. What posture would you expect giraffes to adopt when galloping?

They must hold the head up. As Figure 2.7(b) shows, giraffes gallop with the head held high, not forward as in most other hoofed mammals. They are physiologically incapable of charging predators: kicking, biting or running away are the only forms of defence compatible with their exceptionally long neck and the vascular system that services it. Such physiological constraints determine many other habits and anatomical features:

- Giraffes normally sleep standing up and adopt a unique posture when lying down.
- Rival males 'fight' by swinging their necks against each other with the head held high, and they do not butt or charge each other as most other artiodactyls do.
- They are unable to swim, and their weight on long, slim legs makes them helpless on boggy ground.
- The calves must be tall enough at birth to be able to reach the mother's nipples on the ventral side of her thorax.
- The high blood pressure is associated with structural changes in the major arteries, and numerous modifications of the kidneys and muscles.

These examples are sufficient to illustrate the point that a major adaptation can be associated with a whole syndrome of behavioural and physiological features. It is not necessary to postulate special 'functions' for such features as the small horns and lack of defensive behaviour because they can be understood as a direct consequence of the major adaptation—a long neck.

Sauropod dinosaurs Sauropods (Figure 2.8) were widespread in the Jurassic and early Cretaceous and were some of the largest terrestrial animals that have ever lived. Their habits and habitats have been the subject of lively debate since their fossil skeletons were first discovered in Oxfordshire in the middle of the 19th century. For a long time, sauropods were believed to have lived partly submerged in shallow lakes and rivers, with their feet on the bottom and the head protruding to breathe air and browse on overhanging vegetation. Such a lifestyle is now regarded as unlikely, because when the animal breathed in, the lungs would have had to expand against a pressure of more than 6 m of water. Such a feat would be impossible without exceptionally strong ribs, and no sign of such strengthening has ever been observed in any fossil sauropod thorax. Modern palaeontologists now think that the long-necked sauropods lived on land (as did most other dinosaurs) and fed rather like giraffes (Figure 2.7). Sauropod necks were up to 8 m long, so the pressure required to pump blood to the brain when the head was held up must have been at least twice as great as that of modern giraffes. Calculations suggest maximum blood pressures of up to 717 mm of mercury for a long-necked sauropod such as *Brachiosaurus*. If this interpretation is correct, this dinosaur's heart would have had to generate a blood pressure in the neck arteries at least eight times as great as that generated by the most powerful hearts of living reptiles. They must also have had most of the other adaptations of the vasculature associated with high

Figure 2.8 *(a) The skull of* Camarasaurus, *a Jurassic sauropod.*
(b) Reconstruction of a Lower Cretaceous sauropod (Brachiosaurus).

blood pressure that are found in giraffes. This scenario seems hard to believe, but no more physiologically plausible model has yet been put forward.

The heads of large sauropods were very small relative to their huge bodies. They had few teeth, and the skull (Figure 2.8(a)) does not seem to be well suited to supporting massive chewing muscles. Proponents of the aquatic theory of sauropod habits used these facts to support the suggestion that they ate soft water plants. It is now thought that, instead of chewing as mammals do, sauropods macerated tough food in a muscular gizzard like that of herbivorous birds. The gizzards of ratites (see Figure 2.4), large parrots, swans, geese and other herbivorous birds contain grit or pebbles which act as grinding surfaces. Clumps of rounded pebbles unlike those in the surrounding sediment have been found near fossilized sauropod ribs. This example illustrates the importance of exploring the anatomical, physiological and behavioural implications of important adaptations, and of integrating data from living and extinct species, in assessing the validity of reconstructions of extinct species.

2.2.4 The problems of adaptation

Like most good ideas, the concept of adaptation has sometimes been used inappropriately, leading to confusion and erroneous conclusions. One common source of confusion that arises from extension of the concept of function (see Section 2.1.1) is that *all* features are functional: if it's there, it must be functional, and if functional, it must be adaptive. Such reasoning was ridiculed by Voltaire in the 18th century through his idealistic fictional hero Dr Pangloss, who believed that 'everything was for the best in this best of all possible worlds', quoting such examples as: 'Noses are to support spectacles...'. Such reasoning is not satisfactory because organisms, though often impressively well adapted to their natural environments, are not necessarily perfect. A species is just one stage in ongoing evolution, and retains significant traces of its past (see Section 2.3.2).

Vestigial structures Only four of the five toes of the forepaw of a cat or dog (Figure 2.3) actually support the animal's weight while standing. The other digit, sometimes called the 'dewclaw', seems to be functionless, at least for walking. Its anatomical relations to the other digits and its position on the inside of the leg show that it is homologous with the first digit of five-toed mammals such as hedgehogs. In humans, the first digit is the thumb on the forelimb and the big toe on the hindlimb. Internally, the dewclaws contain bones, muscles and tendons similar to those of the other toes, but these structures are greatly reduced, so the dewclaws are very weak.

In some breeds of dog, the first digit is absent on the forepaws, and greatly distorted on the hindpaws. Walking, running and jumping are not noticeably impaired if the dewclaws are removed surgically. Such structures that have no identifiable function are said to be **vestigial**; they are not demonstrably adaptive and develop only because they were functional in an ancestral form. Some familiar examples in humans are the posterior molar teeth ('wisdom teeth'), the appendix (part of the caecum) and the muscles attached to the lobe of the ear. It is characteristic of vestigial structures that they are inappropriately orientated or too small to be functional (see Figure 2.10(a)) and their degree of development is often very variable in otherwise normal individuals of the same species (see Figure 2.10(b)). For example, in some people, the wisdom teeth are similar to other molars, but in others, they are deformed, misaligned, remain unerupted or do not form at all. The absence of a vestigial structure, whether from incomplete development or by surgical removal, does not detectably impair the function of the organism as a whole. However, one must be wary of designating such structures as vestigial on the anatomy alone, without examining very carefully their function in the living organism. As the example of the aye-aye (Figure 2.1) shows, some apparently distorted or reduced structures prove to be adaptive in the living organism. Very accurate measurements over a long period would be required to show that a feature has *no* effect whatsoever on an organism's survival and reproductive success.

Non-adaptive and maladaptive characters Not all apparently non-adaptive characters can be dismissed as vestigial: some, such as the giraffe's short, weak horns (Section 2.2.3) arise from modifications of other aspects of the organism's physiology and behaviour. Problems arise when the non-adaptive or maladaptive characters are more obvious than the major adaptation with which they are associated. We can interpret the giraffe example quite convincingly because the control of blood pressure has been thoroughly studied (albeit mainly in humans) and suggestions can be tested by direct measurements on the vascular system. However, it is less easy to be precise about more complicated systems; neural processes in particular are not sufficiently well understood for us to do more than guess at the internal mechanisms underlying the many different constellations of structures and behaviours that have evolved.

Handedness and footedness From the age of about two years onwards, most humans prefer to use the right hand to perform intricate tasks such as writing and sewing or powerful movements such as operating saws and screwdrivers. Some people also use one foot or eye more than the other. The same actions performed with the other limb are slow, inefficient and weak, although there are no obvious differences between their muscles or skeleton. Some tasks, such as eating, are almost equally well performed with either hand, although many societies traditionally use mainly the right hand. There is no indication of such laterality in most vertebrates and the evidence for it in higher primates is at best controversial. It is therefore very difficult to see how this situation could be more adaptive than, and hence could have evolved from, the ancestral condition in which either hand or foot could be used for any task. Although humans rely more upon manual tools for food and protection than almost any other species, they are seriously incapacitated by a relatively minor injury to the dominant arm or hand.

Further information about the natural history of dexterity offers some clues to this mystery. About 10% of people — the proportion varies little between racial groups — are left-handed. Such people are completely normal in every other way, and, apart from the inconvenience of using equipment designed for the right-handed majority and the effects of long-standing superstitions, are equally well adapted in an evolutionary sense. Attempts to force children who are naturally left-handed to use the right hand for skilled actions often result in impediments of speech that are very difficult to correct, even with intensive training. This connection probably arises from the organization of the brain. Although we are less aware of it, speech involves temporal and spatial control of the airflow from the lungs, and movements of the vocal cords, tongue and lips that is at least as precise as that involved in writing or handling tools. The neural mechanisms that control these organs are perfected early in life and thereafter are not easily changed: most people who learn a foreign language as adults always have an 'accent', even though their grammar and vocabulary may be perfect; and learning to compensate for injuries to the mouth is very difficult. In early childhood, certain parts of the brain become committed to controlling skilled manual movements and adjoining areas to controlling the speech apparatus. Attempts to 'reprogramme' the brain to right-handed dominance apparently impair the neural control of speech.

Certain parrots and mynah birds are capable of imitating, often with remarkable accuracy, an unusually wide range of sounds including human speech, and have a life-long ability to learn new calls. Parrots perform skilled actions such as shelling nuts using the beak and one foot. The majority, including almost all the best mimics, are left-footed. As in humans, elaborate 'speech' seems to entail

restriction of manual (or pedal) dexterity to one side. It is impossible to argue that such properties are adaptive in either birds or humans. Indeed, there is no evidence that such parrots imitate sounds in the wild, so the whole phenomenon seems to be non-adaptive (see Chapter 17). Possible explanations for the evolution of such abilities include the suggestion that the tendency to imitate sounds is a distorted manifestation of an ability that evolved in relation to other habits—recognizing and responding to other species' calls, or remembering the location of a large number of different trees. Many puzzling human traits, such as the ability to perform complex arithmetical calculations, to remember long passages of music after hearing them only a few times, and to operate (indeed, take pleasure from operating) cars and computers, may have originated in a similar way.

Properties and functions: the central role of adaptation Different kinds of problems arise from the failure to recognize the central importance of adaptation and function in the interpretation of all natural systems. Biochemists, psychologists and livestock producers are more interested in elucidating and manipulating biological structures and processes than in finding out about their role in the organism's natural life. Such people tend to lose sight of the distinction between a 'property' of a molecule, cell or organism, and its natural function. For example, many biological molecules, such as insulin and lysozyme, readily crystallize under certain chemical conditions. Such properties have been very successfully exploited for the analysis of their structure using methods such as X-ray crystallography, but the molecules are biologically active only when in solution: the ability to crystallize is irrelevant to their natural function. Heparin prevents mammalian blood from clotting. It is easily extracted from liver, intestine and several other tissues, and has been widely used in medicine, surgery and research for half a century. The mechanism by which it impedes clotting has been studied in detail, but because we know so little about its role *in vivo*, such research tells us little about how the phenomenon could be adaptive. Heparin cannot be found in normal blood and, although it was first identified nearly 90 years ago, we still know almost nothing about its natural function.

Similar confusion between properties and functions has arisen in some ecological and evolutionary theories. One of the most fashionable is called the Gaia hypothesis (after Gaia, the ancient Greek goddess of the Earth). The whole terrestrial ecosystem, including the oceans, forests and atmosphere, is likened to an organism, with the component species adapted to function together as an integrated unit, and having the capacity for self-adjustment and restoration after perturbations are imposed on the system. However, the resemblance is only partial: there are processes that act to restore the *status quo* after some disturbance—when one species becomes more abundant, its predators also increase, thereby curtailing the numbers of the prey species. But the whole ecosystem does not have a common genetic make-up, nor is it adapted to a common function. Although energy and materials pass from one part of the ecosystem to another, only in a few special circumstances have mechanisms that facilitate such transfer evolved naturally. On the contrary, many features are adapted to preventing predation, herbivory and parasitism, and therefore to thwarting the transmission of material from one species to another (see Section 2.3.2); such organisms cannot be said to be 'adapted' to function together. Although the ecosystem has some superficial similarities to the physiology of whole organisms, its evolution and the relationships between its structures and their functions are quite different, so it is misleading to push the analogy too far.

The perfection of organisms People have long been impressed by the immense complexity and reliability of organisms. Although within the past few hundred years some machines and artificial materials have proved to be as good or better than natural structures and processes, there are many commonplace biological mechanisms, for example photosynthesis, that still cannot be simulated artificially. This situation tended to support the belief that organisms were perfect, both in the sense of being the most efficient and intricate system possible, and being ideally adapted to their environment. The idea that wild organisms were perfect seemed to contrast with the many ailments that afflict people, such as dental abnormalities, short-sightedness and backache.

Careful consideration of the many factors that affect organisms in their natural environment shows that absolute perfection is impossible. Figure 2.9 illustrates some ways in which different kinds of seabirds catch fish from the surface waters. Of course, not all of the species shown occur simultaneously or in the same place, but this example illustrates how prey could be exposed to many different predation strategies. Figure 2.9 shows only avian predators: in nearly all real-life situations, there would be several different kinds of predatory fishes that attack the smaller fish from below and from the side, and perhaps also marine mammals such as seals, dolphins or whales. So jumping out of the water, as 'flying' fish do, avoids swimming predators, but increases the risk from frigate birds and other airborne attackers. Although most prey species can escape from most of their predators for much of the time, no organism can avoid all of its predators all of the time. There are, of course many other constraints on fish behaviour, such as where to find food and lay eggs. Like most species, fish are exposed to a constantly changing constellation of risks and are not perfectly adapted to avoiding any of them. Their structure and habits are a compromise of adaptations to the many different risks to which they are exposed and the activities that they perform.

All multicellular organisms (and many unicellular ones) die, although death cannot be adaptive in that it obviously does not promote the individual's own reproductive success (except sometimes in cases of kin selection, see Chapter 6). Ageing and death seem to be inherent to the way that organisms, particularly higher organisms, work. Natural selection cannot select against degenerative changes that eventually lead to death if they do not become conspicuous until after most of the organism's reproduction is complete. Most elderly mammals suffer from imperfectly healed injuries and broken or septic teeth, and many have signs of arthritis and cardiovascular diseases such as strokes. However, most wild animals die from predation or starvation before these ailments are far advanced. Modern humans have longer lifespans than any other known mammal, so more of the population suffers from diseases of old age for longer before dying.

While some characters and combinations of characters prove to be adaptive, many others are deleterious, and would either fail to develop or could produce an inviable or maladapted offspring. In many species, 'infant mortality' is normally very high (see Figure 2.2(c)), so fetuses or juveniles that die from genetic disorders go unnoticed. Organisms that are obviously suffering from genetic or

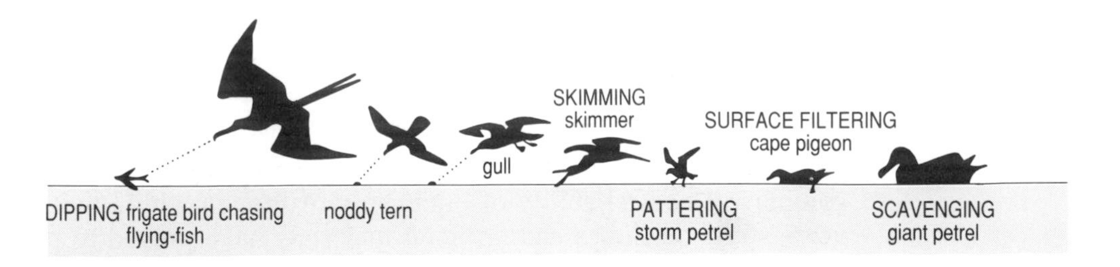

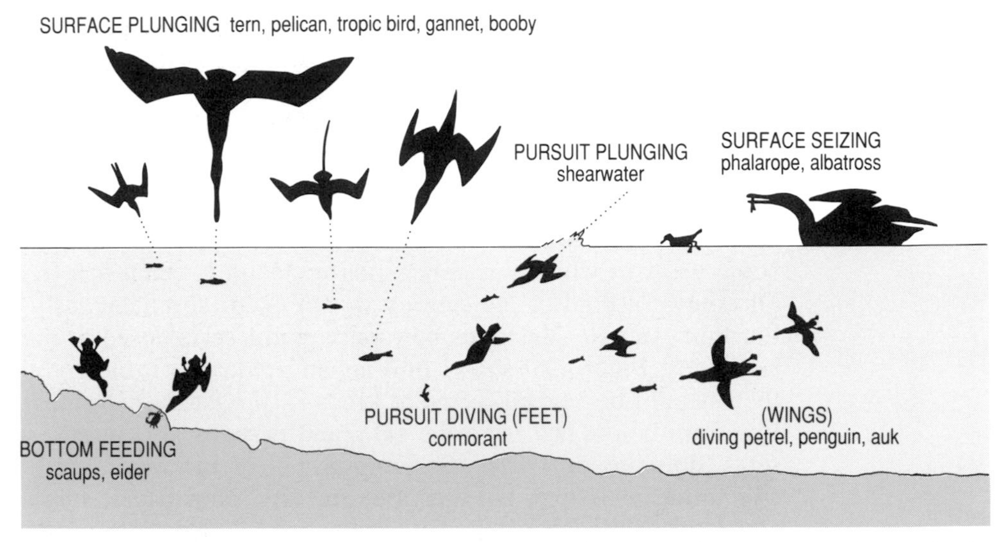

Figure 2.9 *Some ways in which seabirds catch fish and other food swimming or floating in the surface waters.*

developmental abnormalities are therefore not very common. But they do occur even in completely wild populations: Figure 2.10(a) shows the upper jaw of two adult female reindeer shot on Svalbard, an archipelago in the Arctic Ocean (see also Figures 2.17 and 2.18). The specimen on the right has three normal premolar and three normal molar teeth in each jaw, but the third molar of the specimen on the left is at right angles to the other teeth. The homologous tooth on the other jaw of the same specimen is also small and misshapen compared to those of the other specimen. Yet both animals had raised a calf and were apparently living a normal life until collected. As in other Cervidae (deer) and Bovidae (antelopes, cattle, and sheep), the canine teeth are vestigial in reindeer; although visible as small, blunt pegs in the dried skulls (Figure 2.10(a)), they do not erupt through the gums and so never come in contact with the food.

The skull in Figure 2.10(b) is from a wild badger that was killed in a road accident in Milton Keynes, England. The premolar tooth in the left upper jaw just behind the stout, pointed canine is vestigial (see Section 2.2.4)—its small size and simple structure indicate that it could not have been functional. Both the tooth and its socket are absent from the right jaw; they probably never developed. In badgers (and other Mustelidae), the last molar teeth are broad and strong, and play a major role in chewing. In this specimen, the last molar in the left upper jaw is normal but that of the right jaw has almost completely failed to erupt, and most of it remains encased in a bony capsule that protrudes into the eye socket. Nonetheless,

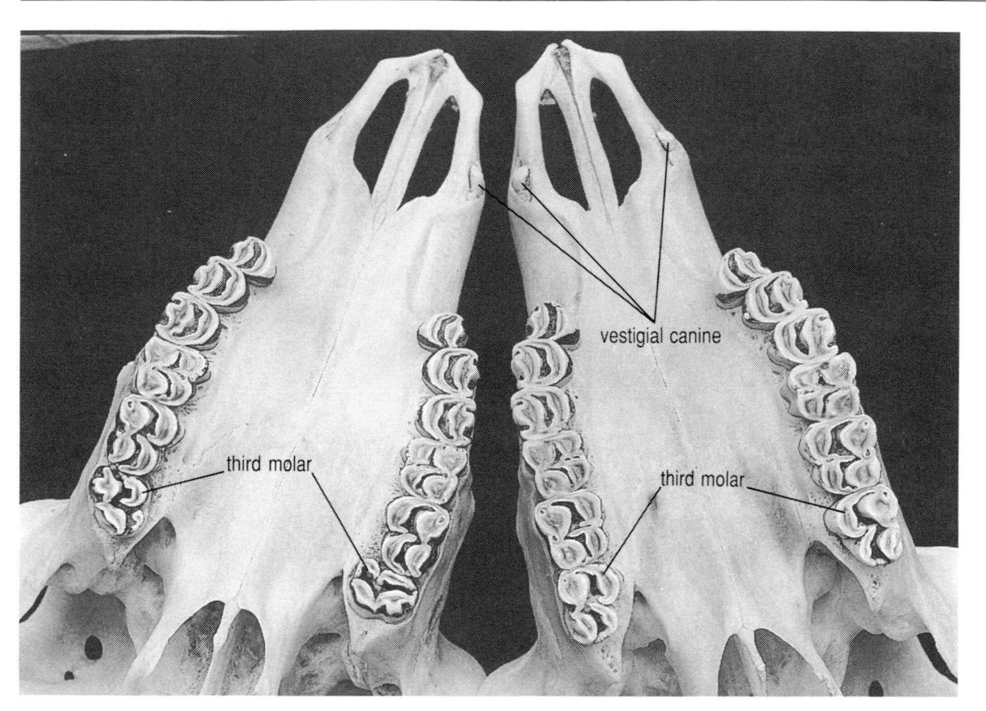

(a)

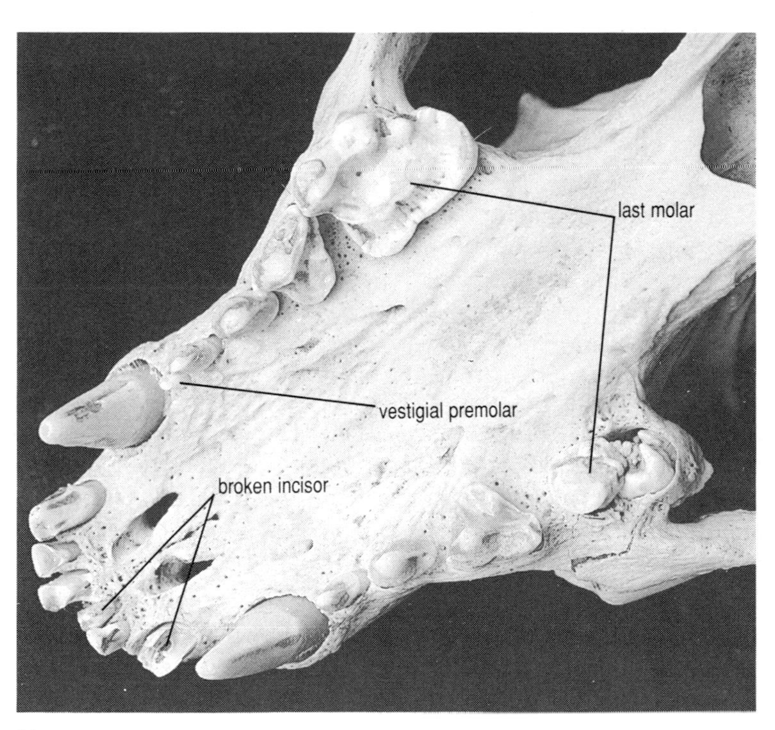

(b)

Figure 2.10 *Ventral view of the palate and upper jaw of (a) two adult Svalbard reindeer* Rangifer tarandus platyrhynchus *and (b) an adult badger* Meles meles.

the animal was moderately fat, and the wear on the other teeth suggests that it was at least three years old. Clearly, it had prospered in the wild in spite of its malformation.

While some people (for instance the small minority who develop inherited diseases such as muscular dystrophy or cystic fibrosis in childhood) clearly have less adaptive sets of genes than others, we all have some deleterious genes that may manifest themselves later in life as susceptibility to disorders such as high blood pressure, heart failure, varicose veins, diabetes and senility. Although such characters are clearly maladaptive, they are not selected against unless they impede successful reproduction.

SUMMARY OF SECTIONS 2.1 AND 2.2

Adaptation is the relationship of the structure and habits of organisms to their natural functions. The concept developed gradually, following improved knowledge of the diversity of species and more detailed study of living organisms in their natural environment, and is important for evolutionary theory because adaptive features promote fitness. The contribution of such features to an organism's fitness can be assessed by investigating how they maximize the numbers of its offspring that reach adulthood and breed themselves. The number of eggs laid per season varies greatly within and between species and is easily measured in wild birds. Comparative and experimental studies have elucidated some of the many factors that determine the relationship between clutch size and the number of offspring that reach maturity. Both fluctuating environmental factors such as weather and food supply, and endogenous factors such as the birds' competence as parents, determine optimal clutch size, which therefore varies between and within populations.

Similar structures often have similar functions, so the habits of extinct species can be deduced from comparisons with living species that have similar adaptations. When adaptations are specific and unique, an extinct species can be successfully reconstructed from a small fragment of the skeleton. When extinct species have features that are not found in living forms, or have a mixture of adaptations, several different reconstructions are possible. Complex adaptations may entail a wide range of modifications of the physiological and biochemical mechanisms that support them, which may influence other features and habits of the organisms. Adaptations must be distinguished from properties; the latter may be striking, and useful for investigating structure or mechanisms, but they cannot be described as adaptive unless they are shown to be functional in the organism's natural habitat.

2.3 ADAPTATION AND THE COURSE OF EVOLUTION

Because adaptation contributes to survival and reproduction, it determines which individuals prosper and breed. However, to contribute to the course of evolution of a species, adaptations must have a genetic basis upon which selection can operate. Identifying the genetic basis of an adaptive feature is not always straightforward, as the following examples show.

2.3.1 The genetic basis of adaptation

Body form and physiological processes of nearly all organisms can change to some extent depending upon the environment; for example, leaf colour and area, the timing and abundance of flowers and fruit and many other features of plants depend upon the conditions under which they are grown (availability of light and water, soil quality, the presence of other plants, etc.). Much, but not all, such plasticity results in adaptive changes and, in many species, such a capacity is itself adaptive. Thus, people who are frequently exposed to bright sunlight develop darker skin (that absorbs harmful radiation, thereby protecting sensitive underlying tissues), and those who often go barefoot develop thicker, tougher soles of the feet. Where these habits are common to all members of the population, these adaptations are universal, but thorough studies are necessary before we can conclude that the features have 'evolved' as a permanent character of the population.

The external ear (pinna) of many mammals, including rabbits and hares (Figure 2.11(a) and (b)), elephants and to a lesser extent humans, has numerous small blood vessels that dilate at high temperature, bringing the blood near to the skin, where it loses heat to the cooler air. Human babies' ears may become bright red when they are too hot or suffering from fever. Rabbits and hares often hold their ears erect, where they are exposed to breezes, and elephants flap their ears when overheated; these activities, combined with the morphological features, are important mechanisms of thermoregulation. The ears of rabbits and hares that are native to hot, dry climates, such as the jack rabbit *Lepus californicus* of south-western USA and northern Mexico, are often relatively very large (Figure 2.11(c)).

Figure 2.11 *(a) The European rabbit* Oryctolagus cuniculus*; note the relatively short ears. (b) The arctic hare* Lepus arcticus *lives in the tundra zone of Alaska, northern Canada and Greenland, and its ears are furry and relatively short. In northern areas, the fur is white throughout the year. More southerly races become grey for a few months in summer. (c) The jack rabbit* Lepus californicus *lives in the semi-desert areas of south-western USA and northern Mexico. Its fur is greyish-brown throughout the year, and its ears are relatively long and erect.*

Similar adaptations have appeared in rabbits from cool climates after they colonize warmer areas, but careful studies are necessary to find out whether the genetic bases of such features are such that they represent permanent evolutionary changes.

In 1859, 24 wild-caught European rabbits *Oryctolagus cuniculus* were shipped from England to Australia and released there. Although the Australian climate is hotter and drier than that of northern Europe, the rabbits multiplied and spread rapidly, colonizing a wide variety of different habitats, and by the middle of the 20th century they were a major pest in many areas. A century-and-a-quarter after rabbits were introduced, biologists compared the body size and body shape of rabbits living in different regions of Australia, and designed experiments to find out how much of the differences arose during the development of the organisms in response to special environmental conditions and how much could be attributed to genetic differences between the populations living in areas with contrasting climates. To obtain the data in Table 2.1, they examined animals trapped in the hot arid deserts of South Australia, the 'Mediterranean' climate of New South Wales and the cooler, wetter, sub-alpine region of the Snowy Mountains.

Table 2.1 *Some average morphometric measurements of wild rabbits caught in three regions of Australia.*

	All males	All females	Desert	New South Wales	Snowy Mountains
No. of specimens	168	155	79	159	85
Body mass/g	951	983	995	957	959
Ear length/mm	69.4	69.1	72.2	69.0	67.2
Ear width/mm	47.5	47.1	48.5	47.3	46.3
Foot length/mm	82.1	82.0	81.6	82.2	82.3

The female rabbits were on average slightly heavier (about 3.4%) than the males, but otherwise there were no sex differences in body dimensions. When data from both sexes were combined, rabbits in deserts of South Australia were 4% heavier than those in the cooler, wetter climates. However, the body shape was also different: the ears of the arid-climate rabbits were 7% longer and 5% wider than those from animals living in the Snowy Mountains, but the length of the foot was almost identical in all three populations.

To investigate the genetic basis of these differences, the biologists caught some specimens from each area and bred them in captivity for several generations at different environmental temperatures. Table 2.2 shows the mean total body mass and the dimensions of the ears of the offspring of rabbits from various locations, bred in captivity under cool (15 °C) or warm (25 °C) conditions.

Table 2.2 *Some average morphometric measurements of the offspring of rabbits from three regions of Australia raised under two different temperatures.*

Temperature/ °C	Measurement	Desert	New South Wales	Snowy Mountains
15	No. of specimens	6	11	4
25		16	16	6
15	Body mass/g	1191	1155	1072
25		1094	1060	985
15	Ear length/mm	75.2	73.9	76.2
25		77.9	76.6	79.0
15	Ear width/mm	50.1	49.8	49.8
25		52.8	52.5	52.5

The captive rabbits were on average 13% heavier than those living in the wild, and this fact must be taken into account when comparing the morphometric measurements in Table 2.2 with those in Table 2.1. At both experimental temperatures, as in the wild, rabbits from the desert were heavier than those from the other two populations. The progeny of all three groups had longer and wider ears when raised at 25 °C, but those of the population from the warmer climates did not consistently become proportionately larger under the experimental conditions than those from the Snowy Mountains. The ears were on average 3.7% longer in all rabbbits raised at 25 °C compared to those raised at 15 °C while the ears of those whose ancestors came from the cool Snowy Mountains were a maximum of 1.4% larger than those descended from desert stock.

▶ What can you conclude about the genetic mechanisms that determine ear size in rabbits?

Adult ear size is determined mainly by the temperature at which they are raised, and not by ancestry. There is no evidence for the evolution of genetic differences that determine ear size in populations that have been exposed to different climates for many generations. These experiments suggest that rabbits in South Australia normally have larger ears because they are all born and raised at higher temperatures. The introduced animals had the ability to produce appropriate responses to environmental conditions. Such conclusions in no way diminish the importance of long ears as an adaptation to dissipating heat, nor do they preclude the possibility that in other populations of this species, or in other species, ear size is determined mainly or entirely by genetic factors. In this case, and in many others, adaptability to climate and habitat is itself adaptive for the long-term evolution of the species. Whether variation of a feature is inherited or environmentally determined does make a difference to the effectiveness of natural selection in bringing about permanent evolutionary change in the population. Jack rabbits (Figure 2.11(c)) of the western desert of America and arctic hares

(Figure 2.11(b)) are species of the genus *Lepus*. The arctic species always has much shorter ears whatever the nurture conditions, so ear length is probably determined genetically in this genus. However, as this example shows, suitable natural variation and thorough, often prolonged, study are necessary to determine whether particular features are determined genetically.

2.3.2 ADAPTATION AND COEVOLUTION

In the previous Sections, adaptations to general features of the environment such as climate, types of terrain, strategies of food collection, etc. were described. This Section is about specific, reciprocal adaptations between a few species, that are believed to have evolved from prolonged, intricate interaction between the species. The evolutionary processes that give rise to such very specific forms of adaptations are often called **coevolution**. A huge variety of often very elaborate adaptations, including those of parasites (parasitic bacteria, tapeworms, roundworms, etc.) and other symbionts to their hosts, herbivores to their host plants (particularly herbivorous insects which only feed on one or a few species of plant), and flowers to their animal pollinators, are believed to have arisen in this way (Chapter 8).

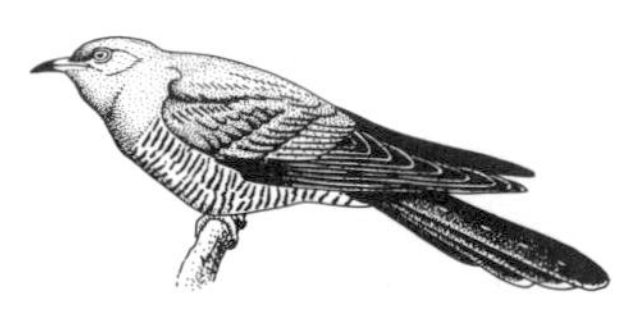

Some birds, including certain members of the family Cuculidae (cuckoos) in Europe, cowbirds in America and certain weaver birds in Africa, exploit the fact that many birds will incubate foreign eggs and feed chicks other than their own (see Section 2.2.1). Instead of expending time and energy building a nest and rearing chicks themselves, such 'brood parasites' lay their eggs in the nests of other species and the foster parents raise their chicks. The rightful occupants of such nests are always disadvantaged by the presence of parasitic chicks: the interlopers take often a large share of the food that the adults bring, and may injure or eject the other nestlings from the nest. Any habits that thwart brood parasites would increase the number and viability of the parents' own offspring.

The European cuckoo *Cuculus canorus* normally lays its eggs in the nests of birds which build nests that are accessible to the female cuckoo and which feed their young on insects and other invertebrates. In Britain in the late 1980s, the species most frequently parasitized were reed warblers, meadow pipits and pied wagtails. Only a small proportion of nests were parasitized: 7% of nests were parasitized in the most severely affected species (reed warblers) in Britain, and less than 4% of nests of most other host species contained cuckoo eggs. The colour and pattern of most birds' eggs are similar within a species but different races of cuckoos, called gentes (singular: gens), produce different-coloured eggs that match those of the host species very closely (Plate 2.2), although the adults of the parasitic and parasitized species are very different. Nick Davies and his colleagues at Cambridge, England, painted eggs to resemble those of cuckoos, and placed them in over 700 nests of a wide range of species breeding in the same areas. Up to 70% of the brooding birds of frequently parasitized species proved able to recognize the artificial cuckoo-like eggs and broke them or ejected them from the nest.

▶ Why are the abilities to recognize foreign eggs, and to respond aggressively to them, adaptive?

Ejecting foreign eggs counteracts parasitism, thereby promoting the hosts' own reproductive success. However, if the birds cannot distinguish accurately between parasitic eggs and their own, they risk destroying their own offspring. These observations suggest that this behaviour evolved as an adaptive response to nest parasitism. However, similar proportions of brooding birds of many of the species (such as chaffinch, blackbird and song thrush) which feed their young similar diets but which are very rarely parasitized by cuckoos, also recognized and ejected or attacked the introduced eggs.

▶ How could you account for this behaviour, which must very rarely occur in the wild?

The response could have become established as an inherited behaviour pattern if the species was formerly used by cuckoos (or other nest parasites), or it has interbred with species that are thus parasitized. This example shows that such adaptive traits evolve even when selection for them is not very strong; more than a dozen different species respond to foreign eggs even though, in nearly all of them, fewer than 1 in 20 breeding pairs is adversely affected by cuckoos at all. Historical data suggest that the proportion of nests parasitized changes rapidly over time. During the past 50 years, more than 73 000 nests of the six most frequently parasitized species have been examined, in many cases by amateur ornithologists, and the incidence of parasitism reported to the British Trust for Ornithology. Analysis of these data shows that the percentage of reed warbler nests parasitized in Britain has more than doubled during this period, while dunnocks, robins and pied wagtails have become less frequently affected. Less reliable reports from the 18th and 19th centuries suggest that 200 years ago, reed warblers were rarely, if ever, parasitized by cuckoos. One explanation is that host species that have long been parasitized have evolved such effective discrimination against exotic eggs that it becomes more adaptive for cuckoos to lay in the nests of alternative hosts. Newly parasitized species, of which reed warblers may be an example in some places, may not yet have evolved a fully effective adaptive strategy that thwarts nest parasites.

Information from other species and populations is also consistent with this conclusion. Cuckoos are absent from Iceland (although it is uncertain whether they formerly bred there, and have recently become extinct, or have never been indigenous) but many of the songbirds that they parasitize breed there. Most of such species rejected experimentally introduced eggs, though less consistently and less vigorously than conspecifics in Britain.

A cuckoo's egg usually hatches a day or two before those of the host. The foster parents start feeding the parasitic chick at once, so it quickly grows bigger and stronger than its nestmates.

▶ Why is being bigger than its nestmates adaptive for the cuckoo chick?

Parent birds give more food to the largest nestling (see Section 2.2.1). The preference for the largest nestling is almost as strong in host species as in those that are never parasitized. The large size and appropriate colouring of the foreign chick stimulate the host parents to feed it at the expense of their own offspring; if food is scarce, the indigenous chicks all die, but the foster parents continue to feed the huge cuckoo chick.

▶ Why does such behaviour persist, when it is obviously maladaptive for the host parents?

It would be adaptive for parents of non-parasitized broods, and normally only a small minority of nests are parasitized. Unless the hosts evolve a means of discriminating between their own and foreign chicks, selection for maximizing the numbers of young fledged (see Section 2.2.1) is stronger than that for avoiding nest parasites.

The nest parasites have also evolved numerous adaptations that promote their effectiveness as parasites. Within hours of hatching, the cuckoo chick tries to lift onto its back, and then to eject, any other object in the nest, whether an egg or another hatchling.

▶ Would you expect nest parasites to lay more or fewer eggs per breeding season than similar non-parasitic species?

Parasites would be expected to lay more eggs: such species do not expend energy or time on incubating their eggs or feeding their young, so they have more physiological resources available for egg production (see Section 2.2.1). Cuckoos can lay up to 25 eggs in a season. Hostile responses from the host parents, as well as predation, accidents and starvation, contribute to higher mortality of both eggs and chicks in parasitic species. The interactions of selective factors are discussed further in Chapter 5.

The evolution of reciprocal adaptations that promote or thwart parasitism (or predation) has been called an 'evolutionary arms race'. The parasite exploits the weak points of the host's defences, such as preferential feeding of the largest chick which is adaptive for non-parasitized birds. As the host's discrimination between its own and exotic eggs becomes more efficient, parasite mimicry becomes more exact. It is important to emphasize that adaptation of host and parasite is never complete or perfect (see Section 2.2.4). The coevolved relationship will eventually fail if either the host or the parasite becomes too efficient. If all the cuckoo chicks are destroyed, the parasite becomes extinct; if the host fails to breed sufficiently to maintain its numbers, both species become extinct (unless the parasite turns to another host).

As already mentioned, different gentes of cuckoos lay eggs of different colours and patterns (Plate 2.2). Egg colour is genetically determined. Successful parasitism depends upon the host's failure to recognize the foreign egg, so it would clearly be maladaptive for a female to produce eggs that were a mixture of different mimetic colours, or to lay her eggs in nests of the 'wrong' species. One way in which females can minimize the chances of such mistakes is illustrated by observations on the paradise whydah *Vidua paradisea*, a finch-like bird that parasitizes more than a dozen different species of finches in tropical Africa. The males learn much of their mating calls from their foster fathers, and the females are selectively attracted to males which sing like the birds that reared them, as well as preferentially depositing their eggs in the nests of such species. These traits together increase the chances that the parasite's eggs resemble closely those of the host, and hence maximize their parents' reproductive success.

▶ Could there be any circumstances under which it would be adaptive for a female to choose a mate or a host nest different from that in which she was raised?

Parasitization of 'new' host species must start by such 'mistakes'. As the cuckoo example shows, one parasitic species uses several different host species, and the proportions of nests parasitized fluctuates significantly over what, on the evolutionary timescale, is a very short period. In the long run, a female that switches to a new host species may leave more offspring than one which follows the strategy to which she is already adapted.

2.3.3 Adaptation and species differences

Most of the differences between the species described in the previous Section could be explained as adaptive. Taxonomists and palaeontologists erect classifications on the basis of consistent, easily recognizable and quantifiable features that may be irrelevant as adaptations to the organisms' natural habitat. Some invertebrate and plant genera include scores, and in the cases of some insects and nematodes hundreds, of species, that are distinguished on the basis of characters, such as the number of bristles on the legs or the colour or pattern of the feathers, that seem to be functionally trivial. On thorough investigation, many such apparently spurious features do indeed turn out to play an essential role in the organisms' natural habits, but it is unwise to push this line of reasoning too far.

Nonetheless, it is important to remember that not all interspecific differences can be convincingly interpreted as adaptations to particular environments. For example, rhinoceroses charge potential predators, rival males and any other intruders, including biologists and busloads of tourists, and the horn(s), when combined with the head-down posture, adds to the effectiveness of this behaviour. Horns are obviously adaptive, but it is not clear that two horns with the anterior one larger, as in the white rhino *Ceratotherium simum* (Figure 2.12(a)) and the black rhino *Diceros bicornis* (Figure 2.12(b)) are most appropriate for the African savannah, but a single horn is best adapted to India, as in *Rhinoceros unicornis* (Figure 2.12(c)), and two horns with the posterior one larger, to Sumatra, as in *Dicerorhinus sumatrensis* (Figure 2.12(d)). The four genera adopt different postures and tactics when fighting each other and with predators, so, as in the case of giraffes (Section 2.2.3), these differences are integrated with other adaptations of the species and are irrelevant to the geographical distribution. Similarly, while camel humps may be interpreted as an adaptation that combines large fat reserves with efficient dissipation of heat from the muscles to the skin, it is not clear that the single large hump of *Camelus dromadarius* is best in Arabia, and the two smaller humps of *C. bactrianus* are best in Mongolia and central Asia. The evolutionary origin of such differences is not well understood; possible mechanisms include the suggestion that they arise from genetic or developmental links between the factors that determine horn or hump shape and (as yet unidentified) adaptive traits.

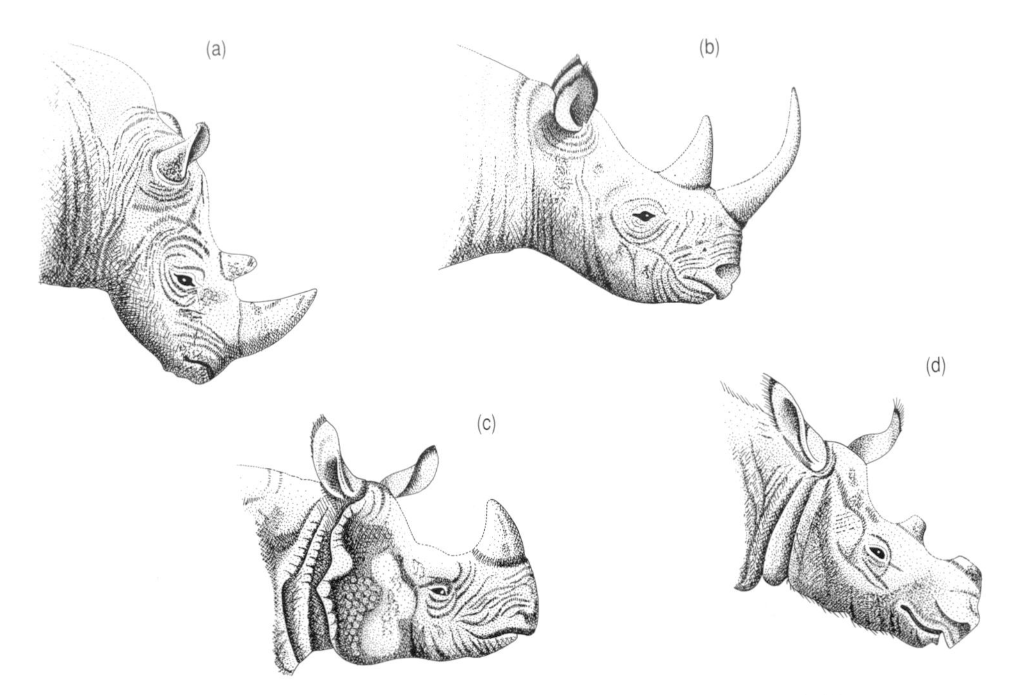

Figure 2.12 *Typical shapes of horns on adult specimens of living species of rhinoceros. (a) African white rhino* Ceratotherium simum. *(b) African black rhino* Diceros bicornis. *(c) Great Indian rhino* Rhinoceros unicornis. *(d) Sumatran dwarf rhino* Dicerorhinus sumatrensis.

Cat-like dogs and dog-like cats When the structure and habits of species in larger taxonomic categories such as families and orders are compared, different combinations of similar adaptations are often found. The families Felidae and Canidae provide some examples. As pointed out in Section 2.2, the structure of their paws and their capacity for sustained exercise adapts cats and dogs to different predatory habits and climbing abilities. You can probably think of many more differences between cats and dogs: cats have relatively short, wide snouts with sharply pointed teeth, and very acute hearing. Felidae have no appetite for fruit or other sweet foods, most species hunt alone and, except when breeding, they are mainly silent and solitary. Dogs have long snouts with several broad, relatively flat back teeth. Their senses of smell and hearing are excellent. They relish sweet and starchy foods, have an elaborate social life involving visual, vocal and olfactory communication, and hunt in packs, cooperating to overcome prey that may be very much larger than themselves.

Foxes are definitely canids—the structure of the skull indicates that they evolved from dog-like ancestors; they can gallop far and fast and they eat fruit and carrion as well as killing live prey. However, in other ways, foxes resemble typical felids. Their paws are soft and semi-retractile, they stalk as well as chase prey and normally hunt alone. Conversely, cheetahs are dog-like felids: their paws are firm and narrow with weakly retractible claws, and their legs and backs are long and strong. Cheetahs do not climb trees but they can run very fast and chase their prey over much longer distances than typical felids can, and they sometimes hunt in pairs or threes. The dewclaws on the forelegs (Section 2.2.4, Figure 2.3) are relatively stout and strong, and seem to have a specific role in the cheetahs' specialized method of catching prey. Lions are dog-like in a different way: although they resemble typical felids in terms of anatomical structure and diet, they live and hunt in large, socially organized groups like wolves and hunting dogs. Both adults and cubs are much more vocal than most other felids, and dominant male lions roar loudly and often, particularly in the breeding season.

These examples illustrate the point that paws, teeth, sensory capacities and social behaviour are readily adaptable and highly adaptive features that can evolve readily and equip species for different habits and habitats. Different combinations of anatomical features and behaviours adapt organisms to their own particular environment. In the examples quoted, anatomical features, as well as genetic properties such as chromosome number and DNA composition, clearly show that foxes are canids and lions and cheetahs are felids. The appearance of such similar adaptations in organisms of different ancestry is called **convergent evolution**. There are many examples of convergence in the fossil record, and between extinct and living forms (see Chapters 11 and 14). Sometimes, convergent adaptations to a specialized diet or habits is so complete that the species' ancestry is obscured. For example, the giant panda *Ailuropoda melanoleuca* is a slow-moving carnivore with unique adaptations of paws and teeth that enable it to live almost entirely on bamboo. Its fossil history is poorly known, and for a long time it was unclear whether the monospecific genus is affiliated to bears (family Ursidae) or raccoons (family Procyonidae): their habit of giving birth to a few relatively very small young suggests the former, but the structure of their teeth and gut, and their occurrence in western China, suggest the latter (see Chapter 11).

2.3.4 The origin of evolutionary novelties

Many complex and physiologically well-integrated structures seem to appear suddenly in fossil lineages. Biologists and palaeontologists have devoted much effort to formulating and verifying the steps by which such **evolutionary novelties** could have evolved, and most such issues are still not fully resolved. The immensely complicated eyes of vertebrates and cephalopods, electric organs in certain tropical fish, mimetic resemblances between distantly related species of insects, and between insects and other arthropods, are among the examples that Darwin and many biologists after him have considered.

The simplest explanation for evolutionary novelties is that a character arose from a new mutation, or a new combination of genes, which produced a structure that was sufficiently adaptive for the organism to be favoured by natural selection. Many such cases have been described in micro-organisms (e.g. the appearance and spread of drug-resistant forms) and some in plants (e.g. pollution-tolerant herbs and grasses), but fewer clear-cut examples are known among vertebrates, mainly because their developmental mechanisms are so complicated that changes to one or a few genes tend to have multiple effects, many of them deleterious.

Albinism, in which the skin (and skin derivatives such as hair, feathers or scales) is white or very pale, is typically due to the lack of the enzyme that catalyses the synthesis of the black pigment, melanin, from tyrosine (a common dietary amino acid), and hence probably arises from the absence or malfunctioning of a single gene. Albino individuals have been recorded in wild populations of many species, including fish, amphibians, snakes, a few birds and many different mammals, including cats, dogs, horses, koala bears and tigers. Albino people occur, albeit rarely, in many parts of the world, including among darker-skinned, dark-haired races such as Chinese and Negroes. Typical albino people and animals have white or pink skin, white hair and pink eyes (because the colour of the blood is not disguised by melanin and other pigments in the iris of the eye). Such characters would normally render the animal vulnerable to sunburn (most albino people in tropical countries cannot do outdoor work such as farming) and conspicuous to predators and prey. But they would not be disadvantageous to those that live in permanent darkness; indeed, in such an environment, albinism may be an advantage because the animal would be able to divert the energy and metabolites normally used in the synthesis of melanin to other physiological processes. Some species that live permanently in burrows or caves, including mole-rats and certain termites, and many cave-dwelling cockroaches, crayfish, fish and amphibians, have most of the features of typical albinos. In such species, pigmentation is no longer adaptive and has become vestigial.

Polar bears evolved from brown bears within the past 0.1 Ma (see Section 2.4.2). The two species are so similar that they occasionally hybridize in captivity (producing khaki-coloured cubs). It is easy to jump to the explanation that polar bears are descended from albino brown bears, but more thorough examination of the facts makes this scenario seem less realistic. Although their fur is creamy white, polar bears have dark eyes and their skin is black, darker than that of most other mammals (Plate 2.3). Clearly they do not lack the biochemical ability to synthesize melanin. There is plenty of melanin in the skin, but it is not incorporated into the fur; it is the site of deposition of pigments, not their synthesis, that has changed. The course of evolution of skin and fur colour in polar bears may have little to do with typical albinism.

Another kind of explanation is that structures that evolved as adaptations to a certain function subsequently became adaptive to other, quite different functions, or they appeared as by-products of other changes. It is often possible to identify with hindsight features in ancestral species that fortuitously adapt them to the environment and habits of a descendent species. Such features are said to be 'preadapted' to the functional role to which they later become adapted. The term causes heated controversy, particularly if 'preadaptations' are mistakenly equated with 'adaptations' that demonstrably promote evolutionary fitness. The problem disappears if the circumstances that give rise to preadaptations and 'functional' adaptations are clearly understood. For example, the ratites (Figure 2.4) are flightless birds, but several features, particularly the distinctive air sacs in their limb bones, suggest that they evolved from flying ancestors (see Section 2.2.2). In the emus (Figure 2.4(c)) and cassowaries, the whole anterior limb is reduced to a single small bone and a few stumpy feathers, suggesting that in these forms the wings are indeed completely vestigial. No trace of the forelimb skeleton has ever been found in moas. In ostriches (Figure 2.4(b)), however, the wings are relatively large, particularly in the adult males; the wings are fully feathered and all the major bones, muscles and nerves are present. Observation of the behaviour of the birds in the wild explains this somewhat surprising situation. During courtship, the male flaps his wings, exposing the shiny black feathers, and both parents extend the wings to shade the eggs and young from the hot African sun. Although vestigial in relation to flight, the wings have become adapted to essential roles in breeding and parental behaviour.

▶ Would you be able to identify this function in a fossil skeleton of an extinct species?

The long, stout legs and greatly reduced sternum would indicate that the birds were flightless, but as the wings were large they were unlikely to be a 'vestigial' structure, although this possibility cannot be excluded. Comparison between living and extinct species might suggest breeding and parental behaviour as an alternative function, but it could not be proved from the skeleton alone. There is more about preadaptation and vestigial structures in Chapter 14.

Many of the most familiar and conspicuous novelties are species-specific structures such as antlers, horns, plumage patterns in birds, skin colour and body shape in fish and insects, etc. Many (but by no means all) of these features evolved under sexual selection, in which the concept of adaptation has a special meaning as will be explained more fully in Chapter 7.

SUMMARY OF SECTION 2.3

Adaptations may be determined genetically, or may arise as adaptive responses to the conditions under which the organisms are living. Only the former are important for the course of evolution in the long term. Where two or a small number of species interact extensively over a prolonged period, specific, reciprocal adaptations may cocvolvc. Some characters, including many of those used for taxonomic identification, are not necessarily adaptive. The evolutionary history of very few complex adaptations is known in detail. Novel structures may arise as a new character from a change in the function of a structure that evolved in connection with some other adaptation.

2.4 THE EVOLUTION OF ADAPTATIONS

It is rarely possible to document in any detail the evolution of adaptations to changes in the environment. Information about the climate and the fauna and flora are also normally incomplete and uncertain. An exception is the evolutionary changes brought about by the Quaternary Ice Age. There have been Ice Ages, involving prolonged cold periods interspersed with warmer interludes, during the Ordovician and in the Carboniferous and Permian, but because they were so long ago, much of the fossil record of these events has been obliterated. On the geological time-scale, the Quaternary Ice Age occurred very recently, so the fossil record is relatively complete. Pollen and seeds are preserved in the anoxic environment of peat bogs, and pits containing tar, brine or asphalt are rich sources of Pleistocene remains. A wide variety of animals and plants, including whole mammoths, rhinoceroses and horses, have been found preserved in the permanently frozen soil of Siberia and Alaska.

2.4.1 The Quaternary Ice Age

During most of the Mesozoic and Tertiary, very little of the world's landmass was cold, and consequently few terrestrial organisms were adapted to arctic conditions. About a million years ago, the Quaternary Ice Age began: the world became significantly colder, and large areas of the Northern Hemisphere, including much of Europe, Asia and North America, became tundra or were covered with permanent ice. There were several successive glacial periods, each lasting about 100 000 years, that were separated by interglacial periods in which the climate was more temperate and much of the ice melted. At least four, and in some places up to eight, such periods can be distinguished. During some of the interglacial periods, the climate in much of the Northern Hemisphere was warmer than it is now. Although the maximum drop in mean annual temperature was only about 5 °C compared to the pre-glacial period, the Ice Age had diverse and profound effects on the fauna and flora.

During glaciations, Europe, Asia and North America became colder and in many places drier. The glaciers and ice-sheets that formed over much of the northern continents were up to 1 km thick in places, turning uplands into highlands. The water thus 'locked up' as ice resulted in a fall in mean sea-level, so that shallow seas, including the English Channel and large areas of the North Sea, dried up. Land bridges formed between continents (such as the Bering Strait between eastern Asia and western North America) and between islands (e.g. most of the British Isles were joined to each other and to the rest of Europe) during glacial periods and were flooded again during the interglacials. Many places, particularly continental areas, were very dry as well as very cold, and the soil was permanently frozen to a depth of several metres, forming permafrost. Primary plant productivity was therefore much lower and in many places probably more seasonal than in the more equable climate that prevailed over most of the world in the Tertiary. There is more about the effects of the glaciations on the evolution of plants in Chapter 13. Many species that could not tolerate the colder climate became extinct, but others adapted to the new conditions in a wide variety of different ways.

The most recent glaciation reached a peak about 20 000 years ago and we are now living in an interglacial period. Except in continental areas like central Canada and Siberia, cold tundra is restricted to high latitudes, where there is continuous darkness in midwinter, and continuous light in midsummer. At lower latitudes, daylength changes much less with the seasons, and plant productivity is limited more by soil conditions, temperature and snow cover than by light. Nonetheless, the conditions under which arctic organisms live are sufficiently similar to those that prevailed over much of the Northern Hemisphere during the glaciations to provide useful indications of how they adapted to the cold climate.

2.4.2 Anatomical and physiological adaptations to cold climate

Alpine shrews, elk, reindeer, musk-ox, tundra voles, lemmings, mammoths, woolly rhinos, arctic foxes and several kinds of bear are among the lineages that adapted to the cooler climate and changed ecological conditions. Some of the species that

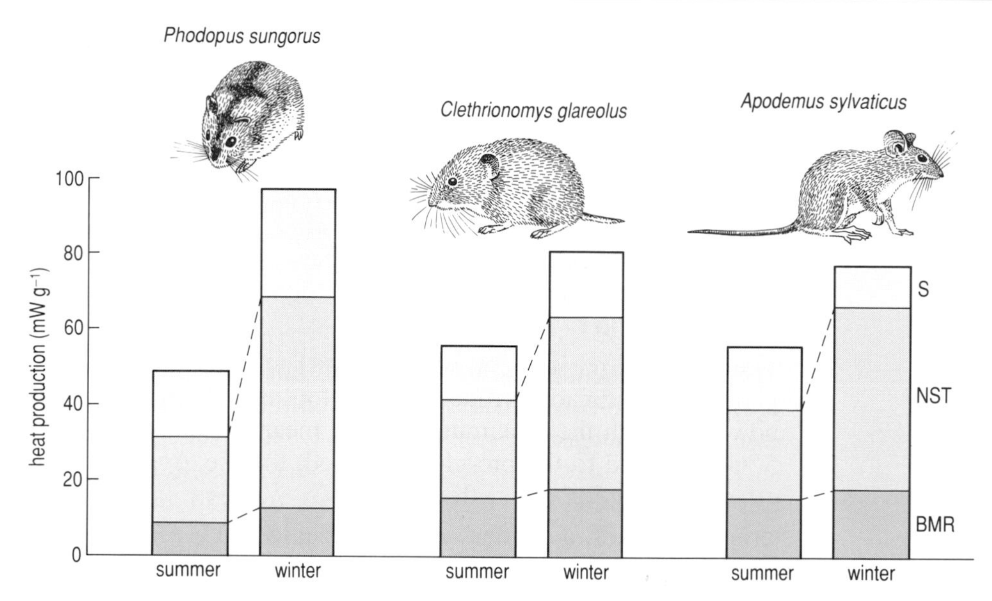

Figure 2.13 *The maximum heat production (mW per gram body mass) arising from the basal metabolic rate (BMR), non-shivering thermogenesis (NST) and shivering (S) in* Phodopus sungorus, Clethrionomys glareolus *and* Apodemus sylvaticus *caught in summer and winter in their natural habitats.*

evolved during the Ice Age are still extant, at least in polar regions, and the fossil record of others is unusually complete. These sources of information can be combined to produce a fairly complete picture of the evolution of the wide range of physiological mechanisms by which vertebrates adapted to the conditions created by the glaciations. Some plants, invertebrates and micro-organisms also occur in the Arctic, and a few species, notably mosquitoes, and lichens are very abundant there. However, their fossil record is more fragmentary and experimental studies of the physiological mechanisms involved are few, so we know much less about how and when they adapted to the extreme cold.

The body temperature of most homeothermic* (warm-blooded) mammals and birds is normally remarkably constant at 37–41 °C. There are several very different ways in which mammals and birds can adapt to cold conditions: more efficient **insulation** of the body surface, physiological **acclimatization** to the cold or increased production of body heat which can involve shivering the muscles, or **non-shivering thermogenesis (NST)**, which is heat production by other biochemical mechanisms, or both. Increased heat production has been most thoroughly studied in small rodents. Figure 2.13 shows some data from three such species of body mass 15–100 g. The bank vole *Clethrionomys glareolus* and the woodmouse *Apodemus sylvaticus* are common throughout Britain and south-west Europe, where winters are mild, but the dwarf hamster *Phodopus sungorus* is native to Siberia and central Asia, where the air temperature is regularly below –30 °C for several months of the year. The investigators caught wild specimens of each species in mid-summer and mid-winter, put them in cold environments for several hours, and measured their basal metabolic rate (BMR), their capacity for NST and their ability to shiver.

* Spelt 'homoiothermic' in some texts.

▶ What are the differences in the mechanism of heat production between summer-caught and winter-caught specimens of these species?

The capacity for NST increases in all three species in winter, but the increase is particularly large in *P. sungorus*, which also shows a larger increase in the capacity for shivering. BMR and shivering increase only slightly in all species studied. Summer-caught *P. sungorus* produce less heat when kept in the cold than the two West European species at the same season, mainly because of lower BMR and capacity for NST.

The physiological basis of NST is still not fully understood, but a tissue called brown adipose tissue (BAT) plays a central role. Proliferation of such tissue is essential to increased capacity for NST. However, growth of an energy-rich tissue such as BAT takes several days, and weather conditions, particularly in Siberia, can change within hours. If the capacity for NST did not increase until the temperature fell, a sudden cold snap could catch *Phodopus* unprepared. The experimental data in Figure 2.14 suggest how *Phodopus* adapts to this feature of the climate. Exposing hamsters to cold temperature causes an increase in NST as expected, but when they are kept on short days (8 h light, 16 h darkness), the capacity for NST also doubles, even if the temperature is constant at 23 °C. When experiencing both cold temperatures and short days, the hamsters' capacity for heat production is even greater. The daylength is a much more reliable cue to seasons than temperature and in arctic mammals and birds many physiological processes, particularly those related to breeding and food selection, are closely linked to daylength. This example shows how adaptation to this extreme climate means adaptation to many different features of the environment, some of which may not be immediately obvious.

Hibernation A completely different way of adapting to cold conditions is **torpor**. This active physiological process is triggered by environmental cues such as falling temperatures, in which the body cools to a few degrees above freezing

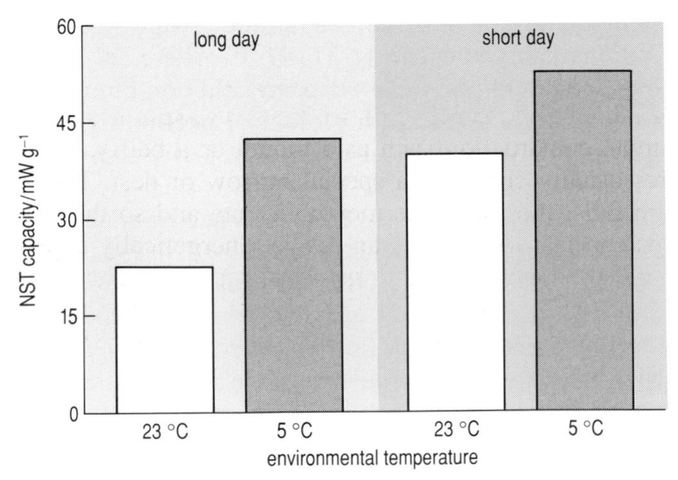

Figure 2.14 *The effect of daylength and environmental temperature on the capacity for non-shivering thermogenesis (NST) in wild-caught* Phodopus sungorus. *Long day = 16 h of light; short day = 8 h of light.*

and the heartbeat, respiration rate, and kidney and brain functions are slowed down. Torpor that is prolonged and occurs during winter is called **hibernation**. Hibernators do not feed (they would be too sluggish to find food and digestion is very slow at low temperatures) but, because energy expenditure is greatly reduced during hibernation, they are able to sustain themselves on body energy stores, usually fat sequestered in adipose tissue. The ability to become torpid has evolved many times, and some form of hibernation is found in several lineages of poikilotherms (cold-blooded animals), such as amphibians, including both anurans (frogs) and urodeles (newts), various reptiles, including many chelonians (tortoises and terrapins), and a few squamates (snakes and lizards), and in several orders of homeothermic mammals. Several species of birds, such as hummingbirds, also become torpid during cold nights, but none hibernates for long periods.

Among mammals, hibernation is most widespread among Insectivora (e.g. shrews, hedgehogs), Rodentia (e.g. dormice, hamsters, ground squirrels) and Chiroptera (bats). Although hibernation is more common among small species, its distribution correlates more with diet than with body size, latitude or environmental temperature. Thus *Phodopus* remains active throughout the winter in Siberia, and indeed increases its capacity to generate body heat (see Figures 2.13 and 2.14), while many larger rodents living in less severe climates hibernate for many months (e.g. marmot). The arctic fox *Alopex lagopus* is no larger than a domestic cat, but it remains active throughout the winter in the high Arctic, often feeding off the scraps of seals killed by polar bears.

▶ What do these facts suggest about the adaptive significance of hibernation in mammals?

They suggest that hibernation is an adaptation to seasonal scarcity of food, rather than to cold. When food is available throughout the year, as in the cases of arctic foxes and *Phodopus*, other physiological and anatomical adaptations enable them to withstand the cold while remaining active. Animals that hibernate are active during the summer and autumn and fatten rapidly if sufficient food is available. The increased appetite and accumulation of fat are an integral part of the hibernation strategy and are controlled by brain mechanisms that respond to environmental cues such as shortening daylength and cooler air temperatures. They seek out a suitable place in which to hibernate: bats often enter a large cave, or a similar human construction such as a tunnel or a belfry, while rodents and insectivores usually construct a special burrow or nest. The lower the body temperature falls, the lower the metabolic rate and so the longer the hibernators' fat reserves can last. Arousal is an active, energetically expensive process, involving the production of large quantities of heat. The exact timing of arousal depends upon both external cues (temperature, daylength, etc.) and internal factors such as the animal's fat stores. The amount of fat stored before hibernation, its average duration and the temperature at which arousal occurs vary greatly between local populations of a species.

Large animals in deep hibernation would need a great deal of energy to raise their body temperature to the normal level, and true hibernation is unknown in mammals larger than about 10 kg adult body mass. Larger mammals may enter sheltered places in the winter and sleep deeply for weeks, even months, without eating or drinking, but as their body temperature and other indicators of metabolic activity such as heart rate are only slightly below normal, they are not true hibernators. The European cave bear *Ursus spelaeus* was massive: adult males probably weighed over 1 000 kg. Literally thousands of fossil skeletons have been found in caves in Europe, northern Asia and, later in the Pleistocene, in North America. Many of them can be assigned to discrete age classes (cubs, one-year-olds, two-year-olds, etc.) from the state of development and maturation of the dentition and skeleton. Many bears seem to have died in the caves, mostly at about the same time of year. These facts, and comparisons with living bears, suggest something about their habits and physiology. Their broad, flat molar teeth suggest that they were mainly, if not entirely, herbivorous, so presumably they fed mainly during the summer and retreated to the caves when plant food became scarce. No living bears hibernate in the same way as modern bats, hedgehogs and dormice do, so the cave bears probably did not hibernate, but they may have spent weeks, perhaps months, sleeping in the caves. Those that failed to lay down sufficient fat during the preceding summer probably starved to death before the end of the winter.

Heat conservation Arctic mammals and birds generally have thicker fur or plumage than temperate zone species of similar size and habits, and areas of skin, such as the feet and the face that are often naked in tropical forms, are often feathered or furred in cold-adapted species. The fur of large arctic mammals such as reindeer is particularly effective in strong winds: the outer hairs are long and fairly stiff, and are not easily blown out of place. The data in Figure 2.15 were obtained by wrapping animal skins over hot-plates and comparing the rates of loss of heat in a wind tunnel. Pelts from winter-caught reindeer and arctic wolves are the best insulators, more than twice as efficient as manufactured cloth. There is also a big difference between the properties of furs from reindeer caught in the summer and winter, showing that moulting is an important part of adaptation to arctic climates.

The efficiency of insulation depends upon the size and shape of the body as well as upon the properties of the fur or feathers. Most arctic mammals are of stocky build with proportionately smaller ears, legs, tails and other extremities that have a relatively large surface area through which heat may be lost (Plates 2.3 and 2.4). Arctic birds have feathered feet and faces and dense, fluffy plumage.

Adipose tissue There are large seasonal changes in both weather and daylength in the Arctic, so most kinds of foods are available for only part of the year. Many arctic animals accumulate energy reserves in storage tissues such as adipose tissue when food is plentiful, and so can fast when food is scarce. In mammals, adipose tissue occurs as many discrete depots that are widely distributed throughout the body, including inside the abdomen and thorax, between the muscles and under the skin. It is widely believed that subcutaneous adipose tissue is an adaptation to cold climates, and indeed most elementary biology texts state that *the* function of subcutaneous adipose tissue in humans is to insulate the warm muscles from the cold skin. Modern humans are often obese and subcutaneous adipose tissue may be thick and extensive. When resting in a cold environment, there is often a thermal gradient across the adipose tissue. However, these facts alone are not sufficient to justify the conclusion that subcutaneous adipose tissue is *adapted* to be an insulator. To confirm this conclusion, it is necessary to demonstrate that, in the

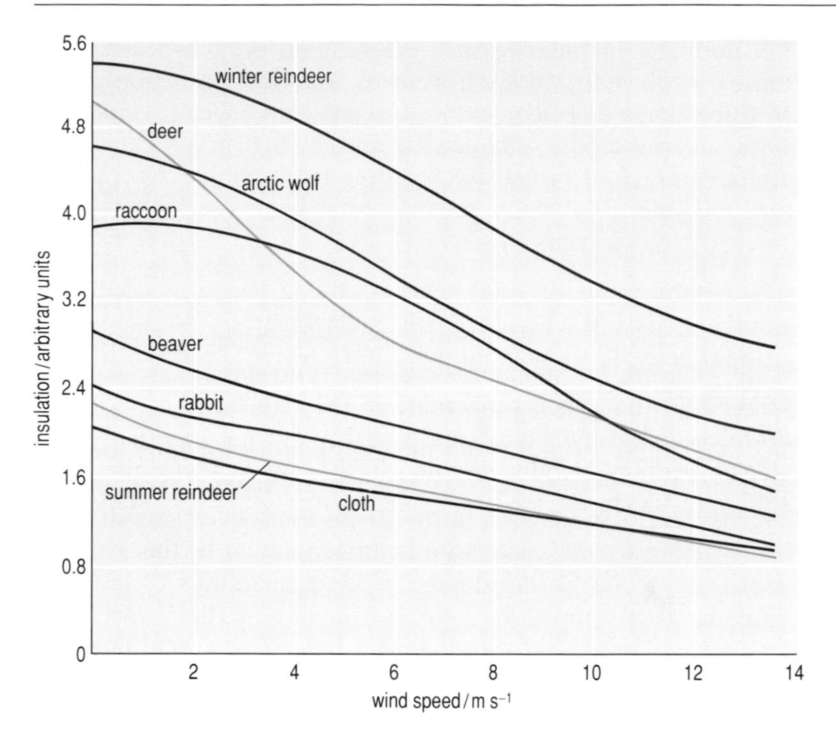

Figure 2.15 *Thermal insulation of the pelts of various mammals and pile fabric.*

evolution of a cold-adapted species from a warm-adapted ancestor, there was a change in the distribution of adipose tissue towards proportionately more 'insulating fat' in the superficial sites. Such a comparison is impossible in the evolution of *Homo sapiens*, because the anatomy of adipose tissue in humans is thought to have been modified by recent changes in diet, and by other factors such as sexual selection, and all other living primate species occur only in tropical or warm temperate areas.

The order Carnivora provides the opportunity for such a comparison. Carnivores are among the most widespread and diverse groups of mammals: the order includes tropical, temperate and arctic species that occur naturally on all continents except Australia and Antarctica. They occur in a wide range of sizes, from weasels (adult body mass about 0.1 kg) to lions and bears (adult body mass more than 500 kg). Modern bears are a homogeneous group: they are all large (adult body mass 50 kg or more) and most are herbivores or omnivores. All but one living species are classified in the genus *Ursus*. The family probably evolved from dog-like ancestors during the Oligocene and the first species to appear as fossils were quite small, about the size of a modern badger, and were probably predators or scavengers. Miocene ursids were bigger, but many of the largest species were more similar to dogs than to modern bears. During the Pliocene, animals similar to the modern genus *Ursus* became larger and more abundant, although almost all were smaller than any of the living species. By the middle of the Pleistocene, bears were moderately abundant, and some species, such as the cave bear, were large. From the late Pleistocene until a few millennia ago, the brown bear was one of the most widespread of all mammals. Their diet varies according to the region, but normally includes berries, grass, small prey, fish and carrion. Body size also varies: adult males in north-western America, particularly on Kodiak Island, are up to three times as heavy as those of the same species in southern Europe. The polar bear, a carnivorous, semi-aquatic species that preys on seals, evolved from a European race of brown bears during the later Ice Age, perhaps during the last 100 000 years. As adults, modern polar bears are larger than all but the largest races of brown bears.

The natural range of polar bears is among the most northerly of any terrestrial mammal. Except when pregnant, polar bears do not enter dens and indeed continue to hunt actively throughout the winter where ice conditions favour seal hunting, even at high latitudes where it is continuously dark for several months. Seals are up to 50% fat, and the bears eat large quantities when hunting is successful, but the availability of prey is erratic and they go for long periods eating little or nothing. For most of the year, polar bears have a layer of adipose tissue between the skin and the muscles, up to 10 cm thick over certain parts of the body. To find out whether its presence there is an adaptation to thermal insulation, or simply represents a site for the storage of large quantities of fat as an adaptation to the irregular food supply, the total masses of superficial and intra-abdominal adipose tissue in 13 adult and subadult wild polar bears were compared with that in 28 other Carnivora native to the tropics or the temperate zone, which were about as fat as the polar bears (Figure 2.16). The scales are logarithmic, so that data from specimens ranging in size from ferrets (body mass less than 1 kg) to large bears of over 400 kg can readily be compared. The lines shown on Figure 2.16 are fitted to all the points *except* those from the polar bears, but they have been extended to the right, into the range of the measurements from the bears. The proportions of adipose tissue that are *predicted* for animals as large as polar bears from the information about smaller, warmer-climate species coincide almost exactly with the measurements from the wild specimens.

▶ What proportion of the total adipose tissue in polar bears is in superficial depots?

There is five to ten times as much superficial as intra-abdominal adipose tissue.

▶ Is there any evidence for *adaptation* of the partitioning of adipose tissue between superficial depots (where it could act as insulation) and intra-abdominal depots (where it could not function as insulation) in the polar bears?

No. The proportion of adipose tissue in superficial depots increases, and that in intra-abdominal depots decreases, with increasing lean body mass in all the species studied. The proportions of intra-abdominal and superficial adipose tissue in arctic bears are not significantly different from that expected for an animal of that size, i.e. the data points from polar bears are near the extended regression lines fitted to the data from tropical and temperate-zone Carnivora. Polar bears have more superficial adipose tissue than smaller carnivores, because they are bigger and usually also fatter than most other carnivores. In spite of their arctic habitat and semi-aquatic habits, there is no evidence for an adaptive change in the distribution of the tissue. Although superficial adipose tissue may help to insulate the body, we

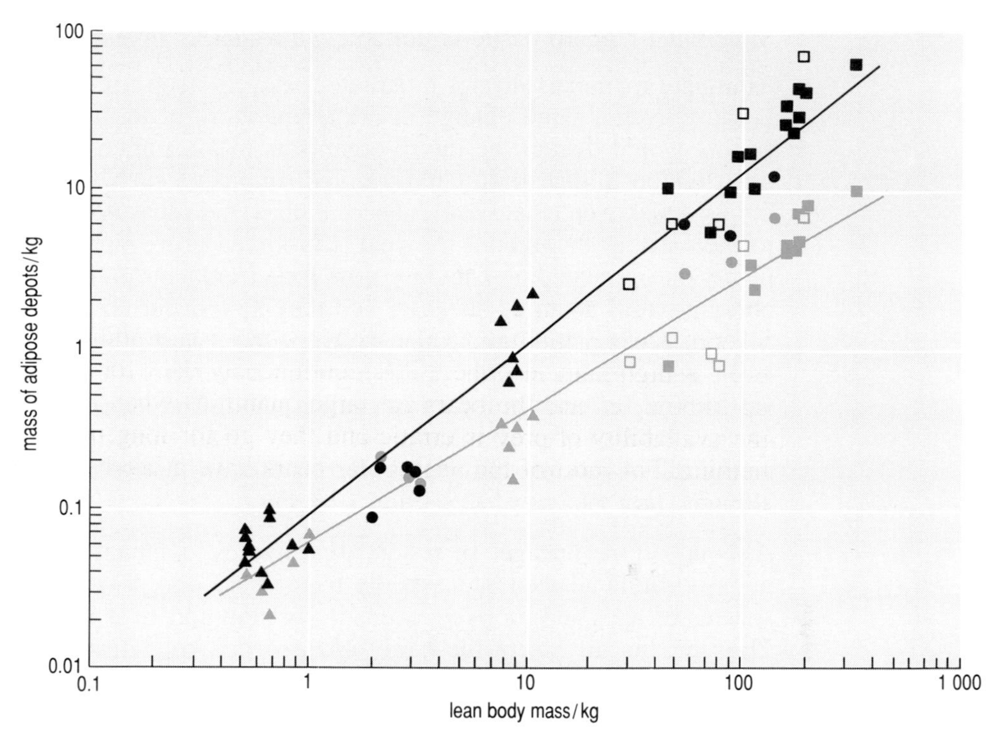

Figure 2.16 *Partitioning of adipose tissue between superficial and intra-abdominal depots in polar bears and various non-arctic Carnivora. The mass of all superficial (black symbols) and all intra-abdominal (green symbols) adipose tissue as a function of lean body mass in 41 adult and subadult Carnivora over 8.4% by weight dissectible adipose tissue. Triangles: Mustelidae (8 ferrets* Mustela erminea, *6 badgers* Meles meles*); circles: Felidae (5 cats* Felis catus*; 1 jaguar* Panthera onca*; 1 tiger* P. tigris*; 1 lion* P. leo*); solid squares: 13 polar bears* Ursus maritimus*; open squares: 6 brown bears* U. arctos. *The regression lines shown are fitted to all data except those from* U. maritimus. *Note that both scales are logarithmic.*

cannot conclude that it is adapted to this role. Thermal insulation must be provided mainly by their thick coat of stiff, woolly fur. Most other large, arctic-adapted mammals are now extinct, but their remains indicate that they also had relatively large quantities of subcutaneous adipose tissue; it was noticeable in a mummified woolly rhinoceros *Coelodonta antiquitatis* and was more than 8 cm thick in a mammoth *Mammuthus primigenius* found frozen in a river bank in Siberia. But, as in polar bears, the large quantities of fat were probably an adaptation to the meagre and highly seasonal food supply and had little or nothing to do with thermal insulation, which was provided by the abundant fur.

In some mammals, notably marine orders such as Cetacea (whales and dolphins) and Pinnipedia (seals, sea-lions and walruses), there is evidence for adaptive changes in the distribution of adipose tissue. Humans have lots of intra-abdominal adipose tissue, and furthermore, as we get fatter, the extra fat accumulates inside the abdomen, even in people who are frequently exposed to the cold.

▶ What do these facts suggest about the role of superficial adipose tissue in thermal insulation of humans?

Superficial adipose tissue is not adapted to act as insulation in people.

Ruminant mammals do not hibernate, or even sleep for long periods, probably because even a small change in the temperature of the abdomen and/or prolonged fasting would disrupt the micro-organisms in the rumen that make an essential contribution to digestion. Cold-adapted ruminants such as reindeer (see Section 2.4.3), muskoxen (Plate 2.4), sheep, llamas and alpacas remain active and exposed to the weather throughout the year. Muskoxen graze lichens and low-growing plants on open tundra. They are now found only in arctic Canada and Alaska, although until about 2 000 years ago they also occurred in northern Europe and Siberia. Their pelt is thick over the ears, snout and other exposed parts of the body, and consists of fine, dense underfur covered with long, strong outer hairs that provide effective protection against wind. The same adaptations make the pelts of such cold-adapted ruminants particularly well suited to spinning and weaving; nearly all wool comes from artiodactyls that are native to high altitudes or high latitudes (see Chapter 17, Section 17.4.1).

Although some lineages of rodents, carnivores, elephants, artiodactyls and perissodactyls adapted successfully to the cold, the primates fared badly. Several species of apes and monkeys were fairly common over much of Europe in the Pliocene, but all except the hardy *Macaca* monkeys disappeared from there with the first major glaciation. Macaque monkeys appeared in parts of southern Europe during the warmer interglacial periods, and briefly spread into southern Britain. The only other primate genus to have established permanent populations in Europe, northern Asia and America during the Quaternary Ice Age was *Homo*. (The 'Barbary Apes' on Gibraltar, a species of *Macaca*, were probably introduced artificially, possibly by the Romans.) Hibernation and torpor are unknown among large primates and most tropical forms are omnivores, eating a wide variety of leaves and fruits. *Homo* was probably able to spread north from Africa, where it first appeared during the Pleistocene, because people became more carnivorous, and had learnt to control fire and to wear animal skins as clothing.

2.4.3 Adaptations to geographical and biological conditions

Many organisms, including many invertebrates and plants and most reptiles and amphibians, became extinct in the colder conditions, although some survived further south, often in isolated populations. During the interglacials and since the end of the last glaciation, these relict populations often recolonized parts of their former range if the climate and habitat were suitable, but changes in sea-level, as well as the climate itself, presented barriers to non-flying terrestrial species. The ice, rather than St Patrick, banished poikilothermic (cold-blooded) terrestrial vertebrates from Ireland. Several species of amphibians recolonized the area when the climate improved, but no reptiles became established before rising sea-levels made the Irish Sea a permanent barrier to colonization by small terrestrial animals. There are still no native amphibians or reptiles in Iceland, much of Scandinavia and Arctic islands such as Svalbard.

Reindeer The absence of reptiles in Ireland makes little obvious difference to the rest of the fauna and flora, but where large predators or large herbivores have failed to recolonize areas, the effects on the rest of the fauna and flora can be very striking. Reindeer (genus *Rangifer*) appear in North American deposits associated with cold-adapted flora dating from early in the first glaciation. Several anatomical features, such as the broad, flat feet, indicate that from an early stage of their evolution they were adapted to walking on deep snow or other soft substrata such as sand. They are the only genus of the deer family (Cervidae) in which both sexes have antlers for part of the year. During the Pleistocene, reindeer (see Figure 2.18) spread over the northern continents and colonized many islands around the Arctic Ocean, and were common on tundra and in boreal forests. The living species, *Rangifer tarandus,* is still the most widespread and abundant wild ungulate in arctic regions of North America (where they are called caribou), Scandinavia and northern Russia. In sheltered areas at lower latitudes, the reindeer browse on bushes and small trees, including birch and rowan, but in open tundra, they graze on grasses, small herbaceous plants such as arctic willow, and on mosses, if necessary digging through snow to reach the dried vegetation underneath. If the snow is too deep, or if periods of thawing and refreezing form layers of hard ice in the snow, the reindeer have difficulty reaching the plants beneath and feeding becomes slow and inefficient. Many populations migrate large distances between feeding grounds.

During the Quaternary Ice Age, the abundance and distribution of reindeer followed the constantly changing pattern of glaciation, with populations becoming isolated on islands and in areas of suitable habitat cut off by changes in climate and sea-level. The distribution of living subspecies of reindeer is shown on Figure 2.17. Most of the differences between the subspecies can be interpreted as adaptations to the climatic and biological conditions.

Throughout the Pleistocene, timber wolves *Canis lupus* were the major predators on ungulates in northern regions. Until this millennium (and in northern Canada and Alaska today), reindeer were the staple diet of wolves in many places and large packs followed the migrations of the herds. There is every reason to believe that reindeer in most areas suffered heavy predation from wolves. There are no wolves on Svalbard (area B on Figure 2.17), or any other large terrestrial predators such as hyenas or humans, and no fossil evidence that any of these species were ever there after the islands were colonized by reindeer. The only other large arctic carnivore—polar bears—can swim across open water and hence readily colonize isolated islands (they breed regularly on several islands of the Svalbard archipelago), but they feed almost entirely on seals and very rarely kill reindeer or any other terrestrial mammal or bird.

Many of the characteristics of Svalbard reindeer *R. t. platyrhynchus* (Figure 2.18(a)) can be attributed to the absence of predators. *R. t. tarandus* (Figure 2.18(b)) forms huge herds, often including thousands of animals, that migrate hundreds of kilometres between feeding areas. They bunch into groups or run away at the first sign of danger. On Svalbard, reindeer occur in small groups of 3–6 animals and adult males are often solitary. They rarely run fast, and when forced to do so they have a lower maximum speed, and become exhausted much more quickly than the mainland subspecies. They often react to intruders more with curiosity than fear. Svalbard reindeer (Figure 2.18(a)) do not have regular times or routes of migration: a group may remain in the same small area for

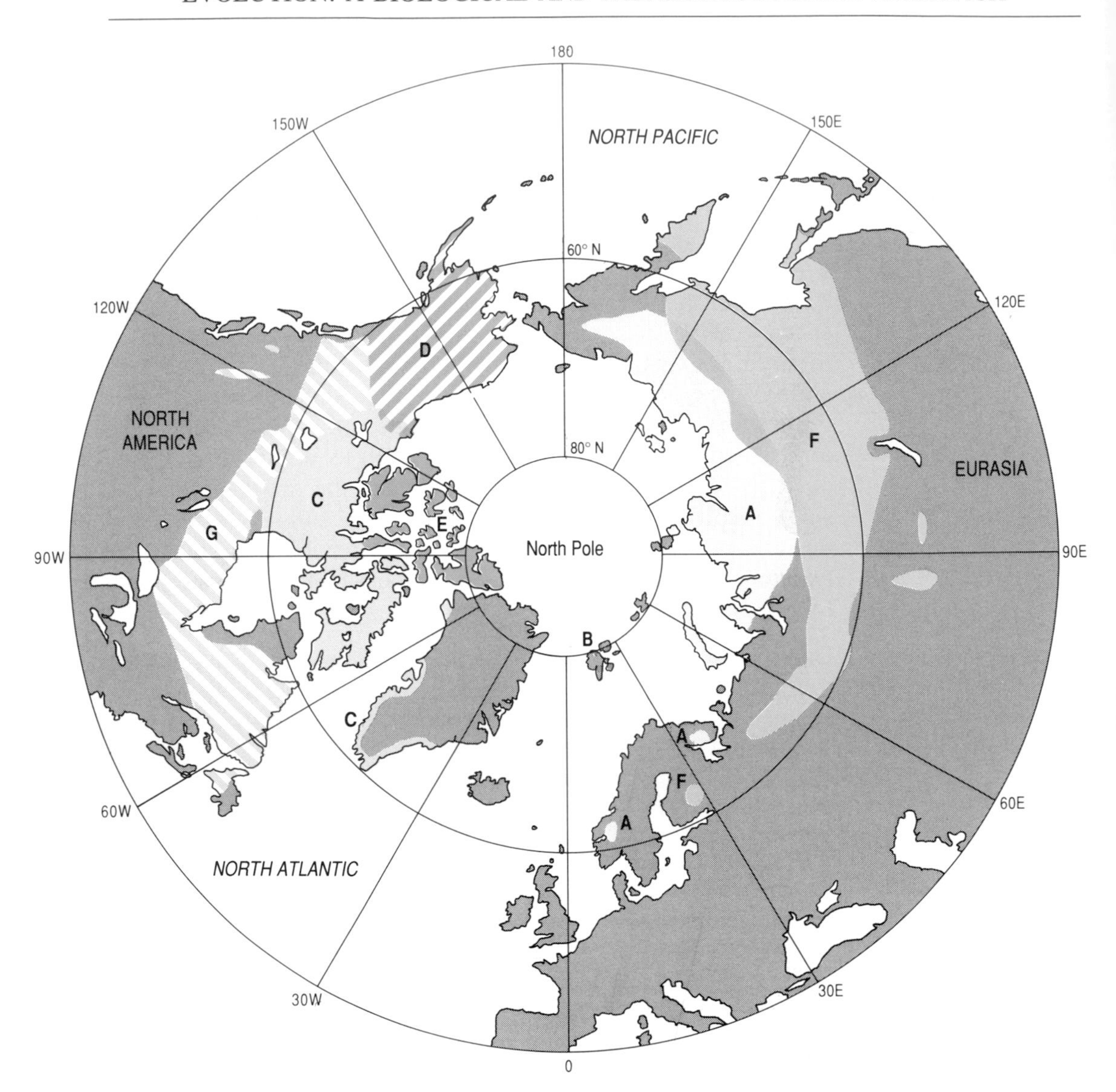

Figure 2.17 *The Arctic, showing the natural distribution of living subspecies of reindeer* Rangifer tarandus. *(A) Mountain or tundra reindeer* R. t. tarandus. *(B) Svalbard reindeer* R. t. platyrhynchus. *(C) Barren-ground caribou* R. t. groenlandicus. *(D) Alaskan caribou* R. t. granti. *(E) Peary caribou* R. t. pearyi. *(F) Forest reindeer* R. t. fennicus. *(G) Woodland caribou* R. t. caribou.

weeks if the grazing is good, moving on only when the food is exhausted. In many areas of Svalbard, the ground is covered with snow for up to nine months of the year, and the reindeer feed almost continuously during the brief period in which the vegetation is readily available. They have about twice as many adipocytes (cells in adipose tissue specialized for storing lipid) as most other wild ruminants of similar size and become very fat by the middle of the winter if food is plentiful. More adipocytes allow the animals to store more energy as fat, enabling them to survive longer when food supplies are exhausted or become inaccessible under deep or hard snow.

Figure 2.18 *Reindeer: (a) Svalbard reindeer* Rangifer tarandus platyrhynchus*; (b) Norwegian reindeer* R. t. tarandus.

Svalbard reindeer *R. t. platyrhynchus* (Figure 2.18(a)) are smaller and have shorter ears, legs and snout. Their fur is longer than that of Norwegian reindeer *R. t. tarandus* (Figure 2.18(b)) and it covers the eyelids, feet and snout much more extensively.

► Are features such as shorter legs in *R. t. platyrhynchus* adaptations to the biological environment or to the climate?

Probably both. Long thin legs lose a lot of heat, so short legs could be an adaptation to the cold climate (see Section 2.4.2). Long legs are an adaptation to fast running, but their growth requires more protein and minerals, so the reduction of this energetically expensive character in a predator-free population is an equally plausible explanation. The reduction in selection for fast running would also facilitate the evolution of a greater capacity for fat storage.

Reindeer also have several less obvious adaptations to the long, cold winter. In all mammals and birds, the air is warmed in the nose so that by the time it reaches the lungs, it is at normal body temperature and fully saturated with water vapour. In species native to warm climates, such as humans, the exhaled air is at about 26 °C and very moist. In very cold conditions, exhaling such warm, moist air represents a substantial loss of both heat and water. There is, of course, no accessible running water on Svalbard for most of the year, and warming snow to body temperature uses quite a lot of heat. All reindeer have elaborately folded nasal turbinals which warm the air as they breathe in and cool it as they breathe out, thereby conserving both heat and water. Those of Svalbard reindeer are especially efficient, and their exhaled air is at about 6 °C in the winter. The kidneys are also relatively large, and the urine is very concentrated. Both these adaptations conserve moisture in an environment where water is present, but energetically expensive to obtain.

Although the range of *R. t. platyrhynchus* is now adjacent to that of *R. t. tarandus*, the island subspecies is anatomically more similar to the Peary caribou *R. t. pearyi* of Ellesmere Island in northern Canada (see Figure 2.17).

▶ What does this fact suggest about how reindeer colonized Svalbard?

The animals came across the southern part of the Arctic Ocean, possibly carried on an eastward-moving ice-floe. The islands were not colonized from the south or east, even though the land is nearer and the sea shallower around those coasts. The west and north coasts of Norway are exposed to a warm current (the Gulf Stream) and the ice probably melted before the climate in Svalbard improved sufficiently for reindeer to become established. Svalbard was completely glaciated in the mid-Pleistocene, and much of the interior is still covered with huge glaciers. During the last glaciation, the climate was dry and cold, rather like Siberia is now, but the coastal areas may have been free of permanent ice for about 130 000 years. There was probably a fairly brief period during which the climate on Svalbard was warm enough and wet enough for plants to become established, thereby providing sufficient food to sustain the reindeer, and the Arctic Ocean was still sufficiently frozen to allow migration of a large terrestrial mammal. Reindeer are by far the largest terrestrial animal on the islands; there are no other ungulates, arctic hares (see Figure 2.11(b)) or rodents (except those introduced by humans), and very few birds are resident there throughout the year. The subspecies *R. t. platyrhynchus* is probably descended from a small founder population, perhaps as few as a dozen animals; even now, when much larger areas are covered with vegetation than was the case in the past, the total population is about only 10 000.

It is not clear why wolves failed to colonize the islands. Wolves flourish in northern Canada and Alaska (see Chapter 17, Section 17.3.1), where the climate is even more severe than in Svalbard, so there seems to be no physiological reason why they could not have adapted to conditions there. Small, isolated populations of prey are not sufficient to sustain breeding populations of large, social predators such as wolves. Perhaps the small numbers of reindeer, and the lack of alternative prey species, prevented wolves from establishing a permanent population. Such factors may often explain the absence of a species from a particular place. When introduced artificially, exotic species often thrive surprisingly well in new habitats (see Sections 2.3.1 and Chapter 17, Section 17.3.3).

SUMMARY OF SECTION 2.4

The Quaternary Ice Age involved prolonged periods in which the climate was significantly colder over much of the Northern Hemisphere than had been the case during the previous 100 Ma, causing widespread extinction and relatively rapid evolutionary change. A wide variety of behavioural, physiological and anatomical adaptations to colder climates and fluctuating food supplies evolved in many different lineages of vertebrates. Descendants of many such species continue to flourish in arctic and subarctic climates, so many of these adaptive mechanisms can be studied experimentally. Comparison between warm-adapted and cold-adapted species is essential to distinguish structural adaptations to cold from features that simply arise as a consequence of differences in body size and shape. Glaciations also produced large changes in sea-level, which linked or isolated areas of land, thereby promoting or preventing colonization by terrestrial animals and plants when the climate became suitable for them. Some modern populations of arctic species have habits and structural features that can be interpreted as adaptations to the depleted fauna and flora.

2.5 CONCLUSION

The concept of biological adaptation is important in three ways: it helps us to understand how organisms live in their natural environment, it can help us to reconstruct the form and habits of extinct organisms from their fossil remains, and it is central to theories about the mechanism of evolutionary change. Correct identification and interpretation of adaptations requires a broad knowledge of the organism's anatomy, physiology, biochemistry and natural history. A major adaptation may entail complex physiological changes that promote or preclude the evolution of other structures and habits. Organisms may adapt to similar environments in contrasting and sometimes unexpected ways. The mechanisms that generate and perpetuate the huge variety of biological structures and processes are the subject of the next Chapter.

REFERENCES FOR CHAPTER 2

Campbell, B. and Lack, E. (1985) *A Dictionary of Birds*, T. and A. D. Poyser, Calton, Staffs.

Chamberlain, J. A. and Westermann, G. E. G. (1976) Hydrodynamic properties of cephalopod shell ornament, *Paleobiology*, **2**, 316–331.

Davies, N. B. and Brooke, M. de L. (1988) Cuckoos versus reed warblers: adaptations and counteradaptations, *Animal Behaviour,* **36**, 262–284.

Heldmaier, G., Klaus, S. and Wiesinger, H. (1990) Seasonal adaptation of thermoregulatory heat production in small mammals, in Bligh, J. and Voigt, K. (eds) (1990) *Thermoreception and Temperature Regulation*, Springer-Verlag.

Lack, D. (1968) *Ecological Adaptations for Breeding in Birds,* Methuen, London.

Moote, I. (1955) The thermal insulation of caribou pelts, *Textile Research Journal,* **25,** 832–837.

Pettifor, R. A., Perrins, C. M. and McCleery, R. H. (1988) Individual optimization of clutch size in great tits, *Nature*, **336**, 160–162.

Pond, C. M. and Ramsay, M. A. (1992) Allometry of the distribution of adipose tissue in Carnivora, *Canadian Journal of Zoology,* **70**, 342–347.

Williams, C. K. and Moore, R. J. (1989) Phenotypic adaptation and natural selection in the wild rabbit *Oryctolagus cuniculus* in Australia, *Journal of Animal Ecology,* **58**, 495–507.

FURTHER READING FOR CHAPTER 2

Alexander, R. McN. (1982) *Optima for Animals*, Edward Arnold, London.

Alexander, R. McN. (1989) *Dynamics of Dinosaurs and Other Extinct Giants*, Columbia University Press, New York.

Bligh, J. and Voigt, K. (eds) (1990) *Thermoreception and Temperature Regulation*, Springer-Verlag.

Davies, N. B. and Brooke, M. (1991) Coevolution of the cuckoo and its hosts, *Scientific American*, **264**(1), 66–73.

Desmond, A. J. (1975) *The Hot-Blooded Dinosaurs*, Blond and Briggs, London.

McLoughlin, J. C. (1979) *Archosauria: A New Look at the Old Dinosaurs,* Allen Lane, London.

Norman, D. (1985) *The Illustrated Encyclopedia of Dinosaurs*, Salamander Books Ltd., London.

SELF-ASSESSMENT QUESTIONS

SAQ 2.1 (*Objective 2.1*) Explain in a few sentences how the following helped in the development of the concept of adaptation in biology:

(a) collecting plants and animals from remote places;

(b) observing wild animals and plants in their natural habitat;

(c) comparing living species with extinct forms.

SAQ 2.2 (*Objective 2.2*) Describe how the following factors influence clutch size in birds:

(a) season;

(b) climate;

(c) irregular differences between one breeding period and the next;

(d) individual differences between pairs.

SAQ 2.3 (*Objective 2.3*) Explain in a few sentences how (a) comparisons with living forms and (b) reconstructions and model building contribute to accurate interpretation of the habits of extinct species.

SAQ 2.4 (*Objective 2.4*) Explain in one or two sentences each how a long neck affects the following features of giraffes:

(a) the heart;

(b) mechanisms that control blood pressure in the head;

(c) response to predators and rivals;

(d) length and thickness of the legs.

SAQ 2.5 (*Objective 2.5*) How does the genetic basis of an adaptive feature influence:

(a) its role in adapting the organism to the conditions in a particular place at a particular time?

(b) its long-term effect on the evolution of the species?

SAQ 2.6 (*Objective 2.6*) Which of the following are characteristic of coevolved adaptations?

(a) They relate to interactions between species that materially affect the reproductive success of both of them.

(b) They relate to all interactions between species.

(c) They relate to all interactions between organisms and their environment.

(d) They evolve in prey or host species, but not in predator or parasite species.

(e) They are reciprocal, involving both prey and predator or both host and parasite.

SAQ 2.7 *(Objective 2.7)* Explain briefly why it is difficult to account for the origin of novel structures using theories of evolution in which adaptation plays a central role.

SAQ 2.8 (*Objective 2.8*) Describe two alternative, physiologically incompatible ways in which mammals adapt to cold winters.

SAQ 2.9 (*Objective 2.8*) Describe in a few sentences the possible causes and consequences of the absence of wolves from Svalbard.

Chapter 3

Heredity and Variation

Prepared for the Course Team by Heather McLannahan

Study Comment *Natural selection operates on variation at the phenotypic level. Evolution can only be brought about by natural selection if variation is heritable. The genetic basis of heredity has been established and observations show that there is genetic variation both within and between species. This Chapter explores the relationship between genotypic and phenotypic variability. It also explains the sources of variation, its maintenance, how it is measured and the evolution of genetic differences between natural populations of the same species in the absence of natural selection.*

Objectives When you have finished reading this Chapter, you should be able to do the following:

3.1 Understand and explain the genetic basis of heredity and the central dogma.

3.2 Distinguish between genotypic and phenotypic variability and explain why there is no simple relationship between the genotype and phenotype of an individual.

3.3 Give an account of the sources of genetic variation.

3.4 Derive the Hardy–Weinberg ratio, citing the assumptions made in its formulation, and show how it accounts for the maintenance of genetic variation in the absence of natural selection.

3.5 Describe methods of measuring genotypic and phenotypic variation, interpret data from such measurements and be able to choose an appropriate method for a given purpose.

3.6 Describe and explain how genetic differences can evolve between natural populations of a species in the absence of natural selection.

3.1 Introduction

Many morphological, physiological and behavioural characters are shared by individuals who are members of the same species, yet no two individuals are ever *exactly* the same. All humans share a great many characteristics and are, as a species, quite different from, say, polar bears. Yet no two humans are exactly alike, not even genetically identical twins. In the latter case differences in personality and appearance arise during development even when both twins are brought up under similar conditions.

The variation of characters may be continuous or discontinuous (i.e. discrete). Where a character shows continuous variation, it may take any value whatsoever—between certain limits—and, using that character, individuals cannot be divided into clearly distinguishable classes. For example, people could probably be found with heights of 1.50 m, 1.51 m or 1.52 m. For each pair of individuals it is possible to find another individual whose height is intermediate between them. Many characters show continuous variation—examples include height, colour, weight and resting metabolic rate, to name but a few.

Where a character shows discrete variation, individuals *can* be divided into clearly distinguishable categories, between which there are no intermediates. People vary discretely in their blood groups. There is, for example, the O, A, B and AB classification—each person has blood of one of these four types. The blood group name reflects the presence or absence of two specific blood glycoproteins (protein molecules with sugar chains covalently attached) and it is necessary to know the blood group of a person requiring a blood transfusion so that they are not supplied with blood that is incompatible with their own. There are no intermediates—one cannot have a blood group halfway between B and O for example.

A population in which discretely different forms (or *morphs*) of a character co-exist is said to be **polymorphic** for that character. If only one form of a particular character is present, the population is said to be **monomorphic**.

3.2 Heredity

Variations of the kinds described above are the raw material on which natural selection acts but, as noted in Chapter 1, the sources of variation puzzled Darwin for he had observed that offspring tend both to inherit features of their parents *and* to possess novel features. Clearly the mechanisms of heredity have to explain both these phenomena. At that time, the pioneering work on heredity of Gregor Mendel (1822–1884) was unknown and Darwin had no satisfactory theory of the mechanism of inheritance. There was a widespread assumption that inheritance was a *blending* process, by which offspring tended to inherit characters intermediate between those of their parents. Thus a tall father and a short mother would be expected to produce children of intermediate height, and indeed this is often the case. If this were always true, however, there would be a tendency for variation between individuals to become less and less over succeeding generations. With Mendel's discovery that some characters, such as seed shape in peas, are usually passed on unchanged to progeny it became apparent that novel features could arise spontaneously, though rarely, by a then unknown process (now known to be due to a change or **mutation** in the structure of the genetic material—see Section 3.3). If such mutations were adaptive in the struggle for existence, they would be passed from one generation to the next without being 'diluted' by blending inheritance. Although Mendel first published his work in 1866, it was largely ignored at the time, to be 'rediscovered' at the turn of the century.

Darwin did in fact accept the inheritance of acquired characters to account for new variation (Chapter 1, Section 1.1.3), but he saw natural selection as much the more important process in directing evolution. Today the former, Lamarckian idea is discounted as an influence in evolution, except in some very limited circumstances (discussed later in Section 3.2.4).

August Weismann (1834–1914), Professor of Zoology at Freiburg, explicitly rejected the Lamarckian view of inheritance. He drew a sharp distinction, at the level of cell division, between two processes that start with the fertilized egg. One leads to the formation of the body or *soma*, the other, via the *germ line*, to the production of **gametes**, the reproductive cells, that go to produce the next generation (Figure 3.1(a)). At each generation, the soma dies, but the germ line is perpetuated and is potentially immortal. The germ line produces the soma but is independent of any changes that may occur in the soma during an individual's life. In the light of modern knowledge about the nature of gene expression, we can replace Weismann's germ line by DNA and the soma by protein (Figure 3.1(b)). The germ line and the soma are not independent entities; the germ line cannot be reproduced without energy and raw materials being provided by the soma, and the soma depends for its existence on the germ line. The crucial role of the germ line is the transmission of *information*. The germ line (DNA) provides the coded information for protein synthesis, but proteins in the soma cannot influence—in any directed sense—the replication of DNA in the germ line producing the gametes. By analogy, a record can generate music in a record player, but we cannot alter the information encoded on a record by making noises into the record player (in the way that we can with a tape-recorder).

Francis Crick (born 1916) realized that the concept of a one-way flow of genetic information (DNA → protein) was fundamental to our thinking about the relation of DNA to phenotype. He referred to this concept as the **central dogma**.

If this central dogma is true, the implications for the evolutionary process are profound. It implies first that evolutionary novelty can arise only through changes in DNA and, secondly, that such changes (mutations) occur through processes that are not directed from outside the germ line. If it can be shown that processes

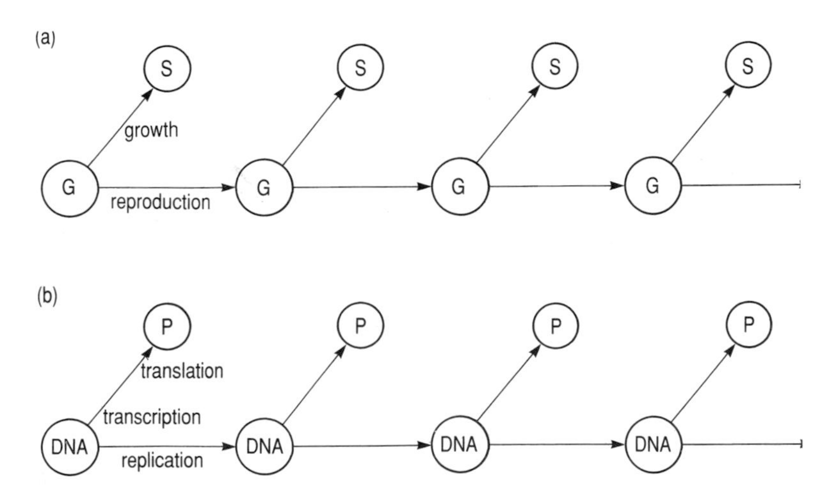

Figure 3.1 *(a) Weismann's scheme (G = germ line; S = soma). (b) The central dogma (P = protein).*

acting in the soma during an organism's life do in fact alter, in a directed way, the genetic material that is passed on to its progeny, then this will undermine the central dogma. There are now various lines of evidence that such processes do exist but they are relatively rare. We shall return to them, and to a discussion of their evolutionary significance, in Section 3.2.4, after we have looked at how the germ line is expressed in producing the soma.

3.2.1 THE FIDELITY OF GENES: FROM MENDEL TO WATSON AND CRICK

From his plant breeding experiments Mendel was able to postulate the following:

1 that there are particles of inheritance;

2 that these particles are present in pairs in the breeding adults;

3 that these particles separate during the formation of gametes, so that each gamete contains only one of each pair;

4 that these particles have alternative *dominant* and *recessive* forms (see Figure 3.2 and Plate 3.1).

The Dane, Wilhelm Johannsen (1857–1927) termed these particles of inheritance *genes*. Following Mendel, the early 20th century geneticists studied the inheritance of *phenotypic characters*. The gene was an abstract concept. Johannsen cautioned against any suggestion that there might be a one-to-one relationship between gene and character:

'The gene is thus to be used as a kind of accounting or calculating unit.... By no means have we the right to define the gene as a morphological structure...Nor have we any right to conceive that each special gene (or a special kind of genes) corresponds to a particular phenotypic unit-character....'

This has turned out to be entirely justified. Not only can genes have more than one effect on the phenotype (**pleiotropy**), but a given phenotypic character is most often the result of the expression of *several* genes (i.e. it is a **polygenic trait**). Additionally, genes do not work in isolation and the expression of a gene (or genes) at one **locus** (that is, the location of the gene on the chromosome) may be affected by genes at other loci. This is known as **epistasis**. Finally, phenotypic characters are affected by the interaction between genes and their environment during the process of development. The relationship between genotype and phenotype is considered further in Section 3.2.3.

Nevertheless, with the identification of the behaviour of chromosomes during cell division toward the end of the 19th century (Section 3.3.1 gives details), it was an inevitable logical step to assume that the gene was a real physical presence embodied within the chromosome. Subsequent research to test this hypothesis culminated about five decades later in the recognition that genes are sequences of DNA. DNA is a long-chain polymer (called a polynucleotide) that consists of units (nucleotides), each containing one of four different nitrogenous bases (see Figure 3.3). This structure fulfils perfectly the necessary attributes of the genetic material, namely that it must be able to direct its own replication and also the synthesis of proteins. Its molecular structure, a double-stranded helix, held together by bonds between the complementary base pairs (Figure 3.3(a) and (b)), immediately

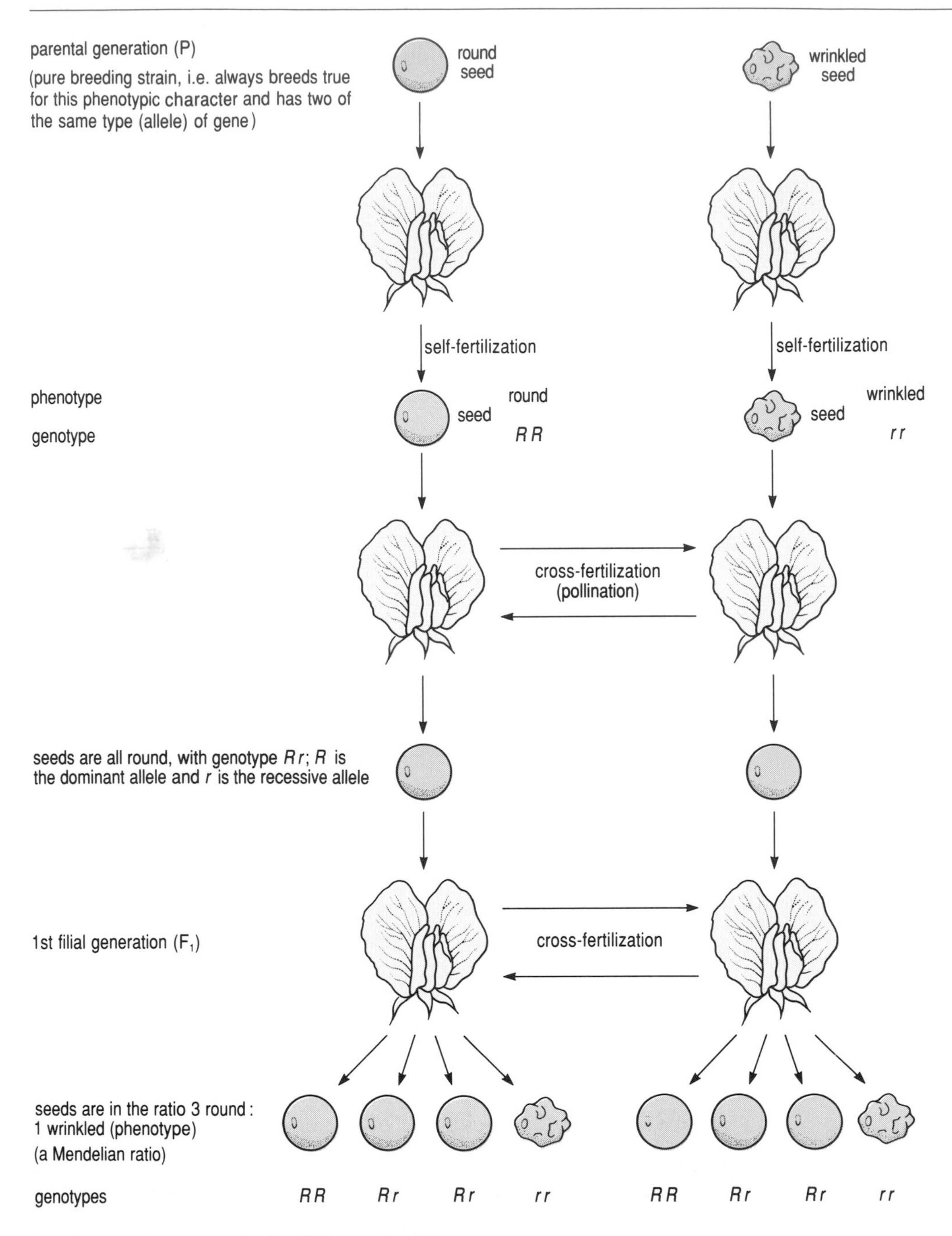

Figure 3.2 *Mendel's breeding experiment with round and wrinkled peas: a summary diagram.*

suggested 'a possible copying mechanism for the genetic material' (Watson and Crick, 1953). Experimental evidence that each strand of the double helix acts as a template came from a classic experiment reported by Meselson and Stahl (1958), which confirmed that the DNA production process proceeds by *semi-conservative replication* (Figure 3.3 (b) and (c)). Thus it is seen that the original strands of the

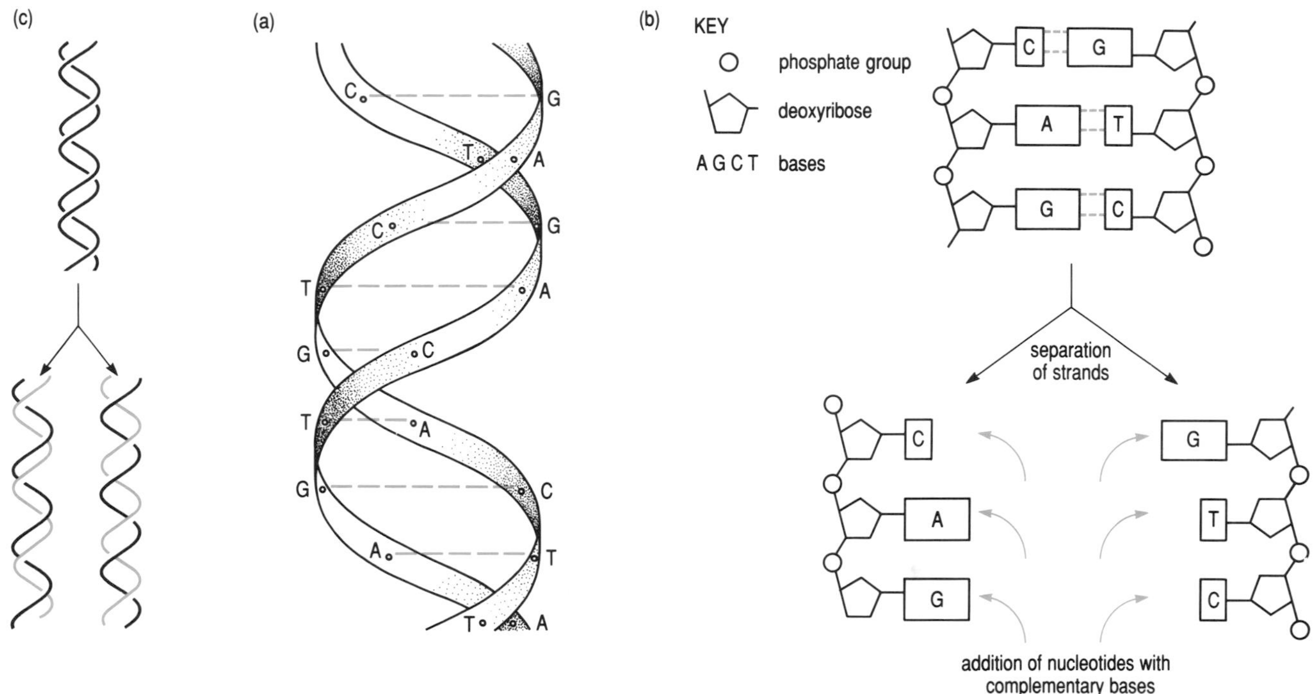

Figure 3.3 *The structure and pattern of replication of DNA. (a) The double helix. (b) Molecular structure. The bases C (cytosine) and G (guanine) always pair together, as do A (adenine) and T (thymine). (b) and (c) both show the semi-conservative replication of DNA. Each existing DNA strand is used as a template for the synthesis of a new complementary strand.*

DNA double helix are conserved throughout the cycles of DNA replication; in other words, information in the genotype is, in general, faithfully copied from one generation to the next. So the findings of molecular biology are consonant with those of Mendelian genetics that genes retain their identity through succeeding generations.

3.2.2 THE GENOME

The link between the chromosome and DNA was made in the previous Section, but the two are not synonymous. In addition to DNA, the chromosome comprises many different proteins, one particularly important group being the *histones*. These are involved in the packaging of the enormously long molecules of DNA.

Eukaryotes are organisms whose cells contain nuclei within which lie the chromosomes. The term **genome** is used to describe all the nuclear DNA of a eukaryote cell. Any other DNA within a cell is termed extranuclear DNA, and this includes the DNA found in organelles such as mitochondria and chloroplasts. **Prokaryotes**, e.g. bacteria, do not have nuclei and the bacterial genome is a closed double-stranded loop of DNA, lying free in the cytoplasm.

The number and constitution of the chromosomes in a eukaryote cell is referred to as its **karyotype**. In multicellular organisms, the cells of the soma (somatic cells), i.e. all the cells except gametes, have the same karyotype and most individuals within a species have the same somatic karyotype. Exceptions occur, for example where males and females have different sex chromosomes. (In humans, females

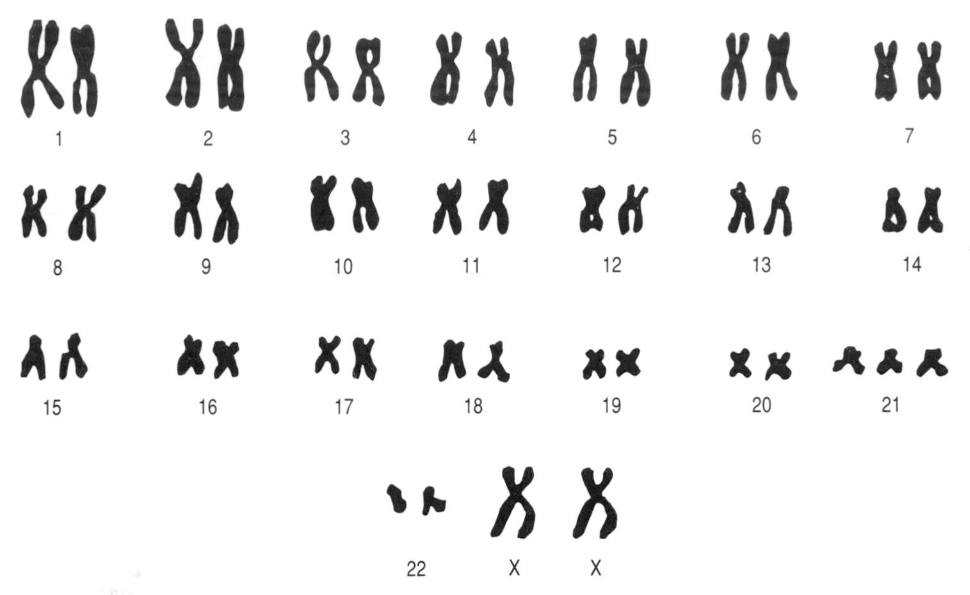

Figure 3.4 *An abnormal human karyotype.*

have two X chromosomes; males have one X accompanied by a smaller Y chromosome.)

Because the somatic cells of most organisms contain paired chromosomes (one derived from each parent) the total number of chromosomes in these is written as $2n$. This is called the **diploid** number. Gametes contain one of each chromosome pair, i.e. n chromosomes in total; this is termed the **haploid** number. In humans $n = 23$ ($2n = 46$).

Homologous chromosomes can be artificially laid out in pairs to show the karyotype of the organism. There are characteristic morphological differences between the chromosome pairs of an organism, and geneticists assign numbers to each chromosome pair.

▶ Figure 3.4 shows an abnormal human karyotype. Study it carefully and say why it is abnormal.

There is an additional chromosome 21. Individuals with an additional chromosome 21 exhibit a range of characteristics classified as Down's syndrome.

▶ Does the karyotype in Figure 3.4 tell you anything else about the phenotype?

Yes. There are two X chromosomes so this is the karyotype of a female.

As expected, the DNA content of the genome seems to be characteristic for each species and varies widely between species, as shown in Figure 3.5. The measure of cellular DNA content is called the ***c*-value**, and is defined as the number of DNA bases in the unreplicated haploid genome of a species.

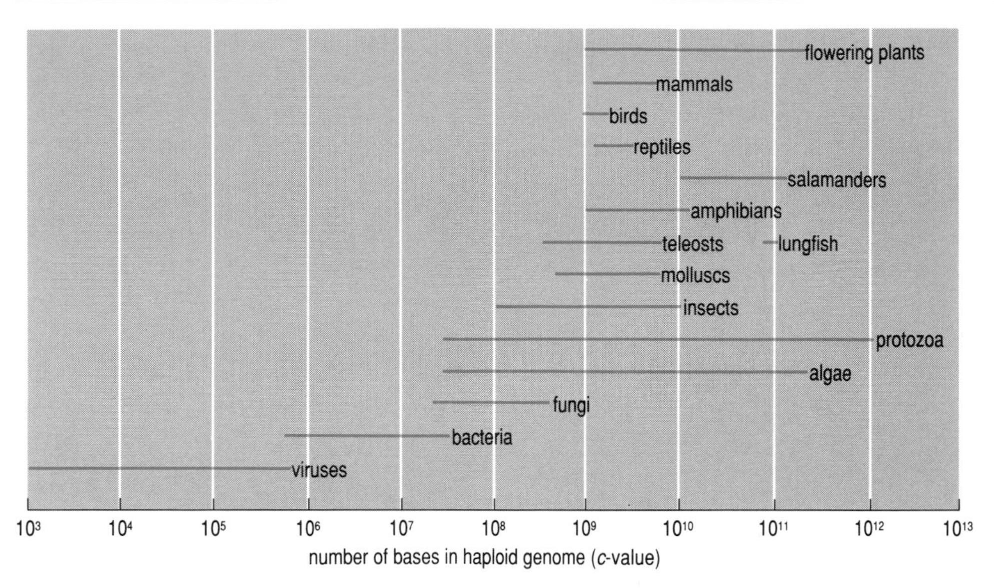

Figure 3.5 *The haploid DNA content* (c-*value) in various groups of organisms.*

▶ Study Figure 3.5 carefully. How do the *c*-values of salamanders and mammals compare?

Salamanders have about 10 to 100 times as much DNA per cell as mammals.

▶ If all the DNA were to code for different proteins, what would be the inference from these *c*-values?

This might imply that salamanders manufacture 10 to 100 times as many different proteins as do mammals.

This seems to be unlikely and to be a paradox, but further investigation reveals that not all of the DNA codes for protein. Figure 3.6 summarizes the process of protein synthesis from a DNA template; notice that some of the DNA must be transcribed into transfer RNA (tRNA) as well as RNA in the ribosomes (rRNA).

Furthermore, much of the DNA in eukaryotes consists of repeated sequences of bases. These may be highly repetitive. For example, a sequence of as few as six bases might be repeated from 10^3 to 10^7 times in one genome. Genetic fingerprints, useful in taxonomic studies, are prepared from this class of DNA (see Section 3.5.2). There are also moderately repetitive sequences of DNA, usually present in tens or hundreds of copies per genome. The latter are longer than the highly repetitive DNA sequences, often consisting of several hundred bases, and include genes coding for histones, rRNA and tRNA. It is in the moderately repetitive DNA sequences that mobile or **transposable elements** are found. These sequences can replicate and then move to new positions within the genome where they may have considerable effects of evolutionary importance, as will be discussed in Section 3.2.4.

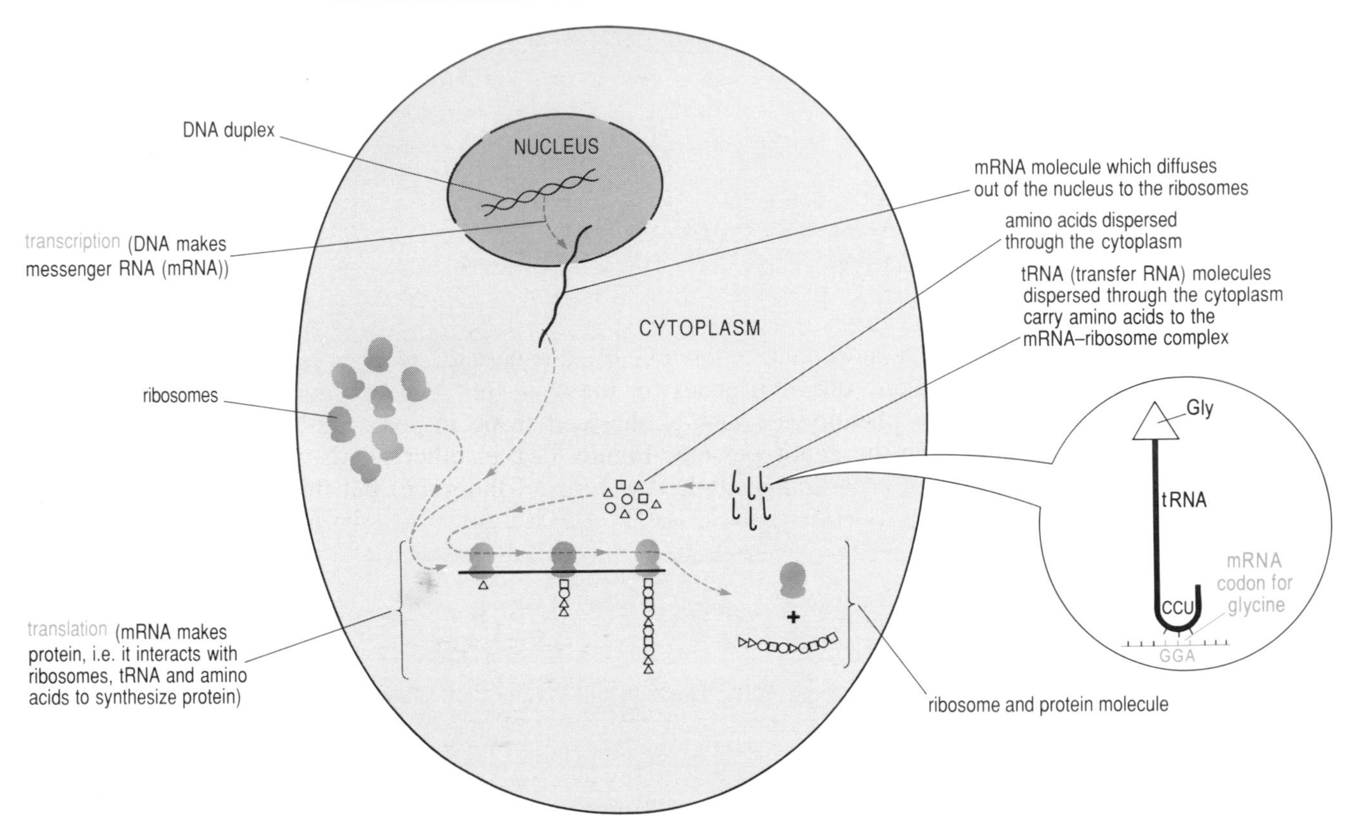

Figure 3.6 *A simplified scheme of protein synthesis showing the two main stages: transcription, 'DNA makes RNA', and translation, 'mRNA makes protein'. Note that this scheme applies to any organism and that the code is universal, e.g. the DNA nucleotide sequence CAA GTA CAA TGC always makes the sequence of amino acids valine–histidine–valine–threonine. In RNA, however, thymine (T) is replaced by another base, uracil (U).*

The *c*-value paradox may thus be explained by differences in the amounts of these repetitive sequences in different groups of organisms. Additionally, the high *c*-values found in some plants are due to **polyploidy** (possession of multiple copies of all the chromosomes), discussed further in Section 3.3.3 and Chapter 9.

It has also been postulated that the cellular DNA content could itself be subject to selection pressure. For example, a low DNA content means that cell volume can be reduced. This gives a high surface area : volume ratio which is favourable for high metabolic rates. It is notable that bats, like birds (see Figure 3.5), have a high metabolic rate and that their genome size is low compared to other mammals and within the range found in birds. For these, and other, reasons it is clear that interspecific comparisons of DNA content need to be treated with caution. They do not, at present, offer definitive information on the relationships, and evolutionary history, of supposedly closely related species.

3.2.3 The expression of genes: from genotype to phenotype

It has already been stated, but it is worth repeating, that there is not a simple one gene–one morphological character relationship. How could there be? DNA is the genetic material and DNA codes for proteins, not for morphological characters. In the previous Section you were reminded that not all the nuclear DNA codes for proteins. In this and the following Section the relationship between the coding DNA and the organism's phenotype (that is, its morphology, physiology and behaviour) will be examined.

Discontinuously varying characters Mendel's original experiments (see Figure 3.2) involved the study of characters that showed discontinuous variation. That is, the characters were quite distinct, e.g. round or wrinkled seeds in peas, between which there were no intermediate forms. This might suggest that the seed shape character was determined by the action of a single gene. However, genetic investigations show that almost every characteristic of an organism requires the activity of many different genes for the character to be expressed. So differences between two phenotypes, such as the seed shape in peas, are the result of differences in the genotypes (see Figure 3.2); in other words, due to the presence or absence of a particular **allele** (i.e. form of the gene) but this is not to say that there is one gene which alone is responsible for the development of the seed shape. For example, in the fruit-fly *Drosophila* some individuals have no eye pigment. It is known that they differ from the pigmented type at one particular locus—the so-called 'white' locus. Alteration of an allele at this 'white' locus results in the developing fly failing to make a protein which is a precursor to eye pigments. However, the formation of eye pigments from the precursor also requires the presence of an enzyme manufactured by activity at a different locus to the 'white' locus. This shows that eye colour in *Drosophila* requires the expression of genes at at least two different loci although a character difference can be brought about by a change at only one of these loci.

But the story is even more complicated. If the allele at the 'white' locus needed for eye pigment is absent it may still be possible for an individual phenotype with pigmented eyes to emerge. This is because the effect of the allele at the 'white' locus can be mimicked by the combined activity of two alleles at two different loci.

Continuously varying characters There are also many characters, particularly those that show continuous variation such as coat colour in the domestic horse *Equus caballus*, that are affected by the cumulative activity of many genes acting together. Genes acting in such a way may not all contribute equally to the final character and individual genes may behave differently depending on the genetic background (i.e. epistasis—Section 3.2.1) and on the environmental conditions. The majority of characteristics are affected in this way; in other words, they are inherited through small effects of numerous loci. This makes it very difficult to assess the genotype of an individual by performing genetic crosses and well nigh impossible to determine allele frequencies at individual loci in a population. The majority of characteristics therefore do not have simple dominant and recessive forms that are inherited in Mendelian ratios. Section 3.5.1 considers how data can be scored for morphological characteristics such as these, and how these data can be used to assess to what extent the considerable variability seen in such characters has a genetic basis.

Pleiotropy Just as one character requires the expression of many genes, so, conversely, one gene usually has more than one effect on the phenotype (pleiotropy—Section 3.2.1). For example, the medical condition of sickle-cell

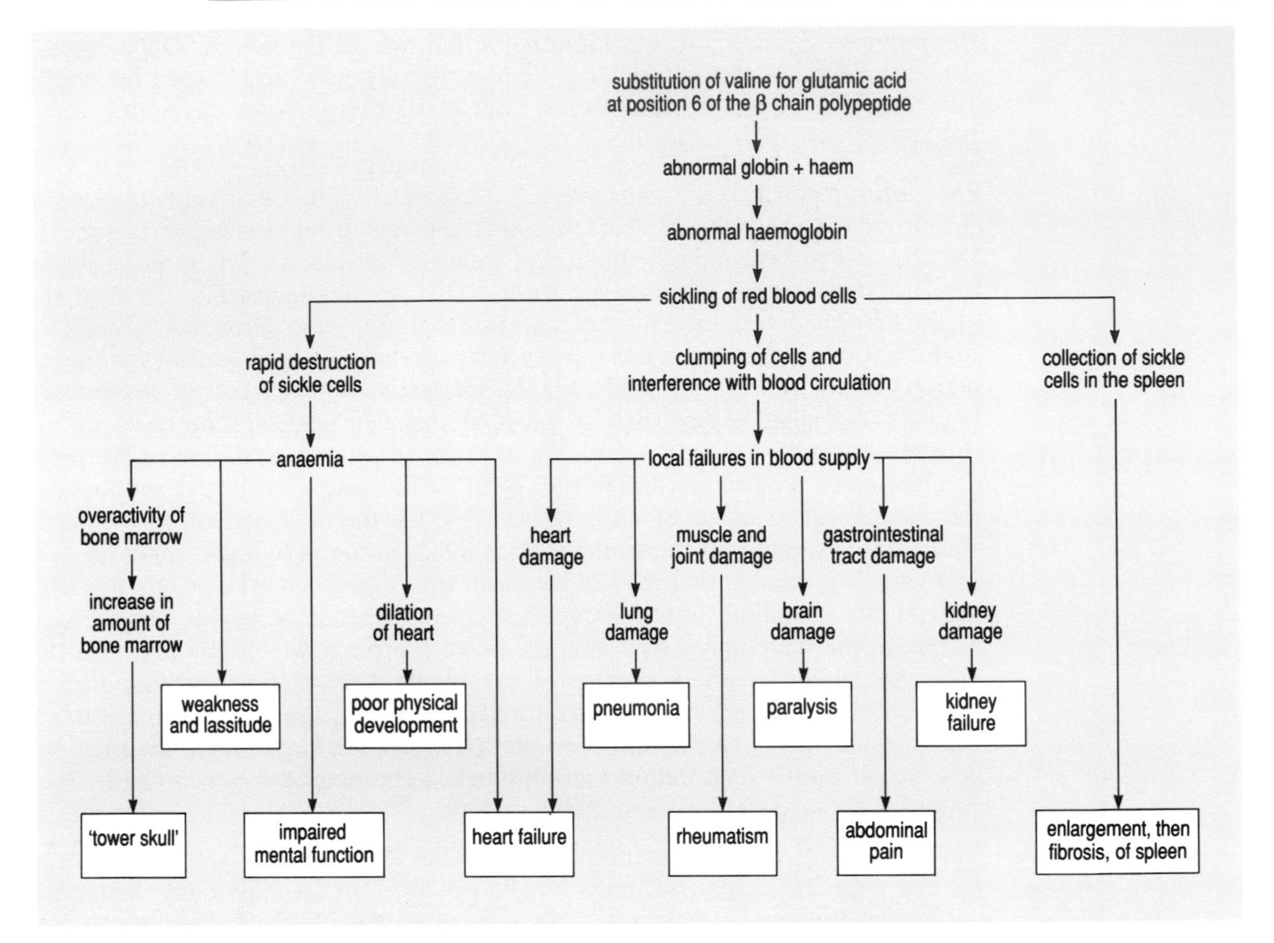

Figure 3.7 *Pleiotropic effects of the amino acid substitution of valine for glutamic acid in a human haemoglobin molecule.*

anaemia is recognized from a multitude of abnormalities in the sufferer. The individual differs from unaffected individuals at one particular locus. The alteration at this locus results in the alteration of one amino acid in the protein that forms a part of the oxygen-carrying molecule, haemoglobin. Figure 3.7 shows the many effects resulting from the alteration of this one amino acid. What exactly has changed in the gene to cause this alteration?

The one altered amino acid is specified by the base sequence of a length of DNA. To be precise, there are three bases forming a triplet code known as a **codon**, that specifies the one amino acid. The four bases found in DNA (adenine, guanine, cytosine and thymine—A, G, C and T in Figure 3.3(b)) can be combined into $4^3 = 64$ different triplets. There are 20 amino acids; some are specified by more than one codon and some codons initiate or terminate polypeptide chains. So to change one amino acid there will, at minimum, have been the substitution of one base for another.

Structural genes Thus, altering one base in a sequence of DNA can produce an altered gene. From the above example it would appear that there can be a simple molecular definition of a gene. A sequence of bases that codes for a protein is

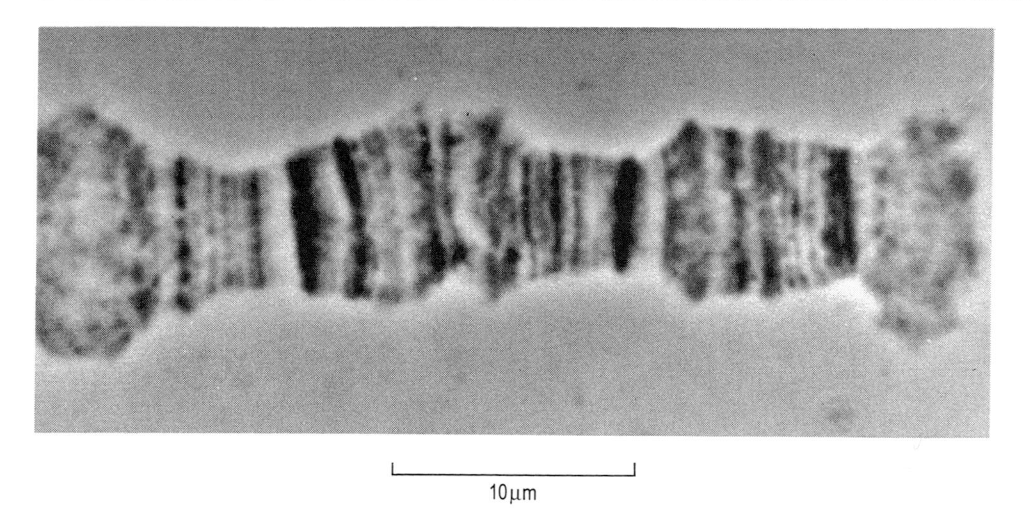

Figure 3.8 *Giant chromosome from the salivary glands of* Drosophila *showing banding.*

called a *structural gene*. Indeed, in certain species, chromosomes may show characteristic banding patterns and there is evidence that each of these bands corresponds to a gene. (Figure 3.8 is a photograph of an unusually large chromosome from *Drosophila melanogaster* showing banding.) But, in 1977, it was discovered that the coding parts of the DNA are often interrupted by sequences of bases that are transcribed but never translated into protein. The coding regions are known as *exons* and the intervening non-coding regions are known as *introns*. Some introns are transcribed into RNA that does not code for proteins but may regulate the transcription or translation of other coding sequences. Figurc 3.9 shows that the introns and exons of an interrupted or split gene form one continuous length of bases yet only the exons code for the protein.

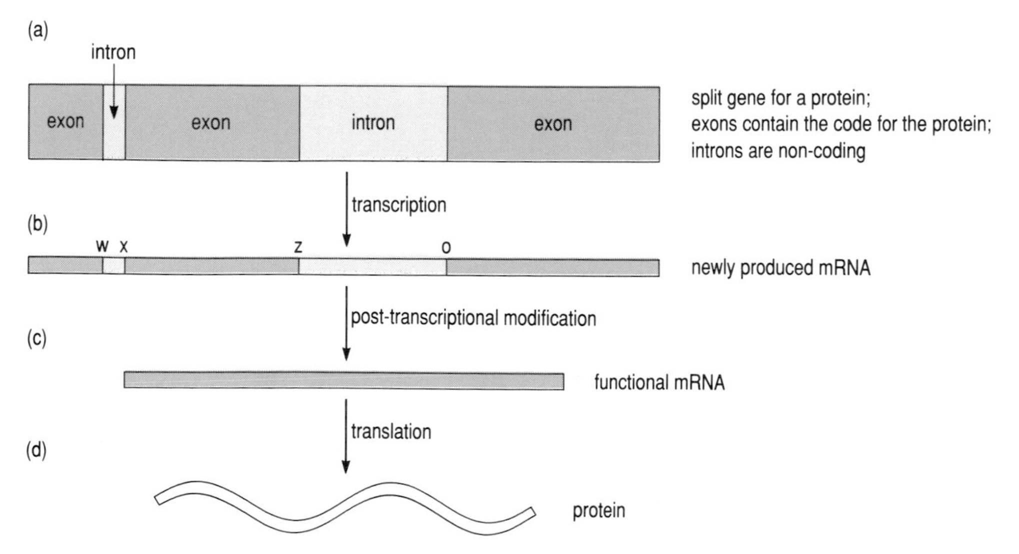

Figure 3.9 *The transcription and subsequent translation of a split gene. Note that both exons and introns are initially transcribed. The newly produced mRNA is then modified to the form in which it is translated.*

So introns are not part of the structural gene. Furthermore, not all structural genes exist as single copies per haploid genome; many exist as repeated units or *multigene families* making up much of the moderately repetitive DNA mentioned earlier (Section 3.2.2). The histone genes are a multigene family. They are arranged in arrays and function simultaneously, making the same histone, so it would be misleading to say that there was a particular gene locus for a particular histone. The human haemoglobin genes also form a complex multigene family but they are responsible for the production of a number of different haemoglobins. This family has been much studied by geneticists and can be used as an example to demonstrate some more of the difficulties in deciding what constitutes a gene.

Regulation of gene expression All the somatic cells of an organism carry the same genome, yet they will not all express the same genes. Haemoglobin is the red pigment that binds to oxygen and transports it within the blood system. A fat cell will not manufacture haemoglobin although it has the necessary genetic information contained in its genome to do so. What regulates the expression of the genome so that different genes are expressed at different times and some genes are never expressed at all? To continue with the human haemoglobin example, during the course of development there are three distinct haemoglobins made—embryonic, foetal and adult. These proteins, though similar, exhibit structural differences which correlate with functioning under slightly different conditions. For example, the foetus does not breathe but obtains oxygen from its mother across the placenta. So the environmental conditions under which foetal haemoglobin loads up with oxygen from the maternal source and releases oxygen to the foetal tissues are different from those for the loading and unloading of maternal (adult) haemoglobin. As development proceeds, some genes must be switched off and others switched on.

Human haemoglobin genes are found in two clusters, on chromosomes 10 and 11. In humans, the order of the haemoglobin genes on the chromosomes corresponds to the order in which they are expressed in development, although the exact control mechanism is unclear. Figure 3.10 shows the arrangement of haemoglobin genes on chromosome 11. Although each gene is predominantly expressed at one stage of development, as indicated in the Figure legend, the total haemoglobin population in the cell at any point in time is invariably a mixture of different haemoglobins. Each of the genes marked in Figure 3.10 consists of three exons and two introns. Work on the control of the expression of these genes suggests that there are regulatory sequences on this chromosome both within the gene cluster and outside of it that are concerned with the temporal control of gene

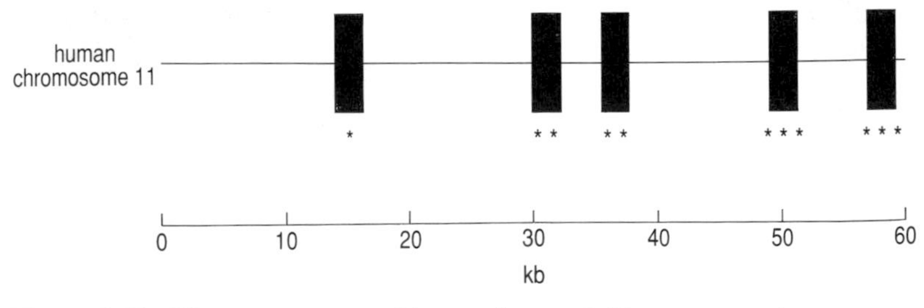

Figure 3.10 *The arrangement of human haemoglobin genes on chromosome 11. Stages of expression: *embryonic; **foetal; ***adult. Scale is in kb (i.e. base pairs* $\times 10^{-3}$*).*

expression. There are also other sites on the same chromosome involved in splicing out the introns correctly. It should be noted that the structure of the entire gene complex is important because the adult protein cannot be manufactured if the embryonic and foetal genes are deleted. The information that controls the developmental sequence must reside mainly within the genome because, once started, this aspect of development is very resistant to interference from changing environmental conditions.

Thus far an attempt has been made to trace the pathway from genotype to phenotype by following the biochemical route from DNA to a particular characteristic of the phenotype, i.e. a particular protein. This proves difficult even when looking at a relatively simple end-product, such as the type of haemoglobin being produced, because from the moment that the zygote begins its development its environment is altering. Genes exert their effects in a changing biochemical environment and the emerging phenotype is a result of genotype–environment interactions. The extent to which the environment affects the gene expression in the developing organism is itself variable. For example, as was stated above, the switching of haemoglobin genes is relatively resistant to environmental influences but there are numerous examples where morphological differences arise as a result of environmental influences during development rather than as a consequence of differences in genotype. For example, butterfly breeders have long known that they can radically alter the appearance of the adults, to the point where they can be mistaken for different species, by keeping larvae and pupae in the fridge (see Figure 3.11). Less well known perhaps is the fact that some vertebrates have two sexes, but no sex chromosomes, and it is environmental factors that initiate sex determination. In the alligator *Alligator mississipiensis*, it is not possible to distinguish the chromosomes of the females from those of a male (compare with Figure 3.4 where the two X chromosomes indicated that you were looking at a female human karyotype). The temperature at which the eggs develop determines whether the young alligators will be male or female. Eggs in nests close to water

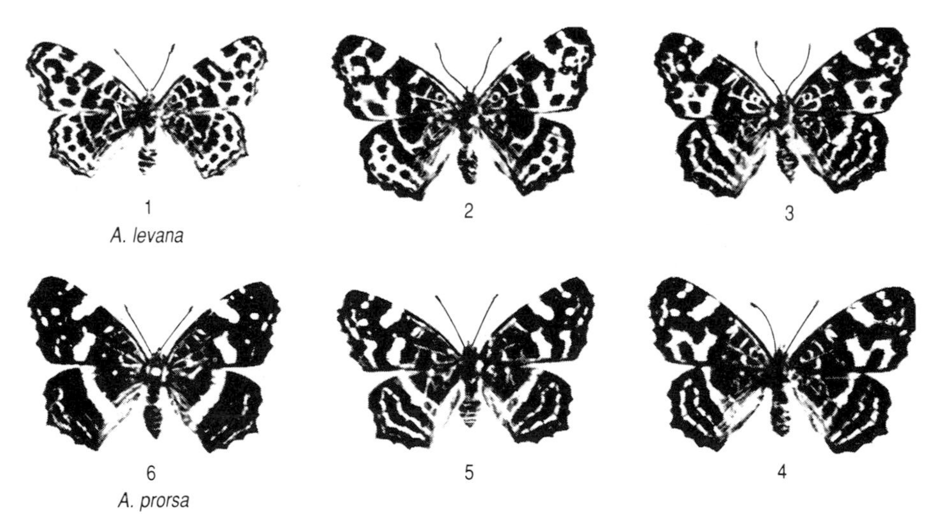

Figure 3.11 *Simulating seasonal polymorphism in* Araschnia levana *(1) and* A. prorsa *(6) connected by experimentally produced intermediates (2–5). These are a single species (the map butterfly).* A. levana *is the spring morph and* A. prorsa *is the form that develops in summer.*

have a mean temperature of 30 °C and the hatchlings are all female, whereas those on drier, level ground have a mean temperature of 33 °C and the hatchlings are all male. The release of pituitary hormones is controlled by the incubation temperature and determines the sex of the developing young. The temperature 'switch' is a total one and there are no young of mixed or indeterminate sex.

3.2.4 THE INFIDELITY OF GENES: THE FLUID GENOME

Look again at Figure 3.9 and note that Figure 3.9(b) shows a length of mRNA not all of which will be translated into protein.

▶ Which portions of the mRNA are not translated into protein?

The portions that were coded for by the intron sequences, i.e. sections WX and ZO.

▶ In manufacturing the functional mRNA shown in Figure 3.9(c) from the mRNA shown in Figure 3.9(b) the portions coded for by introns will be spliced out. In what order would you expect the remaining portions of mRNA to be joined together?

An expectation is that the end of the first sequence (marked W) would join to the second sequence at the point marked X, and that Z would join to O. This may happen but there is clearly the potential for other arrangements. It has in fact been found that for some genes the processing of the newly produced mRNA (Figure 3.9(b)) can proceed in more than one way. In other words, there is not a single unique step from Figure 3.9(b) to 3.9(c) but more than one type of functional mRNA might be produced from the mRNA sequence shown in Figure 3.9(b).

▶ What does this mean in terms of protein production?

It means that different proteins could be produced by one gene. Thus, whilst the gene retains its identity from one generation to the next, it may not always produce the same protein. This happens routinely but it is also a possible mechanism for the evolution of new proteins from novel arrangements of exons.

Genetics of the immune system The genome is flexible in terms of its output in other ways, and this has resulted in the concept of a 'fluid genome'. It transpires that the ordering of genes within the chromosome is not always static. There can be movement and rearrangement of DNA prior to transcription so that genes originally far apart are brought together so as to make a protein. This is known to happen in genes associated with the immune system. The immune system is the mechanism for detecting and protecting an organism against invasion by foreign particles—be they pathogens capable of causing the organism's demise, medical transplants of organs or relatively benign particles such as pollen (to which some people's immune system reacts strongly, resulting in the misery of hay fever). The number of possible substances that could invade is limitless. Certain tissues in the body react by producing **antibodies**, proteins that specifically recognize and bind to foreign substances (which are called **antigens**). Each antigen triggers the production of a unique antibody, and this can only be manufactured if there is an appropriate blueprint in the form of DNA. It turns out that the huge number of different antibodies that can be made by an individual (estimates put this number at 10^6) are not each specified by a unique gene but that the appropriate protein is made as a consequence of DNA rearrangements bringing together genes, previously separate, so as to make the unique protein (antibody). A further aspect of the immune system is that a major source of immunological variation is somatic mutation (i.e. alterations to the DNA structure in body cells, in this case, cells of the immune system). These are detected *after* the DNA rearrangements and occur in the later stages of an immune response.

► Does this challenge the central dogma (Section 3.2.1)?

No. The central dogma is that proteins in the soma cannot influence the replication of DNA in the germ line producing the gametes. It does however shake it slightly to have a mechanism whereby changes in a somatic cell can, apparently, alter the DNA within that somatic cell in a directed fashion.

► If this kind of somatic mutation were found elsewhere (i.e. other than in the immune system), can you think of organisms where these acquired mutations could be passed to progeny?

They could be passed on by plants that propagate by vegetative means or by colonial animals, e.g. corals, that propagate asexually. Somatic mutations provide variability, and selection can then operate through the relative growth of the variants. The key feature of the somatic mutations described in the immune system is that they are *directed*.

There is one example of an environmentally induced *heritable* change that appears to be directed. Flax, *Linum,* changes its appearance in a number of ways in response to high levels of fertilizer and there are associated changes in the genome that are stable through a number of sexual generations in the absence of fertilizer treatment. The morphological changes are said to be adaptive, so this is directed change. The genome change is, however, found to be gene amplification rather than modification or the appearance of new sequences. In other words, the genome acquires more copies of genes rather than new genes. The changes are mainly in the moderately repetitive and highly repetitive sequences and are dispersed throughout the genome. So although the change in total nuclear DNA is large, it is not recognizable by karyotype analysis.

The possession of multiple copies of genes can often be an asset because they enable more of a product to be made at one time. The example of the histone multigene family has already been mentioned (Section 3.2.2).

► When would the ability to produce a lot of histone at one time be an advantage?

When cells are replicating. Histone is associated with DNA and is necessary to enable the DNA to be condensed and packaged into the chromosomal structure. In periods of rapid growth a lot of histone is needed.

Transposons Much of the fluidity of the genome is a consequence of the activity of transposable elements (transposons or jumping genes—Section 3.2.2). These are sequences of DNA that can move within and between chromosomes. Some insert themselves while others replicate and insert copies of themselves into different sites in the genome. In doing this they may disrupt the normal functioning of genes, giving rise to mutations and chromosomal rearrangements. Barbara McClintock, the American geneticist and winner of the Nobel Prize in 1983, was the first to observe chromosomal rearrangements associated with transposition. Although mobile genetic elements occur widely, they often exist in a relatively dormant state. The trigger for transposition can come from disturbances in the external environment. An example of this is when two different populations hybridize, so that the genomic DNA of each comes into contact with the cytoplasm of the other. This could be an important mechanism for producing rapid changes in genome structure and could precipitate rapid evolution, although this idea is speculative at present. It has even been suggested that some genetic material may be transferred between species by a similar means. This is known to occur in bacteria but the extent of this in eukaryotes is quite unknown.

To date the precise mechanisms whereby transpositions cause chromosomal rearrangements and mutations are unknown. The trigger for transposition is also uncertain but the frequency of transposition increases with increased environmental stress. Transposition occurs in germ cells as well as somatic cells so this process could be adaptive by giving rise to more variants some of which may be better able to cope in the new environment.

Notwithstanding the fluidity of the genome it is a remarkable fact that most genes operate in a Mendelian manner, so like begets like. It may be helpful to remember the integrated nature of development. The development of the organism proceeds under the influence of its whole genome. So there is no simple relationship between a gene and a phenotypic character. It is then quite possible to understand that the developmental process may remain stable despite the alteration of a small part of the genome. It requires only that the rest of the system can adjust to the absence or malfunction of a small part. In turn this depends upon a mechanism whereby there can be sophisticated interaction between genes and environment.

▶ Which environmental factors have been suggested to affect the expression of the genome?

Temperature (as in butterflies and alligators), foreign particles (affecting the immune system) and environmental stress (this could include temperature and foreign particles but also shortage or oversupply of nutrients or water) have all been shown to affect the expression of the genome. *However, in the main, the factors have not affected the germ line in a directed fashion and therefore do not challenge the central dogma.*

A question remains as to why this should be so. The evidence from the study of the immune system is that there is a mechanism whereby adaptive changes in the soma can direct changes in the genome and from the flax and fertilizer study there is evidence that Lamarckian inheritance can occur, so why is it so rare?

> 'The answer is that most phenotypic changes (except learnt ones) are not adaptive: they are the result of injury, disease, and old age. A hereditary mechanism that enabled a parent to transmit such changes to its offspring would not be favoured by natural selection.'
>
> J. Maynard Smith, 1989.

3.2.5 NON-GENETIC INHERITANCE AND MATERNAL EFFECTS

The above quotation reminds us that there is also non-genetic inheritance. In a great many species cultural transmission of information, learning from others, is of great importance in securing the animal's survival. This may interact with evolutionary processes.

It is also true that genetic inheritance is not equal. One set of chromosomes is contributed by each parent but virtually all the cytoplasm of the zygote is inherited from the female. This includes organelle DNA and of particular interest is mitochondrial DNA (mtDNA). A given individual usually contains identical copies of mtDNA in all its mitochondria, and these are then passed on to the female gametes unchanged, for reasons that are explained in Section 3.3.1. Mitochondria are very useful when studying relationships in recently diverged lineages (see Chapters 9 and 11). The claim that all existing human mitochondria are probably derived from 'Eve', a female living less than half a million years ago, does not indicate that there was only one breeding pair of humans in existence at that time! There were probably at least 5 000 breeding females in the population. Figure 3.12 should convince you that a group of individuals could have identical mtDNA whilst possessing quite different genomes.

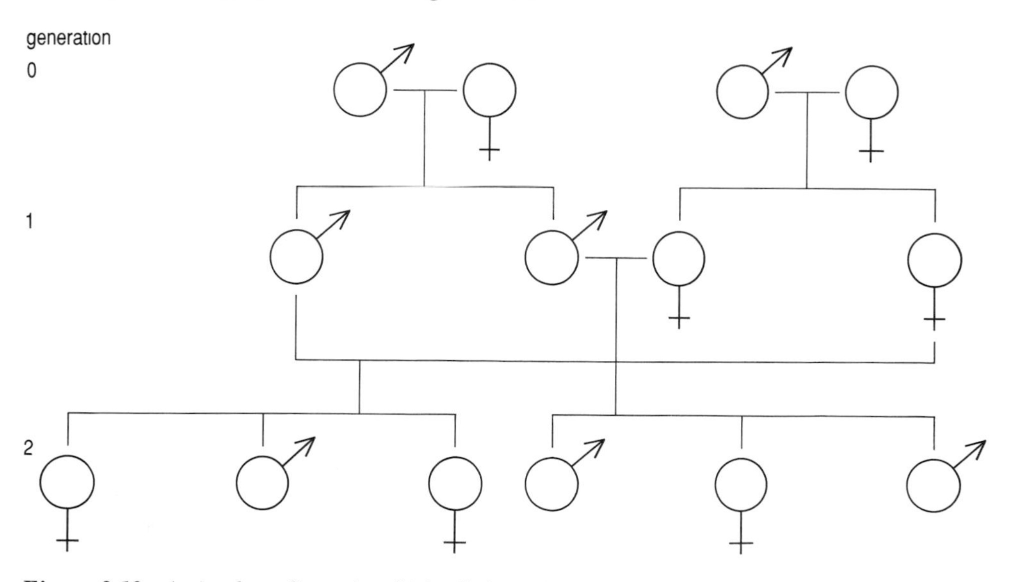

Figure 3.12 *A simple pedigree in which all the individuals in generation 2 trace back, in the female line, to a single female in generation 0, yet all four parents in generation 0 have contributed nuclear genes to all the individuals in generation 2.*

SUMMARY OF SECTION 3.2

The basis of heredity is the passing of (unchanged) genetic material to an organism's progeny. The genetic material is DNA. The central dogma is that information flow from genotype to phenotype is unidirectional. This means that DNA provides the coded information for protein synthesis but proteins in the soma cannot influence the replication of DNA in the gametes in a directed fashion. Eukaryote cells contain nuclear DNA (the genome) and extranuclear DNA such as mitochondrial DNA and chloroplast DNA. The *c*-value and karyotype of a cell are characteristic for each species but the former is of limited taxonomic use because of the many repetitive DNA sequences.

The relationship between genotype and phenotype is not simple. Almost every characteristic of an organism is influenced by more than one locus and genes are usually pleiotropic in their effects.

There are many ways in which the effects of a gene can be modified both from within the genome and by environmental interactions. Nevertheless there is little evidence of the germ-line genome being altered in a directed way.

Examples of non-nuclear inheritance include culture (learning from others) and the inheritance of maternal cytoplasmic factors. e.g. mitochondrial DNA.

This Section has tried to show you that, although DNA is the genetic material that is faithfully transmitted from one generation to the next, its role in shaping the phenotype can be highly variable.

3.3 SOURCES OF GENETIC VARIATION

The previous Section has already indicated how phenotypic variation can arise during development because of the way genes interact with one another and the environment. On the other hand, much phenotypic variation does reflect genetic variability between individuals. Genetic variability is so great that it can be utilized to identify an individual from a small piece of tissue or even a blood stain (so-called genetic or DNA fingerprinting—see Section 3.5.2). The sequences of DNA utilized are unique to individuals, because of the variability (between individuals) of the non-coding DNA regions. Although the latter differences are not expressed in the phenotype, there is a sufficient number of phenotypically expressed genetic differences to merit further consideration. For if the majority of characteristics are adaptive, then a simple expectation is that there will be little variation in phenotypic characteristics when individuals inhabit the same environment, as was described in the study of rabbits' ears in Section 2.1 of Chapter 2. That there is a wealth of phenotypically expressed genetic variability in natural populations is known from experimental data such as were obtained in the studies of cuckoos' eggs (Chapter 2, Section 2.2). A problem with field studies is that it is often difficult to ensure that non-genetic effects such as the influence of a common environment or maternal effects have not confounded the genetic effects. For this reason artificial selection trials are frequently used to reveal genetic variation. The feature to be selected is decided upon and each generation is bred

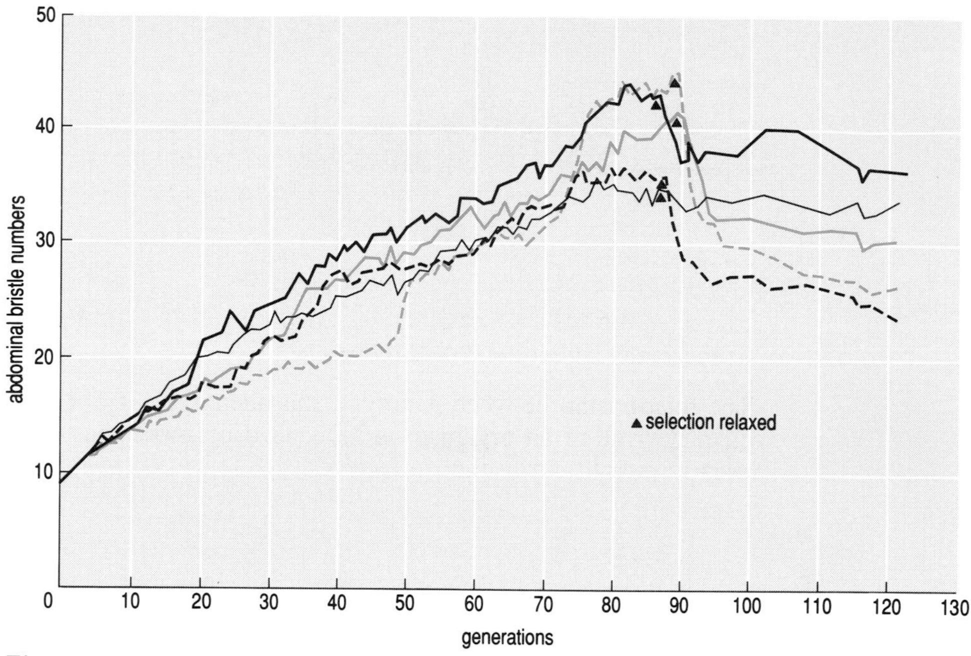

Figure 3.13 *Responses to artificial selection for increased number of abdominal bristles in five laboratory populations of* Drosophila melanogaster. *The bristle number declined after artificial selection was terminated (relaxed) in each population, indicated by the black triangles.*

from individuals showing an extreme of this variation. If the variation has a genetic basis, then selection can alter the expression of the feature in the populations from one generation to another. The fruit-fly, *Drosophila melanogaster,* has a short generation time and is easy to keep in the laboratory so many experiments have used this species. Many features of this species have been altered by artificial selection, including mating speed, tendency to move towards light and abdominal bristle number. Figure 3.13 shows how the number of abdominal bristles increased over 90 generations with each generation bred from flies with the greatest number of bristles.

Thus data from DNA fingerprinting and from artificial selection show that there is much genetic variation in natural populations. The rest of this Section looks at the sources of this variation.

3.3.1 Cell division: mitosis and meiosis

Most individuals start life as a single cell. At its simplest, in asexual reproduction, this involves identical genotypes being produced by the organism dividing into two. Mitosis is the type of cell division that accomplishes this and the cell divisions associated with normal body growth. For multicellular sexually reproducing organisms the single cell from which they derive is the zygote. This is formed by the fusion of two gametes, one derived from each parent. If each gamete contained the complete parental genome then offspring would contain twice as much genetic material as their parents. The process which avoids this is meiosis, a cell division that reduces the genetic material by half in a precise way.

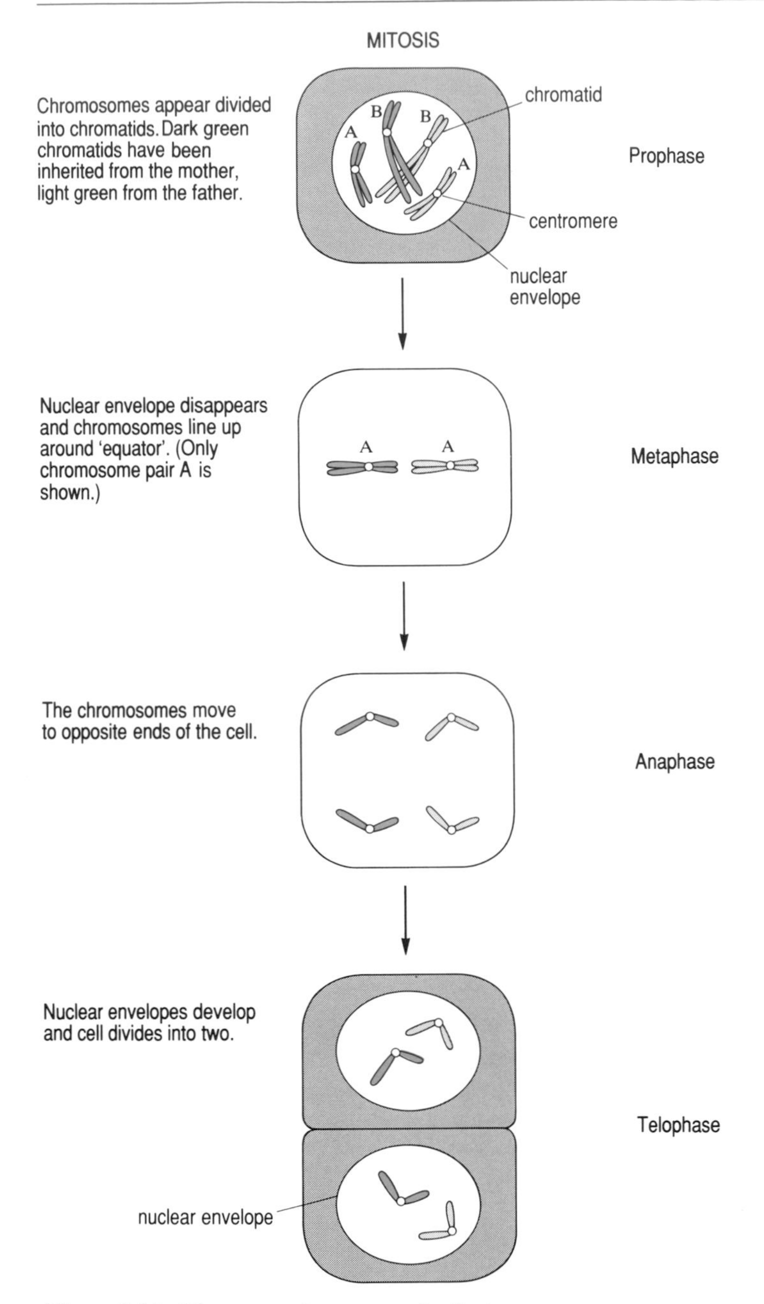

Figure 3.14 *The successive stages of mitosis.*

This Book assumes that you have already studied the phases of meiosis and mitosis but Figures 3.14 and 3.15 provide a summary. One important similarity is that before these processes begin, each chromosome has already replicated and consists of two identical double helices of DNA which become visible as chromatids.

▶ Compare Figure 3.14 with Figure 3.15. What are the main ways in which mitosis differs from meiosis?

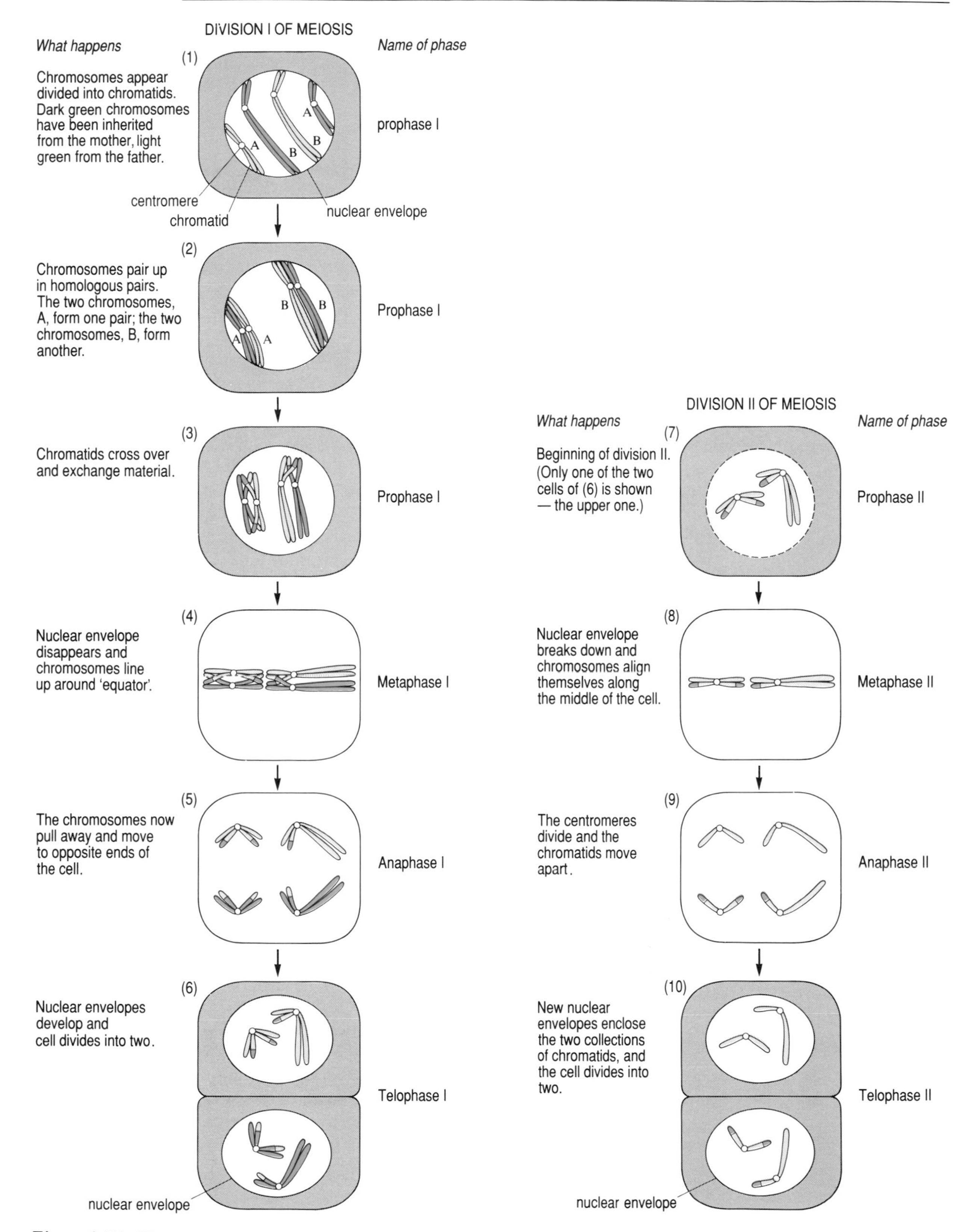

Figure 3.15 *The successive stages of meiosis.*

1 Mitosis involves only one cell division. Meiosis involves two.

2 Mitosis does not involve the pairing of homologous chromosomes. Meiosis does.

3 Mitosis produces cells whose chromosomes are usually exact copies of those of the parent cell. Meiosis does not.

4 Mitosis produces cells with the same number of chromosomes as the parent cell. Meiosis produces cells with only half the number of chromosomes as the parent cell.

Recombination Organisms that are derived from the parent by mitotic division (i.e. by asexual reproduction) contain cells with exactly the same genotype as the parent (point 3 above). The only possible variation between parent and offspring is as a result of mutation. Mutation also makes an important contribution to the amount of variability seen in sexually reproducing organisms but meiosis provides more scope for variability between individuals than does mitosis. This results from *crossing over* of chromatids in prophase I and the consequent exchange of genetic material between homologous chromosomes, and *independent assortment*, that is the random allocation of chromosomes from each pair into the gametes. When new combinations of chromosomes are generated by independent assortment or new combinations of genes within chromosomes are generated by crossing over, the processes are collectively termed **recombination**. Independent assortment can give rise to a total number of different mixes of maternal and paternal chromosomes that is equal to 2^n, where n is the number of chromosome pairs.

If there were three pairs of chromosomes, 2^3 or 8 different types of haploid cell could be produced.

▶ How many different types of cell could be produced as a result of independent assortment during gamete production in (a) humans ($2n = 46$) and (b) *Drosophila* ($2n = 8$)?

(a) 2^{23} (because $n = 23$)—a figure in excess of eight million!

(b) 2^4 (because $n = 4$)—16.

Linkage groups From this, you will readily appreciate that organisms with high chromosome numbers have the capacity to generate a vast amount of genetic variation between individuals by independent assortment alone, provided there are genetic differences between the original maternal and paternal sets of chromosomes. In most instances the cluster of genes found on one chromosome will remain together during meiosis and they are known as a **linkage group**. They can be separated only by crossing over. When you consider that crossing over breaks up and re-forms the combinations of genes on individual chromosomes, there is the potential for even greater variation of the gametes. In humans, this would considerably increase the estimate of over eight million different gametes that it would be possible to produce from one individual! This potential for producing different gametes without introducing any new genetic material explains why—with the exception of identical twins—it is unlikely that two humans would ever have exactly the same genotype.

▶ It was stated in Section 3.2 that mitochondrial DNA is passed on unchanged to gametes. Can you say why this is so?

mtDNA is inherited only from the mother, via the cytoplasm; it is thought that there is no paternal contribution. Unlike the nuclear DNA, mtDNA is not reduced by half during meiosis and so no recombination can occur. Thus mutation is the only source of variation in mtDNA between parent and offspring.

3.3.2 MUTATION

As more became known about the expression of DNA the more difficult it became to define the gene (Section 3.2.3). Similarly 'mutation' is now used in more than one way. In its narrowest sense, it is a change in the base pair sequence of a gene. However, it can also be used more broadly to describe overall changes in the karyotype, and hence is defined as any heritable change brought about by an alteration in the genetic material of an organism.

Inversions and translocations A chromosome can change as a consequence of breaking. This happens naturally but the frequency of occurrence is increased by ionizing radiation and by some chemicals. The result is the loss of part of the chromosome, or a change in the sequence of genes, relative to one another, if the bits of chromosome rejoin in a different order. This kind of change may have no effect if the genes remain intact but in many instances it produces non-viable gametes, particularly when genetic material is lost. Two common structural mutations, *inversions* and *translocations*, and their effects on meiosis are summarized in Figures 3.16 and 3.17. An inversion (Figure 3.16) is a rearrangement within a single chromosome in which a segment has been rotated through 180°. A translocation (Figure 3.17) is a transfer of segments usually by exchange between non-homologous chromosomes.

The importance of such structural changes in evolution is that they often facilitate the reproductive isolation of populations (see Chapter 9, Section 9.2.2). Thus, in an individual heterozygous for a large inversion in which crossing over occurs, meiosis is irregular and about half the gametes are defective (Figure 3.16). Such heterozygotes are therefore less fertile. When the inversion involves only a short segment of the chromosome, the chances of cross-overs *within* the inversion are much reduced so that a high proportion of normal gametes may be produced, and the group of genes within the inversion tends to be conserved. Such a group is known as a *supergene*.

It is known that changes in chromosome number have had both important evolutionary and medical significance. The difference in chromosome number between humans ($2n = 46$) and the ape species, gorilla, chimpanzee and orang-utan ($2n = 48$) appears to have come about as a result of the fusion of two chromosomes which are still separate in the apes. Yet, other species are known where changes in chromosome number resulting from fusion are associated with little or no phenotypic effects. For example, the mouse *Mus musculus* has a normal karyotype of 40 chromosomes, but there is a population in the Central Apennines of Italy that has 22 chromosomes yet shows no obvious phenotypic effects of this apparent fusion of chromosomes.

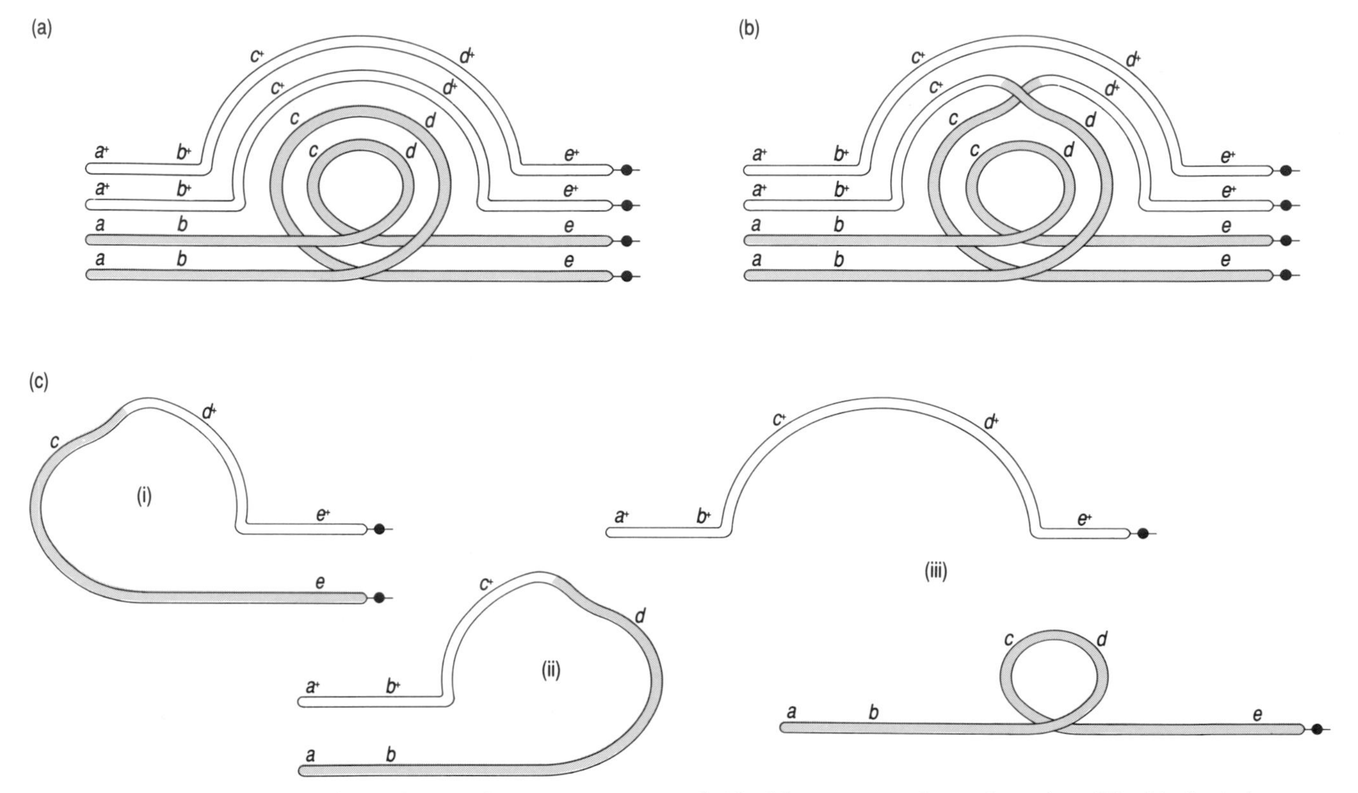

Figure 3.16 *The behaviour of homologous chromosomes in an individual heterozygous for an inversion. (The black circles are the centromeres.) (a) Pairing during meiosis. (b) Crossing over and exchange of genetic material. (c) Separation of chromosomes at anaphase I, showing (i) a chromatid bridge, with two centromeres; (ii) a fragment without a centromere (both (i) and (ii) give non-viable gametes); and (iii) two chromatids with all parts intact (which give viable gametes). Thus half the gametes on average will be defective in the heterozygote. Note that the chromatid pairs are shown separated for clarity.*

Polyploidy The cells of some organisms contain three or more sets of chromosomes. This condition is known as polyploidy (Section 3.2.2) and has been a major factor in plant evolution. Polyploidy is otherwise usually rare in eukaryotes (but is common in earthworms, and is also known in some fishes, reptiles and amphibians). About one-third of all flowering plant species are polyploid. Polyploidy can arise through irregularities in meiosis (or more rarely mitosis), when failure in separation at anaphase leaves some gametes (or somatic cells) with a double number of chromosomes. Progeny arising from fertilization of such gametes, if viable, often grow vigorously but are sterile since they contain an extra set of chromosomes, which cannot divide evenly at the next meiosis. However, many flowering plants can reproduce asexually and so may survive. If subsequently a further doubling of the chromosomes occurs, fertility may be restored and then a viable, sexually reproducing polyploid population can become established, genetically distinct from its parents. Polyploidy is discussed in more detail in Chapter 9, Section 9.3.

Point mutations Although changes in the structure and number of chromosomes have probably been particularly important in plant evolution, the majority of hereditary changes in phenotype have come about as a result of changes to genes. By this is meant alterations in the number of copies of a gene or in its base pair sequence. The latter is known as a *point mutation* and occurs due either to errors in DNA replication or to the inherent slight chemical instability of some of the DNA bases. A point mutation may or may not have a phenotypic effect, depending on the actual alteration made to the base sequence.

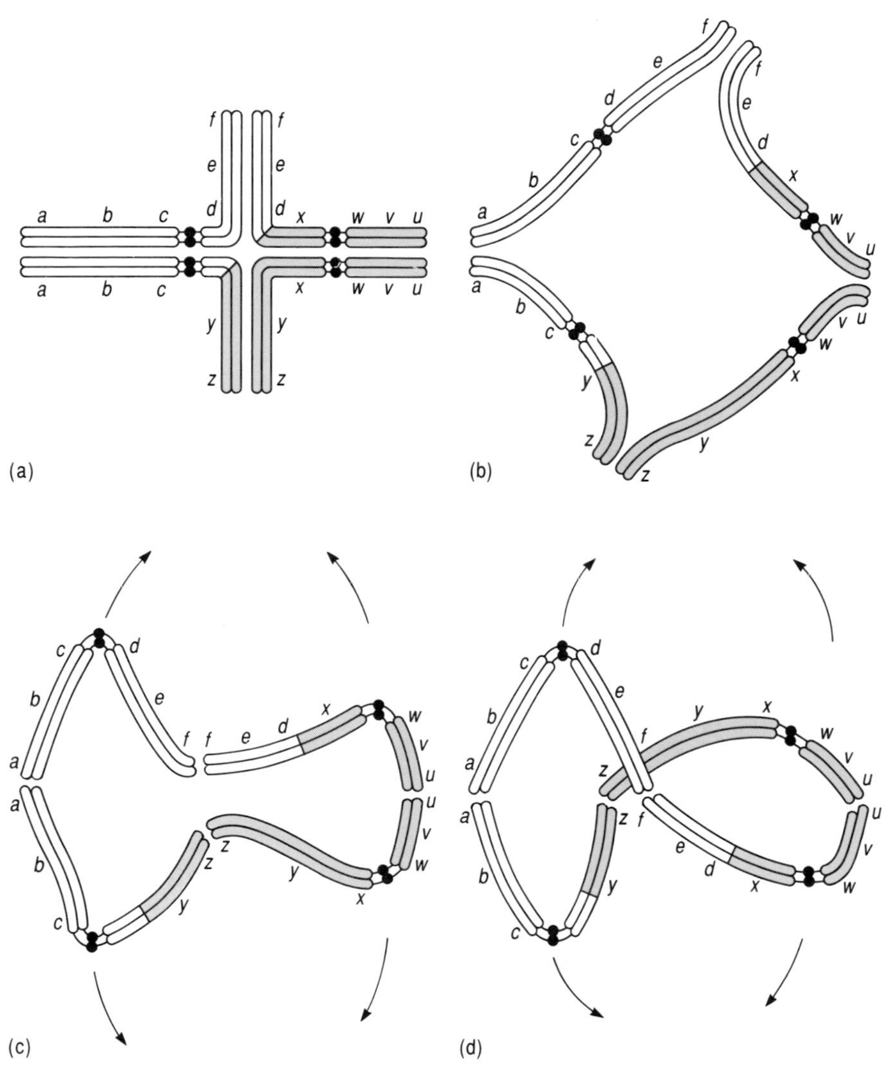

Figure 3.17 *The behaviour of four chromosomes in an individual with a reciprocal translocation. (As in Figure 3.16, the black circles are the centromeres.) (a) Pairing during meiosis. (b) Separation of the chromosomes. (c) and (d) Alternative configurations at anaphase I: (c) gives rise to non-viable gametes because one chromosome segment is duplicated and another is missing in both sets of segregating chromosomes; (d) gives rise to viable gametes as both upper and lower combinations are balanced, i.e. they contain all parts of the two chromosomes. Thus, assuming (c) and (d) are equally likely configurations, on average half the gametes produced will be defective. (The arrows in (c) and (d) indicate movement of chromosomes to opposite ends of the cell.)*

▶ How would the deletion or insertion of a base affect the coding sequence?

It would alter the codon at the point where the insertion or loss occurred and it would shift the 'reading frame' of the subsequent base sequences.

▶ What would be the likely phenotypic effect of this?

The phenotypic effect is likely to be considerable as the amino acid sequence of the protein (beyond the point at which the deletion or insertion occurred) would be altered and the product would probably be non-functional.

Deletions or insertions of bases are known as *frameshift mutations*. The other kind of point mutation involves the substitution of one base for another and is called *base substitution*.

▶ How might base substitution affect the coding sequence?

Just the one codon would be altered.

▶ What would be the likely phenotypic effects of this?

There are several possibilities. As more than one codon may specify a particular amino acid, the alteration of one base in the codon might *not* alter the amino acid and even if the amino acid is altered it might not substantially affect the activity of the protein. In both these instances the mutation would not be observed in the phenotype without detailed molecular analysis of the organism. But altering the amino acid and hence the protein could have marked or lethal effects on the phenotype and there is the further possibility that the new codon might terminate the polypeptide chain prematurely.

Detailed analysis of haemoglobin taken from large numbers of people shows that there are hundreds of variants resulting from hundreds of different kinds of base substitution. Most of them are cryptic or neutral, that is they do not give rise to an observable alteration of the phenotype because they have taken place in parts of the haemoglobin molecule that are not essential to its function. The condition sickle-cell anaemia is an exception: the replacement of glutamic acid by valine affects the functioning of that protein molecule and has considerable knock-on effects, as shown in Figure 3.7.

In general, since organisms are usually well adapted to their environment, phenotypically expressed mutations are more likely to be deleterious than advantageous.

▶ Why should this be so?

There is less scope for improving the function of a molecule than there is for impairing it. The mutation that gives rise to abnormal haemoglobin as shown in

Figure 3.7 produces a molecule that, in its altered behaviour, presents the organism with more costs than benefits. Historically, the term mutation was first used to describe gross morphological changes. Hugo de Vries (1848–1935), the Dutch plant geneticist, was the first to use the term and he used it to refer to new varieties. The initial consensus of opinion that mutations were largely beneficial, giving rise to variants that were fitter than their parents gave way to the opposite view, namely that the majority of mutations were deleterious, as expressed in the previous paragraph.

▶ Is there any evidence that both opinions are wrong?

Yes. If most of the genome has no phenotypic expression then most mutations will not be deleterious; for example, a change in an intron may pass totally unnoticed. The survey of human haemoglobins suggests that many mutations are neutral.

A mutation that neither increases nor decreases an organism's fitness relative to another allele at the same locus is said to be a selectively **neutral mutation**. The extent to which evolution may occur as a result of fluctuations in the frequencies of selectively neutral alleles will be discussed in Chapter 4.

3.3.3 THE RANDOMNESS OF MUTATION AND RATES OF MUTATION

Mutation is described as being random because it is not directed by the environment. In other words, it is not possible for mutations to arise in response to changes in the environment in order to improve the adaptedness of their carriers.

▶ What exceptions are there to this statement?

Some responses of the immune system appear to violate this principle, as does the response of flax to fertilizer (Section 3.2.4). Nevertheless, in the main, mutation is random with respect to its actual or potential usefulness to the organism.

Human genetic disorders Some mutations are readily identifiable because of their expression in the phenotype. The frequencies of occurrence of some human genetic disorders that arise by mutation are shown in Table 3.1.

▶ Look at the information on the genetic defect responsible for each disorder in Table 3.1 and suggest some reasons for the difference in incidence of these disorders.

Down's syndrome and Klinefelter's syndrome are far more common than the other syndromes and they both involve the duplication of chromosomes. You might have suggested that duplication of chromosomes is a more likely event than a point mutation and that chromosomal duplication would always have a phenotypic effect, whereas you have just read that a point mutation can very often occur without an observable phenotypic effect (as in the haemoglobin molecule). A further conclusion from the data in Table 3.1 is that a defect arising from a recessive gene

Table 3.1 *The frequencies of occurrence within the human population of five syndromes that have genetic causes.*

Name	Frequency per 10^5 live births	Genetic defect	Physical manifestation	Life expectancy
Down's syndrome	14	An extra chromosome (3 chromosome 21)	Rounded face with mongoloid eyes; tendency to be overweight; often have cardiovascular problems.	Variable, but the cardiovascular problem may decrease life expectancy.
Klinefelter's syndrome	17	An extra 'sex' chromosome: XXY	Male appearance	Normal
Achondroplasia	4.3	Gene mutation expressed in the heterozygous condition (i.e. dominant) (death *in utero* for the homozygote)	Dwarfism, resulting from very short limbs	Normal
Hutchinson–Gilford syndrome	0.125	Gene mutation expressed in the homozygous condition (i.e. recessive)	Shortness of stature and premature ageing	Not more than 20 years
Cyclopia	rare	Uncertain: recessive gene mutation, deletion of the short arm of chromosome 18	One central eye	Mostly stillborn; at most only minutes

mutation, i.e. one that is expressed only in the homozygous condition, will be rarer than one arising from a dominant one, which can also be expressed in the heterozygous condition.

In fact, in the germ cells of women, prophase I of meiosis is arrested from birth until the menstrual cycle that releases the ovum for fertilization. The 'mistake' that occurs in both Down's and Klinefelter's syndromes is that chromosomes fail to separate at anaphase I. This is a mistake that is more likely to occur the older the woman is, possibly as a result of the long delay in prophase I of meiosis. So the environment could be said to have affected the likelihood of this particular mutation occurring. Such an environmental influence also seems to be involved in achondroplasia (dwarfism resulting from short limbs) where there is evidence that mutations giving rise to this syndrome are more common in the sperm of older men.

In those syndromes that result from recessive mutations (e.g. Hutchinson–Gilford syndrome and cyclopia) it may be that the same symptoms arise by alterations to different genes. Additionally, even if only one base substitution is involved, it can be achieved in more ways in some cases than in others. For example, the amino acid valine is specified by CAA, CAG, CAT or CAC; there are therefore four possible base changes that would result in the triplets coding for alanine—CGA, CGG, CGT and CGC, but only two that would result in glutamic acid—CTT and CTC.

This means that even if each mutation, i.e. base substitution, is equally likely the possible outcomes are not all equally likely. Valine is twice as likely to be replaced by alanine as to be replaced by glutamic acid.

Another factor to consider is that some kinds of mutation have a greater chance of success than others. For example, some produce non-viable gametes, others have a lethal effect early in development.

Cyclopia (characterized by one centrally placed eye) is very rare; most children with this are stillborn and at best they live only minutes. It seems likely that others with this mutation will have aborted long before term. It is not possible to know how often mutations occur that give rise to non-viable gametes but the examples given above give some indication of why certain mutations are observed more frequently than others.

Mutation rates Mutations provide a continuous source of variation and it is useful therefore to know how frequently they arise. This is easier to calculate using bacteria and viruses because they are haploid and have short generation times. In higher organisms recessive mutations are masked by diploidy and generation times are longer, so estimation of mutation rates is more difficult.

The rate of mutation per base per replication is approximately 1×10^{-9}. The mutation rate per gene per cell generation is calculated to be of the order of 1×10^{-5} to 1×10^{-6}, in other words one mutation in 100 000 to 1 000 000 generations, but this is very variable (see Table 3.2). This is very low at the level of the individual but as each individual carries many genes, the chance of it

Table 3.2 *Spontaneous mutation rates of specific genes.*

Species and locus	Mutations per 100 000 cells or gametes
Escherichia coli (bacterium)	
streptomycin resistance	0.000 04
resistance to T1 phage	0.003
arginine independence	0.000 4
Salmonella typhimurium (bacterium)	
tryptophan independence	0.005
Neurospora crassa (fungus)	
adenine independence	0.000 8–0.029
Drosophila melanogaster (fruit-fly)	
yellow body	12
brown eyes	3
eyeless	6
Zea mays (maize)	
sugary seed	0.24
I to *i*	10.60
Homo sapiens (human)	
retinoblastinoma	1.2–2.3
achondroplasia	4.2–14.3
Huntington's disease	0.5
Mus musculus (house mouse)	
a (coat colour)	7.1
c (coat colour)	0.97
d (coat colour)	1.92
In (coat colour)	1.51

carrying a mutation somewhere in the genome is much higher. If we consider the world-wide human population to be around 4×10^9 individuals then at any one time about 80 000 individuals carry a newly arisen mutation within just one gene (4×10^9 individuals $\times$ 2 alleles per gene $\times 10^{-5}$ mutation frequency for the given gene). Chromosomal mutations are more common. Estimates put the level of mutation in a given class of chromosomal mutation, such as reciprocal translocation, at around 1×10^{-4} to 1×10^{-3} per gamete per generation.

So even though by no means all of the genetic variation is phenotypically expressed, there is evidence of a great deal of genetic variation being continuously generated.

There is, however, controversy over the extent to which selection does operate on mutations and the extent to which genetic variation is maintained in populations because it is selectively neutral. The question is therefore whether the amino acid substitutions that become established are due to random processes or to natural selection. The way in which an allele could increase in frequency within a population solely as a result of random processes is discussed in Section 3.6.

SUMMARY OF SECTION 3.3

There is considerable genetic variability found in natural populations of a species. In organisms that reproduce asexually (by mitotic division), variability between parent and offspring can occur only as a result of mutation. However, in sexual reproduction, the process of meiosis results in recombination and this leads to considerable variability between parents and offspring besides any alteration of the genetic material such as occurs in mutation.

Chromosomal mutations include inversions, translocations and polyploidy, each of which can be of considerable evolutionary significance.

Gene mutations (point mutations) are of two kinds: frameshift mutations or base substitutions. The latter are more likely to be cryptic and carried by the phenotype without detriment to the organism.

Mutation is random, in that it does not arise in response to changes in the environment so as to improve the adaptedness of the carrier. The fact that some mutations are more frequently expressed in the phenotype than others is a consequence of such factors as the non-viability of some gametes and the cryptic nature of many mutations. Chromosomal mutations, such as translocations and inversions, are more common than a given point mutation.

3.4 Genes in populations: the maintenance of variation

If there is no selection operating on individuals with a particular character, say flower colour in garden peas, and there is more than one colour in the population, then the relative abundance of each variant will remain constant from one generation to the next provided certain assumptions, to be explained shortly, are fulfilled. This was proved independently by G. H. Hardy and W. Weinberg at the beginning of the 20th century, and their proof shows that variation, i.e. *genotype frequencies*, can remain constant in a population as a consequence of Mendelian ratios (see Figure 3.2).

To understand this, consider an imaginary population of the garden pea *Pisum sativum*, the species in which Mendel studied the inheritance of plant height.

Differences in plant height are determined by a single gene that occurs as one of two alleles, either T (dominant) or t (recessive). Garden peas are diploid, and therefore have two alleles, one on each of the two homologous chromosomes. The possible genotypes for plant height are TT, Tt or tt. The frequency of the T allele in a population of plants is simply the number of plants carrying one copy (Tt) plus twice the number of plants carrying two copies (TT) divided by the total number of alleles at that locus (which equals twice the number of plants, since the plants are diploid). In other words, if the frequency of the T allele is called p, then:

$$p = \frac{\text{number of } Tt \text{ plants} + 2 \times \text{number of } TT \text{ plants}}{2 \times \text{total number of plants}} \tag{3.1}$$

The frequency of the second allele is conventionally represented by q. Because there are just two alleles for plant height in the population, the sum of their frequencies is equal to one, or in other words:

$$p + q = 1. \tag{3.2}$$

Therefore $q = 1 - p$.

If we know the frequency p of an allele in a population, some elementary assumptions allow us to predict the relative numbers of the three possible genotypes present in the population. There are four main assumptions:

1 The mating of individuals is random with respect to choice of partner. A population of such individuals is described as being **panmictic**.

▶ Does it seem likely that a bee or some other pollinator would move entirely randomly amongst a population of flowers?

No. In fact, random movement between plants by pollinating bees is very unusual. Bees usually localize their visits, moving between neighbouring flowers. Pollinator behaviour such as this is one of the factors which make it unlikely that random mating occurs in natural plant populations. Other factors also make it unlikely in animal populations. Non-random mating that involves like pairing with like is said to be **positive assortative**. Where there is a negative correlation between the characters of mating pairs (e.g. short with tall individuals) the mating is termed **negative assortative** (or sometimes *disassortative*). It does, however, simplify the calculations to start with the assumption that mating is random.

In addition, we have to assume that other factors which could affect allele frequencies are ruled out.

2 There are no differences in viability or fecundity of different genotypes.

3 There are no mutations or migration affecting allele frequencies in the population.

4 The size of the population is effectively infinite.

When these conditions are met, the probability of a gamete carrying a T allele uniting with another T gamete to form a TT zygote is the square of the frequency (p) of T, i.e. p^2. On the same grounds, two t alleles will form a zygote with a tt genotype with the frequency q^2. The heterozygote may be formed in two ways, either as Tt or tT, where the first allele derives from the male parent. Therefore, the frequency of the heterozygote is the probability that a Tt pair will form (pq) plus the probability of a tT pair of alleles forming (qp), i.e. $pq + qp = 2pq$. In summary, the probability of all these genotypes forming is shown in Table 3.3.

Table 3.3 *Allele and genotype frequencies for a character showing Mendelian inheritance.*

allele (frequency)	T (p)	t (q)	
T (p)	TT (p^2)	Tt (pq)	genotype (frequency)
t (q)	tT (qp)	tt (q^2)	

The sum of all the frequencies must equal one, so:

$$p^2 + 2pq + q^2 = 1 \tag{3.3}$$

This formula is known as the Hardy–Weinberg equation. The ratio of the genotype frequencies, $p^2 : 2pq : q^2$, is the **Hardy–Weinberg ratio**. Note that it does not matter which allele is dominant as the formula relates allele frequency to genotype frequency. Moreover, the ratio will be maintained in succeeding generations because the frequency of T alleles amongst those of all the progeny is:

$$\frac{2 \times (\text{frequency of } TT \text{ genotype}) + \text{frequency of } Tt \text{ and } tT \text{ genotypes}}{2 \times (\text{frequency of all genotypes})}$$

$$\text{or } \frac{2p^2 + pq + qp}{2 \times (1)}$$

$$= p^2 + pq$$

$$= p(p + q)$$

$$= p \text{ (since } p + q = 1)$$

Thus the phenotypic effect is irrelevant in maintaining the alleles in the population under these conditions.

The Hardy–Weinberg equation is fundamental to the study of evolutionary changes in allele frequencies in populations. It allows us to predict the expected ratio of genotypes from known allele frequencies when a population is at Hardy–Weinberg equilibrium, which is defined as the point when no natural selection is operating on that gene. You will use this equation again in Chapter 4 when examining allele frequencies in the light of possible selection pressures.

In fact, genotype frequencies are quite often very close to the theoretical expectation of the Hardy–Weinberg ratio and you might think of the reason for this as you work through the next example of data on the scarlet tiger-moth *Panaxia dominula* (Table 3.4). E. B. Ford and his colleagues collected 1 612 individuals and categorized them for one character, as follows: 1 469 had white spotting on the forewings, 5 had much reduced spotting and 138 were between the two. Do the genotype frequencies fit the Hardy–Weinberg ratio?

Table 3.4 *Data from a collection of 1612 scarlet tiger-moths.*

Phenotype	No. of individuals	Assumed genotype	No. of S alleles	No. of s alleles
White spotting	1 469	SS	$1\,469 \times 2$	—
Intermediate	138	Ss	138	138
Little spotting	5	ss	—	5×2

The frequency p of the S allele in the population is $(1\,469 \times 2) + 138 = 3\,076$ divided by the total number of alleles in the population, i.e. $2 \times 1\,612$.

So, $p = \dfrac{3\,076}{1\,612 \times 2} = 0.954$

$q = 1 - p$

therefore

$q = 0.046$

The Hardy–Weinberg ratio is $p^2 : 2pq : q^2$

$= 0.910\,1 : 0.087\,8 : 0.002$

For 1 612 individuals the expected ratio is therefore

$(0.910\,1 \times 1\,612) : (0.087\,8 \times 1\,612) : (0.002 \times 1\,612)$

$= 1\,467 : 142 : 3$

The observed ratio is thus very close to the expected ratio. That this is so despite all the constraints or limiting conditions placed on the Hardy–Weinberg ratio is perhaps not altogether surprising. For example, even if mating is non-random with respect to some characters, it may not be for many others. For instance, in a human population some individuals can curl their tongues while others cannot, some can smell freesias, others cannot, and to some the chemical phenylthiocarbamide tastes foul, whilst others find it tasteless. All these are characteristics that are inherited in the simple Mendelian ratios. It is understandable that variation in these characteristics remains constant for it is unlikely that choice of partner will be affected by ability to curl one's tongue, smell freesias or taste phenylthiocarbamide! Where mating is assortative for a given character, however, there will be deviations from the Hardy–Weinberg ratio. Positive assortative mating will lead to an excess of homozygotes and negative assortative mating to an excess of heterozygotes with respect to the Hardy–Weinberg ratio.

SUMMARY OF SECTION 3.4

In the absence of natural selection, and given other assumptions, allele frequencies are held constant in a population and reach equilibrium genotype ratios in a single generation. This is proved mathematically, the ratio of genotypes being known as the Hardy–Weinberg ratio. There are four assumptions. Mating must be panmictic (i.e. random) and there must be no differences in viability or fecundity of different genotypes. The size of the population must be effectively infinite and there should be no mutations or migrations affecting allele frequencies. Genotype frequencies are often close to the Hardy–Weinberg ratio because the conditions are in practice met for the character under investigation.

3.5 MEASURING VARIATION

The ability to measure variation and to make comparisons both within and between species allows suggestions to be made about relationships, leading to hypotheses about evolutionary history and predictions about the possible course of evolution in the future. Most methods of measurement are to some extent indirect measures of heritable variation, so it is important to know something about the way these measures are made in order to assess the worth of the ideas that they generate.

3.5.1 MEASURING VARIATION IN PHENOTYPIC CHARACTERS

In calculating the frequency of alleles in a population of scarlet tiger-moths using the Hardy–Weinberg ratio, Ford and his co-workers first collected and described individuals from a population of these moths. They did not collect every individual but instead took a **random sample**. A random sample is a sample that is supposedly representative of the population as a whole. That is, individuals must be collected without regard to their genotype or location.

The way in which the data are collected is important if the effects of location and genotype are to be eliminated. Setting a moth trap using a sexually receptive female as a lure would obviously bias the sample collected towards sexually active males.

The scoring of information about phenotypic characters can also present difficulties. It is easy enough when variation in a particular character is continuous and exact measurements can be taken, for example, height, weight and the like. These are known as **interval level measurements** and a distinguishing feature is that you can assess the size of the differences between measurements. For example, if you had three pea plants measuring 14 cm, 42 cm and 46 cm, then you could say that the first plant and the second differed in height by 28 cm and that this was seven times the difference between the second and third plant heights. But if the variation in plant height was not continuous across the complete range of possible heights from 14 cm to 46 cm but instead clustered around those two measures then you might feel justified in scoring the plants as either short or tall (and it would certainly save you a lot of time!). If you collect discrete data like this, it is known as **categorical** or **nominal level measurement**. There are many measurements that can be scored only in this way, for example, male and female, horns or no horns, winged or wingless. These categories cannot be arranged in any particular order or rank. The third type of measurement that can be made is **ordinal level measurement**. These are measurements where it is possible to say that one is bigger than another and so to rank them with respect to each other, but where it is impossible to assess the size of the difference. For example, Ford's moths were scored as (1) with spotting, (2) intermediate and (3) little spotting. Now it is possible that with an enormous amount of work these moth data could have been collected using exact interval level measurements, with spotting scored as a percentage of total forewing surface for example, but it was not judged to be a sensible way of representing the data. The point is that once data are scored in this way it is then impossible to assess the size of the difference. Plant height could be scored as (1) tall, (2) intermediate and (3) small. This would be ordinal level measurement, and if you collect data in this way it is impossible to use interval level measurements later, because you have not taken exact measurements in the first place.

In addition to the methods of sampling and of scoring information it is also necessary to consider the effects of deriving information from samples of finite size.

In the worked example given in the previous Section, one of the phenotypes was quite rare.

► What might have happened if Ford had collected only 200 rather than 1 612 moths?

It is possible that there would not have been any individuals showing the rare phenotype with little spotting.

For any given sample size, estimates carried out on different occasions will differ from one another and will also deviate somewhat from the actual population frequency. This is what is known as **sampling error**. Although it is known that the sample will differ from the total population, *inferences* can be made about the population from the sample. Statistical techniques can tell us how confident we can be in drawing certain inferences from a particular set of data.

► Consider the study of the European rabbit in Australia described in Chapter 2, Section 2.3.1. Are the data in Table 2.1 interval, nominal or ordinal level measurements?

They are interval level measurements. Exact measurements have been taken from each rabbit and the values shown in the Table are the averages or **means** (symbol $\bar{x}$). Comparing the means of a particular value, such as body mass, for two different samples is just one way in which the populations can be compared.

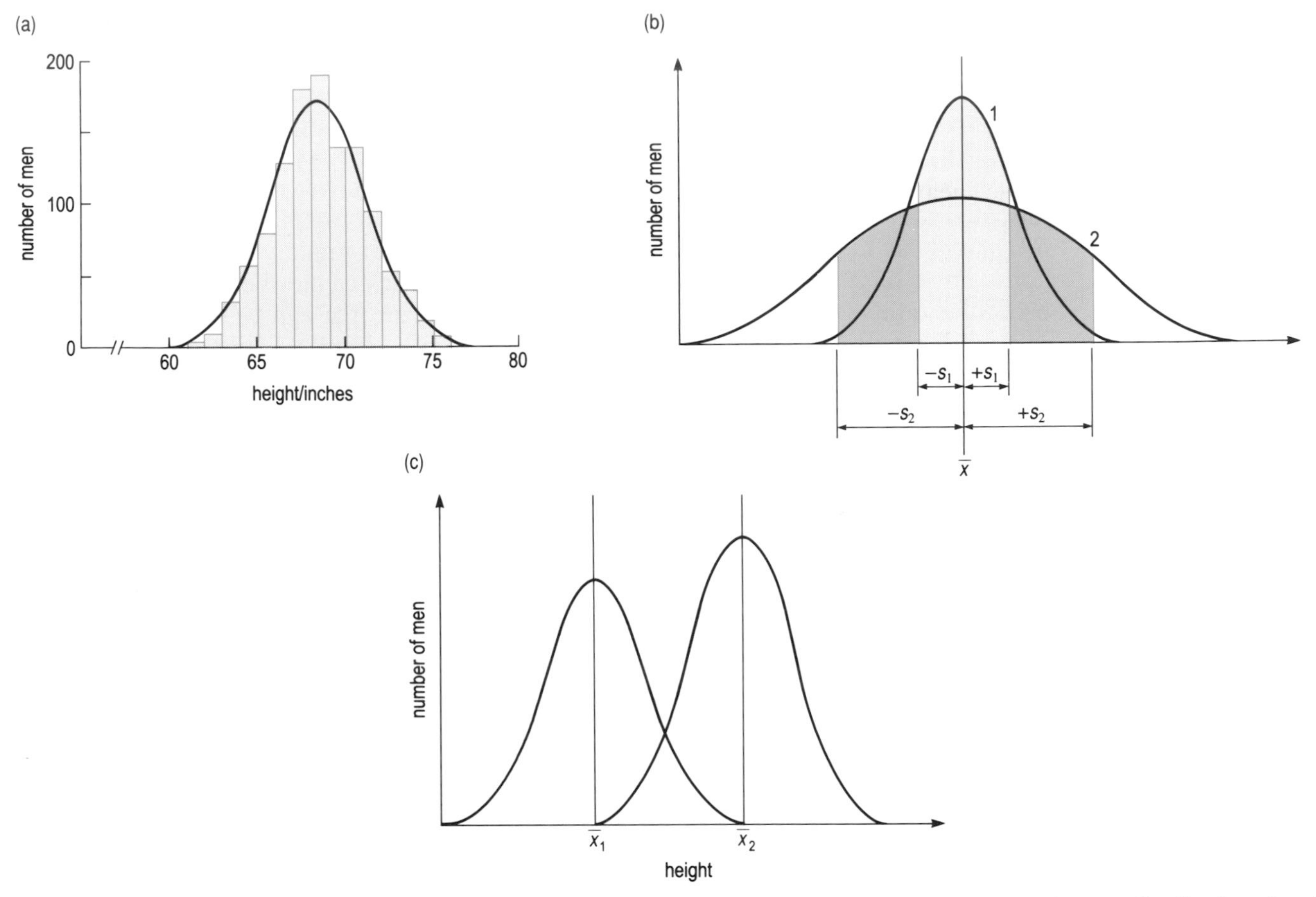

Figure 3.18 *(a) Normal distribution curve fitted to the heights of 1164 British men. (b) Two curves both normally distributed and with the same mean,* $\bar{x}$*, but with a different spread of values around the mean.* s_1 *is one standard deviation for curve 1;* s_2 *is one standard deviation for curve 2. (c) Two curves both normally distributed but with different means,* $\bar{x}_1$ *and* $\bar{x}_2$.

But how reliable is such a comparison? It is possible to select at random two samples of rabbits that have come from the same population and yet find that the samples have different mean values for body mass.

▶ To what would the differences between the two samples be attributed?

Sampling error. To minimize the chance of drawing incorrect conclusions from the data, a more complete comparison between the two sets of data is required. A more complete way of describing sample data can be understood by looking at Figure 3.18.

Figure 3.18(a) shows data on height for 1164 British men. The data, presented as a histogram, show a distribution that is very commonly found in biological measurements. The bell-shaped curve superimposed onto the histogram is known as a **normal distribution**.

Figure 3.18(b) shows the curve for the height data from Figure 3.18(a) superimposed upon a second curve. This second curve has the same mean but there is a much greater spread of values. There are more shorter and more taller men in this population sample than there were in the first one. The curve still shows a normal distribution.

In Figure 3.18(c) both the curves show a normal distribution but differ in their means. A fuller description of the data, describing the spread of values in addition to the means, allows statistical tests to be applied to assess the likelihood that samples differing in their mean value have nevertheless been drawn from the same population (i.e. that the difference is due only to sampling error).

Because biological characters are so often found to be normally distributed in populations, it is generally assumed that there will be a normal distribution unless there is evidence to the contrary. The curve can then be specified by two numbers (correctly known as *parameters* when referring to whole populations or *statistics* when referring to a sample), the mean and the **standard deviation** (see Figure 3.18(b)). This deviation (on both sides of the mean, so it has both a positive and a negative value) contains about two-thirds (in fact 68%) of all the values. Curves with a normal distribution are symmetrical about the mean and approximately 95% of all values will be contained within two standard deviations about the mean. The idea behind a standard deviation of a sample is to find a measure of the distance, or spread, of values from the sample mean. If you were literally to calculate an average of the distances you would find that it was always zero. For example, a sample of three measurements, 8 mm, 9 mm and 10 mm respectively, has a sample mean of 9.0 mm and deviations from this mean of −1.0, 0 and +1.0 mm respectively. The average of these deviations is 0:

$$\frac{-1.0 + 0 + 1.0}{3} = 0$$

This problem can be overcome if each deviation from the mean is squared. The values are now all positive:

$$-1.0^2 = 1.0 \qquad 0^2 = 0 \qquad 1.0^2 = 1.0$$

These squared values are summed and the average value taken to give a measure of the spread of the data around the mean. (In fact, in analysing *samples* it is customary to sum the squares and then divide by one less than the sample size in order to obtain this average value. This compensates for the fact that the smaller a sample is, the more likely it is to *underestimate* the true spread of data around the mean in the population.) Hence:

$$\frac{1.0 + 0 + 1.0}{3} = 0.67$$

for a population (of only three individuals), or:

$$\frac{1.0 + 0 + 1.0}{(3 - 1)} = 1.0$$

for a sample drawn from a larger population.

This figure is called the **variance**. Variance is often used as a measure of the amount of variability found in a population or sample and will be used in Chapter 4 and subsequent Chapters. It is a squared measure—denoted by the symbol s^2—so to get back to a linear measure as represented on the curves shown in Figure 3.18 the square root of the variance is taken. This value is the standard deviation, s.

To return to the original problem (that of making inferences about a population from a sample of the population), it is important to give some thought to the way data are collected because this affects the type of statistical test that can be used. Usually, only interval level measurements show normal distributions, the statistics of which can form the basis of many tests.

Null hypothesis versus alternative hypothesis Before using any statistical test the hypothesis being tested must be stated. Conventionally that hypothesis is the **null hypothesis**. A null hypothesis is a hypothesis of no difference, in other words the expectation of the null hypothesis is that there is no difference between populations being compared. The **alternative hypothesis** is that there *is* a difference.

▶ What would be a possible null hypothesis being tested by the data collected for Table 2.1?

One null hypothesis is that there is no difference in body mass between the populations of rabbits from the three regions of Australia. Another null hypothesis could be that there are no differences between the morphometric measurements of male and female rabbits.

Having formulated the hypotheses to be tested, and having gathered the data, samples can be compared using the statistics derived from the normally distributed curve of each sample, to estimate the probability that two samples come from the same population. Details of the statistical methods used can be found in many textbooks (see 'Further reading' at end of this Chapter).

According to the null hypothesis, any difference between two samples will be due to sampling error alone. But the probability of this being so decreases as the observed difference between samples increases: larger differences between samples make the alternative hypothesis of a real difference between the populations from which the samples came more likely. So at what point do we decide that the probability (P) of the null hypothesis being correct is so low that we can reject it (and so accept the alternative hypothesis)? There is, in fact, no objective criterion for making such a decision. Rather, an arbitrary limit of minimum probability has to be selected at the outset. If the probability of the null hypothesis being correct falls below that level of probability, then, by that preselected criterion, it is rejected. Such a limit of probability is termed a **significance level**, and if any two samples differ to the extent that the null hypothesis may be rejected, they are said to be *significantly different*. It is conventional in much biological and palaeontological work to adopt a probability of 5% ($P = 0.05$) as the significance level for such inferential statistical tests. Sometimes, a more conservative level of 1% ($P = 0.01$) is employed, and sample data differing at this level are said to show highly significant differences. Because significance levels are only an arbitrary criterion, you should always bear in mind that, even if samples *are* significantly different, a small possibility still exists that the null hypothesis might be correct: we choose to reject it only because that probability is so small. Likewise, if no significant difference can be detected between samples, then we cannot necessarily assume that the null hypothesis is correct. Further data might allow us to reject it. Hence all that can be said is that the null hypothesis cannot be rejected from the data collected. The use of inferential statistical tests in morphological analysis will be illustrated in Chapter 10, in relation to the comparison of samples of fossils.

Heritability Where variation in a character results from the contribution of a number of genes and environmental effects, it can be possible to estimate the variance due to individual factors and then sum them. Conversely, a total variance can be partitioned into the variances attributable to a variety of individual factors. For such polygenic characters **heritability** is given as the proportion of the total phenotypic variance that is attributable to genetic causes and it can be calculated from the formula:

$$\text{heritability, } h^2 = \frac{\text{genetic variance}}{\text{total phenotypic variance}} \qquad (3.4)$$

(The convention of denoting heritability by the squared term h^2 indicates that it is based on variance, which is itself a squared term.)

▶ What are the components of the total phenotypic variance?

The phenotypic variance is the sum of the genetic variance and the environmental variance.

▶ What is the range of values that h^2 can take?

h^2 can fall in the range 0 (genetic variance = 0) to 1 (environmental variance = 0).

Clearly for plant and animal breeders it is of practical importance to know what proportion of the phenotypic variance within the population is genetic in origin. This can be discovered by large-scale breeding programmes.

The term heritability is often used to imply resemblance between parents and their offspring. In general the closer the resemblance between parent and offspring the higher the heritability of a particular characteristic.

A mean parent value for a particular characteristic can be compared with the mean offspring value for that same characteristic. If there is a strong association between the two, heritability is high. This method has been used to calculate heritability of several morphological characteristics in a population of Darwin's finches *Geospiza fortis* (see Chapter 1, Figure 1.4), and some values are shown in Table 3.5.

Table 3.5 *Heritability of morphological features in* Geospiza fortis.

Character	Heritability, h^2
Body weight	0.91
Tarsus (digit) length	0.71
Bill length	0.65
Bill width	0.90

The problem with this method is that there may be non-genetic maternal effects as well as environmental effects. In practice it is difficult to eliminate these confounding effects when studying animals in the field, and breeding experiments have to be undertaken.

▶ Can you suggest any manipulation that could reduce maternal effects?

It is possible to use foster mothers, transferring eggs or young (depending on the species). This eliminates some maternal effects and also other environmental effects shared by families.

The problem of measuring heritability in the field will be discussed in more detail in Chapter 4, Section 4.2.

3.5.2 MEASURING GENETIC VARIATION

The inheritance of most phenotypic characters is so complex that we cannot calculate genetic variation by simply measuring phenotypic variability. We have no way of knowing just from the variation found in the height of human beings, for example, how many genetic loci are involved in its inheritance and how genes at any such loci interact with environmental factors (such as nutrition) to determine a person's height.

Gel electrophoresis In the mid-1960s gel **electrophoresis** was first used to detect differences in proteins and polypeptides and thus by inference differences in genes. In simple proteins, minor differences in molecular structure are caused by amino acid changes which frequently alter the electrical charge of the molecule slightly. It is principally differences in charge that are detected by electrophoresis.

Electrophoresis is most often used for detecting genetic variation in enzymes. The different enzyme forms that arise from alternative alleles and have different electrical charges are known as **allozymes**. During electrophoresis the allozymes are separated from each other by their different rates of movement in an electric field. The technique involves placing enzyme extracts obtained from the tissue of different individuals in a jelly-like medium (a starch or acrylamide 'gel', for example) which allows the molecules to migrate from one end of the medium to the other when a high voltage is placed across it.

Once the gel has been run, the positions of enzyme samples from each individual are identified by a stain that reacts with the specific enzyme being tested so that each appears as one or more stained bands as shown in Figure 3.19. Homozygous individuals produce a single band whose position is determined by the enzyme's charge. A heterozygote will show a separate band for each of its alleles, if both produce enzymes. Each diploid individual can show two alleles at a locus, but if there are more than two alleles in the population then this should be revealed if enough individuals are assayed by this method.

Since electrophoretic methods can detect genes with a single allele as well as those with more than one, it is possible to estimate quantitatively the genetic variability of populations. Two measures of this variation are used. The first is the proportion

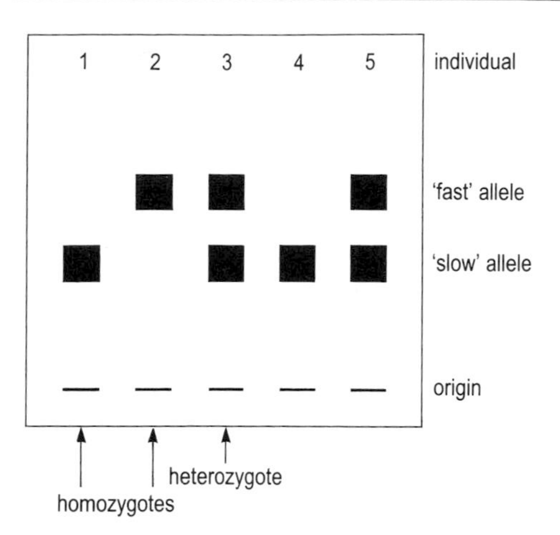

Figure 3.19 *Diagram of a gel showing the separate protein bands produced by electrophoresis.*

of loci with more than one allele (polymorphic loci) in a population, and the second is known as the average heterozygosity per locus, often just called **heterozygosity**, and given the symbol H. This is calculated by working out, for observed allele frequencies, the proportion of genotypes expected to be heterozygous at each locus investigated, adding all these proportions together and then dividing by the number of loci investigated. In mathematical terms this is:

$$H = \frac{\sum_{i=1}^{i=n} \text{proportion of population heterozygous at locus } i}{n} \qquad (3.5)$$

Where n is the number of loci which have been tested and $\sum_{i=1}^{i=n}$ means 'add up the n terms in the expression, from 1 to n'. H can take any value from $H = 1$, indicating that all loci are heterozygous, to $H = 0$, which signifies no heterozygosity.

Some values of polymorphism and heterozygosity obtained by electrophoretic surveys of enzymes in various populations are given in Table 3.6.

▶ Study Table 3.6 and say whether, on average, plants or animals show the greatest genetic variability.

The greatest genetic variability is shown by two plants, maize and the ponderosa pine tree. Both have two-thirds of their loci polymorphic and show average heterozygosity of 0.42 and 0.23, respectively. It should be noted, however, that the number of loci tested is small. It is not known whether the genetic variability found using this method is representative of the overall genetic variability of the species.

Table 3.6 *Some values of polymorphism and heterozygosity found in a variety of animal and plant populations. Polymorphic loci in which one allele has a frequency of less than 0.05 have been considered to be monomorphic.*

Species	Number of loci tested, n	Proportion of loci polymorphic	Heterozygosity per locus, H
Homo sapiens (human)	71	0.28	0.067
Mus musculus-domesticus (house mouse)	41	0.20	0.056
Peromyscus polionotus (white-footed mouse)	32	0.23	0.057
Drosophila melanogaster (fruit-fly)	19	0.42	0.061
Limulus polyphemus (horseshoe crab)	25	0.25	0.061
Oenothera biennis (evening primrose)	10	0.30	0.100
Pinus ponderosa (ponderosa pine tree)	22	0.68	0.230
Zea mays (maize)	11	0.67	0.420
Silene maritima (sea campion)	21	0.29	0.150
Avena barbata (wild oats)	17	0.26	0.040

Immunological techniques Estimates of the degree of similarity between proteins from different species can be obtained by immunological techniques. A protein such as albumin is purified from individuals of a species, say humans, and is injected into those of another species, typically rabbits. The rabbit produces antibody proteins (called immunoglobulins) against the foreign protein (the antigen). These antibodies react with the antigen, combine with it and precipitate it out of solution. Antibodies are very specific to the antigen used (Section 3.2.4), but proteins homologous to the original antigen, such as albumin from related species, will also react with the antibody to a greater or lesser degree depending on their similarity to the original antigen. The degree of dissimilarity between a protein from one species and a homologous protein from another species, as reflected by the strength of the reaction, is expressed as *immunological distance*, and is approximately proportional to the number of amino acid differences between the homologous proteins.

Immunological distances between humans, apes and Old World monkeys are given in Table 3.7. Separate antibodies were prepared against albumin obtained from humans, chimpanzees and gibbons, isolated and reacted with albumins from six species of apes and six species of Old World monkeys. The tests with antibodies prepared against humans show that albumins from the African apes (chimpanzee and gorilla) more closely resemble those from humans than do the albumins of Asiatic apes (orang-utan, siamang and gibbon). The albumins from the Old World monkeys differ most from the human albumin.

Table 3.7 *Immunological distances between albumins of various primates.*

	Antibodies to albumin from:		
Species tested	human	chimpanzee	gibbon
Human	0	3.7	11.1
Chimpanzee	5.7	0	14.6
Gorilla	3.7	6.8	11.7
Orang-utan	8.6	9.3	11.1
Siamang	11.4	9.7	2.9
Gibbon	10.7	9.7	0
Old World monkeys (average of 6 spp.)	38.6	34.6	36.0

▶ Do the tests with antibodies prepared against chimpanzee albumin confirm this result?

Yes. Human and gorilla albumin are more similar to chimpanzee albumin than are the albumins of the Asiatic apes. And again the Old World monkeys are yet more distant.

▶ Using the data in Table 3.7, would you infer that humans are more closely related to chimpanzees or gorillas?

The tests with antibodies prepared against human albumin suggest a closer relationship between human and gorilla than between human and chimpanzee. Yet the distances evidently cannot be interpreted as precise measures of relationship: notice that the distances between pairs of species are not exactly equal when measured either way. For example, human albumin plus antibodies to chimpanzee albumin yields a different result (3.7) to that obtained with chimpanzee albumin plus antibodies to human albumin (5.7).

Another protein, lysozyme, gave results not entirely consistent with those for albumin. On the lysozyme data, the gorilla is further from humans than the orang-utan is. Inconsistencies like this indicate that inferences about phylogenies based on a single protein may be quite misleading, both because of discrepancies in the results from them and because different proteins may have evolved at different rates in different lineages. These sorts of problems in reconstructing relationships will be discussed further in Chapter 11.

Amino acid sequencing A refinement of the method for comparing single homologous proteins is by amino acid sequencing, i.e. determining the position and identity of every amino acid in a protein molecule. Again, consideration of a single protein rarely gives a clear indication of the true phylogenies of the species involved (see Chapter 11, Section 11.3).

DNA sequencing A more direct method of assessing genetic variability uses techniques of DNA sequencing, i.e. the position and identity of each base in a single strand of DNA is determined. These techniques have been available since the 1980s. They complement the earlier methods of using differences in proteins to infer genetic differences because they avoid any assumption that variations in the chemical structure of proteins are due to single gene effects. DNA sequencing also makes accessible the non-coding regions of DNA, and these can show great variability between individuals.

RFLP analysis In studies of variability, sequencing the whole genome is time-consuming and costly. Instead attention has focused on mitochondrial DNA (mtDNA), hypervariable DNA (hvrDNA) and fragments of DNA known as restriction fragment length polymorphisms (RFLPs or 'rifleps'). RFLPs are prepared by treating the (double-stranded) DNA molecules with enzymes known as *restriction enzymes*. Each restriction enzyme recognizes a particular sequence of 4–6 base pairs and cuts at a specific site within this region. For example, the enzyme *Hpa*I recognizes the base sequence:

G T T A A C
C A A T T G

and cuts thus:

G T T   A A C
C A A   T T G

The site where the cut is made is called a restriction site. The fragments of DNA produced after this treatment can be separated on the basis of size using polyacrylamide gel electrophoresis and their lengths can be determined by comparing their positions on the gel against the positions of standard fragments of known length. The fragments are made visible by binding them to radioactively labelled chemicals, a process called autoradiography. A particular fragment (DNA sequence) can be investigated using several restriction enzymes and thus a *restriction map* is built up (Figure 3.20).

Suppose you have a piece of DNA consisting of 10 000 base pairs (10 kb), and a restriction enzyme H cleaves it into two fragments as shown below:

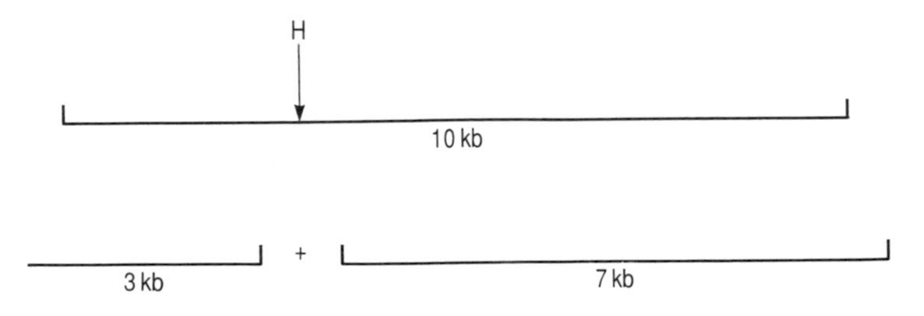

The original piece of DNA is said to have one restriction site for H. Suppose that the original piece of DNA also has one restriction site for the enzyme J, and that this yields fragments of lengths 4 kb and 6 kb.

▶ Does this information allow you to determine the relative positions of the restriction sites for H and J?

No, because there are two possible arrangements:

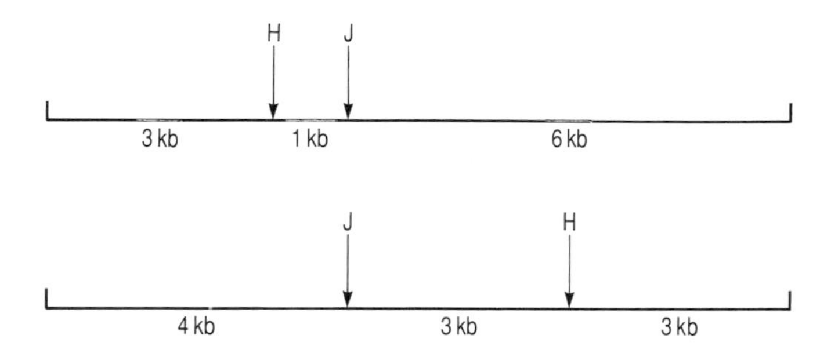

▶ How can you determine the correct sequence?

By using both enzymes together the fragment lengths generated will reveal the relative positions of these two restriction sites.

The use of several enzymes to map restriction sites on short lengths of DNA (about 100 kb) is the prelude to searching for variability between individuals as shown by the pattern of fragments, i.e. the restriction fragment length polymorphism (RFLP).

▶ What kind of variability will be detected?

A change in DNA base sequence that results in the loss of a restriction site will be detected. Also the length of fragments will vary if bases have been lost or gained but it will not be possible to say which base has been affected. Of course, many other kinds of variability will go unnoticed, for example the substitution of a base would not be noticed if it did not occur at a restriction site.

So far RFLPs have been used to assess the degree of relatedness between individuals and, as they focus on polymorphisms at one locus at a time, they can also be useful as markers for hereditary diseases. Another use is shown in Figure 3.20 where restriction enzyme maps of part of the non-transcribed regions within the rRNA genes of humans and apes are given.

▶ Look at Figure 3.20 and say how many restriction sites are identical comparing: (a) humans and chimpanzees; (b) humans and orang-utans; (c) chimpanzees and orang-utans. What does this tell you about relationships between these three species?

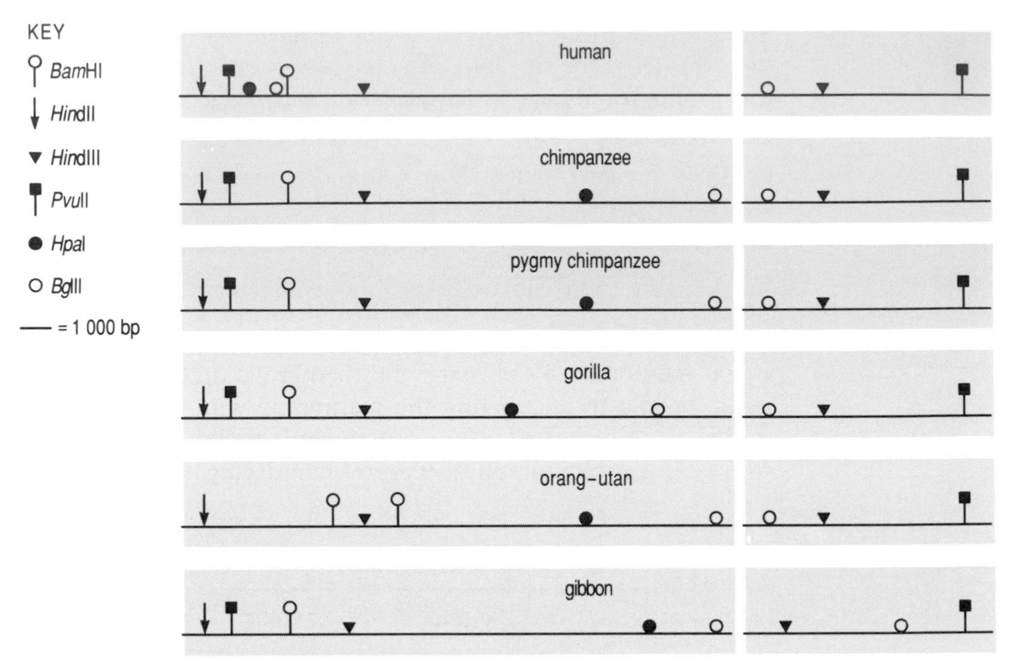

Figure 3.20 *Restriction enzyme maps of the non-transcribed 'spacer' regions in the rRNA genes of human and apes. The sites recognized by seven restriction enzymes are indicated for the non-transcribed region of each of several species. The break in the sequence indicates a portion that has not been studied.*

(a) 7 sites; (b) 5 sites; (c) 7 sites. This might suggest that whilst chimpanzees are equally closely related to humans and orang-utans, the orang-utans are less closely related to humans than are the chimpanzees, but you should realize that this suggestion is based on information from a very small part of the total genomic content of cells from these species. You might wish to compare the data in Figure 3.20 with those in Table 3.7 showing immunological distances.

▶ Is the information on humans, chimpanzees and orang-utans from these consistent?

Not completely. Although humans and chimpanzees are more closely related than humans and orang-utans as judged by both RFLPs and immunological distances of albumins, and chimpanzees are more closely related to humans than to orang-utans using the immunological distances investigated, the restriction enzyme maps by contrast suggest that chimpanzees have equally close relationships with humans and orang-utans. But remember that data using different proteins may also give different results.

DNA fingerprinting A technique that involves procedures that are very similar to those used for RFLP analysis is DNA fingerprinting. This procedure does, however, allow examination and detection of polymorphisms at many loci in one operation. As mentioned earlier (Section 3.2.4) much of the genome in eukaryotes is highly repetitive and it is these sequences that are amenable to this analysis because they contain multiple copies of particular base sequences (of unknown function) dispersed throughout the genome.

First of all, a DNA probe is made. The probe is a short length of DNA containing a repeated core sequence of bases which will bind to complementary sequences within the highly repetitive regions of the DNA under investigation. The probe first has to be cloned, that is it is isolated and inserted into a bacterium so that a large number of identical copies are synthesized as the bacterium repeatedly divides. The cloned probe is then radioactively labelled.

The DNA of interest is cut up into fragments using restriction enzymes and then subjected to electrophoresis to separate the restriction fragments according to length (as described earlier). Thus the different-sized fragments are separated into (invisible) bands within the gel. These bands are subsequently transferred to a nylon membrane by pressing the membrane against the gel. The bands are then made visible by incubating the membrane with the radioactively labelled DNA probe, washing off excess (i.e. unbound) probe and exposing to a photographic film (i.e. autoradiography), thereby identifying the bands on the membrane.

The pattern of bands revealed on the autoradiograph is unique to the individual, hence the term genetic fingerprinting. These bands contain the hypervariable regions (hvrDNA) which are short sequences of DNA (about 20 kb in length) consisting of multiple copies of the core base sequences and they are highly polymorphic because of the loss or gain of these core repeat units. One study of an exceptionally variable locus found 77 different alleles from a sample of 79 humans, with an estimated 97% heterozygosity and mutation rate of 0.003 per gamete.

Genetic fingerprinting is less useful than RFLP analysis for obtaining measures of relatedness because even unrelated individuals can share up to 30% of the bands, but it is a reliable indicator of close relationships and is used in bird as well as human studies, to ascribe or deny paternity to individuals (see Chapter 7) as well as having an important role in forensic sciences. In the future, apart from verifying genetic relationships in long-term field studies, the method could also be used to monitor and maximize the effects of outbreeding projects in populations of zoo animals; for example, with the condor and the Galapagos tortoises in San Diego, California.

Restriction enzymes have also been used to generate fragments of mtDNA. As mtDNA is a homologous structure in animals and shows considerable conservation of gene content, interspecific comparisons can usefully be made. Additionally, animal mtDNA is compact, having no introns and few duplications, and being usually less than 30 kb. So the task of sequencing the whole of the mitochondrial DNA is considerably less than that of sequencing the total genomic content of the cell. Further advantages are that maternal inheritance and the consequent lack of recombination make mtDNA a very useful marker for investigating recent evolutionary events although the absence of a direct link between the rate of evolution of mtDNA and nuclear DNA may introduce a complication. At present the analysis of mtDNA has proved useful in studies of population structure, geographic variation and phylogeny (see Chapters 9 and 11) and seems set to become an increasingly important analytical tool, with the caveat that what is learned about mtDNA in one species will not necessarily be universally applicable to all mtDNA.

DNA hybridization Finally, **DNA hybridization** is a technique that uses the whole genome and exploits the fact that when DNA is dissociated *in vitro* the two complementary strands reassociate more or less completely. This can be used to measure the degree of genetic similarity between two species by placing their dissociated DNAs together in a mixture and then allowing reassociation. The usual procedure is to measure the degree of inhibition of binding of one type of DNA by another, which is present in excess. (The single-strand DNA from one of the species is physically trapped, usually in an agar gel.) Results for human DNA obtained using this technique are shown in Table 3.8. This shows an estimated sequence homology of 100% between human and chimpanzee DNA, thus the method does not give a complete picture of the variability between the two species! But it has been used to establish the relative amount of variation between closely related species. For example, *Drosophila simulans* shows an 80% sequence homology with *D. melanogaster* whilst *D. funebris* shows only a 30% sequence homology with *D. melanogaster*. Note that this technique only tells you the net difference between the compared genomes, not how it is distributed in them. Nor does it indicate the amount of difference between those loci that are different enough not to pair at all, and is thus only a partial measure.

Hybrid DNA can be obtained in varying degrees depending on the amount of similarity (homology) between the base sequences in the DNA from different species. Some strands that associate will not be completely complementary in base sequences and these strands will tend to dissociate again when the temperature is raised. So the degree of mismatch between the nucleotides can be estimated from the difference between the temperature at which hybrid strands come apart and that at which the paired strands from the same species come apart: the smaller the temperature difference, the greater the similarity and hence inferred relatedness of the two species. This method of assessing relationships will be discussed further in Chapter 11.

Table 3.8 *Competition between 200 μg (or more) of DNA of various species and 0.5 μg of ^{14}C-labelled human DNA for binding with 0.5 μg of human DNA held on agar.*

DNA tested	Degree of taxonomic differentiation	% Inhibition of human–human DNA binding
Human	—	100
Chimpanzee	genus	100
Gibbon	family	94
Rhesus monkey	superfamily	88
Capuchin monkey	superfamily	88
Tarsier	suborder	65
Slow loris	suborder	58
Galago	suborder	58
Lemur	suborder	47
Tree shrew	suborder	28
Mouse	order	21
Hedgehog	order	19
Chicken	class	10

Genetic distance The various measures of genetic variability reveal that variability within a species can be extensive; often allele frequencies in one population are quite different from those in another population and to take account of this, measures of genetic similarity or difference have been devised. The one most used is that of Nei and is termed **genetic distance**, *D*. *D* is an estimate of the number of amino acid substitutions that have occurred in proteins that differ between two groups (populations or species) being studied. In other words it is a measure of the average number of allele substitutions accumulated per locus. If it is assumed that the rate of molecular evolution is fairly constant then *D* can be used as an estimate of the time since divergence occurred. This assumption has been useful in studies of hybridization between species (see Chapter 9) although the amount of genetic difference, by itself, is not a useful criterion for determining the evolutionary status of populations of organisms.

SUMMARY OF SECTION 3.5

Data on phenotypic characters can be collected by interval, nominal or ordinal level measurement. The data are often normally distributed, and analysis of variation between populations is amenable to statistical testing. The heritability of a particular trait can be calculated, demonstrating the extent to which variance is due to differences in the genotype.

Variability in genotypes can be inferred from differences between gene products (proteins) and methods mentioned were electrophoresis, immunology and amino acid sequencing. It is now possible to sequence DNA but studies concentrate on sequencing part of the genome because of the cost, both in time and money. The hypervariable regions of the genome, restriction fragment length polymorphisms and mitochondrial DNA have been most used to date (1991). DNA hybridization gives a measure of the extent of the similarity of DNA from different species. Table 3.9 provides a summary of the techniques used to measure genetic variability together with comments on their limitations.

3.6 VARIATION AND POPULATION STRUCTURE

The genetic variation that exists within a single population (known as its **gene pool**) can be a fraction of that which is present in a species over its whole range. This Chapter has demonstrated something of the extent of this variability, its heritability, how it is measured and how it arises. This Section considers the evolution of genetic differences between natural populations of the same species.

Natural selection The best known mechanism is natural selection, which will be examined in detail in Chapter 4. Since populations are always of limited size, selection always acts on a system in which there is some random fluctuation in allele frequency. Sometimes the selection is so strong that the random effects are eclipsed, but sometimes they are important. This Section considers some non-selective ways in which allele frequency is modified.

Table 3.9 *Techniques for measuring genetic variability and their limitations.*

Technique	Material for assay	Mainly used to identify differences between:	Limitations
Gel electrophoresis	protein	individuals populations species	An indirect method. It underestimates the amount of genetic variability. Only amino acid changes involving an alteration in net charge are likely to be identified, i.e. about one-third of all amino acid substitutions. Genes without a protein as a gene product and synonymous codon changes will not be detected.
Immunology	protein	species	An indirect method. Comparison of different proteins from the same two species can give different immunological distances. Reciprocal comparisons of the same protein can give different results.
Amino acid sequencing	protein	species	An indirect method. It underestimates the amount of genetic variability. Genes without a protein as a gene product and synonymous codon changes will not be detected.
DNA sequencing	coding or non-coding DNA	species	Time-consuming and costly.
Genetic fingerprinting	highly variable regions of DNA	individuals	Only uses a small part of the total genome; unrelated individuals can share up to 30% of the bands.
Restriction fragment length polymorphism (RFLP) analysis	(i) DNA from any part of the genome broken into a RFLP	individuals species	Only uses a small part of the total genome; not as reliable as genetic fingerprinting for establishing very close relationships (e.g paternity).
	(ii) mitochondrial DNA (mtDNA)	individuals species	
DNA hybridization	genome (DNA)	species	Gives a single estimate of genetic distance averaged over the whole genome. Does not detect the amount of difference between non-binding loci.

Inbreeding effects The Hardy–Weinberg ratio is a prediction made from known allele frequencies in a population where genotypes are assumed to have equal fitness. Another key assumption is that mating is panmictic (i.e. random), an assumption that is often violated. The most common deviation from panmictic breeding is **inbreeding**, a term applied to mating between close relatives. The most extreme form of inbreeding is self-fertilization (selfing), a breeding system that is common in plants (see Chapter 5). In species with separate sexes, sibling matings or matings between parent and offspring are the most extreme examples of inbreeding. The lack of dispersal of progeny may result in breeding between close relatives being the norm in some species. Inbreeding leads to an increase in homozygosity at all loci because the breeding pairs are initially genetically more similar to one another than would be the case if a pair of individuals had been taken at random from the population. This fact has long been appreciated by plant breeders and is exploited in classical genetical studies where many generations of inbreeding are undertaken to ensure homozygosity for a phenotypic character prior to cross-breeding experiments. (See Figure 3.2, where pure-breeding parental strains of peas were used.) Figure 3.21 shows that the frequency of heterozygotes is halved in each generation when the population is derived by selfing from a heterozygous individual and all individuals continue to self. The rate of loss of heterozygotes is obviously less dramatic in other inbreeding systems but ultimately homozygosity will be achieved if there is no selection or mutation. The frequency of alleles is not altered by inbreeding. It is the proportion of homozygotes that

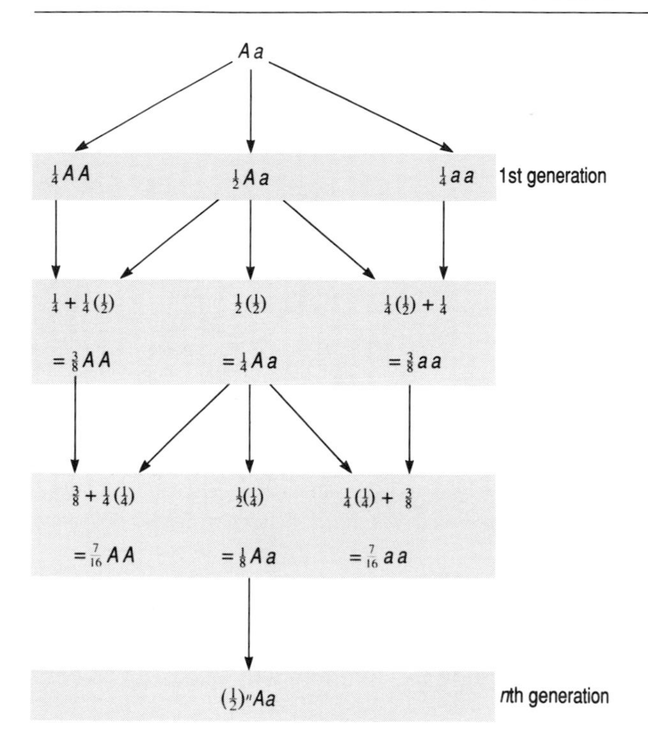

Figure 3.21 *Selfing: the frequency of heterozygotes is halved in each generation.*

increases, whilst the proportion of heterozygotes decreases. The effects that inbreeding can have on evolution will be considered in more detail after the discussion of how homozygosity comes to increase through random events in a panmictic breeding population.

Genetic drift Homozygosity is also achieved in populations by random **fixation** (or stabilization) of alleles. An allele is described as being fixed when it is the only allele at that locus. Fixation can result from **genetic drift**. Genetic drift is a change in the allele frequencies in a population that cannot be ascribed to the action of any selective process. Such differences in allele frequencies arise from the fact that: (i) the gametes produced by an individual contain an assortment of haploid genotypes produced from the germ-line cells at meiosis; (ii) a small random selection from these may form a biased sample of the actual allele frequencies present. In large populations, this gamete sampling error (Section 3.5.1) occurs in all reproducing individuals but produces no net error in the population because biases in one direction in favour of a particular allele in one individual's offspring will be cancelled out by biases in the other direction in other offspring. In small populations, however, there may be too few offspring for these sampling errors to cancel each other out and allele frequencies may alter as sampling errors build up, one generation after another. Moreover, in both sexually and asexually reproducing organisms, there will also be random fluctuations in the survival and reproduction of individuals, which may again lead to drift in small populations. Ultimately, by such random fluctuations in allele frequencies, one allele may become fixed in the populations, all others for the locus in question having been eliminated. The easiest way to visualize this is to look at Figure 3.22 which is drawn from computer simulations of genetic drift in a population of 50 asexually reproducing individuals. Calculations show that fixation of one allele will

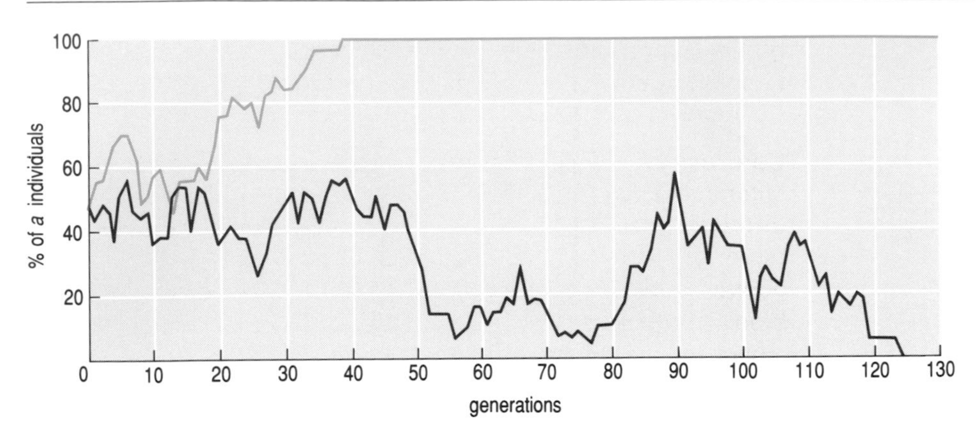

Figure 3.22 *Genetic drift. Two selectively equivalent types,* a *and* A, *reproduce asexually, in a total population of* N = *50. The graphs show the number of* a *individuals in two of the first three simulations carried out. (The third simulation lasted 227 generations before the fixation of* a.*)*

be expected to occur by genetic drift after 2*N* generations on average (where *N* is the number of individuals in the population). If the population consists of diploid sexually reproducing individuals, fixation will occur on average after 4*N* generations.

Population size is often not a good measure when examining the extent of heritable variation within the population, simply because not all individuals breed. If in the example above, shown also in Figure 3.22, there are only 45 breeding individuals out of the total population of 50 individuals then the *effective* population size is 45 and the rate of genetic drift is increased. *N* is now 45 and an allele would be expected to go to fixation on average after 90 generations rather than 100 generations.

▶ Does any aspect of the computer simulation shown in Figure 3.22 appear unrealistic?

The population size is kept constant. Populations tend to increase in size until limited by some external factor, such as lack of food. If the population size increases exponentially then it is less likely that any allele will become fixed.

Once an allele has gone to fixation the other alleles cannot reappear except by mutation or by immigration brought about by interbreeding with another population. This introduction of alleles from another population (termed **gene flow**) can be an important method of re-establishing variability within the gene pool of a population.

Genetic drift and inbreeding both have their greatest effects in small populations, yet many species occur over wide geographic ranges and thus might be expected to be little affected by these processes. However, it is found that despite wide geographic ranges species may be effectively fragmented into quite small local populations genetically isolated from one another, because interbreeding is prevented by simple physical barriers. For example, pollinating insects are not particularly adventurous so the growers of insect-pollinated fruit trees know that it is necessary to have a good source of pollen nearby when growing insect-pollinated varieties that are not capable of self-fertilization. Because the effective population size is often small, genetic drift and inbreeding are probably responsible for more of the differences seen within species than one might initially imagine.

Conclusive evidence that genetic drift does account for some of the variation in naturally occurring populations is difficult to obtain. It has been used to explain differences in the frequency of blood groups found between Icelandic cattle and the original Norwegian population from which they are thought to have been derived over 1000 years ago. In a study of these cattle, allele frequencies at eight blood group loci were examined. The different alleles have no known advantageous effects in the cattle and differences in allele frequencies between the two populations today are small but correspond well to the amount of genetic divergence that could occur from random genetic drift alone.

Although genetic drift is not directed by natural selection, it causes a change in phenotype frequencies and it is very difficult to be sure that a genetic change observed in a population is due to drift and not to natural selection because of the problem of knowing whether a particular trait is adaptive or not. So arguments supporting the theory that genetic drift alone can account for observed rates of genetic change in populations (such as that giving rise to the variations in blood group loci in Icelandic and Norwegian cattle) tend to be circular. Further discussion on genetic drift and natural selection is to be found in Chapter 4.

Founder effect Chance may operate in another way to produce allele frequency changes in small populations when these are isolated from a larger population. When a small population is first formed (e.g. when an isolated island is colonized, or some catastrophic event isolates part of the population), its founders are few in number and their genetic diversity is often low. When founders and their immediate descendants are subject to natural selection, the range of possible outcomes is restricted by the limited variety of alleles present in the population. Furthermore, the actual outcome resulting from the same natural selection acting on two isolated small populations may be very different simply because the two populations *start* with different samples of some larger population's pool of genes.

This effect, called the **founder effect**, has been observed in some experimental populations of *Drosophila*. Ten small populations (20 individuals each) and ten large populations (4000 individuals each) were taken from the same parental population in which there was a 50% frequency of a particular type of chromosomal inversion, called PP, on chromosome 3. The other inversion was AR. This particular mix of chromosomes had been achieved by breeding from two different stocks of flies— one stock contributed the PP chromosome, the other the AR chromosome. Thus the particular frequencies had *not* been determined by natural selection. The PP/AR heterozygote was fitter than either homozygote so the populations were all expected to reach the same balanced frequencies of PP:AR. All 20 populations were then subjected to the same physical environmental conditions and underwent about 19 breeding cycles (during which the populations were allowed to expand). Study the results of this experiment, shown in Figure 3.23.

▶ What changes do the results in Figure 3.23 show that might be due to the founder effect?

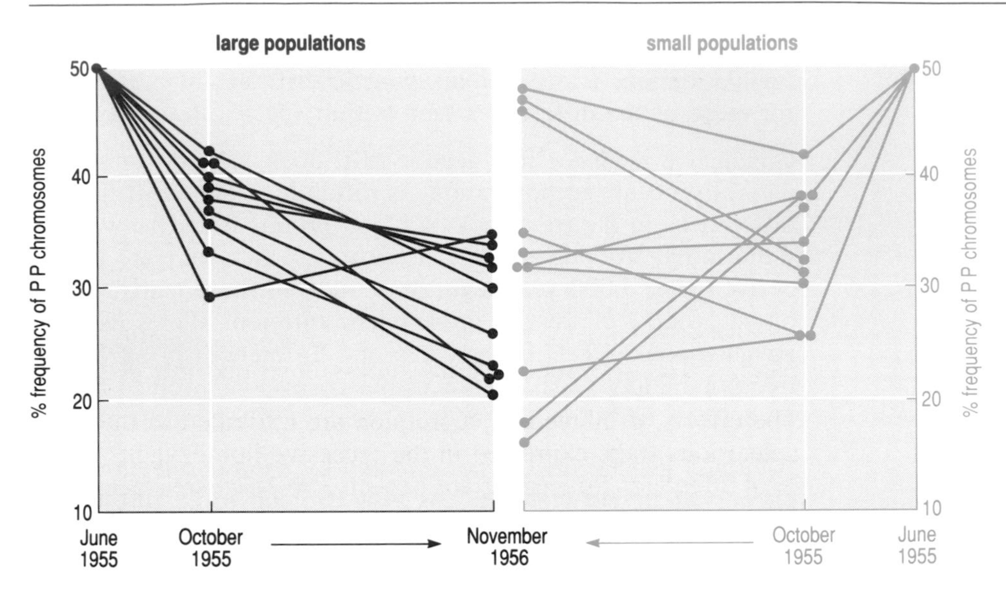

Figure 3.23 Experimental demonstration of the founder effect. The frequency of PP chromosomes in ten initially large (4000) and ten initially small (20) populations of Drosophila pseudoobscura.

The outcome of selection on the small populations shows great variety. This implies that the starting frequencies of genes that influence the fitness of the PP/PP, PP/AR and AR/AR genotypes differed in the founder populations. Natural selection can only work with what is available.

Genetic bottleneck In natural situations a small number of individuals may form an isolated breeding population for some generations and then there may be infiltration, or immigration from other populations, so gene flow between populations can be re-established. There is, however, evidence that there have been occasions when populations have crashed in numbers and then built up again from small founder groups. This is known to have happened in the northern elephant seal *Mirounga angustirostris* as its numbers were reduced to about 20 individuals in the 1890s due to heavy predation (hunting by humans). There are now about 30 000 northern elephant seals and an electrophoretic examination of 24 loci (using the technique illustrated in Figure 3.19) has revealed no genetic variation. (Compare this with the data given in Table 3.6.) When a species shows limited genetic variability thought to be due to a reduction in population size, it is said to have experienced a **genetic bottleneck**. A similar study of the cheetah *Acinonyx jubatus* showed that all 52 loci examined are monomorphic (i.e. there is only one allele for each locus). This suggests that the cheetah experienced a recent genetic bottleneck too, although there is no direct evidence for this. However, the mammalian genome is estimated to have approximately 50 000–150 000 active structural genes, so these studies are based only on a small sample of them.

A small founding population does not necessarily mean that there will be low genetic diversity. Say there are five different alleles for one locus, then a group of 12 individuals, carrying 24 alleles between them, may well carry all five alleles. However, if numbers stay low the loss of an allele through genetic drift would be quite rapid.

The situation is different, however, if a bottleneck leaves only individuals who are closely related because, as discussed above for the northern elephant seal, inbreeding leads to increased homozygosity. After a population crash it may be inevitable that individuals mate with their relatives. There is a great deal of evidence that this is deleterious.

Inbreeding depression In most human societies there is a taboo against incestuous relationships (matings between close relatives), the explanation for which may lie in the observation that the progeny of closely related parents often show poorer survival and lower reproductive success than the progeny of unrelated parents. This lowered offspring fitness is termed **inbreeding depression**.

The effects of inbreeding depression are attributed to the increased frequency of deleterious traits expressed in the recessive homozygous condition. Thus the short-term effects of inbreeding are expected to be unsatisfactory. However, in time these homozygous recessive deleterious genes will be eliminated by natural selection and the inbreeding population will be less variable than an outbreeding population but may be equally viable and fecund.

The effects of inbreeding depression are clearly seen in cultivated plants and the marked improvement in yield obtained by growing F_1 hybrids, in which many of the deleterious recessive alleles are masked, allows the seed merchants to charge a good deal more for the effort of producing such seeds (see Figure 5.5 in Chapter 5). Inbreeding depression is a particular problem for animal breeders in zoos who have only a small population of related animals from which to breed. A comparison of the mortality in inbred and non-inbred young showed that in 15 out of 16 species of captive large mammals (including the Indian elephant, zebra, giraffe, reindeer and Père David's deer), the young born to females mated with a related male had a lower chance of survival than young born to females mated with an unrelated male of the same species. A study of sib (brother–sister) mating in Poland China swine was discontinued after two generations because of the lowered viability of offspring. The results of this study are summarized in Table 3.10, which shows the mean sizes of the litters from initial sib matings (F_1 inbred) and from F_1 sib matings (F_2 inbred) and the mean survivorship of individuals in the various generations.

Table 3.10 *Some effects of inbreeding depression in a herd of Poland China swine.*

	No.	Mean size of litter	% born alive	% raised to 70 days	Sex ratio (males per 100 females)
General herd	694	7.15	97.0	58.1	109.7
F_1 inbred	189	6.75	93.7	41.2	126.1
F_2 inbred	64	4.26	90.6	26.6	156.0

Notice that the ratio of males to females, which was approximately 1.1 : 1.0 in the general herd, has also changed.

▶ Suppose that 20 F_1 sows were mated with their sibs (brothers) to yield one (F_2 inbred) litter each. How many sows would be available as breeding stock in the F_2 generation?

Only eight. Each F_1 female has an average litter size of 4.26 but with only 90% being born alive and 26% surviving to 70 days then on average there is only one survivor in each litter, i.e. 20 piglets. But these are in the approximate ratio of 3:2, male: female, and so only eight sows will be available for breeding in the F_2 generation (assuming no further mortality).

Despite the data in Table 3.10 there is evidence that domesticated animals are better able to resist inbreeding depression than those that are wild. For example, 14 lines of wild mice were sib-mated but only two lines survived after six generations. Sib-matings in *Drosophila* yield similar results. By comparison, 17 out of 35 lines of guinea pigs survived nine years of inbreeding (representing at least 11 generations of sib-mating) and every golden hamster in the Western world is thought to be descended from one pregnant female! The recency of domestication may also affect inbreeding depression. Chicken and turkeys, with a longer history of domestication, show more resistance to inbreeding depression than do Japanese quail and chuckar partridge (see Table 3.11).

Table 3.11 *Effect of inbreeding on the relative performance of four bird species. (Performance of non-inbred lines equals 100.)*

	Performance as percentage of non-inbred birds:			
Trait	Chicken	Turkey	Japanese quail	Chuckar
Hatchability:				
inbred embryo	90.0	83.4	72.2	71.3
inbred hen	97.0	92.1	89.3	89.1
Fertility	99.1	98.8	79.2	71.1
Viability of females	94.3	90.7	81.5	92.1
Egg production	90.4	89.5	83.9	84.1
Total reproduction	74.4	61.6	35.9	34.1

It was suggested earlier that whilst the short-term effects of inbreeding were deleterious, those lines that survived might not suffer any reduction in viability and fecundity. This is discussed further in Chapter 5. However, surviving does not always imply that there has been no reduction in fitness. Even the most vigorous of the surviving lines of guinea pigs, mentioned above, showed a 30% reduction in number of young raised per year when compared with non-inbred controls.

Those populations that have passed through genetic bottlenecks and emerged with low genetic diversity are particularly at risk from infectious diseases. The role of infectious diseases in limiting population growth was recognized by Darwin but their role can go beyond providing a mere check to population growth and they can wipe out a whole population particularly easily when genetic diversity of the population is low.

As noted earlier the cheetah appears to have gone through a population bottleneck, probably at the end of the Pleistocene. Calculations suggest that there was a disastrous decline, maybe to only one breeding pair, and that this was followed by at least 20 further generations of sib-mating. Since then numbers have fluctuated violently and have left the cheetah with little genetic variability. Recently numbers were established at 100 000 but an outbreak of feline infectious peritonitis in the mid-1980s caused mortalities of 50–60% in affected cheetah colonies. The same virus caused only 1% morbidity in domestic cats. Whilst differences in susceptibility to disease do not necessarily implicate inbreeding effects, the susceptibility of the cheetah to the virus is assumed to be related to its genetic uniformity. In 1991 the cheetah population was estimated to be 10 000 and falling fast. Male cheetahs have a high level of abnormal spermatozoa, but the females are very selective in their choice of mate and these factors mean that the effective breeding population may be much lower than the population numbers suggest. Studies such as this emphasize the importance of genetic diversity in maintaining healthy populations.

Both genetic drift and the founder effect may influence allele frequencies, particularly in small populations. Although these changes in allele frequencies do not occur *because* particular genotypes are of higher fitness, it is possible that the fitness of the genotypes involved may be affected by subsequent events and natural selection may take over. It is to natural selection that we turn in Chapter 4.

SUMMARY OF SECTION 3.6

Both inbreeding and genetic drift can lead to an increased level of homozygosity. Inbreeding does not alter the frequency of alleles in a population. Genetic drift is a description of the random fluctuations in allele frequencies that are a consequence of sampling error (i.e. chance). Chance events may lead to the loss of alleles from the gene pool but whether one allele becomes fixed (i.e. the only allele at that locus) and which allele this might be, are also a matter of chance.

It is not possible to provide certain evidence of the occurrence of genetic drift in natural populations, but it is a process that can contribute to evolution independently of natural selection. (See Chapter 4 for a fuller discussion of this.)

After a population crash or where a small population has become isolated, there may be changes in allele frequency because the reduced population, by chance, possesses different allele frequencies to the population from which it derived. This is known as the founder effect. If the founding members of a population are related then there may be a loss of fitness in the progeny, known as inbreeding depression. This frequently leads to the loss of a population in breeding experiments, although some species are less affected than others. Natural populations that have passed through genetic bottlenecks and have retained little genetic variability are particularly susceptible to infectious diseases that can severely reduce, or even eliminate, the population.

3.7 CONCLUSION

Mendelian genetics and molecular biology have established the basis of inheritance. Genes retain their identity through succeeding generations. However, most morphological characters do not segregate according to simple Mendelian ratios; instead they show continuous variation within a population. The inheritance of these characteristics is affected by genes at many loci, each gene having a small effect on the phenotypic expression of the character. For some years many scientists doubted whether there was as great a variability in genotypes as was observed in phenotypes. Modern methods of genetic analysis have demonstrated that, on the contrary, there is enormous variability in genomes, far more than is expressed in the phenotype.

The question that remains is why is there so much variation? Are the repeated sequences of DNA redundant 'junk' or do they have some regulatory role? The only source of new heritable variation is mutation. Rates of mutation can be calculated and the way that the frequency of a mutated allele can change within a population in the absence of a selective agent have been described. But to what extent is variation maintained in this way?

These questions will be addressed in later Chapters of this book.

REFERENCES FOR CHAPTER 3

Meselson, M. and Stahl, F. W. (1958) The replication of DNA in *Escherichia coli*, *Proceedings of the National Academy of Sciences, USA*, **44**, 671–682.

Watson, J. D. and Crick, F. C. (1953) Genetical implications of the structure of deoxyribonucleic acid, *Nature*, **171**, 964.

FURTHER READING FOR CHAPTER 3

Alberto, B. *et al.* (1989) (2nd edn) *Molecular Biology of the Cell*, Garland Publishing Inc., New York & London.

Campbell, R. C. (1989) (3rd edn) *Statistics for Biologists*, Cambridge University Press.

Chalmers, N. and Parker, P. (1989) (2nd edn) *The OU Project Guide*, Field Studies Council Occasional Publication No. 9.

Futuyma, D. J. (1986) (2nd edn) *Evolutionary Biology*, Sinauer Associates, Sunderland, Mass.

Gonick, L. and Wheelis, M. (1983) *The Cartoon Guide to Genetics*, Barnes and Noble Books (a Division of Harper Row Publishers).

Maynard Smith, J. (1989) *Evolutionary Genetics*, Oxford University Press, Oxford.

The Open University (1988) *S102: A Science Foundation Course*, Units 19, 20, 21 and 24.

Sokal, F. R. and Rohlf, J. F. (1987) (2nd edn) *Introduction to Biostatistics*, W. H. Freeman.

SELF-ASSESSMENT QUESTIONS

SAQ 3.1 *(Objectives 3.1 and 3.3)* A mutation is defined as 'any heritable change brought about by an alteration in the genetic material of an organism'. A mutation is acquired during the lifetime of an organism and is inherited by its offspring.

(a) Is this an example of Lamarckian inheritance?

(b) Is this consistent with the central dogma?

SAQ 3.2 *(Objective 3.2)* Explain why it is incorrect to say that there is 'a gene for a particular morphological character.' For example, why is it wrong to talk about a 'gene for red hair'?

SAQ 3.3 *(Objective 3.3)* Both recombination and mutation can contribute to the genetic variability between the individuals of two generations of the same population. Which process is likely to be of greater significance for evolution in the longer term?

SAQ 3.4 *(Objectives 3.4 and 3.6)* The following data were collected from a Brazilian population of *Drosophila polymorpha:*

8 070 flies were collected, of which 3 969 had dark abdomens, 927 had light abdomens and 3 174 had intermediate-coloured abdomens. Abdomen colour is determined by two alleles at one locus. The EE genotype is dark, ee is light and Ee is intermediate.

(a) What frequencies of the three genotypes would be predicted from the Hardy–Weinberg ratio?

(b) Are the observed numbers consistent with the Hardy–Weinberg ratio? If they are not, suggest reasons for this departure.

SAQ 3.5 *(Objective 3.5)* (a) What is the basis for regarding electrophoretic variation in proteins as an estimate of genetic variation?

(b) What are the limitations of this method?

SAQ 3.6 *(Objective 3.6)* Two species that are reported to have low genetic diversity, the northern elephant seal and the cheetah, differ in their present status. The elephant seal is thriving but cheetah population numbers are declining.

(a) Give possible reasons for the low genetic diversity recorded.

(b) Give a possible explanation for the current difference in population numbers of the two species.

CHAPTER 4

NATURAL SELECTION

Prepared for the Course Team by Tim Halliday

Study Comment *The theory of natural selection is the single most important element in contemporary evolutionary theory. This Chapter considers the role of natural selection in producing evolutionary change, in relation to other processes such as mutation and genetic drift. Natural selection will lead to evolutionary change over time, provided that a number of necessary conditions are met. The process of testing the theory consequently involves investigating these conditions, as well as testing the predictions of the theory. Much of this Chapter describes a number of specific studies that have set out to test the theory of natural selection. It concludes with a discussion of the relationship between natural selection and the concepts of fitness and adaptation, with special emphasis on the influence of the genetic systems that underlie phenotypic characters.*

Objectives When you have finished reading this Chapter, you should be able to do the following:

4.1 Discuss how the modern definition of fitness differs from the concept of fitness as used by Darwin.

4.2 Explain the theory of natural selection in such a way that the conditions under which it will occur are differentiated from its evolutionary consequences.

4.3 Recall evidence for, or critically assess observational and experimental data that may provide evidence for, the three conditions that must exist if natural selection is to occur.

4.4 Recall evidence that variation in phenotypic characters is correlated with variation in particular features of the environment, and discuss the limitations of such evidence as a test of the theory of natural selection.

4.5 Recall evidence that experimental and other perturbations of the environment (such as those caused by pollution) lead to adaptive changes in phenotypic characters.

4.6 Derive numerical values for selection coefficients from data on the relative fitnesses of different phenotypes.

4.7 Explain, with examples, why measures of partial fitness can give inadequate or misleading estimates of true fitness.

4.8 Use the Hardy–Weinberg formula to determine whether or not the frequency of an allele may be subject to selection.

4.9 Describe the effects that the frequency of an allele, whether it is dominant or recessive, and its expression in both homozygous and heterozygous forms, may have on the expression of selection on that allele.

4.10 Explain what is meant by frequency-dependent selection and how it can maintain, or lead to a reduction in, variation within a population.

4.11 Describe the relationship between selection on a character, the heritability of that character, and the amount of variation in that character.

4.12 Describe, with examples, how the intensity of natural selection on a given character can be measured.

4.1 THE THEORY OF NATURAL SELECTION

Natural selection is a major part of the theory of evolution; indeed, the terms evolution and natural selection are often viewed (erroneously, as we shall see) as being synonymous. Despite the fact that the theory of evolution by natural selection was proposed over 130 years ago and has remained ever since as the most important single concept in evolutionary biology, there remains much confusion as to exactly what it is and considerable debate as to whether it is a valid explanation of evolution. The first part of this Chapter examines the theory of natural selection, how certain aspects of it have changed since Darwin, and its relationship to evolution. It then goes on to discuss how the theory can be, and has been, empirically tested.

As set out in Chapter 1, Darwin's theory of evolution by natural selection is expressed as four basic premises. Before examining the theory in more detail, it is important to recall what these are.

▶ What are the four components of the theory of evolution by natural selection?

1 More individuals are produced than can survive.

2 There is a struggle for existence, because of the disparity between the number of individuals produced by reproduction and the number that survive.

3 Individuals show variation. Those with advantageous features have a greater chance of survival in the struggle for existence (natural selection).

4 Since selected varieties will tend to produce offspring similar to them (the principle of *inheritance*), these varieties will become more abundant in subsequent generations.

Expressed in this form, the theory of natural selection is a set of logical arguments. It says that, given certain conditions (variation, inheritance and competition), organisms will change over evolutionary time. It does not say that this is the *only* way by which evolution may occur; indeed, the theory of natural selection says nothing about a major component of evolution— the origin of new variations. As expressed by Darwin, natural selection provides an elegant hypothesis that predicts how evolution will occur in the future. It has been argued, however, that it does not provide a valid hypothesis about evolution that has occurred *in the past*, that can be tested scientifically. This has led some philosophers of science to dismiss it as a theory without scientific value. It should perhaps not be called a theory at all, but a law or a syllogism, because what it says is that, if certain conditions are fulfilled, the conclusions necessarily follow. This is an important point because it focuses attention on exactly how natural selection should be approached scientifically. If it is regarded as a law, then the objective must be to test, not its conclusions, but its underlying assumptions. The question to address thus becomes: do the conditions set out by Darwin actually exist in nature?

Later Sections in this Chapter describe a number of studies that have set out to test specific aspects of natural selection theory. Some of these test the assumptions on which the theory is based; where these assumptions are shown to be valid, this increases our confidence that natural selection has been a major factor in evolution that has occurred in the past. Other studies have adopted an experimental approach and make specific predictions about how natural selection will bring about change in the future. These provide a direct test of Darwin's hypothesis and, if the hypothesis is supported, they provide further support for the general applicability of the theory.

Before going further, it is necessary to dwell on a very important concept in evolutionary theory that is implicit in Darwin's third premise, namely the concept of fitness.

4.1.1 FITNESS

Throughout much of *The Origin of Species*, Darwin equated the concept of fitness with 'usefulness', by analogy with the traits selected by animal and plant breeders. As you may recall from Chapter 1, Darwin wrote:

> 'Can it, then, be thought improbable, seeing that variations useful to man have undoubtedly occurred, that other variations useful in some way to each being in the great and complex battle of life, should occur in the course of many successive generations?'
>
> (Chapter 4, 6th edn.)

This raises a central question: useful for *what*? Darwin emphasized qualities that assisted survival as being the criterion of usefulness, but the following quote shows that he meant rather more than this:

> 'If such do occur, can we doubt... that individuals having any advantage, however slight, over others, would have the best chance of surviving and *of procreating their kind*?'
>
> (Our italics)

It was obvious to Darwin that reproduction is an essential aspect of fitness because without it, no feature, however favourable to survival, will be passed on to succeeding generations. His later book, *The Descent of Man and Selection in Relation to Sex* (1871), is devoted largely to the analysis of adaptations that expressly promote reproduction (see Chapter 7). As expressed in *The Origin of Species*, reproduction is an implicit, rather than explicit, aspect of fitness.

The modern view of fitness, however, explicitly emphasizes the **reproductive success** of an individual, defined as the number of that individual's progeny that survive to adulthood relative to other individuals. In fact, this is an inadequate definition of fitness for the following reason. An individual might produce a large number of offspring that survive to reproductive age but these offspring might themselves fail to reproduce, perhaps because they are less healthy than other individuals or because they are the sterile progeny of a hybrid mating. The full definition of fitness therefore incorporates the reproductive success of the progeny; **fitness** is defined as the relative ability of an organism to survive and to leave offspring that themselves can survive and leave offspring.

While this definition is concise, clear and simple, it creates enormous problems for any evolutionary biologist who seeks to obtain accurate measurements of fitness. (It is also, of course, quite inappropriate for palaeontologists!) What it means, in effect, is that to determine the fitness of individual X, a researcher would have to determine the number of grandchildren descended from X and then compare the result with those from other individuals. Imagine trying to do this for a fish that sheds millions of eggs into the ocean, or for a tortoise that lives for 300 years. Some recent studies, particularly of birds and mammals, in which individuals can be identified and followed throughout their lives, have come close to measuring fitness directly and accurately, but for most species it is not possible to do so. Even for organisms such as these, there is a further problem. Certain kinds of social system lead to situations in which individuals are not the genetic progeny of their putative fathers and/or mothers. In such instances, exact measures of fitness can only be obtained by genetic techniques such as DNA fingerprinting (Chapter 3, Section 3.5.2). Examples of studies where DNA fingerprinting have been used in the measurement of fitness will be described in Chapter 6.

In most studies of animals and plants, researchers measure one or more **components of fitness**, such as survival to reproductive age, the number of eggs produced (fecundity), the number of eggs surviving to hatching, etc. These are also referred to as measures of **partial fitness**, recognizing the fact that they are incomplete measures of fitness. It is necessary to be extremely cautious about making the assumption that any measure of partial fitness provides a reasonable approximation to true fitness, as the following example illustrates.

In studying the reproductive success of birds, it is often a relatively simple matter to measure clutch size. As described in Chapter 2, many birds show considerable variation in clutch size and it would be easy to conclude, for example, that a great tit with a clutch of ten eggs is more fit than one with six eggs. It is clear, however, that clutch size is not always a good predictor of the number of young that fledge. In years of poor food supply, parents with small clutches may rear more young than those with large clutches (Chapter 2, Section 2.2.1). If the bird with ten eggs rears only four of them, and that with six eggs rears five, then the second bird apparently has the higher fitness, now that the number of young fledged is used as the measure of partial fitness.

The term 'fitness' has thus undergone a subtle but important shift in meaning from Darwin to the present. To Darwin, it was a *predictive* concept; it described the expected proneness of individuals to survive and reproduce and therefore determined the nature of a species in subsequent generations. In its modern form, fitness is a variable that is measured, and so the term is used in an essentially descriptive, retrospective sense.

The 19th century English philosopher Herbert Spencer coined the phrase 'the survival of the fittest' to encapsulate the theory of natural selection, and Darwin adopted it in later editions of *The Origin of Species*. The phrase is, however, a source of misunderstanding in two ways.

▶ What important aspect of adaptation does Herbert Spencer's phrase ignore?

Reproduction. A major component of fitness, as defined above, is the ability of individuals to pass their characteristics on to succeeding generations. A more subtle problem with the phrase 'the survival of the fittest' is that it seems to contain a circular argument, or tautology, that arises from the modern use of the term 'fitness' as a variable that is actually measured retrospectively. A tautology is a statement that is true by a definition contained within the statement. It has been argued that, because the fittest organisms are, by definition, those that survive, Spencer's phrase says nothing more than 'the survival of the survivors'. The modern view is that organisms that have survived and reproduced are, by definition, fitter than those that have not and so the question shifts to *why* they are fitter. Endler (1986) suggests that, if we must use a catch-phrase to describe natural selection, we should use Oscar Wilde's 'nothing succeeds like excess'.

The concept of fitness will be discussed further in Section 4.4, with particular reference to how it is measured. The next Section returns to the theory of natural selection and looks at it in more detail.

4.1.2 A FORMAL APPROACH TO NATURAL SELECTION

As set out in Darwin's four statements, the conditions for natural selection and its consequences are expressed together. In a book published in 1986, J. A. Endler set out the theory of natural selection in a different, more formal way that separates the conditions from the consequences. He describes natural selection as a *process* in which if, within a population, there is:

(a) variation among individuals in some attribute or trait (*variation*);

(b) a consistent relationship between that trait and a component of fitness, such as the ability to obtain mates, to fertilize eggs, fertility, fecundity, survivorship, etc. (*fitness differences*);

(c) a consistent relationship, for that trait, between parents and their offspring, which is at least partially independent of common environmental effects (*inheritance*);

then:

1 the frequency distribution for that trait will differ among age classes or life-history stages, beyond that expected from variation due to differences between different growth stages;

2 if the population is not at equilibrium, then the frequency distribution for that trait among offspring in the population will be predictably different from that of the parents, beyond that expected from conditions (a) and (c) alone. (By definition, a population at equilibrium with respect to a particular trait shows no change in the frequency distribution of that trait from one generation to the next.)

Items (a) to (c) are the sufficient and necessary conditions for evolution to occur as a result of natural selection; items 1 and 2 are the consequences that logically *may* follow if (a) to (c) are true. Note that there is a proviso here: the conditions are necessary and sufficient for evolutionary change to occur but it does not necessarily follow that evolutionary change *will* occur. This is because, as discussed in the next Section, there is more to evolution than natural selection.

(The phrase 'sufficient and necessary conditions' may require further explanation. To say that (a) to (c) are *sufficient* conditions means that no additional conditions need to be met for natural selection to occur. To say that they are *necessary* means that all three of them must be met if evolution by natural selection is to occur.)

4.1.3 Natural selection, genetic drift and evolution

Natural selection may occur and can lead to phenotypic change from one generation to the next in a population in which there is heritable variation. Phenotypic change may take the form, either of a shift in the mean value of a continuously varying character, or a change in the relative frequencies of different morphs of a polymorphic character. As discussed in Chapter 3 (Section 3.6), a major source of genetic variation between some populations may be genetic drift.

► Can you recall the definition of genetic drift?

Genetic drift is a change in the genetic make-up (frequencies of alleles) of a population occurring by chance. It is particularly likely to occur in small populations that are isolated from the main population of a species. In a population in which genetic drift has occurred, conditions (a) and (c) for natural selection may have been fulfilled, but condition (b), by definition, has not. Because the effect of genetic drift is random with respect to fitness, the range of phenotypes observed in one generation is not consistently related to that in the next. Thus, genetic drift is a process that brings about evolutionary phenotypic change in its own right, independently of natural selection.

Evolution is defined as any net directional or cumulative change in the characteristics of organisms or of populations of organisms over one or more generations. It refers not only to change in the frequency of alleles, variants or traits in a population over time, but also to the *origin* of new alleles, variants or traits. Thus, mutation, which gives rise to new characteristics, may also be a cause of evolutionary change.

The relationship between evolution, genetic drift and natural selection is depicted diagrammatically in Figure 4.1 (overleaf). This shows the three phenomena partially overlapping one another. What this means is that evolution may occur as

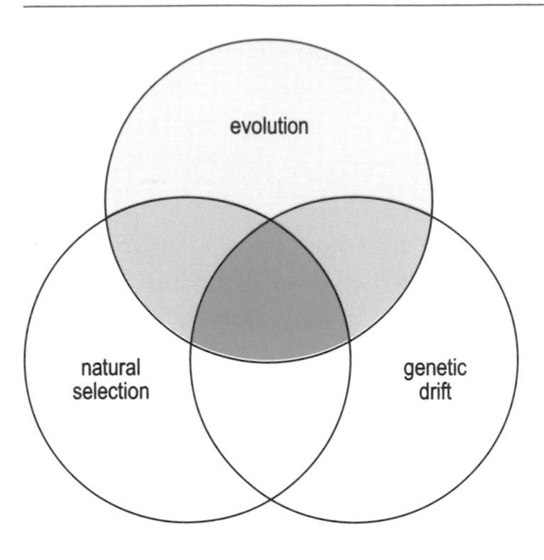

Figure 4.1 *The relationship between evolution, natural selection and genetic drift.*

a result of natural selection, of genetic drift, or of a combination of the two. It can also occur independently of either; mutation falls within that part of the evolution circle that does not overlap with either natural selection or genetic drift.

The relative sizes of the three circles in Figure 4.1, and the extent to which they overlap, are of no significance and do not imply anything about the relative importance of the three phenomena. The diagram simply symbolizes their interactive nature.

Figure 4.1 suggests that natural selection can occur in the absence of evolution; but how can this be? Consider a population that is at equilibrium, that is, its characteristics do not change from one generation to another. It is, by definition, not evolving. Nonetheless, conditions (a) to (c) may still apply and the stability of the population may be maintained by natural selection, by genetic drift, or by a combination of the two. To take an extreme example, having six legs is a character of insects that has not changed over a very long period of evolutionary time—it is not evolving. This does not mean, however, that there may not be variation in this character or that natural selection does not normally eliminate individuals that have more or less than six legs.

It is important to emphasize that the scheme shown in Figure 4.1 is only one view of evolution. Many evolutionists use the term 'evolution' in a looser way, to cover all three processes. Some confusion has also arisen as a result of the definition of evolution used by population geneticists, namely that evolution is a change in allele frequencies between generations. This definition includes natural selection and genetic drift but has led some biologists to underemphasize or to give no consideration at all of the origin of new alleles, an important factor in evolution. *The crucial point to remember here is that natural selection is not synonymous with evolution; it is a cause of evolution, but it is not the only cause and, when it happens, it does not necessarily cause evolutionary change to occur.* Indeed, natural selection can have a powerful stabilizing influence (see Section 4.4.3).

4.2 TESTING THE THEORY OF NATURAL SELECTION

In the previous Section, the theory of natural selection was set out in such a way that three conditions ((a) to (c)) were differentiated from two consequences (1 and 2). Investigations of natural selection in nature may seek to obtain evidence for the conditions, the consequences, or both. We now first consider evidence that the three necessary conditions exist and, secondly, describe a number of studies that have sought to provide evidence that the predicted consequences of natural selection have occurred.

4.2.1 INVESTIGATING THE CONDITIONS FOR NATURAL SELECTION

This Section concentrates on a single character, adult body size, in a particular group of animals, the anurans (frogs and toads), and describes evidence that conditions (a) to (c) exist. First, a few facts about anurans. For much of their lives, most frogs and toads are very secretive in their habits, being mainly active at night and when conditions are warm and wet. They become most apparent when they gather, often in huge numbers, to breed in ponds. It is common that males are more numerous in breeding aggregations, for reasons that will be discussed in Chapter 7. Males either call to attract females, most notably in tree frogs, or engage one another in protracted struggles to attain a position (called amplexus) in which they hold onto a female's back. The eggs are deposited by the female and, at the same time, the male in amplexus with her at that moment sheds sperm onto them (external fertilization).

Condition (a): variation in body size If you visit a breeding population of common frogs *Rana temporaria* or common toads *Bufo bufo*, you will readily see that there is considerable variation in body size in both sexes (Figure 4.2, overleaf)

▶ From the data presented in Figure 4.2, what conclusion do you draw about the form/shape of the distribution of body size in male and female toads?

Body size in both sexes shows a distribution that approximates to a normal distribution (Chapter 3, Section 3.5.1).

▶ What other conclusion do you draw from these data?

Female toads are, on average, substantially larger than males. The fact that the same character, body size, shows a difference between the two sexes means that, assuming it is a character that is subject to natural selection, natural selection must be affecting the two sexes differently. It is necessary, therefore, that, in the rest of this analysis, males and females arc considered separately.

The variation in body size shown in Figure 4.2 is fairly typical for anurans. The sex difference in body size is also common but by no means universal. In some

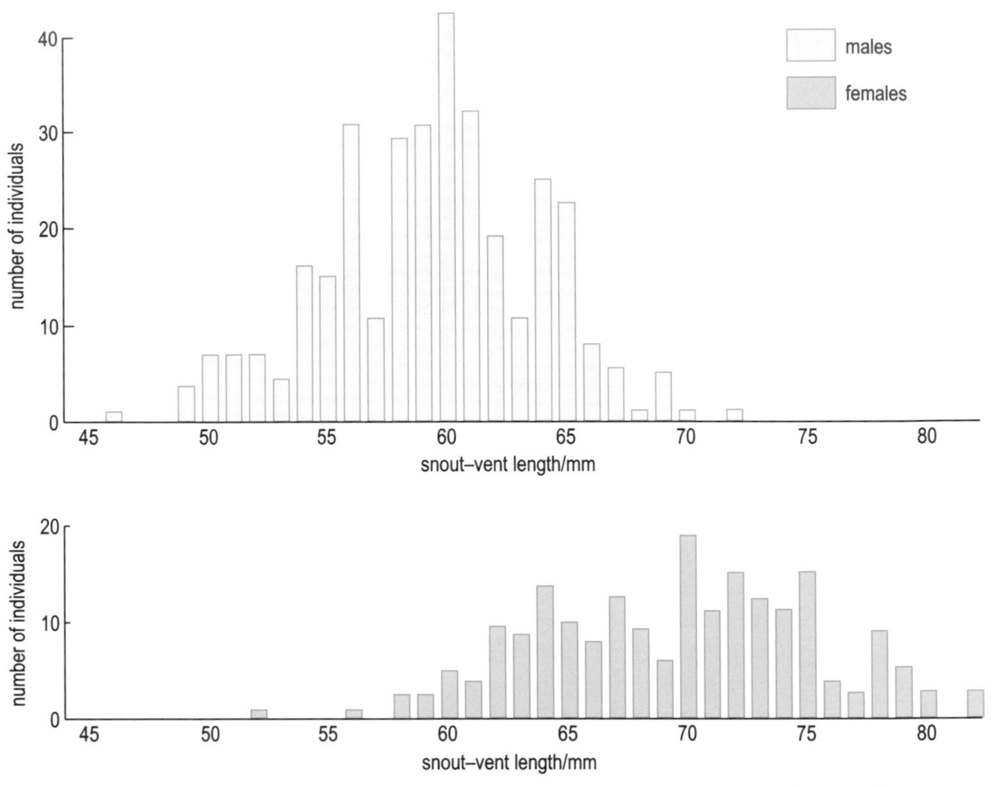

Figure 4.2 *Size–frequency histograms of male and female common toads* Bufo bufo *from a pond in southern England. Size is measured as snout–vent length (the distance between the tip of a toad's snout and the hindmost part of its body— its cloaca or vent). Sample sizes: 336 males, 193 females; mean male size = 59.4 mm; mean female size = 69.4 mm.*

species, males and females are of comparable size; in others, males are larger, on average, than females. Some of the variation in body size in anurans is due to age because, like most fish, amphibians and reptiles, they show indeterminate growth, i.e. they continue to grow, albeit slowly, throughout their adult life. Thus, there is a positive correlation between body size and age in toads, but this correlation is quite weak and accounts for only a small proportion of the total variation in body size. Put another way, body size is a poor indicator of age; a male of average size (60 mm long) could be anything from three to seven years old. Most of the variation in adult body size is thought to be the result of variation in growth rate during the pre-reproductive (juvenile) phases of life, when it is more rapid than during adulthood. The difference in size between male and female toads shown in Figure 4.2 is related to the fact that, whereas males typically breed for the first time when two or three years old, females begin to breed later, at three or four years old. These sources of variation are important, because they have to be borne in mind when considering the influence of natural selection on body size.

Condition (b): fitness differences Are size differences between individual toads correlated with components of fitness? Considering females first, the relationship between body size and one component of fitness, i.e. fecundity (or number of eggs), is shown in Figure 4.3.

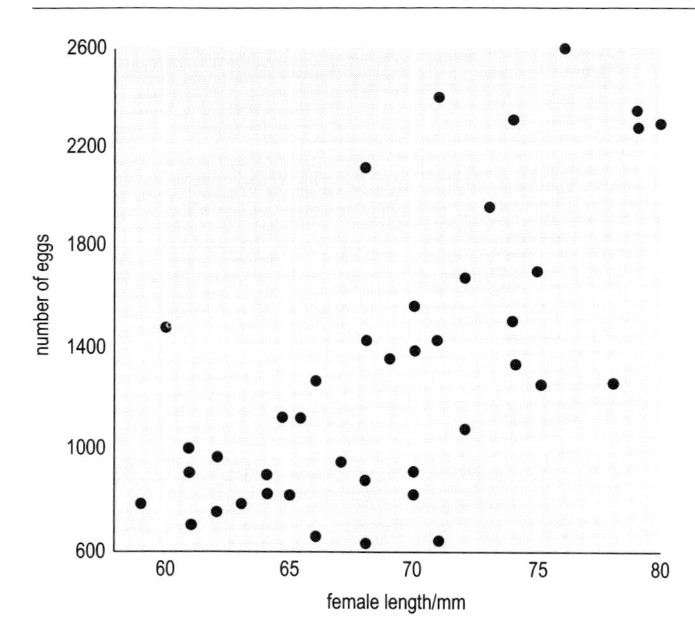

Figure 4.3 *The relationship between female size and number of eggs in a sample of female toads* Bufo bufo. *The correlation coefficient* (r = *0.669: see Chapter 10) is highly significant* (P$<$*0.001).*

▶ Do the data presented in Figure 4.3 indicate that variation in female body size is correlated with this component of fitness?

Yes, they do. Female body size is correlated with fecundity, which is one component of fitness, but, although the correlation is quite strong, there is considerable variation, suggesting that other, unknown, factors contribute to fecundity.

Turning to males, a comparison of the size of males that are measured while spawning with females, with that of unpaired males, reveals a difference (Table 4.1). (In toads, males outnumber females by about 3.5 to 1, so that only a small proportion of males will mate in a given season. Of those that do mate, the great majority mate only once in a season.)

Table 4.1 *Comparison of the size of those male toads* Bufo bufo *that mate and those that do not, over seven seasons. (Figures in brackets give the numbers of animals measured.)*

Year	Mean length of mating males/mm	Mean length of non-mating males/mm
1977	62.2 (16)	58.4 (216)
1978	63.4 (59)	60.6 (781)
1979	61.8 (108)	60.5 (410)
1980	64.1 (57)	62.3 (539)
1981	63.8 (8)	60.7 (405)
1982	64.5 (55)	63.1 (334)
1983	65.0 (131)	64.2 (584)

▶ Do the data presented in Table 4.1 indicate that variation in male body size is correlated with fitness?

Yes; in all years, mating males are larger than unmated males, though this effect is more marked in some years than in others. The size difference between mating and non-mating males is statistically significant in some years, but not in others. This example illustrates an important point: the strength or intensity of selection may vary from time to time—a topic to which we return later in this Chapter. It may also vary from one locality to another; studies of toad mating populations elsewhere in Europe (e.g. in Sweden) have not revealed any relationship between male size and mating success. The data in Table 4.1 provide evidence that large size can confer an advantage on males during mating. This is because larger males are more likely to win fights for the possession of females.

Taking up a point made in Section 4.1.1, it is important to emphasize that mating success represents only one component of fitness. It is a reasonable assumption that those males who have an advantage in obtaining mates will leave more progeny, but it remains an open question whether or not those progeny are in any way superior to those of other males. Selection for characters that enhance mating success is called sexual selection and is discussed at length in Chapter 7.

Condition (c): is body size heritable? This is the most difficult of the three conditions to verify and, for that reason, is the one that is least frequently established. The problem is that, for most organisms, it is very difficult to establish in their natural environment which individuals are the progeny of which parents, an essential ingredient in determining whether a character is inherited. Exact parentage can be established more easily in certain kinds of animal, such as birds and mammals, in which there is parental care, although, as discussed in Chapters 6 and 7, it may be necessary also to have genetic data to establish genetic relationships precisely. Most commonly, the heritability of a character must be established by means of a breeding experiment. This has not been done for the European toad but has for an American species, the wood frog *Rana sylvatica*, by K. A. Berven, in an experiment based on a 'half-sib' design (sib = sibling). In this, each of a number of males is mated with two females, of differing size. Thus, the progeny of a given male all have the same father, but half of them have a different mother to the other half; hence they are called half-sibs, as opposed to full-sibs.

Berven had previously found that there are pronounced differences in growth rate and size at metamorphosis (the point at which tadpoles transform into froglets and leave the water) between different localities. At high elevation sites, tadpoles grow more slowly but metamorphose at a larger size than at low elevation sites. When tadpoles from high and low elevation sites are reared in the laboratory under identical conditions, these differences persist, indicating that they have a genetic component. This is further supported by the observation that hybrid tadpoles (the parents of which came from sites at different elevations) show intermediate growth rates and size at metamorphosis. Berven carried out the same experiments on animals from a high elevation and a low elevation site.

The half-sib design makes it possible to compare the size at metamorphosis of a large number of young frogs, some of which are full-sibs (same mother and father), some half-sibs (same father, different mother) and some non-sibs (different mother and father). By correlating the size of each froglet with the size of each of its parents, it is possible to estimate the heritability of body size for each parent independently (see Chapter 3, Section 3.5.1).

Berven found that heritability was high for body size, with respect to both parents, at the high elevation site, but not at the low elevation site. Thus, Berven's study provides evidence that body size is heritable in wood frogs, from both the male and female, but only in populations at high elevation. This again makes the important point that, within a species, natural selection may act differently at different localities. Heritability measures must be interpreted with caution. First, a given heritability score is applicable only under the conditions in which that score is obtained; a character may show high heritability in one environment, low heritability in another. Thus, Berven's results show that body size is heritable in the laboratory but this does not necessarily mean that it is equally heritable in nature. Secondly, the high correlation, for high elevation frogs, between female size and froglet size may not be genetic; it could be due to what is called a **maternal effect**. Larger female frogs produce larger eggs which, because they contain more yolk, could give rise to faster-growing tadpoles. Other studies by Berven suggest, however, that for the wood frog this is not an important factor, though it has been shown to be so in other anuran species. In the fire-bellied toad *Bombina orientalis*, for example, the size of eggs produced by individual females varies from one clutch to another, probably reflecting variation in food supply. This leads to variations in fitness because larger eggs produce larger tadpoles (Kaplan, 1989).

SUMMARY OF SECTION 4.2

From this analysis of body size in toads, and of the extent to which this character fulfils the three conditions for natural selection to occur, a number of points emerge, many of which are of general relevance to characters that may or may not be subject to natural selection:

1 Variation in phenotypic characters is commonly observed; it is particularly marked in continuously varying characters such as body size.

2 Variation in the character under consideration can be correlated with fitness differences but, typically, much of the variance in that character remains unaccounted for, indicating that other factors are involved. On the basis of the data presented above, you might well ask: if body size is so strongly correlated with fitness in both male and female toads, why have toads not evolved to be much larger than they actually are? There are at least three possible answers to this question:

(a) The character could be subject to other selective factors that a simple correlational approach has not taken into account. It must be emphasized that the data presented above for toads, were collected over a comparatively very brief period of their lives, i.e. during mating, and that many other factors may impose selection at other times in their lives. It might well be, for example, that larger toads, because of their greater food requirements, are more susceptible to starvation at times of food shortage, with the result that there is then a negative correlation between body size and survival. Such a negative correlation would also arise if larger toads had greater difficulty in hiding or escaping from predators. In other words, there may be some factor acting counter to that which has been identified by simple correlational analysis.

(b) The character may be subject to constraints that prevent it evolving in a given direction. Body size is the product of growth and, all other things being equal, it takes longer to grow to a large size than to a small one. Toads are continuously subject to various hazards and the size distribution that we observe may simply reflect the fact that they die before they can grow any bigger. (This is not to say that life history variables, such as growth rate and body size at maturity, are not subject to natural selection—see Chapter 8).

(c) The character may not be heritable. If body size in toads is not heritable, then no matter how strong the correlation between body size and fitness, evolution by natural selection favouring larger body size will not occur.

3 Of the three conditions, heritability of a character is the most difficult feature to establish. In the field, it requires a considerable effort, often over several years, to measure the character both in parents and their offspring. This has been done in a few instances; one is clutch size in great tits *Parus major*, a character that has proved to be heritable (Chapter 2, Section 2.2.1). Heritability is more commonly established in the laboratory, using a breeding experiment of the kind described above.

4.3 Evidence for natural selection

In his comprehensive and critical book on the evidence for natural selection, Endler (1986) lists no less than ten distinct methods that can and have been used to detect natural selection in natural populations. No one method is capable of testing all aspects of the process and we only have space to consider a few of these methods.

4.3.1 Correlation with environmental factors

The rationale of this approach is that, if a character has evolved by natural selection in response to some feature of the environment, then, if that feature shows variation, for example across a geographical range, the character in question should show correlated variation. The following examples illustrate this approach.

Defensive adaptations in sticklebacks Three-spined sticklebacks *Gasterosteus aculeatus* possess three dorsal spines and a pair of pelvic spines. There is geographical variation in the size and number of these spines, in the size and number of a series of hard plates called lateral scutes along the flanks, and in the strength of the pelvic bones to which the pelvic spines are attached. For example, on the island of South Uist off the west coast of Scotland, sticklebacks are

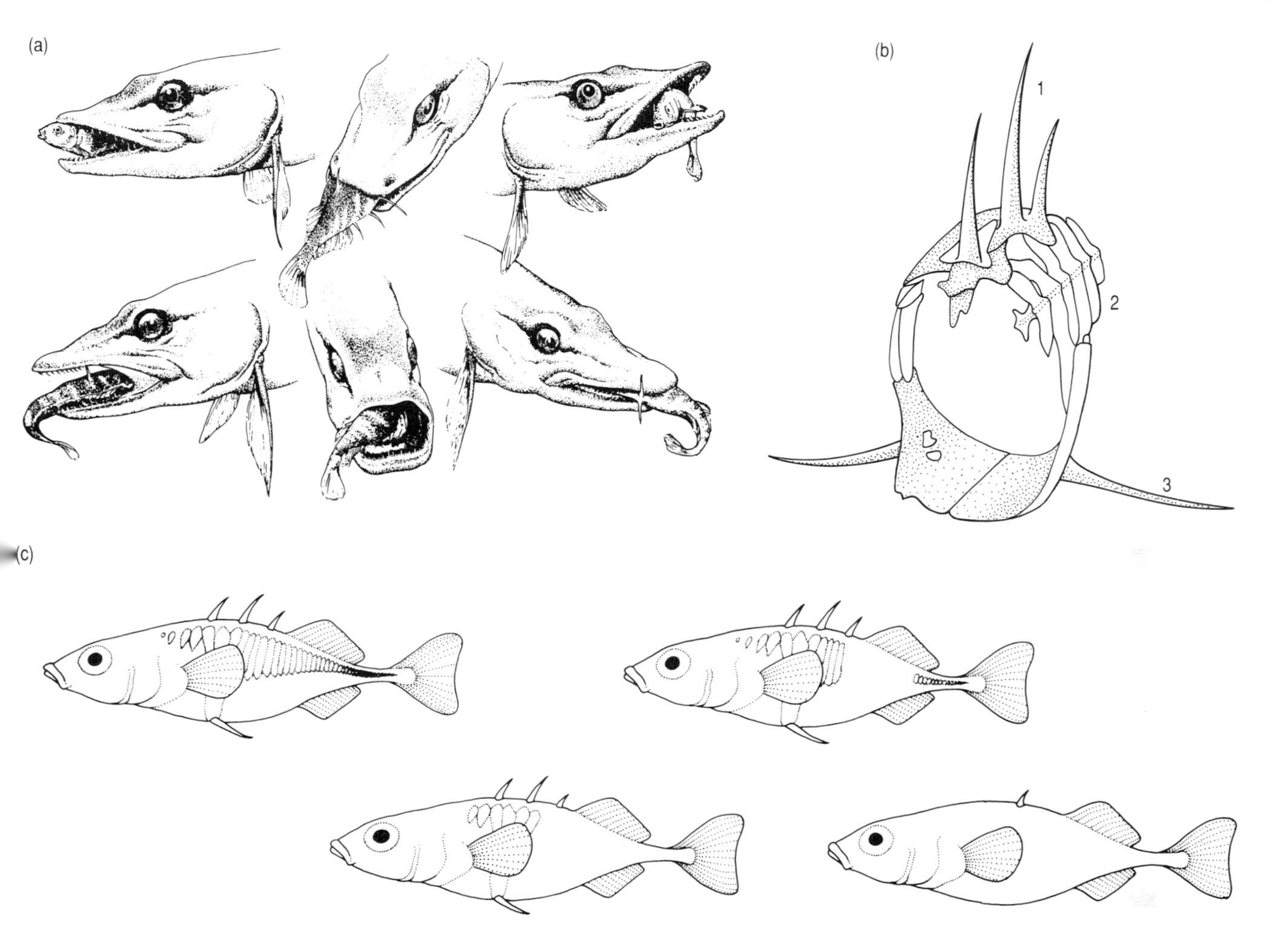

igure 4.4 *(a) Various positions of a stickleback in a pike's mouth. (b) The relationship between (1) the three dorsal spines, (2) some f the lateral scutes, and (3) the pelvic bones and pelvic spines of a stickleback. (c) Four examples of sticklebacks showing variation in pine size and in lateral scute number.*

virtually spineless. The spines have long been assumed to fulfil a defensive function; a predatory fish that catches a stickleback finds the spines sticking into its mouth and spits its prey out, still alive (Figure 4.4(a)). Figure 4.4(b) shows the spines, some of the lateral scutes, and their relationship to the pelvic bones; and Figure 4.4(c) shows representative individuals, illustrating some of the variation that occurs in these characters.

Within the species *Gasterosteus aculeatus*, two morphs are distinguished: a completely plated morph (*G. a. trachurus*, Figure 4.4(c), top left) and a low-plated morph (*G. a. leiurus,* Figure 4.4(c), bottom right). The *trachurus* form is predominantly marine but also occurs in the lower reaches of many rivers and predominates in freshwater habitats in eastern Europe. The *leiurus* form is typical of freshwater habitats. Where the ranges of the two forms meet, intermediates (*G. a. semiarmatus*) are also found.

The hypothesis that these characters evolved as a defensive adaptation as a result of natural selection exerted by predators has been tested by H. P. Gross (1978),

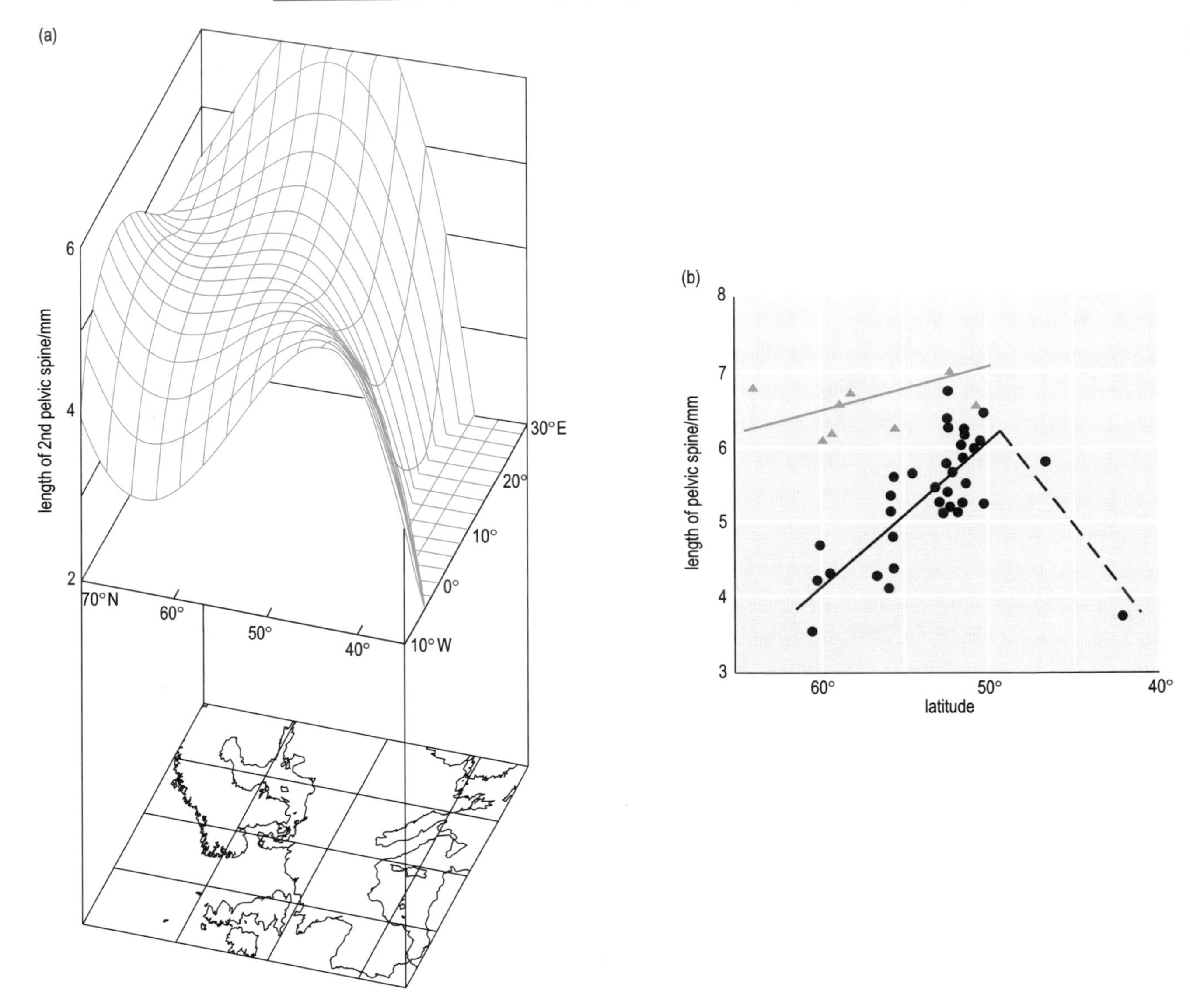

Figure 4.5 Latitudinal variation in defensive apparatus in three-spined sticklebacks. (a) Length of the second dorsal spine in populations of the completely plated form (G. a. trachurus) *from north-western Europe. (b) Length of the pelvic spine for both completely plated (▲) and low-plated* (G. a. leiurus) *(●) forms.*

who set out to correlate geographical variation in defensive characters with variation in the number of predators at different localities. If the hypothesis were correct, he expected to find that the most heavily defended sticklebacks would be found in localities where predation was most intense. He found that in all marine localities that he examined, predators were numerous and that the defensive apparatus of sticklebacks was well developed. In freshwater habitats, predators were most abundant in the middle of the geographical range of sticklebacks and became less numerous (especially pike and perch) towards the north and south extremes. Various measures of the defensive characters, such as the length of dorsal and pelvic spines and number of lateral scutes, tended also to peak in the middle of the range and declined to north and south (Figure 4.5).

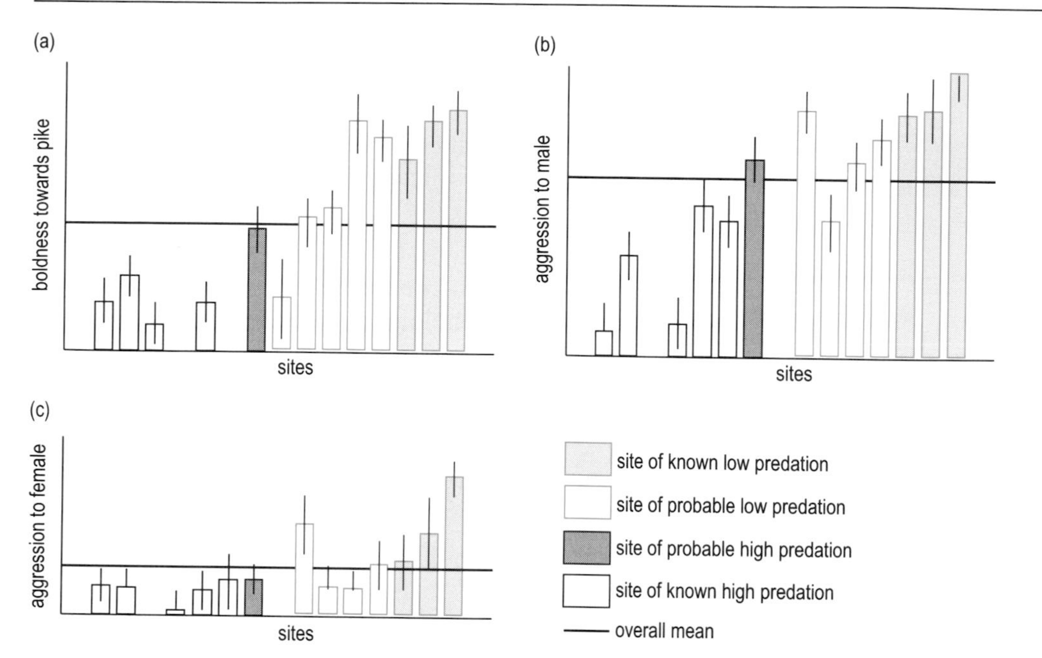

Figure 4.6 *Mean values (with standard errors) for three behavioural scores in sticklebacks collected from 15 sites at which the abundance of predators varied. (a) Boldness towards a pike; (b) aggression towards a male stickleback; (c) aggression towards a female stickleback.*

Other studies of geographical variation in sticklebacks suggest that natural selection due to predators influences behaviour as well as morphology. F. A. Huntingford (1982) collected sticklebacks from 15 sites in Britain which varied in terms of the abundance of predatory fish, and tested their behavioural responses in the laboratory to a hunting pike *Esox lucius* and to a stickleback in breeding condition. From her observations, Huntingford produced a score of each stickleback's 'boldness' towards the pike and a score of its aggressiveness towards another stickleback. The results are shown in Figure 4.6.

▶ Do the results shown in Figure 4.6 support the hypothesis that the behaviour of sticklebacks varies according to the intensity of predation at their collection sites?

Yes, they do. Sticklebacks from high predation sites are markedly less bold towards the pike than those from low predation sites.

▶ How do you suppose that natural selection would have influenced the behaviour of sticklebacks?

Individual sticklebacks that were bold in the presence of a predator would be more likely to fall prey to that predator. Natural selection would thus tend to eliminate bolder sticklebacks in populations where predators were abundant. It is possible, of course, that the responses of sticklebacks to predators are learned; fish in high predation sites have more opportunities to learn to avoid predators than those at low predation sites. The evidence, however, is that their response to predators has a substantial inherited component.

Interestingly, Huntingford found that less bold sticklebacks are also less aggressive to both male and female sticklebacks. This suggests that that there may be a genetic correlation between intraspecific aggression and behaviour towards predators. If such a correlation exists, it makes the task of identifying the effects of natural selection more difficult. Does selection act primarily on boldness towards predators, or on aggression between conspecifics?

These two studies of sticklebacks have demonstrated that there is a correlation between variation in an environmental factor—predation— and variation in defensive characters, morphology and behaviour, that is consistent with the hypothesis that natural selection by predation has favoured the evolution of those characters. While the results are consistent with natural selection theory, they do not, however, confirm that the characters in question have evolved by natural selection. This is because correlation between two variables does not necessarily mean that there is a causal relationship between them: *correlation does not prove causation*. Correlational evidence does, however, show that a hypothesis of causation is probable and it also enables researchers to reject a null hypothesis that there is *no* link between two variables. Correlational evidence provides only an indirect, rather than a direct, test of the theory of natural selection. This point is well-illustrated by yet another study of sticklebacks, this time by T. E. Reimchen in western North America.

Reimchen (1989) investigated geographical variation in the extent of the red ventral coloration that male sticklebacks develop in the breeding season. One hypothesis, proposed by previous researchers, to explain this variation is that brightly coloured males will be more conspicuous and will therefore suffer increased predation, leading to the prediction that the degree of coloration should be negatively correlated with the abundance of predators. Reimchen, however, also considered an alternative hypothesis, that the extent of red coloration is related to the clarity of the water. Data from 31 localities showed no correlation between coloration and the presence or absence of predators, but there was a strong effect whereby red pigment was most developed in habitats with high water clarity, but was lost where the water was heavily stained. This association could be explained by two hypotheses. First, male sticklebacks develop red colour only in habitats where such colour is clearly visible to females. Secondly, male sticklebacks can develop red coloration only in habitats where the nutrients essential for the synthesis of red colour (called carotenoids) are available in their diet. These two hypotheses may not be mutually exclusive; it could be that clear water selects for red coloration, but it can only develop in localities where the critical nutrients are present. Neither hypothesis has yet been tested. The first hypothesis suggests that there is a relationship between the characters in question and an environmental factor, but that the relevant factor is visibility to females, not predation by other fish. The second hypothesis suggests that the variation in colour may be due, not to selection at all, but to an environmental effect; male coloration simply reflects the availability of a particular nutrient in the environment. Such an effect can easily mislead an investigator into thinking that selection is occurring when it is not, a point amplified by the next example.

Variation in land snails The land snail *Cerion uva* occurs on islands in the West Indies and has been extensively studied by S. J. Gould (1984). There is a great deal of variation in size and shape among *Cerion* from different localities

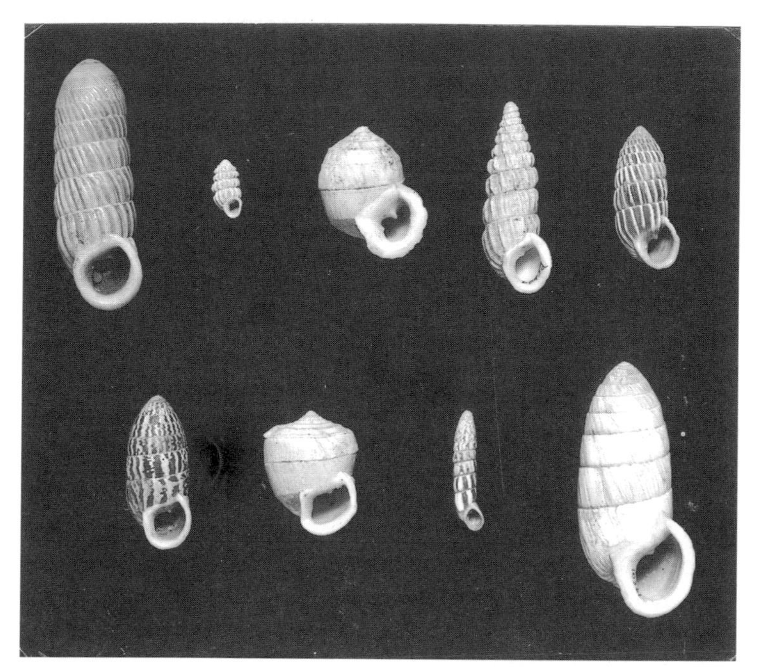

Figure 4.7 *Various* Cerion *shells from the Bahamas and Cuba, showing the diversity of form within the genus.*

(Figure 4.7), to the extent that early naturalists described several hundred distinct species. Gould found, however, that most of these forms could readily interbreed when brought together and thus that they were not genetically distinct and were not true species at all. Rather, the various forms reflect differences in the rate at which the snails grow at different stages in their lives. Certain shell sizes and shapes occur in particular kinds of environment, consistent with the hypothesis that they are the product of selection, but Gould's analysis reveals that this is not so. In places where conditions for the snails are good, they grow faster and larger, giving rise to differences in shape that are due to varying patterns of growth rather than selection. This example shows that correlation between variation in a character and environmental variables, if not combined with a knowledge of genetics and development, can give rise to a spurious assumption that that character is the product of natural selection.

Heavy metal tolerance in plants Environments vary, not only in space, but also in time, and one important way of documenting the action of natural selection is to investigate how organisms respond to changes in the environment. The generally damaging effects of human activities on the environment have provided numerous opportunities for such studies. Pollution from industrial sites often has a devastating effect on vegetation near such sites. However, it is observed frequently that after a time certain plants recover and begin to flourish again in polluted areas. Experimental analysis reveals that plants that are able to grow in polluted soil are tolerant to the particular pollutant involved. An example is provided by the wind-pollinated grass *Agrostis capillaris* in which a copper-tolerant form has been identified. Figure 4.8 (overleaf) shows the results from a study of this grass along a transect that crossed a copper mine. Two sets of data were collected: from adult plants growing naturally at various points along the transect, and from plants

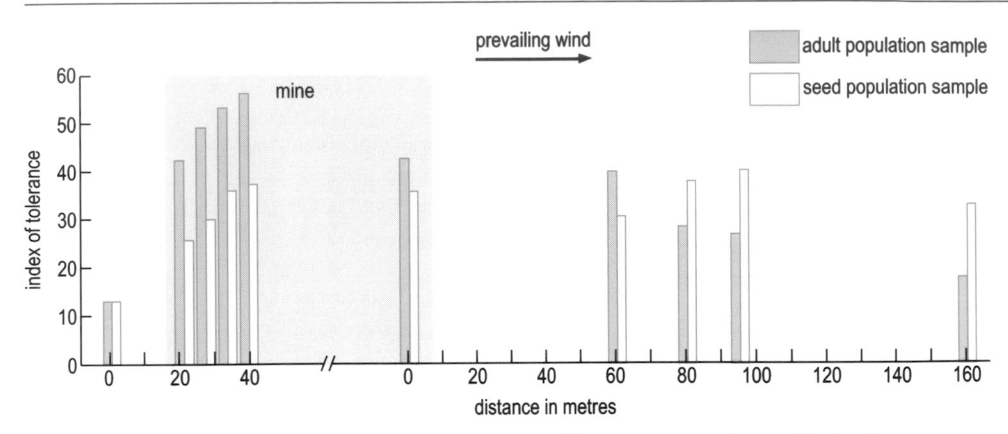

Figure 4.8 *Variation in average copper tolerance of the grass* Agrostis capillaris *along a transect through the edge of a copper mine. The green shaded area represents the copper-impregnated part of the transect. Green bars indicate average copper tolerance scores for adult plants; white bars for plants grown from a random seed sample.*

grown from a sample of seed selected at random from plants over a wide area. The growth of each plant was scored to provide an index of tolerance that reflected how vigorously it had grown. At the mine, and for a distance of 60 m downwind from the mine, the adult plants showed a higher tolerance index than those grown from the seed. Away from the mine, in a downwind direction, the seed-grown plants had an increasingly higher tolerance index.

▶ How do you account for the presence, in the mine area, of adult plants that have higher tolerance scores than the plants grown from the random seed sample?

Natural selection has favoured copper tolerance in the mine area, and has eliminated non-tolerant genotypes that are present in the random seed sample.

▶ How do you account for the fact that, at distances greater than 60 m from the mine, adult plants have lower tolerance scores than the plants grown from the random seed sample?

In this area, natural selection has reduced the frequency of tolerant genotypes below that which occurs in the random seed sample. This suggests that copper tolerance, while being of advantage in a polluted area, imposes a cost that makes it disadvantageous in non-polluted areas.

▶ Why do you suppose that copper-tolerant plants occur at all outside the mine area?

The grass is wind-pollinated and the wind will have carried pollen from tolerant plants in the mine area to plants growing downwind from the mine. As a result, alleles from copper-tolerant plants have spread into the non-polluted area. Some copper-tolerant plants will naturally occur in unpolluted areas, if only at very low frequency; the fact that this character evolves in polluted areas indicates that alleles for copper tolerance are present in the population.

4.3.2 Comparison of populations subject to selection over a long period

The rationale of this method for detecting natural selection is that, if a population is subject to natural selection, we should expect certain characteristics of individuals in that population to show change in a predictable direction over time.

▶ Can you think of a situation in which this would not be true?

If the population is at equilibrium at the start of the period of study, we would detect no change in the population, even if natural selection were operating. This method is therefore most revealing when the environment, and thus the selective regime acting on the population being studied, is known to have changed. Pollution (as discussed in the previous Section) provides such a situation. Alternatively, the investigator may introduce some kind of perturbation into the environment. This was the method used in the following example.

Coloration in guppies Guppies *Poecilia reticulata* (Plate 4.1) are small freshwater fish native to Central America and are a popular aquarium fish because they breed very rapidly. The males are substantially smaller than females, and their bodies are decorated with a mosaic of spots and patches that vary, between individuals, in colour, size, position and reflectivity. Endler (1980) studied guppies in a number of streams in Trinidad, where the details of male colour patterns vary markedly from one stream to another. The most colourful fish were found in streams where there were few predators, suggesting that predation may impose natural selection on the male colour patterns. These patterns are most fully developed around the time of mating, suggesting that they have a function related to sexual behaviour. Endler set up a long-term experiment, using artificial ponds, to test the hypothesis that two kinds of selection are involved in the evolution of male coloration:

1 Selection imposed by predators favouring colour patterns that make males more cryptic (camouflaged) within the environment.

2 Selection imposed by females favouring colour patterns that make males more conspicuous and attractive to females.

The selection for crypsis hypothesis was tested by lining the experimental ponds with gravel of one of two grain sizes: large (7–15 mm) and small (2–3 mm). This gravel was made up of grains of several colours (black, white, green, blue, red and yellow), mixed in proportions that matched those found in natural streams. The assumption was that fish with smaller spots would be more conspicuous against the larger-grained background, and fish with larger spots would be more conspicuous against the small-grained background. Some of the ponds contained a single voracious predator of guppies, the cichlid fish *Crenicichla alta*; others contained six individuals of the relatively innocuous (but still predatory) cyprinodont fish *Rivulus hartii*; while others contained no larger fish at all. Endler set up a total of ten ponds, arranged such that all combinations of gravel size and predation regime were included (Table 4.2).

Table 4.2 *Design of Endler's pond experiment with guppies. C1 to C4 and R1 to R4 denote ponds containing different predator regimes and K = no predators.*

Background	Severe predation (*Crenicichla alta*)	Mild predation (*Rivulus hartii*)	No predation
Fine gravel	C1, C3	R2, R4	K1
Coarse gravel	C2, C4	R1, R3	K2

Each pond was first stocked with a population of male and female guppies containing fish derived from several localities, and with fish from all collection sites included in all the ponds. As fish were placed in ponds, each male was scored individually for the number, size, position and colour intensity of his spots. The ponds were then left for six months, at which point the colour patterns of males were scored again and the predators were added. After a period of a further five months, a first 'census' was conducted, in which each individual male guppy was scored for the same measures as at the start of the experiment. At this point, the guppy populations had passed through three to four generations since the introduction of predators. Finally, the ponds were left for another nine months, allowing another five generations to pass, before a second census was conducted, with the same measures being taken. Thus, Endler was able to measure a wide range of male characters at the beginning and end of a period representing nine to ten generations during which guppies had been or had not been exposed to predation.

Figure 4.9 shows the results for one measure—the total number of spots per individual male. Note that, because the guppies were initially taken from a wide range of populations, there was a very large amount of variation in the number of spots per male in the initial pond populations.

▶ What effect did the presence or absence of predators have on this measure of male coloration?

In all ponds, the variation in spot number decreased dramatically over the course of the experiment, particularly over the first six months. In the ponds in which there was the dangerous predator (C), the average number of spots per fish declined rapidly over the 14 months of the experiment. In those ponds in which there was either no predator (K) or the innocuous predator (R), however, the average number of spots per fish increased.

▶ Do the results shown in Figure 4.9 support the hypothesis that the colour patterns of male guppies are subject to natural selection through predation?

Yes, they do. There is a clear divergence in the number of spots per fish between populations with and without dangerous predators over the course of the experiment.

Experiments such as these are always open to the charge that they are highly artificial and that conditions in artificial environments do not accurately mimic those in nature. To counter this problem, Endler also conducted a parallel

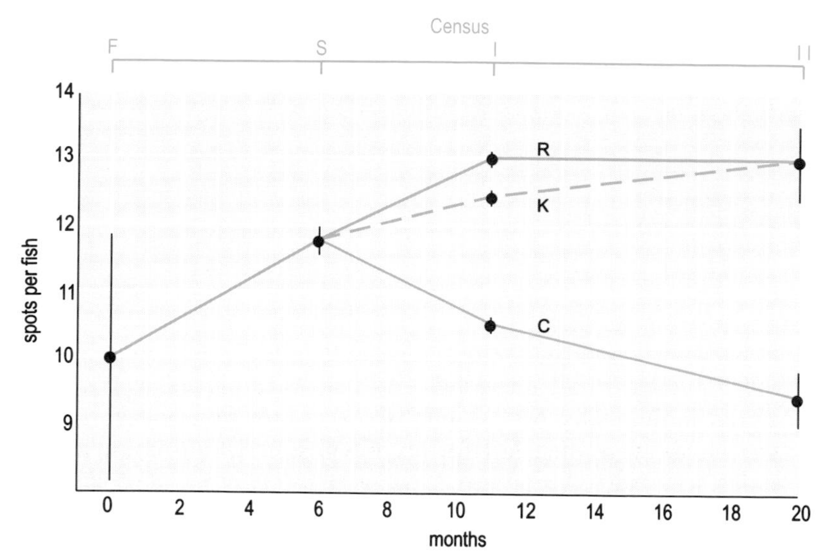

Figure 4.9 *Changes in the mean number of spots per fish during the course of Endler's experiments with pond populations of guppies. R: ponds with the innocuous predator (6* Rivulus hartii *per pond); K: ponds with no predators; C: ponds with the dangerous predator (1* Crenicichla alta *per pond); F: foundation population of guppies (no predators); S: start of the experiment (predators introduced); I: first census; II: second census. The points show mean scores; and small vertical lines are two standard errors.*

experiment in the field. He transferred 200 guppies from a natural stream that contained the dangerous predator *Crenicichla alta* to one that contained only the innocuous predator *Rivulus hartii*. The transfer was made in July 1976 and the two sites were sampled and the guppies scored in May 1978, sufficient time for 15 generations to have elapsed. The results are shown in Figure 4.10.

▶ Are the results shown in Figure 4.10 consistent with those from the artificial ponds?

Yes, they are. As in the artificial ponds, there was a marked reduction in the mean number of spots per fish in the site where the dangerous predator was present (C). Consequently, it is clear that Endler's ponds did satisfactorily mimic the selection regime that guppies encounter in nature.

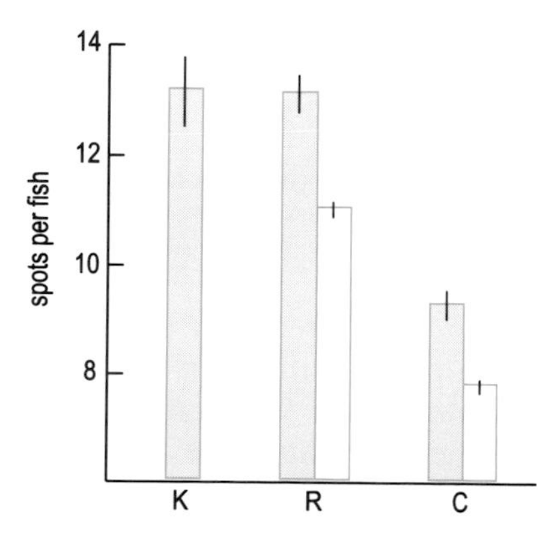

Figure 4.10 *Mean number of spots per fish at Census II in experimental ponds (green bars) and in the field experiment (white bars). K: ponds with no predators; R: ponds/stream with the innocuous predator; C: ponds/stream with the dangerous predator. The bars show mean scores; and small vertical lines are two standard errors.*

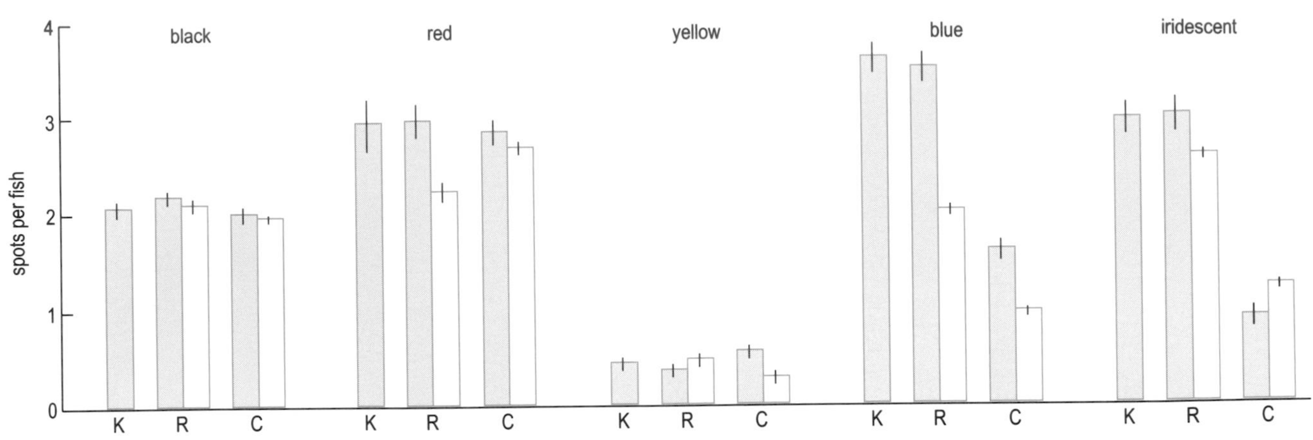

Figure 4.11 *Number of spots of different colours per fish at Census II. Green bars: data from experimental ponds; white bars: data from the field experiment. The bars show mean scores; and small vertical lines are two standard errors. Letters denote different regimes, as in Figure 4.10.*

The data presented in Figure 4.9 are for the overall number of spots and do not reveal the details of how selection exerted by predators influenced male coloration. Now examine Figure 4.11.

▶ What conclusion do you reach from the data presented in Figure 4.11?

Under conditions of severe predation, there was a reduction in the number of blue and iridescent spots, but black, red and yellow spots were not affected. Thus, the reduction in the overall number of spots shown in Figure 4.9 is mostly due to the marked reduction in blue and iridescent spots.

▶ Are the results from the field experiment consistent with those from the artificial ponds?

Yes, they are, but with the exception that guppies subject to predation by the relatively innocuous *Rivulus hartii* show a reduction in the number of blue spots, and in the total number of spots that is not apparent in the pond data.

Turning now to the effect of the background on the colour patterns of guppies in the artificial ponds, Figure 4.12 shows data for the length of spots of all colours, measured at Census II.

▶ Do the data presented in Figure 4.12 support the hypothesis that background influences the size of colour spots?

Yes, they do. In ponds in which there was predation (either weak (R) or strong (C)), spots were larger on fish from ponds with coarse gravel, and smaller on fish from ponds with fine gravel. In contrast, fish from ponds with no predators (K) had larger spots in ponds with fine gravel than those from ponds with coarse gravel.

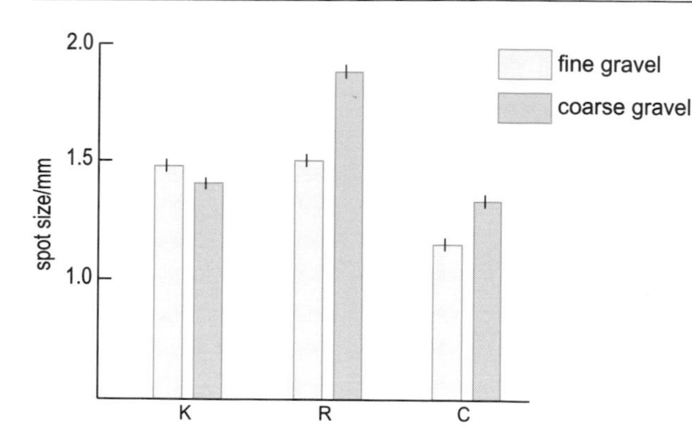

Figure 4.12 *Size of spots at Census II in the experimental ponds. The bars show mean scores; small vertical lines are two standard errors.*

▶ What would these differences in spot size mean in terms of the relative conspicuousness of fish in the two kinds of pond?

In ponds with fine gravel, fish with larger spots would be more conspicuous than those with smaller spots; in ponds with coarse gravel, fish with smaller spots would be more conspicuous.

▶ Do the results shown in Figure 4.12 support the hypothesis that predation influences the size of spots?

Yes, they do. In all ponds where there were predators (R and C), spot size tended to match the degree of coarseness of the background gravel, but in ponds where there were no predators (K), spot size contrasted with the background gravel. Data from the field experiment also show an effect on spot size, though here there was no experimental manipulation of the texture of the background. Recall that fish were transferred from a site with high predation to one with low predation. The results for spot size are shown in Figure 4.13 (overleaf).

▶ How do you interpret the data presented in Figure 4.13?

For spots of all four colours, spots were larger in the low-predation sites than they were at the original, high-predation site. They were very similar in size to those at the control low-predation site, with the exception of blue spots. These results thus confirm the conclusion that, in the absence of severe predation, guppies have larger spots and are, therefore, more conspicuous.

In summary, a number of conclusions can be drawn from these results. First, the hypothesis that the colour patterns of male guppies are subject to two kinds of selection, cryptic with respect to predators and conspicuous with respect to females, is strongly supported. In the presence of dangerous predators, males evolved fewer, smaller spots that tended to match the grain size of the environment. Secondly, the fact that most of the effects seen in experimental ponds were replicated in the field experiment indicates that this effect is not an artefact of the experimental conditions. Thirdly, marked changes in spot patterns in response to predation occurred over a relatively short time, about nine generations

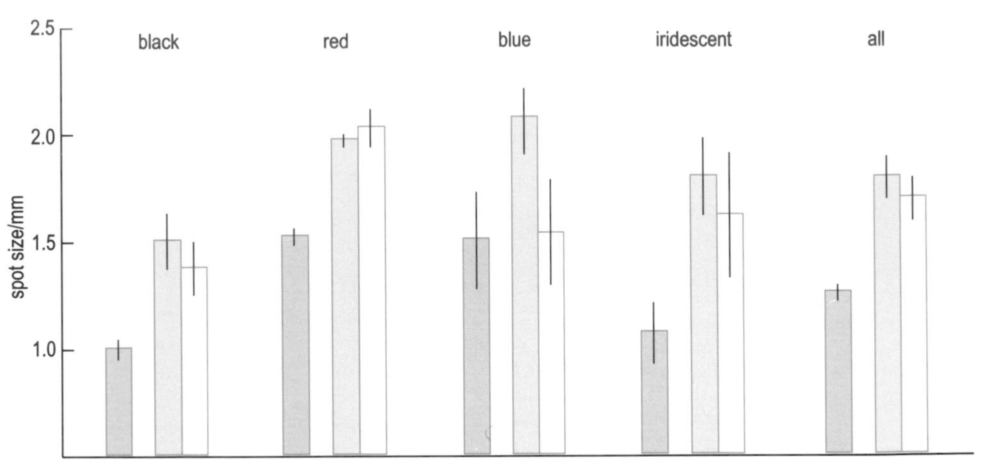

Figure 4.13 *Size of spots of different colours at Census II in the field experiment. Measures were made at the original stream, where predation was severe (dark green), at the stream to which fish were transferred where predation was mild (light green), and at a control stream where predation was mild (white). The bars show mean scores; and small vertical lines are two standard errors.*

in the experimental ponds and about 15 in the field. This indicates that selection exerted by predators can be very strong and can lead to a very rapid change in phenotypic characters.

The data for spot number (Figure 4.11) showed that certain spots, blue and iridescent, were influenced by selection, but that those of other colours were not. The data for spot size (Figure 4.13), however, showed that, with respect to their size, spots of all colours were affected by selection in the same way.

► What do these results suggest to you about the genetic control of spot characters?

It is clearly complex, and must involve alleles at many loci. The numbers of blue and iridescent spots must be controlled by genes different from those that control other colours. Also, for at least these two colours, the number and size of the spots must be controlled by separate genes.

Endler's experiments examined the influence of predation on male colour patterns. They leave unanswered, however, the important question of why male guppies possess bright colours at all. Endler assumed that bright colours are important for attracting females but his experiments did not set out to test this hypothesis. Other experiments, by Endler and other researchers, have tested this hypothesis and, briefly, have come to the following conclusions. When given a choice of males differing in the size and number of their spots, females are more likely to respond positively to a male with numerous, large red spots. This female preference for red coloration in males is more apparent in females from populations where males generally have well-developed red coloration. Red coloration is partly environmentally controlled; males with access to carotenoids in their diet develop brighter red spots. Finally, males can vary their behaviour according to their immediate environmental conditions; they display less vigorously to females at times of day when there is bright light, and they are consequently more conspicuous, and they display less vigorously in the presence of predatory fish. The role of female preferences in the evolution of male characters such as coloration is discussed extensively in Chapter 7.

The methodology that Endler used to test the theory of natural selection had two important components. First, he measured the phenotypic characters of different generations in the same populations at different times, to test the prediction that selection will cause phenotypic change over successive generations. Secondly, he manipulated what he assumed to be a principal agent of selection—predation. This methodology provides a much stronger test of natural selection than the correlation approach described in the previous Section and, because it involved experimental manipulation of a selective agent, it is a direct, rather than an indirect test of the theory.

Melanism in the peppered moth Changes in the frequency of a particular phenotype have been recorded, in varying degrees of detail, over a period of more than 100 years for the peppered moth *Biston betularia* in Britain. Prior to 1848 the peppered moth occurred in a single form that is speckled grey in colour, but in 1848 a completely black individual was reported from Manchester. This form, called the *carbonaria* form, rapidly increased in frequency until, by 1900, it formed 95% of populations in industrial areas, though it remained rare in most rural areas. The relationship between the relative frequency of dark and pale forms and pollution was attributed to their differing degrees of being cryptic against the tree trunks on which they rest during the day. In industrial areas, sulphur dioxide pollution kills the pale lichens that grow on trees so that tree trunks are much darker than in rural environments. The hypothesis was that birds would find and eat conspicuous moths more readily than cryptic ones so that the typical pale form would be selected against in polluted areas, the black form in rural areas. This hypothesis was tested by H. B. D. Kettlewell who caught, marked and released a large number of peppered moths and then scored the number of marked individuals that he recaptured in moth traps. The results are shown in Table 4.3.

Table 4.3 *The numbers of peppered moths released and recaptured at two localities.*

Locality	*Carbonaria* form	Typical form	Total
Birmingham (urban):			
Released	154	64	218
Recaptured	82	16	98
% Recaptured	53.2	25.0	
Dorset (rural):			
Released	473	496	969
Recaptured	30	62	92
% Recaptured	6.3	12.5	

▶ Do the data shown in Table 4.3 support the hypothesis that there is differential predation of pale and black forms in different habitats?

Yes, they do: the pale form suffers higher mortality in the industrial area, the black form in the rural area. It must be emphasized, however, that these results support the hypothesis but do not confirm it, because the actual form of selection that is assumed to be operating—selective predation by birds—has been observed, but has not been quantified, nor has it been manipulated. Other research on peppered moths indicates that predation is not the only factor that determines the relative survival of the different forms. It has been found, for example, that their viabilities differ even in the absence of predation, for reasons that are not understood.

The genetic basis of the colour polymorphism in *Biston betularia* is simple and well understood. The black *carbonaria* form is controlled by a single dominant allele, so that the heterozygous form is black. As discussed below (Section 4.4.3), this results in the black allele spreading very rapidly when it is favoured by selection.

In Britain, industrial pollution is now strongly controlled by legislation with the result that, by the early 1970s, sulphur dioxide levels in the environment had fallen significantly. The impact of this change has been monitored by Open University students taking the Science Foundation Course. By 1985, they had collected 1 825 peppered moths from 190 localities. This survey has documented a steady decline in the black form across Britain such that, by 1985, it existed at very high frequency only in the extreme north-east of England (Figure 4.14).

The change in the frequency of *carbonaria* is also apparent in data from a single locality (Figure 4.15). These data provide a good fit to a mathematical model that assumes that the *carbonaria* form has a 12% disadvantage relative to the typical form.

There remain several features of melanism in the peppered moth that have yet to be explained. For example, the frequency of the melanic form in East Anglia, UK, is as high as 80% in some places, despite this being an area that has never

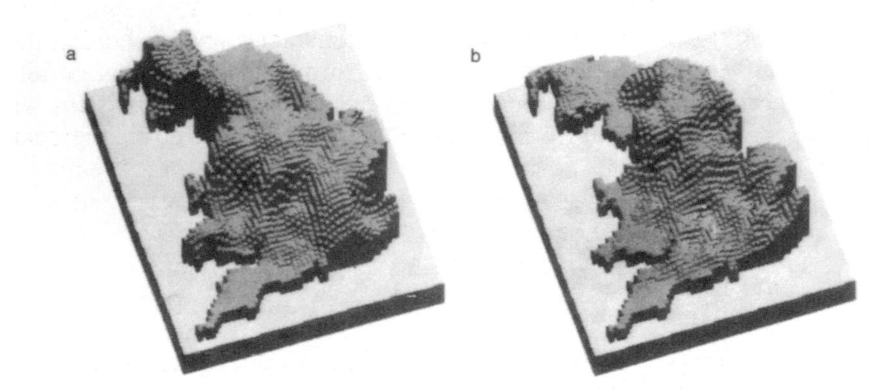

Figure 4.14 *Computer-generated three-dimensional surfaces showing the relative frequency of the* carbonaria *form of the peppered moth* Biston betularia *in Britain in (a) 1970 and (b) 1984.*

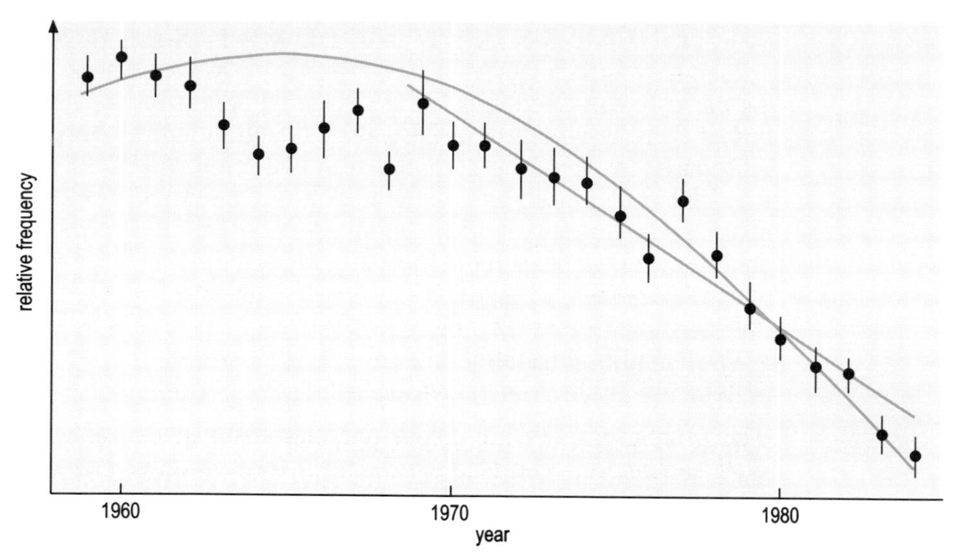

Figure 4.15 *Change in the relative frequency of the* carbonaria *form of the peppered moth* Biston betularia *at Caldy, the Wirral (north-west England), between 1959 and 1984. The curve shows the expected frequency based on a model that assumes a direct relationship with sulphur dioxide levels in the locality. The straight line shows the frequency to be expected if the* carbonaria *form has suffered a constant 12% disadvantage since 1969.*

received significant industrial pollution (Jones, 1982). In some places, anomalies can be explained by dispersal of melanic individuals from polluted into unpolluted areas, but this does not explain the East Anglian situation because individuals migrate by a maximum of about only 2.5 km per generation. Selection on melanism in the peppered moth is clearly more complex than can be accounted for purely by selective predation by birds.

SUMMARY OF SECTION 4.3

Two frequently used methods for testing the hypothesis that a particular character has evolved as a result of, and is maintained by, natural selection have been discussed: correlation between the character and an environmental factor, and change in the character over time in response to a change in the environment. The first method yields correlational evidence that is consistent with the hypothesis but it does not confirm a causal relationship. Variation in the character in question could be the result of selection related to some other feature of the environment, or it could be due to a purely environmental effect. Analysis of change in a character over time provides more direct evidence for natural selection, when the environment is undergoing some sustained change, such as that caused by pollution. Alternatively, environmental changes may be made by the experimenter, as in Endler's studies of guppies. The results from studies that involve measuring changes in characters over time indicate that quite marked changes in phenotype can occur in the course of only a few generations.

4.4 FITNESS AND NATURAL SELECTION

As discussed in Section 4.1.1, an essential condition for natural selection to occur is that there is variation in the ability of individual organisms to survive and reproduce. In the modern approach to evolution, fitness is used in a descriptive way and is also used as a variable that can be quantified (Section 4.1.1). In the neo-Darwinian approach to natural selection that incorporates consideration of genetics, fitness is attributed to particular genotypes. The genotype that leaves the most descendants is ascribed the fitness value $W = 1$, and all other genotypes have fitnesses, relative to this, that are less than 1. It is important to note that fitness is a *relative*, not an absolute, measure; what is important are differences between individuals in terms of their survival and reproduction.

Fitness measures the relative evolutionary *advantage* of one genotype over another, but it is often important also to measure the relative *penalties* incurred by different genotypes subject to natural selection. This relative penalty is the corollary of fitness and is referred to by the term **selection coefficient**. It is given the symbol s and is simply calculated by subtracting the fitness from 1, so that:

$$s = 1 - W \qquad (4.1)$$

Table 4.4 gives some hypothetical values for three fruit-fly genotypes in a situation in which two alleles determine body colour.

In this example, the fittest genotype, *EE*, has a fitness of $W = 1$, by definition, and therefore a selection coefficient of $s = 0$. This does not mean, however, that it is not subject to selection. s measures the relative penalties incurred by different genotypes so that the genotype with $s = 0$ incurs no relative penalty; in other words, it is the genotype most strongly *favoured* by selection. It is important to note that values for W and s are applicable only in the environmental conditions in which they are measured; in another environment the relative fitness of the different genotypes may be different.

▶ Can you recall a specific example, given earlier in this Chapter, of where this is true?

In the peppered moth, the pale form has the higher fitness in rural areas, the black form in polluted urban areas. In the guppy, bright coloration is favoured in the absence of predators, but not where predators are present.

Table 4.4 *Partial fitness values and selection coefficients for a hypothetical population of fruit-flies (drosophilids) with two alleles for body colour.*

Genotype	Phenotype	No. of eggs per female	No. of eggs surviving	Proportion of eggs surviving	Fitness (W)	Selection coefficient (s)
EE	orange	250	200	200/250 = 0.80	200/200 = 1.00	1 – 1.00 = 0.00
Ee	orange	250	170	170/250 = 0.68	170/200 = 0.85	1 – 0.85 = 0.15
ee	normal	250	140	140/250 = 0.56	140/200 = 0.70	1 – 0.70 = 0.30

4.4.1 The measurement of fitness

Fitness has two components—survival and reproductive success—and so measurement of either one of these components will provide an incomplete estimate of total fitness. There are strong theoretical reasons for assuming that the relationship between survival and reproductive success will be quite complex in nature. As discussed in later Chapters, there is an inherent 'trade-off' between survival and reproduction; an organism that puts all the resources available to it into survival will have none to put into reproduction, and one that puts too much into reproduction may fail to survive as a result. Thus, at a given time, selection will favour an optimum apportionment of energy, nutrients and other resources to survival and reproduction. The picture becomes more complex when long-lived organisms that breed many times in their lives are considered. An individual that puts a lot of effort into breeding in one year may have less to sustain its breeding effort in the following year; there is thus a trade-off between current and future reproductive success (see Chapter 8).

A number of studies have set out to measure survival and reproductive success in the field and to test for the presence of such trade-offs. The results are rarely clear and vary considerably from one species to another. For example, J. N. M. Smith (1981) sought to test the hypothesis that individual Canadian song sparrows *Melospiza melodia* that produce a large clutch in one year will be less likely to survive to breed in the following year than those that produce a small clutch. In fact, he found no such effect; females that were most successful at reproducing in a given year were just as likely to survive to the next as those with low fecundity. Smith concluded that females vary considerably in fitness and that the fitter ones are both capable of higher reproductive output and are better able to survive. In this example, therefore, the fittest birds were superior in terms of both survival and reproduction, but it cannot be assumed that this will always be true. Negative results, like those obtained by Smith, do not disprove the hypothesis that there is a trade-off between survival and reproductive success, but may reflect the fact that comparison between individuals is not a valid way of testing the hypothesis. The analysis of life history factors and of the relationship between them is examined in more detail in Chapter 8.

Measuring the components of fitness in the field is no easy task and it is generally the case that it is easier to measure some variables than others. For example, a researcher studying a bird can generally be more confident that a female's clutch size is an accurate measure of her reproductive success than that the number of females with which a male has been observed to mate is an accurate assessment of his reproductive success. As birds can rarely be followed all the time, it is very likely that a male will engage in matings that are not recorded by the researcher. Consequently, measures of fitness obtained in the field are referred to as measures of partial fitness, in recognition of the fact that they are incomplete measures of fitness (Section 4.1.1). Depending on the good fortune or judgement of the researcher, a measure of partial fitness will represent a substantial proportion of total fitness but it is important to recognize the limitations of any single measure of partial fitness. For example, the data for the reproductive success of female toads discussed earlier (Figure 4.3) were collected in a single season. Toads can breed for several successive seasons, and reproductive success measured in one year may or may not correlate well with that achieved in other years. Because of

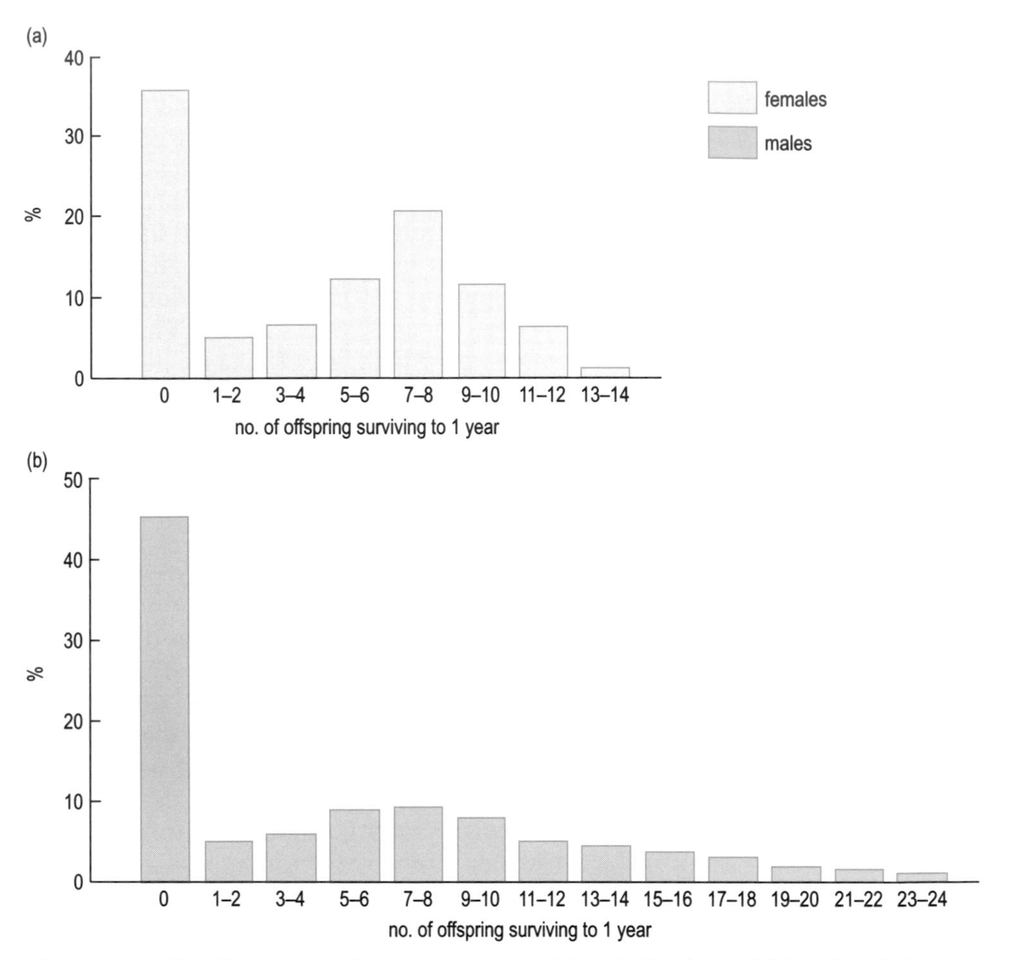

Figure 4.16 *The lifetime reproductive success of (a) female (hinds) and (b) male red deer (stags) on the island of Rhum, off western Scotland. The data are based on the known reproductive history of a large number of individuals.*

these problems, biologists are increasingly seeking to measure the lifetime reproductive success of individuals by following them over an extended period and measuring their reproductive success over several breeding episodes. This is of course much easier for an organism like an insect that may breed for only a few days (a few hours in the case of mayflies) than for long-lived animals such as large mammals. The data in Figure 4.16 show the estimated lifetime reproductive success of red deer *Cervus elaphus* and are the result of about 20 years' research.

The data in Figure 4.16 serve to illustrate how estimates of fitness can vary over different time-scales. In a single breeding season, a breeding hind has, at most, a single calf which either does or does not survive until the following year. Consequently, data from a single season would suggest that the reproductive success of hinds is either 0 or 1. There is considerable variation, however, in the longevity of red deer and in the number of years over which hinds continue to breed, giving rise to the data in Figure 4.16(a). (Hinds generally start to breed at age 3 or 4 and live for up to 20 years, though few live for more than about 14 years.) Many hinds have one or more barren years during their reproductive lives. The most successful hind produced 13 surviving calves; the average lifetime reproductive success of hinds was 4.5, and 35% of hinds had no surviving calves at all during their lives. There is thus much greater variation in the reproductive success of hinds over a lifetime than would be apparent from a single season, due primarily to variation in the duration of their reproductive life.

Turning to stags, their reproductive success is highly variable within a season because only a small proportion of them secure most of the matings, a small number of stags successfully defending a group of hinds, so that a successful stag may father as many as five calves in a single season. Even the most successful stags, however, rarely breed for more than four or five consecutive years so that their reproductive life is considerably shorter than that of females. Nonetheless, there is greater variation in the lifetime reproductive success of stags than of hinds, with the most successful stag fathering 24 surviving calves in this study. In contrast to females, the lifetime reproductive success of stags is determined primarily, not by their longevity, but by their size and strength, which determine their ability to compete for and defend hinds.

▶ For both stags and hinds, average lifetime reproductive success is 4.5 calves. Express this, for each sex, as a measure of relative fitness.

For hinds, the highest value for lifetime reproductive success is 13; giving this a fitness value of $W = 1$, W for an average female = 4.5/13 = 0.35. A similar calculation for stags gives W for an average male = 4.5/24 = 0.19. This difference in the average fitness of individuals and in the variation of fitness within each sex will be taken up again later in this Chapter.

4.4.2 Selection and genetics

As discussed in Chapter 3, Section 3.4, the Hardy–Weinberg formula provides a null hypothesis for the relative frequencies of genotypes against which the effect of selection on those genotypes can be determined. This is because one of the assumptions of the Hardy–Weinberg formula is that no selection is operating. Consider the following example.

A sample of 692 moths is collected: 326 are black, 9 are white and 357 are intermediate and are known to be heterozygotes. Is it possible to deduce from these figures whether selection is acting on any of the three forms? First, it is necessary to calculate the actual frequency of the two alleles underlying these characters, using the procedure set out in Chapter 3 (Table 3.4):

Table 4.5 *Data for a sample of 692 moths.*

Phenotype	No. of individuals	Genotype	No. of B alleles	No. of b alleles
Black	326	BB	652	—
Intermediate	357	Bb	357	357
White	9	bb	—	18
Total	692		1 009	375

The relative frequency of the rarer allele (b), $q = 375/(1\,009 + 375) = 375/1\,384 = 0.271$.

Therefore, the relative frequency of the commoner allele (B), $p = 1 - 0.271 = 0.729$.

From these values, the Hardy–Weinberg ratio can be determined as:

$p^2 = 0.531 : 2pq = 0.395 : q^2 = 0.073$

For a population of 692 individuals, this gives an expected ratio of the three forms: 368 : 273 : 51.

▶ How does the observed ratio of the three forms differ from that to be expected from the Hardy–Weinberg formula?

The white form is much rarer than expected, and the intermediate form somewhat commoner. This suggests that, in its homozygous form, the allele for white coloration is subject to strong selection, and that the heterozygote is at some advantage (assuming random mating and no immigration of genotypes).

The effect that selection will have on the spread of a character will vary considerably depending on the genetic basis of that character. This can be illustrated by a simple hypothetical case, in which the character in question is controlled by two alleles, A and a, at a single locus. Suppose that A confers higher fitness than a and that A is fully dominant, so that the fitness of AA (W_{AA}) is equal to that of the heterozygote (W_{Aa}). If the frequencies of the three genotypes, AA, Aa and aa, are p^2, $2pq$ and q^2 and their relative fitnesses are 1, 1 and $1 - s$, then the change in frequency of A after one generation of selection, Δp, is given by the equation:

$$\Delta p = spq^2/(1 - sq^2) \qquad (4.2)$$

This equation is derived as follows:

The frequencies of the two alleles, A and a, are p and q, such that $p + q = 1$ (Chapter 3). The Hardy–Weinberg formula gives the frequencies of the three genotypes at the locus, AA, Aa and aa as: $p^2 + 2pq + q^2 = 1$ (Equation 3.3 in Chapter 3).

You should remember, however, that one of the assumptions of the Hardy–Weinberg formula is that there is no differential selection on the three genotypes. If there *is* selection, the three genotypes have different fitnesses. Under the conditions set out above, these are:

$AA = 1$; $Aa = 1$; $aa = (1 - s)$

The frequencies of the three genotypes after selection will then be:

$AA = p^2$; $Aa = 2pq$; $aa = q^2(1 - s)$

The last item can be rewritten as: $q^2 - sq^2$.

The three genotypes will yield the two types of gamete, A and a, as follows:

A: $2p^2 + 2pq$

a: $2pq + 2(q^2 - sq^2)$

The twos cancel out throughout so that:

A: $p^2 + pq = p\ (p + q) = p$ (since $p + q = 1$)

a: $pq + (q^2 - sq^2) = q\ (p + q - sq) = q\ (1 - sq) = q - sq^2$

Thus, after selection, the new frequency, p', of A in the next generation is given by the equation:

$$p' = p/(p + q - sq^2) = p/(1 - sq^2)$$

Therefore, the change in the frequency of A, Δp, is given by the equation:

$$\Delta p = p' - p = (p/(1 - sq^2)) - p$$

$$= (p - p\ (1 - sq^2))/(1 - sq^2) = psq^2/(1 - sq^2)$$

Or: $\Delta p = spq^2/(1 - sq^2)$

From Equation 4.2, it can be seen that the change in the frequency of A (Δp) depends on the strength of selection (denoted by the coefficient of selection, s) and on the frequencies, p and q, of the two alleles; if any of these three values is small, Δp will be small. This illustrates an important general principle, i.e. that the rate of genetic change is great only if both alleles are common in the population.

If allele A were to arise by mutation in a population containing only a alleles, its initial frequency in the population (p) would be very small and its spread through the population would be very slow, as shown below. This contrasts with a situation whereby an allele that was previously not at a selective advantage, but which is already at quite a high frequency in the population, assumes higher fitness because of some environmental change. Such an allele will spread quite rapidly.

In the hypothetical example, A will spread at the expense of a, but a will not be completely eliminated because when it occurs in the heterozygous condition it confers no selective disadvantage. Thus, when the frequency of a (q) becomes very small, the value of Δp becomes correspondingly reduced (Equation 4.2). In general, disadvantageous recessive alleles are rarely eliminated by selection because of this effect and, for this reason, natural populations contain many deleterious recessive alleles at low frequency. As in this hypothetical case, new alleles do arise by mutation continuously in natural populations. In contrast, deleterious alleles that are dominant are not 'protected' in the heterozygous form and are very rare in natural populations.

Equation 4.2 and related equations can be used in computer simulations of natural populations to see what will happen when the values of the different variables in the equation are changed. For example, it is possible to simulate the spread of advantageous alleles that are recessive rather than dominant. The results of some such simulations are shown in Figure 4.17 (overleaf).

Figure 4.17 simulates the spread of an allele (A) under various conditions, with respect to two factors: the initial frequency of the allele and whether or not it is recessive or dominant. In Figure 4.17(a), A is initially at the very low frequency of $p = 0.01$; in (b), $p = 0.1$. In each case, there are three conditions: A dominant, A recessive and A intermediate. For these three conditions, fitness values were assigned to the various genotypes as in Table 4.6.

Table 4.6 *Fitness values assigned to the various genotypes included in the simulation shown in Figure 4.17.*

Genotypes:	*AA*	*Aa*	*aa*
Condition 1—*A* dominant	1.0	1.0	0.8
Condition 2—*A* intermediate	1.0	0.9	0.8
Condition 3—*A* recessive	1.0	0.8	0.8

First consider Figure 4.17(a). When *A* is dominant and initially very rare, its rate of increase is slow at first, but it accelerates after about 10 generations. After about 40 generations, it begins to spread less rapidly and levels off to a plateau, without reaching fixation (a frequency of 100%). This is because the allele that it is displacing, *a*, remains present at low frequency because of its 'protected' status in heterozygous individuals. When *A* is neither fully dominant nor fully recessive, but is intermediate, the phenotype of the heterozygote will be different from that of either homozygote. Consequently, *A* spreads less rapidly but does, after about 60 generations, go to fixation. This occurs because, in this condition, heterozygous individuals are less fit than *AA* individuals (see Table 4.6). Finally, when *A* is recessive and initially very rare, it barely increases in frequency at all within the timescale shown. In fact, if the simulation continues for many more generations (about 1 000), it will eventually go to fixation. From Figure 4.17(b), it can be seen that when the initial frequency of the advantageous allele *A* is reasonably high, it spreads much more rapidly under all three conditions.

It is important to remember, when considering the effect of selection on gene frequencies, that selection does not act directly on genes but on their respective phenotypes. Although selection operates directly on the phenotype, there is nevertheless a net effect on the frequencies of alleles which can be used as a convenient abstraction for simple models of evolutionary change in allele frequencies as a result of natural selection.

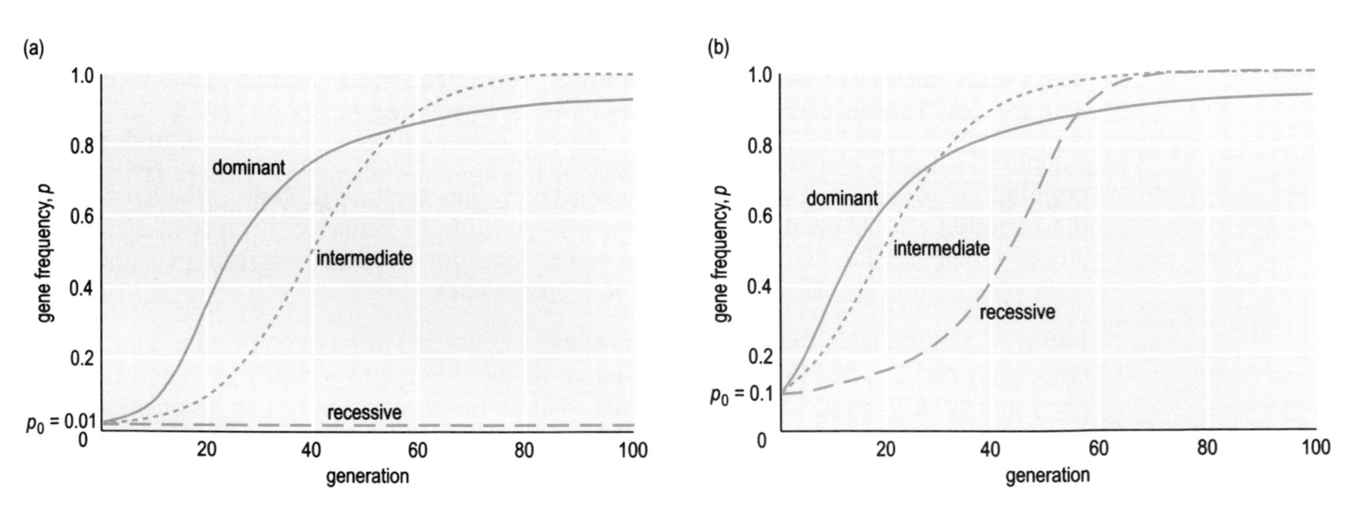

Figure 4.17 *The results of computer simulations of the spread of an advantageous allele,* A, *under a variety of conditions. See text for explanation.*

In the example discussed above, deleterious recessive genes are not eliminated by selection under certain conditions because they are not exposed to selection in the heterozygote. This effect is especially strong when there is **heterozygous advantage**, i.e the heterozygote has higher fitness than either homozygote. However, every heterozygous locus that is maintained in this way in a population incurs a cost in terms of fitness to any individual carrying it, due to the death or reduced fitness of the homozygous offspring that are frequently produced whenever heterozygotes reproduce. The combined effect of all these effects for all heterozygous loci maintained by selection is called the **genetic load** of the population.

An extreme example of genetic load is provided by the European great crested and marbled newts *Triturus cristatus* and *T. marmoratus*. These species possess a pair of chromosomes that are heteromorphic (differing in size and shape), with one being much larger than the other. Any individual that is homozygous for this chromosome pair (both short or both long) dies at a specific early stage of development. As a result, 50% of all progeny die in these species, a genetic load of 50%. This example involves whole chromosomes rather than gene loci but it serves to illustrate the principle of genetic load.

4.4.3 Modes of selection

As discussed in Section 4.1, natural selection may produce phenotypic change over time if three conditions hold: (1) that there is phenotypic variation, (2) that this variation correlates with variation in fitness, and (3) that the variation has a heritable component. The direction in which the phenotypic change will occur depends on the nature of the relationship between (1) and (2), i.e. exactly how variation relates to fitness. When there is a simple positive or negative correlation between a given character and fitness, such that individuals showing that character to the greatest extent have the highest or the lowest relative fitness, selection will produce change in the direction of the fitter individuals. This is called **directional selection**. Figure 4.18 (overleaf) illustrates directional selection and compares it with two other modes of selection. Two kinds of variation are shown: continuous variation; and discontinuous, polymorphic variation (Chapter 3, Section 3.1). If selection favours intermediate phenotypes at the expense of individuals at both extremes of the phenotype distribution, there is **stabilizing selection** and a decrease in the phenotypic variance of the population. If selection favours both the extremes at the expense of individuals with intermediate phenotypes, there is **disruptive selection**. This will tend to make the population bimodal and then, through recombination of the genotypes, to have an increased phenotypic variance in later generations. If there is also positive assortative mating (Chapter 3, Section 3.4) between extreme phenotypes, the bimodal distribution may even split into two groups (Chapter 9, Section 9.6.1).

The direction and rate of phenotypic change due to natural selection will depend on the relative fitnesses of different phenotypes within a population. It cannot be assumed, however, that these relative fitnesses remain constant over time. Environmental changes may cause one phenotype to acquire greater or smaller relative fitness, as in the examples discussed in Section 4.3 of environmental change due to pollution. Less obviously, the relative fitnesses of different

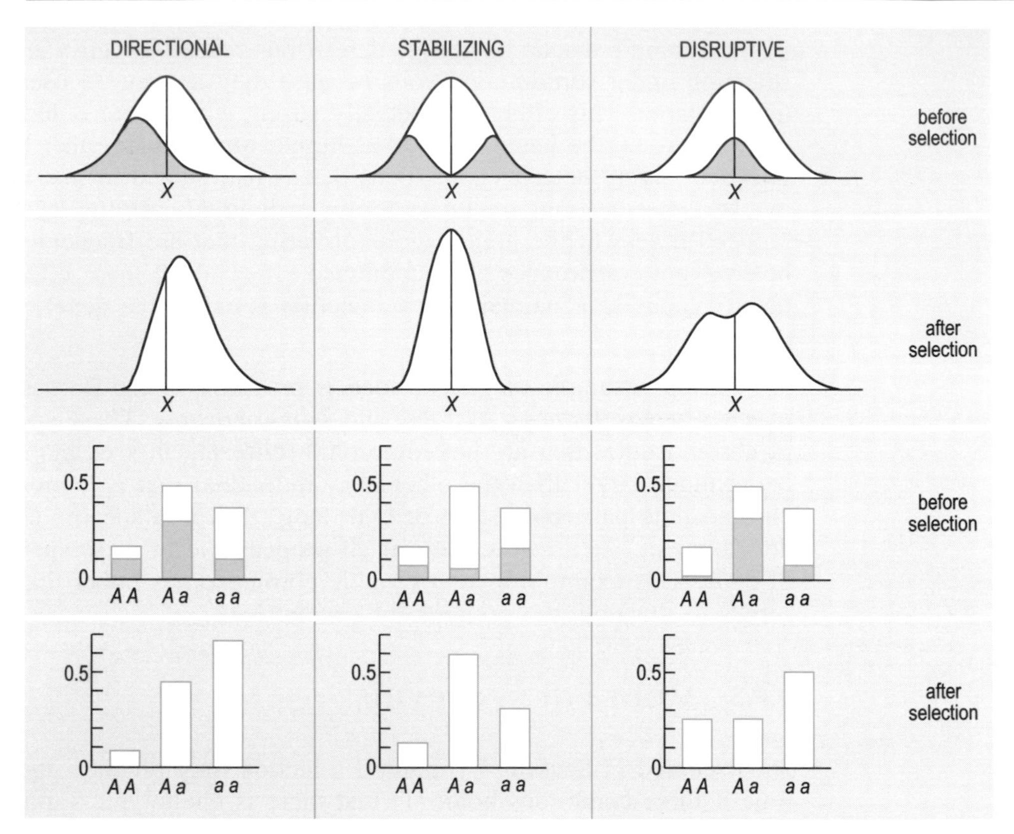

Figure 4.18 Three modes of selection. Upper rows: *effect of selection on a continuously varying character;* lower rows: *effect on a polymorphic character. In each case, the vertical axis is the proportion of individuals, and the horizontal axis is the value of the character. The shaded portions represent individuals that are at a relative disadvantage.* X *indicates the mean value of the character before selection.* AA, Aa *and* aa *are the three genotypes at the polymorphic locus.*

phenotypes within a population may change in relation to their relative frequencies, such that a particular phenotype has a different fitness when it is rare than when it is common. This form of selection is called **frequency-dependent selection**, and its two possible forms are shown in Figure 4.19.

In positive frequency-dependent selection, the fitness of a particular phenotype increases as its frequency in the population increases. As a result, one phenotype will generally increase at the expense of others very rapidly until the others are eliminated from the population. Should there exist an equilibrium, as indicated in Figure 4.19(a), it is highly unstable because any small increase in the frequency of one of the phenotypes will lead to its rapid increase at the expense of the other. Therefore, this is not a process that is likely to be observed operating in nature but is one that may have produced changes in the evolutionary past.

In negative frequency-dependent selection, the fitness of a particular phenotype increases as its frequency in the population decreases. In other words, rarer phenotypes have higher fitness than common ones. This is a process of considerable evolutionary importance and interest because of its possible role in maintaining polymorphisms within populations. Any departure from a stable

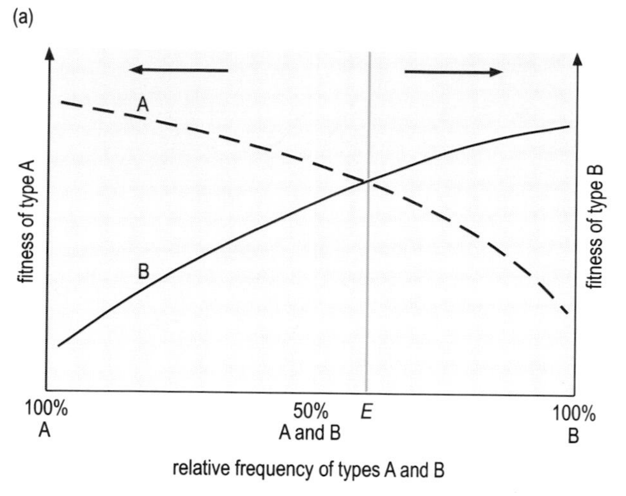

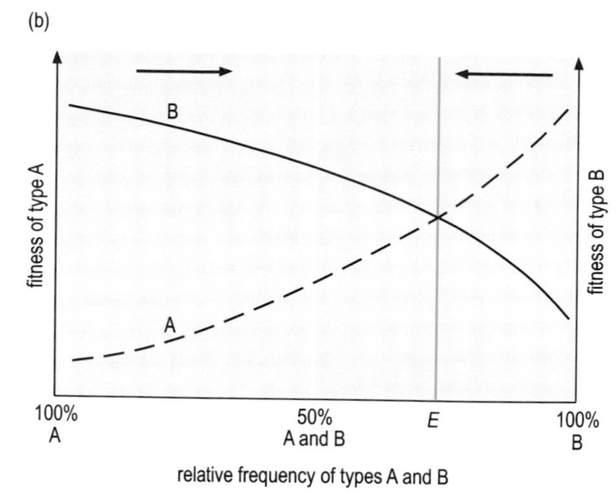

Figure 4.19 *Graphical representations of the two kinds of frequency-dependent selection. (a) In positive frequency-dependent selection, the fitness of two types, A and B, increases as they become relatively more common. This leads to an unstable equilibrium ratio of the two types at* E. *Any departure from this point will lead to selection (indicated by arrows) that reinforces that departure. (b) In negative frequency-dependent selection, the fitness of each type decreases as it becomes relatively more common. The equilibrium ratio* E *is stable in this case as any departure from equilibrium is opposed by selection.*

polymorphic equilibrium tends to be corrected by selection acting against the commoner phenotype, as shown in Figure 4.19(b). There are several contexts in which negative frequency-dependent selection is believed to play an important role in evolution. The following are just three examples:

1 Maintenance of a 50:50 sex ratio. Should a situation arise in which the sex ratio in a population departs from 50:50, some members of the more common sex will fail to obtain mates, whereas all members of the rarer sex will typically obtain mates. Consequently, the average fitness of the rarer sex will be greater than that of the commoner sex, favouring an increase in the frequency of the rarer sex which will tend towards a restoration of the 50:50 sex ratio.

2 Maintenance of polymorphism in prey species. There is considerable evidence that predators, such as birds hunting for cryptic insects and for the snail *Cepaea nemoralis*, are more likely to find and eat common prey types and that they tend to overlook rare types (see also Figure 4.20). As a result, rare morphs in a polymorphic prey species are at a selective advantage, with the result that the polymorphism tends to be maintained.

3 The rare male effect. There is some evidence, mostly from laboratory animals such as fruit-flies, that females within a polymorphic species prefer to mate with males who belong to a rarer phenotype. One study, of the two-spot ladybird *Adalia bipunctata*, has found evidence for such an effect in nature. This species basically exists in two forms, a normal form (red with black spots) and a melanic form (black with red spots), and females are found more often than expected on the basis of random mating to be paired with the rarer type of male in their locality. The adaptive value of a rare-male preference is obscure but, as in 2, its effect is to maintain polymorphism within a population.

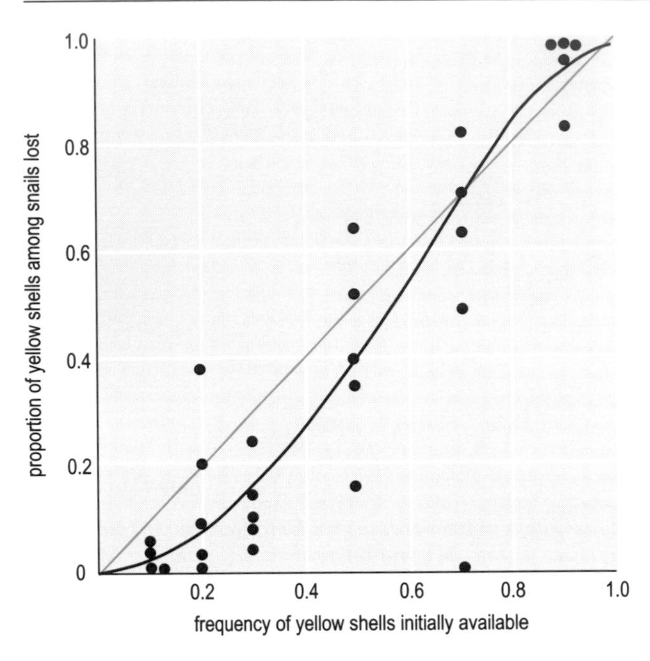

Figure 4.20 *Negative frequency-dependent selection. Predation by unknown predators on the yellow (as opposed to pink) form of the mangrove snail* Littoraria filosa. *The green line shows the proportion that would be taken if predators took the snail morphs in direct proportion to their frequency; the solid black line shows the proportion that was actually taken. The yellow morph is taken less often than expected when it is relatively rare, more often when it is very common.*

4.4.4 THE RATE OF EVOLUTION

The rate at which natural selection will produce phenotypic change will depend on a number of properties of a population. These are illustrated graphically, for a continuously varying character, in Figure 4.21, which shows the frequency distribution of a character in two successive generations. In the first, parental, generation, the mean value for the character among those individuals that survive to reproduce differs from the mean value for the whole population by a value *S*; this is called the **selection differential** (*S* must not be confused with *s*, the selection coefficient (Section 4.4)). In the second, offspring generation, the mean value for the character differs from that in the parental population by a value *R*; this is called the response to selection. Note that *R* is smaller than *S*.

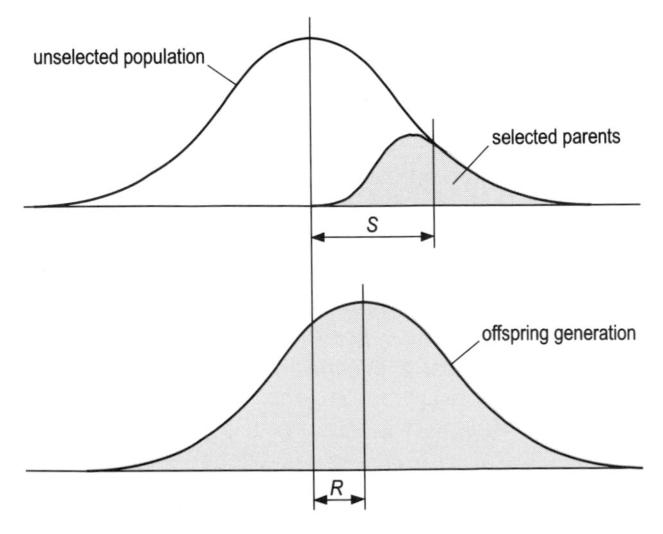

Figure 4.21 *The relationship between the selection differential* S *and the response to selection,* R. *See text for explanation.*

▶ Why do you suppose that this is so?

Because only a proportion of the variation in the character in the first generation is heritable. R and S are related to one another by heritability (h^2, defined in Chapter 3, Section 3.5.1) in the following way:

$$R = h^2S \tag{4.3}$$

▶ How will the value of h^2 affect the relationship between R and S? (Note that the maximum value of h^2 is 1, and the minimum is 0.)

When h^2 is large, R will approach the value of S. When h^2 is small, R will be small relative to S. R is a measure of the rate of evolution. When it is large, phenotypic change from one generation to the next will be pronounced and evolutionary change will therefore be rapid. What, then, influences the value of R?

Consider, first, the relationship with R of the selection differential, S. The value of S depends largely on the amount of variation present in the parental population: if there is a lot of variation, S can be large; if variation is slight, S must be relatively small. This effect was expressed by R. A. Fisher (1930) in what has come to be known as **Fisher's fundamental theorem**, which states that: *the rate of increase in fitness is equal to the additive genetic variance in fitness*. Note that Fisher's theorem applies not to the total variation in a character but to the genetic variation in fitness of that character.

When selection on a character is strong, variation in that character tends to decrease rapidly from one generation to the next (e.g. see Figure 4.9). This effect is well-known from artificial breeding programmes, in which strong directional selection on a character leads, after a few generations, to a situation in which there is no longer any heritable variation in that character. Strong selection thus leads to reduced variation; S consequently becomes smaller and the rate of evolutionary change is reduced. This raises an important question: how can directional evolutionary change be maintained over a long period of time? One possible answer is that, as selection reduces variation in a character, at least some of that variation is restored by continual mutation or by immigration of individuals from other populations subject to different patterns of selection. In fact, the situation in which strong directional selection is maintained over many generations may be relatively rare. In nature, selection pressures typically vary as the environment fluctuates. For example, environmental variation from one year to another favours large clutch size in birds in some years, small clutch size in other years (Chapter 2, Section 2.2).

Turning to heritability, natural selection tends to reduce heritability, just as artificial selection reduces variation in a selected trait, simply because selection leads to increasing homogeneity in genotype over successive generations, as less fit genotypes are eliminated. This leads to the general conclusion, which is somewhat surprising at first sight, that heritability is lower for traits that are strongly related to fitness than it is for other traits. Data illustrating this effect in *Drosophila melanogaster* are shown in Table 4.7. A character that is obviously strongly related to fitness, egg production, has a value of h^2 of 0.2, whereas

Table 4.7 *Approximate values of heritability* (h^2) *for various traits in the fruit-fly* Drosophila melanogaster.

Trait	h^2
Abdominal bristle number	0.5
Body size (thorax length)	0.4
Ovary size	0.3
Egg production	0.2

abdominal bristle number, which apparently has little or no fitness consequences, has a value of 0.5. A low value of h^2 means that there is little variation in the character and that what variation there is is attributable to environmental rather than genetic effects.

To conclude, strong selection can produce a marked phenotypic change from one generation to the next provided, first, that there is a lot of variation in the character in question and, secondly, that the heritability of that character is high. Over time, however, variation in the character, and its heritability, tend to decline, so that the rate of evolution of that character will tend to slow down. This does not mean, however, that change will necessarily cease altogether. Genetic variation for the character may be maintained, to some degree, by mutation, by immigration of individuals from other populations, or by variations in the intensity or direction of selection from one generation to another as a result of environmental fluctuations.

4.4.5 THE INTENSITY OF SELECTION

In Chapter 2, it was emphasized that the presence of a particular character may have several consequences for an organism; in other words, it may have a variety of effects in fitness terms. Returning to the example of body size in anurans (Section 4.2.1), large body size enables a female frog or toad to have high fecundity. However, it may also involve certain costs; larger individuals may be subject to higher stress at times of food shortage and it generally takes them longer to reach sexual maturity. To understand fully how natural selection influences the evolution of a character such as body size, it is necessary to measure the intensity or strength of selection on that character that results from these different factors. S. J. Arnold and M. J. Wade (1984) have developed a technique for doing this, based on Fisher's fundamental theorem (Section 4.4.4), which can be illustrated by studies of anuran body size.

In the American bullfrog *Rana catesbeiana*, males defend territories around the edges of ponds. They call from their territories and females, attracted by the calls, approach a particular male, mate and lay their eggs in the chosen male's territory. The eggs develop and hatch within the territory in which they are laid and detailed studies of bullfrogs by R. D. Howard (1979) have shown that females select males according to features of their territories. They prefer parts of the pond where there is a low incidence of a predatory leech that eats bullfrog eggs and where the water temperature is high, such that the eggs develop rapidly.

Howard made intensive studies of the mating dynamics of bullfrogs and found that there was considerable variation in both male body size and male reproductive success, measured as the total number of zygotes hatching per male in a season (Figure 4.22(a)). A regression analysis of male reproductive success in relation to male body size gives a regression line, called a selection gradient; the steep, positive slope of this reflects the fact that larger males have much higher reproductive success than smaller males. The Figure also includes two mean values: the mean size of all males in the breeding population (z) and the mean size of those males that achieved a reproductive success greater than zero (z_1). The disparity between these two values represents the difference in size between all males in one generation and those that contributed to the next generation.

▶ What is the difference between z_1 and z called?

The selection differential (S) (see Figure 4.21).

The fact that S has a positive value means that there is directional selection, favouring larger male body size. Exactly how does this arise? Why do larger males have higher reproductive success? Howard's data enabled Arnold and Wade to consider three possible factors:

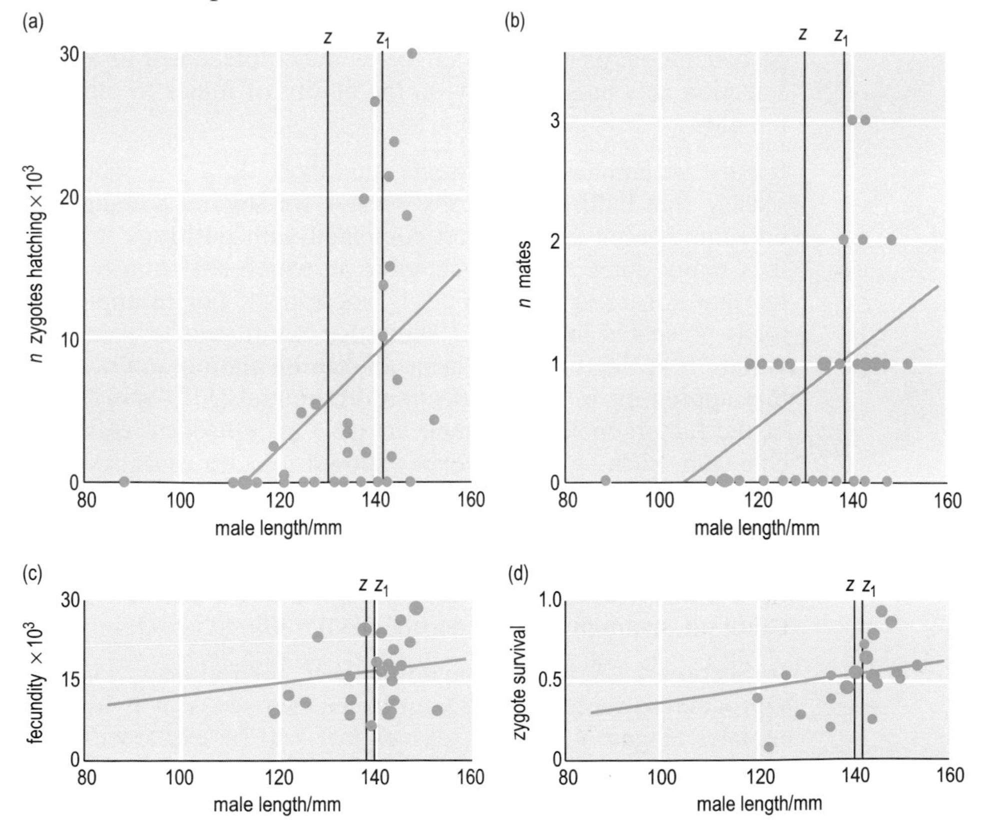

Figure 4.22 An analysis of the intensity of selection on body size in male bullfrogs. The scales of all the y-axes have been standardized against each other according to the variation in fitness established during each episode of selection, so (a)–(d) are directly comparable. See text for explanation.

(i) Larger males obtain more mates because of their superior competitive ability.

(ii) Larger males mate with larger females, which produce larger numbers of eggs.

(iii) Larger males defend territories with fewer predatory leeches and a better temperature for egg development, so that a higher proportion of the eggs laid in their territories survive to hatching.

Factors (i) to (iii) could all be making a contribution to the variation in male reproductive success shown in Figure 4.22(a). Arnold and Wade carried out the same kind of analysis for each of the three factors separately; these are shown in Figure 4.22(b) to (d).

▶ From the results presented in Figure 4.22, what conclusion do you reach about the relative strength of selection in relation to factors (i), (ii) and (iii)?

The difference between z_1 and z is large for factor (i), but is small for factors (ii) and (iii). This suggests that factor (i) makes a much larger contribution to the variation shown in Figure 4.22(a) than factors (ii) and (iii). Likewise, the slope of the selection gradient, though positive for all three factors, is steeper for (i) than for (ii) and (iii). In other words, the intensity of selection on male body size in relation to mating success is much greater than that related to the fecundity of their mates or to the hatching success of eggs laid in their territories. Thus, it can be concluded, on the basis of the evidence presented in Figure 4.22, that natural selection acts most strongly on the ability of males to attract and mate with females.

It must be emphasized, however, that these data were obtained in a single breeding season, that bullfrogs typically survive for several breeding seasons, and that each breeding season is very short compared with bullfrogs' total lifespans. There are thus many other potential situations in which selection may act on their body size that are not taken into account by these data. For example, the data for European toads presented in Table 4.1 show that, from year to year, there is considerable variation in the difference in mean size of mating and non-mating males, a value that approximates to the selection differential (S). Arnold and Wade refer to each of the factors included in their analysis as 'episodes' of selection. A single character, such as body size, is exposed to a series of such episodes, each relating to a different kind of selection pressure, so that whether or not it evolves in a particular direction will depend on the additive effect of all these episodes. A researcher may reasonably expect, for a given character, to be able to collect data for a number of selection episodes, but it is very unlikely that they will be able to study all such episodes in an organism's life.

In Section 4.2.2, it was pointed out that, for red deer, variation in lifetime reproductive success is different in the two sexes; it is higher in males than in females (Figure 4.16). For reasons that will be explained in Chapter 7, it is commonly true that male fitness is more variable than female fitness. Body size is a character common to both sexes and this gives rise to the question, with reference to anurans, which are commonly sexually dimorphic for body size, of whether natural selection acts on the same character in the same way in the two sexes. This could be addressed by analysing the intensity of selection at different episodes in the lives of the two sexes.

4.4.6 Genetic variation and natural selection

A major problem in the theory of evolution by natural selection concerns the relationship between selection and genetic variation. Selection can only lead to evolutionary change in characters for which there is genetic variation, but selection tends to reduce such variation by favouring some alleles and eliminating others. How, then, can selection produce sustained evolutionary change, if it tends to remove the basis of that change? Before the 1960s there was much debate about how much genetic variation actually exists in natural populations. With the advent of molecular techniques such as gel electrophoresis (Chapter 3, Section 3.5.2), it became clear that there is an enormous amount of genetic variation. The question then became: how important is this variation? Does it provide the 'raw material' for evolutionary change or is it irrelevant? One school of thought was that many of the mutations that give rise to genetic variation are selectively neutral and that their frequency is therefore not affected by selection; rather, they increase or decrease in frequency as a result of mutation and genetic drift.

Many of the genes identified by gel electrophoresis code for enzymes and, it appears, many point mutations may occur in such genes that have no favourable or adverse effect on the action of those enzymes; i.e their effect is selectively neutral (Chapter 3, Section 3.3.2). This is not to say that enzymes are not subject to natural selection, however. For example, closely related fishes occurring on each side of the Panama isthmus have differences in the enzyme muscle lactate dehydrogenase that are clearly adaptive. Species living on the Pacific side have enzymes that function best at the cooler temperatures at which these species live; the equivalent enzymes in species on the Atlantic side function best at slightly higher temperatures. For many enzyme differences between populations, however, there appear to be no adaptive consequences.

Whereas electrophoresis reveals variation in enzymes and other biochemical characters, it tells us very little about variation in genes controlling other characters, such as body size, morphology and coloration. Thus, rather little is known about the amount of genetic variation underlying the kind of characters that many evolutionary biologists study in nature, such as those described in Sections 4.2 and 4.3. Much of the variation in the genome occurs in that large proportion of the DNA that is never expressed (Chapter 3, Section 3.2.3), and so is never subjected to natural selection. For these reasons, the evolutionary significance of the high level of variation that exists in natural populations at the genetic level remains somewhat obscure. There are, however, a number of phenomena that are known to sustain this variation.

The frequency of an allele is not independent of that of other alleles. Rather, specific alleles often show a strong degree of linkage with one another. Strong linkage occurs between alleles at loci that are close together on the same chromosome and between which crossing-over never or very rarely occurs. A consequence of strong linkage between alleles is that an allele that is selectively neutral, or even deleterious, may be maintained at high frequency in a population because it is linked to an allele that is strongly favoured by selection, a phenomenon called 'hitch-hiking'.

While it is important to identify sources of the genetic variation on which natural selection acts, it is also important to remember that much genetic variation is itself a consequence of natural selection. There are a number of processes, two of which have been discussed earlier in this Section, that tend to maintain variation:

1 Heterozygous advantage (Section 4.4.2).

▶ How will heterozygous advantage contribute to the maintenance of genetic variation?

It favours a combination of two alleles at a given locus, preventing either going to fixation or extinction. The frequencies of alleles that, in the homozygous condition, have an adverse effect on fitness thus become balanced.

2 Negative frequency-dependent selection (Section 4.4.3).

▶ How will negative frequency-dependent selection contribute to the maintenance of genetic variation?

Again, it favours more than one allele at the same locus because, as either of them gets rare, its relative fitness increases. As a result, neither allele will be eliminated by selection.

Finally, genetic variation is maintained in natural populations by changes in the environment. From one generation to the next, the environment may shift slightly, such that a phenotype that has high relative fitness in one generation may not do so in the next.

▶ Can you recall an example of this effect?

Two examples have been mentioned earlier in this Book. The optimal clutch size of birds may differ from one year to the next (Chapter 2, Section 2.2.1) and, in toads, larger males have a mating advantage in some years but not in others (Section 4.2.1). Environmental changes will tend to promote genetic variation because particular alleles will increase in frequency when the environment is favourable for them but will not be eliminated by a change to a less favourable environment unless that change is sustained over a very long period of time.

SUMMARY OF SECTION 4.4

To study evolution quantitatively, it is necessary to give the fitness of a phenotype a numerical value. In the context of natural selection, what is important is the relative fitnesses of different phenotypes. The fittest phenotype in a population is given a fitness value of $W = 1$ and other, less fit phenotypes have values of $W < 1$. In practice, fitness is very difficult to measure, especially in long-lived organisms that breed several times during their lives.

The extent and direction of changes in the frequency of a character as a result of natural selection depend critically on the genetic basis of that character. The Hardy–Weinberg formula, which assumes no selection, provides a null hypothesis against which the impact of selection on gene frequencies can be deduced. The rate of change in the frequency of a given allele in a population depends on: the fitness consequences of that gene (in both homozygous and heterozygous form), its frequency in the population, and whether or not it is dominant or recessive.

With respect to changes in the expression of a given character over time, natural selection may be directional, stabilizing or disruptive. The relative fitness of a given phenotype is not constant but can change with the frequency of that phenotype, if there is frequency-dependent selection. Negative frequency-dependent selection, in which phenotypes have higher fitness when they are rarer, tends to maintain variation in a population.

The rate at which selection leads to phenotypic change depends on three factors: the strength of selection (i.e. the magnitude of fitness differences between phenotypes), the heritability of the character in question, and the amount of variation in fitness in that character (Fisher's fundamental theorem). A consequence of sustained directional selection on a character is that, over time, both variation in, and the heritability of, that character tend to decline. The intensity of selection on a given character, such as body size, can be measured by relating the expression of that character to its fitness consequences.

Genetic variation is very high in natural populations, but the evolutionary significance of this variation is not clear because much of it may not be subject to selection. Some genetic variation is a consequence of certain aspects of natural selection and also of fluctuations in the environment.

4.5 FITNESS AND ADAPTATION

Natural selection and fitness are very important explanatory concepts in evolutionary theory but, to gain a complete picture of the evolutionary process, they need to be related to, and integrated with, the equally important concept of adaptation which was discussed in Chapter 2. In this Section, a conceptual model is developed which relates changes in allele frequencies to the fitnesses of individuals in a population. This model can be a powerful device for predicting the patterns of adaptive change in populations of differing size and structure.

4.5.1 MEAN FITNESS

Before addressing the question of how selection relates to adaptation, it is necessary to introduce the variable **mean fitness ($\bar{W}$)** of a population. It is important not to misunderstand this term. It is *not* intended to imply that the population has some corporate fitness. Rather, as its name suggests, it simply refers to the mean value of the fitnesses of all the individuals in a population. It is readily calculated as the sum of the fitnesses of each genotype multiplied by the frequency of that genotype. Suppose that there are two alleles, X and x, at a single locus, that XX individuals have a fitness of 1, xx individuals $1 - s$ and that Xx individuals have intermediate fitness, e.g. $1 - s/2$. These values can be plotted graphically (Figure 4.23). If all the individuals in the population are homozygous for X (i.e. $p = 1$) then both the frequency of genotype XX and the fitness of all

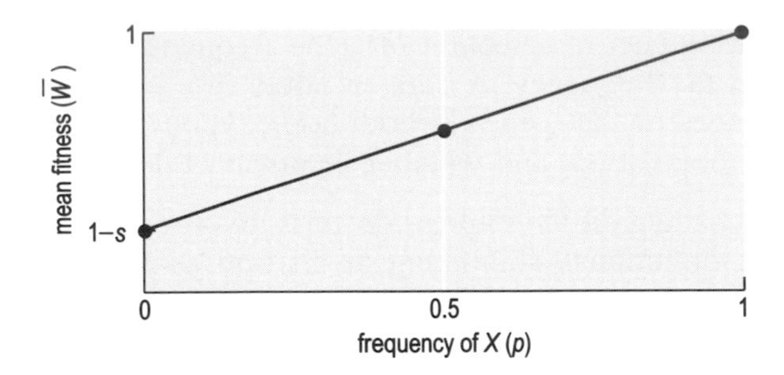

Figure 4.23 *The relationship between the mean fitness ($\overline{W}$) of a population and the frequency of an allele X at a single locus. See text for explanation.*

individuals is 1. Therefore, $\overline{W} = 1 \times 1 = 1$ (extreme right of the graph in Figure 4.23). If all individuals are homozygous for x (i.e. $p = 0$), the frequency of the genotype xx is 1 and the fitness of all individuals is $1 - s$. Therefore, $\overline{W} = 1 \times (1 - s) = 1 - s$ (extreme left of the graph in Figure 4.23).

Suppose, finally, that the frequency of A is 0.5. According to the Hardy–Weinberg ratio, the frequencies of the three genotypes, XX, Xx and xx (before selection) are 0.25. 0.50 and 0.25, respectively. Then:

$$\begin{aligned}\overline{W} &= (0.25 \times 1) + (0.50 \times (1 - s/2)) + (0.25 \times (1 - s)) \\ &= (0.25) + (0.50 - 0.50 \times s/2) + (0.25 - 0.25s) \\ &= (0.25 + 0.50 + 0.25) - (0.50 \times s/2 + 0.25s) \\ &= 1 - 0.5s\end{aligned}$$

This is represented by the middle point in Figure 4.23.

This hypothetical example shows that the mean fitness of a population is affected by the relative frequencies of alternative alleles at a given locus in that population. Any population may thus be represented as a point on the line in Figure 4.23, which shows the mean fitness for given frequencies of the allele. The occurrence of less adaptive alternative alleles at a locus reduces mean fitness, an effect described in Section 4.4.2 as genetic load.

▶ What effect will natural selection have on the frequency of X in the population shown in Figure 4.23, if its initial frequency is less than 1?

As all the less fit genotypes are eliminated, the frequency of X will increase to fixation, i.e. x will be eliminated. On the graph in Figure 4.23, this is represented by the point moving up the line towards the right. Thus, natural selection tends to increase the mean fitness of a population. This is always true where the fitnesses of different genotypes remain constant. Where fitness is frequency-dependent, however, it is possible for selection in favour of an allele to lead to a *decrease* in mean fitness. As this only occurs in rather special circumstances, we will not consider such exceptional cases further.

4.5.2 The adaptive landscape

The adaptive landscape is a way of visualizing the combined effect of variation at several loci on fitness. In a simple case, there are two loci (Figure 4.24). At one locus (Figure 4.24(a)) there is heterozygous advantage such that $\overline{W}$ is highest when

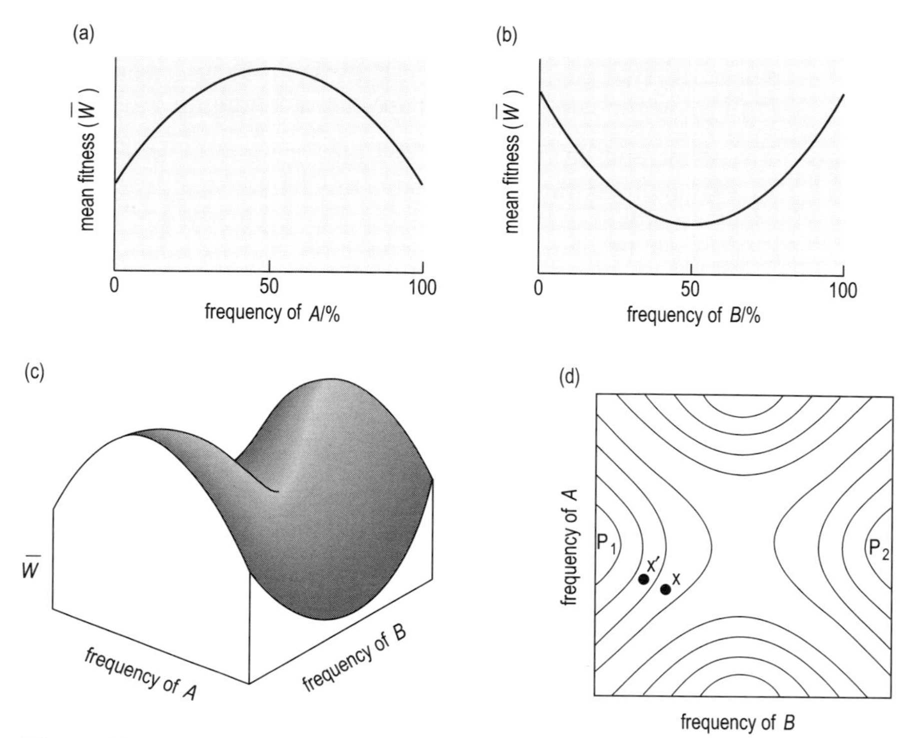

Figure 4.24 *A simple 'adaptive landscape', based on two genetic loci. See text for explanation.*

allele A occurs at a frequency of 50%. At the other locus (Figure 4.24(b)) there is heterozygous disadvantage such that $\overline{W}$ is lowest when allele B occurs at a frequency of 50%. If these two loci form two axes of a three-dimensional plot in which $\overline{W}$ is the third (vertical) axis, we get the picture shown in Figure 4.24(c) which translates into a contour map as in Figure 4.24(d). In the example represented here, there are two areas where mean fitness is very high, called 'adaptive peaks', P_1 and P_2.

▶ Suppose that at locus B there is heterozygous advantage, as there is at locus A in Figure 4.24(a). What shape would the adaptive landscape be in this case?

It would form a single peak, occupying the centre of Figure 4.24(d).

Consider, in Figure 4.24(d) a population for which, at a given time, the relative frequencies of alleles A and B is such that it occupies position x. Natural selection will act on that population, altering the frequencies of A and B such that the value of $\overline{W}$ will tend to increase.

▶ If the value of $\overline{W}$ increases for the species at position x, how will this affect its position on the map in Figure 4.24(d)?

It will move 'up the hill' towards peak P_1, to position x′. By favouring the fitter genotypes, natural selection will act so as to increase mean fitness (except in the rare circumstances mentioned earlier). The population cannot move by natural selection from position x towards P_2, because this would mean moving across a 'valley' in the adaptive landscape, i.e. undergoing a decrease in mean fitness.

The shape and distribution of the fitness contours in Figure 4.24(d) are determined by the nature of the environment. In Chapter 2, it was emphasized that environments in which species live are, typically, not constant from generation to generation. In other words, the position, steepness and height of an adaptive peak in the landscape may vary over time; nature is constantly 'shifting the goalposts'. This would mean that, over time, the population is moving about over the landscape, tending always to move towards the nearest peak, but never quite getting there. This has been expressed by Van Valen (1973) as the **Red Queen's hypothesis**. In Lewis Carroll's *Through the Looking Glass*, the Red Queen tells Alice 'Now here, you see, it takes all the running you can do, to keep in the same place'.

The adaptive landscape shown in Figure 4.24 is grossly simplistic in that it depicts variation at only two loci. Moreover, it has been assumed that the loci have independent effects on fitness. In reality, epistatic interactions between loci may cause the shape of the curve in Figure 4.24(a) to vary as the frequency of allele *B* varies, and that of Figure 4.24(b) to vary as that of *A* varies, yielding a more complex landscape. Furthermore, there are many thousands of loci, many with complex and interacting effects on fitness, so that a more realistic representation would have many thousands of axes, yielding an inconceivably complex landscape. Nevertheless, this simple version, which can readily be visualized, is a useful device for thinking about genetic change and adaptation in populations.

The adaptive landscape model can be used to make another important point about evolutionary change brought about by natural selection. Suppose that, as in Figure 4.25, P_1 is higher than P_2; that is, the mean fitness of the population located at z would be higher if it could acquire the gene frequencies that define P_1. Such a move would be called a *peak shift*. If the adaptive landscape remains static over generations, there is no way that, through a process of adaptive change, the population can move from P_2 to P_1.

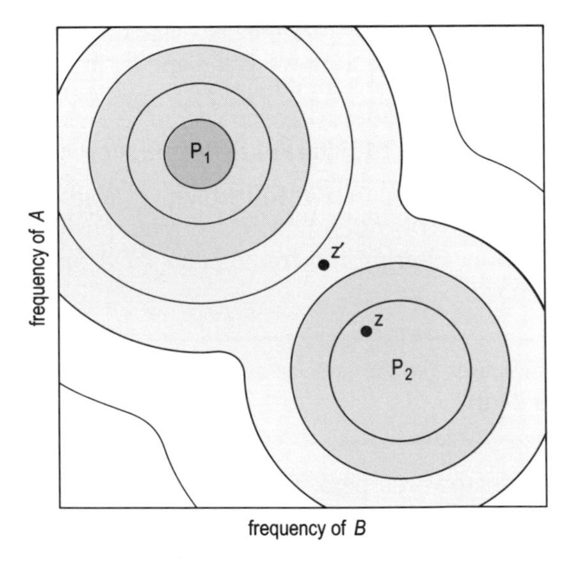

Figure 4.25 *An adaptive landscape with two adaptive peaks of different height. See text for explanation.*

► Can you think of any way by which the population could move to P_1?

Peak shifts might occur through genetic drift. Suppose that the population located at position z in Figure 4.25 becomes greatly reduced in terms of the number of individuals. A property of small populations is that gene frequencies may change as a result of genetic drift (Chapter 3, Section 3.6). Thus, while selection will tend to move the population from z towards P_2, genetic drift may move it in any direction across the landscape. Suppose that drift moves the population downhill towards the position z′.

► In what direction will natural selection now tend to move the population?

It will tend to move the population up the opposite slope of the 'valley' towards P_1. In this way, genetic drift might move a population about over the adaptive landscape and may thus bring about a peak shift.

The American evolutionary biologist Sewall Wright suggested that many species consist of a number of populations, of varying size, between which there may be little migration of individuals and, therefore, very limited gene flow. Through the combined effect of genetic drift and natural selection, each population could be at or near a slightly different adaptive peak, of which some are higher than others. If a particular population reaches a higher adaptive peak than others, it may start to disperse more, fitter migrants and colonists than other populations, thus spreading its new adaptation to the rest of the species. In this way, Wright suggested that more rapid evolutionary change could occur in a species than is envisaged on the basis of selection operating over the species as a whole. This is known as the *shifting balance* model of evolution.

The adaptive landscape model illustrates the fact that the capacity of natural selection to bring about adaptive change in a population is constrained by the genetic structure of that population. This point can be illustrated by an analogy. The design of aircraft can be said to have 'evolved'; numerous features of aircraft, such as their shape, size, speed, etc., have diversified and undergone directional changes over the years. In the course of this evolution, there was a dramatic change with the invention of the jet engine. This is fundamentally different from the propeller engine, involving quite different design principles, and, if aircraft design were directly comparable to natural selection, such a change is very unlikely to have occurred; it is analogous to a large peak shift. One of the major issues that has concerned evolutionary theorists is how major changes in evolution, such as those from fish to terrestrial vertebrates, or from reptiles to mammals, took place. Did they involve major 'jumps' or can they be accounted for by the accumulation of very many, small changes. This issue is discussed further in Chapter 14.

SUMMARY OF SECTION 4.5

1 At a given time, a population has a given value of mean fitness which is a function of the fitnesses of all genotypes in that population and of the relative frequencies of those genotypes.

2 Natural selection tends (with rare exceptions) to increase mean fitness by reducing the frequency of, or eliminating, alleles that reduce individual fitness.

3 The range of potential values of mean fitness can be plotted as an adaptive landscape on which a population will tend to move about, natural selection driving it towards a local adaptive peak. The shape of the landscape changes as the environment changes so that a population will typically never be at an adaptive peak.

4 Natural selection cannot move a population from one adaptive peak to another but, if the population is small, genetic drift may lead to such a peak shift.

REFERENCES FOR CHAPTER 4

Arnold, S. J. and Wade, M. J. (1984) On the measurement of natural and sexual selection: theory, *Evolution*, **38**, 709–79.

Arnold, S. J. and Wade, M. J. (1984) On the measurement of natural and sexual selection: applications, *Evolution*, **38**, 709–79.

Berven, K. A. (1987) The heritable basis of variation in larval development patterns within populations of the wood frog (*Rana sylvatica*), *Evolution*, **41**, 1088–1097.

Clutton-Brock., T. H., Guinness, F. E. and Albon, S. D. (1982) *Red Deer: The Behaviour and Ecology of Two Sexes*, Chicago University Press, Chicago.

Cook, L. M., Mani, G. S. and Varley, M. E. (1986) Postindustrial melanism in the peppered moth, *Science*, **231**, 611–613.

Darwin, C. (1859) *On the Origin of Species*, John Murray, London.

Darwin, C. (1871) *The Descent of Man and Selection in Relation to Sex*, John Murray.

Davies, N. B. and Halliday, T. R. (1977) Optimal mate selection in the toad *Bufo bufo, Nature*, **269**, 56–58.

Davies, N. B. and Halliday, T. R. (1979) Competitive mate searching in male common toads *Bufo bufo*, *Animal Behaviour*, **27**, 1253–1267.

Endler, J. A. (1980) Natural selection on color patterns in *Poecilia reticulata, Evolution*, **34**, 76–91.

Endler, J. A. (1986) *Natural Selection in the Wild,* Princeton University Press.

Fisher, R. A. (1930) *The Genetical Theory of Natural Selection*, Oxford University Press.

Gould, S. J. (1984) Covariance sets and ordered geographic variation in *Cerion*, from Aruba, Bonaire and Curaçao: a way of studying non-adaptation, *Systematic Zoology*, **33**, 217–237.

Gross, H. P. (1978) Natural selection by predators on the defensive apparatus of the three-spined stickleback, *Gasterosteus aculeatus* L., *Can. J. Zool*, **56**, 398–413.

Howard, R. D. (1979) Estimating reproductive success in natural populations, *Am. Nat.*, **114**, 221–231.

Huntingford, F. A. (1982) Do inter- and intraspecific aggression vary in relation to predation pressure in sticklebacks?, *Animal Behaviour*, **30**, 909–916.

Jones, J. S. (1982) More to Mendelism than meets the eye, *Nature (Lond.)*, **300**, 109–110.

Kaplan, R. H. (1989) Ovum size plasticity and maternal effects on the early development of the frog, *Bombina orientalis* Boulenger, in a field population in Korea, *Functional Ecology*, **3**, 597–604.

Reimchen, T. E. (1989) Loss of nuptial color in threespined sticklebacks (*Gasterosteus aculeatus*), *Evolution*, **43**, 450–460.

Smith, J. N. M. (1981) Does high fecundity reduce survival in song sparrows?, *Evolution*, **35**, 1142–1148.

Van Valen, L. (1973) A new evolutionary law, *Evolutionary Theory*, **1**, 1–30.

SELF-ASSESSMENT QUESTIONS

SAQ 4.1 (*Objective 4.1*) What is the most important difference between the modern definition of fitness and Darwin's use of the term?

SAQ 4.2 (*Objective 4.2*) For each of the following observations (i) to (iii), decide whether it is (a) a necessary condition for natural selection to occur, or (b) a consequence of natural selection.

(i) A comparison of two successive generations and their respective progeny shows a statistically significant difference between them in the frequency distribution of body size.

(ii) Larger individuals have higher mating success.

(iii) Body size is heritable.

SAQ 4.3 (*Objective 4.3*) For each of the following statements (i) to (iv), decide whether it represents evidence for one of the three conditions necessary for natural selection to occur. If your answer is 'yes', state to which of the following three conditions it relates:

(a) variation, (b) fitness differences, or (c) inheritance.

(i) Adult birds defending breeding territories that are rich in food produce larger fledglings that are more likely to survive to adulthood.

(ii) In a population of birds, some males spend more of their time singing than do other males.

(iii) Within a species of fish, the more colourful males tend to engage in more matings.

(iv) The habit shown by small birds, notably blue and great tits, of pecking into the tops of milk bottles on doorsteps, has increased markedly in frequency during the past 30 years.

SAQ 4.4 (*Objective 4.4*) Give two examples of phenotypic characters in which geographical variation is correlated with features of the environment that appear to exert a selection pressure on those characters.

SAQ 4.5 (*Objective 4.5*) Give two examples of phenotypic characters in which variation has undergone adaptive change over time as a result of some perturbation of the environment.

SAQ 4.6 (*Objective 4.6*) Suppose that, in a species of frog, 100 individuals of each sex are marked in one breeding season. In the following year, 85 marked males and 70 marked females are recaptured. Calculate the coefficient of selection on (a) males, (b) females from these data.

SAQ 4.7 (*Objective 4.7*) Why might the following measures of fitness be inadequate or misleading estimates of true fitness?

(i) Measurements made of toads at a pond in 1977 reveal that larger males are more likely to mate than smaller ones (see Table 4.1).

(ii) Observations of a population of blue tits reveal that some females lay more eggs than other females.

SAQ 4.8 (*Objective 4.8*) In a sample of 130 moths, 44 are black (genotype BB), 16 are white (bb) and 70 are heterozygous intermediates. Use the Hardy–Weinberg formula to determine whether the population from which the sample was taken is subject to natural selection. (Assume that mating is random and that there is no immigration or emigration.)

SAQ 4.9 (*Objective 4.9*) Does selection cause a more rapid increase in the frequency of a favourable allele if:

(i) that allele is dominant rather than recessive?

(ii) that allele is very rare rather than common?

SAQ 4.10 (*Objective 4.10*) Under what conditions does frequency-dependent selection lead to (a) an increase, (b) a decrease in the amount of genetic variation within a population?

SAQ 4.11 (*Objective 4.11*) How are variation in a character, selection on that character, and the heritability of that character related to one another?

SAQ 4.12 (*Objective 4.12*) What measure can show the intensity of selection on a given character, and how is that measure obtained?

CHAPTER 5
REPRODUCTIVE PATTERNS

Prepared for the Course Team by Tim Halliday

Study Comment *It is a characteristic feature of all organisms that they reproduce. Sexual reproduction involving the fusion of gametes from two individuals of different gender is a very widespread pattern, but is by no means the only mode of reproduction found among plants and animals. In this Chapter, we review briefly a variety of modes of reproduction, concentrating on the relative advantages and disadvantages of sexual and asexual reproduction. In so doing, we will see that the mode of reproduction shown by a species has consequences for the long-term survival of the species as well as for the immediate reproductive success of the individual. This raises the possibility that, in this context, selection at a level above the individual may be involved in evolution. We look also at the evolution of gender, raising the question of why males and females evolved. We shall see that, where there are gender differences, different selection pressures operate on the two sexes leading to a variety of secondary aspects of reproductive modes. This is an area of evolutionary theory in which there are many questions, ideas and hypotheses, but few clear answers; it is a source of continual lively debate among evolutionary biologists. Be prepared for many questions to be raised in your mind but do not expect most of them to be unequivocally answered.*

Objectives When you have finished reading this Chapter, you should be able to do the following:

5.1 Recall and describe the genetic consequences of sexual and asexual reproduction and evaluate them according to whether they represent costs or benefits to the individual or the species.

5.2 Recall and describe the essential features of the sib-competition model for the evolution of sexual reproduction.

5.3 Recall the arguments for the evolution of sex as an adaptation for competition with other species and as an outcome of coevolution with other species.

5.4 Outline the genetic and fitness consequences of inbreeding and outbreeding and describe some of the adaptations of plants and animals that ensure a particular balance between the two processes is maintained.

5.5 Recall and describe the disruptive selection hypothesis for the evolution of anisogamy.

5.6 Describe what is meant by an evolutionarily stable strategy and relate this concept to the evolution of the sex ratio and to sequential hermaphoditism in fishes.

5.1 INTRODUCTION

Because humans and the domestic and farm animals with which we are familiar reproduce sexually, we tend to regard reproduction and sex as synonymous processes. In fact, sex of the kind practised by humans and other mammals is only one of a very diverse array of reproductive patterns found among plants and animals. Many organisms sometimes reproduce by mechanisms that do not involve sex, called **asexual reproduction**. For example, some organisms reproduce by simply dividing into two; others show **parthenogenesis** (meaning 'virgin birth'), the eggs developing into young without being fertilized. Some organisms are hermaphrodites that fertilize their eggs with their own sperm, and others exchange sperm with another individual. In this Chapter, we will not attempt to describe all the reproductive patterns that exist, but will concentrate on sexual and asexual reproduction and discuss the factors that have been involved in the evolution of sex and of asexual reproduction and their long-term evolutionary consequences.

Consideration of the evolution of a particular form of reproduction, such as sex, raises two principal questions: why did it evolve in the first place, and why is it maintained among living animals and plants in the presence of alternative modes of reproduction? Although the first of these questions is clearly important, it is not addressed in this Chapter, which instead concentrates on the second question. There are a number of reasons for this. First, the fossil record tells us virtually nothing about the mode of reproduction of extinct species and so it is possible only to speculate about the evolutionary origins of reproductive patterns. Secondly, whereas plausible theories can be made about the origins of sex by, for example, comparing living organisms (see Chapter 16), such theories are very difficult to test experimentally. In contrast, it is often possible to test hypotheses about how sex is maintained in a species. This is especially true of species that show both sexual and asexual forms of reproduction, as will be described later in this Chapter.

5.1.1 THE LIFE CYCLE OF APHIDS

If we wish to understand sexual and asexual reproduction as evolutionary adaptations to particular aspects of the environment, a good place to start is to consider organisms that have life cycles which incorporate both modes of reproduction. Such species should give us clues to the ecological context in which each mode is used. One such species is the lettuce-root aphid *Pemphigus bursarius* (see Figure 5.1).

In the early spring, lettuce-root aphids exist as eggs that have overwintered in crevices in the bark of poplar trees. The eggs hatch into immature founder females that crawl out and find newly growing poplar leaves; at this stage, males do not exist. A female injects a leaf with growth stimulators, causing it to form a protective and nutritive gall around her. As the female grows, so does the gall, until it reaches a size of about 10×15 mm, and she then gives birth, parthenogenetically, to 100–250 winged daughters, but no sons. After about nine weeks, the young fly off, already containing embryos of the next generation, and land on lettuce leaves. There they reproduce, again parthenogenetically, feeding initially on lettuce leaves and later on the lettuce roots, building up very large populations of wingless aphids over several generations.

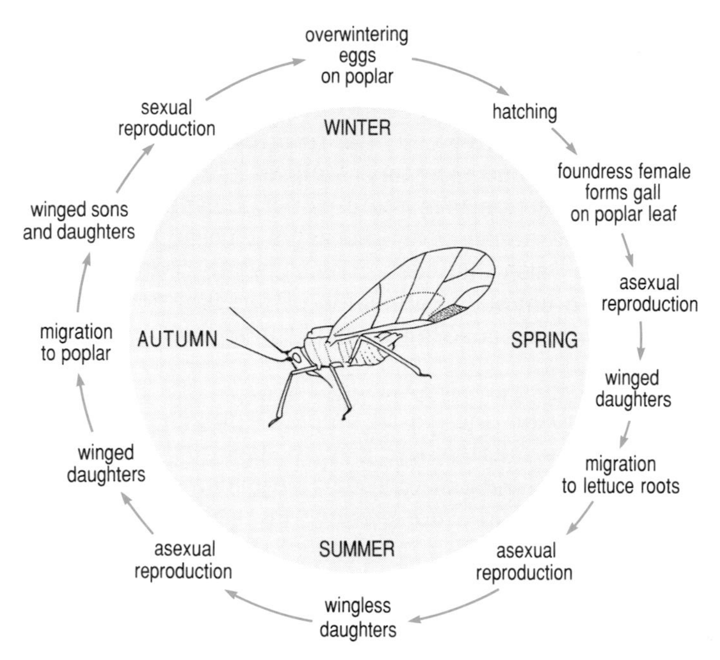

Figure 5.1 *The annual cycle of the lettuce-root aphid* Pemphigus bursarius, *showing the alternation of sexual and asexual reproduction at different times of year.*

In August, the lettuce growing season is drawing to a close. The flightless aphids living in the roots produce a generation of females that crawl up to the leaves and develop wings; these fly to poplar trees where they produce, parthenogenetically, both male and female offspring. This is the only phase in the life cycle at which males exist in this species. These sexual aphids lack mouthparts and cannot feed. Each female consists of little more than a single large egg with legs; once she is fertilized by a male, she deposits her egg in a bark crevice where, the following spring, it will hatch into a founder female. The annual cycle of the aphid is summarized in Figure 5.1.

In this example, we see that asexual, parthenogenetic reproduction is characteristic of the spring and summer period, when the aphid's food supply is abundant and increasing. It enables the aphids to reproduce very rapidly; an aphid on a lettuce in spring contains several daughters about to be born and, inside each of these, another generation of young is already developing. In late summer, by contrast, the food supply is beginning to decline and aphids face the prospect of a long winter of dormancy without any food. At this stage they switch to sexual reproduction; this produces the large eggs that found a new population the following year. This association of asexual reproduction with favourable environmental conditions, and sexual reproduction with declining conditions and environmental uncertainty, is a common one and provides clues to the adaptive significance of the two kinds of reproduction. To fully understand sexual and asexual reproduction as adaptations to particular kinds of environment, we must first consider the different genetic consequences of these and other modes of reproduction.

5.1.2 Genetic consequences of different modes of reproduction

Sexual reproduction has two essential features that distinguish it from asexual reproduction: meiosis and syngamy. Meiosis is the process of cell division that is intrinsic to the production of haploid gametes; syngamy is the fusion of two haploid gametes to produce a diploid zygote. During meiosis the genotype of the parent is 'shuffled' such that each gamete, while containing half of the parental genotype, carries a unique combination of parental genes. This shuffling arises through recombination (crossing over and independent assortment—Chapter 3, Section 3.3.1).

In asexual reproduction, the parent produces progeny that are exact genetic replicas of itself. There is no intermediate gamete stage and thus no meiosis and no syngamy. As a result, asexual reproduction does not involve either assortment or crossing over and thus does not produce the variety of new genotypes that is produced by sexual reproduction (Figure 5.2). Let us suppose that the two genes, *A* and *B*, in Figure 5.2 are deleterious. The asexual parents inevitably pass them on to their progeny but the sexual parents produce some progeny that lack both of them; sexual reproduction thus has the capacity to eliminate deleterious mutations. Now let us suppose that *A* and *B* are advantageous. Again, they are conserved in asexual progeny but some of the sexual progeny contain both genes; sexual reproduction thus has the capacity to create new combinations of favourable mutations. With asexual reproduction, *A* and *B* can arise in the same individual only if both mutations happen in the same lineage. For example, a successful lineage may arise that carries *A*, in which mutation *B* arises later. Despite this theoretical prediction that sex should produce greater genetic diversity, genetic studies of asexual species have revealed surprisingly high levels of heterozygosity, for reasons that are not well understood.

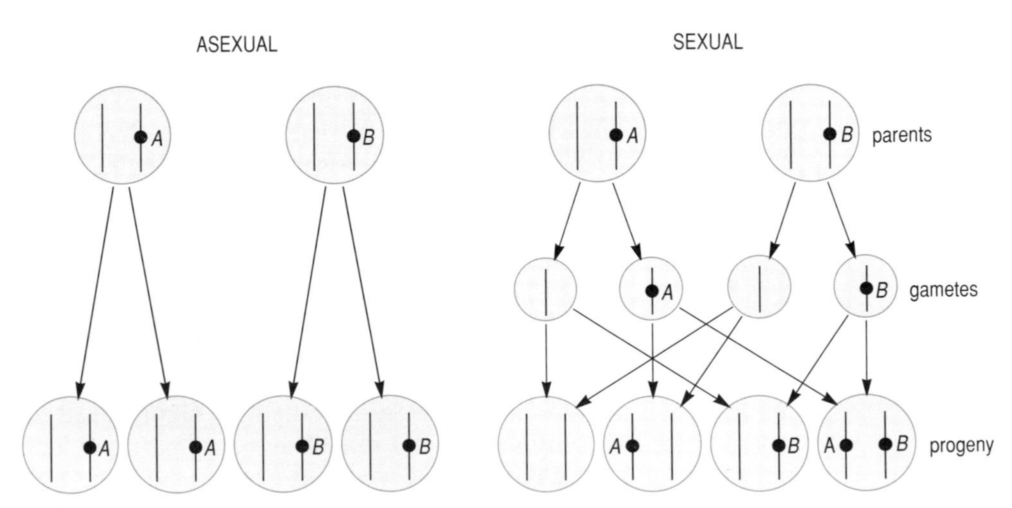

Figure 5.2 *The genetic consequences of asexual (left) and sexual reproduction (right). Asexual reproduction produces progeny that are exact replicas of the parents, carrying the same genes* (A, B)*. Sexual reproduction produces genetically variable progeny that carry genes from two parents in different combinations.*

Reproduction can be likened to a lottery in which tickets represent progeny in the next generation. To win a lottery (leave some surviving offspring), one should buy as many tickets as possible (maximize reproductive success). In asexual reproduction, all the 'tickets' have the same number (same genotype); in sexual reproduction, each 'ticket' has a different number. Thus, asexual reproduction is likely to be favoured when the environment to be faced by the next generation is the same as that experienced by the parents. When the environments faced by parents and progeny differ, however, the capacity of sexual reproduction to produce a range of different genotypes will be favoured. This association between mode of reproduction and predictability of the environment fits the example of lettuce-root aphids rather well.

As described above, sexual reproduction has the capacity to eliminate deleterious mutations; a corollary of this is that asexual reproduction does not have this capacity. Because asexual species cannot eliminate harmful mutations, they tend to accumulate them over many generations, a process that represents a major cost of asexual reproduction. Turning to sex, while a major advantage of sexual reproduction is its capacity to produce new combinations of genes through recombination, the same process may also represent a cost.

▶ Why might this be so?

Recombination will tend to break up favourable combinations of genes, by the very same process that brings them together. This is referred to as the cost of recombination. The strength of this effect depends critically on the relationships between particular genes. As described in Chapter 3, Section 3.3.1, a number of genes may be linked through being close together on the same chromosome to form a linkage group. Recombination during sexual reproduction will be less likely to break up combinations of genes that are strongly linked than it will others.

Asexual and sexual reproduction, as depicted in Figure 5.2, represent only two of a variety of modes of reproduction. Each mode has different genetic consequences in terms, first of the degree of genetic variation between parents and progeny and, secondly, the level of heterozygosity present in the population. We will now describe briefly a number of modes of reproduction, indicating for each their genetic consequences. (It is not necessary that you remember the details of these different reproductive modes; the important point is that there is great diversity among organisms, and that different modes of reproduction yield a continuum of genetic diversity among the progeny.)

- *Mitotic parthenogenesis* Female parent produces eggs by mitosis. Progeny are genetically identical to parent and all individuals are female. Heterozygosity is preserved.
- *Sexual parthenogenesis* Female parent produces eggs by meiosis that can develop without uniting with a male gamete. (The diploid state is restored, either by a cell division that doubles the number of chromosomes, or by the fusion of the egg nucleus with another maternal nucleus.) Progeny will differ genetically from the parent to a small degree, depending on the level of heterozygosity in the parent. (At any heterozygous locus, there are two different alleles, so that meiosis will produce eggs that have different alleles at that locus. Eggs will be identical at

homozygous loci.) Note that, in plants, parthenogenesis, with or without meiosis, is called **apomixis**.

- *Self-fertilizing hermaphroditism* Female parent produces male and female gametes meiotically; these unite to form progeny. Progeny will differ genetically from the parent to a small degree, depending on the level of heterozygosity in the parent. Heterozygosity will decline.
- *Sex with polyembryony* Females and males produce gametes that unite to form a single fertilized egg; this divides mitotically to produce offspring that are genetically identical to each other but which differ from both parents. (This process produces identical twins in mammals. More rarely, the zygote divides into four, producing identical quads, as in armadillos.)
- *Inbreeding sex* Females and males that are closely related to each other produce gametes that unite to form fertilized eggs. Progeny differ genetically from their parents, but less so than in outbreeding sex. Heterozygosity will decline over successive generations.
- *Outbreeding sex* Females and males that are not closely related to each other produce gametes that unite to form fertilized eggs. Progeny differ genetically from their parents to a high degree. Heterozygosity will be maintained over successive generations.

This list provides a very brief and simplistic overview of just a few of the many modes of reproduction that exist among animals and plants. The details of the mechanisms involved are beyond the scope of this Book. What is important is an appreciation that there is much diversity in reproductive patterns among organisms and that, as a result, different species can, during reproduction, produce progeny that differ genetically from themselves to a greater or lesser extent. A general hypothesis is that each mode of reproduction, and the level of genetic diversity that it generates, is an adaptation to a particular set of environmental conditions.

The fact that some species have inbreeding sex is worthy of mention, as it is common knowledge that, at least for humans and many domesticated animals, inbreeding can have genetically harmful effects, as discussed in Chapter 3, Section 3.6. We will return to this issue in Section 5.1.6.

5.1.3 Costs and benefits of sexual and asexual reproduction

In the previous Section, we saw that sexual reproduction can produce individuals with new combinations of favourable mutations and individuals that have lost unfavourable mutations, in a manner that asexual reproduction cannot achieve. Thus, sex can be regarded as a mechanism for the elimination of deleterious mutations and the production of genetic novelty whereas, in an asexual population, deleterious mutations will tend to accumulate and new, favourable gene combinations will tend to arise at a low rate. As a result, a sexual population will be able to evolve more rapidly than an asexual one. If a sexual and an asexual population coexist in an environment that changes from one generation to the next, the result may be the eventual extinction of the asexual population. Therefore, a major benefit of sexual reproduction is that it enables sexual species to compete successfully with asexual species, especially in fluctuating environments.

There are two problems with this argument as an explanation for the original evolution of sex. First, asexual reproduction does exist in a very large number of species, either as the sole means of reproduction or as a mode of reproduction that is used under certain conditions, as in aphids and many plants. This suggests that there are benefits to asexual reproduction, otherwise it would have been eliminated earlier in the evolution of life. Secondly, the argument made above is in terms of a benefit accruing to the species; it is a long-term explanation in which the evolutionary consequence is the survival or extinction of a population or an entire species. Such an explanation is framed in terms of what is known as *group selection* (see Chapter 6).

▶ Why might an argument that considers adaptation in terms of the species pose a problem for the theory of natural selection?

The theory of natural selection is expressed in terms not of species but of individuals; the struggle for survival is between individual organisms, not between populations or species (see Chapter 6). It is essential to consider the costs and benefits of alternative modes of reproduction from the individual's point of view.

Consider a female in a sexual species that produces several offspring. When she reproduces, she allocates 50% of her genes to each of her progeny, the other 50% being provided by one or more males. If she abandoned sex and produced young by parthenogenesis, each of them would contain 100% of her genes. Thus, an asexual mutant in a sexual population has a two-fold reproductive advantage over other females; she passes her genotype on to the next generation twice as effectively. This leads us to the conclusion that, at the individual level, asexual reproduction should be favoured by natural selection over sexual reproduction. This cost of sex is often referred to as the cost of producing sons.

Some species are able to vary the proportion of sons to daughters that they produce. For example, females of the water flea *Daphnia magna* reproduce parthenogenetically (producing only females) at some times, sexually at others. Experiments have shown that the environmental stimulus which triggers the production of males is water-borne; individuals exposed to water taken from a container in which fleas are at high density increase their production of sons even when they are living at low density. Over a season, females produce about seven broods of young, with an overall sex ratio among their progeny of one son to two daughters. The evolution of sex ratios is discussed further in Section 5.4.

A number of other costs of sex have been identified. One of these is the cost of recombination. As pointed out in the previous Section (Figure 5.2), an advantage of sexual reproduction is that it produces offspring with novel, and sometimes favourable, combinations of genes. This genetic variation results from recombination during meiosis and from syngamy. The very same process, however, will also break up favourable gene combinations. So, genotypes that have proved successful in one generation tend not to be conserved in the next generation. This could be a severe cost if the environmental conditions experienced by the next generation are very similar to those experienced by the parental generation. The recombination that is inherent in sexual reproduction appears to be a process that reduces the chance that adaptive genotypes will be passed on.

Another cost of sex is the cost of mating. To reproduce sexually, an individual must find a mate and then engage in a mating act, activities which, in animals, can be time-consuming and hazardous. Sexual reproduction has other attendant risks, such as failing to find a mate at all, and mating with a member of another species and so producing no offspring or hybrid offspring of low fitness.

An additional advantage of asexual over sexual reproduction is that it saves time. Mitosis takes less time than meiosis and, especially for unicellular organisms, asexual reproduction leads to a faster rate of reproduction under good conditions. In animals, asexual reproduction can reduce the time and energetic costs involved in finding and competing for a mate, as well as those involved in the mating act itself.

It is not yet clear what advantages might offset these various costs in sexually reproducing organisms. It is possible that, for many species, the net benefits to the individual of asexual reproduction are in fact greater than those for sexual reproduction. Why then does sexual reproduction exist at all? One possible evolutionary scenario is that asexual mutants arise within sexual species quite frequently. Because of their two-fold reproductive advantage, they become established and spread at the expense of their sexual competitors. An asexual population, or clone, will, however, lack the high degree of genetic diversity existing among the offspring of sexual individuals and so will never be able to displace the sexual population entirely. Furthermore, as the environment changes, it will be less able to adapt and will eventually become extinct in the face of the capacity of the sexual population to adapt more rapidly. Note that, in this argument, the cost of asexual reproduction is borne not by the individual organism but by the population; it persists for fewer generations than a sexual population.

Certain features of the distribution of asexual reproduction among organisms provide some support for this scenario. First, particular asexual species typically are closely related to particular sexual species, suggesting that they are of recent origin. Secondly, with the exception of the bdelloid rotifers (an invertebrate group in which males do not exist), no large taxonomic group (family, order or class) consists entirely of asexual species, suggesting that their capacity to diversify and persist for long periods of evolutionary time is indeed very limited.

So far, we have suggested that sex is disadvantageous at the individual level. An alternative approach to the problem is to consider whether sex may be adaptive for individuals. As G. C. Williams has pointed out, the existence of both sexual and asexual reproduction in the life cycle of a single species (as in aphids) suggests that there may be a balance of costs and benefits. At certain times of year the balance shifts in favour of sex, at others it shifts in favour of asexual reproduction. The aphid example suggested that asexual reproduction is favoured when the environment is favourable for rapid reproduction and population growth, but that sex becomes more adaptive when the dispersal phase of the life cycle is being produced.

▶ Why might there be a link between dispersal and sexual reproduction in aphids?

The dispersal phase of aphids has to survive the winter and found a new population under conditions that may be very different than those in a current season. Because dispersal involves progeny moving to new, less predictable environments, greater genetic diversity among progeny may give sexual reproducers an adaptive advantage over asexual reproduction in this context.

5.1.4 The sib-competition model

Williams (1975) proposed a model for the maintenance of sexual reproduction, in a population containing both sexual and asexual genotypes, based on the hypothesis that sexual individuals will have an advantage because they produce progeny with a variety of genotypes. An essential feature of this model is that each individual of this hypothetical organism produces a large number of offspring which must compete with one another for limited resources to reproduce; the model therefore later came to be called the sib-competition model.

Suppose that the hypothetical organism is an annual plant that sheds its seeds in large numbers into its immediate environment. The environment consists of a limited number of patches suitable for seed germination, none of which is large enough to support more than a limited number of mature plants; we will assume that this number is one. If several seeds land on a particular patch, there will be density-dependent mortality among the sibling seedlings that grow from them; i.e. the more dense they are, the higher the mortality rate. Thus, as only one seedling can survive, the proportion of siblings which die increases with the number of them that germinates. This is shown diagrammatically in Figure 5.3(a) (overleaf).

Let us now suppose that the patches are not uniform but differ in a variety of ways such that there is variation among them in the probability that a seedling with a particular genotype will survive to reproduce. The offspring of sexual parents will have varied genotypes and, as a result, progeny from a single parent will be able to survive successfully and reproduce in several kinds of patch (Figure 5.3(b)). Those of asexual parents, in contrast, having genotypes identical to their parents, will survive and reproduce only in patches that are identical to the parental patch (Figure 5.3(c)). So, although asexual plants may produce more seeds than sexual plants, they may leave fewer offspring that survive to reproduce. Whether or not sexual plants are more successful at reproducing in this hypothetical system depends critically on the heterogeneity of the habitat; the more variable are the patches, the more likely is it that sexual reproduction will be more successful than asexual reproduction. In essence, this model expresses the lottery effect in a formal way.

Suppose that each patch has a number and that seedlings also have corresponding numbers; the closer a seedling's number to the patch number, the more likely it is to survive in that patch. Consider what happens if three progeny (i.e. sibs) from each of two parents, one sexual, one asexual, land on a patch with the number 56. The numbers of the three sexual progeny are 23, 54 and 89, those of the three asexual progeny are 47, 47 and 47. In this case, one of the sexual progeny (54) will survive, but it would have been one of the asexual progeny if the patch number had been 43. In this situation, the sexual progeny have three times the chance of 'winning the lottery' than the asexual progeny because they have three times as many numbers among them. The British evolutionary biologist John Maynard Smith has modelled this kind of situation and has shown the conditions under which sex is and is not advantageous. Sex does not win if progeny from

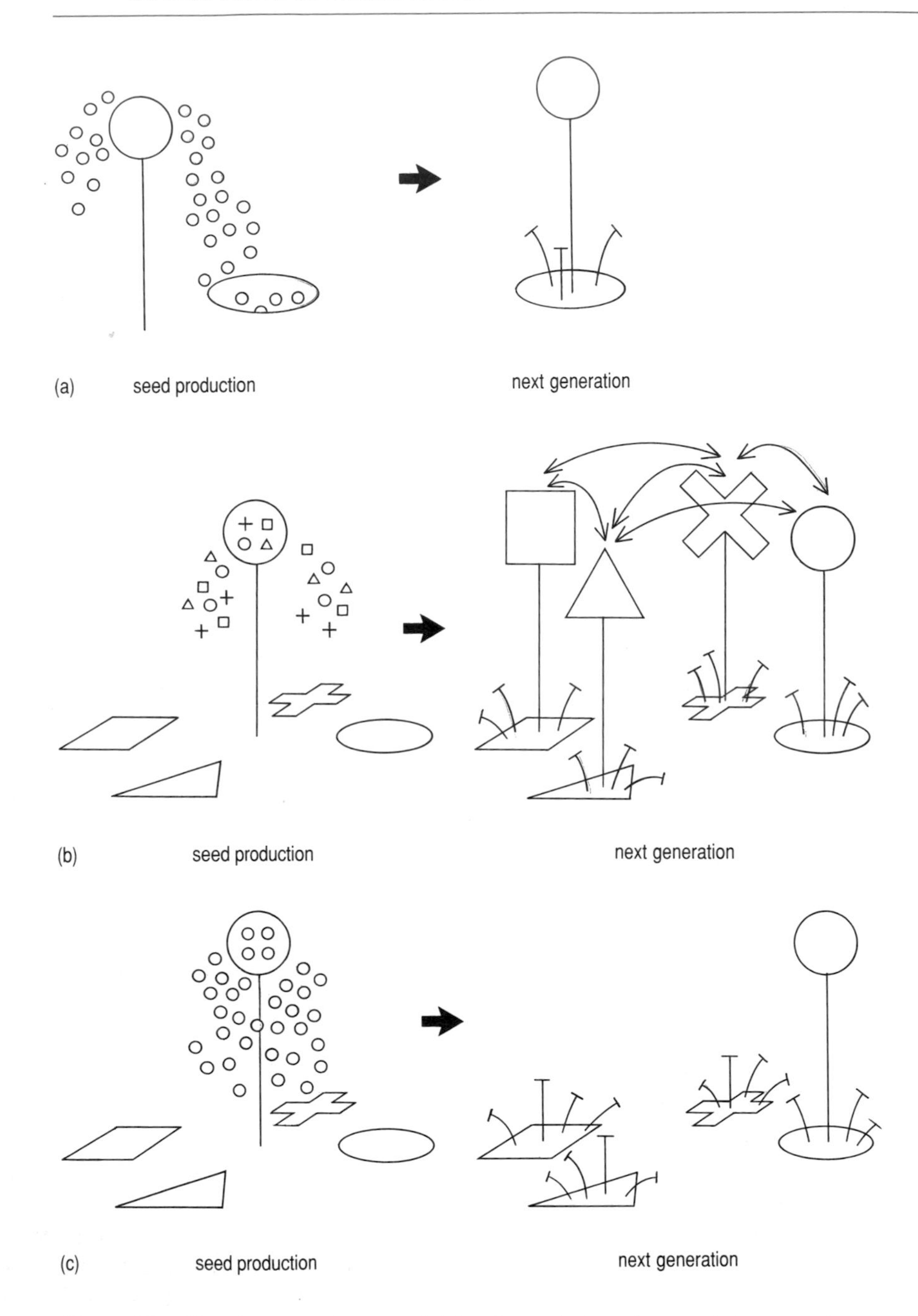

Figure 5.3 *The sib-competition model for the evolution of sex. (a) An annual plant sheds its seeds into a patch. However many seeds enter the patch, only one can develop into a mature plant. As a result, there is mortality among seedlings and this is density-dependent. (b) An annual plant reproduces sexually, producing genetically variable progeny that are shed into variable patches. Selection operates on these offspring, with the result that one of each type of seed survives and develops in the patch to which it is genetically suited. (c) An annual plant reproduces asexually, producing genetically identical progeny that are shed into variable patches. Only those seeds that are adapted to the same patch type as the parent can survive and develop.*

only one parent enter a patch, however many of them there are. More importantly, sex does not win if only one progeny per parent enters a patch; it is critical that there be competition between the progeny of a given parent, hence the name of the model. Indeed, the advantage of sex increases with the number of progeny supplied per parent. Extending the numerical example given here, if five siblings per parent enter a patch, the sexual progeny have a five-fold advantage over asexual progeny.

Computer simulation of the sib-competition model thus suggests that sexual genotypes can indeed have higher fitness in some circumstances. This begs the question, however, of whether the assumptions of the model are biologically realistic. The assumption that there is intense competition between siblings is realistic for many plants, the seeds of which, despite the existence of a variety of dispersal mechanisms, typically settle close to their parents. Seeds are, however, usually more mobile than the asexual progeny produced by plants, which are commonly attached to their parents, as in the stolons of strawberries. As we saw in Section 5.1.1, a similar relationship between mode of reproduction and mobility occurs in the aphid. Progeny produced parthenogenetically are wingless, whereas sexual females and males have wings that enable them to disperse from their summer food plant to their overwintering sites. Habitat patches that are remote from the parent are more likely to differ from those in which asexual progeny will grow, again suggesting a link between sexuality and varying or unpredictable environments.

This link can be tested, observationally and experimentally, by examining the kinds of organisms that inhabit frequently disturbed, and therefore unpredictable, habitats. Among plants, for example, weed species colonize recently cleared patches of soil.

▶ Would you predict that weed species should reproduce sexually or asexually?

The prediction, on the basis of the sib-competition model, is that weeds should reproduce sexually, since they will typically reproduce by invading new patches that are likely to differ in a number of ways from places in which the parent plants grow. In fact, weeds are rarely fully sexual, contrary to this prediction; they are more commonly parthenogenetic or self-fertilizing. It is now thought that what may be more important for weed species than the physical nature of their environment is their biotic environment, specifically the presence or absence of competitors. Weeds thrive in cleared patches of ground where there are few or no competitors. In such a context, selection apparently favours the rapid production of as many progeny as possible rather than of offspring that are genetically diverse.

In plants, sexual reproduction is more characteristic of species living in saturated habitats, i.e. those in which there are many other plant species competing for the same resources, such as water, nutrients and light, as well as established populations of animal herbivores and a variety of parasites and pathogens (disease-causing organisms such as viruses and bacteria). Sex is advantageous in saturated environments because it generates genetic diversity, enabling plants continually to evolve effective defences against competitors and other enemies that are themselves constantly evolving more effective means of attacking them.

This concept that different species of organisms are engaged in a constant evolutionary struggle with one another is central to the 'Red Queen's hypothesis' (see Chapter 4, Section 4.5.2) which was put forward primarily to account for the observation that, over evolutionary time, species are continually becoming extinct and being replaced by other species. No matter how long a species has survived and how well it is adapted to its environment at any particular time, its environment can be said to be constantly deteriorating as far as its adaptations are concerned. At all times, its fitness is determined by a variety of biological and physical pressures, such as predation, competition for resources, parasites, etc., but the relative importance of these factors will vary in time. If a species acquires a character that is adaptive in response to one kind of pressure, this may be deleterious for its adaptations to other pressures which, at that time, are relatively weak. As a result, the individuals will suffer a decline in fitness when the balance of pressures changes. Consider, for example, a plant growing in your garden. One year, a major selective pressure may be drought favouring the ability to retain water; the next year, water may be abundant but there could be a plague of aphids. The balance of selective pressures to which a species must respond is constantly changing, not least because of the continuous adaptive responses that different species living in the same habitat make to one another. Thus, any temporary adaptive advantage gained by individuals of one species over others represents a worsening of the environment experienced by the others. As a result, the adaptiveness of a species relative to others will fluctuate with time.

Metaphorically, each species within an ecological community is engaged in a permanent race with the other species; hence the reference to the Red Queen. Sexual reproduction, because of its capacity to produce genetically varied offspring, may have the effect of enabling species to keep up with their competitors in this race. Insect pests provide good examples of the evolutionary advantages of sex. The mite *Tetranychus mcdanieli*, which infests apple orchards, evolved resistance to four different pesticides, one after the other, in under 20 years so that finally no effective spray was available for its control. Such rapid evolution would be unlikely in an asexual population of these animals.

One possible way to test the relative advantages and disadvantages of sexual and asexual reproduction is to find very similar, closely related species that differ in their mode of reproduction. Such an approach has been used by T. J. Case (1990) on two lizards with overlapping ranges in south-western USA and Mexico, the sexual *Cnemidophorus tigris* and the asexual *C. sonorae*. Case could find no evidence for the asexual species having an advantage over the sexual species in terms of numbers or the extent of its range. He did, however, find evidence that the sexual species is more diverse in terms of its exploitation of food. There was greater variation, between individuals, in terms of the size and diversity (number of different species) of their prey, in the sexual than in the asexual species. Suggestive as these results are in supporting the hypothesis that sexual reproduction leads to greater within-species diversity, Case cautions that his study could not entirely control for the fact that lizards living in different sites might have had different prey species available to them.

5.1.5 Sex, parasites and disease

Another argument for the adaptive value of sexual reproduction again relates to its capacity to produce genetically varied progeny. Most organisms are subject to infestation by pathogens and parasites. These may have a profound effect on individual fitness by causing the death of, or a reduction in the reproductive success of, their hosts. Hosts have evolved a variety of defences against pathogens and parasites, notably a complex immune system that recognizes and attacks alien organisms. On the other hand, pathogens and parasites evolve mechanisms that circumvent or counteract their hosts' defence mechanisms. There is thus coevolution between an organism and its enemies (Chapter 2, Section 2.3.2).

In this coevolutionary process, pathogens and parasites have a distinct advantage. They typically have a very short generation time and/or a very high reproductive rate. A bacterium that infects a human, for example, may generate several dozen generations within a single day. (The generation time of the bacterium *E. coli*, for example, is 20 minutes under optimal conditions.) As a result, pathogens and parasites can evolve, thus becoming better adapted, within the lifetime of their host. If the host reproduces asexually, it produces progeny that are genetically identical to itself and for which there are, therefore, populations of pathogens and parasites that are already adapted to infect it. This leads to the hypothesis that sexual reproduction, because it produces genetically variable progeny, is an adaptation against pathogens and parasites. It ensures that at least some progeny will be less susceptible to attack by those organisms that have evolved during the lifetime of their parents.

Evidence that parasites may play a role in determining the mode of reproduction of their hosts comes from studies of a freshwater snail *Potamopyrgus antipodarum* in lakes in New Zealand (Lively, 1987). Individuals in this species are either male or female, unlike many snails that are hermaphrodites, but females are capable of parthenogenesis. Some populations consist entirely of females and so must reproduce parthenogenetically; others contain as many as 40% male individuals, indicating that they have the potential for sexual reproduction. In a comparison of nearly 50 sites, Lively found a strong tendency for males to be more frequent in localities where snails were heavily infested by various trematode parasites (flatworms). This finding supports the hypothesis that sex is maintained in populations where parasites impose a selection pressure favouring the production of genetically diverse progeny.

It is common in both the scientific and the popular literature for an analogy to be drawn between the coevolution of pathogens and their hosts and the human **arms race**. The two processes have in common the fact that each innovation by one side leads to some kind of response from the other side. The value of this analogy is, however, strictly limited. In the human arms race, there has been a tendency towards the development of greater destructive power; weapons become more powerful and more numerous. In the analogous evolutionary process, this kind of escalation need not occur and, indeed, it is clear that in many instances it does not. Consider a pathogenic organism. If it evolves to become more virulent, it may so reduce the numbers of its host that it threatens its own continued existence; escalation could lead to the extinction of both sides. If it becomes less virulent, however, its host will survive in greater numbers and so the pathogen is assured of a population of hosts that is adequate for its support. There is good evidence from long-term studies of diseases that the direction of evolution is often towards reduced, not increased, virulence in the pathogen. It appears, for example, that the disease myxomatosis became less virulent over a number of years after its first introduction into rabbit populations in both the UK and Australia, resulting in a steady increase in rabbit populations over several years following an initial sudden decrease (see Chapters 8 and 17). At the same time, resistance to the disease evolved in the rabbits.

5.1.6 Inbreeding and outbreeding in sexual species

As described in Section 5.1.2, there is a great diversity of modes of reproduction, such that the category 'sexual reproduction' embraces a variety of breeding systems. An important variable to be considered here is the degree of inbreeding that occurs in a given system. It is common knowledge that inbreeding sometimes has genetically harmful effects, especially when continued over several generations (Chapter 3, Section 3.6); we all know what happened to various royal families that practised incest (mating between siblings) or frequent matings among cousins. Repeated inbreeding in humans leads to increased frequency of cleft palate, mental retardation, albinism and other physical abnormalities in the offspring. It is commonly assumed that organisms should avoid mating with close relatives and animal breeders generally ensure that such matings are avoided, but rather little is known about their frequency in nature. Models in population genetics, such as the Hardy–Weinberg model (Chapter 3, Section 3.4), commonly assume that the population is panmictic; that is, mating is random and any individual is equally likely to mate with any other member of the population. In nature, it is often very unlikely that this will be true. Plants are sedentary and the probability that one particular plant will fertilize another must be higher if they are close together than if they are far apart (Figure 5.4). Many animals show some degree of **philopatry**, a tendency by an individual to return to breed in the locality where it was born. Thus, for both plants and animals there is good reason to suppose that, in natural populations, mating will not be random and that inbreeding may occur to some extent.

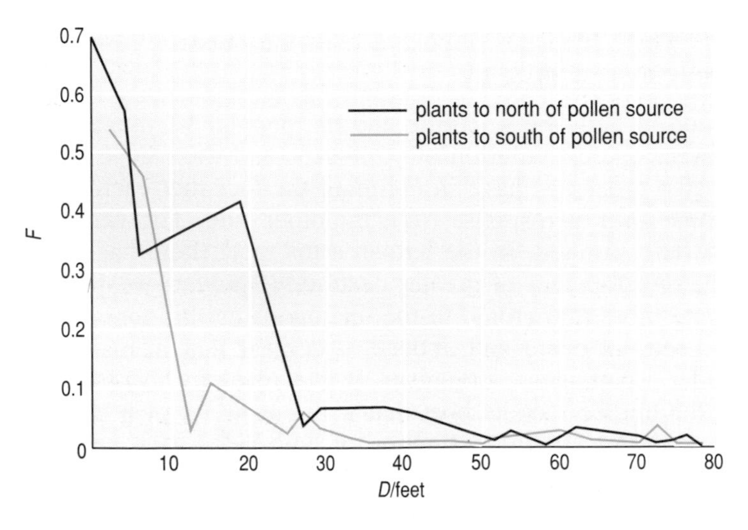

Figure 5.4 *Effect of distance on the probability that one plant will fertilize another.* D = *distance between corn plants of one genetic strain and plants of a different strain;* F = *proportion of seeds fertilized by a plant at a particular point. Here, the prevailing wind was from the south. A particular plant is more likely to fertilize plants to the north than to the south of it, but fertilization rates of plants more than 10 m (~30 feet) away in any direction is very low.*

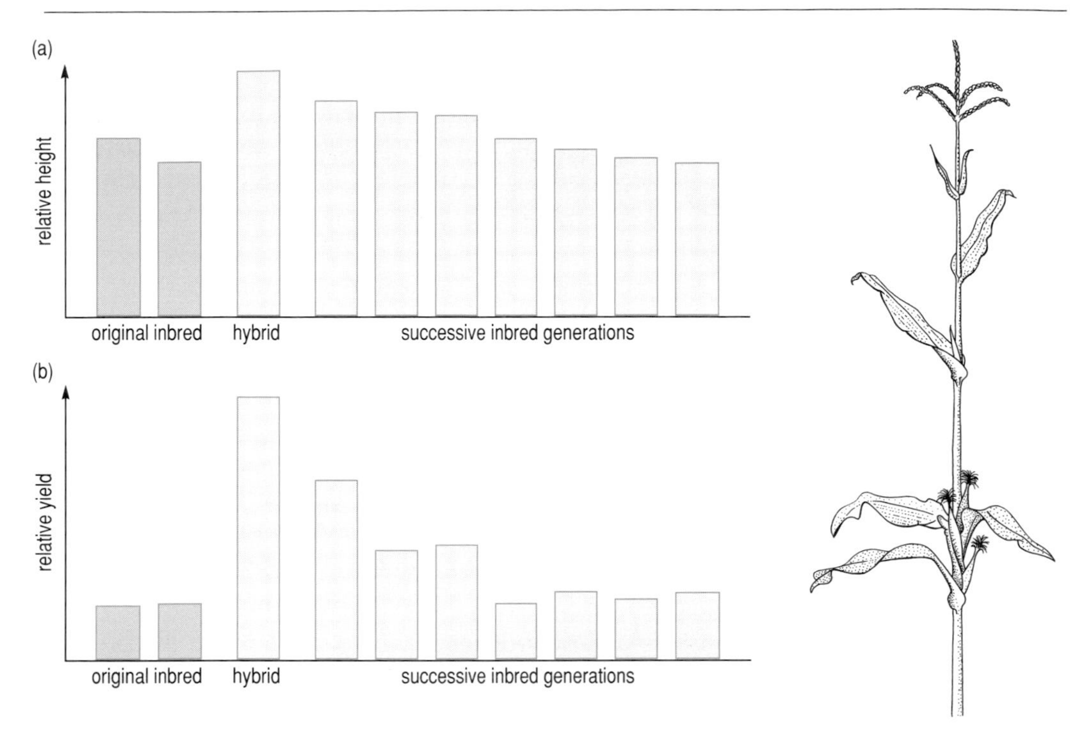

Figure 5.5 *Histograms showing the phenotypic effect of inbreeding and outbreeding in corn plants: (a) height of plant; (b) yield of grain. In each case, the first two columns on the left are from two original inbred strains. To their right are the hybrids (outbred) between them, followed by successive generations of self-fertilized (inbred) progeny of the hybrid.*

The best documented consequence of inbreeding is lowered offspring fitness, or inbreeding depression (Chapter 3, Section 3.6). This is a well-known phenomenon in laboratory and domestic animals (see Chapter 3, Table 3.10) and in crop plants (Figure 5.5), but there is somewhat contradictory evidence for it in natural populations. For example, a field study carried out in Britain found that nestling mortality among great tits *Parus major* is up to 70% higher among the offspring of related than of unrelated birds. A comparable study of great tits in the Netherlands did not, however, find such an effect. The explanation for inbreeding depression, in genetic terms, lies in the presence of deleterious recessive mutations in populations. In humans, for example, it has been estimated that each of us, on average, carries the equivalent of three to five lethal recessive alleles. These are not normally expressed because they are in the heterozygous condition.

▶ What are the likely genetic and phenotypic consequences for the offspring resulting from a mating between two close relatives?

Close relatives are more likely than non-relatives to carry the same deleterious or lethal alleles, because of their common ancestry, with the result that inbreeding leads to an increased occurrence of such alleles in the homozygous condition, and therefore their phenotypic expression in the offspring. The offspring will thus tend to express recessive alleles that are not expressed in their parents.

If inbreeding decreases fitness, we would expect animals and plants to have evolved mechanisms that reduce its occurrence. A common mechanism in animals is differential dispersal by the two sexes. In mammals, males generally disperse further than females from their place of birth before breeding; in birds, it is more commonly females that disperse further. For example, the mean dispersal distance from natal site to breeding site in Belding's ground squirrel *Spermophilus beldingi* is 450 m for males, 47 m for females. The equivalent figures for great tits are males 558 m, females 879 m.

▶ How does differential dispersal reduce inbreeding?

If brothers and sisters breed in different localities, it is very unlikely that they will mate with one another. There is also accumulating evidence that animals can recognize and avoid mating with their close kin. Often this is achieved simply by their not mating with those individuals with which they were reared, animals preferring to mate with unfamiliar rather than familiar partners. There is evidence, however, for a variety of groups, including mammals, amphibians and arthropods, that kin recognition can occur without individuals having prior experience of one another.

Plants possess a variety of mechanisms that reduce the probability of inbreeding. The most extreme form of inbreeding is self-fertilization, the fusion of gametes from the same individual. A feature of many plants that prevents or reduces the likelihood of self-fertilization is that they possess separate flowers producing male and female gametes. A plant with both male and female flowers on the same individual is called **monoecious** (from the Greek, meaning 'one house', e.g hazel *Corylus avellana*); those in which individuals exclusively bear either male or female flowers are **dioecious** ('two houses', e.g holly *Ilex europaeus*). Monoecious plants, and those in which male and female functions occur in the same flower (plants with 'perfect' flowers), are both forms of hermaphrodite and have evolved a variety of adaptations that prevent self-fertilization. In many monoecious plants (e.g lords and ladies *Arum maculatum*) male and female flowers mature at different times; in other hermaphrodites with perfect flowers, the male and female parts of a flower are physically separated in such a way that pollen cannot, or is unlikely to, reach the stigma of the flower in which it is produced (Figure 5.6). A widespread mechanism in flowering plants is genetic self-incompatibility: the pollen from a given individual fails to grow down the stigma of that individual or, less commonly, the embryos resulting from self-fertilization fail to develop.

Although the available evidence suggests, first, that inbreeding can cause a reduction in fitness and, secondly, that animals and plants may possess adaptations that reduce its occurrence, we cannot assume that outbreeding is necessarily always a more adaptive mating pattern.

▶ From what you read earlier about the costs of sex, what can you suggest might be a deleterious effect of outbreeding?

Mating with unrelated individuals will tend to break up adaptive combinations of genes, whereas mating with relatives will tend to conserve such combinations in the progeny. This consideration has lead to the idea that there may be an optimum level of inbreeding. Very close relatives and totally unrelated individuals should be

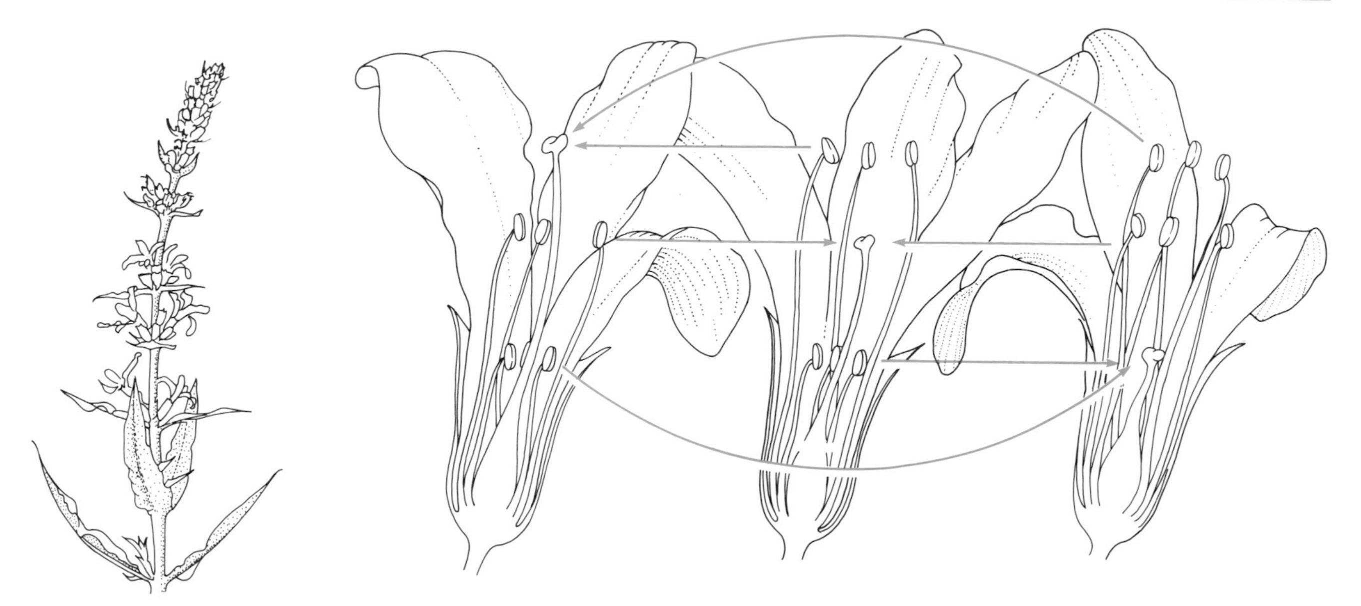

Figure 5.6 *Prevention of self-fertilization by means of heterostyly in purple loosestrife* Lythrum salicaria. *Each flower possesses 12 stamens (male parts), arranged in 2 groups of 6, that come in 3 sizes—short, medium and long. Each flower contains stamens of 2 lengths only. The single pistil (female part) can also be long, medium or short and is always of a different length to the stamens. Each plant possesses flowers with only one of the 3 possible combinations of stamen and pistil length. Visiting insects will pick up pollen from 2 positions in the flower that differ from the position of the pistil, making self-fertilization unlikely, but increasing the probability that fertilization of a flower with a different arrangement will occur.*

avoided as mates, but medium to close relatives should be preferred. Reduction of fitness due to mating with unrelated individuals is called **outbreeding depression**. Evidence that both inbreeding and outbreeding lead to reduced fitness is provided by a study of an insect-pollinated plant, the delphinium *Delphinium nelsoni*, by Price and Waser (1979). They identified a number of 'donor' plants and, taking their pollen, hand-pollinated many other plants at varying distances from the donors. The donor plants were also self-pollinated. This procedure was based on the assumption that plants close to a donor would be more closely related to it than those growing further away. They then scored the fitness of all the hand-pollinated plants by counting the number of seeds produced per plant. The results are shown in Figure 5.7.

▶ Do these data support the hypothesis that there is an optimum level of in/outbreeding?

Yes they do. Fertilization by donors of plants about 10 m distant from them generally yields higher numbers of seeds than self-fertilization, fertilization with near-neighbours or fertilization with plants further away than 10 m. (There is, however, a lot of variation within and between samples and some samples do not fit the general pattern, suggesting that other factors influence the number of seeds per flower.) This species therefore shows evidence of both inbreeding and outbreeding depression. Interestingly, the same study also found that the average distance between donors and recipients of pollen carried naturally by insects is only about 1 m, considerably less than what appears to be the optimum distance for this species.

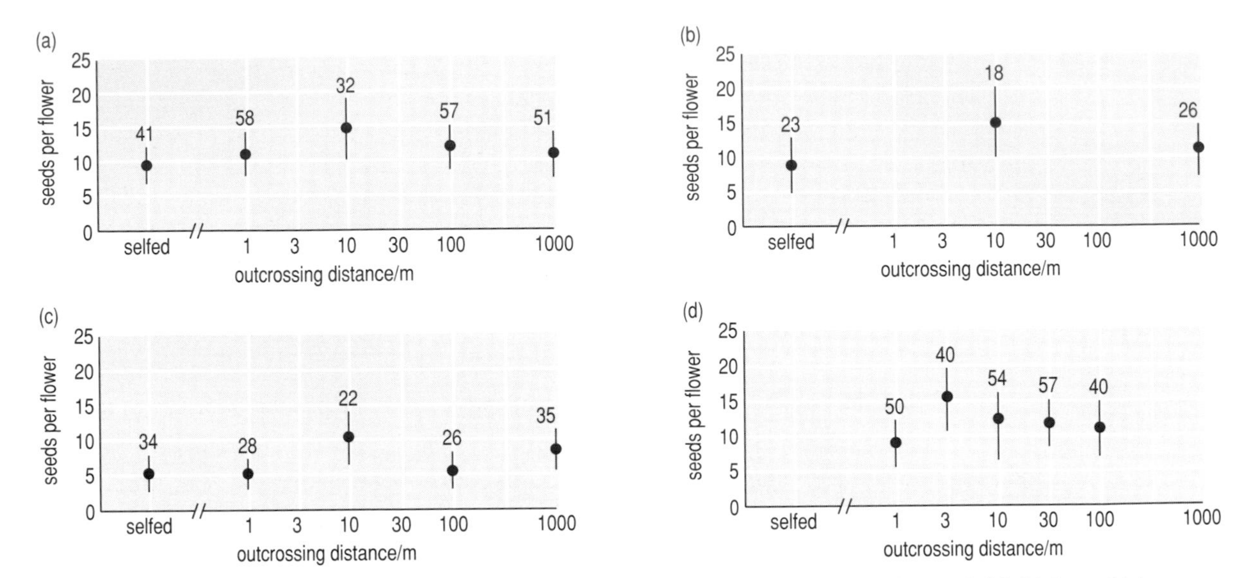

Figure 5.7 *Effects of outcrossing distance on number of seeds produced by hand-pollinated delphiniums. Values are means with 95% confidence limits, with sample sizes shown above. (a), (b) and (d) are different samples, taken in different years at the same site, (c) is a sample taken in a single year at another site.*

Waser and Price (1989) have conducted a more recent study on a long-lived plant, *Ipomopsis aggregata*, that is pollinated by humming-birds. Using the mean number of seeds produced per flower as a measure of fitness, they found that individuals crossed with plants 1–10 m away had higher fitness than self-pollinated plants or those crossed with plants 100 m away. Furthermore, the offspring of 10 m crosses had higher survival, were more likely to flower, and flowered earlier than those of 1 m or 100 m crosses. The relative lifetime fitnesses of the offspring of 1 m, 10 m and 100 m crosses were 0.47, 1.00 and 0.68 respectively. In the same study, offspring were planted at varying distances, over a range of 1–30 m, from their parents and a decline in offspring fitness was detected at greater distances.

▶ What do you conclude from the latter result?

It suggests that the cause of outbreeding depression observed in 100 m crosses is that the progeny are not as well adapted to the local habitat far from their parents as they are to that in which their parents grew.

Among animals, indirect evidence for this idea comes from laboratory studies of Japanese quail *Coturnix coturnix japonica*, in which individuals prefer to mate with first cousins rather than with their siblings on the one hand, or with unrelated birds on the other (Bateson, 1980). A field study of great tits has shown that females tend to be paired with males whose songs differ slightly, but not very much, from the songs of the females' own fathers. Great tit males learn their songs from their fathers and so sons tend to sing much more like their fathers than like unrelated males.

A particular situation in which a degree of inbreeding will be advantageous is where local populations of organisms are specifically adapted to local ecological conditions. Such locally adapted populations are called **ecotypes** and are discussed further in Chapter 9. Mating with individuals belonging to the same ecotype will tend to ensure that the offspring inherit the same adaptations to local conditions. The Canadian three-spined stickleback *Gasterosteus aculeatus* exists in two forms—an exclusively freshwater ecotype and an anadromous ecotype—which breeds in freshwater but otherwise lives in the sea. (The term *anadromous* means ascending rivers and streams from the sea in order to breed.) Hay and McPhail (1975) conducted mate choice tests and found that individual sticklebacks, of both sexes, preferred a partner of the same ecotype. This tendency for mating to occur non-randomly, with individuals pairing with partners of a similar type, is an example of positive assortative mating (Chapter 3, Section 3.4).

▶ What, in this example, could be the adaptive value of positive assortative mating?

It could ensure that progeny will be adapted to one habitat or the other. Progeny resulting from mixed pairings would probably be less well-adapted to either the freshwater or the anadromous life history than pure-bred progeny.

The related phenomena of philopatry and assortative mating among ecotypes have important implications in the formation of new species, and are discussed further in Chapter 9.

Studies of the degree to which organisms are inbred or outbred, and of the fitness consequences of different degrees of inbreeding, are few at present, but this is an area of research that is developing and expanding rapidly. A major development in this area is the technique of DNA fingerprinting (Chapter 3, Section 3.5.2) which will eventually make it possible to establish, in a natural population, precisely which individuals mated successfully with one another and which descendants they produced. Such studies will make it possible to determine, first, how closely related mating partners are in nature and, secondly, what are the fitness consequences of inbred and outbred pairings. Only then will it be possible to test the hypothesis that, for a given species, there is indeed an optimal level of inbreeding.

5.1.7 CHANGING FROM ONE MODE OF REPRODUCTION TO ANOTHER

As indicated in Section 5.1.3, asexual reproduction appears to have evolved many times during the course of evolution, with the result that, at the present time, asexual species are known within a diverse array of plant and animal groups. If this is so, why has it not evolved more often, given that it appears to confer an immediate selective advantage in certain circumstances? One possible answer is that such circumstances do not arise very commonly. Another is that many asexual species have evolved but most have become extinct. It may also be the case, however, that the number of asexual species is limited by the possibility that it may be very costly or, indeed, not possible to change from one mode of reproduction to another. Evidence for this comes from asexual species that retain certain aspects of sexual reproduction.

A small fish, the Amazon molly *Poecilia formosa*, is a parthenogenetic species consisting entirely of females. However, before an egg can develop it must be penetrated by a sperm, which triggers cell division but does not fuse with the cell nucleus. To obtain sperm, female mollies must elicit matings from males of a related, sexual species. A similar situation exists among some European frogs (genus *Rana*). In many lizards, females require stimulation from males, in the form of elaborate courtship behaviour, if their eggs are to develop fully. Some species of lizard, such as the whiptail lizard *Cnemidophorus inornatus*, have become parthenogenetic but, though they are emancipated from the need to obtain sperm, they still require 'male' stimulation. This is provided by each lizard in a population spending about half the time that it devotes to mating behaving like the males of an ancestral, sexual species.

What these examples suggest is that, once a species has evolved a particular mode of reproduction, it acquires a range of adaptations—morphological, physiological and behavioural—that enhance the efficiency of that mode of reproduction. Although it might be a relatively simple step to switch from producing eggs meiotically to producing them mitotically, or to producing eggs that do not need to be fertilized, effective asexual reproduction may be prevented by a legacy of adaptations for sexuality. This line of argument raises an interesting possibility. It may be that, for a number of existing sexual species, it might actually be more adaptive, at least in the short term, if females could switch to asexual reproduction, but this option may be closed to them because it is not possible to switch from one mode to another without going through a series of steps that would be less adaptive than sexual reproduction. In terms of the adaptive landscape model presented in Chapter 4, Section 4.5, a species on a sexual 'peak' might be unable to move to a higher asexual 'peak' because of the reduced fitness engendered by crossing the intervening 'valley'. In other words, there may be severe constraints on the extent to which a species can evolve from one state to another.

Some organisms, like the lettuce-leaf aphid, employ both sexual and asexual reproduction, switching between them at different phases in the life cycle (Section 5.1.1). A few can switch between reproductive modes facultatively, such as *Hydra*, a small freshwater cnidarian (Figure 5.8). When prevailing conditions are good, *Hydra* reproduces asexually, by growing buds that detach as small replicas of their parents. However, when its habitat begins to deteriorate as it dries up in the

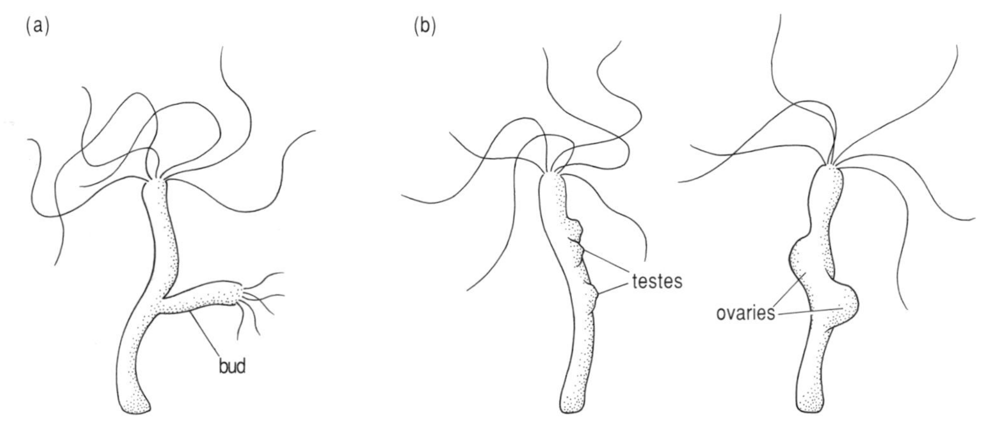

Figure 5.8 *Asexual and sexual reproduction in* Hydra. *(a) An individual reproducing asexually by growing a bud that will eventually become detached. (b) Individuals bearing testes (left) and ovaries (right).*

summer, *Hydra* develops testes and ovaries producing male and female gametes that fuse to form zygotes which can survive for a long period until conditions become favourable again.

Organisms like aphids and *Hydra* appear to enjoy the best of both worlds, being able to switch between sexual and asexual modes of reproduction as environmental conditions change. Why it should be that there are so many other species that do not enjoy this facility is one of the many mysteries about the evolution of reproductive patterns that have yet to be explained.

5.1.8 Empirical tests of aspects of reproduction using plants

The discussion so far has taken an essentially theoretical approach to the relative costs and benefits of sexual and asexual reproduction. However, a number of studies have been carried out to try to test some of the assumptions and predictions of various aspects of theory relating to the evolution of different modes of reproduction. The great majority of these studies have involved plants, because they have a number of advantages over animals in this context. First, and most important, many plants reproduce sexually and asexually at the same time, making it possible to carry out direct comparisons of the two modes of reproduction. Secondly, plants do not move about, making it much easier to monitor the survival and reproductive success of individuals and their progeny. Thirdly, plants can often be pollinated by hand, making it possible to control exactly which individual(s) a particular plant 'mates with'. Finally, plant progeny can be planted out in different localities and at different densities, so that their environment can be very precisely controlled.

One assumption of the theory for the evolution of reproductive modes is that producing male progeny represents a cost of sex. This hypothesis has been examined by McKone (1987) in brome grasses (genus *Bromus*). He studied five species, of which one, *B. inermis*, is a fully outcrossing, sexual species; another, *B. tectorum*, is entirely self-fertilizing. The remaining three species are partially outcrossed, partially self-fertilized. For each species, McKone measured the expenditure of energy and five nutrients (nitrogen, phosphorus, potassium, magnesium and calcium) that went into different components of reproduction. In the fully outcrossing species, 50% of the energy and nutrients allocated to reproduction went into producing pollen, whereas in the fully self-fertilizing species the comparable figure was only 2%. The three intermediate species allocated between 5% and 11% of energy and nutrients to pollen production. These results support the argument that outcrossing sexual reproduction incurs a high cost in terms of the allocation of resources to male, as opposed to female, progeny.

Another important assumption is that sexual reproduction confers higher fitness by producing genetically more variable progeny than asexual reproduction. Bierzychudek (1989) compared progeny of the plant *Antennaria parvifolia* that were produced by sexual reproduction with those produced by apomixis. Both kinds of progeny were grown in a series of six growth chambers that provided a wide diversity of conditions in terms of both temperature and moisture. Contrary to expectation, apomictic progeny exceeded sexual progeny in terms of mean survival, mean number of flower heads produced, and in mean biomass in all six chambers, prompting the question: why does sexual reproduction persist in this species? One limitation of this kind of experiment is that it may not have replicated exactly the environmental conditions that progeny are exposed to in nature. Greenhouse and field experiments have yielded conflicting results in a series of studies of another plant, *Impatiens capensis*. In a greenhouse study, Schmitt and Ehrhardt (1987) found no differences between outcrossed and self-pollinated seedlings. McCall, Mitchell-Olds and Waller (1989) obtained similar results with this species in greenhouse-reared progeny, but among young plants planted out in the field, they found that, whereas self-fertilized and outcrossed progeny grew equally well when planted close to related progeny, outcrossed progeny grew larger, and self-fertilized progeny were smaller, when they were each planted out with non-relatives. These varying results suggest that the exact environment faced by progeny, including their genetic relatedness to their neighbours, may determine whether sexual reproduction is or is not more adaptive than asexual reproduction.

The sib-competition model for the maintenance of sexual reproduction has been tested in a number of botanical studies. A key feature of this model is that the genetic diversity among progeny that arises from sex means that sexual progeny will be more likely to survive in competitive growing conditions. This part of the model can be tested by comparing the growth of progeny from a plant that has been pollinated by a single father with that of progeny from a plant pollinated by several fathers. This has been done for the wild radish *Raphanus sativus* by Karron and Marshall (1990). In the wild, individual flowers of this species receive pollen from six to eight flowers, and the seeds from a single flower are fathered by between one and four fathers. The seeds fall directly to the ground so that, in this species, they are particularly likely to be in competition with full-sibs and half-sibs (progeny with the same mother but different fathers). Karron and Marshall set up pots containing four full-sibs, four half-sibs, four unrelated individuals or one individual seedling. After 14 weeks, all the plants were dried and weighed. Plants grown individually had grown to three times the size of those grown in groups of four, indicating that, in the four-plant pots, there was intense competition. In terms of mean weight, there were no significant differences between plants grown with full-sibs, half-sibs and unrelated individuals. However, there was greater variation in size between individual plants in groups of four non-relatives than in groups of half-sibs, and likewise in groups of half-sibs than in groups of full-sibs. This result supports the crucial assumption of the sib-competition model, that genetic variability among progeny is adaptive in competitive conditions.

Finally, a number of studies have been carried out on the grass *Anthoxanthum odoratum*. Kelley (1989a, b) found that sexually produced progeny had higher reproductive rates than asexually produced progeny when both were planted out in a natural grassland habitat, but not when they were planted in a greenhouse. Schmitt and Antonovics (1986) planted out young plants, some with four unrelated neighbours, some with four sibs, and some without neighbours. In this study, some 45% of all the plants became infested with aphids. Among aphid-infested plants, those planted with sibs survived less well than those planted with non-relatives, but this effect was not apparent among non-infested plants. This result supports the hypothesis that a major advantage of sex is that it can produce new genotypes that are resistant to parasites and diseases.

SUMMARY OF SECTION 5.1

In species that reproduce both sexually and asexually within their life cycle, sex is typically associated with a deteriorating environment and dispersal of propagules to new environments, asexual reproduction with a favourable environment. This appears to be related to the genetic consequences of the two modes of reproduction. Sexual reproduction generates genetically diverse progeny; it can eliminate harmful mutations; it increases the rate at which new, favourable genotypes are generated, but it also breaks up favourable genotypes. Asexual reproduction generates genetically homogeneous progeny; harmful mutations tend to be conserved and to accumulate; new genotypes are not generated except by mutation, but favourable genotypes are conserved.

These genetic consequences represent costs and benefits that accrue at the individual and the species level. At the individual level, sexual reproduction appears to be adaptive in saturated environments in which individuals face intense competition and are in a coevolutionary relationship with predators, parasites and pathogens. Asexual reproduction appears to confer greater fitness in species that colonize new, unsaturated environments, such as those colonized by weeds. At the species level, the genetic variation generated by sex confers a long-term advantage, enabling sexual species to adapt to changing environmental conditions; asexual species appear to become extinct more rapidly than sexual species. Because asexual reproduction can be more adaptive than sex at the individual level, it has arisen many times during evolution, but, because sexual reproduction confers a long-term advantage on the species, sexual reproduction is more widespread, particularly among animals. Asexual reproduction may not have evolved in many species because of an evolutionary constraint on switching from one reproductive mode to another, resulting from the evolution of a variety of adaptations associated with sex.

Sexual species have evolved a number of adaptations that reduce the costs of sex. These include optimal in/outbreeding, which reduces the cost of recombination; biased sex ratios, which reduces the cost of producing sons; and hermaphroditism, which reduces the cost of producing sons and the cost of mating.

5.2 DIVERSITY IN LIFE CYCLES AMONG PLANTS

Most algae, ferns and mosses have two separate phases in their life cycles: one diploid and one haploid. One generation consists of diploid plants called **sporophytes**. Some of their cells undergo meiosis to produce haploid spores that develop directly into haploid plants called **gametophytes**. These produce gametes, by mitosis, that may or may not differ in appearance and be differentiated as ‘male’ or ‘female’. Gametes fuse to form a diploid zygote that develops into a diploid sporophyte plant. This pattern, called the **alternation of generations**, occurs not only in algae but also in the more recently evolved vascular plants.

In the sea lettuce *Ulva lactuca*, sporophyte and gametophyte plants are virtually indistinguishable (Figure 5.9). In this species the gametes produced by the gametophyte are all the same in appearance, and cannot be differentiated into male

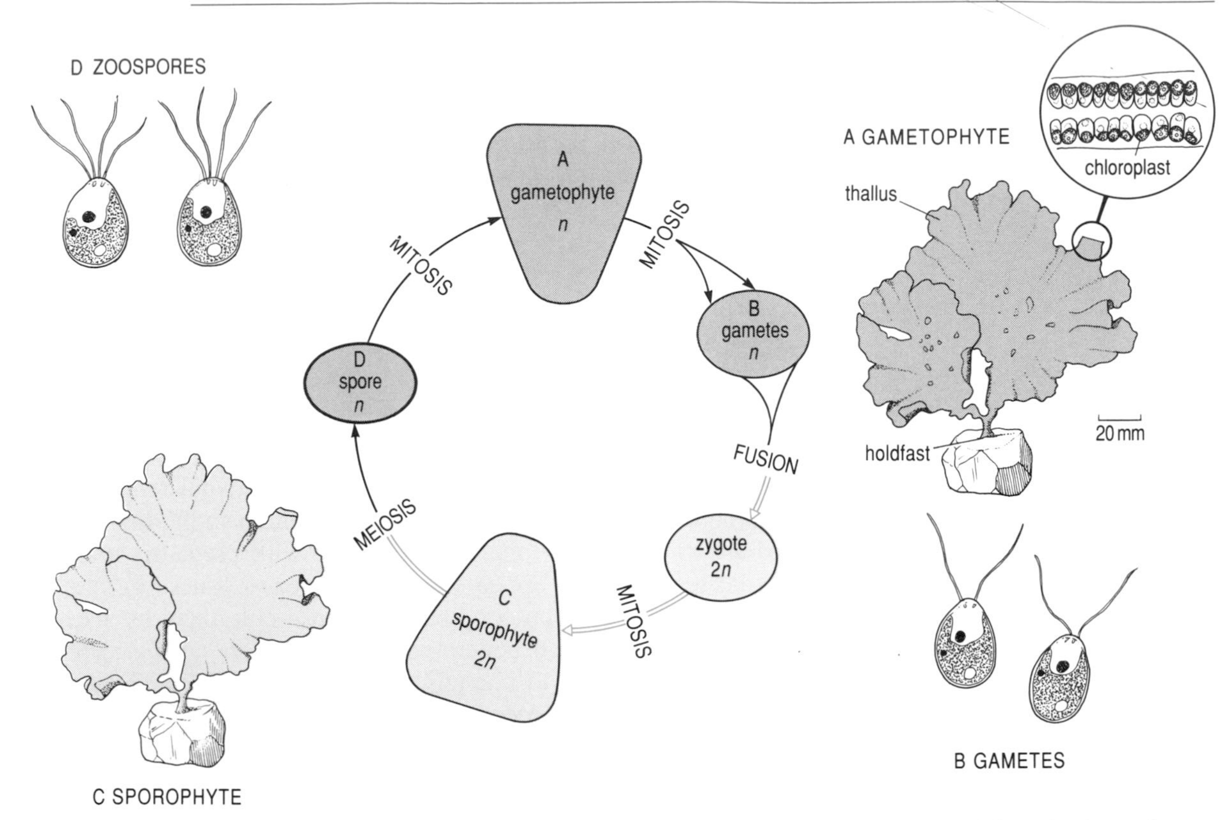

Figure 5.9 *The life cycle of the sea lettuce* Ulva lactuca. *Diploid stages are shown in green. There is alternation between haploid* (n) *and diploid* (2n) *phases.*

and female — a condition called **isogamy** ('iso' meaning 'same'). They fuse with gametes from other plants, but not with gametes from the same plant, to form a new sporophyte. In this life cycle, two essential features of sexual reproduction occur at separate phases. Meiosis occurs when the diploid sporophyte produces haploid spores; syngamy occurs when gametes from two gametophyte plants fuse to form a new sporophyte.

In the oarweed *Laminaria*, a similar cycle exists but with two important differences (Figure 5.10). First, sporophyte and gametophyte are very different in form. The sporophyte is the well-known large brown seaweed (kelp), whereas the gametophyte consists of small, branched filaments, only a few cells in length, that form a felt-like covering on rock surfaces. Secondly, gametophyte plants exist as two sexes, males producing mobile, biflagellate spermatozooids, females producing oogonia, each containing a larger, immobile egg cell. This condition is called **anisogamy**.

In the larger seaweeds like *Laminaria*, the sporophyte forms the larger, more well-developed and complex plant. However, in the most primitive group of true land plants, the bryophytes (liverworts and mosses), the gametophyte is exceptionally the dominant generation (Figure 5.11). The sporophyte generation consists of capsules that are attached to and are dependent for their nutrition on their gametophyte parent.

The more advanced vascular plants consist of two main groups, the Pteridophyta (including ferns, horsetails and clubmosses) which are dispersed by spores, and the

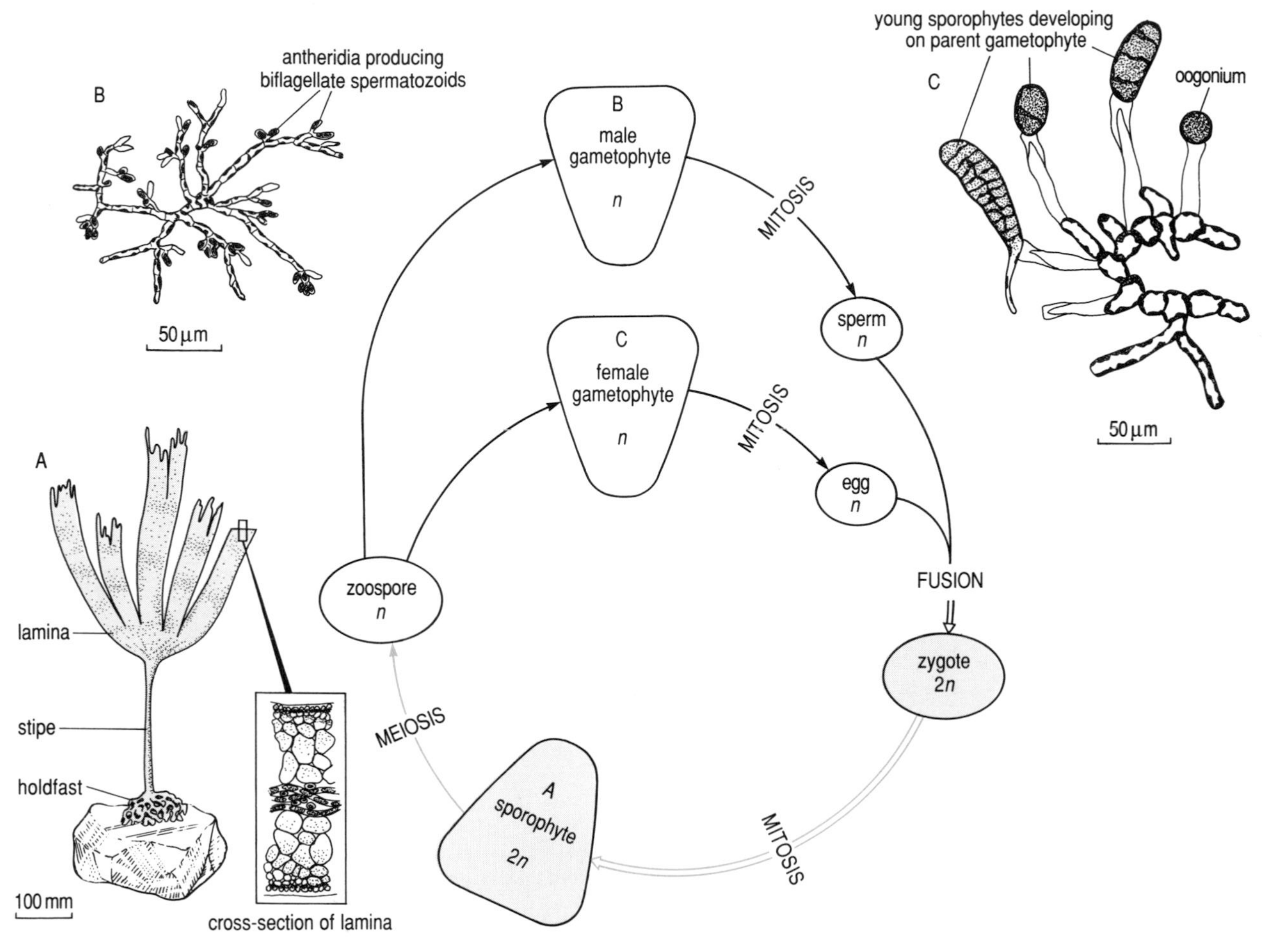

Figure 5.10 *The life cycle of the oarweed* Laminaria. *Diploid stages are shown in green. A, the large seaweed sporophyte; B, a male gametophyte bearing antheridia that produce spermatozoids; C, a female gametophyte bearing oogonia.*

Spermatophyta (including conifers and flowering plants) which, of course, produce seeds. Their life cycles are rather different, but in both groups the sporophyte is the familiar predominant form. Pteridophytes have an independent gametophyte stage but these are insignificant structures, in ferns looking like small, featureless lobes of green tissue that contain both male and female reproductive organs. The male reproductive organs, called antheridia, produce flagellate spermatozooids, which fertilize egg cells within the female reproductive structure, the archegonium.

In the seed plants, the gametophyte stage has no independent existence. The female gametophyte consists of an embryo sac containing eight nuclei; the male gametophyte is contained within the pollen grain and contains three nuclei. When a pollen grain lands on a female flower, a pollen tube grows out of the grain towards the embryo sac, where fusion of one of the pollen nuclei with one of the embryo sac nuclei produces a zygote.

Spermatophytes differ from all other plants, including pteridophytes, in that their gametes are not free-swimming. Pollen is carried to female flowers by a variety of vectors, such as wind, insects, and other animals; there the male gamete is carried

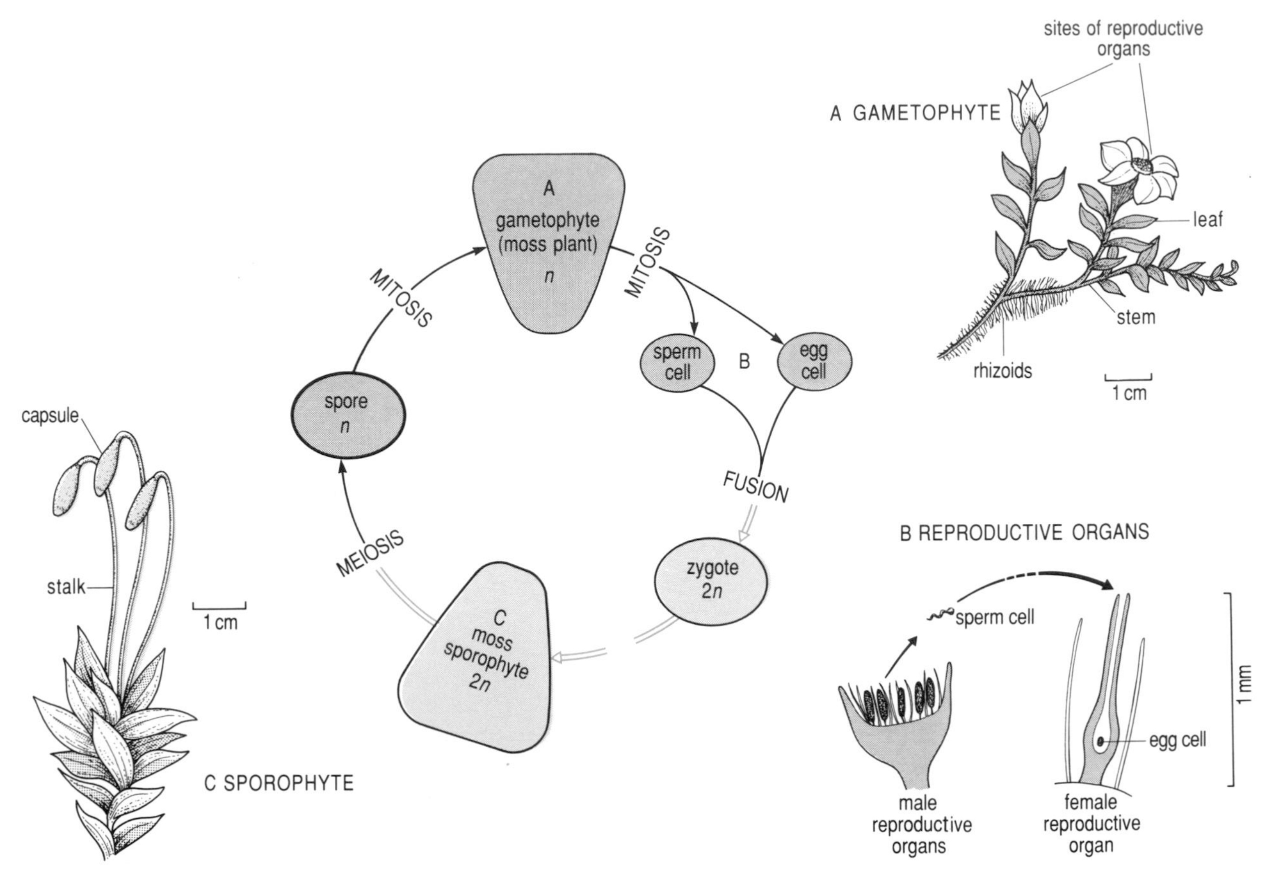

Figure 5.11 *The life cycle of the moss* Mnium hornum. *Diploid stages are shown in green.*

to the female gamete by the pollen tube. Much of the evolution of reproductive cycles in plants is linked to their dependence, or non-dependence, on water for reproduction. The lower plants, with free-swimming gametes, are dependent on a watery medium in which their gametes can move. Higher plants are emancipated from this constraint and, partly as a result, have been able to colonize the land.

5.3 THE EVOLUTION OF ANISOGAMY

An essential feature of sexual reproduction is the fusion of two haploid gametes to form a diploid zygote. In the great majority of organisms, there is anisogamy, with two kinds of gamete being produced: relatively large ova and tiny spermatozoa. The morphology of gametes has no bearing on the evolutionary basis of sexual reproduction *per se*; the genetic consequences of sex are the same, whether gametes are of the same or of different forms. It is important, however, to understand why anisogamy evolved because, as we shall see in this and subsequent Chapters, it provides the basis for many differences that exist between males and females and which are of great evolutionary significance.

While anisogamy is associated with there being two distinct sexes, this is not a universal phenomenon. In bacteria and single-celled organisms such as *Paramecium*, no gametes are produced but there is a form of sexual reproduction in which nuclei from two individuals fuse. Individuals belong to one of several 'mating types' and fusion only occurs between two individuals that belong to different mating types.

It is generally assumed that the primitive condition is isogamy and that anisogamy evolved from it. In the isogamous condition, all gametes are similar, though we can reasonably assume that there is some degree of variation in their size and form. Each gamete contains a certain quantity of nutrients stored in its cytoplasm and possesses some means of locomotion, such as one or more tail-like flagella (e.g. the isogametes of sea lettuce in Figure 5.9). The nutrients contained in a gamete not only sustain it until it fuses with another gamete, but also help to sustain the zygote that results from their fusion in the early stages of its development. To be successful, each gamete must fuse with another which involves swimming actively before its reserves are used up and it dies. There are alternative ways by which individual gametes may maximize the probability of fusion. They may increase their mobility by having more powerful flagella and smaller bodies, so increasing the amount of space they can search in a unit time. Alternatively, they can sacrifice mobility in favour of larger size, expanding their storage capacity and thus increasing their longevity. What will happen when gametes of these two types meet? Two of the smaller, more mobile type will tend to find one another very quickly but, if they fuse, they will produce a small zygote with very limited reserves to sustain it. A fusion of two of the large, less mobile type will produce a very large zygote, but this will be a rare event that is not likely to occur until the two gametes are old and have used up most of their reserves. Fusions between a small and large gamete tend to occur quickly, because of the mobility of the smaller one, and will yield a viable zygote because of the cytoplasmic reserves of the larger one. Such a zygote will, therefore, be more likely to survive than those formed by the fusion of two gametes of the same type.

Once two distinct types of gamete have evolved, initially with a small morphological difference between them, disruptive selection will operate on them (Chapter 4, Section 4.4.3). That is to say, selection will favour the evolution of smaller, more mobile gametes, provided there are large ones with which they can fuse, and *vice versa*. Furthermore, selection will favour the capacity of each type of gamete to fuse preferentially with one of the opposite type.

Selection operates on gametes in a number of ways. First, as they have a finite life and may be existing in an environment that is only temporarily hospitable, there will be selection favouring those that find another gamete as quickly as possible. Secondly, as gametes are produced in very large numbers, selection will favour those that fuse to form a large, viable zygote. Thirdly, once a degree of anisogamy exists, there will be competition among the smaller, more mobile ones, creating selection that favours those that reach the larger, less mobile ones first. The first two of these selection pressures are conflicting and will yield either a single type of gamete that is a compromise between conflicting requirements (isogamy), or two distinct types each of which can exist only if the other type exists (anisogamy).

Sperm, being small, are metabolically cheap to manufacture and typically are produced in vast numbers. They tend to be short-lived, though in some species they may live for some time in a nutritious external medium. In some animals, this medium is provided by the female, who stores sperm, sometimes for a year or more, between the time of mating and when the eggs are fertilized. Eggs are much more expensive to make and tend to be produced in relatively smaller numbers. The differences between eggs and sperm form the basis for many other differences between males and females. Most importantly, the reproductive success of females is limited by the number of eggs they can produce and (in some species) care for until they become self-sufficient. In contrast, the reproductive success of males is not generally limited by the ability to produce gametes, but by the ability to obtain matings.

The relationship between egg and sperm is an interesting one because it embodies, at the cellular level, the essential relationship between male and female organisms. Each is dependent on the other if they are to survive to form a viable zygote; their relationship is thus an example of mutualism, in which each benefits from the particular characteristics of the other. Their contribution to the zygote is, however, unequal because, while each contributes equally to the genotype of the zygote, it is the ovum that contributes virtually all of the cytoplasm; in this respect the sperm can be said to be exploiting or parasitizing the egg. As we shall see in Chapter 7, sexual reproduction, especially in animals, involves elements of both cooperation and conflict of interests between males and females. The basis of these aspects of sexual reproduction is the phenomenon of anisogamy.

Dimorphism in gametes represents the basis of a diverse array of differences between males and females that are referred to collectively as **gender**. Males, by definition, produce sperm, females produce eggs. Gender is not, however, as clear a concept as might at first appear. In everyday language it refers to a range of characteristic anatomical and behavioural differences between men and women. For some biologists, it refers to chromosomal or other genetic differences between males and females, for others to gonadal differences. Consider, however, the unisexual lizards discussed in Section 5.1.7. In terms of their morphology and their gonads, they are female, but in their behaviour they can show the full repertoire of male as well as of female behaviour. Hermaphrodites (see Section 5.5) function gonadally and behaviourally as both males and females; for them, gender cannot be ascribed to the individual organism but to certain aspects of its reproductive physiology and behaviour.

These conceptual difficulties arise from the fact that, in biological terms, sex and gender are not equivalent or synonymous concepts. Species which, like the sea lettuce, produce isogametes, have sexual reproduction involving meiosis and syngamy, but they have no gender; all such individuals have similar morphology, reproductive organs and sexual behaviour. The existence of such systems indicates that anisogamy and gender are not obligatory features of sexual reproduction; rather they have evolved secondarily, as a consequence of the existence of sexual reproduction.

5.4 EVOLUTION OF THE SEX RATIO

In the great majority of sexually reproducing animals and plants that exist as separate sexes (a condition called **gonochoristic** in animals, dioecious in plants), sons and daughters are produced in roughly equal numbers, giving a 1 : 1 sex ratio. In genetic terms this ratio is readily explained by the chromosomal mechanism of sex determination in which one sex (males in mammals, females in birds) is heterogametic (*XY*), the other is homogametic (*XX*). Such a system would appear to make it inevitable that, at least at conception, sons and daughters will be produced in equal numbers. There are, however, examples of species in which the sex ratio is biased towards one sex, suggesting that selection can operate on the sex ratio in some circumstances. A more general reason why we should not accept a 1 : 1 sex ratio as inevitable is that, for a number of reasons, it may not be adaptive for the sexes to be produced in equal numbers.

Let us consider a specific example, the elephant seal *Mirounga angustirostris*. Female elephant seals produce one pup each year and they make a massive contribution to the rearing of their pups in the form of large quantities of very rich, highly nutritious milk. Males play no part in parental care but fight in a particularly violent way for sexual access to females. Only the largest and most powerful males mate with females; they establish the status of 'harem masters' with exclusive access to a large number of females. The outcome of this competition among males is that, whereas all females that survive to breeding age produce progeny, only 10% of males breed during their lifetime. From the point of view of the species, or of a population, a 1 : 1 sex ratio thus seems maladaptive. Many more young would be recruited into the population if males and females were produced in the ratio 1 male : 9 females. No effort would then be wasted in producing sons that fail to breed. Natural selection does not, however, act on species or on populations, but on individuals, and so we must seek to explain the evolution of the sex ratio in terms of whether it is adaptive for individuals to produce more sons or more daughters. Such an explanation was formulated by Fisher (1930).

Suppose that a situation arises in a population of elephant seals in which females are twice as numerous as males. A parent that produces x daughters will have xy grandchildren, where y is the average expected number of young produced in a female's lifetime (in elephant seals y = approximately 10—one pup per year for up to 10 years). A parent that produces x number of sons will have $2xy$ grandchildren, because each son will, on average, inseminate two females. As a result, it is more adaptive to produce sons than daughters in a female-biased population, and the frequency of sons will tend to increase as a result. The fact that only 10% of sons actually reproduce in the elephant seal mating system is irrelevant; it is the average reproductive success of males that is important. Likewise, it will be more adaptive to produce daughters than sons in a male-biased population. Selection thus acts against any departure from a 1 : 1 sex ratio; put another way, a 1 : 1 sex ratio is said to be evolutionarily stable, a biased sex ratio is not. The concept of an **evolutionarily stable strategy (ESS)** was developed by J. Maynard Smith and G. R. Price (1973), and is defined as a pattern of behaviour or some other phenotypic character which, when adopted by all members of a population, cannot be bettered by an alternative strategy.

This argument holds true only if the effort required of a parent to produce a son is the same as that required to produce a daughter. In elephant seals, this is the case; young of each sex are the same size at weaning and require the same amount of parental care. If, however, female progeny 'cost' a parent twice as much to produce as sons (for example), that parent has the option (assuming that the sex ratio is under genetic control), given that the resources it can allocate to reproduction are finite, of producing n number of daughters or $2n$ sons, or some equivalent combination of sons and daughters. If the population sex ratio is 1 : 1, a parent producing $2n$ sons would have, on average, twice as many grandchildren as one producing n daughters and so would have higher fitness. In this situation, a 1 : 1 sex ratio is therefore not evolutionarily stable and would evolve towards one in which males are more numerous than females. The evolutionarily stable sex ratio would be 1 female : 2 males. Thus, a more accurate statement of Fisher's argument is that parents should expend equal amounts of effort on producing sons and daughters.

If the *XX*/*XY* mechanism of sex determination always produces sons and daughters in a 1 : 1 ratio, there will be no variance on which selection could act, even if it were adaptive for parents to produce more progeny of one sex than the other. It is not, at present, clear to what extent genetic determination of sex is an important constraint on the evolution of the sex ratio (see below). There is, however, increasing evidence that biased sex ratios do occur in nature and that these arise by other than genetic mechanisms. In an American bird, the common grackle *Quiscalus quiscula*, male and female eggs are produced in roughly equal numbers but, by the time they fledge, females outnumber males by 1.6 : 1 (Howe, 1977). In this species, males, at fledging, are approximately 20% heavier than females; consequently, they require a great energetic investment on the part of their parents. It appears that the parents devote roughly equal amounts of parental care to male and female progeny with the result that mortality among male nestlings, which have a greater energetic requirement, is higher than among females.

In an experimental study of wood rats *Neotoma floridana*, McClure (1981) kept a control group of mothers on an unrestricted diet and an experimental group on a diet that provided only 70 to 90% of the nourishment required to maintain the body weight of non-reproductive females of comparable size. In both groups the sex ratio at birth was 1 : 1 but, whereas this was maintained until the time of weaning in the fully fed control group, the sex ratio in the food-restricted group fell steadily until, by 20 days after birth, it was three males to seven females. This change resulted from the food-restricted mothers allowing their sons less opportunity to suckle than their daughters. The young of mothers kept on the restricted diet were smaller at all ages, including full maturity, than those of fully fed mothers. The small adult size of daughters did not reduce their eventual reproductive success, as measured by litter size, but small sons were less able than the sons of fully fed mothers to obtain and defend territories and so suffered reduced reproductive success in comparison with the sons of fully fed mothers. The adaptive explanation for the discrimination shown by food-deprived mothers to their male progeny is that, because their sons have poor reproductive prospects, mothers on a restricted diet devote less parental care to them.

Baboons live in matriarchal societies, in which the females of a given age within a group are socially dominant to males of the same age. Within each sex, there is also a hierarchy, with larger individuals usually being dominant over smaller ones. As a result, larger females have priority of access to high quality food over smaller females or males and so have larger progeny that have a greater chance of surviving. Individual young, as a result of these size differences, tend to 'inherit' (because of this phenotypic effect, not because of a genetic effect) the status of their mother; high-ranking females tend to have progeny that subsequently themselves attain high rank. In addition, high-ranking females produce more daughters than sons; low-ranking females produce more sons. The mechanism underlying this sex ratio bias, in a species that has genetic sex determination, is not known.

It is often assumed that, in animals such as birds and mammals in which sex is determined chromosomally, a 1 : 1 sex ratio is inevitable, at least at conception. In fact this is not so. In many mammals, including humans, more than 50% of zygotes are male and, for reasons that are poorly understood, this proportion falls between conception and birth. In the elk *Cervus canadensis*, for example, over 60% of foetuses are male at conception, but this proportion has fallen to around 50% at birth.

The examples of the grackles and the wood rats suggest that, although there may be no departure from a 1 : 1 sex ratio at birth (the primary sex ratio), the sex ratio in a population can be distorted by unequal allocation by parents of their parental care to sons and daughters. One group of animals in which the primary sex ratio is manipulated, though in a quite different way, is the hymenopteran insects, such as bees. Workers determine the sex of the queen's progeny by controlling whether or not her eggs are fertilized. Unfertilized eggs become haploid males, fertilized eggs become diploid females. Many Hymenoptera show strongly biased sex ratios with a much higher proportion of females than males; in honeybees for example, the sex ratio within a hive is one male to three females though the great majority of females are sterile.

It can be argued that, in a variable environment, a female should manipulate the sex ratio among her progeny according to whether her sons or her daughters are more likely to maximize their reproductive potential under the prevailing conditions. The basis of this prediction is that, if females do this, they will maximize the number of their grand-progeny, i.e. maximize their fitness. Such a model has been developed by Charnov *et al.* (1981) and has been applied by them specifically to parasitic wasps, in which a female lays single eggs on the paralysed bodies of hosts which vary in size and on which the growing wasp larva feeds. The size of the young wasp that eventually develops from each egg depends on the size of the host on which it is laid. Because eggs are large and energetically expensive to produce, the fecundity of a female is likely to be a function of her size; the larger she is, the more eggs she will produce. This dependency of fecundity on body size will generally be less marked in males; even the smallest male will produce enough sperm to fertilize several females.

▶ If these assumptions are correct, how should female wasps apportion male and female eggs among hosts of varying size?

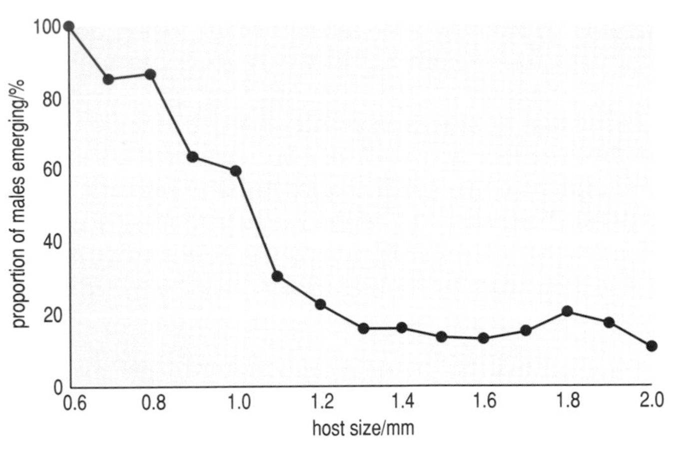

Figure 5.12 *Graph showing the relationship between size of insect hosts and sex of young parasitic wasps (*Lariophagus*) laid on those hosts. Male progeny are laid on smaller hosts, females on larger ones.*

They should lay female eggs on larger hosts, male eggs on smaller hosts. In a laboratory experiment, this prediction was tested using the parasitic wasp *Lariophagus distinguendus*, which attacks granary weevils and in which progeny that were laid on large hosts are larger than those laid on small hosts. The results are shown in Figure 5.12.

▶ Do these results support the prediction?

Yes, they do; females laid a greater proportion of female eggs when the available prey were large than when they were small.

In the solitary bee *Osmia lignaria propinqua* females produce equal numbers of each sex over the course of a breeding season, but produce a preponderance of daughters early in the season and switch to producing sons later. This appears to be related to a greater availability of flowers that provide food early in the season. When food is abundant, progeny can attain large size and will achieve high fecundity if they are female. Later, when food is scarce, they will be small and can be more successful as males.

Perhaps the most puzzling form of departure from a 1 : 1 sex ratio is that which results from environmental effects. This **environmental sex determination** is best known among reptiles, in several species of which the gender of an individual is determined by the temperature at which it undergoes embryological development in the egg (Chapter 3, Section 3.2.3). In the leopard gecko *Eublepharis macularius*, for example, eggs reared at a temperature of 32 °C yield progeny of which 80% are male, 20% are female. In contrast, eggs reared at 26 °C result in 100% female offspring. A 1 : 1 sex ratio results if the eggs are reared at 29 °C (Figure 5.13). These results are due to a real effect of temperature on development, the nature of which is not known, and cannot be attributed to differential mortality of males and females at different temperatures. Whatever the temperature at which it is reared, an individual gecko is unequivocally male or female; hermaphrodites or individuals of indeterminate sex are never found and there are no physiological differences

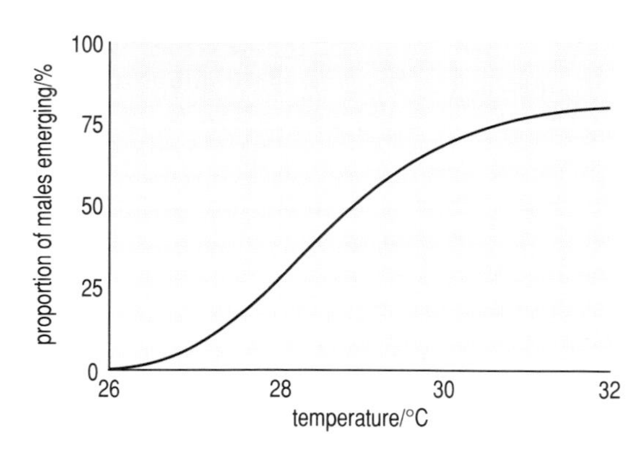

Figure 5.13 *The effect of temperature during development on the primary sex ratio in the leopard gecko* Eublepharis macularius.

between 'hot' females, reared at temperatures that produce a preponderance of males, and 'cold' females. It appears that temperature during development activates some kind of switch mechanism that determines whether an individual will become male or female. In another group of reptiles, the chelonians (turtles and tortoises), in contrast, high temperatures produce females, low temperatures males.

▶ How does this kind of departure from a 1 : 1 sex ratio differ from the other examples discussed earlier (with the exception of the social hymenopterans)?

In those reptiles that show environmental sex determination, the effect is on the primary sex ratio, i.e. that among progeny at birth. In the other examples discussed, females manipulated the sex ratio of their progeny in a variety of ways to produce a biased ratio, but (for most) in the context of a primary sex ratio that is 1 : 1.

The ecological and therefore evolutionary significance of environmental sex determination is not yet clear. The phenomenon has been studied intensively in the laboratory, where temperature can be controlled, but there are rather few data on sex ratios in natural populations. It is known, however, that the laboratory results found in *Alligator mississippiensis* are consistent with what happens in nature. Crocodiles and alligators lay their eggs in large nests consisting of mounds of vegetation; the temperature within nests varies from one to another, hotter nests produce a higher proportion of males than cooler nests, and, in natural populations there is a preponderance of females, suggesting that the majority of nests are cool. It remains to be seen whether female crocodiles, and other reptiles, can adaptively manipulate the sex ratio among their progeny by controlling the temperature within their nests.

SUMMARY OF SECTION 5.4

For many but not all species, genetic mechanisms tend to produce a 1 : 1 primary sex ratio. There exists, however, a variety of mechanisms by which particular species can alter their behaviour towards their progeny to produce a population sex ratio that is biased towards sons or daughters. There is some evidence, furthermore, that such manipulation can be adaptive, inasmuch as individuals can produce a larger proportion of offspring of the sex that will yield them the greater fitness. These effects are interesting because they suggest that some species can reduce one of the costs of sex identified in Section 5.1.3, i.e. the cost of producing sons.

5.5 HERMAPHRODITES AND ANIMALS THAT CHANGE SEX

Hermaphrodites are animals that are capable of producing both male and female gametes either at the same time or at different stages in their life cycle. Simultaneous hermaphrodites are those that can produce both at the same time, a condition found in a wide variety of animals, including platyhelminths, molluscs, crustaceans and fishes. The most likely selection pressure favouring hermaphroditism is difficulty in finding a mate. Many hermaphrodites are sessile animals (that cannot move about in search of mates), animals that live at very low population densities (and have a low probability of finding a mate), and internal parasites (in which individuals typically live in an environment that is very isolated from conspecifics). For such animals, the ability to fertilize one's own eggs in the absence of a mating partner is a great advantage.

► In terms of the costs of sex discussed in Section 5.1.3, what additional gain may self-fertilizing hermaphrodites gain?

They do not bear the costs of mating, since they do not have to seek or compete for mates and are not exposed to the risks that attend mating.

► What cost are self-fertilizing hermaphrodites likely to bear?

The cost of inbreeding, of which self-fertilization is the most extreme form. Self-fertilization will only be adaptive for hermaphrodites that are self-fertile and which do not carry a large number of lethal recessive genes. A further possible cost is that imposed by individuals having to develop and maintain both male and female reproductive systems.

Even if hermaphrodites are not self-fertile, they are at an advantage over gonochoristic animals that live at low densities, because any two individuals that meet one another will be able to mate (each produces both male and female gametes). Mating in many hermaphrodites, such as earthworms, involves a mutual exchange of gametes, each individual passing sperm to the other.

Among plants, as we have seen (Section 5.3), many adaptations exist that prevent self-fertilization. There are, however, a number of plants, such as the violet, that possess **cleistogamous** flowers that never open up, ensuring that self-fertilization occurs. Much less is known about the incidence of self-fertilization in animals. Some gastropod snails, such as *Helix* and *Cepaea*, are completely self-sterile, whereas another, *Rumina*, typically fertilizes itself in nature. The white-lipped land snail *Triodopsis albolabris* will never fertilize itself if kept in pairs, but will do so after being kept in individual isolation for several months. The reproductive success of cross-fertilizing pairs is 86 times greater than that of self-fertilizing individuals, suggesting that, for this species at least, self-fertilization incurs severe fitness costs that will make it adaptive only when no partners are available.

There are many hermaphrodites that are not sessile or parasitic and which do not live at low densities, most notably several species of coral reef fish. One species, the black hamlet *Hypoplectrus nigricans*, employs its dual sexuality in a remarkable way. Each individual has a large ovary and a relatively minute testis. When spawning, they form pairs which last for a day or more and the two partners take turns to adopt male and female roles. One fish produces a batch of eggs which its partner fertilizes externally. The other fish then produces a batch of eggs which the first fish fertilizes and this alternating pattern is repeated many times until both fish have exhausted their egg supply. The fertilized eggs are released into the plankton. The two partners in effect 'trade' eggs with one another, in return for having them fertilized by their mate. This system greatly reduces one of the major costs of sex — the production of sons — as all individuals are fully capable of functioning as females. Furthermore, because each fish needs only a small testis to produce sufficient sperm to fertilize the eggs of its mate, the allocation of resources to the production of male capability is much less than it is in gonochoristic species, in which, as discussed in Section 5.4, it is typically equal to the allocation to female capability. The black hamlet, by egg-trading and cross-fertilizing with another fish, gains all the benefits of sexual reproduction, but bears very few of the costs. It is very curious that this reproductive pattern is known for only a very few species.

Several species of fish consist of individuals that are capable of producing both male and female gametes, but not at the same time, a condition called sequential hermaphroditism. Some are **protandrous**, beginning life as males and becoming female later in life; others are **protogynous** and change sex in the reverse direction. (Many plants are protandrous, with pollen production preceding the maturation of the ovules.) The existence of these two patterns raises the question: under what conditions is it better to be male, or to be female, first? In both protandrous and protogynous species, sex change typically occurs when an individual attains a certain size, suggesting that body size is an important determining factor. The exact role of body size in determining the direction of sex change depends on the **mating system** of the species. (Mating system is a term that describes the pattern of mating relationships within a population, such as whether individuals have one mate or several, and is discussed more fully in Chapter 7; with reference to plants, the term *breeding system* is generally used.) Size-dependent sex change also occurs in a perennial plant, the jack-in-the-pulpit *Arisaema triphyllum*. Small individuals do not flower, larger ones reproduce as males, and the largest as females. As an individual plant increases or decreases in size, so it may change sex.

Fishes typically continue to grow after they have begun to reproduce. For both sexes, an individual's fecundity typically increases with its size because its gonads will grow along with the rest of its body. Thus, larger males will produce more sperm than smaller individuals, and larger females more eggs. For females, because they produce relatively very large gametes, fecundity is strictly limited

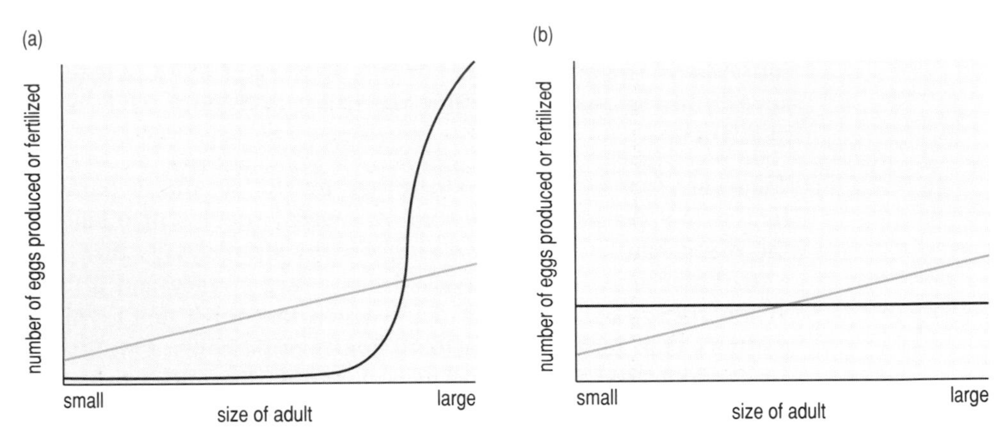

Figure 5.14 *The expected reproductive success, in terms of number of eggs produced or fertilized, for male and female fishes in two mating systems. In both systems, female reproductive success increases linearly with body size (green line). In (a), males compete for access to females and only very large males compete successfully. In (b), pairing is random with males of all sizes being equally likely to gain access to females.*

and will typically show a linear increase with body size (Figure 5.14). For males, however, sperm can be produced very cheaply and in huge numbers so that even a tiny male has a very large reproductive potential, in terms of the number of eggs he can fertilize. If, however, the mating system of a particular species of fish is such that larger males dominate and prevent smaller ones from mating — a common pattern — only larger males will be able to realize their reproductive potential (Figure 5.14(a)). In this situation, it is adaptive for an individual to start breeding as a female and to continue to do so until it has reached a critical size at which it can compete effectively with other males. This pattern is called protogyny and is characterized by males being larger and less numerous than females.

If mating is random, however, and there is no advantage to being a large male, small males can produce enough sperm to fertilize many eggs, many more in fact than they could produce eggs. In this situation, what limits a male's reproductive success is his ability to find females to mate with and variation among males in terms of mating success is not related to their size (Figure 5.14(b)).

▶ In this situation, should an individual start its reproductive life as a female or as a male?

As a male; it can expect to fertilize more eggs than it could produce as a female of comparable size. It will change to being a female when it has grown large enough to be able to produce large numbers of eggs. This is protandry, in which males are typically smaller than females.

There is another factor to be considered here, and that is the sex ratio in the population. Here we can apply the same kind of argument that was developed in Section 5.4, based on the idea that, in a given population, there will be a particular sex ratio that is evolutionarily stable. If, in a population of protandrous fishes, males are very abundant relative to females, average male reproductive success will be relatively low, because there are few eggs available for them to fertilize. It will then be adaptive for one or more individuals to change into females, because they will then produce more progeny than they would if they remained male. Likewise, in a protogynous species in which females become relatively very common, an individual may gain higher reproductive success by changing into a male. This hypothesis has been tested experimentally in a number of protogynous species. If one or more large males are removed from a population, one or more of the larger females typically turn into males and take their place. An example of a protogynous fish is the bluehead wrasse *Thalassoma bifasciatum*, found in the western Atlantic and the Gulf of Mexico.

In sequential hermaphrodites, each individual has the option of being male or female. Which sex it is more adaptive for it to be, i.e. which yields the higher fitness benefit, depends not only on properties intrinsic to the individual, such as its size, but also on what other individuals are doing. Thus, the sex ratio in a population which, at first sight, may seem to be an attribute best understood at the population level, is a consequence of selection operating on the behaviour of individuals. We shall return to the question of what entities natural selection acts on in the next Chapter.

The capacity to change sex during an individual's lifetime is relatively common among fishes, probably because most have external fertilization. Animals with internal fertilization have evolved complex, sexually dimorphic genitalia and specialized female organs in which the fertilized eggs or young develop. The costs of changing such structures makes sex change in species with internal fertilization much less likely to evolve.

Summary of sections 5.2–5.5

Many plants have separate and distinct haploid and diploid phases in their life cycle, with alternation of generations between them. In others, and in all animals, only the gametes are haploid. There is inherent competition among gametes that has resulted, by disruptive selection, in the evolution of anisogamy, in which gametes occur in two very different forms. While there are commonly genetic mechanisms that predispose organisms to produce male and female offspring in equal numbers, there are a number of factors that lead to distorted sex ratios. These include differential investment by parents in sons and daughters and environmental effects. Some organisms are hermaphrodite (sequentially or simultaneously) and provide opportunities to assess the relative advantages of being male or female under different environmental conditions.

5.6 Conclusion

The evolution of sex has been described as one of the outstanding unanswered questions in evolutionary biology. There are a number of reasons why this is so. In an ideal world, biologists would be able to associate each mode of reproduction that occurs in nature with a specific set of ecological parameters, and thereby determine precisely how each mode is adaptive. The mode of reproduction of a given species will be influenced by many features of its environment and, as discussed in Chapter 2, it is often very difficult to identify and quantify what these factors are.

First, there is the question of whether we should be looking at sex as an adaptation at the individual or at the species level. As discussed in this Chapter, there are good reasons for believing that sexually reproducing species persist for longer over evolutionary time than asexual species. The only way to test this hypothesis directly would be to make observations, over millions of years, of comparable sexual and asexual species to determine which became extinct first. Such information could only be inferred indirectly from the fossil record of the relative durations of species (Chapter 15). To test the hypothesis that sex is adaptive at the individual level, we would ideally, as with other characters, study the fitness consequences of variation in sexuality. Among animals, however, there is typically no such variation; members of a given species are sexual or asexual and they do not show varying degrees of sexuality. There are species, like aphids and *Hydra*, where both sexual and asexual reproduction occur, but they rarely, if ever, occur at the same time and in comparable ecological conditions, so that meaningful comparisons of their fitness consequences cannot be made. Many plants, in contrast, reproduce sexually and asexually at the same time, offering better opportunities to assess the relative costs and benefits of different modes of reproduction.

Another problem is that sexual reproduction typically involves a number of features that are secondary aspects of sexuality, notably anisogamy and gender. These aspects generate fitness variations on their own account. As we have seen, for example, an individual may have higher fitness by producing more sons than daughters, or *vice versa*. It is very difficult then to separate such fitness consequences from those due to sex *per se*.

These and other problems make the evolution of sex a particularly problematic area of research. There are, however, a number of conclusions that can be drawn from the evidence presently available. Most current theories about the evolution of sex focus on the degree of genetic, and therefore phenotypic, diversity that each kind of reproductive mode generates among progeny. Sexual reproduction, because it involves a particular form of cell division—meoisis—that involves genetic recombination, produces a very high degree of variation. Indeed, there is reason to believe that, in terms of individual fitness, it generates too much; hence the cost of recombination. As a result, selection may favour asexual forms that arise within sexual species. The phylogenetic distribution of asexual species provides some support for the hypothesis that asexual forms have arisen quite frequently from sexual species, that they flourish for a relatively short period of evolutionary time, but that they become extinct more quickly than sexual species. This results from the fact that asexual reproduction does not generate sufficient genetic variation to enable a species to adapt to changing environments.

There is, however, an alternative direction in which a sexually reproducing species may evolve. As far as the generation of genetic diversity is concerned, asexual reproduction and outbreeding sex represent extremes on a continuum. There are other modes of reproduction that produce intermediate levels of variation. Perhaps the most important factor to be considered here is the degree to which a sexual species shows inbreeding or outbreeding, which is also a continuum. Inbreeding reduces the cost of recombination but may incur costs in the form of inbreeding depression. The balance of these costs and benefits may be such, however, that some degree of inbreeding sex may be more adaptive than either asexual reproduction or outbreeding sex, because it yields an intermediate, optimal level of genetic variation.

REFERENCES FOR CHAPTER 5

Bateson, P. (1980) Optimal outbreeding and the development of sexual preferences in Japanese quail, *Zeitschrift fur Tierpsychologie,* **53**, 231–244.

Bierzychudek, P. (1989) Environmental sensitivity of sexual and apomictic *Antennaria*: do apomicts have general-purpose genotypes?, *Evolution,* **43**, 1456–1466.

Case, T. J. (1990) Patterns of coexistence in sexual and asexual species of *Cnemidophorus* lizards, *Oecologia,* **83**, 220–227.

Charnov, E. L., Los-den Hartogh, R. L., Jones, W. T. and van den Assem, J. (1981) Sex ratio evolution in a variable environment, *Nature,* **289**, 27–33.

Fisher, R. A. (1930) *The Genetical Theory of Natural Selection,* Clarendon Press, Oxford.

Hay, D. E. and McPhail, J. D. (1975) Mate selection in threespine sticklebacks (*Gasterosteus*), *Canadian Journal of Zoology,* **53**, 441–450.

Howe, H. F. (1977) Sex ratio adjustment in the common grackle, *Science,* **198**, 744–746.

Karron, J. D. and Marshall, D. L. (1990) Fitness consequences of multiple paternity in wild radish *Raphanus sativus*, *Evolution,* **44**, 260–268.

Kelley, S. E. (1989a) Experimental studies of the evolutionary significance of sexual reproduction. V: A field test of the sib-competition lottery hypothesis, *Evolution,* **43**, 1054–1065.

Kelley, S. E. (1989b) Experimental studies of the evolutionary significance of sexual reproduction. VI: A greenhouse test of the sib-competition hypothesis, *Evolution,* **43**, 1066–1074.

Lively, C. M. (1987) Evidence from a New Zealand snail for the maintenance of sex by parasitism, *Nature,* **328**, 519–521.

Maynard Smith, J. and Price, G. R. (1973) The logic of animal conflict, *Nature*, **246**, 15–18.

McCall, C., Mitchell-Olds, T. and Waller, D. M. (1989) Fitness consequences of outcrossing in *Impatiens capensis*: tests of the frequency-dependent and sib-competition models, *Evolution,* **43**, 1075–1084.

McClure, P. A. (1981) Sex-biased litter reduction in food-restricted wood rats (*Neotoma floridana*), *Science,* **211**, 1058–1060.

McKone, M. J. (1987) Sex allocation and outcrossing rate: a test of theoretical predictions using brome grasses (*Bromus*), *Evolution,* **41**, 591–598.

Price, M. V. and Waser, N. M. (1979) Pollen dispersal and optimal outcrossing in *Delphinium nelsoni*, *Nature,* **277**, 294–297.

Schmitt, J. and Antonovics, J. (1986) Experimental studies of the evolutionary significance of sexual reproduction. IV, *Evolution,* **40**, 830–846.

Schmitt, J. and Ehrhardt, D. W. (1987) A test of the sib-competition hypothesis for outcrossing advantage in *Impatiens capensis*, *Evolution,* **41**, 579–590.

Waser, N. M. and Price, M. V. (1989) Optimal outcrossing in *Ipomopsis aggregata*: seed set and offspring fitness, *Evolution,* **43**, 1097–1109.

Williams, G. C. (1975) *Sex and Evolution,* Princeton University Press.

FURTHER READING FOR CHAPTER 5

Maynard Smith, J. (1984) The ecology of sex. In: J. R. Krebs and N. B. Davies (eds) *Behavioural Ecology. An Evolutionary Approach* (2nd edn), pp. 201–221, Blackwell Scientific Publications, Oxford.

Stearns, S. C. (ed.) (1987) *The Evolution of Sex and its Consequences*, Birkhauser, Basel.

SELF-ASSESSMENT QUESTIONS

SAQ 5.1 (*Objective 5.1*) In Table 5.1, List 1 gives a number of possible genetic consequences of reproduction, A to F. For each, categorize it according to which item, 1 to 4, in List 2 is most appropriate.

Table 5.1 *For use with SAQ 5.1.*

List 1	List 2
A Genetic recombination leading to genetically diverse progeny	1 A benefit of asexual reproduction
B Elimination of deleterious mutations	2 A cost of asexual reproduction
C Accumulation of deleterious mutations	3 A benefit of sexual reproduction
D Genetic recombination leading to break-up of favourable combinations of genes	4 A cost of sexual reproduction
E Faster production of progeny	
F Production of sons	

SAQ 5.2 (*Objective 5.2*) Why is it an essential feature of the sib-competition model for the maintenance of sexual reproduction that individuals produce a large number of progeny?

SAQ 5.3 (*Objective 5.3*) In what way can sexual reproduction be an adaptation that enables one species to compete successfully with others in the same habitat?

SAQ 5.4 (*Objective 5.4*) (a) Why does inbreeding sometimes lead to a reduction in fitness?

(b) In a species in which there is differential dispersal of the two sexes, why might it not be advantageous for the average dispersal distances of the two sexes to differ to a very large extent?

SAQ 5.5 (*Objective 5.5*) In what way does the evolution of anisogamy from isogamy represent an example of disruptive selection?

SAQ 5.6 (*Objective 5.6*) Suppose that, in a species in which male and female progeny are equally costly to produce, there is a sex ratio of 45 males to 55 females. What does it mean to say that this sex ratio is not evolutionarily stable?

CHAPTER 6
THE FOCUS OF SELECTION

Prepared for the Course Team by Tim Halliday

Study Comment *This Chapter is almost entirely about social behaviour in higher vertebrate animals. This is because certain kinds of behavioural interaction between animals, specifically those that involve one individual conferring a fitness benefit on another, pose problems to evolutionary theory. They are difficult or impossible to explain in terms of classical natural selection theory. Detailed analysis of these social systems enables us to explain their evolution in terms of a precise, modified theory of natural selection, called kin selection. Kin selection is important, not just because of its capacity to explain these particular kinds of behaviour, but because it has modified the way that we view natural selection and makes very explicit the way in which selection acts upon individual animals.*

Objectives When you have finished reading this Chapter, you should be able to do the following:

6.1 Calculate the values of the coefficient of relatedness between individuals and various categories of their relatives.

6.2 Explain the role of ecological factors, individual interests and the interests of relatives in the evolution of cooperative breeding.

6.3 Recall Hamilton's rule and apply it in simple calculations of the conditions under which kin selection will operate.

6.4 Discuss critically the relative merits of theories based on group, kin and individual selection as explanations for the evolution of altruistic behaviour.

6.1 THE PROBLEM: ON WHAT DOES SELECTION ACT?

The innumerable organisms that inhabit the Earth can be hierarchically organized. Individuals belong to a particular species, a number of species are classified together within a single genus, genera are classified together into families, and so on through the higher taxonomic categories: orders, classes and phyla. At all levels, we can observe, in historical or in geological time, that essentially similar 'demographic' phenomena occur. Each category of organization has a definable origin in time, it flourishes for a while and the number of its constituent units increases, it then declines and becomes extinct. Similar phenomena occur at other than the individual level. A particular character, such as resistance to a disease or melanism in moths, arises at a particular time and spreads through a population; some characters become fixed in a population, others become rarer and eventually disappear. In many instances, such characters can be attributed to the actions of specific genes and we can describe the origin, proliferation and decline of those genes. Thus, at each level of biological organization, some kind of *'sorting process'* is going on; the number of constituent units at a given level at any one time may be greater or less than the number of units at a previous and at a later time. The question we address here is whether or not the same kind of sorting process is involved at each level of biological organization. This is a very important question in understanding evolution, because natural selection is a major cause of sorting and it is very important that we establish at what level of organization natural selection acts. Does selection act on genes, on characters, on individuals, on species, or on any other level in the hierarchy of living organisms?

As we have seen in earlier Chapters, Darwin made it quite clear that natural selection acts on individuals. Those individuals that survive and reproduce successfully in one generation pass on their characteristics to subsequent generations; those that die or fail to reproduce do not. Darwin had no knowledge, however, of the genetic mechanisms of inheritance. In the neo-Darwinian view of evolution, individual genotypes are seen as being passed on, from one generation to another. In fact during sexual reproduction, these genotypes are broken up by recombination, but the genes making up those genotypes are passed on largely unaltered. Within a population, a mutation may give rise to a particular character and, if it increases fitness, it may spread through the population. Likewise the mutant allele will spread through the gene pool of the population. As a result, population geneticists see evolution as change in allele frequencies, and see the same kind of 'sorting process' going on at the level of the genes. Genes appear, spread, decline and become extinct over time within a population. This raises the question of whether the sorting process taking place at the level of the gene is natural selection.

The Oxford biologist Richard Dawkins has made a human analogy between genes and individual organisms on the one hand, and oarsmen and boat crews on the other. Crews of eight oarsmen each represent individuals; each oarsman is a gene, interacting with seven others to determine the speed of a given boat. Suppose that there were a large number of oarsmen, from which we wished to select the best possible crew of eight. In the human context there are two ways of doing this. One is to assess each oarsman's ability as an individual, rowing a boat on his own; this is not analogous to nature because genes do not exist as independent entities. Another would be to watch several crews of eight and to pick from these the eight individual oarsmen who seem to be the best rowers. Again, the natural situation is not like this, because genes are only passed on from one generation to the next as part of a genotype. What would be analogous to nature would be to stage a large number of races, each representing a generation, and, between each race, to assign the crew members of winning crews to new crews (i.e. reassortment). At the end of a series of races, we would have a single crew consisting of eight oarsmen who would not necessarily be the eight best individual oarsmen, but who have successfully survived selection to interact successfully with others. In nature, there is no process equivalent to the selection of oarsmen as individuals; genes cannot selectively be taken from various individuals and brought together in a single individual. Whether or not a gene survives from one generation to the next depends on whether individuals that carry it survive and reproduce.

The critical point to consider here is whether, at a given level of organization, there are 'agents' of selection. At the individual level, there are numerous factors that determine whether individuals die or survive, and whether or not they reproduce. These include predators, diseases, parasites, climatic factors and the competitive abilities of individuals. At the gene level, there are no such agents; no natural process can selectively eliminate one gene from a given genotype and leave its other genes intact. There is, however, one area where we have to consider the effect of natural selection on the spread of specific genes, and that is where individual organisms behave in ways that benefit other individuals, enhancing the latter's fitness, at the expense of their own fitness. We shall return to this question in Sections 6.3 and 6.4.

Turning to the species level, is there a selective agent that can account for the sorting process that we observe among species? The answer again is a qualified 'Yes'. 'Yes' because, as will be discussed in Chapter 15, there is a sense in which processes that eliminate some species and favour the continued existence of others are selective. In the previous Chapter, for example, we saw that asexual species may have an initial reproductive advantage over sexual species, but that, over evolutionary time, this advantage may be transitory. The qualification is that this process is not natural selection, as defined by Darwin, because that relates only to selection among individual organisms.

6.2 Individuals and genes

Darwin's theory of natural selection incorporated, as an essential feature, the phenomenon of inheritance of characters from one generation to the next. The modern neo-Darwinian approach to natural selection incorporates our understanding of genetics and of the genetic material, DNA (Chapter 3). One of the most important books of recent years that explores the relationship between individuals and genes in the evolutionary process is *The Selfish Gene* by Richard Dawkins (1976).

Dawkins emphasizes a number of features of genes:

1 They reproduce themselves; Dawkins describes them as 'replicators'. Genes can produce numerous copies of themselves. Replication of genetic material typically involves the production of numerous accurate copies but, occasionally, mutation gives rise to variation in the products of replication. Thus, at the gene level, three essential features of the theory of natural selection are apparent: reproduction, multiplicity of replicates and variation among those replicates.

2 Genes may survive in a population for much longer than the individuals that carry them; they are, potentially, immortal, because they are passed on from one

individual to another. For example, an allele for haemophilia appeared in Queen Victoria and was inherited by a number of her descendants. In this instance, the allele is detrimental to an individual that carries it.

▶ Only males with this allele suffer from haemophilia; heterozygous females do not express the allele and are phenotypically normal. Why is it detrimental to them to carry the allele?

The allele reduces their fitness because it affects the survival of their male descendants. Such alleles would, in nature, eventually be eliminated by selection acting on those individuals, whereas alleles that have a favourable effect will spread through a population from generation to generation and may persist within a species for many millions of years. Indeed, genes may persist from one species to another through geological time.

3 Genes cannot survive and replicate in isolation. A particular gene's survival and replication depends on whether, in combination with other genes, it produces an individual that survives and reproduces, so that copies of it are passed to the next generation.

Dawkins describes individual organisms as 'survival machines', whose function is to produce progeny that carry copies of the genes that they contain. Thus, Dawkins' answer to the chicken and egg conundrum is that an individual organism is a gene's way of making more genes.

The Selfish Gene created a considerable stir on its publication. While some evolutionary biologists hailed it as the clearest exposition of neo-Darwinism, others attacked it, for a variety of reasons. A major area of criticism concerned the implication that all features of organisms, notably their behaviour, are genetically determined and therefore fixed, a view that disregards environmental effects, such as learning, that affect an individual's phenotype. Another kind of criticism concerned Dawkins' treatment of genes as if they were discrete entities with clearly definable effects, whereas we know that genes interact in a variety of complex ways. An organism is an expression of its entire genotype and cannot be viewed as a set of discrete phenotypic characters, each of which is the expression of a single gene. Dawkins has responded to these and other criticisms in his subsequent writings but they are not of immediate concern to the issues that we shall discuss in the rest of this Chapter.

One of the most important features of the selfish gene approach is the notion that natural selection acts on genes, since this appears to be a radical departure from Darwinism, in which selection is seen as acting on the individual. In subsequent writings, Dawkins clarified his views on this issue. The agents of selection, such as predation, starvation, disease and reproductive success, act only on individuals. There is no direct natural agent that can selectively take a particular gene out of one genotype and put it into another. There are, however, certain characters of individuals, the adaptiveness of which depends on their fitness consequences, not only for the individual showing them, but also for other individuals. For example, in several birds certain individuals assist others in breeding, apparently to the detriment of their own breeding effort (Section 6.3.4). If such characters have a genetic basis, we cannot understand their evolution by considering their effects on the survival and reproductive success of individuals in isolation. These characters relate to certain aspects of social behaviour in animals and, in the remainder of this Chapter, we look at a number of examples of such behaviour and consider the extent to which an understanding of their genetic basis is essential to understanding their evolution.

Before we look in detail at some examples of social behaviour, it is important to emphasize that the phenomena that will be discussed are very specific and occur in only a very small proportion of animals. You may feel that the discussion is somewhat esoteric in relation to evolution as a whole. In a sense this is true, but the issues raised are important outside the specific context of social behaviour, for two reasons. First, the material that follows addresses the question of how certain patterns of behaviour evolved and how they are maintained; that is, in what way are they *adaptive*? This Chapter thus provides a 'case study' in the analysis of adaptation and shows just how much detail, in terms of measuring costs and benefits, is required to establish that a given character is or is not adaptive. Secondly, the question of whether natural selection acts on individuals or on genes is obviously a fundamental issue and is central to the neo-Darwinian approach to evolution.

Summary of Sections 6.1 and 6.2

At all levels of biological organization, from genes, through individual organisms, populations and species, to taxonomic groups, there is a sorting process over evolutionary time, such that some units at each level persist for longer than others. The theory of natural selection provides a mechanism that brings about sorting at the level of individuals. An important question is whether the same, comparable, or different processes occur at other levels. This Chapter is concerned with evolutionary change at the level of genes and individuals and is focused on particular forms of social behaviour in which behaviour shown by one individual has fitness consequences, not only for that individual but also for other individuals.

6.3 Kin selection

In this Section, we consider particular patterns of social behaviour, the evolution of which can be understood only in terms of the genetic relationships between individuals engaged in that behaviour. All the examples we will consider relate to reproductive behaviour. The general phenomenon we are seeking to explain is called **altruism**. Altruism is defined as a pattern of behaviour performed by an individual that decreases the fitness (i.e. the survival and reproductive success) of that individual, but which increases the fitness of one or more other individuals. (It is important to note that altruism is defined purely in terms of its consequences; there is no implication that an animal performing altruistic behaviour is conscious of being altruistic or that it has an intention to be so.) As so defined, altruism poses a clear challenge to the theory of natural selection: if such behaviour *decreases* an individual's fitness, how can it evolve?

The great British geneticist J. B. S. Haldane is said to have spent some time in a pub one evening making calculations on an envelope, eventually announcing that he would be prepared to lay down his life for the sake of two brothers or eight cousins. The calculations he was making and the conclusion that he reached were his attempt to explain altruistic behaviour in terms of what has since come to be called *kin selection.*

6.3.1 PARENTAL CARE

The most obvious and widespread example of behaviour in which one animal gives aid to another is parental care. In many species, one or both parents defend, feed and generally care for their offspring. There is abundant evidence that by doing so they are taking risks, and are using up energy and resources which they might otherwise use for their own survival. Thus parental care decreases the survival of individual parents. It is also obvious, however, that, by caring for their offspring, parents are increasing the survival prospects of their progeny. On the face of it, the evolution of parental care poses no challenge to the theory of natural selection, as long as we bear in mind that the fitness of an individual is measured not simply in terms of that individual's survival, but also in terms of the number of its offspring that survive. If parental care is necessary for offspring survival, it will be favoured by natural selection. Obvious as this may seem, it does not provide a sufficient explanation for the evolution of parental care. Such an explanation must account for three aspects of care-giving behaviour:

1 Individuals are typically selective in terms of the young in which they invest care. In most species, they care only for their own offspring. In the herring gull (*Larus argentatus*), for example, adults care for their own chicks but eat other chicks whenever they get the opportunity.

2 Individuals in certain species care for young that are *not* their own progeny.

3 The way that individuals allocate care to offspring can vary and, in consequence, their fitness can vary. As shown in Chapter 2, there will be an optimum number of young to rear; to attempt to rear more or fewer could lead to reduced fitness. This point may seem obscure at present but will become clear in Section 6.3.2 below.

6.3.2 THE COEFFICIENT OF RELATEDNESS

Each individual in a sexually reproducing species obtains half its alleles from its father and half from its mother. Thus the probability that an individual and one or other of its parents will have any given allele that is identical by descent is 50%, or 0.50. This value is called the **coefficient of relatedness**, abbreviated to ***r***. The value of r can also be calculated using the general equation:

$$r = n(0.5)^L \qquad (6.1)$$

where n is the number of 'routes' between the related individuals along which the particular allele can be passed (inherited); while L is the number of meioses, or generation links, between them. In the simplest case, of a parent and its offspring, $n = 1$ and $L = 1$, so $r = 1 \times (0.5)^1 = 0.5$ (Figure 6.1(a)). For a grandparent and its grandchild (Figure 6.1(b)), $n = 1$ and $L = 2$, so $r = 1 \times (0.5)^2 = 0.25$. Brothers and sisters (siblings) all inherit 50% of their alleles from each of their parents, but each sibling will not receive the same 50%. The probability that two siblings will inherit a particular allele from their father is $0.50 \times 0.50 = 0.25$. Likewise, the probability that they will both inherit a particular maternal allele is 0.25. By simply adding the paternal and maternal contributions, we find that the probability that two siblings will have any given allele in common by descent is $0.25 + 0.25 = 0.50$; thus $r = 0.50$. Alternatively, r can be calculated using Equation 6.1: $n = 2$ and $L = 2$, so $r = 2 \times (0.5)^2 = 0.50$ (Figure 6.1(c)). For an individual and each of its nephews and nieces, $n = 2$ and $L = 3$, so $r = 2 \times (0.5)^3 = 0.25$ (Figure 6.1(d)). (These calculations assume that there is no inbreeding. The offspring from matings between siblings, for example, would have an $r = 0.75$, but this is a complication that need not be considered further.)

▶ What will be the value of r for two half-sibs, e.g. individuals that have the same mother but different fathers?

$r = 0.25$. The probability that they will both inherit the same allele from their mother is 0.25, but the probability of sharing a paternal allele is 0. This value can also be arrived at using the general equation: here $n = 1$ and $L = 2$, so $r = 1 \times (0.5)^2 = 0.25$ (Figure 6.1(e)).

Now let us apply the coefficient of relatedness to the evolution of parental behaviour. Imagine an extreme case in which an individual parent defends its young against a predatory attack, as a result of which it dies and they survive. We

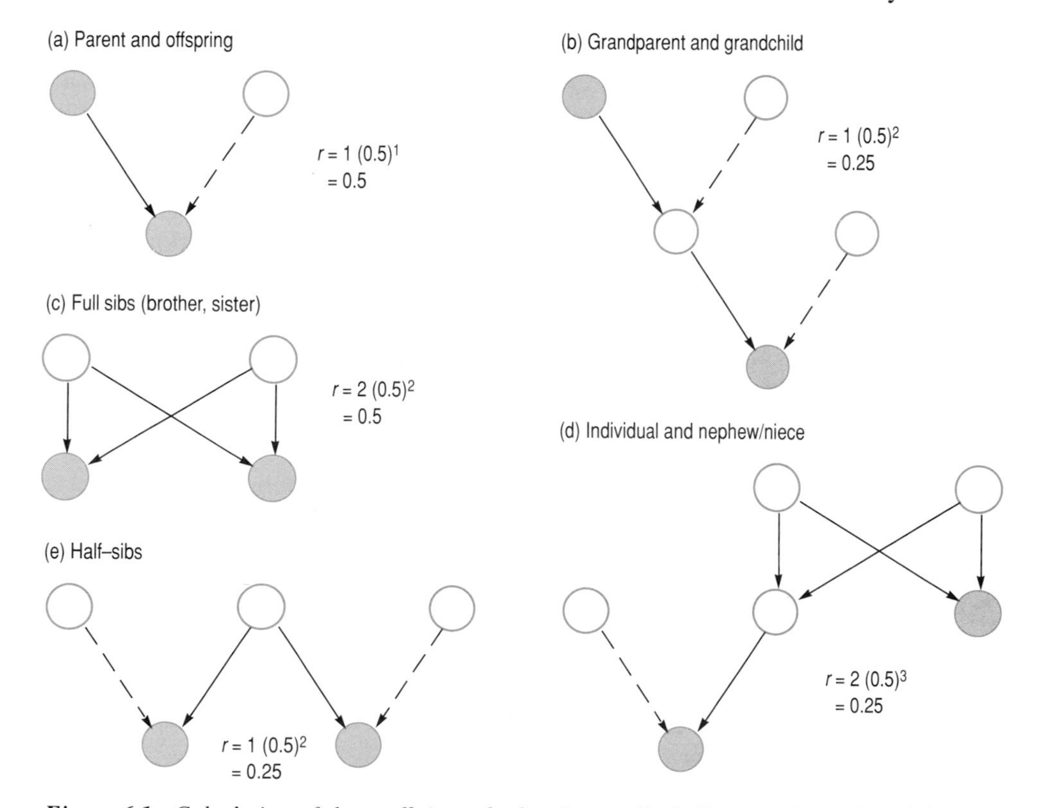

Figure 6.1 *Calculation of the coefficient of relatedness* r. *Each diagram shows (in solid green) the two individuals for which* r *is being calculated. Open circles denote other relatives. Solid lines are generation links used in the calculation, dashed lines are other links in the pedigree.*

need to define the conditions under which natural selection will favour individuals expressing an allele *A* that promotes such behaviour, rather than individuals expressing allele *a* that does not. When *A* individuals are attacked, they defend their progeny but die. The probability that the *A* allele will be passed on is 0.50 for each of its progeny, since this is the probability that each offspring will carry that allele. When *a* individuals are attacked, they do not defend their offspring; they survive but their progeny die. The probability that the *a* allele will survive is 1.0 in the parent and 0 in its offspring. We can now draw up a Table as shown in Table 6.1.

Table 6.1 *The spread of an allele favouring altruism in relation to the number of offspring produced.*

	Probable number of surviving alleles *A* or *a* if:	
Number of offspring	parent is altruistic (expresses *A*)	parent is not altruistic (expresses *a*)
1	$0.5 \times 1 = 0.5$	1.0
2	$0.5 \times 2 = 1.0$	1.0
3	$0.5 \times 3 = 1.5$	1.0
4	$0.5 \times 4 = 2.0$	1.0
	etc.	etc.

When there is only one offspring, the likelihood of *A* surviving is less than the likelihood of *a* surviving. Thus natural selection would favour *a* over *A*. When there are two offspring, *A* and *a* are equally likely to be passed on, and selection will not favour one over the other. However, for more than two offspring, *A* is at an advantage over *a* in terms of its chances of being passed on, and so will tend to increase in frequency as a result of natural selection.

This hypothetical example has been grossly simplified to illustrate the principle of **kin selection**, namely, that natural selection can favour behaviour that confers a disadvantage on the individual if it confers a benefit on that individual's kin. The strength of this effect depends on the closeness of the kin relationship. Extending the above example, if the beneficiaries of the altruistic act are grandchildren of an individual carrying *A* (in which case $r = 0.25$), *A* will be at a selective advantage over *a* only if more than four individuals benefit. This is a very important point. It shows how natural selection, which can readily explain the evolution of parental care, can be extended to explain behaviour directed towards more distant relatives.

Furthermore, the model is simplistic in that it expresses advantages and disadvantages in terms of certain survival or certain death. In nature, the costs and benefits of altruistic behaviour will not be so absolute. Altruism will change the chances of survival and breeding (fitness) of altruist and beneficiaries in a probabilistic way. This does not alter the principle of the argument; it merely complicates the mathematics.

It must be emphasized that this theoretical discussion should not be taken to imply that altruistic behaviour, or parental care, is determined by a single gene. A major criticism of Dawkins' *The Selfish Gene* is his frequent use of language that implies that there are 'genes for characters'. In reality, a complex pattern of behaviour such as caring for young will be controlled by many genes. The model that we have just developed should be seen as relating to alleles that increase an individual's propensity to behave altruistically.

Let us return to the three features of parental care that were listed at the end of Section 6.3.1, and see how our perspective on these has been changed by thinking in terms of genes:

1 That individuals are typically selective in terms of the young in which they invest care is readily explicable. Individuals carrying genes that cause them to care for the progeny of others, to which they are not related, will not, by so doing, pass those genes on to the next generation.

2 That individuals in certain species care for young that are not their own progeny is explicable, if, as in the case of individuals and their grandchildren, the beneficiaries are genetically related to them. We shall return to this topic in Sections 6.3.3 to 6.3.5.

3 The evolution of caring for progeny clearly depends on the nature of the benefits to the young, the cost to the care-giver and the number of progeny involved. In the example given, the cost to the carer is high, and the benefit to the beneficiary is high and more than two progeny must benefit if such behaviour is to evolve.

► Suppose that the benefit to the progeny is reduced, so that self-sacrifice by a parent in the face of a predator only statistically increases the survival of the progeny (i.e. the effect occurs on some occasions but not on others). What do you suppose this would do to the number of progeny that must benefit if the behaviour is to be favoured by natural selection?

If the survival of the progeny becomes probabilistic, then their survival prospects become less than 1. As a result, the critical number of offspring that must benefit (more than 2 in the above example) will increase. Likewise, if we reduce the risk to the parent, by assuming that it faces a risk of death rather than certain death, *A* becomes more likely to be favoured at the expense of *a*.

6.3.3 TASMANIAN NATIVE HENS

The Tasmanian native hen (*Tribonyx mortierii*) is a species of bird in which males outnumber females by about three to two. Breeding groups may consist of pairs, trios and, in a few instances, quartets and quintets. Whatever the group size, only one bird in a group is a female. All members of a breeding group participate in defending a territory against neighbouring groups, in incubating the eggs and in caring for the young. In groups with several males, one male is clearly dominant over the others, but he does not prevent them from mating with the female. This raises the question: why does the dominant male not drive out the other males and so ensure that he alone mates with the female?

At least two factors are involved. First, the presence of one or more subordinate males enables a group to defend a larger territory. As a result, a female in a multi-male group has a higher reproductive success than one in a pair. The second, crucial, factor is that, where two or more males share a female, they are brothers. Thus, though some of the young produced in a multi-male group are not the direct offspring of the dominant male, they are closely related to him and thus will inherit some proportion of his genes.

We can appreciate the significance of these two factors more clearly by means of a simple arithmetic exercise. In this we consider the question: under what conditions will an allele favouring tolerance of other males in a breeding group be selected for? Suppose that there are two males, A and B, and the following outcomes:

Male A carries allele *t*. He excludes other males.	Male B carries allele *T*. He tolerates the presence of a subordinate male, but allows him to perform only a proportion of all the matings with the female. (In nature, a single subordinate male achieves about one-third of all the matings with the female.)
Male A and his single female defend a relatively small territory.	Male B, his subordinate male and their shared female defend a larger territory than male A.
The female raises 4 chicks.	The female raises 6 chicks, of which 4 are fathered by male B, 2 by the subordinate male.

To calculate the probability with which males A and B will pass on their respective alleles, *t* and *T*, we need to incorporate the coefficient of relatedness between parents and their offspring.

▶ What is the value of r for parents and offspring?

$r = 0.5$. Let us first assume that the subordinate male tolerated by male B is *not* related to him ($r = 0$). Then:

The probable number of offspring that will inherit allele *t* from male A $= 4 \times 0.5 = 2.0$.	The probable number of offspring that will inherit allele *T* from male B $= (4 \times 0.5) + (2 \times 0) = 2.0$.

▶ Is one allele more likely to be selected for than the other?

No; the probability of inheritance of each allele is the same.

▶ Continuing under the assumption that the subordinate male is *not* related to male B, can you think of two conditions under which *T* would be more likely to be passed on than *t*?

The two conditions are:

1 The trio (male B, subordinate male and shared female) produces more than 6 chicks so that, even if male B still only fathers two-thirds of them, he fathers more than 4.

2 Male B restricts the subordinate male's sexual access to the female to less than one-third of the matings, so that he fathers more than 4 of the possible 6 chicks.

Now let us keep the original conditions the same, but suppose that the subordinate male in the trio is a brother of male B. Thus, the offspring that are fathered by the subordinate male will be nephews and nieces of male B.

You will recall that the value of r between an individual and its nephews and nieces is 0.25. We can now recalculate the probable number of offspring carrying the different alleles in the two breeding groups:

The probable number of offspring that will inherit allele t from male A = $4 \times 0.5 = 2.0$.

The probable number of offspring of male B that will inherit allele $T = (4 \times 0.5) + (2 \times 0.25) = 2.5$.

▶ Is one allele now more likely to be selected for than the other?

Yes; allele T is passed on with greater frequency than allele t. This example enables us to define clearly the conditions under which tolerance of another male in a breeding group will be favoured by natural selection:

1 The presence of another male must enhance the breeding success of the female, and thus of the dominant male.

2 If males are unrelated, only a proportion of the increase in breeding success can be allowed to accrue to the additional male.

3 The strength of this second condition is relaxed if males sharing breeding access to a female are genetically related.

So far, our analysis of the breeding system of the Tasmanian native hen has considered only the interests of the dominant male, and has examined the conditions under which he should tolerate a subordinate. We must also consider, however, the subordinate male and consider the conditions under which it will be adaptive for him to join a dominant male. As stated earlier, males outnumber females in the Tasmanian native hen population, so that it is not possible for all males to have exclusive access to a female. A subordinate male therefore has two options: to join a dominant male and form a trio, or not to breed at all. By joining a dominant male, he does gain some sexual access to a female and fathers some progeny of his own. This is sufficient benefit for his behaviour to be favoured by natural selection.

▶ What additional factor will make his behaviour adaptive?

If he is the brother of the male that he joins, and is therefore likely to share with him an allele that promotes female sharing, there is a reasonable probability that the progeny that are fathered, not only by him but also by the dominant male, will also carry that allele.

It is important to note that, in considering the genetic basis of kin selection, we are concerned only with those genes, such as the hypothetical tolerance gene T, that are involved in the particular behaviour with which we are concerned, in this case tolerance. Being brothers, male B and his subordinate companion will have many alleles in common, but these are not relevant to the argument.

Kin relationships are also important in the evolution of the social behaviour of other animals. A pride of lions, for example, consists of a number of females and their cubs, defended against rivals by a small group of males. From time to time a group of pride-holding males will be ousted by a coalition of males that have previously not held a pride. Such coalitions often consist of full brothers, or at least half-brothers. Like Tasmanian native hens, they share in the defence of their pride and also share sexual access to the females in the pride. It is significant in the context of kinship that, when a group of males take over a pride, they kill all the cubs that are still being suckled by the females. These are unrelated to the males and their death hastens the time at which the females will bear cubs that are the offspring of the new males. Breeding coalitions among close relatives that bear some similarity to those observed in native hens and other animals also occur in some human societies.

6.3.4 COOPERATIVE BREEDING

Section 6.3.3 analysed a social system in which individuals, in that case adult males, shared mates. We saw also that, because the males are genetically related, such a system can be explained by kin selection. In a number of other birds, the situation is more complex, because it involves cooperation in breeding by several individuals, at least some of which are not genetically related, either to the adults with which they cooperate, or to the progeny that are reared by their efforts. The analysis of such cooperative breeding systems focuses on certain individuals called 'helpers at the nest', a pattern of behaviour that has been described in over 150 species of bird. What are the selection pressures that have led to the evolution of such behaviour?

Example 1 Grey-crowned babblers The grey-crowned babbler (*Pomatostomus temporalis*) is an Australian bird that has been studied by J. L. Brown and his co-workers (1978). Breeding pairs may be assisted by anything up to eight helpers, or by none at all. This variation between nests in the number of helpers makes it possible to measure quantitatively the role played by helpers in the total breeding effort at nests. A group at a nest consists of two parents, their chicks and a variable number of helpers. The parents and the helpers all bring food to the nestlings. To understand the evolution of this system, we have to analyse, separately, the benefits that accrue to three categories of individuals: the parents, the nestlings and the helpers.

▶ Which of these categories do you suppose most obviously benefits from the helpers, and how?

The nestlings appear to be the most likely beneficiaries since, if they are fed by several birds, they should receive more food. The parents should also benefit because the effort that they individually have to invest in feeding the young will be enhanced by the contribution of the helpers; as a result, they may be able to rear more young.

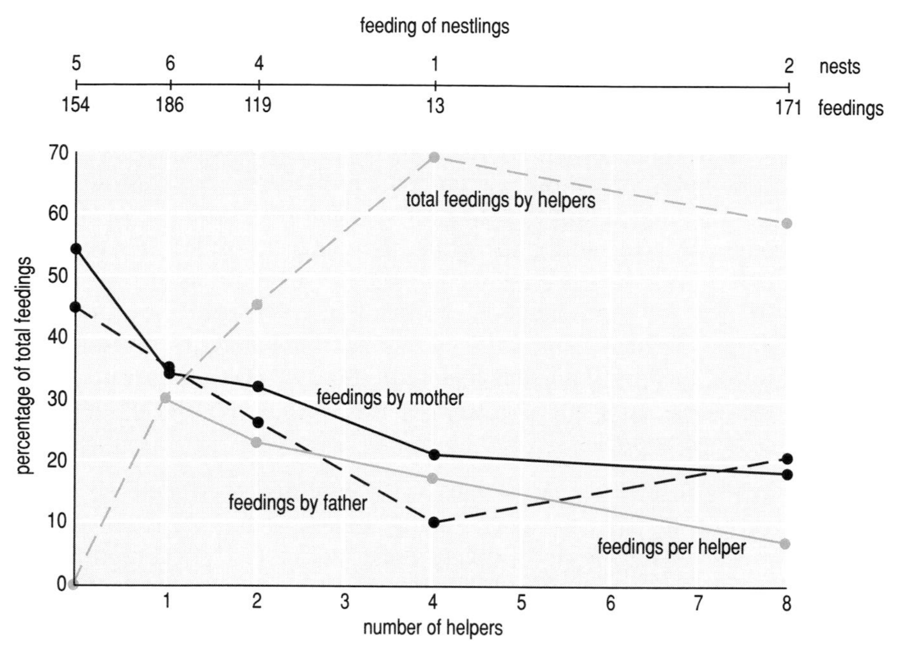

Figure 6.2 *Feeding of nestling grey-crowned babblers by helpers, the father and the mother. The number of nests watched and the total number of feedings at those nests are shown on the scales at the top of the Figure, corresponding to the number of helpers (bottom scale).*

In fact, detailed observations of the quantity of food that is brought by each adult to a nest reveal that the first of these suppositions is incorrect. All broods receive more or less the same amount of food, regardless of the number of adults involved in feeding them. The major benefit of having helpers falls to the parents, which have to bear a smaller proportion of the total feeding effort when they have helpers. Study Figure 6.2 carefully.

▶ What proportion of the total feeding effort is carried out by the two parents at nests where there are (a) no helpers and (b) four helpers?

Where there are (a) no helpers, the male carries out 45% and the female about 55% of the feeding effort. Where there are (b) four helpers, the male carries out 10%, the female just over 20% of the feeding effort. At nests where there are one or two helpers, the feeding effort of the parents is reduced by smaller amounts. It is thus clear that parents benefit considerably by having helpers, since they have a reduced feeding load.

▶ Do (a) the parents and (b) the helpers benefit when there are eight, as compared to four helpers at a nest?

For the parents (a), no, there is no benefit; the female brings roughly the same amount of food, the male a lot more. For the helpers (b), yes, there is considerable benefit; at nests with eight helpers, each carries out 7% of the total feeding effort; at nests with four helpers, each contributes 17%.

Having considered the data, we can return to the point made earlier, that it is necessary to consider separately the benefits gained in a cooperative breeding system that accrue to all categories of individual involved. In the case of the grey-crowned babbler, one category of individual, parents, clearly benefits to varying extents, depending on group size. Parents benefit from the activities of helpers, by being able to reduce the effort they expend in feeding their young. The time and energy that they save enables pairs to start a second brood, greatly increasing their fitness. Another category, the nestlings, apparently do not benefit at all. Individual helpers benefit from the presence of other helpers, in that they have to bear a lower feeding load, but this begs the question of why it should be adaptive for them to help at all. We shall return to the question of why helping may have evolved shortly.

While nestlings, on the basis of the evidence considered so far, apparently receive no benefit from the presence of helpers, it is important to point out that the data presented in Figure 6.2 were collected in just one breeding season. It is possible that in other seasons food might be less abundant, in which case helpers might significantly increase the total amount of food brought to a nest. The question of whether helpers can actually make a significant contribution to the breeding success of a group was addressed experimentally by Brown *et al.* (1982). They selected 20 groups of approximately the same size and occupying comparable territories. They then caught and removed all but one helper from 9 groups, leaving the other 11 intact. The intact groups contained four, five or six helpers. The breeding success of all groups was then monitored (Figure 6.3).

▶ Do these results provide evidence that helpers contribute to the reproductive success of a breeding group?

Yes, they do. The average number of chicks ($\bar{x}$) reared by the intact groups is three times that of the groups from which helpers had been removed. The most likely reason for this is that groups containing many helpers are more effective at detecting and driving off predators than are smaller groups.

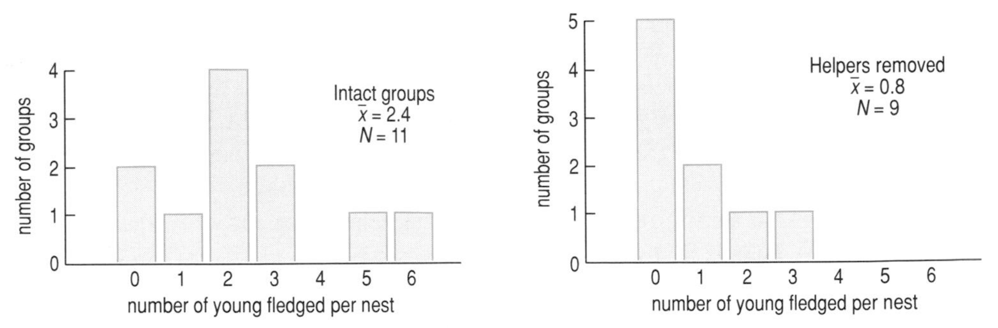

Figure 6.3 *The effect of the removal of helpers on the reproductive success of breeding groups of grey-crowned babblers. There were 11 intact groups (left) and 9 groups with helpers removed (right)*

This brings us to the puzzling question of how the helpers might benefit. On the face of it, they seem to be putting themselves at a disadvantage by expending time and energy caring for young that are not their own offspring.

▶ What factor might explain the behaviour of helpers?

It will benefit individuals to help at nests where the nestlings are genetically related to them. In other words, helping at the nest may have evolved through kin selection. In the case of the grey-crowned babbler, the degree of kinship between parents and their helpers is known in many instances; many helpers are the earlier offspring of the two parents. In most bird species in which nest-helping has been studied in detail, it has been found that parents and helpers are close relatives. In the European moorhen (*Gallinula chloropus*), for example, a pair will often produce two, occasionally three, broods during a breeding season. The chicks from the first brood will help their parents to rear the second brood, primarily by feeding them. Essentially, most groups that show cooperative breeding behaviour are extended families.

While the fact that helpers are usually genetically related to the individual breeding birds that they help suggests that helping has evolved through kin selection, there are several advantages that may accrue to helpers which are not dependent on the closeness of their genetic relationship to the breeding birds. First, helpers are usually young birds that will become breeders later in life, and their experience as helpers may enable them to acquire skills that will subsequently make them more effective parents. Secondly, in most communally breeding species there is a dominance hierarchy among members of a group, with the breeding pair being dominant over the helpers. When one member of a breeding pair dies, its position is typically taken by the next most dominant individual of the same sex. Thus, to become a breeding bird, a young individual may have to work its way up a hierarchy of helpers. Thirdly, the presence of helpers increases the breeding success of a group so that an individual that acts as a helper may be contributing to the success of a group in which it may eventually become a dominant, breeding bird. The larger the group when it becomes a breeder, the more offspring it may produce. Factors such as these will reinforce the action of kin selection in the evolution of patterns of cooperative breeding behaviour.

It is important in considering the various factors involved in cooperative breeding to make a distinction, in relation to the behaviour of individuals, between *staying* in a group and *helping*. The fact that birds that have belonged to a group for some time may eventually acquire breeding status makes it adaptive for them to stay in a group rather than attempting to breed independently, but it does not necessarily make it adaptive for them to help. Two factors may be involved here. First, kin selection will favour helping if birds that stay are related to the young. Secondly, birds that stay, but do not help, are, in some species, not tolerated by other birds in the group and are driven away. Thus there is a 'payment principle' by which birds are 'allowed to' stay in a group in return for their helping efforts.

The various factors involved in the evolution of cooperative breeding can be considered under two headings: genetic predispositions and ecological constraints. Genetic predispositions refers to the argument, developed above, that the principle that it is adaptive for individuals to care for their immediate progeny can be extended to other categories of relative. Ecological constraints refers to certain environmental features that are commonly associated with cooperative breeding. The most important of these is that, in a population, there are limited opportunities to breed. This may be because there is a shortage of suitable nest sites, because predation on nests is very intense, or because there is a shortage of food. In such a situation, an individual may face two options: to join and help an established breeding pair, or to seek a mate and set up a breeding group of its own. If conditions for breeding are very hostile, if prior experience of rearing young significantly increases breeding success, or if staying in a group eventually leads to acquiring breeding status within an established group, it may be more advantageous to an individual to stay in a group than to breed independently, whether or not it is genetically related to the young that it helps to rear. When a high-ranking breeding individual dies, for example through predation, its position is generally filled by one of the helpers within the group.

It is important to note that these two factors, genetic predispositions and ecological constraints, do not represent alternative or mutually exclusive explanations for cooperative breeding. Rather, their influence is additive. Thus, if ecological conditions are such that it is more adaptive for an individual to stay than to leave and attempt to breed independently, it will further increase its fitness if, while it stays, it joins a group in which it helps its relatives. The interplay between these various factors is illustrated by the next example.

Example 2 Florida scrub jays The Florida scrub jay (*Aphelocoma coerulescens*), studied extensively by Glen Woolfenden and John Fitzpatrick (1984), lives in areas where habitat suitable for breeding is scarce and patchily distributed. These are just the kinds of conditions that may make it impossible for all individuals to breed, an example of an ecological constraint favouring cooperative breeding. Breeding pairs defend a territory throughout the year; about 50% of pairs have helpers, on average 1.8 per pair. The helpers, generally birds of one or two years of age, contribute to the breeding effort of their group in two ways. First, they warn the parents and the chicks of the presence of predators and they attack such predators. This is the major cause of the enhancement in breeding success shown by pairs that have helpers (see below). Secondly, they help to feed the young, providing up to 30% of the food brought to the nest. This does not contribute to an improvement in breeding success because, as in the previous example (Figure 6.2), it does not necessarily lead to an increase in the overall amount of food brought to the nest. Rather, because it decreases the feeding load borne by the parents, it enhances their survival; parents at nests without helpers have a 77% probability of surviving to the next year, parents at nests with helpers, 85%. The data collected by Woolfenden and Fitzpatrick enabled them to answer a number of critical questions:

1 Are helpers genetically related to the individuals that they help? The great majority of helpers are genetically related to the breeding pair and, therefore, to their chicks. Of 165 helpers whose kinship was known, 64% were the earlier progeny of the breeding pair, and thus full siblings of the nestlings. 24% were the progeny of one of the breeding pair, the other parent having died and been replaced, and so were half-siblings of the nestlings. Only 4% were not related to the parents and nestlings. Woolfenden and Fitzpatrick point out an interesting feature of the scrub jay system, that helpers derive genetic benefits in two ways.

First, by helping to rear their own kin (siblings in this case) and, secondly, by enhancing the survival, and thus fitness, of their parents.

2 Do breeders benefit from having helpers? Data relating to this question are presented in Table 6.2.

Table 6.2 *The number of young successfully reared per nest at various categories of nest in the Florida scrub jay.*

	Nests without helpers	Nests with helpers
Inexperienced pairs	1.24	2.20
Experienced pairs	1.80	2.38

On average inexperienced pairs (those breeding for the first time) had 1.7 helpers, experienced pairs 1.9.

▶ What conclusions do you draw from these data in terms of the benefits derived from having helpers, and the previous breeding experience of pairs?

Helpers appear to increase considerably the breeding success of pairs, especially that of inexperienced pairs. Breeding experience clearly enhances reproductive success; both with and without helpers, experienced birds rear more young. To confirm that helpers do enhance the breeding success of a pair, it is necesary to perform an experiment in which some helpers are removed, because some or all of the effects shown in Table 6.2 could be due to a confounding variable, e.g. variation in the quality of the parents. Better parents will have more progeny and, if progeny become helpers, they will have more helpers.

3 Is helping a better option than being a breeder? By making various assumptions about the breeding success that might be expected for birds breeding for the first time, and using their knowledge of the genetic relationships between breeders and helpers, Woolfenden and Fitzpatrick calculated the expected genetic benefits of helping an established pair and of breeding. They concluded that it would be substantially more adaptive, in reproductive terms, for an individual to breed than to help. Why then did individuals help rather than breed?

▶ What possible answer can you give to this question?

An obvious possibility is that there are insufficient nest sites available, so that the breeding option is not actually available. Support for this argument came from observations made during the period 1979–1980, when nearly half the adult population died of an unidentified disease. In the following year, most of the young birds set up nests in the territories vacated by the deceased breeding birds and very few remained at their natal nests as helpers. This suggests that, if the option is open to them, young birds will breed rather than be helpers.

4 Does helping enhance an individual's future breeding success? In the scrub jay, the size of the breeding territory defended by a breeding group is positively correlated with the size of that group. This results, in part, from a positive feedback process; helpers enhance breeding success which produces more helpers,

and so on. Eventually a large group acquires such a large territory that it becomes possible for a second nest site to be established within it; eventually this becomes the focus of a territory in its own right. This new nest and territory is defended by a breeding pair, in which the male was previously a helper in the original group. Thus, being a helper in a successful group can eventually lead to the acquisition of breeding status.

Before we leave the Florida scrub jay, there is a further aspect of its breeding biology that deserves mention and which relates to a topic discussed in Chapter 5. The fact that helpers remain at their natal nest and frequently become breeding birds there could have important genetic consequences.

▶ Can you identify what these are?

If breeding birds at a particular nest were themselves bred at that nest, they are likely to be close relatives, and so inbreeding might occur. In fact, in the Florida scrub jay, this does not occur because there is differential dispersal from the natal nest by males and females. Young males generally remain at their natal nest, as helpers, for longer than females and may inherit that nest as a breeding bird. Young females leave their natal nest earlier, either to become a non-related helper at another nest or to replace a breeding female that has died.

The relative importance of genetic predispositions and ecological constraints will be discussed in Section 6.3.7.

6.3.5 COOPERATIVE BREEDING IN SOME OTHER SPECIES

Cooperative breeding is not restricted to birds, but occurs also in some mammals and fishes and it occurs in its most elaborate form in some social insects, such as bees and ants. (In bees, breeding females (queens) are diploid, but males (drones) are haploid (Chapter 5, Section 5.4). As a result, the female offspring of a drone and a queen inherit identical paternal alleles (0.5×1) but a typical complement of maternal alleles (0.5×0.5). Therefore, $r = 0.5 + 0.25 = 0.75$.) In the black-backed jackal (*Canis mesomelas*), helpers are close kin of breeding individuals, but in another African mammal, the dwarf mongoose (*Helogale paerula*), some helpers are kin, others are not. Non-related helpers in dwarf mongoose groups provide just as much help as related helpers do, and detailed observations of this species have revealed three factors that maintain this system:

1 Individuals that do not provide help are evicted from groups by dominant, breeding individuals.

2 Helpers, whether related or unrelated, frequently eventually become breeding members of those groups in which they are helpers.

3 The young animals that are helped by helpers subsequently help those helpers when they become breeders.

In essence, what appears to happen in the dwarf mongoose is that unrelated individuals are accepted into groups if they provide help. A similar pattern has been observed in the anemone fish *Amphiprion akallopisos* and in some African kingfishers.

6.3.6 Conflicts of interest within cooperatively breeding groups

The examples discussed so far have emphasized the cooperative aspects of certain features of social behaviour and how these can be explained in terms of kin selection. It is important to remember, however, that selection acts upon individuals and that, within a social group, each individual will be subject to particular selection pressures relating to the role that it plays within the group. It will often be the case that there is a conflict of interest between individuals. For example, a helper that assists a breeding bird might, potentially, achieve higher fitness if, instead of helping, it could somehow take the place of the breeding bird. Conflicts of interest are most apparent in species in which, unlike the examples discussed so far, several females breed and share the same nest.

Example 1 Groove-billed anis The groove-billed ani (*Crotophaga sulcirostris*) is a Central American member of the cuckoo family that has been extensively studied by Sandra Vehrencamp (1978). A breeding group consists of one to four monogamous pairs that cooperatively defend a territory, lay their eggs in a single nest, and, to varying extents, share in the incubation of the eggs. Within this apparently cooperative system, competition between females is intense and takes the form of destroying eggs by tossing them out of the nest. The females in a breeding group show a dominance hierarchy. The most subordinate female is the first to start laying her eggs, visiting the nest at intervals of one or more days to add another egg. After a time, the next female up the hierarchy starts to visit the nest to lay eggs. At this time the dominant female does not lay eggs but does visit the nest periodically, frequently tossing out one of her subordinates' eggs. Eventually she starts to lay, at which time she no longer ejects eggs from the nest, thus ensuring that she does not destroy any of her own. Once all the eggs are laid, incubation begins, with all members of the group, male and female, making a contribution.

The effect of the dominant female's destruction of the subordinate females' eggs is to limit the size of the clutch that will eventually be incubated by the whole group. If a clutch is too big, it cannot be incubated efficiently and some eggs will fail to develop normally. By throwing out eggs laid by other females, the dominant female ensures both that the clutch will be of a reasonable size and that it will consist largely of her own eggs. What happens to the eggs laid by the various females in ani groups is summarized in Table 6.3.

Table 6.3 *The fate of eggs laid by female groove-billed anis, according to their status within breeding groups.*

	Order of laying		
	First female (lowest rank)	Second female	Third female (dominant)
Mean no. of eggs laid	7.0	6.3	5.8
Mean no. of eggs tossed out	4.0	2.5	0
Mean no. of eggs incubated	3.0	3.8	5.8

▶ Do female groove-billed anis in a group derive equal benefit from breeding in a group?

No, they do not. Low-ranking females expend a greater effort in eggs, but less than 50% of those eggs survive to be incubated; dominant females lay fewer eggs and all of them are incubated. The net result is that the dominant female has nearly twice as many eggs in the communal clutch as the lowest-ranking female. In view of the loss in reproductive potential suffered by subordinate females, the question arises as to why it should be adaptive for them to join groups at all; they might do better to breed elsewhere as part of a pair. The answer seems to relate to the fact that ani nests are subject to heavy predation and that a group of birds defends a nest much more effectively than a pair does. Detailed analysis has shown that, despite the loss of many of their eggs, subordinate females have slightly higher reproductive success in groups than they do if they breed monogamously. Thus, while dominant females enhance their own reproductive success at the expense of subordinates, they do not do so to such an extent that the group might split up.

Example 2 Acorn woodpeckers The breeding system of the acorn woodpecker (*Melanerpes formicivorus*) has been studied extensively in Carmel Valley, California by Walter Koenig and his co-workers (1987). A group of up to 15 birds defends a communal territory and feeds the young. A typical group contains two breeding birds of each sex and their one-year-old offspring. The breeding males in a group are brothers and the breeding females are sisters, but birds of each sex come from a different family and so are unrelated. In at least some groups the eggs of a given female are fathered by different males. As in the anis, females start laying at different times and some of the eggs of the first female to lay are removed by the female that starts later, despite their close kinship. The eggs, however, are not tossed away but are eaten in a communal feast in which even the female whose egg it is joins in. What this example shows is that, even when females are related, there can be conflicts of interest between them.

A major factor that maintains breeding groups and mitigates against individuals leaving to breed independently is that group members cooperate to construct and defend huge food stores. These consist of acorns and other seeds, collected in the autumn, and pushed into innumerable holes drilled in a tree (Figure 6.4). These communal 'larders' provide food for the woodpeckers at times of year when food becomes scarce.

6.3.7 Is kinship really important in the evolution of helping?

In all the examples discussed so far, emphasis has been placed on two factors in the evolution of helping behaviour: ecological constraints and genetic relationships between helpers, breeding birds and their progeny. The emphasis that has been placed on kinship has been criticized by Jamieson and Craig (1987), who suggest an alternative explanation. They suggest that individuals have evolved a 'provisioning rule' that causes them to respond to any chick within their territory that begs for food. Such a rule would be adaptive because it would ensure that birds would feed their own chicks, normally the only chicks in the territory. If the

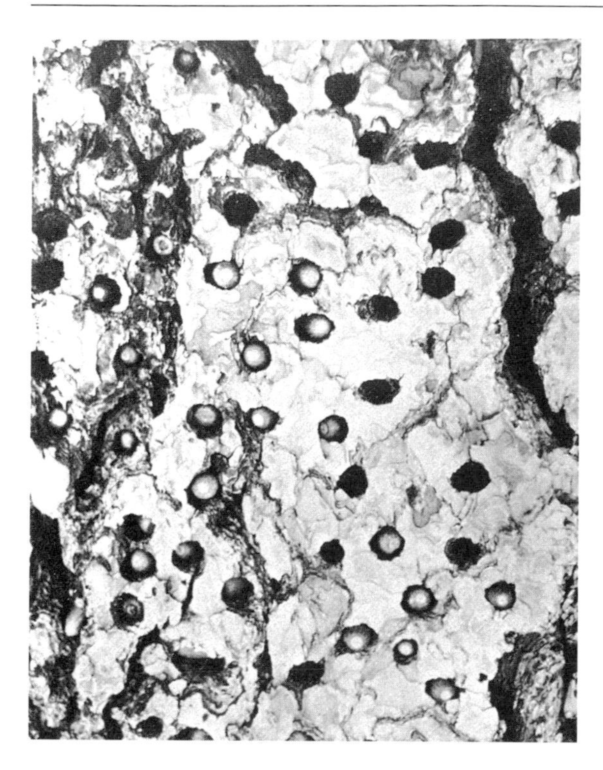

Figure 6.4 *Part of the communal larder of an acorn woodpecker group, consisting of numerous holes drilled in a tree, most of which are stuffed with acorns.*

environment is such that younger birds have no opportunity to breed, they stay in their home territory where this provisioning rule will cause them to feed their parents' chicks. This argument thus sees the helping of kin as a by-product of ecological factors and a response that has evolved in connection with parental care, and rejects the argument that helping has evolved through kin selection.

This alternative view raises an important issue that is of very general relevance in the study of evolution, the question of identifying characters or traits that are subject to selection. The kin selection approach sees helping kin as the trait whose evolution has to be explained. Jamieson and Craig see provisioning young as the important trait, not helping kin. They suggest that, by identifing helping kin as a functional unit of behaviour, we may be making the unwarranted assumption that there exists a mechanism by which birds preferentially help their kin. Their hypothesis makes no such assumption.

This alternative explanation for the evolution of helping has been tested, at least for some species, by certain observational and experimental data. Among white-fronted bee-eaters (*Merops bullockoides*), studied in Kenya by S. T. Emlen, non-breeding birds help at the nests of breeders, but individuals belonging to colonies in which they have no close kin show no helping behaviour.

▶ Does this observation support the Jamieson and Craig hypothesis?

No, it does not. It suggests that helping behaviour is performed specifically in response to the presence of kin. An experimental test of the Jamieson and Craig hypothesis has been conducted on acorn woodpeckers by Koenig and Mumme

(1987). They temporarily removed the dominant male from a number of breeding groups. In one set of trials they removed him during the egg-laying period, so that he could not father any of the chicks in his group, and returned him during the incubation period. On his return, the dominant male destroyed the entire clutch, causing the females to become sexually active again and to lay a new clutch of eggs. In another set of trials, the dominant male was removed during incubation; he did not destroy any eggs on his return.

▶ Do these results support the Jamieson and Craig hypothesis?

No, they do not. The dominant male's behaviour on his return to his group was dependent on whether or not the young in that group could be his own. If they could not, he destroyed the eggs. Koenig and Mumme carried out the same procedures with subordinate males. Regardless of the time at which they were removed, they did not destroy eggs on their return but helped to rear the young. Their behaviour appears, therefore, to be consistent with the Jamieson and Craig hypothesis. Why dominant and subordinate males should differ in this respect is not clear. It may be that the subordinate males are deterred from destroying eggs by the dominant male. Another possible explanation is that, because they are related to the dominant male, kin selection may favour subordinate males not destroying eggs even though none are likely to be their direct descendants.

SUMMARY AND CONCLUSION OF SECTION 6.3

From the information examined in Section 6.3, we can reach the following general conclusions abour cooperative breeding:

1 Helping other individuals to breed is commonly an option adopted by individuals that are unable to breed themselves, either because of ecological factors (scarcity of breeding resources such as food or nest sites) or social factors (a lack of available mates or low social status).

2 Helping behaviour benefits breeding individuals, most commonly by reducing the effort that they have to expend in parental care. As a result it may enhance their future reproductive success.

3 Helping behaviour does not always benefit the young directly.

4 Helpers may derive non-genetic benefits, such as gaining experience of breeding-related activities and acquiring a social position in a group that will enable them eventually to acquire breeding status.

5 Helpers are commonly, but not always, genetically related to the individuals that they help. Most commonly, helpers are previous progeny and thus full siblings of the young that they help to rear. Where helpers are related, they will derive genetic benefits.

6 Conflicts of interest may exist between individuals in cooperatively breeding groups. These conflicts are most marked where more than one female uses the same nest.

These conclusions have a critical bearing on the general question posed in this Chapter: what is the focus of selection? We have examined the factors that favour the evolution of a particular character shown by individual animals, i.e. helping others to breed. This character has fitness consequences, not only for those individuals that possess it, but also for other individuals. Because it incurs costs for the performer and imparts benefits to the recipent, helping is, at first sight, difficult to explain in terms of classical natural selection theory. As we have seen, however, the performer also derives benefits. These are of two kinds, genetic and non-genetic, that combine to determine whether or not helping will evolve. The concept, central to the theory of natural selection, that it is the individual on which selection acts, must thus be subtly changed in this context. Selection can favour behaviour by an individual that reduces the fitness of that individual, provided that it increases the fitness of a number of individuals that are closely related to it. Such selection is called kin selection.

6.4 KIN SELECTION EXPRESSED FORMALLY: HAMILTON'S RULE

The concept that selection will favour characters that are costly to the individual, provided that they confer some benefit on genetic relatives, was anticipated by R. A. Fisher, in the 1930s, and J. B. S. Haldane, in the 1950s. The theory of kin selection was developed and expressed formally by W. D. Hamilton (1964). He defined precisely the conditions under which an allele promoting an altruistic act will spread, as follows.

Suppose that we observe an interaction between an altruistic donor and a recipient, and that we can measure the costs and benefits accruing to each in terms of fitness (survival and reproduction). If the donor sustains cost C, and the receiver gains a benefit B as a result of the altruism, then an allele that promotes the altruistic act in the donor will spread in the population if

$$B/C > 1/r \qquad (6.2)$$

where r is the coefficient of relatedness between donor and recipient (Section 6.3.2). (The symbol > means 'is greater than'.) This formula can be re-written as:

$$rB - C > 0 \qquad (6.3)$$

This is called **Hamilton's rule**. It is essentially the same calculation that Haldane made to enliven his pub conversation. In that case, the gene in question is one that promotes self-sacrifice to enable relatives to live. The benefit B gained by each recipient is that they survive, and is therefore 1. The cost C to the donor is certain death, also 1. If the relatives are cousins, and because Haldane stipulated 8 of them, the net benefit is 8. To complete the calculation we need to know the value of r. This can be calculated by applying Equation 6.1 (Section 6.3.2), and is found to be 0.125. The net benefit is thus $(0.125 \times 8) - 1 = 1 - 1 = 0$. In other words, by sacrificing itself for eight cousins, an individual achieves exactly the same probability of passing on an allele that promotes self-sacrifice as it would if it did not perform that behaviour. Thus, to satisfy Hamilton's rule nine cousins would have to survive.

▶ Suppose that the relatives are nephews and nieces. How many of them would have to survive to satisfy Hamilton's rule?

The answer is more than 4, and is derived as follows. For nephews and nieces, $r = 0.25$. Thus the net benefit for 4 is $(0.25 \times 4) - 1 = 0$.

▶ Suppose that the effect of the self-sacrifice is not to ensure the survival of the recipient, but to increase its probability of survival, say to 50%. How does this effect the number of nephews and nieces that must benefit?

It increases it to more than 8. Making the benefit *B* probabilistic decreases its value, in this case to 0.5 for each recipient: hence $(0.25 \times (0.5 \times 8)) - 1 = 0$.

Self-sacrifice in order that others may live is a rather extreme form of altruism and one that is uncommon in nature. Much more widespread is the kind of phenomenon that we discussed in Section 6.3, in which an individual helps others to reproduce and, as a result, reduces its own reproductive success. To apply Hamilton's rule to these situations, it is convenient to express it in a different form:

$$B/C > r_{\text{donor to own offspring}} \,/\, r_{\text{donor to recipient's offspring}} \qquad (6.4)$$

Suppose that an individual has two options: to rear offspring of its own or to help its mother to rear offspring. The individual's own offspring and those of its mother both have $r = 0.5$ (assuming that the offspring and its mother's offspring are full siblings, which they will be, provided the mother has not changed her mate). Then Formula 6.4 becomes $B/C > 1$. Thus, kin selection will favour genes for helping the mother provided that, as a result of that help, the mother produces more additional offspring than the helper has lost through not breeding itself.

Hamilton's rule provides a very simple formula by which we can define the conditions under which kin selection will favour altruistic behaviour. It requires us to obtain values for just three variables: *B*, *C*, and *r*. To establish the value of *r* between two individuals requires the collection of very accurate genealogical data, so that we can be sure of the genetic relationship between two individuals. This is often not easy to do, especially where, for example, a female mates with more than one male so that we cannot determine the paternity of her offspring from social interactions alone. This problem is now being alleviated by the widespread application of the technique of DNA fingerprinting, by which the parentage of an individual can be determined precisely (Chapter 3, Section 3.5.2). To measure benefits (*B*) and costs (*C*) is not easy either, because they should be measured in terms of fitness. To measure fitness accurately requires data on the lifetime reproductive success of individuals. Such data, especially for long-lived species that breed many times in their lives, are not easy to obtain. In practice, we have to use measures of partial fitness (Chapter 4, Section 4.1.1), such as the number of young reared in a season, on the assumption that these make a significant contribution to overall fitness. This was the case in the worked examples in Section 6.3.4.

6.5 Communication in social groups

Some animals live in stable social groups in which individuals cooperate, for example in obtaining food, in defence against predators or, as we have seen in Section 6.4, in breeding. Such cooperation requires the exchange of information between individuals, and it is in social species that we find the most complex and sophisticated forms of animal communication. It is surely no coincidence that the species with perhaps the most highly ordered social systems, the honey-bee and humans, have two of the most complex communication systems known in the animal kingdom.

When considering communication in social species, we often find that it involves an exchange of information which brings benefit to the group as a whole. This is true of the honey-bee's dances, by which individuals inform the swarm as a whole about the location of rich food sources. This raises a problem when considering the evolution of such communication systems: why do individuals give information to others when they might, to their own advantage, keep it to themselves? The following discussion concerns a simple form of social communication and a number of theories that have been proposed to explain its evolution.

6.5.1 The hawk alarm call

Many small birds, like chaffinches and tits, spend part of the year in flocks, moving around looking for food. These are not stable, cohesive social groups, but may change in size and membership from hour to hour or day to day. It can be shown that individuals benefit from being in these flocks in at least two ways. First, individuals find food more readily, largely by being able to observe where other individuals find food. Secondly, flocks are quicker to detect predators, partly because several pairs of eyes are more likely to detect a predator early than one pair, and partly because members of a flock may collectively mob a predator. In addition, individuals that see a predator warn other members of a flock. When a bird sees a hawk flying nearby, it immediately flies into cover and gives a characteristic call. Other birds in a flock respond by flying into the nearest cover where they too may give the call. Figure 6.5 (overleaf) shows recordings of this call from five different bird species.

▶ What do you notice when you compare these five alarm calls?

They are remarkably similar in form. In all five species, the call lasts about 0.5 s and has a frequency of around 7 kHz. This similarity between species is evidence for powerful convergent evolution among these small birds, in contrast with the radiation that has occurred in their other songs, which are highly species-specific. These species are readily distinguished on the basis of their territorial songs, but even a trained ear cannot tell one hawk alarm call from another.

The hawk alarm call has a particular property. For purely acoustic reasons, it is very difficult to locate the source of a simple high frequency sound like the alarm call. This makes good sense in terms of adaptation. A hawk may hear the call but

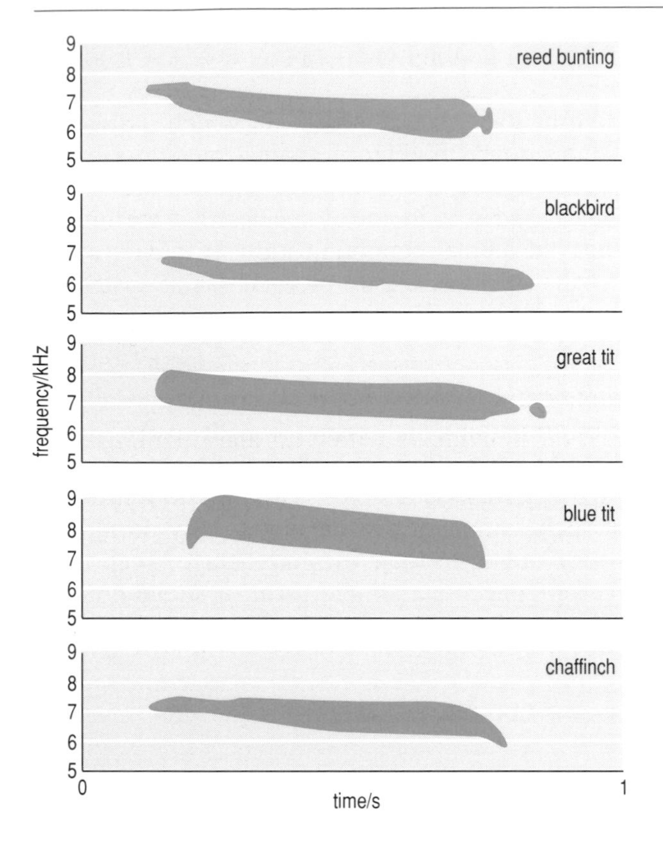

Figure 6.5 The hawk alarm call of five species of birds, shown as a plot of frequency ('pitch') against time.

will not be able to locate the bird that gave it. What appears to have happened during evolution is that several species, subject to the same selection pressure, have evolved similar, unlocatable alarm calls.

▶ How would selection have led to the evolution of unlocatable alarm calls?

By eliminating individuals that gave locatable calls; such individuals were, presumably, at risk of being caught by hawks. This suggests that giving alarm calls is potentially costly to the individual and poses the question that we have to answer: why do individuals show behaviour that is costly to themselves and which is beneficial to others? The behaviour is apparently altruistic. Three theories that have been proposed for the evolution of altruistic behaviour are now considered, using the hawk alarm call as an example.

6.5.2 Group selection

The theory of **group selection** recognizes the problem that the evolution of certain types of behaviour, such as altruism, is difficult to explain in terms of natural selection acting on individuals. Instead, it explains the behaviour in terms of advantages that accrue to a group of organisms. It is a necessary condition for group selection to operate that the species in question lives in small, isolated groups and that there is minimal or no gene flow between groups. If all individuals within a particular group show an altruistic behaviour pattern like the alarm call then, although each individual may be at a slight disadvantage in comparison with members of other, non-altruistic groups, this may be exceeded by the advantage that accrues to the group as a whole. Thus a group in which the alarm call is a feature may suffer a lower loss of members through predation than groups that do not give the alarm call. Groups in which alarm calling occurs will tend to increase in numbers faster than groups where it is absent, and so alarm calling will eventually become a feature of the population as a whole.

On the face of it, this argument seems quite plausible. It depends, however, on two important assumptions:

1 It requires that *all* individuals within an alarm calling group show that behaviour. Any individual that does not show the behaviour will not suffer whatever costs are incurred by calling, and so will be favoured by natural selection at the expense of those members of the group that do. Its progeny will thus tend to survive at the expense of the progeny of altruists.

If the alarm calling behaviour is based on a complex of genes, then it is necessary for all members of a group to acquire the same gene complex at the same time. It is highly improbable that this would happen. It could come about in exceptional circumstances (for example, when a small population consists of the progeny of a single female) but such events are probably very rare in nature.

Suppose that this condition has, somehow, come about. A population or group of altruistic animals is always open to invasion by individuals that are not altruistic, because the latter will have greater fitness. Non-altruists could arise within the group by mutation. If a population consists of individuals that behave in such a way that the population is open to invasion, through mutation or immigration, by individuals that behave in a different way, it is said not to be an *evolutionarily stable strategy* (Chapter 5, Section 5.4). We shall return to this in Chapter 7.

2 It requires that there is no gene flow between groups. Suppose that a situation has arisen in which all members of a group exhibit the behaviour. Any movement of individuals from elsewhere into that group will lead to the inclusion of non-altruists which, as argued above, will tend to flourish at the expense of the altruists. We know that it is a feature of group-living animals that there is continuous interchange of individuals between groups. Indeed, such movements are often interpreted as an adaptation that minimizes the deleterious effects of inbreeding as in the Florida scrub jay (Section 6.3.4). The takeover of prides by groups of male lions (Section 6.3.3) ensures that males and females in such prides are not closely related, so that inbreeding does not occur.

Thus, even if conditions do exist in which all members of a group share the same behaviour, this situation is not likely to last for long if there is movement of individuals between groups.

There is fairly wide agreement that group selection could theoretically operate, given that the two special conditions described above exist. However, because it is recognized that these conditions are very improbable in nature, group selection is a discredited theory, at least as an explanation for the evolution of altruism.

6.5.3 Kin selection

The basis of the theory of kin selection, as discussed in Sections 6.3 and 6.4, is the concept that an altruistic act may be favoured by natural selection if there is some resulting benefit to genetically close relatives of an individual performing that act. This is because close relatives are likely to share any genes that promote such behaviour, through common descent. The strength of this effect depends on how closely related individuals are, a variable that can be expressed precisely as the coefficient of relatedness r (Section 6.3.2).

Let us apply the theory of kin selection to the evolution of the hawk alarm call. If some individuals in a flock of birds are genetically related to the individual that gives the call, this will favour the evolution of the alarm call if the overall benefit (increase in fitness) to the kin, multiplied by the appropriate value of r in each case, is greater than the cost (reduction in fitness) incurred by the caller.

It is one thing to propose a process by which altruistic behaviour could have evolved. It is quite another to demonstrate that that process occurs, or has occurred, in nature. Evidence in support of the kin selection theory for the evolution of the alarm call would be if flocks consisted of closely related individuals. If members of flocks of chaffinches, great tits etc. were always unrelated, the kin selection theory would be untenable. Unfortunately, we know very little about the kin relationships within bird flocks, though such evidence as there is does suggest that some, but not all, members of flocks are genetically related. Data bearing on this question have, however, been obtained for certain mammals.

Belding's ground squirrel (*Spermophilus beldingi*) has been extensively studied in the USA by Paul Sherman. He found that individual females were more likely to give an alarm call in the presence of a predator if they had close relatives nearby. The best evidence for this kind of effect comes from a study of another colonial rodent, the American black-tailed prairie dog (*Cynomys ludovicianus*), by John Hoogland. Prairie dogs live in social groups called coteries that typically consist of one adult male, three or four adult females, and their offspring. Females remain in their natal coterie throughout their life; males leave to join another coterie in their second year. At a given time, a particular individual within a coterie may be unrelated to any other individual, or it may be a close relative of other individuals; those relatives can be grouped into two categories, descendent kin (offspring) and non-descendent kin (siblings, half-siblings, cousins etc.).

▶ Consider a two-year-old male that has just left his natal coterie and moved into a new one. What kind of genetic relationship will he have with other members of the coterie? How will this change as he gets older?

Initially, he will be unrelated to other members of the coterie. As he gets older and begins to breed, some members of the coterie will be his offspring, i.e. his descendent kin.

▶ Consider a two-year-old female. What kind of genetic relationship will she have with other members of the coterie? How will this change as she gets older?

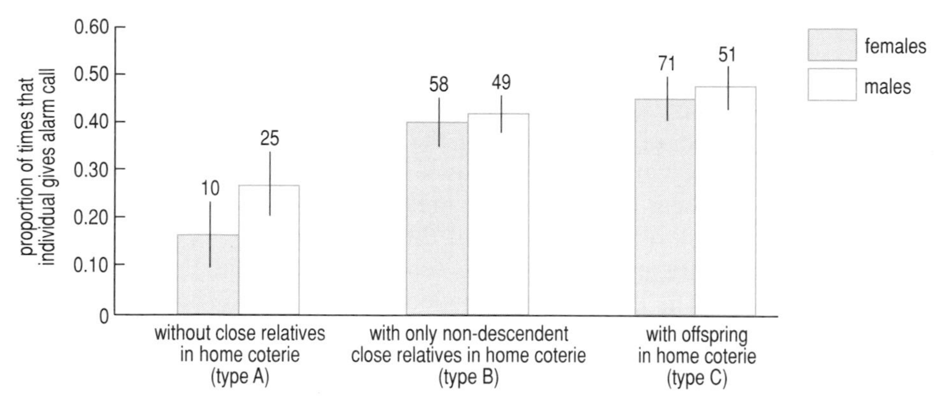

Figure 6.6 *The occurrence of alarm calling by black-tailed prairie dogs to a stuffed badger. The differences between type A and type B, and between type A and type C individuals are statistically significant; the differences between types B and C are not. The histograms give mean scores with standard errors (vertical bars) and sample sizes (numbers above error bars).*

Initially, the other females in the coterie will be her sisters and other non-descendent kin. As she gets older and begins to breed, some members of the coterie will be her offspring, i.e. her descendent kin.

Hoogland measured the responses of individual prairie dogs to a stuffed badger, badgers being a natural predator. The results of over 700 trials are summarized in Figure 6.6.

▶ Do these data support the hypothesis that alarm calling could have evolved because it benefits kin?

Yes, but to a limited extent. Individuals, especially females, call only rarely when they belong to coteries in which they have no kin (type A), but they call on about 40% of occasions when their coterie includes either descendent or non-descendent kin (types C and B respectively). The frequent occurrence of calling when there are only non-descendent kin is significant; it shows that individuals do direct behaviour towards relatives, just as Hamilton's rule predicts, and that warning relatives is not just a by-product of warning offspring. The fact that they call at all when there are no kin suggests that factors other than kin selection are involved in the evolution of alarm calling in this species. Hoogland suggests that there are direct advantages to an individual that alarm-calls (see Section 6.5.4).

6.5.4 INDIVIDUAL SELECTION

Faced with the problems that arise from attempts to explain altruistic behaviour such as the hawk alarm call in terms of group or kin selection, evolutionary explanations have been sought in terms of **individual selection**. The essence of this kind of argument is that an individual that gives an alarm call is, in fact, increasing its own fitness and is not at any selective disadvantage.

An animal that sees a hawk is in possession of two items of information: that there is a hawk in the area, and where that hawk is. When it makes an alarm call, it conveys only one of these items (that a hawk is nearby) to its fellow flock members. It does not pass on any information about the hawk's location. The response of birds that hear the alarm call is to fly into the nearest available cover. Not knowing where the hawk is, they will not necessarily fly in a direction that takes them away from the area of greatest danger. The result is that the birds in the flock all fly in different directions. It is a well established observation that predators are less likely to capture a prey animal when it is part of a group than when it is alone. This is especially true if the members of the group are moving in different directions. The predator is distracted and unable to concentrate its attentions on any one prey individual. This confusion effect has been demonstrated for a wide variety of predators, including fish, birds and mammals.

The argument for individual selection is that a bird that sees a hawk passes on only part of the information that it possesses and thereby creates a situation that is confusing to the hawk. It uses this to cover its own retreat which, since it alone knows where the hawk is, can be in a direction that takes it away from the danger area. Some of its fellows may actually be at a greater risk of capture because they may fly towards the predator.

The individual selection argument thus does not interpret the hawk alarm call as an altruistic act. Instead, it argues that individuals manipulate the behaviour of their fellow flock members and so decrease their own risk, relative to that of the rest of the flock.

Another possible explanation for alarm calling has been suggested; this sees the call as communication, not only between potential prey individuals, but also between the caller and the predator. A predator's best chance of making a successful attack is to catch its prey, such as a prairie dog, unawares. The alarm call effectively signals to the predator that it has been detected by at least one potential victim and that its attack is less likely to be successful. As a result, the predator may move away and seek a meal elsewhere.

6.5.5 WHICH THEORY IS CORRECT?

We have discussed three theories for the evolution of alarm calling. Can we now say which theory accounts best for the phenomenon? In trying to answer this question we are at a severe disadvantage, one that applies to many evolutionary questions and especially those that concern behaviour: there is no fossil evidence of ancestral stages in the evolution of the character we are seeking to explain. Our task, therefore, is to explain how selection will act to *maintain* such a character in an extant population. What we can do is to compare the three theories and ask what assumptions each makes and whether or not these assumptions are valid. The theory that makes the fewest unverified assumptions is, arguably, the one we should accept.

The group selection theory rests on two important assumptions: that animals live in genetically isolated groups, and that all individuals within a group are genetically identical with respect to the behaviour pattern in question. As we have already argued, both conditions are likely to occur only in very exceptional circumstances. For this reason, this theory is generally rejected as an explanation for the evolution of cooperative behaviour. (In Chapter 5, Section 5.1.3, however, we saw that group selection *is* entertained as an explanation for the maintenance of sexual reproduction.)

▶ What major assumption does the kin selection theory make?

The major assumption is that at least some of the individuals in a flock are relatives of a bird that gives the alarm call. Before we could accept the kin selection theory, we would need to know whether or not individuals have relatives in their flock and, if so, that these relatives are sufficiently numerous for any benefits accruing to them to offset any survival cost to the caller. We would also need to know two more things. As we have seen, the strength of kin selection will decline as the coefficient of relatedness between altruist and beneficiary decreases. We therefore need to know *how* closely related flock members are. In addition, we need to know the relative magnitude of the benefits to the beneficiaries and the cost to the caller. Without going into mathematical models, it should be obvious that, if the cost to the caller is large compared to the benefits to relatives, kin selection will be less powerful than if the cost to the caller is small relative to the benefits to the kin.

The kin selection theory gives a plausible explanation for the evolution of certain kinds of social interaction, those that apparently involve altruism. However, before we can accept it, we must recognize that it depends on detailed supporting evidence of the kinds outlined above. Bear in mind the fact that, as in the analysis of cooperative breeding in Section 6.3, detailed study of social systems that appear initially to involve altruism turn out to be largely explicable in terms of the costs and benefits that accrue to individuals.

▶ What assumption does the individual selection theory make?

It makes three assumptions:

1 The caller does not pass on all the information in its possession. This seems to be consistent with what is observed in flocks under attack by hawks.

2 The predator is confused by many birds flying in different directions. This has been confirmed for many predators.

3 The benefits accruing to the caller are greater than the costs to it. This has not been tested.

▶ How might you test assumption 3?

One way would be to observe many hawk attacks on bird flocks and to record whether an alarm caller or the non-callers are more likely to be caught. The individual selection theory predicts that callers should be less likely to be caught than birds responding to the call.

It would be wrong to imply that the kin and individual selection theories are mutually exclusive. As shown in the discussion of cooperative breeding (Section 6.3.4), apparently altruistic behaviour may benefit the individual as well as conferring a benefit on genetic relatives. In effect, individual selection and kin selection may act in the same direction. Nor should we assume that the factors promoting such behaviour are the same for all species; individual benefits may be most important in one species, kinship effects in another.

6.5.6 Conclusion of Section 6.5

This discussion of alarm calling and of the three theories that have been proposed to explain its evolution may have left you wondering what is the point of the exercise, in view of our failure to arrive at a firm conclusion. When people first realized that the evolution of certain types of social behaviour, notably altruism, is difficult to explain in terms of traditional natural selection, they frequently asserted loosely that such behaviour patterns evolved 'for the good of the species' or of the social group. When this argument is analysed closely and expressed formally as the theory of group selection, it becomes apparent that it is untenable because of the assumptions it makes. Kin selection, on the other hand, provides a plausible explanation of such behaviour and is, in fact, no more than a formal extension of Darwin's argument that natural selection favours characters that benefit the individual *and* its progeny.

The general importance of kin selection theory is that it alters our view of what it is that natural selection actually acts upon. According to the theory, selection does not simply favour characters that increase individual fitness; it favours characters that increase the net fitness of individuals *and* of their kin. What is favoured by natural selection is not individual fitness but **inclusive fitness**; that is to say, fitness should be seen as a function, not of the individual and its immediate progeny, but of the individual and that component of its genetic make-up that it shares with its kin. Inclusive fitness is a measure of the reproductive success, not of the individual in isolation, but of the individual and its close relatives.

A recognition of the importance of kinship in the evolution of social behaviour has generated a great deal of research into mechanisms by which individuals recognize their kin. The kin selection argument for the evolution of altruism requires that individuals discriminate between kin and non-kin; it then becomes important to understand how they do this. Many studies have established that animals belonging to a wide range of taxonomic groups, including insects, fishes, amphibians, birds and mammals, can recognize their kin. In some instances this ability is based on experience; in essence, individuals recognize other individuals with which they were reared as their kin. Some animals can, however, recognize their kin without having had previous experience of them, suggesting a genetically determined ability to recognize kin.

Kin recognition has considerable importance outside the context of altruistic behaviour. As discussed in Chapter 5, the reproductive success of an individual can be profoundly influenced by the genetic relationship between it and those individuals with which it mates. Both extreme inbreeding and extreme outbreeding can lead to reduced reproductive success; if they are to be avoided, organisms must be able to discriminate between potential mates on the basis of kinship. Thus kin recognition in the context of breeding is important, not just to birds and mammals, but to all animals.

Before leaving this topic, it is important to sound a note of caution. The discussion of altruism has assumed that such behaviour has a genetic basis. This may well be true of helping behaviour and alarm calling in birds, but it should not be assumed that the many forms of altruism observed in humans and other animals necessarily have a genetic basis. Kin selection theory provides an explanation for the evolution of altruism in terms of genes for such behaviour. It does not follow that all forms of altruism have a genetic basis, and under artificial selection apparently non-adaptive behaviour can become established.

SUMMARY OF SECTIONS 6.4 AND 6.5

Hamilton's rule provides a formula for determining whether, in a given situation, altruistic behaviour will be maintained by natural selection. Whether altruism is adaptive depends on the magnitude of both the fitness benefit and the fitness cost to the performer, and of the closeness of its genetic relationship to the beneficiary. Alarm calling may be a form of altruism in which an animal that detects a predator has the option of whether or not to pass that information on to other individuals. The theory of group selection is untenable because it is based on unrealistic assumptions about the distribution of alleles in a population. Evidence that alarm calling is maintained by kin selection comes from ground squirrels, in which individuals show more calling if either descendent or non-descendent kin are present in their social group. Individual selection may also be important in the evolution of altruistic behaviour such as alarm calling, because individuals may gain an individual advantage by calling. All three hypotheses for the evolution of altruism are based on assumptions, and make specific predictions, that can be tested in nature. In the context of altruistic behaviour, natural selection is seen to act, not simply on the fitness of an individual in isolation, but on its inclusive fitness, a measure of fitness that includes its descendent and non-descendent relatives.

6.6 CONCLUSION

Understanding the evolution of altruistic behaviour, and the role of kin selection in its evolution, is important for a number of reasons:

1 We are able to explain the evolution, through natural selection, of behaviour which, at first sight, appears difficult to explain without recourse to other evolutionary mechanisms.

2 Natural selection, combined with kin selection (itself an extension of natural selection) provides sufficient explanation for phenomena that had been attributed to a quite different process, group selection, which envisaged characters evolving 'for the good of the species'. Group selection is shown to be based on unrealistic assumptions.

3 The detailed analysis of the evolution of altruism has very general implications for understanding evolution because it refines the concept of adaptation. Feeding young, for example, is an adaptation. At first sight, its adaptive value is obvious; it promotes the survival of the young. The fact that individuals may feed young that are not their progeny causes us to question the way that such behaviour is adaptive. The examples discussed in this Chapter have shown that helping, and other forms of altruism, involve costs and benefits, to both the performer and the beneficiary. The adaptive value of helping can only be understood by drawing up a detailed balance sheet of these costs and benefits.

The study of kin selection is severely limited by procedural and methodological constraints. As this Chapter has shown, to demonstrate whether or not a particular pattern of behaviour is adaptive, it is necessary to measure the reproductive success of individuals in great detail. With great persistence and devotion, biologists have managed to do this for a few birds and mammals, but there are many animals for which, for purely practical reasons, such analyses are impossible.

The cost–benefit approach to adaptation is continued in the next Chapter. There we look at certain characters, typically shown by males, that appear to be costly to individuals that possess them. Their evolution, and their adaptive value, can only be understood if we correctly identify the benefits that they confer.

REFERENCES FOR CHAPTER 6

Brown, J.L., Dow, D. D., Brown, E.R. and Brown, S.D. (1978) Effects of helpers on feeding of nestlings in the grey-crowned babbler, *Pomatostomus temporalis*, *Behavioural Ecology and Sociobiology*, **4**, 43–60.

Brown, J.L., Brown, E.R., Brown, S.D. and Dow, D.D. (1982) Helpers: effects of experimental removal on reproductive success, *Science*, **215**, 421–422.

Dawkins, R. (1976) *The Selfish Gene*, Oxford University Press.

Emlen, S. T. (1984) Cooperative breeding in birds and mammals, in J. R. Krebs and N. B. Davies (eds) *Behavioural Ecology. An Evolutionary Approach* (2nd edn), pp. 305–339, Blackwell Scientific Publications, Oxford.

Hamilton, W. D. (1964) The genetical evolution of social behaviour, *Journal of Theoretical Biology*, **7**, 1–52.

Hoogland, J.L. (1983) Nepotism and alarm calling in the black-tailed prairie dog, *Cynomys ludovicianus*, *Animal Behaviour*, **31**, 472–479.

Jamieson, I. G. and Craig, J. L. (1987) Critique of helping behaviour in birds: a departure from functional explanations, in P. Bateson and P. Klopfer (eds), *Perspectives in Ethology*, vol. 7, pp. 79–98.

Koenig, W. D. and Mumme, R.L. (1987) *Population Ecology of the Cooperatively Breeding Acorn Woodpecker*, Princeton University Press, Princeton, New Jersey.

Koenig, W. D. (1990) Opportunity of parentage and nest destruction in polygynandrous acorn woodpeckers, *Melanerpes formicivorus*, *Behavioral Ecology*, **1**, 55–61.

Maynard Smith, J. and Ridpath, M.G. (1972) Wife sharing in the Tasmanian native hen, *Tribonyx mortierii*: a case of kin selection? *American Naturalist*, **106**, 447–452.

Sherman, P.W. (1977) Nepotism and the evolution of alarm calls, *Science*, **197**, 1246–1253.

Vehrencamp, S.L. (1978) The adaptive significance of communal nesting in groove-billed anis, *Crotophaga sulcirostris*, *Behavioural Ecology and Sociobiology*, **4**, 1–33.

Woolfenden, G.E. and Fitzpatrick, J.W. (1984) *The Florida Scrub Jay*, Princeton University Press, Princeton, New Jersey.

FURTHER READING FOR CHAPTER 6

Emlen, S. T. (1991) Evolution of cooperative breeding in birds and mammals, in J. R. Krebs and N. B. Davies (eds) *Behavioural Ecology. An Evolutionary Approach* (3rd edn), pp. 301–337, Blackwell Scientific Publications, Oxford.

Sober, E. (1984) *The Nature of Selection. Evolutionary Theory in Philosophical Focus*, MIT Press, Cambridge, Mass.

Stacey, P. B. and Koenig, W. D. (eds) (1990) *Cooperative Breeding in Birds: Long-term Studies of Ecology and Behaviour*, Cambridge University Press, Cambridge.

SELF-ASSESSMENT QUESTIONS

SAQ 6.1 (*Objective 6.1*) Draw a diagram similar to those in Figure 6.1 and use it to calculate the coefficient of relatedness, *r*, for two cousins.

SAQ 6.2 (*Objective 6.2*) For each of the following statements, decide whether it is true or false. Give a brief reason for your answer.

(a) Helping behaviour will evolve only in situations in which parents cannot rear young on their own.

(b) Helping behaviour will evolve only if the helpers are genetically related to the young birds that they help to rear.

SAQ 6.3 (*Objective 6.3*) Suppose that a juvenile bird has the option of helping its mother to rear her next clutch of chicks or of breeding itself. Assume that any female can rear two chicks on her own, five if she has help. Will helping evolve through kin selection if (a) the mother has kept her previous mate and (b) the mother has changed her mate since the previous clutch?

SAQ 6.4 (*Objective 6.4*) Are theories for the evolution of alarm calling incompatible if they are based on (a) individual as opposed to group selection and (b) individual as opposed to kin selection?

Chapter 7

Sexual selection and mating systems

Prepared for the Course Team by Tim Halliday

Study Comment *This Chapter discusses a particular form of natural selection which produces marked differences in morphology and behaviour between males and females, and so is sometimes regarded as a distinct process. This process, called sexual selection, has attracted a great deal of attention, largely because it leads to the evolution of characters that are apparently maladaptive. The Chapter discusses how such characters could evolve and, in so doing, further develops the concept of adaptation that was discussed in Chapter 2.*

Objectives When you have finished reading this Chapter, you should be able to do the following:

7.1 Discuss how anisogamy leads (i) to differences between males and females in the way that reproductive effort is allocated to various aspects of reproduction, and (ii) to different degrees of variation in reproductive success in the two sexes. Discuss, too, how patterns of resource allocation in the two sexes are influenced by the nature of the environment and by the form of parental care in a particular species.

7.2 Explain what is meant by the mating system of a species and discuss how it is influenced by a number of features of the environment. Explain why monogamy is more common among birds than any other animal group.

7.3 Describe the factors that may cause one individual in a population to be polygamous, and another to be monogamous.

7.4 Explain what is meant by certainty of paternity and discuss how it can influence the mating system of a species.

7.5 Using examples, discuss how the mating system observed in a population represents a compromise between selection acting separately on males and on females.

7.6 Describe how elaborate male characters may have evolved through the effect of female choice and recall evidence that selection arising through male–male competition and through female choice has led to the evolution of conspicuous male characters.

7.7 Analyse and interpret data on variation in reproductive success among individuals in terms of how such variation has led to the evolution of specific sex-related characters.

7.8 Explain, with examples, what is meant by an alternative mating strategy and discuss how such a strategy evolves and is maintained in a population by natural selection.

7.1 INTRODUCTION

Charles Darwin was very aware that his theory of natural selection, as set out in *The Origin of Species*, failed to account for the evolution of an important group of characters, namely extreme characters present in only one sex, usually males. The male peacock's train in Plate 7.1 is one such example. Such characters seem to defy explanation in terms of the essentially utilitarian concept of fitness developed in his book, because they apparently are a serious encumbrance to males and are therefore likely to reduce their survival. To overcome this problem he proposed the theory of sexual selection, set out in *The Origin of Species* and developed more fully in his later book, *The Descent of Man and Selection in Relation to Sex* (1871). Sexual selection has an importance in the study of evolution beyond being an explanation for a particular kind of character, for a number of reasons. First, differences between the sexes are very widespread among animals and it is important to understand their evolutionary basis. Where sex differences are extreme, they can lead to the misidentification of individual specimens, particularly fossils, with their being placed in different species. Secondly, there are good reasons to believe that natural selection in relation to sex can lead to very rapid evolutionary change, and so sexual selection has relevance to questions about the rate at which evolution occurs. Thirdly, it has been suggested that sexual selection may lead to the formation of new species, a topic discussed in Chapter 9. Finally, because characters that evolve through sexual selection may *appear* to be maladaptive, it has been suggested that they could have led to the extinction of certain species, although there are grounds for questioning such an assertion.

The characters that will be discussed in this Chapter belong to a category called **secondary sexual characters**. These are characters in which the two sexes of a species differ, but they exclude the gonads, their ducts and associated glands (the primary sexual characters). In humans, secondary sexual characters include the mammary glands, the pitch of the voice and the distribution of hair over the body.

In many animal species there are marked morphological differences between the sexes, a phenomenon called **sexual dimorphism**. In some species, sexual dimorphism is very striking. Male elephant seals are about three times heavier than females while some male spiders have a body mass less than one hundredth that of the female. Size is not the only character that may differ between the sexes; in many species one sex, usually the male, is more brightly coloured, and has more elaborate plumage or specialized weapons, such as horns, that are less well developed or absent in the other sex. Figure 7.1 shows sexual dimorphism in a

Figure 7.1 *Sexual dimorphism in the ruff* Philomachus pugnax. *The female (a) is considerably smaller than the male. The male's plumage is similar to the female's in the winter (b), but becomes more elaborate, with the addition of head and neck plumes, in the breeding season (c).*

bird, the ruff, and illustrates two common features of sexual dimorphism. First, one sex, in this instance the male, is larger than the other and, secondly, the male possesses elaborate plumage that is present only during the breeding season.

▶ What does the seasonal nature of the male's elaborate plumage suggest about its function and its evolution?

The association with the breeding season suggests that it is related to reproduction. The fact that it is absent outside the breeding season suggests that it may impose some cost on males and so is developed only when it has a function to serve.

The evolution of sexually dimorphic characters that are related to sexual behaviour has been the subject of heated debate in evolutionary biology ever since Darwin addressed the issue. At the centre of this debate is the assumption that very elaborate and extravagant male characters, such as the peacock's train and the huge antlers of some deer, must impose a severe cost on males, in terms of their survival. Many of these characters make males very conspicuous and, in some instances, somewhat clumsy. It has been suggested, for example, that the Irish elk *Megaloceros giganteus* became extinct because its antlers became so large they weighed the animal down, although this is now considered unlikely (see Chapter 14).

Early proponents of Darwin's theory of natural selection saw the process as one by which characters that promote individual *survival* would be favoured and would change over successive generations. This was an essentially utilitarian view, summed up in Herbert Spencer's unfortunate phrase 'the survival of the fittest' (see Chapter 4). With such a view of evolution, it was obviously difficult for evolutionists to understand how characters, such as the peacock's train, could possibly evolve if they tended to reduce an individual's chances of survival. Darwin's solution to the problem, as set out in his books in 1859 and 1871, was to suggest that characters promoting an individual's reproductive success would evolve even though they incurred a cost in terms of survival. He envisaged sexual selection as occurring in one or other of two ways. In species in which males fight for the possession of females, characters that enhance their fighting ability such as large size and specialized weapons will evolve. In species in which males must attract females by courting them, elaborate plumage and display behaviour will evolve. (We shall return to the validity and utility of this distinction later in this Chapter.) While most evolutionary biologists accepted the argument that fighting among males for females could lead to the evolution of morphological characters used in aggression, most rejected the notion that characters could evolve simply because they made males attractive to females, largely because it depended on females preferring certain males to others. As discussed later in this Chapter, this idea has now gained widespread acceptance and there is considerable evidence to support it.

Darwin's legacy has been the notion that there are two kinds of selection, typically operating in opposition to one another: sexual selection favours elaborate sexually dimorphic characters; natural selection opposes it. This is an odd state of affairs in view of the fact that Darwin made it abundantly clear that he regarded reproduction, rather than simply survival, as the critical factor in natural selection. As discussed in Chapter 4, the modern view of fitness places very heavy emphasis on individual reproductive success. When fitness is seen in these terms, it becomes illogical to regard natural and sexual selection as distinct and opposing processes. A character that promotes reproductive success will increase fitness, even if it imposes a survival cost, provided that the cost is not so great that it prevents the individual attaining breeding age. Sexual selection should be envisaged not as a distinct process but, to use Darwin's own words, as 'selection in relation to sex'.

The rest of this Chapter will consider a number of aspects of natural selection as it affects characters involved in reproduction, particularly mating. It will identify a variety of selective forces that are generated by sex, their consequences in terms of differences and relationships between males and females, and how the evolution of male characters through female choice can be explained. It will also review empirical evidence that sex-related characters like the peacock's train are adaptive, and discuss the biological significance of alternative mating strategies. Finally, it will return to the question of whether the distinction between natural selection and sexual selection is a valid one and will discuss, very briefly, the extent to which sexual selection occurs in plants.

7.2 Sexual competition and conflict

From a human perspective, sexual reproduction is widely regarded as a cooperative enterprise between males and females. Furthermore, the assumption is often made, implicitly or explicitly, that the 'goal' of reproduction is the perpetuation of the species. A cursory consideration of some patterns of behaviour associated with reproduction reveals that both these assumptions are false. In birds that breed in dense colonies, such as gulls, any nest that is left unguarded by both parents for just a few minutes is quickly raided by neighbours and the eggs destroyed. In elephant seal breeding groups, or in rookeries, males engage in vicious fighting that often results in serious injury. These are not the kinds of behaviour to be expected from animals seeking to perpetuate their species. Among praying mantids, it is not uncommon (at least in captivity) for the female to eat her mate as he copulates with her; this does not suggest that they are acting cooperatively. To understand the evolution of sexual behaviour, it is essential to focus on the individual animal, and on what is most adaptive for it, not for the pair or for the species. This Section looks at sexual behaviour from the perspective of the male and of the female, and considers to what extent their interests are similar and to what extent they are in conflict. This analysis reveals that the patterns of sexual relationship that are observed among animals are the evolutionary product of selection pressures acting separately on males and females, sometimes in the same direction, sometimes in different directions. We begin by considering a fundamental difference between the sexes, anisogamy, the evolutionary origins of which were discussed in Chapter 5, Section 5.3.

Males typically produce sperm in enormous numbers, each sperm being energetically very cheap to make, whereas females produce relatively fewer eggs, each of which is relatively expensive in energetic terms. (This generalization is less apparent in some animal groups than others; many fish, for example, produce huge numbers of eggs, yet they still produce fewer eggs than males produce sperm.) As a result, males inherently have a much greater reproductive potential than females. If a male mates with just one female, his reproductive success (in terms of number of progeny) is the same as hers and is much less than his potential reproductive success. He can realize a greater proportion of his reproductive potential if he mates with several females. Among males, selection will thus tend to favour those that mate with the most females, since they will leave the most progeny. Males do not, however, have an infinite capacity to produce sperm, and their ability to inseminate females may vary in the short term as their sperm supplies are depleted and replenished. For the female, however, mating with several males does not increase the number of her progeny; she can enhance her reproductive success, relative to that of other females, only by improving the survival prospects of her limited number of offspring. This she may do in a number of ways, particularly by caring for her offspring. She may also improve the quality of her offspring by mating with particular males rather than others, depending on the exact nature of the male's role in reproduction, as we shall see later. Thus, anisogamy leads to a fundamental difference between males and females. All other things being equal, males are under strong selection to mate with as many females as possible. Since females are limited in number this generates intense competition among males for access to females. Females, on the other hand, are under strong selection to care for their progeny and to be selective in terms of which males father their progeny. The fact that a pattern of this kind occurs in only some species, and that there are many species that depart from it, indicates that all other things are not equal. The sources of variation between species in terms of male and female roles is discussed in the next Section.

7.2.1 VARIATION IN REPRODUCTIVE SUCCESS: BATEMAN'S PRINCIPLE

The present era of intense interest in sexual selection was largely initiated by a classic experiment on fruit-flies *Drosophila melanogaster* by A. J. Bateman (1948). To compare the variation in the reproductive success of males and females, Bateman placed 3 to 5 virgin flies of each sex in laboratory containers in which they could breed. Each female thus had a choice of several males and each male had several competitors. Each adult fly carried distinctive genetic markers so that the progeny produced in each container could be attributed to specific parents. The results are shown in Figure 7.2.

▶ What do the results indicate about the variation, within each sex, in the number of mating partners (Figure 7.2(a))? In which sex is it the greater?

It is greater between males. Several males obtained no matings while several had 2, 3 or 4. In contrast, most females mated once or twice; very few did not mate at all.

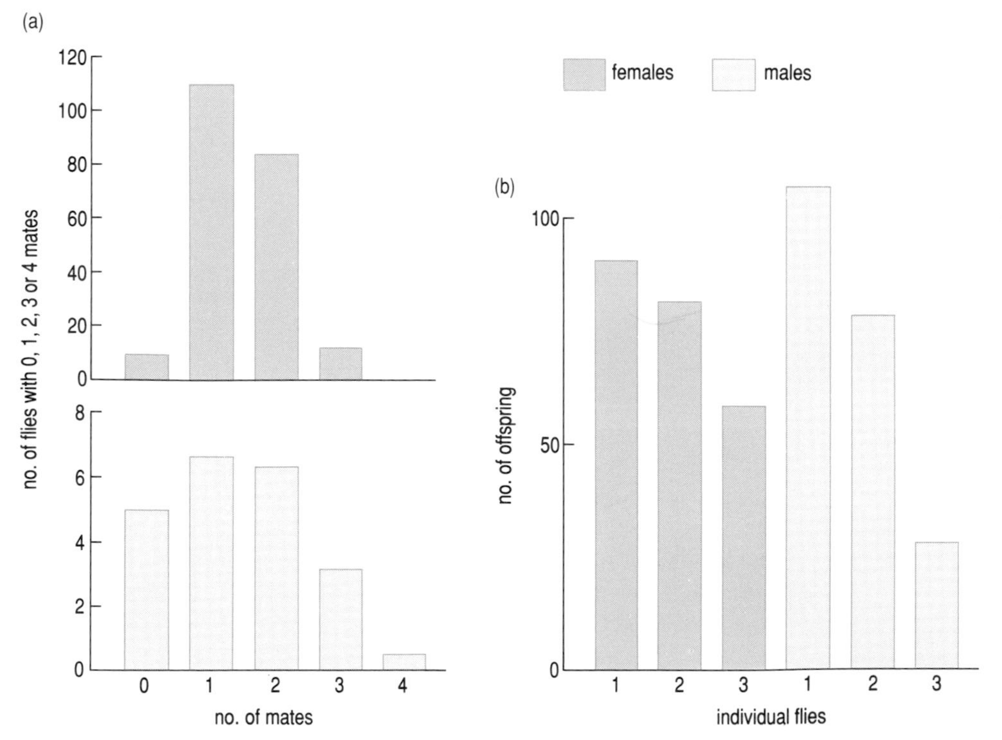

Figure 7.2 *Variation in the reproductive success of male and female fruit-flies* Drosophila melanogaster. *(a) Frequency of mating: data from experiments in which 3 to 5 flies of each sex were caged together. (b) Number of offspring: data from experiments in which 3 flies of each sex (1 to 3 on the* x *axis) were caged together. Females and males are ranked left to right in order of decreasing fecundity.*

▶ What do the results indicate about the variation within each sex in terms of reproductive success (number of progeny) (Figure 7.2(b))? In which sex is it the greater?

It is greater between the males. The reproductive success of females varies between about 55 and 90 progeny per female, that of males between about 30 and 110 per male. On the basis of these results, Bateman argued that, because of the fundamental difference between male and female gametes, eggs represent a limited resource for which males must compete. As a result, males are selected for the ability to mate with many females. In fruit-flies, males must stimulate females by means of elaborate displays and there is little direct interaction between males. Females are not selected to mate with several males, but to be selective about the males with which they mate.

Bateman's results were obtained in a highly artificial laboratory setting. Since then, however, numerous studies of animals in the field have revealed a similar effect, for example in red deer (Chapter 4, Section 4.4.1). Variation in male mating success is particularly marked in animals with a **lek** mating system, in which males cluster in dense groups visited by females for the sole purpose of mating. At a lek, males typically defend very small territories which do not contain food or cover but simply provide a space where mating occurs. Individual females generally mate only once, but some males mate very often, others not at all (Figure 7.3). These differences between the sexes in terms of the variation in mating success have a profound effect on the strength of selection acting on them. Because some males sire many progeny while others sire none at all, selection for those characters that promote male mating success will be very strong and will lead to a rapid change over generations in those characters. Selection on females will not only act on quite different characters but will be much less intense.

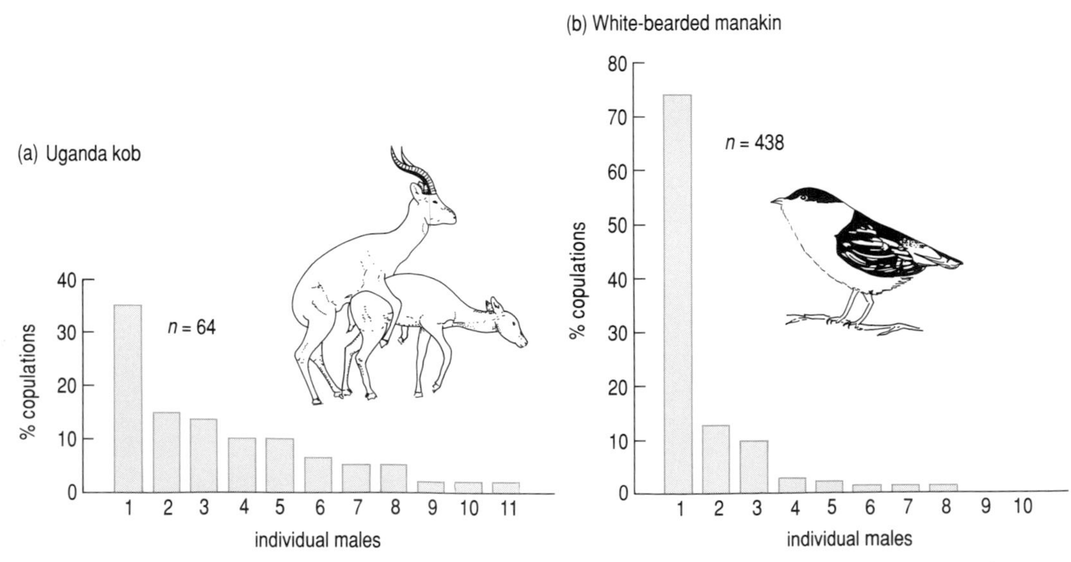

Figure 7.3 *Variation in male mating success among (a) Uganda kob* Adenota kob thomasi *and (b) White-bearded manakins* Manacus m. trinitatis. *In both species, males gather in compact groups, called leks, which females visit in order to mate. In each histogram, males are ranked left to right in order of mating success. Thus in (b) the male manakin ranked 1 obtained 74% of 438 matings observed whereas male 10 obtained none.*

7.2.2 ALLOCATION OF REPRODUCTIVE EFFORT

In the course of their lives, animals acquire resources in the form of nutrients and energy, which they then allocate to various activities that promote their survival, growth and reproduction. With the exception of resources that go into energy reserves such as fat, resources allocated to one activity cannot be subsequently used again for another activity. As a result, an organism's life-history can be regarded as a continuum of alternative possibilities as to how resources should be allocated so that fitness is maximized. This introduces the important concept of a life-history **trade-off** whereby any benefit derived by allocating resources to one activity will incur a cost in terms of a reduced capacity to engage in other activities. This concept is illustrated in Figure 7.4.

The fundamental alternatives for resource allocation upon which selection may act lie between survival and growth, on one hand, and reproduction on the other. Resources that are allocated to survival and growth are referred to as **somatic effort**. As young animals develop and grow, they typically allocate very little to reproduction; once they reach reproductive age, however, their gonads mature and, in many species, animals cease to grow. In species that reproduce only once in their lives (*semelparous reproduction*), the shift of resources to reproduction may

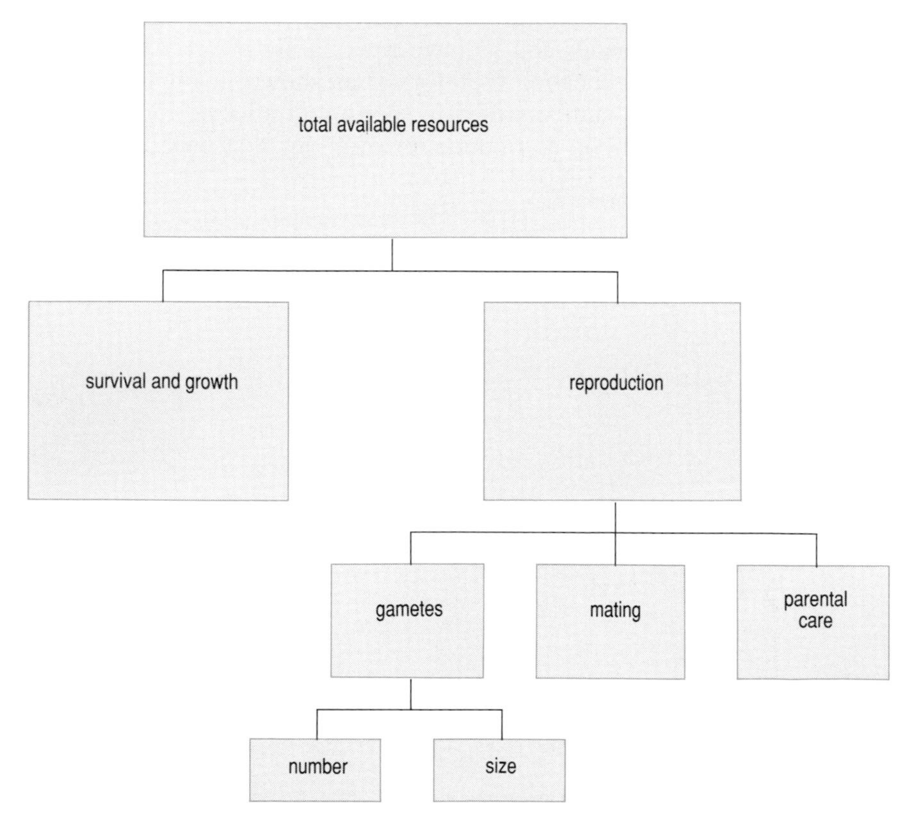

Figure 7.4 *Schematic diagram to show the principle of trade-offs in resource allocation to different components of lifetime reproductive success. At each level, resources are allocated to two or more aspects. The amount allocated is indicated by the size of each box; at each level the combined size of boxes is equal to that of the appropriate box at the preceding level.*

be almost total and the individual dies after breeding. In species that breed several times during their lives (*iteroparous reproduction*), there must be a trade-off between current reproduction and ensuring survival until the next breeding period. We shall return to these overall life-history patterns in Chapter 8.

Resources that are allocated to reproduction are referred to as **reproductive effort** and must be further allocated to different aspects of reproduction (Figure 7.4). For animals, there are three principal components of reproductive effort: the production of gametes, mating, and caring for the progeny. Gamete production involves a trade-off between gamete size and number; thus, a female has a finite allocation of reproductive effort with which she can produce a few large eggs or many small ones. Mating effort includes a great variety of activities apart from the mating act itself, such as developing secondary sexual characters that attract mates, finding a mate, courtship displays, and competition with rivals to secure matings. Since all these activities require resources, there must be trade-offs between them too but, to keep the picture simple, they are not represented in Figure 7.4.

To see the significance of patterns of allocation of reproductive effort to different activities in the context of sexual selection, a number of factors must be considered, the first of which is the fundamental difference between the sexes arising from anisogamy. As argued above, females typically produce relatively few, large eggs in which they invest parental care; they gain little or nothing by mating with several males and so have low mating effort. Their typical pattern of reproductive allocation therefore is as shown in Figure 7.5(a). Males, by contrast, produce large numbers of tiny sperm and allocate much of their reproductive effort to the competition and courtship activities that enable them to secure matings with several females (Figure 7.5(b)).

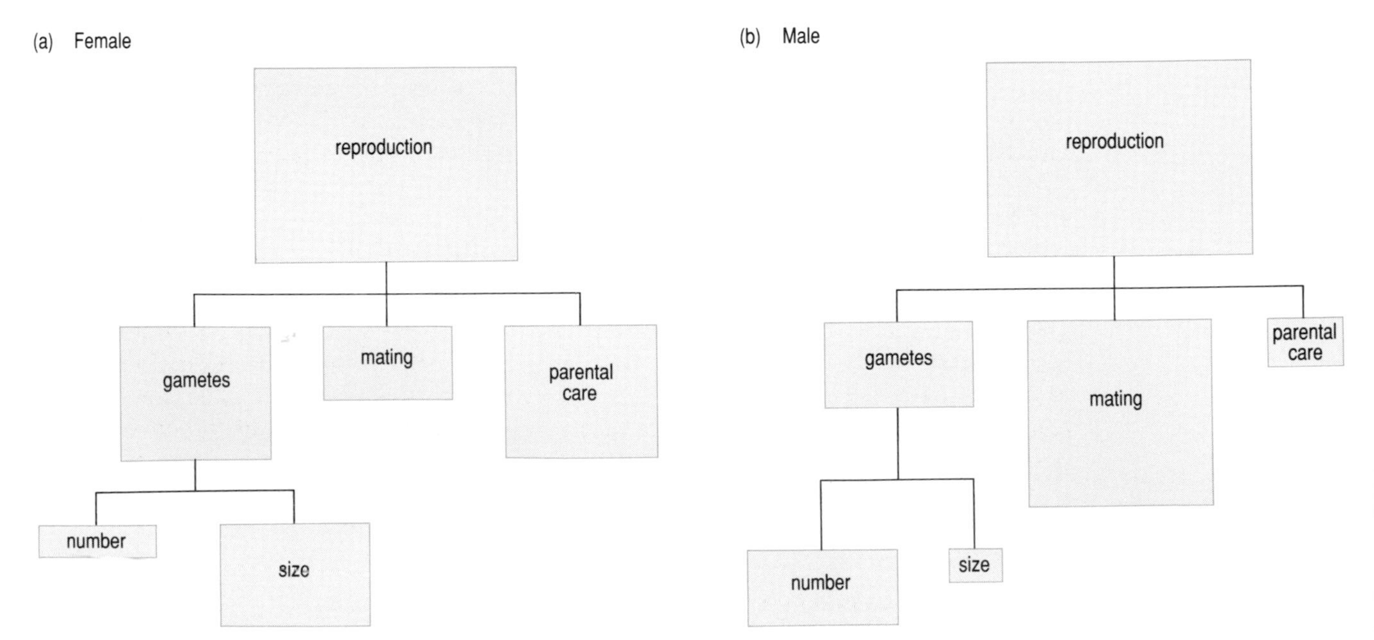

Figure 7.5 *Differential resource allocation in males and females. (Conventions as in Figure 7.4.) (a) Females allocate most of their reproductive effort to producing relatively few, large gametes which receive substantial parental care. (b) Males allocate most of their reproductive effort to producing large numbers of small gametes and to seeking a large number of matings.*

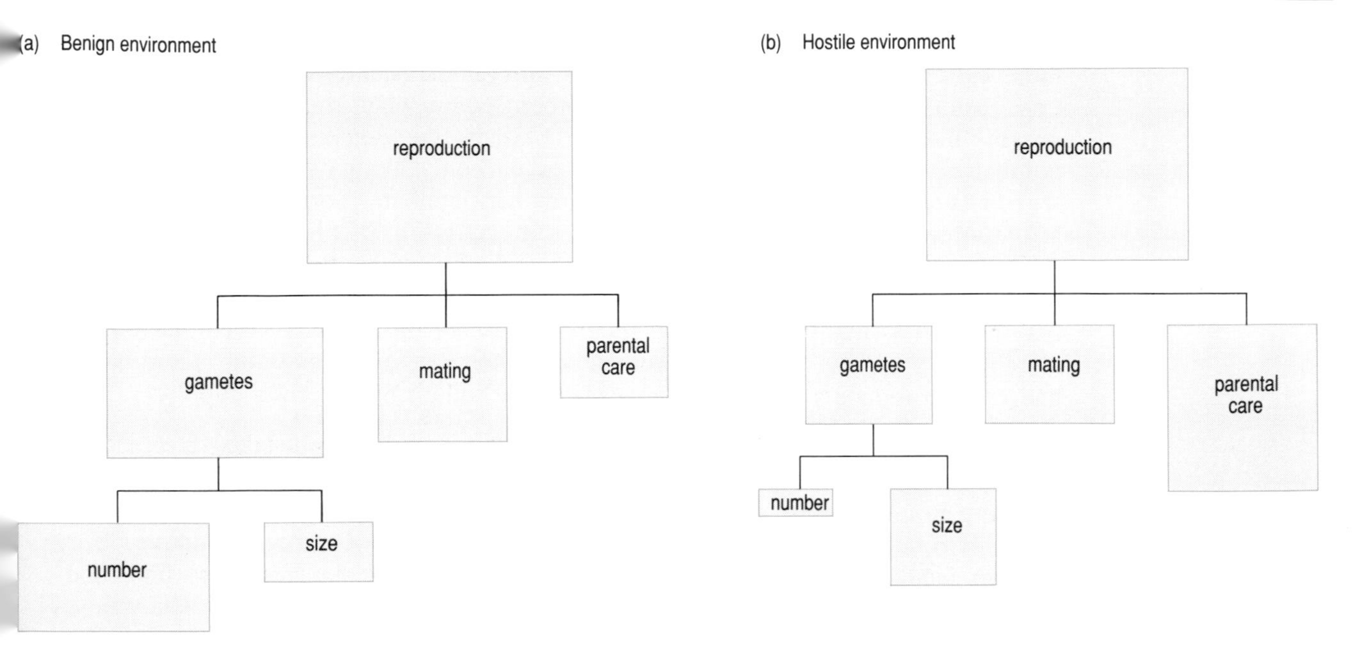

Figure 7.6 *Resource allocation influenced by the environment. (Conventions as in Figure 7.4.) (a) In a biologically benign environment favourable to the survival of the young, an individual female allocates reproductive effort primarily to producing a large number of young (gametes). (b) In a biologically hostile environment in which the young cannot survive unattended, individual females produce relatively few, large eggs, which receive substantial parental care.*

These patterns are, however, profoundly influenced by the nature of the environment in which an animal has to breed. Let us consider just one aspect of reproduction, the survival of the young, and see how the environment can affect the reproductive allocation of one sex, the female, to parental care in different environments. If progeny are born into a biologically benign environment where predators and competitors are relatively few and food is abundant then their chances of surviving with little or no parental care may be quite high. This promotes the evolution of a pattern in which females invest heavily in producing a large number of relatively small eggs but invest very little in parental care (Figure 7.6(a)). In an environment in which progeny can only survive if they are protected and fed by a parent, females will tend to produce relatively few, large eggs in which they invest a great deal of parental care (Figure 7.6(b)).

The environment is not the only factor that we must consider here, for different kinds of animal are subject to different ecological and physiological constraints that will affect their patterns of reproductive effort allocation. The production and care of eggs by reptiles and birds serves to illustrate this point. Reptiles are ectothermic (their body temperature is determined largely by heat absorbed from the environment) and they, and their eggs, can tolerate quite large changes of temperature. Birds are endothermic (they maintain an almost constant body temperature by generating heat through internal metabolism) and for them and, to a lesser extent, their eggs, a prolonged substantial drop in core body temperature is usually lethal. A reptile can therefore leave its eggs buried in the sand where they are reasonably safe from predators; they can be heated by the Sun during the day and they will tolerate a fall in temperature at night. Free of the need to care for her eggs, a female reptile can produce many eggs. Some large turtles lay over a thousand eggs per breeding season. Birds, in contrast, nearly always incubate their eggs, using their own body heat to keep them warm when the weather is cold, and provide them with shade when it is hot. They are thus constrained in terms of the number of eggs that they can produce; they can only incubate as many eggs as will fit beneath their bodies. Very few birds can incubate more than a dozen eggs at one time and many incubate only one. (In one group of birds, the megapodes of Australasia, eggs are laid in large mounds of rotting vegetation so that they are 'incubated' by heat generated by decomposition in the mound. These mound nests may contain as many as 35 eggs.) Incubation is not the primary constraint on clutch size for some birds, however. In some species, such as certain game birds and owls, a female's clutch size appears to be limited by her food supply.

So far, we have considered animals as individuals and have not taken into account the influence that the behaviour of one sex may have on the other. Consider a bird breeding in a temperate climate in which it may be frequently cold and food may be in short supply. A female will have to allocate much of her reproductive effort towards incubating her eggs and guarding her chicks, but she must also find time to go out and feed herself and her young. If these demands are heavy, she will only be able to rear a few young successfully. If, however, a male assists her in incubating and feeding the eggs and chicks, she will be able to rear more progeny. Clearly, it is in the interests of a female to receive help from a male. For a male, there are two options. He may devote his reproductive effort to securing as many matings as possible, or he may mate with only one female and then remain with her and share parental duties with her. In severe environments in which females can produce only a few offspring on their own, or none at all, males will gain little or nothing in terms of reproductive success by seeking multiple matings and it is better for them to allocate reproductive effort to parental care rather than mating effort. There is thus a trade-off for males between mating with several females, each of which has a low probability of reproductive success, or mating with one which, with his help, has a high chance of reproductive success. As we have seen, the reproductive biology of birds imposes particularly severe demands on parents and it is among birds that monogamy, in which both parents care for the young and establish a long-lasting pair bond, is most common.

Before going on to consider in more detail monogamy and other patterns of relationship between males and females, we must consider evidence for a critical assumption in the arguments presented in this Section which is that there is a trade-off between one component of reproduction and another. Table 7.1 (overleaf) shows, for three animal groups, data on offspring number and offspring size within species. In all three groups, species with large offspring produce relatively few of them at a time, suggesting that there is indeed a trade-off between progeny number and progeny size.

Figure 7.7 (overleaf) presents data for two male reproductive characters in several species of primates (monkeys, apes and humans). The characters are: relative size of canines (the extent to which males have larger canine teeth than females) and relative testes size (corrected for overall size variations within the species). The species are divided into three categories: those that breed in monogamous pairs, those that breed in groups consisting of one male and several females, and those that live in groups consisting of several males and several females.

Table 7.1 Number and size of progeny in pairs of similarly sized species

Species	No. of eggs per breeding episode	Size of eggs
Salamanders:		(Egg diameter/mm)
Ambystoma talpoideum	100–460	1.25
Desmognathus fuscus	9–16	2.5–3.0
Birds:		(Egg length/mm)
Blue tit *Parus caeruleus*	7–16	15.3
Great tit *Parus major*	6–14	17.3
Frogs:		(Egg volume/mm^3)
Bombina bombina	32.5	3.4
Bombina variegata	17.4	8.2

▶ Bearing in mind that canine teeth are used extensively in fighting among males, what do the data in Figure 7.7 suggest about the relative allocation of reproductive effort to gametes and mating competition in the three categories?

In species such as gibbons where males and females form pairs, males have small canines and small testes, suggesting that they allocate relatively little to competition with rivals or to gamete production. In species with single-male groups, such as the gelada baboon *Theropithecus gelada*, they have large canines and small testes, indicating a large allocation to fighting. In species with multi-

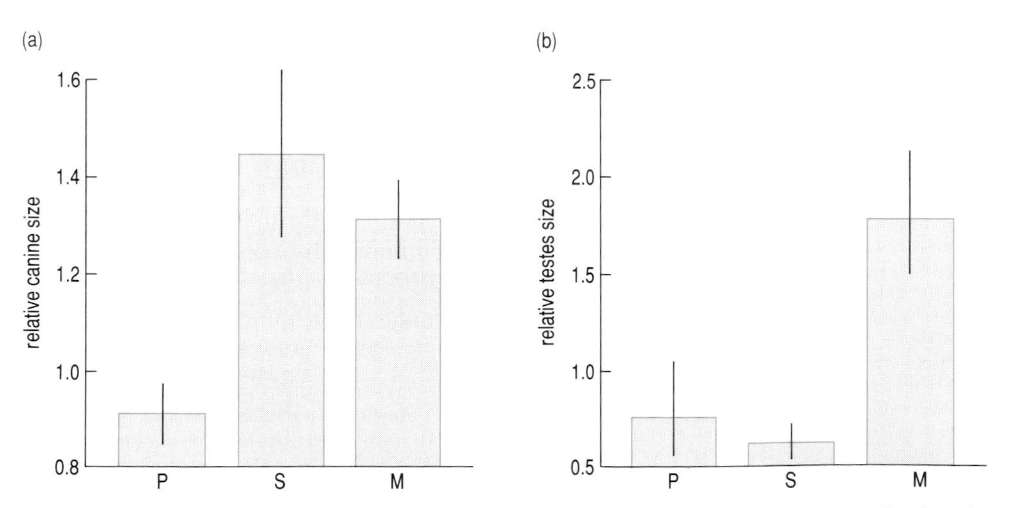

Figure 7.7 *Variation in the size of two male sexual characters among primates. (a) Canine size, relative to that of females. (b) Relative testes size, allowing for the effect of overall body size. P = species in which males live in monogamous pairs; S = species in which males live in groups containing a single male and several females; M = species in which males live in groups containing several males and several females. Histograms show mean values, vertical lines show standard errors.*

male groups, such as the olive baboon *Papio anubis*, they have large canines and testes, suggesting a large allocation both to fighting and to gamete production.

Clearly, there is variation among primate species in the pattern of allocation of reproductive effort to different aspects of reproduction that is related to their different social systems. The nature and evolutionary origins of this variation are analysed in the next Section.

7.3 ANIMAL MATING SYSTEMS

There are two basic categories of mating system; **monogamy** (meaning 'one mate'), in which an individual has one mating partner, and **polygamy** (meaning 'many mates'), in which an individual has several partners. Polygamy is subdivided into **polygyny**, in which a male mates with several females, and **polyandry**, in which a female mates with several males. (Some classifications recognize a further category, promiscuity, in which both sexes have several partners. However, this term has obvious human connotations that are inappropriate for animals so it is no longer generally used; polygamy is an adequate term to describe such systems in animals.) The various terms may be applied to species, to populations or to individual animals.

These categories of mating system are not as tidy or discrete as they might at first appear. The distinctions between categories become blurred, depending on the level of analysis (individual, population, or species) and on the time-scale. For example, individuals in some species breed with a succession of individuals, changing their partners between breeding episodes, a system called **serial monogamy**. In terms of the demography of males and females at any given time, individuals are monogamous, but in terms of the progeny that they produce over the course of a lifetime, they are polygamous; the consequences in terms of genetic diversity among progeny are the same as if individuals had several mates at the same time. In species described as polygynous, it is self-evident that, if there is a 50:50 sex ratio, not all males can enjoy long-term polygynous relationships; some will have several mates, some will be monogamous, and some will have no mates at all (there is no technical term for this celibate condition!).

The term 'mating system' has the unfortunate connotation that it is somehow enforced on a population of animals by some external influence. Rather, in analysing the evolution of different mating systems, we are seeking to explain them as adaptations to particular environmental conditions. Since natural selection acts on individuals, not on groups or populations, we must seek to explain mating systems in terms of the reproductive consequences that accrue to individuals. Furthermore, as we saw in Chapter 5, there is a basic conflict of interests between the sexes arising from anisogamy. This means that our analysis of the evolution of mating systems must consider the interests of each sex separately.

7.3.1 DETERMINANTS OF MATING SYSTEMS

A number of ecological, social and other factors interact to determine the kind of mating system that evolves in a particular species. We will consider a number of these factors separately before going on to see how they interact. In this Section we will draw all our examples from birds, a group that has been more thoroughly studied than most in terms of mating systems.

1 Quantity of resources In a given environment, resources essential for reproduction, such as food for the young and suitable nest sites may be abundant or scarce. Where food is abundant, females may be able to feed the young adequately on their own; consequently their brood size will be limited by the number of eggs they can lay and/or incubate, not by their capacity to feed nestlings. Males may gain more by mating with several females than by sharing parental care with one female. Rich environments thus tend to favour polygyny, impoverished environments monogamy.

2 Distribution of resources Resources essential for reproduction may be evenly or patchily distributed in the environment. Where resources are patchy, it may be possible for individual males to monopolize a patch, defending it against rival males and thereby gaining access to those females that must use the patch. Patchy distribution of resources may thus favour polygyny.

3 Predation In environments in which predators are numerous, predation may pose a serious threat to an individual's reproductive success. It may be the adults that are most at risk, or the young when the nest is left unattended for example. Monogamy may yield a higher return to the male than polygyny (in terms of offspring raised) if a male and female, by staying together, can defend the brood more effectively than a female on her own. In this context, it is not only heterospecific predators (predators of a different species) that are important. In many colonial-breeding birds such as gulls, individuals readily eat one another's eggs and chicks if they are left unguarded for a moment.

4 Other determinants of social behaviour Animals may form groups for a variety of adaptive reasons other than reproduction. Living in a group may enhance an individual's feeding efficiency and may afford protection against predators. If a male can drive all the other males out of a social group, he may enjoy exclusive access to several females. This is a relatively unimportant factor among birds, but it is the basis of polygyny in many ungulates and primates.

5 Female availability in time Within a species, females may show varying degrees of synchrony in terms of when they mate and breed. If females are highly synchronized, all mating at the same time, then individual males may be severely limited in the number of females that they can mate with and defend in the limited time available. If females are asynchronous, however, an individual male may be able to obtain exclusive access to several in succession. Thus, asynchronous breeding among females tends to favour polygyny.

6 Requirements of the young Species show much variation in the amount and kind of care that the young require. Among many birds that nest in trees, the (*altricial*) young hatch at an early stage in development; they are helpless, naked and blind and require a great deal of parental care to keep them warm and fed. If two parents can rear substantially more chicks than one parent alone, males may achieve higher reproductive success by being monogamous. In contrast, among many birds that nest on the ground the (*precocial*) young hatch at an advanced stage of development ; they are feathered, have open eyes, are able to run around and feed themselves very soon after hatching. In such species, the young require much less parental care so males tend to be polygynous.

In considering these various factors, we have largely concentrated on the interests of the male and whether it is to his advantage, under particular conditions, to be monogamous or polygamous. What of the female's interests? As discussed earlier, females typically have a very limited reproductive potential compared to the males, and cannot increase the number of their progeny by mating with several males. However, if a female's progeny require parental care to survive then her reproductive success may be considerably lower, in terms of number of offspring, than her reproductive potential. In many birds, for example, a female's clutch size is limited mainly by the number of eggs she can incubate and/or by the number of chicks she can feed which, for most species, is considerably fewer than the number of eggs she can lay. Therefore, if she can somehow obtain assistance in the caring for her young, either from a male or from other individuals, she can devote less of her own reproductive effort to parental care and more to producing eggs, so enjoying higher reproductive success. We will discuss below certain adaptations by females that enable them to secure parental care from males.

In many species, females do in fact mate with several males, even in species that are apparently monogamous. There are a number of possible adaptive explanations for this. Mating with more than one male may be an insurance against the possibility that any individual male is sterile or has a low fertility. Multiple paternity of her offspring will increase their genetic diversity and perhaps increase the proportion of them that will survive (refer to the arguments for the evolution of sex in Chapter 5).

7.3.2 Examples of mating systems

In this Section, we shall look at various mating systems, describe specific examples of each, and discuss the ecological and social factors that influence their evolution.

Monogamy Monogamy is more common among birds than any other group of animals. It is estimated that about 90% of bird species are monogamous whilst it is very rare among other groups of vertebrates. The prevalence of monogamy among birds is probably due, as we have already discussed, to the fact that they produce young which, both as eggs and chicks, require considerable parental care if they are to survive. Eggs require constant incubation and defence against predators; chicks require warmth, protection and, in many species, frequent parental feeding. Only in habitats where the climate is generally warm, food is abundant and predators are few can one parent rear altricial young alone. If females are assisted by a male, they can realize a greater proportion of their reproductive potential. If males do engage in the parental care of the young they have fathered, they gain greater reproductive success than if they do not.

In monogamous birds parental duties are shared between males and females to varying extents depending on the species. In some birds, such as gulls, both sexes share in incubating and feeding the young and in defending the nest and brood, each sex spending roughly equal amounts of time in each activity. By contrast some hornbills, such as *Tockus flavirostris,* divide parental duties between males and females in a rather extreme and bizarre way. Their nest is built in a deep hollow in a tree where the female hornbill walls herself in with mud collected by the male, leaving a hole just large enough for him to pass food to her. She thus carries out all the incubation and feeding of the young and he does all the foraging for food for his mate and their brood. She does not break out of the nest until the chicks are so large that there is no room for her. The chicks are then walled in again and are fed by both parents until they are ready to leave the nest.

Another unusual form of shared parental effort occurs in pigeons and doves. The young (called squabs) are fed on a milk-like secretion from the crop which is regurgitated into the squabs' mouths. This crop milk is produced by both the male and female during the early days of the squabs' life. Contrast this to the situation in mammals where only the female produces milk for the young. (Crop milk is also produced by both parents in flamingoes and some seabirds.)

Some birds maintain their pair-bonds over several years. In the kittiwake *Rissa tridactyla* the breeding success of individuals and pairs in a colony in northern England has been carefully monitored over several years. It has been possible to show that the longer a pair stays together the higher is their reproductive success. This advantage provides an additional adaptive reason for individuals to maintain a monogamous relationship.

In recent years, it has become apparent that monogamous relationships among birds are not quite what they seem to be. The technique of DNA fingerprinting (Chapter 3, Section 3.5.2), by which it is possible to establish precisely the maternity and paternity of individual young, has revealed that, in several species, young birds are not always the progeny of their putative fathers and/or mothers. There are two principal reasons for this. First, females and males may engage in 'extra-pair copulations' with partners other than their mates.

▶ On the basis of what you have read in this Chapter, suggest why it might be adaptive for females to engage in extra-pair copulations.

It increases the genetic diversity of their offspring (Section 7.3.1), and also ensures against the possibility that their mate is sterile.

▶ Why might it be adaptive for males to do so?

It increases the number of their progeny, increases the genetic diversity of their offspring, and insures against the possibility that their mate is sterile.

The second major reason why young birds may not be the offspring of their apparent parents is that in many species, females, if they find a nest that is unguarded, will lay eggs in nests other than their own, a phenomenon called 'egg-dumping'. In doing this, a female is essentially parasitizing another female of her own species, in a manner comparable to the way cuckoos parasitize females of other species (Chapter 2, Section 2.3.2).

▶ Why might it be adaptive for females to dump eggs?

As we have seen, a female bird's reproductive success is typically limited by her ability to incubate eggs or to feed chicks, not by the number of eggs she can lay. Egg-dumping enables her to produce more young, to be cared for by other birds, than she is able to care for herself.

The phenomenon of extra-pair copulation amplifies a point that was made earlier which was that, in using terms such as monogamy and polygamy, it is essential to be quite clear as to the level of analysis at which the terms are being applied. In terms of general social structure and in their pattern of caring for young a species may be strictly monogamous, but individuals that engage in extra-pair copulations are, in terms of the genotypes of their progeny, polygamous.

Resource-based polygyny This type of mating system occurs where males control access by females to resources, such as food or nest sites, that are essential to the reproductive effort of females, and in which males compete to do so in such a way that the resources are divided unequally among males in a population. As a result, males that successfully defend a large share of resources attract and mate with many females; those that hold a small share obtain few or no matings.

▶ What feature of a resource is most likely to result in it providing the basis of polygyny?

The resource should be patchily distributed (see Section 7.3.1). If the resource occurs in patches, it is much more likely that there will be a pattern in which some males hold territories that are rich in the resource, while others will hold resource-deficient territories. A particularly clear example of resource-based polygyny is provided by the orange-rumped honeyguide *Indicator xanthonotus* of Nepal. In this species, beeswax obtained by raiding the nests of giant honeybees is an essential component of the diet of both sexes. Bee nests are scarce; they occur only on cliff faces and so are patchily distributed. Males do not show any parental care but focus all their reproductive effort around the bee nests where a small proportion of males successfully establish territories. Only females and non-breeding immature birds are allowed into these territories. The males copulate with all mature females entering their territories to eat wax. Females are unresponsive to the courtship displays of non-territorial males.

In a species showing resource-based polygyny, we would expect there to be a positive correlation between male mating success and the quantity or quality of the resources in their territories. Figure 7.8 (overleaf) shows data for a bird and a mammal, obtained in studies in which particular ecological aspects of males' territories were measured. Both show the positive correlation we would expect.

The distribution of resources and the mating success of individuals are variables that can be measured reasonably accurately in the field. It is therefore possible to make specific quantitative predictions about how females should mate with males in a resource-based system, and to test these predictions against actual data. The

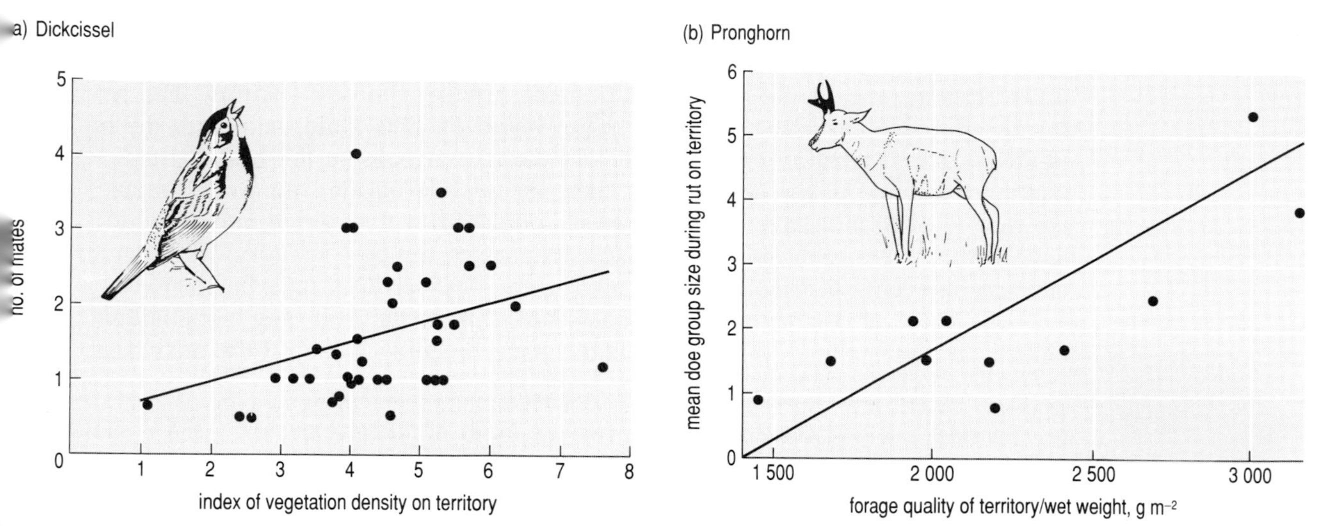

Figure 7.8 Positive correlation between males' territory quality and mating success. (a) In the dickcissel Spiza americana, *males with territories containing denser vegetation attract more females. (b) In the pronghorn* Antilocapra americana, *males with territories containing good grass attract more females during the breeding period.*

polygyny-threshold model (Figure 7.9(a)) provides the theoretical basis for such a test. This model is based on three assumptions: that resources required by females for reproduction are unevenly distributed across a number of male territories; that a female's fitness (in terms of the number of offspring reared) will be correlated with the quantity of resources to which she gains access; and that females choose which males to mate with on the basis of the resources contained in their territories.

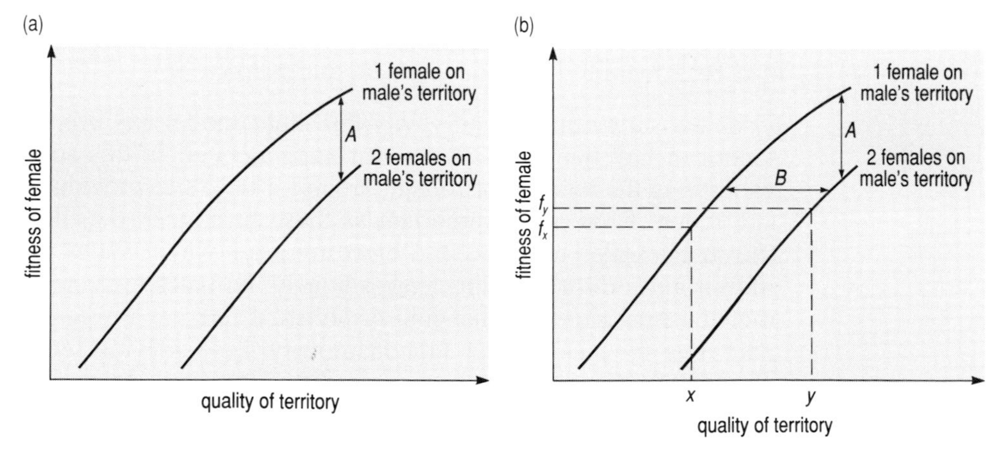

***Figure 7.9** The polygyny-threshold model. (a) It is assumed that a female's fitness (in terms of the number of young successfully reared) is correlated with the quality of resources in the territory of the male with which she mates. If two females occupy a given territory, the fitness of the second female is reduced by an amount* A, *compared with her expected fitness if she bred alone in that territory. (b) A female has a choice between an unoccupied poor territory* (x) *and a good territory that already contains a female* (y). *If she chooses* x, *her expected fitness is* f_x; *if she chooses* y *it is* f_y. *When the difference in quality between an already occupied territory and an empty one exceeds a certain value* (B), *a female gains higher fitness if she chooses the occupied one;* B *is the polygyny-threshold.*

Consider an area in which a number of males have established territories; some rich in resources, some poor. What we would expect the first female that arrives to do is obvious; she should mate with the male holding the best territory. The next female to arrive has a less obvious choice, however. She could mate with the male in the next-best territory or she could join the first female on the best territory, making that male polygynous. If she takes the second option, she (and perhaps the first female) will have reduced fitness compared to that which either would have if they were each the sole female occupant of the best territory. If resources are very unevenly distributed among males, the best territory may be so much better than the next-best that two females breeding in the best may, nonetheless, both have higher fitness than either would have breeding alone on the next best.

▶ Consider a female choosing between two territories of differing quality, x and y, shown in Figure 7.9(b). The better territory (y) already contains a female; x is empty. Extrapolate from the horizontal axis to the vertical axis (fitness) to determine which option the female should choose.

She should choose to join the female on y, rather than enter x (Figure 7.9(b)). Consider now the first female, already established on territory y.

▶ How would you expect her to respond to the arrival of the second female?

She should try to exclude the second female, since the latter's occupancy of the territory may reduce the first female's fitness. In fact, there is some observational evidence, for example from American red-winged blackbirds *Agelaius phoeniceus*, that females are aggressive towards one another during the early part of the breeding season when they are settling to breed in male territories. Humans tend to stereotype males as the aggressive sex; this example shows that aggression can also be a feature of female reproductive behaviour.

Evidence supporting the polygyny-threshold model has been obtained for the American lark bunting *Calamospiza melanocorys*. In this species the critical resource is the amount of foliage around a nest site providing shade for the nest and young. Food is not important as these birds feed outside breeding territories. The first females to arrive in a breeding area, where males have already established territories, pair monogamously. Females arriving about 5 or 6 days after the first arrivals either join established females in bigamous groups (one male, two females) or pair monogamously. Late-arriving females pair monogamously. About 20% of males attracted no females. In 23 out of 35 comparisons between bigamous and monogamous groups, each of the females in bigamous groups reared as many or more young than monogamous females. A detailed study of the habitat in the various territories revealed that bigamous groups occurred in territories with the most shade.

Female-defence polygyny In some species males compete for females directly (not by competing for resources) and successful males establish a group or 'harem' of females. The formation of harems is facilitated if females show a predisposition to gather in groups for some reason other than mating. Female seals leave the water once a year, hauling themselves out onto a beach to give birth to their pups. Because suitable beaches are few and far between, large numbers of females tend to gather in confined areas. Mating occurs at the same site immediately after a female has given birth. The site thus becomes the focus of intense competition among males to gather groups of females. In the northern elephant seal *Mirounga angustirostris*, fights among males are extremely violent and only the largest and most powerful males become 'harem masters'. This intense selection for fighting ability has favoured large size in males, the largest of which are three times heavier than females. In a single season only a third of males mate at all and fewer than 10% of the males fertilize nearly 90% of the females. Fighting incurs high costs for males; some die immediately after a season in which they have been harem masters and few achieve this status for more than three years, whereas females may breed for up to 10 years in succession. Most males never mate during their lives.

A male's success in maintaining a harem depends not only on his ability to compete with rival males, but also on his ability to control the movements of females. Red deer *Cervus elaphus* stags spend all their time in the rut (breeding period) competing with rivals by roaring, frequently fighting, and chivvying the hinds like a sheep-dog to keep them in a compact group; even so, females frequently wander away and join other harems. In the marine isopod crustacean *Paragnathia formica,* the male captures females and takes them into a burrow, from which he can prevent them leaving (Figure 7.10). The rarity of this kind of polygyny among birds is probably partly due to the fact that they are highly mobile making it difficult for males to control the movements of females.

Leks A quite different form of polygyny occurs when males do not defend or control either resources or females. Instead, males gather in dense clusters called leks where they defend very small territories (Section 7.2.1). These contain no resources of importance to females but simply provide sites on which males display to and mate with females that visit the lek. Males display vigorously to females and it is typical of lek species that males have very elaborate plumage; the ruff is a typical example (Figure 7.1). Lek species have been studied very intensively, largely because they provide opportunities to study female choice

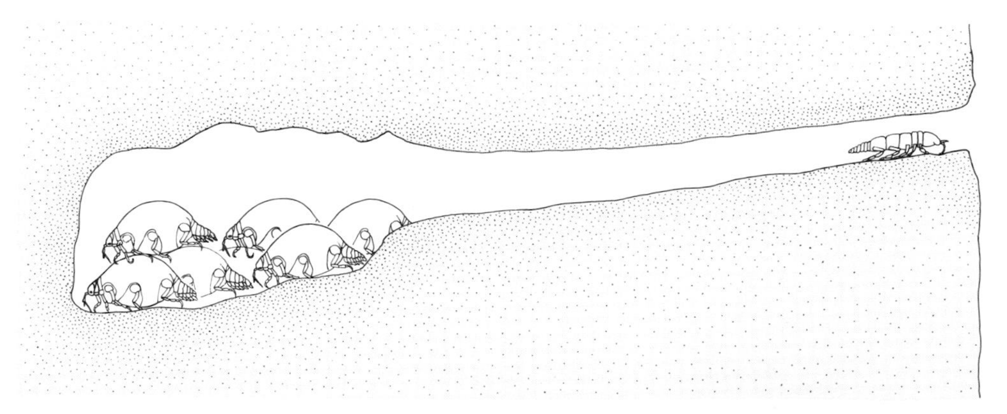

Figure 7.10 *Harem defence in the marine isopod* Paragnathia formica. *The male digs a burrow into which he drags as many as 25 females. In his position at the mouth of the burrow he can prevent females leaving and also prevent rival males entering. This animal is about 5mm long.*

which, as we have seen, has been a controversial aspect of sexual selection. Females could, potentially, choose males on a variety of criteria including their plumage, their behaviour and their position within a lek (central or peripheral). In several studies of leks the approach has been to score the mating success of individual males in order to try to correlate this with one or more aspects of male morphology and behaviour. Which aspect or aspects is important varies from one species to another (Table 7.2).

Table 7.2 *Correlates of male mating success in a number of lek-breeding birds.*

Species	Male character(s) positively correlated with frequency of mating
Sage grouse *(Centrocercus urophasianus)*	Rate and duration of display
Ruff *(Philomachus pugnax)*	Central position of territory Rate of display
Peacock *(Pavo cristatus)*	Number of 'eyespots' in male's train
Jackson's widowbird *(Euplectes jacksoni)*	Rate of display Length of tail

Polyandry Polyandry, in which individual females simultaneously have long-lasting mating relationships with several males, is a very rare mating system. For this reason and because it contradicts the general proposition that males are expected to be the polygamous sex, it has been intensively studied. It is exemplified by a water bird, the American jacana or lily-trotter *Jacana spinosa*. Jacanas live on lily-covered lakes, where their greatly elongated toes enable them to walk on floating vegetation. Each female defends a large territory within which there are several smaller territories defended by males. Each smaller territory contains a floating nest that is attended and defended by a particular male. The female moves around her territory mating from time to time with all the males and laying eggs in each nest. Each male defends a nest and carries out the complete process of incubation of the eggs in it. The female is thus emancipated from parental care and allocates all her reproductive effort to mating, egg-laying and defending her territory against other females. She is considerably larger than the males (50–75% heavier), is dominant over them and will stop any fights that break out between them.

The critical factor in the evolution of polyandry is the nature of the breeding habitat. Suitable breeding sites are scarce and nests are subject to heavy predation. These conditions favour the ability in females to lay large numbers of eggs, both to exploit such suitable habitats as do exist and to replace eggs that are lost to predators. Selection has thus favoured two aspects of female reproductive biology that enhance their egg-laying capacity: large body-size and a shift in resource allocation away from parental care and towards egg production. Selection for large body-size is accentuated by the advantage that it gives females in competition with other females. It has also led to females being larger than males and thus being able to dominate them.

This mating system is clearly highly advantageous to the female since she is able to achieve much higher reproductive success than if she had to care for her young. What are the selective pressures acting on males that have led to their behaving as they do? As discussed under monogamy, an environment that is hostile to egg survival will generally favour parental care by males; polyandry most probably evolved from a monogamous mating pattern. A high level of predation requires the ability to replace lost eggs; only females can do this so the allocation of parental care shifts towards males and large size in females is favoured. When females are large enough to dominate and control males, the options open to the latter become restricted. For the male to seek further matings he would have to leave his nest, exposing it to predation and himself to attack by the female.

There remains the question of why males within a female's territory tolerate one another. This is related to the pattern of mating and egg-laying shown by the female and introduces a critical factor in the evolution of this and other mating systems: *certainty of paternity*. The female mates with all the males in her group before and during egg-laying. Each male thus guards and incubates a clutch of eggs, only some of which are likely to have been fertilized by him; on the other hand, some eggs that he has fertilized are being cared for by other males in the group. Natural selection will favour males that allocate paternal care only to those progeny that they themselves have fathered and will select against those that care for progeny that do not carry their genes (Chapter 6). In the jacana, however, males cannot distinguish between eggs that they have fathered and those of other males because particular episodes of mating by the female are not linked to particular episodes of egg-laying. Selection thus favours the males caring for a clutch of eggs (in which some eggs are probably their own) and not disrupting the breeding effort of other males that are caring for other clutches of eggs (in which there also may be some of their eggs).

Synthesis These accounts of various mating systems are, of necessity, rather superficial and do not provide a complete picture of either the full range and diversity of different mating systems or of the multiplicity and complexity of factors that determine their evolution. For each species that has been analysed in detail the balance of selective forces acting on the various components of male and female reproductive effort has produced a particular mating system. What the above accounts have sought to do is to illustrate three general points:

1 The nature of the mating system found within a given species is profoundly influenced by certain aspects of the environment in which it lives, especially food supply, predation and the availability of nest sites.

2 The way that one sex allocates reproductive effort to different aspects of reproduction affects the allocation pattern of the other sex.

3 Mating systems must be understood as being the result of selection acting separately on individual males and females.

The fact that selection operates differently on the two sexes can lead to situations in which the best interests of males and females are not the same. This raises the question of whether the mating systems we observe in nature are a compromise between male and female interests or whether one sex is at an advantage over the other. On the face of it polygyny appears to be a system in which males, at least those that are successful, gain at the expense of females whereas polyandry seems to represent the opposite situation. This conflict of interests between the sexes is well illustrated by our final example.

The dunnock: a variable mating system Despite its drab appearance, the common dunnock or hedge sparrow *Prunella modularis* has a far from drab sex life. Meticulous studies of a population of dunnocks in the Cambridge Botanical Gardens (U.K.) by N. B. Davies (Davies and Houston, 1986) have shown that there is considerable variation in the pattern of sexual relationships among males and females. Some males are polygynous and have a territory containing two nests with a female and a brood in each; they provide some parental care to both broods. Some males are monogamous and devote all their parental care to a single brood. Others are polyandrous and care exclusively for a brood whilst the female of the brood also cares for another brood that is attended by another male. Davies recorded the number of young successfully reared from a large number of nests, according to whether their parents were polygynous, monogamous or polyandrous (Table 7.3). He also recorded the frequency with which individual males and females copulated with one another (individuals were marked with leg rings), so that he could estimate the certainty of paternity of males in polyandrous groups.

Table 7.3 *Reproductive success of male and female dunnocks in different mating combinations within a single population.*

Mating system	No. of adults caring for young in each nest	Reproductive success (no. of young fledged per season)	
		Per female	Per male
Polygyny (1 male, 2 females)	Female + part-time male	3.8	7.6
Monogamy (1 male, 1 female)	Female + full-time male	5.0	5.0
Polyandry (2 males, 1 female)	Part-time female + full-time male	6.7	*a* 4.0* *b* 2.7

* In polyandrous groups, females mate more often with one male *a* than the other *b*; paternity of *a* and *b* estimated according to their relative frequency of copulation.

▶ From the data in Table 7.3, which mating system is most advantageous for females?

Polyandry. In polyandrous system females produced, on average, 6.7 fledglings.

▶ Which mating system is most advantageous for males?

Polygyny. On average each polygynous male produced 7.6 fledglings.

▶ For each sex, how advantageous is monogamy, compared to the two other alternatives?

For both sexes, it yields an average of 5.0 fledglings. For females this success rate is better than polygyny, not as good as polyandry. For males it is better than polyandry but not as good as polygyny.

This example shows very clearly that there is indeed a conflict of interests between the sexes and that, within a single population, there can be a variety of mating systems in which the balance of interests can tip towards one sex or the other. In the dunnock, the critical determinant of which pattern any one individual will be involved in depends on his or her competitive ability. If a female can dominate males, she achieves polyandry; a very dominant male will achieve polygyny. Birds of intermediate dominance that are well matched will become monogamous.

In Davies' original study it was possible only to estimate the reproductive success of males according to the frequency with which they were observed copulating with particular females. It was assumed that the probability of their being the father of a particular female's young was proportional to the relative frequency with which they were seen to mate with her. Therefore if a particular male was the only one observed to mate with a particular female he was assumed to be the father of all her offspring; if two males mated equally frequently with a female each male was assumed to be the father of 50% of her young.

The development of the technique of DNA fingerprinting (Chapter 3, Section 3.5.2) makes it possible to determine paternity unambiguously. DNA obtained from blood samples from chicks was compared with that from all possible fathers and the paternity of each chick was determined. The results obtained for Davies' dunnock population are shown in Table 7.4.

Table 7.4 *Paternity of dunnock chicks, determined by DNA fingerprinting.*

Mating system	Sample sizes:		Percentage (%) young fathered by:	
	nests	chicks	alpha male	beta male
Monogamy	15	49	100	0
Polyandry	11	34	52.9	44.1*
Polygynandry **	19	50	72.0	28.0

* One chick was fathered by neither the alpha nor the beta male but by a male in a neighbouring group.

** Polygynandry describes a group in which there is more than one male and more than one female.

▶ Do the data in Table 7.4 support the assumption that the relative frequency with which a male copulated with a female provides a good estimate of his reproductive success?

Yes, they do. In monogamous relationships the male fathered all of a female's young. In relationships involving more than one male paternity is shared with the

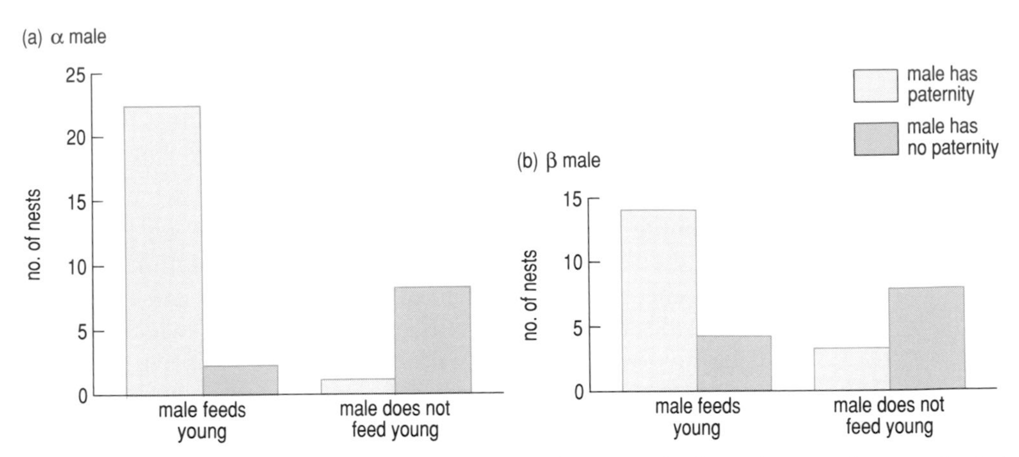

Figure 7.11 *The behaviour of (a) alpha male and (b) beta male dunnocks towards young in their nests according to their status within a breeding group (alpha or beta) and to whether or not they have paternity of at least some of the young in a nest.*

alpha male (the one that mated most often) having higher reproductive success than the lower-ranked males.

These paternity data enabled Davies to test further hypotheses about how males at polyandrous nests should behave towards young in those nests. For example, males should allocate time to caring for young in proportion to their certainty of paternity of those young (see under *polyandry* above). There are two possibilities:

1 Males should care only for those chicks that they themselves have fathered. To do this males would have to be able to recognize their own chicks.

2 Males should spend time providing care at a nest in proportion to their certainty of paternity of the young in that nest.

Davies found no evidence to support possibility 1. If males brought food to a nest, they fed all the chicks in that nest with equal frequency. Data relating to possibility 2 are shown in Figure 7.11.

▶ Do the data in Figure 7.11 support the hypothesis that males allocate care to young in proportion to their certainty of paternity of those young?

Yes, they do. Males (both alpha and beta males) at nests where they had fathered young were more likely to bring food to the nest, than males at nests where they had not fathered any young. Also, alpha males carried out more feeding than beta males; alpha males had mated with the female more often than had beta males.

7.4 SEXUAL SELECTION IN ACTION

Darwin's theory of sexual selection is based on the observation that males commonly compete with one another to mate with females. As described in Section 7.2.1, this results in a greater variation in mating success among males than among females. A consequence of this high variation in male mating success is that any character contributing to the reproductive success of individual males will be strongly favoured by selection. Darwin drew a distinction between two kinds of competition among males:

1 Males fight with each other; the winners claim the disputed females. This form of selection favours the evolution of greater male strength and of weapons such as horns; it is commonly called **intrasexual selection**.

2 Males do not fight but compete to attract the attention and sexual response of females. This will favour the evolution of conspicuous colours, structures and displays that attract females. This is commonly called **intersexual** or **epigamic selection**.

(The validity and usefulness of this distinction will be discussed later in Section 7.4.6.)

7.4.1 SELECTION ARISING THROUGH MALE AGGRESSION

Male aggression associated with mating provides a satisfactory explanation for the evolution of the antlers of male deer or the horns of stag beetles which are used in fighting. It also helps to explain why, in many species, males are bigger than females; a larger male is more likely to be successful in fighting than a small male. Figure 7.12 shows the relationship between harem size and the degree of disparity in size between the sexes for a number of species of pinniped mammals (sea-lions, seals and walruses).

▶ How can you explain the strong positive correlation between harem size and the degree of sexual dimorphism in size?

The larger the harem size the smaller the proportion of males that actually mate. (This conclusion assumes that there are as many males as females in the breeding population which is true for these species.) The smaller the proportion of males that mate, the more intense will be selection for large male size.

There is abundant evidence from several species such as elephant seals that larger males have much higher mating success than smaller males. Consequently there has been no dispute with Darwin's argument that selection arising from male

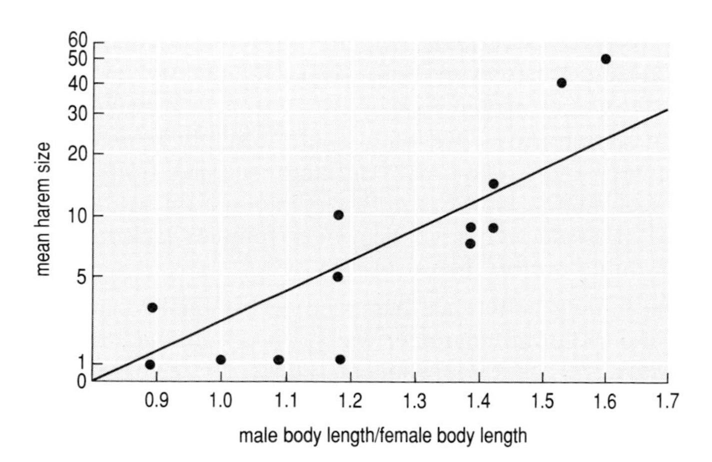

Figure 7.12 *The relationship between mean harem size (number of females) and sexual dimorphism in body-size in 13 species of pinniped mammals (sea-lions, seals and walruses). Note that mean harem size is plotted on a logarithmic scale.*

fighting has led to the evolution of both sexual dimorphism in size and of male weapons.

7.4.2 Selection arising through female choice

The argument that male characters have evolved through the action of female choice has long been, and still remains, controversial because it depends on the assumption that females 'prefer' to mate with certain males rather than others. Those males whose appearance or behaviour matches the female preference will be most successful in mating. Consequently they will pass on to their sons those characters that females prefer, assuming that those characters are heritable. As a result elaborate plumage or sexual displays have evolved in males. This argument begs an important and fundamental question: why should females have preferences for certain male characters?

Before we look in detail at this question we must narrow down the range of phenomena we are seeking to explain. As discussed in Section 7.3 the mating systems of many species are based on resources that are defended by males. In such species we would expect females to choose males on the basis of the resources they hold (as in the polygyny-threshold model), and there is much evidence that they do; this kind of female choice is not controversial. Where males provide parental care we might expect females to choose males according to their parental abilities; there is some evidence for this too. For example, in some birds such as the common tern *Sterna hirundo,* the male shows courtship-feeding by presenting the female with fish prior to mating. Females show a preference for males that bring them a lot of fish and a male's performance in courtship-feeding is positively correlated with the rate at which he subsequently brings food to chicks. What is difficult to explain is female choice in those species in which a male's sole contribution to reproduction is his sperm, as in lek species.

Darwin provided no answer to the question of why females might prefer certain males to others; he simply assumed that female preferences exist. As a result his theory of sexual selection through female choice was widely rejected until R. A. Fisher (1930) closed the loophole in the theory. Fisher argued that, for selection to lead to the evolution of a male character preferred by females, some selective advantage must accrue to females that show that preference. He envisaged the following two-stage evolutionary process:

1 Males possessing a certain character attractive to females have some other heritable advantage, however slight, over other males. In effect, the preferred character is a marker of some other factor that increases male fitness. For example, an ancestral peacock with a train slightly larger than that of other males might have been more adept at flying and escaping from predators. Females preferring such males would pass this advantage on to their sons; they would therefore be at a selective advantage over females lacking the preference since their sons would be more likely to survive.

2 Once the female preference has become established in the population, males possessing the preferred character are at an advantage simply because they are preferred; they will, as a consequence, attract more mates. More importantly, females preferring them will be at an advantage as they will pass the preferred character on to their sons. Their sons will be more attractive than other males and,

in their turn, will be more likely to attract mates. In other words, females that prefer attractive males are at a selective advantage because they will tend to have more descendants. Therefore male attractiveness is selected for its own sake. Fisher showed that this would lead to a *runaway process* in which ever greater development of an attractive male character such as a long tail would be favoured by selection. At this stage, it is no longer necessary that the preferred male character be associated with some other adaptive male attribute. Indeed, the preferred character will continue to be selected for even if, in other respects, it becomes a disadvantage to the male possessing it. The process will stop, however, when the advantage in terms of increasing attractiveness balances the disadvantage in terms of male survival. There can be little doubt that the peacock's train imposes a considerable burden, if only in terms of the energy required to grow it, but it has been selected for because a peacock's fitness is increased if his train enables him to attract more females.

In recent years several models of the process envisaged by Fisher have been developed. These models are very complex; an exposition of their detail is beyond the scope of this Book, but, in essence, they model the spread of genes for a male character and for a female preference through a population over successive generations. In general the models support Fisher's theory, particularly the argument that a male character that is costly in survival terms but which makes males attractive spreads through a population and becomes more exaggerated if preference genes for that character are present in females. The models also support the concept of a runaway process leading to rapid evolution of a male character. In genetic terms, this process leads to a phenomenon called **linkage disequilibrium** because females possessing 'choice genes' will non-randomly mate with males possessing 'character genes'. Provided that the respective genes are expressed only in the relevant sex then the sons of such females will inherit the character genes, the daughters the choice genes.

Genetic models of selection through female choice also reveal a problem that any theory of sexual selection must address. If variation in male mating success is so high that only those males possessing a preferred character to an extreme degree are successful then variation in the development of that character decreases within a very few generations (Chapter 4, Section 4.4.4). This effect is well-known in artificial breeding programmes. If, for example, one seeks to select for high milk-yield in cows and does so by breeding from only the very few, highest-yielding individuals in a population, then all the progeny will have high milk-yields after a few generations but there will be less variation on which further selective breeding can be based. In such a situation, some degree of variation is always maintained by mutation. For most genetic models of sexual selection to produce realistic results, it is necessary to assume rather unrealistically high levels of mutation.

This problem is largely overcome by one particular model, the *condition-dependent* or *viability-indicator* model, which assumes that males, according to their condition, vary in the extent to which they develop those characters that attract females. Therefore, very healthy, well-nourished males develop larger characters than those that are sick, weak or poorly-nourished. Because condition is a character that will be the expression of a male's entire genotype, not just a few specific genes, variation is likely to remain high, even under intense selection, compared with variation in a character that is the expression of only a few genes.

Another feature of this model is that females which choose males with large characters will mate with the most vigorous males; if male vigour has a heritable component, such males have more vigorous progeny.

There is little direct evidence to support the condition-dependent model though there is good circumstantial evidence. For example, the size of the male peacocks' train (Plate 7.1), shows considerable age-related variation; each year a male, after moulting, grows a slightly larger, more elaborate train than in the previous year. Males do not develop a full-length train until they are about four years old, after which time the train acquires additional 'eyespots' each year. This concept is consistent with the idea that it takes several years for males to develop their foraging skills to a level that enables them to sustain the energetic cost of a large train. Only males with fully-developed trains attract females and, among them, those with the most eyespots obtain the most matings.

More direct evidence comes from a study by J. M. R. Baker at the Open University of the great crested newt *Triturus cristatus* (Baker 1992). He captured male newts as they entered a breeding pond, weighed and measured them to obtain a score of their general condition, expressed as weight divided by length. Some days later he captured the same animals in the water when they had developed the dorsal crest that is displayed prominently during courtship. There was a positive correlation between condition and crest-height; males that were in good condition on arrival at the pond, i.e. heavy for their size, developed larger crests.

7.4.3 Evidence for selection through male fighting

In the European common toad *Bufo bufo*, breeding takes place over a few days in spring with males and females arriving, often in large numbers, at suitable ponds. At most breeding sites males outnumber females by three or four to one, so there is intense competition among males to clasp females in a position called *amplexus* which is maintained until the female spawns. Most females are intercepted by males before they reach the pond so that they arrive already in amplexus; they spawn some two to four days after entering the water. Figure 7.13 shows the relationship between male body size and the probability of being in amplexus with a female.

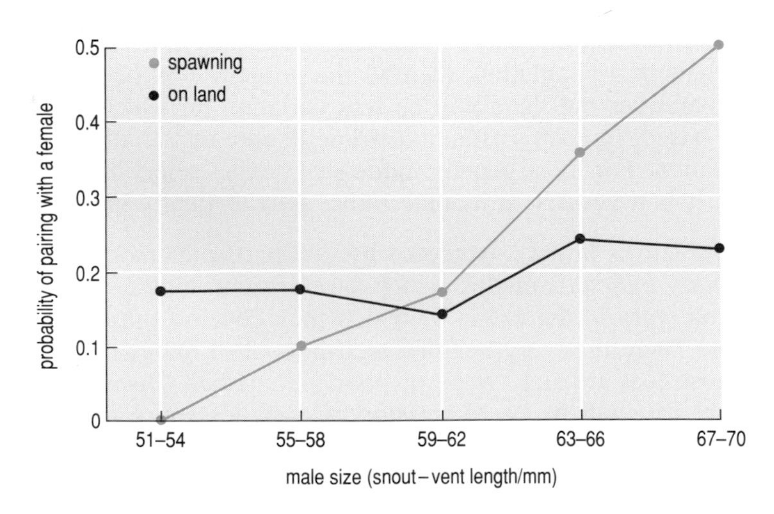

Figure 7.13 *The probability of being in amplexus with a female for five body-size classes of male toads* Bufo bufo *at two points in the breeding season: on land as animals approach a pond, and in the pond as females spawn.*

▶ For animals approaching the pond, is there any relationship between male size and their probability of being in amplexus?

No. Males in the five body-size classes have roughly equal probability (approximately 0.2 = 20%) of being in amplexus.

▶ Is there a relationship among the males when spawning is taking place?

Yes. The largest male body-size class is much more likely (0.5) to be in amplexus than smaller males; males in the smallest body-size class have zero chance of being in amplexus.

▶ How can you explain the difference in male probability of being in amplexus between the two stages of mating?

Clearly those smaller males that succeed in gaining amplexus with females as they approach the pond tend to get displaced by larger males.

These data provide clear evidence that large body-size confers a mating advantage on males. They suggest that this arises because of a variation among male body-size classes in the ease with which they can be displaced by larger rivals. This theory was tested in an experiment in which three toads (one female, one small and one large male) were placed in a bucket and left to see which male assumed amplexus with the female. Forty-one replicates of this experiment were conducted; the results are summarized in Figure 7.14.

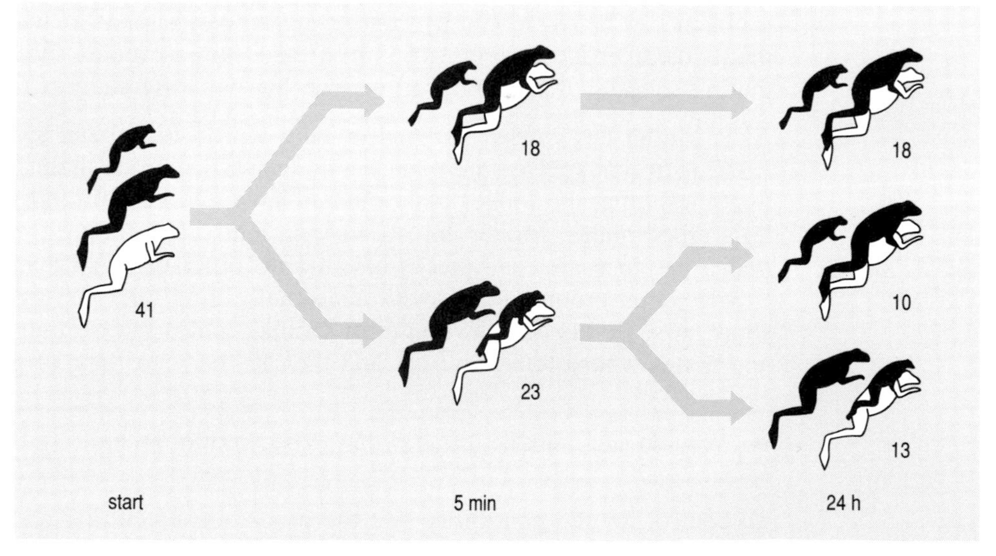

Figure 7.14 *Results of 41 replicates of an experiment in which two male toads* Bufo bufo, *one large and one small, were placed in a bucket with a single female. Within five minutes, one of the males has clasped the female. After 24 hours the amplexed male has been displaced by the other in some of the replicates.*

▶ Do the results shown in Figure 7.14 suggest any male body-size advantage in terms of initially capturing the female?

No. Small males were equally likely (if anything more likely) than large males to be first to clasp the female.

▶ Does the pattern of pairing observed after 24 hours support the hypothesis that large males can displace smaller ones from the backs of females?

Yes. In all buckets in which the larger male amplexed the female first he was still in amplexus 24 hours later. In 10 out of 23 buckets in which the small male was first to clasp the female he had been displaced by the larger one. Toads provide clear evidence, both observational and experimental, supporting the hypothesis that large body-size is adaptive in mating competition. Despite this, toads provide a paradox because males are considerably smaller on average than females, a point to which we will return shortly.

7.4.4 EVIDENCE FOR SELECTION THROUGH FEMALE CHOICE

The theory of sexual selection predicts that where males vary in the size of characters that are attractive to females, those with larger characters should attract more females than those with smaller characters. This prediction can be tested observationally by measuring the size of males' display characters and recording their mating success. In many instances the prediction has been supported. Observational data are not, however, sufficient to test the prediction thoroughly. It could be that males with large display characters have some other attribute, such as being more competitive, that makes them more successful at mating. This problem can be overcome by conducting an experiment in which the size of a male character is manipulated; if it is made larger, males should enjoy higher mating success; if it is reduced in size, their mating success should decrease.

A. P. Møller has carried out such a study on the European swallow *Hirundo rustica*. In this species the male has longer tail plumes than the female. To test the hypothesis that the longer male plumes have evolved through female choice for long-tailed males, Møller cut off varying lengths of plume from the tails of several males. He then attached pieces of tail onto other birds so that he had two experimental groups; males with abnormally long tails and males with abnormally short tails. He also had two control groups; males whose tails were left untouched and males whose tails were cut and then re-made to their original length.

▶ Why do you think he included this second control group?

To control for the possibility that the procedure of cutting and remodelling tails might influence the birds' behaviour and reproductive success. After treatment the males were marked and released into the wild. Møller then monitored their reproductive behaviour and gathered data on three aspects of their reproductive success: how soon they acquired a female; whether they (and their female)

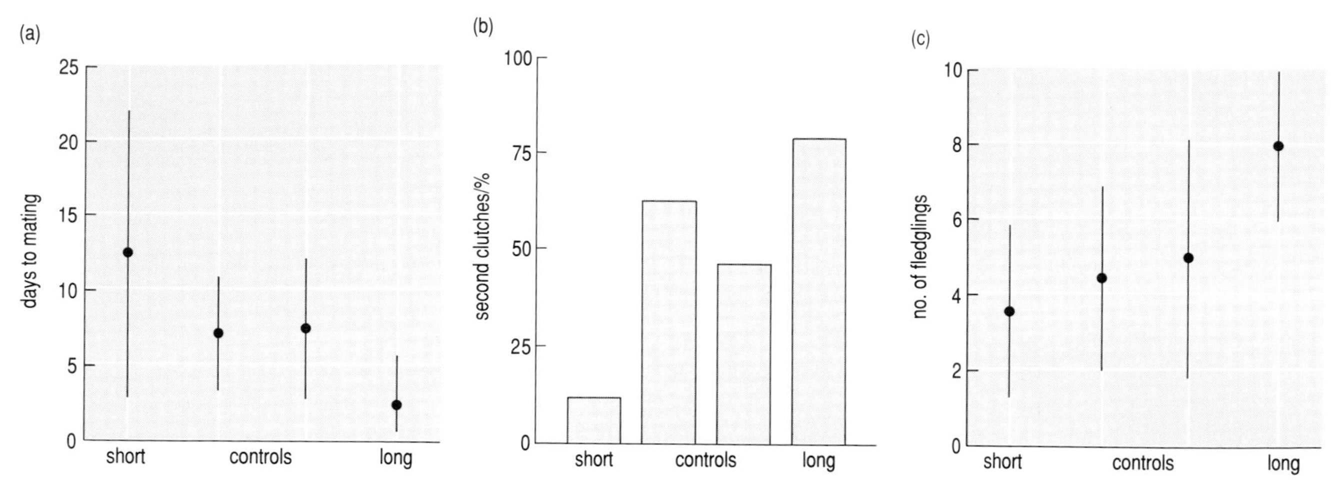

Figure 7.15 *Three measures of the reproductive success of pairs of swallows* Hirundo rustica, *in which males had lengthened, shortened or normal-length tails. (a) Length of pre-mating period measured in days from arrival at the breeding site to pair formation. Points are mean scores; vertical lines show standard deviations. (b) Proportion (%) of males in which the pair produced a second clutch. (c) Number of fledglings reared per pair. Points are mean scores; vertical lines show standard deviations.*

produced a second clutch in the season; and the number of fledglings they successfully reared. The results are shown in Figure 7.15.

▶ Do the data in Figure 7.15 support the hypothesis that males with longer tails are more attractive to females?

Yes. Males with lengthened tails attracted females more quickly than those with normal-length or shortened tails. Males with shortened tails attracted females later than those with normal tails.

▶ What conclusion, about the females' fitness, do you draw from the other two scores?

The preference for longer-tailed males is adaptive. Females mating with longer-tailed males were more likely to have a second clutch and reared more fledglings.

Møller also monitored all the males after the breeding season and into the following season. He found that males with artificially-lengthened tails gradually accumulated more signs of wear and tear in their plumes. He recorded the size of prey that males caught and found that the long-tailed experimental males caught, on average, smaller prey items (Figure 7.16). Swallows catch insects in flight and the longer-tailed males appeared to be less agile in the air when chasing prey. In the following season, after their annual moult, males whose tails had, in the previous year, been short or of normal length grew tails slightly longer than in the first year. In contrast, males whose tails had been lengthened in the first year grew a shorter tail in the second year than they had grown naturally in the previous year (Figure 7.16).

▶ What conclusion do you draw from the data in Figure 7.16?

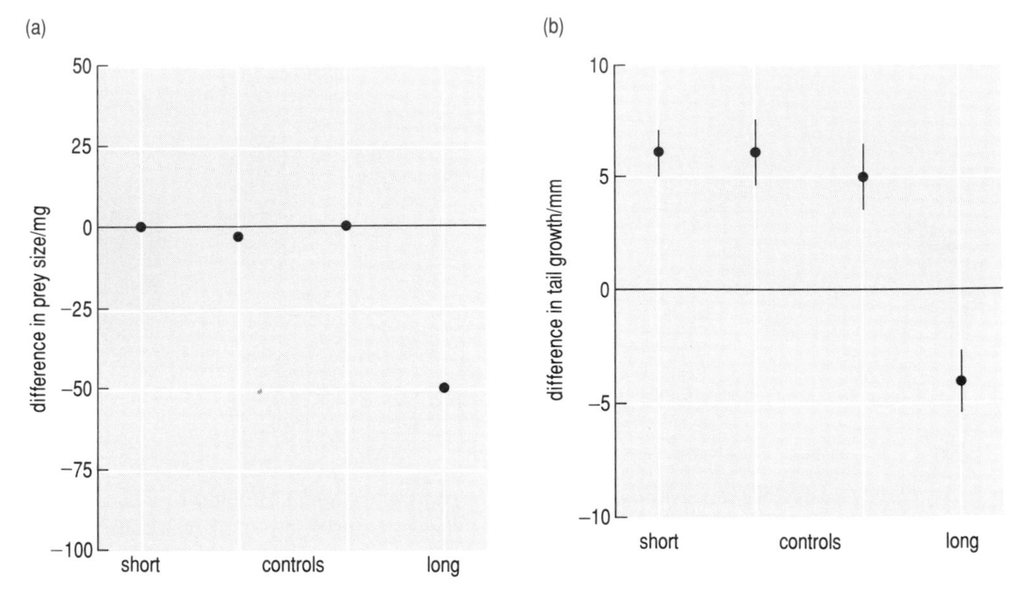

Figure 7.16 *Two measures of male reproductive effort according to whether males had lengthened, shortened or normal-length tails. (a) Median values of the difference in the size of prey brought to the nest by a male and his female partner. Points are median values. (b) The change in the length of males' tails from one year to the next as a result of a manipulation of their tail in the first year. Points are mean values; vertical lines show standard deviations.*

That a long tail imposes fitness costs on males. They are less able to catch large prey, they lose condition and, in the following year, they grow a shorter tail making them less attractive to females.

This elegant experiment provides empirical support for a number of assumptions in the theory of selection for male characters through female choice. Females do prefer males with extreme characters. This preference is adaptive for females but extreme characters impose fitness costs on those males that possess them.

7.4.5 Sexual selection and sexual dimorphism

It is commonly assumed that a sex difference in size or appearance is *a priori* evidence that sexual selection has acted on one sex. This assumption is unsafe. The recently extinct Huia *Heterolocha acutirostris* of New Zealand showed marked sexual dimorphism in beak shape (Figure 7.17). The male had a short, stout beak and found insect food by hammering into branches, whereas the female used her longer, curved beak to probe into crevices. As a result of this dimorphism, male and female obtained their food from different parts of a tree and so avoided competition for the same food sources. Similarly, in many birds of prey the female is slightly larger than the male and takes slightly larger prey (Chapter 8).

Differences in body size between the sexes can also be the result of life-history differences between the sexes. As we saw in the previous Section male common toads are smaller than females despite there being strong selection favouring large male size in competition for females. To understand this paradoxical effect we have to consider the life history of male and female toads. Toads are long-lived

Figure 7.17 *The extinct Huia of New Zealand. The male is on the left.*

animals that breed several times during their lives; they are *iteroparous*. They continue to grow throughout life although the rate at which they grow slows down markedly when they breed for the first time. Females start to breed one or two years later than males. For females, fecundity (in terms of number of eggs) is strongly correlated with body size; larger females lay more eggs. Females achieve large body size and consequently greater reproductive success due to delay in their first breeding relative to males. As a result fewer females than males survive to breeding age which accounts for the fact that males greatly outnumber females in toad breeding populations. For males, as we saw in the previous Section, there is an advantage to being large but this effect is not seen at all localities and does not occur in all seasons at a given locality. It appears that selection favouring breeding early in life is stronger than selection for large body size in males. Therefore sexual dimorphism in body size in toads is not explained by the action of sexual selection. Sexual selection is clearly acting on males but its effect is not as strong as that favouring large body size in females. In many animals, notably fishes and insects, female fecundity is strongly correlated with body size and, as a result, females are larger on average than males.

It is important to realize that sexual dimorphism is not a character upon which selection can act directly. Natural selection will act directly on both male and female body size but the difference between them is a consequence of these separate selection pressures acting in different ways on the two sexes.

7.4.6 Intrasexual and intersexual selection

The distinction between intrasexual and intersexual selection focuses our attention on the fact that competition among males for females can take different forms. Nature is not simply red in tooth and claw, it is also a source of beauty. The dichotomy is, however, misleading in certain ways. First, the two processes are not alternatives and there are numerous examples of mating systems in which male aggression and female choice for male characters are both involved as, for

Figure 7.18 *Female elephant seals* Mirounga angustirostris *can influence fighting among males. Females may 'protest' loudly when a male attempts to mount them. (a) If they protest, the male's mounting attempt is more likely to be interrupted by a dominant, high-ranking male, especially if the mounting male is of low rank. (b) Females are more likely to protest when a young, low-ranking male attempts to mount them than when a dominant male does.*

example, in elephant seals (Figure 7.18). Second, both kinds of selection can operate on a single male character. For example, the black throat patch of the European house sparrow *Passer domesticus* varies in size according to a male's dominance status. Males with large throat patches are more likely to win aggressive interactions and so have higher status; they are also preferred as mates by females. Intrasexual and intersexual selection should be seen as different manifestations of a single evolutionary process: competition among males to obtain mates. In some species males compete by fighting, in others they compete by being attractive to females.

7.4.7 Sperm competition

So far, this Chapter has focused on sexual competition between males as individuals and has considered very obvious behavioural and morphological manifestations of this competition such as fighting and weapons. There are, however, more subtle ways by which males compete—a very widespread example is sperm competition. Sperm competition occurs when one or more males produce sperm such that those sperm compete to fertilize an individual female's eggs. In many fish species in which the eggs are fertilized outside the body (external fertilization), a single male will dart towards a pair engaged in mating and will ejaculate over the eggs at the same time as the mating male. In the bluegill sunfish *Lepomis macrochirus*, small, single males 'sneak' into the nests of larger males by mimicking female behaviour; once there they shed sperm onto the eggs just after they are laid. In animal species in which internal fertilization takes place, the occurrence and nature of sperm competition depends on the female showing two features in her reproductive biology. The first is that she stores sperm for a period of time between mating and the fertilization of the eggs and the second is that she mates with more than one male during this period. When ejaculates from two or more males are stored within a female's reproductive tract one of the following patterns may occur:

1 The ejaculates mix such that the male that delivered the most sperm fertilizes the highest proportion of her eggs.

2 The first male to mate with the female fertilizes a higher proportion of her eggs than later males (first male sperm-precedence).

3 The last male to mate with the female fertilizes a higher proportion of her eggs than previous males (last male sperm-precedence).

Sperm competition has been described in a wide variety of animals, notably insects and birds, and patterns 1 and 3 are both relatively common whereas pattern 2 is rather rare.

The existence of sperm competition has led to the evolution of a wide variety of sexual adaptations in males. As described in Section 7.3.2, certainty of paternity is an important determinant of some mating systems. If a male can prevent sperm competition by ensuring that he alone fertilizes a female's eggs then adaptations that enable him to do so will be favoured by selection. In some insects, snakes, and mammals such as rats, males leave a gelatinous secretion in the female's reproductive tract after mating; this acts as a mating plug that hinders or prevents matings by other males. In the damselfly *Calopteryx maculata,* the male has a special, scoop-shaped organ at the end of his abdomen with which he removes sperm from a female's reproductive tract before mating with her himself. A common feature of damselflies and closely-related dragonflies is a prolonged period of 'mate-guarding' during which the male remains attached to the female between the time of mating and when she lays her eggs so warding off the attentions of other males. In some insects, mate-guarding may last for many hours or even several days.

► What effect will mate-guarding have on a male's ability to mate with several females?

It severely limits it. For males there is a trade-off between guarding a female with which they have already mated and seeking new females. If they seek new matings they are likely to have their reproductive success from previous matings 'diluted' by the effect of sperm competition. Among birds, males cannot guard females by physically holding onto them as many insects can, instead they may protect their paternity of the female's eggs by repeatedly mating with her and/or by driving other males away. In the European house sparrow *Passer domesticus,* a pair mate as often as 40 times a day during the time that a female is laying her clutch of eggs.

In general, the effect of sperm competition is to force males to invest their available resources in certain components of reproductive effort such as mate-guarding, sperm production and repeated mating. As a result they have fewer resources to allocate to other activities, particularly searching for and attracting other females.

7.5 ALTERNATIVE MATING STRATEGIES

In mating systems where there is intense competition among males, typically only a certain proportion of males will achieve any mating success. This creates a situation in which selection will favour any pattern of behaviour in less competitive males that enables them to achieve a degree of mating success greater than they would otherwise achieve. These behaviour patterns, which differ from those shown by the successful males, are called **alternative mating strategies**. The nest-building and 'sneaky' mating behaviour patterns of large and small bluegill sunfish (Section 7.4.7) provide one example.

Many species of frogs gather in large groups, called choruses, in which males call loudly to attract females. In many instances it is observed that at any one time only a proportion of males are calling. Non-calling males tend to aggregate around calling males and show what is called *satellite* behaviour when they attempt to intercept and mate with females that approach the calling males (Figure 7.19). Calling and being a satellite are both alternative mating strategies. In some species such as the American bullfrog *Rana catesbeiana,* larger males call more loudly and more persistently than smaller males. Calling is energetically very expensive for frogs and it appears that smaller males simply cannot sustain the same level of calling activity as larger males. In this species, larger males tend to be callers and smaller males adopt the alternative, satellite strategy. If calling males detect the presence of satellite males they attack them and drive them away.

In some frogs, however, calling and satellite strategies are not related to size. In the North American green treefrog *Hyla cinerea,* there is no difference in the mean size of callers and satellites and, when individuals are monitored from night to night, it is clear that any individual male is a caller on some nights, a satellite on others. Which strategy should a male adopt on any given night? If all males call at once, the probability that any one of them will attract one or more of the relatively few receptive females available on that night is very low; the best option for an

Figure 7.19 *Callers and satellite males in a frog chorus.*

individual is to stop calling, conserve his energy and adopt satellite behaviour. If all males are silent, no females will be attracted into the chorus. This kind of reasoning leads to the conclusion that males should adopt calling or satellite behaviour according to the relative mating success achieved by each strategy and that a balance will occur when each strategy yields the same return. If callers are experiencing higher mating success than satellites, an individual satellite should switch to calling. If satellites are being more successful, an individual caller should shut up and become a satellite. The ratio of 'time spent calling' to 'time spent as a satellite' that achieves greatest mating success is called the evolutionarily stable strategy or ESS (Chapter 5, Section 5.4) for that individual.

In a caller-satellite system, the mating success of each type depends on the frequency of that type relative to the frequency of the other type.

▶ What kind of selection is operating in such a situation?

Negative frequency-dependent selection (Chapter 4, Section 4.4.3)

The green treefrog system has been experimentally investigated by Perrill *et al.* (1978, 1982). They first observed choruses to determine the relative numbers of callers and satellites and found that, of 1 237 calling males observed, 207 (16%) had satellites close to them. They also observed 30 matings of which 17 were obtained by calling males, 13 by satellites. Thus, satellites achieved a mating success nearly equal to that of callers, suggesting that a caller to satellite ratio of 84 to 16 is close to an ESS. They then tested the hypothesis that male green treefrogs are sensitive to what other males are doing and alter their behaviour accordingly. They removed the calling male from 19 caller-satellite associations and found that the satellite soon began to call on 11 occasions. In a further 10 trials, a calling male was replaced by a loudspeaker playing green treefrog calls; on all occasions the satellite male moved and sat near the speaker. Finally, they played recorded calls to 14 calling males and found that in 11 trials the calling male stopped calling and adopted a satellite position near the speaker. These observations support the hypothesis that frogs alter their behaviour according to the behaviour of their immediate rivals and by so doing maximize their chances of obtaining mates while minimizing the physiological costs of calling.

7.6 IS SEXUAL SELECTION A DISTINCT PROCESS?

Having looked at various aspects of sexual behaviour and at the evolution of characters related to sex, it is time to return to an issue raised at the beginning of the Chapter: is sexual selection different from natural selection? The fact that natural selection and sexual selection are often treated as distinct processes undoubtedly owes much to the fact that Darwin discussed them in separate books. Also, by the time the later book in which he set out his theory of sexual selection was published, biologists had established a view of natural selection that emphasized survival. However it must be emphasized that Darwin, in his earlier book, stressed the importance of reproduction in natural selection and in it he did allude to sexual selection. Some authors, such as Ernst Mayr (1972), have suggested that what makes sexually-selected characters different is that, while they confer an advantage in terms of the reproductive success of individual males, they are detrimental to the fitness of the species. This notion contains a fundamental misconception about natural selection, departing, as it does, from Darwin's clear view that natural selection acts on *individuals*, not on species. (This issue was discussed in Chapter 6.)

The neo-Darwinian view of evolution places greater emphasis on the reproductive success rather than the survival of individuals. This is especially true in population genetics where evolution tends to be seen as the spread of alleles through populations as a result of the individuals carrying those alleles reproducing more successfully than other individuals. The neo-Darwinian view sees exaggerated male characters as the evolutionary result of a trade-off, not between individual fitness and species fitness, but between individual survival and individual reproductive success. According to this, the evolution of extreme male characters involves a trade-off between reproductive and somatic effort directly analagous to that made by females.

Detailed analysis of the evolution of sex-related characters gives us a more balanced perspective on the concepts of adaptation and function because it is so apparent that such characters can be costly in survival terms. For such a character it is necessary to recognize that it confers both costs and benefits. Its function is the net balance between costs and benefits. It can be said to be an adaptation if the benefits exceed the costs. In this respect, sex-related characters are not as different from other characters as they might at first appear. As discussed in Chapter 2 for example, many animals that have evolved a cave-living habit have lost their colour and many island-dwelling birds have reduced or vestigial wings. Colour and wings are clearly adaptive features in most environments but the fact that they are lost in new contexts suggests that they incur costs. The guppy, discussed in Chapter 4, provides a good example of how colour pattern, a sex-related character, is the product of opposing selective factors.

Before leaving this question, it is important to point out that the assumption that extravagant male characters are costly has only rarely been tested. There is some evidence from certain birds that, in highly sexually-dimorphic species, the larger and/or brighter male has reduced longevity compared to the female, though the reasons for this are not known. This is true also of the polar bear, in which males are larger than females. In one instance, an exaggerated male character is positively advantageous in survival terms. The male fiddler crab *Endocimus albus* has one greatly enlarged claw with which he displays to females; the female fiddler crabs mate preferentially with males with larger claws. This claw also affords enhanced protection against predation by birds such as ibises.

Therefore, although sexual selection is widely referred to in the literature as if it were a process distinct from natural selection, it should be regarded as a form of natural selection; in Darwin's words it is 'selection in relation to sex'. It does not involve any process that is inherently different to those involved in the evolution of characters that are not related to sex.

7.7 Sexual selection in plants

So far this Chapter has dealt exclusively with animals and has ignored plants. Are there any phenomena in plants that can be attributed to the action of sexual selection? At first sight, the answer would appear to be 'no'. Plants cannot move about and so cannot fight over mates, nor, apparently, can they actively choose their mates. Furthermore, many plants are bisexual and carry male and female reproductive organs on the same individual; and it seems illogical to think of a plant competing with itself. There are, however, certain features of plants where sexual selection, or a comparable evolutionary process, is clearly important.

The brightly coloured and highly elaborate flowers of many plants appear to be the botanical equivalent of male display structures such as the peacock's train in that they are very conspicuous. In many plants, however, female flowers may be as conspicuous or more conspicuous than male flowers, and in many others the functions of both sexes are performed by the same flowers. In most species with conspicuous flowers it is clear that the flowers have evolved to be attractive but not to other plants; they serve to attract insects, birds and other animals that visit flowers and carry pollen from one flower to another. In this context the underlying basis of selection for attractiveness is likely to be, not only competition between one plant and another, but the fact that plants may be widely separated from one another and so must attract pollinators over a considerable distance.

A more direct comparison between plants and animals concerns competition between gametes rather than between adult individuals. In animals this takes the form of sperm competition (Section 7.4.7); in plants it is seen as competition between pollen grains. In flowering plants female flowers, or the female parts of bisexual flowers, include a stigma on which pollen grains settle, brought there either by the wind or by a pollinator, such as a bee (Chapter 5, Section 5.2). If a bee has previously visited several flowers it may deposit pollen grains from several individual plants on the same stigma. From each pollen grain, a pollen tube grows down the stigma in what amounts to a race to reach the female gametes. Each pollen tube contains a male cell, in effect a sperm, that will fuse with a female gamete. To grow down the stigma the pollen tube must interact with the tissues of the female flower; it uses nutrients in the stigma to sustain its growth. Experiments have shown that pollen tube growth rate is a heritable character controlled by genes in the pollen grain. A feature of the stigmas of many flowers is that they are very long relative to the size of the gametes so that pollen tubes have a long way to grow. It has been suggested that female flowers have evolved long stigmas as a mechanism that selects for 'superior' male genes. Those pollen genotypes that interact most effectively with the stigma grow fastest and will thus fertilize the female gametes. There is rather little evidence to support this theory at present and most evidence that does exist relates to inbreeding and outbreeding (Chapter 5, Section 5.1.6). In some plants the pollen of very closely related individuals do not grow as fast down the stigma as those from unrelated individuals. In other words, the interaction between stigma and pollen grain may be such as to favour maximum complementarity between male and female genotypes.

SUMMARY OF CHAPTER 7

A character difference between the sexes (sexual dimorphism) suggests that natural selection operates differently on the two sexes. When the character in question is related to sexual behaviour, the process by which it has evolved is called sexual selection. Of particular interest are extravagant sex-related characters in males such as very large weapons and elaborate plumage.

The basis of morphological and behavioural differences between the sexes is anisogamy, the existence of two kinds of gamete, eggs and sperm. Because these gametes are typically produced in very different numbers, there is a conflict of interests between the sexes in terms of how each may maximize its reproductive success. Data from both laboratory and field confirm that, as expected, variation in reproductive success is often higher among males than among females.

For individuals of either sex, reproduction involves a trade-off between allocation of energy to reproductive effort and to somatic effort. There are also trade-offs between various categories of reproductive effort such as gamete production, competition for mates, mating and parental care. For a given species the optimal pattern of trade-offs is often different for the two sexes because of anisogamy; it is also strongly influenced by environmental factors.

The pattern of mating between males and females is called the mating system. The principal mating systems are monogamy, polygyny and polyandry. The mating system within a species is determined largely by environmental factors, particularly the abundance and distribution of resources and the intensity of predation. Also important are the specific requirements of the young such as the stage of development at which they are born. The mating system of a species is the result of selection acting separately on males and females; there may be much variation in mating behaviour within a species, as observed in the dunnock.

Sexual reproduction commonly involves competition among males for access to females. This competition may be expressed in terms of fighting and may lead to selection favouring large body size and weapons; it may consist of competition to be attractive to females leading to the evolution of elaborate plumage and sexual behaviour. These two processes often occur together and the distinction between them is not a useful one. There is considerable evidence that particular male characters confer an advantage in both forms of mating competition. Not all instances of sexual dimorphism are attributable to sexual selection; some are related to ecological differences between the sexes.

Intense mating competition may lead to the evolution of alternative mating strategies, callers and satellites among male frogs for example. Where two strategies exist, a balance will be established such that the two strategies will occur at those frequencies at which they achieve roughly equal pay-offs. This balance is called an evolutionarily stable strategy (ESS).

Sexual selection should be regarded, not as a separate process from, but, as a particular form of natural selection.

REFERENCES FOR CHAPTER 7

Baker, J. M. R. (1992) Body condition and tail height in crested newts, *Triturus cristatus, Animal Behaviour*, in press.

Bateman, A. J. (1948) Intra-sexual selection in *Drosophila, Heredity,* **2**, 349–368.

Burke, T., Davies, N. B., Bruford, W. W. and Hatchwell, B. J. (1989) Parental care and mating behaviour of polyandrous dunnocks *Prunella modularis* related to paternity by DNA fingerprinting, *Nature,* **338**, 249–251.

Darwin, C. (1871) *The Descent of Man and Selection in Relation to Sex,* John Murray, London.

Davies, N. B. and Halliday, T. R. (1977) Optimal mate selection in the toad *Bufo bufo, Nature,* **269**, 56–58.

Davies, N. B. and Halliday, T. R. (1979) Competitive mate searching in male common toads, *Bufo bufo, Animal Behaviour,* **27**, 1253–1267.

Davies, N. B. and Houston, A. I. (1986) Reproductive success of dunnocks *Prunella modularis* in a variable mating system. II. Conflicts of interest among breeding adults, *Journal of Animal Ecology,* **55**, 139–154.

Fisher, R. A. (1930) *The Genetical Theory of Natural Selection,* Clarendon Press, Oxford.

Mayr, E. (1972) Sexual selection and natural selection, in B. Campbell (ed.) *Sexual Selection and the Descent of Man, 1871–1971*, 87–104, Heinemann, London.

Møller, A. P. (1988) Female choice selects for male sexual ornaments in the monogamous swallow, *Nature,* **332,** 640–642.

Møller, A. P. (1989) Viability costs of male tail ornaments in a swallow, *Nature,* **339**, 132–135.

Perrill, S. A., Gerhardt, H. C. and Daniel, R. E. (1978) Sexual parasitism in the green tree frog (*Hyla cinerea*), *Science,* **200**, 1179–1180.

Perrill, S. A., Gerhardt, H. C. and Daniel, R. E. (1982) Mating strategy shifts in male green treefrogs (*Hyla cinerea)*: an experimental study, *Animal Behaviour,* **30**, 43–48.

FURTHER READING FOR CHAPTER 7

Emlen, S. T. and Oring, L. W. (1977) Ecology, sexual selection, and the evolution of mating systems, *Science,* **197**, 215–223.

Gould, J. L. and Gould, C. G. (1989) *Sexual Selection,* Scientific American Library, New York.

Halliday, T. R. (1980) *Sexual Strategy,* Oxford University Press.

Harvey, P. H. and Bradbury, J. W. (1991) Sexual selection, in J. R. Krebs and N. B. Davies (eds) *Behavioural Ecology. An Evolutionary Approach* (3rd edn), 203–233, Blackwell Scientific Publications, Oxford.

Thornhill, R. and Alcock, J. (1983) *The Evolution of Insect Mating Systems,* Harvard University Press.

Trivers, R. L. (1972) Parental investment and sexual selection, in B. Campbell (ed.) *Sexual Selection and the Descent of Man, 1871–1971,* 136–179, Heinemann, London.

SELF-ASSESSMENT QUESTIONS

SAQ 7.1 (*Objectives 7.1 and 7.8*) Define the following terms:

(a) anisogamy;

(b) polygyny;

(c) somatic effort;

(d) satellite male.

SAQ 7.2 (*Objective 7.1*) Figures 7.5(a) and (b), and Figures 7.6(a) and (b) show different possible patterns of reproductive resource allocation. From these four possible scenarios, select the diagram that best fits the situation of each of the following:

(a) female turtles;

(b) modern Western human males.

SAQ 7.3 (*Objectives 7.2 and 7.5*) What is it about the nature of parental care in mammals and birds that contributes to the greater frequency of monogamy found among birds compared with that found among mammals? (*Hint*: pigeons differ in this respect from mammals but are still monogamous.)

SAQ 7.4 (*Objective 7.3*) Use Figure 7.9 to answer this question.

(a) Consider the situation where there is one female on territory *y* and where there is an additional territory *z* which is exactly intermediate in quality between *x* and *y*. If both *x* and *z* are empty, predict the choice that will be made by the second and third females to arrive in the area.

(b) Suggest what resources might determine territory quality.

(c) Figure 7.9 does not show what would happen if a third female chose territory *y*. Based on your answer to (b), suggest some possible outcomes.

SAQ 7.5 (*Objective 7.4*) Give two ways in which possible uncertainty of paternity may limit the range of reproductive options open to a male.

SAQ 7.6 (*Objective 7.5*) Explain why polyandry is a rare mating system.

SAQ 7.7 (*Objectives 7.6 and 7.7*) Jackson's widowbird *Euplectes jacksonii* is a lek-breeding bird. Adult males show variation in the length of their tails, and it is said that long tails have evolved through the effect of female choice. What evidence would you require to support this assertion?

CHAPTER 8

EVOLUTIONARY ECOLOGY

Prepared for the Course Team by Tim Halliday

Study Comment *This Chapter discusses a number of ecological factors that have a major impact on the evolution of plants and animals. Much of the Chapter is concerned with ecological interactions between species, such as competition for resources, predation and parasitism, and it looks at the way the different organisms involved in such interactions influence each other's evolution. The final Section looks at a number of features of the life history of organisms, such as their lifespan and pattern of breeding, and discusses how these have evolved in response to the nature of the environment in which organisms live.*

Objectives When you have finished reading this Chapter, you should be able to do the following:

8.1 Compare and contrast the evolutionary consequences of intraspecific and interspecific competition.

8.2 Explain what is meant by a cost–benefit approach to aggression in the context of competition between animals.

8.3 Interpret information concerning ecological differences between species in terms of whether or not it provides evidence that competition actually or potentially occurs between them.

8.4 Categorize different kinds of interaction between pairs of species according to whether each species derives a fitness benefit, incurs a fitness cost, or whether the interaction has no impact in fitness terms.

8.5 Describe adaptations of prey species that may evolve as an evolutionary response to predation.

8.6 Explain, with relevant examples, why parasite–host interactions may often evolve towards less exploitative, more commensal forms of interaction.

8.7 Distinguish between Batesian, Müllerian and aggressive mimicry.

8.8 Present general arguments concerning selective factors that may affect the timing and frequency of breeding within a given species.

8.9 Distinguish betweem semelparity and iteroparity and compare the ecological conditions under which each is likely to evolve.

8.10 Recall the nature of life-history variables that are characteristic of r- and K-strategist species.

8.1 INTRODUCTION

Ecology is the scientific study of the interrelationships among organisms, and between organisms, and all aspects, living and non-living, of their environment. The word is derived from the Greek *oikos*, meaning 'house' or 'dwelling place'. Because ecology is concerned with the interaction between organisms and their environment, an understanding of ecology is fundamental to understanding adaptation. As emphasized in Chapter 2, the detailed analysis of adaptations requires that we know a great deal about how those features of organisms that we call adaptations enable plants and animals to survive and reproduce in their natural environment. The long finger of the aye-aye enables it to find its prey (Chapter 2, Figure 2.1), the clutch size of birds is related to the available food supply, egg recognition by birds enables them to eject the eggs of cuckoos, the ability to lay down adipose tissue and to hibernate enables a variety of mammals to survive the harsh winter. All these adaptations are understood in terms of how animals survive and reproduce in the face of demands imposed by their physical environment (climate etc.) and their biological environment (members of the same or other species).

In this Chapter, we look at a number of features of the ecology of plants and animals and consider how those features shape the course of evolution. This is a huge subject that has filled several large textbooks and what follows is, of necessity, a very restricted selection from a vast array of topics. Because this Book combines biological and palaeontological approaches to evolution, and because the principal common ground between the two disciplines is morphology, greater emphasis is placed on the relationship between ecology and morphology than on other features of organisms.

8.2 COMPETITION

The theory of natural selection, as formulated by Darwin and Wallace, expressly assumed that there is a *struggle for existence* between individuals within a species (Chapter 1); this struggle for existence is now referred to as *competition*. Competition refers to interactions between individual organisms in which one individual consumes a resource that would otherwise have been available to, and might have been consumed by, another individual. The acquisition of resources, such as oxygen, light, food and mates, is a prerequisite for the survival and reproductive success (i.e. the fitness) of the individual. Because competition involves the acquisition of resources by one individual at the expense of another, it leads to variation in fitness. Variation in fitness provides the driving force for the evolution of adaptations by natural selection.

Competition may be defined as an interaction between individuals, brought about by a shared requirement for a resource in limited supply, and leading to a reduction in the survivorship, growth and/or reproduction of the competing individuals concerned.

▶ What implication does this definition have for the concept of *fitness*?

The second part of the definition states that competition leads to a reduction in one or more components of fitness. As discussed in the rest of this Chapter, what is important in evolutionary terms is the extent to which individuals show *variation* in the effects of competition. A distinction is made between **intraspecific competition**, in which the individuals belong to the same species, and **interspecific competition**, in which they belong to different species. The two kinds of competition are discussed separately below.

Consider a very simple, hypothetical, community of plants and animals, a field seeded with grass. As the seeds germinate and the plants begin to grow, they have abundant water and nutrients in the soil and unlimited access to light, but as they get larger, their roots and foliage come ever closer until individual plants will be competing for these resources. The intensity of this competition will depend on two things: how many seeds were sown and whether or not all of them germinated at the same time. If there were many seeds, competition will be more intense; if they germinate at different times, the earlier plants will be larger and so have a competitive advantage over the later ones. Competition is thus influenced both by the absolute numbers of individuals and by differences in the behaviour of those individuals. Suppose the field is visited by rabbits that eat the grass. If they eat a lot of grass, this will affect both the grass and the rabbits. If grazing by rabbits is intense, late germinating seeds face less intense competition from established plants than they otherwise would. The rabbits continually have to move on from one already-grazed plant to another, using up more energy so that, ultimately, they have less energy for reproduction and produce smaller litters of young. If the farmer lets sheep into the field, there are additional elements of competition; the sheep are competing with one another and the rabbits are competing with the sheep. There are many possible results of this competition, for grass, rabbits and sheep. For example, rabbits and sheep could so over-graze the grass that it was wiped out, the rabbits might leave the field to feed elsewhere, or all three species might coexist. This is a three-species community; if we consider a real community, in which there are many plant species, several herbivores, and predators and parasites of those herbivores, it is clear that competitive interactions within and between species will become very complex and very difficult to unravel. Such is the task of the ecologist.

8.2.1 The nature and consequences of competition

Competition between individuals, of the same or of different species, is commonly indirect; an individual acquires a large share of some resource, leaving another individual with a smaller share or none at all. There is no direct physical interaction between the competitors. This is called **exploitation competition** and it is contrasted with **interference competition**, which involves some kind of direct interaction between competing individuals. Interference competition may take the form of overt fighting or some less damaging form of aggression (see Section 8.2.4). Many sessile animals living on hard surfaces, such as limpets, barnacles and corals, interfere directly with each other; occupation of a space by an individual denies that space to other individuals. In extreme cases, such animals may develop bizarre shapes as a result of such competition (Figure 8.1). Some plants secrete substances into the soil around them that prevents seeds germinating, so ensuring the plants exclusive access to the soil.

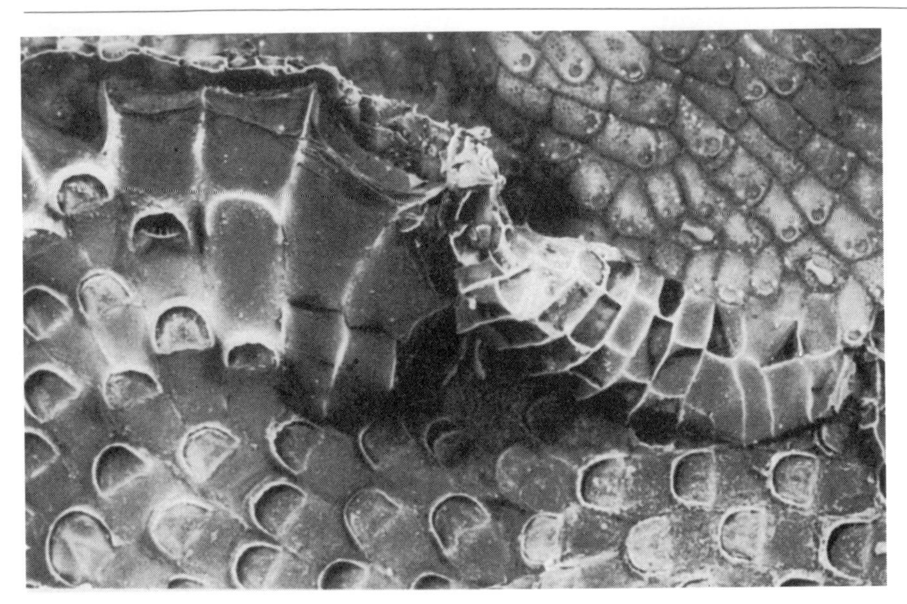

Figure 8.1 *Interference competition in bryozoans (see Chapter 10).* Stylopoma *(top) is overgrowing* Steginoporella *(bottom) in a flank attack to the right while the* Steginoporella *has raised its growing edge and has prevented overgrowth in a frontal encounter at the left.*

When two individuals compete, it is possible, in theory, that their effect on one another will be equivalent and reciprocal such that the fitness of each is reduced by the same amount. In nature, such reciprocity is rare because there is typically some kind of asymmetry between any two individuals. Such an asymmetry could be entirely due to chance (extrinsic) and be no reflection of any intrinsic properties of the individuals concerned, as when one barnacle larva just happens be first to settle on a bare patch of rock. Other asymmetries, however, are intrinsic to the individuals concerned. Age, for example, is a common source of such asymmetry. Young plants growing from seeds that germinated late in the season may have to compete with older, larger plants. Among birds of prey and several other birds, the eggs in a clutch hatch at different times so that the first chick is one or two days older than the second, and so on (Chapter 2, Section 2.2). The oldest, largest chick competes aggressively with the younger ones, often killing them or ejecting them from the nest. This effect is especially marked when the parents are unable to find sufficient food to feed the whole clutch adequately. This rather gruesome behaviour, called *siblicide*, has a valuable spin-off for conservation. The active conservation of a number of endangered bird species involves removing and artificially rearing the otherwise-doomed second and third chicks in incubators, or with foster parents of another species.

The most important kind of asymmetry between competing individuals, from the perspective of evolution, is that due to genotype, since only inherited asymmetries can lead to evolutionary change. If individuals with a particular genotype have an advantage in competition with conspecifics, that genotype will increase in frequency and the species will show evolutionary change. In other words, competition produces much of the variation in fitness on which natural selection acts (Chapter 4).

8.2.2 Intraspecific competition

Competition, as described above, leads to increased variation in fitness among members of a population. As a result, the summed reproductive output of all the members of a population will fall short of its full potential; some individuals do not reproduce as successfully as they would if there were no competition. In some instances, such failure can mean that the overall reproductive output of the population is reduced. Whether or not this happens depends on the effect of competition on mortality and fecundity. A crucial point here is that the effect of competition is **density-dependent**, i.e. the more competitors there are, the greater the deleterious effects of competition. The following examples illustrate this.

Density, mortality and fecundity As the hypothetical example of grass, rabbits and sheep showed, natural communities of plants and animals are so complex that it is very difficult to analyse the effects of competition within one species, in isolation from the influence of interactions with other species. This problem can be circumvented by studying population dynamics in simple, artificial communities. Figure 8.2 shows the results from a series of experiments in which populations of the flour beetle *Tribolium confusum* were reared in glass tubes containing a mixture of flour and yeast. Different numbers of *Tribolium* eggs, up to about 140, were placed in the tubes which were later sampled to record the number of beetles that had died, the mortality (Figure 8.2(a)), and the number that survived (Figure 8.2(b)). Obviously, the more beetles that are present in a population at the start, the more will die, but Figure 8.2(c) shows that initial population density also affected mortality *rate*.

The curves plotted in Figure 8.2 are divided into regions 1 to 3, corresponding to increasing initial densities (low, medium and high) of beetles.

► What happens to mortality rate in region 1 (representing initial populations of 0 to 40 beetles per tube) as population density increases, and how does this affect population size at the end of the experiment (Figure 8.2(b))?

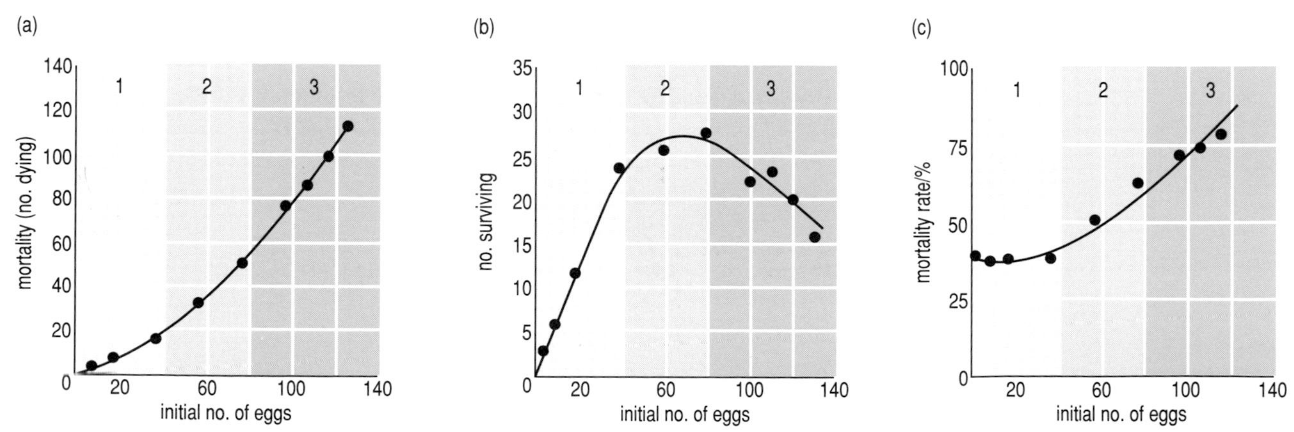

Figure 8.2 *The effect of initial population density on (a) mortality (b) survival, and (c) mortality rate in artificial populations of the flour beetle* Tribolium confusum.

In region 1, mortality rate remains constant and the final population size rises sharply as initial population size increases.

▶ Compare Figure 8.2(c) with Figure 8.2(b). What effect does mortality rate have on final population size in region 2, as compared with region 3?

In region 2, final population size rises with initial population size, but much less sharply than in region 1, because of the increased mortality rate in region 2 (Figure 8.2(c)). In region 3, mortality rate increases further and the final population size falls sharply as the initial population density increases. Figure 8.2(c) shows that mortality rate is density-dependent; as density increases above a certain point, so does mortality rate. The effect of population density on mortality rate is called *compensation*. In region 2, increased mortality rate reduces the rate of increase in population size but does not stabilize it; this is called undercompensation. In region 3, increased mortality rate causes a reduction in population size; this is called overcompensation. The term 'compensation' might suggest to you, because of its everyday connotations, a process that is adaptive in an evolutionary sense. This is erroneous, because the effect operates at the level of the population, not the individual. To regard compensation as adaptive would be to invoke group selection (Chapter 6, Section 6.5.2).

Figure 8.3 shows comparable data derived from a study on an artificial population of trout.

▶ How do these results compare with those for flour beetles?

At a certain initial population density, the population stabilizes at about one fish per square metre. In this instance, the density-dependent effect of mortality rate exactly compensates for variation in the initial density.

The *Tribolium* and trout studies considered the effect of population density on mortality rate. What of the other major component of fitness, i.e. fecundity? The density-dependent effects of population size on fecundity are, in essence, a mirror image of those on mortality. Figure 8.4 shows three examples of the relationship between fecundity and population density.

▶ Is fecundity density-dependent or density-independent in longhorn cattle (Figure 8.4(a))?

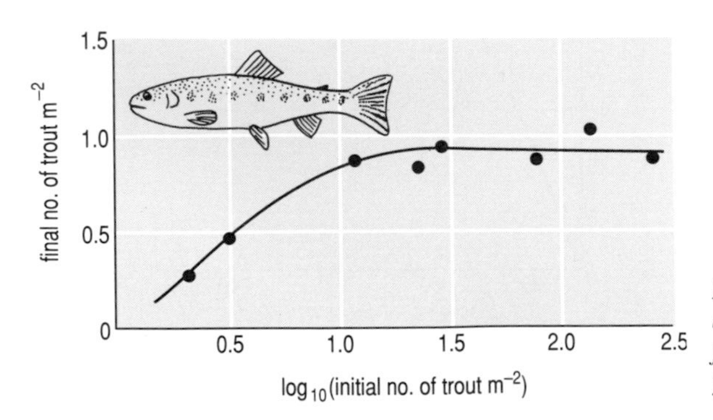

Figure 8.3 *The effect of initial population density (measured at constant depth) on final population size in a population of young trout.*

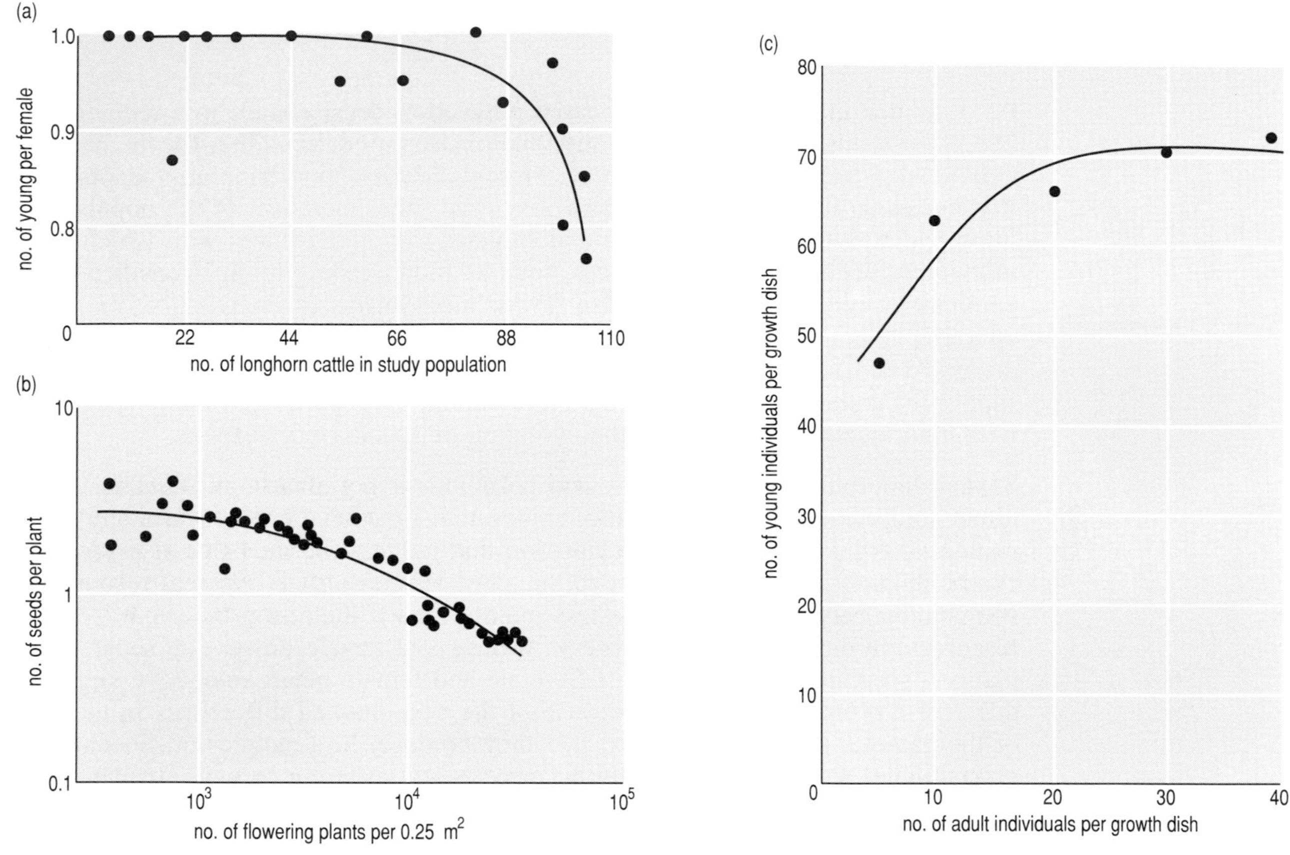

Figure 8.4 *The relationship between fecundity (number of progeny per adult individual) and population density in (a) longhorn cattle, (b) the annual dune plant* Vulpia fasciculata, *and (c) the fingernail clam* Musculium securis, *in culture dishes (here total numbers of offspring per dish, instead of fecundity, are shown).*

Up to a population size of about 60 cattle, it is density-independent; at higher population densities it shows a density-dependent decline.

► What kind of compensatory influence will the density-dependent effect of population size on fecundity have on the eventual population size of longhorn cattle?

It has an overcompensation effect. As fecundity falls, the cattle, on average, produce less than one calf per female. Thus the eventual population size must inevitably fall.

► What kind of compensation occurs in *Vulpia* (Figure 8.4(b)) and in the clams (Figure 8.4(c))?

Vulpia shows undercompensation below about 10^4 plants per $0.25\,m^2$; fecundity falls at high densities but is still higher than one seed per adult plant, and so population size increases. At densities higher than 10^4 plants per $0.25\,m^2$ there is overcompensation. The clams show exact compensation; large population sizes become stabilized, implying a density-dependent decline in fecundity.

8.2.3 Evolutionary consequences of intraspecific competition

As described in Section 8.2.2, increased population density leads to a reduction in fitness, in terms of increased mortality and/or decreased fecundity, for the average individual. Clearly, this effect imposes strong selection, favouring any adaptation that makes an individual more competitive than other members of the population. Such selection favours adaptations that enhance the effectiveness with which individuals utilize essential resources, such as food, leading to an intensification of exploitative competition. It may also favour mechanisms for more effective interference competition, such as fighting and other forms of aggression. As discussed in Chapter 7, females are a vital resource that is typically in short supply for male animals, and sexual reproduction is associated in many species with high levels of fighting and the evolution of specialized weapons.

Sexual dimorphism in morphology and behaviour is not always attributable to direct competition related to mating, however; it may have evolved in many instances, at least in part, as an adaptation that reduces the intensity of exploitative competition. If males and females of the same species utilize different resources, competition between them will be less intense than if they used the same resources. In dioecious plants (those with male and female flowers on separate plants—see Chapter 5, Section 5.1.6), male and female plants may have slightly different resource requirements because of the very marked differences in the form of the gametes (pollen and ovules) that they produce. In a number of dioecious plants, males and females typically occupy slightly different habitats. In the meadow rue (*Thalictrum fendleri*), for example, males occupy drier, sunnier and more elevated positions than females (Figure 8.5). Among animals, there are numerous examples of ecological differences between the sexes. In the malarial mosquito, females suck mammalian blood; males suck juices from plants. Many birds are sexually dimorphic in body size and in most birds of prey the female is larger than the male. In an extreme case, male peregrine falcons (*Falco peregrinus*) weigh 500 to 800 g, as compared to 750 to 1 200 g for females. The two sexes take very different prey, with males concentrating on small birds and

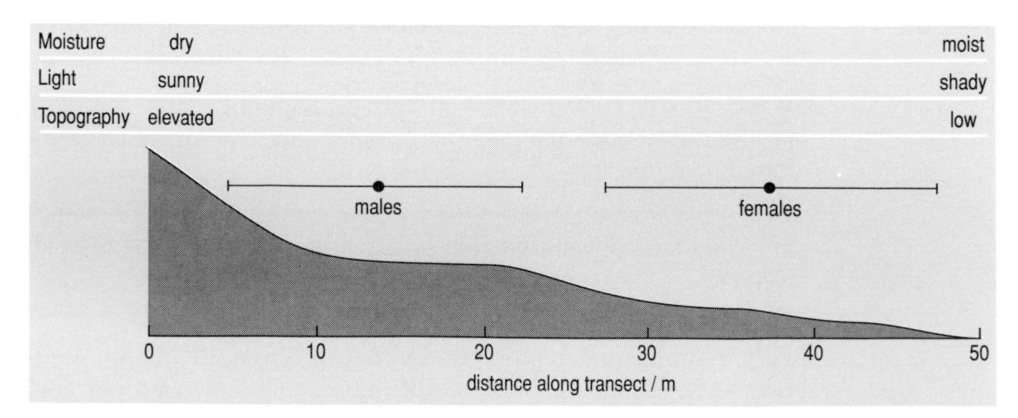

Figure 8.5 *Ecological segregation of male and female plants in meadow rue (*Thalictrum fendleri*). The diagram shows the mean and standard deviation of the distribution of each sex along a topograpical gradient that varies also in terms of moisture and light conditions.*

females on larger ones. In general, however, sexual dimorphism leading to ecological differences between the sexes is the exception, not the rule.

The relationship between morphological sexual dimorphism and sex differences in ecological habits is a complex one, as the following examples show. In the North American downy woodpecker (*Picoides pubescens*), males and females forage in different parts of trees, but there is no sexual dimorphism in body size or other morphological characters. Males feed on smaller branches high up in trees; females feed lower down, mostly on larger branches and tree trunks. W. D. Peters and T. C. Grubb (1983) set out to determine if this ecological difference between the sexes is due to their having differing food requirements or whether one sex simply displaces the other from feeding sites preferred by both sexes. In one experimental area males were removed, in another females were removed. Where males were removed, females adopted male-like foraging habits, but where females were removed, males did not change their foraging habits.

▶ Do these results support the hypothesis that male and female downy woodpeckers have intrinsically different ecological requirements?

No, they do not. The fact that females adopted male-like habits where males were absent suggests that the difference between the sexes is a social phenomenon and is not related to their having different ecological requirements. In this species, males are socially dominant to females and displace them from the feeding habitat that is preferred by both sexes.

An ecological difference associated with a morphological difference between the sexes was particularly striking in the recently extinct huia (*Heterolocha acutirostris*) of New Zealand (see Chapter 7, Figure 7.17). The male had a short, stout beak and found insect food by hammering into branches, whereas the female used her longer, curved beak to probe into crevices. As a result of this dimorphism, male and female avoided competition for the same food sources.

When the two sexes of a species differ in size or some other aspect of their morphology, the question arises: to what extent is this due to selection related to sexual behaviour and to what extent to the sexes having different ecological requirements? Did sexual competition lead to size differences that then meant that males and females, as an inevitable consequence, had different food requirements, or did selection pressure to reduce competition lead to sex differences that predisposed the sexes to behave differently in relation to sex? Questions about 'what came first?' in evolution are often sterile, because we cannot 're-play' evolution to see what happened, but they do raise important issues about how we define and measure exactly how a particular character is adaptive, because a single character has many biological consequences (Chapter 2). An attempt to tease the various factors apart has been made by T. D. Price (1984) for the sexually dimorphic species, the medium ground-finch *Geospiza fortis* on the Galápagos Islands (one of the group known as Darwin's finches— see Chapter 1, Section 1.3.1), by describing their morphology and aspects of their ecology.

Sexual dimorphism in this species is slight but is statistically significant; males are, on average, 5% heavier than females, and have correspondingly larger beaks (Figure 8.6(a)). Price studied the feeding behaviour of both sexes and recorded the

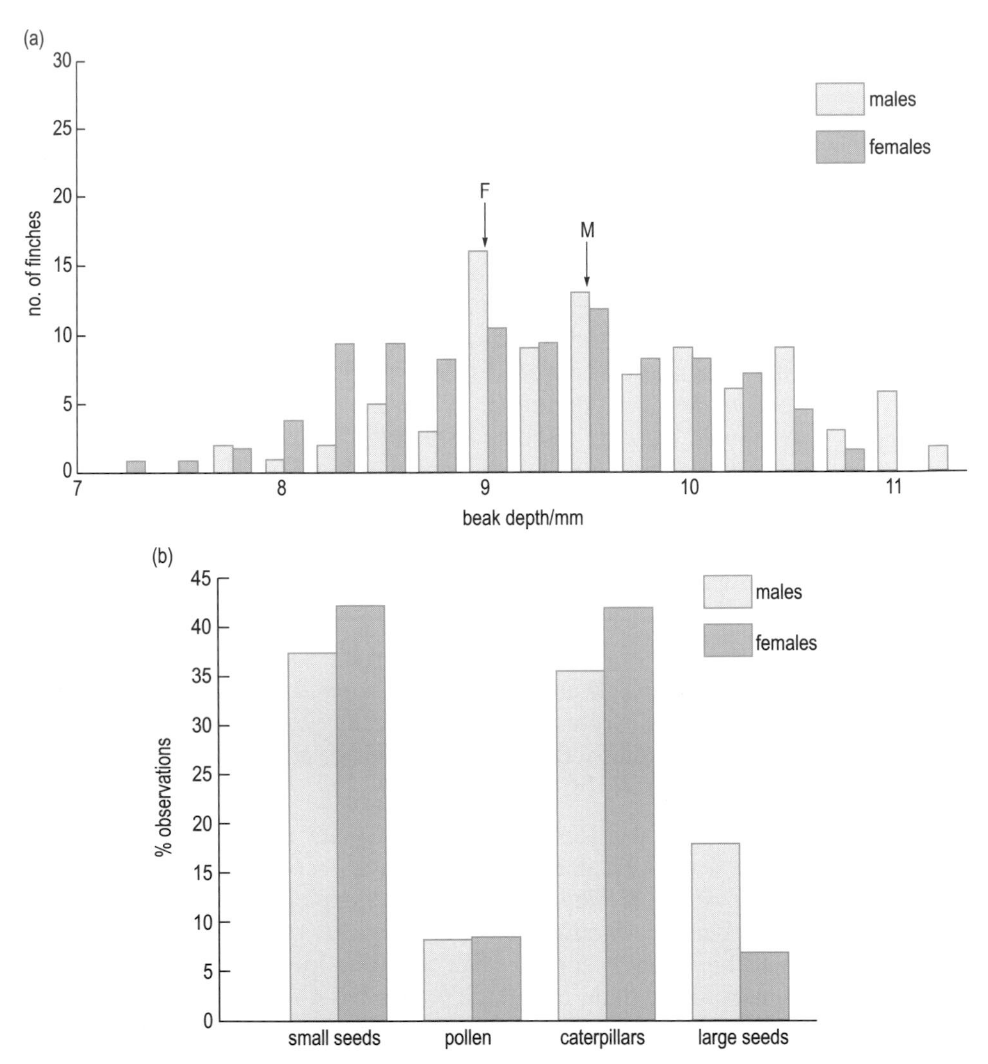

Figure 8.6 *(a) Frequency of beak depth in 93 males (light green) and 93 females (dark green) of Darwin's medium ground-finch. F and M = mean beak depth for females and males respectively. (b) Differences in the foraging behaviour of males (light green) and females (dark green) of Darwin's medium ground-finch. The only difference between the sexes that is statistically significant is that for large seeds.*

reproductive success of a large number of breeding pairs. He found differences in the food taken by the two sexes, but these were slight (Figure 8.6(b)). Males, with their larger beaks, fed slightly more on large seeds than did females but, because large seeds were not a major component of their diet in the breeding season, Price concluded that dietary differences between the sexes were unimportant in the evolution of sexual dimorphism in this species.

Price found that there was great variation in the reproductive success of individual breeding pairs, resulting from a high rate of clutch loss due to apparently random factors. However, two significant correlations did emerge from an analysis similar to that proposed by Wade and Arnold (Chapter 4, Section 4.4.5):

1 Smaller females showed a strong tendency to breed one year earlier than larger females. As a result they produced more clutches in a lifetime and so had higher fitness. This effect appeared to be due to smaller females having lower energetic requirements than larger females, and so being ready to breed at a younger age.

2 Larger males had higher reproductive success than smaller males. In Price's populations there was a skewed sex ratio, with males more numerous than females. In the resultant competition for females, larger males were more succcessful.

▶ Given effects 1 and 2, what would you expect to be their evolutionary result, in terms of sexual dimorphism, in this species?

Males should be larger than females, much more so than the 5% difference that actually exists (Figure 8.6(a)). The rather surprising lack of marked sexual dimorphism in this species is explained by another result that Price obtained in his study. He weighed both adult birds and their chicks and found high positive correlations between the weight of each of the parents and the weight of both their male and female chicks. In other words, heavier fathers had heavier sons and daughters, as did heavier mothers. Assuming that body mass is at least partly heritable, these data indicate a high degree of between-sex genetic correlation in body size, a feature that will act counter to selection pressures favouring divergence between the sexes. For effects 1 and 2 to lead to marked sexual dimorphism, the favoured type in each sex (larger males, smaller females) would need to pass its characters on to its same-sex offspring, but *not* to its opposite-sex offspring. The fact that offspring inherit size from the opposite-sex parent can act as a genetic constraint on the evolution of sexual dimorphism, as in this case.

This study, although it involves a species in which sexual dimorphism is relatively slight, points to some important conclusions concerning the relationship between sexual dimorphism in morphology and sex differences in ecology:

1 The ecological sex difference observed (males eating larger seeds) is related to a morphological difference (larger beak size), but appears to be a consequence of that difference, not its ecological cause. Because males have larger beaks, they exploit seeds that are too large for females to handle.

2 There exist strong selection pressures operating in different directions on the two sexes, but this has not led to marked sexual dimorphism, because of a genetic constraint. This conclusion reinforces the important point, made in Chapter 4 (Section 4.4), that the outcome of natural selection depends critically on the genetic basis of a character.

As discussed above, competition between members of a species may be particularly deleterious for certain categories of individual where some intrinsic asymmetry between individuals exists. Sex is just one example of such an asymmetry. A widespread form of asymmetry is age; younger individuals are typically smaller, weaker, less experienced and generally less competitive than adults. Competition between adults and young will be reduced if they have different ecological requirements. In lizards, individuals typically take prey of a size that roughly matches their jaw size, so that younger, smaller individuals concentrate on small prey items; older, larger animals take larger ones. This effect is very well illustrated by an analysis of the diet of Nile crocodiles (*Crocodilus*

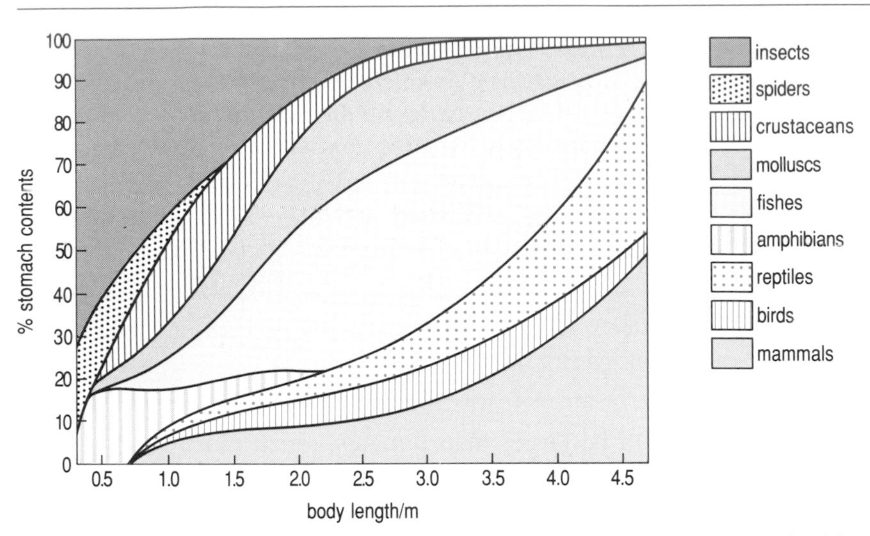

Figure 8.7 *The proportions of different food items found in the stomachs of wild-caught crocodiles of various body lengths.*

niloticus) of different ages (Figure 8.7). Young, small individuals primarily eat insects, and large adults mostly eat mammals and reptiles. Fishes form a very small proportion of the diet of full-grown crocodiles and are absent in the diet of very young individuals, but for crocodiles of intermediate size, they represent a major component of the diet. A difference in the diet of older and younger life stages is shown in its most extreme form in animals that have a larval stage, such as insects and amphibians. In butterflies, for example, the caterpillars feed on the leafy parts of plants, while adults feed on nectar or pollen. In many frogs, the tadpoles are herbivorous whereas the adults are carnivorous.

While these dietary differences between age classes have the *effect* of reducing competition for food between them, it does not necessarily follow that competition has been the only factor in their evolution. A degree of dietary change will be inevitable for many animals as they grow, simply because smaller animals can eat only smaller prey.

In a population of animals, competition between individuals will be reduced if different individuals feed on slightly different diets. In some birds, such as oystercatchers and herring gulls, detailed research on the diet of individuals has revealed a high degree of individual specialization. Among oystercatchers, one individual may feed primarily on mussels, another on cockles, another on oysters. In herring gulls, there are specialists feeding on starfish, crabs and mussels. These individual specializations develop through learning in these species and the exact nature of a given individual's specialization is probably determined largely by the particular feeding habits of its parents, because it is through being fed by their parents as chicks that such birds acquire their first experience of food. In dogwhelks (*Nucella lapillus*), individuals habitually feed either on barnacles or on mussels, but switch to the alternative food type when their favoured food becomes very scarce.

Competition between individuals within a population can lead to a situation in which there is ecological polymorphism (i.e. the existence of two or more distinct ecological morphs that differ genetically) which, in turn, can lead to the formation of new species. If there is intense competition for resources, two or more forms may coexist within a species, each of which has slightly different ecological requirements. For example, the three-spined stickleback (*Gasterosteus aculeatus*) in Canada exists in two forms, an exclusively freshwater type and an *anadromous* type, meaning that it breeds in freshwater but spends the rest of its life in the sea. Mate choice experiments have been conducted in which individual sticklebacks were offered a choice between a freshwater and an anadromous mate; 62% chose a mate of the same type, i.e mating was positive assortative (Chapter 3, Section 3.4). A consequence of positive assortative mating, if it is a strong effect, is that the polymorphism is maintained, the different types may become genetically distinct, and these may eventually become true species. Other examples of this process, called *competitive speciation*, are discussed in Chapter 9 (Section 9.6.2).

8.2.4 Aggression, fighting and territoriality

Competition between members of an animal species is commonly expressed in the form of aggressive behaviour. The term *aggression* is most obviously applicable to animals that are engaged in fighting, i.e. actually or potentially inflicting harm on their opponent, but it really includes a wide range of behaviour patterns, such as scent-marking by mammals and singing by insects, frogs and birds which, at first sight, do not appear to be aggressive at all, at least in the everyday sense of the word. In the early days of **ethology** (the scientific study of animal behaviour), aggression was defined in terms of the *form* of the behaviour and so was limited to activities that actually or potentially inflict injury. More recently, the emphasis has been placed on the *consequences* of the behaviour, so that **aggression** is now defined as behaviour which results in one individual acquiring a contested resource at the expense of another individual. Thus, inasmuch as its behaviour serves to keep rivals out of its territory, a mammal that leaves scent marks around the boundaries of its home range is showing aggression. The link between aggression and contested resources intimately links the behaviour of animals and their ecology. This Section explores this link, with particular emphasis on the evolution of morphological characters that are involved in aggressive behaviour.

The evolution of aggressive behaviour One of the founders of ethology, Konrad Lorenz, propounded the view that aggression is a spontaneous 'urge' within all animals that builds up if it is not expressed. The modern view is very different and makes a crucial link between the occurrence and intensity of aggression and the nature of the resources over which the aggression is taking place. As stated above, aggression is a behavioural mechanism by which one individual obtains a resource at the expense of another individual. There is abundant evidence, from a wide variety of animals, that aggression becomes more frequent and more intense (i.e. more dangerous to the contestants) as resources such as food, space and mates become more limited. The nature of aggressive behaviour is thus related to the *benefits* that animals derive by being aggressive. Another important factor affecting the nature of aggression concerns the *costs* that animals may incur by fighting. A general prediction that can be made is that natural selection will favour forms of aggressive behaviour in which the costs of aggression do not exceed the benefits that individual animals stand to gain as a result of that behaviour. In this kind of analysis, costs and benefits are measured or estimated in the 'common currency' of fitness.

This cost–benefit approach to aggressive behaviour has been developed using what is known as 'game theory' to analyse the various ways in which animals compete for resources. In particular, it is used to explain why the majority of competing animals do not conform to the stereotype of 'nature red in tooth and claw' but resolve their interactions by means of ritualized, non-damaging behaviour. For example, many animals possess lethal weapons that are adaptations for defence against predators or for killing their prey, but, in interactions with conspecifics, they show aggressive behaviour patterns in which such weapons play no part (Figure 8.8).

As discussed earlier, competing animals will typically show some kind of intrinsic asymmetry, such as a difference in size or age. Such asymmetries make it likely that, in the context of aggression between two individuals, one is more likely to win a contest than the other. There is a considerable body of evidence, from a variety of species, that morphological asymmetries between competing animals are used to settle contests quickly and without recourse to fighting. For example, male European house sparrows (*Passer domesticus*) and American Harris sparrows (*Zontrichia querula*) have a black throat patch, the size of which correlates with the probability that they will win disputes with other males (Figure 8.9). In winter flocks, disputes between two male sparrows, for example over food or a perch, are

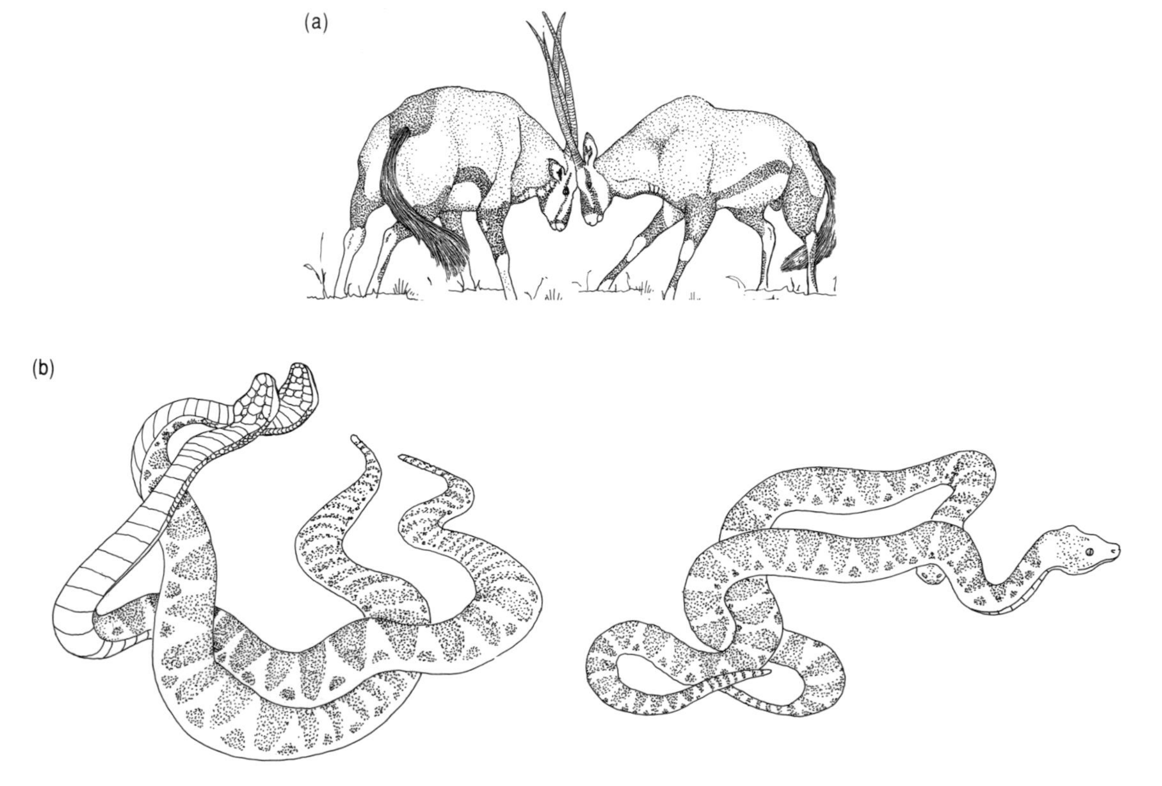

Figure 8.8 *Non-lethal fighting between animals that possess lethal weapons. (a) Oryxes have long, sharp horns that are used to impale predators such as lions but, in fighting, two males push against one another using the base of the horns. (b) Rattlesnakes have a lethal bite for killing prey and predators but, in fighting, they wrestle; the winner is the one that succeeds in pressing its opponent's head against the ground.*

*Figure 8.9 Plumage variation in male Harris sparrows (*Zontrichia querula*). The bird on the far right is the most dominant and wins most fights.*

generally quickly resolved, with the individual with the larger throat patch winning; only when two individuals have patches of comparable size are disputes protracted and likely to involve fighting (Møller, 1987).

Characters such as the sparrow's throat patch are referred to as 'badges of status' and, in some species, they are morphological characters, rather than colour patterns. Both sexes of the moorhen (*Gallinula chloropus*) possess a red frontal shield on the forehead (Figure 8.10) which increases in size during the breeding season. The size of this shield correlates, among individuals, with body size, such that larger, more powerful individuals have larger shields. In fact, the relationship between body size and shield size shows positive allometry, meaning that the shield of a large bird is not only larger in absolute terms than that of a smaller bird, but is larger in proportion to body size (see Chapter 10, Section 10.2.3). A positively allometric relationship between male body size and the size of structures used in threat or fighting has been found in a number of animals, including the antlers of several species of deer. The same effect emerges when different species of deer are compared; stags of larger species have larger antlers, in relation to their body size, than those of smaller species (Figure 8.11). The extreme expression of this effect was shown by the extinct Irish elk *Megaloceros* (see Chapter 14).

The evolution of large and elaborate weapons has perhaps reached its most extreme expression in various groups of horned beetles. Horns may be outgrowths from the head or greatly enlarged jaws (mandibles). They are used in fighting to

Figure 8.10 *Measurement of the frontal shield of a moorhen (*Gallinula chloropus*).*

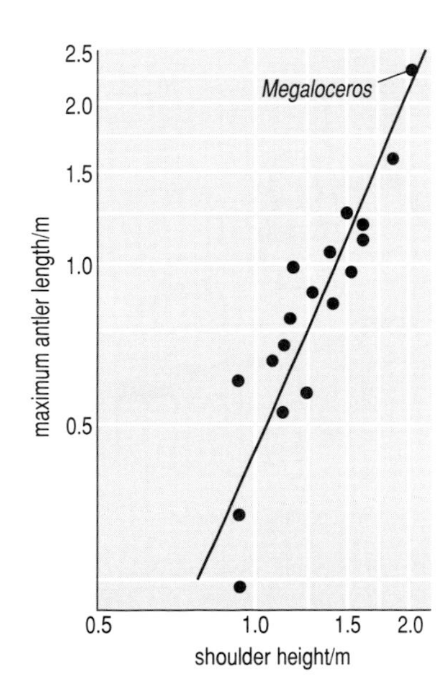

Figure 8.11 *The relationship between body size (expressed as shoulder height) and antler size in 18 species of deer, both plotted on log scales.*

push, lift or turn a rival over and some studies have found a positive correlation among males between horn size and mating success. In most instances, horns are sexually dimorphic, being absent in the female (A and B in Figure 8.12(a)). In some other species of beetles, the male differs from the female in having small spines on its abdomen that he uses to lock himself into his burrow when defending it against rivals (C, Figure 8.12(a)). Figure 8.12(b) shows collections of males of two different horned beetle species, showing how horn size, in relation to the rest of the body, changes during growth. In some species, horns grow in proportion to the rest of the body; in others they show positive allometry and grow at a faster rate than the rest of the body. There are also some species of horned beetle in which there is polymorphism among males; one form develops horns, the other does not but tends to have a larger body size because resources do not have to be allocated to horns. There is thus an enormous amount of variation within some horned beetle species, both between the sexes and among males, that becomes apparent only when a large sample is available.

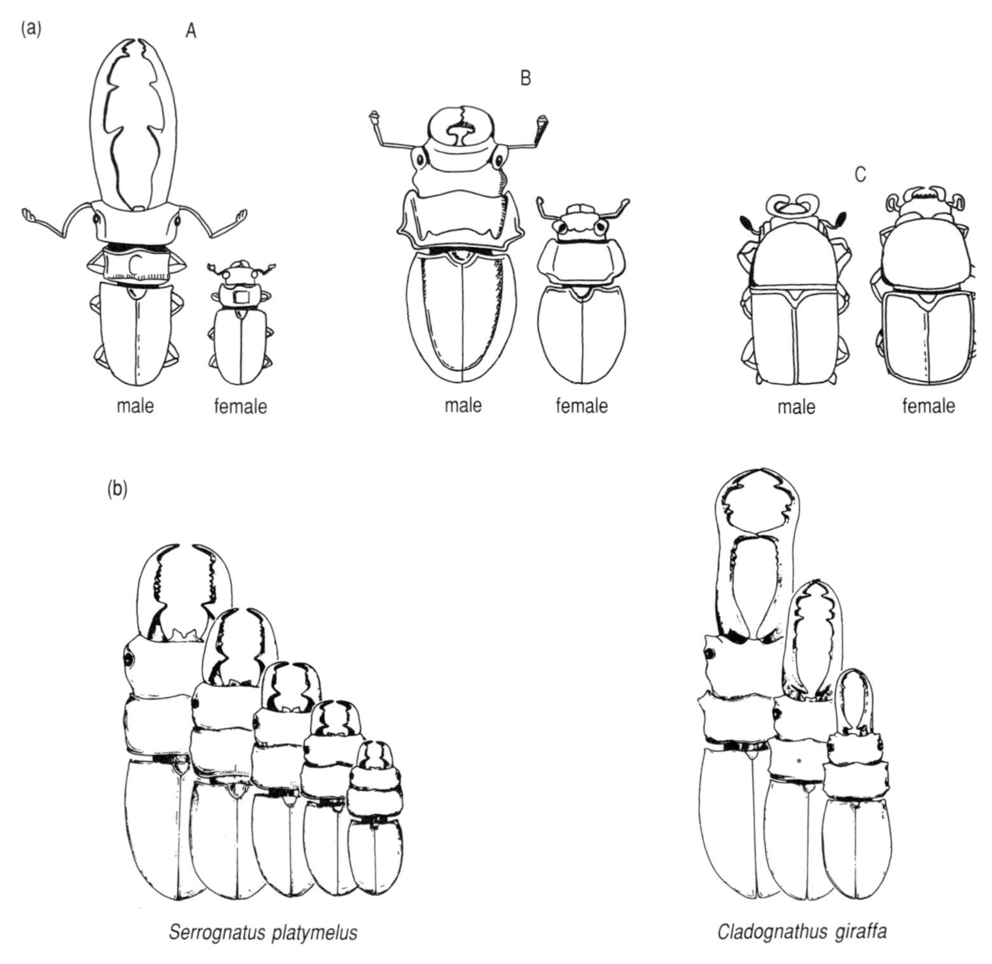

Figure 8.12 *(a) Sexual dimorphism in beetles. In species A and B, males have much larger jaws than females; in species C, the sexes are similar except that the male has two small abdominal spines. (b) Changes in body and jaw size during growth in males of two species of beetle,* Serrognatus platymelus *and* Cladognathus giraffa.

Territorial behaviour and population regulation In many species, competition for resources takes the form of territorial behaviour in which individuals defend an area of habitat against rivals. Like the word 'aggression', the way that territoriality is defined has undergone a change over the years. Previously, a territory was defined as 'a defended area', with the emphasis on the form of the behaviour involved. More recently, the emphasis has shifted to the ecological consequences of the behaviour, in recognition of the fact that territorial behaviour often does not involve overt aggression. **Territorial behaviour** is now defined as behaviour that leads to individual animals or groups of animals being spaced out more than would be expected from a random occupation of suitable habitats. Spacing between individuals is typically assessed by measuring the distance between individuals in a population and their nearest neighbour, and then comparing the frequency distribution of nearest-neighbour distances with the distribution to be expected if the same number of individuals were distributed at random in the same area (Figure 8.13).

► In what ways do the actual distributions of nearest-neighbour distances in the two examples shown in Figure 8.13 differ from those to be expected if the animals were distributed at random?

The average observed nearest-neighbour distance is greater than the expected average. Put another way, the frequency of large nearest-neighbour distances is greater and the frequency of small nearest-neighbour distances is less than the comparable frequencies expected by chance. The two examples shown in Figure 8.13 represent contrasting behavioural mechanisms by which spacing is brought about. Individual male great tits hold territories around a nest which they defend by singing, visual threat displays and occasional fights. Buffaloes form groups and a group defends a territory, occasionally engaging in massed attacks against other groups if they intrude.

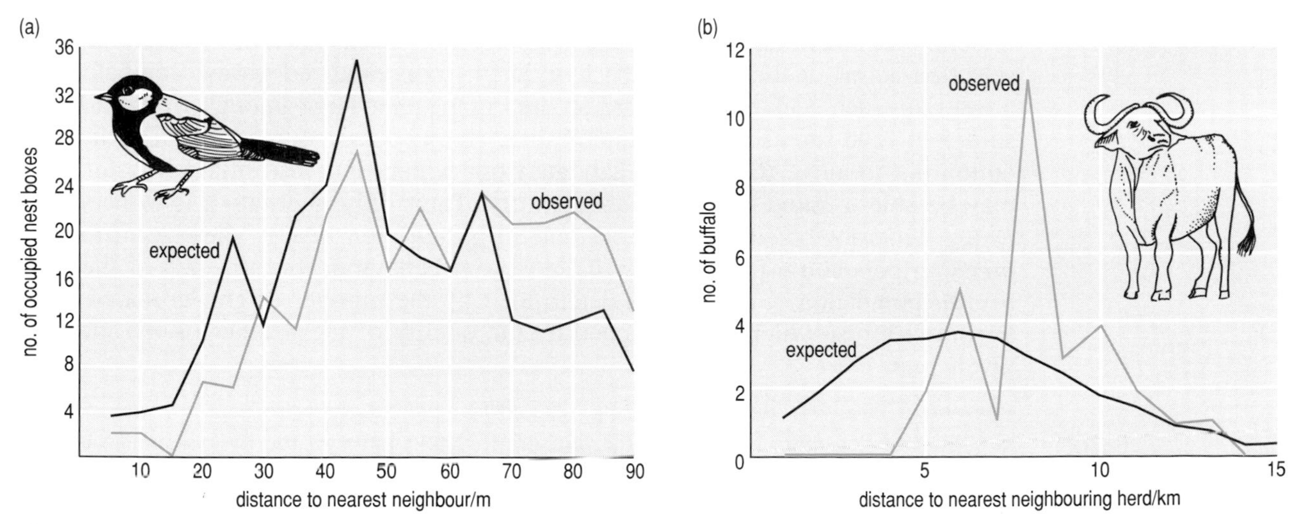

Figure 8.13 *Comparison of actual nearest-neighbour distances with those to be expected if animals occupy habitat at random in (a) male great tits (*Parus major*) in Wytham Wood, Oxford, and (b) herds of buffalo (*Syncerus caffer*) in the Serengeti.*

Because territorial behaviour leads to greater spacing between individuals or groups than would arise if they were distributed randomly, it follows that, in many instances, some individuals have no territory and are thus excluded from the resources on which the territorial system is based. If a male great tit is caught and removed from his territory, he is replaced, sometimes within minutes, by another male from a 'floating' population of non-territorial males. Excluded from breeding habitat, floating males will not attract females and cannot breed. Thus territorial behaviour limits the size of a breeding population and may therefore have a profound influence on the overall population dynamics of a species. The relationship between territorial behaviour and population size has led to a vigorous debate about the function of territorial behaviour, a debate that has focused on the red grouse (*Lagopus lagopus scoticus*).

Red grouse derive 90% of their diet from heather, particularly young shoots, and pairs set up breeding territories on areas of grouse moors in the autumn. Birds that are unable to obtain a territory are forced into marginal habitats and most of them fail to survive the winter through lack of suitable food. Consequently, the territorial pattern in the autumn has the *effect* of limiting the size of the breeding population in the following spring. V. C. Wynne Edwards suggested that population regulation was the *function* of territorial behaviour in grouse. He argued that territorial behaviour evolved for the benefit of the species because it would prevent grouse from over-exploiting an area of grouse moor; the non-territorial birds, by not forcing their way into the breeding population, were behaving in the best interests of the population as a whole.

▶ What kind of selective process does this argument invoke (refer back to Chapter 6)?

It is an argument expressed in terms of group selection, i.e. one that explains the adaptive value of a behaviour pattern in terms of benefits accruing to the population rather than to the individual (Chapter 6, Section 6.5.2).

The modern interpretation of the red grouse territorial system rejects the group selection argument and explains the phenomenon strictly in terms of natural selection acting at the level of the individual. Individual red grouse compete for territories, each seeking to acquire an area of heather large enough to provide sufficient food for itself, its mate and their chicks. Less competitive birds have no option but to move to marginal habitats from which, if a territorial bird dies, they may be able to move into a vacant territory. Their only alternative is to persist in losing fights for breeding territories and so hasten their death. The size of the *breeding* population is limited by its territorial behaviour, but the size of the *overall* population is not; that is determined by the number of chicks reared by the territorial birds, which is determined by the quality of the heather on the moor in a given year.

▶ What evolutionary effect *does* the territorial system have on the overall red grouse population?

It affects the genetic composition of the population, because only a proportion of the total population in any one year breeds and thus contributes genes to the next generation. It thus means that only part of the gene pool of the population is passed on from one generation to the next.

This example illustrates an important point about adaptation, made in Chapter 2, Section 2.2.4. We cannot assume that an *effect* of a character is necessarily the evolutionary *function* of that character. Territorial behaviour has the effect of limiting the size of a breeding population, but its function is to ensure that individual birds have sufficient food to rear their young. It also underlines the fact that natural selection acts on individuals, not at the level of the population.

Alternative competitive strategies Intense competition among members of a population for a scarce resource can lead to the evolution of alternative strategies, most commonly seen in the context of competition among males for females (Chapter 7, Section 7.4). The relative frequency of different types in a population may be determined by frequency-dependent selection (Chapter 4, Section 4.4.3). The more common one type becomes, the lower is the average fitness of individuals of that type and the higher is the average fitness of an alternative type. These conditions will lead to a stable polymorphism. In species in which there is a polymorphism among males in terms of morphological characters used in fighting, such as some horned beetles, natural selection favours two types of male in a population. One type has a 'fast and furious' life history; it allocates a large proportion of its resources to growing horns, it engages in frequent fighting over mates, but it lives for only a few seasons. The other type is 'slow but sure'; it does not grow horns, it breeds at a lower rate and lives for a much longer time.

In a fish, the swordtail (*Xiphophorus nigrensis*), studied by M. J. Ryan and B. A. Causey (1989), individual males belong to one of three relatively distinct size classes, small, medium and large, that also show behavioural differences. Larger males can dominate smaller males and are preferred as mates by females. Large males perform courtship displays to females; small males adopt 'chasing' behaviour and seek to intrude into the courtship interactions of large males to achieve 'sneaky' matings. Medium-sized males may either display or chase, depending on the relative size of males with which they are in immediate competition. This polymorphism has a genetic basis; the morphological and behavioural differences between large and small morphs are due to alternative alleles at a single locus, but the genetic make-up of medium-sized males is uncertain.

8.2.5 Interspecific competition

Animals and plants compete for resources, not only with members of their own species but also with members of other species. The greater the similarity in the resources required by two coexisting species, the greater will be the potential for competition between them. Many birds defend their territories, not only against conspecifics but also against heterospecifics (members of other species) and the signals that determine the outcome of fights between individuals are often recognized across species as well as within them. For example, male chaffinches (*Fringilla coelebs*) and great tits (*Parus major*) have mutually exclusive territories on the island of Eigg off Scotland, although birds in mainland populations have overlapping territories. Experiments using tape-recorded songs showed that, whereas mainland birds of both species respond aggressively only to conspecific song, chaffinches and great tits on Eigg both respond to both conspecific and heterospecific song (Reed, 1982). When male chaffinches were removed from an area of woodland on Eigg, male great tits quickly moved in to replace them.

The nature and evolutionary consequences of interspecific competition are discussed more fully in Section 8.3.

Summary of Section 8.2

Competition for resources between individuals is an integral component of natural selection; it is what Darwin referred to as 'the struggle for existence'. Competition leads to a reduction in the fitness of individuals, because it reduces survival, reproductive success, or both. These effects are density-dependent: the higher the density of a population, the more intense the competition and the more marked is the reduction in average fitness. Competition may be indirect, with one individual making resources unavailable to another (exploitative competition), or may involve direct interactions between individuals, such as fighting (interference competition). Some sexually dimorphic characters appear to be adaptations that reduce resource competition between males and females but this effect may, as in the ground finch, be only one factor among several underlying the evolution of sexual dimorphism. Competition may lead to variation in the use of resources by individuals that differ in size and age, to the development of individual differences in resource use, and to the evolution of ecological polymorphism. Thus, competition is a phenomenon that can promote variation within a population. Many species of animal have evolved forms of aggression that minimize the risk of injury or death. Competition and aggression are involved in territorial behaviour, which involves the unequal distribution of resources among individuals. Territoriality has the effect of limiting the size of a breeding population, but it is erroneous to argue that this is the adaptive value of territorial behaviour.

8.3 Ecological niches

Whereas the previous Section considered competition between individuals within a species, this Section is concerned with competition between species and with its evolutionary consequences. In nature, every species 'has its place'; it has a range of adaptations that equip it to survive and reproduce within a particular habitat, and with particular relationships with a number of other species. The Section begins with a discussion of the term that is used in ecology to define the specific place that a species has in an ecological community, its *niche*.

8.3.1 The concept of a niche

Every species of plant or animal is typically found in a particular habitat and it has particular requirements in terms of the resources that it needs to grow, survive and reproduce. Any good gardener knows that a particular plant thrives best in places with particular properties; some plants do best in warm, sunny spots, others must be planted in cool, shady places. Each species is specifically adapted for life in a particular kind of habitat; in other words, it has evolved to occupy a specific ecological **niche**. This term has been part of the ecological vocabulary for over 50 years but, until recently, its meaning has been rather vague. It is not a term that lends itself to a simple concise definition and its meaning is best illustrated by examples.

In the lower reaches of British rivers are found three species of crustacean of the genus *Gammarus*. One species, *G. pulex*, occurs only in freshwater; another, *G. locusta* only in water near the sea where salinity is high, and the third, *G. zaddachi*, occurs in stretches of rivers where salinity is intermediate (Figure 8.14).

In this example, the three species occupy different ecological niches, but this is the simplest possible case, in which the different niches are arrayed along a single dimension, salinity. For most species, it is necessary to consider two, three or more dimensions. Furthermore, the *Gammarus* example is much simpler than most, because it implies that the niche of a given species has sharply defined boundaries. Figure 8.15 shows two examples of niches, of a fruit-fly and a shrimp, that are defined in two dimensions. In each example it is clear that there are combinations of values for the two dimensions at which these species thrive better than at others.

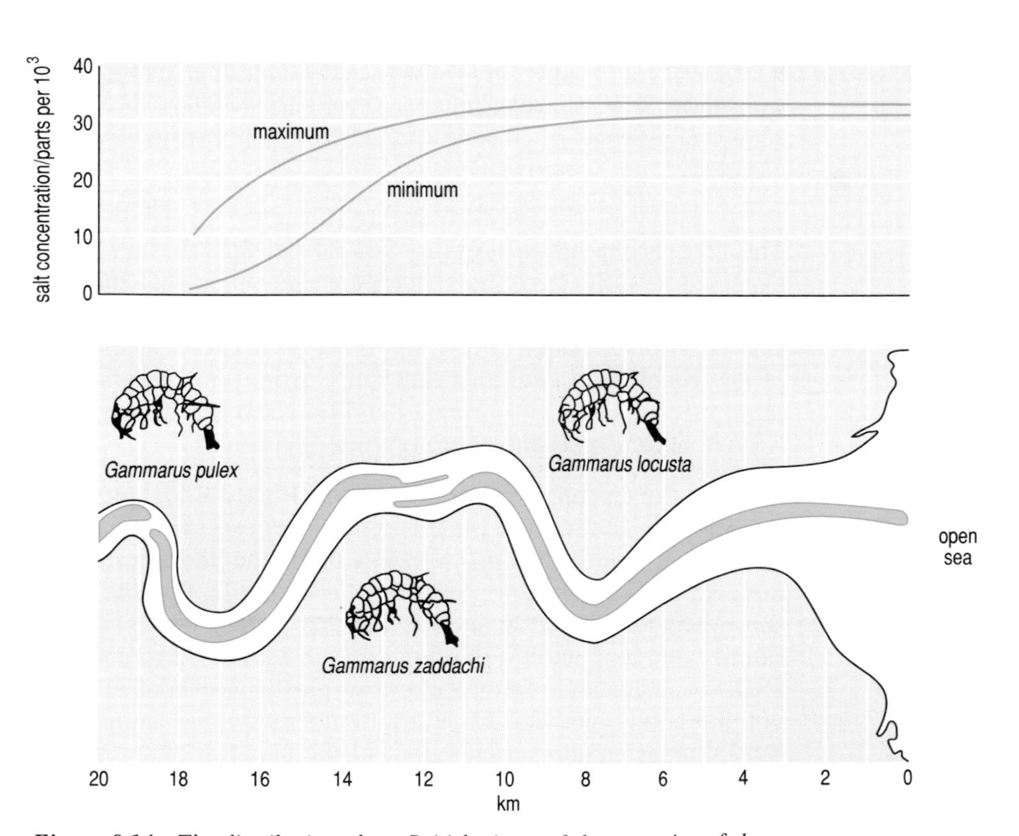

Figure 8.14 *The distribution along British rivers of three species of the crustacean* Gammarus *(green tone areas, below), in relation to the salinity of the water (above).*

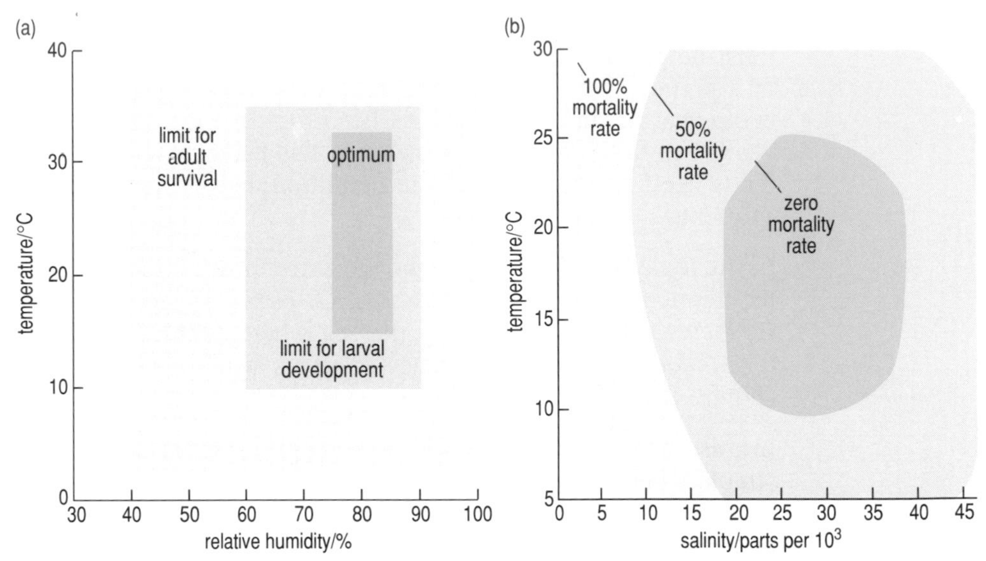

Figure 8.15 *Niches defined in two dimensions. (a) The distribution of a Mediterranean fruit-fly* Ceratitis capitata *with respect to temperature and humidity. The inner rectangle defines conditions optimal for growth, the middle rectangle conditions suitable for larval development, the outer rectangle conditions in which individuals can survive to adulthood. (b) The distribution of the sand shrimp* Crangon septemspinosa *with respect to temperature and salinity. The shaded areas represent contours of mortality rate in cultures of differing temperature and salinity.*

▶ Consider, in Figure 8.15(b), two sand shrimps, one living in a habitat in which temperature is 10 °C and salinity is 20 parts per 10^3, the other living at a temperature of 15 °C and a salinity of 30 parts per 10^3. How will the fitnesses of the two shrimps compare?

The first shrimp has a mortality risk of between 50 and 100%; the second is expected to suffer zero mortality. Note, however, that these data are from laboratory cultures, in which there was no mortality due to such factors as starvation or predation. Thus, the niche of a given species includes combinations of environmental conditions that range from those in which it survives and reproduces maximally (optimal conditions) to those in which individuals can barely survive and/or reproduce (sub-optimal conditions).

The two-dimensional niches depicted in Figure 8.15 are obviously oversimplified in that they leave out other dimensions of niches in nature, such as food supply. The addition of a third dimension makes the niche concept more conceptually complex but also more interesting. Suppose, for example, that the food supply of sand shrimps is distributed in such a way that it does not occur in those conditions of salinity and temperature that are optimal for the shrimp's survival. The shrimp must feed to survive and reproduce and so must live where its food is, that is, in sub-optimal conditions with respect to temperature and salinity. In other words, adding a third dimension alters the 'shape' of the niche in three-dimensional space. If there is also a predator of the shrimp, the optimal niche for the shrimp may shift again, depending on where the predator is most abundant, because it will be adaptive for shrimps to minimize predation. While it is possible to depict, as in Figure 8.15, a two-dimensional niche on paper, a three-dimensional niche is difficult, and niches with more than three dimensions are impossible to visualize graphically. In reality, niches involve many more than three dimensions, hence the modern definition of a niche: the *n*-dimensional 'hypervolume' within which a species can maintain a viable population. A hypervolume is an abstract concept; hence, what appears at first to be a rather simple idea has become a very complex, abstract one.

The *n* dimensions of a niche include both environmental conditions, such as temperature, humidity, salinity etc., and the various resources that organisms require to sustain growth, survival and reproduction, including solar radiation, water and minerals for plants, and food, nest sites, etc. for animals. Provided that a location provides all the requirements, defined by the *n* dimensions, within acceptable limits for a given species, that species can, potentially, survive and persist in that location. Two further conditions must be met, however. First, the species must be able to reach the location; this ability depends both on the remoteness of the site and on the dispersal and colonization capabilities of the species in question. Secondly, the capacity of the species to thrive in the location may be reduced or precluded altogether by the presence of members of other species that interact with it in various ways, such as by competition, predation or parasitism. These interactions with other species form the subject matter of much of the rest of this Chapter.

The fact that a species may be prevented by other species from living in certain localities that meet the requirements of its niche leads to a distinction between the *fundamental niche* and the *realized niche* of that species. The **fundamental niche** is that niche which can be occupied by a species in the absence of competitors, predators and other species that adversely affect it. The **realized niche** refers to a more limited set of conditions and resources within which the species actually maintains a viable population.

8.3.2 NICHE OVERLAP AND RESOURCE PARTITIONING BETWEEN SPECIES

If two or more species with similar niches coexist in a given habitat, they are likely to compete for one or more resources. Competition, as defined in Section 8.2, entails a reduction in fitness. It follows, therefore, that natural selection will tend to favour adaptations that reduce the extent to which the niche of one species overlaps with that of another occupying the same habitat. All plants growing in an area of soil are potentially in competition for water in the soil, but they can reduce the adverse effects of such competition by **resource partitioning**, i.e. by the possession of adaptations that enable two species to coexist by exploiting a common resource in different ways. For example, plants vary in the form of their roots such that they extract water from different depths in the soil (Figure 8.16).

Among animals too, several species may exploit a particular environmental resource in different ways; a group of such species is called a **guild**. Among birds, for example, two major guilds are seed eaters and insect eaters. The various species belonging to a particular guild may partition their use of the common resource in such a way that competition between them is reduced. For example, warblers of the genus *Dendroica* gather insects from different parts of spruce trees in Maine (Figure 8.17).

Figure 8.16 *The root morphologies of (a) burdock (*Arctium lappa*) and (b) mullein (*Verbascum thapsus*).*

Because niches are *n*-dimensional, considering resource partitioning along only one dimension may give a very false impression of the extent to which two or more species show niche overlap. Species may overlap considerably along one dimension but, when one or more additional dimensions are considered, niche overlap may prove to be less than expected. This concept is illustrated in Figure 8.18 by a hypothetical example.

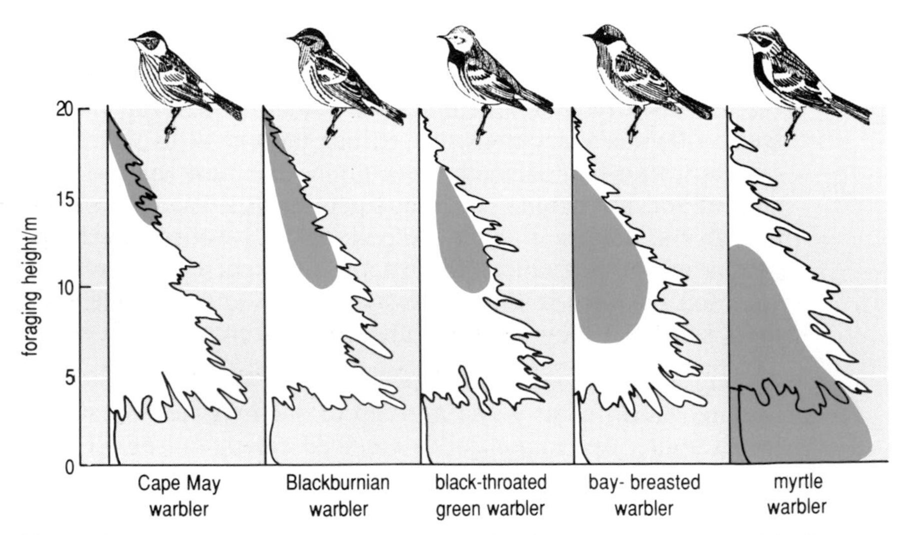

Figure 8.17 *Typical feeding locations (identified by green shading) used by five species of warbler (genus* Dendroica*) in spruce forests in Maine.*

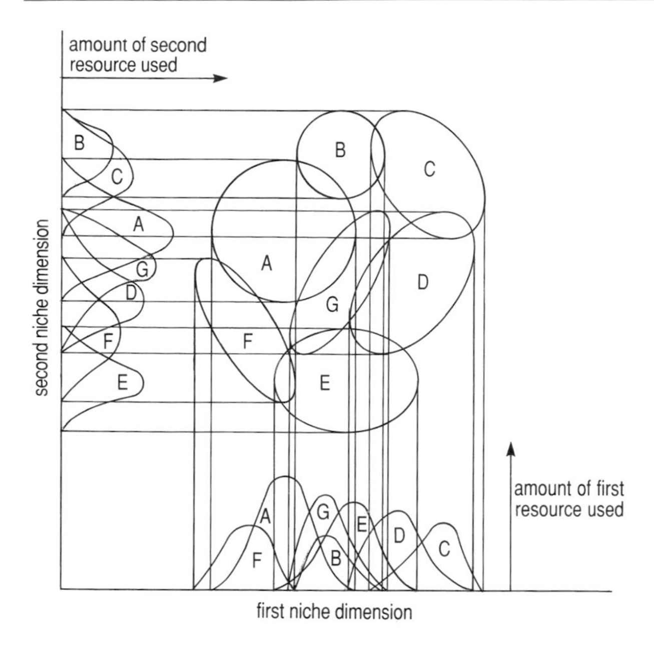

Figure 8.18 *A two-dimensional plot of the niche space of seven hypothetical species A–G.*

▶ Suppose that the species in Figure 8.18 are fishes, that the first niche dimension is salinity, and that the second is invertebrate prey size. To what extent do species A and D, and species B and E show niche overlap?

There is no niche overlap between either pair. Species A and D overlap in terms of the size of prey that they eat but they would be found in different parts of a stream, since they do not overlap on dimension 1 (salinity). Species B and E are found in the same parts of a stream but they take prey of very different sizes.

8.3.3 INTERSPECIFIC COMPETITION

When two or more species occur in the same habitat and exploit the same resources, and if individuals of at least one of the species suffer reduced survival and/or reproductive success as a result, then the species are in competition. If there is no reduction in fitness, there is no competition between them. Thus coexistence within the same habitat is not in itself evidence of competition between species, though it is usually a necessary condition for competition. Species may coexist, and exploit similar resources, but are not in competition if the numbers of each species are maintained at a low level by other factors such as predators or parasites. Predators may keep down the population sizes of two coexisting species such that the resources they require in common never become limiting.

A recent study by R. A. Griffiths (1986) of two similar-sized species of newt, the smooth newt (*Triturus vulgaris*) and the palmate newt (*T. helveticus*), which frequently coexist in ponds in Britain, looked for differences in the diet taken by adults by examining their gut contents. The two species were found to take a virtually identical array of food types, and in very much the same proportions (Figure 8.19).

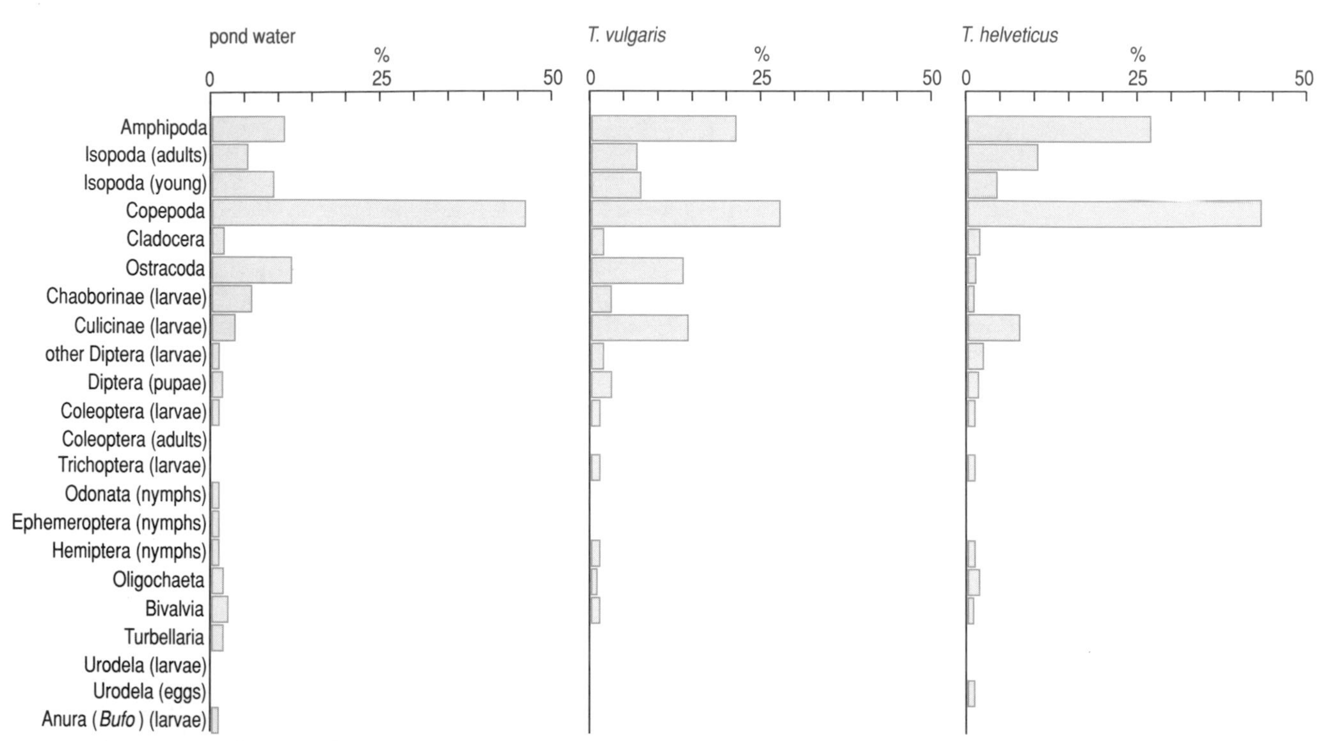

Figure 8.19 *The diet of adult smooth (*Triturus vulgaris*) and palmate newts (*T. helveticus*), measured as number of each prey type taken. (Note that a very similar pattern is obtained when the results are expressed in terms of the volume of each prey type consumed.)*

▶ Can you suggest two reasons why there is apparently total niche overlap between these two species?

1 The number of adults of the two species in a pond is kept below a level at which competition for food might occur, by some other ecological factor.

2 The species differ in some other, unidentified, niche dimension such that niche overlap is illusory (as in Figure 8.18).

Under 1, a likely factor is predation, either of adult newts or, more likely, of their eggs and larvae; adult newts may simply not be sufficiently abundant to be in competition for food. Under 2, no niche dimension that separates smooth and palmate newts within the same pond has been identified; both species occupy similar parts of a pond and they seem to be identical in their habits. The data presented above do not prove that there is no competition between these two species, however. It is possible, for example, that there is competition between their larvae, which are produced in large numbers.

The example of the newts shows just one possible outcome when two species with similar niches occupy the same habitat; they coexist without apparently competing. The following three examples illustrate other possible outcomes.

1 *American woodland salamanders*. Two terrestrial salamanders, *Plethodon jordani* and *P. glutinosus*, live in the southern Appalachian mountains of the USA. In general, *P. jordani* lives at higher altitudes than *P. glutinosus*, but in some

localities they occur together. Nelson Hairston (1980) studied seven experimental plots over a six-year period; from two plots he removed all the *P. jordani* he could find, from another two he removed all the *P. glutinosus*. The remaining three plots served as controls from which no salamanders were removed. Each year the plots were scored for the number and age of all salamanders found within them. In the control plots, *P. jordani* was much the commoner species. In plots from which *P. jordani* was removed, there was a significant increase in the number of *P. glutinosus*; in those from which *P. glutinosus* was removed there was no change in the abundance of *P. jordani*, but there was an increase in the proportion of *P. jordani* individuals that were one or two years old.

► Do these results provide evidence that *P. jordani* and *P. glutinosus* compete with one another?

Yes. When *P. jordani* is removed, *P. glutinosus* increases in numbers. When *P. glutinosus* is removed, there is an increase in the abundance of younger *P. jordani* individuals, suggesting a decrease in juvenile mortality. These two species are thus in competition, but nonetheless coexist.

2 *Bedstraws*. In a classic experiment, one of the pioneers of plant ecology, A. G. Tansley, studied two species of bedstraw, *Galium hercynicum* which is found on acidic soils and *G. pumilum* which occurs on calcareous soils. Both species would thrive in either kind of soil provided they were grown separately but, if both species were grown together, only *G. hercynicum* grew successfully in acidic soil and only *G. pumilum* in calcareous soil.

► In terms of coexistence and competition, how would you characterize the relationship between these two species?

They compete with one another and cannot coexist. Which species survives in a given habitat depends on the nature of the soil in that habitat. When, as in this example, one species prevents another from maintaining a viable population, **competitive exclusion** is said to occur.

3 *Barnacles*. In the intertidal zone of rocky shores around north-western Europe there occur two barnacles, *Chthamalus stellatus* and *Balanus balanoides*. Adult *Chthamalus* generally occur higher up the shore than adult *Balanus*, although young *Chthamalus* settle and start to develop on rocks throughout the intertidal zone. Connell (1961) monitored the settlement, growth and survival of young *Chthamalus* and also maintained some experimental areas of the lower intertidal zone free of adult *Balanus*. He found that young *Chthamalus* survived well in the lower zone, indicating that the usual absence of *Chthamalus* in this area is the result of mortality due to competition with *Balanus*, which smothers, undercuts and crushes *Chthamalus*. *Balanus*, on the other hand, cannot survive in the upper parts of the intertidal zone, because it is less resistant than *Chthamalus* to desiccation during periods when the tide is out.

► Is this an example of resource partitioning or competition?

Elements of both phenomena are involved. *Chthamalus* occupies parts of the shore where, on the dimension of desiccation, *Balanus* cannot survive. *Chthamalus* can also live lower down on the shore but is there competitively excluded by *Balanus*. Thus, if one regards the intertidal zone as a single habitat, these two barnacles coexist but are spatially separated as a result of competition.

The results of the four examples above are summarized in Table 8.1.

Table 8.1 *Variation in the outcome of coexistence within a habitat of two species that occupy similar niches.*

Species	Competition?	Coexistence?
Newts (*Triturus vulgaris, T. helveticus*)	No	Yes
Salamanders (*Plethodon jordani, P. glutinosus*)	Yes	Yes
Barnacles (*Chthamalus stellatus, Balanus balanoides*)	Yes	Yes (with spatial separation)
Bedstraws (*Galium hercynicum, G. pumilum*)	Yes	No

Theoretically, it is to be expected that, when two species share very similar niches and show extensive overlap in their use of resources, it is virtually certain that one of the species has at least a slight advantage over the other. This fundamental rule of ecology, the **competitive exclusion principle**, states that ecologically identical species cannot coexist in the same habitat. This principle makes good sense at the theoretical level and there is much evidence to support it, such as the barnacle and bedstraw examples discussed above. Other evidence does not support it, however; in the salamander and newt examples, two species appear to coexist. The problem is that, in these two examples, it is quite possible that some critical ecological factor has been overlooked; their niches may not be as identical as it appears. The two salamanders may show some kind of niche differentiation that has not been detected. The two newts apparently do not compete when they are breeding in ponds in spring, but competition may occur at some other stage in their life cycle. The competitive exclusion principle is difficult enough to *confirm*; it requires experimental manipulation, as in the salamander, bedstraw and barnacle examples, to conclusively demonstrate that one species competes with another. It is even more difficult, if not impossible, to *disprove*; because niches have so many dimensions, it is always possible that competitive exclusion is based on some dimension that has been overlooked, however detailed and thorough the study.

The evolutionary consequences of interspecific competition It follows from the competitive exclusion principle that, if two species with similar niches occur in the same habitat, they can coexist in the long term only if they evolve niche differentiation. Thus, a general prediction is that natural selection favours adaptations that enable a given species to occupy an exclusive niche and which separate it ecologically from other, similar, species. Some specific examples of such adaptations are discussed in Section 8.3.4.

It is frequently observed that **sympatric** (occupying the same geographical area), closely related species often differ in various features, such as body size or beak size in birds, for example, and such differences have often been interpreted as adaptations that reduce interspecific competition by minimizing niche overlap. Recently, however, an alternative explanation has been proposed to explain such differences between species sharing a particular habitat. A group of sympatric species may be regarded as a sample from a larger 'pool' of species that might potentially occupy that habitat, and the morphological or other differences between those species may be no greater than would be expected if species were taken at random from the pool. This idea provides a 'null' hypothesis (see Chapter 3, Section 3.5.1) against which the hypothesis of adaptive niche differentiation can be tested.

Birds of prey come in a wide range of sizes, from small kestrels to large eagles. In general, there is a positive correlation between body size and prey size; smaller hawks take smaller prey than larger hawks. The theory of competitive exclusion predicts that two hawks of similar size cannot coexist in a given habitat, because they will tend to compete for the same prey. Tom Schoener (1984) has tested the hypothesis that hawks occurring together in a given area tend to differ in size, against a random model, for 47 species of the bird-eating genus *Accipiter* found throughout the world. Members of this genus that occur in Europe are the sparrowhawk (*A. nisus*) and the goshawk (*A. gentilis*). Schoener calculated the probability with which any two *Accipiter* species, differing in size by a given ratio, would occur sympatrically if they were drawn at random from the pool of 47 species. He then compared the expected frequency of size ratios between sympatric species pairs generated by this model with the actual size ratio of pairs of sympatric *Accipiter* species found across the world (Figure 8.20).

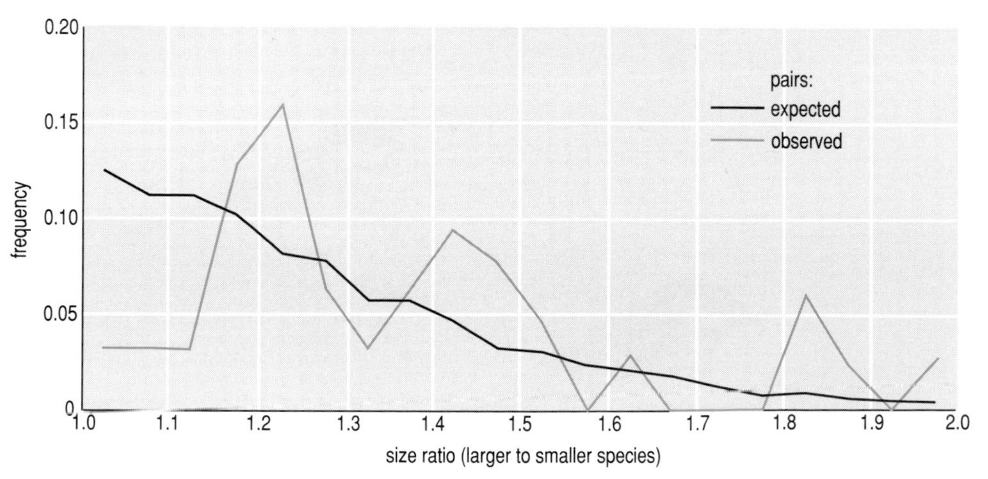

Figure 8.20 *The observed frequency of sympatric pairs of hawks (genus* Accipiter*) in relation to the ratio in their body sizes, compared with the frequency expected if the world's 47* Accipiter *species occurred at random.*

▶ How does the observed distribution of size ratios differ from that expected on the random model?

Pairs of species of very similar size (small size ratio) are much rarer than expected. It is only at size ratios of 1.15 : 1 and greater that species pairs are statistically as common or more common than expected. Schoener's study thus refutes the random model and supports the hypothesis that birds of prey are distributed, with respect to body size, such that competition between them is reduced.

Determining the role of interspecific competition in the evolution of differences between species is highly problematic, as the following example shows. Five species of tit (genus *Parus*) commonly occur together in woodland in Britain (Figure 8.21). There are small size differences between the species, with the great tit (*P. major*) appreciably larger than the other four, the marsh tit (*P. palustris*), the blue tit (*P. caeruleus*), the coal tit (*P. ater*) and the willow tit (*P. montanus*). All five species exploit two main, seasonally abundant food sources: leaf-eating caterpillars in spring and beech mast and other seeds in winter. Both these foods are so abundant during their respective seasons, that it is unlikely that the five tit species will be in serious competition for them. The five species do, however, show niche differentiation. For example, great tits forage mostly on the ground, blue tits on smaller twigs and leaves high up in trees. The species differ also in the size of the insects and in the hardness of the seeds that they eat.

There are three possible ways in which these differences can be explained:

1 *Current competition.* The five tit species are in competition for resources but coexist through niche differentiation. They could each expand their realized niche if one or more of their competitors were absent.

Figure 8.21 *The five tit species that may occur together in British woodland.*

2 *Competition in the past.* The species competed in the past but natural selection has favoured individuals of each species that possess specialized habits such that competition with other species was reduced. This explanation invokes what has been called 'the ghost of competition past'.

3 *Independent evolution.* The five species have responded to natural selection in different and independent ways. They are distinct species, with different morphologies and habits. They do not compete now and never have done; they are simply *different.*

Although explanation 1 might be tested experimentally, testing of 2 and 3 would require the sort of inferential test employed in the hawk example earlier.

8.3.4 Character release and character displacement

The hypothesis that interspecific competition influences the evolution of morphological and other characters can, potentially, be tested by comparing the nature of such characters in localities where there is competition and in localities where there is not. If species A, which is normally sympatric and in competition with species B, spreads into a habitat where B is absent, it may come to occupy some or all of the niche that had previously been occupied by species B and so evolve characters similar to those of B; this is called **character release** (Figure 8.22(a)). Conversely, if species A and B are **allopatric** (occupying separate geographical areas) over most of their ranges, there may be some degree of character shift in areas of sympatry (i.e. where they come into contact), such that competition between them is reduced; this effect is called **character displacement** (Figure 8.22(b)).

Both processes lead to the same phenomenon in nature, namely that two ecologically similar species show distinct differences where their ranges overlap, but show little or no differentiation where they do not. The essential difference between character release and character displacement is simply the order of events. Character release occurs as a result of sympatric populations becoming allopatric; character displacement occurs as a result of allopatric populations becoming sympatric. Thus, without an accurate knowledge of the history of an ecological interaction between two species, it would be impossible to attribute any ecological

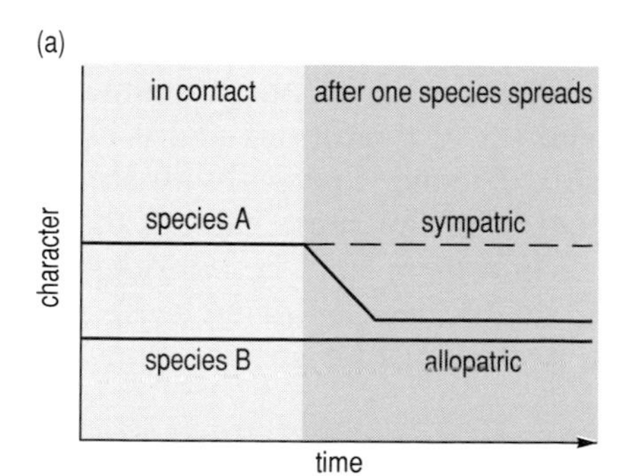

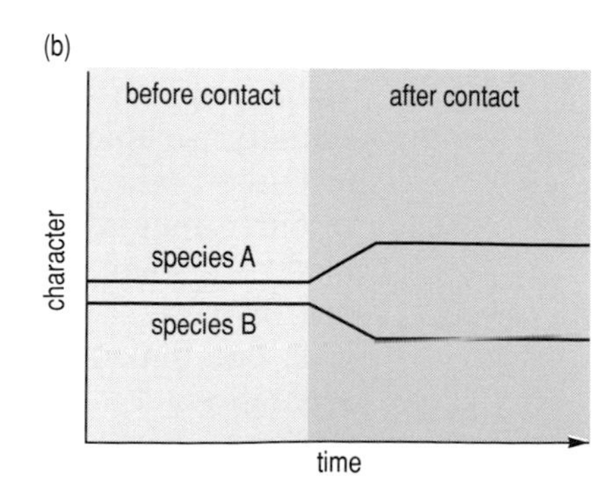

Figure 8.22 *Diagrammatic representation of (a) character release and (b) character displacement between two species A and B, with similar ecological niches.*

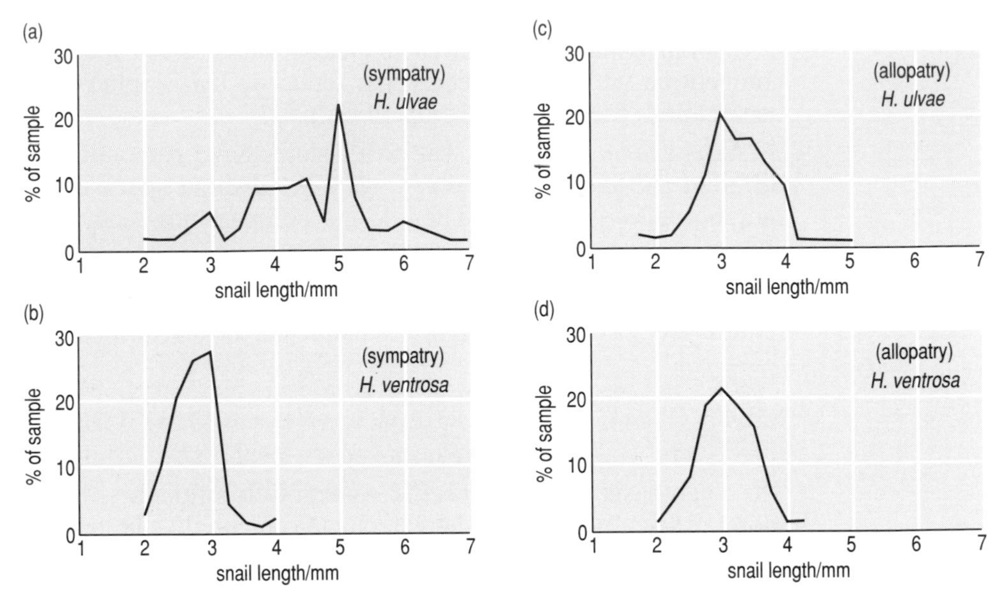

Figure 8.23 *Frequency distributions of the lengths of the shells of the snails* Hydrobia ulvae *and* H. ventrosa, *in samples taken from an area of sympatry, (a) and (b), and from areas of allopatry, (c) and (d).*

differences between them reliably to one process or the other. An example of probable character displacement occurs in very narrow zones of overlap that occur between the otherwise allopatric marine snails *Hydrobia ulvae* and *H. ventrosa* (Figure 8.23). These species are typically very similar in size, but where they coexist, one species (*H. ulvae*) is larger than usual while the other is abnormally small.

SUMMARY OF SECTION 8.3

Each species of plant or animal occupies a specific niche, defined as an *n*-dimensional hyperspace, for which the *n* dimensions represent all the environmental conditions and resources which that species requires for growth, survival and reproduction. A given species potentially occupies a fundamental niche that is defined, in *n* dimensions, by the conditions in which it can maintain a viable population. Because of the presence of other, competing species, however, it actually occupies a more limited set of conditions and uses a more limited set of resources, which define its realized niche. Species with similar niches (i.e. similar resource requirements) may coexist in the same habitat; they generally show some kind of resource partitioning. While there may be considerable overlap between one or more species along any one dimension of niche space, there is typically at least one dimension on which two coexisting species show some degree of niche separation.

Two or more species with similar niches may coexist in the same habitat without there being any competition between them. In many instances, the populations of potentially competing species are maintained, for example by predation, at levels below those at which competition would have a measurable impact on their abundance or distribution. Species with similar niches may coexist in the same habitat even if there is competition between them; though in many instances, such competing species show some form of spatial separation within the common habitat. However, some species cannot coexist in the same habitat because of competition between them; one species competitively excludes the others.

Ecological differences between species with similar niches occupying the same habitat are often attributed to natural selection having brought about niche differentiation, but this is only one possible evolutionary interpretation. Such differences may be due to competition operating in the present, to competition having operated in the evolutionary past, or they may have arisen because the species evolved independently and in divergent ways. Where two species are sympatric over part of their ranges, they may show divergence in one or more characters in the area of overlap. Such divergence may be due to character release in allopatry, or to character displacement in sympatry; which mechanism is involved in a given instance reflects the history of the spatial interaction between the two species concerned.

8.4 INTERACTIONS AMONG SPECIES

Competition is only one of many kinds of interaction that may occur between species. In essence it results from two or more species exploiting the same resource in the same place. In this Section, we turn to interactions between species of a fundamentally different kind, those in which one species is itself a resource that is exploited by another. Such interactions will clearly be a very powerful force in evolution because the exploiting species will typically gain at the expense of the other. Exploitative interactions will thus tend to favour the evolution of adaptations in the exploiting species that increase its effectiveness as an exploiter, and adaptations in the exploited species that enhance its ability to avoid or minimize the impact of that exploitation. They thus provide an evolutionary pattern of adaptation and counter-adaptation. For example, African antelopes are prey for a number of predators, including leopards and cheetahs. These two species hunt in fundamentally different ways. Leopards are 'sit-and-wait' predators that hide for long periods in dense cover and take unwary prey by surprise; they lack the speed and endurance necessary for pursuing prey in the open. Cheetahs are 'pursuit' predators and are anatomically and physiologically adapted for chasing prey at high speeds over moderately long distances (Chapter 2, Section 2.3.3). Antelopes have adapted to predation in a variety of ways. Species like impala and gazelles are relatively small and are equipped for high speed and endurance; they can escape cheetahs by being able to run faster or for longer, and can escape leopards by being alert and wary near cover. Other species, such as water buck are large and rather slow, but they have large horns that can kill a predator. Other important behavioural adaptations are involved in predator–prey interactions. Many antelopes live in groups that provide defence against predators because several sets of eyes (or ears and noses) are better than one. Some predators, notably lions and wolves, hunt in groups (Chapter 2, Section 2.3.3); by behaving cooperatively they can compensate for their lack of speed by 'ambushing' prey that can run faster than them, and they can also kill animals, such as buffaloes, that are much larger than they are.

The pattern of adaptation and counter-adaptation in predation and other kinds of interaction between species has led many biologists to use the analogy of an 'arms race' (Chapter 5, Section 5.1.5), in which any adaptation by one species must be countered by an adaptation in the other. What follows in this Section will show that interactions between species are typically very complex, are often very subtle, and that the arms-race analogy is actually rather a poor way of thinking about about how species interact in evolution.

8.4.1 A CLASSIFICATION OF INTERSPECIFIC INTERACTIONS

Five principal kinds of interaction between species are generally recognized: competition, predation, parasitism, mutualism and detritivory (all these terms are explained below). There is a trend among ecologists to use the term predation to include both the consumption of animals by other animals (carnivory) and the consumption of plants by animals (herbivory), because the two kinds of interaction have many features in common. These various interactions can be classified according to whether each of two interacting partners derive a gain (denoted by a +), incur a cost (–) or whether there is no impact in fitness terms (0). Thus a predator–prey interaction would be classified as + –. The various kinds of interaction are summarized in these terms in Table 8.2.

Table 8.2 *A classification of interactions between species.*

Interaction	Consequences for species A and B	Comments
Competition	0 –	Competition in which A competitively excludes B.
	or	
	– –	Competition in which A and B coexist.
Predation	+ –	Carnivory: A typically kills B.
		Herbivory: A typically browses on but does not kill B.
Parasitism	+ –	A is parasite; B is host.
Commensalism	+ 0	
Detritivory	+ 0	B is dead prior to interaction, so incurs no cost.
Mutualism	+ +	
Neutralism	0 0	

Some of the terms listed in Table 8.2 require formal definition and further comment:

Parasitism is an interaction in which one organism lives in (*endoparasite*) or on (*ectoparasite*) another organism, its *host*, obtaining nourishment at the latter's expense. Typically, the host is not killed by the parasite.

Commensalism is an interaction in which one organism lives in or on another organism, its host, at no detriment to the host.

Mutualism is an interaction in which two organisms live in a close association, to the benefit of both.

Parasitism, commensalism and mutualism are all forms of **symbiosis**, defined as the living together in permanent or prolonged association of members *(symbionts)* of two different species with beneficial or deleterious consequences for at least one of the parties.

Detritivory is the consumption by a variety of organisms, including animals, plants, fungi and bacteria, of organic debris derived from dead organisms or from the waste products of living organisms.

Neutralism is a rarely used term, but is included in Table 8.2 to complete the list of all possible kinds of interaction. It refers to species that coexist without having either adverse or beneficial effects on the other.

Distinctions between many of the categories of interaction listed in Table 8.2 are rarely clear-cut. For example, parasites commonly hasten the death of their hosts, but are not regarded as predators. Herbivores browse on, but generally do not kill, plants, but are not regarded as parasites. Their impact on plants is akin to that of mosquitoes or vampire bats that 'browse on' their animal hosts. Note that, in Table 8.2, parasitism and predation have the same notation (+ –). What differentiates parasitism from other exploitative (+ –) interactions is that a parasite typically lives in an intimate, long-lasting association with its host, whereas the relationship between an individual predator and its prey is transitory. One particular kind of parasite, called a **parasitoid**, is an insect that lays its eggs inside the eggs, larvae or pupae of another insect species. In most cases only one parasitoid egg is laid in each host. As the parasitoid larva grows and develops, it consumes its living host, finally killing the host as it emerges as an adult.

Humans living in affluent modern societies generally fail to appreciate the biological importance of parasites, because they are so rarely encountered in everyday life. In fact, on a world-wide scale, parasites are a major cause of human mortality and many millions of people are killed each year by parasitic diseases such as schistosomiasis and elephantiasis. It is reasonable to assume that, for most animals and plants, parasites are a major cause of mortality and reduced fecundity. Parasites are also extremely diverse and numerous in terms of number of species. Any free-living organism typically harbours several individuals of a number of different parasite species. Furthermore, most parasites are normally specific to one host or to a very narrow range of host species. It follows from these two facts that more than 50% of the species on the Earth must be parasites. It is estimated that about 25% of all insect species are parasitoids, some of which are parasitoids of other parasitoids.

8.4.2 Predator–prey interactions

There are innumerable examples of adaptations that enhance the effectiveness with which predators find, capture and devour their prey. Many species show specific morphological adaptations related to the food that they eat; for example, the many shapes of birds' bills are related to their diet (Figure 8.24). Some predators are generalists and can feed on a variety of prey; others are specialists and are specifically adapted to one prey type. One such species is the snail kite (*Rostrhamus sociabilis*) which is endangered in the Everglades swamps of south-eastern USA. It has a bill that is highly adapted for extracting a particular species of snail from its shell (see margin). The marked decline of this species in many parts of its range is attributable to its specialist habits; where its prey declines as a result of ecological deterioration, a viable population of kites can no longer be maintained.

A common feature of generalist predators is that they take a disproportionately large number of commoner prey types and relatively few of rarer types (Figure 8.25). Consequently, the various prey types are subject to frequency-dependent selection (Chapter 4, Section 4.4.3), such that rare food types are at an advantage over commoner ones, leading to an increase in their relative frequency until a balance is achieved.

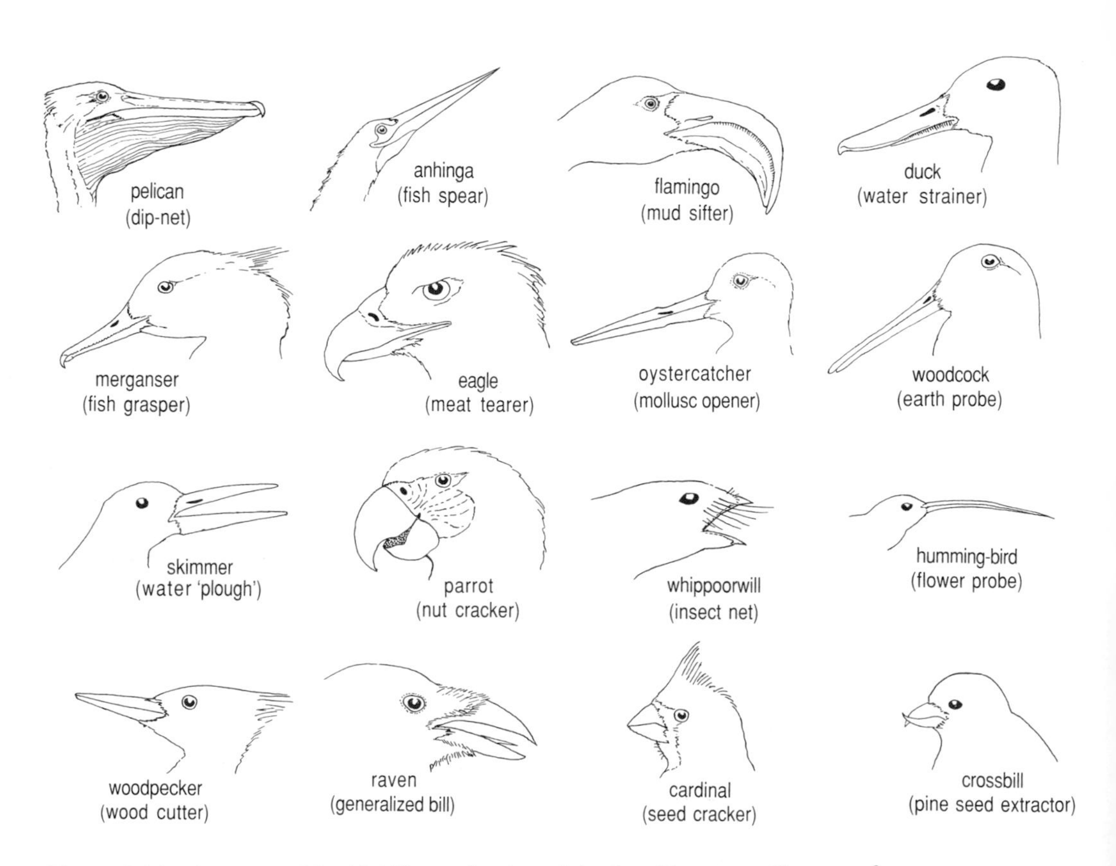

Figure 8.24 *A variety of birds' bills, each adapted for handling a specific type of prey.*

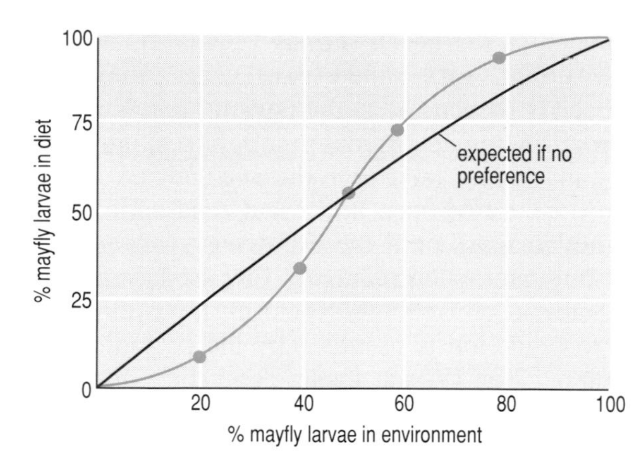

Figure 8.25 *The consumption of prey by water boatmen (*Notonecta*) in an experimental environment in which they have varying proportions of two prey, mayfly larvae and hog slaters (freshwater crustaceans). The black line shows the proportion of mayfly larvae that would be eaten if water boatmen ate the two prey in direct proportion to their relative frequencies; the green line shows the proportion that was actually taken.*

► What consequence will frequency-dependent predation have on the diversity of prey species in a particular habitat?

It will tend to maintain diversity because, when any one prey species becomes rare as a result of heavy predation, selection will be relaxed as predators switch to the more common food types. As a result, the rarer species tends to become more common once again.

Equally numerous are examples of adaptations that reduce the risk of predation in prey species. As described in Chapter 4 (Section 4.3.1), the spines and lateral armour of sticklebacks protect them against predation and their development varies geographically according to the local abundance of predators. Many plants possess morphological features such as thorns and stinging hairs that discourage herbivores from eating them; others produce chemicals that render them distasteful or poisonous to potential browsers. Some plants can increase the effectiveness of such defences in immediate response to predation. Pines attacked by insects show increased production of chemical defences, and dewberry plants grazed by cattle have longer and sharper prickles than those on ungrazed plants nearby. A less direct form of defence shown by some plants is polymorphism in leaf shape, with different individuals of the same species having leaves of different shape. Herbivores typically form a short-term 'search image' for leaves of a particular shape and thus overlook plants with leaves of a different shape. As discussed in Section 8.2.3, such polymorphism favours individuals with rare leaf shapes.

The role of herbivores in the evolution of defensive adaptations in plants is also revealed by the fact that such adaptations are often lost in populations of plants that have colonized islands where herbivores are absent. Examples include nettles that do not sting, raspberries that lack prickles and mint that has no smell.

One adaptation by which a number of prey species reduce the impact of predation is to synchronize their activities such that they 'swamp' their predators at times of greatest risk. If a large number of potential prey are gathered together in one place and at the same time, the risk to any one individual from the limited number of predators present in the area will be reduced. Many birds that breed colonially breed synchronously, so all the eggs in a colony hatch within a few days of each other. Some trees, such as the beech (*Fagus sylvatica*), show synchronized peaks of reproduction; in certain years, called 'mast years', they produce exceptionally large quantities of seeds (beech mast) that swamp resident populations of seed-eating birds and small mammals. Predators might respond to occasional periods of food abundance by increasing their reproductive success, but this response is always delayed. In the case of mast peaks and seed-eating birds, the herbivore population would not increase until the following year, and mast years are typically followed by years when relatively few seeds are produced. Perhaps the most remarkable example of breeding synchrony that may have evolved as an adaptive response to predation is provided by the periodic cicadas. These insects emerge as adults on a 17 or 19-year cycle, depending on the species. Note that these intervals are large prime numbers, which means that no predator that cycled on a 3, 4 or 5-year cycle, for example, could keep track of the availability of its prey. Only predators with 17 or 19-year cycles could do so, and no such species appear to have evolved.

Amongst terrestrial organisms, a carnivore typically kills its prey but herbivores generally browse on a plant, taking only part of it and leaving much of the plant alive and intact. The fitness consequences of predation for animal and plant prey are therefore not the same and the evolutionary consequences are different. Animal prey are selected to survive by avoiding predation altogether; plant prey can restore a loss of fitness by compensating for the effects of browsing and by deterring it. Many plants compensate for the loss of foliage to browsers by increased growth. Familiar to gardeners is the effect of pruning; plants can be encouraged to grow new foliage by being pruned at appropriate points. Grasses are plants that have become highly adapted to repeated grazing, a fact that makes them ideal for lawns. Their aerial growth tissue (the apical meristem) is situated close to the ground where it is unlikely to be removed by herbivores (or lawn mowers). Grazing can drastically alter the reproductive pattern of plants. The grass *Poa annua*, as its Latin name implies, naturally flowers and dies within a year of germination. When this species is repeatedly mown, as on a bowling green, it neither flowers nor dies but reproduces asexually (by vegetative growth) and is, in effect, immortal.

An analogous situation exists in marine environments in relation to certain animals. Many colonial benthic animals, such as corals, bryozoans, tunicates etc., have an asexual pattern of growth similar to that of grasses. Such a habit enables them to occupy areas of rocky substrate, just as grasses cover the soil. These animals are 'grazed' by predators, often without lethal consequences.

Many plants possess adaptations that turn the activity of herbivores to their advantage. Sweet fruits attract animals to collect and eat them; the remains, containing seeds, are discarded at a distance from the parent plant so that the herbivore enhances the plant's dispersal. The seeds of some plants are commonly contained in brightly coloured fruits that are attractive to animals and germinate only if they have passed through the gut of a herbivore; deposited in a pile of dung, they get a well fertilized start in life. No individuals of a tree called *Calvaria major*, endemic to the island of Mauritius, are known that are less than about 300 years old. It has been suggested that the seeds of this tree needed to be processed by the dodo which became extinct some 300 years ago. It is perhaps more likely, however, that introduced seed predators, such as rats and mice, eat the seeds, and that introduced sheep and rabbits eat any seedlings that do germinate.

Carnivores do not have a wholly detrimental effect on populations of their prey. Predators are more likely to be able to catch individual prey that are already weakened by old age or disease. A detailed study of musk-rats (*Ondatra zibethicus*) showed that individuals that fell prey to mink (*Mustela vison*) were mostly those that could not obtain a territory or that were injured in fights. Such individuals were unlikely to have bred, so that the impact of predation by mink at the level of the population was probably minimal.

An experiment to investigate means of controlling pigeons (*Columba palumbus*), a serious agricultural pest, showed that shooting large numbers failed to control population size in the long term.

▶ Can you suggest why this might have been so?

The elimination of many birds by shooting reduced intraspecific competition for food among the survivors. Consequently, they had more food to eat and enjoyed higher reproductive success. (This provides an example of compensation, as discussed in Section 8.2.2.)

As suggested earlier, predator–prey interactions are not static over evolutionary time but involve adaptation and counter-adaptation. For example, the evolution of chemical defences in plants may provide comprehensive protection from herbivores for only a short time in evolutionary terms.

▶ What benefit will accrue to a herbivore that evolves tolerance to the chemical defences of a plant species?

It will have access to a food resource for which it will face no competition from other herbivore species. As a result, some herbivores have become specialized to feed on particular plant species that are protected against other herbivores. Many herbivores, especially insects, have evolved tolerance to plant toxins and have, furthermore, exploited them to their own advantage. The milkweeds (*Aschapias*) contain compounds that are toxic to vertebrates and many insects, causing abnormal activity of the heart. The caterpillars of the monarch butterfly (*Danaus plexippus*) feed on milkweed and sequester the plant's toxins in their own bodies. Consequently, they and the adult butterflies into which they develop are protected from bird predators, that vomit if they eat them. Both caterpillars and adults of the monarch butterfly are brightly coloured. The association of distastefulness with bright coloration is called **aposematic coloration**, and will be discussed in more detail later in this Chapter.

A similar phenomenon can be seen in some marine organisms. For example, sea slugs that graze on corals are unaffected by their prey's nematocysts (stinging cells), and indeed transfer them, 'unfired', through their body to the brightly coloured dorsal tentacles, where they serve to protect the sea slug from its potential predators.

8.4.3 Parasite–host interactions

Parasites are enormously important in evolutionary biology, for three reasons. First, as mentioned above, they are very numerous in terms of number of species. Secondly, as a category of exploitative (+ – in Table 8.2) interactions between species, they often have a seriously detrimental effect on the fitness of their hosts, and as a result may, in extreme cases, severely reduce the host population size. Thirdly, they possess a number of features that have made them a very potent force in the evolution of many species, as will be discussed later.

Parasites are extremely diverse in terms of how they live and reproduce and how they affect their hosts. Figure 8.26 illustrates this diversity; it shows for a single host, a representative bird, a range of parasites belonging to 20 different taxonomic groups, that can be found in 30 distinct parts of birds. In the limited space available here it is possible to describe only a few examples of parasite–host interactions, from which a number of general principles can be drawn.

There are three key elements in a parasite–host relationship that are of concern to an evolutionary biologist:

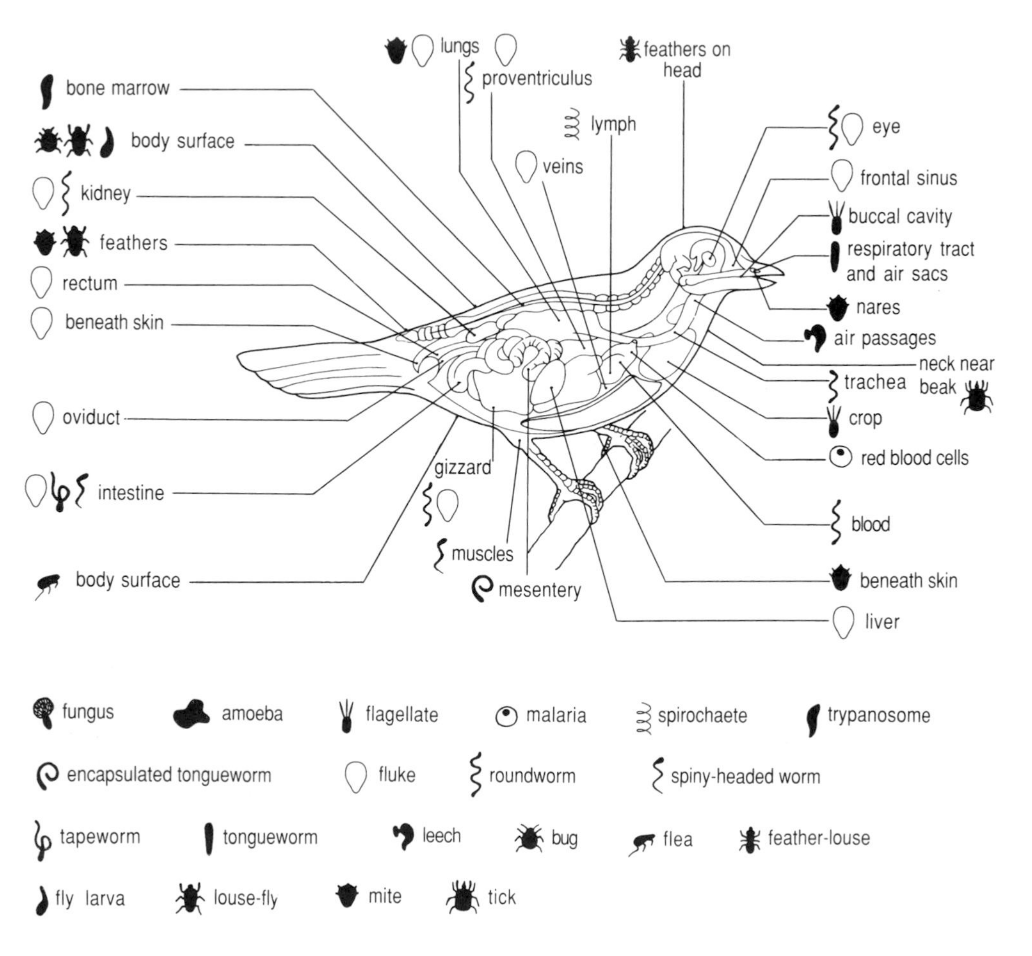

Figure 8.26 A typical bird illustrating the main groups of bird parasites and their sites of infection.

1 The association is often very intimate. Parasites typically possess a number of adaptations that enable them to maintain a position in or on their host and to resist most of the host's efforts to remove them. For example, different species of mammals have fleas with grasping limbs specifically adapted to clasp the hair of their host species. Thus cat fleas resist grooming by cats, but cannot maintain a secure grip on human hair.

2 The parasite is largely or wholly dependent on its host for the resources that it requires for survival, growth and reproduction.

3 The activities of the parasite, to varying degrees, reduce the fitness of the host by limiting its growth and/or reducing its reproductive success, and often make it more vulnerable to predation. In extreme instances, a parasite may be the sole cause of its host's death.

The harm caused by parasites to their hosts shows much variation, both within and between species. Within a host species, there are typically individuals that are totally unaffected by a particular parasite, others that have slight infestations that cause little or no impact on their fitness, and some that are seriously debilitated or killed. Between parasite species, there is a spectrum of effects, from organisms that are generally benign to those that are often lethal.

▶ If a symbiont is wholly benign, that is it has no detrimental effect on the fitness of its host, what would that interaction be called?

Commensalism (Table 8.2). There are strong theoretical reasons for expecting that, over evolutionary time, there is a dynamic relationship between parasitism and commensalism. As discussed later, it will often be in a parasite's best interests not to kill its host or increase the latter's vulnerability to predation; as a result, many commensal relationships have probably evolved from parasitic ones.

Such is the diversity of parasites that there are many ways of categorizing and classifying them. One major distinction is between micro- and macroparasites. Microparasites are those that reproduce and multiply within their host, usually within its cells, such as the many kinds of viruses and bacteria that infect plants and animals; they are often referred to as **pathogens** rather than parasites. Macroparasites grow either in their host, usually in body cavities, or on the outside of the host, and, when reproducing, release infective stages into the outside world. These include, for animals, lice, fleas, tapeworms and flukes, and for plants, leaf miners and gall-forming insects. Another important distinction concerns the mode of transmission from one host to another. For some parasites and pathogens, transmission is direct; individual parasites move from one host to another when individual hosts come into close contact, e.g. rabbit fleas and human body lice and the pathogens causing the common cold and venereal diseases. While some parasites and pathogens can survive only a few minutes outside a host, others can survive for a very long time in a dormant form before infecting a new host. For example, the small island of Gruinard off the Scottish coast was uninhabitable for more than 40 years after if was used for testing the anthrax bacillus as a biological weapon in World War 2. Other parasites have an **intermediate host** or **vector** that carries the parasite, or its infective stage, between host individuals. Mosquitoes carry malaria and sleeping sickness between mammal hosts; aphids pass viruses

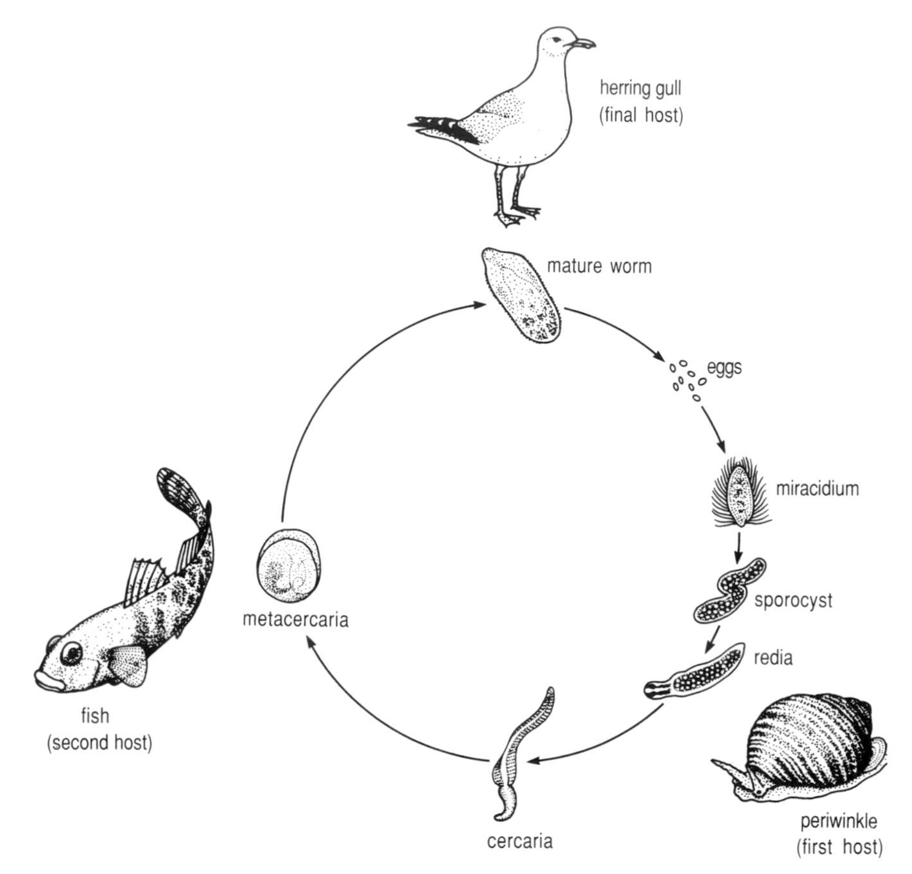

Figure 8.27 *The life cycle of the gull fluke* Cryptocotyle lingua.

between one plant and another. Some parasites have highly complex life cycles involving more than one intermediate host. The gull fluke, for example, passes through a snail and a fish on its way from one final host to another (Figure 8.27).

A common feature of parasite–host interactions is that parasites are distributed among individual hosts such that the number of parasites per host individual does not show a normal distribution. Rather, their distribution is highly clumped, such that many host individuals carry no parasites, some carry a few, and a small proportion are heavily infested (Figure 8.28). There are a number of possible reasons for this distribution pattern, each of which may apply more or less in different cases. It may, for example, reflect the fact that it is difficult for parasites to disperse from one host to another, with the result that infections affect only a restricted number of individuals living in a confined area. Another common factor is that parasites may reproduce prolifically only in individual hosts that are already weakened by other factors such as malnutrition or another kind of infection. Finally, individual hosts vary in their resistance to parasites.

Parasites and pathogens can severely reduce the fitness of individual hosts and, as a consequence, can adversely affect their host at the population level. Figure 8.29(a)–(d) shows four measures of partial fitness for the water bug *Hydrometra myrae*, in relation to the intensity of infection (number of parasites per individual

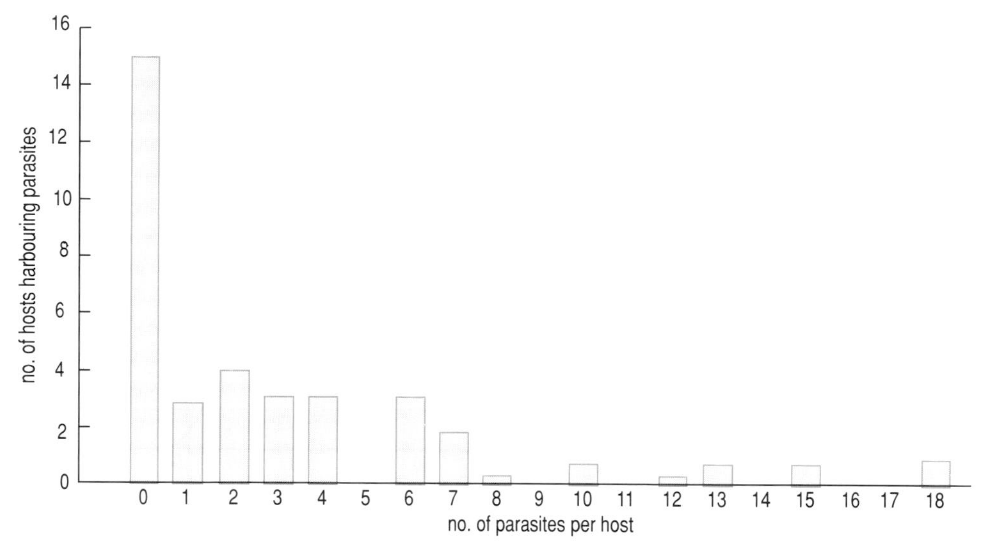

Figure 8.28 *An example of a parasite that shows a highly aggregated distribution:* Toxicara canis, *a nematode worm that infects the guts of foxes and dogs. (Ingestion of the faeces of an infected dog can cause blindness in humans.)*

host) by the parasitic mite *Hydraphantes tenuabilis*. Heavy infestation with the mite causes individual bugs to delay breeding (a), to produce slightly fewer eggs per day (b), and to have reduced survival (c) and a lower life expectancy (d).

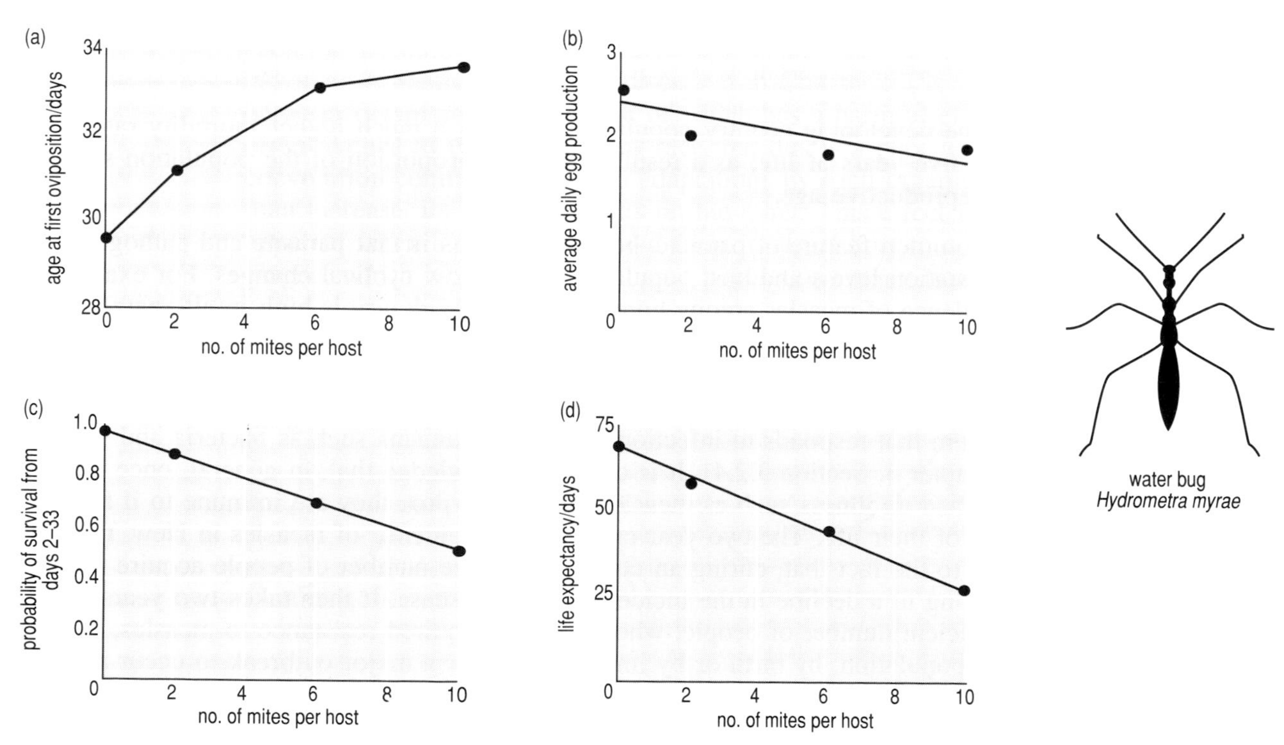

Figure 8.29 *Four measures of partial fitness for the water bug* Hydrometra myrae, *in relation to the intensity of infection (number of parasites per individual host) by the parasitic mite* Hydraphantes tenuabilis.

history, is contracted from rats, in which it is a far less lethal disease. Other examples of diseases that are more harmful to humans than their natural hosts are rabies (foxes), yellow fever (monkeys) and brucellosis (cattle). These and other examples suggest that parasite–host relationships tend to evolve towards a situation in which both organisms can coexist without the host being affected as severely as it was when the parasitic relationship first began. Is this effect due to evolutionary changes in the host, the parasite or both?

The history of the disease myxomatosis (a form of fibrous skin cancer) bears on this question. The *Myxoma* virus, which causes myxomatosis, is indigenous to South America, and rabbits there are prone to infection, but do not usually die from it. The virus was unknown in other parts of the world until it was deliberately introduced into Australia in 1950 in an attempt to control the rabbit population, which threatened the livelihood of sheep farmers. Rabbits had themselves been introduced to Australia from Europe, without most of their natural parasites, in the 19th century and had undergone a population explosion (see Chapter 17). Carried by mosquitoes, the *Myxoma* virus caused an epidemic that killed 99.8% of all the rabbits. A year later, the virus killed only 90% of the rabbit population, and then seven years after that, it killed less than 30%. This rapid decline in mortality due to myxomatosis was due both to an increased immunity to the virus in the rabbits and to reduced virulence of the virus.

▶ Why do you suppose that reduced virulence might be adaptive for the *Myxoma* virus?

A pathogen that does not kill its host has greater opportunity itself to survive, reproduce and infect new hosts than one that kills its host quickly (Chapter 5, Section 5.1.5). It appears to have been a common feature of parasite–host and pathogen–host relationships that, during evolution, they have changed from being one of true parasitism (+ – in Table 8.2) towards one of commensalism (+ 0).

As noted in Chapter 5 (Section 5.1.5), parasites and pathogens typically have a marked evolutionary advantage over their hosts because they have a much shorter generation time. As a result, a parasite can evolve faster than its host species. For this reason, it has been suggested that parasites and pathogens may be an important factor in the evolution and maintenance of sexual reproduction. Another important role for pathogens and parasites has been identified in the context of sexual selection in animals (Chapter 7). W. D. Hamilton and Marlene Zuk published a highly influential paper in 1982 in which they suggested that male secondary sexual characters such as bright coloration, elaborate visual displays, courtship songs, mating calls and courtship odours often act as 'revealing handicaps'. Their argument was, first, that animals are subject to debilitating parasites; secondly, that resistance to such parasites is heritable; and thirdly, that only individuals that are resistant to parasites are able to express fully their secondary sexual characters. Consequently, female choice for males with the best-developed sexual characters would result in offspring that are likely to inherit resistance to parasites from their father.

This important theory, which has generated an enormous amount of research in recent years, offers a possible solution to two serious problems in the theory of sexual selection, both of which were noted in Chapter 7. First, there is the question of why it should be adaptive for females to prefer males possessing particular characters over other males; the parasite theory provides a possible adaptive explanation for female choice for highly decorated males. Secondly, the strong, directional selection that is apparently involved in the evolution of extreme male secondary sexual characters is expected, on theoretical grounds, to be a transitory effect because heritable genetic variation in those characters will decrease over relatively few generations in response to strong selection. Because parasites and pathogens have so much shorter generation times than their hosts, the latter are having constantly to adapt to new varieties, strains and species of parasite or pathogen, so that different alleles will be favoured by selection in different generations. This theory thus provides a potential mechanism by which heritable genetic variation in male characters may be maintained despite there being strong directional selection on those characters.

8.4.4 Mimicry

So far, this Section has discussed interactions between species that have been direct, that is, one species has a direct effect on the fitness of another. We now turn to an indirect interaction between species, one in which the evolution of one species is influenced by the existence of another, but in which the two species do not interact in any direct way. Rather, the interaction is effected through a third species. **Mimicry** is the resemblance of one organism (called the *mimic*) to another organism (the *model*), such that the two organisms are confused by a third organism. Mimicry takes many forms, of which three will be discussed here.

Batesian mimicry Batesian mimicry refers to a situation in which the model species is poisonous or distasteful and warns potential predators of its unpalatability by some conspicuous feature such as aposematic coloration (Section 8.4.2). The mimic has evolved similar features but is neither distasteful nor poisonous. This form of mimicry relies on predators learning, through aversive experience, that the model is not palatable, so that they also avoid eating the mimic. The interaction is thus advantageous to the mimic but is of no advantage or disadvantage to the model. Batesian mimicry is essentially equivalent to camouflage; just as a stick insect, for example, resembles a non-palatable stick, so a mimic resembles an unpalatable model. Selection favours any adaptation in the mimic that improves its resemblance to the model. Some Batesian mimetic resemblances are remarkably precise, and they are particularly abundant among insects. For example, the harmless grasshoppers of the Bornean genus *Condylodera* so closely resemble tiger beetles (genus *Tricondyla*), both in morphology and in the way that they move, that they have been mistakenly put together in museum collections (Figure 8.33).

There are innumerable examples of presumed Batesian mimicry in textbooks of natural history and evolution, but the great majority of these should be regarded as hypothetical. Simple resemblance between two species is insufficient evidence for the existence of a mimetic relationship between them. A number of supposed instances of mimicry have, on close investigation, proved to be quite complex. For example, some supposed models are palatable, and some supposed mimics are

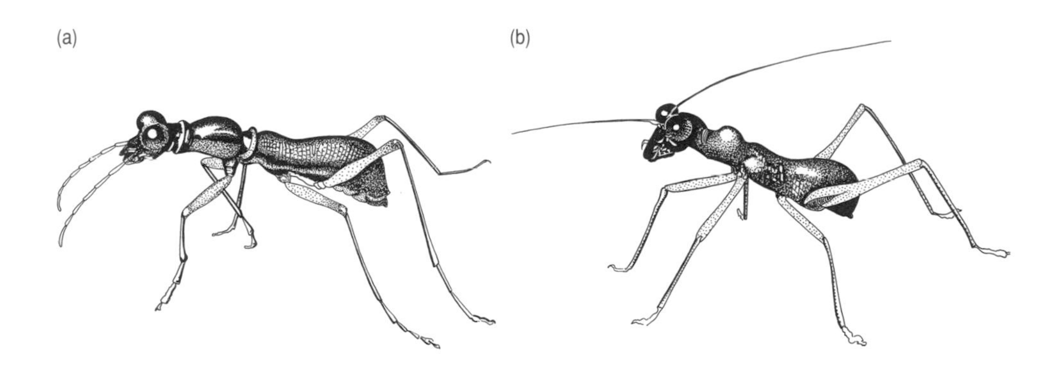

Figure 8.33 Batesian mimicry between (a) a tiger beetle (genus Tricondyla*) and (b) a grasshopper (genus* Condylodera*).*

unpalatable to some predators but not others. Furthermore, there are other selection pressures that may cause two species to resemble one another, as described below. One way to test the hypothesis of Batesian mimicry between two species is to demonstrate that a population of mimics suffers greater predation in the absence of the supposed model than when the model is present. One way to do this is to compare the response to the mimic of captive-reared predators that have or have not been exposed to the model. Another way is to exploit the fact that, in nature, some models and mimics have geographical ranges that only partially overlap (Figure 8.34).

The aposematic pipe-vine swallowtail butterfly (*Battus philenor*) occurs in the southern half of North America. Over this region, another butterfly, *Limenitis*

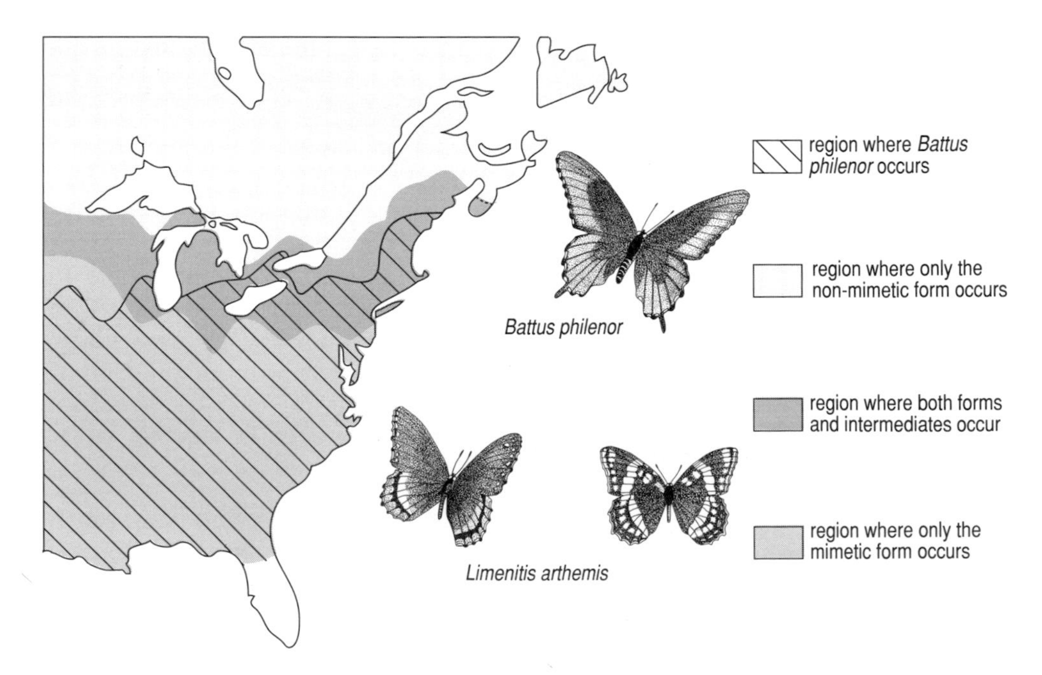

Figure 8.34 The geographic ranges of the aposematic butterfly Battus philenor *and two forms of* Limenitis arthemis.

arthemis, occurs extensively in a form that, like *B. philenor*, is conspicuous and almost entirely black. However, in Canada, where *B. philenor* is absent, *L. arthemis* is cryptically coloured black and white (i.e. it is camouflaged). There is also a region, around the northern limit of the range of *B. philenor*, where the mimic species occurs as the pure-black (mimetic) form, the black-and-white (cryptic) form and as intermediates.

▶ Do these distribution patterns support the hypothesis that *Limenitis arthemis* mimics *Battus philenor*?

Yes, they do. Where both species occur, *L. arthemis* occurs in a conspicuous form that resembles the aposematic *Battus philenor*, but where they do not occur together, such coloration would not protect *L. arthemis* and it must rely on camouflage as a defence against predation.

A theoretical prediction concerning Batesian mimicry is that selection only favours the mimic if it is less common than the model.

▶ Why do you suppose that mimics would not be protected if they were more common than models?

Because predators would learn to associate their conspicuous colour pattern with palatable food and so would tend to take them at a high rate. Some mimetic butterflies have adapted to this constraint on their evolution by being polymorphic, existing in two or more distinct forms, each of which resembles a different model species. In others, such as the tiger swallowtail butterfly (*Papilio glaucus*), another mimic of *Battus philenor*, females are mimetic but males are not. Another factor, however, relates to how toxic and distasteful is the model. In some associations in which the model is highly distasteful, there is evidence that their mimics can be relatively abundant, so strong is the tendency of predators to avoid their shared colour pattern.

Müllerian mimicry In this form of mimicry, a number of aposematic, distasteful species have evolved the same or similar warning signals. A familiar example from the UK includes social wasps (genus *Vespula*), solitary digger wasps (*Bembex*) and the caterpillars of the cinnabar moth (*Callimorpha jacobaea*), all of which have black and yellow striped bodies. In Müllerian systems, all the species derive benefit because any predator that has learned to avoid one of them will generalize its experience and avoid the others. Typically, the degree of resemblance between Müllerian mimics is not as great as that between Batesian mimics and their models.

Aggressive mimicry As the name implies, aggressive mimicry involves deception of a third participant, the victim. One animal (the mimic) is able to prey on or otherwise exploit the victim because it mimics a signal that is attractive to, or is not avoided by, the victim. Some praying mantids mimic flowers and devour insects that come to feed from them. On Pacific coral reefs the cleaner wrasse or sea swallow (*Labroides dimidiatus*) lives in close association with much larger fish

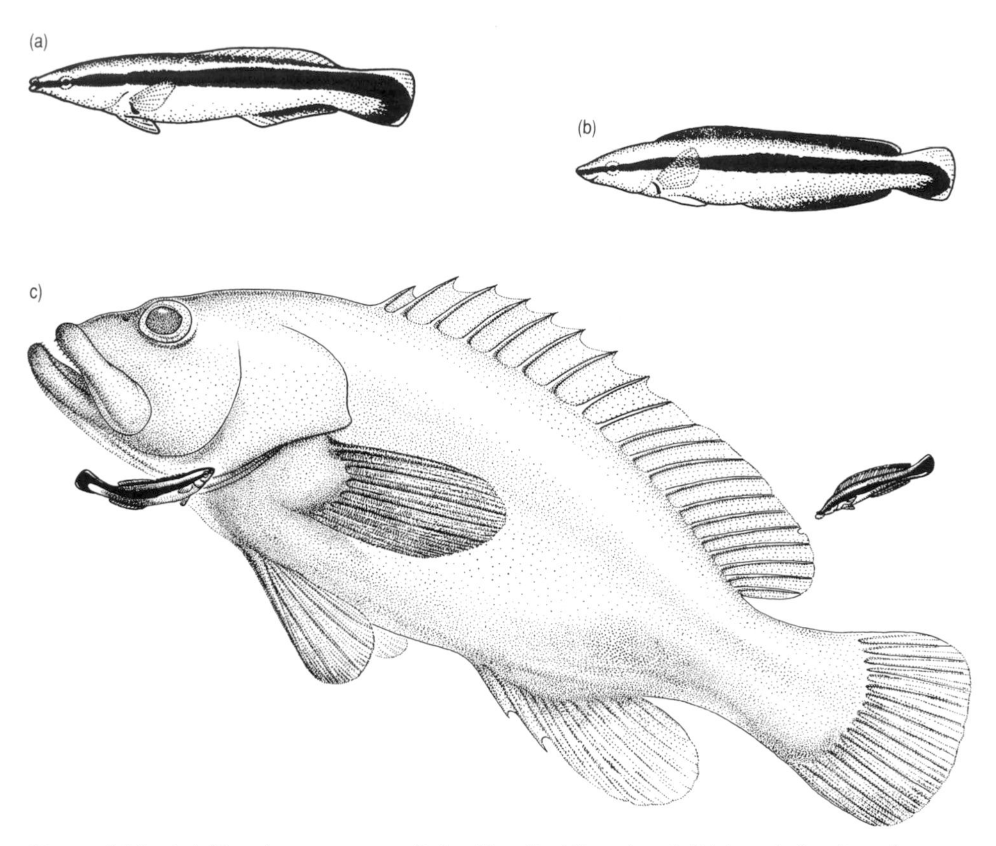

Figure 8.35 *(a) The cleaner wrasse (*Labroides dimidiatus*) and (b) its mimic, the sabre-toothed blenny (*Aspidontus taeniatus*), which have very different kinds of relationship with larger fish (c). (See text for details.)*

(its 'clients'), from which it removes parasites and dead skin. The cleaner wrasse avoids being eaten by having a distinctive colour pattern and a particular way of moving as it approaches its client. The sabre-toothed blenny (*Aspidontus taeniatus*) mimics both the appearance and behaviour of the cleaner wrasse but, instead of cleaning a client fish, it bites chunks out of its fins (Figure 8.35).

Another example of aggressive mimicry, discussed in detail in Chapter 2 (Section 2.3.2), is egg mimicry in cuckoos.

8.4.5 DIVERSITY, INTERACTIONS AND THE PHYSICAL ENVIRONMENT

So far, this Chapter has considered biological interactions between organisms, emphasizing the point that each species evolves in an environment that it shares with many other species and with which it interacts in a variety of ways. It is important to remember, however, that the environment of any given species also has important non-biological aspects, such as climatic, chemical and geological features, that also have a profound effect on its evolution. These factors are important when comparisons are made between the major ecosystems that exist in the world, such as tropical forests, small oceanic islands, and arctic tundra.

(**Ecosystem** is a term that refers to an ecological unit usually comprising several habitats and their many communities of organisms, together with their physico-chemical environment.)

Tropical rainforests and coral reefs are both ecosystems which typically contain very large numbers of species, of both plants and animals. Interactions between these species are often highly complex and include numerous parasitic associations, in some of which one parasite has multiple hosts, elaborate commensalisms, and very elaborate reproductive systems that involve interactions between plants and a variety of animals. A striking feature of tropical forests, for example, is the large number of plant species (e.g. orchids and ferns) that live as epiphytes (attached to) or parasites of other plants. These tropical habitats are long-established, highly productive, and relatively infrequently disturbed by fluctuations in climate, nutrient supply etc.; large fluctuations in the population sizes of individual species are unusual. By contrast, recently formed habitats, such as volcanic islands, and those at high altitude or high latitude, with sharply fluctuating conditions, tend to be much simpler and less stable. They contain far fewer species and complex interactions between species are a less obvious feature than they are in tropical ecosystems. Populations of many species are subject to dramatic fluctuations, as a result of variations in food supply or due to disease or parasite epidemics, such as those shown by lemmings and arctic foxes. As discussed later in the Course (Chapters 13 and 15), these variations between ecosystems in terms of their complexity and stability, have a profound influence on the rate at which extinctions occur in different kinds of environment.

SUMMARY OF SECTION 8.4

Interactions between organisms, whether plant–plant, animal–animal, or animal–plant can be classified according to whether each of two interacting species derives a gain in fitness, suffers a fitness cost, or is unaffected in terms of fitness. In predator–prey and herbivore–plant interactions, selection favours adaptations in the exploited partner that tend to reduce the effect of the predator or herbivore. Such adaptations favour the evolution of counter-adaptations in the predator or herbivore that enhance the effectiveness with which they obtain food. Parasite–host and pathogen–host interactions involve a similar coevolutionary pattern of adaptation and counter-adaptation, but evolution commonly leads towards a less exploitative, more commensal kind of interaction. Parasites and pathogens are very important, both because they make up a large proportion of species and because they have a profound effect on the evolution of their hosts. Not only can they have an effect on the fitness of individual hosts, and on the population size of their hosts, but they may be involved in the maintenance of sex and the evolution of male secondary sexual characters. Mimicry differs from other interactions in that it is not direct, but is mediated through a third species. Complex interactions between species are particularly common in the stable, complex ecosystems that occur in the tropics, such as rainforests and coral reefs.

8.5 THE EVOLUTION OF LIFE HISTORIES

In the life of an individual organism there are two fundamental events, reproduction and death. These relate to the two major components of fitness, survival and reproductive success, survival being essentially the delaying of death. The life history of an organism comprises the characteristics and timing of its reproduction and its death. There is considerable variation between individuals within a species in these two parameters and it is on this variation that natural selection acts (Chapters 4 and 7). This Section is primarily concerned with variation in life history between species and examines such questions as why the individuals of some species are short-lived, and others long-lived, why some reproduce only once in their lives, and others several times, and how different kinds of life history are related, in an adaptive sense, to features of the environment.

8.5.1 LIFE, REPRODUCTION AND DEATH

There is immense variation in the life expectancy of multicellular organisms. Amongst plants, many species, such as groundsel (*Senecio vulgaris*), are **annuals** and complete a life cycle, from seed germination, to seed production, to death, within a single year. Others, called **biennials**, such as the foxglove (*Digitalis purpurea*) take two years to complete a life cycle, typically growing leaves in the first year and flowering and setting seed in the second. Some plants that live for more than two years, called **perennials**, have very high longevities; the North American bristlecone pine (*Pinus cristata*), for example, can live for up to 4000 years. Among animals there is also considerable, but narrower, variation in longevity, from some crustaceans with life histories of a few days, to a few weeks for some insects, to more than 100 years for some of the larger tortoises.

There is also wide variation among species in the frequency of reproduction. Among fishes, for example, the Pacific sockeye salmon (*Onchorhynchus nerka*) lives for seven years, spawns once and then dies, whereas the Pacific sardine (*Sardinops caerulea*) spawns an average of three times in its ten-year life. There are innumerable plants that reproduce repeatedly, irrespective of their longevity, from the kidney vetch (*Anthyllis vulneria*), which has a life expectancy of three years to the sessile oak (*Quercus petraea*) which commonly lives to over 150 years. A long life does not necessarily mean that a species reproduces many times; a species of bamboo, *Phyllotrachys bambusoides*, regularly lives to an age of about 120 years without flowering, reproduces at the end of this period and then dies. This case is somewhat exceptional and probably evolved as a means of avoiding seed predation (Section 8.4.2). As discussed in Section 8.5.2, most long-lived plants and animals reproduce several times in their lives.

High longevity within a species obviously implies an ability on the part of individuals of that species to survive. It is to be expected, therefore, that there will be an inverse relationship between mortality rate, that is, the probability of dying within a given year, and life expectancy. Table 8.3 shows data for a variety of birds that show this effect.

Table 8.3 *Annual mortality rate and life expectancy among birds.*

Bird	Mean annual mortality rate among adults/%	Mean expectation of further adult life/years
Small passerine*	70	0.9
Wood pigeon (*Columba palumbus*)	40	2.0
Swift (*Apus apus*)	20	4.5
Yellow-eyed penguin (*Megadyptes antipodes*)	10	9.5
Royal albatross (*Diomedea epomophora*)	3	32.8

*The term 'small passerine' covers a wide range of familiar bird species, such as sparrows, finches, tits etc.

The data in Table 8.3 show that, within a taxonomic group, birds in this instance, there can be enormous diversity in such life-history parameters as annual mortality rate and longevity. There is also variation in the rate at which birds reproduce. A small passerine, such as the great tit (*Parus major*) produces one, occasionally two broods, of eight to ten eggs, in its short life; a royal albatross may not breed until it is 15 years old and then produces a single egg once every two years. If all the birds in the Table were to breed at comparable rates, the world would be overrun with albatrosses!

▶ What other pattern of variation can you detect in Table 8.3?

There is a tendency for the low mortality rate, greater longevity species to be large birds, and for the high mortality rate, short-lived birds to be small. The swift, however, is an exception to this pattern, illustrating the point that, as we shall see, few if any generalizations about life-history patterns are true for all species.

How are these diverse kinds of life history to be reconciled with the theory of natural selection, which might appear to suggest that the fittest organisms should be those that produce the greatest number of offspring in the shortest possible time? This question is the basis of much current theoretical and experimental investigation and the rest of this Section will examine just a few aspects of it.

A naive interpretation of the theory of evolution by natural selection is that it would be expected to produce an organism that has *all* the attributes that confer high fitness. Such an organism would reproduce soon after birth, would produce large numbers of progeny, in which it would invest effective parental care, and it would continue to do these things throughout an infinitely long life. To do this it would have to out-compete any competitors, avoid predation and be itself hugely efficient at obtaining food. Such an organism does not, of course, exist, and nor could it do so, for a number of reasons, of which we shall consider two, one specific, the other general.

Consider, first, the specific question of why organisms die. Why has natural selection not produced an organism that is immortal? The answer to this question is that natural selection cannot prevent senescence and death because it cannot eliminate certain kinds of alleles. Alleles that cause death *before* an organism has reproduced are, obviously, not passed on to progeny and so are eliminated by selection. However, those that cause death *after* an organism has begun to reproduce are inevitably passed on to progeny and thus persist in the gene pool of a population. This question of the 'evolution of death' makes a number of important points about evolution. First, as emphasized in Chapter 4, natural selection operates on organisms within constraints imposed by the processes of genetic inheritance. Secondly, death is an *effect* of natural selection, but is not a character that is favoured by natural selection; it is important to remember that one cannot assume that everything that is identified as a character is an adaptation (Chapter 2). Thirdly, and most relevant in the present context, the range of possible options in terms of life-history characters that can evolve by natural selection is limited by the inevitability of death.

The second, and more general, reason why an organism that maximizes all components of fitness cannot exist is that, as discussed in Chapters 4 and 7, there are trade-offs between one fitness component and another. For example, an increased allocation of resources to survival and growth typically reduces the amount of resources that can be allocated to reproduction, and vice versa. Trade-offs arise because resources are not in infinite supply but are limited, an effect exacerbated by the fact that individuals must compete with one another for such resources as are available. As suggested in Chapter 4, however, obtaining empirical evidence of life-history trade-offs is often not as straightforward as might be expected. For example, if there is a trade-off between reproductive effort and survival, one might expect, by measuring the reproductive output and survival of a large number of individuals, to find a negative correlation between the two measures. In fact, in many instances a positive correlation is found (Figure 8.36). What data like these suggest is that, as in the case of Canadian song sparrows (Chapter 4, Section 4.4.1), there is a wide variation in individual fitness and that individuals of high fitness invest heavily in both survival and reproduction. This conclusion does not negate the theoretical argument that each individual makes a trade-off between the two, rather it shows that this correlational approach does not provide evidence for such a trade-off.

Evidence for life-history trade-offs can be obtained by means of experiments and we will consider one simple and elegant example of an experimental approach.

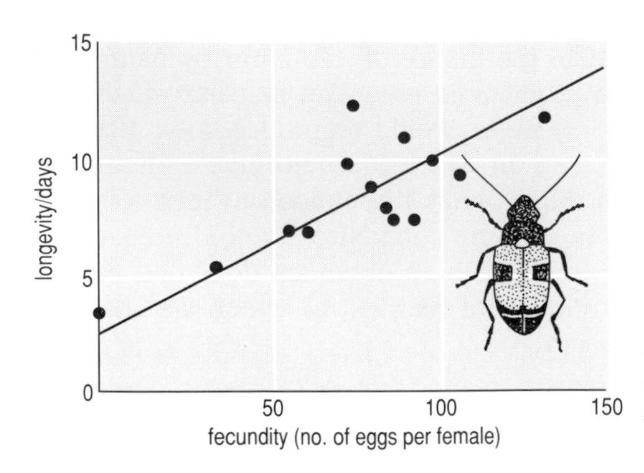

Figure 8.36 *The relationship between longevity and fecundity (number of eggs produced) for female beetles (*Callosobruchus maculatus*).*

Kate Lessells (1986) studied the Canada goose (*Branta canadensis*), a species which normally lays an average of six eggs per clutch. She added a single egg to the clutches of a number of experimental females and compared various components of their fitness with those of a control group of birds that had not been given an additional egg. The effect of the extra egg was to force pairs of birds to invest more heavily than they otherwise would in parental care. At the end of the breeding season, the experimental birds showed reduced body weight and condition (a score that measures an animal's weight per unit body size). In the following year, both the female and the male from experimental pairs bred later in the season than control birds.

▶ Do these results support the hypothesis that there is a trade-off between reproduction and survival?

Yes, they do. The reduced weight and condition of experimental birds suggests that rearing an extra young reduced the amount of resources that they could allocate to somatic, as opposed to reproductive, effort (Chapter 7, Section 7.2.2). Lessell's results show an additional effect, that an increase in reproductive effort in one year led to a reduction in reproductive effort the following year, as suggested by the fact that experimental birds bred later in the following season. We shall return to this effect shortly.

Life-history theory seeks to explain, for a given species, how a number of variables relate to one another and to the nature of that species' environment. The most important of these variables are: birth rate, death rate, longevity, body size and the efficiency with which individuals exploit their environment. In the rest of this Section we will look at some of these variables and will then integrate them by means of some general theories about life-history evolution.

8.5.2 Reproductive strategies

As we have seen, there is variation between species, in terms of how reproductive effort is expended during an individual's life. In some species individuals breed once early in life, in others once late in life, and in others they breed repeatedly during the course of their life. An individual's lifetime reproductive success depends both on its survival and on the way it distributes its reproductive effort. The two effects are combined into a single value, for an individual organism, called its **reproductive value**, $\boldsymbol{V_i}$. Reproductive value varies over the life of an individual and so is calculated for a particular age interval, denoted by the subscript i. V_i is calculated as follows:

Reproductive value at age i	=	contemporary reproductive output	+	residual reproductive value

$$V_i = m_i + \sum_{t=i+1}^{t=\infty} (l_x/l_i)m_x \qquad (8.1)$$

where m_i is the number of offspring produced up to the present interval i, l_i is the survivorship to the beginning of the interval i, l_x is the survivorship to the

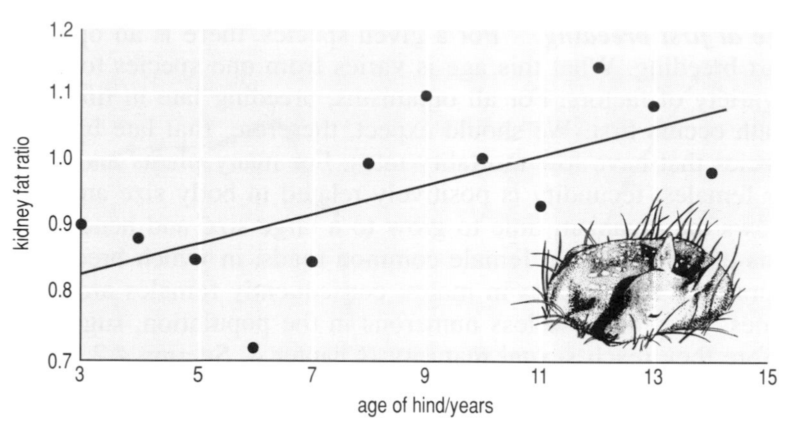

Figure 8.38 *The ratio of kidney fat in red deer calves to that in hinds, plotted against hind age.*

future reproductive value, increases with age, especially towards the end of her life. These concepts lead to the prediction that an iteroparous animal will invest relatively more effort in each of its later progeny than it did in its earlier progeny. This prediction has been tested for the red deer (*Cervus elephas*) by Tim Clutton-Brock (Figure 8.38). Clutton-Brock measured the amount of fat around the kidneys in hinds of different ages and in calves born to hinds of different ages and found that the ratio of calf fat to maternal fat increased with maternal age. (A ratio greater than one means that a calf contains proportionately more kidney fat than its mother).

▶ Do the data in Figure 8.38 support the hypothesis that red deer hinds invest more reproductive effort in their later offspring?

Yes, they do. As they get older, hinds store a smaller proportion of their available resources as fat and invest more in feeding their calves. As a result, calves born later in a hind's life have higher fat reserves, relative to those of their mother.

Number of progeny As discussed at some length in Chapter 7, Section 7.2.2, there is a trade-off between the number of young that an individual organism produces and the proportion of its reproductive effort that goes into the care of those young. To recapitulate briefly, in environments in which young are able from an early stage to survive without parental care, individuals produce large numbers of progeny and invest relatively little effort in parental care. In environments in which the young can survive only if they receive protection, care and feeding by the parents, few young are produced at each breeding episode. There is marked variation in the number of young produced per breeding episode between different animal groups, because different taxonomic groups are subject to different constraints. For example, the clutch size of birds is constrained by the number of eggs that they can incubate at one time (Chapter 2, Section 2.2.1) and the litter size of female mammals is constrained by the number of foetuses that they can carry in the uterus without impairing their own mobility and by the needs of lactation.

8.5.3 LIFE-HISTORY STRATEGIES

A number of attempts have been made to develop a theoretical framework in which the various life-history variables discussed above can be related, both to one another, and to the nature of the environment in which a species lives. The most influential of these has been the dichotomy between what are called *r* and *K* life-history strategies. This dichotomy is based on two parameters that relate to phenomena at the population level, which are illustrated in Figure 8.39.

A population that colonizes a vacant area of habitat will show an initially slow, but then rapidly accelerating increase in numbers, as a result of there being no competition. The maximum rate of growth of the population, that is, its rate of growth when there is zero mortality, is defined as the **intrinsic rate of natural increase, *r***. As the population gets larger, the habitat becomes crowded and competition for resources becomes increasingly severe. As a result, population growth slows down until the population reaches a stable size. This size is determined by the nature of the habitat and is called its **carrying capacity, *K***.

As described in Section 8.5.2, the optimal reproductive strategy for an individual, that is, the strategy that maximizes its reproductive value, varies according to the population density. At low densities it will most adaptive to reproduce at as high a rate as possible; at high densities it will be more adaptive to invest in survival, to produce few young and to invest heavily in each of them. This, in essence, is the basis of the *r*- and *K*-strategy dichotomy. An ***r*-strategist** is an organism, the life history of which is adapted to maximize its reproductive rate. A ***K*-strategist** is an organism, the life history of which is adapted to maximize its competitiveness and adult survival. There is considerable potential for confusion here, which is one reason why the *r*/*K* dichotomy has not found favour with all ecologists. The variables *r* and *K* relate to *populations*, as in Figure 8.39; yet life-history strategies are the product of natural selection acting on *individuals*.

The *r*-strategy is associated with organisms living in unstable, frequently disturbed or harsh habitats, the *K*-strategy with those living in stable environments that are favourable to survival and growth. Table 8.4 lists the life-history characteristics that are associated with the two kinds of strategy.

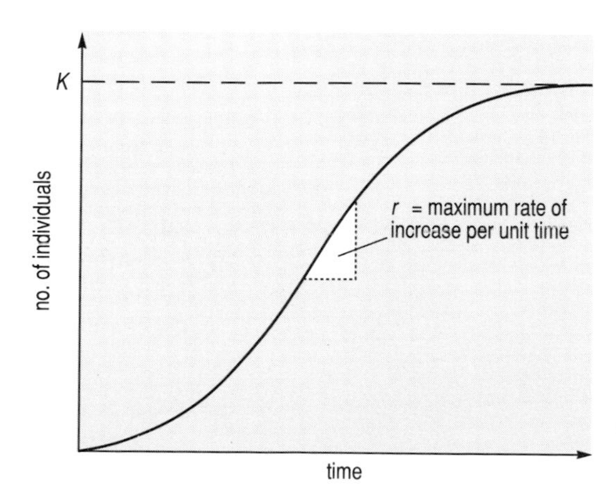

Figure 8.39 *The relationship between time and the population size of an expanding population of organisms, showing the meaning of the variables* r *and* K. *See text for full explanation.*

Table 8.4 *Characteristics of* r- *and* K-*strategists.*

K-strategy	*r*-strategy
Habitat: stable, favourable to growth and survival	Habitat: unstable, harsh
Slow development	Rapid development
Large adult body size	Small adult body size
Late first reproduction	Early first reproduction
Iteroparity	Semelparity
Few progeny per breeding episode	Many progeny per breeding episode
Long lifespan	Short lifespan
Long generation time	Short generation time

In addition to the features listed in Table 8.4, there is a tendency among plants for *r*-strategists to reproduce asexually, and *K*-strategists sexually.

The *r*/*K* dichotomy has proved of value as a heuristic device, helping ecologists to bring some order to the enormous diversity that exists among life histories. It is, however, one that should be used with caution. While it is possible to identify certain species that exemplify the *r*/*K* dichotomy rather well, the great majority of species show an intermediate condition and may combine features associated with each extreme. The bamboo (mentioned in Section 8.5.1) is one such example. In delaying breeding until it is 120 years old, it behaves like a *K*-strategist, but in being semelparous and spectacularly fecund, it shows two elements of the *r*-strategy. The *r*- and *K*-strategies should be seen as the extremes of a continuum, along which the majority of species will be clustered nearer the middle than the ends. Another problem with the *r*-strategy is that it is associated with two features of environments, instability and severity, that do not necessarily go together. For example, a desert is a severe habitat that is also highly stable. J. P. Grime has devised a three-way classification of habitats and plant life histories, known as Grime's triangle (Figure 8.40).

This classification has two axes, one relating to the stability or frequency of disturbance of the habitat, the other to its severity as a habitat in which plants

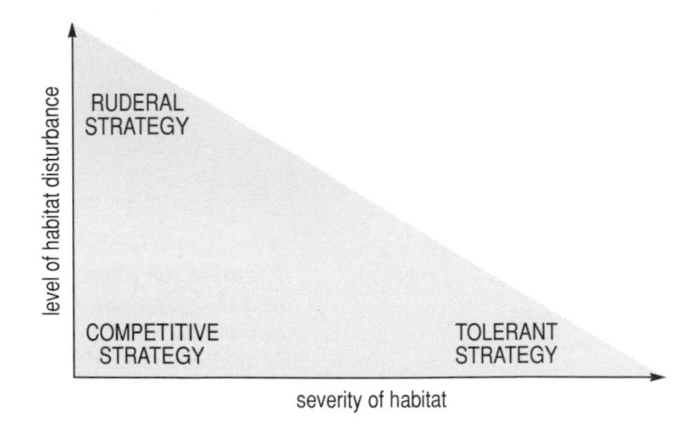

Figure 8.40 *Grime's triangle. (See text for explanation.)*

must survive. Habitats that are stable and benign (lower left in the Figure) select for life histories that maximize competitive ability in the adult, essentially the characteristics of K-strategists. Benign but unstable habitats (top left) favour life histories that maximize reproductive rate and which are characteristic of weed species; this is called the *ruderal* strategy. Habitats that are stable and severe (lower right), such as deserts, favour a tolerant strategy. This includes adaptations for resisting drought over long periods, but also for being able to reproduce abundantly at unpredictable times when conditions temporarily become favourable, for example after rain.

SUMMARY OF SECTION 8.5

Different species of plants and animals vary enormously in a number of life-history parameters, such as longevity, frequency of reproduction, age at first reproduction and body size. The evolution of these characters involves many trade-offs between one activity and another, because organisms are strictly limited in terms of the quantity of resources that they can allocate to different activities. During the course of its life, an individual's reproductive value changes; it always reaches a maximum and then declines with age, but the precise pattern of this change varies from one species to another. The lifetime reproductive output of an organism is determined by three factors: age at first breeding, frequency of breeding and the number of progeny produced per breeding episode. The many factors relating to the survival and reproduction of individuals in a particular species of animal or plant combine to make up the life-history strategy of that species. One way of classifying life-history stategies is on a single-axis continuum, from r-strategist to K-strategist. Another recognizes two principal axes, habitat stability and habitat severity. All such classifications are very simplistic and, in practice, it is rarely possible to ascribe a particular species unequivocally to a particular life-history category.

REFERENCES FOR CHAPTER 8

Connell, J. H. (1961) The influence of interspecific competition and other factors on the distribution of the barnacle *Chthamalus stellatus, Ecology*, **42**, 710–723.

Griffiths, R. A. (1986) Feeding niche overlap and food selection in smooth and palmate newts, *Triturus vulgaris* and *T. helveticus*, at a pond in mid-Wales, *Journal of Animal Ecology*, **55**, 201–214.

Hairston, N. G. (1980) The experimental test of an analysis of field distributions: competition in terrestrial salamanders, *Ecology*, **61**, 817–826.

Hamilton, W. D. and Zuk, M. (1982) Heritable true fitness and bright birds: a role for parasites, *Science*, **218**, 384–387.

Lessells, C. M. (1986) Brood size in Canada geese: a manipulation experiment, *Journal of Animal Ecology*, **55**, 669–689.

Møller, A. P. (1987) Variation in badge size in male house sparrows *Passer domesticus*: evidence for status signalling, *Animal Behaviour*, **35,** 1 637–1 644.

Peters, W. D. and Grubb, T. C. (1983) An experimental analysis of sex-specific foraging in the downy woodpecker, *Picoides pubescens*, *Ecology*, **64**, 1437–1443.

Price, T. D. (1984) The evolution of sexual dimorphism in Darwin's finches, *American Naturalist*, **123**, 500–518.

Reed, T. M. (1982) Interspecific territoriality in the chaffinch and the great tit on islands and the mainland of Scotland: playback and removal experiments, *Animal Behaviour*, **30**, 171–181.

Ryan, M. J. and Causey, B. A. (1989) 'Alternative' mating behavior in the swordtails *Xiphophorus nigrensis* and *Xiphophorus pygmaeus* (Pisces: Poeciliidae), *Behavioural Ecology and Sociobiology*, **24**, 341–348.

Schoener, T. W. (1984) Size differences among sympatric bird-eating hawks: a worldwide survey, in D. Simberloff, L. G. Abele and A. B. Thistle (eds) *Ecological Communities: Conceptual Issues and the Evidence*, pp. 254–281, Princeton Univ. Press.

FURTHER READING FOR CHAPTER 8

Begon, M., Harper, J. L. and Townsend, C. R. (1990) *Ecology* (2nd edn), Blackwell Scientific Publications, Oxford.

Cockburn, A. (1991) *An Introduction to Evolutionary Ecology*, Blackwell Scientific Publications, Oxford.

SELF-ASSESSMENT QUESTIONS

SAQ 8.1 *(Objective 8.1)* For each of the features of organisms (a) to (e), state whether it is more likely to have evolved as a consequence either of intraspecific or of interspecific competition, whether it could be the result of either, or whether it has nothing to do with either form of competition.

(a) Territorial behaviour.

(b) Niche overlap.

(c) Sexual dimorphism in feeding habits.

(d) Character displacement.

(e) Differences in diet and/or feeding habits between age classes.

SAQ 8.2 *(Objective 8.2)* In what circumstances, in terms of fitness costs and benefits, would you expect animals to settle competitive disputes over resources (a) by fighting to the death, and (b) by means of ritualized behaviour?

SAQ 8.3 *(Objective 8.3)* Section 8.3.3 quoted the example of five tit species in Britain and the niche differentiation that is observed among them, and suggested three hypotheses for its evolution: (i) current competition, (ii) past competition, and (iii) independent evolution. With which of these are the hypothetical observations (a) to (c) compatible?

(a) When great tits are experimentally removed from an area of woodland, blue tits spend more time feeding on the ground.

(b) Where one of the smaller tit species occurs alone, it has a larger body size than in localities where it is sympatric with larger tit species.

(c) When great tits are experimentally removed from an area of woodland, blue tits show no change in their behaviour.

SAQ 8.4 *(Objective 8.4)* For each of the interactions (a) to (c), categorize it according to whether it has a positive (+), a negative (–) or no (0) impact on each of the two species involved in that interaction.

(a) Parasitism.

(b) Commensalism.

(c) Mutualism.

SAQ 8.5 *(Objective 8.5)* Explain briefly how predation can favour the evolution of polymorphism within a prey species. What is the particular kind of selection that is involved in this process?

SAQ 8.6 *(Objective 8.6)* Give two general reasons, one related to the host, the other to the parasite, why parasite–host interactions will tend to evolve towards commensalism.

SAQ 8.7 *(Objective 8.7)* For each of the statements (a) to (d), state whether it relates to an example of Batesian, Müllerian or aggressive mimicry.

(a) An organism is deceived by the mimic.

(b) A palatable prey species resembles an unpalatable, aposematic species.

(c) The mimic species possesses a signal that is attractive to its victim.

(d) A number of distasteful species share a similar colour pattern.

SAQ 8.8 *(Objective 8.8)* For a female toad, what factors will select in favour of (a) breeding early in life, and (b) breeding late in life? In what ways are these factors similar or different for male toads?

SAQ 8.9 *(Objective 8.9)* For each of the ecological factors (a) to (c), state whether it tends to promote the evolution of semelparity or iteroparity.

(a) High adult mortality.

(b) Decreasing intraspecific competition.

(c) An unstable environment in which new habitats are created frequently.

SAQ 8.10 *(Objective 8.10)* For each of the life-history factors (a) to (e), state whether it is characteristic of an r-strategist or a K-strategist.

(a) First reproduction late in life.

(b) Semelparity.

(c) Many progeny per breeding episode.

(d) A long generation time.

(e) Large adult body size.

Part II
Evolution of Species

CHAPTER 9

SPECIES, SPECIATION AND EXTINCTION

Prepared for the Course Team by Marion Hall

Study Comment *There are problems in defining species, and this Chapter considers the advantages and disadvantages of three common definitions. A variety of hypotheses have been put forward to explain how new species arise and why species become extinct. Various biological models of speciation are described and evaluated in relation to each other and to the available evidence; and the possible causes of extinction are discussed.*

Objectives When you have finished reading this Chapter, you should be able to do the following:

9.1 Distinguish between the evolutionary biological and recognition species concepts, and evaluate each species concept in terms of its applicability to living or fossil organisms.

9.2 Describe the types of reproductive barrier that genetically isolate sexually reproducing species, and give or recognize examples of each type.

9.3 Describe the ways in which species vary geographically, and give or recognize examples of species that show such geographical variation.

9.4 Describe and distinguish between allopatric, sympatric and parapatric models of speciation.

9.5 Describe and assess the evidence for and against allopatric, sympatric and parapatric models of speciation.

9.6 Understand the theoretical difficulties inherent in models of sympatric and parapatric speciation.

9.7 Describe the possible causes of population extinction, and how the extinction of populations relates to the extinction of species.

9.1 INTRODUCTION

No-one knows how many species there are living in the world today, but some of the most recent estimates put the figure as high as 30 million or more, of which only a few million have been discovered and about only 1.5 million described in detail. Of course, this is but a fraction of the total number of species that have existed throughout the Earth's history. During evolution, new species arise while others cease to exist—i.e. **speciation** and **extinction** take place. It is easy to see how a species may become extinct—if more individuals die than are replaced by reproduction, the species must eventually die out. Extinction will be discussed in Section 9.8; but first, how are new species formed?

We can define speciation as the multiplication of species, i.e. the division of one species during evolution into two or more separate species. (The gradual evolution, without such division, of one species into a form sufficiently different to be classified as a different species—the two forms are said to be different **chronospecies**—is not included in this definition and will not be discussed further here. However, you will return to the idea of chronospecies in Chapter 10.)

Taking a very simple view, speciation can happen in one of two ways: 'splitting' or 'budding' (Figure 9.1). A species could split fairly equally into two, the two halves evolving differently until they eventually become separate species. Alternatively, a small part of the species population could 'bud off' from the main part and evolve rapidly (in terms of geological time-scales) to form a new species while leaving most of the original species population relatively unchanged. If the new species is successful, it could rapidly increase in numbers to become as abundant as its parent species.

It is simple to define speciation as the division of one species into two—but how does that division come about? And how do we identify the point at which a species has changed enough to be classified as a new species? In other words, how is a species defined? As you will see, there are many problems in trying to answer these questions.

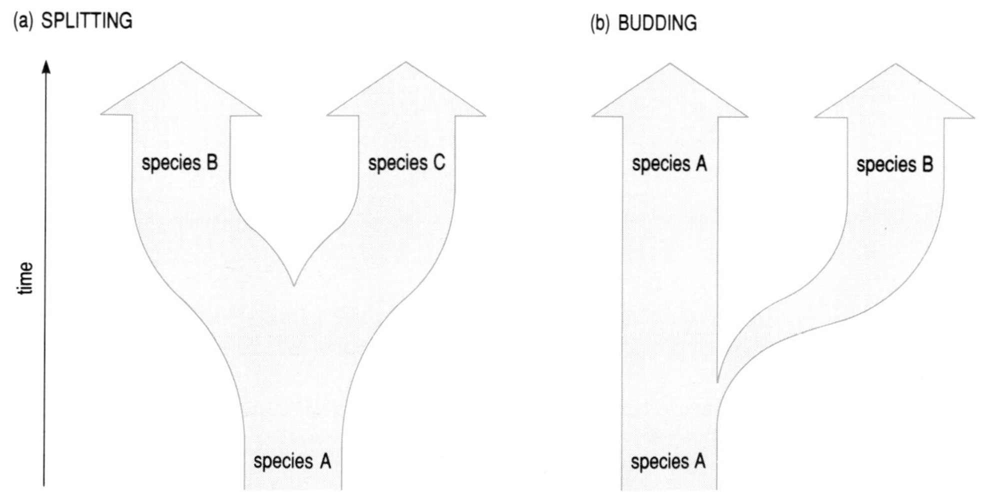

Figure 9.1 *Speciation as a process of (a) splitting or (b) budding. Degree of differentiation is represented across the page, so that the farther apart two lines are, the greater is the differentiation between them; thickness of each line represents population size.*

9.2 WHAT IS A SPECIES?

Taxonomists—scientists who, amongst other things, describe and classify living organisms and decide what to call them—divide living things into discrete units called *species*. Classification into recognized groups is necessary for the purposes of discussion of and communication about those groups, but are species *real* entities? That is, do they exist as discrete units in nature or are they just convenient divisions invented by taxonomists? The study of natural populations shows that living things do tend to occur in units that are distinctly different in form, ecology, genotype and evolutionary history, and within which individuals resemble each other and between which there are generally no intermediate forms (but see Section 9.3.1 for exceptions). Thus, species must be considered as real entities but there remains the problem of definition.

Three commonly used definitions—the Evolutionary Species Concept, the Biological Species Concept and the Recognition Species Concept—are discussed below.

9.2.1 THE EVOLUTIONARY SPECIES CONCEPT

According to this concept, a species consists of all those individuals that share a common evolutionary history. This means that an evolutionary species consists of a single lineage of ancestral–descendant populations, which remains distinct from other such lineages; it evolves roughly as a unit and independently of other such units. This definition is useful because it comes close to that used in practice by most taxonomists and palaeontologists (see Section 9.2.4 and Chapter 10), who usually decide where the limits of a particular species lie on the basis of similarity of phenotype within the species and discontinuity between it and other related species.

The concept has two main problems. First, how do we judge what constitutes a common evolutionary history? The individuals within a species are never completely identical, so they cannot have followed an identical evolutionary path. Also, two chronospecies are part of the same lineage yet are considered to be different species. How much variation in phenotype is 'allowed' within a single species? Secondly, although speciation is seen to occur as a result of different populations within a species evolving differently, the concept says nothing about the mechanisms by which species are maintained or by which new species evolve.

9.2.2 THE BIOLOGICAL SPECIES CONCEPT

Until very recently, the generally accepted modern definition of a species was that provided by Ernst Mayr in 1940:

> 'Species are groups of actually or potentially interbreeding natural populations which are reproductively isolated from other such groups.' (Mayr, 1970.)

Thus, according to this Biological Species Concept (also known as the Isolation Species Concept), an individual belongs to the species with whose members it can successfully reproduce.

► Using the above definition of species, are the various breeds of domestic cat all members of a single species or of different species?

They all belong to a single species. Any of the different breeds of cat can reproduce successfully with each other, if allowed to do so!

According to the Biological Species Concept, separate species are prevented from merging with each other by a number of **isolating mechanisms** which form barriers to gene flow (Chapter 3, Section 3.5.2) between them. Speciation is thus seen in terms of the evolution of isolating mechanisms and is said to be complete when reproductive barriers are sufficient to prevent gene flow between the two new species. Isolating mechanisms are often divided into **pre-zygotic** and **post-zygotic mechanisms** (or into **pre-mating** and **post-mating mechanisms**), depending on whether or not fertilization (or mating) is able to take place.

Pre-zygotic mechanisms include:

1 **Ecological** (or **habitat**) **isolation**: the species occupy different habitats in the same geographical area, and so are unlikely ever to meet. For example, the larvae of the European mosquitoes *Anopheles labranchiae* and *A. atroparvus* live in brackish water, those of *A. maculipennis* in running freshwater, and those of *A. melanoon* and *A. messeae* in stagnant freshwater.

2 **Seasonal** (or **temporal**) **isolation**: species have different mating or flowering seasons or times of day, or become sexually mature at different times of the year. Closely related plants growing in the same geographical area are often seasonally isolated. The pines *Pinus radiata* and *P. muricata* are both found on the Monterey Peninsula in California. *P. radiata* sheds its pollen early in February while *P. muricata* sheds its pollen in April. Hybrids are relatively rare and are less vigorous and fertile than either of the parental species.

3 **Ethological** (or **sexual**) **isolation**: the sexual attraction between males and females of different animal species is reduced or absent due to a mismatch in behaviour or physiology. Ethological isolation occurs in many groups of animals, and is often the most powerful reproductive barrier between closely related species in the same geographical area. Many insects and vertebrates have species-specific courtship displays, which may combine visual, auditory, tactile and olfactory stimuli. For example, the courtship displays of different *Drosophila* species often involve different visual stimuli—the speed and style of their courtship 'dance'—and different auditory stimuli—the sound frequencies making up the courtship 'song' and the rate at which the song is delivered; olfactory stimuli may also be involved.

4 **Mechanical isolation**: the mechanical structure of the reproductive organs or genitalia impedes or prevents the transfer of gametes between species. Mechanical isolation is frequently found among flowering plants. Some species, because of the structure of their flowers, can be pollinated only by a few or even a single species of insect. The flowers of some orchids, for example, attract sexually active male bees of a particular species by their resemblance in colour and shape to female bees of the same species. The males copulate with the flowers and, in so doing, pollinate them. Different orchid species resemble female bees of different species, so cross-pollination is unlikely.

5 **Gametic isolation**: gamete transfer takes place but fertilization does not occur, either because the male and female gametes of different species fail to attract or unite with each other, or because male gametes are inviable in the reproductive organ of the female of another species. Many species of *Drosophila*, for example, show an 'insemination reaction', in which the walls of the female's vagina swell enormously in the presence of sperm from another species, killing the sperm. In flowering plants, the pollen of one species may fail to germinate on the stigma (that part of the female reproductive organ which receives the pollen) of another species. And even if it does, the pollen tube (an extension from the pollen grain that grows down to the ovary so that fertilization can take place) may grow more slowly, or it may start to grow but then stop.

Post-zygotic isolating mechanisms include:

1 **Hybrid inviability**: the egg is fertilized but does not develop, or the development of the embryo becomes arrested at some stage, or the hybrid dies before it reaches sexual maturity. In crosses between sheep and goats, for example, the embryos die very early on in their development.

2 **Hybrid sterility**: the hybrid survives but fails to produce functioning sex cells or gametes. The mule—the product of a cross between a female domestic horse (*Equus caballus*) and a male domestic ass (*E. asinus*)—is the classic example of hybrid sterility. Sometimes, sterility may be limited to one sex, usually the male, as in many interspecific crosses in *Drosophila*.

3 **Hybrid breakdown**: the hybrid is fully viable and fertile but the viability and/or fertility of its offspring are reduced. Hybrid breakdown is probably due to the formation of less-fit genotypes by recombination (Chapter 3, Section 3.3.1) between the genotypes of parental species. Recombination occurs at meiosis, so that, in a first-generation hybrid offspring between two different species, recombination between the two parental genotypes will not occur until the hybrid produces gametes of its own by meiosis.

The lack of gene flow between any pair of species may be due to one or more of the above reproductive barriers. For example, the American toads *Bufo americanus* and *B. fowleri* are isolated both seasonally and by habitat preference: *B. americanus* lives in forested areas and breeds earlier in the spring than *B. fowleri*, which lives in grasslands.

A major problem with the Biological Species Concept, particularly in relation to evolutionary studies, is that it is not applicable to all species.

▶ What is the most obvious group of species to which the Biological Species Concept cannot be applied?

It cannot be applied to species that reproduce asexually.

There are also practical problems in applying the Biological Species Concept. It is impossible to apply it to fossil species, for example, since they no longer mate! Thus, all extinct organisms plus the relatively large number of organisms that are asexual or exclusively self-fertilizing are excluded from the definition. Yet these organisms can be divided into species on the basis of the same pattern of similarity in phenotype within species and discontinuity between species that is shown by living sexual species.

Even for extant sexual species, the criterion of reproductive isolation cannot be used to identify species that are geographically separated. And, where they do live in the same geographical area, the Biological Species Concept does not provide a practical definition because, for most species, it is not possible to observe mating in the wild. For example, in the case of many plants where pollen is carried to the female by wind or insects, or marine organisms that simply shed their gametes into the water, it is almost impossible to check which individuals are mating with which. And successful breeding in the unnatural conditions of the laboratory or zoo is no guarantee that it actually happens in nature.

Another major problem is that the concept of a reproductively isolated group frequently breaks down. Hybridization between species does occur in nature, creating many problems in the delimitation of species. Closely related species often occupy different geographical areas, but some overlap at the edge of their ranges is common, and in these areas it is not unusual for hybridization to occur. Hybridization is especially common in plants. Plant taxonomists frequently group species in larger units called *syngameons*, within which natural hybridization may take place. Yet the species within a syngameon remain separate species. How much reproductive isolation must there be before a group is recognized as a species?

This problem is illustrated by two American tree species from the genus *Populus*, the balsam poplar and the cottonwood. The fossil record shows that they have been distinct morphologically for at least 12 million years, yet they have apparently formed hybrids throughout this time. Such hybrids today are widespread and fertile, but the balsam poplar and the cottonwood have still maintained their identity within their own type in terms of genotype, phenotype and ecology. Thus, the two species are real biological units. Similar situations occur in animals. For example, coyotes and wolves hybridize in the wild, yet they are distinct from each other and have evolved as separate lineages for at least 500 000 years.

Lastly, there is considerable lack of agreement between the degree of reproductive isolation and the amount of morphological or genetic divergence shown between two species (Section 9.2.5).

9.2.3 THE RECOGNITION SPECIES CONCEPT

Since the mid-1980s, H. E. H. Paterson has argued that some of the problems of the Biological Species Concept can be avoided by looking at reproductive isolation mechanisms from a different angle. He defines the species as '... that inclusive population of individual biparental organisms which share a common fertilization system'.

The fertilization system consists of all the components, such as courtship behaviour, genital structure, or attractiveness of the ovum to the sperm or pollen, that contribute to the ultimate function of bringing about fertilization with another individual having the same fertilization system. Speciation thus occurs when a different fertilization system evolves, and reproductive isolation is merely a by-product of a change in the fertilization system, rather than an active part of the process of speciation. Paterson is thus emphasizing the factors that keep a species together—mating and reproduction—rather than those that keep species apart.

▶ Which of the problems associated with the Biological Species Concept also apply to the Recognition Species Concept?

The Recognition Species Concept cannot be applied to asexual species, and the existence of hybrids creates the same difficulties as for the Biological Species Concept. The Recognition Species Concept may be more meaningful biologically than the Biological Species Concept for species separated in space and time, as fertilization systems can be compared to some extent between species even if they are geographically separated or fossilized—e.g. by looking at the morphology of genitalia—but it is just as difficult to test because mating often cannot be observed in nature. The problem also remains of how similar fertilization systems must be in order to define two groups as members of the same species.

9.2.4 The identification of species in practice

If there are problems with all of the species definitions given above, how are species identified in practice? Generally, as we have already seen, organisms are divided into species on the basis of similarity in phenotype and genotype within species and discontinuity between species. Traditionally, organisms were classified mainly on the basis of morphological differences by taxonomists working on museum specimens, and this is still the way in which most species are distinguished. Where a group of organisms requires more-detailed investigation in order to identify species, however, modern taxonomists have several additional methods of studying variation among organisms and of working out their evolutionary relationships with each other. These methods include studying the behaviour of the organisms concerned, and/or the differences among them in chromosome number and banding pattern or in their molecular genetics, e.g. by electrophoresis of proteins or DNA fragments, DNA hybridization or measurement of immunological distance (Chapter 3, Section 3.5.2). They have frequently also revealed the existence of **sibling species**—two groups of organisms that are morphologically identical but which are actually two separate species, being behaviourally, genetically and/or karyotypically different, and with limited or no hybridization between them. Palaeontologists, of course, have to rely almost entirely on morphological features to assign their fossil specimens to species. The problems that this creates will be discussed further in Chapter 10.

Even where several methods of assessing differences are available to the taxonomist, things are still not cut and dried—there is the perennial problem of how different something has to be before it can be classified as a separate species. Hybrids and organisms intermediate in form between two species also present taxonomists with great problems (though, as you will read in this Chapter, they are of great interest to evolutionary biologists). Taxonomists frequently revise the classification of related groups of species as more information becomes available.

In practice, biologists and palaeontologists generally adopt the definition of species that is most suited to the type of animal or plant on which they are working. Thus, for biologists working on living, sexually reproducing species in the field, the Biological Species definition is most commonly used; and the main criterion for deciding whether two forms are members of the same species is whether they generally interbreed in nature. Taxonomists working on specimens in museums rely more on morphological criteria, as by necessity do palaeontologists studying fossils.

9.2.5 THE RELATIONSHIP BETWEEN PHENOTYPIC AND GENETIC DIFFERENCES BETWEEN SPECIES

We know almost nothing about the relationship between DNA and the organism's morphology, physiology, ecology or behaviour. Measurements of genetic and karyotypic variation have shown that there is very little relationship between the amount of morphological difference or the degree of reproductive isolation between two species and the amount of genetic and/or karyotypic difference between them. Some species appear to differ by relatively few genes yet rarely hybridize, while others show very many genetic and/or karyotypic differences and hybridize frequently where they come into contact with each other.

For example, electrophoretic studies of proteins in the Hawaiian picture-wing *Drosophila* (so called because of their ornately patterned wings) have shown that there is an extremely high degree of genetic similarity within groups of related species. For *Drosophila* species elsewhere, indices of genetic similarity (see Chapter 3) between populations of the same species in different geographical areas are generally greater than 0.95, those between sibling species are approximately 0.56, while those between morphologically distinct species average about 0.35. On these criteria, most Hawaiian picture-wing species would not be considered as separate species because similarity indices can be as high as 0.90, yet these flies are highly differentiated morphologically and do not interbreed.

This major problem in the study of evolution was highlighted by A.C. Wilson. He compared the rate of evolution in two groups of vertebrates—frogs and placental mammals (Table 9.1). Placental mammals have experienced much faster phenotypic evolution than frogs. There are thousands of frog species living today but they are all very similar morphologically, whereas the placental mammals vary enormously in morphology—just think about bats, whales, horses and cats!

Table 9.1 *Rates of evolution in frogs and placental mammals.*

	frogs	placental mammals
number of living species	3 050	4 600
minimum age of the group/Ma	242	79
rate of phenotypic evolution	slow	fast
rate of albumin* evolution	standard	standard
rate of change of chromosome number	slow	fast

*A protein found in the blood.

In terms of immunological distances estimated by Wilson from the albumins of hundreds of species of frogs and mammals, two closely related species of frog can differ by as much as do a bat and a whale. Similarly, one part of the haemoglobin molecule (the β-polypeptide) from two species of the genus *Rana* differs by 29 amino acid substitutions, a greater difference than is found between any two orders of placental mammal, and the differences between the nucleotide sequences of DNA in two subspecies of the clawed toad, *Xenopus laevis laevis* and *X. l. borealis*, are greater than those between humans and New World monkeys. Yet frogs with widely different proteins show little difficulty in hybridizing, whereas even closely related and similar species of mammals often fail to produce viable hybrids.

In contrast to their rates of genetic change, placental mammals have experienced a much faster rate of karyotypic change than frogs. This is illustrated in Figure 9.2, which shows the percentage of pairs of species having the same chromosome number plotted against the immunological distance between the albumins of the pairs, for both frogs and mammals.

▶ Study Figure 9.2. Which of the two groups—frogs or placental mammals—shows greater variation in (a) immunological distance and (b) chromosome number?

Frogs show a greater range of immunological distance but much less variety in chromosome number—a much higher percentage of frog pairs (100% over most of the range of immunological distances shown by placental mammals) have identical chromosome numbers.

The variation in morphology shown by mammals is not necessarily due to the variation in karyotypes, however. Many species and species groups show a considerable amount of karyotypic variation without a corresponding variation in

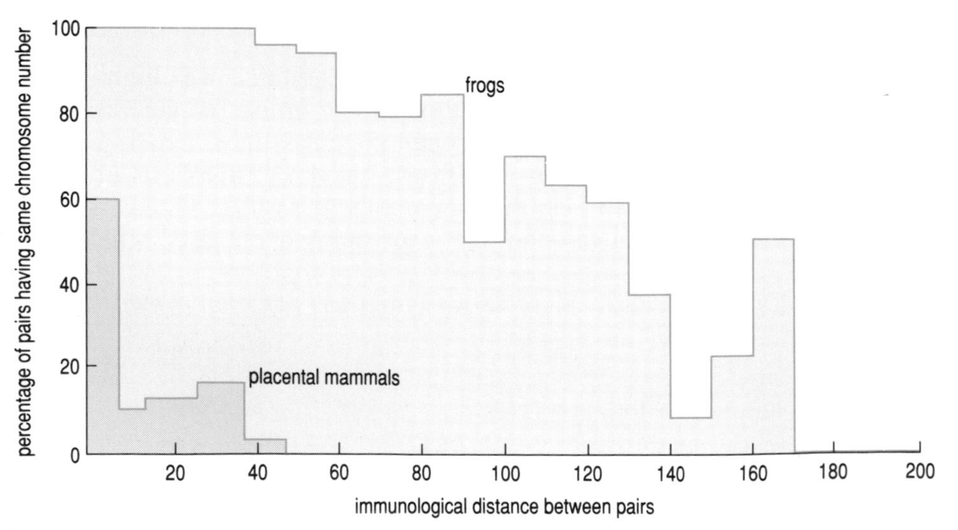

Figure 9.2 *Percentage of species pairs with identical chromosome number as a function of the immunological distance between the albumins of the pairs. The light green histogram summarizes the results for 373 different pairs of frog species, and the dark green one summarizes the results for 318 pairs of placental mammal species.*

morphology. Conversely, species that are very different morphologically may have very similar karyotypes. For example, chimpanzees (*Pan troglodytes*), gorillas (*Gorilla gorilla*), orangutans (*Pongo pygmaeus*), and humans (*Homo sapiens*), are very different morphologically and physiologically, but they have very similar karyotypes. In terms of their proteins, humans and chimpanzees appear to be as close as sibling species genetically, yet they are traditionally classified in different families on morphological and physiological grounds.

If genetic or chromosomal change is not directly related to phenotypic change, then we cannot explain evolution simply in terms of the substitution of certain alleles by others. Indices of genetic distance or similarity are based mostly on genes that code for specific enzymes or structural proteins, but it is possible that much of phenotypic evolution may be a result not of changes in these genes but of changes in *regulatory* genes. Such regulatory genes may control, for example, stages in development or the activity of other genes; a small genetic change in a regulatory gene could give rise to a large phenotypic change (see Chapter 14, Section 14.6.3). Some possible regulatory genes are now beginning to be identified, so it is possible that indices of genetic distance based on regulatory genes may soon be developed.

9.2.6 Are species differences adaptive?

Many evolutionists, including Darwin himself, have emphasized that not all evolutionary change can be understood in terms of adaptation. For example, a North American species of flowering plant, *Clarkia lingulata*, which has evolved recently, lives in a habitat which is marginal for its parent species, *C. biloba*, i.e. a habitat in which it is only just possible for *C. biloba* to grow. (Presumably, *C. lingulata* does not occupy a habitat that is *optimal* for *C. biloba* either because it has not yet had enough time to spread into it or because *C. biloba* is better able to compete in such habitat and can exclude *C. lingulata*.) The adaptedness of *C. lingulata* compared with that of *C. biloba* was investigated by means of extensive field, garden and laboratory experiments. Within the limits of these experiments, no adaptive superiority of the new species in its native environment could be demonstrated—both species did equally well in the marginal habitat. Similarly, some of the Hawaiian *Drosophila* species (see Section 9.4.5) are reproductively isolated but live in identical habitats in the same geographical areas, and differences between them do not appear to be adaptations to their habitat. Such apparently non-adaptive differences between species might be the result of genetic drift or founder effects (Chapter 3, Section 3.6). Many genes also have pleiotropic effects (Chapter 3, Section 3.2.1), so that selection favouring one effect of a gene will often carry along other effects of the same gene that may not necessarily have any adaptive value. *Speciation may not necessarily involve adaptation therefore and, conversely of course, adaptation may occur without speciation.* Thus, differences between species may or may not be adaptive. Even where species *are* adapted to their own environments, different species may merely represent alternative solutions to the same problems of survival.

Summary of Section 9.2

Living things can be divided into recognizable groups on the basis of similarities within and discontinuities between groups. The most commonly used definitions of species — the Evolutionary Species Concept, the Biological Species Concept and the Recognition Species Concept — all have advantages and disadvantages.

Only the Evolutionary Species Concept, according to which species consist of those individuals that share the same evolutionary history, can be applied to all species. It suffers from the problem of how similar evolutionary paths must be to be considered the 'same'.

The Biological Species Concept, according to which a species consists of those individuals that can reproduce with each other, cannot be applied to fossil species, species that reproduce asexually or species that are geographically separated. It is frequently not testable because mating often cannot be observed in nature. Biological species are seen as being kept apart by reproductive isolating mechanisms which prevent gene flow between species because either no hybrids are produced (pre-zygotic isolation) or the hybrids produced are highly unfit (post-zygotic isolation). The criterion of reproductive isolation frequently breaks down because apparently fit hybrids may be formed between different species in nature.

The Recognition Species Concept, according to which a species consists of those individuals with the same fertilization system, has similar problems to the Biological Species Concept except that it can sometimes be applied to fossil species and to species that are geographically separated. It suffers from the problem of how similar fertilization systems must be to be considered the 'same'.

In practice, taxonomists usually identify species on the basis of morphological characteristics, with the help of more-detailed genetic and chromosomal analysis if necessary, even though the relationship between degree of morphological difference, degree of reproductive isolation, and amount of genetic or chromosomal difference is extremely poor. For sexually reproducing species, the ultimate criterion for deciding if two forms are separate species is whether or not they normally breed in the wild. Differences between species may or may not be adaptive: speciation may not necessarily involve adaptation; and adaptation may occur without speciation.

9.3 Geographical variation within species

As you have already seen in Chapter 3, within a particular population of a particular species there is usually variation in genotype and phenotype. A considerable amount of variation is often also found between populations in different parts of the species' range. Species may show great uniformity of phenotype over wide areas where the species' range is continuous, but isolated populations can be distinctively different both from the main population and from each other.

Some of the patterns of geographical variation commonly shown by natural populations are illustrated for a hypothetical species in Figure 9.3. This imaginary

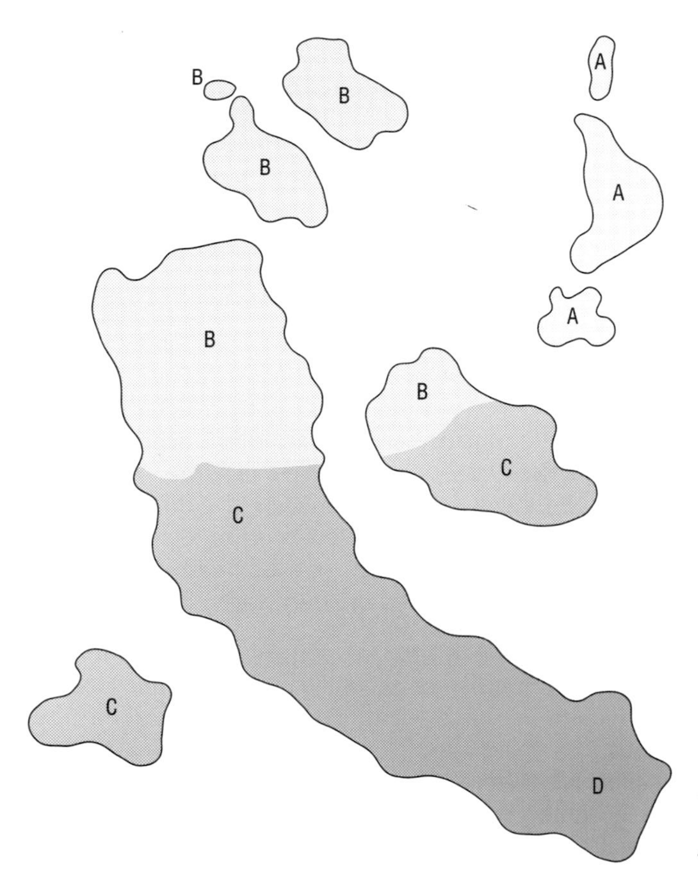

Figure 9.3 *The distribution of a hypothetical species divided into four types, A–D.*

species is divided into a number of **allopatric**—i.e. geographically completely separate—populations. Where there is no geographical separation between populations, they are said to be **sympatric**. Within the species, four geographical races or subspecies, A–D, can be identified on the basis of morphological, genetic, karyotypic, ecological or behavioural differences.

There are three populations of A, all of which are allopatric to each other and to the populations making up the rest of the species. Virtually all species contain some such isolated populations, especially near the periphery of the species' range (where they are called *peripheral isolates* or **peripatric** populations). Sometimes, such an insular distribution pattern for the species reflects geographical or ecological conditions, e.g. the populations could be on islands separated by water.

► What other geographical or ecological conditions could lead to an island-like distribution pattern?

Conditions that can lead to an insular distribution include not only oceanic islands but also ecological 'islands' such as mountain tops, lakes or patches of forest surrounded by grassland.

There are five populations of B. Three of these are allopatric to each other and to the rest of the species. The other two are allopatric to other populations of B but share a boundary with populations of C and are said to be **parapatric** to C. B and C may be adapted to different environmental conditions on each side of the boundary or may have become differentiated from each other as a result of genetic drift. The boundary may be an abrupt one, caused perhaps by some physical barrier such as a river, which reduces the possibility of B and C meeting and interbreeding and, therefore, restricts gene flow between them. Alternatively, the boundary may be more gradual, with some interbreeding between B and C leading to the formation of an area between pure B and pure C populations where hybrids are produced.

9.3.1 HYBRID ZONES

Two closely related parapatric species are often separated by an area of hybridization (Section 9.2.2) called a **hybrid zone**. Similarly, species themselves are frequently subdivided into a patchwork of subspecies or races by intraspecific hybrid zones. Hybrid zones have often been recognized on morphological criteria, because these are the most obvious, but they can involve behavioural, ecological, karyotypic or genetic differences or, frequently, any combination of these.

The differentiation between the two populations on either side of a hybrid zone could have arisen primarily or secondarily. In **primary contact**, the two populations remain in contact with each other throughout the period of differentiation. A hybrid zone forms between the two differentiating populations, and this disappears if speciation becomes complete. In **secondary contact**, the two populations become geographically separated at some point in their evolutionary history; they differentiate and then, at some later time, as a result of changes in their range, they come into contact again and may or may not form a hybrid zone in the area of overlap (see Figure 9.4). As the end result is the same, it is impossible to be certain on the basis of present distributions which of these two possibilities is the correct one, and in most cases the fossil record provides inadequate data to make a decision. However, on the basis of known climatic changes, such as those during the last ice age, and deductions about the resulting changes in species' ranges, many hybrid zones are thought to have originated from secondary contact between differentiated populations (though the differentiation may not necessarily have arisen entirely in allopatry). Explanations for the formation of hybrid zones by secondary contact have been put forward for many temperate species of mammals, birds, amphibians and invertebrates.

Hybrid zones can be a temporary phenomenon, resulting from the secondary contact of two populations that eventually merge together completely; but many of them appear to be very stable and have been maintained for a long time. For example, the crow *Corvus corone* is divided into two subspecies, the carrion crow *Corvus corone corone* and the hooded crow *C. c. cornix*. The all-black carrion crow inhabits western Europe including Scotland and eastern Ireland; the hooded crow is grey with a black head, wings and tail, and inhabits central and eastern Europe, northern Scotland and western Ireland. Where the two subspecies meet, there is a zone of hybridization varying in width from 24 to 170 km (Figure 9.5), which has probably been in existence since the two subspecies came into

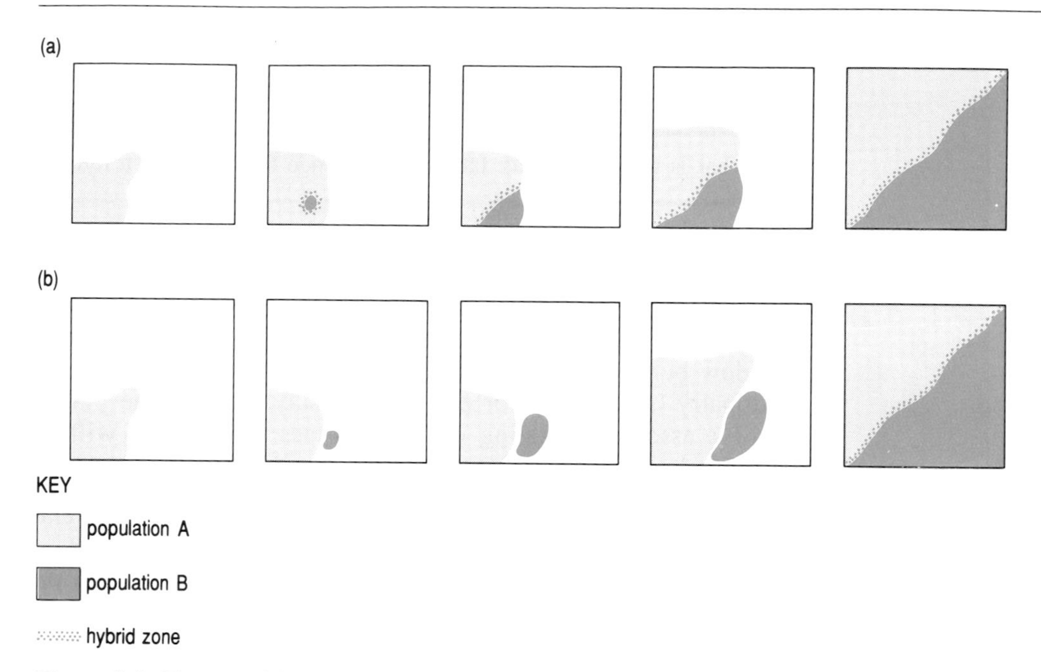

Figure 9.4 *Two possible scenarios for the origin of a hybrid zone, each eventually producing an identical pattern of distribution. (a) Primary contact: differentiation between two types occurs in sympatry and the two types then extend their range, forming an extended hybrid zone between them. (b) Secondary contact: differentiation occurs in allopatry and subsequent spread of the two populations also leads to an extensive hybrid zone between them.*

secondary contact after the last ice age. Mating within the zone appears to be random and every conceivable combination of parental characters and all degrees of hybridity are present. Hybrids are rarely found outside the hybrid zone.

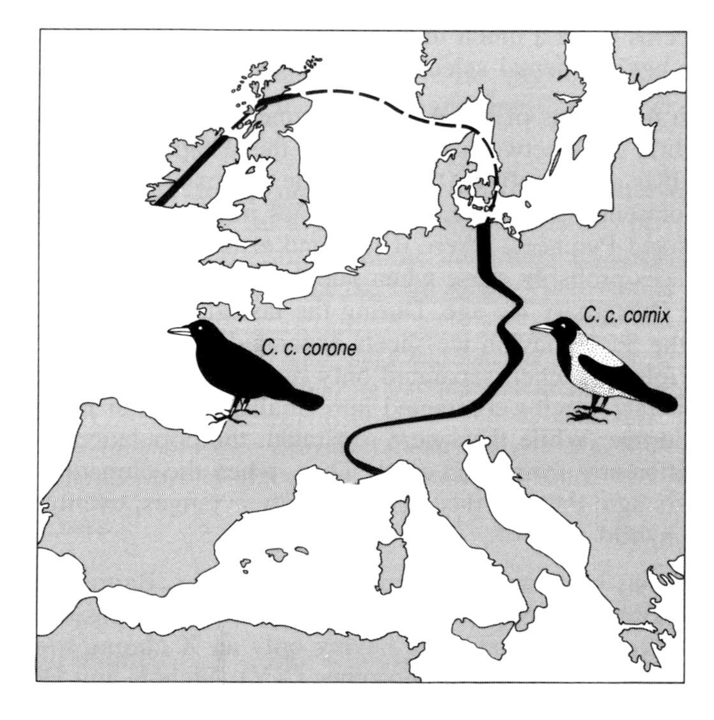

Figure 9.5 *Position of the hybrid zone between the carrion crow* Corvus corone corone *and the hooded crow* C. c. cornix *in western Europe.*

▶ From the data in Table 9.2, can you tell whether there is selection against hybrids between X0 and XY races, and, if so, when it is operating?

Selection against X0–XY hybrids does appear to be operating. About 50% more non-hybrid eggs hatched and survived compared with hybrid eggs (53% compared with 34%), partly because more hybrid eggs failed to develop (UD = 31% for hybrids compared with 17% for non-hybrids) and partly because more of the hybrid eggs that did hatch died soon after hatching (R = 20% for hybrids compared with 15% for non-hybrids).

The two races of *Podisma* interbreed freely in the laboratory. As they do not seem to be adapted to different habitats on either side of their boundary, without selection against hybrids the two races should have mingled extensively as a result of the unimpeded gene flow between them. This appears to be prevented and the hybrid zone kept to a narrow area by selection operating against hybrids and by the short dispersal distance of the grasshoppers.

A hybrid zone maintained by selection against hybrids is often called a **tension zone**. Because a tension zone is not necessarily the result of adaptation to different habitats, its position can change. It has been shown, however, that a mobile tension zone will tend to come to rest in a region of low population density. This is because, if more parental types from one population are dispersing in from their side of the hybrid zone than are parental types from the other side, the centre of the zone will move towards the area with the lower density.

In Figure 9.7(a), the population of parental type A is at a greater density next to the tension zone than is that of parental type B. As a result, more A genotypes are likely to disperse into the tension zone than B genotypes, so some of the dispersing As will mate with other dispersing As, producing pure A offspring rather than hybrids. Fewer hybrids are therefore produced on the A side of the zone. The reverse is true on the B side of the zone because the few dispersing Bs are less likely to mate with each other. As a result, the zone will gradually move towards the B population until the rate of dispersal of As into the tension zone is the same as the rate of dispersal of Bs. When the tension zone arrives at the area of lowest population density (Figure 9.7(b)), the rate of dispersal of A into the hybrid zone equals that of B and an equilibrium position is reached.

If an allele arises that is advantageous over another, it will gradually replace the disadvantageous allele, with a hybrid zone between the two genotypes (with and without the new advantageous allele) moving out like a ripple from the superior gene's point of origin. Such moving hybrid zones will be observed very rarely, however—the process can be very rapid, taking only a few generations in a small population of a fast-breeding species if the advantage is large, so in most cases the exchange will already be complete. Most recorded examples involve species where there has been some kind of human interference, such as the spread across Britain, probably in less than 50 generations, of resistance to the poison Warfarin in rats.

9.3.2 Clines

The term **cline** was first coined by Julian Huxley to describe a continuous gradation in some measurable character, whether morphological, physiological,

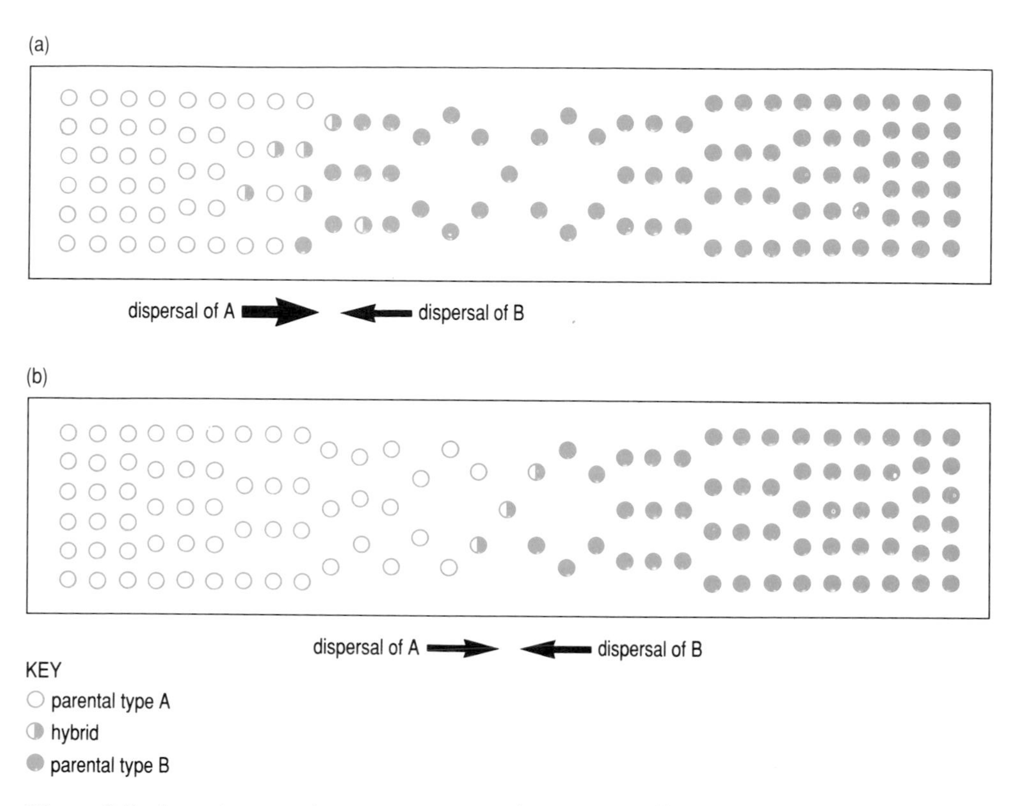

Figure 9.7 *A tension zone between two parental genotypes will move to an area of low population density. The thickness of the arrows is related to the rate of dispersal.*

genetic, chromosomal or behavioural, across the range of a species. Many species of Australian birds, for example, show clinal variation in size, gradually becoming smaller from Tasmania northwards to the Torres Strait.

The cline for any one character is, in theory, independent of any other character. Many of the Australian birds that show the north–south size cline also show a cline for intensity of colour, but this time it follows a gradient from the humid periphery of Australia to the arid interior.

A cline is often associated with some environmental gradient such as climate, though it is not always easy to prove that it represents an adaptation of the species to that environmental gradient. It is generally impossible to distinguish between an adaptive cline formed primarily (i.e. where all parts of the clinal population remain in contact with each other but there is adaptation to local conditions) and one formed after two allopatric populations that have drifted apart genetically come into secondary contact and form a wide hybrid zone. Such a wide hybrid zone, resulting from long-distance dispersal and/or negligible selection over many generations against hybrids or pure types outside their own environment, would show a very gradual change from one parental type to the other through a series of intermediates. In other words, it would look the same as an adaptive cline but would not necessarily be related to ecological differences in the environment. Because of this, the term cline is now often used to describe any hybrid zone, whether between species or between subspecific types, where there is a gradual change from one type to another, even where the zone is narrow. Where the change from one type to another is particularly rapid in the centre of the hybrid zone, it is also called a **step cline**. Figure 9.3 shows clinal variation between the types C and D of a hypothetical species.

9.3.3 Mosaic hybrid zones

Where hybrid zones correspond with a habitat change, they do not always show a gradual change from one 'pure' type to another. Rather than changing gradually between neighbouring habitat types, the environment may go through a transitional zone consisting of a mosaic of patches of the two habitats (Figure 9.8). For example, as you move from an area that is completely forested to a neighbouring area that is all open grassland, the forest may not become open grassland gradually, as the trees get more and more thinly but still evenly spread. Instead, continuous forest may change to patches of forest interspersed with open areas; the forest patches become smaller and farther apart until, eventually, the country is completely treeless. A species might consist of two types, one adapted to forest and the other adapted to open grassland, and they might hybridize where the two habitats meet. In such a **mosaic hybrid zone**, the hybrids most like the forest type might occupy the forest patches, while the hybrids most like the grassland type might occupy the open patches.

R. G. Harrison and his colleagues have studied such a hybrid zone between two species of field cricket, *Gryllus pennsylvanicus* and *G. firmus*, in Connecticut, USA. *G. pennsylvanicus* is found inland and *G. firmus* along the coast. The two species differ to some extent in appearance and calling song, and electrophoretic studies have shown differences in allele frequencies at several loci. But there is no evidence that different alleles at any locus have become fixed in the two species, and the genetic distance between them (Chapter 3) is very small. The two species have, however, become fixed for different mitochondrial DNA (mtDNA) variants. Pre-mating and post-mating barriers both operate: both species show strong positive assortative mating and, while crosses between *G. firmus* males and *G. pennsylvanicus* females produce fertile and viable offspring, crosses between *G. firmus* females and *G. pennsylvanicus* males fail to produce any offspring at all. Because most adults of both species are flightless, dispersal distances are short—probably less than 50 m on average per generation.

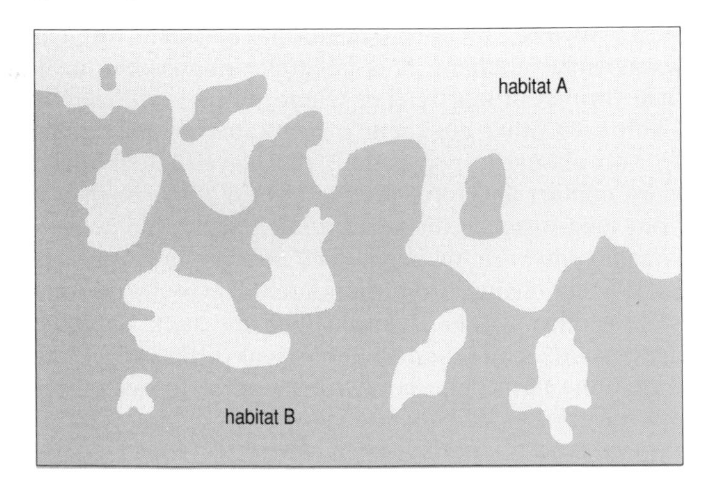

Figure 9.8 *Mosaic environment with a transition between two habitat types, A and B.*

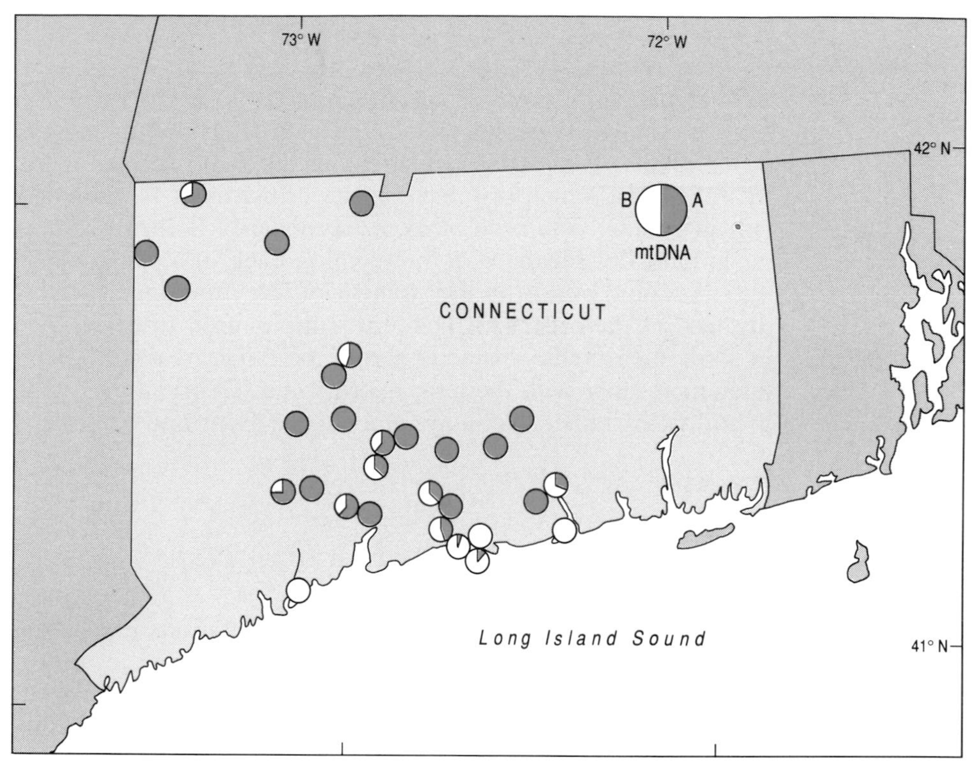

Figure 9.9 *A mosaic hybrid zone between* Gryllus pennsylvanicus *and* G. firmus *in Connecticut. The frequencies of* G. pennsylvanicus *(A) and* G. firmus *(B) mtDNA genotypes found within each population sampled are represented by pie-charts.*

Sampling of a series of populations along a line starting at the coast and moving inland showed a gradual transition from 'pure' *G. firmus* to 'pure' *G. pennsylvanicus*, suggesting a simple hybrid zone. But when more extensive sampling throughout the whole hybrid zone was carried out, the researchers found that the situation is not quite so simple. The hybrid zone does not represent a straight transition from 'pure' *G. firmus* along the coast to 'pure' *G. pennsylvanicus* inland—instead there is a mosaic of populations (Figure 9.9) which seems to be related to the patchy distribution of soil types in the area. *G. firmus* is found on sandy well-drained soils, *G. pennsylvanicus* on loam soils. Within the hybrid zone, hybrids most like *G. firmus* tend to be found on sandy soil patches, while those most like *G. pennsylvanicus* tend to be found on loam soil patches.

▶ In what two ways might this association between hybrid type and soil type come about?

The association could be a result of (1) each hybrid type choosing to live in a particular habitat and/or (2) differential selection operating in the two habitats, so that hybrids most like *G. firmus* are better able to survive and breed on sandy soils while hybrids most like *G. pennsylvanicus* are fitter on loam soils.

SUMMARY OF SECTION 9.3

Individuals from the same species can vary to a large extent, especially if they are from a population geographically isolated from the main species range. Between differentiated parapatric populations of the same species there may be an area of hybridization, which may have arisen primarily if the populations diverged while still in contact with each other, or secondarily if the populations diverged while geographically separated before coming back into contact. The width of a such a hybrid zone depends on the strength of selection against hybrids and the dispersal distance of the organism. The term cline is used to describe continuous gradation in some measurable character across the range of a species but it is also used interchangeably with the term hybrid zone. Hybrid zones may show a mosaic of hybrid types rather than a gradual change from one parental type to another.

9.4 ALLOPATRIC SPECIATION

The recognition that spatially separated populations can diverge from each other and that differences between varieties, races, subspecies and full species are to some extent a matter of degree, led Mayr to develop the theory of **allopatric speciation**. Allopatric speciation takes place when populations become geographically separated. Like the other models of speciation discussed later in this Chapter, it is usually applied to sexually reproducing species (for a discussion of speciation in asexual species, see Chapter 5). In its basic form, it is still the most widely accepted model of speciation for animal populations, though there is controversy about the actual mechanisms involved—some evolutionists believe that adaptation plays the major role in the process of divergence while others think that chance factors are much more important.

According to Mayr's version of allopatric speciation, which he originally put forward in 1957 (see Mayr, 1970), two populations of the same species become geographically separated, either by some physical barrier or by elimination of the intervening populations. Gene flow between the two populations, which would maintain homogeneity between them, is thus reduced, and this allows them to diverge from each other. Divergence can be the result of adaptation to local conditions—the physical structure of the habitat, the climate, or other ecological conditions such as the other species present or the density of members of the same species—and/or random genetic drift, though this is likely to be important only in very small populations. If the range of one or both of the populations later expands so that the two types come into secondary contact, Mayr argues that one of three things can happen:

1 The two populations have become so different that they do not interbreed at all. They are therefore separate species and speciation has occurred.

2 The two populations have differentiated to only a certain extent so that limited interbreeding is still possible. However, as long as at least one isolation mechanism has evolved which is sufficient to prevent hybridization in nearly all cases, natural selection will be able to act to improve reproductive isolation by favouring the acquisition of additional isolating mechanisms. This is because, when

hybrids *are* formed, they will usually be greatly inferior as a result of behavioural, ecological or genetic incompatibilities; such hybrids will reproduce very poorly, and there will be no breakdown in the species barrier.

3 The amount of differentiation is small, the two populations are still able to interbreed, hybrids are almost or equally as fit as parental types, and the two populations mingle completely and remain as a single species.

9.4.1 Geographical isolation

Geographical isolation is defined by Mayr as 'the division of a single gene pool into two by strictly extrinsic factors'. It is seen by Mayr and other proponents of allopatric speciation as an essential prerequisite to speciation, as they believe this is the only way of preventing gene flow between the two populations, so that a genome adapted to local conditions can evolve. The 'strictly extrinsic factors' can include the following:

1 The sea. The drosophilid flies of the Hawaiian islands (see Section 9.4.5) are a classic example of rapid and profuse speciation through isolation on islands.

2 Bodies of freshwater. Just as the sea separates land islands, so lakes, streams, etc., are 'islands' of water separated by land.

3 Mountain ranges, especially if they separate climatic zones as do the Himalayas between India and Tibet.

4 Valleys. Just as mountains are barriers for lowland organisms, valleys may be barriers for mountain species.

5 Glacial masses. The Quaternary ice-masses of the northern continents were among the most effective barriers in recent geological time (see Section 9.3.1).

6 Habitat zones. Types of vegetation or physical habitat may be sharply divided into zones, forming effective geographical barriers for many species. For example, along a coast, rocky shores may form 'islands' separated from each other by different habitats such as sandy or muddy shores. Habitat can be divided into zones vertically as well as horizontally—species occupying different depths of the ocean may be as isolated as species occupying different vegetation types on land.

9.4.2 Genetic reorganization and divergence in isolated populations

Mayr argues that as soon as two populations are isolated from each other, even if initially they are genetically identical and occupy identical environments, they will drift apart in their genetic make-up. He sets out a number of reasons for this:

1 The chance that the same mutations will occur in both populations is very small.

2 Each mutation alters the genetic background of the population (i.e. the genetic make-up of the population as a whole), and thus affects the intensity of selection acting on all the alleles in that population and on all subsequent mutations.

3 Recombination will produce different genotypes in the two populations and this can affect allele frequencies, because the intensity of selection operating on a particular allele may be different when it is part of different genotypes.

selection against hybrids. For reinforcement to occur, therefore, natural selection would have to act to favour positive assortative mating, not between the same parental types, but between compatible genotypes. Moreover, hybrids that happen to be formed between compatible genotypes will be at an advantage compared to those hybrids formed between incompatible genotypes, which will be selected against because they will be less fit. Selection may therefore be more likely to operate to produce more viable or more fertile hybrids, i.e. post-mating isolation becomes less strong. As a result the parental gene pools gradually become less incompatible and so selection in favour of positive assortative mating becomes less and less intense. It can be calculated that, even if the average fitness of all hybrid genotypes is only a quarter of that of their pure parents, hybrid genotypes will still make up 99.9% of the population within eight generations after the two parental types come into contact. Two populations will remain distinct only if, when they first meet, the rate of hybridization is extremely low, and/or the fitness of hybrids is extremely low. This fits more with the views of Mayr than with those of Dobzhansky.

A computer simulation of reinforcement in a hybrid zone between two races of plants was devised by Jack Crosby. The plant races had differences at eight unlinked loci affecting fertility such that F_1 hybrids were only 25% as fertile as the parental types. Both races also varied at three unlinked loci affecting the time flowering started and at another two unlinked loci affecting the duration of the flowering period. There was no selection against hybrids for flowering time or flowering duration. At the start of the simulation, the two races had the same duration of flowering but slightly different dates for the start of flowering. Seed and pollen dispersal were limited and plants could be pollinated only by other plants in flower at the same time.

The simulation started with the two races' ranges coming into contact for the first time. Within a few generations, a narrow hybrid zone formed at the boundary between the two races, but gradually, over successive generations, the frequencies of the two parental combinations of fertility alleles increased in the hybrid zone, the overlap between these types increased, and the frequency of hybrid combinations of fertility alleles decreased. This happened because alleles controlling fertility, flowering time and duration came to be associated, so that individuals with one parental combination of fertility alleles had early flowering alleles and those with the other parental combination of fertility alleles had late flowering alleles. Duration of flowering decreased in both races. There was no direct selection on flowering time or duration—the changes observed were the result of selection operating only against hybrids for fertility alleles.

▶ How does this simulation provide support for the theory of reinforcement?

Selection against hybrids brought about increased reproductive isolation. The computer simulation was a genuine model of reinforcement because hybrids were only partially sterile.

On the other hand, the simulation showed how weak reinforcement may be. The changes described required a relatively large number of generations. The model is also not applicable to all situations because there is evidence that, in some natural populations, stabilizing selection operates on mate recognition systems, so that 'unusual' types have lower reproductive success. When Crosby built only a very small amount of stabilizing selection, e.g. selection against shorter flowering duration, into his model, no divergence took place at all.

Also, in Crosby's simulation, a mere eight loci generated a selection against F_1 hybrids of 75%—i.e. almost 10% selection per locus. Data from actual hybrid zones suggest that, although selection against hybrids may be strong—you will remember from Section 9.3.1 for example that selection is of the order of 36% against hybrids between the chromosomal races of *Podisma pedestris*—many genes are involved (about 150 in *Podisma*) and each individual locus experiences only weak selection (less than 1%). Reinforcement is much less likely when selection at each locus is weak and many loci are involved (certainly there is no evidence that reinforcement is taking place in the *Podisma* hybrid zone). This is because the most common hybrids within a hybrid zone will be those 'halfway' between parental types. They will have about 50% of their alleles from each parental race, though which particular alleles come from which type of parent will vary. Even if they mate with an individual with a similar phenotype, i.e. mate assortatively, their offspring are still likely to be heterozygous at a relatively large number of loci and so they will still be subject to relatively strong selection. The potential benefits of mating positive assortatively will thus decrease as the number of loci on which selection is operating increases.

To summarize, reinforcement is *favoured* where selection against hybrids is very strong, where few loci are involved and where there is close linkage between the genes causing positive assortative mating and the loci causing the hybrid disadvantage. It is *opposed* by stabilizing selection on the mate recognition system, gene flow into the hybrid zone, and recombination. These theoretical considerations indicate that reinforcement, although possible, requires conditions that are likely to be very rare in nature. It is difficult, however, to get evidence from the field as it is impossible to distinguish between reinforcement and reproductive character displacement *once the process is completed.*

Butlin has pointed out that, for any study to provide convincing evidence of the occurrence of reinforcement, it must demonstrate:

1 That gene flow occurs between the populations in question, or at least that it did when they first met.

2 That aspects of the mate recognition system have diverged in the area of contact between the two populations in the time since contact was first established and (ideally) that this divergence is not a result of other selection pressures on the mate recognition system.

3 That these changes in the mate recognition system are sufficient to decrease the frequency of production of hybrids.

Several studies that purport to provide evidence of the occurrence of reinforcement involve the demonstration of greater pre-mating isolation in areas where two species' ranges overlap, but where in fact there is no gene flow between them.

▶ Unless it can be shown that there was gene flow at some time in the past, what is a more likely explanation than reinforcement in such cases?

Reproductive character displacement, because there is no gene flow between the two species. For example, Dobzhansky found greater pre-mating isolation between sympatric than between allopatric populations of two races of *Drosophila paulistorum*, and used this result as evidence of reinforcement. But the study was carried out in the laboratory, and evidence that hybrids can be produced in the laboratory and that they are viable and fertile is not evidence of gene flow in the wild, because laboratory conditions are inevitably unnatural.

The snail *Partula suturalis* on the island of Moorea, near Tahiti, possibly provides the most convincing example of reinforcement at present. In this snail, the shell coils either to the right (dextral) or to the left (sinistral) such that it is always in the opposite direction to that of other sympatric *Partula* species (Plate 9.1). Snails with opposite coil directions experience mechanical difficulties in mating, and coil types mate positive assortatively, but even so there is gene flow between species in some areas. There is still an alternative explanation to reinforcement, however. This is that the present situation could have arisen from the fixation of alternative coil directions in allopatric populations of *Partula suturalis*, followed by increases in range so that *P. suturalis* came into contact with other *Partula* species; it was only able to carry on living in sympatry with another species, however, where it differed from that other species in coil direction.

Several other studies of natural populations involve hybrid zones or clines associated with environmental gradients, so gene flow does occur, but the evidence for reinforcement is generally weak or ambiguous.

There have been several laboratory tests of reinforcement but many of these imposed a fitness of zero on hybrids and so were tests of reproductive character displacement rather than reinforcement. One exception is an experiment carried out by Thoday and Gibson in 1962. They kept a population of *Drosophila melanogaster* in the laboratory and, in each generation, selected those individuals with either a high or a low number of chaetae (bristles) on the abdomen as parents for the next generation (in other words, individuals with an intermediate number of chaetae were removed). The number of chaetae an individual has is controlled by several genes. Since individuals with high and low chaetae number were kept together and allowed to breed freely and not all hybrids were removed, gene flow continued to be high. After twelve generations there was a marked increase in pre-mating isolation between types with high and low chaetae number, though this never became absolute. This experiment demonstrated that reinforcement is possible but unfortunately, despite several attempts, no-one else has ever managed to repeat the result.

Evidence that many hybrid zones are stable and have been in existence since the two populations came into contact, without the evolution of pre-mating barriers, runs counter to the idea of reinforcement. Some indirect evidence against reinforcement also comes from a study by Coyne and Orr in 1989. They used indices of genetic distance to determine the time since divergence for 119 pairs of closely related species of *Drosophila* from all over the world. They also measured the degree of pre-zygotic isolation (measured as the number of fertilized matings that were between species as a proportion of all fertilized matings, when given the choice of mating with the same species or a different species), and the degree of post-zygotic isolation (measured as the degree of sterility or inviability of hybrids) in laboratory crosses between each pair of species. They found:

1 Both pre-zygotic and post-zygotic isolation increase with the time since two species became separate.

2 Among species that have separated most recently (genetic distance is small), presently sympatric species tend to show strong pre-zygotic isolation more often than strong post-zygotic isolation.

3 Among recently separated species, presently allopatric species show similar degrees of pre-zygotic and post-zygotic isolation.

If reinforcement takes place, post-zygotic barriers should evolve first, in allopatry. Then, once post-zygotic barriers have evolved, they should favour the evolution of pre-zygotic barriers between species that come into secondary contact.

► If reinforcement is taking place, which do you predict that presently sympatric species should show more often—pre-zygotic or post-zygotic barriers?

Post-zygotic barriers. You would expect post-zygotic barriers to have evolved first, so they should have evolved more frequently than pre-zygotic barriers.

► Do Coyne and Orr's results support this prediction?

No—pre-zygotic isolation was more common in sympatric species than post-zygotic isolation. Their results show that some species have evolved strong pre-zygotic isolation *without* evolving strong post-zygotic isolation, a situation which is incompatible with the theory of reinforcement.

► If pre-zygotic barriers evolve as a result of reinforcement, which would you expect to be more common in allopatric species—pre-zygotic or post-zygotic isolation?

Post-zygotic barriers, because pre-zygotic barriers are favoured by selection only once populations become sympatric.

► Do Coyne and Orr's results support this prediction?

No—Coyne and Orr found that presently allopatric species show similar degrees of pre-zygotic and post-zygotic isolation. This result also cannot be explained by the theory of reinforcement.

In summary, the evidence for reinforcement is weak and ambiguous and there is some evidence against it, though this is not conclusive. If the process does occur in nature, it probably does so under very rare conditions. This too remains a controversial area in evolutionary biology.

9.4.4 ALLOPATRIC SPECIATION IN THE CICHLID FISHES OF LAKE VICTORIA

A group of closely related species can sometimes be confined to a small geographical area, with no close relatives anywhere else. Detailed study of such **species flocks** or **swarms**, involving the geological and biotic history of the environment, and the biology, life histories and genetic relationships of the organisms concerned, can allow the possible sequences and mechanisms of speciation to be reconstructed. Two groups of organisms where profuse multiplication of species appears to have been the result of allopatric speciation are discussed in this Chapter. This Section considers the cichlid fishes of Lake Victoria; the next Section looks at the Hawaiian drosophilids.

Until very recently, a single family of small fish—the cichlids—dominated the fish fauna of Lake Victoria, both ecologically and in terms of number of species. There may have been as many as 170 species of the cichlid genus *Haplochromis* alone. As a result probably of overfishing and the artificial introduction in the 1960s of the Nile perch *Perca nilotica* (a large edible species but a voracious predator on other fish) in an attempt to improve the local fisheries, however, many of the cichlids are now rapidly becoming extinct. This is a classic example of human interference upsetting the natural ecological balance and leading to large-scale extinctions (see Chapter 17).

All of these *Haplochromis* species evolved after the lake's origin in the mid-Pleistocene, about 750 000 years ago. Lake Victoria is a large but shallow drainage basin fed by several large rivers (Figure 9.11). Although it is connected to Lake Kioga in the north by its only outflow, the Victoria Nile, its fauna has always been isolated, originally by the Ripon Falls on the Victoria Nile, more recently by a dam that replaced and submerged the Falls.

The lake basin began as an area crossed by several westward-flowing rivers which drained the eastern highlands of Kenya. As a result of geological changes to the west of what is now Lake Victoria, a two-way drainage system was established, eastwards into the developing Victoria basin and westwards into several protolakes that are now Lake Mobutu Sese Seko (once Lake Albert), Lake George, Lake Edward and Lake Kivu. As the eastward-flowing river valleys gradually filled, a series of shallow lakes formed, which eventually joined together as the water levels rose to form a single body of water covering a larger area than Lake Victoria does today. The geology of the lake basin was not stable, however, and there were several periods when water levels dropped, so that several small peripheral lakes separated off from the main lake, interspersed with periods when water levels were high and the peripheral lakes were reincorporated.

Greenwood (1974) has studied the *Haplochromis* species of Lake Victoria since the 1950s and has found circumstantial evidence that all of them may have originated from a single ancestral species or a group of related species which resemble the living species *H. bloyeti* (Figure 9.12). The evidence is as follows:

1 Lakes Victoria, Mobutu Sese Seko, Edward, George and Kivu were originally fed by the same East African drainage system. The species in these lakes were similar to each other in many respects, but they did not have certain features found in species occurring in other East African lakes, such as Lake Malawi,

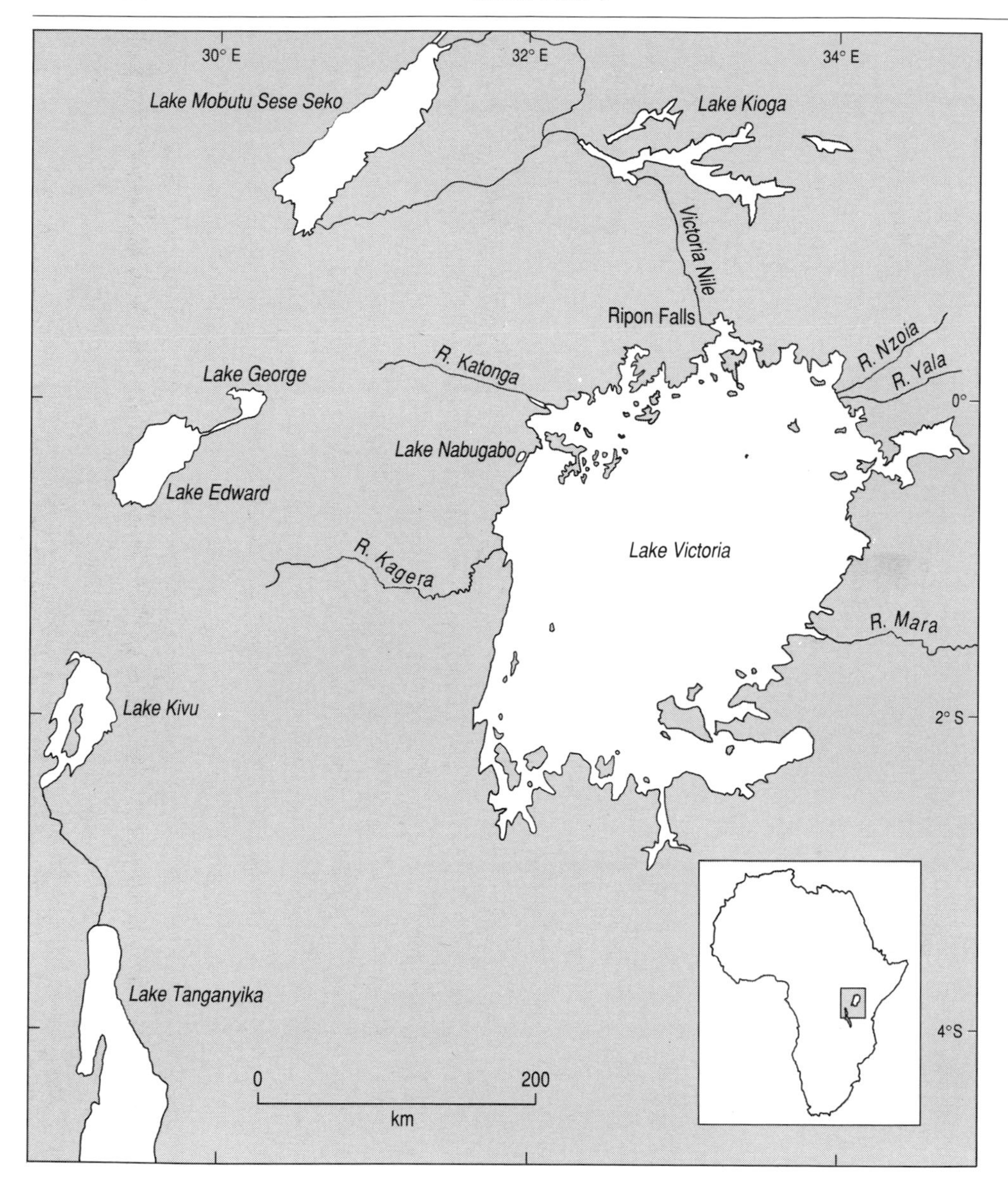

Figure 9.11 *The geography of the Lake Victoria region. Lake Nabugabo is separated from Lake Victoria by only a narrow sand-bar.*

which never shared the same drainage system as Lake Victoria.

2 All the species of *Haplochromis* occurring in Lakes Victoria and Nabugabo are endemic (i.e. found nowhere else) except one.

3 All the lakes that have at some time shared a drainage system with Lake Victoria are populated by *H. bloyeti*, or at least by a number of species that resemble *H. bloyeti* very closely. But *H. bloyeti* does not occur in lakes that have not shared a drainage system with Lake Victoria.

4 All the species in the Lake Victoria species flock resemble each other and *H. bloyeti* relatively closely, and intermediate forms exist between *H. bloyeti* and the most differentiated species.

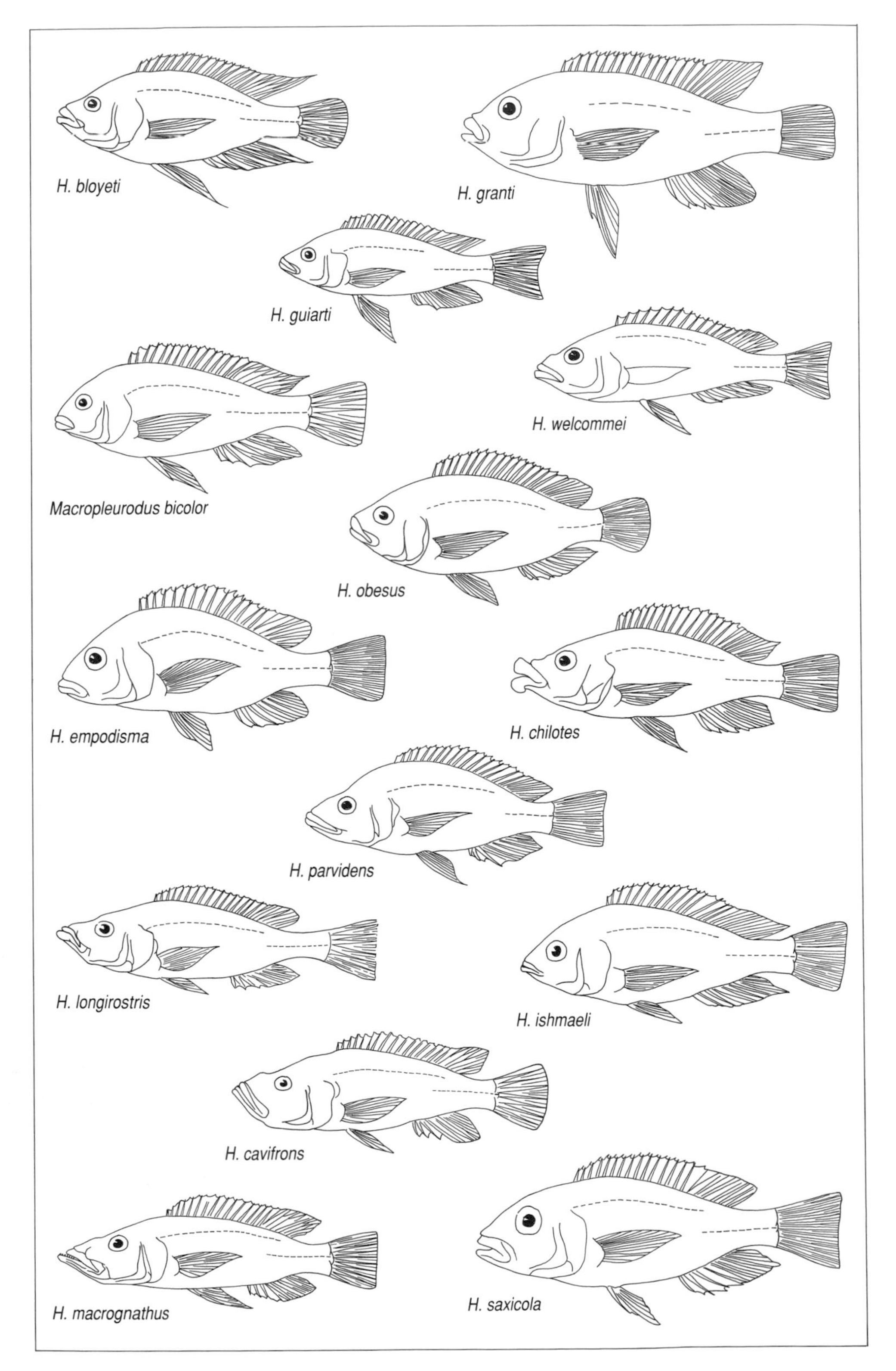

Figure 9.12 *Some Lake Victoria cichlid species (not to scale).*

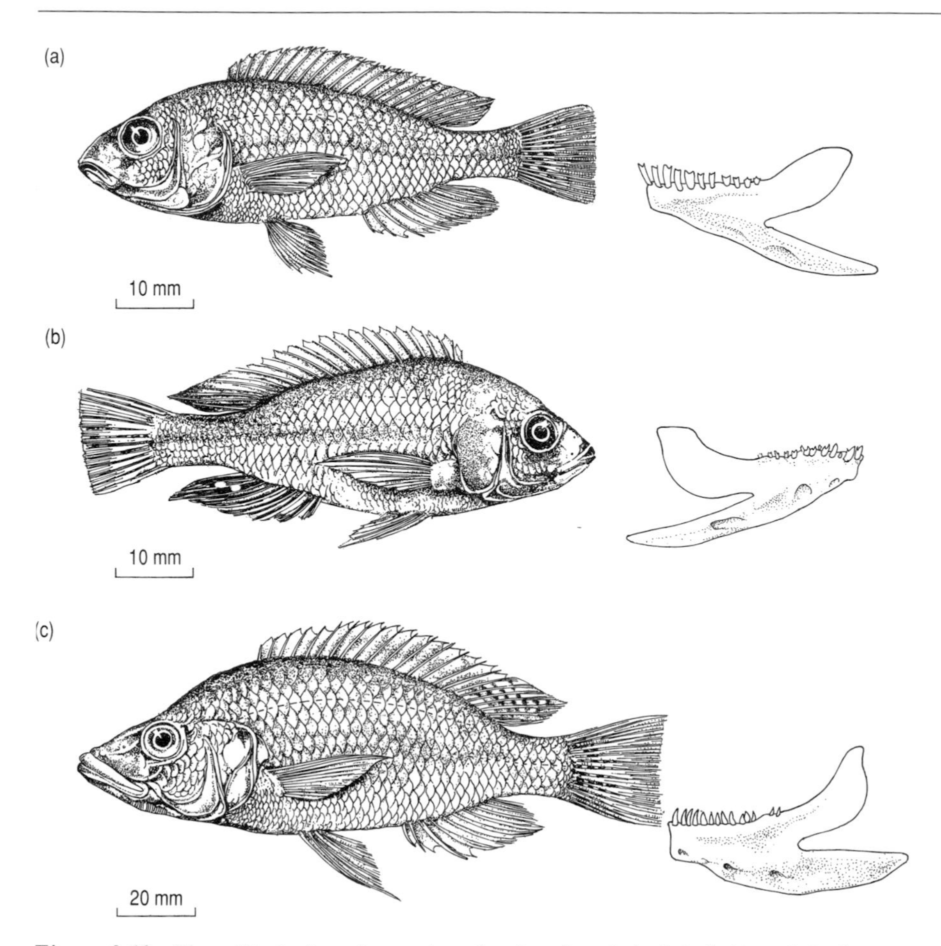

Figure 9.13 *Three* Haplochromis *species, showing the whole fish (left) and the lower jaw (right). (a)* H. pallidus *(diet: detritus and insects); (b)* H. ishmaeli *(diet: molluscs); (c)* H. parvidens *(diet: fish eggs and larvae).*

The various *Haplochromis* species show differences in morphology, mainly in the shape and size of the jaws and the bones around the throat, in dental patterns and in the shape of the skull correlated with the dental differences (Figure 9.13). These morphological features are connected with the feeding habits of the fish. Every major food source in the lake, except zooplankton, is exploited by one or more *Haplochromis* species (Figure 9.14). Although there is considerable overlap between species in feeding habits and in the type of habitat they occupy, groups that are basically insectivores, crustacean eaters, plant eaters, mollusc eaters or paedophages (species that prey on the larvae of other fish, mainly other *Haplochromis*) can be distinguished. One species, *H. welcommei*, has the peculiar habit of feeding on fish scales, which it scrapes off the tails of other fish. Within each of these groups, there are several species which show a gradient of increasing specialization in the morphological features associated with the food they eat.

Within species, geographical differentiation in morphology, if any, is very slight, probably because there is little geographical restriction on the distribution of any species within the lake, so gene flow is not inhibited.

Probably all the *Haplochromis* species in Lake Victoria are mouth brooders, the eggs being fertilized and incubated in the mouth of the female. All known

Figure 9.14 *Food sources exploited by different* Haplochromis *species in Lake Victoria.*

Haplochromis species are sexually dimorphic, with the male having brighter and more colourful skin patterns than the female, even during non-breeding periods. Males from different species all have different colour patterns from each other (though in some cases the difference is fairly subtle) and different ocelli (coloured spots that resemble eggs) on the anal fin (Plate 9.2). Reproductive isolation between species is very marked, with hybrids being found very rarely, and male coloration may be a major factor in species recognition during mating. The egg spots, which become more prominent when the male is sexually active, play a particularly important role during mating. After the female lays her eggs, she takes them into her mouth. The male then displays the ocelli on his anal fin and the female attempts to pick them up, as if they were eggs. As soon as she does this, the male discharges sperm which are drawn into the female's mouth, fertilizing the eggs.

What factors could have contributed to the profuse speciation of *Haplochromis* in Lake Victoria? Specialization in diet and/or habitat could have contributed, despite the large amount of overlap between species, for example if the degree of specialization increases when competition increases. It is also possible that populations within the lake might have been effectively isolated, for example if they were separated by the deeply indented bays of the lake.

▶ From what you have learned so far, why is geographical separation between populations within the lake unlikely to have been a major factor in speciation in *Haplochromis*?

There is little morphological differentiation between fish in different parts of the lake and no species is restricted in its geographical distribution within the lake. (Though this does not preclude the possibility, of course, that populations were previously geographically restricted and/or that they may be differentiated genetically and/or karyotypically.)

Greenwood suggests that each time peripheral populations were isolated from the main body of the lake as water levels fell, they diverged as a result of genetic drift, founder effects and adaptive ecological shifts, i.e. they underwent allopatric speciation according to Mayr's model. There would have been no gene flow at all between populations in each lake because *Haplochromis* does not migrate through rivers connecting lakes. Once populations became allopatric, any specialization in diet and habitat would have speeded up morphological and genetic divergence. Many *Haplochromis* species are opportunistic feeders, eating whatever is available, and the primary factor encouraging such an adaptive shift may have been the lack of competition in the new habitat (ecological character release—see Chapter 8).

Reproductive isolation could have arisen as the result of only very slight changes in male coloration or in behaviour during courtship. As soon as any divergence in mating behaviour or breeding colour occurred, the sexual behaviour and mouth-brooding habit of *Haplochromis* could have helped to bring about reproductive isolation and maintain it when the separated populations became sympatric again. For example, differential selection pressures in separated populations could have brought about changes in egg spots. Perhaps in an area with a high density of visually hunting predators, ocelli might have become smaller, fewer and duller and so less conspicuous to predators, while in an area with few predators, they might become more conspicuous so as to be more attractive to females (as in the guppies

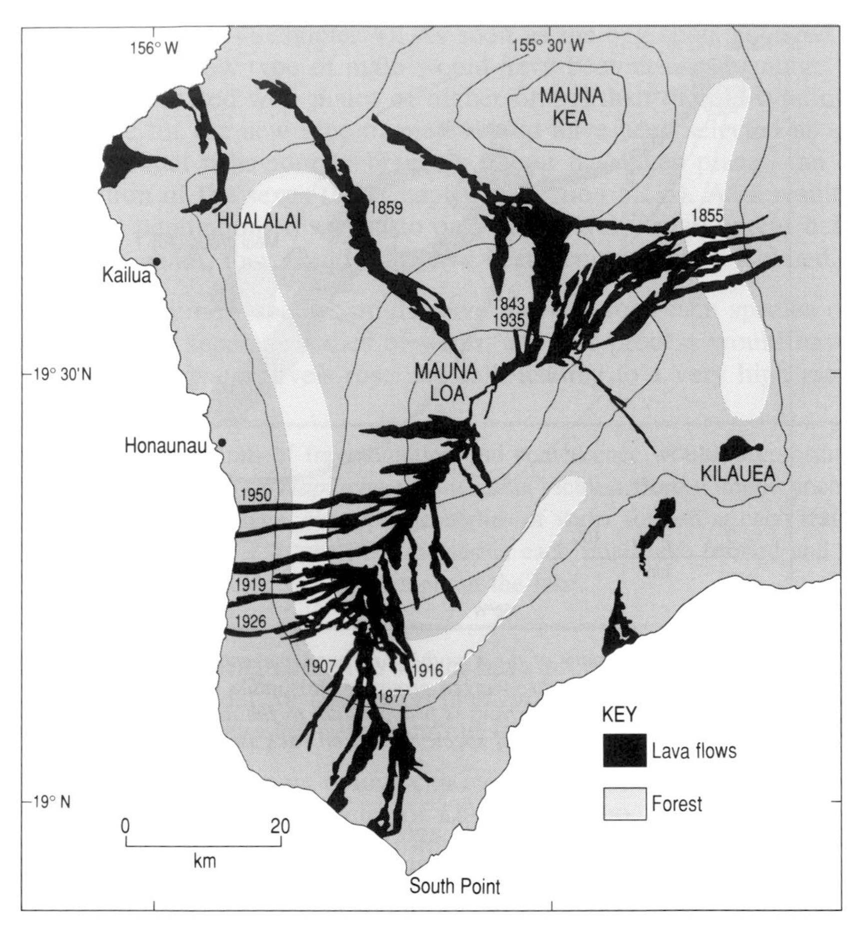

Figure 9.16 *The southern two-thirds of the island of Hawaii showing division of* Drosophila silvestris *habitat (rainforest between altitudes of 1000 and 1700m, shown by the green shaded area) by lava flows recorded over the past 150 years.*

of their results, they suggest that all of the endemic drosophilid species may be descended from a single successful fertilized female colonizing the oldest island, making this one of the most spectacular examples of explosive speciation known.

The islands contain a wide diversity of habitats, from deep rainforest to dry alpine forests, seashore to mountains 4000 metres high, lush vegetation to recently colonized lava flows. The habitats are very patchy and have been repeatedly split up by lava flows which frequently cut wide swathes through the vegetation, making large areas totally barren, and dividing the vegetation into islands within islands (see Figure 9.16). Colonization of these barren areas takes place from adjacent habitats, though the eventual composition of the vegetation may be different from the original. It is clear that founder events could have occurred frequently during the islands' evolutionary history.

Not only is there an enormous number of Hawaiian drosophilid species, but there is also a huge diversity in morphology, ecology and sexual behaviour between Hawaiian species compared with that between other members of the Drosophilidae. The body and wings are often ornately patterned and unusual modifications of the

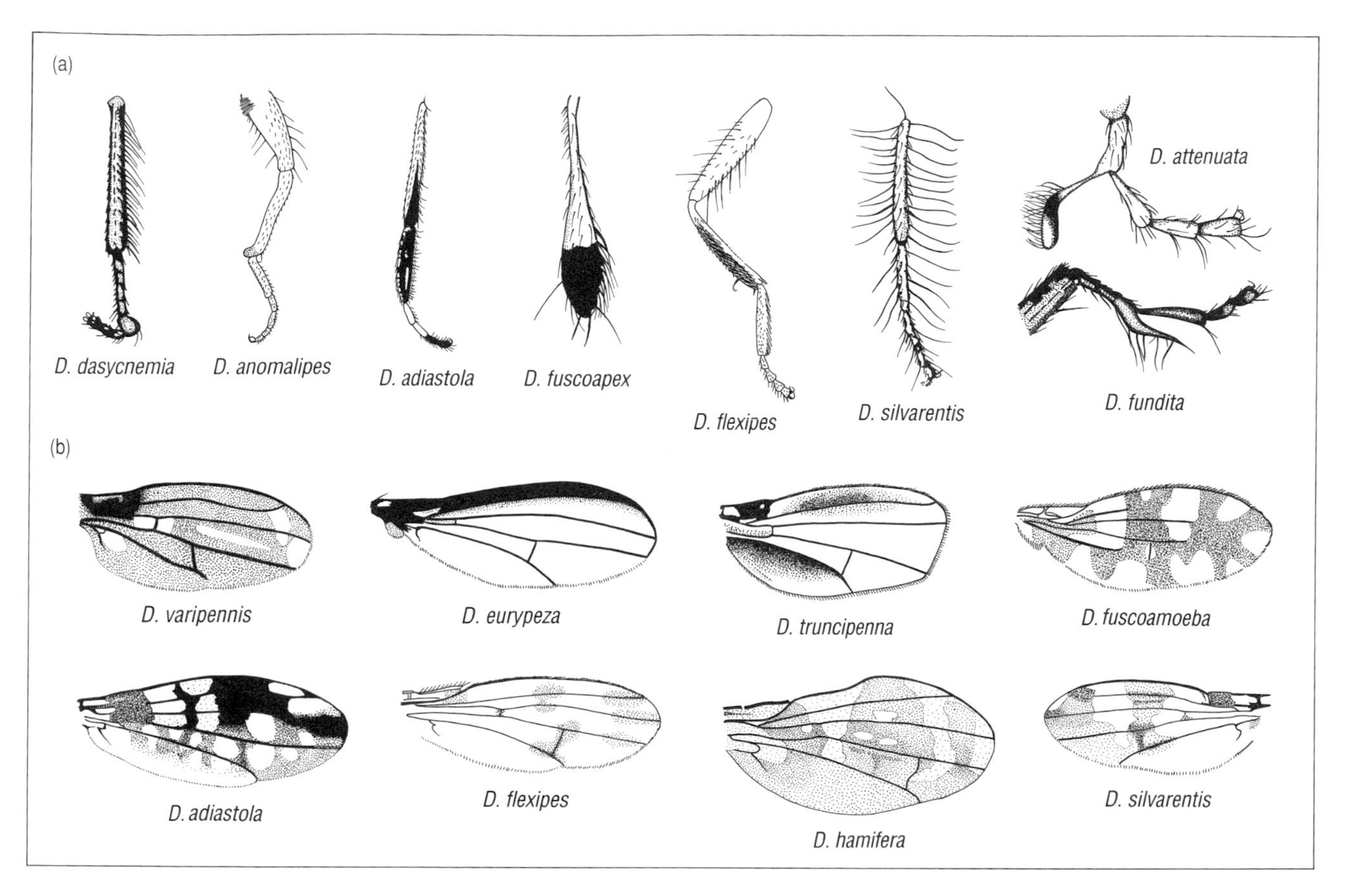

Figure 9.17 *Morphological diversity in Hawaiian drosophilid species: (a) legs and (b) wings from the males of various species of* Drosophila.

mouthparts and legs of the males in different species frequently occur (Figure 9.17), though females of different species are often very similar. Nearly all the species are saprophytic, i.e. the larvae feed as scavengers on decaying bark, leaves, flowers or fruits, on slime fluxes (sugary, fermenting sap that exudes from a number of endemic trees as the result of a fungal infection), or on fleshy fungi. Most species are highly specialized, the female laying her eggs on only one or two species of native plant.

Chromosomal analyses have shown that there is very little variation in the pattern of chromosome banding or the number of inversions (see Chapter 3, Section 3.3.2) among the Hawaiian drosophilids compared with that among drosophilid species elsewhere in the world, even though the degree of morphological differentiation among the Hawaiian species is generally much greater. Evolutionarily older species outside Hawaii show large karyotypic differences that correlate with evolutionary distance between species. This suggests that, in drosophilids at least, chromosomal mutations are not especially important in the initial stages of species formation and that chromosomal structure changes gradually after speciation.

As pointed out in Section 9.2.5, electrophoretic studies of proteins in the Hawaiian drosophilids revealed that there was also an extremely high degree of genetic similarity within groups of related species. More recently, the detailed structure of satisfy their courtship requirements (because the population is so small) and so may never mate. Less-discriminating females will be more likely to mate and leave offspring. Thus, there may be a shift in the distribution of female mating types within the population towards females that are less discriminating. Kaneshiro argues that, during the genetic revolution that accompanies founder events, such a shift in mating patterns may become fixed in the new population. On the other hand, whatever the size of the population, selection will always favour males with superior mating abilities, i.e. those that are 'good' at courting females. In Kaneshiro's model, rather than runaway sexual selection being balanced by selection for viability (see Chapter 7, Section 7.4.2), sexual selection for males with high courtship ability is balanced by sexual selection for less-discriminating females (who will generally mate with males with low courtship ability and will therefore produce low-ability sons).

Some support for this idea has been provided by mate preference experiments carried out by Arita and Kaneshiro in which two virgin females, one from each of two laboratory strains of the Hawaiian species *Drosophila adiastola*, were placed with a single male from the same strain as one of the females. One strain had been kept in the laboratory for about six years, during which time it had gone through several population crashes. In the other strain, which had been kept for about one year, numbers had always been maintained at a high level.

▶ Assuming Kaneshiro's suggestion is correct, predict the result of mate preference experiments between the two strains, i.e. which females would be less discriminating.

Females from the strain subjected to repeated bottlenecks, i.e. the 6-year-old strain.

Table 9.3 *Results of mate preference experiments in two laboratory strains of* Drosophila adiastola.

Strain of male	Number of matings	Mates with female of same strain	Mates with female of different strain
6-year-old	48	39	9
1-year-old	50	24	26

▶ Study Arita and Kaneshiro's results, given in Table 9.3. Do they support your prediction?

Yes. Females from the 1-year-old strain discriminated against males of the other strain (only 9 out of 48 matings by the male from the 6-year-old strain were with a female from the 1-year-old strain). Females from the 6-year-old strain, however, had a lowered receptivity threshold and were just as ready to accept the courtship of a male from the 1-year-old strain as was a female from the 1-year-old strain (males of the 1-year-old strain mated randomly with females of either strain).

SUMMARY OF SECTION 9.4

According to the basic model of allopatric speciation, two populations of the same species become geographically separated; they adapt to different local conditions and thereby diverge to such an extent that, even if they later come back into contact, they are unable to interbreed and have, therefore, become separate species. Mayr argued that divergence is likely to take place more rapidly in a small isolated population because chance factors such as founder effects and genetic drift will play a more important role than they would in a large population, so new species are more likely to arise from small, peripheral, isolated populations than from large populations. Some biologists such as Mayr envisage a decrease in genetic variability resulting from such chance factors; this facilitates a reorganization of the genome which then allows the new species to occupy a new niche. Low genetic variability lasts until the new species is able to expand into this niche. They thus emphasize the role of chance factors. Other evolutionists such as Carson see founder effect speciation occurring with little loss of genetic variability, which allows the founder population to respond to the different selection pressures operating in its new environment. They thus emphasize the role of adaptation. There is evidence for both viewpoints, but none of it is conclusive.

There is controversy about the origin of reproductive isolation. Mayr and his supporters think it comes about in allopatry as a by-product of genetic divergence. Dobzhansky and his supporters think that speciation can only be completed once two diverging populations come into secondary contact. Post-zygotic isolation resulting from genetic divergence in allopatry creates a selection pressure on the unfit hybrids produced, which leads to the evolution of pre-zygotic isolation—the process of reinforcement. Most of the evidence for reinforcement is, however, weak or ambiguous and it is likely to occur only rarely in nature. Carson, Kaneshiro and their supporters believe that sexual selection plays an important role in speciation, because any very slight change in allopatric populations could lead to rapid divergence in reproductive morphology and behaviour as the result of runaway sexual selection. This could explain the lack of correlation between reproductive isolation and genetic divergence in many species.

There is good evidence that geographical separation has been important in the speciation of a number of groups of organisms, though how frequently allopatric speciation is likely to take place is still disputed. Two examples are given here: the cichlids of Lake Victoria and the drosophilids of Hawaii. Changes in mating behaviour may have played a major part in the speciation of both of these groups, lending support to the idea that sexual selection is frequently involved in the speciation process.

9.5 NON-ALLOPATRIC MODELS OF SPECIATION

No one disputes that allopatric speciation has happened. Even where two closely related species are now completely sympatric, this can be ascribed to secondary invasion of one species' range by the other species. Some biologists doubt, however, that allopatric speciation could have happened frequently enough to produce the number of species present in the world today—in which case non-allopatric speciation must sometimes have taken place. Others think that, with all the geographical barriers that have constantly divided and redivided populations and the changes in climate that have led to repeated contractions and expansions of species' ranges throughout the evolutionary history of this planet, there ought to be *more* species than there actually are.

Non-allopatric speciation can be of two types:

1 **Sympatric speciation**—where there is no geographical separation between the speciating populations and all individuals are, in theory, physically able to meet each other during the speciation process.

2 **Parapatric speciation** —where the speciating populations are parapatric and so there is only partial geographical separation between them. The populations are contiguous and are not separated by extrinsic factors, so some individuals on each side are able to meet across the common boundary during the speciation process.

► Study Figure 9.19. Which of the three situations (a–c) represents (1) allopatric speciation; (2) sympatric speciation; and (3) parapatric speciation?

Allopatric speciation is represented by (a) in Figure 9.19, sympatric speciation by (c) and parapatric speciation by (b).

According to models of sympatric and parapatric speciation, divergence can take place between two forms (the parental type and the novel type) without total

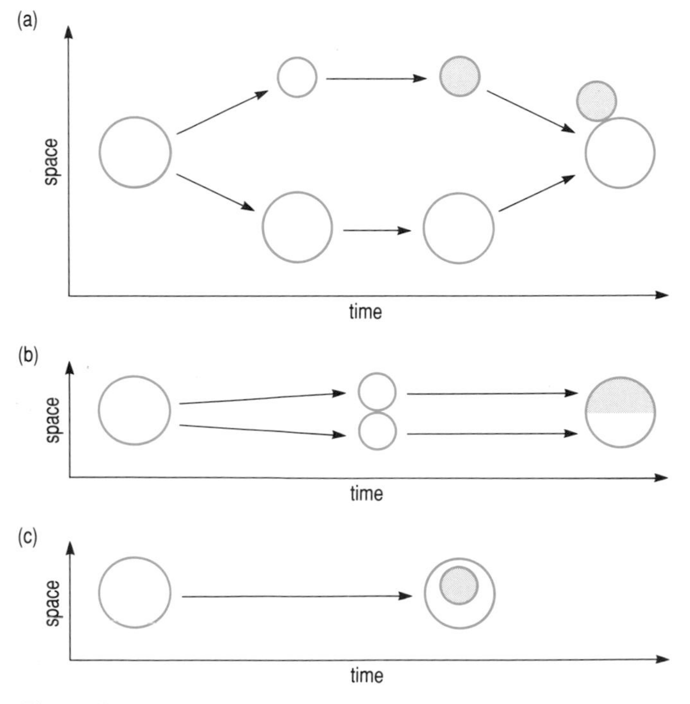

Figure 9.19 *A comparison of allopatric, parapatric and sympatric modes of speciation. Open circles represent populations or species and shaded circles represent a new species.*

geographical separation as long as gene flow between them is reduced sufficiently and for long enough to allow the novel type to adapt differentially without being swamped by parental genes. In this respect, these models share the same basic premise as models of allopatric speciation—the only difference is that gene flow in allopatric models is reduced by geographical separation, whereas gene flow in sympatric or parapatric models is reduced by other means.

9.6 SYMPATRIC SPECIATION

Several models of sympatric speciation have been proposed. Most of them involve a change in host preference, food preference or habitat preference, or the partitioning of an essential but limiting resource. Three of the major models are discussed here: speciation by disruptive selection, competitive speciation and polyploidy.

9.6.1 SPECIATION AS A RESULT OF DISRUPTIVE SELECTION

After Mayr's theories about allopatric speciation were published, one of the first models of sympatric speciation was produced by John Maynard Smith in 1962 and elaborated in 1966. He showed (Maynard Smith, 1966) that the existence of a stable polymorphism within a single population in a heterogeneous environment could, in theory, lead to speciation, and that such a stable polymorphism could result from disruptive selection (Chapter 4).

▶ From what you read in Chapter 4, Section 4.4.5, list three ways in which a stable polymorphism between two alleles at a particular locus may be established.

A stable polymorphism may result if the heterozygote is fitter than either of the homozygotes, if one allele is neutral or even deleterious but is tightly linked to an allele at another locus and this is favoured by selection, or if selection is negatively frequency-dependent, i.e. the fitness of one genotype increases as its frequency in the population decreases (as, for example, where predators are more likely to prey on the common type).

Maynard Smith showed mathematically that disruptive selection favouring two different phenotypes controlled by a single gene could also lead to a stable polymorphism, whether heterozygotes have the same fitness as one of the homozygotes (dominance is operating) or whether heterozygotes are intermediate in phenotype and less fit than either homozygote.

For Maynard Smith's model to work, however, certain very stringent conditions must be fulfilled:

1 The two types favoured by disruptive selection must be adapted to different ecological niches (e.g. different food plants or different types of nesting place).

2 The numbers of individuals occupying each niche must remain approximately constant.

3 The selective advantage of each type in its own niche must be large.

These conditions may have been met inadvertently in a number of laboratory experiments where disruptive selection has led to a stable polymorphism, e.g. in those where a fixed number (condition 2) of 'high' and 'low' individuals (condition 1) were chosen as parents for the next generation and where — as is mostly thc case in laboratory studies — the selection against intermediates was strong (condition 3). Thoday and Gibson's 1962 experiment which involved *Drosophila* with different numbers of abdominal bristles (Section 9.4.3) is one such example. However, it is difficult to judge how often such conditions will be met in natural populations. For example, suppose predators took only medium-sized individuals from a population, perhaps because the bigger ones were too big for them to handle and the smaller ones escaped their notice. Such disruptive selection would normally lead to a population consisting entirely of small individuals or entirely of large individuals. It could lead to a stable polymorphism with a balance between large and small types only if the population contained a fixed number of large individuals and a fixed number of small individuals (condition 2), with large and small individuals each adapted to a different niche (condition 1) and if a lot of intermediate individuals were eaten by predators (condition 3). Condition 2 is extremely unlikely in the wild even if conditions 1 and 3 are satisfied.

The model does not make any assumptions about **habitat selection** (individuals showing habitat selection end up in the niche to which they are adapted). Maynard Smith showed that if habitat selection is included, i.e. if individuals choose to live in or lay their eggs in the habitat to which they are adapted, then a stable polymorphism will be more likely.

Habitat selection could arise either:

1 as a pleiotropic effect of the gene or genes that determine the habitat to which individuals are adapted; or

2 because individuals choose the habitat in which they themselves grew up or hatched out.

In the latter case, individuals are said to be *conditioned* to choose a particular habitat. The gene or genes controlling habitat choice 'instruct(s)' the individual to choose the same habitat in which it grew up rather than specifying a particular habitat as in (1) above.

The first possibility seems unlikely but there are many examples of organisms that show habitat selection seemingly as a result of such conditioning. The sawfly *Pontania salicis* is an insect that lays its eggs on willow trees, causing the tree to produce a growth called a gall around the developing larvae. There are several biological races of *P. salicis* in North America, each forming galls on a different species of willow. When a race which normally formed galls on *Salix andersoniana* was reared in the laboratory on *S. rubra*, most of the larvae in the first generation died, but this heavy mortality became much less in subsequent generations. After four generations confined to *S. rubra*, three further generations were given a choice of willow species on which to lay their eggs, but the sawflies continued to lay eggs on the new host plant.

▶ What is the most likely explanation for the shift in egg laying to the new host plant?

Females are conditioned to lay eggs on the plant species on which they themselves were reared, i.e. the gene or genes controlling egg-laying behaviour cause the female to lay her eggs on the same type of plant as that on which she was raised rather than specifying a particular plant species. Genetic change was thus unnecessary—forcing the sawflies to live on a different plant was sufficient to change the females' egg-laying behaviour.

▶ Why is a change in the gene or genes controlling egg-laying behaviour an unlikely explanation?

For a genetic change to have taken place, an allele that caused females to lay eggs on *S. rubra* would have had to be present in the original laboratory population (unlikely, since the caterpillars were collected from *Salix andersoniana*), or to have arisen by mutation after the transfer to *S. rubra*, and this allele would have had to become fixed in the laboratory population, all within the space of only four generations.

There is also some evidence for specific control of habitat preference in relation to adaptation to a particular habitat. Partridge has shown that two species of tit, the coal tit *Parus ater* which mainly lives in coniferous woodland, and the blue tit *Parus caeruleus* which mainly lives in broad-leaved woodland, differ not only in their feeding skills but also in their habitat preferences. Blue tits tend to feed while hanging upside down, forcefully excavating insect food from the vegetation. Coal tits feed upright, taking small camouflaged insect prey from the surface of the vegetation. These differences are related to the availability of prey in the two habitats—many of the insects on broad-leaved trees are found on the undersides of leaves or buried in the bark, while those on conifer trees are found on the surface of the needles and are accessible from above. Given a choice between a conifer branch and a broad-leaved branch, coal tits tend to choose the conifer and blue tits the broad-leaved branch. These differences are shown to a large extent even if the birds are hand-reared under identical conditions, without their parents and without access to tree branches.

▶ What does the fact that hand-reared birds show similar differences to wild birds tell you about the basis of the differences?

It shows that the differences are not due entirely to learning or conditioning—as the hand-reared birds can only have inherited the differences, there must be some genetic basis to them.

Maynard Smith showed that, once the most difficult step—establishment of a stable polymorphism—has been achieved, natural selection will favour the evolution of reproductive isolation between the two types and hence speciation. This can happen in four ways:

1 By habitat selection. As soon as a polymorphism for niche adaptation becomes stable, selection will favour genotypes showing habitat selection even if habitat

selection was not necessary to establish the polymorphism in the first place. Individuals laying eggs or rearing offspring in the same habitat as that in which they themselves were raised (and survived to breed) will be at an advantage because their offspring, being similar to their parents, are likely to do better in that niche than in any other. Individuals will also be at an advantage if they mate in the habitat to which they are adapted because then they are more likely to find mates adapted to the same habitat and thus to produce offspring adapted to that niche rather than less well-adapted hybrids. If mating and rearing of offspring takes place entirely within the habitat to which each type is adapted, the populations are reproductively isolated and speciation is complete.

2 By a pleiotropic effect of the gene causing the adaptation to a particular niche, such that it also happens to cause positive assortative mating.

3 By modifier genes, which in this context means genes that determine whether or not mating will take place between two particular individuals. The modifier gene causes mating to take place only between individuals with the same genotype as each other at the locus affecting adaptation to a particular niche. The modifier gene does not itself affect adaptation to the particular niche and is at a different locus from the gene that does. Maynard Smith showed that reproductive isolation is unlikely to evolve under these circumstances, however, unless there is also habitat selection.

4 By positive 'assortative mating' genes. A second gene causes positive assortative mating between individuals carrying the same positive assortative mating alleles, regardless of the genotype at the locus affecting adaptation to a particular niche. Unless, however, the gene causing positive assortative mating and that causing adaptation to a particular niche are very closely linked genetically—which is possible but unlikely—an association between the appropriate alleles at each locus can never become complete because the genotypes will be continually split up by recombination. And without such an association, there will be no reproductive isolation. Maynard Smith showed that the two genes could become associated without close linkage, however, if individuals always breed within the niche to which they are adapted, i.e. if there is habitat selection.

Thus, the most important step in sympatric speciation according to Maynard Smith's model is the establishment of a stable polymorphism in an environment consisting of more than one habitat type. Unless reproductive isolation arises as the result of a pleiotropic effect of the gene controlling adaptation to the niche, however, some form of habitat selection is probably crucial in the speciation process, either in the development of the stable polymorphism or in the evolution of reproductive isolation.

There are several ways in which reproductive isolation could arise as a pleiotropic effect of adaptation to a particular niche, for instance if the timing of mating or development of eggs is dependent on the life cycle of a specific food plant. A group of insect species called treehoppers, all from the genus *Enchenopa*, provide a good example. There are at least six closely related species, each of which utilizes a different species of host plant within the same geographical area. Mating takes place once a year over only a very short time period, and females lay their eggs in the twigs of the host plant. The eggs remain dormant over the winter and hatch only when sap begins to flow through the plant in spring. Different host plants differ in the time their sap starts to flow, so that the life cycles of the various *Enchenopa* species are asynchronous and populations on different host plants mate at different times. As a result, individuals can mate only with others from the same host plant, so the populations on different host plants are reproductively isolated. Thus, in theory, the six species could differ by only one gene controlling the species of host plant utilized, with the times of mating being determined automatically by the host plants themselves.

It is also possible to see from Maynard Smith's model how sympatric speciation could take place if habitat selection operates. For example, if for any reason a female from an insect species which has adapted to feeding on only one species of plant were to lay her eggs on an unusual plant, two things might follow. First, the larvae might be exposed to new conditions and may therefore be under intense selection pressures—like the sawfly larvae on *S. rubra* that suffered heavy mortality in the first few generations. Adoption of the new host may have selective advantages, though—allowing the insect to escape from its predators, pathogens or competitors, both intra- and interspecific.

Secondly, the females hatching out from those eggs might tend to lay their eggs on the same plant, not because they differ genetically from the rest of their species but because they have been conditioned to do so. If the species were also one that tends to mate on or near its host plant, then females would tend to mate with males that emerged from eggs on the new host plant. In this way a population could arise that is sufficiently reproductively isolated from its parent population for further divergence to take place, and the greater the degree of divergence, the more selection will favour habitat selection and assortative mating, eventually leading to complete speciation.

A lot of insect species have life histories that fulfil several of these conditions, so it is not surprising that many of the examples where it is thought that sympatric speciation has occurred (or is occurring) in nature are among the insects. (If sympatric speciation is common in insects, it might also explain why there are so many insect species in existence.) Closely related sympatric insect species usually use different host plants while closely related allopatric species use identical or similar plants (though this could be interpreted as resulting from a secondary invasion by one species of another species' range, which would only be possible in those species using different plants). Insects often have very specific habitat preferences, e.g. parasites that have only a single host species, herbivores that feed on only one species of plant and predators that have only one prey species. They frequently mate within or after dispersal to their preferred habitat, so disruptive selection on genes that determine habitat or host selection will lead pleiotropically to positive assortative mating, and they frequently lay their eggs in their preferred habitat, often as a result of conditioning.

For example, soon after emerging from the pupa, adult *Drosophila* disperse—i.e. move some distance before settling in one place. They tend to mate only after dispersing, so a gene causing selection of a particular habitat in which to settle would automatically have the pleiotropic effect of producing positive assortative mating. Sympatric speciation could therefore have played a part in the spread of the Hawaiian *Drosophila*, despite the primary role probably played by allopatric speciation. Thus, some closely related sympatric species use different habitats or food: the fully sympatric species *D. silvarentis* and *D. heedi* feed exclusively on

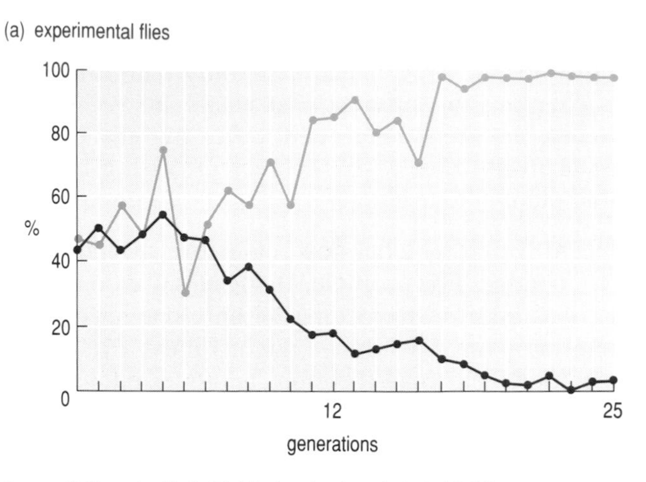

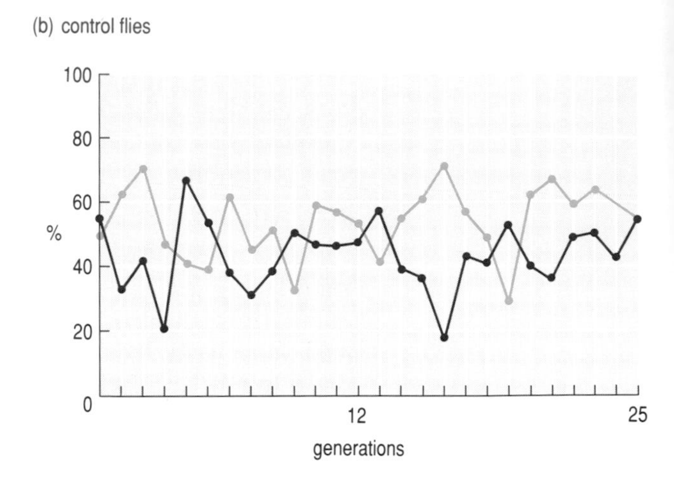

flies raised in habitat A choosing to mate in habitat B

flies raised in habitat A choosing to mate in habitat A

Figure 9.20 *The percentage of flies raised in habitat A choosing habitats A and B for (a) experimental flies and (b) control flies. Similar results were obtained for flies raised in habitat B.*

the slime flux of a single species of tree, but in different places—one on the trunk well above the ground, the other in the soil.

W. R. Rice and G. W. Salt (1988) applied disruptive selection to habitat preference in a laboratory population of *D. melanogaster*. The laboratory population was set up such that males and females mixed freely before selecting their habitat from a choice of eight types. Mating, egg laying and the growth of the resulting larvae took place only in the selected habitat. In one experiment, strong disruptive selection was imposed by taking only those flies which had chosen either of the two extreme habitat types, A and B, as parents of the next generation. In the control population, parents were selected at random across all habitats. Rice and Salt's results are shown in Figure 9.20.

▶ What percentage of experimental flies were switching between habitats by the end of the experiment, i.e. how many were moving from the habitat which their parents had selected to mate in the other habitat?

After 25 generations, only 1–2% of flies reared in one of the two habitats were choosing to mate in the other.

▶ What percentage of control flies were switching between habitats after 25 generations?

Approximately 50%.

▶ What do Rice and Salt's results indicate about the effect of disruptive selection on their laboratory population?

As a result of the disruptive selection imposed, the experimental population divided into two types, each preferring one of the two habitats. Mating was almost completely positively assortative, with very little gene flow between the two sub-populations.

Selection in Rice and Salt's experiment was much stronger than might be expected in the wild, but there is some evidence that natural populations may be undergoing a similar process to that seen in their *Drosophila* population. For example, caterpillars of the ermine moth *Yponomeuta padellus* (a small British moth with white front wings spotted with black and grey hind wings) feed on apple and hawthorn trees. There is variation in colour among adult moths but caterpillars developing into different coloured adults are found on both tree types and the adults show no other differences in morphology. About 80% of females raised on apple, however, lay their eggs on apple while about 90% of females raised on hawthorn lay their eggs on hawthorn. There is also a tendency for moths to mate with individuals raised on the same food plant and caterpillars show a strong preference for the food plant on which their mothers were raised. The 'apple' and 'hawthorn' types are not completely isolated as some interbreeding probably takes place in the wild. The present situation may represent an intermediate point in the ongoing process of sympatric speciation, and some investigators even consider that the two types are essentially two species.

Another possible example of sympatric speciation according to the Maynard Smith model, this time one where speciation is already complete, has been proposed by C. A. Tauber and M. J. Tauber. They studied two North American species of lacewing (a common insect with delicately veined wings): *Chrysoperla carnea* and *Chrysoperla downesi*. The range of *C. downesi* is completely within that of *C. carnea* and there is no evidence that the two species have ever been geographically isolated. *C. carnea* is brown in the autumn and light green in the spring and summer, while *C. downesi* is dark green all year round. The colour differences are related to differences in the species' habitat: *C. carnea* lives in grassland in summer and migrates to deciduous trees in autumn; *C. downesi* lives on coniferous trees. Each species is thus cryptically coloured in its own habitat. The larvae of both species prey on a variety of soft-bodied arthropods such as greenfly, and the adults eat nectar and pollen.

No hybrids are found in nature, probably because the two species mate at different times of year: *C. carnea* breeds in winter and again in summer; *C. downesi* breeds only once, in the spring. *C. downesi* adults also tend to overwinter in conspecific aggregations and then mate before dispersing, decreasing the likelihood of their mating with a member of the other species.

In the laboratory, however, the two species readily interbreed to produce fully viable and fertile hybrids. Tauber and Tauber have carried out extensive hybridization studies which show that the differences in colour and breeding time between the two species are controlled by only three genes. One gene controls colour and has two alleles, one producing the dark green colour when homozygous and the other producing the light green/brown colour when homozygous. The heterozygous hybrid has an intermediate colour. The other two genes control the time of breeding, again with two alleles for each gene, one recessive and one dominant. *C. downesi* has two recessive alleles at both loci and *C. carnea* has two dominant alleles at both loci. Heterozygotes (i.e. all first-generation hybrids) breed in the summer and the winter like *C. carnea*.

habitat selection. This would eventually lead to the separation of the ancestral species into two separate species, with the process then being repeated in one of them to produce the three separate species seen today. There is some evidence that a pleiotropic effect may be responsible for positive assortative mating. In *Megarhyssa*, mating generally takes place before the females emerge from the log. The male inserts most of his body into the female emergence tunnel and as a result large males tend to mate with those females found deep in the log, i.e. the offspring of females with long ovipositors. Body size and ovipositor length are probably correlated and both may be pleiotropic effects of the same gene or genes.

It is possible that the division of a single gene pool into two or more adaptive types as a result of density-dependent selection may be more common in nature than is generally thought. T. J. Ehlinger studied the bluegill sunfish *Lepomis macrochirus*, a common fish species in lakes and ponds of North America. Shortly after hatching, the fish move into open water and feed on zooplankton. Very young fish are relatively safe in open water because their small size, inconspicuous colour and quick escape reactions make it difficult for fish predators to detect and catch them. Juveniles become more susceptible to predation and eventually move into the cover of aquatic vegetation near the shore, even though there may be less food there. Once the fish are longer than about 80 mm, however, they become relatively immune to predation, which enables them to move freely between open water and vegetation, wherever food is most abundant. All mating takes place in the vegetation.

Until Ehlinger studied this fish in detail, it was assumed that populations within a single lake were monomorphic (i.e. not polymorphic). He found, however, that the fish were divided into vegetation and open water foraging specialists. The two habitat types require different foraging techniques if the maximum rate of prey capture is to be achieved—fish in open water hover for a short time and move quickly between hovers in order to catch the dispersed and relatively conspicuous

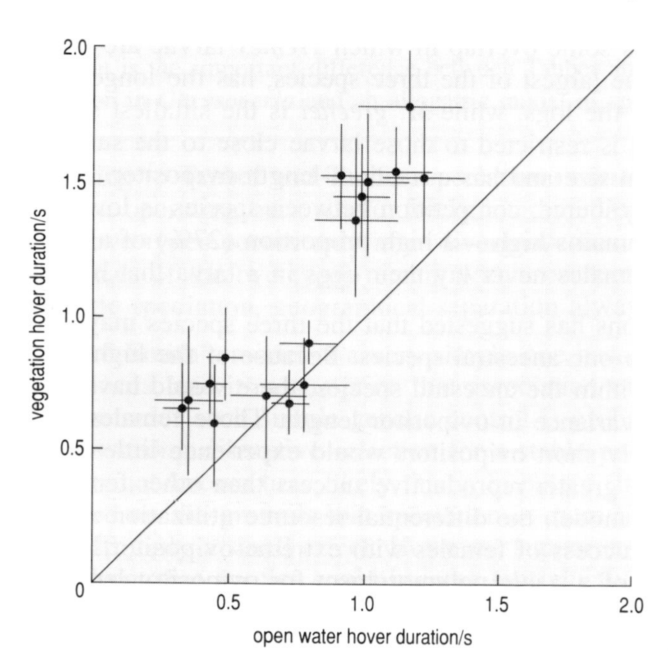

Figure 9.21 *Behavioural polymorphism in the bluegill sunfish* Lepomis macrochirus. *Hover durations of individual fish were measured in both open water and in vegetation. Dots represent the average hover duration, vertical and horizontal lines represent standard errors. Where a dot lies above the diagonal, it means that the individual hovers longer in vegetation than it does in open water. The fact that the dots are divided into two clusters means that inflexible differences exist between fish, which can be divided into 'long-hover' and 'short-hover' forms.*

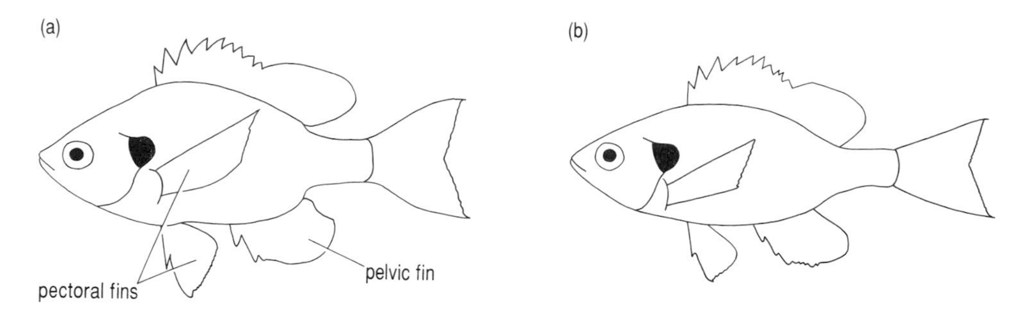

Figure 9.22 Line drawings illustrating differences in the morphology of bluegill sunfish collected from a Michigan lake in (a) vegetation and (b) open water habitat. Compared with open water fish, fish from the vegetation have deeper bodies, longer pectoral and pelvic fins, and have their pectoral fins positioned farther back and up on their bodies.

prey found there, while fish in vegetation hover for a long time and show slow stealthy movements between hovers in order to lie in wait for cryptic prey hidden among the weeds (Figure 9.21).

The fish also showed differences in morphology, in both sexes and in all sizes, that can be related to differences in their foraging technique. Fish caught in vegetation are deeper bodied, have longer pectoral and pelvic fins and the pectoral fins are further back on the body than fish caught in open water (Figure 9.22). These features would help vegetation fish to swim better amongst vegetation and open water fish to swim better in habitat without obstacles.

The two types also differ in the kinds of parasites that they carry. Fish caught in vegetation are infested mainly by a parasitic flatworm whose intermediate host is a snail found on the vegetation, while fish caught in open water carry a tapeworm whose intermediate host is a small crustacean usually found in open water. This implies that the two forms spend most of their time in their respective habitats.

Thus, bluegill sunfish of the same size are not found in one habitat or the other but segregate into the habitat to which they are best adapted by morphology and behaviour. Density-dependent selection will maintain the polymorphism because the fitness of each type decreases as the strength of competition for food increases, i.e. as the density of individuals of the same type increases. It is possible that the two foraging types will eventually become reproductively isolated, so making them two separate species.

If density-dependent or frequency-dependent selection often cause a randomly mating population to separate into ecologically distinct forms whose intermediates are relatively unfit, then sympatric speciation may have taken place more frequently than previously thought.

9.6.3 Polyploidy

Polyploidy involves the multiplication of whole sets of chromosomes (each set being the haploid number n—see Chapter 3). Polyploids occur frequently in plants and in some groups of animals such as earthworms and rotifers (a type of microscopic aquatic animal). There are two types: **autopolyploids** and **allopolyploids**.

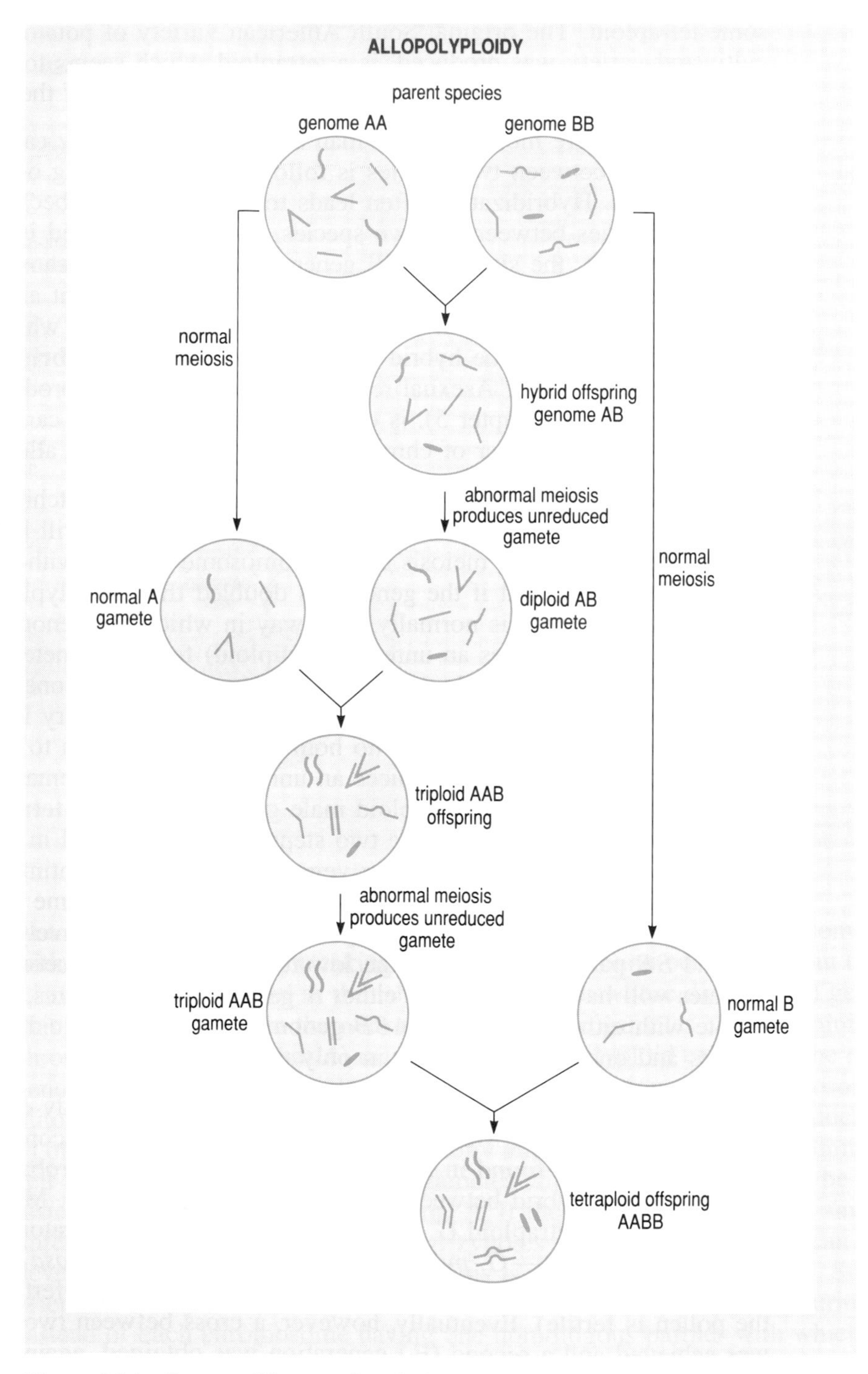

Figure 9.24 *One possible route by which an allopolyploid may be formed.*

can be crossed easily and produce fully fertile offspring, providing further evidence that *G. tetrahit* probably arose in this way in the wild.

Polyploid plants are often found to differ from diploids in having larger cells (because the nucleus has to contain twice as many chromosomes and is therefore

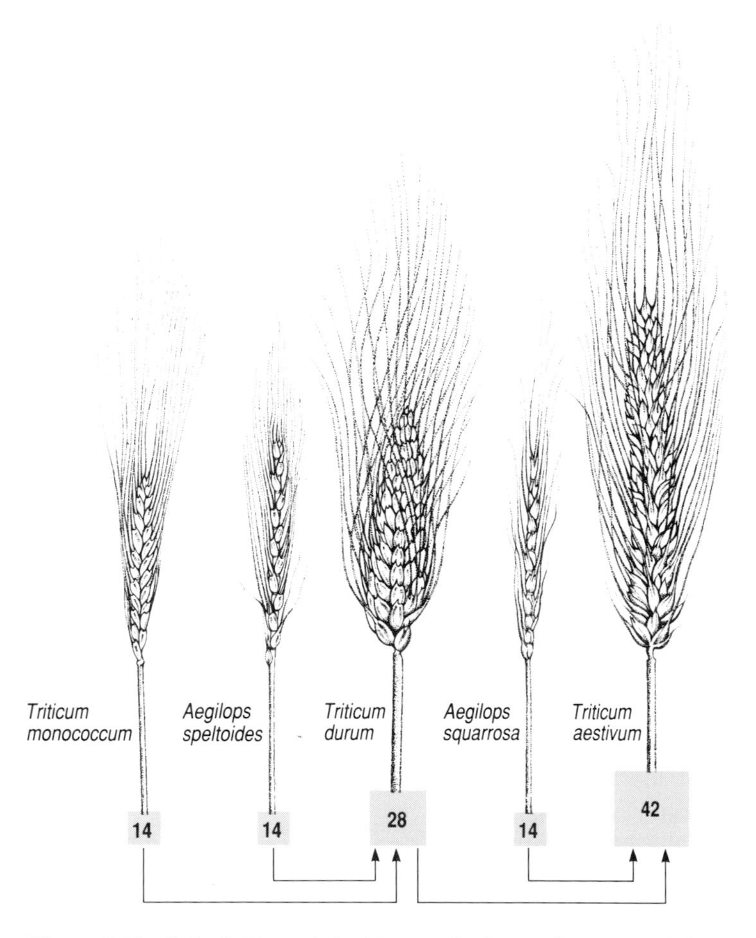

Figure 9.25 *Polyploidy and the history of wheat—the origin of durum wheat* Triticum durum *and bread wheat* T. aestivum *by chromosome doubling in two different hybrids. Diploid chromosome numbers are shown in the boxes.*

larger), thicker, fleshier leaves and larger flowers and fruits. The last two features are very desirable in plants cultivated for food, so it is not surprising that many of our common food plants are polyploids. Wheat provides a good example (see Figure 9.25). The einkorn wheat first grown by humans, and still found growing wild in the Near East, was a diploid species, *Triticum monococcum*, with 14 chromosomes. Modern cultivated durum wheat *Triticum durum*, which is used to make pasta, is a tetraploid species with 28 chromosomes. It probably arose by chromosome doubling in a hybrid between *T. durum* and a species of grass that grows as a weed in cultivated fields in the Near East, *Aegilops speltoides*, which also has 14 chromosomes. The tetraploid wheat may have been found growing wild by early humans, or may have appeared in an early cultivated wheat field and been selected for breeding because of its larger size. The wheat now grown for bread, *T. aestivum*, is a **hexaploid** species, with $6n$, i.e. 42 chromosomes. This probably arose by chromosome doubling in a second hybrid, this time between *T. durum* and another grass species with 14 chromosomes, *Aegilops squarrosa*.

Polyploidy is of necessity a sympatric process and is probably the only model of sympatric speciation accepted without question. It has occurred most frequently in the speciation of plants. The chromosome numbers of related plant species are often multiples of some common basic number (different species of *Chrysanthemum* for example have $2n$ = 18, 36, 54, 72 and 90, which are all multiples of 18) and some estimates suggest that as many as 80% of flowering plants originated as polyploids. This is probably due to several factors, including the ability of many plants (unlike most animals) to reproduce vegetatively or to fertilize themselves. Vegetative reproduction avoids any problems in chromosome pairing during meiosis and self-fertilization increases the chances of two unreduced gametes from a plant that has a tendency to produce them fusing with each other to form an autopolyploid. Where polyploidy has taken place in animals, it is generally in species that are hermaphrodite or asexual, though a few polyploid fish and amphibian species have arisen from sexual species.

The main limiting factor on this type of speciation is ecological—the polyploid must be able to coexist with its parent species or at least survive long enough to expand into a new range. In fact, polyploidy itself often induces many ecologically significant changes, making some degree of ecological divergence between the polyploid and its parent species almost automatic.

9.7 Parapatric speciation

Many of the problems associated with sympatric speciation also apply to parapatric speciation. As a proportion of the two neighbouring populations are, at least in theory, able to meet each other as potential mates, gene flow between the two populations must be reduced by some means if speciation is to take place.

Endler (1977) has developed in detail the idea of parapatric speciation as a result of the formation of a cline (Section 9.3.2). According to Endler's model, an ancestral species spreads over a spatially variable area and this leads to geographical differentiation of the sort described in Section 9.3. The range of the species is continuous, however, and all populations remain in contact. Geographical differentiation leads to the formation of a cline, which acts as a barrier to gene flow so that further divergence can take place in the populations on each side of the cline.

▶ From what you read in Section 9.3.1, what factors will tend to reduce gene flow through a hybrid zone?

Short dispersal distances, selection against hybrids and selection against 'pure' types either in the hybrid zone or on the 'wrong' side of the hybrid zone.

▶ What likely effect will an increase in the divergence between two populations have on a hybrid zone between them?

The hybrid zone will become narrower. Selection against hybrids and against pure types, either within or on the wrong side of the hybrid zone, will be increased because the two parental genotypes will be even less compatible in hybrid form and pure types will be even less fit outside their own environment. Gene flow through the hybrid zone will therefore be reduced and hence the hybrid zone will become narrower.

As a result of greater divergence, the cline thus becomes steeper and so presents a stronger barrier to gene flow, leading to an even steeper cline, until eventually a very narrow hybrid zone is formed, with highly differentiated populations on either side. Pre-mating reproductive isolation evolves as a by-product of divergence, i.e. as a pleiotropic effect of the alleles that control adaptation to the environment, rather than as a result of reinforcement (Section 9.4.3). If it becomes complete, the hybrid zone disappears and the two new species occupy contiguous areas.

In this model, isolation is 'by distance' rather than by geographical or ecological separation, though these may also play a part. For example, differential adaptation may take place in two populations on either side of a boundary where ecological conditions change abruptly (called an **ecotone**), with a narrow hybrid zone forming between them.

Some evidence for Endler's idea of isolation by distance is provided by the existence of **ring species**. Where a species covers a large geographical area, individuals at the extreme ends of its distribution can be very different. Yet they will usually be considered to be members of the same biological species if the species forms a continuous interbreeding population or if neighbouring populations interbreed, even if it is not possible to tell if the two extremes would interbreed in nature because they are too far apart ever to meet for mating. But in ring species, the two geographical extremes do meet because the species is distributed in a circle.

For example, the distribution of the gull known in Britain as the herring gull *Larus argentatus*, forms a complete circle in the Northern Hemisphere (Figure 9.26). Moving westwards from Britain to North America, the gulls are recognizable as herring gulls, although they gradually become somewhat different from the British form. Their appearance continues to change to the west until in Siberia, they begin to look more like the gull that in Britain is called the lesser black-backed gull *Larus fuscus*. From Siberia, through Russia, to northern Europe, the gulls gradually become more and more like the British lesser black-backed gull. The ring overlaps in Europe—the two geographical extremes meet and appear to be two perfectly good biological species, very different in appearance and normally with no interbreeding between them.

Proponents of parapatric speciation take the view that, in a ring species, the two ends are far enough apart to have become differentiated into separate species despite the gene flow through the population which tends to homogenize it. Opponents of the idea of isolation by distance point out, however, that the current distributions of ring species are often not completely continuous and that the species' range may also have been split up at some time during its evolutionary history, allowing differentiation to take place according to the traditional allopatric model. Thus, in the *L. fuscus–L. argentatus* ring, there are now pronounced gaps between some of the populations in western Europe and eastern Asia and populations must also have been split into a number of refuges during recent ice ages. Unfortunately, there is no way of actually telling, on the basis of present distributions, whether *L. fuscus* and *L. argentatus* differentiated allopatrically or parapatrically.

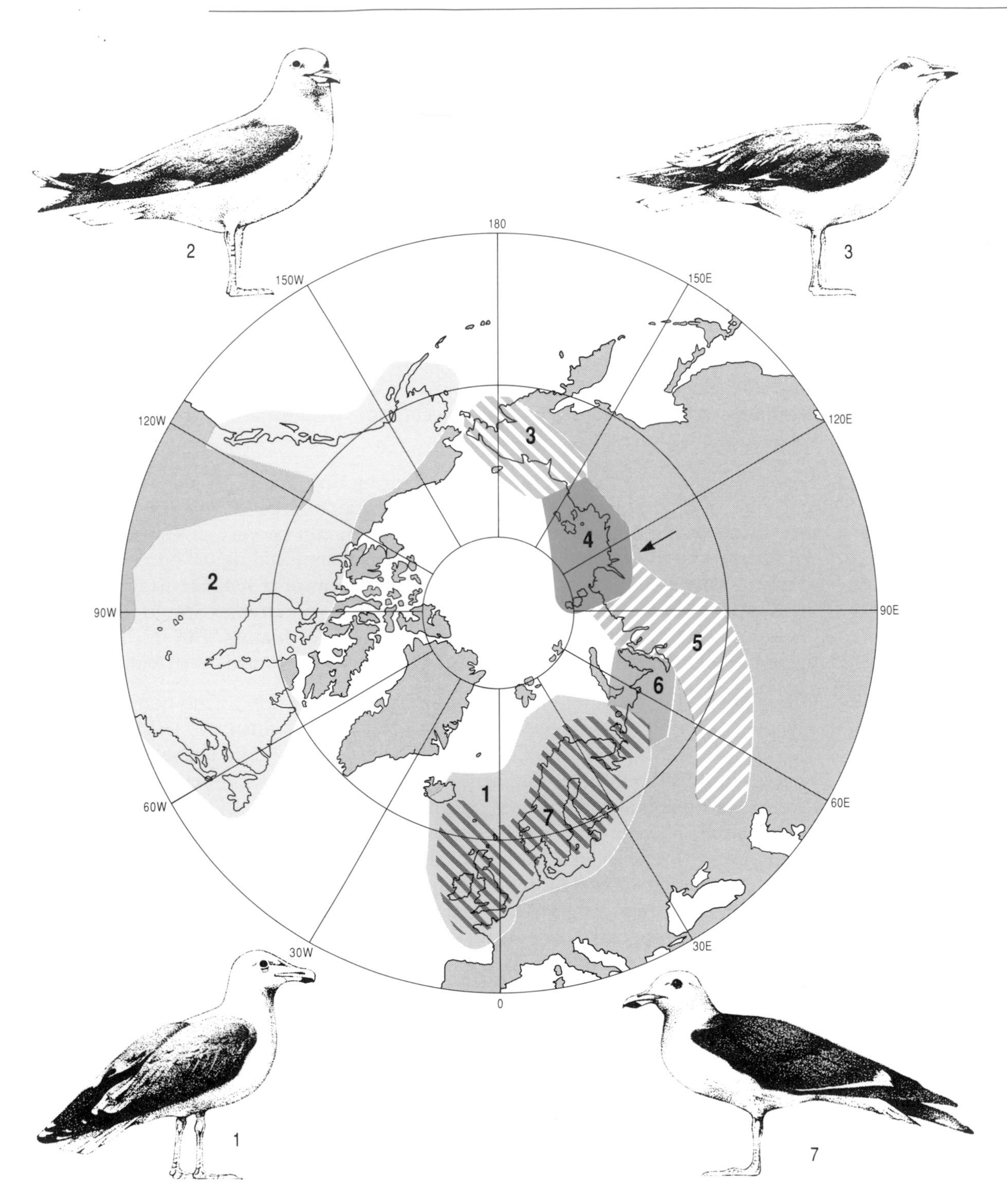

Figure 9.26 *Map showing the circumpolar distribution of the ring species formed by the herring gull* Larus argentatus *and the lesser black-backed gull* L. fuscus, *and the areas occupied by some of their subspecies. The toned area is the overlap between the two species; the arrow marks the boundary between them. 1, herring gull* Larus argentatus argentatus*; 2, American herring gull* L. argentatus smithsonianus*; 3, Vega herring gull* L. argentatus vegae*; 4, Birula's gull* L. argentatus birulae*; 5, Heuglin's gull* L. fuscus heuglini*; 6, Siberian lesser black-backed gull* L. fuscus antelius*; 7, lesser black-backed gull* L. fuscus fuscus.

Endler has shown mathematically that a narrow hybrid zone (or step cline) can form between two parapatric populations even in the face of dispersal between them. Barton and Hewitt, on the other hand, have calculated that, because of recombination, neutral alleles (Chapter 3, Section 3.3.2) will diffuse through a hybrid zone, taking about only 10 000 generations to spread 100 times the dispersal distance of the species (i.e. the average distance individuals move each generation), while an allele that is even only slightly favourable will move through with little hindrance because, as soon as a few genes penetrate to the other side of the barrier, they will increase very rapidly and spread through the population. Clines for different characters will not necessarily coincide unless they are all responding to the same environmental gradient or unless, as frequently happens, they are tension zones that have all come to rest in a region of low population density. But if they do coincide, the strength of the barrier to gene flow will be increased considerably. Barrier strength thus depends on the number of coincident clines, mean fitness, recombination rate and the steepness of the cline.

Endler has pointed out that, within single species, there are more differentiated parapatric populations than there are differentiated allopatric populations. He argues that this is what we would expect to see if populations are more likely to differentiate in parapatry than in allopatry and that therefore parapatric speciation happens more often than allopatric speciation.

▶ How can the fact that there are more differentiated parapatric populations and fewer differentiated allopatric populations also be used as evidence *against* the idea that parapatric speciation happens more often than allopatric speciation?

It can be argued that there are more differentiated parapatric populations because speciation is less likely to be completed in parapatry and so the populations never get beyond the differentiation stage; on the other hand, there are fewer differentiated allopatric populations because speciation is more likely to become complete in allopatry and what we see is the end result—separate species, not differentiated races.

The existence of apparently stable hybrid zones in the face of free dispersal between the populations shows that gene flow need not destroy spatial divergence. However, as pointed out in Section 9.3.1, it is generally not possible to be absolutely certain whether hybrid zones have formed from primary or from secondary contact, though many researchers are of the opinion that most hybrid zones are formed by secondary contact. It is usually not possible, therefore, to distinguish definitely between parapatric speciation and speciation by reinforcement (Section 9.4.3). There are a few cases, however, where there is good circumstantial evidence for parapatric speciation.

One such example involves the common bent grass *Agrostis capillaris*, already mentioned in Chapter 4 (Section 4.3.1), which is able to grow on spoil heaps such as those thrown up by the mining industry, despite the high concentrations of metals like copper, lead, zinc and tin in the soil which would normally prevent plants growing there at all. It also grows on naturally occurring areas of soil containing high metal concentrations, such as those covering veins of ore. Experiments have shown that there are different types of *A. capillaris*: the type growing on spoil heaps is more tolerant of metals than is the 'normal' type, which is a common species in grassland, and that the higher tolerance is due to genetic differences. On the spoil heaps, only the tolerant grass can survive; but, in the surrounding areas, tolerance to high metal concentrations is not necessary and is associated with physiological characteristics that are actually a disadvantage. Disruptive selection against the tolerant type outside the spoil heaps and against the normal type on the spoil heaps is therefore intense. The boundary between contaminated and uncontaminated soil is usually very sharp and there is normally only a narrow hybrid zone between the two types of grass. How far the hybrid zone extends into each habitat type depends on whether it is upwind or downwind. Hybrids are found only a metre or two on the upwind side of the edge of a spoil heap but extend for about 150 metres on the downwind side.

▶ Why does the wind have this effect?

You will remember from Chapter 4 that *A. capillaris*, like other grasses, is wind-pollinated, so the width of the hybrid zone on either side of the edge of a spoil heap probably reflects the dispersal distance of the pollen, either helped or hindered by the wind.

The difference in the width of the hybrid zone upwind and downwind shows that gene flow between tolerant and non-tolerant populations of *A. capillaris* can be considerable. Yet the two types are diverging and already (spoil heaps are not a long-standing phenomenon in evolutionary time-scales!) show signs of incipient reproductive isolation—they flower at slightly different times and the tolerant type self-fertilizes to a greater extent than the non-tolerant type. The differences in flowering time are genetically determined and seem to be an adaptation to local conditions. However, there is no evidence of incompatibility between the two genotypes—seeds are produced in equal numbers and germinate equally well regardless of whether they are formed by crosses between the same or different types.

It is likely that the tolerant type of *A. capillaris* emerged from the population in the area surrounding a spoil heap, probably developing independently several times at different spoil heaps, and that the different populations have always been continuous. For the tolerant type to evolve in allopatry it would have had to develop on a spoil heap not surrounded by normal *A. capillaris*, and then 'jump over' the normal population to colonize the spoil heaps where it presently grows!

SUMMARY OF SECTIONS 9.5–9.7

It is now accepted that speciation can take place either when there is no geographical separation whatsoever and two diverging populations share exactly the same range—sympatric speciation—or when geograpical separation is not complete and two diverging populations share a common boundary—parapatric speciation.

Sympatric speciation can take place only when some factor other than geographical separation reduces gene flow between two sections of the population. This can happen when there is a change in host preference, food preference or habitat preference by some members of the original population. Maynard Smith showed mathematically that, once a stable polymorphism exists within a sympatric population such that the two types are each adapted to a different ecological niche, then natural selection will favour the evolution of reproductive isolation between the two types and hence speciation. Reproductive isolation can be brought about by pleiotropic effects of the gene causing adaptation to the habitat or by habitat selection, e.g. if individuals mate or lay eggs in the habitat to which they are adapted. The stable polymorphism could arise in the first place as the result of disruptive selection whereby each type has a large selective advantage in its own niche. Disruptive selection is more likely to lead to a stable polymorphism if there is habitat selection. Habitat selection is thus a key factor in both stages of this sympatric speciation model. There is some evidence from nature and from laboratory experiments in support of Maynard Smith's model. Many insect species have life histories that fulfil its conditions and this may explain why many of the examples where it is thought sympatric speciation has taken place or is occurring are among the insects.

Sympatric speciation can also take place when different members of the populations divide an essential but limiting resource between them or as the result of polyploidy, in which an individual is produced that has one or more extra sets of chromosomes. Such polyploids often constitute a new species because they cannot generally interbreed successfully with normal members of the population—speciation can thus take place in as little as one generation.

Parapatric speciation also requires a reduction in gene flow between two adjacent populations. This can occur where a cline forms in a large population as a result of adaptation to local conditions. The cline will act as a barrier to gene flow if there is selection against hybrids or against individuals moving outside the area to which they are adapted. The greater the divergence between the two populations, the stronger the barrier to gene flow presented by the cline between them, which in turn allows even more divergence until reproductive isolation becomes complete and the two populations become separate species.

9.8 THE EXTINCTION OF SPECIES

Taking the whole of evolutionary history, extinction of species has been only very slightly less common than speciation. There has been a constant turnover in species, with speciation and extinction taking place continually. The vast majority of all the species that have ever existed, therefore, have become extinct—even if there are as many as 30 million species living today, this is only a small proportion of the thousands of millions that have probably existed over the whole of evolutionary time.

Despite the prevalence of extinction, very little is known about its causes. A *population* will eventually die out if more individuals are lost by mortality and emigration than are gained by reproduction and immigration. What factors tend to lead to the extinction of a population? A population can die out as the result of simple chance fluctuations in birth and death rates. Small populations in particular have a high risk of dying out as a result of such 'demographic accidents'—if one year the birth rate happens to be low and the mortality rate high and/or if, by chance, many more males are born than females, then there is a strong possibility of the population becoming extinct, especially if it has similar 'bad luck' in the next breeding season. Small populations are also vulnerable to the dangers of inbreeding—this can lead for example to low fertility. Island populations are frequently small in number and often become extinct. Likewise, founder populations, being not only small in number but also in novel and often adverse conditions, are highly likely to die out. Extinction rather than speciation has probably been the fate of most peripheral isolates.

All other things being equal, such as rates of reproduction, the risk of dying out is higher for populations of species of short-lived individuals than for populations of species of long-lived individuals.

▶ From what you have read about *r*-selected and *K*-selected species in Chapter 8, why is it unlikely that all other things *will* be equal?

Longevity and intrinsic rate of increase are both related to size: larger organisms tend to be longer-lived and reproduce more slowly than smaller organisms. Reproduction rates are therefore unlikely to be the same for a species of short-lived organisms and a species of long-lived organisms. For the moment, however, assume that they *are* equal. Short-lived organisms, if their numbers are reduced either by chance or by a temporary change in the environment, have less time in which to recover than long-lived organisms. Consider two populations, one each of two imaginary species. Individuals of one species have a maximum lifespan of two years (species A) and those of the other have a maximum lifespan of ten years (species B). Otherwise, their rates of reproduction and of juvenile or seedling mortality are the same. One year, the populations crash—large numbers of individuals of both species die because of adverse weather conditions, and both populations are reduced to ten individuals. Over the next two years, either as a result of continuing adverse conditions or because of chance, the mortality rate among the offspring of both species is 100%. After that the situation improves and juvenile or seedling mortality decreases.

▶ How many individuals will be left in the population of species A?

None. No offspring will have survived during the past two years and, as individuals cannot live longer than two years, the ten individuals that survived the population crash will all be dead too. Some of the adults of the population of species B, however, will probably have survived (as their lifespan can be as much as ten years) and, when the situation improves, some of their offspring will also survive and numbers may increase again. The population of species B survives because individuals can 'wait out' the bad times while that of species A goes extinct because individuals do not have the time to do this.

All other things being equal, the risk of population extinction is also greater for species with a low intrinsic rate of increase (Chapter 8), i.e. those species where individuals produce only a few young, at infrequent intervals, and the young take a long time to grow and reach breeding condition themselves. A classic example is the elephant: the female does not breed until she is about ten years old and cannot produce offspring more frequently than once about every three years. Populations of such species can only recover very slowly if their numbers are drastically reduced and so remain longer at risk from demographic accidents. In contrast, populations of species such as greenfly, which are able to produce large numbers of quickly maturing offspring, can increase in numbers very rapidly.

Finally, the risk of dying out is higher for those populations whose environment varies greatly through time, leading to large fluctuations in numbers. Many insects, for example, show frequent population crashes and flushes.

▶ Why should fluctuating numbers mean that a population is more likely to die out?

The reason is that, at times of low numbers, populations face the risk of extinction because of demographic accidents.

To summarize, the populations most likely to die out are small ones, those whose members are short-lived and slowly reproducing, and those whose numbers vary greatly over time and which inhabit a fluctuating environment. Since longevity and intrinsic rate of increase are both related to body size, so that larger organisms tend to be longer-lived and to reproduce more slowly, these two factors tend to cancel each other out. Stuart Pimm and his colleagues (1988) have calculated, however, that for an extremely small population (fewer than seven breeding pairs), the risk of it dying out is greater for small, fast-growing, short-lived species than for large, slow-growing, long-lived ones. For larger populations, the reverse is true. This is an important consideration for conservationists and wildlife managers, concerned with saving populations of rare species.

What selective forces are likely to cause the numbers of a population to decline and thereby become vulnerable to extinction? Selection pressures can be due to any aspect of an organism's environment: physical or biotic. Even in a constant physical environment, competitors, predators, prey, parasites and disease-producing organisms can constitute powerful selective forces. A species may evolve to become more skilled at capturing its prey, for example, or able to run faster to better escape its predators. As each species evolves, this in itself creates a change in the biotic environment of other species (the Red Queen's hypothesis—see Chapters 5 and 8). So, a reduction in numbers in a population of a particular species is likely to be due to a disadvantageous change in its biotic or physical circumstances that takes place faster than the species can adapt to it. Human activities, for example (discussed in detail in Chapter 17), have led to some very rapid changes in the environment of many species and have probably been responsible for many extinctions as a result. Those populations that are able to respond quickly to new selection pressures by evolving rapidly are less likely to become extinct, but rapid evolution is not always possible. Species that reproduce asexually may not be able to evolve as quickly (in terms of amount of change per generation) as sexual species, because they lack the possibility of genetic recombination in each generation (Chapter 5). Developmental constraints may also prevent a species evolving in a particular direction, as will be described in Chapter 14. Remember too that the characteristics selected during evolution are not very little part, while others think the opposite. Again, this cannot be resolved on present evidence. We do not know enough, for example, about the strength of natural selection in wild populations. In fact, there is unlikely to be a clear-cut difference between adaptive radiation and founder-effect speciation. The two processes probably form the extremes of a continuum, with the relative importance of adaptation and chance varying, depending on the particular circumstances.

Certainly, chance plays some part in speciation, as when one population happens to become separated from the rest of its species by some extrinsic factor. That separated population could become a new species, join up with the rest of its species again if the extrinsic factor disappears, or it could become extinct. It is impossible to know how many incipient species have followed the last two courses, but extinction must have been a frequent occurrence, especially while the population was small and therefore particularly vulnerable.

Very little is known about the actual causes of extinction or about how species extinction relates to population extinction. Most extinctions are probably due to several factors acting together, not just to a single cause.

REFERENCES FOR CHAPTER 9

Coyne, J. A. and Orr, H. A. (1989) Two rules of speciation, in D. Otte, and J. A. Endler (eds), *Speciation and its Consequences,* Sinauer Associates, Sunderland, Massachusetts.

Dobzhansky, T., Ayala, F. J., Stebbins, G. L. and Valentine, J. W. (1977) *Evolution*, W. H. Freeman, San Francisco, California.

Endler, J. A. (1977) *Geographic Variation, Speciation and Clines*, Princeton University Press, Princeton, New Jersey.

Greenwood, P. H. (1974) *Cichlid fishes of Lake Victoria, East Africa: the biology and evolution of a species flock*, Trustees of the British Museum (Natural History), London.

Hewitt, G. M. and Barton, N. H. (1980) The structure and maintenance of hybrid zones as exemplified by *Podisma pedestris*, in R. L. Blackman, G. M. Hewitt and M. Ashburner (eds), *Insect Cytogenetics: Symposia of the Royal Entomological Society (London)*, **10**, pp.149–169, Blackwell Scientific Publications, Oxford.

Maynard Smith, J. (1966) Sympatric speciation, *American Naturalist*, **100**, 637–650.

Mayr, E. (1970) *Populations, Species and Evolution: An Abridgment of Animal Species and Evolution*, Belknap Press of Harvard University Press, Cambridge, Massachusetts.

Pimm, S. L., Jones, H. L. and Diamond, J. (1988) On the risk of extinction, *American Naturalist*, **132**, 757–785.

Rice, W. R. and Salt, G. W. (1988) Speciation via disruptive selection on habitat preference: experimental evidence, *American Naturalist*, **131**, 911–917.

FURTHER READING FOR CHAPTER 9

Carson, H. L. and Templeton, A. R. (1984) Genetic revolutions in relation to speciation phenomena: the founding of new populations, *Annual Review of Ecology and Systematics*, **15**, 97–131.

Giddings, L. V., Kaneshiro, K. Y. and Anderson, W. W. (eds) (1989) *Genetics, Speciation and the Founder Principle*, Oxford University Press, Oxford.

Grant, V. (1981) *Plant Speciation* (2nd edn), Columbia University Press, New York.

Hewitt, G. M. (1988) Hybrid zones—natural laboratories for evolutionary studies, *Trends in Ecology and Evolution*, **3**, 158–167.

Kaneshiro, K. Y. and Boake, C. R. B. (1987) Sexual selection and speciation: issues raised by Hawaiian drosophilids, *Trends in Ecology and Evolution*, **2**, 207–212.

Otte, D. and Endler, J. A. (eds) (1989) *Speciation and its Consequences*, Sinauer Associates Inc., Sunderland, Massachusetts.

SELF-ASSESSMENT QUESTIONS

SAQ 9.1 (*General Objective*) Define (a) speciation and (b) extinction.

SAQ 9.2 (*Objective 9.1*) What is the major advantage of the Recognition Species Concept (RSC) compared with the Biological Species Concept (BSC)? What are the major disadvantages that apply to both of them?

SAQ 9.3 (*Objective 9.2*) Which of the following are examples of pre-zygotic isolation mechanisms, post-zygotic isolation mechanisms, or neither? (a) Offspring of hybrids have reduced fertility when they come to breed themselves. (b) Snails of two species are physically unable to mate because their shells coil in opposite directions. (c) Birds of one species feed on nectar while those of another species feed on grass seeds. (d) Fish of one species live solely in salt water while those of another species live solely in freshwater.

SAQ 9.4 (*Objective 9.3*) What is the difference between a cline and a mosaic hybrid zone?

SAQ 9.5 (*Objective 9.4*) Which of the following are examples of allopatric, sympatric, parapatric speciation, or none of these? (a) A particular species of worm is found all over England and Scotland. Then the Scottish worms become adapted to cooler temperatures, with a short, thick, dark-coloured body, while the English worms become adapted to warmer temperatures, with a long, thin, light-coloured body. Offspring of matings between Scottish and English worms are adapted to neither climate and rarely survive to adulthood. (b) A particular bird species, previously found only in Britain, colonizes France. After a large number of generations, birds from France that disperse as far as Britain are unable to breed with British birds. (c) In a particular species of cuckoo, a breeding male will remain close to a particular nest built by a member of its host species. He will mate only with a female that approaches this nest, and she then lays her eggs in that nest. Individual cuckoos always choose a nest built by the same species as the one by which they were reared. One year, as the result of a series of extremely unusual circumstances, a mated female lays her eggs in the nest of a different host. All her offspring survive to breed themselves.

SAQ 9.6 (*Objectives 9.5 and 9.6*) What is the main requirement, common to all models of speciation, if divergence between two populations, leading ultimately to speciation, is to take place? Why does this requirement present difficulties for sympatric speciation?

SAQ 9.7 (*Objective 9.7*) Which of the following situations are likely to make a population of the species in question more liable to extinction, less liable to extinction, or to have no effect on the likelihood it will become extinct? (a) A bird eats any kind of seed or insect food. (b) A particular species of *Drosophila* feeds only on the sticky sap exuded by a

10.7 Carry out simple graphical and statistical analyses of morphometric data to discriminate groupings among, or to detect differences between, samples of fossils of similar morphology.

10.8 Analyse forms of simple *allometry* from *lines of relative growth* drawn on bivariate plots of age-dependent variates, and explain the implications of allometry for the comparison of morphometric data from different samples of fossil specimens.

10.9 Describe three palaeontological models of the evolution of species, based upon the 'classical' and the 'neo-Darwinian' versions of *phyletic gradualism*, and upon *punctuated equilibria*, respectively, and explain the differences between them.

10.10 Given suitable data, calculate rates of morphological evolution measured in darwins, and explain how such rates are related to the time-span considered.

10.11 Using relevant skills from Objectives 10.1–10.8 and 10.10, assess the implications of data on samples of fossils for the various models of species evolution listed in Objective 10.9.

10.1 THE FOSSIL RECORD

Biologists have to look at living populations at various stages of separation when trying to work out how speciation takes place, since the whole of that process is usually far too slow for continuous human observation. Although the models they produce, illustrated in the previous Chapter, highlight the various ecological, behavioural and genetic factors that may be involved, they usually have little to say about rates of evolutionary change. In some cases, estimates for the timing of evolutionary changes may be constrained by circumstantial evidence. In Section 9.4.4 of Chapter 9, for example, it was stated that the age of the sand bar now separating Lake Nabugabo from Lake Victoria, in East Africa, means that several species of the cichlid fish genus *Haplochromis* which evolved in Lake Nabugabo must have done so within the last 4 000 years. Yet the rates of their morphological evolution remain unknown. The species could conceivably have evolved in the first 500 years and then remained static thereafter. For an historical record of the longer-term changes in populations, it is necessary to consider the evidence of fossils. This Chapter explores the use of fossils in charting such microevolutionary histories.

10.1.1 FOSSILS AND FOSSILIZATION

Fossils are the original or altered remains of organisms (**body fossils**) or traces of their activity, such as burrows, footprints and droppings (**trace fossils**), which have been preserved usually by burial in sediment. Most body fossils are of only the resistant hard parts of organisms, such as shells, bones, teeth and scales. Some toughened, chemically resistant organic components, such as the woody tissues of land plants and the more robust parts of insect cuticles, are also common in some sedimentary deposits. Otherwise, soft (organic) tissues are only fossilized in exceptional circumstances (Chapter 12).

The original mineral components of the hard parts (e.g. calcium carbonate, $CaCO_3$, in most shells, and calcium hydroxyapatite, Ca_{10} $(PO_4)_6.2OH$, in bone) are sometimes preserved, either with their original texture, or recrystallized. Alternatively, water percolating through the surrounding porous rock may have dissolved the skeletal material away, leaving only hollow spaces bounded by surfaces of the rock matrix (**moulds**). The later filling of these cavities by crystalline mineral *cement* (like the 'fur' in a kettle), precipitated from the circulating water, can produce replicas, or **casts** of the original hard parts (Figure 10.1). Artificial casts can also be made by filling the moulds with some suitable substance, such as latex rubber.

Some people might not consider Recent remains, such as a snail shell dug up in the back garden, as real fossils, and archaeologists sometimes talk about these as *sub-fossils*. Yet it is not really useful to have an arbitrary minimum geological age for recognizing fossils. Fossilization may involve many overlapping processes, such as biological and chemical attack and physical transport, as well as those following burial in sediment (Chapter 12). These processes operate to different extents in different environments. Thus a shell buried in one setting may be completely altered in composition after only a few hundred years, while another, in a different geological setting, may retain its original form and composition for millions of years. To call the latter a fossil and the former a sub-fossil on some arbitrary age criterion would hardly seem compatible with their different states of preservation, and so no such distinction will be made in this Book.

A remarkably small proportion of organisms end up as fossils. Biological breakdown, through the activities of consumers and bacteria, as well as physical and chemical degradation of any remaining components, consign the vast majority of dead organisms to oblivion within a short time. Fossils preserved in sedimentary rocks are the residue of accidental exceptions to this fate. What sense can we hope to make of such an incomplete record?

Despite the general improbability of fossilization, some fossils may be very common. The remains of certain kinds of organisms, dwelling in particular habitats, are preferentially accumulated and the fossil record therefore shows characteristic patterns of bias. By understanding these patterns, the palaeontologist

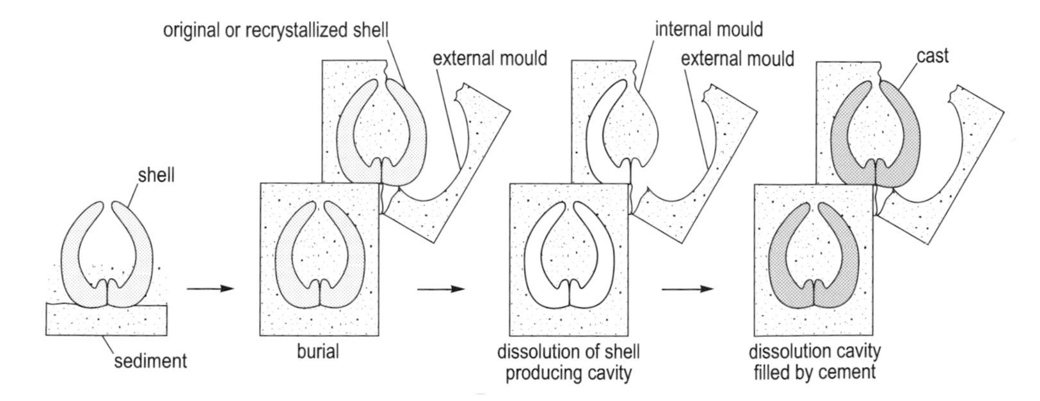

Figure 10.1 *Diagram illustrating the formation of moulds and casts. Bottom: the sequence of events that may follow burial of a shell (shown in section). Above: what is revealed if the rock is broken open at different stages.*

Figure 10.2 *A sample of fossil oysters (*Catinula*) collected from a single bed of Middle Jurassic shale in SW England, illustrating individual variability in a species.*

can prospect for suitable material for study in rather the same way that a petroleum geologist prospects for likely reservoirs of oil. Reconstructing microevolutionary history requires information on changing patterns of variability within populations. It is therefore desirable to have numerous and abundant samples of specimens. As the next Section will explain, the focus of such studies tends to be upon the readily fossilized hard parts of organisms which dwelt in seas, shallow oceanic areas and lakes (e.g. Figure 10.2).

Other kinds of fossils, including fossilized soft parts and trace fossils, usually have less to offer for microevolutionary investigations, the former because of their rarity, and the latter because of their relative lack of information on the detailed morphology of their makers. They are nevertheless of considerable importance for understanding patterns of macroevolution, and will therefore be surveyed in more detail in Chapter 12.

10.1.2 THE SEDIMENTARY RECORD

The fossil record can reveal only those patterns of change which took longer to unfold than the time represented by each layer of sediment sampled for fossils. It would be worthless, for example, to attempt to detect changes century by century if each sample of fossils took a thousand years to accumulate. When reconstructing microevolutionary history from samples of fossils, it is therefore crucial to know both the period of time over which all the individuals in any given sample lived, and the time that elapsed in the intervals between successive samples. To estimate such time-spans, it is necessary to consider patterns of sediment accumulation.

Sediment is ultimately derived from the weathering and erosion of exposed rock. The physically and chemically degraded products of weathering are transported from their sites of origin by a variety of agents (e.g. wind, flowing water and ice, and gravitational mass movement such as landslides), either as fragmental sedimentary grains, or in solution. Material in solution eventually re-assumes solid form as a result of precipitation, especially in organisms with mineralized hard parts (e.g. shells). The breakdown of the latter then yields further sedimentary grains.

Eventually the sediment is deposited somewhere in a specific environment, where successive layers, or **beds**, forming a **sedimentary sequence**, may incorporate the remains of organisms destined to become fossils. As the sediment accumulates, compaction and the growth of cement in the pore spaces between the grains hardens, or *lithifies,* the sediment, to form sedimentary rock. These processes constitute the sedimentary part of the *rock cycle* (Figure 10.3), whereby the materials of the Earth's crust are endlessly transformed from one state to another in response to the energy supplied by the Earth's hot interior and by the Sun.

Different environments of sediment deposition (e.g. river systems, deltas, shallow seas and ocean basins) tend to show characteristic types of sedimentary sequences. The make-up of these is controlled not only by the nature of the sediment brought into them, but by the conditions at the site of deposition. Thus the frequency and strength of currents, for example, determine both the variety of deposited grain sizes and the character of the sedimentary bedding (e.g. flat-bedded or cross-bedded). Any organisms present may both contribute skeletal grains to the

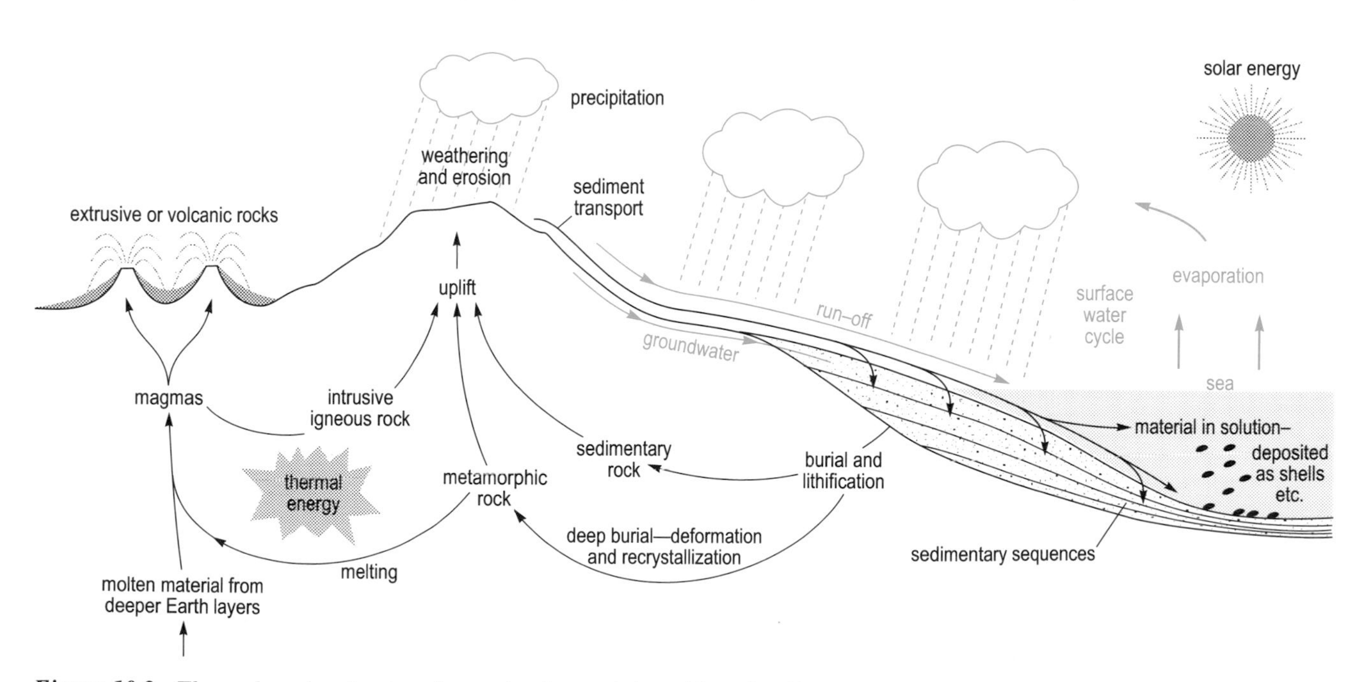

Figure 10.3 *The rock cycle, showing the production and deposition of sediments.*

sediment and affect its later history. Burrowing organisms, for example, churn the sediment (**bioturbation**). Sessile, surface-dwelling organisms, in contrast, may stabilize areas of sediment against subsequent erosive removal (e.g. Marram grass on coastal sand dunes, and colonial corals in reefs). Careful analysis of the composition and structure of sedimentary sequences allows the original depositional environment to be identified in most cases with some confidence. This is the concern of that branch of geology known as **sedimentology**.

Rates of sedimentation Nowhere is the deposition of sediment a continuous, let alone uniform process. Short-term fluctuations in the strength of winds, river flows, waves, tides and other transporters of sediment mean that these may deposit some of their load one moment, at any given spot, only to set it in motion again and carry it away later. Even in what might be thought of as relatively quiet settings such as in deep lakes and on ocean floors, there is similarly the coming and going of sediment. Scouring gravitational flows of dense, sediment-laden water and, in the oceans, wandering deep currents such as the northward flow of dense cold bottom water in the Atlantic, derived from the Antarctic icecap, may erode previously deposited sediment. 'Here today, gone tomorrow' aptly describes the short-term behaviour of most sediment.

Any sedimentary sequence is thus only the net accumulation of sediment deposition and removal. The sedimentary record is therefore characteristically incomplete. The longer the period of time considered, the more this is so. For, in addition to the ephemeral fluctuations mentioned above, sediment accumulation at any given point may, over a longer period, reach a limit, or *base level*, where erosion balances deposition, as on the crest of a sand bank or a delta margin. These hiatuses cause larger scale gaps in the long-term record. Ultimately, the only reason that any sequence is ever preserved at all is because space for its accumulation was created over an extended period of time. **Sedimentary basins** are depressions produced by relative subsidence of the Earth's crust on a local or regional scale, within which thick sedimentary sequences have accumulated. Alternatively, a rise in sea-level may have enlarged the space available for sediment accumulation on the continental margins by deepening the water above them. The episodic pattern of these vastly slower processes causes yet more interruption of the record in the long term.

The combined effects of these various factors result in a broadly inverse relationship between measured (*net*) rates of sediment accumulation and the time-span to which the measurements relate: the longer the period of time considered, the more incomplete the record, and so the lower the net rate of sedimentation.

This relationship and its importance for the interpretation of evolutionary lineages from fossils is best illustrated with a hypothetical example (Figure 10.4). Let us suppose that an unnaturally long-lived geologist has had the opportunity to study the pattern of sedimentation in a shallow sea, at a point initially offshore from an advancing delta, over a period of some 7 500 years. We will assume that waves (rather than tides or river flows) were the main agents of sediment distribution around the front of the delta. The overall sedimentary sequence built up during the study by the progressive advance of the deltaic sediments might look something like that shown diagrammatically on the left in Figure 10.4(a) on page 452. In the lower part of the sequence (i) is clay, deposited in deeper water, where wave currents would rarely have impinged on the sea-floor. Storms would occasionally have washed in sand and silt from shallower areas, to form isolated beds in the clayey sequence. With shallowing, as the sediment accumulated, (ii), the sea-floor would have become subject to more or less continuous wave action. Some of the plentiful influxes of sand would have accumulated, though there would also have been frequent erosion, and most clay and silt, in particular, would have been carried off in suspension (to be deposited farther offshore). Finally, as the shoreline itself reached the study area, an intense interplay of sand deposition and erosion would have left only a condensed unit, (iii), of delta crest sands at the top of the sequence.

This sequence would be but a budgetary residue from a long history of sediment deposition and erosion. Assume that our geologist, from the outset, had fixed a vertical ruler on the sea-floor (at 0 m) so that sediment accumulation and removal could be monitored. After each 100 years, the net rise or fall of the sediment surface at that point could be measured and the results plotted as a graph of sediment level against time. This graph might look something like that shown on the right in Figure 10.4(a) (assume, for simplicity, that the sediment undergoes no further compaction after each century of accumulation). Episodes of net sediment accumulation are shown as thick black vertical bars, net stillstands (where erosion balanced deposition) by thin horizontal lines, and episodes of net erosion by depressed lines, indicated by arrows.

The three units, (i)–(iii), show rather different patterns of sedimentation. The overall net rate of accumulation for each unit can readily be calculated by dividing the total thickness of that unit by the time elapsed during its deposition. Values for net rates of accumulation are usually expressed in a standardized form, as metres (m) per 1 000 (10^3) years.

► Calculate the net rate of sediment accumulation for each unit, listing these in declining order.

Unit (ii) shows the highest net rate, with 12 m accumulating over 1 500 years (= 8 m/10^3 years). Next comes Unit (i), with 12 m accumulating over 3 000 years (= 4 m/10^3 years). Finally comes Unit (iii) with only 2 m in 3 000 years (= 0.67 m/10^3 years).

When one looks at rates of sediment accumulation over shorter intervals, however, a different picture emerges. We are interested here only in those 100-year intervals during which there was positive net sediment accumulation, because, by definition, only these are recorded in the sequence. The heights of the black bars show how much sediment accumulated for each of these.

► Considering *only* the episodes of positive net sediment accumulation, list the three units in declining order of mean 100-year rates of accumulation.

Unit (ii) again shows the highest rates of sediment accumulation, since the black bars here are, on average, taller than those of the other two units. Unit (iii) comes next, followed by Unit (i), in which the 100-year rates are all relatively small.

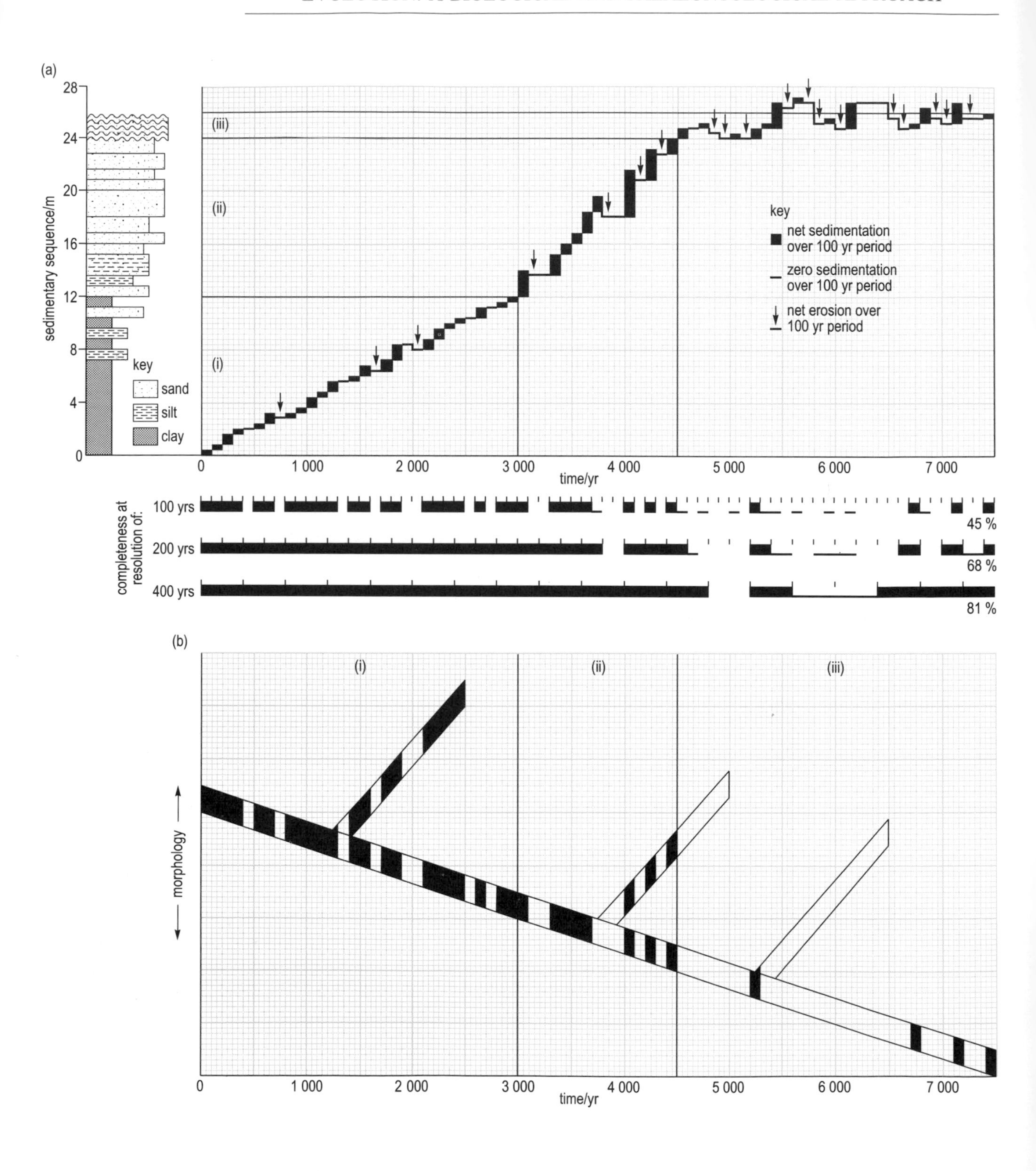
(a)
sedimentary sequence/m
key
sand
silt
clay
(i)
(ii)
(iii)
key
net sedimentation over 100 yr period
zero sedimentation over 100 yr period
net erosion over 100 yr period
time/yr
completeness at resolution of:
100 yrs
200 yrs
400 yrs
45 %
68 %
81 %
(b)
morphology
time/yr

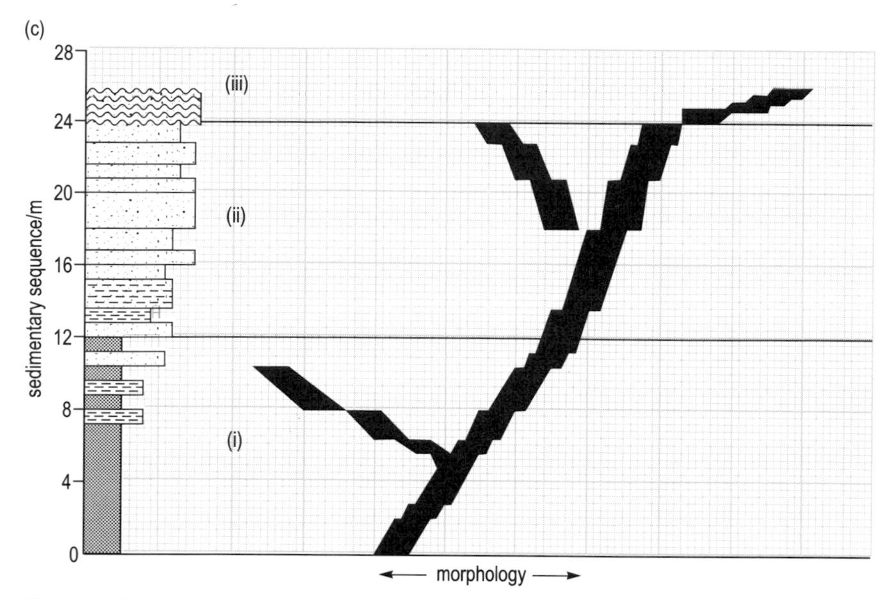

Figure 10.4 *A hypothetical sedimentary sequence and the effect of its deposition upon the fossil record of evolving populations. (a) Left: the sequence with beds projecting according to their grain sizes (see key). Right: net sediment gain or loss in successive 100-year intervals. (b) Change in the morphology of some evolving populations during the same period of time. (c) The same sequence as in (a), with the preserved fossil record of the evolving populations in (b) shown alongside. See text for further details.*

In order to compare these shorter-term rates with the overall net rates for each of the units, they should again be standardized to metres/10^3 years. Thus, from Figure 10.4(a), the maximum 100-year rate in Unit (ii) (that immediately after 4000 years) is 3.6 m/100 years = 36 m/10^3 years. Likewise, that for Unit (iii) is 2 m/100 years = 20 m/10^3 years and that for Unit (i), 1.2 m/100 years = 12 m/10^3 years. Note that all these values are markedly higher than the overall net rates for the units, calculated earlier. This is because the latter also incorporate those 100-year episodes when sediment accumulation was nil, or even negative. Had observations been made over yet shorter intervals than 100 years the pattern would have been repeated: positive net rates of accumulation measured over, say, yearly intervals, would tend, again, to have much greater values than the net rates for the 100-year intervals. So the positive net rates of sediment accumulation are inversely related to time-span. Moreover, the nature of this inverse relationship clearly varies according to the environment of deposition. The difference between the shorter and longer term rates is least for Unit (i), in which intervals of stillstand or erosion were infrequent, and greatest for Unit (iii), where erosion almost counterbalanced deposition, overall.

A major consequence of this variation in the difference between shorter and longer term rates is that the completeness of any given sequence, in terms of its record of successive time intervals, will vary according to the environment of deposition. This is illustrated along the bottom of the graph in Figure 10.4(a). Here, intervals of 100 years, 200 years and 400 years, respectively, have been blacked-in where a sedimentary record of them remains. Absence of a record for any given interval may be due either to subsequent erosion (shown by a thin horizontal line) or to lack of net positive accumulation during the interval in question (shown by a blank). Completeness can simply be calculated as the percentage of intervals which are blacked-in (shown at right). Two important points should be noticed. First, the completeness depends upon the length of the time interval chosen as a limit of resolution. Longer intervals give higher values of completeness, because they become more and more inclusive of episodes of deposition. Thus, whereas only 45% of the 100-year intervals have a sedimentary record in the sequence as a whole, 68% of 200-year intervals have a record, and 81% of 400-year intervals have a record. An interval as long as the period of deposition of the sequence itself of course has a completeness of 100%. Secondly, as noted above, completeness clearly varies markedly between the three units, especially at shorter limits of resolution.

▶ Calculate the percentage of completeness for each of the three units, (i)–(iii), at the 100-year limit of resolution.

In Unit (i), 22 out of 30 one hundred year intervals are blacked in, giving 73% completeness. For Unit (ii) the value is $8/15 \times 100 = 53\%$, and for Unit (iii), $4/30 \times 100 = 13\%$. Clearly, any historical patterns of change which might be captured in the sedimentary record (such as evolutionary change recorded by fossils) have a far greater probability of being captured in Unit (i) than in, say, Unit (iii).

Sedimentation and the fossil record What effects do these patterns of sedimentation have on the fossil record of evolving populations? In Figure 10.4(b), a hypothetical evolutionary tree is shown. The *y* axis represents, in a generalized way, the morphology of individuals in the evolving populations, and the *x* axis, time, as in Figure 10.4(a). One major population is shown as evolving throughout the entire period with a consistent morphological trend, while three new populations successively branch off from it, each evolving for a time with a contrary trend, but then becoming extinct. Let us assume that these were abundant organisms, preserved as fossils in all the sediments in the sequence (perhaps some shelly free-swimming organism). As above, the 100-year intervals for which sedimentary (and therefore fossil) records have been preserved are blacked-in. The disparity of completeness of preservation of the evolutionary history in the three units is again evident, with Unit (i) preserving quite a good record, but Unit (iii) a very poor one. Indeed, no record whatever of the third population to branch off from the main one is preserved in Unit (iii); its entire history has simply been lost in a gap in the sequence.

Figure 10.4(c) translates the recorded parts of the evolutionary history in Figure 10.4(b) back alongside the original sequence (at left), and thus shows the morphological changes seen in the fossils, passing up the sequence. This translation produces some intriguing effects. The third daughter population, as noted above, is entirely missing. The second daughter population, again because of a sedimentary break, shows a much more abrupt pattern of separation from the main population than does the first daughter population. But these gross patterns are not the only effects seen. The rates of morphological change in the populations through *time* are constant, as indicated by the straight slopes in Figure 10.4(b). Yet the rates of change *per unit thickness of rock*, as in (c), are decidedly variable. The abrupt sideways shifts are readily attributable to the gaps in the sequence. As one might expect these are most pronounced in Unit (iii), with the later part of the main lineage becoming chopped into a succession of fossil assemblages with widely differing morphologies. However, another, more subtle effect is also apparent: within the continuous segments of the sequence (i.e. those without any gaps of 100 years or more), the rate of morphological change per unit thickness of rock is also variable (as indicated by the changes in slope of the segments in Figure 10.4(c)).

▶ Explain why the rate of *continuous* morphological change in the main population per unit thickness of rock is less in Unit (ii) (say, between 13.6 m and 18 m) than it is in Unit (i) (say, between 8 m and 10.4 m).

The difference in slope is due to the difference in the short-term positive net rates of accumulation between the two units. Where sediment accumulated continuously over several centuries in Unit (ii), it did so relatively rapidly, and so the fossil record of the evolving population became stretched out over a sizeable thickness of sequence. In Unit (i), in contrast, short-term rates of accumulation were slow, and so the record was squeezed into a thinner part of the sequence. Thus the 4.4 m of rock between 13.6 m and 18 m in Unit (ii), and the 2.4 m of rock between 8 m and 10.4 m in Unit (i) both accumulated over periods of 400 years. Since the morphological change through time was constant, this means that the change per unit of rock thickness (in continuous segments) was less in Unit (ii) than in Unit (i).

Analysing the record So far we have been assuming the truth of the history of sedimentation shown in the graph in Figure 10.4(a), and of the evolutionary history in Figure 10.4(b), because of the convenient fiction of our unnaturally long-lived geologist who was able to observe these things. Sadly, not even the healthy pursuit of field geology confers longevity on this scale and so we must now consider what we could realistically expect to know of this hypothetical example. The total information that would be available to us would be that shown in Figure 10.4(c)—just the sedimentary sequence itself and the sequence of fossil morphologies shown alongside. In the majority of cases it would not be possible to calibrate the sequence accurately with an absolute time-scale (say, by radiometric dating methods) because the errors on the dates calculated for successive beds might well even exceed the entire 7500 years of the sequence. (Exceptions to this problem will be referred to in Section 10.3.2). So the representation of time in the sequence must be gauged by other means. From the earlier discussion, it is evident that it would be misleading simply to assume deposition at a constant rate. That would lead us (incorrectly) to interpret the pattern of morphological change shown in Figure 10.4(c) literally, as a pattern of change through time. One way to gauge, if only approximately, how continuously time might be represented through such a sequence is to estimate the probable depositional completeness of Units (i)–(iii). For each unit this could be done by comparing an assumed value for the short-term rate of accumulation of sediment, derived from equivalent environments today, with its overall net rate of accumulation, as explained below.

Figure 10.5 shows how the mean values of measured rates of sediment accumulation vary with time-span considered, for several different types of depositional environment. The graph is based on a large number of measurements of net rates of sedimentation, ranging from those of hourly or daily accumulation to those calculated by dividing the thicknesses of sequences by their time-spans derived from various dating methods. Although the raw data (not shown here) inevitably have some scatter, the average trend lines shown in the Figure provide useful approximations for estimating completeness.

Steeper slopes on Figure 10.5 are associated with the deposits of more energetic, or frequently disturbed environments, such as rivers, coastal wetlands and shallow carbonate platforms and reefs.

▶ Think back to the factors affecting net rates of sediment accumulation and decide why steeper slopes characterize the data for these environments.

Because these environments are more frequently subjected to fast flowing currents, they tend to experience both rapid influxes of sediment (giving high short-term rates of sediment accumulation) *and* greater erosive activity (lowering net rates of accumulation in the longer term). Hence net rates decline more sharply with increasing time-span. This was the situation with Units (ii) and (iii) in Figure 10.4.

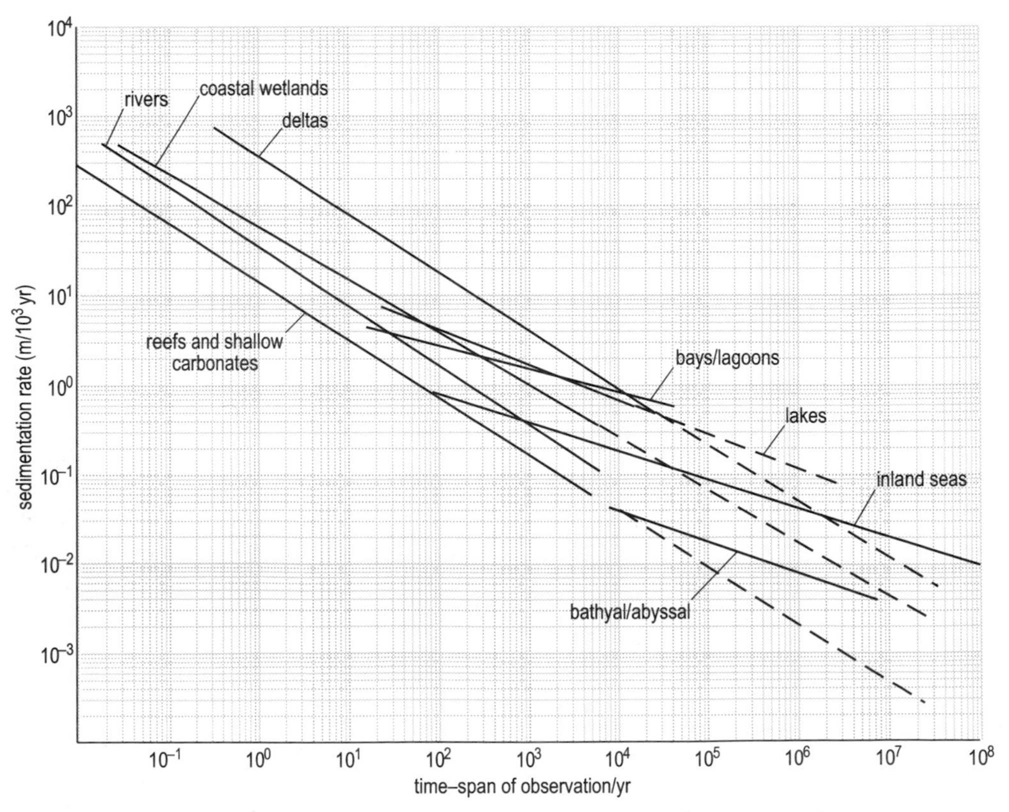

Figure 10.5 *Summary trend lines for the deposits of different sedimentary environments on a plot of net sediment accumulation rate vs. time-span considered. Note that axes are logarithmic.*

Gentler slopes, by contrast, are associated with the deposits of generally quieter settings, such as lakes, small basinal seas and abyssal areas. However, some secondary steepening of their trends is introduced by sediment compaction, which is particularly pronounced in these generally muddier sediments. The compaction has the effect of reducing the measured rate of sediment accumulation as the sediment is squashed under greater thicknesses of sediment.

Using these data, it is possible to estimate how time is represented in various thicknesses of a sedimentary sequence and so tackle the sorts of question raised at the beginning of this Section concerning the likelihood of being able to resolve patterns of evolutionary change. This is termed **resolution analysis** (Schindel, 1982). Three factors need to be determined:

1 **Temporal scope** (T) is the total time elapsed during deposition of a sequence of thickness X, based on the dating of its upper and lower limits.

2 **Microstratigraphical acuity** (m) is an estimate of the time elapsed during the deposition of an individual sedimentary interval in the sequence. This may be an interval sampled for fossil specimens, or one separating different sampling intervals. It is important that there are no obvious sedimentary discontinuities within the interval considered, since these will correspond to a period of non-deposition or erosion of unknown duration. Two estimates may be made for each interval. The *longer* estimate (*minimum acuity*, m_{min}) assumes that deposition has always been continuous and uniform, so that the interval accumulated at the net rate for the sequence. It is calculated by dividing the interval thickness (i) by the net rate of sedimentation for the sequence:

$$m_{min} = \frac{iT}{X} \qquad (10.1)$$

The value of m_{min} for any 1 m interval of Unit (ii) of Figure 10.4, for example, would be

$$\frac{1\,\text{m} \times 1\,500 \text{ years}}{12\,\text{m}} = 125 \text{ years}$$

The *shorter* estimate (*maximum acuity*, m_{max}) is the amount of time, on average, it would have taken such a thickness to have accumulated in an equivalent environment today, at some shorter-term rate of accumulation. The latter depends on the limit of time resolution arbitrarily selected for the investigation. In this calculation it is assumed that there are gaps in other parts of the sequence, but that the interval in question accumulated continuously, at least to the selected limit of time resolution. As noted earlier, no deposition is entirely continuous, so continuous deposition at a limit of resolution of, say, 100 years would mean that there had been some net accumulation for each 100 years in the interval; any shorter-term gaps are ignored. The limit of resolution chosen should be relatively short in relation to the time-span of the evolutionary pattern to be investigated, so that the ignored gaps below the limit of resolution are trivial compared with the evolutionary time-span. The short-term net rate of accumulation (s) used in this calculation is read off from the line corresponding to the equivalent environment in Figure 10.5, at the time-span value selected as the limit of resolution. Thus, for example, if we were dealing with a reef sequence, and chose a limit of resolution of 100 ($=10^2$) years, then the value of s would be about 0.7 (just under$10^0=1$)m/ 1 000 years. The estimate, m_{max}, is calculated by first multiplying the interval thickness by a suitable factor (c) to allow for it having been shortened by sediment compaction, and then dividing the result by the short-term net rate of sedimentation obtained above:

$$m_{max} = \frac{ic}{s} \tag{10.2}$$

The value of c depends upon the type of sediment. Muddy sediments tend to undergo much greater compaction than sandy sediments. This is because the initially randomly orientated clay flakes in the mud tend to flatten down as more sediment accumulates. The more equidimensional grains of sandy sediments, in contrast, merely become somewhat more closely packed. A value of c for any given sedimentary rock can be estimated by studying the relative flattening of the less robust fossils (or other objects of known original shape), or that of the sedimentary laminations around solid objects (Figure 10.6).

▶ Estimate c in Figure 10.6(a).

In Figure 10.6(a), $c = l_0/l_1 = 3$. Values of c up to about 2 are common in sandy sediments, while those for muddy sediments may run up to about 10.

Calculation of m_{max} for a 1 m interval in Unit (ii) of Figure 10.4 would be as follows: c is 1 in this hypothetical example, since it is assumed that there is no

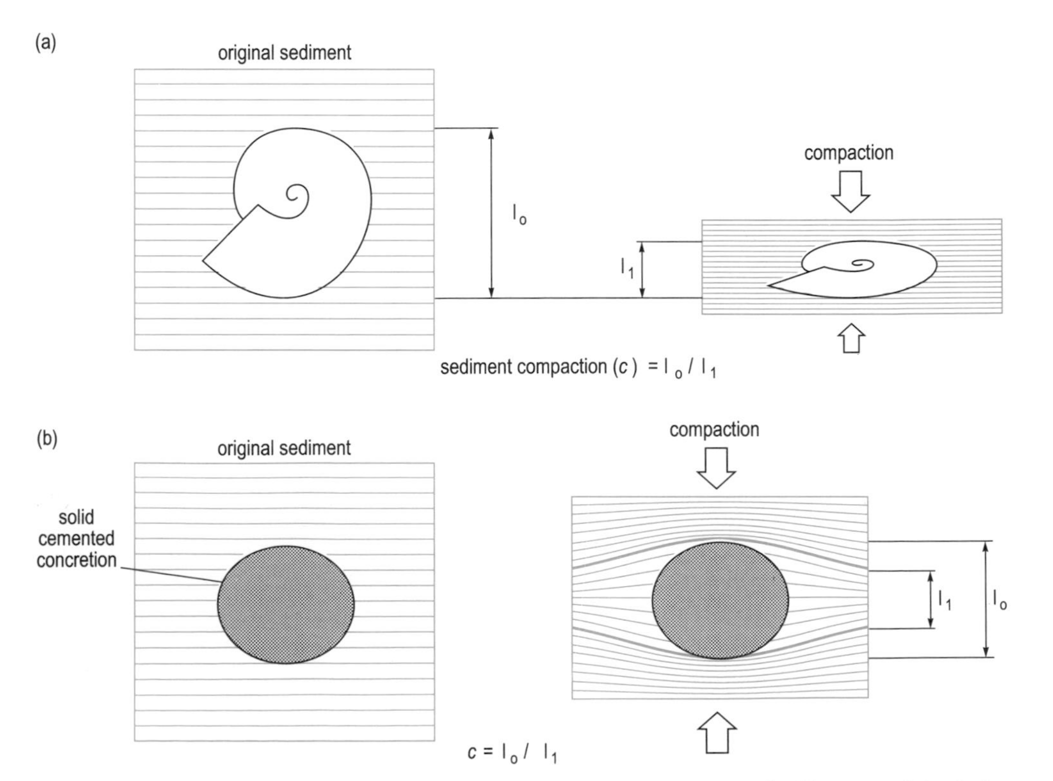

Figure 10.6 *Estimating sediment compaction from (a) a deformed fossil of known original shape; and (b) from the deformation of sedimentary laminations around a concretion formed by the growth of cement in the uncompacted sediment.*

further compaction after each century of deposition. If a limit of time resolution of 100 years is selected, then $s \doteq 17$ (just under 2×10^1) m/10^3 years, from the line for deltaic sediments in Figure 10.5. Hence:

$$m_{max} = \frac{1\,\text{m}\times 1}{17\,\text{m}/10^3\ \text{years}} = 58.8\ \text{years}$$

3 **Stratigraphical completeness** (*S.C.*) is the proportion of the temporal scope which is physically represented by strata, as opposed to that lost in discontinuities. It is expressed as a percentage. It therefore represents the amount of time it actually took for the sediment that *is* present to be deposited, as a proportion of the temporal scope of the sequence. Again, it is given to a stated limit of time resolution, since, for example, a sequence would have a greater completeness to a resolution of 1 000 years than to that of 100 years, as noted earlier.

$$S.C. = \frac{Xc}{sT}\times 100\% \qquad (10.3)$$

Thus for Unit (ii) of Figure 10.4, again at a limit of resolution of 100 years,

$$S.C. = \frac{12\,\text{m}\times 1}{(17\,\text{m}/10^3\,\text{years})\times 1\,500\,\text{years}}\times 100\% = 47\%$$

Note that this estimate approximates to the *real* percentage completeness for this Unit (53%), calculated earlier from Figure 10.4(a). (With unusually complete sequences, the *estimate* of *S.C.* may exceed 100%. In such instances *S.C.* is treated as being 100%.)

Together, these estimates can give the palaeontologist some idea of the time represented by given intervals in a sequence, and over what kind of time-scale any observed evolutionary changes have occurred.

To illustrate the use of resolution analysis, let us suppose that a 25 metre sequence of sandy lake sediments has been dated and shown to have been laid down over a period of 100 000 years. The net rate of accumulation for the sequence ($X/T = 25$ m/10^5 years $= 0.25$ m/10^3 years) can be seen, from Figure 10.5, to be close to that expected for lake deposits with a temporal scope of 10^5 years. Now let us further suppose that fossils of a species of pond snail are common throughout the sequence, so a palaeontologist decides to sample these from the lower 25 cm of each successive metre (i.e. the top of each 25 cm sampling interval is separated by 75 cm from the base of the next one up).

▶ What are the longer estimates of microstratigraphical acuity (m_{min}) for each of the sampling intervals and the intervening intervals?

Using Equation 10.1, the results are:

$$m_{min}\ \text{(sampling interval)} = \frac{0.25\,\text{m}\times 100\,000\ \text{years}}{25\,\text{m}} = 1\,000\ \text{years}$$

$$m_{min}\ \text{(intervening interval)} = \frac{0.75\,\text{m}\times 100\,000\ \text{years}}{25\,\text{m}} = 3\,000\ \text{years}$$

10.2.1 Species concepts in palaeontology

There are both practical and logical reasons why the biological species concept and others based on reproductive recognition are problematical for palaeontologists. On the practical front, it is obvious that one cannot directly test for reproductive isolation between co-occurring populations which are now represented only by fossils, although this may sometimes be inferred by indirect means. In theory, the *recognition species concept* (Chapter 9, Section 9.2.3) might be tested by study of the genitalia of those species in which internal fertilization is effected with complex copulatory structures (as in many insects, for example). But such structures were either originally absent, or have failed to be preserved in the great majority of fossil organisms. Palaeontological species therefore have to be recognized almost entirely using the criterion of morphological similarity. The relationship between these *morphospecies* and biological species is often unclear. In fact the problem is by no means limited to fossils, since, as noted in Chapter 9, Section 9.2.4, the vast majority of known species of living organisms are themselves only founded upon the description of morphological characters of collected and preserved specimens. As one commentator noted, 'They might as well be fossils!'.

▶ How might some collections of morphologically similar individuals contain two or more biological species?

The genetic analysis of living populations has increasingly uncovered groups of *sibling species* (Chapter 9, Section 9.2.4), which, though genetically distinct, show little or no mutual morphological divergence. The detection of sibling species amongst fossil populations is probably one of the most intractable problems of evolutionary palaeontology: most, it seems, are doomed to remain clumped together in morphologically indivisible fossil 'species'.

In order to investigate the extent of this problem in a group of living organisms with an abundant fossil record, Jackson and Cheetham (1990) carried out a series of breeding experiments and protein analyses of several morphologically recognized species of bryozoans. These are colonial marine organisms which produce mat-like, or upright skeletal structures on the sea-floor or on the surfaces of other organisms (Figure 8.1 in Chapter 8). Each individual in the colony occupies a tiny box-like compartment, from which it filter-feeds with a protrusible feeding organ furnished with tentacles. The skeletal compartments, which in many species are readily fossilized, usually possess numerous intricate morphological features by which the species have been recognized. Jackson and Cheetham found that the *genetic distances* (Chapter 3, Section 3.5.2) between the morphologically defined species were consistently much greater than those between populations of the same morphological species. The morphological characters used in recognizing the species were also found to be highly heritable. Moreover, the different local populations of the same morphological species were found to have very similar allele frequencies, with no allele unique to any one population. Jackson and Cheetham therefore concluded that in these organisms, at least, there is a close correspondence between the species recognized on morphological grounds and the biological species inferred from genetic and breeding criteria.

On the other hand, sibling species have certainly been detected among what were previously considered widespread single species, as for example in marine mussels, fiddler crabs and even some other bryozoans with unmineralized skeletons. Much further work therefore needs to be done on the genetics of living wild populations to establish the frequencies of sibling species amongst different groups of organisms, so that the reliability of species recognized from fossils can be assessed in biological terms.

A more important problem, however, is the logical inapplicability of the biological species concept in the time dimension. The reproductive criterion makes sense only when applied to contemporaneous populations. We cannot mate with our medieval ancestors simply because they have long been dead. It would clearly be silly to propose that we and they should therefore be considered separate species. This might be countered by the proposition that had we and they lived together there would probably have been no reproductive barrier. This may well be true but the hypothesis is untestable. Only in the special case of allopolyploid hybrid species (Chapter 9, Section 9.6.3) will successive generations become divided from one another by genetic incompatibility. Another criterion is therefore needed to draw a line between ancestral and descendent species in other cases.

Let us suppose that large collections of fossils have been made from each of six successive sampling intervals in a sedimentary sequence (Figure 10.7). In this hypothetical example, the collection in the first interval falls within one (morpho) species population, P_1. In each of intervals 2 to 4, however, the collections naturally fall into two groupings, taken to represent two distinct descendent lineages, P_{2A}–P_{3A}–P_{4A} and P_{2B}–P_{3B}–P_{4B}. The first lineage continues in the last two intervals (P_{5A} and P_{6A}), but the second shows a further subdivision, into P_{5B}–P_{6B} and P_{5C}–P_{6C}. Since the discrete populations in each interval were co-occurring with no intermediate forms, we will assume that they were distinct biological species.

Two alternative criteria for separating ancestral from descendent species can be applied to Figure 10.7. One criterion is to use the branching points to mark the boundaries between ancestral and descendent species, and the other is to use considerable morphological change as the key.

▶ How might the various populations, P_1, P_{2A}, P_{2B}, etc. in Figure 10.7 be grouped into species, using each of the criteria given above?

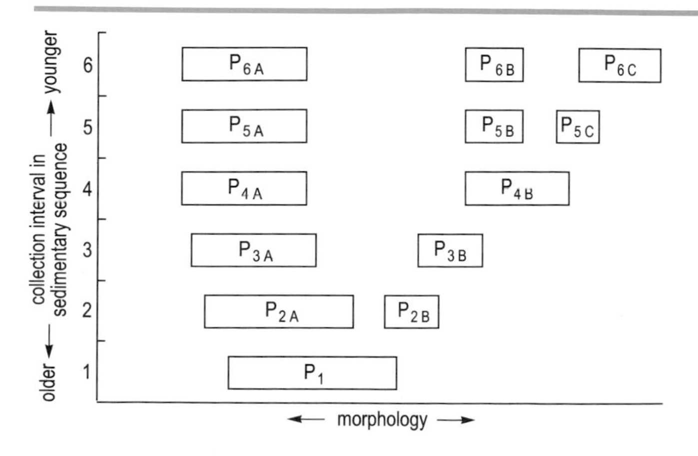

Figure 10.7 Hypothetical samples of fossils from six successive sampling intervals in a sedimentary sequence. Range of overall morphological variation of each sample is indicated by width of rectangle.

Adopting the criterion of the branching points would yield five species, I–V (Figure 10.8(a)), where P_1 is species I, P_{2A} to P_{6A} comprise species II, P_{2B} to P_{4B} make up species III, P_{5B} to P_{6B} form species IV, and P_{5C} to P_{6C} form species V. This corresponds to the *evolutionary species concept*, introduced in Section 9.2.1 of Chapter 9, since each species recognized here comprises an evolutionary lineage of individuals sharing a 'common evolutionary fate through time'. The ancestor/descendent boundaries are drawn each time populations diverge to create separate lineages with independent histories.

According to the second criterion, morphological change, populations P_{2A}–P_{6A} change little from the ancestral condition in P_1, and so P_1–P_{6A} could be placed in one species, α. P_{2B} then marks the inception of a distinct species, β (which, of course, must be separate from P_{2A}). But then, so great is the change from P_{2B} to P_{4B} that there is no overlap between these two. *If they had been contemporaneous populations, we would have felt bound to recognize them as distinct species.* So a boundary should be placed somewhere between P_{2B} and P_{4B} (either above or below P_{3B}), separating species β from γ. Such divisions within a single lineage are called **chronospecies** (Chapter 9, Section 9.1), each successive one entailing the arbitrary termination, or **pseudoextinction**, of its predecessor. Thus the chronospecies γ is here shown to commence between levels 3 and 4, meaning that chronospecies β becomes pseudoextinct there (Figure 10.8(b)). P_{5B} and P_{6B} substantially overlap morphologically with P_{4B}, and so can also be included in species γ (this is analogous to the union of the P_{2A} lineage and P_1 in species α). P_{5C} then marks the inception of another species, δ.

Figure 10.7 shows morphological variation along a single axis for the sake of clarity. In using morphology to separate chronospecies, it is advisable in practice to consider as many traits as possible, but particularly avoiding those subject to environmentally induced developmental variation. Single features may change inconsistently among contemporaneous populations of the same species, and so be unreliable for recognizing the same chronospecies in different sequences. As the genetic controls on most morphological variations are poorly understood, use of a combination of many characters reduces the risk of error from environmentally induced phenotypic variations.

The allocation of the populations among species dictated by the two methods is thus very different. Indeed, even the numbers of species recognized may differ, as in Figure 10.8(a) and (b).

In theory, at least, the branching criterion would seem to be preferable, because it appears objective and consistent. The morphological criterion, with its uncertain genetic significance and arbitrary separation of chronospecies, at first seems less precise. In practice, however, it is the latter which yields a scheme less prone to alteration following new fossil discoveries. This is because of the incompleteness of the fossil record. It is often hard, if not impossible, to decide exactly when related species have split from each other, and new branching points might also have to be accommodated with the discovery of new fossil species. Any system of recognizing species relying solely upon the branching criterion is thus likely to need the constant renaming of known fossil populations, as ideas about the numbers and timing of branching points change. Morphological characterization of

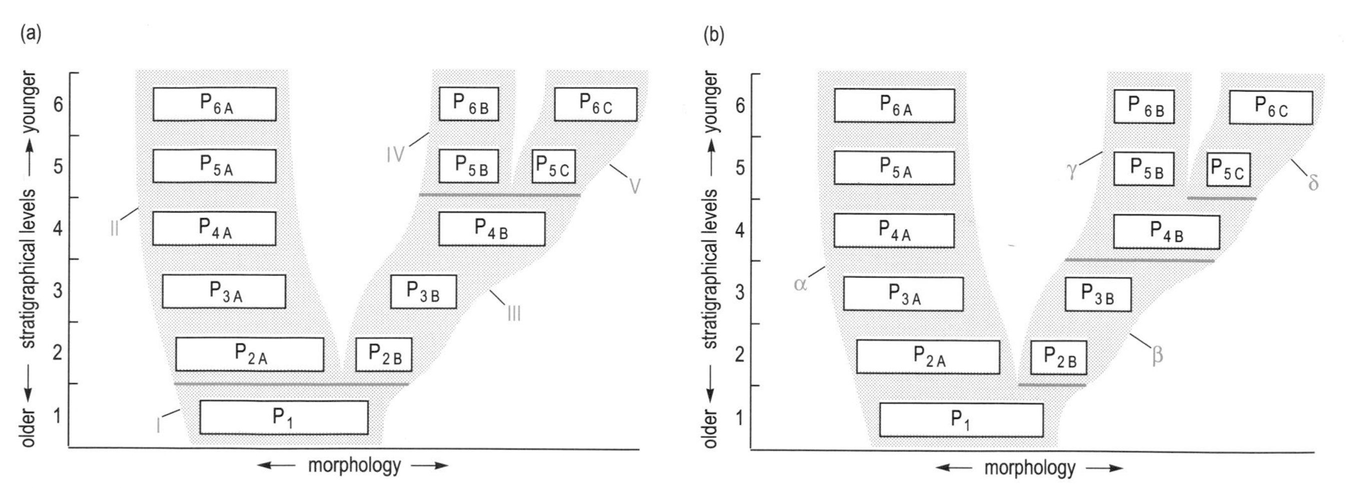

Figure 10.8 *Ways of dividing Figure 10.7 into species: (a) according to branching events; (b) according to morphological change (defining chronospecies).*

the species, in contrast, should simply become more secure as further information accrues. Morphology is, after all, what can be seen; branching points are merely inferred.

Ultimately, therefore, the palaeontologist is forced, by practical considerations, to delimit fossil species in time, as well as in space, on the basis of morphology.

However, besides the problem of sibling species discussed earlier, there are several other pitfalls to be guarded against in using morphology to recognize species, as the next Section will illustrate.

10.2.2 Morphological variation and palaeontological species

Individuals may differ because of genetic differences, or because they are at different stages of development, or because of the effects on their development of differences in the environment (Chapter 3, Section 3.2.3).

Ecophenotypy A serious problem in the use of fossils for microevolutionary studies is the difficulty of distinguishing the genetic from the non-genetic components of variation between individuals. Phenotypic divergence of local populations due solely to environmentally induced developmental changes (yielding distinct **ecophenotypes**), for example, might be erroneously interpreted as representing speciation. Even cuttings from a single house-plant, which are likely to be genetically identical, may grow leaves somewhat different in shape, size or colour according to how much light, water and nutrients they get. If certain morphological types of fossil are found to be consistently associated with sedimentary deposits which indicate particular environmental conditions, and these occurrences are widely separated from one another both in space and time, then it may be suspected that they represent local ecophenotypes of some more widely distributed species. Alternatively, environmentally induced developmental variations in living relatives might provide clues for recognizing ecophenotypy in fossils.

Smith and Paul (1985) interpreted a stratigraphical sequence of changes in shape in the small echinoid, *Discoides subucula*, in the Upper Cretaceous of SW England, as having been more probably ecophenotypic than evolutionary. They found that the ratio of height to diameter of the echinoid shell, or *test*, showed a pattern first of decrease (i.e. flattening of the test), and then of increase, in successive intervals in the sequence studied by them (Figure 10.9). This pattern was closely matched by changes in the mud-grade (silt and clay) content of the sandy to chalky sediments of the sequence (at left on Figure 10.9): the higher the mud content, the more steeply conical were the echinoid tests. By analogy with living echinoids of this shape, it is probable that this species burrowed into the sediment and kept a narrow space open above the test in which water could circulate. A more steeply sloping top to the test would have facilitated the clearance of any mud-grade sediment that collapsed from the burrow roof, so keeping the zone of water circulation clear. Smith and Paul argued that the changes in test shape through the sequence were more probably ecophenotypic (in this case produced as a developmental adaptive response) in origin, than due to genetic differences between the populations. Similar variations in the test, correlated with habitat differences, can be seen in living burrowing echinoids. Since these have widely dispersed, free-swimming larvae, which may settle on a wide range of sediment types, such variations are more likely to be due to adaptive developmental plasticity in individuals (Chapter 2, Section 2.3.1) than to fixed genetic differences between individuals.

Such arguments are at best only suggestive, not conclusive. Further studies (on the actual genetic variability of the living analogues and on the relative ability of individuals of different shape to clear mud) would be desirable. Unfortunately, it is often the case that appropriate background information on the living analogues is not available. Moreover, many fossil groups quite simply lack living relatives showing similar kinds of variation, which might serve as analogues. The distinction between ecophenotypic and genetic variability in many fossil organisms therefore remains unclear.

Deficiencies in sampling Another false trail that may lead to the erection of spurious species is insufficient sampling. In order that fossil species may be characterized as completely as possible, their description should be based on large numbers of specimens. Only then can the natural variability of the original species be recognized. Unfortunately, however, preservational or logistical constraints frequently permit the collection only of small samples. Consequently, mere variants from what is in fact a morphological continuum may appear to stand out on their own through random sampling, and so be mistakenly separated as distinct species. Refer back to the assemblage of fossil oysters shown in Figure 10.2, and imagine taking just four specimens at random. This could result in your having some smooth shells and some completely ribbed shells. Confronted just with these specimens alone, one might be tempted to judge them, erroneously, to belong to different species. Subjective judgements of this sort, usually based on very small samples, plague much of the older palaeontological and, indeed, also biological taxonomic literature, leaving us with a still awesome legacy of superfluous and biologically meaningless species names. These clearly have to be 'mopped up' into more objectively identified morphological groupings before any serious study of microevolution in fossil species populations can commence.

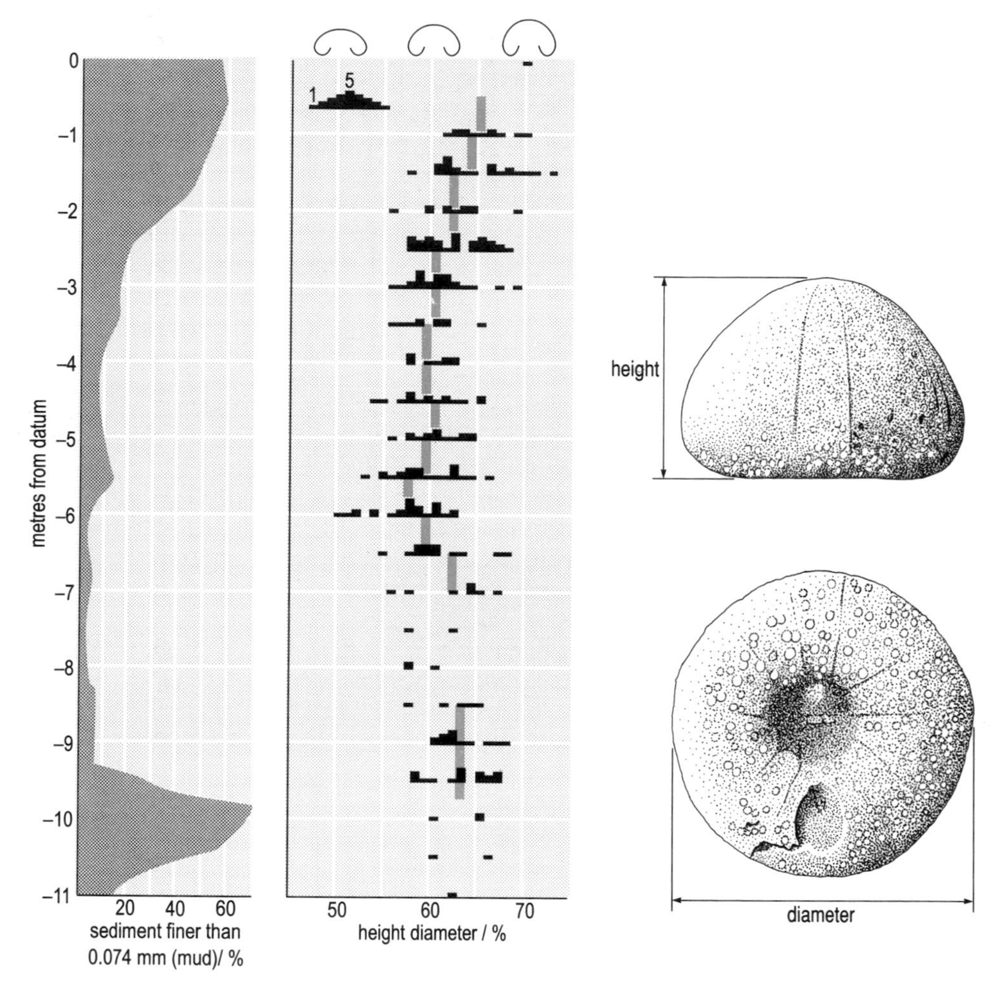

Figure 10.9 Changes in the height/diameter ratio of the test of the irregular echinoid Discoides subucula *in an Upper Cretaceous sequence in SW England (test shape is shown diagrammatically in section above). Percentages of mud-grade sediment are shown to the left. Histograms show nos. of specimens in each height/diameter class (scale at top left). Green bars in some histograms show the mean values.*

Morphological groupings within species Even when large samples of fossils have been collected and statistically sorted into natural groupings, we cannot always consider each to represent one species. This is an extremely important point because many spurious species have also been created through failure to recognize the connections between discrete morphological groupings.

▶ Apart from ecophenotypy which was discussed earlier, what other kinds of variation seen in biological species might give rise to morphologically distinct fossils?

There are three more kinds of marked variation in living organisms which could result in distinct morphological groupings in the fossil record. First, species may be divided into geographically separated races, or even subspecies of distinct appearance (Chapter 9, Section 9.3). Secondly, species may be *polymorphic*, with two or more distinct morphological types of individual (Chapter 3, Section 3.1). A common type of polymorphism is *sexual dimorphism* (Chapter 7, Section 7.1). The markedly different gametophyte and sporophyte generations of some plants (Chapter 5, Section 5.2) are another case in point. Finally, individuals of the species may have morphologically distinct devclopmental stages, which would look different as fossils, as with the larvae, pupae and adults of insects.

Though presenting the palaeontologist with quite a headache, such kinds of variation in fossil populations can be identified. In all but the example of geographical variation, the different morphological types of any given species can be expected to have co-occurred geographically (in broad terms at least, since there may have been some local separation of, say, different developmental stages). They should also have shared the same stratigraphical range. Such co-occurrence can be detected by close inspection of the fossil record in cases where this is complete enough. Moreover, in many shelly organisms, the shell is grown throughout life by accretion, and so sexual dimorphs may be recognized because they share identical immature stages of the shell. Several pairs of ammonite dimorphs, which were previously considered separate species (Figure 10.10) have been matched up using these methods.

Examples of geographical variation may be harder to resolve, as they often are even for species of living organisms. But even steep local clines (Chapter 9, Section 9.3.2) have been detected in some fossil populations, and so some examples of allopatric divergence within species can be detected.

Another, quite different reason why a single species may be represented by two or more discrete morphological groupings of fossils is geological, rather than

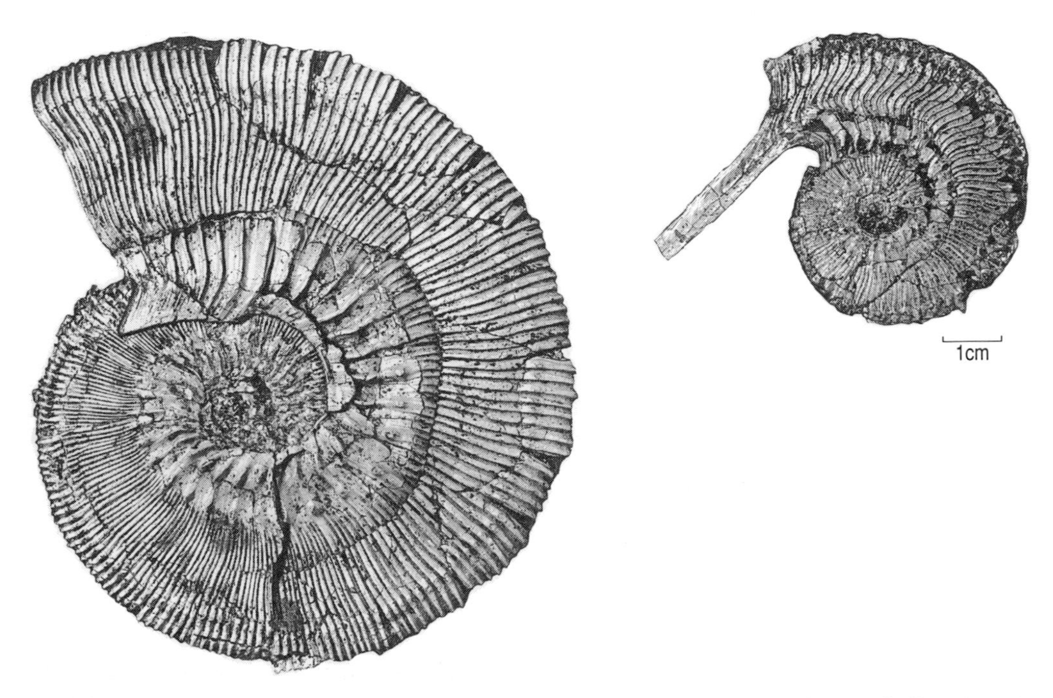

Figure 10.10 *Sexual dimorphs of the Jurassic ammonite,* Kosmoceras. *The larger shell is thought to be that of a female.*

biological, and has to do with the vagaries of fossilization (e.g. the different styles of fossilization illustrated in Figure 10.1). A particular problem arises with organisms which had more than one skeletal element in life (as with vertebrate skeletons), or which readily became broken into distinctive parts: post-mortem dissociation of these might have led to palaeontologists giving them different taxonomic names simply because comparable living organisms, or complete fossils of the original organisms, are not available to show how the components fitted together.

Palaeontological species A final point, which has already been referred to in Section 10.2.1, is that the recognition of palaeontological species should be based upon an assessment of variation in many aspects of morphology rather than just one or two. Single features may vary enormously within a species. You need only think of some of the extreme modifications of single features, such as snout length and body height, produced by artificial selection in breeds of dogs. Yet all of these breeds belong to one species, *Canis familiaris*. While studies of variation and change in single features tell us about evolution within a species, the decision on where to draw the line between species must be based on an assessment of overall morphology, as gauged by the measurement of many features. This involves a subjective element of judgement in relation to chronospecies in evolving lineages. But in the case of co-occurring species, the natural divisions are likely to be evident in most cases, provided that a wide enough range of morphological features is investigated (as explained in the next Section).

When palaeontological species are sensibly diagnosed, with due allowance being made for the splitting or lumping of morphological groupings discussed above, then they may serve as a bridge between the raw variability of fossil assemblages and the *real* evolutionary species of the past. Of course, some mistakes are inevitable, but with perseverance and critical testing of interpretations many of the pitfalls can be avoided. The way can thus be opened for studying morphological variation within the species of the past.

10.2.3 Morphometrics and patterns of growth

The quantitative description of morphology in organisms—**morphometrics**—was introduced in Chapter 3. It is a fundamental aspect of evolutionary studies, both of living and fossil organisms. In studies of living organisms, however, it takes its place alongside other approaches, particularly the direct analysis of genetic variation in populations, as earlier chapters have demonstrated. But in the study of fossils, morphological analysis is usually all that is possible, and so evolutionary palaeontology lays particular emphasis on morphometrics for microevolutionary investigations.

▶ Recall from Chapter 3 the three kinds of measurement scale which are generally employed for describing the morphology of organisms.

They are the categorical, ordinal and interval scales (Chapter 3, Section 3.5.1). Categorical differences may exist between closely related species, and even within species populations (e.g. eye colour in humans), but quantifiable differences on ordinal and interval scales tend to be predominant. Most fossils are morphologically complex enough to present a potentially bewildering array of recognizably variable features, or **variates** for possible quantification. How is one to choose what to count or to measure?

Variates Variates need to be defined such that, together, they allow consistent and accurate discrimination between specimens. The features being measured must be *homologous* (Chapter 2, Section 2.2.2) in the specimens that are being compared. Appreciation of the biological organisation of the original organisms, such as their planes and/or axes of symmetry and their anatomical orientations (front to back, and top to bottom), must, therefore, first be obtained from a qualitative study of the specimens. Only then can biologically equivalent traits be identified and measured in different specimens. Measurements between points which can be easily identified (e.g. the width of the upper jaw between the outer margins of the sockets for the canine teeth) ensure good repeatability of the measurements.

When analysing morphometric data, it is also important to remember that fossilized remains were originally the products of organic growth. As an organism develops, not only its size, but usually also its shape changes. Failure to take account of the relationships between size, shape and age may result in confusion between evolutionary differences among fossil organisms, and those merely reflecting their different stages of growth.

Most features show some increase in size or number during development: their measurements are thus said to be **age-dependent**. The shells of bivalves, such as cockles, oysters and mussels, for example, continue to grow in height throughout the lives of their makers, though with a marked slowing down in adults. Human height, by contrast, is only partially age-dependent. A histogram of any such single variate in a sample could be difficult to interpret, and a comparison between samples confusing, because some of the variation would be due to phenotypic differences between individuals of the same age and some would merely reflect age differences in the organisms sampled.

▶ How might you eliminate the effects of the age differences?

If individuals of the same age were selected for measurement, then the age-dependent component of variation could be removed from consideration. It is usually difficult, if not impossible at times, however, to assess the exact age (at death) of fossil specimens. Sometimes a periodicity of growth is expressed in skeletal components, and annual (or other periodic) increments may be recognized. Bivalve shells, for example, grow larger by the addition of increments of shell material around the shell margin, and this leaves distinct **growth lines** on the outside of the shell, running parallel with the margin (these are visible on the fossil oyster shells shown in Figure 10.2). In many bivalves, growth lines show distinct annual patterns, from which the age of the shell's maker can be estimated. Growth rings in some bones and tree rings are examples of a similar phenomenon. In such cases, there is scope for eliminating the age-related component of variation between individuals, by selecting individuals of the same age for measurement.

Another approach to the problem is to seek out variates in which age-dependence is limited. Some features become permanently established at an early stage of growth. A notable illustration of this is provided by the adult teeth of most mammals. These are usually fully formed early in development, prior to eruption from the gums. Following the earliest stages of growth, their dimensions are therefore independent of the age of their possessors (except for the heights of the crowns, which typically become secondarily reduced through wear). If incompletely grown juvenile specimens are eliminated (e.g. by ignoring any completely unworn and hence unerupted teeth), then the differences between specimens (apart from crown height) reflect only the phenotypic differences between individuals. In any given population such continuous phenotypic variation commonly shows a *normal distribution*, as explained in Chapter 3, Section 3.5.1. The expectation of normal distributions in such variates allows palaeontologists to distinguish morphological groupings among collections of similar fossils. Sample data that do not show a normal distribution may be suspected of comprising a mixture of different morphological groupings. Sometimes the deviation from a normal distribution is obvious, but where it is ambiguous, and might result from sampling error, inferential statistical tests may be applied (e.g. Chi Square, or, for smaller samples, Kolmogorov–Smirnov tests, explanations of which can be found in elementary statistical textbooks).

Discrimination of morphological groupings The discrimination of groupings in such data is well illustrated by one of the classic studies of the 1950s by the Finnish palaeontologist Björn Kurtén (reprinted in 1988), on fossil mammalian teeth. Being highly resistant, these teeth are relatively abundant as fossils in certain deposits, and, as noted above, most of their measurements may be considered age-independent. One of Kurtén's studies was of a large collection of fossil hyaenas from the Pliocene of China. Figure 10.11 shows a histogram of measurements of one of the variates studied, the length from front to back of the crown in the upper fourth premolar (cheek) tooth (P^4). Taken as a whole, the data in this histogram

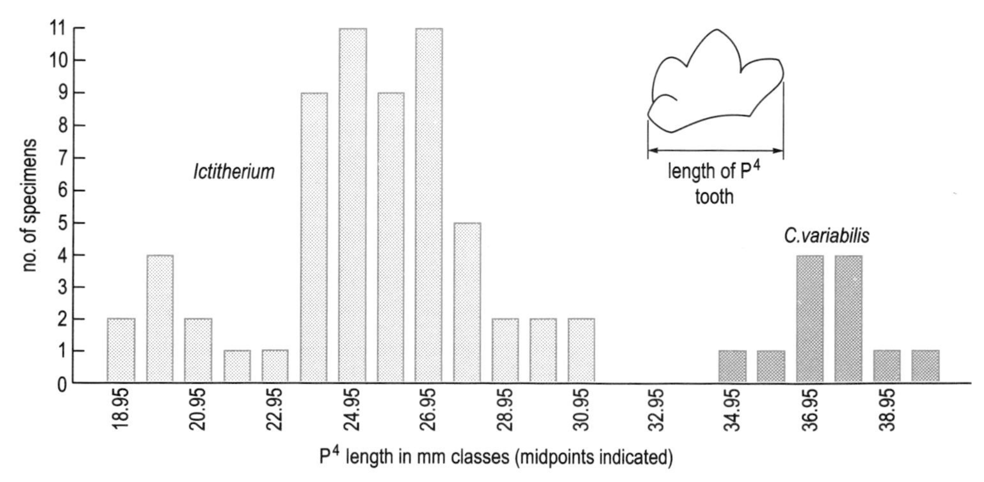

Figure 10.11 *Histogram of measurements of the length of the upper fourth premolar (P^4) for all specimens of the Chinese Pliocene hyaenas studied by Kurtén. The data are grouped in mm intervals, the mid-points of which are shown on the x axis. The clusters of data for* Crocuta variabilis *and* Ictitherium *are indicated. Further details discussed in text.*

deviate markedly from a normal distribution since those for one species, *Crocuta variabilis* (which is also distinguished on other characteristics), stand out as a distinct grouping at the right of the histogram. The remaining data, which relate to forms placed in the genus *Ictitherium*, are harder to interpret.

Although the *Ictitherium* data do not show a perfect normal distribution, random sampling error might explain the deviation (the deviation is not, in fact, statistically significant). However, another aspect stands out:

▶ How does the distribution of the *Ictitherium* data differ most from that for *Crocuta variabilis*?

The range of variation of the *Ictitherium* data is much greater than that for *Crocuta variabilis*. This led Kurtén to ask if there might not be an amalgam of morphological groupings, to explain the broader spread of the *Ictitherium* data (although he conceded that this need not be so). To see if any such groupings could be discriminated, he combined these measurements with those of another variate, on a graph with one variate on each axis—a **bivariate scatter**. The width of the upper second molar (back cheek) tooth (M^2) was chosen as the second variate, and the results are shown in Figure 10.12. (Note that not all the *Ictitherium* data from Figure 10.11 are shown here because corresponding measurements of M^2 width were unobtainable for some.) Although a few specimens are intermediate, the majority now lie within more clearly discriminated clusters. Inspection of yet further variates might confirm the clustering, and so resolve the borderline cases. Such techniques of *multivariate analysis*, which allow discrimination of clusters of specimen data by considering several variates together, are widely used in morphometric work today, thanks to the facility with which computers can handle such operations. But the object here is more to illustrate the basic principles involved in discriminant analysis than to provide a

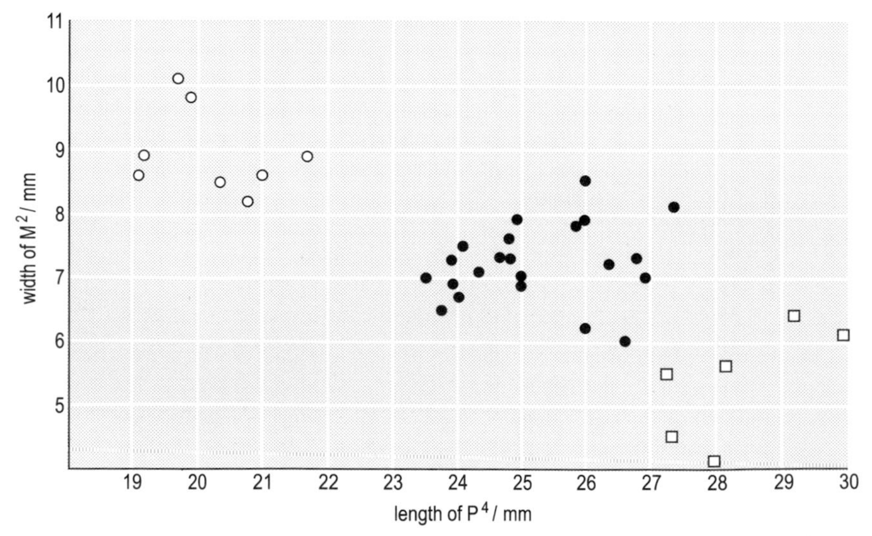

Figure 10.12 *Bivariate scatter of two age-independent variates (length of P^4 and width of M^2) in fossil specimens of* Ictitherium, *showing discrimination of three species.* I. wongii *is represented by the black circles in the middle.*

comprehensive dossier of techniques; those of multivariate analysis go beyond the technical scope of this Book, and will not be considered further.

Kurtén interpreted the morphological groupings shown in Figure 10.12 as three separate species. Fossils of the three groupings do not share identical geographical distributions, so it is unlikely that any two of the groupings merely represent the sexes of one species. Yet all three co-occur at some localities, without intermediates (in multivariate terms), implying that they were indeed distinct biological species. Moreover, possible ecological differences between the three were also noted. The larger species have more robust premolar cheek teeth, and relatively reduced hind molar teeth. This is consistent with a gradation from primitive meat-shearing habits to the bone-crushing habits typical of modern hyaenas.

If all the values of P^4 length for the most abundant species (*I. wongii*: the central cluster in Figure 10.12) are now separately plotted as a histogram (Figure 10.13) their distribution can be seen to be approximately normal in form, as expected. The deviation from perfect normality can readily be attributed to the small size of the sample. Although in this instance the different morphological groupings were interpreted as separate species, this need not always be the case, as the previous Section made clear. Other considerations must always be taken into account, as in this example, in making such a judgement.

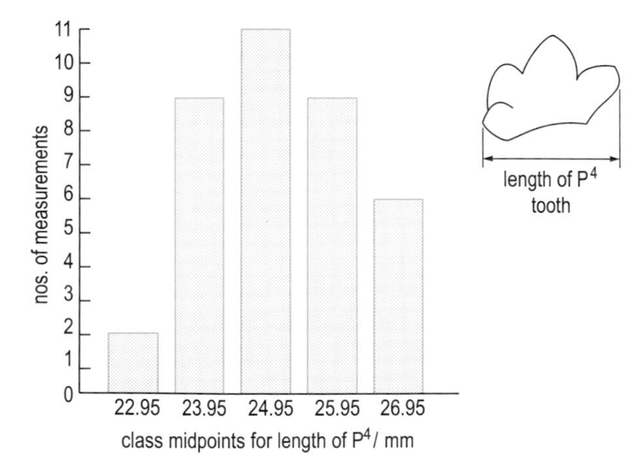

Figure 10.13 *Histogram of measurements of the length of the upper fourth premolar (P^4) of the fossil hyaena,* Ictitherium wongii, *arranged in mm classes around the mid-points indicated on the* x *axis.*

Detecting change in populations Another way in which normal distributions of age-independent variates can aid morphometric analysis is in the detection of changes in stratigraphically successive populations of a single lineage. Figure 10.14(b) shows a pair of histograms for measurements of a single variate in samples of a fossil trilobite species, *Onnia superba*, collected from successive intervals in a sedimentary sequence of Ordovician age in western central England. The variate in question is the number of small pits around half of the outer fringe of the headshield (cephalon) of the trilobite (Figure 10.14(a)). Numbers of pits were apparently established early in life, and so can be treated as virtually age-independent. Here the data are, strictly speaking, ordinal rather than interval scale measurements (i.e. numbers of items rather than dimensions). But it is nevertheless possible to calculate their mean values and standard deviations (Chapter 3, Section 3.5.1), because the pits are very numerous and so can be treated as if they were the units of continuous measurements. The calculated statistics of the samples are indicated alongside the histograms in Figure 10.14(b). Note that for the purposes of statistical comparison the values of the means and standard deviations are recorded exactly as calculated, and are *not* rounded off as whole numbers.

The question arises: Can the differences between the successive samples be taken to mean that there had been a genuine change in morphology in the populations from which they were drawn, or might the differences merely reflect random sampling from unchanging (i.e. identical) populations? Note, for example, the range of variability in the two samples, and the extent of their overlap: the answer to the question above is by no means immediately clear. The question is tackled along the lines explained in Section 3.5.1 of Chapter 3.

At the outset, the null hypothesis that the differences in the means and standard deviations of the two samples are due to sampling error alone (from identical populations), is proposed. The probability of this being the case can be calculated from the sample statistics (as shown below). If that probability is found to fall below some chosen level of significance (here 5%, or 0.05, will be used), then the null hypothesis is rejected, and the alternative hypothesis, that the samples come from populations which were different, is accepted. Full explanations of the methods outlined here may be found in textbooks on statistics (see 'Further Reading' at the end of this Chapter). The following is merely a brief introduction to the procedures used.

The first step in such a comparison is to compare the variances of the samples, and this is done by means of an '***F* test**'. The statistic, F, is calculated as the ratio of the larger value of variance over the smaller (recall from Chapter 3 that variance is simply the square of standard deviation, s):

$$F = \frac{s_1^2}{s_2^2} \tag{10.4}$$

where s_1^2 and s_2^2 are the larger and smaller values, respectively, of the sample variances. If the samples had come from identical populations, then you would expect most values of F to be close, if not equal, to 1. Progressively larger values could be regarded as a decreasingly probable result from sampling error alone. The calculated value of F is therefore compared with Tables (given in statistical texts) that show the critical values of F which correspond to various levels of significance. These critical values themselves vary according to sample size, since smaller samples show greater random variation, and hence yield a larger expected spread of values of F consistent with the null hypothesis. The appropriate critical value of F for the chosen level of significance is therefore read off from the position in the Table of values corresponding to $(n_1 - 1)$ and $(n_2 - 1)$, where n_1 and n_2 are the respective sample sizes. Note that $(n - 1)$, termed the *degrees of freedom*, is used for each sample, rather than n, as in the calculation of variance (Chapter 3, Section 3.5.1). If the value of F is found to exceed the critical value for the chosen level of significance, then the null hypothesis is rejected, and the alternative hypothesis, that the populations from which the samples were drawn had different variances, is accepted. A point you should note in making such calculations is that the values of the statistics should *not* be rounded off to only a few decimal places. Successive approximations can lead to a multiplication of errors, yielding an inaccurate result.

▶ What is the value of F for the two samples shown in Figure 10.14?

$F = (1.9863)^2/(1.6952)^2 = 1.3729$. Tables of values of F show that, for a level of significance of 0.05, and for the two sample sizes considered here, F must be greater than about 1.60 for rejection of the null hypothesis. In this case, then, the null hypothesis is not rejected, and the differences in the variances of the two samples are said to be *not significant* (at the 0.05 level).

If there is therefore no reason to suppose that the populations from which the samples came differed in variance, can the same be said of the differences in the means of the samples? The null hypothesis in this case is that the samples came from identical populations, and that their different mean values are due to sampling error. Again, if the probability of the null hypothesis is found to be less than the chosen level of significance, then that hypothesis is rejected in favour of the alternative hypothesis of difference between the populations. The suitable test in this case is known as a ***t* test**, and it depends, as before, on the calculation of a statistic, t. This statistic is the difference between the two mean values, divided by the *standard error* for difference in means (which is the expected standard deviation, around a mean of zero, of values of differences in the means of pairs of samples randomly drawn from the same population):

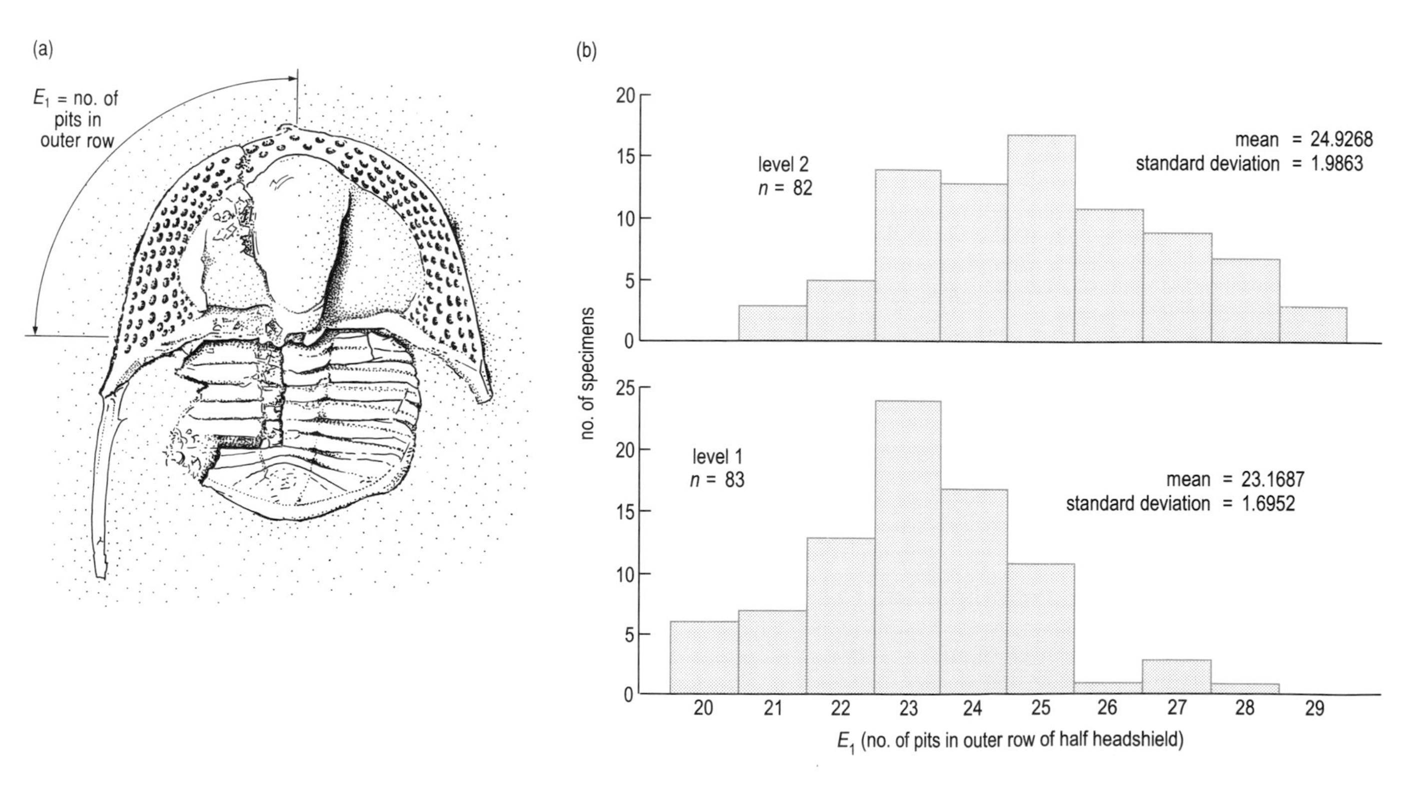

Figure 10.14 *(a) The trilobite* Onnia superba *from a sequence of Ordovician shales in western central England. Note the rows of small pits around the fringe of the headshield. (b) Histograms showing the variation in the number of pits in the outer row for half of the headshield in specimens from two successive intervals in the sequence. The sample means and standard deviations are indicated to the right.*

$$t = \frac{\bar{x}_1 - \bar{x}_2}{s_e} \qquad (10.5)$$

where $\bar{x}_1$ and $\bar{x}_2$ are the means of the two samples, and s_e is the standard error for the difference of the means. It is necessary to calculate s_e from the statistics of the two samples:

$$s_e = s_p \sqrt{\frac{1}{n_1} + \frac{1}{n_2}} \qquad (10.6)$$

where s_p is the *pooled estimate of the standard deviation* for both samples, and n_1 and n_2 are the sample sizes. s_p, itself, is calculated by combining the standard deviations of the two samples, thus:

$$s_p = \sqrt{\frac{(n_1 - 1)\, s_1^2 + (n_2 - 1)\, s_2^2}{n_1 + n_2 - 2}} \qquad (10.7)$$

where s_1 and s_2 are the standard deviations of the two samples.

All that is ultimately needed, then, for the calculation of t, *are the means, standard deviations and sizes of the two samples being compared.* The procedure is to calculate s_p first (Equation 10.7), then s_e (Equation 10.6), and finally, t itself (Equation 10.5).

Most pairs of samples from identical populations could be expected to yield values of t close, if not equal, to 0. Progressively greater deviations from 0 are less and less probable in the null hypothesis. As before, then, the calculated value of t is compared with a Table of critical values of t for various significance levels, in order to see if the difference in means may be regarded as significant. As in the previous test, the critical values are themselves dependent upon the degrees of freedom involved. The appropriate critical value of t in this case is read from the position in the Table corresponding to $(n_1 + n_2 - 2)$ degrees of freedom. In this test, t may be positive or negative in value, and so a significant deviation may be in either direction (constituting what is called a *two-tailed* test, in contrast with *one-tailed* tests, such as the F test, in which there are deviations in only one direction).

▶ Calculate t for the two samples shown in Figure 10.14.

From Equation 10.7:

$$s_p = \sqrt{\frac{(82-1)(1.9863)^2 + (83-1)(1.6952)^2}{82+83-2}} = 1.8456$$

From Equation 10.6:

$$s_e = 1.8456 \sqrt{\frac{1}{82} + \frac{1}{83}} = 0.2874$$

Hence, from Equation 10.5,

$$t = \frac{24.9268 - 23.1687}{0.2874} = 6.117$$

Tables show that, for a significance level of 0.05 (i.e. for $P = 0.025$ on either side of 0, because this is a two-tailed test), and for samples of these sizes, values of t

of magnitude exceeding ±1.980 (greater than 1.980 or more negative than −1.980) indicate a significant difference in the means.

▶ Can we conclude that there had been a change in the populations from which the two samples shown in Figure 10.14 came?

Yes, at the 0.05 level of significance, the null hypothesis of equivalence of the populations has been (very well) rejected, and so it may be surmised that the two populations differed.

The biological interpretation of this difference is quite another matter, however, as previous Sections should have made clear: there are many possible reasons for differences in morphology, not all of which need involve evolutionary change. These need to be carefully considered before it can be concluded that there has been real evolutionary change in the lineage, and the issue will be discussed further in Section 10.3.2. The purpose of the present exercise is simply to show how decisions may be made concerning whether differences between samples may be considered to reflect differences in the original populations rather than being due to random sampling effects.

Growth and form So far we have concentrated on the analysis of single variates which reflect phenotypic differences. In order to analyse changes in shape, or the growth relationships of different parts of the body, at least two measurements need to be compared. Changes in the shape of the sea-urchins in Figure 10.9, for example, were studied by comparing the height of the test with its diameter in specimens from successive levels in the sequence. Likewise, the disproportionate growth of the human head with respect to the body could be monitored by comparing head height with that of the whole body at different stages of growth. As in the latter example, the size relations of many features vary with body size. It is therefore usually best to avoid simply calculating the ratio of any two such measures, since this procedure loses track of absolute size. Instead, it is better to plot a *bivariate scatter*, to reveal the pattern of relative growth of the two variates.

On a bivariate scatter, measurement of any two body dimensions taken from a single individual at different stages of its growth will fall along some form of line (Figure 10.15). This is called a **line of relative growth**. A similar line can also be derived from the scatter of points in a plot of measurements taken from many individuals in a single population, provided that a wide range of individual ages is represented in the sample (Figure 10.16). Because of slight differences in relative growth between individuals, a plot of relative growth derived from several specimens inevitably shows some spread of data points. A line therefore has to be fitted to the scatter. There are several ways of fitting lines to scatters of data, explanations of which can be found in statistical texts (see Further Reading). Where two variates are expected to be mutually correlated along a straight line (as discussed below), the most appropriate method is to plot a **Reduced Major Axis** (**RMA**) for the bivariate scatter. In effect, this is the line which, for a given scatter of data, minimizes the sum of the triangular areas formed on the graph by drawing lines parallel with the x and y axes from each data point to the line itself (see Figure 10.16).

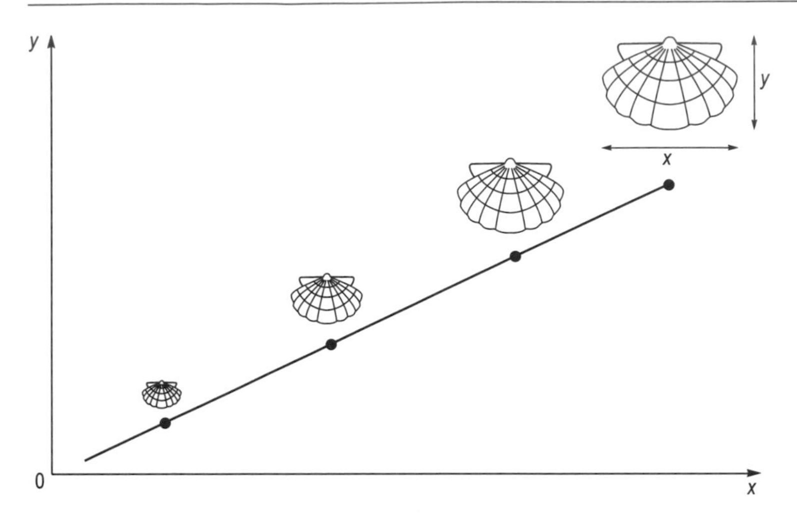

Figure 10.15 *Hypothetical line of relative growth of two variates,* x *and* y*, based on successive growth stages of a single individual (illustrated diagrammatically at each data point).*

Construction of an RMA is relatively simple. The ratio, s_y/s_x, gives the slope of the line, where s_y and s_x are the standard deviations for the two variates which are plotted respectively on the y and x axes. A line of this slope is drawn through the point on the graph corresponding to the mean values for the two variates, $\overline{y}$ and $\overline{x}$, to give the RMA. It should be stressed that, in drawing such a line, it is assumed (as is usually the case) that there is mutual dependence in the growth of the variates, or, at least, that both variates are mutually dependent on some third factor (e.g. if one were to compare the lengths of the left and right legs of people). If one variate grows independently, but the size of the other variate is dependent upon it (e.g. if one were to plot the dependence of trouser length on leg length in humans), then another method of line-fitting (regression) would be required, which will not be discussed here. Figure 10.16, for example, shows a plot of measurements of height (y) versus length (x) of the shells of a species of ostracod (a small crustacean with an enclosing carapace consisting of two matching parts of the shell, or *valves*, hinged along the back) from the Lower Carboniferous of the French Pyrenees, with a reduced major axis fitted to the scatter.

An important point to note is that age itself is not represented in such graphs: each pair of measurements simply relates the size of one variate to that of the other in a single specimen, no matter what its age. Countless studies have shown that no matter how absolute rates of growth vary, the *relative* rates of growth of dimensional variates do tend to show remarkably regular relationships. It is not hard to see why. As Chapter 2, Section 2.2.3, stressed, the major adaptations of organisms are supported by anatomical and physiological features that place constraints on the form of other structures and habits. Thus the growth in one dimension of a feature is usually matched in some regular way by that in other dimensions. This is not to say that age is irrelevant. Long-term evolutionary changes in the ages at which given sizes and shapes are attained by individuals form the basis of much macroevolutionary change. The latter theme will be taken up in some detail in Chapter 14. In the present context, however, we are only

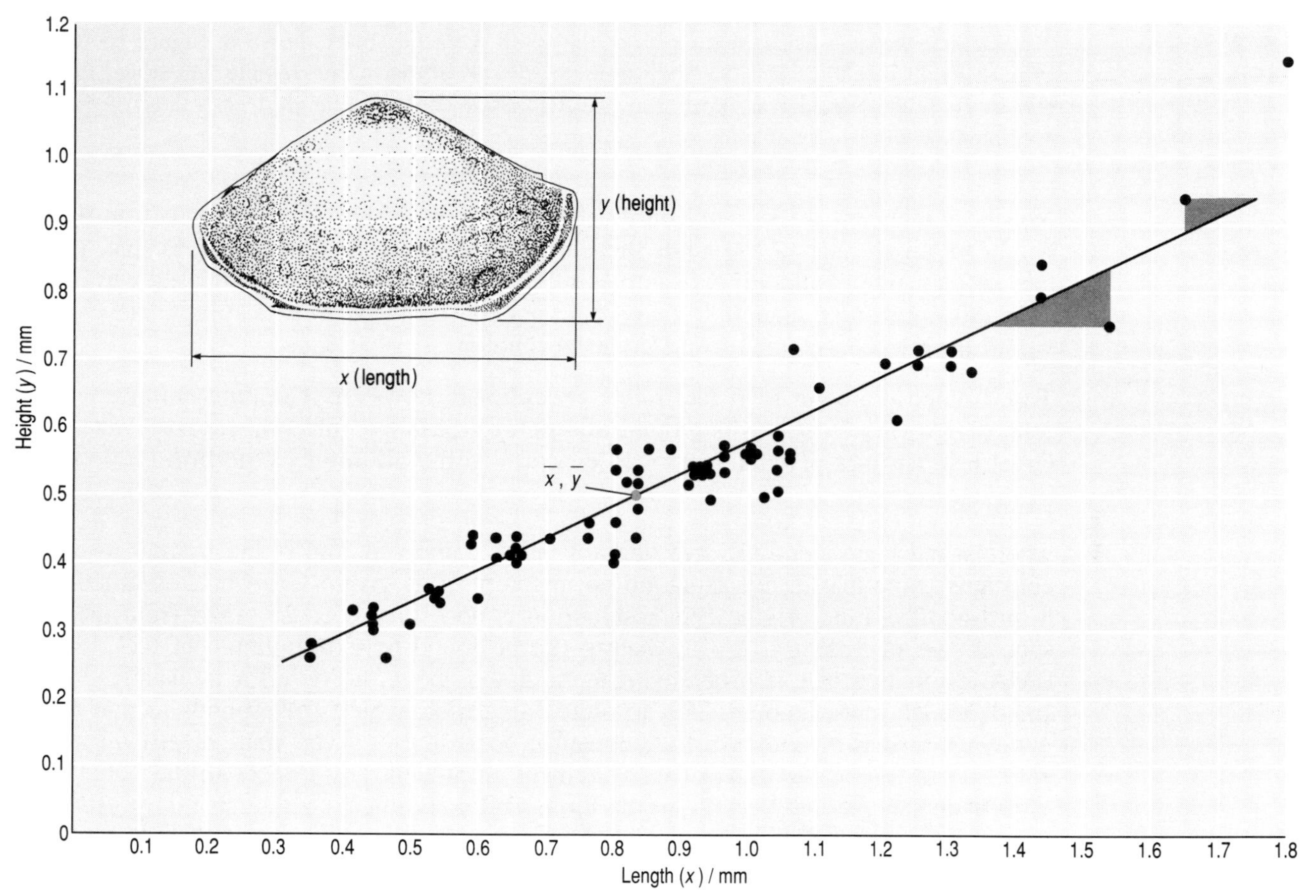

Figure 10.16 *Bivariate scatter of measurements of height versus length of the shells of an early Carboniferous species of ostracod,* Bairdia perretae *(inset), from the French Pyrenees. A line of best fit for the data (RMA) is also shown passing through the joint means,* $\bar{y}$ *and* $\bar{x}$*, represented as a green dot in the Figure. The triangular areas between the points and the line, the sum of which is minimized in the construction of an RMA, are shown for two of the points at the right of the scatter.*

concerned with individual growth within species, and can therefore restrict attention to the relationships between size and shape, for the time being.

The straight line relationship of the two variates in Figure 10.16 shows that their growth was proportionate: in other words there was a constant ratio between the two rates at all sizes. Such a pattern of growth is said to show **isometry** (meaning 'same measure' in Greek). As in this example, the line usually extrapolates just to one side or the other of the origin because the initial proportions of the earliest (larval) stages of growth are commonly unlike those of the adult. In effect, the isometric growth commences from a false origin on the graph established by the growth of the larva. In such a case, isometric growth conforms to the equation:

$$y = bx + c, \qquad (10.8)$$

where b is the constant of proportional growth (i.e. the slope) for the variates x and y, and c is the intersect on the y axis. Because c is usually small compared with adult values of x and y, changes of shape, expressed as changes in the ratio of y and x, are minor when growth is isometric.

▶ Suppose that you had two samples of the fossil ostracod shells shown in Figure 10.16, from different localities, and that the mean size of specimens in one sample was larger than that in the other. Would you expect much difference in the mean height/length ratio of the specimens in the two samples?

No: the y intercept (c) is about only 0.1 mm. This has relatively little effect on the height to length ratio of the adult shells. The importance of this finding is that, in this species, one would expect samples of specimens of different mean sizes to have closely similar shapes. If shape did vary markedly between samples, they would have to be assigned to different morphological groupings.

True isometric growth is in fact rather exceptional, however. The *differential* relative growth of variates is termed **allometry** (meaning 'different measure' in Greek). Most lines of relative growth conform to the **allometry equation**:

$$y = bx^a, \qquad (10.9)$$

where y and x are again the two variates being compared, and b and a are constants. Note that in this equation no intercept, c, is given. Growth is assumed to proceed from the origin. This assumption is necessary because, apart from the special case of isometry ($a = 1$) where c can be found by extrapolation of the RMA, estimation of c is problematical. As noted above, however, it is usually small enough to be ignored. The absolute value of b, which corresponds to that of y when $x = 1$, has no biological significance, because on curved lines of growth it varies according to the units of measurement employed. Differences in the value of b between samples, however, may be biologically significant although rarely used for discrimination of species.

The value of a, the **allometric exponent**, in contrast, is of considerable importance. Significant change in shape with growth occurs when a is not equal to 1. Here the line of relative growth is curved, in exponential form (Figure 10.17). Allometry is said to be *positive* when a is greater than 1 (yielding a line of relative growth which steepens upwards as in Figure 10.17), and *negative* when a is less than 1 (yielding a line which decreases in slope upwards). Relative growth of the variates is thus disproportionate, though in a regular fashion, as in the growth of the human head in relation to the body mentioned earlier. There are numerous possible reasons for disproportionate growth of this kind. The effect of scale when anatomical surfaces are functionally linked with volumes is one very common factor, and Figure 10.17 is an example of just this, as explained below.

Many marine animals feed by filtering fine particles of food from the surrounding water with some form of net or mesh. As they grow, their food requirements tend to increase in scale with the mass of their soft tissue, and hence its volume (i.e. with the *cube* of their linear body dimensions). Yet, if their filter-feeding net grew isometrically, its capacity for food-entrapment would increase in scale only with its area (i.e. with the *square* of its linear dimensions). Isometric growth would therefore result in there being less food per gram of body mass as the animal became bigger. This discrepancy is overcome by many different kinds of adaptation of growth pattern in nature. One of the commonest of these is disproportionate growth of the filter-feeding system such that the *area* of the filtering net keeps pace with increasing body mass (*volume*). In Figure 10.17, the

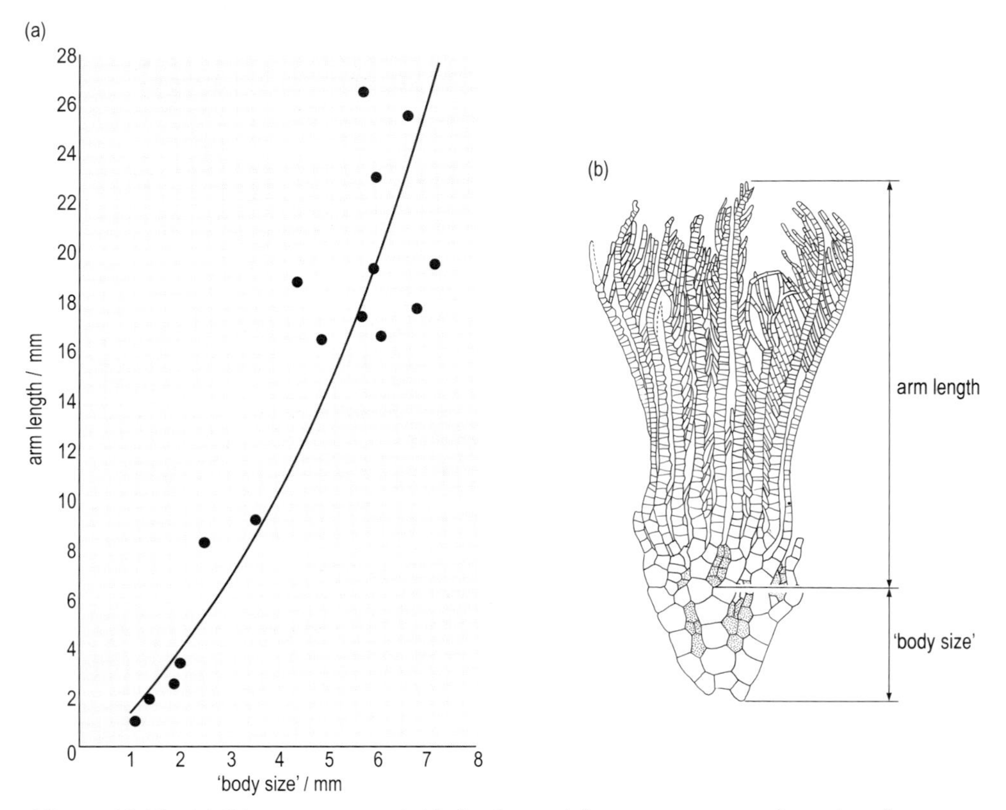

Figure 10.17 *(a) Bivariate scatter (with fitted curve) for measurements of arm length versus a linear measure of 'body size' (variates shown in (b)), for a species of crinoid,* Alisocrinus tetrarmatus, *from the Ordovician of Illinois, USA.*

arm length, a linear measure of the filter-feeding net of an Ordovician species of crinoid, *Alisocrinus tetrarmatus,* is plotted, on the *y* axis, against another linear measurement representative of body size, on the *x* axis. Examples of different growth stages in this species are illustrated in Figure 10.18. In life the animal would have been rooted to the sea-floor with a long stalk (not shown in these specimen drawings), and the fan of frond-like arms around the main body, or crown of the animal would have spread out to form an umbrella-shaped filter (top left diagram in Figure 10.18).

▶ How is the allometric growth of the crinoid's filter system mainly expressed in terms of morphological change?

The allometry is mainly expressed by a disproportionate increase in the lengths of the arms, relative to that of the crown, during growth. It is true that the numbers of sub-branches from the arms (*pinnules*) also increased with growth, but these are set at regular intervals along the arms, and so would have increased even with isometric growth. Only because of the disproportionate elongation of the arms has the number of pinnules expanded beyond what would have been expected from isometric growth. It should perhaps be noted that, though common, this form of allometry is not universal in crinoids. Other modifications to growth are also known.

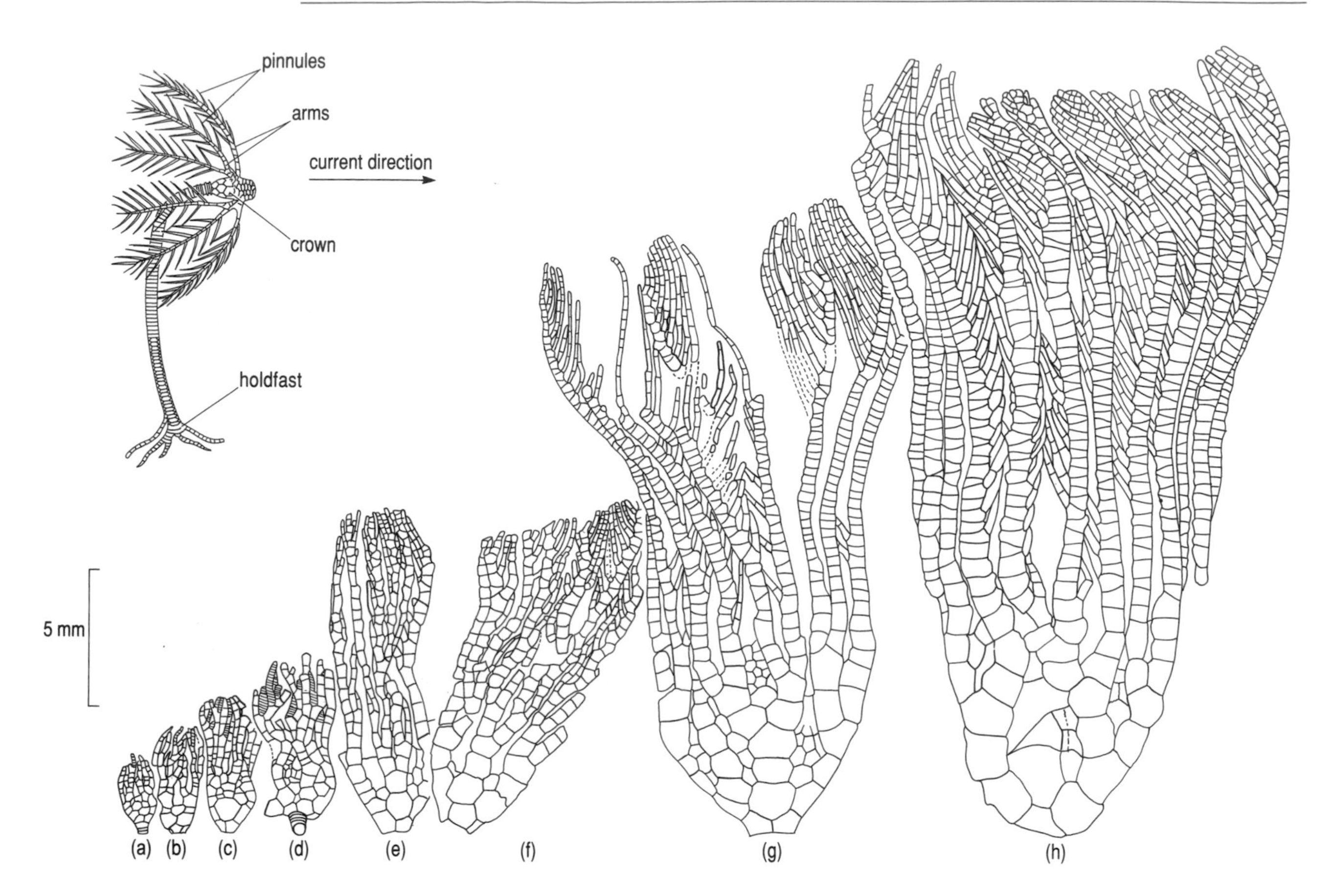

Figure 10.18 *Specimens of the crinoid* Alisocrinus tetrarmatus *(see Figure 10.17) showing different stages of growth. Diagram at top left shows reconstructed feeding posture.*

► Now suppose that, as with the earlier question on the ostracods, you had two samples of specimens of this crinoid species, with different mean body sizes, from different localities. Would you still expect the specimens of the two samples to have similar average shapes?

In this case, the answer is no, because of the allometric growth. The relative proportions of the filter system and the body differ markedly with size, as the graph in Figure 10.17, and the examples of specimens of different sizes in Figure 10.18 show. Thus, in order to see whether or not two such samples might be assigned to the same morphological grouping, it would be necessary to see if they shared similar values of the constants a and b in the allometry equation, notwithstanding their size-related differences in shape.

It is difficult, however, to fit a line to a curved scatter of data, like that in Figure 10.17, with accuracy, and to derive values of a and b from the graph. A simple way around this problem is to convert the measurements of the variates into logarithms and to plot these. From Equation 10.9, it follows that:

$$\log y = \log b + a \log x \tag{10.10}$$

A plot of log y against log x will yield a straight line of slope a, and an intercept of log b on the y axis, and so the constants a and b can readily be derived. Figure 10.19 shows a plot of the data in Figure 10.17, converted to logarithms, with a straight line fitted. The line is an RMA, as in Figure 10.16, only here it is calculated from the logarithms of the data for y and x. The value of a is also shown on the graph.

Using such graphical techniques, it is possible to recognize members of a single morphological grouping, characterized by a common pattern of allometric growth, despite the obvious differences in appearance of specimens of different sizes. However, it should also be pointed out that adult size may itself show evolutionary change (Chapter 14).

Sometimes, a bivariate scatter may appear to show a straight (i.e. isometric) trend, though with a sizeable intercept, c. This appearance may be misleading, however, if the range of measurements taken in fact falls along the gently curved part of an exponential line. If the latter (derived from the logarithmic plot) is drawn on the arithmetic plot, it will usually be found to fit the scatter, though curving sharply in towards the origin (as on Figure 10.17). If measurements from smaller individuals can be obtained, the appropriateness of the line can then be checked. An alternative explanation for a straight trend with a large intercept is that there *was* a marked change in the pattern of growth during post-larval development, following

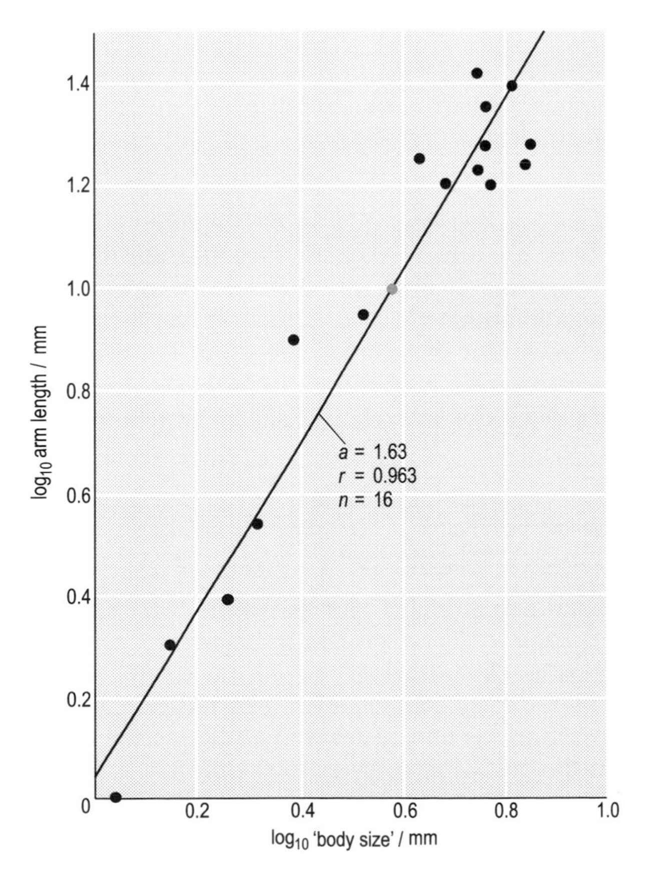

Figure 10.19 *Plot of logarithms of the data shown in Figure 10.17. Values of* a, *in the allometric equation,* r, *the correlation coefficient, and* n, *the sample size, are indicated.*

which growth became isometric. Such a change, however, would normally be quite evident from the specimens studied. A complete range of possibilities also exists between these two alternatives. To pick up a point made earlier, this uncertainty is why it is difficult to estimate c without measurements of very small individuals. In most instances, however, it may be assumed that c reflects the larval stage of growth and is thus negligibly small.

Inferential testing of allometry As with the histograms of single variates discussed earlier, bivariate scatters can be statistically analysed. Consider the scatter of data points either side of an RMA, as in Figures 10.16 and 10.19. It seems intuitively obvious that the more tightly the points cluster towards the line, the greater the strength of the correlation between the two variates that is implied. This measure of the strength of correlation can be represented statistically by the **correlation coefficient, *r***, which may vary between +1 and −1. When r is +1, or −1, there is a perfect positive, or negative, correlation (i.e. all the data points fall exactly on a straight line which either rises, or descends, from left to right on the graph) respectively. As r approaches 0, so the cluster of points becomes increasingly dispersed. The calculation of r is explained in statistics texts (see Further Reading), but the value of r is usually also directly available for keyed-in bivariate data in scientific calculators. Not only is r useful in its own right, as a measure of the strength of correlation, but it also allows a standard error for the allometric exponent, a, to be calculated:

$$s_e = a\sqrt{\frac{1-r^2}{n}} \qquad (10.11)$$

where s_e is the standard error of a (i.e. in this case, the expected standard deviation of sample values of a around the *true* value of a for the population); r is the correlation coefficient of the RMA for a plot of the logarithms of the two variates (yielding a as the slope: see Equation 10.10); and n is the sample size.

As in the t test described earlier, s_e is useful for inferential testing. In particular, it is important to know at an early stage of investigation if a is significantly different from 1, or not. If a is not significantly different from 1, then the analysis may proceed with a simple arithmetic plot of the two variates being studied, on which an RMA may be drawn for the relationship $y=bx+c$ (Equation 10.8) as in Figure 10.16. However, if a significantly differs from 1, then the logarithms of the variates must be plotted, for the relationship $y=bx^a$ (Equation 10.9).

▶ What null hypothesis should thus be proposed to test for deviation from isometric growth in a sample of specimens?

The appropriate null hypothesis is that the value of a for the population from which the sample came is 1 (i.e. growth is isometric), and that any deviation is due to sampling error. The probability of a, in a sample, deviating from 1 by more than a given amount can be estimated from a statistic called the **Z statistic** (although the probability is underestimated with smaller sample sizes, especially below $n=30$). In this test, Z is calculated as follows:

$$Z = \frac{a-1}{s_e} \qquad (10.12)$$

Values of Z have a normal distribution around a mean of 0 and a standard deviation of ±1. Hence, from the known distribution of data in a normal distribution (Chapter 3, Section 3.5.1), we expect to find values of Z to fall within two standard deviations approximately 95% of the time (more accurately, 1.96 standard deviations correspond to 95% of values). Thus, if the magnitude of Z from Equation 10.12 is found to exceed ±1.96, then the null hypothesis can be rejected at the 0.05 level of significance. This, again, is a two-tailed test.

Looking at Figure 10.17, for example, you may consider it questionable whether relative growth really was allometric, with a curved line of relative growth, as shown. It might seem as if a straight, i.e. isometric, line could just as easily pass through the scatter of points. This question can now be settled:

▶ Given the values of a and r shown in the logarithmic plot in Figure 10.19, and employing the Z test described above, decide whether a significantly differs from 1, or not (at a significance level of 0.05).

Your calculations should have shown that a does differ significantly from 1. From Equation 10.11:

$$s_e = 1.63\sqrt{\frac{1-(0.963)^2}{16}} = 0.110$$

Hence, the value of Z for the difference of the slope (a) from 1 (Equation 10.12) is:

$$Z = \frac{1.63-1}{0.110} = 5.73$$

The magnitude of Z (positive in this instance) is thus considerably greater than ±1.96, and so the null hypothesis may be rejected. (Strictly, the Z test underestimates random deviation in this instance because of the small sample size, but still not sufficiently to account for so high a value of Z.)

▶ What type of allometric growth is now confirmed for arm length vs. 'body size' in *Alisocrinus tetrarmatus*?

Positive allometry: a has been shown by the test above to have been significantly greater than 1.

The various graphical methods (histograms and bivariate scatters) and statistical tests described above illustrate how morphological groupings may be discriminated in assemblages of similar fossils, and how morphological changes in populations can be detected from samples collected from successive intervals in a sedimentary sequence. In the next Section the evolutionary significance of such studies will be discussed.

SUMMARY OF SECTION 10.2

Reproductive isolation cannot be directly recognized from fossils. Palaeontological species must therefore be recognized from morphology. Most sibling species remain undetected in the fossil record, so their past frequency is unknown. Studies of living groups yield differing estimates. Morphological change, rather than (inferred) branching, provides the most practical criterion for separating ancestral from descendent species. Single lineages may thus be divided into successive chronospecies.

Many specimens must be studied in order to recognize morphological groupings of fossils. Different groupings may be included within a single species if there is evidence for ecophenotypy, geographical variation, polymorphism or differentiation of developmental stages. Fossilized juvenile stages, and patterns of distribution, can yield such evidence. Fossils may also differ for geological reasons.

Variates selected for morphometric analysis must be homologous in the compared organisms. Single variates can reveal only phenotypic rather than age-related differences either if specimens of the same age are studied (which may be difficult), or if age-independent variates are selected. The latter usually show normal distributions of values in single populations, enabling groupings to be discriminated. Bivariate or multivariate analysis can further improve discrimination. Morphometric differences can be assessed using inferential statistics based on the means and standard deviations of samples.

Bivariate scatters are used for the analysis of the relative growth of variates. Values for a pair of age-dependent variates, taken from single individuals at various stages of development, or from many individuals of different ages, may be plotted to yield a line of relative growth, which shows how the two variates increase relative to each other. Most relative growth conforms with the allometry equation: $y=bx^a$ where x and y are the two variates, and b and a (the allometric exponent), constants. If $a=1$, growth is isometric (proportionate), although there may be a small intercept on the y axis (c), such that $y=bx+c$. If a does not equal 1 (allometry), in contrast, the line of growth has an exponential form, and shape varies with size. Allometry is positive or negative when a is greater than, or less than, 1, respectively. A common reason for non-isometric growth is a functional relationship between some anatomical surface and body volume. Values of a and b can be derived from a plot of log y versus log x, which yields a straight line of slope a and intercept (on the y axis), log b. It is necessary to analyse allometry in samples of fossil specimens in order to see if any differences in shape between them might be size-related. Inferential statistical tests can be applied to lines of relative growth to detect deviations from isometric growth.

10.3 MICROEVOLUTION IN THE FOSSIL RECORD

We have now considered the nature of the fossil record and the resolution it offers (in Section 10.1), as well as how the morphological characteristics of fossils may be used as a basis for recognizing species and their variability (in Section 10.2). We are now in a position to return to the issue raised at the beginning of the

Chapter—the patterns of change actually observed within species lineages, **phyletic evolution**, and in speciation. It will be useful to start by outlining some of the different patterns for the evolution of species which palaeontologists have variously proposed, based partly upon what they believed they saw in the fossil record and partly upon how they thought microevolutionary change might be expressed in it.

10.3.1 PALAEONTOLOGICAL MODELS FOR THE EVOLUTION OF SPECIES

What might be described as the classical view of microevolution in the fossil record is that of **phyletic gradualism**. Where a complete enough fossil record could be found, it was argued, it would show very gradual morphological change through time in lineages of populations, and where lineages became split, gradual divergence of species (Figure 10.20(a)).

▶ To which of the patterns of speciation shown in Figure 9.1 of Chapter 9 does this model relate?

It is based on the *splitting pattern*, illustrated as Figure 9.1(a). This model was derived from Darwin's own emphasis upon the slow, preferential accumulation of slight variations within populations (see Chapter 1, Section 1.4). He attributed the geologically abrupt appearance of most fossil species to the disruption of gradually changing lineages by frequent gaps in sedimentary sequences, concluding, with regret, that:

> 'The noble science of Geology loses glory from the extreme imperfection of the record. The crust of the earth with its embedded remains must not be looked at as a well-filled museum, but as a poor collection made at hazard and at rare intervals.'
>
> C. R. Darwin (1859)

Despite Darwin's acknowledgement that rates of evolution might, nevertheless, also have been highly variable, his general insistence on gradual change soon became widely adopted by palaeontologists as an article of faith. Reconstructed phylogenies commonly took a form like that in Figure 10.20(a).

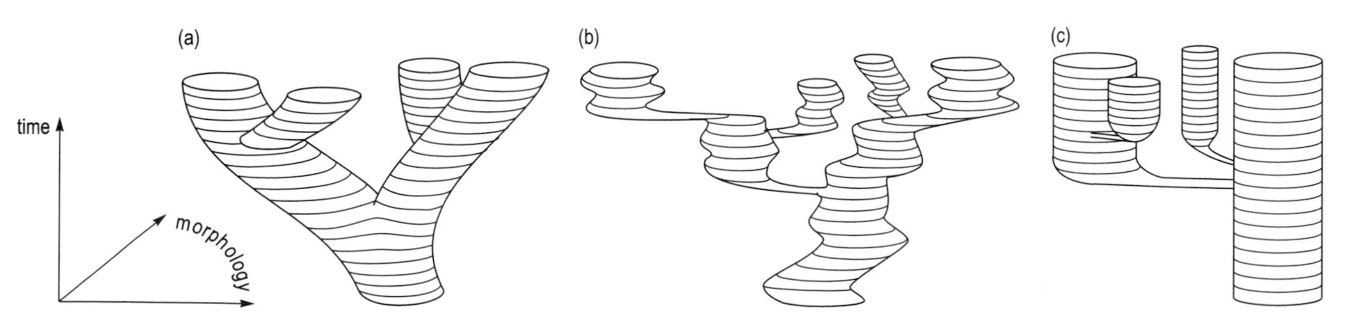

Figure 10.20 *Palaeontological models for the evolution of species: (a) 'classical' phyletic gradualism; (b) revised neo-Darwinian version of phyletic gradualism (punctuated gradualism); (c) punctuated equilibria. Morphological state indicated on horizontal axes, and the succession of populations through time on the vertical axis.*

In fact, many such evolutionary diagrams involved an important confusion of scale. Darwin's own notion of *gradual* change was intended to refer to the microevolution of populations, when perceived from the limited human time-scale of generations, or even of years. In many of the reconstructed phylogenies, however, it was longer-term, macroevolutionary change, spread across millions of years, which was being portrayed in a gradualistic manner, implying an apparent constancy of rates never really intended by Darwin. In some instances, however, the authors of these schemes also had other, distinctly non-Darwinian reasons for showing such exaggeratedly gradual patterns. Some palaeontologists of the late 19th and early 20th centuries were sceptical of the power of natural selection to explain major evolutionary changes. They toyed with the idea of innate and inexorable evolutionary tendencies within lineages, which were divorced from the influence of selection—**orthogenesis**. It was believed, for example, that many mammalian groups showed continuously sustained evolutionary trends for increasing body size and the elaboration of such features as horns and antlers (a theme discussed in Chapter 14).

Orthogenesis and other such non-Darwinian ideas were rejected when the fundamental role of natural selection in microevolution was re-established in the 1940s and 1950s. The American palaeontologist G. G. Simpson and others showed, for example, that the fossil evidence of horse evolution pointed to a complex history of divergently branching lineages showing differing patterns (see Chapter 15). Some lineages showed various modifications evidently linked with major changes in habits, such as the evolution of high-crowned teeth (adapted for grazing on grass), but others did not show these changes. Some even showed reversals of common tendencies, such as that for increasing body size. These complex evolutionary patterns, though yielding some accumulative net trends, contradicted any idea of orthogenesis divorced from adaptation to environmental changes.

Nor, then, was there any remaining justification for the exaggeratedly gradual schemes proposed by some of the earlier authors. Instead, Simpson returned to Darwin's surmise that rates of evolution might have been highly variable. By then, however, it had become possible to derive realistic estimates of likely rates of evolution by natural selection, both from direct observation and from theoretical models used in population genetics (as discussed in Chapter 4). These sources showed that rates of morphological change, even at low levels of selection, could be expected to achieve relatively rapid transformations when perceived on a geological time-scale. Simpson therefore interpreted geologically abrupt changes in morphology in fossil lineages as representing bouts of pronounced directional selection. Long periods of little or no change, in contrast, could then be interpreted as implying very weak or no directional selection. This neo-Darwinian, revised version of phyletic gradualism, which some authors have referred to as *punctuated gradualism*, may be characterized by Figure 10.20(b), which should be contrasted with Figure 10.20(a).

An alternative to phyletic gradualism was proposed by Eldredge and Gould (1972), in a major paper which explored the pattern that should be expected in the fossil record from the peripatric model of speciation (Chapter 9, Section 9.4.2), which Mayr and others believed to be overwhelmingly dominant in nature. Eldredge and

Gould termed their alternative the **punctuated equilibrium** model (Figure 10.20(c)), and argued that this most closely portrays what the fossil record shows. The pattern envisaged was along the lines of the budding scheme of speciation, illustrated in Figure 9.1(b) of Chapter 9. They proposed that the morphological change accompanying the speciation of a peripherally isolated population would, even if thousands of generations were allowed, be accomplished in a relative moment of geological time, and also within a small area. Such a local process, involving only a small population, would, in the great majority of cases, escape registration in the fossil record. Most new species could thus be expected to make stratigraphically sudden or punctuational appearances in the record, due to their migration into any given area from their unknown, local area of origin. However, once a species had become well established, the evolutionary inertia of large populations predicted by Mayr (Chapter 9, Section 9.4.2) should then be reflected, Eldredge and Gould argued, by a geologically long period of negligible morphological change—**stasis**, or morphological equilibrium. According to the punctuated equilibrium model, then, virtually all morphological change is associated with rapid and localized speciation events.

▶ What scope does each of the three models in Figure 10.20 provide for the recognition of chronospecies?

Both versions of phyletic gradualism show considerable morphological change within their lineages, and thus scope for chronospecies, in addition to branching speciation. In the revised, neo-Darwinian version (Figure 10.20(b)), the task is facilitated by the episodes of rapid evolution, which could be treated as natural morphological boundaries separating chronospecies.

In contrast, the characteristic stasis of established species in the punctuated equilibrium model, Figure 10.20(c), leaves virtually no scope for chronospecies. Note that although every species is here initiated by a speciation event, the situation is not quite equivalent to the use of branching events to delineate species, discussed in Section 10.2.1. In that scheme *both* the daughter lineages would be distinguished as separate species from the parent lineage preceding speciation.

The next Section critically examines the evidence for and against the three models outlined above.

10.3.2 THE RECORD OF CHANGE IN FOSSIL SPECIES

Recall from Section 10.1.2 which kinds of sedimentary sequences are likely to yield the best record of changes in populations.

▶ Which slopes in Figure 10.5 are most likely to promote stratigraphical completeness in the record?

A minimum contrast between the sedimentation rates for short and long time-spans is desirable in order to preserve stratigraphical completeness; gaps created over the longer term, by erosion or non-deposition, must be avoided as much as possible.

Hence lines of low gradient on the graph should be sought. If you refer back to Equation 10.3 in Section 10.1.2, you will see that this is equivalent to minimizing the difference between cX/T and s, so that completeness approaches 100%.

▶ Which environments indicated on Figure 10.5 satisfy this criterion?

Low gradients are shown by the lines for the deposits of bays/lagoons, lakes, inland seas and ocean floors. As noted earlier, although sedimentation may be interrupted even in these settings, such disturbances generally occur less frequently than in the other environments shown.

However, this is not the only consideration since it is also desirable to minimize the amount of time represented by each interval sampled, so that as precise as possible a record of a population at particular times may be made.

▶ Which aspect of sedimentation rate favours sampling from short time intervals?

The greater the short-term rate of sediment accumulation, the better the microstratigraphical acuity (see Equation 10.2). Thus the lines higher up on the y axis of the graph are those which favour the best values.

Unfortunately, then, precisely those depositional environments which have just been identified as tending to be associated with low stratigraphical completeness, also tend, by contrast, to show the best microstratigraphical acuity. One may either have a good short-term record but poorer long-term record, or *vice versa*, but usually not both together. The simple reason for this is that the combination of good completeness and microstratigraphical acuity would require the sustained accumulation of exceptionally thick sequences and this in turn would require areas of rapid but long-term subsidence—an unusual, though not impossible, geological circumstance.

Among those environments which commonly preserve high stratigraphical completeness, Figure 10.5 suggests that the best microstratigraphical acuity tends to be associated with bays/lagoons and lakes. However, it is rare to find examples of deposits formed in these environments with temporal scopes of more than a few thousands of years (hence the dashed extrapolations of the lines in Figure 10.5); these are, geologically speaking, relatively ephemeral environments. Nevertheless, they would appear to offer some of the best prospects for good microevolutionary records, and it is indeed the case that lake deposits, in particular, have furnished some of the most outstanding examples. Especially favourable sites for extensive accumulation of the latter are continental *rift valleys*. These are subsident linear belts, bounded variously on one or both sides by normal faults, caused by stretching of the continental crust.

The Pliocene–Pleistocene lake deposits around Lake Turkana, in northern Kenya, formed in part of the East African Rift Valley, were studied by Williamson (1981), who documented the changes in their shelly fauna. Both the stratigraphical correlations of the sequences here, as well as absolute dating of them, were based upon a number of widespread *tuff* (volcanic ash) beds. A part of Williamson's study particularly focused upon changes around one of these levels, the Suregai

Tuff (Figure 10.21). This lies near the base of a stratigraphical unit (the Lower Member of the Koobi Fora Formation), 110 m thick, which has a temporal scope (based on radiometric dates from tuffs) of about 100 000 years. Williamson's samples were taken from discrete sandy fossiliferous beds, each mostly less than 25 cm thick.

▶ Calculate the net rate of sediment accumulation in this sequence and decide whether this is typical for lake deposits (for the time-span of the sequence), and then calculate the stratigraphical completeness of the sequence, at the 100-year level of resolution. Assume a compaction factor of 2.

The net rate of sediment accumulation is 110 m/100 000 years = 1.1 m/1 000 years. Figure 10.5 indicates that the average rate for lake deposits, over a 100 000-year time-span, is about 0.3 m/1 000 years. (Remember, in reading off the value from the graph, that the scales are logarithmic.) Therefore, the rate recorded in this example is exceptionally high. This presumably reflects the sustained subsidence associated with the rifting, allowing sediment to accumulate with relatively little subsequent erosion.

Stratigraphical completeness (Equation 10.3) at the level of resolution of one century is:

$$\frac{110\,\text{m} \times 2}{(4\,\text{m}/10^3\ \text{years}) \times 100\,000\ \text{years}} \times 100\% = 55\%$$

We may therefore expect at least every other century to have some sedimentary record in the sequence—an impressive degree of completeness.

The fossils, however, are not spread evenly throughout the sequence, but are restricted to the shell beds mentioned above. It is difficult to assess their individual microstratigraphical acuities, because abnormal conditions may have led to their formation. Nevertheless, they are frequent throughout the sequence, and so the overall acuity of a given interval containing a number of shell beds can be assessed.

Five shell beds were distributed through the 2 m of sequence immediately beneath the Suregai Tuff, which was followed by a barren interval of almost 2 m, terminating in another shell bed. The thin layer of tuff itself, which presumably represents a single ash fall (or cluster of ash falls), may be ignored in the calculation, as a geologically instantaneous deposit. The time represented by the 4 m of sequence sampled (calculated on the same basis as microstratigraphical acuity) is estimated to be between 2000 and 3 636 years. The lower five samples are therefore probably separated from one another by intervals of a few hundred years, and the barren interval above the tuff may represent some 1 000 to 1 800 years, before the final shell bed.

The morphological changes in the shells over this interval are striking (Figure 10.21). Williamson undertook a multivariate morphometric study of them and found that although little change was evident in the the sequence below the 4 m described here, there were marked changes in several characters, in all the species, through the 4 m sequence itself. The differences between specimens of each species in the final shell bed and those in earlier shell beds lower in the sequence were considered by Williamson to be equivalent to those seen between the living species of the same genera. Higher up, following a further barren interval of nearly 3 m (about 1 500 to 2 700 years) the original morphologies return, and the modified forms disappear. The latter replacement is associated with evidence for a marked rise in the water-level of the original lake (Figure 10.21).

Williamson interpreted these findings as evidence for speciation according to the punctuated equilibrium model. The samples below the 4 m sequence were judged to exhibit stasis, while the changes in the interval around the tuff were interpreted as punctuational speciations of the local populations, now isolated from populations elsewhere because of the low level, and consequent fragmentation of the lake (rather like the speciation of the cichlid fishes discussed in Section 9.4.4 of Chapter 9). Williamson argued that these new species did not survive when the rise in the lake level allowed the return of the unaltered parent species from an unknown area. That the latter had therefore presumably persisted elsewhere was taken to suggest true speciation (i.e. branching evolution) of the locally modified populations.

There has been much discussion concerning whether or not these changes really did represent speciation, a particularly favoured alternative explanation being that the changes were ecophenotypic, associated with environmental stress in the then retreating lake. However, in the present context, what is of undoubted importance

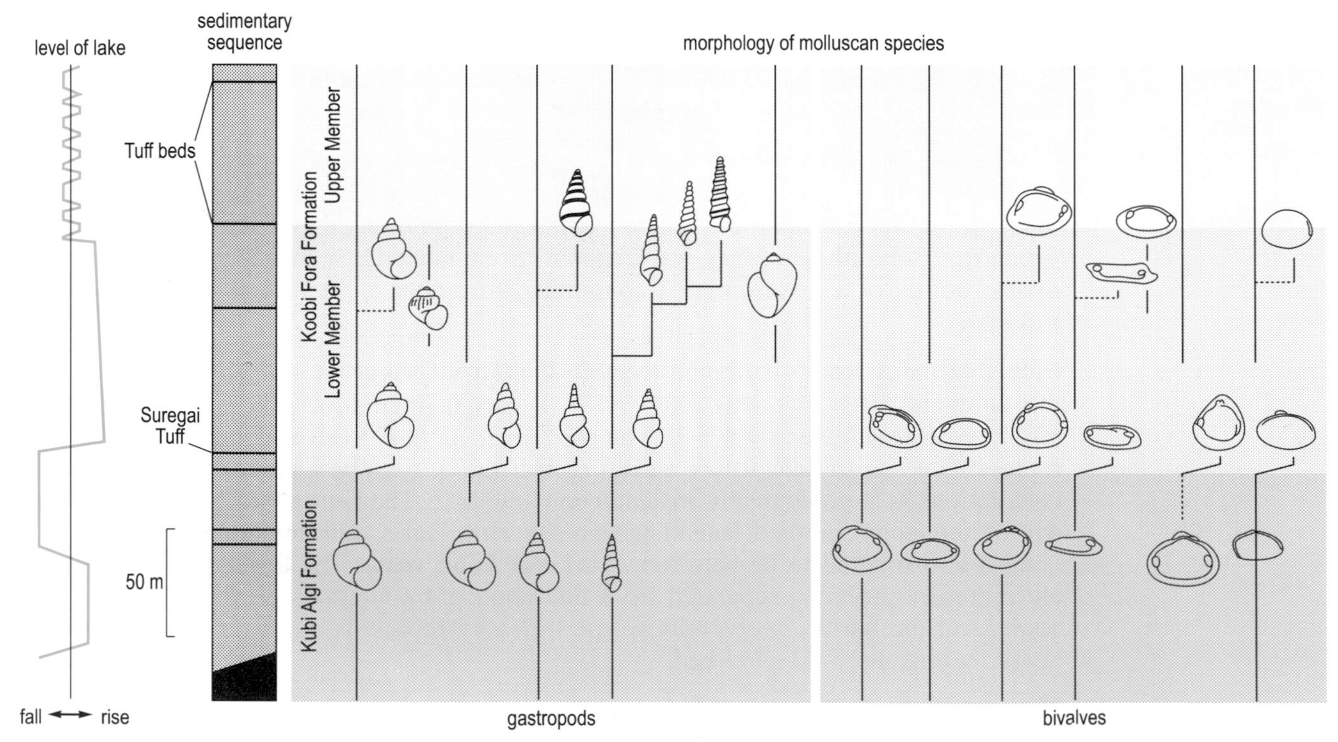

Figure 10.21 *Diagrammatic summary of patterns of morphological change in several species of gastropod and bivalve molluscs around the Suregai Tuff in the Plio–Pleistocene of Lake Turkana, Kenya. The sequence of lake sediments, with a diagrammatic curve indicating rises and falls in the level of the lake, is shown at left. Typical morphologies of specimens at various levels are indicated. Sideways shifts of the vertical lines for the species represent episodes of rapid morphological change (see text for discussion).*

is the demonstration of the possibility, *given the right geological circumstances*, of being able to sample changes in some fossil populations over intervals representing only a few hundred to a few thousand years. Where the divergence of populations is spread over thousands, or even tens to hundreds of thousands, of years, then such a record would certainly be adequate to monitor at least the broad pattern of morphological change in whichever population dwelt in the area of study.

Nevertheless, one virtually insurmountable problem for the complete documentation of speciation remains, even where the record in a sequence is as good as that discussed above.

► What other major aspect of morphological variability should be monitored in order to provide a record of the changes involved in the divergence of species?

Geographical variation: as allopatric modes of speciation are likely to be important in nature (Chapter 9), it is necessary that the geographical variability of populations also be monitored. Remember that the interpretation of the Lake Turkana shells as manifesting allopatric speciation *assumed* the continued presence of the unaltered populations elsewhere, beyond the study area.

There are, however, two distinct problems in monitoring geographical variability. First, there is the geographical incompleteness of the sedimentary record—if anything, a much more severe problem than stratigraphical incompleteness—and secondly there are the considerable difficulties attached to correlating between sequences in widely separated areas.

The lack of deposition of any sediment, or subsequent erosion, may result in no sedimentary record whatever in some (perhaps even most) parts of the original geographical range of a species. Even where there is a record, most of it will be practically inaccessible to the palaeontologist, who can sample only from surface exposures or from drilled cores. Moreover, all manner of logistical, environmental and political problems may allow only a fraction of what is accessible to be studied. Palaeontologists are therefore usually restricted to monitoring the geographical variability of fossil species on the basis of a limited number of widely separated geological sections. Given the usual stratigraphical incompleteness of most of these, the probability of their co-incidentally preserving samples of populations that fall even within the same 1 000-year interval, for example, becomes quite small. And the probabilities, in turn, of two or more isolated sections preserving samples from several successive intervals become miniscule. All this presupposes, in any case, that one would be able to correlate samples, from widely separated sections, within a few thousand years of each other.

Up until the 1960s, the correlation over large distances of sedimentary strata relied almost entirely upon fossils. The basic premise of such correlations is that any given species persists through only a single span of time, from its origin to its extinction (or pseudoextinction). Therefore, where its fossils may be correctly recognized, and it is possible to exclude the possibility of their having been eroded from older strata and redeposited, or, in the case of microscopic fossils, carried down into older strata during, for example, drilling operations, then the strata containing them can be correlated at least to within whatever is the time-span of

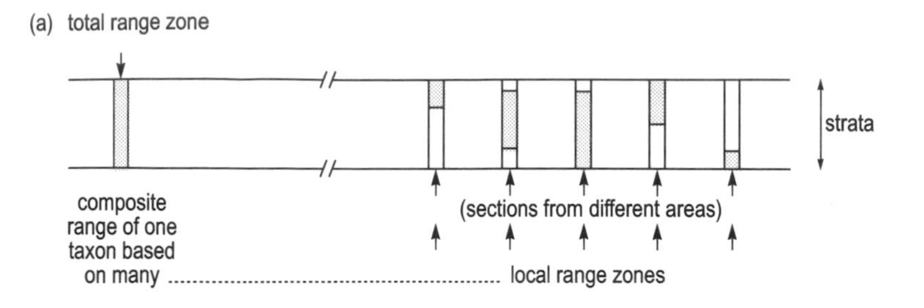

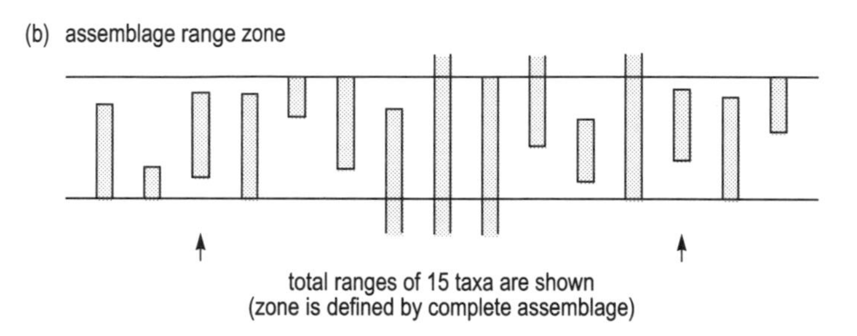

Figure 10.22 *Synopsis of the ways in which different kinds of biostratigraphical zone are recognized. (a) The total range for a given species is composited from several local ranges. (b) The ranges of several species can define an assemblage range zone.*

that species. More precise correlation can be obtained by identifying the more limited ranges of overlap of two or more species within sequences (Figure 10.22). Stratigraphical ranges recognized by such methods are termed **biostratigraphical zones**.

▶ Why are zonal methods of correlation unlikely to offer a satisfactory degree of resolution to clarify the geographical aspects of microevolution in most species?

The problem with zonal schemes of correlation is that they are based on the ranges of whole species, or of assemblages of species. These ranges can therefore resolve only the relative time relationships of the geographically separated populations of very slowly evolving species.

Since the 1960s, alternative, non-zonal methods of biostratigraphy (e.g. graphic correlation) have been developed, which offer a considerable improvement on the crude resolution of classical zonal methods. These have been joined by a variety of other, independent means of correlation, based upon the recognition of the physical or chemical signatures of global events such as reversals in the Earth's magnetic field. The details of these methods may be found in modern stratigraphical texts. These methods offer greater promise for high resolution studies of the geographical aspects of microevolutionary change in species. Nevertheless, examples are still likely to be scarce, in view of the problem of incomplete geographical coverage, and to remain largely limited to organisms with already restricted distributions (in single lake complexes for example) or those associated with unusually widespread deposits such as oceanic oozes. Planktonic microfossils in the latter show much promise in this respect. Figure 10.23, for instance, shows an example of morphological divergence following allopatric speciation in fossil radiolarians (protists with siliceous skeletons) in Pliocene to Pleistocene oceanic sediments in the North Pacific. The morphological changes shown here in the lineages of *Eucyrtidium* were detected in a single core, but other cores which have been correlated with it show that *E. matuyamai* initially evolved from a population of *E. calvertense* that was isolated for a short time about 1.9 Ma ago in a restricted area of the ocean. The majority of cases of allopatric speciation, however, are likely to remain incompletely recorded, if at all, for the reasons discussed above.

The best prospect which the fossil record has to offer, then, is sampling of the patterns of microevolution within *single* sequences, at a resolution of perhaps thousands, or in some cases even hundreds, of years, as already illustrated by the Lake Turkana study. Other examples have emerged, especially in response to the punctuated equilibrium/gradualism debate. Examples of everything from punctuational evolution and gradualism to stasis have now been extensively reported, and variously interpreted.

A particularly detailed study of microevolution in a lineage is that of Bell *et al.* (1985), on a species of stickleback fish, *Gasterosteus doryssus* (Figure 10.24), from the Miocene of Nevada, USA. Again, the sequence studied had been deposited in a lake, though apparently a saline one. A particular aid to study was the presence of annual sedimentary laminations, or *varves*, caused by seasonal

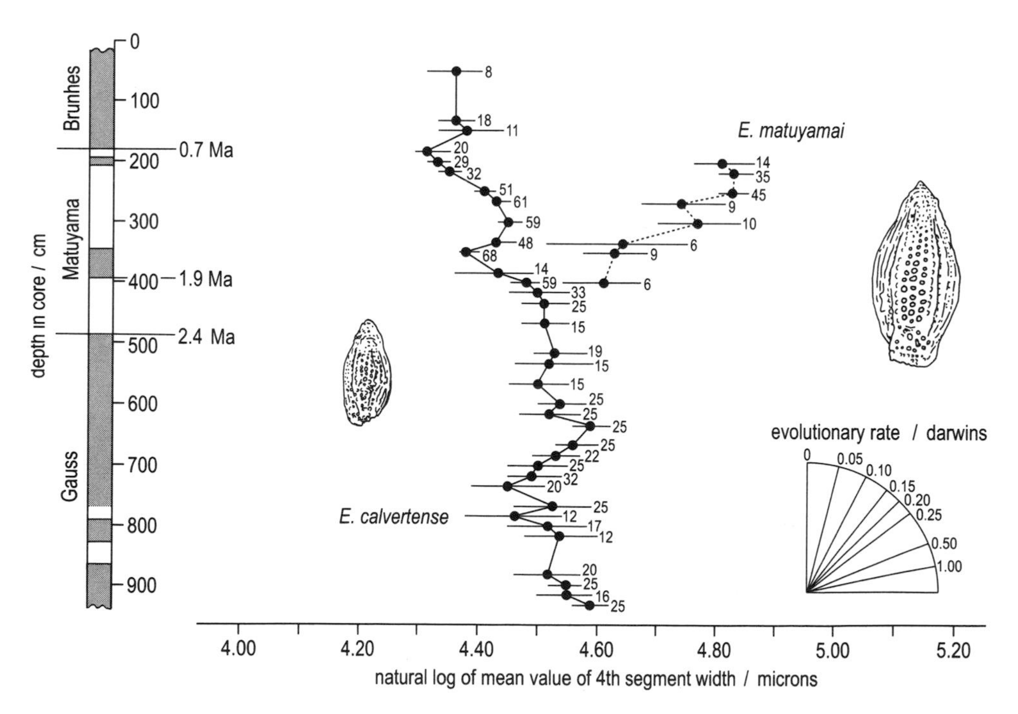

Figure 10.23 *Gradual morphological divergence of secondarily sympatric populations of the Plio–Pleistocene radiolarian,* Eucyrtidium, *from the northern Pacific Ocean, based on a core drilled from the ocean floor sediments. Depth in the core is shown at left, with a time-scale based on reversals of the Earth's magnetic field (normal field is shown in black, reversals in white). The variate shown on the* x *axis is the logarithm of the width of the skeleton. Means, 95% confidence intervals and sample sizes are shown.*

fluctuations in sediment supply. These allowed good estimates to be made of the time intervals between successive samples, removing the need for the sort of calculated estimates discussed earlier. Samples were collected at levels mostly corresponding to approximately 5 000-year intervals. A number of variates were studied, and the mean values of four of these (the variates indicated in Figure 10.24) for successive samples, are plotted in Figure 10.25.

Except for standard length, the variates selected were based on features established early in development, and so can be considered age-independent. In order to tackle the problem of age-dependence of the standard length, mean values of the largest 20% of specimens in each sample were plotted separately (the upper line in Figure 10.25(d)). The similarity between this and the line for all specimens (below) suggests that, despite the inclusion of juvenile specimens, mean standard length here reflects phenotypic size differences quite well.

▶ From what you read about living sticklebacks in Chapter 4, Section 4.3.1, which of the four features labelled in Figure 10.24 might you expect to have been particularly subject to evolution by natural selection?

The dorsal spines (DS) and the pelvic structure (PS), which supports other spines, are important in the living sticklebacks for defence against predators, and so might be expected to have been subject to strong selection.

Sample sizes were generally much greater than 30 specimens (and some more than 100), and so errors in the values of the means due to the effects of random sampling are likely to be relatively small. Statistical tests applied by Bell *et al.*, showed that some (at least) of the changes observed can be said to signify changes through time in the populations from which the samples were drawn.

The variates investigated do show slight developmental variation in living sticklebacks, according to temperature, but this is much less than that shown between some of the samples of the fossil specimens. The (significant) changes in Figure 10.25 can therefore be attributed with some confidence to genetic

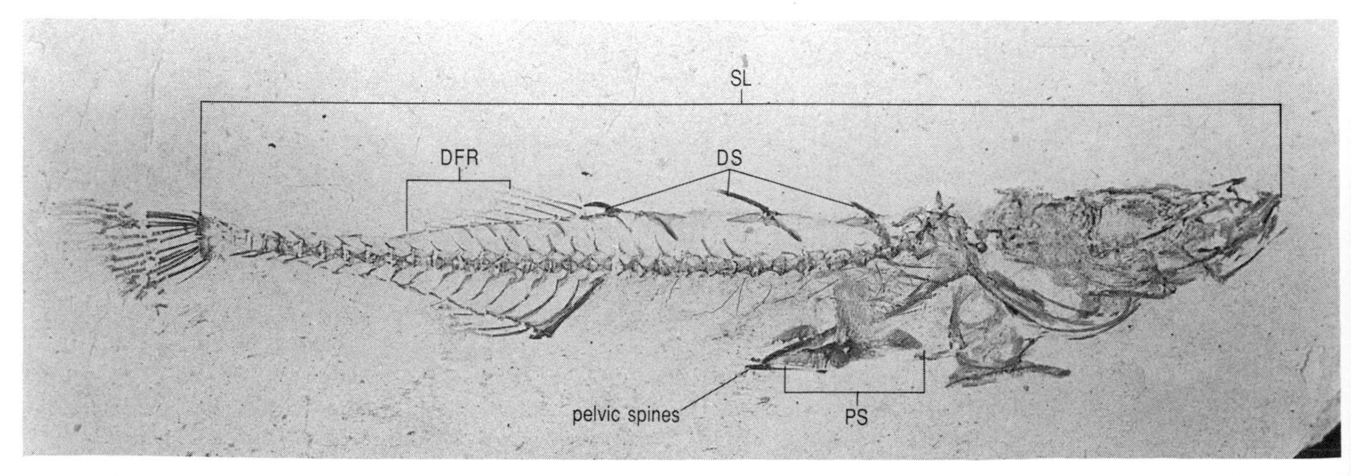

Figure 10.24 *Specimen of the fossil Miocene stickleback,* Gasterosteus doryssus, *showing four of the variates measured by Bell* et al. *(1985): DFR, dorsal fin rays; DS, dorsal spines; PS, pelvic girdle structure; and SL, standard body length. PS was recorded on an ordinal scale of numbered progressive modifications to the bone structure.*

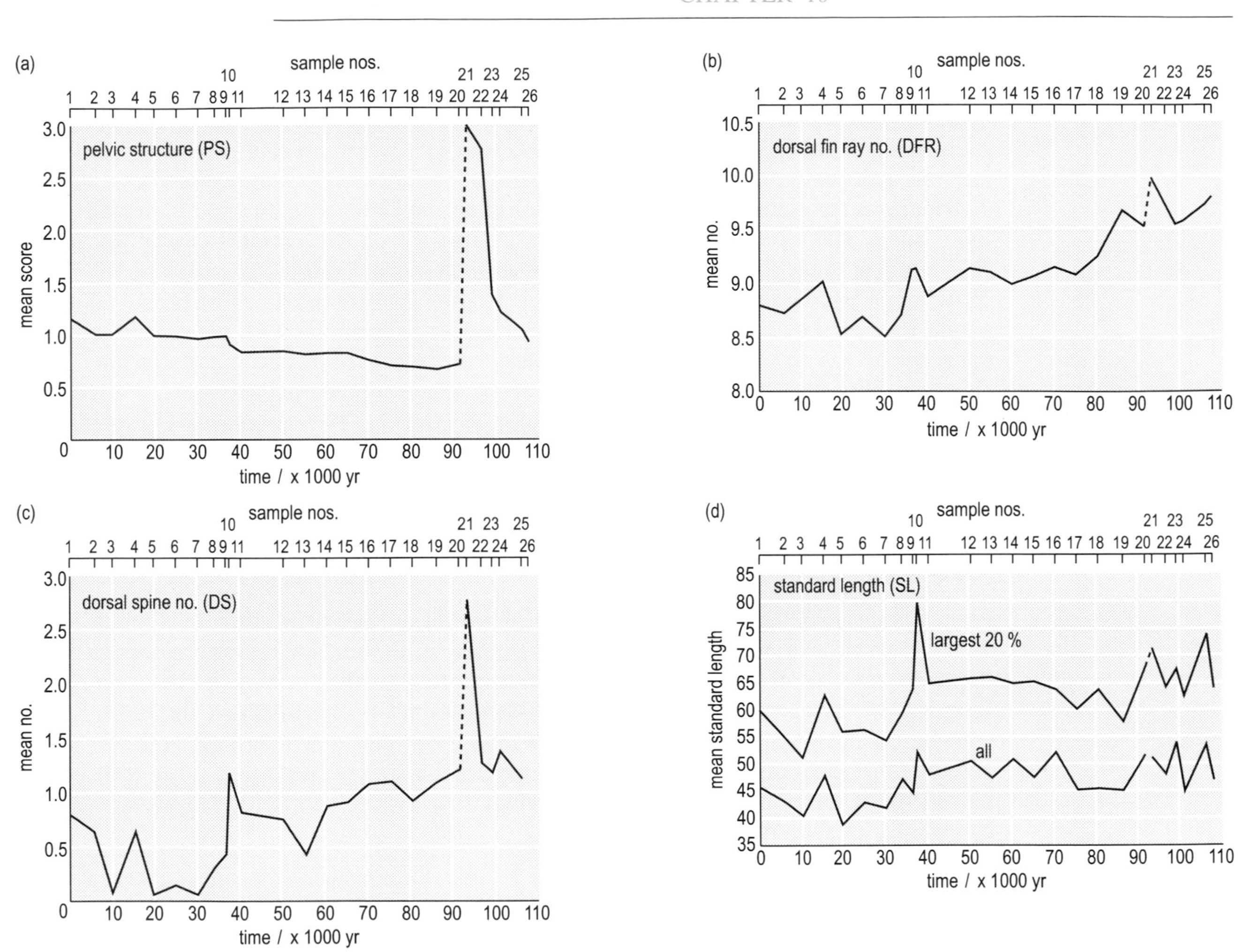

Figure 10.25 *Plots of mean values for the variates shown in Figure 10.24 through successive samples of* Gasterosteus doryssus *in a Miocene sequence, in Nevada, USA. See text for discussion of details.*

differences in the populations from which the samples were drawn. Particularly dramatic, and highly significant, is the sequence of changes in pelvic structure and the dorsal spine number, after sample 20. There is a sharp discontinuity between samples 20 and 21, followed by a rapid change in subsequent samples, so as to return to values similar to those before the discontinuity. How may these changes be interpreted?

Again, the major impediment to interpretation is the lack of knowledge of the geographical variability of the populations. It is open to question whether the morphological changes arose because of genuine microevolutionary change in essentially the same lineage, or whether the differences reflect the shifting to and fro in parts of the lake (or lake complex) of different populations. In the latter case, there might have been little or no evolution in the populations themselves; the changes seen in the study area could simply reflect movements of the different populations, into and out of the area.

Bell *et al.* indeed felt that the latter explanation was possible for the sharp change between samples 20 and 21, noting that the co-existence of two (living) species of *Gasterosteus* within a single lake has been observed. However, they were against such an explanation for the other changes. Any sympatric mixing of two species, they argued, would tend to yield histograms of variates with non-normal distributions (as with the hyaena teeth in Figure 10.11)—a feature not detected in their study. The more or less normal distributions of all the variates except, in a few samples, length (which, as an age-dependent variate might be expected to show some deviation) found by Bell *et al.* in their samples, argues in favour of a single species population in each (as explained in Section 10.2.3). Movement of a cline across the site might be posed as an alternative hypothesis to explain the changes. This, too, was rejected by Bell *et al.* They argued that the slow, consistent change in many characters, over several consecutive 5 000-year intervals (especially for the variate pelvic structure), would have required a quite implausibly steady and sustained migration of a cline within the confines of a single lake. Their interpretation, that the changes were evolutionary (except perhaps for those between samples 20 and 21), thus seems to be the most plausible hypothesis. If this is accepted, what then are the implications of the patterns of change detected?

First, it was noted that the close spacing of the samples had picked up a much more dynamic pattern of change than that noted in earlier studies at the same site. *Apparent* rates of change therefore varied according to the time-spans between samples. If, for example, only every fifth sample had been collected, the entire episode of change associated with sample 21 could have been missed, and a pattern of gradual change would have emerged. This very important point will be picked up again later, as it has major implications for palaeontological ideas on patterns of evolution.

Although the change from samples 20 to 21 may represent a discontinuity between different populations, the series of changes over the next few samples (i.e. over a time-span of only thousands to tens of thousands of years, according to the variate looked at), appears to document continuous, but relatively rapid evolutionary change. In a less complete sequence (or sampling programme), a change at this rate might well be missed, and if the initial and final stages were preserved, would then be registered as a single, punctuational shift in morphology. This therefore has all the requirements for Simpson's rapid phyletic evolution, unusually 'caught in the act', so to speak. The reasons for the changes, however, remain unknown. Although dorsal spine number and pelvic structure did show some marked changes, it is not known at present what part selective predation might have played in these. At present, evidence for predation on the sticklebacks by other fish is wanting. Yet a point worth noting is that the changes in the pelvic structure here are equivalent to the differences used by taxonomists to separate some families of bony fishes. A *gradual* morphological transformation, larger than the difference between species, can here be seen to have occurred over a period that would normally be considered *punctuational* by palaeontologists. Whether such changes are called gradual or punctuational thus depends on the scale of perception.

This result in turn allows one argument that has occasionally been advanced to explain punctuations to be rejected. Some evolutionary scientists have stressed that individual development is constrained, or *canalized* by the interactions of

Plate 2.1 Ammonite *Pleuroceras solare* (Lower Jurassic) from Alderton Hill, Gloucestershire, UK. *(Courtesy Neville Hollingworth)*

Plate 2.3 Polar bear *Ursus maritimus* has white fur but black eyes and black skin. *(Courtesy Caroline M. Pond)*

Plate 2.4 Muskox *Ovibos moschatus*, an arctic bovid with adaptations similar to those of reindeer. *(Courtesy Caroline M. Pond)*

Plate 2.2 Host, cuckoo and experimental model eggs. *Top row:* model eggs; *centre row:* real cuckoos' eggs; *bottom row:* hosts' eggs. *Left to right:* robin, pied wagtail, dunnock, reed warbler, meadow pipit, great reed warbler. *(Courtesy Nick Davies)*

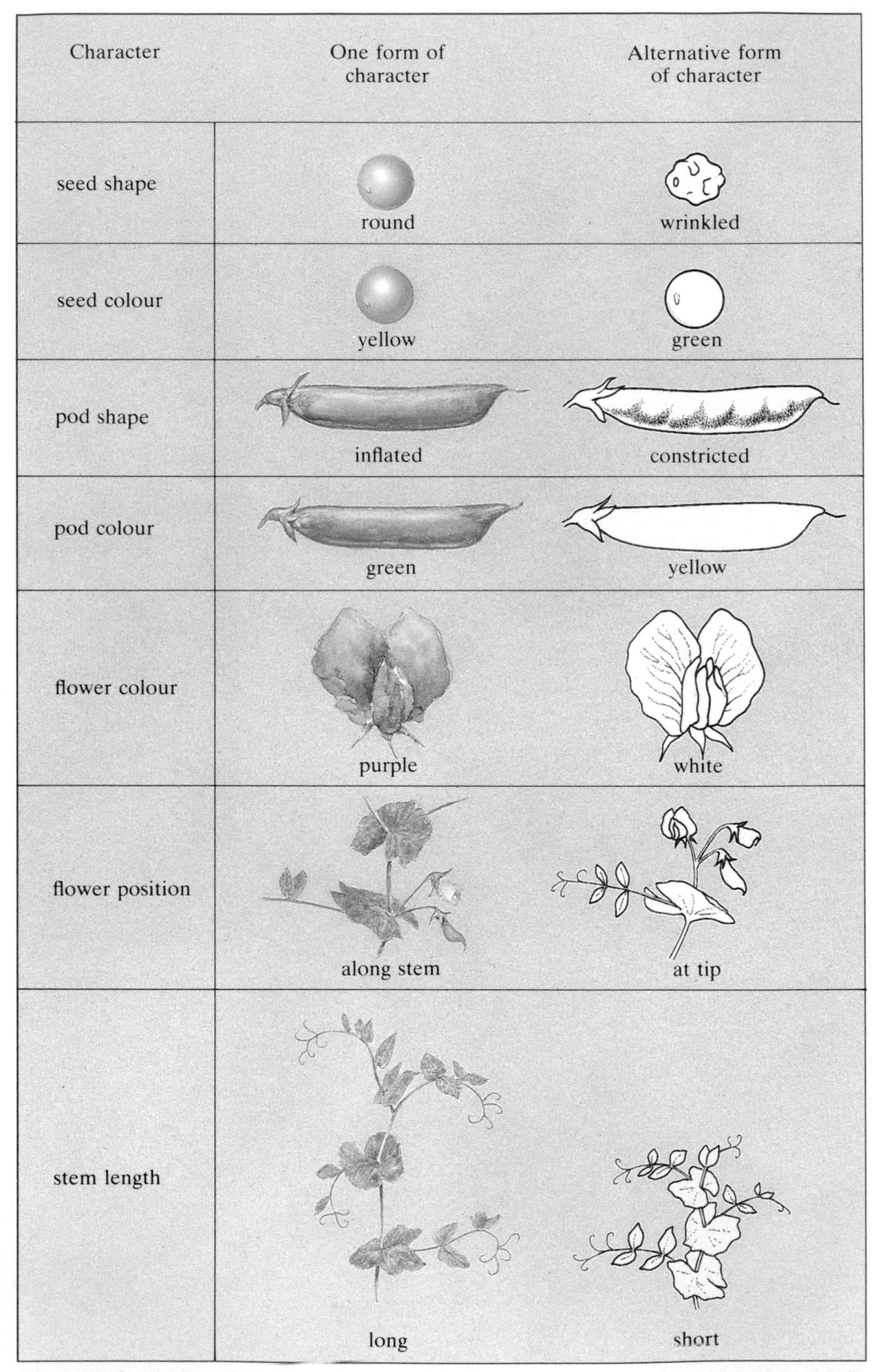

Plate 3.1 Pairs of contrasting characters in peas investigated in Mendel's breeding experiments. For each pair of characters, the dominant form is shown on the left and the recessive form on the right.

Plate 4.1 Guppies *Poecilia reticulata*: (above) female, (below) male.

Plate 7.1 Male peacock *Pavo cristatus* displaying to female. *(Courtesy Marion Petrie)*

Plate 9.1 The snail *Partula suturalis*: (a) dextral type — the shell coils to the right; (b) sinistral type — the shell coils to the left. (*Courtesy Prof. B. C. Clarke)*

Plate 9.2 Breeding coloration in *Haplochromis*. Above: *Haplochromis riponianus*, sexually active male. Note the prominent egg spots on the anal fin. Below: *Haplochromis brownae*, adult male, showing almost complete development of breeding coloration. *(Courtesy Natural History Museum)*

Plate 10.1 Fossil trilobites from the Middle Ordovician of central Wales: (a) *Cnemidopyge*; (b) *Platycalymene*, moulted specimen. *(Courtesy of Peter Sheldon)*

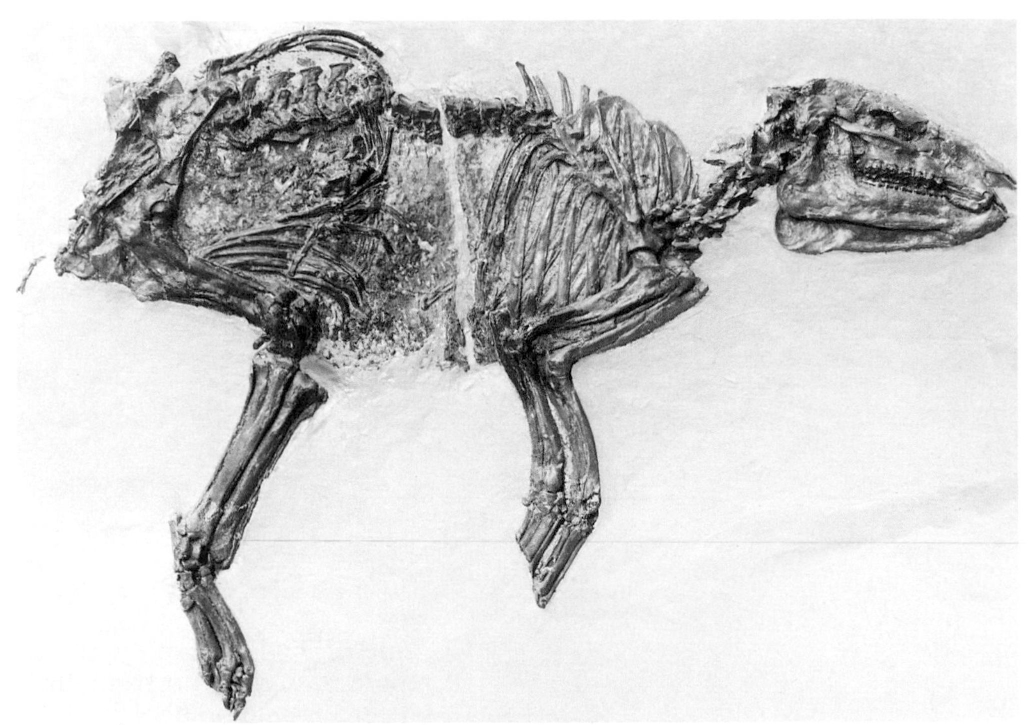

Plate 12.1 The primitive fossil horse *Propalaeotherium* from the Middle Eocene of Messel, south-western Germany. It was approximately the size of a cocker spaniel dog. *(Courtesy of David Amy, BBC, with permission of the Hessisches Landesmuseum, Darmstadt, Germany)*

Plate 12.2 Fossil magnolia leaf from the Miocene Clarkia fossil beds of northern Idaho, USA, containing preserved DNA. *(Courtesy of Edward M. Golenberg and David Gianassi)*

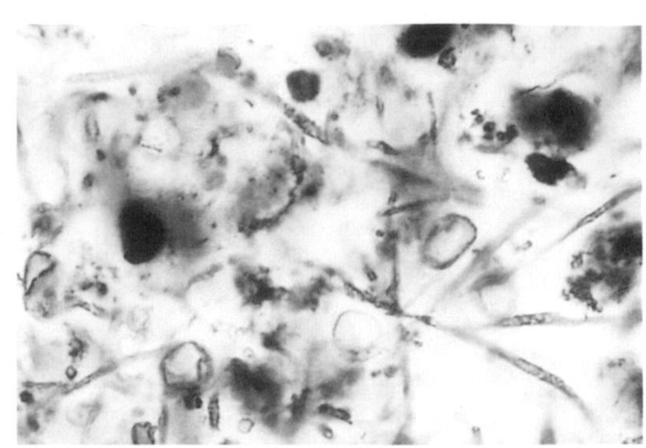

Plate 12.4 Photomicrograph of prokaryotic cells in chert from the Proterozoic Gunflint Iron Formation, Ontario, Canada. Spheres are *c*. 10 μm across. *(Courtesy of Andrew Knoll)*

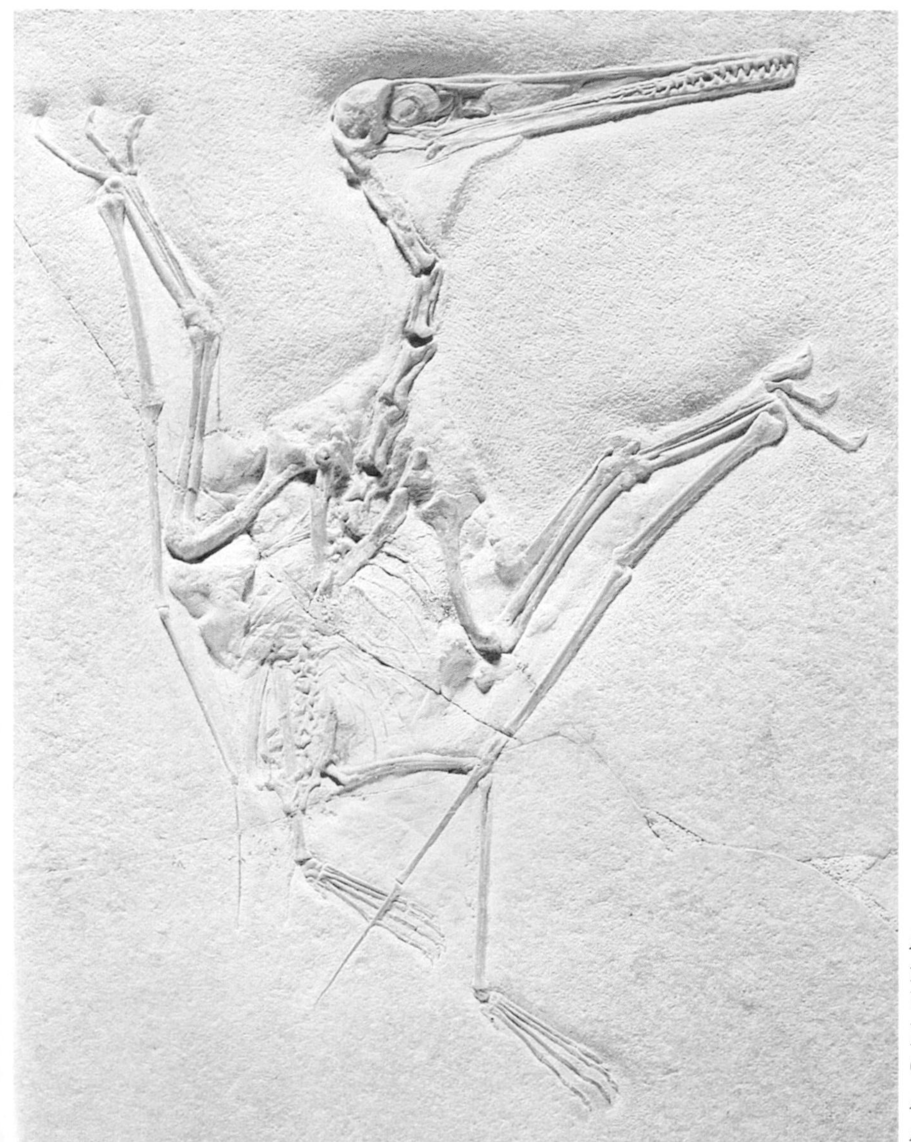

Plate 12.3 Fossil of the flying reptile *Pterodactylus* from the Upper Jurassic Solnhofen Limestone of southern Germany. *(Staatsamlung für Palaeontologie und Historische, Munich)*

Plate 14.1 Pearly nautilus. *(Courtesy of P. Laboute/Jacanda)*

Plate 14.3 *Hexagonaria* or Petoskey Stone, a Lower–Middle Devonian fossilized coral. *(Courtesy of Townsend P. Dickinson)*

Plate 14.2 Fossil of *Archaeopteryx* from the Upper Jurassic Solnhofen Limestone of southern Germany. *(Dr P. Wellnhofer, Munich)*

(a)

(b)

(c)

(d)

(e)

Plate 17.2 Domesticated breeds of sheep *Ovis aries*: (a) Soay; (b) Southdown; (c) Longwool (Wensleydale); (d) Black-faced. *(Courtesy of Caroline M. Pond)*

Plate 17.1 The giant toad *Bufo marinus* (*c*. 20 cm long) in its native Panama, Central America. *(Courtesy of Professor Tim Halliday)*

Plate 17.3 Père David's deer *Elaphurus davidianus* at Woburn Abbey, UK. *(Courtesy of Caroline M. Pond)*

regulatory genes, acting, so to speak, like a complex mesh of checks and balances to the pattern of growth. They believe that the only way changes on this scale can occur is through drastic *macromutations*, which divert development from its canalized pattern so as to produce novel 'hopeful monsters' in single, abrupt steps. This was not, it should be stressed, a part of the original formulation of the punctuated equilibrium model, which was, instead, based upon Mayr's peripatric model of speciation, involving rapid, but continuous changes in small, isolated populations (Section 10.3.1). Some confusion did arise later (principally in the late 1970s and early 1980s), simply because observed punctuation in the fossil record might, in theory, encompass both kinds of change; the nature of the record is usually such that it cannot distinguish between the two possibilities. The particular value of the highly resolved record of Bell *et al.*, is that the detection of continuity in what might elsewhere be seen as a punctuational change rules out the macromutational hypothesis, in this case at least.

In this instance there is no evidence for branching evolution. The rapid morphological change therefore cannot be definitely associated with speciation. The retention of all the samples within a single species, by the authors, indicates that they chose not to recognize chronospecies, despite the magnitude of the changes. Figure 10.25 shows this to have been a sound decision. Had sample 21 been recognized as a distinct species from earlier samples, then the return in later samples to pretty much the same morphology would have made it difficult to avoid the taxonomic *faux pas* of a species with a divided stratigraphical range. In fact, this is not so much a problem for evolutionary theory, but rather, it highlights one of the shortcomings of the (pre-evolutionary) taxonomic system of discrete category names applied to changeable populations. The problem is best avoided (as in this case) by recognizing some fossil species on the basis of a rather broader range of variation than that found in living species, so that the sorts of fluctuations seen in Figure 10.25 can be accommodated. Chronospecies can then be limited to when a lineage eventually and irreversibly evolves into a new and different morphological range (for several variates). You may recognize this problem, of continuous variability extending beyond normal species limits, as being analogous, in the time dimension, to that noted, in the geographical dimension, in *ring species* (see Chapter 9, Section 9.7).

Another apparent casualty of this study is the hypothesis of universal stasis in species—one of the essential tenets of the punctuated equilibrium model (Section 10.3.1). It is clear that this species did not remain morphologically static.

Although stasis is therefore not universal in established species, examples of it have nevertheless been convincingly documented. Stanley and Yang (1987), for example, undertook a multivariate study of 19 burrowing bivalve lineages in late Tertiary and Quaternary deposits of the North Atlantic and Mediterranean (Italy). They found that, over the last 4 Ma, virtually all the changes in the variates studied in the fossil samples of each lineage remained more or less within the scope of the differences seen between different populations of the corresponding living species. Thus, although there had indeed been some small-scale zig-zag patterns of change between successive samples, each of the lineages had remained essentially confined within the same limits of variability recognized in the living species. The same results were even established for three of the species going back some 17 Ma.

Stasis has also been demonstrated in Cheetham's (1986) study of Miocene and Pliocene species of a Caribbean bryozoan genus (see Section 10.2.1), called *Metrarabdotos*. Multivariate analysis revealed no significant *overall* change (according to t tests) in the morphology of any of the species investigated. Moreover, the variance of short-term rates of change within established species was found to be significantly less (in F tests) than that of minimum inferred rates of change during the evolution of nine new species from their putative ancestors. In all cases, the related species overlapped in time and so the new species must have arisen by branching. Thus, the morphological changes involved in speciation were significantly greater than those recorded within established species lineages.

▶ With which model of species evolution in Figure 10.20 is this example in best agreement?

The pattern of evolution in *Metrarabdotos* species is most consistent with that expected by the punctuated equilibrium model (Figure 10.20(c)). The close correspondence between morphospecies and biological species established among similar living bryozoans (Section 10.2.1) suggests that, in this case, the fossil record is a faithful reflection of the pattern of biological speciation.

If rapid phyletic evolution, punctuational speciation and stasis have thus all been adequately demonstrated in the fossil record, so, too, has longer-term *gradualistic* change. An outstanding example is that documented by Sheldon (1987), who studied the evolution of eight lineages of trilobites in an Ordovician sequence in central Wales which had a temporal scope of some 3 Ma (Figure 10.26).

The sequence studied by Sheldon was deposited in a marine shelf-basin, with good stratigraphical completeness, as expected from Figure 10.5. The principal variate

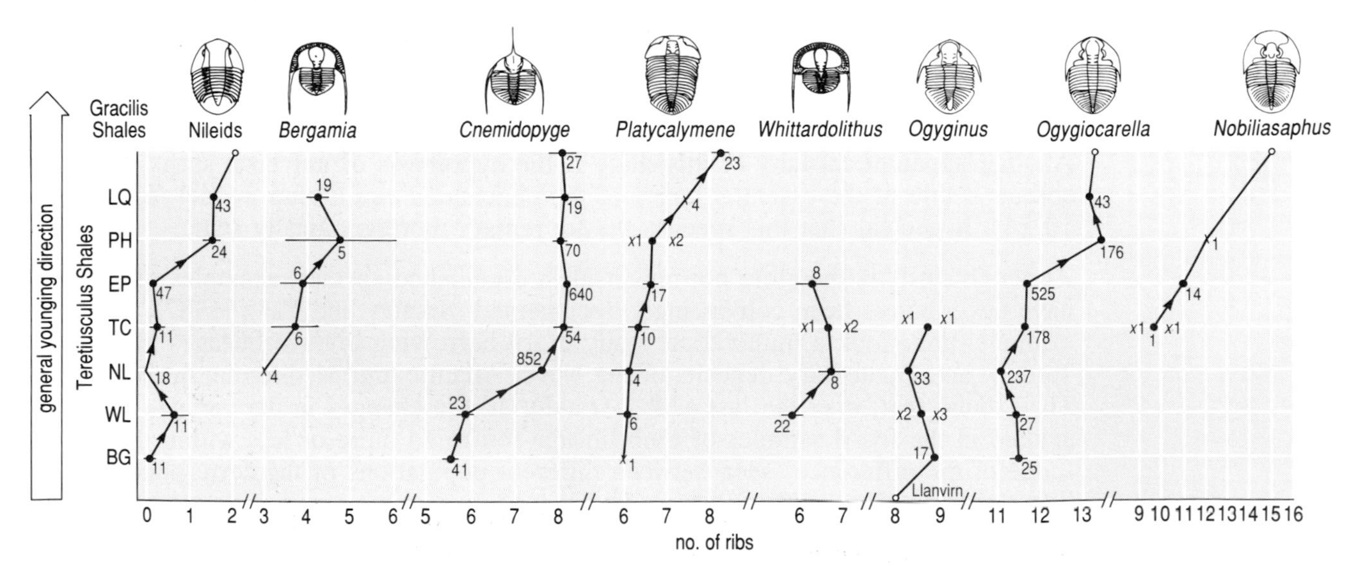

Figure 10.26 *Mean values (with 95% confidence intervals) of the numbers of pygidial ribs in successive samples from eight trilobite lineages in an Ordovician sequence in central Wales. Small arrow-heads indicate significant changes. Sample sizes are indicated next to the data points. The letters at left denote (pooled) sampling intervals, in chronological sequence.*

studied was the number of radial ribs on the tailpiece or *pygidium* of the trilobites (Plate 10.1). These counts, like those of the facial pits in *Onnia* (see Section 10.2.3), are almost age-independent, except in the most juvenile stages.

Sheldon found that all eight lineages showed a slow net increase in the rib counts, though there was little short-term parallelism in the changes. In particular, a number of reversals were also noted within some of the lineages, which were not, however, synchronized in the different lineages. These changes are summarized in Figure 10.26, in which significant changes in each lineage are marked with an arrow-head. Sheldon interpreted the changes as evolutionary, rather than ecophenotypic, since the latter kind of variation might have been expected to show a far greater degree of parallelism in the different lineages.

The reversals are of particular interest. Had the record been poorer, or the sampling less thorough, these might have been missed, and so would have been subsumed in much slower *apparent* rates of change in the lineages. Indeed, Sheldon's data points shown in Figure 10.26 are the pooled values for whole stratigraphical sections, and involved some aggregation of consecutive samples. He has noted, however, that when the original single mean values are plotted out, an increasingly zig-zag pattern (with even more reversals) emerges at finer time-scales. As in the other studies described, then, it is evident that, with increasing resolution, the fossil record tends to reveal an increasingly dynamic pattern of evolutionary rates. This provides a vital clue to what has for a long time been considered a paradox by many evolutionists: many of the *gradualistic* trends documented from fossils *seem* to have been extraordinarily slow, when compared with the rates of change that may be observed, as a consequence of natural selection, in living populations in the laboratory, or the field. If some of these fossil rates were taken at face value, it has been calculated that only about one selective death per 100 000 individuals per generation would be needed to account for them — a rate that even random drift could be expected to swamp! With reversals built into the record, however, the paradox evaporates: the apparent slowness of gradualism is just a product of poor resolution.

This conclusion, evident from all the highly resolved records discussed in this Section, should lead us to expect an inverse relationship between measured rates of evolution and the time-spans considered, precisely analogous to that found for rates of sedimentation. Exactly such a relationship has indeed been noted by Gingerich (1983), from a broad survey of measured rates (Figure 10.27).

In order to compare rates of evolution in quite different features from different organisms, Gingerich used a standardized unit of morphological change called a **darwin**. This is a proportional increase of 2.718 in any linear measurement per million years. The number 2.718, or 'e', has some rather useful mathematical properties when used as a base for logarithms, which need not concern us here. A logarithm to the base of e, or **natural logarithm (ln)** of any given number (x) is the power to which e must be raised to equal that number:

$$e^{\ln x} = x \qquad (10.13)$$

The natural logarithms of numbers can be readily obtained on scientific calculators, or from standard Tables. An evolutionary rate in darwins can thus be calculated for any linear variate according to the formula:

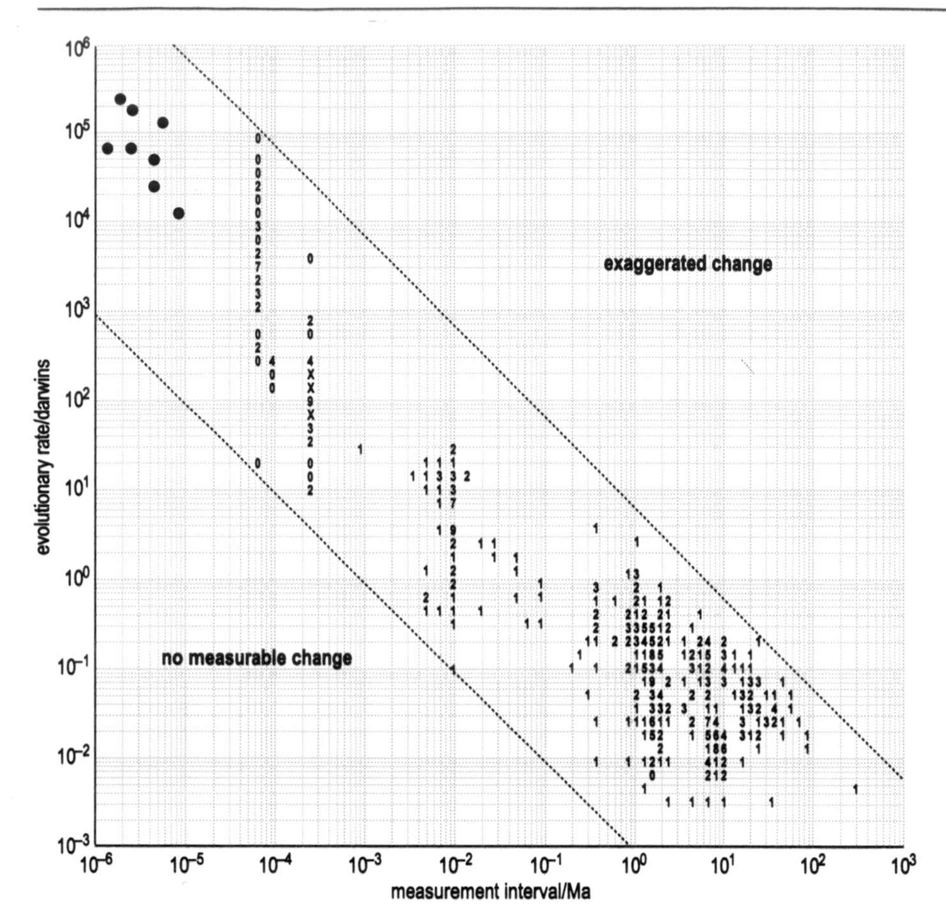

Figure 10.27 *Double logarithmic plot of measured (net) rates of evolution (in darwins) against time-span (measurement interval) considered. Dots, zeros and crosses are single data points from different sources; other nos. refer to nos. of data points.*

$$\text{Rate of change in darwins} = \frac{\ln \bar{x}_2 - \ln \bar{x}_1}{t_2 - t_1} \quad (10.14)$$

where $\bar{x}_1$ and $\bar{x}_2$ are mean values for the variate in question at two successive times, t_1 and t_2, measured in millions of years. Notice from Figure 10.27 that logarithmic scales are used (as in the plot for sedimentation in Figure 10.5).

Putting the results of these various examples of documented evolution in fossil lineages together, rather an interesting pattern emerges. From the relatively coarse time-scale of observation usually permitted by the fossil record, it is evident that punctuational evolution, stasis and gradualism may all be observed in the fossil record. Yet it is clear that these are all gross patterns, and that, in every case, a common pattern of dynamic, small scale change is evident at finer levels of resolution. These rapid changes are much more compatible with rates expected from the action of natural selection, and so the gross patterns should be seen not so much as direct expressions of microevolutionary rates, but rather as the different possible *net* effects of fluctuations over long time intervals, shown in an imperfect record. Looking at microevolution in the fossil record is thus like trying to trace an intricate pattern through frosted glass: the main outlines of the pattern can be seen, but its finer details are all but lost.

10.3.3 Implications for the evolution of species

In Section 10.3.1, three palaeontological models for the pattern of evolution of species were outlined, two of them versions of *phyletic gradualism* and the third, that of *punctuated equilibria*. How do these now stand, in light of the findings presented in Section 10.3.2?

The 'classical' gradualist model, in its most exaggerated form (Figure 10.20(a)), can be immediately rejected (as it had been, in effect, by G. G. Simpson: see Section 10.3.1). The appearance of extremely slow, gradual change can now be explained as being an artefact of poor resolution, as explained above.

The dynamic pattern of rapid fluctuation detected in the best resolved records confirms that the microevolution familiar to population geneticists is simply too fast to be registered in the vast majority of sequences. This observation is certainly consistent with both abrupt change in single lineages (Figure 10.20(b)) and the punctuated speciation of the punctuated equilibria model (Figure 10.20(c)). We now need to consider if punctuational change is limited to small, peripherally isolated populations, hence usually involving speciation (as in the latter model), or whether it may affect entire species populations.

The examples discussed in Section 10.3.2 show that, although stasis exists, it is not ubiquitous among fossil species. Also, it is apparently only a gross pattern, perhaps no less dynamic than any other when looked at with fine enough resolution. These findings contradict the idea of stasis through inertia, of the theory of punctuated equilibria (Figure 10.20(c)). The Simpsonian or *punctuated gradualistic model* (Figure 10.20(b)) remains as the most satisfactory general model, encompassing the variety of patterns reviewed in Section 10.3.2. Punctuation need not always coincide with speciation and stasis reflects only an absence of *net* change. Nor should we forget the biological evidence for sibling species. These show that speciation need not immediately entail morphological divergence. A wide spectrum of patterns is possible.

That is by no means the end of the story, however, because explanations for the *gross* patterns of stasis and gradualism are still needed. It is evident from the discussion above that these two patterns are really at opposite ends of a continuum. Indeed, individual lineages may show different patterns at different times, and different characters may evolve at different rates (Figure 10.25 for example).

If the finer-scale fluctuations are, most plausibly, attributed to the interplay of natural selection, genetic drift and other such conventional principles of neo-Darwinian theory, then the gross patterns must reflect the longer-term context within which the finer-scale mechanisms operate. One argument that has steadily been gaining ground is that stasis reflects net stabilizing selection, effectively confining any population to a more or less static *adaptive peak* (Chapter 4, Section 4.5).

Given the environmental changes that occur all the time, it might at first seem rather a preposterous proposal that the selective regime, and hence adaptive landscape, of so many species might have remained so conservative for millions, if not tens of millions, of years. However, one possibility is that these species may effectively track particular habitats through time and space, such that they then remain subject to conservative regimes of selection, notwithstanding the kaleidoscopic change of environments. There are two ways in which this tracking of habitats may come about. First, many organisms actively seek out suitable habitats for which they are adapted, a behaviour pattern which is itself adaptive (Chapter 9, Section 9.6.1). Juvenile mussels, for example, settle twice during their development, the first time as larvae, *away* from adult clusters (in which there is a high risk of their being drawn in on the adults' inhalent feeding currents), and the second time as slightly larger, juvenile mussels, amongst established adult clusters, to which they are now directly attracted. Secondly, even where there is no active habitat choice, the preferential proliferation of individuals which happen to have dispersed into favourable habitats has the same effect on the eventual distribution of populations.

Accordingly, populations may be seen as being largely confined, by conservative stabilizing selection, to adaptive peaks, which are themselves either static, producing stasis in the long term, or slowly changing through time, producing (gross) gradualistic change. In both cases, the busy fluctuations of populations at the finer scale might reflect either the effects of small, ephemeral changes in the adaptive landscape itself, or genetic drift within adaptive peaks, in populations of smaller effective size (Chapter 3, Section 3.6). Some punctuational change might, however, reflect the effects of a peak shift to a higher adaptive peak with ensuing shifting balance (Chapter 4, Section 4.5).

Why, in turn, some adaptive peaks may remain static for long periods of time, while others change, is largely a matter for speculation at present. The question is likely to be a focus of much interest in future years.

SUMMARY OF SECTION 10.3

Three models for the pattern of microevolution in species have been proposed by palaeontologists. The 'classical' version of phyletic gradualism is gradual morphological change within lineages and speciation by gradual divergence. A revised version, punctuational gradualism, involves greatly varying rates of morphological change in lineages, ranging from abrupt change (reflecting directional selection) to negligible change (reflecting little or no net directional selection). Rapid morphological evolution is necessarily associated with speciation in the model of punctuated equilibria, based on Mayr's peripatric theory of speciation. In this model the budding speciation of small, localized populations is postulated to produce punctuational changes in morphology in the fossil record, while large, established species populations, in contrast, are believed to show morphological stasis.

The suitability of sedimentary sequences to test between these models depends upon high stratigraphical completeness and good microstratigraphical acuity. These two factors tend to be mutually exclusive, but may be associated in areas of sustained rapid geological subsidence, such as rift valleys. In some lake sediments morphological changes in single lineages at levels of resolution of hundreds of years can be studied.

It is very difficult to monitor geographical variation in fossil species, for two reasons. First, there is the geographical incompleteness of the sedimentary record, and secondly, conventional methods of correlation often cannot resolve the relative ages of geographically separated populations of a single species. Newer methods of correlation may go some way towards easing this problem. The fossil record best reveals the patterns of change in lineages in single sequences.

The fossil record provides some examples of rapid phyletic evolution, at a rate which would appear punctuational in a less complete sequence, or programme of sampling. At a larger scale, both stasis and gradualistic change can also be demonstrated, and seem only to be end members of a continuum. High resolution studies of these, however, show that both tend to comprise a finer-scale pattern of rapid fluctuations of morphology, including frequent reversals. There is consequently an inverse relationship between measured net rates of morphological change and the time-spans considered. This explains the apparently paradoxical discrepancy between rates of evolution determined from the fossil record and those measured in living populations.

The exaggeratedly smooth version of phyletic gradualism and punctuated equilibria may be rejected as universal models. Morphology may show anything from rapid, or punctuational, evolution to stasis in lineages, and there is no exclusive link between morphological change and speciation. Nevertheless, the gross patterns of stasis and gradualism still need to be explained. Stasis may result from net stabilizing selection, effectively confining a population to a stable adaptive peak, and gradualism may result from a changing adaptive landscape. Long-term stabilizing selection may be promoted by habitat-tracking.

10.4 Conclusion

Within its limits of resolution, and when interpreted with due caution, the fossil record can throw some light on the tempo and mode of microevolutionary change, including that associated with speciation. A wide variety of phyletic patterns has now been documented, including *punctuational* to *gradualistic* morphological change, and *stasis*. Although allopatric speciation can be inferred in a few cases, the original mode of most speciations is obscured by both geographical and stratigraphical gaps in the record. Moreover, the existence of sibling species today suggests that the morphological divergence of some fossil lineages may have post-dated their biological speciation.

None of these patterns is inconsistent with the expectations of neo-Darwinian theory, as explained in earlier Chapters, though the variety of the patterns suggests a corresponding variety of processes in nature: no one explanation is universally applicable. Nevertheless, some of the larger-scale patterns, especially stasis and long-term gradualistic trends (both superimposed on more dynamic patterns at finer scales of resolution), would not necessarily have been predicted by neo-Darwinian theory. The explanation of these larger-scale patterns poses a challenge for future detailed palaeontological studies.

REFERENCES FOR CHAPTER 10

Bell, M. A, Baumgartner, J. V. and Olsen, E. C. (1985) Patterns of temporal change in single morphological characters of a Miocene stickleback fish, *Paleobiology*, **11**, 258–271.

Cheetham, A. H. (1986) Tempo of evolution in a Neogene bryozoan: rates of morphologic change within and across species boundaries, *Paleobiology*, **12**, 190–202.

Darwin, C. R. (1859) *On the Origin of Species by Means of Natural Selection, or The Preservation of Favoured Races in the Struggle for Life*, John Murray, London.

Eldredge, N. and Gould, S. J. (1972) Punctuated equilibria: an alternative to phyletic gradualism, in T. J. Schopf (ed.), *Models in Paleobiology*, 82–115, Freeman, Cooper, San Francisco, California.

Gingerich, P. D. (1983) Rates of evolution: effects of time and temporal scaling, *Science*, **222**, 159–161.

Jackson, J. B. C. and Cheetham, A. H. (1990) Evolutionary significance of morphospecies: a test with cheilostome Bryozoa, *Science*, **248**, 579–583.

Kurtén, B. (1988) *On Evolution and Fossil Mammals*, Columbia University Press, New York.

Schindel, D. E. (1982) Resolution analysis: a new approach to the gaps in the fossil record, *Paleobiology*, **8**, 340–353.

Sheldon, P. R. (1987) Parallel gradualistic evolution of Ordovician trilobites, *Nature*, **330**, 561–563.

Smith, A. B. and Paul, C. R. C. (1985) Variation in the irregular echinoid *Discoides* during the early Cenomanian, *Special Papers in Palaeontology*, **33**, 29–37.

Stanley, S. M. and Yang, X. (1987) Approximate evolutionary stasis for bivalve morphology over millions of years: a multivariate, multilineage study, *Paleobiology*, **13**, 113–139.

Williamson, P. G. (1981) Palaeontological documentation of speciation in Cenozoic molluscs from Turkana Basin, *Nature*, **293**, 437–443.

FURTHER READING FOR CHAPTER 10

Briggs, D. E. G. and Crowther, P. R. (eds.) (1990) *Palaeobiology: A Synthesis*, Blackwell Scientific Publications, Oxford.

Cope, J. C. W. and Skelton, P. W. (eds.) (1985) Evolutionary case histories from the fossil record, *Special Papers in Palaeontology*, **33**, 203.

Davis, J. C. (1986) *Statistics and Data Analysis in Geology*, 2nd edn, John Wiley & Sons, New York.

Gingerich, P. D. (1985) Species in the fossil record: concepts, trends and transitions, *Paleobiology*, **11**, 27–41.

Harper, D. A. T. and Ryan, P. D. (1990) Towards a statistical system for palaeontologists, *Journal of the Geological Society*, **147**, 935–948.

Lande, R. (1986) The dynamics of peak shifts and the pattern of morphological evolution, *Paleobiology*, **12**, 343–354.

Levinton, J. (1988) *Genetics, Paleontology, and Macroevolution*, Cambridge University Press, Cambridge.

McKinney, F. K. (1991) *Exercises in Invertebrate Paleontology*, Blackwell Scientific Publications, Oxford.

Sadler, P. M. (1981) Sediment accumulation rates and the completeness of stratigraphic sections, *Journal of Geology*, **89**, 569–584.

SELF-ASSESSMENT QUESTIONS

SAQ 10.1 (*Objectives 10.1 and 10.5*) Imagine that a palaeontologist wishes to compare some fossils, which have suffered geological alteration, with unaltered and intact shells in a museum collection. By what practical means might the problems of preservational differences be overcome, to allow morphological comparison, in the following cases:

(a) The fossils comprise only moulds of bivalve shells in a sedimentary rock matrix.

(b) The fossils are of ammonites that have been deformed by compaction, as in Figure 10.6(a), in a deposit of shale with concretions like that in Figure 10.6(b).

SAQ 10.2 (*Objectives 10.2 and 10.3*) Assume that a geologist is studying the deposits of an ancient inland sea and those of some coastal wetlands which originally formed a part of its margin. In both deposits the dominant sediments are sandy mudstones containing concretions, as in Figure 10.6(b), from which a value of c can be estimated for both deposits. A section of thickness 10 m is studied in each deposit, and both sections are found to have about the same temporal scope, approximately 100 000 years.

Using the data in Figure 10.5, estimate:

(a) The stratigraphical completeness of each section, at a limit of time resolution of 100 years;

(b) The minimum and maximum values of microstratigraphical acuity of a 10 cm interval in each section (assuming no obvious sedimentary changes within the sequences).

Which section yields the more favourable value of (a), and of (b), for the purposes of palaeontological sampling, and how might these differences be explained?

SAQ 10.3 (*Objective 10.4*) Why is the *evolutionary species concept*, if literally applied, impractical for palaeontological purposes?

SAQ 10.4 (*Objectives 10.5 and 10.6*) Study the fossil oysters shown in Figure 10.2. Each shows an indented *scar* at its beak, marking where the shell was originally attached to the substrate.

(a) Would you consider the size of this scar to be an age-dependent variate?

(b) Is the size of the scar likely to be a suitable variate in distinguishing separate species among such fossil oysters?

SAQ 10.5 (*Objectives 10.6 and 10.7*) In view of three of the variates in Figure 10.25 showing marked and significant changes in mean values from sample 20 to sample 21, it is of interest to inspect the fourth variate, *standard length*, SL, to see if any such change can be detected there. Assuming that, in this instance, mean SL values can be taken to reflect normally distributed adult size differences (Section 10.3.2), F and t tests can be applied. The basic statistics for the two samples are shown in Table 10.1.

Table 10.1 *Statistics of SL values for samples 20 and 21 of the fossil stickleback* Gasterosteus doryssus, *studied by Bell* et al. *(1985).*

	n	Mean SL/mm	Standard deviation/mm
Sample 20	68	51.51	12.312
Sample 21	62	50.45	14.564

For a pair of samples of these sizes, the critical values of F and t for rejecting the null hypothesis of equivalence, at a level of significance of 0.05, are +1.65 and ±1.98, respectively. Now decide:

(a) Do the samples differ significantly, either in variance, or in mean value?

(b) What is the implication for the interpretation of the changes seen in the other variates?

SAQ 10.6 (*Objective 10.8*) Why might it be misleading to compare the ratios of growth-dependent variates (e.g. shell length/shell width) in samples of specimens, and in which circumstances is such a comparison justified?

SAQ 10.7 (*Objectives 10.9, 10.10 and 10.11*) Study Figure 10.23, which plots changes in the mean values of a linear measurement (*note*: already converted to natural logarithms), on the x axis, against depth in the core (and time), on the y axis, for the radiolarian *Eucyrtidium*.

(a) Calculate the *net* rate of evolution of the measured variate in *E. matuyamai*, in darwins, between its first and last mean values, dated at 1.9 Ma and 0.7 Ma ago, respectively. You can check your result against the diagram at bottom right of the Figure as follows. Since the x axis is already expressed in natural logarithms, evolutionary rates in darwins are directly shown by the slope of a line. By finding the slope which corresponds to the net change described above, you can confirm the rate from the diagram.

(b) Again using the rate diagram, estimate the highest rate of change shown in the record of *E. matuyamai* (just below 300 cm depth in the core), and the time-span over which that episode of evolution is shown as having taken place.

(c) Now plot the two sets of results (rate vs. time-span) found in (a) and (b) on the graph shown in Figure 10.27. Does the difference between them correspond with the general trend for measured rates of evolution shown in Figure 10.27?

(d) With which of the three models of evolutionary change shown in Figure 10.20 do your findings best agree? Explain how the results specifically disagree with the expectations of the other two models.

Part III
Macroevolution

CHAPTER 11

EVOLUTIONARY RELATIONSHIPS AND HISTORY

Prepared for the Course Team by Peter Skelton

Study Comment *Evolution beyond the species level—macroevolution—is the concern of Chapters 11–15. Central to all studies of macroevolution is the analysis of evolutionary relationships between both living and extinct species, and so the methods by which these are analysed are particularly emphasized in this Chapter. It starts, however, by introducing the main sources of information and the methods of study used to investigate macroevolution.*

Objectives When you have finished reading this Chapter, you should be able to do the following:

11.1 Explain how, in principle, macroevolutionary hypotheses may be tested.

11.2 Determine whether given higher taxa are monophyletic, paraphyletic or polyphyletic from information concerning their phylogenetic histories.

11.3 Outline the three main methods for classifying species in a taxonomic hierarchy and for analysing their evolutionary relationships (evolutionary systematics, phenetics and cladistics), and apply them to given examples.

11.4 Decide which of the differing character states in a group of organisms are likely to be primitive, and which are probably derived, given suitable information about the distribution of the character states among these and other organisms, the patterns of individual development of the organisms, and/or their fossil record.

11.5 Given information on the character states found in a number of taxa, construct a cladogram for the taxa, and analyse their possible phylogenetic relationships.

11.6 Explain the principle of the 'molecular clock' and its role in phylogenetic reconstruction.

11.7 Interpret the phylogenetic implications of phenograms and cladograms based on molecular data.

11.1 THE STUDY OF MACROEVOLUTION

The emphasis so far in this Book has been on microevolution. Chapters 11–15 deal with evolution above the species level, or **macroevolution**. This concerns changes both in the form and in the diversity of organisms established over long periods of time through the accumulated evolution and extinction of many species. The study of macroevolution is therefore necessarily historical, and ideas about what happened, and how, can only be inferred from the legacy of the remaining evidence.

Evolutionary relationships are inferred from comparative studies of species, both living and fossil, and in practice such studies largely focus upon their morphological (and/or physiological and behavioural) features on the one hand, and their molecular features (genes and gene products) on the other. These are topics addressed in later Sections of this Chapter.

As already noted in Chapter 10, fossils provide the only tangible evidence of past life, and indeed the only evidence at all of extinct organisms. In order to assess what light fossils can shed on macroevolution, Chapter 12 delves further into how fossilization takes place, and the retrieval of information from the fossil record.

The evolutionary history of organisms is also reflected in their geographical distributions, both today and in the past. The study of these is taken up in Chapter 13, along with a brief review of those aspects of the Earth's physical development which seem to have had an important influence on the distributions of species.

The patterns of macroevolution reconstructed from these various lines of evidence, and the interpretation of how they are likely to have come about, are explored in Chapters 14 and 15—the former dealing with the evolution of form in organisms, and the latter with patterns of diversification and extinction.

Macroevolution, like any other area of historical analysis, poses an interesting problem for scientific methodology.

▶ How is the scientific investigation of hypotheses concerning macroevolution limited, in terms of the method of study?

The testing of hypotheses by experiment is clearly ruled out because the events considered generally unfolded over vastly greater periods of time than that available to human scientific observation. In contrast, some hypotheses about fundamental microevolutionary processes, such as natural selection itself, can be experimentally tested, as Chapter 4 illustrated. This limitation does not mean, however, that hypotheses concerning macroevolution cannot be tested at all (and are therefore unscientific, as some people have mistakenly supposed). Rather, a different kind of testing is required. A given historical hypothesis generally allows one to predict certain other necessary consequences besides the known evidence upon which the idea was originally founded. Different associated consequences may be predicted from alternative hypotheses. If such additional lines of evidence are confirmed by subsequent observation, then the hypothesis in question may be considered increasingly probable. In some instances, such probabilities may even be assessed statistically. This procedure constitutes **retrospective testing**, because it relies upon the observation of the already existing, but perhaps hitherto unrecorded, outcomes of past events. You might think of it, in effect, as predicting the outcome of a natural experiment that has already been run. A similar approach is adopted, for example, in astronomy: the 'Big Bang' theory for the origin of the Universe is hardly amenable to experimental repetition, yet detection of the theoretically expected 'background radiation' makes it a highly plausible scientific theory.

11.1.1 INVESTIGATING PATTERN AND PROCESS IN MACROEVOLUTION

The sorts of questions that are addressed in macroevolutionary studies can be illustrated with reference to Figure 11.1. This shows a reconstructed evolutionary tree, or **phylogeny**, of the reptiles, birds and mammals, of a kind commonly encountered in evolutionary texts.

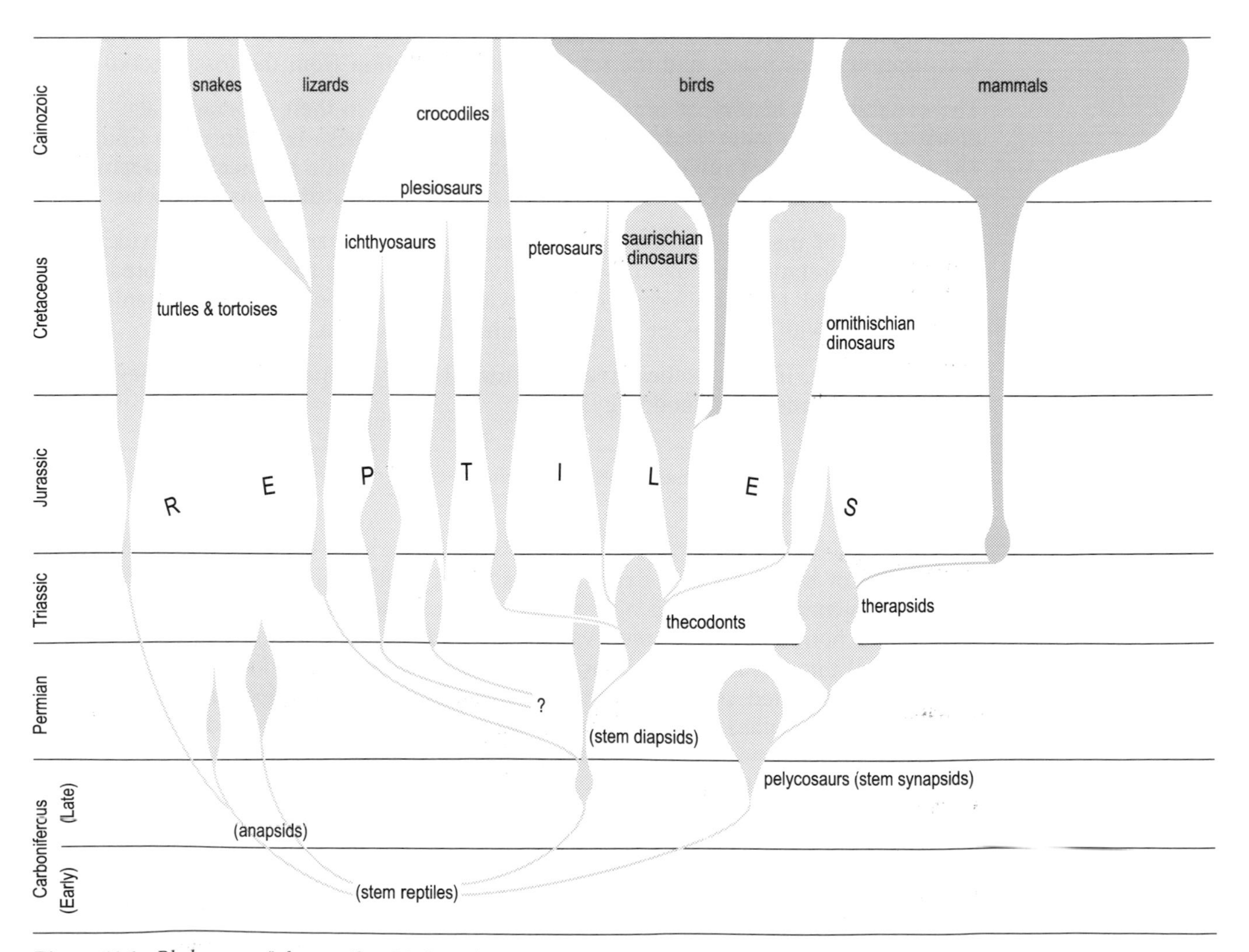

Figure 11.1 *Phylogeny of the reptiles, birds and mammals.*

The diagram has the appearance of a branching tree or bush. In this case, the branches do not represent individual species lineages, though they might be shown in other phylogenies. When considering such large numbers of species, it is more convenient to group related species together into a few higher taxa (such as orders or classes: Chapter 1, Section 1.3.3), and it is these that the branches on the diagram represent. The pattern of branching shows the inferred evolutionary relationships of these taxa through time, which is indicated by the vertical axis. Branches that reach the top contain extant members, while the rest of the tree shows the life of the past. Several of the branches are extinct, terminating below the top of the diagram.

The width of each branch is intended to show its **taxonomic diversity**, i.e. the numbers of constituent taxa (such as species, genera or families) at any given time. Ideally, it would be preferable always to show numbers of contemporaneous species. Usually, however, reliable estimates of their numbers cannot be obtained because of the incompleteness of the fossil record, insufficient taxonomic documentation of both living and fossil species, and uncertainties arising from the practical difficulties involved in identifying them, as explained in Chapters 9 and 10. For these various reasons, numbers of genera or even higher taxa are commonly estimated instead, as proxies for the relative numbers of species. Because a genus may contain several species, its fossil record is likely to be relatively more complete than that of any individual species: any fossil from *any* constituent species is sufficient to register the existence of a genus in a given time interval. For the same reason, the fossil record of families is yet more complete. The higher the taxonomic category, the more likely a taxon is to be represented by some fossils. Moreover, specimens can usually be allocated to higher taxa even when individual species have not been discriminated. Some of the differences which distinguish members of the cat family (Felidae) from that of the dogs (Canidae) were discussed in Chapter 2, Section 2.2.2: allocating a fossil specimen to one or other family could readily be accomplished without having to determine to which species it belonged.

There is nevertheless a problem in using higher taxa instead of species for measurements of taxonomic diversity. Contemporaneous species can be considered as naturally discrete entities, even if there may be some practical problems in recognizing them (Chapter 9, Section 9.2).

▶ Can higher taxa likewise be thought of as *naturally discrete* entities?

No. Although different species may be grouped together on the basis of similarity (morphological, genetic or otherwise) to form natural clusters, there is no consistent criterion for deciding how inclusive each cluster should be. Humans, chimpanzees and gorillas, for example, certainly form a natural cluster, recognized today as the family Hominidae, but the decision to draw the line there, and exclude the more distantly related orang-utan (see Chapter 3, Section 3.5.2), is arbitrary. So higher taxa are arbitrarily defined in terms of their limits, although taxonomists endeavour to ensure that only evolutionary relatives are included within those limits. The donkey, for example, could not be included as a natural member of the Hominidae. In other words, the phylogenetic 'tree of life' must be chopped into separate parts by the taxonomist, to create the higher taxa.

▶ What problem does this arbitrariness pose for basing taxonomic diversity on numbers of higher taxa such as genera or families?

The measurements of diversity will depend upon the scheme of classification used, which may in turn render comparisons between groups misleading. Taxonomists have traditionally tended to assemble species in higher taxa between which there appear, subjectively, to be marked differences. So the taxonomic categories recognized in different groups of organisms are bound to differ in scope. For example, is a dinosaur family, on average, equivalent in scale to a mammal family, or genus, or even order? This question could be answered by counting numbers of species in several of the better known families of the two groups to see if they do show any systematic differences. In general, when higher taxa are employed for estimates of taxonomic diversity, it is assumed either that variation in the numbers of species per family (or genus, or whatever) yields comparable *average* values for the major groups being compared, or that the higher taxa are themselves distinct adaptive groupings (e.g. the mammalian order Carnivora), which allow an assessment of adaptive diversity. These assumptions, and the need to test them, should always be borne in mind when studying patterns of evolution such as that shown in Figure 11.1, and particularly when investigating their causes, as some aspects of the pattern may be no more than taxonomic artefact.

Two broad kinds of question may be asked when considering a scheme such as Figure 11.1. These concern pattern and process, respectively.

Pattern The pattern questions have to do with the construction of the diagram itself. In addition to the problems in gauging relative taxonomic diversity mentioned above, there may be difficulties in reconstructing the phylogenetic relationships of the various groups shown. For example, nobody was around, of course, to document the evolution of the earliest mammals from some therapsid ('mammal-like' reptile) ancestor in the late Triassic, at first hand. This and other phylogenetic relationships have had to be inferred from comparisons between fossil and extant species. Ideas about these relationships may indeed vary according to which features of the species are studied (see Section 11.2). There has been some argument, for example, as to whether all the mammals evolved from a single ancestral species: the celebrated vertebrate palaeontologist George Gaylord Simpson suspected that the monotremes (the egg-laying platypus and echidna of Australia) evolved independently of all other mammals from different therapsid ancestors. Later detailed morphological comparisons and molecular work, however, have strongly supported the theory of common ancestry for the mammals. Here, then, the molecular data served as a retrospective test of the hypotheses of relationship based upon morphological information.

Phylogenies such as that shown in Figure 11.1 are thus only theoretical reconstructions of history, which, like any other scientific theory, must be tested against alternatives.

Process With the acceptance of any given phylogenetic pattern, the focus of interest can shift to the processes that caused it. One might ask, for example: was the waxing and waning of the various groups the random outcome of myriads of independently caused speciations and extinctions — simply the 'luck of the draw'—or were there more widely embracing causal factors affecting large numbers of species together? Look, for example, at the timing of the extinctions shown in Figure 11.1.

▶ Does the hypothesis of groups becoming extinct merely through random extinction seem likely, given the pattern of extinction of the groups shown in Figure 11.1?

It does not seem likely: of the twelve groups which are shown to have become extinct, five did so at or near the end of the Cretaceous Period. Remember, also, that each group comprised many species. Such a coincidence could not reasonably be expected had the species extinctions been purely random.

Such episodes of large-scale extinction, affecting many different groups within a short interval, are known as **mass extinctions**, and they have long been recognized as important features of macroevolution. The search for a cause naturally turns to some large-scale environmental change, of catastrophic proportions. Physical evidence for such a change at the time can again be sought by way of retrospective testing. This particular example, the Cretaceous–Tertiary boundary extinction, is perhaps the best known, because the available evidence can be interpreted as implicating a major asteroid impact (Chapter 15).

The timing of episodes of expansion of certain groups likewise invites explanation. It is remarkable, for example, that the mammals (shown to the right in Figure 11.1) underwent such a rapid diversification in the early Tertiary, despite having been around at much lower levels of diversity since the late Triassic. This particular episode of rapid, sustained increase in numbers of taxa also involved the spectacular diversification, over some 15 Ma, of adaptive types within the class, ranging from bats to whales. This episode of large–scale **adaptive radiation** might be attributed to ecological release, resulting from the earlier demise of the dinosaurs and the flying and swimming reptiles. Some other groups which came through the mass extinction, such as birds, lizards and snakes, likewise underwent concurrent radiations. These are the kinds of issue discussed in Chapter 15.

Another sort of question is *how* did the birds evolve? What changes in the morphology, behaviour and/or circumstances of their reptilian ancestors brought into being the novel features, such as feathers and wings, which have allowed them to exploit so many aerial (and other) habitats? The origins of such major innovations and the circumstances which allow them to evolve are discussed in Chapter 14.

SUMMARY OF SECTION 11.1

Because of the time-scale involved, macroevolutionary hypotheses can be tested only retrospectively, i.e. by seeking evidence for additional necessary consequences not expected from competing hypotheses. Macroevolutionary studies can be broadly divided into those concerning pattern and those concerning process. Pattern questions involve the evolutionary relationships of species (their phylogeny), their systematic arrangement in higher taxa, and assessments of the taxonomic diversity of the latter. Process questions probe the causes behind the patterns of origination and extinction of taxa. Major patterns of particular interest are mass extinctions, adaptive radiations and the appearance of evolutionary innovations.

11.2 Systematics and the reconstruction of phylogeny

To the lay person, it might seem surprising that there is any problem with the recognition of higher taxa. The very existence of long-established vernacular names for inclusive groupings of species (e.g. finches, thrushes, parrots and hawks as distinct groups of birds) suggests that higher taxa are self-evident. Accordingly, the task of the taxonomist might seem merely to consist of recognizing these groupings and assembling them in a hierarchy of increasingly inclusive categories.

Indeed, taxonomists had been at work on this task long before Darwin's time. The original intention of most earlier taxonomists had been to reveal the 'divine plan of Creation': each higher taxon was seen to be united by a common basic design, and the differences between its constituent species to be derived from the specific adaptations of the design for different places in nature. The relative 'affinities' (similarities of basic design) between species were thus widely held to reflect a natural order before Darwin and Wallace re-interpreted them as indicating phylogenetic relationships (Chapter 1, Section 1.3.3). In the light of evolution, the hierarchy of taxa could now be seen as reflecting the phylogenetic tree of life.

If the phylogeny of life were itself known, there would be little problem in recognizing higher taxa: the only arbitrary part of the exercise, of deciding where to 'sever' the branches to delimit the higher taxa, could simply be a matter for consensus. The problems arise because, as noted earlier, the true phylogeny is unknown, and can only be inferred from the available evidence. In practice, then, organisms are grouped according to criteria deemed to reflect relationship, and phylogeny is construed from these groupings. Conclusions may vary not only according to the characteristics of the organisms which are investigated, but also according to how they are analysed.

11.2.1 Taxa and relationships

Until the mid-20th century, inferences about evolutionary relationships between species were generally based upon as wide a range of evidence as could be mustered. **Evolutionary systematics** is the name given to this eclectic approach, because of its explicit focus on evolutionary conclusions. The disparate nature of the evidence used (ranging from the taxonomic attributes and geographical distribution of living organisms to the stratigraphical distribution of fossils) meant that there was no single underlying method of analysis, and so the conclusions were reached by a variety of lines of reasoning. Consequently, the discipline became notoriously the domain of widely experienced experts, who tended to acquire an unfortunate reputation in the popular imagination as a sort of unassailable priesthood. Frustration with the lack of a consistent method of analysis, and hence with the ultimately subjective nature of evolutionary systematics, led, in the 1950s, to the development of two new approaches to systematics both of which claimed to be more objective: in **phenetics**, species are clustered according to their overall morphometric similarities; in **cladistics**, relationships are inferred from the extent to which different species share evolutionarily modified features apparently derived from common ancestors. Yet neither new approach proved to be without its problems, and so all three continue to be practised today. The methods and the advantages and disadvantages of each approach will be discussed in the following Sections. Nevertheless, cladistics has emerged in recent years as the most powerful and widely used method of phylogenetic analysis in most instances, and so most emphasis will be given here to this approach.

Before considering the different approaches to phylogenetic analysis, some general points need to be made concerning evolutionary relationships.

▶ Using the idea of blood relationships in people as an analogy, can you think of two distinct types of relationship between species?

One is the relationship of descent—as in the parent–child relationship—and the other is that of shared parentage, as with brothers and sisters. Similarity between species may reflect either of these two kinds of relationship. It would be misleading to push the analogy too far, though, for two reasons. First, with the exception of allopolyploid hybrid species (Chapter 9, Section 9.6.3), new species are derived from single parental species. Secondly, a newly evolving species derives its characteristics directly from those currently present in the ancestral population: there is nothing in a species population that corresponds to the unchanging germ line of individual parent organisms (Chapter 3, Section 3.2). Two fundamental patterns of change in a phylogeny thus give rise to the differences between related species: **anagenesis** refers to descent with modification, within any given single lineage; and **cladogenesis** refers to the evolutionary division of lineages causing a proliferation of species. The grouping of species to form higher taxa can emphasize either or both of these components of phylogeny (Figure 11.2).

If species are grouped together because they show a similar extent of accumulated anagenetic change with respect to their ancestors, then the taxa so formed constitute **grades**. In Figure 11.2, morphological change is represented along the horizontal axis. The three columns show grades of anagenetic modification, with parts of the phylogeny occupying each grade. Grades are easy to recognize, because they are based upon raw similarities between species, but they may be misleading as far as the reconstruction of phylogeny is concerned.

▶ Why may some grades contain more than one branch of a phylogenetic tree (as in the central column of Figure 11.2)?

Convergent features (Chapter 2, Section 2.3.3) may have evolved independently in separate lineages. A grade grouping of such species based on these features (as in the central column of Figure 11.2) would thus exclude their latest common ancestor (which remains in the left-hand column in Figure 11.2). Thus the central grade grouping in Figure 11.2 does not comprise a single branch from a

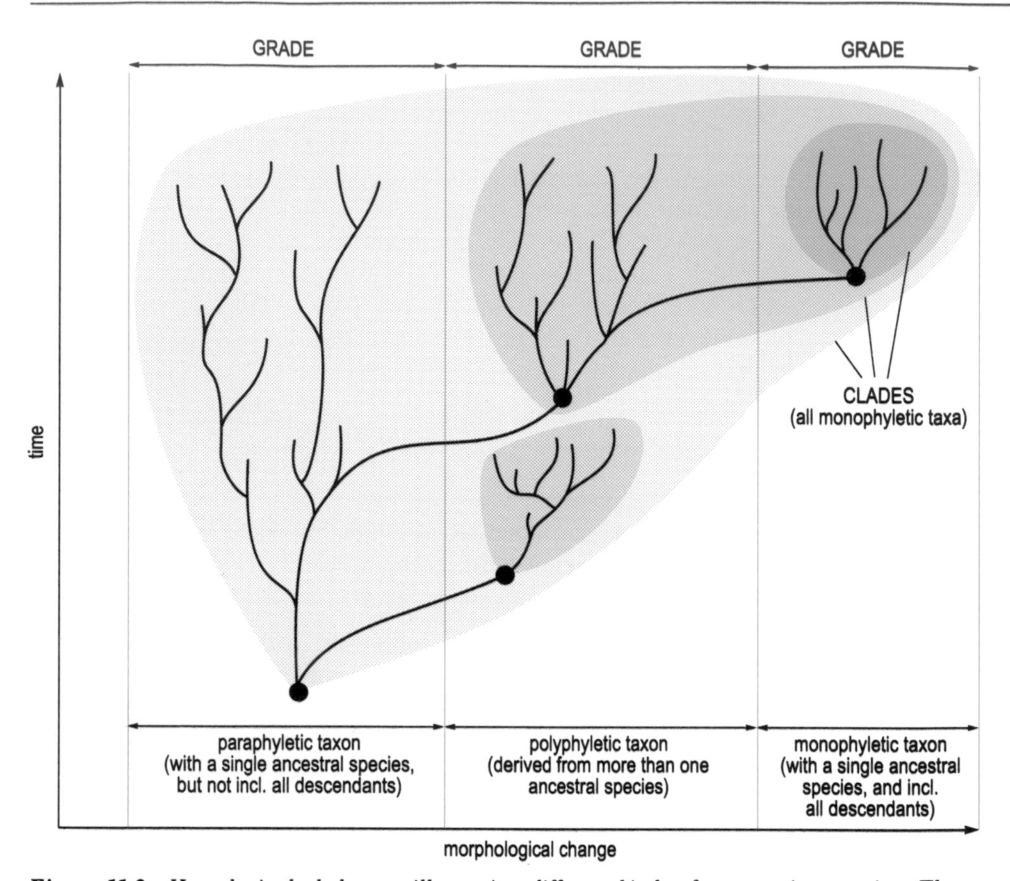

Figure 11.2 *Hypothetical phylogeny illustrating different kinds of taxonomic grouping. The three columns represent morphological grades.*

phylogenetic tree. A grouping which assembles species with independently evolved similarities is said to be **polyphyletic**. Some early Victorian naturalists, for example, grouped elephants and rhinoceroses together as 'pachyderms', because of their thick skins (which is what the name means in Greek). However, it has long since been recognized that this feature is convergent in these animals, and so the name is no longer used for systematic purposes. Modern taxonomists attempt to avoid using polyphyletic taxa because they are misleading in phylogenetic reconstruction. To continue to talk about 'pachyderms', for example, might create the false impression that elephants and rhinoceroses are more closely related to each other than either is to, say, horses. Numerous other lines of evidence indicate, rather, that horses and rhinoceroses are the more closely related pair.

Not all grades are polyphyletic, however, and grade groupings which do include their common ancestor (as in the left and right columns in Figure 11.2) make up modern classifications. In Figure 11.1, for example, birds and mammals represent groupings considered to differ sufficiently from their reptilian ancestors to be recognized as distinct grades.

If, alternatively, the pattern of cladogenesis (i.e. shared ancestry) is taken as the sole criterion for recognizing higher taxa, then *all* the descendants of a common ancestral species must be grouped together (along with the ancestral species itself), to form a **clade** (Figure 11.2: various clades are enclosed by different shades of green). A clade represents a single whole branch from a phylogeny, and, because it is derived from a single common ancestor, it is said to be a **monophyletic** grouping.

▶ Are the mammals a clade?

Yes, despite Simpson's earlier reservations about their possible polyphyletic origins, morphological and molecular data now strongly suggest that they are all indeed derived from a single ancestral mammalian species (Section 11.1.1). So the mammals are both a clade *and* a grade grouping (as are also the birds).

If the taxonomic hierarchy is to give an unambiguous reflection of phylogenetic relationships, then the recognition of clades is the most desirable objective of systematics: members of any taxon, so recognized, will be more closely related to each other than to any member of any other taxon, by definition. A classification based purely on a hierarchy of clades is the objective of cladistics.

▶ What aspect of evolutionary pattern is missing from such a scheme of classification?

Because cladistic hierarchies reflect only increasing levels of inclusiveness of the branchings in a phylogeny, they cannot reflect the different amounts of evolutionary change between ancestral and descendent organisms. In other words, they ignore the anagenetic component of pattern, upon which grades are based.

▶ Are the reptiles a proper clade (Figure 11.1)?

No, because despite the reptiles being derived from a common ancestor, two descendent groups—the birds and the mammals—have been removed from them. The reptiles therefore do not include *all* the descendants of the primordial reptile species and so are not a complete monophyletic taxon. A taxon which thus comprises a single branch from which one or more clades have been removed is called a **paraphyletic** taxon (as in the left column of Figure 11.2): the reptiles are therefore paraphyletic. In cladistic classifications, paraphyletic taxa are not recognized, and so the 'reptiles' would not be accepted in such a scheme.

So much for the nature of higher taxa, but how are the constituent species grouped together in the first place? The short answer is 'through comparison of their characters', but this begs the question of what is a 'character'. Because organisms are so complex and so highly integrated, the identification of separate aspects to be treated as taxonomic characters has to be arbitrary, and is thus a subjective issue which presents problems whatever the approach to phylogenetic reconstruction. A particular problem with most morphological characters is that the genetic controls on their development are complex and often poorly understood (Chapter 3, Section 3.2.1). As with the selection of variates (i.e. shared but variable characters) for the morphometric description and discrimination of closely related species (Chapter 10, Section 10.2.3), the choice of characters for phylogenetic analysis is pragmatic. Those features which are reasonably consistent within each species, but which differ sufficiently in expression from species to species so as to permit degrees of similarity between species to be noted, tend to be used. Problems that arise over the definition of characters will be discussed later (Section 11.2.4), but one fundamental consideration must be mentioned here. As noted earlier in reference to the 'pachyderms', the similarities of some features will be misleading as an indicator of relationship if they result from evolutionary convergence. Features showing similarities due to convergence are said to be **analogous** (Chapter 2, Section 2.2.2), and a prime objective of modern systematic methods is to avoid the confusion they can cause in classification. Other features, in contrast, are interpreted as being of similar construction because they have been inherited from a common ancestor. These features are said to be **homologous** (see Chapter 2, Section 2.2.2).

If homologies could be recognized as such, then the relationships between species could be inferred from their shared homologies. Unfortunately, however, homologies and analogies cannot always be unambiguously distinguished in practice. The risk of confusion is especially great when closely related species are compared, because similarities in their morphology and ecology make the parallel evolution of analogous features in separate lineages quite likely. As with other statements concerning history, homologies must themselves be inferred. In some cases, this may seem easy enough. Figure 11.3 shows the structure of a human arm, a bird's wing and an insect's wing. We readily recognize the first pairing as being homologous and the second as being analogous, but why?

▶ What aspect of the human arm and the bird's wing would suggest that they are homologous?

In spite of the differences of their superficial form, they share the same basic construction: corresponding bones, with the same spatial relationships, though with

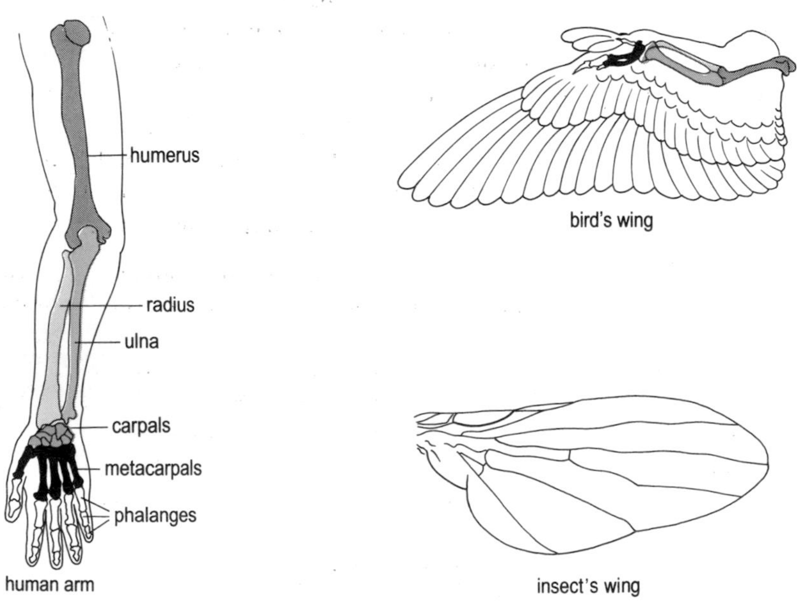

Figure 11.3 *Diagram of the structure of a human arm, a bird's wing and an insect's wing.*

differing proportions, may be recognized (indicated by different shadings in the Figure). The similarity of the wings of the bird and the insect, in contrast, is only superficial (reflecting their common adaptation to flight); they are of markedly different construction, the insect's wing, of course, having no bones at all.

Thus the mode of construction seems to offer a clue, and a useful concept in this respect is that of the information content of features. The more numerous the points of resemblance between structures being compared, in terms of the elements making them up, their positional relationships with respect to each other and their pattern of development, the more likely they are to be homologous. In other words, there would be an improbably large amount of detailed similarity to explain away as coincidental convergence. The appendages in Figure 11.3 are all features of high information content, and so the numerous structural similarities of the first pairing strongly imply homology, while the lack of them for the second pairing implies analogy.

Many other examples are less easy to resolve, however, and continue to create systematic problems to this day. This is a common problem with many fossil taxa. Many of the morphological features of simple fossil shells, for example, have a very low information content, and so homology and analogy are readily confused.

Other approaches to the problem of distinguishing homology from analogy can be adopted, but these vary according to the different systematic methods, which must now be considered.

11.2.2 Evolutionary systematics

Whereas Darwin and Wallace initially cited the hierarchical arrangement of 'natural' taxa, already recognized by their contemporaries, as evidence for evolution, their successors soon turned the link around and looked to evolution as the practical key to taxonomy. To this end all available forms of evidence for evolutionary relationship were exploited. The methods used were thus *ad hoc*: no single consistent basis underlay phylogenetic reconstruction. At one extreme, extinct groups with a rich fossil record, but whose fossils reveal only rather few, simple characters (as with many shelly marine invertebrates), might be analysed using mere similarity and the stratigraphical succession of forms as criteria for relationship. At the other extreme, complex living organisms, with numerous features of high information content but with a relatively poor fossil record (e.g. songbirds), would better lend themselves to being grouped on the basis of features singled out as shared homologies.

The higher taxa recognized by evolutionary systematists combine the qualities of both clade and grade groupings (Section 11.2.1): both paraphyletic and monophyletic taxa are accepted.

► Would both the reptiles and the birds be recognized as valid higher taxa by an evolutionary systematist?

Yes, they would. The reptiles are paraphyletic and the birds are monophyletic (Section 11.2.1). Polyphyletic taxa, however, are rejected from evolutionary systematics because they are not founded on phylogenetic branches, and do not therefore reflect pathways of evolution.

The major taxa shown on Figure 11.1, which are still widely used for classification, are the legacy of traditional evolutionary systematics, including both paraphyletic and monophyletic taxa, but excluding polyphyletic taxa. They have been recognized as such using many different lines of evidence. Analogous features, for example, have been identified by various means and excluded from consideration in the scheme of classification. A common criterion for recognizing them is their incongruence with comparisons based upon other features deemed to be homologous. *Endothermy* (internal heat generation, especially by the liver and muscles, or 'warm-bloodedness') is found in both birds and mammals, for example, but is believed by most authorities to be analogous in the two groups. Hence, it is not used as a justification for uniting them (although a few have proposed exactly that, erecting the group 'Haemothermia' to receive the birds and the mammals—Section 11.2.4). The fossil record for the origins of the mammals is now reasonably well documented, and so it is taken more or less at face value as evidence for their independent descent from the therapsids, unlike the birds, as shown in Figure 11.1.

Evolutionary systematics thus allows all available data to be exploited. The higher taxa so identified also tend to correspond closely with those recognized by common sense, and regarded as natural groupings. However, there are also disadvantages. The lack of any consistent method of analysis means that there is no unambiguous criterion for deciding between competing phylogenetic hypotheses. For example, the comparative morphology of living species may suggest one taxonomic arrangement, but their fossil record may suggest another. The taxonomist can only decide between these possibilities by a subjective assessment of where the weight of the evidence lies, doubtless backed up by personal preferences over the relative reliability of different kinds of evidence. Taxonomic decisions thus tend to depend as much on the personal views of the taxonomist as upon objective criteria. Where the evidence is scarce, phylogenetic schemes tend also to be very sensitive to new discoveries. Though intensively studied, the fossil record of human evolution, for example, is relatively sparse (comprising mostly isolated bones or teeth, with rare whole or partial skeletons from fewer than a hundred localities worldwide). For many years, the story of human evolution gained a certain notoriety for being rewritten every time a new fossil hominid was described. This situation was doubtless also exacerbated by the natural tendency for many of the discoverers of the new finds to see them as being of central importance to the story, which they accordingly re-arranged: 'ancestor-hunting' took on a positively competitive spirit. Latterly, however, the introduction of cladistic methods of systematic analysis (explained in Section 11.2.4), and a broadening of the evidence considered, particularly to include molecular data, has considerably stabilized current hypotheses on human evolution, although plenty of scope for debate remains.

Frustration with the subjectivity of evolutionary systematics, and the consequent circularity of many of the arguments over which features were homologous and which analogous, led to a quest, from the 1950s, for more rigorous methods of systematic analysis.

11.2.3 PHENETICS

A pragmatic approach to the problem of subjectivity in evolutionary systematics is to give up altogether trying to base classification upon *ad hoc* inferences concerning relationships, and to aim, instead, for a scheme based solely upon the seemingly more objective criterion of raw phenotypic similarity between species. The relative levels of similarity must themselves be quantified, being derived from as many measurements of characters (e.g. limb length, numbers of teeth, etc.) as can practicably be obtained. This is the approach known as phenetics, from the Greek 'to show' (or, because of its quantitative methods, sometimes referred to as *numerical taxonomy*). Although evolutionary relationships are no longer explicitly sought after, it is nevertheless supposed that they tend to be reflected in phenetic classifications because the misleading indications of analogous features are usually swamped by the contrasting indications of other features.

As no special (evolutionary) significance is attached to particular characters, all are given an equal weighting in the assessment of similarity between species. For each measured variate (Chapter 10, Section 10.2.3), the difference between the mean values for any two given species defines a 'morphological distance' between them. Thus one might measure, for example, the difference in mean adult ear length between the American jack rabbit *Lepus californicus* and the European rabbit *Oryctolagus cuniculus*, discussed in Chapter 2 (Section 2.3.1). The distances for all variates studied are then aggregated to yield an overall distance between each of the species being compared. Usually, dozens or even hundreds of characters are measured and the calculation of overall distances is a complex computation, but the essential point can be made here by reference to just two character measurements (Figure 11.4(a)).

Mean values of the character measurements for five hypothetical species are plotted as a bivariate scatter in Figure 11.4(a). The direct distances on the graph between the various points can then be used to construct a hierarchy of distances between them, by means of some convention for grouping the points called a *cluster statistic*. If, for example, the points are progressively grouped according to the distances between nearest neighbours, then the hierarchy shown in Figure

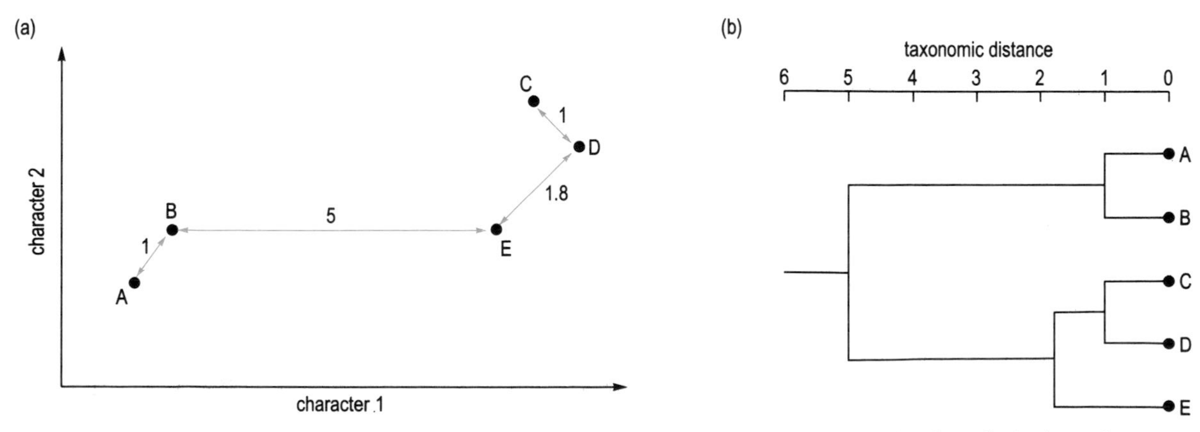

Figure 11.4 *(a) Bivariate scatter of mean values of morphometric measurements from five hypothetical species (A–E). Distances between nearest neighbours are indicated. (b) Phenetic classification (phenogram) of the species in (a), using a nearest neighbour cluster statistic.*

11.4(b) emerges. Such a diagram is known as a **phenogram**. The order of the species at the right of the phenogram is arbitrary, and merely follows the pattern of branching. The lengths of the branches, indicated by the horizontal scale, are explained below.

The distances between species A and B and between species C and D in Figure 11.4(a) are the same; let us call this distance 1 unit of taxonomic distance. On the phenogram (Figure 11.4(b)), each of these pairs of species is therefore associated at the level of 1 unit on the taxonomic distance scale. Likewise, the nearest neighbour of E is D, at a distance of 1.8 units, and so it becomes associated with the C, D pairing at that level. Finally, the two clusters (A, B and C, D, E) are associated at the level of 5 units, being the distance between their closest members, B and E.

It is important to reiterate that a phenogram is no more than a hierarchy of relative phenotypic similarity of a set of species. It is objective, in the sense that any two taxonomists should achieve similar results, only provided that the same characters are measured and that the same cluster statistic is used. It is not intended as an accurate portrayal of phylogenetic relationships, although, as noted earlier, it may commonly reflect them if enough characters are measured. Even setting aside the possible errors introduced by convergence, however, phenograms may still be misleading if treated literally as equivalent to phylogenies.

▶ What assumption concerning rates of evolution would have to be made if the relative distances between pairs of species in a phenogram were to be interpreted as indicating the relative times of the mutual divergence of their ancestors?

It would have to be assumed that the rates of evolution for the characters measured were more or less constant. Had that indeed been so for the characters upon which Figure 11.4(b) is based, and had there been no problem with convergence, then the greater proximity to each other of C and D than of either to E, for example, could be taken to mean that C and D had diverged from one another *after* E had already split from their common ancestor (Figure 11.5(a)). It is possible, however, that the rates of evolution had been uneven, and that E had diverged very rapidly from D *following* the split between C and D (Figure 11.5(b)). Had that been the case, then the phenogram in Figure 11.4(b), though perfectly correct as a representation of relative phenotypic similarity, would be misleading as a phylogenetic scheme: the paraphyletic grouping of C and D, based on raw phenetic similarity, would obscure the more recent common ancestry of D and E.

▶ From what you read about rates of morphological evolution in Chapter 10, is this likely to be a common problem in the use of phenograms for reconstructing phylogenies?

Yes, unfortunately it is. The study of microevolution in sticklebacks (Section 10.3.2), for example, showed large variations in the rate of change both of single variates at different times and between different variates. Phenetic analysis of morphological characters therefore has to be interpreted with caution in reconstructing phylogeny.

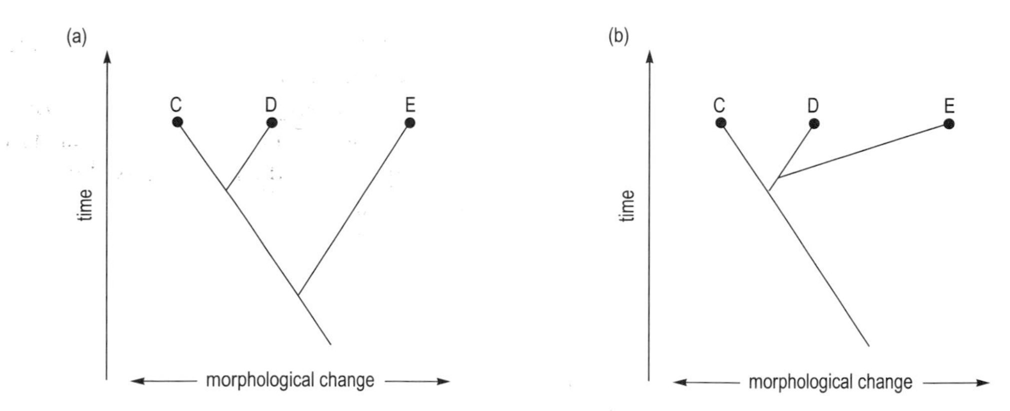

Figure 11.5 *(a) A phylogeny for C, D and E in Figure 11.4(a) consistent with the phenogram in Figure 11.4(b). (b) A possible phylogeny for C, D and E in Figure 11.4(a) which differs from the branching pattern shown by the phenogram.*

Another problem with phenetics, however, is that it is not after all as objective an approach as it first seems to be. The approach has been strongly criticized, ironically, because of the very weakness in evolutionary systematics that it was intended to remedy—that of subjectivity of procedure. In addition to the obvious problem of the arbitrary (or selective) choice of characters to be measured, there are in fact also many possible cluster statistics, which yield different classifications from exactly the same data. Figure 11.6(a) shows another bivariate scatter, of the mean values of two variates for seven hypothetical species, and Figure 11.6(b) and (c) shows two possible phenograms based on different cluster statistics. Figure 11.6(b) employs the nearest neighbour statistic (as in Figure 11.4(b)). Figure 11.6(c), in contrast, is based on an average neighbour cluster statistic, where distances are measured between each point and the midpoint or average of its

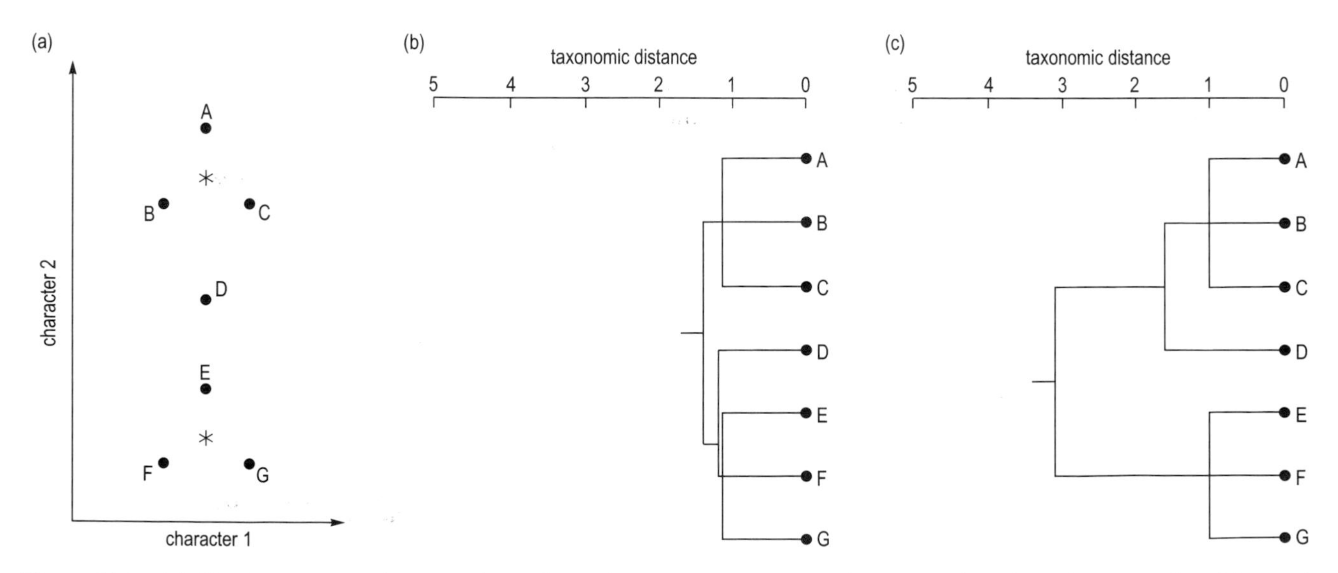

Figure 11.6 *(a) Bivariate scatter of mean values of morphometric measurements from seven hypothetical species. Asterisks mark the centres of the two clusters, ABC and EFG. Note that E is the closest point to D, but that the central asterisk for ABC is closer to D than that for EFG. (b), (c) Phenetic classifications of the species in (a), using (b) a nearest neighbour cluster statistic, and (c) an average neighbour cluster statistic.*

nearest cluster of points (shown by an asterisk for each of the clusters ABC and EFG). Notice, in particular, that species D associates with the cluster EFG in Figure 11.6(b), because E is its nearest neighbour, but with ABC in Figure 11.6(c) because the midpoint between A, B and C is closer to D than is that for E, F and G. As there is no guiding criterion for deciding which is the best cluster statistic, the choice between them must be arbitrary.

For these various reasons, pure phenetic methods have rather lost favour for phylogenetic studies today, although they remain useful for comparing subspecies and races within species, and more especially for molecular data, as Section 11.3 will explain. For the phylogenetic analysis of morphology, the spotlight has moved to quite a different approach—that of cladistics.

11.2.4 CLADISTICS

While phenetics abandoned the inference of relationships as the basis for systematics, alongside it developed a school of thought with the opposite aim of refining that approach. Its founder, the German entomologist Willi Hennig (1913–1976), accepted the premise that a natural taxonomy should reflect phylogeny. He sought to avoid the subjectivity of conventional evolutionary systematics by establishing a single criterion for grouping related species—strict monophyly. Thus, if the three species in Figure 11.5(b) had evolved as shown, with D and E sharing a common ancestor which post-dated the split with C, then only D and E could be put together to form a subordinate higher taxon within the clade C + D + E. The taxa (C) and (D + E) would be said to comprise *sister groups*.

▶ Does this arrangement make any concession to the relative amounts of anagenetic change (Section 11.2.1) shown by the three species?

No, because E underwent considerably greater change than did either D or C, both of which remained fairly conservative. Yet E and D are here grouped together, and separated from C, because of their more recent common ancestry. The grouping is thus purely cladistic, and the paraphyletic grade grouping of C + D is not recognized.

In both evolutionary systematics and phenetics, in contrast, relative anagenetic change *is* incorporated in the classification: in this instance, the grouping together of C and D (as in Figure 11.4(b)) would reflect their conservative grade of evolution with respect to E. Cladistics rejects such paraphyletic taxa (as well as polyphyletic taxa) for two reasons. First, they obscure the pattern of phylogenetic branching, as noted in Section 11.2.3. Secondly, their limits are subjectively defined by the taxonomist, who decides which subordinate grade groups to remove in order to create them. Monophyletic taxa are judged to be less arbitrary in the sense that they include the entire issue of any given phylogenetic branch.

Character states and cladograms In common with the other approaches, cladistics nevertheless still has to group taxa by reference to some aspect of similarity. So the question arises, which features can be expected to reveal the pattern of cladogenesis alone? The principles and terminology of cladistics can be illustrated by reference to Figure 11.7.

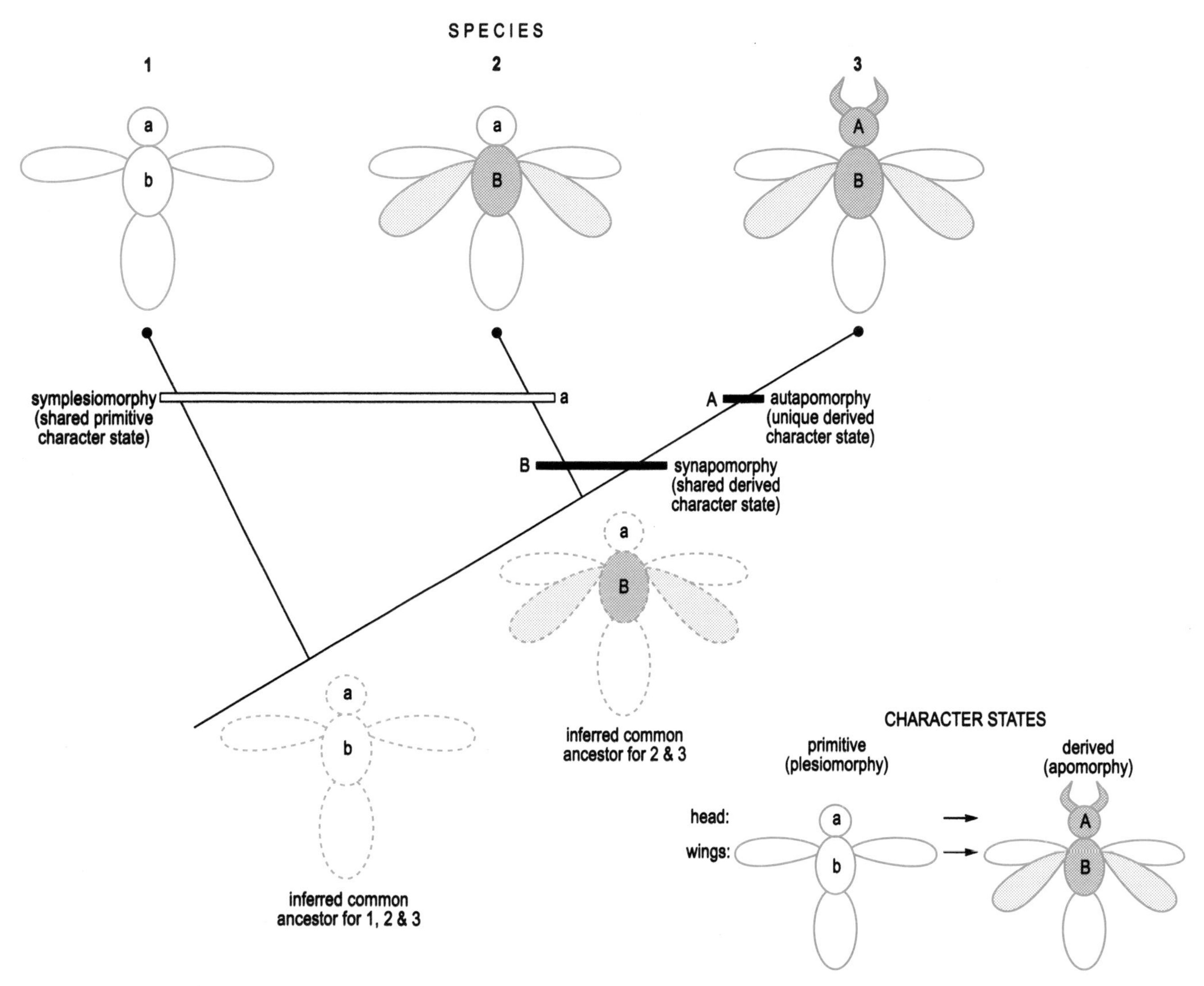

Figure 11.7 *Cladogram for three hypothetical species of insect-like organisms. The primitive and derived states of two of their characters, the form of the head (a and A) and the number of pairs of wings (b and B), are indicated below. The distribution of derived character states and of the shared primitive character state are indicated on the cladogram by black and white bars, respectively. The construction of the cladogram is explained in the text.*

Along the top of Figure 11.7 are shown, in diagrammatic form, the morphologies of three hypothetical insect-like animals, labelled 1, 2 and 3. Note that all three share the basic homologies of a three-part body, with wings attached to the central part. We will concentrate on just two characters—the form of the head (shown at the top of each diagram), and the number of pairs of wings. Each of these characters exists in one of two states: the head is either simple (state a in 1 and 2), or it has prominent antennae (state A in 3); and there is either one pair of wings (state b in 1), or two (state B in 2 and 3). For each character, methods exist for inferring whether a given character state is either **primitive** for the animals considered (present in the latest common ancestor to *all* the species) or **derived** (subsequently evolved in one or more members of the group). Primitive character states are termed **plesiomorphies** (sing. plesiomorphy, meaning 'near-shape' in Greek), and derived character states are termed **apomorphies** (sing. apomorphy, meaning 'from-shape' in Greek). Methods for distinguishing primitive and derived character states will be explained later in this Section, but for present purposes suppose that the distinction has already been established in this case: the simple head (a) and possession of one pair of wings (b) are inferred to be primitive, and the head with antennae (A) and possession of two pairs of wings (B) are judged to be derived (as shown at the bottom of Figure 11.7).

The cladistic analysis may now proceed as follows. Species 2 and 3 share the derived state of the wings (B). There are two possible explanations: either both species inherited the state from a common ancestor in which B had already evolved (as shown where the branches for 2 and 3 join in Figure 11.7), or they independently evolved that character state. The first explanation is simpler than the second, however, as it requires B to have evolved only once (prior to the splitting of the ancestral lineage), while the second requires B to have evolved twice, once in each separated lineage. In the absence of any evidence to the contrary, it is conventionally considered preferable to accept the simpler explanation, because that requires the fewest conjectured events. This convention is termed the principle of **parsimony**, and, as later discussion will show, it plays an important part in phylogenetic inference. It must be stressed, however, that the most parsimonious explanation is not necessarily the correct one. There is no reason why evolution should always have followed the simplest pattern that is consistent with the available evidence. Rather, parsimony is merely employed as a procedural guide, because it leaves least to be filled in by speculation, for want of evidence. New evidence may indeed overturn a parsimonious explanation in favour of another, which is less so, but which remains consistent with the new findings.

Accepting, then, that 2 and 3 in Figure 11.7 inherited B from a common ancestor, it may be inferred that the lineage leading to 1 (with the primitive character state b) had already branched off from a yet earlier ancestor, common to all three species, which likewise lacked B (shown alongside the bottom branching point in Figure 11.7). Species 2 and 3 are thus inferred to be more closely related to each other (i.e. to share a more recent common ancestor) than either is to species 1. Taxa (1) and (2 + 3) may thus be designated as sister groups.

▶ Another possibility, however, would be that species 1 branched off from a common ancestor with either of 2 or 3 alone, and subsequently lost its second pair of wings, reverting to character state b. Why do you think this possibility is rejected in this analysis?

Though possible, this explanation (involving both the gain and the loss of a derived character state) is less parsimonious than that given previously (which requires only the gain of the derived character state).

The relationships between the three species inferred above are indicated by a diagram composed of successive branches, called a **cladogram**, drawn in beneath the three animals in Figure 11.7. The shared derived character state (or **synapomorphy**, meaning 'apomorphy-together' in Greek), B, which unites 2 and 3, is shown by a black bar, labelled B, joining the branches for these two species.

Other bars are also shown. Only 3 has the derived character state A, which is thus termed an **autapomorphy** (meaning 'self-apomorphy' in Greek), and this is shown as a black bar on the branch for 3 alone. 1 and 2, in contrast, share the primitive character state a, which is thus termed a **symplesiomorphy** (meaning 'plesiomorphy-together' in Greek), and this is shown as a white bar, labelled a, joining their branches.

▶ Which of the different kinds of bar (patterns of character state distribution) yields the information on cladistic relationships indicated by the cladogram?

Only the synapomorphy allows one to infer that (1) is the sister group to (2 + 3), as shown. The autapomorphy is unique and so gives no clue as to whether 1 or 2 is more closely related to 3. The symplesiomorphy merely tells us that 1 and 2 have retained the primitive state of the head, but again gives no clue as to which animal is more closely related to 3.

▶ What kind of taxon would you call (1 + 2), if united on the basis of their primitive head form?

It would be a paraphyletic taxon (Section 11.2.1), representing a primitive grade grouping. Such a taxon would be rejected in a cladistic classification because it does not group together those species which share the most recent common ancestor (in this case, 2 and 3). *Cladistic analysis is thus based solely upon the identification of synapomorphies.*

Before leaving the simple example of Figure 11.7, a further feature of the cladistic method should be pointed out. Note that the cladogram presupposes a sequence only of dichotomies (splitting of lineages into *two*). Again, this is only a methodological assumption, the biological validity of which will be critically examined at the end of this Section.

An example of cladistic analysis: lizard, crocodile, duck and dog An example using some familiar animals will now serve to introduce how primitive and derived character states are distinguished, and employed in systematic analysis, and how cases of convergence of derived character states can be detected.

Table 11.1 shows a selection of the character states observed in four well-known kinds of vertebrate animal. The following discussion of inferences concerning the evolutionary relationships of these animals is based on the selected characters only for the sake of simplicity, in order to highlight the methods of analysis. These characters are by no means the only evidence for current ideas on the phylogeny of these and other related animals, which are based rather on much larger numbers of characters. Some controversy nevertheless still remains over the conclusions reached, so it is as well to remember that phylogenies are only theoretical constructions (Section 11.1.1): you are always entitled to disagree, provided that you have good grounds for doing so!

Before getting on to the cladistic analysis itself, consider the implications of an uncritical comparison of raw similarity (i.e. adopting a phenetic approach).

Table 11.1 *Matrix of selected characters in four vertebrates.*

	State of character in:			
Character	Lizard	Crocodile	Duck	Dog
(a) Non-living (keratinous) outgrowths from the skin covering the body	Scales	Scales	Feathers	Hair
(b) Body temperature regulation	Ectothermic ('cold-blooded')	Ectothermic	Endothermic ('warm-blooded')	Endothermic
(c) Mode of embryonic development	In shelled egg	In shelled egg	In shelled egg	In uterus, with placental connection to mother
(d) Skull type*	Diapsid	Diapsid	Diapsid	Synapsid
(e) Openings at sides of palate* (palatine fossae)	Present	Present	Present (in modified form)	Absent
(f) Aperture in skull in front of the eye socket* (antorbital fenestra)	Absent	Present in extinct forms, but secondarily lost in living forms	Present	Absent

*See Figures 11.8 and 11.9.

► Study Table 11.1 and for each of the possible pairings of the animals count the characters for which they show *differing* states (e.g., the duck–dog pair differ in five of the six characters, sharing the same state only for character b). Treat the aperture in character f as being present in the crocodile. You will find it helpful to record your answers in a grid showing the four animals listed along the top, and again down the side. Only one triangular half of the grid need be filled in, of course, as the other half will be symmetrical to it.

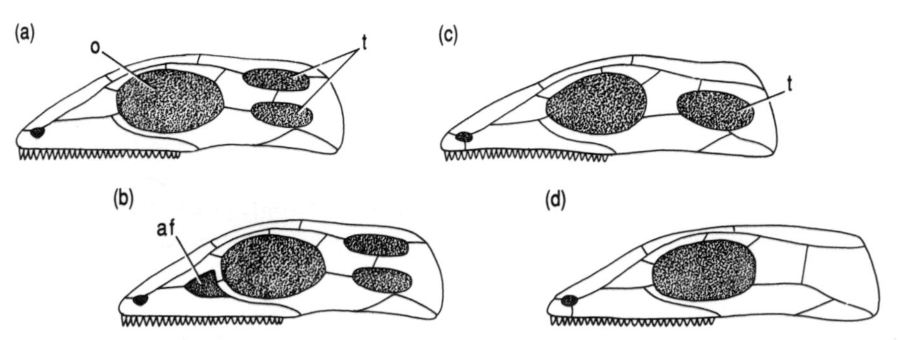

Figure 11.8 *Diagrams of various types of skull. Thin lines indicate the sutures between the various bones of the skull. Openings in the skull are shaded: o, eye socket; t, temporal openings behind the eye socket; af, antorbital fenestra, an opening in front of the eye socket. (a) Diapsid skull (two temporal openings). (b) Diapsid skull with an antorbital fenestra. (c) Synapsid skull (one temporal opening). (d) Anapsid skull (no temporal openings).*

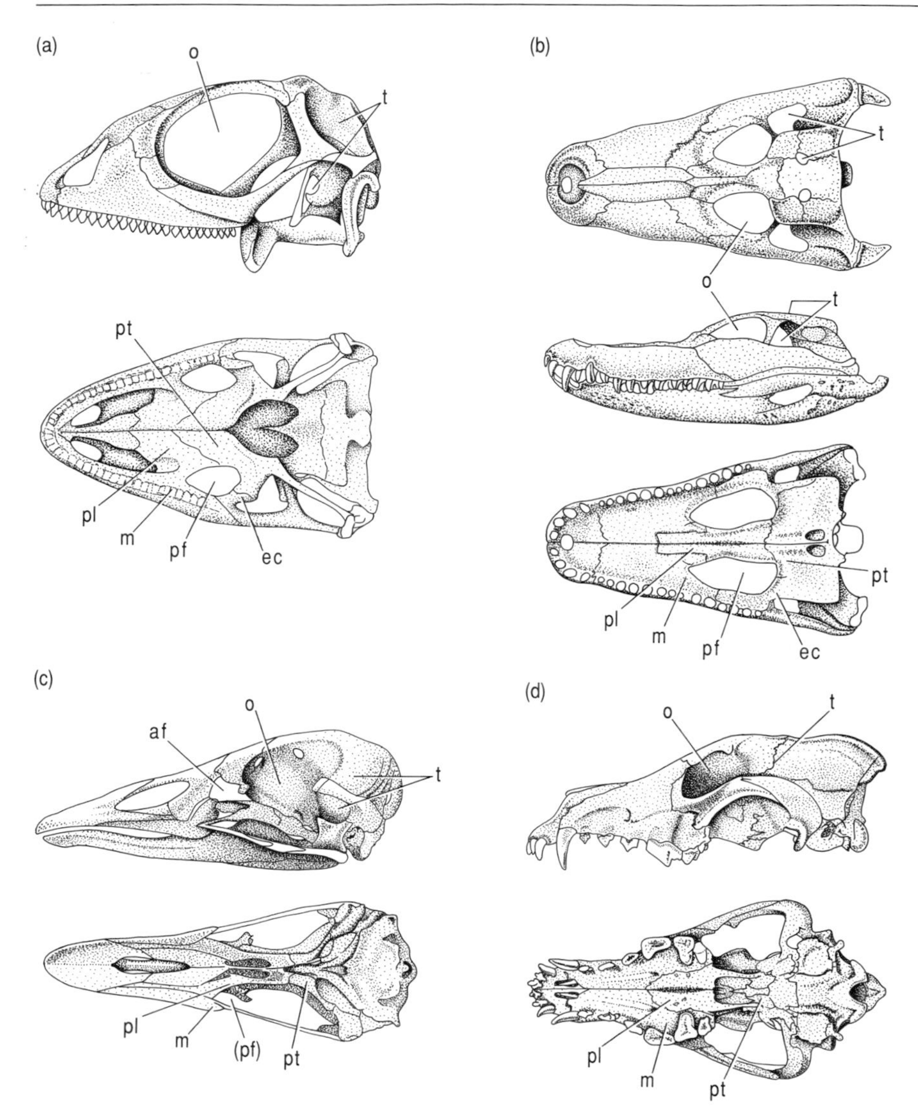

Figure 11.9 *Skulls of the animals in Table 11.1. (a) Lizard (*Iguana*), in side view and seen from below. Note that the bony bar beneath the lower temporal opening is absent. Well developed palatine fossae (pf) visible on the underside, each surrounded by the maxilla (m), palatine (pl), pterygoid (pt) and ectopterygoid (ec) bones. (b) Crocodile (*Crocodilus*), seen from above, from the side and from below. Temporal openings somewhat reduced. Antorbital fenestra lost here, though present in certain fossil crocodiles (cf. Figure 11.11). Symbols as in (a) and in Figure 11.8. (c) Duck (*Anas*), in side view and from below. The bony bars separating the eye socket from the temporal openings have been lost, but a short spur of bone separates the upper and lower temporal openings. The bony margins of the antorbital fenestra are likewise reduced. Loss of the ectopterygoids means that the palatine fossae are joined to the apertures behind (as explained in the text). Symbols as in (a) and Figure 11.8. (d) Dog (*Canis*), in side view and from below. The bony bar separating the eye socket from the single temporal opening has been lost so that the two are connected. Symbols as in (a) and Figure 11.8.*

The six possible comparisons are shown in Table 11.2. If each of the character differences in Table 11.2 is treated as one unit of difference, a crude phenogram can be constructed from the relative scores for phenetic distance between the pairs of animals (Figure 11.10).

Table 11.2 *Grid showing the numbers of character state differences between the pairings of animals listed in Table 11.1. Entries show, for each pair, the number of character state differences and (in brackets) the characters concerned, from Table 11.1.*

	Lizard	Crocodile	Duck	Dog
Lizard	—	1(f)	3(a, b, f)	5(a, b, c, d, e)
Crocodile		—	2(a, b)	6(a, b, c, d, e, f)
Duck			—	5(a, c, d, e, f)
Dog				—

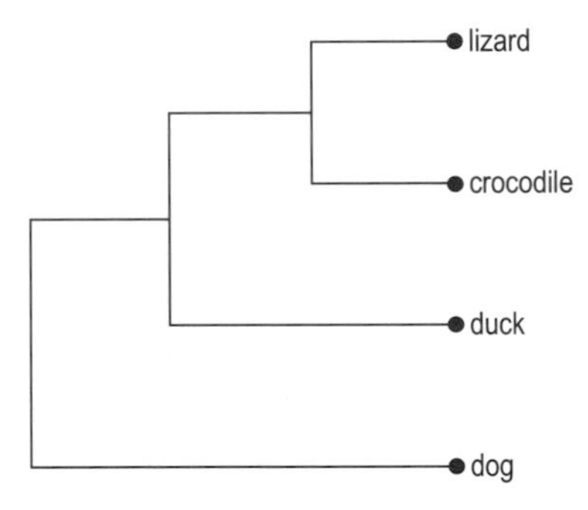

Figure 11.10 *Phenogram for the four species in Table 11.1, based on the scores for character differences listed in Table 11.2.*

▶ Can we infer from Figure 11.10 that the lizard and the crocodile are the two closest relatives, in the sense of sharing a more recent common ancestor, than either does with any of the other animals?

No, we cannot, because the scores are based only on raw comparisons. Some of the similarities between the lizard and the crocodile, which differ only in one character, may merely reflect their joint retention of primitive character states, compared with the other two animals (as with animals 1 and 2 in Figure 11.7).

▶ If this were so, what sort of grouping would the lizard–crocodile pair constitute?

They would constitute a paraphyletic taxon, which, as noted earlier, would obscure the pattern of cladogenesis.

To tackle this problem, it is necessary to determine whether any of the alternative character states can be considered primitive or derived with respect to each other. In other words, is it possible to detect character *polarity* from primitive to derived?

The first point of similarity between the lizard and the crocodile is the possession of scales. Certainly with respect to the feathers of the bird, scales appear to be primitive. The mode of development of feathers suggests that they are highly modified scales; indeed, unaltered scales, like those in the reptiles, persist on the legs of most birds. Here, then, the polarity is detected by embryological comparison. Behind this lies a useful generalization concerning development, first enunciated by the German biologist von Baer, in 1828 (Chapter 1, Section 1.3.2):

'The general features of a large group of animals appear earlier in the embryo than the special features.'

(Translated by S. J. Gould, 1977.)

So feathers, being a developmental modification of scales, are judged to be derived, and the scales primitive. There are certainly exceptions to von Baer's law, but apparently not so frequent as to undermine its general usefulness in suggesting polarity, particularly in conjunction with other criteria which can serve to test it. The presence of scales in the lizard and the crocodile results only from their shared retention of a primitive character.

► What is the term for this condition?

A shared primitive character is a symplesiomorphy (Figure 11.7). A similar argument may be applied to the comparison of the two reptiles with the dog as far as character a is concerned, although hairs are outgrowths of highly modified, keratinized skin, rather than modified scales, as such. Nevertheless, some mammals do retain reduced scales, as on the tails of shrews and rats. So character a in the dog shows another derived state.

The lizard and crocodile also share the character state b of ectothermy (the body temperature is mainly regulated by external sources of heat; i.e. they are 'cold-blooded' in popular, though inaccurate, speech). Again, this state appears to be primitive with respect to the endothermic condition in the duck and the dog, in which body temperature regulation relies heavily on internal heat production, especially from the liver and muscles. The polarity in this instance is confirmed by another widely used means for detecting polarity, termed *outgroup comparison*. This comparison involves looking to see which (if any) of the character state alternatives exists in another organism generally regarded to lie outside the clade that includes all the organisms being compared. In this case an obvious candidate would be an amphibian, such as a newt. This too is ectothermic, so ectothermy is likely to be the primitive condition for the clade which includes all the animals in Table 11.1. The justification for this inference is simply that it is more parsimonious to suppose that lizards and crocodiles inherited ectothermy from their latest common ancestor shared with the newt, rather than that the latest common ancestor of the lizard, crocodile, duck and dog was already endothermic, but that the first two then reverted to ectothermy. Such an argument is certainly not infallible, but like that based on embryology, it can be a useful rule of thumb. It would now be tempting to designate endothermy as a synapomorphy for the duck and the dog. The alternative is that it evolved independently (as convergent autapomorphies) in the birds and mammals. We will return to this issue after considering the other characters.

▶ Now decide by outgroup comparison with amphibians whether development within an egg or in a uterus is the more derived state of character c in these taxa.

This comparison suggests that development within an egg is primitive (for the sake of simplicity here, we ignore the fact that the shelled egg of the lizard, crocodile and duck is itself greatly modified, i.e. derived, with respect to that of amphibians). The derived condition of development in a uterus is thus an autapomorphy for the dog.

▶ A further clue that development in the uterus is a derived condition is provided by certain other mammals discussed earlier. Can you recall which these are?

Section 11.1.1 mentioned the monotremes: in these, the primitive egg-laying state has been retained.

So far in our consideration of the characters shown in Table 11.1, only symplesiomorphies (scales and ectothermy in the lizard and crocodile, and development in a shelled egg in these animals plus the duck) and autapomorphies (feathers in the duck, and hair and uterine development in the dog) have been unambiguously detected. The identification of cladistic relationships, however, relies upon synapomorphies, as noted earlier in the Section. As yet, endothermy in the duck and the dog is the only possible candidate, and, if no further evidence were available, a clade uniting the duck and dog would be the most parsimonious conclusion. The further evidence of the three remaining characters, however, suggests otherwise.

Despite individual differences in the structure of the skull (character d), the lizard, crocodile and duck all share the same basic pattern of *diapsid* skull form (Figure 11.8(a)). This is characterized by the possession of two openings in the skull behind each eye socket. Among other things, these openings lighten the skull and allow more secure attachment of the jaw-closing muscles. Outgroup comparison shows this to be a derived condition. However, the primitive, *anapsid* condition (in which there are no such openings in the skull, Figure 11.8(d)) is not that seen in the dog. The latter has an alternative derived condition—*synapsid* skull form—where there is but one hole in the skull behind each eye socket (Figure 11.8(c)). Note, therefore, that the primitive-derived polarity for this character is not exhibited within the four animals being compared here, but, rather, between alternative derived states in these, and the presumed primitive state in outgroup animals. The best candidates to serve as the latter are some early fossil reptiles which share a few, though not all, of the synapomorphies which unite all other reptiles, birds and mammals. They also illustrate a third criterion for establishing polarity, that of palaeontological antiquity (the oldest fossils tending to show the most primitive character states). However, this criterion is fraught with difficulties because of the incompleteness of the fossil record. Organisms with more derived character states may, by chance, be caught in the fossil record before those with primitive character states, especially when the fossil record is as patchy as it is for early vertebrates. It tends to be used only as a last resort, and then with some caution.

In addition to the diapsid openings behind the eye sockets, the lizard, crocodile and duck also share the character state e of possessing prominent openings on the underside of the skull, to either side of the palate. The *palatine fossa* (pf), as the opening on each side is called, is clearly visible in the skulls of the lizard and crocodile (Figure 11.9(a) and (b), respectively), separating the palatine (pl) bone, in the roof of the palate, from the rear flank of the upper jaw, or *maxilla* (m). The fossa is closed off from another opening behind by a bridge of bone running from the *pterygoid* (pt) bone, behind the palatine, to the rear of the maxilla, which is called the *ectopterygoid* (ec). The arrangement in the skull of the duck (and other birds) is a little different, as the ectopterygoid has been lost. Nevertheless, the palatine fossa is still recognizable, flanked by elongate extensions of the palatine and maxilla, but now forming the front part of a single large opening beside the bones of the palate and lower braincase (Figure 11.9(c)). In contrast to these three animals, the dog lacks palatine fossae, the palatine directly adjoining the rear extension of the maxilla on each side (Figure 11.9(d)). Outgroup comparison, even with certain other living reptiles such as turtles and tortoises, suggests that absence of the palatine fossae in the dog is primitive, and its presence in the lizard, crocodile and duck, derived. This finding highlights an important point, that it is character states, not whole organisms, which are judged to be relatively primitive or derived. Although most of the dog's features might seem relatively derived (as with the previous four characters), here is one respect in which it seems to have conserved a primitive trait which is modified in the other three animals.

Characters d and e thus show a congruence of presumed synapomorphies (diapsid skull, with palatine fossae), suggesting that the lizard, the crocodile and the bird share a more recent common ancestor than any does with the dog. In other words, it would appear that ancestors leading to the dog (the mammal-like reptiles) split off early in the evolution of the reptiles, before the splits giving rise to those groups which include each of the three other animals. The clade which thus contains the lizard, the crocodile and the duck has been given the name Diapsida, after the form of the skull. It contradicts the bird–mammal grouping suggested by character b, and it is upon such evidence that endothermy may be considered merely convergent in the duck and the dog.

The conclusion reached so far, that the diapsid animals and the mammals are sister groups, now allows us to use the latter as an outgroup for resolving the polarity of variable character states within the diapsids.

▶ Using this approach with reference to character f in Table 11.1, would you consider the presence of an aperture in the skull in front of the eye socket (as indicated in Figure 11.8(b)) to be a derived or primitive condition within the diapsids?

Its absence in the dog, which here serves as the outgroup, suggests that it is relatively derived within the diapsids.

▶ What is the implication of this answer for sister groupings within the three diapsid animals?

The presence of such an aperture appears to be a synapomorphy uniting the crocodile and the duck. However, this statement needs careful qualification. The justification for grouping the crocodile with the duck depends upon the assertion in Table 11.1 that the aperture in front of the eye socket (*antorbital fenestra*, af in Figure 11.8(b)) *was* present in extinct crocodiles, but has been lost in living forms. Figure 11.9(b) confirms that it is indeed absent in the living example, and so, on the evidence of that example alone, the crocodile–duck grouping seems false. The problem is resolved by fossils. Triassic and Jurassic crocodiles (Figure 11.11) still retain the antorbital fenestra: evidently this opening has become closed off with the evolution of increasing robustness of the skull in later crocodiles. You might well ask how fossil species such as that shown in Figure 11.11 can themselves be identified as crocodiles. The answer is that they possess other, highly characteristic crocodilian autapomorphies (such as a distinctive type of ankle joint). This little detour in the analysis of Table 11.1 raises an important point. Some enthusiasts of cladistics, impressed by the vast amount of information on characters that can be obtained from living organisms, have asserted that fossils supply too little information to be of much value: it has been claimed that fossils can only be fitted into, but cannot alter, a cladogram based on living taxa. The crocodile example discussed above shows that, on the contrary, fossils which comprise a blend of derived and primitive character states can bring crucial information to a cladistic analysis. The derived states allow the fossil to be grouped with a particular living taxon (e.g. the crocodiles), while the *relatively* primitive traits (e.g. retention of the antorbital fenestra in early crocodiles) give information on character states no longer seen in the living taxa. In the present instance, the crocodile and the duck can thus be united within a clade (termed the Archosauria), to which the lizard (and its relatives, such as snakes) form a sister group.

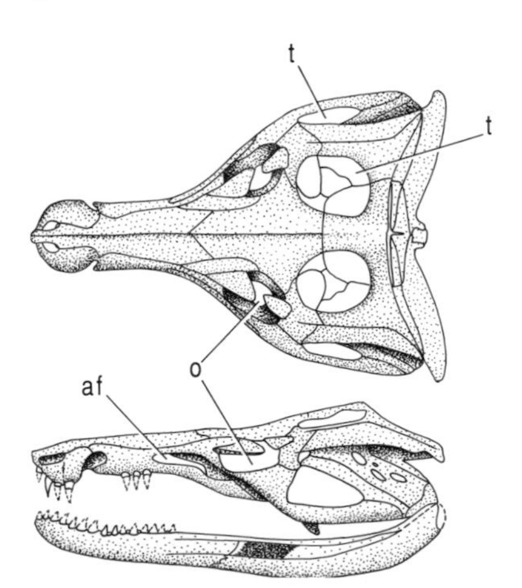

Figure 11.11 *Skull of the early Jurassic crocodile* Nothochampsa, *seen from above and in side view. Symbols as in Figures 11.8 and 11.9. Note the antorbital fenestra (cf. Figure 11.9(b)).*

The conclusions reached above can now be summarized in a cladogram (like that in Figure 11.7) illustrating the inferred phylogenetic relationships between the four animals (Figure 11.12). Here, only apomorphies are shown (by black bars), arranged on each branch in an arbitrary sequence. The sequence of the animals along the top is also arbitrary, being arranged so as to suit the branching pattern of the cladogram.

By ignoring the symplesiomorphies which led to the lizard and the crocodile being the most closely associated pair in the phenogram (Figure 11.10), and by

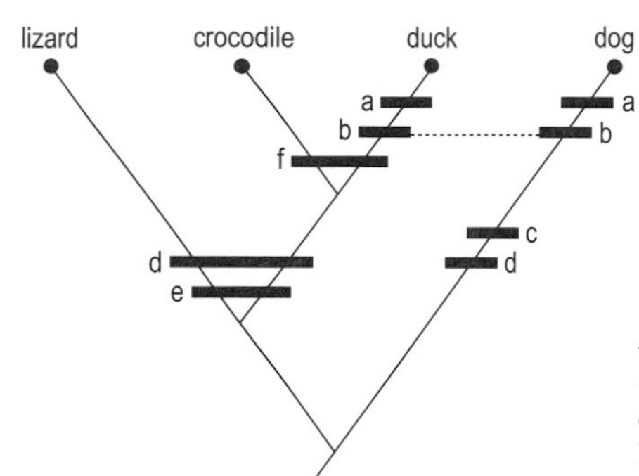

Figure 11.12 Cladogram for the four species in Table 11.1, based on the character states listed there and discussed in the text. Black bars show the apomorphies. The sequence of the latter on each branch is arbitrary. Dotted line shows inferred convergence.

concentrating only on the synapomorphies, the cladogram shows the most plausible arrangement of relative recency of common ancestors for the taxa considered. Note how the crocodile is now more closely associated with the duck than the lizard: evidently, the line that led to the lizard group split off before the common ancestor of the crocodile and the birds had evolved antorbital fenestrae. As suggested earlier, this suggests that the phenetic lizard–crocodile pairing is only paraphyletic, and so obscures their respective evolutionary relationships with the birds. In a nutshell, this obscuring of relationships is why a cladist would not accept the taxon 'Reptilia', as conceived by the evolutionary systematist, and as shown in Figure 11.1.

An important feature of this cladogram is that it represents the most parsimonious analysis of the evolution of the derived character states. Other arrangements are possible, but they would require a greater number of evolutionary changes to bring about the combinations of character states observed in the taxa being compared. Suppose, for example, that the endothermy (character b) of the duck and the dog were judged to be an overriding consideration (Figure 11.13). To bring about the known character state distributions in the animals, it would now be necessary to suppose that no less than three derived character states (for characters d, e and f) evolved convergently in the lizard–crocodile group and in that containing the duck. The number of discrete evolutionary 'events' required to yield the derived states in each cladogram can be obtained simply by adding up all the black bars shown.

▶ How many such discrete evolutionary 'events' are required in each of the cladograms?

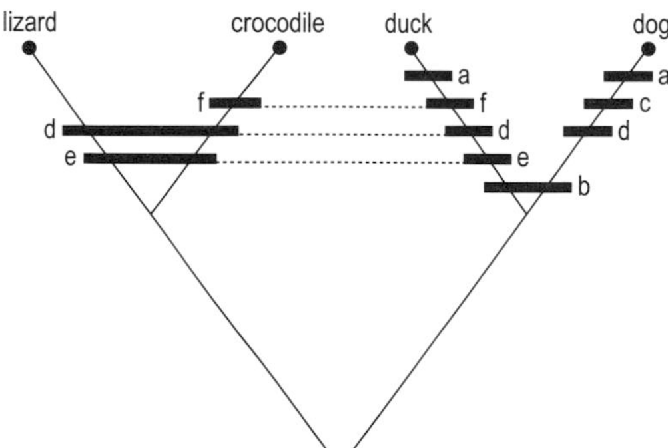

Figure 11.13 Alternative cladogram for the species in Table 11.1 (cf. Figure 11.12), requiring a larger number of apomorphies.

The first (Figure 11.12) requires nine, while the second (Figure 11.13) requires 11. The first is therefore the simpler, or more parsimonious analysis. As noted earlier, we cannot say that the second has to be wrong, but merely that, if we accept parsimony as a guide, it is the less likely explanation. It is on this basis that the scheme shown in Figure 11.12 is accepted by most authorities, with its implication that endothermy has evolved convergently in birds and mammals. However, it should be reiterated that only a small part of the evidence has been discussed here, for the sake of simplicity, and the full debate concerns many more characters. Indeed, supporters of the 'Haemothermia' grouping (Section 11.2.2) cite 28 'synapomorphies' for birds and mammals. Yet a counter-argument is that many of these have to do with the physiological constraints of endothermy itself, so that it is questionable whether they should be counted as separate synapomorphies. So cladistics, too, has its Achilles' heel of subjectivity—the selection and treatment of characters for analysis. Its objectivity lies in the consistency of the method of analysis once the characters have been decided upon.

Parsimony Determining the most parsimonious cladogram for a given array of taxa is itself an elaborate task, which, in practice, is usually accomplished with the aid of a computer (several software programs exist for this purpose). The total number of possible alternative cladograms rises startlingly with the number of taxa considered. While only three possibilities exist for three taxa (Figure 11.14(a)), there are in fact 15 possible cladograms for four taxa (Figure 11.14(b)), 105 for five, and over 2 million for nine! In practice, when confronted by such a plethora of possible cladograms, the systematist does not always blindly accept the most parsimonious one, as implied by the simplified example discussed above. Several different cladograms may differ little, if at all, in parsimony, and often one of these may easily supplant another with only a small change in the choice, or treatment, of characters. Consequently, the top five per cent, say, of the most parsimonious trees may be generated, and a consensus tree (in which some of the dichotomous branches may be combined to form multiple branches) constructed from these. The consensus approach at least picks out the stable parts of the tree, leaving the more problematical parts unresolved.

Unrooted trees In the illustration given above, the likely polarity of the character states was independently assessed from embryological, outgroup and palaeontological considerations. Sometimes, however, it may not be possible to gauge polarities at the outset. Branching diagrams showing the different possible patterns of linkage between the taxa being compared may still be constructed (Figure 11.15 shows the three possible trees for linking the four taxa in Table 11.1). These are known as **unrooted trees**, because they lack any stem presumed to connect with other outgroup taxa showing primitive character states. On these trees can be marked the boundaries between the different character states found in the taxa (shown as bars on the branches in Figure 11.15: on any one branch, the sequence of bars is arbitrary). Thus, for example, the boundary for character b in Figure 11.15(a) separates the (ectothermic) lizard and crocodile, on the left, from the (endothermic) duck and dog on the right. In Figure 11.15(b) and (c), in contrast, the character b boundary appears twice, in order to separate the different

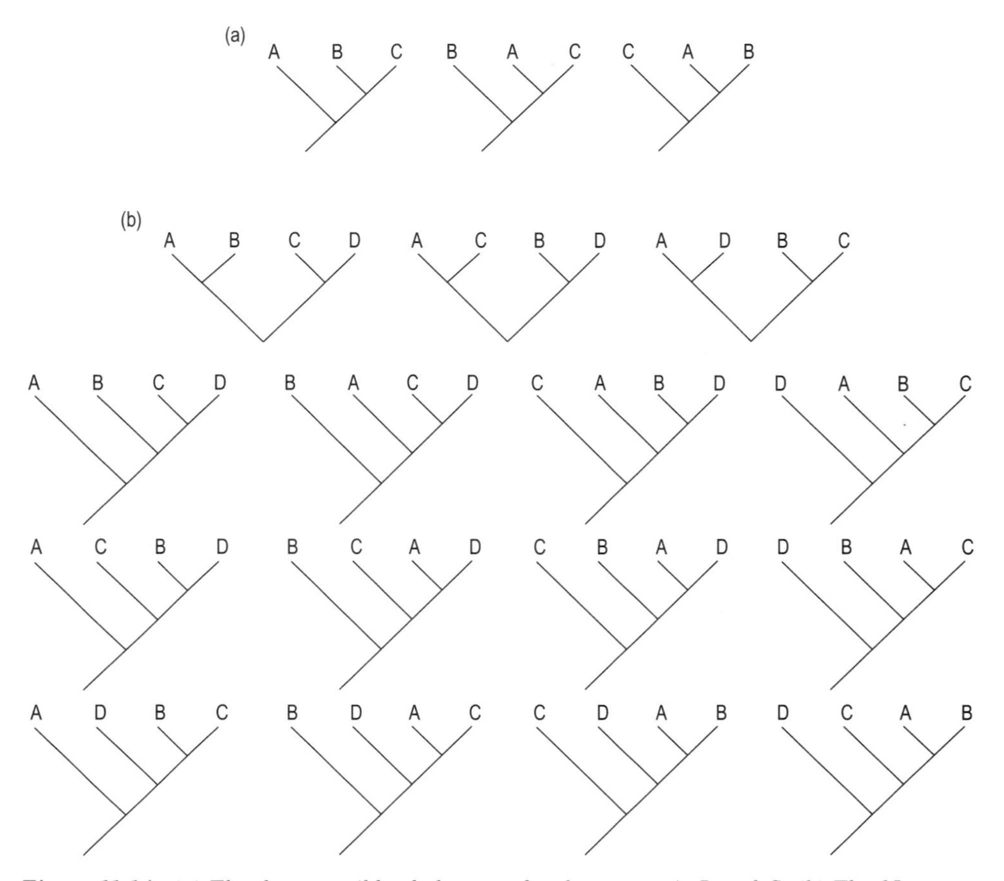

__Figure 11.14__ (a) The three possible cladograms for three taxa, A, B and C. (b) The 15 possible cladograms for four taxa, A, B, C and D.

character states in the animals. Two boundaries for character a appear in all the diagrams because of the unique states of that character in the duck and dog, respectively.

The principle of parsimony can again be applied to unrooted trees, despite the lack of knowledge of polarity of the character states, by simply comparing the number of boundaries shown on alternative trees.

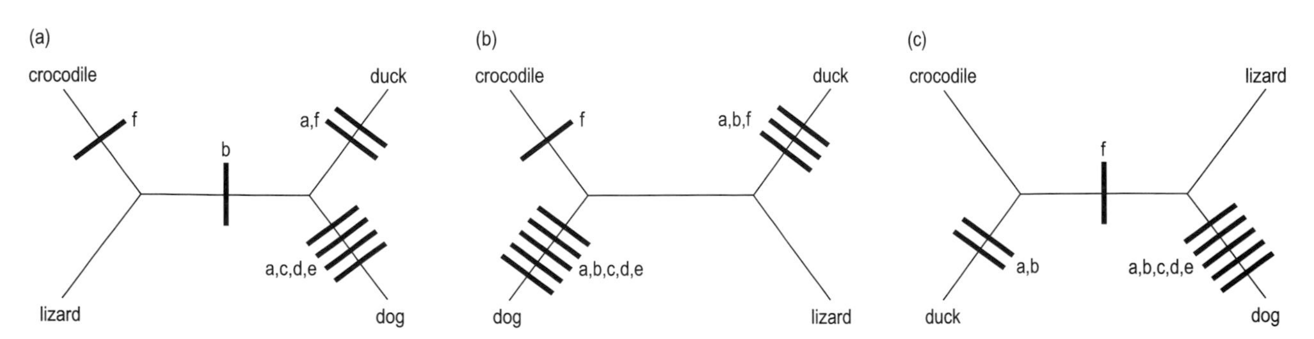

__Figure 11.15__ The three possible unrooted trees for the four species in Table 11.1. The bars mark the boundaries between the different states of the characters. The sequence of the bars on each line is arbitrary.

▶ Can any of the trees shown in Figure 11.15 be rejected on the basis of parsimony?

Yes, the middle tree (Figure 11.15(b)) shows nine boundaries, compared with eight in the other two, and so it may be rejected.

Note that Figure 11.15 shows that only three alternative unrooted trees are possible for four taxa (switching the crocodile and lizard around in Figure 11.15(a), for example, would not alter the relative positions of the boundaries). Each unrooted tree thus corresponds to five of the fifteen possible cladograms that can be constructed for four taxa (Figure 11.14(b)). Rejection of Figure 11.15(b) on the basis of parsimony therefore usefully reduces the field of possible cladograms to ten. In order to 'root' such a tree, imagine pulling down any one of the five branch lines, so that the four taxa flip up into a line, with the now rooted cladogram beneath: Figure 11.16, for example, shows how rooting Figure 11.15(c) on the dog branch generates the cladogram shown in Figure 11.12 in this manner. (The two versions on the right in Figure 11.16 are merely alternative ways of showing the same sequence of branching.)

▶ Which tree in Figure 11.15 should be rooted, and from which branch, in order to generate the cladogram of Figure 11.13?

Figure 11.13 associates the lizard and crocodile, and the duck and dog as discrete pairs, and so Figure 11.15(a) is the appropriate tree. This should be rooted from the central branch to yield the pattern shown in the cladogram.

Any evidence for one of the branches of a tree being an outgroup with respect to the others allows the tree to be rooted. Evidence for the polarity of one or more characters may serve this purpose, and simultaneously imply the polarities of the other characters. Thus, the recognition that character e (absence of palatine fossae) is primitive in the dog would allow that to be considered the outgroup to the other three. Rooting the dog branch in Figure 11.15(c) yields the cladogram shown in Figure 11.12, as noted in Figure 11.16. Rooting the dog branch on the tree shown in Figure 11.15(a), however, would yield some rather bizarre implications for polarity of the other characters.

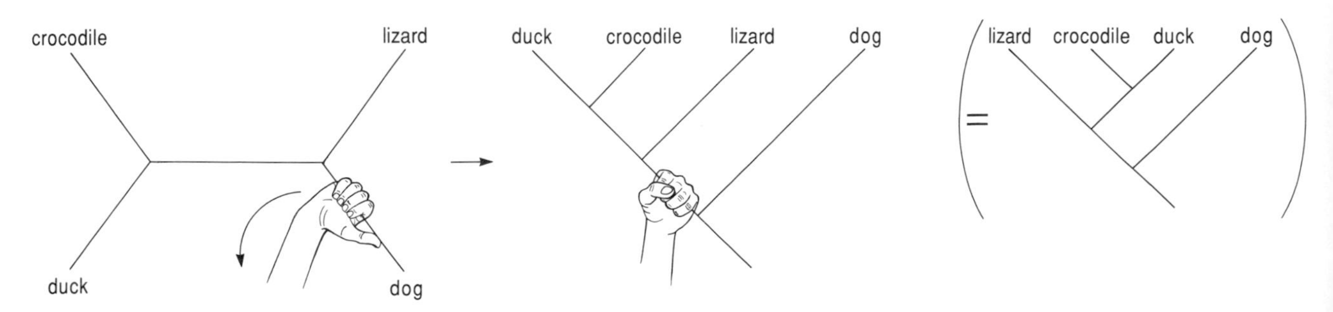

Figure 11.16 *Rooting the tree shown in Figure 11.15(c) on the dog branch line (left) yields the cladogram (right) shown in Figure 11.12.*

▶ What would be the implication for the polarity of character b (body temperature regulation), if Figure 11.15(a) were rooted at the dog branch?

The implication would be that ectothermy, in the lizard and crocodile, was derived. Because of that, Figure 11.15(c) is the more plausible tree, and Figure 11.12 becomes the preferred cladogram. Unrooted trees thus allow testable hypotheses to be postulated for the polarities of certain character states, on the evidence of others. By such a process of elimination of alternatives, the most plausible cladogram can eventually be arrived at.

Cladograms and phylogenies An important point to note is that a cladogram shows only the hierarchical arrangement of increasingly inclusive synapomorphies observed in a set of compared taxa. The latter are invariably arranged at the branch terminations. The branching points, or *nodes*, in the diagram thus indicate only successive sister groupings of the taxa, indicating the inferred relative recency of their latest common ancestry. A cladogram is therefore not precisely equivalent to a phylogeny, because none of the compared taxa is recognized as an ancestor. If all the taxa being compared are living, their phylogeny will have the same form as the cladogram, because the taxa can only be evolutionary sisters and cousins to each other, so to speak. In that case, the nodes can then be treated as representing unknown ancestors (as in Figure 11.7) and this is how Hennig originally conceived the diagrams.

The distinction between a cladogram and a phylogeny becomes important when fossil taxa are considered, because it is conceivable that one or more of these may indeed be ancestral to some of the other taxa. Consider, for example, a simple hypothetical three-taxon cladogram for three fossil species, A, B and C (Figure 11.17(a)). In the manner of cladograms, the three taxa are set in a single row and the cladogram constructed on the basis of their synapomorphies, as shown. If the three species appear in succession in the fossil record, any of the four phylogenies shown in Figure 11.17(b) would be consistent with the cladogram. All that is known from the cladogram is that A is more primitive than B and C. It could either have branched off from their stem, or it could itself have been ancestral to them. Likewise, B could have been ancestral to C. Combining these options gives

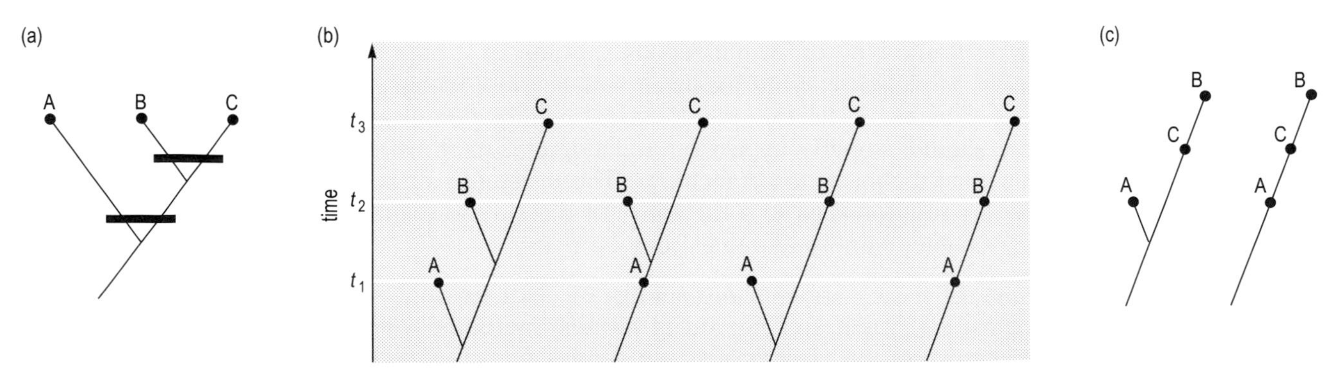

Figure 11.17 *(a) A hypothetical cladogram for three species, A, B and C; and (b) four possible phylogenies, which are consistent with it given the occurrence of the three species in three successive time intervals,* t_1, t_2 *and* t_3. *(c) The two further phylogenies which would be possible if the relative ages of the taxa were unknown.*

the four possible phylogenies in Figure 11.17(b). Moreover, if nothing were known about the relative ages of the fossil species, another option would be C ancestral to B, giving a further two possibilities (Figure 11.17(c)).

▶ What additional information on the cladogram in Figure 11.17(a) might allow you to reject the likelihood of any given taxon being ancestral?

Were any of the species to show an autapomorphy, the possibility of it being ancestral to another in the cladogram is lessened on the grounds of parsimony: its ancestral status would require the additional evolutionary 'event' of loss of the autapomorphy in the other species. Autapomorphies thus allow the hypothesis of a fossil species being a direct ancestor to be rejected (or at least considered improbable). *A major problem with the cladistic analysis of fossil taxa, however, is that an ancestral taxon cannot be positively identified as such.* The absence of autapomorphies merely allows the possibility of a species being directly ancestral to another. Subsequent discovery of autapomorphies, in more detailed studies, could always reject that hypothesis. Some cladists therefore argue that a cladogram is as far as one can go in analysing relationships, and that it is not possible to choose between the different phylogenies which may be compatible with a given cladogram. This has inspired a lively and at times vituperative debate. Palaeontologists who work with organisms that have a relatively complete fossil record, in particular, argue that the stratigraphical distribution of fossil taxa can allow one to reject some of the possible phylogenies. To deny such conclusions, because cladistic analysis alone is mute on the issue, simply highlights a limitation of the latter, rather than of what it is possible to discover.

Another methodological problem with cladistics, mentioned earlier, is its assumption of dichotomous branching of lineages.

▶ Think back to the various models of speciation discussed in Chapter 9. Does the above assumption seem reasonable in relation to these?

No. Models of speciation are frequently represented, for the sake of simplicity, as yielding pairs of daughter species (e.g. Chapter 9, Figure 9.1). However, there is nothing, in principle, to prevent most of the models of speciation discussed in Chapter 9 allowing the simultaneous splitting of a population into several potential species. Models involving localized isolates, in particular, such as the peripatric model (Chapter 9, Section 9.4.2), might readily permit such multiple splitting. These should properly appear on a cladogram, then, as multiple (*polytomous*), rather than dichotomous branchings. The insistence of some purist cladists upon resolving relationships to a hierarchy of dichotomous branchings alone is therefore unreasonable: some polytomies may be real.

In spite of these various shortcomings of cladistic analysis, it remains undoubtedly the most consistent and rigorously testable method for reconstructing phylogeny in most cases. At the very least it can furnish well-founded initial hypotheses on phylogeny, which can then be tested by other forms of evidence (such as the stratigraphical sequence of fossil taxa).

11.2.5 Comparison of the methods

Phenetics and cladistics both developed in response to the perceived subjectiveness of evolutionary systematics, each attempting to provide a more consistent and therefore objective method of analysis. Phenetics eschews the quest for the unknowable phylogeny, and concentrates instead on raw phenotypic similarities, quantified by multivariate analysis, for clustering taxa. Cladistics, in contrast, attempts to refine the reconstruction of phylogeny by grouping taxa purely according to the relative recency of latest common ancestry, inferred from synapomorphies. Despite offering considerable improvements in the consistency of approach, both of the new schools still have shortcomings.

Cladistics is the least ambiguous means for reconstructing phylogeny, in most cases. It is particularly effective for living organisms which show numerous characters with well-understood polarities. If the group being investigated has a relatively complete fossil record, the evolutionary systematist's approach of using stratigraphical information for resolving questions of possible ancestry is also needed. Where the fossils are rather similar, with features of relatively low information content, the polarity of character states may be difficult to determine, and convergence and evolutionary reversals of single features may be fairly common. In such a case, a combination of detailed stratigraphical and phenetic analysis (sometimes referred to as *stratophenetics*) may yield more information than cladistics can offer. The case studies discussed in Chapter 10, Section 10.3.2, illustrated this approach. The evidence for phylogenetic history that remains today varies enormously between different groups of organisms, according to the availability of living examples and the vagaries of the fossil record. The systematist therefore cannot afford to be doctrinaire in the face of such variable evidence, and must tailor the methods used for interpreting phylogeny accordingly.

In all cases, however, it should never be forgotten that phylogenies are hypotheses concerning historical patterns, and that, as such, they must be repeatedly tested with other data and by alternative means of analysis.

Summary of section 11.2

Evolutionary systematics, the classical approach, is eclectic, exploiting all forms of evidence for both anagenetic change and cladogenesis. A major objective is to distinguish homology from analogy, so as to exclude the latter from the classification. This distinction is clearest in features that have a high content of information concerning the number and arrangement of components. Both monophyletic and paraphyletic taxa are recognized; polyphyletic taxa are rejected. The lack of a consistent method, however, promotes subjectivity in the treatment of evidence.

From the 1950s–1960s, two opposing systematic schools sought to remedy these problems. Phenetics groups taxa according to overall phenotypic similarity. Large numbers of variates are measured and the taxa clustered in a hierarchy (phenogram) according to the morphological distances between them. The phenogram is considered likely to reflect phylogeny, as the misleading similarities due to analogy will tend to be 'swamped' in the comparisons by contrasting homologous similarities, if enough variates are studied.

Cladistics groups taxa according to inferred relative recency of latest common ancestry, based upon shared derived character states (synapomorphies). Thus only monophyletic taxa are recognized, which unambiguously reflect the pattern of cladogenesis. The primitive-derived polarity of character states can be recognized from embryological, outgroup comparison, and palaeontological criteria. The relationships between taxa are inferred from that hierarchy of sister groupings (cladogram) which shows the most parsimonious arrangement of shared derived character states. Analogous derived character states are detected by their lack of congruence with the other synapomorphies in the preferred cladogram. Where polarities are not evident at the outset, unrooted trees may be constructed for the taxa in question, and unparsimonious trees rejected. Evidence for any of the taxa being an outgroup with respect to the others can then be used to root the tree and so turn it into a cladogram.

With the inclusion of fossil taxa, several alternative phylogenies may be consistent with one cladogram. In these phylogenies, the fossil taxa may be variously interpreted as being directly ancestral to others. The hypothesis of a fossil species being a direct ancestor may be rejected if it possesses autapomorphies, but cladistic analysis alone cannot positively identify an ancestor as such. Stratigraphical considerations can help to resolve this issue.

Cladistics offers the most consistent and unambiguous method for reconstructing phylogeny in most, though not all, cases. If a good fossil record is available, stratigraphical information is also needed to resolve questions of ancestry. Where the compared taxa are somewhat similar with features of low information content, combined stratigraphical and phenetic (stratophenetic) analysis may be more informative than cladistics.

As phylogenies are themselves only hypotheses, they should be repeatedly tested against the results derived from other forms of evidence and methods of analysis.

11.3 PHYLOGENETIC ANALYSIS FROM MOLECULAR DATA

Since the elucidation of the molecular basis of heredity in the 1950s (Chapter 3, Section 3.2.1), there has been a rapid growth in the techniques for directly assaying the genetic make-up of organisms. As Chapter 3, Section 3.5.2, showed, these techniques initially concentrated upon gene products (either amino acid frequencies or their sequences in polypeptides), but later moved on to DNA (or RNA) itself. Both the phenetic and the cladistic methods of analysis, which were discussed in relation to morphology in Section 11.2, can also be applied to such molecular data from living, and even a few extinct, organisms.

In comparative morphology, the characters to be compared are usually self-evident by virtue of their anatomical position and structure. For example, birds' wings are recognized as being homologous with human arms, and not legs, because of their anatomical positions as forelimbs. With molecular data, in contrast, there is an immediate problem of what to compare with what, in order to recognize homology.

One criterion for identifying some homologous proteins is equivalence of function, because the particular biochemical functions of certain proteins seem to have been highly conservative in the evolution of a wide diversity of organisms. One of the first proteins to be investigated in detail from an evolutionary point of view, for example, was cytochrome *c*. This large, complex molecule is present in all aerobic organisms, where it plays a precise role in the electron transport chain in cellular respiration. Its structure is very constrained, varying little between organisms, as small changes can upset its vital function. Its structure is well known and it has been studied in a wide range of organisms. The few differences that have evolved in cytochrome *c* thus give valuable clues to phylogeny. Another conservative type of molecule, whose specific function constrains its structure, is ribosomal RNA (rRNA). Sequences from rRNA are widely used to investigate the phylogenies of disparate organisms.

Recognition of homology at the finer level of specific sequences of amino acids, or of the nucleotide bases of DNA (Chapter 3, Section 3.2.3), involves an elaborate statistical matching exercise. The logic of this method is best illustrated with a much simplified hypothetical example.

Imagine that an organism has within its DNA the sequence of 12 bases shown at the top of Figure 11.18(a). The exactly corresponding part of the DNA of another, genetically identical, individual can be detected by making a series of comparisons of the two DNA sequences, shifting the second base sequence along by one position each time. As the lower part of Figure 11.18(a) shows, when the two sequences are exactly aligned (marked by the arrow), the match is, of course, 100 per cent. There is nevertheless a variable amount of chance matching even when the two sequences are misaligned, as the other comparisons in the Figure show.

▶ What is the *average* percentage of chance matching you would expect in a long sequence of DNA, in which the four bases were present in equal frequencies?

With equal frequencies of the bases, one would expect, on average, some 25 per cent of the bases to be matched, by chance alone, in each comparison. This is because, for any given base on one sequence, the probability of the same base being present in the compared position in the other sequence is one in four. However, greater degrees of matching by chance alone become less and less probable.

▶ What is the probability of a 100 per cent match in a sequence of twelve bases arising by chance alone?

With the probability of a match at each site being 0.25, the probability of matches at all 12 sites will be $0.25 \times 0.25 \times 0.25$.... etc., 12 times over, or $(0.25)^{12}$, which is about 6×10^{-8}, or six-millionths of 1 per cent. The null hypothesis of the 100 per cent match in Figure 11.18(a) being due to chance alone could safely be rejected, and the alternative hypothesis of a common origin for the sequence (i.e. homology) accepted.

(a)

	AGG ATG AAT CCC	proportion of matching bases
GCAAGG	ATG AAT CCC CAT	4/12 = 33%
GCAAG	GAT GAA TCC CCA	T 2/12 = 17%
GCAA	GGA TGA ATC CCC	AT 5/12 = 42%
→ GCA	AGG ATG AAT CCC	CAT 12/12 = 100%
GC	AAG GAT GAA TCC	CCAT 5/12 = 42%
G	CAA GGA TGA ATC	CCCAT 1/12 = 8%
	GCA AGG ATG AAT	CCCCAT 3/12 = 25%

(b)

	AGG ATG AAT CCC	proportion of matching bases
GCAAGG	GTG AAA CCC CAT	3/12 = 25%
GCAAG	GGT GAA ACC CCA	T 4/12 = 33%
GCAA	GGG TGA AAC CCC	AT 7/12 = 58%
→ GCA	AGG GTG AAA CCC	CAT 10/12 = 83%
GC	AAG GGT GAA ACC	CCAT 5/12 = 42%
G	CAA GGG TGA AAC	CCCAT 2/12 = 17%
	GCA AGG GTG AAA	CCCCAT 2/12 = 17%

Figure 11.18 *(a) Successive comparisons of the DNA base sequence for one organism (below the line) with that of 12 bases in another, genetically identical organism (above the line). (b) As above, though with two base substitutions (green boxes) in the compared organism. In both cases, matching bases are shown in green type, while the proportion of matching bases is shown to the right. Note that homologous zones can be detected by a sharp rise in the proportion of matches (arrowed).*

In comparing different organisms, and especially from different species, however, mutation is likely to have caused divergence of the homologous sequences. Figure 11.18(b) repeats the comparisons of Figure 11.18(a), but with the DNA of the second organism shown to differ by two bases in the homologous sequence (boxed). Note that, despite these differences, the two homologous sequences still show a significantly greater percentage of base matching than that arising by chance alone. By this means, homologous sequences can be identified statistically, notwithstanding mutational differences between them. The same logic can be applied to recognizing homologies in the amino acid sequences of proteins. The recognizability of homology by this means naturally fades with time, as increasing divergence renders the distinction from chance matching less clear. Hence, the importance of using conservative sequences (such as that for cytochrome *c*) when comparing distantly related organisms.

As noted at the outset, the explanation of this method for recognizing homologies outlined above has been greatly simplified. Complications include the possibility of mutational deletions or additions in some sequences, the fact that the different possible amino acids or nucleotide bases are not present in equal frequencies, and the differing probabilities of forward and reverse mutations between one amino acid or base, and another (Chapter 3, Section 3.3.2). Allowances can be made for all of these problems, but their details are beyond the scope of this Book. Suffice it to say that the complex comparisons are now usually carried out by using computers.

Nor are the problems at an end here, because of the 'fluidity' of the genome, which was discussed in Section 3.2 of Chapter 3. Two important sources of confusion in the recognition of molecular homologies may arise as a result of structural changes in chromosomes. Translocation (Chapter 3, Section 3.3.3) may lead to *partial homology*, whereby only a part of a given sequence in one organism is homologous with that in another. An abrupt change in the consistency of statistical matching of molecular sequences can expose such cases of partial homology.

Another source of confusion is sequence duplication (Chapter 3, Section 3.2.2). This process is important because it requires the recognition of a distinction between two kinds of molecular homology. Homology between corresponding sequences in *different* organisms is termed **orthology** (equivalent to the conventional form of homology in comparative morphology). Homology between corresponding sequences in the *same* organism, which have arisen by duplication of a single original sequence, by contrast, is termed **paralogy**. The potential for confusion is that paralogous sequences can diverge just as orthologous sequences do, yet only the latter divergence coincides with cladogenetic divergence; divergence of paralogous sequences may ensue from the time of sequence duplication within a single evolutionary lineage. Therefore, if a sequence from one species is mistakenly compared, not with its orthologous counterpart in another, but with a paralogous sequence in the second species, resulting from a duplication in some common ancestor, then the genetic divergence will reflect the time elapsed since the duplication event, rather than that since the mutual divergence of ancestors of the two species. In practice, the distinction is often rather obvious because the duplication event was (evidently) sufficiently long ago for the extent of paralogous divergence to be inconsistent with other evidence (e.g. from the fossil record) for the timing of cladogenesis.

One of the better known examples of paralogy is provided by the group of proteins known as globins, which function as oxygen carriers. Successive duplications of genes, long ago in evolutionary history, have given rise to a number of paralogous forms which co-occur in the majority of vertebrates. Among these forms are myoglobin and the two types of protein chain, α and β, that combine to give adult haemoglobin. Myoglobin is a single protein that occurs within red muscle fibres (giving them their red colour). Haemoglobin is the oxygen carrier of blood (again, giving the red colour). In some vertebrates, including humans, further paralogous sequences provide different versions of the haemoglobin chains at different stages of development (Chapter 3, Section 3.2.3). When paralogy can be clearly distinguished from orthology, as with the globins, then a tree may be constructed to show both duplications and inferred cladogenetic relationships. Figure 11.19 shows a simplified version of one constructed from globin sequence data by Goodman *et al.* (1987).

Even in simplified form, such diagrams can seem quite daunting, and you may wish to spend a little time studying it.

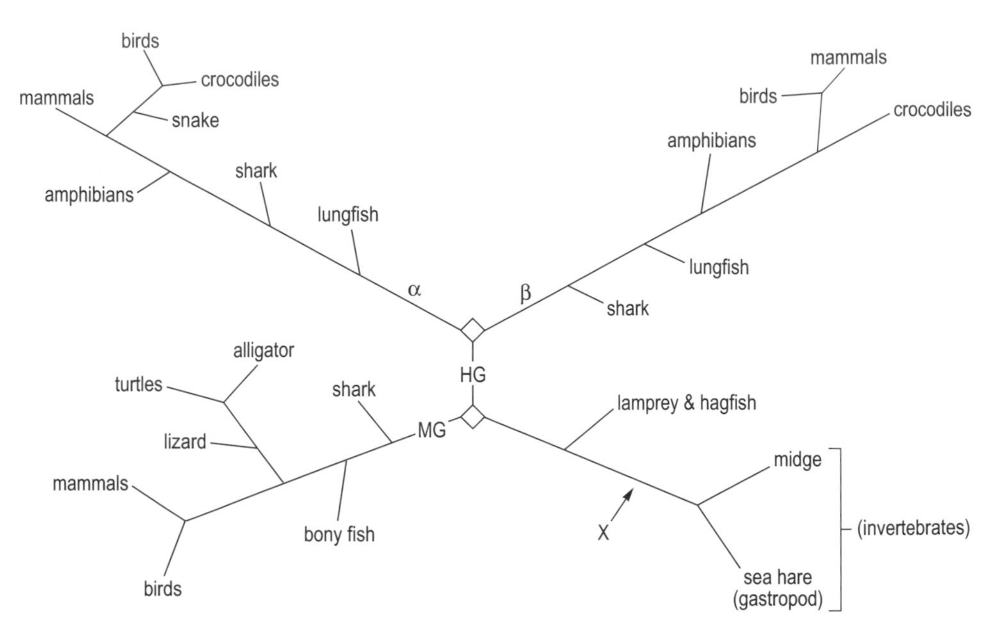

Figure 11.19 *Simplified globin tree for selected vertebrate taxa, based on 218 sequences, derived from Goodman* et al. *(1987). Duplications are shown by diamonds, giving rise to the paralogous proteins, myoglobin (MG), and the α and β chains of haemoglobin (HG). The position X is discussed in the text.*

▶ Why do you suppose the same animals appear more than once, on different branches in the tree?

They appear more than once because of the gene duplications: the different main branches of the tree represent the different paralogous globins (as labelled on the diagram). As most of the animals possess all or most of the different globins, a tree of relationships between the animals can be built up for each globin molecule. The 'diamonds' in the tree thus represent the duplications which gave rise to the various paralogous genes. The ordinary branch points, in contrast, show where orthologous genes have diverged, with the splitting of lineages.

The ways in which such trees may be constructed will be discussed in the following Sections, but first it is worth noting certain aspects of Figure 11.19 in relation to earlier discussions.

▶ Does Figure 11.19 show a rooted or an unrooted tree?

It is an unrooted tree (Section 11.2.4), as no branch is left free to connect with possible outgroups. It is not difficult to see where it should be rooted, although the reasons for so doing rely upon other conventional taxonomic data, rather than on the molecular evidence itself. The small cluster of invertebrates at the bottom of the tree may confidently be treated as an outgroup for all the vertebrates, and so the root should be placed on the branch leading to them (marked X on the Figure).

Consider again the duplication events marked by diamonds.

▶ The lampreys and hagfishes are primitive jawless fishes. Did any of the globin duplications occur in their ancestors (shared with all the other vertebrates)?

Apparently not: the first diamond, representing the duplication that yielded myoglobin and haemoglobin, comes *after* the branch for the jawless fishes. In other words, the presence of the two paralogous globins is a synapomorphy (Section 11.2.4) which unites all other vertebrates, to which the jawless fishes are therefore a sister group.

▶ Recall from Section 11.2.4 that it was concluded, from morphological data, that crocodiles and birds share a more recent common ancestor than either does with the mammals. Study each of the trees for myoglobin and the haemoglobin α and β chains in Figure 11.17, and decide whether or not they are consistent with that conclusion.

Only the tree for the α chain of haemoglobin yields the same result: here the crocodile and bird clades branch from each other after the mammal clade has already branched off. The β chain of haemoglobin, in contrast, unites the mammal and bird clades, with the crocodile clade as the sister group. Myoglobin does likewise. The discrepancies between the results from these proteins means that they cannot all be right (as a representation of phylogeny). Phylogenies based on molecular data, too, are fallible.

The data yielded by some of the methods of molecular assay may constrain the approach to phylogenetic reconstruction. In this respect, molecular data fall into two categories:

1 Measures of net differences between homologous proteins, DNA, or RNA, though without information on the exact sequences involved.

2 Comparative sequence data, again, either of amino acids in polypeptides or of DNA (or RNA) bases.

▶ Study Table 3.9 in Chapter 3 again and decide to which of categories 1 and 2 above data from the following techniques should be assigned: (a) electrophoresis of allozymes; (b) immunological comparison of proteins; (c) DNA sequencing; and (d) DNA hybridization.

From this list, only (c), DNA sequencing, falls into the second category (as does, also, amino acid sequencing). The other three only provide measures of net similarity (or dissimilarity), and so fall into the first category. The distinction between the two categories is important because only for data in the second category is it known exactly where the differences between the compared sequences lie, and therefore cladistic analysis can only be applied to such data.

With category 1 data, in contrast, cladistic inferences cannot be made, because the exact locations of the differences are not known. With only raw distances between compared taxa available, only phenetic analysis can be employed.

11.3.1 Phenetic analysis of molecular data

In Section 11.2.3, attention was drawn to certain disadvantages of phenetic analysis for the reconstruction of phylogeny. A major problem is the false implication of cladogenetic relationships that paraphyletic taxa give.

▶ What aspect of morphological evolution leads to the recognition of paraphyletic taxa?

Paraphyletic taxa reflect marked differences in the rates of anagenetic change: two taxa may be similar merely because they have not departed greatly from the ancestral condition, while a third, which may have shared a more recent common ancestor with one of them, may nevertheless differ more because of rapid anagenetic evolution (Figure 11.5(b)). If morphological divergence between taxa were to arise at a constant rate, then the phenetic distances between them would directly reflect the relative recency of their common ancestry. In such a case, paraphyletic taxa would not emerge as apparently natural groupings. The question thus arises: how variable are rates of molecular evolution? This beguilingly simple question opens up a complex and widely ranging debate, which can only be touched on here. A reconsideration of some basic principles will help clarify the general issues.

▶ What is the ultimate source of new alleles?

Mutation is the usual source of new alleles in a population; recombination merely shuffles alleles (or parts of alleles) into different arrangements (Chapter 3, Section 3.3.1). Transfer of transposable elements between species (Chapter 3, Section 3.2.4) and hybridization (Chapter 9, Section 9.3.1) may also introduce new alleles in some species, but mutation remains the dominant process.

Mutations themselves occur randomly, with characteristic (but not equal) frequencies at given loci (Chapter 3, Section 3.3.3). However, they do not necessarily spread through populations at corresponding rates.

▶ Why not?

Natural selection occurs, such that alleles associated with lower fitness are either kept to low frequencies, or are eliminated (Chapter 4). Hence a deleterious mutation, which may arise with a certain frequency in a population, is prevented from spreading, and the *status quo* of allele frequencies preserved. Conversely, new mutations that increase fitness may spread rapidly, giving rise to rapid genetic evolution of the population, and hence divergence from other populations. Genes subject to selection can thus be expected to evolve at different rates.

Should there be any mutations with no influence on fitness (i.e. not affected by natural selection), however, then the frequencies of these may fluctuate in a population, and in time become fixed. Some genetic differences between populations may thus arise simply through a combination of mutation and drift (Chapter 9, Section 9.2.6).

▶ What would you call such mutations, and how might they arise?

Mutations with no effect on fitness are termed neutral mutations (Chapter 3, Section 3.3.2). There are various means by which they might arise. Changes in the amino acid sequences of parts of some proteins can leave the function of the protein unaffected. Alternatively, a point mutation may merely provide a synonymous codon (Chapter 3, Section 3.3.4). Although such a change may affect the amount of an enzyme that is produced, it would not alter its structure, and so may be of neutral effect. Much DNA is not, apparently, even transcribed in the first place (Chapter 3, Section 3.2.2), and so mutations in this DNA are phenotypically mute.

Neutral mutations spread purely according to their frequency of appearance and to genetic drift. Although these effects cannot be expected to yield literally constant rates of molecular change in populations, they might be expected to yield constant *average* rates of change (known as 'stochastically constant' rates). Such rates are the basis of the **molecular clock theory**, which relates the average rate of fixation of new alleles in an equilibrium population (as illustrated by hypothetical examples in Chapter 3, Figure 3.22) to the rate of mutation. According to the theoretical prediction of the Japanese geneticist M. Kimura, neutral mutations will become substituted at loci at an average rate equal to the rate of mutation per generation. On this basis, genetic distances between species, insofar as they involve neutral mutations, can be expected to reflect the time elapsed since their latest common ancestor. Even where selection does operate, an averaging of selection coefficients over long periods could also provide a molecular clock for times of divergence. A phenogram of such distances might thus indeed be interpreted as a phylogeny.

There is much debate about how generally the molecular clock theory can be applied, and it is certainly acknowledged that rates do seem to vary quite widely between different proteins and even for different parts of the same protein. Differing functional constraints on structure (Section 11.3) are one major source of differences in rates. Generation time (which affects the rate of meiotic production of gametes (Chapter 3, Section 3.3.3) and hence the rate of some mutations) also seems to exert some influence on rates of substitution.

Estimates of the average rates of molecular change for given proteins can be obtained with the aid of data from the fossil record. For example, in 1967 Allan Wilson and Vincent Sarich of the University of California, Berkeley, used the fossil evidence for the origin of Old World monkeys some 30 Ma ago to calibrate a molecular clock based on blood proteins (serum albumins). Their immunological distance data gave a value for the distance between humans and the African apes approximately one-sixth of that found between humans and various Old World monkeys (a ratio only slightly differing from that in Chapter 3, Table 3.7). They concluded that the human–African ape split therefore dated from about only 5 Ma ago. Their claim has been countered by the assertion that the molecular clock in this case (as indeed in many others) can be expected to have 'slowed down' over time, for a variety of reasons including that of increasing generation time.

A simple but effective way of testing the molecular clock theory is called the *relative rate test*. Suppose that for three species, two (let us call them A and B) are known by cladistic analysis to be more closely related to each other than to a third (C). If the rate of molecular evolution had been effectively constant, then, by simple logic, we should expect the distance between A and C to be equal to that between B and C. This is because A and B should have continued to diverge from C at equal rates, following their own mutual separation. Where more than three species are involved, there are multiple opportunities for applying the rate test.

Table 11.3 shows some data on mean thermal stability of hybrid DNA (Chapter 3, Section 3.5.2) made from a variety of members of the mammalian order Carnivora. The data have been extracted from a study conducted by O'Brien *et al.* (1985) that aimed to elucidate the relationships of the giant panda and the lesser panda (see Chapter 2, Section 2.3.3). Here, the mean difference between the temperature of dissociation of hybrid DNA and that of pure DNA for each pair of species will be used as a measure of the genetic distance between them.

Table 11.3 *DNA hybridization data for the pandas and other selected Carnivora, modified from O'Brien* et al. *(1985). Values are of the difference between the mean temperature of dissociation of the hybrid DNA and that of conspecific DNA for each pair compared. The values for each pairwise comparison and its converse (e.g. brown bear against giant panda and giant panda against brown bear) have been averaged to simplify the analysis.*

	Brown bear	Giant panda	Lesser panda	Raccoon	Dog
Brown bear	—	4.6	14.0	14.4	18.7
Giant panda		—	14.3	14.5	18.3
Lesser panda			—	14.1	18.9
Raccoon				—	18.5
Dog					—

▶ From the values given in the Table, which of the five species indicated is probably the most distantly related to the other four, and why?

The dog is the most likely outgroup species, because it shows the greatest genetic distances from all the other species.

▶ Study the distance values between the dog and each of the other species. Are they consistent with the molecular clock theory in this instance?

Yes, the values are all fairly close to one another, implying no great overall differences in the rates of DNA change in the various lineages that arose after their latest common ancestor. The *relative rate test* has therefore been approximately satisfied in this instance.

A molecular phenogram (Figure 11.20(a)) can now be constructed from the data in Table 11.3, and it is likely also to reflect the phylogeny, in view of the conclusion given above. It is here constructed employing an average neighbour cluster statistic (Section 11.2.3). Note how the distance between any given pair of species has been apportioned equally along the branch lines running to each species, in

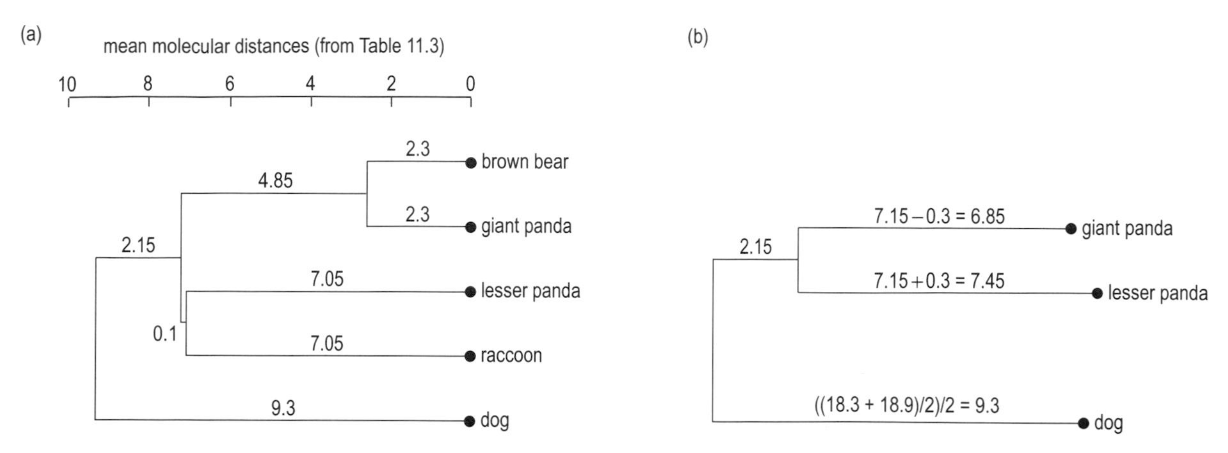

Figure 11.20 *(a) Phenogram for the data in Table 11.3, using an average neighbour cluster statistic. Distances indicated on each branch are based on an assumption of the molecular clock theory. (b) Phenogram for the dog and the two pandas in Table 11.3 using a cluster statistic that allows for variation in the rates of molecular evolution.*

accordance with the assumption of the molecular clock. Thus, the 4.6 between the brown bear and the giant panda is shown as 2.3 along each branch line; and the *averaged* distance between either the brown bear or the giant panda, and either the lesser panda or the raccoon ((14.0 + 14.4 + 14.3 + 14.5)/4 = 14.3), is split into a total of 7.15 along the branch lines running to each.

▶ To return to the question of the relationships of the giant panda raised in Chapter 2, Section 2.3.3, does the phenogram in Figure 11.20(a) weigh in favour of including it in the bear family (Ursidae) or with the raccoons (Procyonidae)?

The giant panda is clearly more closely associated with the bears, and might therefore be included in the Ursidae. The lesser panda, by contrast, is more closely allied with the raccoons: the similarities of the two pandas thus appears to be a case of convergence. O'Brien *et al.* also studied electrophoretic data on isozymes and immunological data, as well as karyological evidence from the pandas, with the same conclusions. Such fossil evidence as there is is also consistent with this conclusion.

Notwithstanding the consistency of these conclusions, the earlier warning about the fallibility of molecular indications must be reiterated. Recall from Section 11.2.3, first of all, that different cluster statistics may give differing results. Secondly, the relative rate test does show up some discrepancies in molecular rates in many cases. This should hardly come as a surprise: many allele frequencies clearly are modulated by natural selection. However, more sophisticated cluster statistics can cater for such discrepancies, provided that they are not excessive: phenograms may be generated to show different amounts of genetic distance along different branch lines (Figure 11.20(b)). The underlying logic of phenograms of this kind is as follows (using just the data for the dog and the two pandas in Table 11.3). As before, it is assumed that the dog is the outgroup species, because the distances between it and each of the two pandas are greater than that between the latter pair. Yet the distance between the dog and the lesser panda (18.9) is greater than that between the dog and the giant panda (18.3). Evidently, the rate of molecular evolution in the lesser panda lineage was slightly greater than that in the giant panda lineage, and so the distance between them may be split unequally along their respective branch lines.

11.3.2 Cladistic analysis of molecular data

Sequenced molecular data allow precise homologous matching, and thus the recognition of differing molecular character states. As already shown in Figure 11.19, unrooted trees can be constructed from such data. However, the recognition of primitive-derived polarity for molecular characters is more difficult than for most morphological characters. First of all, the embryological criterion (Section 11.2.4) cannot be used at all, because the genome is fixed at birth: it has no development. Secondly, the palaeontological criterion offers exceedingly little scope, fossilized DNA being very rare. All is not lost in that respect, however, as this Section will later illustrate. In effect, though, only outgroup comparison remains for gauging polarity.

A prominent danger with outgroup comparison of molecular data arises from the probability that some mutations (e.g. base substitutions) may be repeated and others reversed.

▶ How might such effects confuse the assessment of polarity?

Base substitutions at a given position may be independently repeated in different lineages, giving rise to analogous derived states. Reversals, by contrast, may bring about the appearance of plesiomorphy in lineages which in fact are doubly apomorphic (with a derived substitution followed by a reverse substitution). All these sources of potential confusion have their characteristic probabilities, and so the way around them is to assay long sequences (or aggregates of many short sequences). The product of these probabilities of error is reduced with increasing quantities of data.

The main advantage of cladistic analysis of molecular data (where possible) over phenetic analysis for establishing relationships, is that the molecular clock need not be assumed (Section 11.3.2). All that matters is the recognition of molecular synapomorphies (e.g. derived base substitutions), from which cladograms may be constructed (Section 11.2.4).

To illustrate a molecular cladogram, an example is shown which has the unusual merit of incorporating data from an extinct animal. The 'marsupial wolf' *Thylacinus cynocephalus* was once widespread in Australia, but rapidly declined following settlement by humans and the dogs that they introduced. The last known specimen died in Hobart Zoo in 1933 (although rumours of wild sightings persist). For some time, there has been debate about whether it was more closely related to similar-looking, extinct carnivorous South American marsupials (known as borhyaenids), or to other Australian carnivorous marsupials. If the latter, it would then only have been convergent with the borhyaenids. Thomas *et al.* (1989) were able to extract minute amounts of intact DNA and RNA from museum specimens and archaeological remains, from which larger samples were synthesized by cloning (Chapter 3, Section 3.5.2). The sequences from the thylacine were then

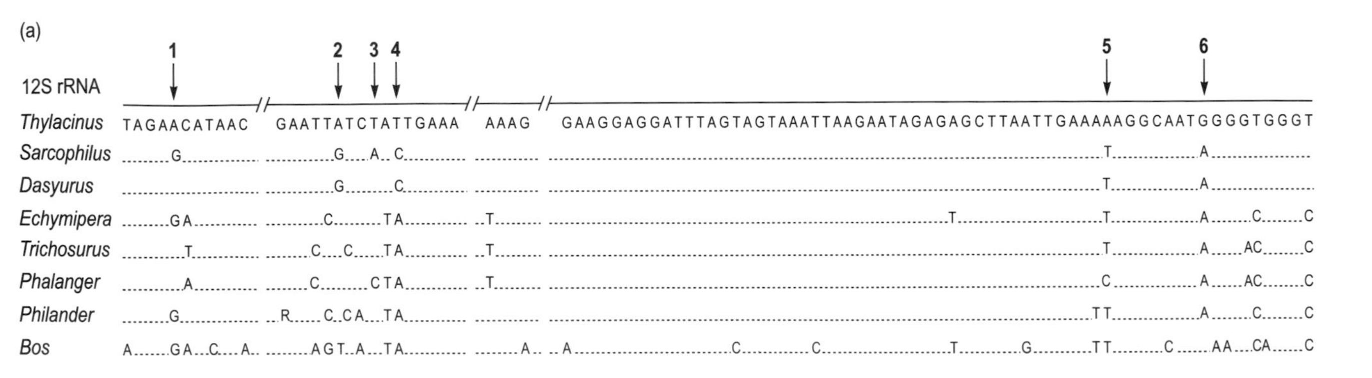

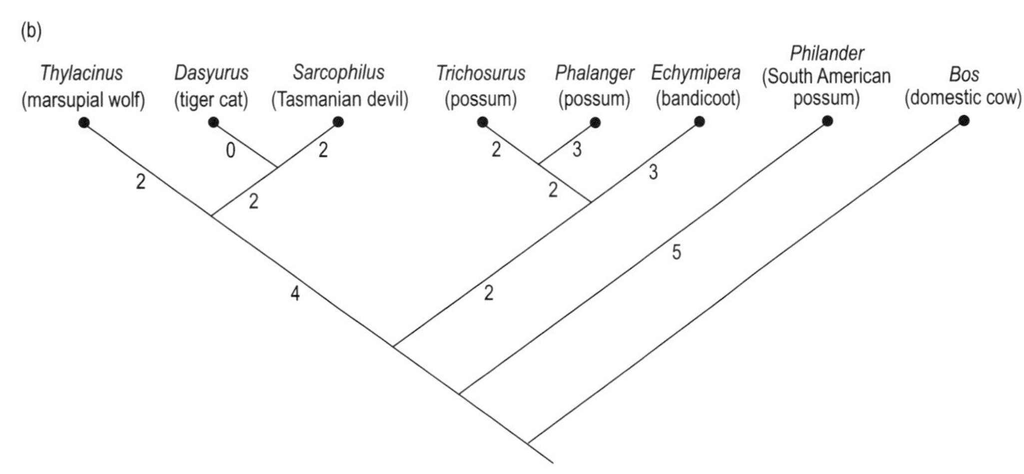

Figure 11.21 *(a) Base sequence of part of the 12S rRNA gene from the marsupial wolf, six other marsupials and a cow (as outgroup). Bases in the other animals identical to those in the thylacine are indicated by dotted lines. Cladistic analysis was conducted for the sites covered by the solid lines at the top (intervening sequences not shown). Positions 1–6 indicate sites where there are analysed differences between* Thylacinus, Sarcophilus *and* Dasyurus. *(b) Cladogram based on (a), with numbers of inferred base substitutions (= apomorphies) indicated on branches.*

compared with those from other taxa, including both Australian carnivorous marsupials (the Tasmanian devil *Sarcophilus harrisii* and the tiger cat *Dasyurus maculatus*—see marginal illustrations) together with other Australasian marsupials (the bandicoot *Echymipera kalubu* and two possums: *Trichosurus vulpecula* and *Phalanger sericeus*), and the South American possum *Philander opossum*. The domestic cow *Bos* is included as an outgroup to the marsupials. One of the sequences studied, that for some ribosomal RNA (12S rRNA), is shown in Figure 11.21(a), and a cladogram derived from these data is given in Figure 11.21(b).

The cladogram is based only on parts of the sequences, indicated by the solid lines at the top of the diagram. Numbers of apomorphies are indicated along each branch. For example, *Sarcophilus* is shown to have two autapomorphies and *Dasyurus* none; the two share two synapomorphies; *Thylacinus* has two autapomorphies, while all three species share four synapomorphies; and so on. The character state polarity of the alternative bases at homologous positions is established as follows. The six positions where differences in bases (beneath the solid lines) exist between the three species are indicated on Figure 11.21(a). In position 1, *Sarcophilus* has G, but *Dasyurus* has A. *Thylacinus*, the sister group to the pair, has A, implying (by outgroup comparison) that the A in *Dasyurus* is primitive, and the G in *Sarcophilus*, derived.

▶ At which of the other positions is the second autapomorphy of *Sarcophilus* situated?

Position 3 shows T for *Thylacinus* and *Dasyurus*, but A for *Sarcophilus*. The latter is therefore judged to be derived by the outgroup comparison with *Thylacinus*.

Position 2 shows *Dasyurus* and *Sarcophilus* sharing G, and *Thylacinus* with A. In this instance, the sister group to these three species is that comprising *Trichosurus*, *Phalanger* and *Echymipera*. As the latter three all share A at position 2, the G shared by *Dasyurus* and *Sarcophilus* is judged by this outgroup comparison to be derived, and hence a synapomorphy.

The other synapomorphy (at position 4) for *Dasyurus* and *Sarcophilus* is more difficult to resolve, because neither these species (with C) nor *Thylacinus* (with T) share(s) the A shown here by the outgroup. The decision to treat the T as the primitive state for the three species in question and the C as derived in *Dasyurus* and *Sarcophilus* depends upon the relative probabilities of the sequence of base substitutions involved.

Positions 5 and 6 show the bases in *Thylacinus* to be derived, by outgroup comparison with the sister group taxa.

▶ What term would you use to describe the shared bases found here in *Dasyurus* and *Sarcophilus* (T in position 5, and A in position 6)?

These shared bases are examples of symplesiomorphy (Section 11.2.4).

▶ What does the cladogram suggest about the likely affinities of the extinct thylacine?

Its close linkage with the two other Australian carnivorous marsupials suggests that it was more closely related to them than to any other marsupials (including the borhyaenids).

Cladistic analysis thus offers considerable scope for reconstructing phylogenies from molecular data, with the extra advantage over phenetic analysis of not relying upon the molecular clock theory.

SUMMARY OF SECTION 11.3

Some protein or nucleotide sequence homologies can be identified because of their relative structural conservatism, linked with some basic vital function. Such molecules are especially useful for establishing the phylogeny of distantly related organisms. Homologies at the sequence level, whereby point mutational differences may be detected, can be identified by the statistical matching of sequences.

There are two kinds of molecular homology: that between corresponding sequences in different organisms (orthology); and that between duplicated sequences in the same organism (paralogy). Trees showing both kinds of homology may be constructed, with the same organisms repeated on different paralogous branches.

Phenetic analysis is possible where only overall molecular differences are known (e.g. immunological distances and DNA hybridization data). Molecular phenograms are likely to match phylogenies where rates of molecular evolution have been more or less constant. Molecular clock theory predicts stochastically constant rates for the fixation of neutral mutations. The theory is controversial, however, and wide variations in apparent rates are attributed to the effects of natural selection, variation in generation time and other factors. Molecular clocks may be calibrated by reference to the fossil record, and tested by the relative rate test, whereby members of a given monophyletic taxon are predicted to show equal molecular distances from an outgroup taxon.

Cladistic analysis requires sequenced molecular data, so that particular nucleotide base (or amino acid) substitutions may be recognized as derived character states for the construction of cladograms. Outgroup comparison of homologous sequences is used to determine the primitive-derived polarity of alternative bases (or amino acids) at a given position in the sequences.

11.4 MOLECULES AND MORPHOLOGY IN PHYLOGENETIC ANALYSIS: A CONCLUSION

Throughout the discussion of both morphological and molecular evidence, it has been stressed that there are potential pitfalls: no database is wholly reliable. Moreover, no method of analysis (evolutionary systematics, phenetics and cladistics) is wholly objective; they all have skeletons of subjectivity in their cupboards. The selection of characters is a particularly difficult problem. As more comes to be understood about the behaviour of genes in shaping phenotypic attributes, it may be possible to start defining characters in terms of the genes which regulate them, thereby providing a common currency, so to speak, in the analysis of characters. Such a development could help to reduce the subjectivity involved in comparing many disparate morphological characters.

Although cladistic analysis usually offers the most rigorous route to the reconstruction of phylogeny, the various methods are differently suited to different situations. The seeker after phylogeny cannot afford to be doctrinaire, and must choose suitable tools for each job. At the end of the day, it is not a case of competition between techniques for acquiring data and between methods for analysing them, because none is foolproof. Rather, you should recall at all times what was asserted at the outset, that phylogenies are only hypotheses, and must therefore be repeatedly subjected to testing. The more forms of data that are available, and the greater the range of methods used to analyse them, the more frequently phylogenetic hypotheses can be independently tested. Multiple corroboration is as close to the truth as we shall ever get.

REFERENCES FOR CHAPTER 11

Goodman, M., Miyamoto, M. M. and Czelusniak, J. (1987) Pattern and process in vertebrate phylogeny revealed by coevolution of molecules and morphologies, in Patterson, C. (ed.), *Molecules and Morphology in Evolution: Conflict or Compromise?*, Cambridge University Press, Cambridge, UK, pp. 141–176.

Gould, S. J. (1977) *Ontogeny and Phylogeny*, Belknap Press of Harvard University Press, Cambridge, Mass., USA.

O'Brien, S. J., Nash, W. G., Wildt, D. E., Bush, M. E. and Benveniste, R. E. (1985) A molecular solution to the riddle of the giant panda's phylogeny, *Nature*, **317**, 140–144.

Thomas, R. H., Schaffner, W., Wilson, A. C. and Pääbo, S. (1989) DNA phylogeny of the extinct marsupial wolf, *Nature*, **340**, 465–467.

FURTHER READING FOR CHAPTER 11

Ax, P. (1987) *The Phylogenetic System. The Systematization of Organisms on the Basis of their Phylogenesis*, English edn trans. by Jefferies, R. P. S., John Wiley & Sons, Chichester, UK.

Benton, M. J. (1990a) *Vertebrate Palaeontology*, Unwin Hyman, London, UK.

Benton, M. J. (1990b) Phylogeny of the major tetrapod groups: morphological data and divergence dates, *Journal of Molecular Evolution*, **30**, 409–424.

Carroll, R. L. (1988) *Vertebrate Palaeontology and Evolution*, W.H. Freeman & Co., New York.

Patterson, C. (ed.) (1987) *Molecules and Morphology in Evolution: Conflict or Compromise?*, Cambridge University Press, Cambridge, UK.

Ridley, M. (1986) *Evolution and Classification—The Reformation of Cladism*, Longman, London.

SELF-ASSESSMENT QUESTIONS

SAQ 11.1 *(Objectives 11.2–11.5)* Table 11.4 shows the states of just two morphological characters in humans, chimpanzees and gibbons: (i) the extent of body hair; and (ii) the number of wrist bones (carpals) found in the adult.

Table 11.4 *Selected character states in humans, chimpanzees and gibbons.*

	State of character in:		
Character	Human	Chimpanzee	Gibbon
(i) Body hair	Localized	Extensive	Extensive
(ii) No. of carpals	8	8	9

The embryos of humans and chimps have nine carpals, but two of them fuse during development, to give eight in the adult.

(a) Construct a cladogram from the data in Table 11.4. First decide on the primitive-derived polarity of the character states shown in the Table, on the basis of information given here or from general knowledge. Then draw the most parsimonious cladogram for the three species, marking apomorphies as bars on the appropriate branch lines (as in Figure 11.7).

(b) Suppose it were put to you that the chimps and gibbons should be classified together as apes, because their relative hairiness implies close relationship, but that humans should be excluded from the group because of their relative hairlessness. What kind of taxon would this 'ape' grouping constitute? And, in the light of your cladogram, comment on the usefulness of that proposal for establishing phylogenetic relationships.

SAQ 11.2 *(Objectives 11.1, 11.3, 11.6 and 11.7)* (a) Construct a phenogram from the immunological distance data on humans, chimpanzees and gibbons shown in Table 3.7 of Chapter 3. First, construct a grid for the three species (like that in Table 11.3) to show the mean immunological distances between each of the species pairs. To simplify matters, take the average value for each pairwise comparison and its converse (e.g. the average of chimp albumin against antibodies to human albumin, and human albumin against antibodies to chimp albumin), and thus fill out only half the grid. Then construct a phenogram from these values, using an average neighbour cluster statistic (as shown in Figure 11.20(a)).

(b) What assumption must be made if the molecular phenogram is to serve as a reflection of phylogeny? Is this assumption justified in this instance?

(c) What is the implication of this phenogram for the discussion of phylogenetic relationships in part (b) of SAQ 11.1?

(d) If two of the three species were to be grouped together as a monophyletic taxon, which two should they be?

CHAPTER 12
FOSSILIZATION AND THE RECORD OF PAST LIFE

Prepared for the Course Team by Dave Martill and Peter Skelton

Study Comment *This Chapter returns to the study of fossils, this time as a source of inferences on macroevolutionary patterns. An understanding of how fossils form is indispensable to interpreting the fossil record. This topic is therefore covered in some detail, starting with the processes involved in fossilization, passing on to how different kinds of assemblages of fossils may form and what can be learnt from them, and finally considering the circumstances in which exceptional assemblages have come about. The latter yield invaluable privileged insights into evolutionary history. A short, final Section briefly reviews the retrieval of information on macroevolutionary patterns, considering first some sources of bias, and then outlining how data are recorded for the purposes of further analysis (as illustrated in subsequent Chapters).*

Objectives When you have finished reading this Chapter, you should be able to do the following:

12.1 Describe the different processes which may operate on the remains or traces of organisms, identify their potentially destructive or conservative effects, and, given information on particular fossil specimens, interpret their history of preservation.

12.2 From information on a given fossil assemblage, classify the probable autochthonous, parautochthonous and allochthonous components, and identify the exotic or the derived fossils among the latter.

12.3 Explain how different original environments can influence the nature of fossil assemblages in different ways, citing or recognizing examples which illustrate these effects.

12.4 Describe or recognize fossil evidence for interactions between species, and suggest how the evolution of such relationships may be investigated.

12.5 Explain how preservational bias and time-averaged accumulation distort the fossil record of past communities of organisms, how mass mortality assemblages may avoid at least some of these distortions, and apply such insights to a given fossil assemblage.

12.6 Describe how different kinds of fossil-Lagerstätten may form, citing or recognizing appropriate examples, and explain how they provide special insights into evolutionary history.

12.7 From information given on the stratigraphical distributions of fossils, construct range charts and spindle diagrams, and outline the sources of bias in such compilations.

12.1 Fossils and macroevolution

You have met fossils in Chapter 10 in the context of microevolutionary studies. Here we examine the nature of fossils and the processes of fossilization more broadly, in relation to macroevolution.

In textbooks on evolution, it is customary to bemoan the incompleteness of the fossil record. It would be a mistake, however, to conclude that it is thus a poor record of evolution, for two reasons. First, completeness of evidence is a relative concept: it depends upon the question to be answered. Detecting patterns of microevolution, for example, does indeed require the sorts of close sampling and detailed quantitative analysis illustrated in Chapter 10. Yet if the question concerned only, say, whether dinosaurs reached Antarctica, then a single identifiable specimen from there, showing evidence for burial close to where the original animal lived, would settle the issue: further material would be redundant (indeed this question was answered in 1986, when dinosaur remains were discovered for the first time from Upper Cretaceous rocks of the Antarctic Peninsula). There is an important asymmetry, however, in the possibilities for answering such questions: the *absence* of evidence is much harder to interpret. Inferences may be based on the pattern of distribution of taxa elsewhere (as explained in Section 12.4.2 and in Chapter 13) together with an assessment of the likelihood of failure either of preservation or of sufficient study. As with the testing of a null hypothesis (Chapter 3, Section 3.5.1), a hypothesis of original absence may readily be rejected (e.g. in the case of the Antarctic dinosaurs), but cannot be confirmed. Nevertheless, many macroevolutionary questions, relating to longer-term changes in the anatomy, taxonomic diversity and distribution of major groups of organisms, can be answered satisfactorily with a relatively sparse record. Indeed, this circumstance is not peculiar to palaeontology. The fact that only a few per cent of the estimated numbers of living species have been described (Chapter 9, Section 9.1) has not prevented a great deal being discovered about the patterns and processes of evolution, as earlier Chapters have shown. The key to evolutionary palaeontology, then, is understanding the causes and nature of the record's incompleteness, and thus its patterns of bias, so that sensible questions can be asked with some prospect of being satisfactorily answered. This Chapter therefore places much emphasis on how fossilization may occur and the nature of the record that is left behind.

The second reason for not being dismissive of the fossil record is that its incompleteness is remarkably variable. Body fossils of some hard parts and trace fossils can be exceedingly common, especially in marine deposits. Non-skeletal body fossils on the other hand are relatively rare, but can be abundant in certain types of deposit, where they may offer exceptionally fine anatomical detail. **Chemical fossils** (organic molecules or their breakdown products, specific to certain taxa) can be abundant, and they may offer information on rates of production of biomass, for example; in some instances, they may yield the only available evidence for the first appearance of taxa (as Chapter 16 will illustrate).

The main importance of fossils is that they offer the only direct record of the evolutionary history of organisms over long time-spans. Fossils thus provide evidence for past adaptive radiations and other changes in clade diversity, including mass extinctions (Chapter 11, Section 11.1, and Chapter 15), and when fossiliferous rock sequences are accurately dated, they allow rates of morphological change to be calculated.

Figure 12.1 and Plates 12.1 and 12.2 show four different types of fossil, which illustrate the range of evolutionary inferences that can be made. The Ordovician fossil in Figure 12.1(a) is the internal mould of a bizarrely shaped animal which lived on the sea-floor. It was originally surrounded by a complex skeleton of calcite plates, as shown in the reconstruction alongside. Indications of the internal anatomy revealed by the moulds allowed R. P. S. Jefferies (1990) to infer a curious combination of features. As in the echinoderms (which also have skeletons composed of calcite plates: e.g. Figure 10.18 in Chapter 10), the animal appears to have possessed a *water vascular system* (a system of water vessels from which feeding tentacles (*tube feet*) arise). Yet the internal structure of the tail (lower part of Figure 12.1(a)) is comparable with that of chordates. This species, and related animals, thus constitute a paraphyletic 'stem group' (the Soluta), which combines primitive character states of the echinoderms and of the chordates, each secondarily lost in one or the other of those groups. It thus provides insights on the likely nature of the latest common ancestor of these two phyla.

Figure 12.1(b), by contrast, shows trace fossils—the footprints of dinosaurs. Although it is usually impossible to determine to which species the animal belonged, such prints can often be assigned to some higher taxon within the dinosaurs. Using reasonable estimates of the size of the animals (e.g. based on that of the footprints), and comparing the stride length with that of living animals of similar size, it is possible to infer the likely speed at which the animals moved. From such clues, an idea of dinosaur life habits may be built up (see also Section 2.2.2 in Chapter 2).

The articulated skeleton of a mammal called *Propalaeotherium* (Plate 12.1) contains a preserved hind-gut full of wads of leaves and even grape pips. This is direct evidence of the diet of *Propalaeotherium* and provides a retrospective test of the postulated diet of such early horses previously inferred only from the form of the teeth. Finally, the fossil leaf (Plate 12.2), although some 17–20 Ma old, yielded some preserved DNA (Golenberg *et al.*, 1990), from which relationships with living taxa and rates of molecular evolution may be inferred (Chapter 11, Section 11.3.1).

However, before considering the usefulness of fossils to macroevolutionary studies, we need to examine how they are preserved, and to consider the biasing effects on the fossil record.

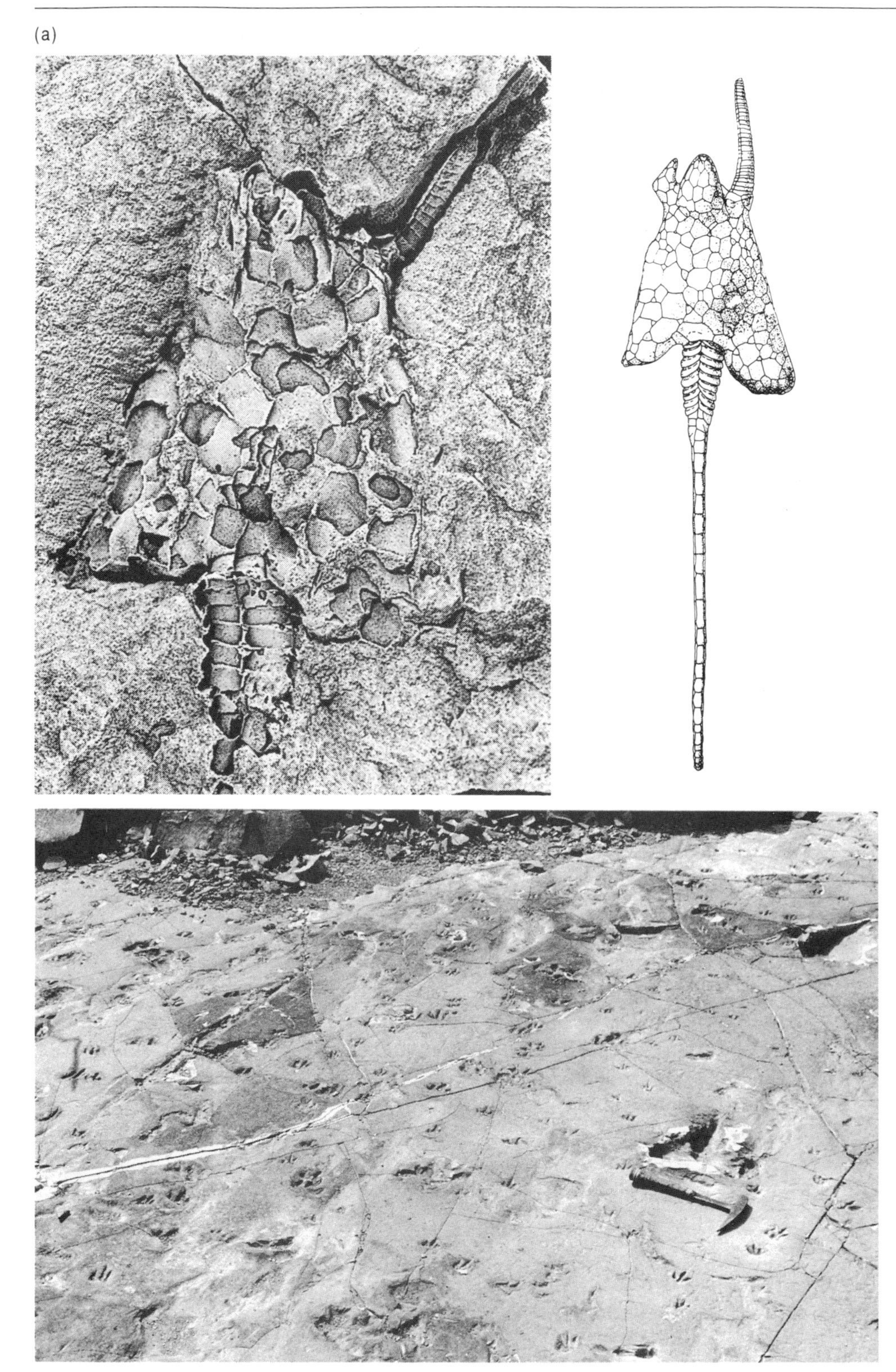

Figure 12.1 *(a)* Dendrocystoides scoticus, *an extinct member of the Soluta, a group possibly embracing the common ancestor of chordates and echinoderms (further discussion in text). (b) Footprints of three-toed dinosaurs preserved on a layer of mud (from the Cretaceous of Lark Hill, Australia).*

12.2 Taphonomy: the processes of fossilization

The study of those processes that act on the remains or traces of organisms is called **taphonomy** (from the Greek *taphos*, a tomb, and *nomos*, a law). Such processes include predation, scavenging, bacterial decomposition, transportation in, say, streams and rivers, weathering, burial by sediment, and a host of microbial and geochemical processes within the sediment. In general, the effect of many of these processes is to break the organism down and to recycle its constituents within the environment. However, some of these processes may arrest breakdown and/or transform the materials in a way that permits preservation. Burial, and some of the biochemical and geochemical reactions that take place within sediment, can thus assist in the fossilization process, which is why we have any fossil record at all!

12.2.1 Getting into the fossil record

Many biological, chemical and physical processes fall under the general heading of taphonomy. In the following Sections we examine how these processes affect the fossil record.

Death and predation Death is a common starting point, though not a prerequisite for entry into the fossil record.

▶ What fossils might be derived from living organisms?

Although death is the usual gateway for body fossils, moults or teeth shed by living organisms may become fossilized. Trace fossils are also made while the organism is alive. Chemical fossils may result from the death of an individual, but they may also be a result of metabolic processes.

Common causes of death include predation, starvation, sickness, accident, and rarely, old age. It is sometimes possible to establish the cause of death from a fossil organism. Predation may be recognized, for example, by the presence of bite marks, or other characteristic damage produced by animals with a piercing or crushing apparatus. Where particular kinds of predator can be diagnosed from such traces, evolutionary changes in predator–prey interactions may be detected. Today, several predatory gastropods gain access to their bivalve prey by drilling distinctive holes in their shells (Figure 12.2). The marked proliferation of such drill-holes in fossil bivalves from around the middle of the Cretaceous testifies to the rise of these gastropod predators at that time.

Predation is probably the most important agency of death and biological recycling of animals. It is usually totally destructive. However, many skeletal components can survive predation, most notably shells, bones and teeth, and other remains which are difficult to digest, such as beetle carapaces. Plants are less frequently killed by their predators (herbivores), but some of the components that are eaten may likewise resist being digested (e.g. seeds and fibrous tissues). These items may pass through animal guts almost unaffected, and are important elements in the

Figure 12.2 *A boring made in a Recent bivalve shell by a predatory gastropod.*

fossil record. Some animals have been preserved with such gut contents (Plate 12.1), so giving clues to ancient feeding relationships.

Sickness, including osteoarthritis and syphilis, has been recognized from diseased bones, but most diseased animals become prey before they die. Cases of accidental death by rapid burial, volcanic eruption, and other physical processes have been documented also (discussed in Section 12.3.1).

Scavenging Carcases provide an important resource for a host of animals, which can rapidly reduce a carcase to a skeleton, often damaging it in the process and scattering it widely. Generally, carcases are spared from scavenging only when their remains are buried rapidly or sink into waters devoid of oxygen. Some scavengers may assist fossilization, however, by removing carcases to lairs where their chances of preservation are increased, as for example in caves.

Decomposition and chemical degradation In general the soft tissues of animals (e.g. muscles, lungs, gills, alimentary tract etc.) and plants are decomposed very rapidly by bacteria and fungi, which utilize them as a food source, and by enzymes released from organelles within the cells of the tissues themselves. In contrast, though, bacteria may ultimately also assist in the preservation of soft tissues by favouring the growth of minerals or, in exceptional circumstances, by themselves become mineralized (see Section 12.3.3).

There are a few environments in which microbial breakdown may be slowed down, or even inhibited completely; the frigid, but dry arctic wastes, hot dry deserts, and highly saline lakes are environments in which carcases of animals have been found in a well preserved condition.

Skeletons, both internal and external, are a common feature of many animals and plants, and collectively have a remarkable variety of composition. Many animal skeletons are composites of minerals, such as calcium carbonate or calcium hydroxyapatite, and organic material (Chapter 10, Section 10.1.1). Other skeletons are composed of very large organic molecules, such as lignin in plants and chitin

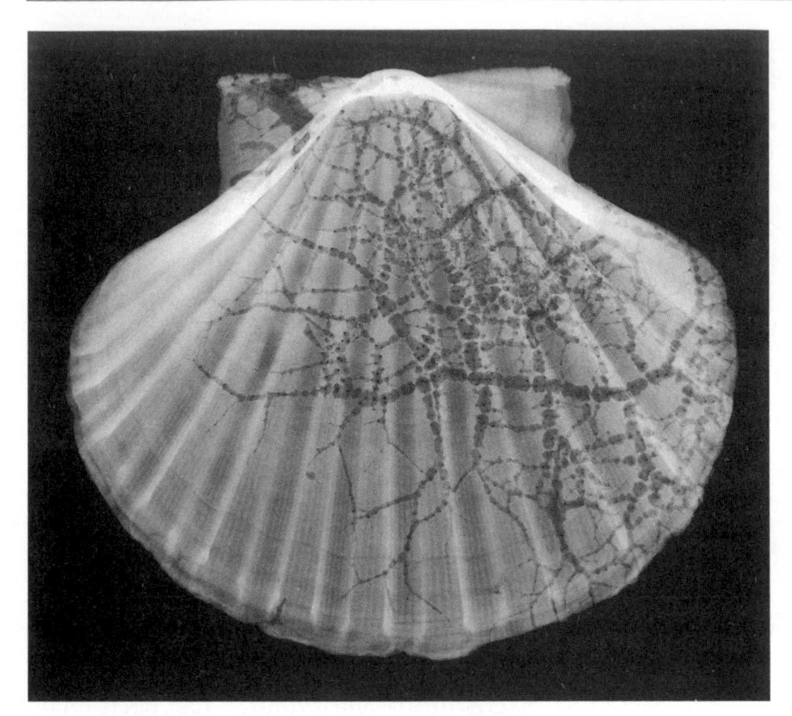

Figure 12.3 *Borings made by sponges in a scallop shell (shown by X-radiography).*

in insects. Even skeletons composed of mineralized tissues may be subject to decomposition by bacteria, which attack the organic matrix. Likewise, many organisms bore into bones and shells, some to feed on the organic matrix or even to feed on the other boring organisms, and some merely to take refuge from their own predators (Figure 12.3). These invasions may result in the total disintegration of bones and shells, if they are exposed for prolonged periods. However, the organic component is often so intimately linked with the mineralized component that it may be difficult to get at. Thus bones, teeth and shells stand a much greater chance of surviving long enough to become buried, and so of entering the fossil record, than do soft tissues.

Most of the minerals in skeletons are reasonably hard and insoluble in the short term. They are relatively stable at normal surface temperatures and pressures, and do not break down unless subjected to sustained exposure to acidic or alkaline conditions, or prolonged physical abrasion.

Physical transport Besides transport due to predators and scavengers, water currents may move animal and plant remains from their place of death to a site where they may become buried. Wind too is an important transport agency, especially for microscopic items such as spores and pollen. Localized accumulations of transported remains may result (Figure 12.4).

Skeletons of animals and plants are commonly composed of a number of components. Some of these may be fused together, as with the bones making up the adult human skull; others are merely held together by organic tissues, as with the limb bones. Post-mortem separation of the various parts of organisms frequently occurs during transport. Indeed many fossils represent only isolated components.

(a)

(b)

Figure 12.4 *An accumulation of transported Recent shells on a beach in Venezuela. (a) General view. (b) Close-up.*

The reconstruction of a Carboniferous tree in Figure 12.5 is based on fossil remains of the various components of the tree. In the past, individual parts were given separate generic names. This was the result of separation of the various components, and their incorporation into the fossil record as discrete entities. Sometimes it is not possible to confirm the integral nature of the various elements until sufficient fossils demonstrating unequivocal relationships have been found (Chapter 10, Section 10.2.2). Similar situations have arisen with vertebrate skeletons, where fossil remains from the limbs and the skulls, for example, have been given distinct names. In the past this resulted in an artificial picture of diversity due to the creation of numerous redundant taxonomic names (*synonyms*).

In fast-flowing streams and rivers there is much abrasion by sedimentary particles, which may result in physical wear and destruction of biological materials. On low gradients the abrasive power of rivers is reduced. Here skeletal components may be said to have a higher **preservation potential**, i.e. they stand a higher chance of becoming incorporated into the fossil record. The term preservation potential is very useful and can be applied not only in the context of different environments, but also in relation to different kinds of organisms or even their constituent parts.

▶ Which component of the mammalian skeleton has the greatest preservation potential (Chapter 10)?

It was noted in Section 10.2.3 of Chapter 10 that the teeth of mammals are particularly resistant to destruction, and are thus an important source of information on the evolution of the group.

Another contributory factor is the number of skeletal components, as discussed above. A gastropod shell, consisting of one component, stands a better change of remaining intact than does that of a bivalve or brachiopod, consisting of two components which may become separated on death; and multi-component skeletons (e.g. the crinoid in Figure 10.18 of Chapter 10) are likely to retain their integrity

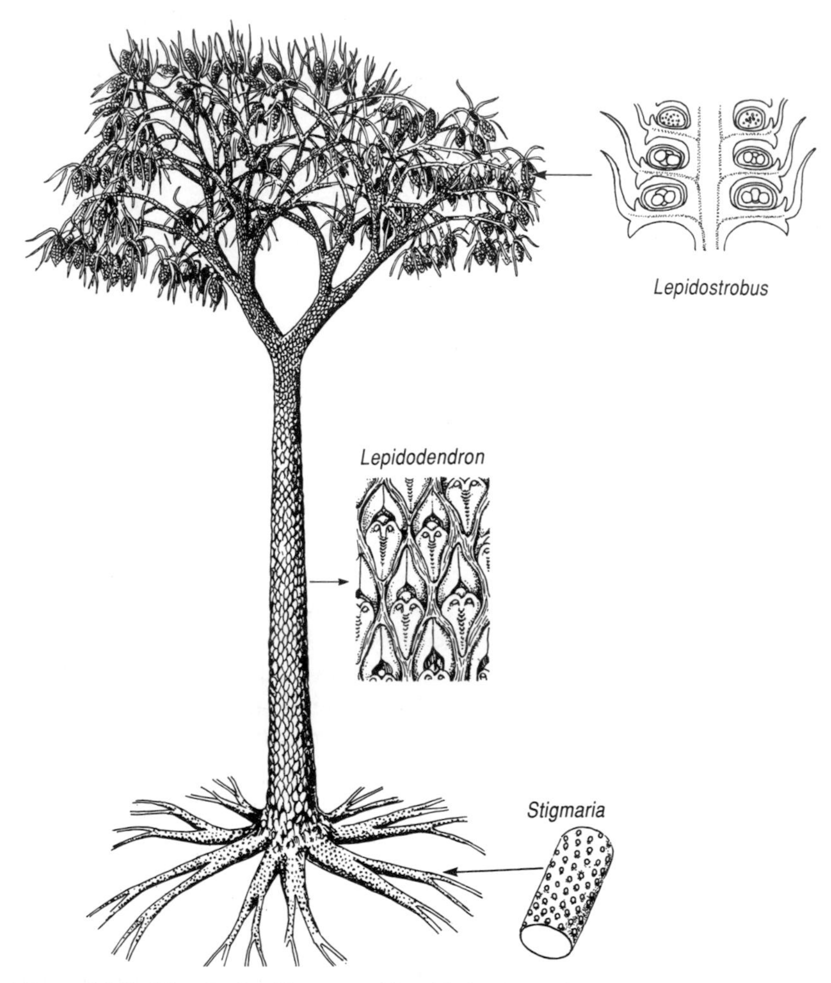

Figure 12.5 *The Carboniferous tree* Lepidodendron. *At previous times separate taxonomic names were given to the roots (*Stigmaria*), trunk (*Lepidodendron*) and reproductive bodies (*Lepidostrobus*), although all these parts originally belonged to a single type of plant. The height of the tree was about 15 metres.*

only under special circumstances, such as rapid burial of the live animal. Mode of growth and life habits are also important. Animals that moult their skeletons (e.g. arthropods, including the extinct trilobites: see Plate 10.1) have more opportunity, so to speak, to invest their remains in the fossil record than do those that grow a single skeleton throughout life. Relative abundance is another obvious factor: animals from high up in the food chain (i.e. carnivores) tend to be much less common than those from lower down the chain (e.g. herbivores), and are correspondingly rarer as fossils.

Some skeletal elements are more resistant to abrasion than others; for example, teeth are harder than calcitic shells, and persistently turbulent environments, e.g. beaches, may selectively destroy relatively softer materials, artificially elevating the frequency of harder elements. Sustained current activity may also sort skeletal elements according to size, shape and density, again yielding highly biased local accumulations. Thus, where there is sedimentological evidence for accumulation in such settings, the relative abundances of different taxa are likely to be highly distorted with respect to the original frequencies of the living organisms.

Burial Biological materials that are not destroyed by the processes discussed above may eventually become buried by sediment. Once buried, the carcase or skeleton is protected to some degree from many destructive processes (though repeated episodes of burial and reworking may precede the final burial of some specimens).

Rates of burial vary greatly, as discussed in Section 10.1.2 of Chapter 10. Fossil preservation is enhanced when short-term sedimentation rates are high, and when burial occurs soon after death. Rapid burial reduces the time that a carcase spends exposed on the sea or lake floor.

▶ Why might exposure on the sea-floor aid breakdown?

During periods of extended exposure (perhaps hundreds of years), organic components may rot and some skeletal elements may dissolve. Such degradation is usually enhanced, especially in shallow waters, by organisms that bore into the remains. Nevertheless, in marine environments, exposed shells may also become encrusted by other shelly organisms, such as corals, bryozoans, and calcareous algae which may afford some protection against dissolution.

Many sediments are host to burrowing animals, especially annelids, crustaceans and molluscs, which may disturb the sediment so intensely that all the original sedimentary layering is destroyed (see Chapter 10, Section 10.1.2). Such bioturbation may also cause the disarticulation of delicate skeletons and disturb any shells buried in the position in which they lived. The degree of bioturbation varies with both the number and mobility of the burrowing organisms, with the rate of sedimentation and with the oxygen content of the water. Where sedimentation rates are low, burrowers have more time to plough through the sediment; where they are high, the burrowers move upwards to keep pace with sedimentation, and may not have time to process all the sediment (Figure 12.6).

Figure 12.6 *Two successive beds of silty sand originally deposited rapidly, by storm currents, in an otherwise normally quiet offshore area. The lower part of each bed still retains its original sedimentary structure of hummocky cross-stratification, but that has been destroyed in the upper part of each by bioturbation. Lower Jurassic of Dorset, UK. (The lens cap is about 5 cm across.)*

Burrowing is also inhibited if the overlying water becomes stagnant, since few burrowing organisms can survive very low oxygen levels.

▶ How might the evolution of organisms capable of burrowing have affected the record of fossils found as impressions in sediment?

Before the burrowing mode of behaviour became common, sediments deposited below the level affected by storm and wave currents would have remained relatively undisturbed. Any impressions produced in these sediments would have stood a reasonable chance of being preserved. Thereafter, only when burrowing organisms were excluded because of rapid sedimentation rates, oxygen deficiency or hypersalinity, could sediments escape biological reworking. As you will see in Section 12.3.3, this changeover occurred in latest Precambrian times, and is marked by a proliferation of trace fossils.

In the face of so many frequently destructive influences, often acting together, it might appear that biological materials, including mineralized tissues, normally stand very little chance of becoming fossils. Yet fossils can be very common, and in some cases they are the major components of sedimentary rocks, especially limestones. The reason why fossils can be so abundant is both because of the vast amounts of time that are available for their accumulation and because many processes that act on buried remains serve to preserve, albeit by modification, rather than to destroy. How this occurs will be discussed in the next Section.

12.2.2 DIAGENESIS: FROM BONE TO STONE

Diagenesis refers to those physical and chemical changes that take place within sediment. Especially important are geochemical reactions, many microbially mediated, which produce minerals that cement sediment to form rock. Many biological components, from whole animals and plants to isolated bones, teeth and shells, are affected by diagenetic reactions. Importantly, many of these reactions conserve biological remains, either by replacing them by new, perhaps more stable minerals, or by forming a protective concretionary coating around their surface (Figure 10.6(b) in Chapter 10).

Many different chemical reactions take place within sediments. Some are specific to certain sediment types, while others are specific to certain depths within the sediment. The major controls are usually pH and the redox (reduction/oxidation) potential E (the potential to take up or donate electrons), but the availability of elements dissolved in the water within the sediment (or coming up from compacted sediment beneath) is also very important. There is not space here to discuss these controls in detail, but simply to point out some of their consequences where they are of importance to the fossilization of animal and plant remains.

Diagenetic mineral growth and replacement Most sediments contain some detrital organic material which is fed upon and decomposed by diverse microscopic organisms, usually bacteria, protozoa, fungi etc. In sediments with a good oxygen supply, both the detritus and the decomposers may themselves be fed upon by larger organisms (detritivores), e.g. annelid worms and molluscs, which return excretory products and faecal material. The sediment may thus become enriched in nutrients containing such essential elements as phosphorus, nitrogen and sulphur, as the organic matter is broken down (as in a garden compost heap). With increasing depth of burial, the supply of oxygen to the sediment from above decreases. This depletion is due to the oxygen being used up, largely by aerobic (oxygen-requiring) bacteria living in the top few millimetres of subaqueous sediments (but to greater depths in aerated soils), and also by the oxidation of gases (such as H_2S) diffusing upwards from below. Below this level, conditions are referred to as **anoxic**, i.e. lacking oxygen. Anoxia does not inhibit all life: anoxic sediments frequently contain bacteria capable of respiring by reducing compounds other than molecular oxygen (e.g. sulphate ions, SO_4^{2-}), as, indeed, the earliest forms of life probably did (Chapter 16) The insoluble by-products (diagenetic minerals) that can be produced by these reactions form microscopic crystals in the spaces between sediment particles, sometimes filling the spaces and cementing the sediment together. For example, the reduction of sulphate may eventually lead to the production of sulphide ions (S^{2-}), and these may be rapidly precipitated by contact with free ferrous ions (Fe^{2+}) to form the insoluble FeS. The latter is usually later converted to the mineral pyrite (FeS_2). More importantly from a palaeontological perspective, such new minerals may also fill void spaces in biological tissues or replace them altogether to produce a fossil (Figure 12.7).

Many shells, and occasionally bones, are progressively replaced by other minerals during diagenesis. Calcite ($CaCO_3$) shells, for example, may be replaced by silica (SiO_2), by pyrite (FeS_2) or, less commonly, by a host of other minerals. The process occurs when the original mineral of the shell is out of chemical equilibrium with the water in the pores of the sediment. If a different mineral can precipitate under the conditions that cause the original mineral to dissolve, and

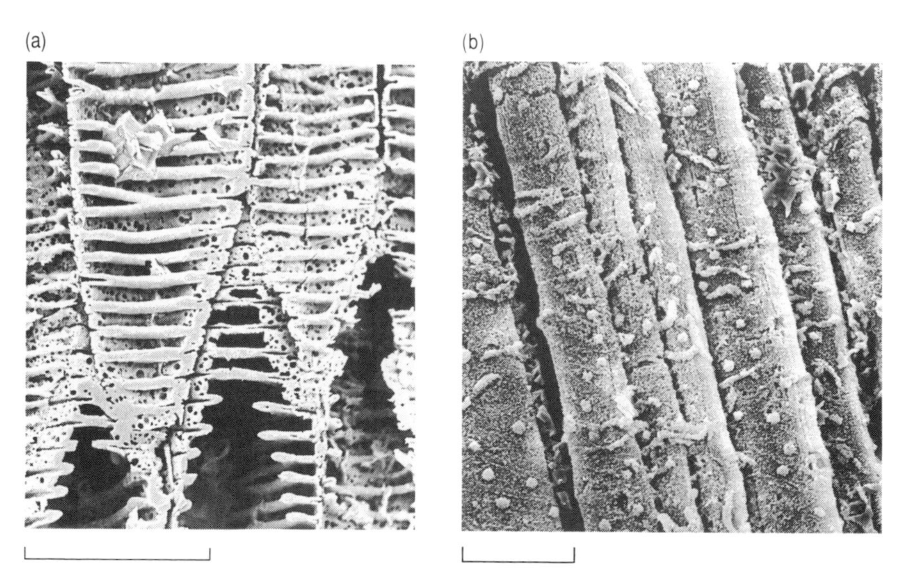

Figure 12.7 *(a) Scanning electron micrograph of some lignified water-conducting cells of the early land plant,* Gosslingia, *which have been protected from compaction by the growth of pyrite within the cell walls. Here, the pyrite has been removed with nitric acid to reveal the cell walls. Lower Devonian of Wales. (b) Scanning electron micrograph of muscle fibres replaced by apatite, in a fish from the Lower Cretaceous of Brazil. The small round objects are believed to be fossilized nuclei and the transverse branching structures, capillaries. In both (a) and (b) scale bar = 50 μm.*

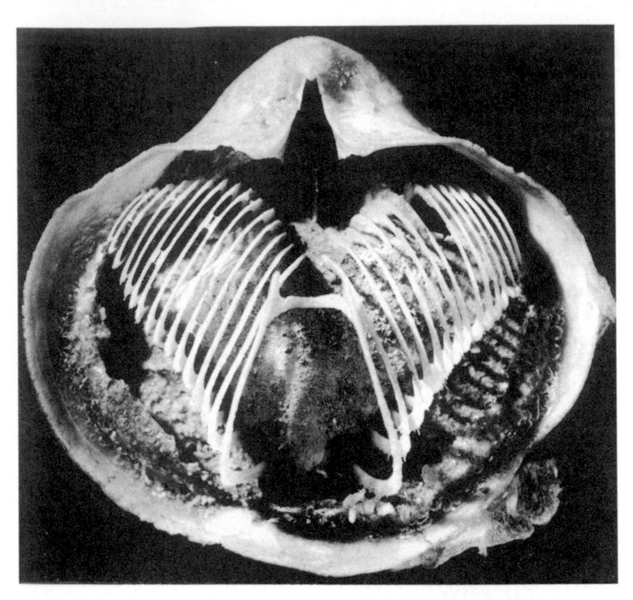

Figure 12.8 *Silicified brachiopod shell (*Spiriferina*) etched from its limestone matrix with acid. The upper (dorsal) valve has been removed to show the large spiral structures which supported the feeding tentacles. The specimen is 30mm wide.*

reactants are widely available, the new mineral may precipitate in the sites where the original mineral was present. Replacement can occur along the interfaces between growing and dissolving crystals, in which case the original structure can be retained in exceptional detail. Where calcitic brachiopod shells have been selectively replaced by silica, the limestone matrix can be dissolved in dilute acid to reveal the superbly preserved fossils (Figure 12.8). Such exquisitely detailed preservation allows accurate morphometric analysis of the abundant fossils for taxonomic and other purposes. Replacement may also occur by complete dissolution of the original mineral, resulting in a void, or mould, which later becomes filled with a new mineral, such that only the gross morphology is retained, as a cast (Chapter 10, Section 10.1.1).

A special type of replacement, known as *neomorphism*, occurs when an unstable mineral is converted to a more stable form with the same chemical composition. For example, many shells (e.g. those of most molluscs) are composed of the relatively unstable form of $CaCO_3$ called aragonite. When not dissolved away, these shells commonly become neomorphosed to calcite, which is the more stable form of $CaCO_3$ at surface (or near-surface) temperatures and pressures.

Dissolution Some skeletal components are more soluble in water than others. Their solubilities also vary with water chemistry and temperature. Dissolution of shells rarely occurs at the surface in shallow marine waters (which are commonly saturated with respect to $CaCO_3$), but it does occur in deep, cold waters in the oceans, and in freshwater. In general, aragonitic shells are the most soluble, the calcite of some mollusc shells, such as oysters, being less soluble than the aragonite. Hydroxyapatite, the phosphatic mineral component of bones and teeth, is the least soluble of the common minerals forming hard parts of organisms. As the pH of the water decreases (i.e. becomes more acidic), so dissolution of most mineralized skeletal components increases. Organic structural tissues, such as wood, however, tend not to break down under slightly acidic conditions; rather they tend to be preserved, as in the pickling process for preserving foods.

Dissolution of shells may take place on the sea-floor when reduced sedimentation rates leave skeletal material exposed for long periods, or it may occur within the sediment, where pore water pH levels can fluctuate due to the microbially mediated reactions discussed above. This can result in a **chemical lag** deposit, where a fossil accumulation originally containing skeletal material of a range of compositions has had the more soluble elements removed by dissolution.

▶ Ammonites had shells composed of aragonite, but many had jaws composed of calcite. Which component would have been the first to be leached away if the sediment were to be flushed by groundwater which was not saturated with respect to $CaCO_3$?

The shells of the ammonites would have been less resistant as they were composed of the more soluble aragonite. A number of deposits which contain the jaws of ammonites, but which lack ammonite shells, have been interpreted as chemical lag deposits.

Occasionally pore waters may become sufficiently acidic (due to a build-up of organic acids resulting from bacterial activity) that even bones and teeth will be dissolved. Alternatively the waters may become alkaline due to the release of ammonia from the decomposition of proteins. Any items that consist of, or may have already been converted to, acid-resistant minerals, such as silica, will then be dissolved.

▶ How might differential dissolution affect the perception of faunal diversity within a given fossil assemblage?

Differential dissolution will remove some components and may thus artificially reduce the diversity of the fossil assemblage, enhancing the relative abundances of the remaining taxa.

Compaction and metamorphism The result of sedimentation may be to crush a fossil, so flattening it (e.g. Figure 10.6(a) in Chapter 10). However, a large number of fossils are cemented well before the pressure due to burial is high enough to crush them. If the spaces within the fossil have been filled with minerals it may resist compaction. So, even in highly compacted sediment, such as shale, it is possible to find more or less undeformed fossils from which good morphometric data can be obtained. Likewise, concretions often form before compaction crushes the fossils. Because concretions frequently nucleate around organic remains, again as a result of the locally intense bacterial activity, they often contain the very best-preserved fossils. The work of diagenesis in preserving fossils may eventually be undone, however, if a sedimentary sequence becomes deeply buried and/or deformed. When a sedimentary rock becomes metamorphosed, the action of high temperature, high pressure and chemical activity may obliterate completely any fossils it once had.

12.2.3 THE EFFECTS OF TAPHONOMIC PROCESSES ON THE RECORD

The brief review of taphonomy in the preceding two Sections highlights the wide variety of interacting processes that may operate upon organic remains and lead to the preservation of some as fossils. Some of the processes are of destructive, and some of conservative effect. Whether or not a fossil results in any particular instance depends crucially upon the sequence of such influences: clearly, one or more of the conservative factors must intervene at an early stage, if complete degradation is not to ensue. Preservation by this means may come about in many ways, and there are consequently many different styles of fossilization.

▶ How might the following circumstances contribute to the chances of fossilization of the remains concerned? (a) Burial of plant fragments in muddy marine sediments with pore water becoming anoxic just below the sediment surface; (b) sinking of a fish carcase to an area of sea-floor overlain by stagnant, anoxic water; and (c) rapid burial of a crinoid (like that in Figure 10.18 in Chapter 10) by a one-metre-thick layer of sediment during a storm.

In (a), the anoxic pore water would arrest decay caused by aerobic bacteria and favour the early diagenetic growth of minerals such as pyrite, which might permeate and thus protect the plant tissues (as in Figure 12.7(a)). The anoxic bottom conditions in (b) would exclude scavengers and burrowing organisms, so permitting, perhaps, the eventual burial of the undisturbed carcase. The depth of sediment overlying the crinoid in (c), if not subsequently eroded, would prevent the disturbance of the intact skeleton by burrowing organisms.

The extraordinary quality of preservation that may result in certain exceptionally favourable circumstances will be explored later (Section 12.3.2). But what of the taxonomic representativeness of the fossil record? Preservation potential varies enormously between different kinds of organisms and between different environments. How comprehensive in its coverage of taxonomic diversity is the fossil record capable of being?

All palaeontologists agree that there is a heavy bias towards marine organisms with mineralized or resistant skeletons. Indeed, the fossil record of some animals with mineralized skeletons may be exceedingly good. In 1989, J. W. Valentine attempted to test the taxonomic coverage of the fossil record by comparing modern shelf sea molluscan (bivalve and gastropod) faunas of southern California with fossil molluscan faunas from Pleistocene rocks in the same region—the time difference between the living and fossil faunas being no more than one million years (Table 12.1). He collected his fossil assemblages from sediments originally laid down in areas of shelf sea, now exposed on land, while the modern faunal list was based on samples from all over the continental shelf, down to 180 metres depth. A small number of unique, dubious or taxonomically poorly understood records were excluded from the study.

Table 12.1 *Numbers of living benthic molluscan taxa from the continental shelf, and their Pleistocene fossil record from shelf deposits, of the Californian Province, recorded by Valentine (1989).*

Taxa	Total living taxa	Living taxa also known as fossils	% of living taxa known as fossils
Bivalvia			
Families	53	48	90.6
Genera	132	111	84.1
Species	230	182	79.1
Gastropoda			
Families	60	53	88.3
Genera	159	130	81.8
Species	468	354	75.6
Total			
Families	113	101	89.4
Genera	291	241	82.8
Species	698	536	76.8

Table 12.1 shows that, in this case, the taxonomic completeness of the fossil record is impressive: at least three out of every four living species have fossil representatives, the bivalves being slightly better represented than the gastropods.

► What trend is evident for the percentage completeness values going from genera to families?

The completeness values increase: some nine out of every ten families have fossil representatives. This example thus illustrates the point made in Chapter 11, Section 11.1.1, that the coverage of the fossil record improves as the taxonomic hierarchy is ascended, because of the more inclusive nature of higher taxa.

When Valentine examined the 162 (i.e. 698 – 536) living species without fossil counterparts, he found that their members are either very rare today, live largely outside the area, or are very small with delicate shells. The last group he supposed might have had a low preservation potential, whereas some of the forms that are rare in the area today might not have lived there at all during the Pleistocene.

► What other factors might account for some of the absences of the rarer living taxa from the fossil list?

Some rarer taxa are quite likely to have been missed as fossils simply through sampling failure. Although about 400 fossiliferous localities were sampled, with an average of 50 species being recorded at each, Valentine estimated that a doubling of the sampling effort might have yielded some 10% more fossil species.

In addition to those with living representatives, Valentine recorded 211 Pleistocene fossil species not living in the area today.

▶ What might have become of these?

Some species might have become extinct, while others may now live elsewhere. In fact 98 of these species are believed to be extinct. Of the others, 72 species are known to the south of the area, and 41 to the north. Climatic shifts are the most likely cause of these movements.

▶ What therefore is the main reason for the total number of fossil species sampled (211 + 536 = 747) being *higher* than the total number of living species sampled (698)?

The main reason for the higher number of fossil species is that the accumulated numbers from up to a million years' worth of migrations into the area have been pooled together, in contrast to the few years' worth of the living sample. Note that the extinct species alone (98) are insufficient to bring the numbers of fossil species still living in the region today (536) up to the total of living species (698). The fossil sample is thus said to show the effect of **time-averaging**—the pooling together of data from an extended period of time. This is an effect of great importance when individual fossil assemblages are considered, as the next Section will show.

12.3 FOSSIL ASSEMBLAGES

In the conclusion to the previous Section, the taxonomic completeness of the fossil record was briefly considered on a regional scale, at least for organisms with good preservation potential. In this Section, the emphasis is upon the interpretation of ecology from fossils—**palaeoecology**—and, in particular, what assemblages of fossils from single beds can reveal about past communities of organisms. You might have supposed that little evidence could possibly remain of the often ephemeral interactions of living organisms with one another and with their physical environments. Yet, given the right circumstances, a surprising amount can sometimes be inferred concerning, say, life habits, feeding relationships and other such interactions between particular species, the diversity of species co-existing in some past communities, and even in some instances aspects of the population dynamics of species. Most such inferences are of more strictly ecological, rather than evolutionary, interest. However, it is important to remember that other species and the nature of the physical environment provide the context for evolution by natural selection, as was stressed in Chapter 8. So it is worth briefly reviewing what can be learnt from fossil assemblages concerning this context. As you will see, the key to such understanding is careful taphonomic analysis.

12.3.1 THE COMPOSITION OF FOSSIL ASSEMBLAGES

When studying a fossil assemblage, five considerations are particularly relevant to palaeoecological interpretation:

1 the original situation of each of the organisms in life (e.g. whether or not they dwelt at the site of burial);

2 the physical character of their original habitat(s);

3 any evidence for consistent interactions between species (e.g. predation, parasitism etc.);

4 the taxonomic completeness of the assemblage, as a record of the original community or communities involved;

5 the temporal distribution of the organisms represented in the assemblage (e.g. whether they lived at the same time, or whether their remains accumulated over a long period).

Original situation of fossilized organisms The eventual burial and fossilization of the remains of an organism may take place where it lived, or only after it has somehow been removed from its original habitat. There is a continuous range of possibilities, as outlined below.

When preserved at the site, and in the original orientation of life, a fossilized organism is said to be in **life position**; such fossils are termed **autochthonous** (from the Greek, 'self-interred').

Sometimes the orientation has been disturbed post-mortem, but the fossil remains at or near the original site of life; such fossils are termed **parautochthonous** (from the Greek, 'beside-self-interred').

Finally, the remains may have been introduced from elsewhere by some means; such fossils are termed **allochthonous** (from the Greek, 'from-elsewhere-interred').

▶ Cast your mind back to the taphonomic processes discussed in Section 12.2.1: which among these might give rise to allochthonous fossils?

The most obvious process leading to fossils being allochthonous is that of physical transport of the original remains by currents, although another possibility is the removal of carcases to distant lairs by predators and scavengers. Mere localized scattering of remains, in contrast, would yield parautochthonous fossils.

▶ Can you think of another possible common cause of physical disturbance of remains *following* burial?

Disturbance within sediment commonly results from bioturbation. So extensive is this in certain deposits that entire fossil assemblages may become parautochthonous.

▶ What pattern of sediment deposition over time (discussed in Chapter 10, Section 10.1.2) would be likely to inhibit extensive bioturbation, and so favour the preservation of fossils in life position?

High *short-term* rates of sedimentation could be expected to lead to the rapid burial of many specimens by thick layers of sediment, the lower parts of which burrowing organisms would have little opportunity to rework (Figure 12.6).

Another important consideration in this respect is that of the life habits of the organisms themselves. Clearly, only organisms living on or within preservable substrata can be preserved in life position, so the fossil record of ecological relationships is biased in favour of organisms that are **benthic** (i.e. bottom-dwelling marine or freshwater) or ground-dwelling. **Pelagic** (i.e. free-swimming or floating) organisms, or, in terrestrial environments, those that fly, can only be preserved as allochthonous elements in sediments. The interpretation of their ecological relationships is correspondingly more tenuous.

▶ Even if it could be shown that an assemblage of fossils of pelagic organisms (e.g. fish and ammonites) comprised contemporaneous individuals, how might it be possible for the different species concerned never to have had overlapping habitats in life.

If members of the different species characteristically lived at different depths, some near the surface and others near the sea-floor, then there might never have been any overlap between them in life. An assemblage of fossils from the remains of such organisms, which had sunk to the sea-floor, would thus comprise a 'cocktail' of allochthonous specimens from different habitats.

Those allochthonous elements from a fossil assemblage which are thus more or less contemporaneous with any autochthonous or parautochthonous elements, but representative of other habitats elsewhere, are said to be **exotic fossils**.

Derived fossils, in contrast, are fossils that originally formed in rocks older than those in which they are found. This process occurs commonly when cliffs composed of fossiliferous sediments are eroded and the fossils are washed out, and incorporated as pebbles in recent sediments. Obviously they were not a part of the biota alive at the time the sediment was deposited. It is important that such fossils be recognized as derived, and they should be discounted when assessing taxonomic diversity. A continuum of possibilities exists, from fossils which have been derived from very much older strata to those reworked through the erosion and redeposition of sediments laid down earlier in the same environmental setting. The former rarely present problems, as their mode of preservation and original environmental associations are usually strongly contrasted with those of other fossils in an assemblage. Fossils reworked from only slightly older sediments, however, may be harder to identify, since both their mode of fossilization and their original environmental associations may closely match those of the younger fossils with which they are found.

▶ Casting your mind back to the discussion of sedimentation in Section 10.1.2 in Chapter 10, are such nearly contemporaneous derived fossils likely to be a common problem?

Yes, unfortunately, they are, because of the frequency of erosion of sediments in virtually all environments (as illustrated in the hypothetical example of Figure 10.4 in Chapter 10). The problem tends to be most severe in relatively sediment-starved environments in which any sediment that is reworked is likely to have been deposited a good many years earlier. Offshore shallow marine environments are particularly sensitive in this respect to fluctuations in sea-level (as, for example, during the Quaternary Ice Age: Chapter 2, Section 2.4.1). Many of the shells that may be dredged from the sea-floor around parts of Britain and the United States today and that may be picked up along some beaches, are reworked from Pleistocene deposits and may be tens to hundreds of thousands of years old.

If you are unfamiliar with fossils in the field, you may be beginning to get the uneasy feeling from the discussion above that all fossil assemblages are likely to be hopelessly jumbled mixtures of autochthonous, parautochthonous and exotic and derived allochthonous elements. Fortunately, however, this is far from the case: there is enormous variety in the constitution of fossil assemblages and some may offer a relatively pristine record of organisms that were associated in life, as later Sections will show. In analysing any given assemblage, the status of each constituent species needs to be considered individually, and different criteria must be employed for different kinds of organisms. Those that are habitually fixed to a particular spot in life (**sessile** organisms), such as trees and shrubs on land, and corals, oysters and barnacles in the sea, can be categorized readily as autochthonous when their fossils are found still rooted or attached to a preserved surface (Figure 12.9(a)). **Vagile** benthic organisms (those that move around on or in the sediment) also can be preserved as autochthonous fossils, though in this case identification as such relies upon the orientation of the fossils. For example, when a burrowing clam dies, the muscles that draw the two halves of the shell together rapidly rot away, and the horny ligament between them causes the shell to gape wide open, unless constrained by surrounding sediment. Fossil shells of such clams, if found closed and in the usual burrowing position (Figure 12.9(b)) are thus likely to be autochthonous.

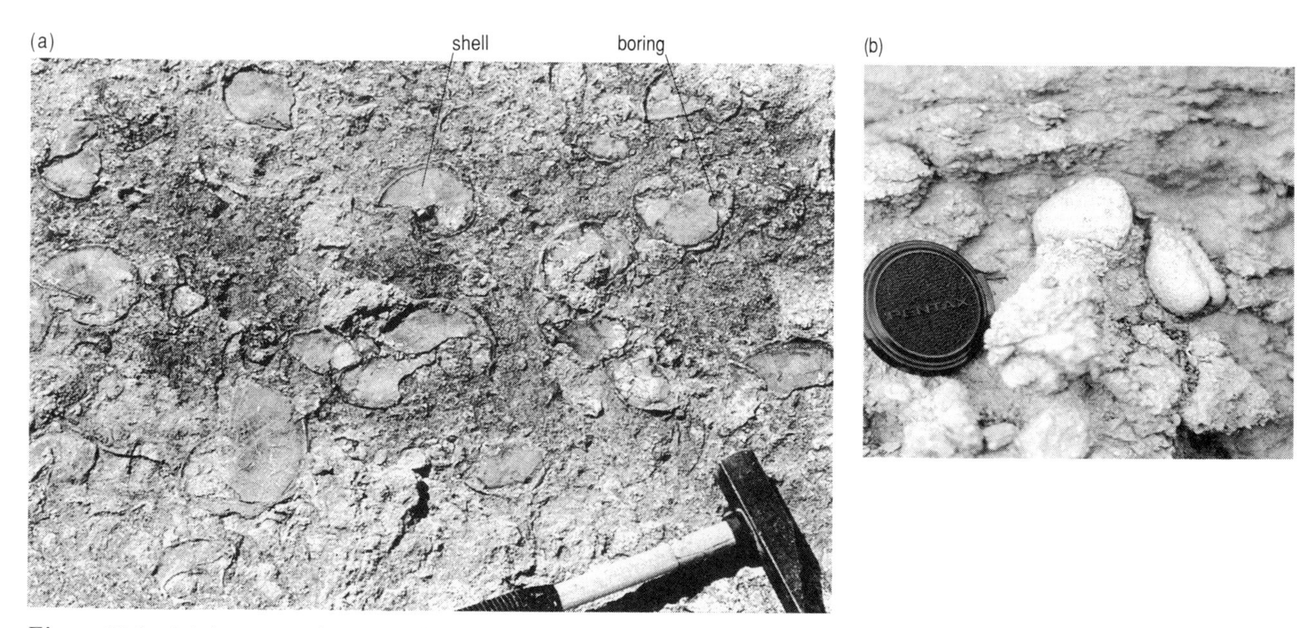

Figure 12.9 *(a) A preserved cemented surface (hardground) showing fossil evidence both of encrusting and of boring organisms. (b) Fossil burrowing clams preserved in life position in a bed of fine-grained limestone. Both from the Lower Cretaceous of western central Portugal.*

The extent of displacement and transport of remains can be assessed using various criteria. Remains that are buried where the organisms could not have lived are obviously exotic, no matter how intact they may be (e.g. Plate 12.1). Those that are neither in life position nor clearly exotic are more problematical to interpret.

▶ What do you suppose might be the effects of prolonged current transport upon compound skeletons, such as those of trilobites (Plate 10.1) and crinoids (Figure 10.18)?

Effects that readily come to mind are break-up and wear (or even complete destruction), but a third, more subtle effect is the *sorting* of components of different size and shape, through differential transport, as they are carried along by currents (Section 12.2.1). However, the extent of these effects is not simply correlated with the amount of net displacement: other factors have to be considered. Shells on a beach may suffer very considerable wear as waves and tides repeatedly wash them to and fro, and yet the net transport distance may be small (although in other instances, zig-zag transport up and down the shore can lead to extensive transport *along* it). In contrast, remains that are caught up in slumps of sediment on submarine slopes may be carried long distances (perhaps for miles), yet be protected from damage and sorting by the viscous slurry, or even stiff sediment, around them. Careful sedimentological analysis is thus a prerequisite for interpreting the extent of transport of specimens in an assemblage. Moreover, not all breakage of shells or bones is attributable to transport or to the impact of objects carried along by currents.

▶ What other important means of breakage must be considered?

Predation and scavenging were also identified in Section 12.2.1 as destructive processes. Whether or not there is any associated transport depends entirely upon the habits of the predator or scavenger.

Various observations would imply a relative lack of differential transport or destruction of specimens in an assemblage. These might include a wide range of specimen sizes, particularly with a mixture of juveniles and adults, and relative frequencies of different skeletal components (such as the two valves of bivalves and brachiopods, or the different parts of trilobites, or of plants), such as might be expected from living populations.

Physical character of original habitats The kind of environment in which a fossil assemblage originally formed may be interpreted by sedimentological analysis. Sedimentology is a huge field, covered in detail elsewhere (see Further Reading): for present purposes, an extremely generalized diagram (Figure 12.10) suffices to illustrate some of the characteristic structures and associated fossils of sediments deposited in different kinds of environments. As in Figure 10.4(a) in Chapter 10, the sequences of sediments are shown in diagrammatic profile.

A palaeoecological objective of particular relevance to evolutionary studies is the identification of those settings in which organisms that dwelt together in the same habitat are likely to have been been preserved together in fossil assemblages. Evolutionary changes in the interactions of species both with each other and with

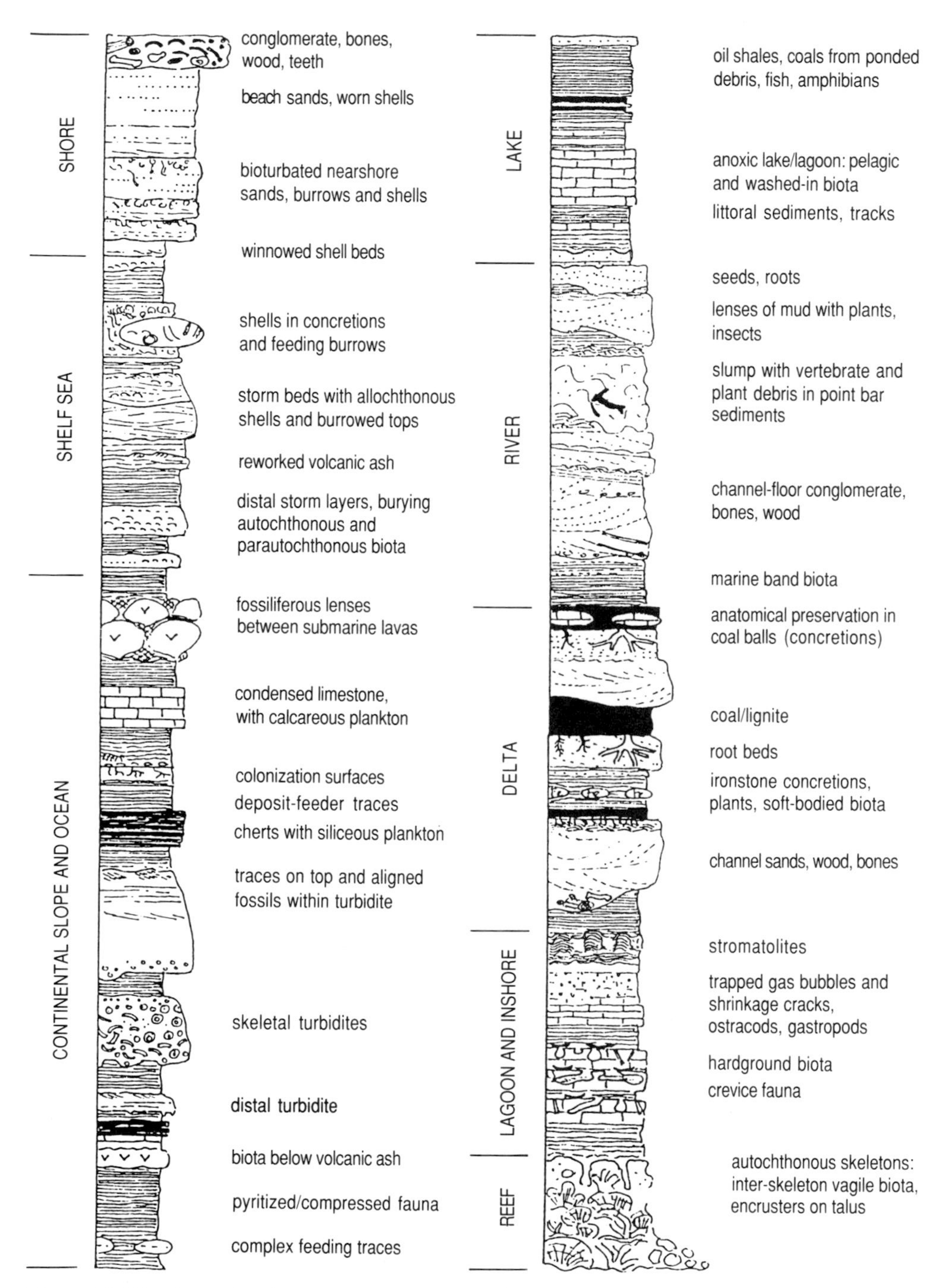

Figure 12.10 *Synopsis of sedimentary sequences deposited in different types of environment, with typical associated fossil assemblages. The beds are shown in diagrammatic profile, those projecting furthest having the greatest relative grain size.*

their physical environment may then be monitored. Figure 12.10 shows how the constitution of fossil assemblages may be expected to vary in this respect between different environments. A primary consideration is the physical character of the substratum: some types of substratum are especially susceptible to preserving an imprint of certain elements of the communities of organisms that occupy them. You should by now be equipped to surmise some of these patterns of preservation from what you have already read in this, and the previous Section.

▶ Study the upper part of the column on the left in Figure 12.10, where some characteristics of shallow marine deposits are summarized. Which type of fossil is most likely to be both common and autochthonously preserved in such deposits?

The loose, sandy sediments of such environments, overlain by well oxygenated water, are likely to be host to a rich fauna of burrowing organisms (e.g. worms and burrowing molluscs and shrimps), which will leave an abundance of trace fossils. As mere rearrangements of sediment, these traces are unlikely to survive disturbance without being destroyed, so those that remain are essentially autochthonous. Of the burrowers themselves, in contrast, only skeletal elements are likely to remain as body fossils, and frequent current reworking of the sediment will ensure that most of these are parautochthonous or even allochthonous. Thus not only do the assemblages of trace fossils record the presence of soft-bodied organisms (e.g. worms) not represented by body fossils, but they also more faithfully document the co-occurrence in life of many organisms.

▶ What is the main drawback of trace fossils for detecting the presence of particular species?

It is usually not possible to tell to which biological species a trace-maker belonged (Chapter 10, Section 10.1.1), although the general type (e.g. clam, worm or shrimp) and size of organism are often identifiable. By judiciously combining observations on trace fossil assemblages with those on suitable examples of body fossils (e.g. bivalve shells in life position: Figure 12.9(b)), some idea of the diversity of organisms originally present, and even some idea of their feeding interactions may be gained.

▶ Refer again to Figure 12.10, and decide which settings might be particularly favourable for the preservation of autochthonous body fossils.

There are several answers you might have given to this question. The possibility of sudden burial under a blanket of sediment arises frequently in several settings (e.g. beneath layers of sediment washed by storm currents into offshore areas, volcanic ash-falls (*tuffs*) or distal *turbidites* (deposited from mud clouds which have flowed downslope from sedimentary slumps) in the sea and in lakes, and mudflows and flood deposits on land). Such events can yield exceptionally well preserved assemblages, which will be discussed in more detail in Section 12.3.2. Another possibility on land is that of fossilized soil beds, in which plant roots and some soil fauna may be preserved. The early growth of concretions in swamp peats (*coal balls*) may further enhance their preservation.

In the sea, a major opportunity for the preservation of organisms in life position is provided by hard, or at least relatively firm, substrata. The clearest example in Figure 12.10 (bottom right) is that of fossil reefs. Exposed hard surfaces, either of

(a)

(b)

Figure 12.11 *Fossils of bivalves which have bored into corals. (a) In section. (b) From the outside. Upper Jurassic of central France. (The coin is about 1 cm across.)*

dead coral and other such organisms or of cemented sediment, are common in reefs and there is much competition between sessile organisms to colonize these surfaces. Colonial growth, involving asexual budding, is a common adaptation for such competitive occupation of space (Chapter 8, Section 8.2.1). Fossil reefs thus tend to preserve a rich record of attached shelly colonial organisms. That is not all, however, for the trace fossils of boring organisms (Figure 12.11) and the body fossils of many other sessile forms, as well as vagile crevice-dwellers, may also be preserved.

Information from fossil assemblages of the sorts briefly outlined above can be sufficient to monitor some broad patterns of change in the make-up of certain types of community, and hence the context for evolutionary change in their constituents. This point is well illustrated by the record of another kind of hard-surface community—that occupying *hardgrounds* (lower right in Figure 12.10). Hardgrounds form in areas of condensed (i.e. slow and sporadic) sedimentation, when sediment near to the surface becomes cemented at an early stage (usually by calcium carbonate) and then exposed by erosive currents. If exposed for long enough, these surfaces may be colonized by a rich biota of encrusting and boring organisms, a high proportion of which are likely to leave fossil evidence of their presence if the hardground itself becomes buried and preserved (Figure 12.9(a)). Fossil hardgrounds are moderately common in marine limestone sequences

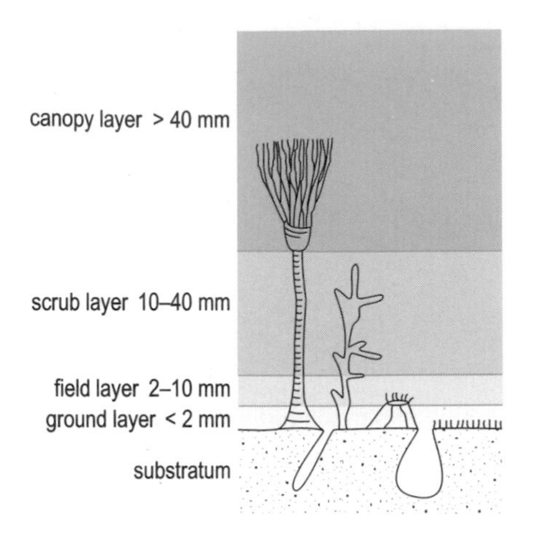

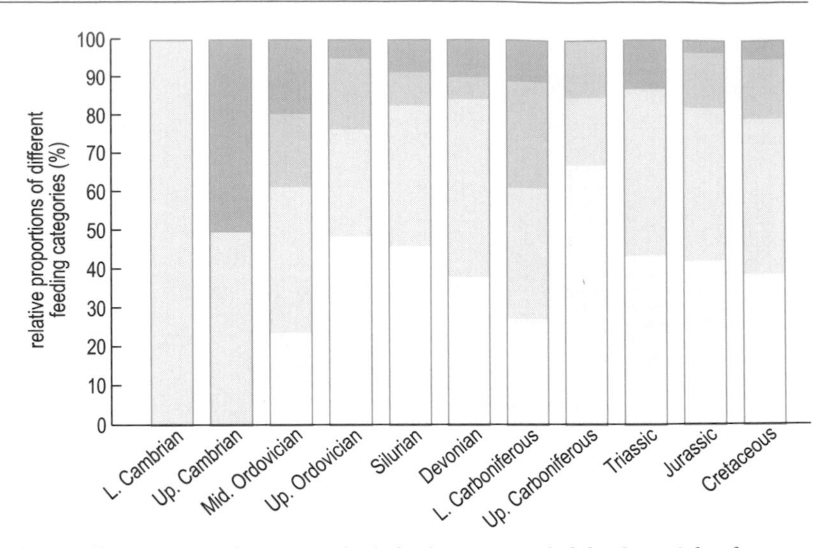

Figure 12.12 *Relative proportions, on hardgrounds of different ages, of species which feed on suspended food particles from different levels above the substratum.*

throughout most of the Phanerozoic, so T. J. Palmer (1982) conducted a comparison of their fossil assemblages, in order to detect any patterns of change (Figure 12.12).

Several quite clear patterns did emerge from Palmer's study. One was that the partitioning of niches by the filter-feeders (e.g. Figure 10.18 in Chapter 10), in terms of the height above the surface from which different species fed, was established quite early on, by mid-Ordovician times. The earliest, Cambrian examples, in contrast, show only a low diversity of borings, which were probably associated with low-level feeding. Thus the main phase of development of such communities, from initial colonization of the habitat to diverse exploitation, as found today, was limited to the Cambrian to mid-Ordovician interval. Nevertheless, important changes in the characteristics of the constituent organisms did take place later on. In particular, there was a marked increase in the frequency of well protected shelly forms among the exposed sessile taxa (not shown in Figure 12.12). The significance of these and other, related patterns will be discussed further in Chapter 15.

Interactions between species The majority of interactions between organisms either leave so little preservable evidence (e.g. many forms of behaviour) or are so destructive (e.g. much predation) that the fossil record of ecological relationships between species can be only partial. It would be in vain to attempt to resurrect from fossil assemblages all the details of relationships such as can be observed in living communities. All is not lost, however, for the reason that should by now be familiar to you: the record is *predictably* biased and some kinds of interaction can thus be reliably monitored from fossil assemblages. To study correlated evolutionary changes in the participants in such relationships is thus a realistic objective.

▶ Give an example of a specific predator–prey interaction, mentioned earlier in this Chapter, that has a good preservation potential.

In Section 12.2.1 the shell-drilling behaviour of certain predatory gastropods was discussed. The good preservation potential of the shells of their victims ensures a correspondingly good record of such predatory encounters (e.g. Figure 12.2). Both the time of inception of the habit and its effects on the 'menu' of prey species can thus be closely monitored from fossils. Using this approach, Dudley and Vermeij (1978), for example, traced changes in the intensity of drilling into the shells of a common sand-dwelling gastropod, *Turritella*, and found that there was an initial sharp rise in late Cretaceous times, to reach levels in the Eocene similar to those of today .

Another form of interaction, interspecific competition (Chapter 8, Section 8.2.5), is often elusive, and is thus a source of considerable debate concerning its role in shaping patterns of macroevolution (see Chapter 15). Again, however, reliable subjects for study can be identified. An example of a potentially fruitful topic for investigation is the evolution of competitive hierarchies in the overgrowth of neighbours by encrusting shelly organisms (as illustrated in Figure 8.1 of Chapter 8).

Evidence for some of the other forms of relationship that are listed in Table 8.2 of Chapter 8 can also be detected from fossils. Several examples of parasitism, for example, have been documented (e.g. Figure 12.13). However, only isolated instances have been studied to date, so the elucidation of the evolutionary origins and history of such associations remains a tantalizing prospect for future work.

The evolution of organisms engaged in some ecological relationships can be investigated by an approach that combines both experimental and retrospective testing of hypotheses. For example, in a study of the evolution of the cemented habit in bivalves (e.g. as in oysters), E. M. Harper (1991) explored the possibility that the habit evolved largely as a defensive adaptation against predators. The majority of sessile bivalves attach either by means of organic threads secreted by the animal's foot (e.g. the *byssus* or 'beard' of mussels), or by cementation of the shell. In most instances, cladistic analysis (Chapter 11, Section 11.2.4) shows the former condition to be the primitive character state relative to the latter, derived condition. In order to test the hypothesis that cemented forms were at a selective advantage over byssate forms when attacked by certain known major predators (seastars and crabs), Harper set up seawater tanks containing such predators, and added to each tank equal numbers of randomly distributed byssate and cemented

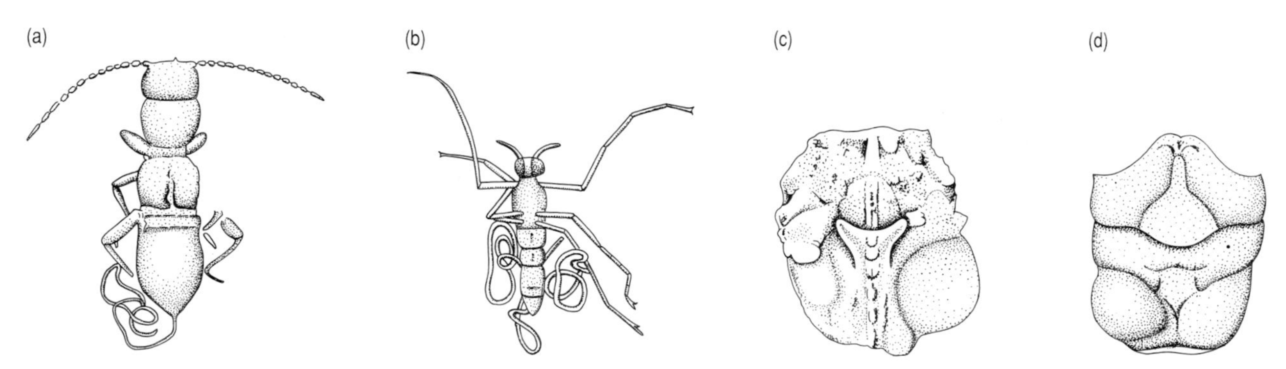

Figure 12.13 *Fossils of parasites. (a) and (b) Parasitic nematodes associated with a beetle from Oligocene lignite (a) and with a fly preserved in Oligocene amber (b). (c) and (d) Crab carapaces from the Lower Cretaceous of England (c) and the Upper Jurassic of Poland (d), showing swellings (asymmetrical bulges) due to internal parasites.*

prey. Frequencies were maintained by restocking during the experiment. Any prior preferences of the predators for particular prey types were overcome by employing the same species (the common mussel *Mytilus edulis*) for both types of prey: cementation was simulated using epoxy cement. (The possibility of the latter putting predators off, e.g. by being distasteful, was tested, and rejected, by daubing some of the byssate individuals with the epoxy cement and monitoring the reaction of the predators.) The predators encountered the two prey types with similar frequencies.

▶ What is the appropriate null hypothesis for this experiment?

The two prey types should be equally vulnerable to successful attack, according to the null hypothesis, so no significant difference is expected between the numbers of each that are eaten. Harper's results are shown in Table 12.2.

Table 12.2 *Numbers of byssate and 'cemented' mussels successfully eaten by selected predators.*

Predator	Total no. of prey taken	No. of each type expected	No. of byssate eaten	No. of 'cemented' eaten in place	No. of 'cemented' pulled free and then eaten	*P* (Chi Square test)
Asterias rubens (seastar)	121	60.5	95	11	15	< 0.001
Cancer pagurus (crab)	132	66	96	7	29	< 0.001
Carcinus maenas (crab)	27	13.5	19	4	4	< 0.05

The numbers of cemented individuals that were pulled free and then eaten were included with the numbers of cemented individuals eaten in place. Probability under the null hypothesis was calculated using a Chi Square test (as explained in statistical texts cited in the Further Reading for Chapters 3 and 10).

▶ What do you conclude from the results in Table 12.2?

The null hypothesis may be rejected for all three predators. The alternative hypothesis, that cementation confers an advantage on prey individuals, can thus be accepted.

In parallel with her experiment, Harper conducted a retrospective test, surveying the first fossil appearances of the various clades of cemented bivalves to compare with the record of known mollusc-eating predators.

▶ What is the expected pattern from Harper's evolutionary hypothesis?

A strong correlation between the times of appearance of the predators and the cemented clades should be expected.

The comparison gave a clear result. From early Mesozoic times onwards there was a marked proliferation of predators specially equipped to deal with shelly prey. At least 20 clades of cemented bivalves have evolved, yet only one or two of these arose in the Palaeozoic. The rest first appear in post-Palaeozoic rocks, with many making their debut in the early Mesozoic—in parallel with the predators.

The results from both the experimental and the retrospective tests thus strongly support the hypothesis that increased predation pressure from early Mesozoic times was a major cause of the evolutionary proliferation of cemented bivalves. In fact this pattern of change appears to have been but one aspect of an episode of considerable re-organization in benthic marine communities that Vermeij (1977) termed the **Mesozoic Marine Revolution (MMR)**.

▶ Can you think of another correlated change in certain benthic communities, discussed earlier, which would be consistent with the hypothesis of increased predation pressure?

Hardground assemblages show a marked increase in the frequency of sessile taxa with robust shells. The possible causes and effects of the MMR will be explored further in Chapter 15.

Taxonomic completeness of assemblages Valentine's (1989) study of taxonomic completeness, which was discussed in Section 12.2.3, focused deliberately on benthic molluscs, because the objective was to assess the quality of the record for organisms with good preservation potential. It should be obvious from the earlier discussion of taphonomic processes that the majority of fossil assemblages are likely to be entirely bereft of any body fossils of those organisms that were either soft-bodied, or only had lightly mineralized skeletons. Trace fossils rarely compensate adequately, for the purpose of cataloguing of species, because of the problems of associating them with specific trace-makers. Fossil assemblages of the kind normally encountered (i.e. those consisting merely of hard parts and trace fossils) thus provide only a biased perspective on past life, although the judicious selection for comparative study of particular types of assemblage (e.g. those of reefs and hardgrounds), or of specified components of the biota, can minimize this handicap.

An alternative approach is to prospect for those localities where exceptional combinations of taphonomic circumstances have resulted in the preservation of soft tissues as well as skeletal elements (e.g. Plate 12.1). So important are these as sources of information, that discussion of them will be deferred to a separate Section (Section 12.3.2).

Temporal distribution of specimens Most assemblages show some extent of time-averaging (Section 12.2.3). Not only individuals, but entire local populations may have only a fleeting existence relative to the time-spans over which fossiliferous beds form. In shallow sands off the coast of Britain today, for example, the living shelly burrowing fauna in any one local area may be overwhelmingly dominated by one species. That population may be virtually annihilated by a passing storm after only a few years, to be replaced by another, perhaps of a different species, seeded by whichever cloud of planktonic larvae next drifted into the area. Over a century, say, populations of several species might come and go in this manner.

▶ Which factors might lead to mixing of the shells of these different populations in the resulting sedimentary sequence?

The most important factor in such an environment would be current erosion: this might move considerable volumes of sand, and leave a parautochthonous sheet of shells behind (to form a *shell bed*: see Figure 12.10, upper left column). Another mixing agent could be bioturbation. The resulting assemblages would thus contain a greater diversity of species than ever co-existed there at any time.

Palaeoecological interpretation of fossil assemblages thus relies crucially upon analysis of the taphonomic circumstances in which they formed: assemblages can rarely be treated merely as single fossilized communities (although, regrettably, naive interpretations of this kind have frequently found their way into the literature). Nevertheless, valuable exceptions may be provided by some instances of **mass mortality**. Periodically, large numbers of individual organisms may be killed in a single event. There may be many causes, both biological and physical. Such events can affect a single species or many species. Usually they do not wipe out whole species, but may severely reduce local populations to such an extent that it can take many years for populations to be restored to pre-event levels.

A mass mortality event taking place over a period of a few hours, days or weeks may result in a large concentration of dead animals and/or plants in a single place, and on a single **bedding plane**, the interface between two layers of sediment. Figure 12.14 shows a bedding plane with a monospecific mass mortality assemblage of fossil fish, possibly caused by an increase in salinity or temperature.

Figure 12.14 *A bedding plane from a sequence of fine-grained limestone showing a mass mortality assemblage of fish fossils. From the Eocene of Wyoming, USA.*

The outstanding value of mass mortality assemblages is that they sample populations of coexisting individuals, and so may be used, for example, to assess the range of morphological variation in a population at a given time. They may also provide suites of specimens showing all stages of development from young to old.

▶ What is the value of such deposits in studying microevolutionary change as discussed in Chapter 10?.

They allow morphometric analysis of populations at 'instants' of time—hence there is no blurring of the record by mixing of populations from different times.

There are many possible causes of mass mortality, and sometimes it is difficult to determine which has operated. Recent mass mortalities have been caused by sudden burial, fluctuations in oxygen levels, toxification of water by micro-organisms, and rapid fluctuations in salinity and temperature. Cases of mass death by volcanic eruption, shock waves from earthquakes, and even poisoning by the rapid escape of gas from sediment have also been reported.

12.3.2 Fossil-Lagerstätten: exceptional fossil assemblages

The term **fossil-Lagerstätten** (sing. fossil-Lagerstätte, a 'fossil bonanza' in German) was introduced by the German palaeontologist Adolf Seilacher to describe sedimentary deposits that are an especially rich source of fossil material. He recognized two broad categories. **Conservation Lagerstätten** are deposits in which fossils, although not necessarily common, are exceptionally well preserved. **Concentration Lagerstätten**, on the other hand, are large accumulations of fossils, such as shell beds, in which it is possible to obtain large numbers of individuals. The two types thus emphasize, respectively, quality and quantity of preservation. There are some sedimentary environments that can concentrate remains to high abundances *and* preserve them exceptionally well. The two types of fossil-Lagerstätte are not therefore mutually exclusive.

Examples may vary in size from very small, as in the case of amber formed from the resin of pine trees, where a deposit may be only tens of millimetres across, to the deposits of restricted marine basins which may cover several tens of square kilometres. Figure 12.15 (overleaf) shows some of the environmental settings in which concentration and conservation Lagerstätten are known to accumulate.

Fossil-Lagerstätten are of significant value in macroevolutionary studies. First, because of the abundance and/or high quality of fossils, they have been the targets of much collecting activity. More importantly, the exceptional quality of the fossils in conservation Lagerstätten allows for fuller and more accurate interpretations of past life.

12.3.3 Conservation Lagerstätten

The key to many conservation Lagerstätten is the felicitous combination of taphonomic circumstances whereby processes that normally destroy tissues are delayed, and/or diagenetic mineral growth is accelerated, such that the two overlap.

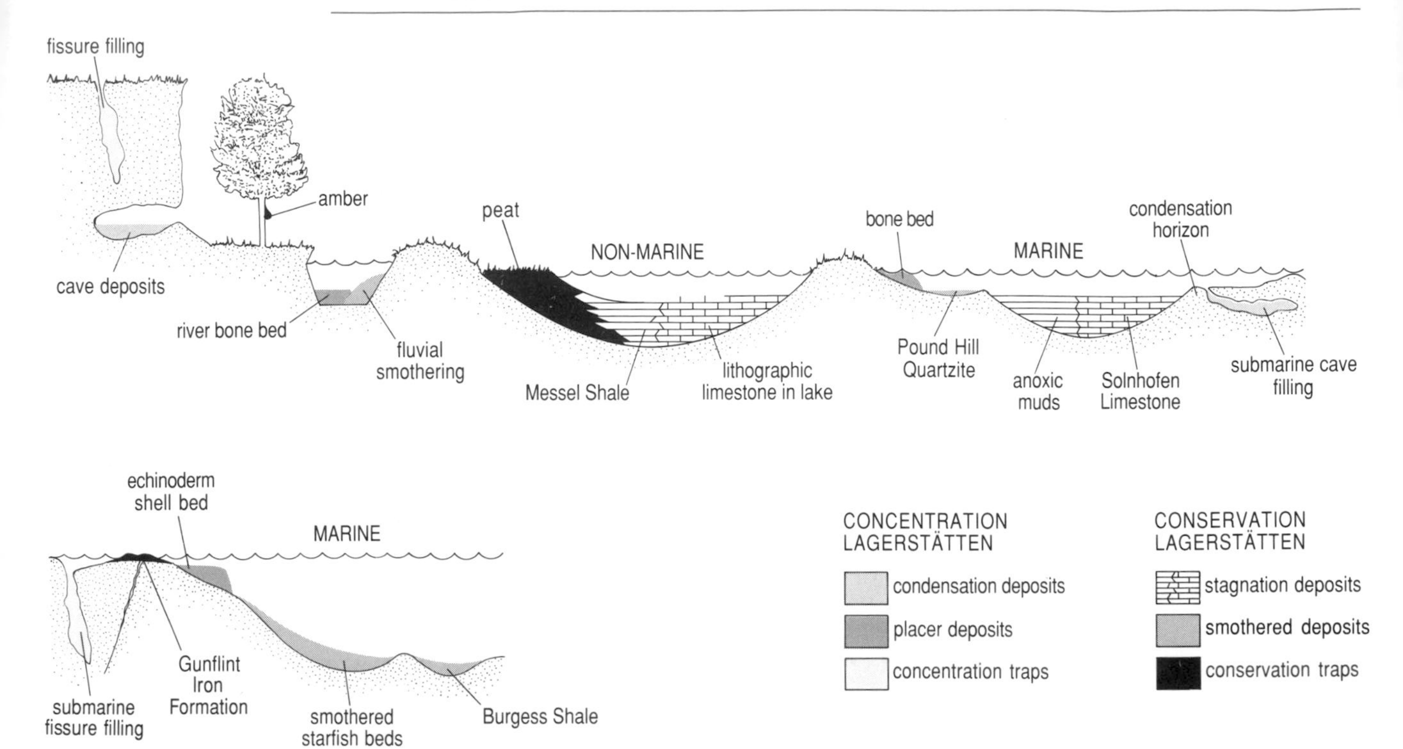

Figure 12.15 *Environmental settings for the formation of fossil-Lagerstätten.*

The consequence may be *petrifaction* (replication in stone) of tissues that would otherwise be lost to the record.

▶ What processes might delay or prevent the destruction of soft tissues?

Referring back to Section 12.2.1, there are several possibilities. Either rapid sedimentation or anoxic bottom conditions could inhibit scavenging and bioturbation. Bacterial decomposition may be limited by unfavourable water chemistry (as in pickling). Such conditions are usually accompanied by rapid diagenesis, occurring before or during microbial decomposition of soft tissues.

In some instances, diagenetic mineral growth is absent but extreme conditions have arrested decay. Such is the case with desiccated mummies, frozen carcases and insects trapped in amber.

Conservation Lagerstätten thus have a wide variety of origins, and form in many different settings (Figure 12.15). Some more notable examples are briefly introduced below.

In the Late Precambrian Pound Hill Quartzite in the Flinders Range, near Ediacara, South Australia, rapid sedimentation of sand in a shallow sea buried large numbers of soft-bodied, bottom-dwelling animals. These organisms occur as fossils at a number of localities around the world in rocks of similar age. There is a debate about their affinities because it is difficult to relate them to existing phyla (see Chapter 14, Section 14.1.2 and Figure 14.3).

Why are these fossils found at all? Sedimentological evidence indicates that they were buried on the sea-floor by rapidly migrating sand. In addition, the sediments show no evidence of burrowing, and indeed it is doubtful if the ability to burrow deeply had been acquired by any organisms at this time. So the conservation of the fauna is due, at least in part, to the absence of destructive bioturbation—providing, so to speak, a taphonomic window that was to be usually closed in later times. However, there may also have been an important structural factor involved. It has been suggested that these organisms may have had a relatively tough body wall, that resisted decomposition until the sediment had become lithified. Thus the Ediacaran assemblages provide a privileged glimpse at late Precambrian life, at a time that is crucial for our understanding of the early evolution of animals.

The Burgess Shale in southern British Columbia, Canada, is another conservation Lagerstätte, of Middle Cambrian age. Though deposited in a deeper-water setting than the Pound Hill Quartzite, its exceptional preservation is also largely attributable to the rapid burial of organisms, mostly forms with no mineralized skeleton, but including some, such as trilobites and brachiopods, that did have mineralized skeletons (Chapter 14, Section 14.1.4 and Figure 14.4). Burial in this case involved the periodic slumping downslope of fine sediment, with its associated fauna, at the foot of a reefal escarpment. The bottom water in the basin where the slumped sediments accumulated is thought to have been anoxic, so excluding bottom-living organisms which might otherwise have disturbed the washed-in carcases, by scavenging or bioturbation.

▶ Casting your mind back to Section 12.3, how would you classify the Burgess Shale fossil assemblage?

It is an allochthonous mass mortality assemblage, formed by the episodic dumping of associations of more or less contemporaneous organisms in an area away from where they lived. The assemblage nevertheless consists of organisms which largely co-existed. The Burgess Shale fauna is exceedingly diverse, comprising many of the phyla that are present throughout the rest of the Phanerozoic.

Together these two conservation Lagerstätten, although separated by only a few tens of millions of years, demonstrate a marked change in the types of animal that inhabited ancient seas (Chapter 14, Section 14.1).

The concretions produced by diagenetic reactions, discussed in Section 12.2.2, may also be considered as conservation Lagerstätten, especially when formed during or before decomposition of soft tissues. For example, Lower Cretaceous concretions from the Santana Formation of Brazil contain fossil fish with mineralized muscles, gills and tissues of the alimentary tract preserved uncompacted. Accurate comparisons are possible, between the soft tissue morphology of fossil and living fish. Such preservation is even more important when soft tissues are preserved in organisms that are now extinct, providing unparalleled insights into their anatomy.

Among the most spectacular conservation Lagerstätten are the various deposits of amber, of late Cretaceous and Tertiary age. These are accumulations of resins from ancient pine trees, and they occasionally contain insects. The insects became trapped, probably while trying to feed on the resin. Many are preserved in excellent detail, and show delicate features such as fine hairs, and in at least one instance two flies have been preserved in the act of mating. The exceptional preservation here is thought to be due in part to the inhibition of bacterial decay by chemicals within the resin, accompanied by rapid sealing by the amber, as well as dehydration by osmosis.

Also spectacularly preserved are the rare frozen mammals from the tundra. Although these are relatively young on a geological time-scale (usually less than 100 000 years old), they retain original soft tissues, and even contain well preserved macromolecules such as proteins and polynucleotides.

Conservation Lagerstätten do, however, present palaeontologists with a dilemma for the documentation of taxonomic diversity. It is highlighted by Figure 12.16, which shows the numbers of families of fossil birds recorded from various stratigraphical stages in the Jurassic and Cretaceous.

The first record shown in Figure 12.16, in the last stage (Tithonian) of the Jurassic, is due to a conservation Lagerstätte in southern Germany, known as the Solnhofen Limestone. This is a very fine-grained limestone (used for lithographic printing) that allowed for the preservation of exquisite detail in such delicate organisms as pterosaurs (Plate 12.3), as well as the earliest-known indisputable birds. Since the discovery, in 1860, of an isolated feather, six partial to almost complete fossil skeletons, some also furnished with feathers, have been recovered to date (1992) from the Solnhofen Limestone, and all have been placed in the

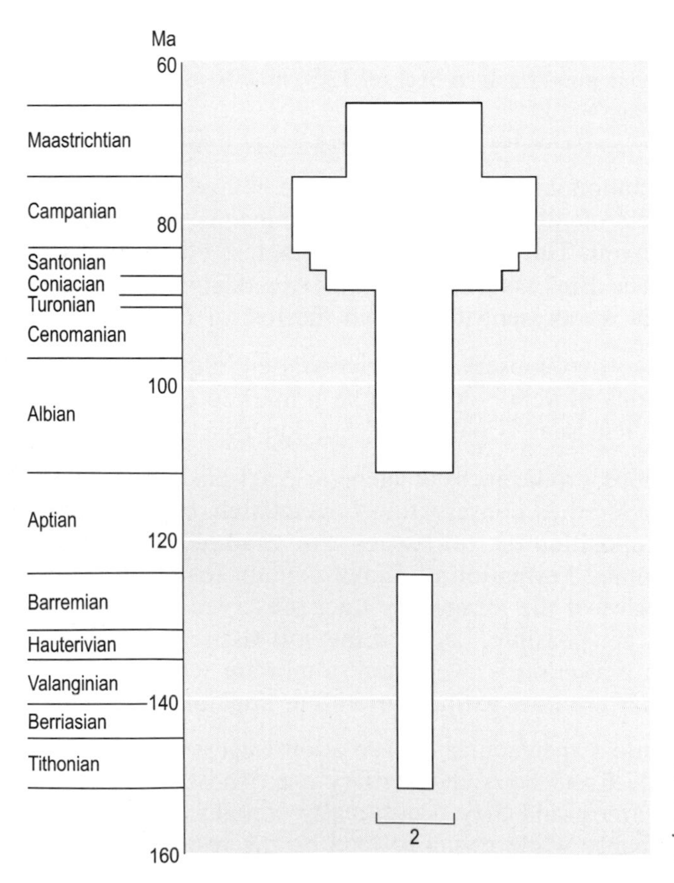

Figure 12.16 *Changes in the numbers of families of birds, known from fossils, through the Upper Jurassic and Cretaceous.*

species *Archaeopteryx lithographica* (see Plate 14.1). These are the only known fossil birds from the Tithonian Stage. The next-oldest specimens comprise some more questionable remains from the lowest Cretaceous (Berriasian Stage) of Roumania, followed by a specimen of a much more advanced genus (*Sinornis*), reported as recently as 1992, from the Valanginian Stage (see Figure 12.16) of China. Thus gaps of several million years separate these early records. Had the Solnhofen *Archaeopteryx* not been preserved, or discovered, the diversity diagram would commence only in the Berriasian, or, but for the Roumanian specimens, in the Valanginian. First occurrences from conservation Lagerstätten such as this must therefore be treated as providing only a minimum limit to the possible age of the taxon in question, and the discovery of older fossils is distinctly possible. The picture is less bleak, however, for organisms with high preservation potentials, and thus more complete fossil records.

Some idea of the extent to which a first record might misrepresent the true inception of a given taxon (or, indeed, the last record, its extinction) can be obtained from the relative completeness of the record over the already known stratigraphical range of the taxon. It stands to reason that the more frequently fossils of the taxon in question have been recorded over that range, the less probable is it that their absence beyond its limits is due to preservational failure alone. In other words, it ought to be possible to place some form of margin of error on the first and last records of a particular taxon. Obvious though this principle is, in theory, its transformation into a rigorously quantitative method is difficult, because certain complicating assumptions, such as the effects of changes in geographical range on the probability of preservation, would need to be applied in particular to the inception and demise of each taxon.

We now examine two conservation Lagerstätten in more detail, particularly highlighting some of their evolutionary implications.

The Gunflint Iron Formation The fossil record offers few clues on the nature of early life, examples of which have only been preserved in exceptional conditions. It is likely that the first forms of life were prokaryotes (see Chapter 3, Section 3.2.2), as will be explained in Chapter 16. Most prokaryotes lack a mineralized skeleton, although a few have stages in which they encyst in structures with resistant organic walls. A few also produce banded columnar stony structures known as **stromatolites** (see Chapter 16). These are built up as layers of fine-grained carbonate sediment become trapped by mats of filamentous organisms such as blue–green algae (a group of prokaryotes, also termed cyanobacteria). Stromatolites may be several metres across, and one or more metres high. They are among the first easily recognized fossils of life on Earth.

The Gunflint Iron Formation crops out around the shores of Lake Superior in Minnesota (USA) and Ontario (Canada). The Gunflint fossils occur within stromatolites which have been replaced by *chert* (a form of silica, SiO_2). They are some of the oldest well preserved fossil cells known (Plate 12.4): radiometric dates provide an age of approximately 2 000 Ma.

The fossils comprise microscopic spheres and unbranched filaments. There are also many objects which are difficult to interpret, but may be poorly preserved microscopic fossils. Judging by their small size and simple structure (see

Chapter 16), the microfossils probably represent the remains of various forms of bacteria or cyanobacteria. As yet, no eukaryote microfossils have been identified in the Gunflint chert.

Special conditions are required for the fossilization of such organisms. Accordingly, it is not known if the microfossil assemblage is typical of the Earth's early biota or whether it was peculiar to the environment in which this type of preservation can occur. This is a common problem with conservation Lagerstätten. However, the circumstantial evidence suggests that a widespread environment (and hence biota) is represented, and thus that it was the preservation that was exceptional.

It was presumably a shallow-water biota: cyanobacteria are photosynthetic, so the stromatolite-forming micro-organisms would have required sunlight for photosynthesis, but they were also benthic, requiring a substratum on which to grow (many other photosynthetic micro-organisms float freely in surface waters). The presence of stromatolites is therefore an indication of water no deeper than the **photic zone** (the zone to which light will penetrate water with sufficient intensity for photosynthesis to occur). Modern stromatolites are largely restricted to intertidal or confined subtidal areas where salinities are often elevated with respect to normal seawater, such as at Shark Bay, in eastern Australia. Yet this restriction, even within the photic zone, is due in part to the presence in normal marine environments of organisms which graze on the mats. As there is no evidence for grazing organisms in the early Proterozoic, stromatolites probably had a wider distribution in the sea than at present.

The rocks of the Gunflint chert are thought to have been originally deposited as limestones. This is based on the observation that all modern stromatolites are constructed of calcium carbonate. So it is reasonable to suppose that they were subjected to diagenetic replacement.

▶ Is the diagenetic alteration of the stromatolites likely to have been a geologically early or late event?

It is somewhat surprising that such delicate organic microfossils have survived at all. It must be assumed that the diagenesis was very early, before the organisms had any significant opportunity to decay.

Rare though such preservation is, that seen in the Gunflint chert is of immense importance due to the exceptional preservation of its fossils, combined with its extreme antiquity. Here a conservation Lagerstätte tells us about the existence of a variety of single-celled organisms some 2000 Ma ago, and it allows the minimum grade of organization of life forms to be assessed in relation to the age of the planet.

The Messel Oil Shale Messel is a small village near Darmstadt, in Germany. It is famous for a deposit of oil shale, a shale with a high organic carbon content, which contains many remarkably well preserved fossils.

The shales are of Eocene age, around 50 million years old, and were deposited in a rift valley lake (rift valleys and lake sediments are discussed in Chapter 10,

Section 10.3.2). Rivers flowed into the lake, bringing in large quantities of dead plant and animal material washed in from the surrounding forests. The lake itself was deep and stagnant, and dissolved oxygen in the water column was depleted by the bacterial decay of seasonal blooms of planktonic algae to such a degree that the bottom water became completely anoxic. In short, the lake acted as a huge settling tank.

The bottom water became toxic with dissolved gases such as hydrogen sulphide, methane, ammonia and carbon dioxide which built up to high concentrations. Their presence was important on two counts. First, in the water they probably slowed the rate of decomposition of carcases by inhibiting decay to just a few bacterial species. Secondly, upwellings of poisonous gas may occasionally have erupted from the lake, or spread over it as a result of nearby volcanic activity. The pall of gas hanging over the lake was sufficient to kill many birds and bats flying over its surface: bats and birds are particularly common at the Messel site compared with other fossil sites of equivalent age elsewhere. Many other animals living in and around the lake were also killed and accumulated in it. A recent parallel, with tragic human consequences, occurred at a lake in Cameroon in the mid-1980s.

The Messel vertebrate fauna is dominated by fish, but besides the birds and bats there are also many snakes, lizards, crocodiles, and mammals including small horses (Plate 12.1) and some early primates. The invertebrate fauna is dominated by flying insects. There is a rich flora derived from numerous broad-leaved trees and conifers. Microfossils are also common, and include pollen, spores, fungal hyphae, and even fossil bacteria, as well as the algae.

The style of preservation is remarkable. Most of the skeletons are fully articulated, although somewhat flattened due to compaction, and many of them show the outlines of the soft tissue to a high level of detail. Insect wing cases usually show the original colour when first revealed in the rock, although this rapidly disappears on exposure to the air.

The soft tissues at Messel are preserved in an unusual manner, as replacements by what are interpreted as lithified bacteria (Figure 12.17). It appears that the bacteria spread through the soft tissues and were themselves fossilized there (quite how the bacteria become fossilized is unclear, but it was perhaps due to the rapid precipitation of minerals onto the bacterial cell membranes, as a consequence of metabolism in bizarre conditions). The process has been referred to as a sort of

Figure 12.17 *Fossilized bacteria in the soft tissue body outline of a bat (from the Eocene of Messel).*

bacterially mediated 'photography'. Again, it is attributable to very early diagenetic mineralization, as in the case of the Gunflint chert.

The early Tertiary was a time of rapid diversification of many of the groups that had survived the mass extinction at the end of the Cretaceous, especially the mammals (see Figure 11.1 in Chapter 11). The Messel fossils offer an unparalleled opportunity to observe closely the anatomy of many species in the early stages of these radiations. The crucial importance of the little fossil horses, for example, for the interpretation of the early evolution of that group has been mentioned in Section 12.1. So well preserved, also, are some of the many bats, that it has proved possible to identify the types of insects on which they were feeding, from the preserved contents of their guts. Preliminary studies suggest that resource partitioning (Chapter 8, Section 8.3.2) already existed between the different species of bats, which seem to have shown some specialized preferences in their prey.

In conclusion, although such deposits are rare, it is often the circumstances of preservation, rather than the associations of organisms, which have been unusual. The resulting fossil assemblages can provide invaluable glimpses of past life, from which much can be learnt about aspects of macroevolution that would be hard, if not impossible, to glean from more normal assemblages.

12.3.4 Concentration Lagerstätten

Concentration Lagerstätten are fossil assemblages containing large abundances of fossils. They are not necessarily well preserved, and diversity may be low or high. The agencies of concentration can be physical, chemical or biological (Section 12.2.1). Physical means of concentration are common, and include water current action, and wind and wave action. These processes may concentrate remains by sorting on the basis of hydrodynamic properties (such as size, shape and density), or by the removal of sediment to leave behind *lags* of resistant skeletal materials.

Reduced sedimentation rates may also allow concentrations of fossils to build up over prolonged time periods. Chemical concentrations are rare, but when sediments weather, concentrations of fossils consisting of stable minerals may remain. This is sometimes the case for bone beds dominated by teeth, where highly resistant elements are left behind as a by-product of both physical and chemical removal of sediment. Biological agencies for concentration have already been mentioned in Section 12.2.1. Many cave-dwelling predators and scavengers concentrate remains, often to exceptionally high levels, especially when occupancy of a cave extends over several generations. A curious combination of phenomena evidently led to the accumulation of many vertebrate remains in a number of caves in South Africa. Moist air tends to issue from the mouths of the caves, and this favours the local growth of trees (in an otherwise sparsely wooded landscape) above them. Large cats, such as leopards, often haul the carcases of their prey up into the trees, and devour them there. As a consequence, falling pieces of the carcases tend to accumulate in the caves below. Some of the South African Pleistocene cave assemblages are an important source of early hominid material accumulated in this manner (although some of the caves also show evidence of occupation by the hominids).

Concentration Lagerstätten can provide a wealth of specimens for morphometric analysis.

► What advantage would a mass mortality concentration Lagerstätte (such as an accumulation from a herd of animals caught and drowned in a flood) have over a time-averaged concentration Lagerstätte for assessing morphometric variation?

A mass mortality assemblage would sample the morphological variation within a population of individuals alive at the same time. A concentration Lagerstätte formed by condensation from several populations will show the overall range of variation of these aggregated populations over a prolonged time period (Section 12.3.1). This mixing may mask evolutionary changes in morphology; e.g. a real shift in the mean size of an organism may be recorded only as a broad range of sizes in a time-averaged assemblage.

As was repeatedly stressed in Section 12.3, then, it is essential to interpret the taphonomy of any such fossil assemblage with great care, before drawing inferences from the fossils themselves concerning their biological characteristics.

12.4 INFORMATION RETRIEVAL

Having examined the processes of fossil preservation, we can now go on to examine the retrieval of information on past diversity from fossils. Caution is needed as taphonomic processes are highly selective, so the fossil record can be heavily biased.

12.4.1 BIAS IN THE FOSSIL RECORD

In Chapter 10, Section 10.1.1, it was stated that the fossil record is highly biased, in terms of abundance, in favour of certain types of organisms and certain environments. These are naturally occurring biases which limit the information obtainable from the fossil record. Nevertheless, as demonstrated in the previous Section, the fossil record can sometimes be extremely good in terms of quality, and you have seen how this fact can also be used to advantage in studying evolution.

In addition, an important bias is introduced by time. The surface of the Earth is constantly subject to geological activity, as illustrated by the concept of the rock cycle (Figure 10.3 in Chapter 10). In the long term this cycle destroys fossils. As a consequence, the record becomes patchier the further back in time one looks. But this bias must be properly understood. The record does not become uniformly worse in quality with age: deposits such as the Burgess Shale and the Gunflint Iron Formation disprove that. Rather, the amount of available record decreases. This bias needs to be borne in mind when global diversity values are assessed (Chapter 15).

The fossil record as it occurs in rocks is the primary source of information for evolutionary palaeontology, but it is specimens in collections that form the basis of many such studies. And here another very important bias, artificial bias, can become important; for example, many more fossil taxa are known from Europe and North America than from Africa, Asia and South America. This discrepancy undoubtedly does not reflect real differences in diversity, but merely the concentrated activity of palaeontologists in the first two continents for more than 200 years.

12.4.2 PRESENTATION OF DATA ON THE DISTRIBUTION IN TIME OF FOSSIL TAXA

The *stratigraphical range* of a fossil species is the span on the geological column lying between the oldest and the youngest record of the species (although either or both of these limits may be arbitrary if chronospecies are involved: see Chapter 10, Section 10.2.1).

▶ Apart from chronospecies, why might the stratigraphical range of a fossil species not correspond to the true duration of an evolutionary species?

There are two main possibilities. First, because of the incompleteness of the record, the duration is almost invariably under-represented (Section 12.3.3). In contrast, only when derived fossils are not recognized as such, or where fossils are artificially displaced to lower levels (e.g. by drilling operations) can the apparent range become over-extended. Secondly, where speciation at first involved only sibling species, followed later by morphological divergence, only the latter would be registered in the fossil record, with the appearance of distinct morphospecies (Chapter 10, Section 10.2.1). As discussed in Section 12.3.3, the probable extent of the first problem can be estimated from the relative completeness of the record within the stratigraphical range of a species. The second problem is more difficult to tackle, though studies of the patterns of speciation in comparable living organisms may provide some clues (Chapter 10, Section 10.2.1). The evolutionary palaeontologist has little choice but to use stratigraphical ranges as proxies for species durations in analysing patterns of macroevolution. However, the problems outlined above always need to be borne in mind as possible sources of error.

Range charts are compilations of the stratigraphical ranges of several taxa, shown alongside a geological time-scale (Figure 12.18). A range chart can be constructed at any taxonomic level, e.g. for species, genera or families.

The horizontal axis of the range chart need not be defined (i.e. the sequence of taxa may be arbitrary), but it is possible to arrange the taxa in some sort of morphological order, or in order of first appearance, or in geographical groupings, according to the issue being addressed. This arrangement may reveal patterns in the stratigraphical distributions of fossil taxa, such as numerous first appearances clustered in time or space, or coincident extinctions.

Another useful graphic representation of stratigraphical range data is the **spindle diagram**. This is a modification of the range chart based on the numerical abundance of fossil taxa within discrete time slots (as shown in Figure 11.1 of Chapter 11). The vertical scale again represents geological time. Against this are

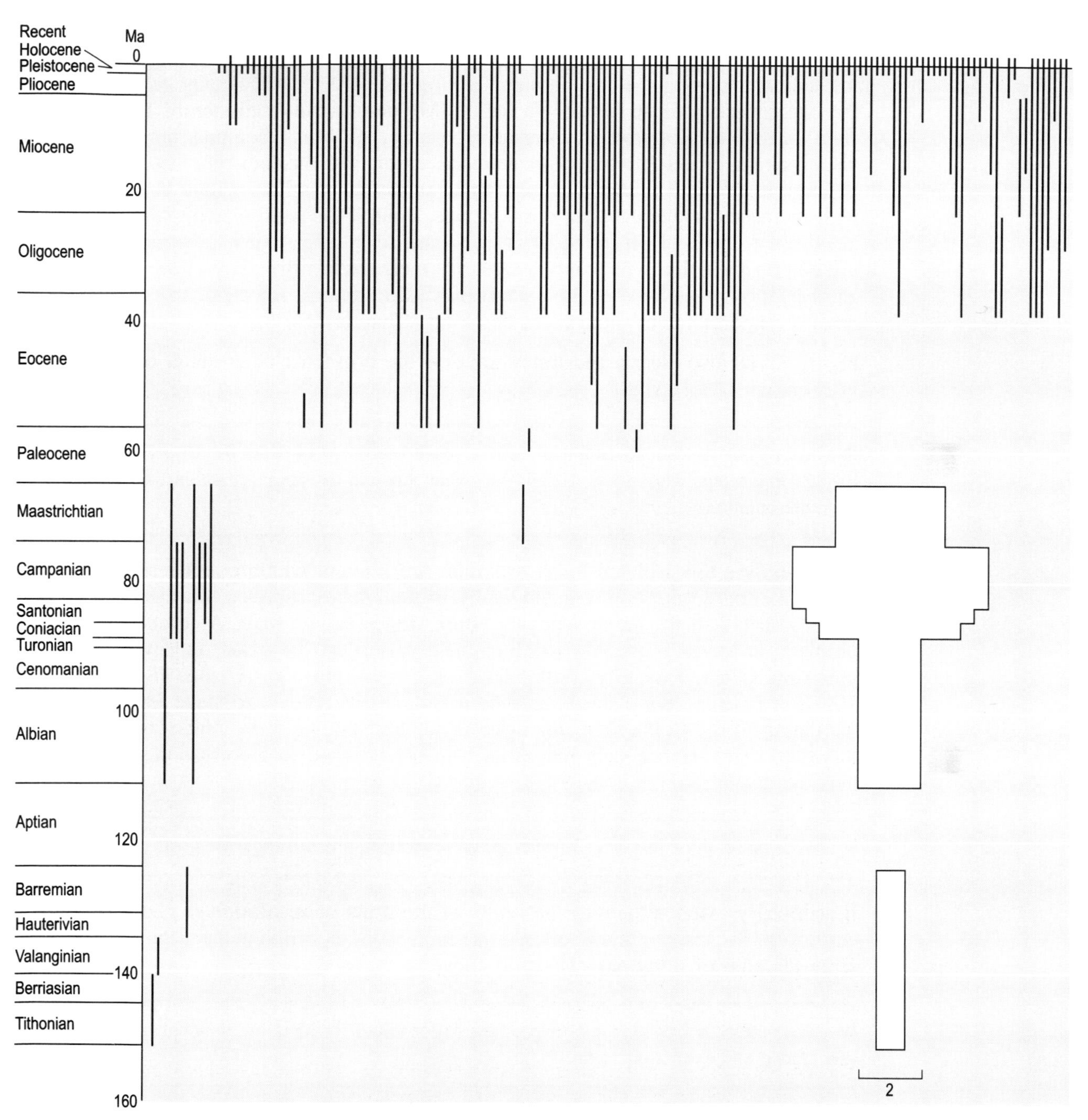

Figure 12.18 *A range chart for families of birds represented by fossils (based on Unwin, 1988). Extensions at the top of the Figure denote families that are extant. In the lower right of the Figure is a spindle diagram for the Mesozoic ranges shown in the range chart. (See also Figure 12.16.)*

plotted the numbers of taxa (e.g. genera or species) in each higher taxon through successive time intervals. Each spindle is a histogram drawn on its side, but instead of arranging the bars in a row sitting along the vertical axis, traditionally the bars are centralized (as shown for the Mesozoic data in Figure 12.18 (lower right), which also appeared earlier as Figure 12.16). Such diagrams provide a rapid visual means of assessing relative abundances of taxa through time. In this way it is possible to see if increases or decreases in diversity of one group correspond with increases or decreases in other groups, though the biases mentioned above need also to be borne in mind. It is important to note that such patterns alone do *not* demonstrate that one group necessarily increased in diversity at the expense of another, though analysis of the pattern may allow one to reject, or to corroborate other arguments for hypotheses of this kind (Chapter 15).

Figure 12.18 also clearly illustrates some of the problems of artificial bias inherent in diagrams of this kind.

▶ From what you already know about the fossil record of birds in the Upper Jurassic and Lower Cretaceous, from Section 12.3.3, in what sense is the range chart shown here a misrepresentation?

In fact only a few isolated fossil examples are known, which ought really to appear as dots (representing geologically instantaneous samples) on the range chart, rather than the bars occupying entire stages shown here. A common problem, however, is that the localities in question can be dated only with limited precision. Often, a locality can be dated only to within a fossil zone (Section 10.3.2 of Chapter 10) or even, as shown here, more crudely to stage level (though the dating of the Solnhofen Limestone is actually known more precisely than is indicated here). Hence such ranges have to be recorded to a correspondingly broad level of resolution.

▶ What effect does this have on the *apparent* stratigraphical ranges of taxa?

It artificially extends them to the limits of the units of resolution (e.g. stages) employed. In practice this is often a yet more serious problem than the under-representation of durations discussed above.

▶ What effect is this bias due to poor resolution likely to have on apparent patterns of appearance and extinction of taxa?

Because ranges are systematically extended to stratigraphical boundaries (e.g. between zones, stages etc.), an artificially high degree of co-incidence in timing is suggested. Bias of this sort is obviously a major problem for any attempt to detect episodes of rapid evolutionary radiation or mass extinction, requiring a careful assessment of the limits of resolution of the data available for the study.

A further bias evident in Figure 12.18 is what has been termed *the pull of the Recent*. Notice that, as the Recent is approached, increasing numbers of ranges extending from earlier levels up to the present are shown. Much of this apparent increase in diversity is attributable to the discrepancy in relative completeness between the living and the fossil record. If a taxon is extinct, at least two fossil occurrences are required to yield an extended range by interpolation, and that is likely to be a considerable underestimate of the true duration if the fossil record of the taxon is poor. If a taxon has living representatives, however, then a single fossil occurrence will suffice to interpolate a range at least from that level to the present. Thus, there is a pronounced bias in favour of showing more of the original durations of extant taxa, so that *recorded* diversity appears to increase towards the present. Of course, this problem does not arise with wholly extinct taxa (e.g. trilobites), and is of minor effect on those with a good fossil record (e.g. many groups of shelly marine organisms). However, it is an important source of bias that needs to be taken into account when a group with a relatively sparse or patchy fossil record (as in Figure 12.18) is considered.

Range charts and spindle diagrams are the basis for much interpretive work, with due allowance made for the problems discussed above, but it should be remembered that they do not supply the reasons for extinction or radiation. Nevertheless, the quantitative analysis of such patterns may permit some explanatory hypotheses to be rejected, and this approach will be explored in Chapter 15.

Range charts and spindle diagrams frequently also reveal biases in the fossil record. Ranges may have gaps showing time periods which lack fossils, and spindles may show rapid co-incident increases in diversity followed by equally rapid decreases over small time periods.

▶ How might the latter pattern arise as a consequence of preservational bias?

It may arise when the time period being examined includes a conservation Lagerstätte.

It should be remembered also that range charts and spindle diagrams are statements of the current state of knowledge. Ranges might be—and often are—extended in both directions as a result of new discoveries. As such, range charts are usually more reliable for organisms with high preservation potential, than they are for organisms with low preservation potential.

SUMMARY OF CHAPTER 12

Taphonomy is the study of the processes that may affect the dead or shed remains of an organism (or traces of its activity) and perhaps lead to eventual fossilization. Pre-burial processes include predation, scavenging, decomposition, transport and physical and/or chemical degradation: these are generally of destructive effect, though in some instances may promote preservation. Following eventual burial, further disruption may be caused by bioturbation. Diagenetic mineral growth, in contrast, frequently assists fossilization.

Fossil assemblages may contain elements that are preserved in life position (autochthonous), disturbed post-mortem but still at the original site of life (parautochthonous), or brought in from elsewhere (allochthonous), either from other contemporaneous environments (exotic fossils) or reworked from older strata (derived fossils). The status of a specimen in this respect is affected both by the life habits of the original organism and by taphonomic factors. The environment(s) in which the original organisms lived may be inferred from sedimentological study. The original environmental conditions, especially the nature of the substratum and the pattern of sediment accumulation, strongly influence the composition and completeness of any associated fossil assemblages. Fossil reefs and hardgrounds, for example, commonly offer a good record of their sessile benthic biota. Some ecological relationships between organisms have preservation potential (e.g. shell-drilling predation, competition for space among sessile organisms and some symbioses), and so evolutionary changes in these can be monitored. Fossil assemblages are usually highly incomplete records of original communities, though many are also time-averaged mixtures of specimens from successive populations which may never have overlapped in life. Mass mortality assemblages provide particularly valuable insights on contemporaneous populations.

Fossil-Lagerstätten are exceptional fossil deposits. Suppression of destructive taphonomic processes and acceleration of diagenetic mineralization can result in conservation Lagerstätten, in which even soft tissues may be fossilized. These deposits provide privileged insights into the biology of past organisms and their diversity. Concentration Lagerstätten furnish abundant fossils, which can be useful for morphometric analysis, especially if resulting from mass mortality.

Apart from preservational biases, the record is also subject to biased collection and study. Estimates of the distribution of taxa in time can be represented as stratigraphical ranges on range charts, or on spindle diagrams, though errors may arise from the biases in the record and in the methods of representing it.

12.5 CONCLUSION

This Chapter has examined the way organisms can be incorporated into the fossil record, and it has shown that the processes involved select certain organisms, or certain parts of organisms, and operate with varying effect in different environments and at different periods of the Earth's history. Notwithstanding the selectivity of fossilization, it is also clear that parts of the fossil record are able to provide abundant biological information, and in some cases the quality of preservation is such that even soft tissue anatomy can be examined. The key to the storehouse of information potentially available from the fossil record is therefore a thorough grasp of taphonomy.

Exceptionally well preserved fossils offer the greatest opportunity for palaeontologists to reconstruct the biology of extinct organisms, though even isolated fossil bones and shells can contribute significantly to our knowledge of, say, the geographical and temporal distributions of taxa. Moreover, although the fossil record contains numerous gaps, it is possible to use these gaps themselves to determine the completeness of the fossil record so that the error on ranges can be assessed, on at least a semi-quantitative basis. Progress in evolutionary palaeontology thus relies upon asking sensible questions, tailored to the strengths of the fossil record, and avoiding speculations for which it can offer no test.

REFERENCES FOR CHAPTER 12

Dudley, E. C. and Vermeij, G. J. (1978) Predation in time and space: drilling in the gastropod *Turritella*, *Paleobiology*, **4**, 436–441.

Golenberg, E. M., Giannasi, D. E., Clegg, M.T., Smiley, C. J., Durbin, M., Henderson, D. and Zurawski, G. (1990) Chloroplast DNA sequence from a Miocene *Magnolia* species, *Nature*, **344**, 656–658.

Harper, E. M. (1991) The role of predation in the evolution of cementation in bivalves, *Palaeontology*, **34**, 455–460.

Jefferies, R. P. S. (1990) The solute *Dendrocystoides scoticus* from the Upper Ordovician of Scotland and the ancestry of chordates and echinoderms, *Palaeontology*, **33**, 631–679.

Palmer, T. J. (1982) Cambrian to Cretaceous changes in hardground communities, *Lethaia*, **15**, 309–323.

Unwin, D. M. (1988) Extinction and survival in birds, in G. P. Larwood (ed.) *Extinction and Survival in the Fossil Record, Systematics Association Special Volume*, **34**, 295–318, Clarendon Press, Oxford.

Valentine, J. W. (1989) How good was the fossil record? Clues from the Californian Pleistocene, *Paleobiology*, **15**, 83–94.

Vermeij, G. J. (1977) The Mesozoic marine revolution: evidence from snails, predators and grazers, *Paleobiology*, **3**, 245–258.

FURTHER READING FOR CHAPTER 12

Allison, P. A. and Briggs, D. E. G. (eds) (1991) Taphonomy: releasing the data locked in the fossil record, *Topics in Geobiology*, vol. 9, Plenum Press, New York.

Briggs, D. E. G. and Crowther, P. R. (eds) (1990) *Palaeobiology: A Synthesis*, Blackwell Scientific Publications, Oxford.

Donovan, S. K. (ed.) (1991) *Fossilization: the Process of Taphonomy*, Belhaven Press, London.

Goldring, R. (1991) *Fossils in the Field—Information Potential and Analysis*, Longman Scientific and Technical, Harlow, UK and John Wiley & Sons, New York, USA.

The Open University (1983) *S236: Geology, Block 4 Surface Processes,* and *Block 5 Fossils.*

Martill, D. M. (1988) Preservation of fishes in the Cretaceous of Brazil, *Palaeontology*, **31**, 1–18.

Storch, G. (1992) The mammals of island Europe, *Scientific American*, February 1992, pp. 42–47.

Taylor, J. D. (1981) The evolution of predators in the late Cretaceous and their ecological significance, in P. L. Forey (ed.) *The Evolving Biosphere*, 229–240, British Museum (Natural History) and Cambridge University Press, London.

Whittington, H. B. and Conway Morris, S. (eds) (1985) Extraordinary fossil biotas: their ecology and evolutionary significance, *Philosophical Transactions of the Royal Society of London*, 192 pp.

SELF-ASSESSMENT QUESTIONS

SAQ 12.1 (*Objectives 12.1, 12.2 and 12.3*) Below are listed the main constituents of five fossil assemblages ((a)–(e)), four taphonomic factors (A–D), and four categories of fossil (1–4). For each of the assemblages, select the single most likely taphonomic factor to have led to their preservation, and the appropriate category in which the fossils should be classified.

Assemblages

(a) Open burrow systems preserved in a bed of limestone, with an encrusting fauna lining their interior walls, and borings cutting across them.

(b) Abundant intact skeletons of pelagic fish.

(c) Worn mammoths' teeth in a Recent beach gravel.

(d) An assortment of trace fossils reflecting frequent vertical migration of burrowing organisms through sediment.

(e) Intact skeletons of vagile marine invertebrates, composed of many pieces, which lived on the surface of offshore muds.

Taphonomic factors

A Anoxic bottom water.

B Vigorous stirring of shallow waters by seasonal storms.

C Highly condensed sediment accumulation in warm seawater in an arid climate.

D Low levels of predation.

Fossil categories

1 Autochthonous

2 Parautochthonous to autochthonous

3 Exotic allochthonous

4 Derived allochthonous

SAQ 12.2 (*Objectives 12.4 and 12.5*) Suppose that a palaeontologist decides to investigate changes through time in the relative frequency of mortality in mussels due to predation by shell-drilling gastropods. What preservational biases would need to be taken into account, if the study were based upon either of the following kinds of fossil assemblage?

(a) Assemblages of shells buried beneath thick beds of sediment rapidly introduced during storms.

(b) Assemblages of parautochthonous shells formed by long-term accumulation in areas with low net rates of sedimentation.

SAQ 12.3 (*Objectives 12.6 and 12.7*) (a) From the range chart in Figure 12.18, plot a spindle diagram for the Cainozoic families of birds, with a horizontal scale of 20 families per cm (i.e. one-tenth of that used in the spindle diagram in Figure 12.18).

(b) What is the most likely explanation for the sharpest increase in diversity shown in the spindle diagram, and from where might further information be sought to test your interpretation?

CHAPTER 13
GEOGRAPHY AND MACROEVOLUTION

Prepared for the Course Team by Charles Turner

Study Comment *This Chapter considers how geographical factors, both present and past, influence or have influenced the pattern of macroevolution. It takes a broader view of geological processes than that discussed in Chapter 12, considering changes that affect the fate not just of individuals or particular species but of whole ecosystems and the faunal and floral assemblages they contain and support.*

Objectives When you have finished reading this Chapter, you should be able to do the following:

13.1 Discuss hypotheses concerning latitudinal and environmental variation in species diversity.

13.2 Describe, giving examples, how the distribution of taxa may be limited by ecological factors or by different kinds of geographical barriers, giving rise to the concept of biogeographical regions.

13.3 Discuss critically the differences of approach between dispersal and vicariance biogeography.

13.4 From suitable data, construct an area cladogram for a group of organisms.

13.5 Explain how continental drift and other plate tectonic processes have affected the diversity, composition and course of evolution of both marine and continental biota.

13.6 Discuss the processes of migration, replacement and extinction of terrestrial faunas as a result of plate movements.

13.7 From suitable data, calculate S, the Simpson coefficient of similarity.

13.8 Explain, giving examples, how climatic and sea-level changes during the Quaternary have affected the distribution patterns of fauna and flora, and discuss the possible evolutionary effects of such changes.

13.1 INTRODUCTION

Chapters 10 and 12 have emphasized the importance of the fossil record in evolutionary studies, at both the species and macroevolutionary levels. But clearly, evolutionary processes must be considered in a framework not only of time but also of space. This is underlined by the importance that many evolutionary biologists give to allopatric models of speciation. In fact, both the present geography of the Earth's surface and past geographical changes have been important influences on patterns of evolution.

At a macroevolutionary level, two important issues need to be considered. The first relates to the patterns of species diversity found in different ecosystems and particularly at different latitudes, both on land and in the oceans. A great deal of debate surrounds factors which might encourage or discourage species diversity in any given ecosystem.

The second issue concerns geographical distribution patterns, not only of related species, but also of groups of higher taxa, which are presumed to have a common ancestor. An example of this kind of problem is illustrated by the distribution of flightless birds, the group known collectively as ratites, which has isolated species in many parts of the Southern Hemisphere. Several of these birds have already been discussed in Section 2.2.2 of Chapter 2. The ostrich *Struthio camelus* (Figure 2.4(b)) is, or rather was, widespread as a fast-running bird of open country in Africa and formerly even spread into the adjacent deserts of Arabia and Syria. In South America, the same kind of habitat is occupied by similar large birds, the rheas, of which two species survive. In Australia, there are emus *Dromaius novaehollandiae* (Figure 2.4(c)) and cassowaries; and in New Zealand kiwis *Apteryx owenii* (Figure 2.4(d)). To these can also be added species extinct within historical time—the ostrich-like *Aepyornis* and the smaller *Mullerornis* (the elephant birds from Madagascar), and, from New Zealand again, the even larger moas *Dinornis*, up to 3 m in height (Figure 2.4(a)), as well as several smaller relatives, all now lost. Structural features suggest that these genera are descended from a common, flightless ancestor. How, therefore, did birds without the power of flight become so widely distributed in areas separated by vast expanses of ocean?

Investigations of such questions can only be attempted, though not always with success, with a firm understanding both of patterns and of processes. Biological processes, already discussed in earlier Chapters, are important, but in this Chapter the *geological* processes which lead to palaeoclimatic and palaeogeographical change are shown to form an indispensable background to any evolutionary analysis. Hypotheses resulting from biogeographical data can also be subjected to more rigorous analysis, along the lines of the cladistic analyses described in Chapter 11.

13.1.1 BIOGEOGRAPHY

Biogeography is the study of geographical patterns in the distribution of organisms. There are two main branches of modern biogeographical studies—**ecological biogeography** and **historical biogeography**. Ecological biogeographers seek to explain present-day distribution patterns in terms of environmental factors

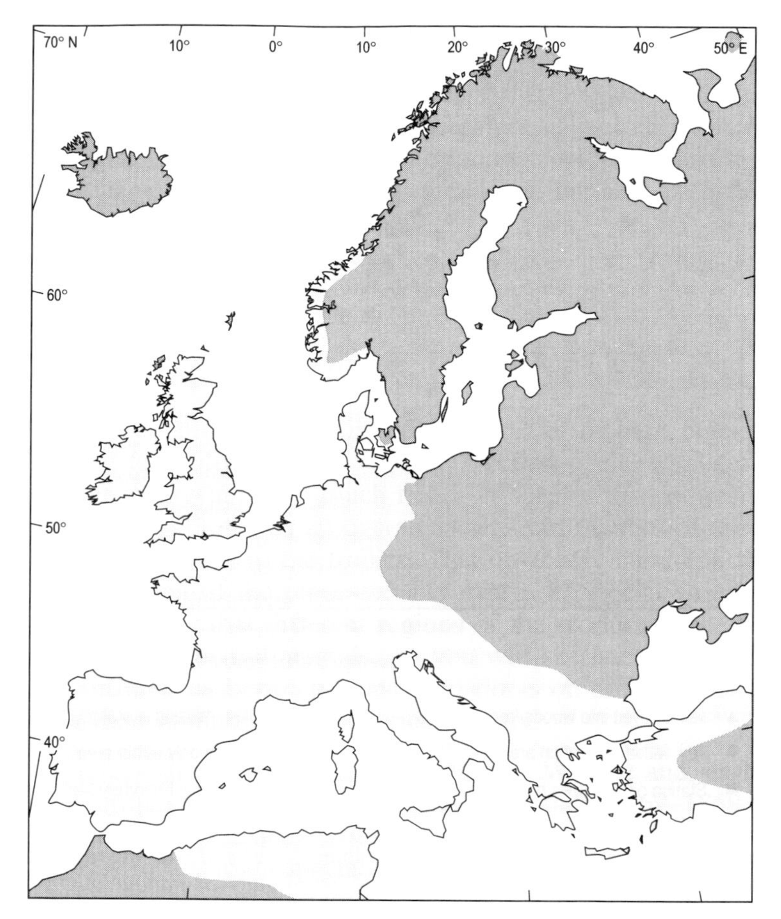

Figure 13.1 *Distribution of the holly* Ilex aquifolium *in Europe.*

and interactions with other species. Environmental factors include a variety of climatic parameters and, particularly for plants, soil conditions. Competition between species is important (Chapter 8, Section 8.2), especially where the presence of one species leads to the exclusion of another. For animals, foodplant and predator–prey relationships may also control or restrict their distribution (Chapter 8, Section 8.4).

Indeed, ecological biogeographers are able to use observational (including experimental) data to identify limiting factors, or more often combinations of such factors, that may determine boundaries of species distributions. The holly *Ilex aquifolium* is at the north-eastern edge of its distribution range in Denmark (Figure 13.1), and it was noted that in some areas of the country this shrub suffered much damage during the harsh winters of 1939–42. Figure 13.2 records data on the average temperature of the warmest and coldest months for 58 meteorological stations in Denmark. A survey was also carried out to see whether holly was growing within an area of 20 km radius around each site and within an altitudinal range of 40 m.

▶ What can you conclude about the plant's climatic tolerances from these data?

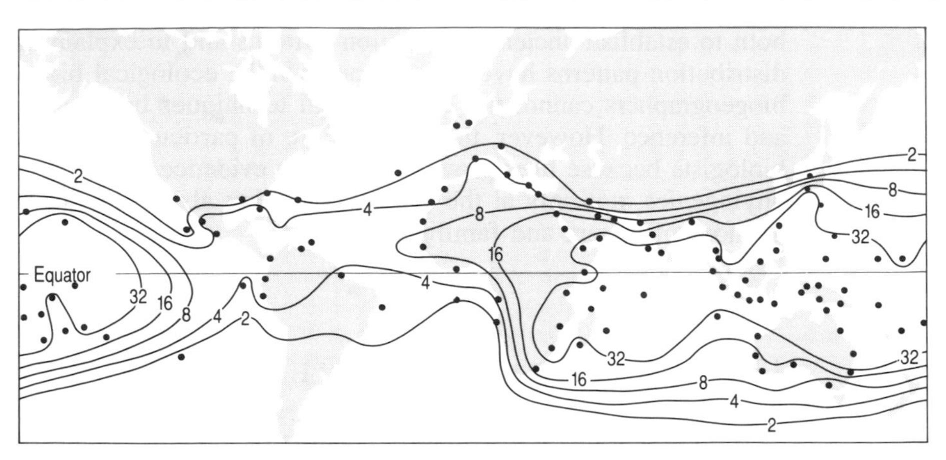

Figure 13.4 *Distribution of living cowrie genera, showing greatest diversity on tropical continental shelves. Note that the contours have been extrapolated across deep oceans and land, where no cowries live.*

This trend is also evident for whole ecosystems, such as grasslands or forests. There are, of course, exceptions to this generalization. Coniferous trees, for example, are both more abundant and more diverse in the forests of the temperate zones than in tropical forests. Other groups have quite specialized ecological preferences or restricted geographical ranges. Thus, all but one isolated penguin species (on the Galápagos) prefer the subpolar southern oceans, while the Cactaceae (cacti family) are virtually restricted as native plants to North and South America, but have a diversity of species which are adapted to environments characterized by a shortage of water—semi-desert areas of the continental interior, coastal dunes and the high, dry mountain slopes of the Andes. Nevertheless, the sheer numbers of species in tropical environments covering the whole spectrum of the plant and animal kingdoms ensures that the phenomenon cannot be ignored or denied.

The diversity of terrestrial organisms is much greater than that of aquatic ones: only about 20% of named species occur in the oceans. This may in part be explained by the virtual absence from the oceans of representatives of two very diverse groups—the insects and the flowering plants—but it should be remembered that the oceans remain poorly explored biologically and undoubtedly contain very many undescribed taxa.

Many hypotheses and arguments have been put forward to explain the occurrence of latitudinal diversity gradients and this very high species diversity of tropical areas. The problem can be summarized by two questions.

1 Which factors explain why more species can live together within an ecosystem in certain areas than in others, and particularly in tropical as compared to temperate ecosystems?

2 Why do tropical areas apparently encourage speciation or alternatively accumulate more species (i.e. by immigration of species that have originated elsewhere and/or through lower rates of extinction) than non-tropical areas?

Similar questions can, of course, be asked about latitudinal diversity gradients across polar and temperate areas and about any particularly species-rich ecosystem.

Note that the first question relates to biotic and environmental factors, whereas the second is concerned with historical factors and processes of speciation. A few of the arguments and hypotheses that have been used to illuminate, if not solve, these questions are given below:

(a) Environmental stability It is generally suggested that stable environments are likely to have a more constant and dependable supply of resources than unstable environments and, therefore, they are able to support a greater diversity of species of plants and animals. Stability of an environment is a difficult concept to define, but the main feature is predictability. Important factors might be that vital variables, such as temperature, rainfall or salinity, show few fluctuations and that catastrophic events, such as fires or hurricanes, are rare. In tropical environments, the relatively even daylength throughout the year might also be significant.

Terrestrial ecosystems where apparently stable environments do occur, for example tropical forests, are usually highly partitioned, with organisms occupying narrow niches (Chapter 8, Section 8.3), and depending on a regular and predictable food source. Such ecosystems contain a large number of specialist species which frequently implies a long history of co-evolution between plants and animals, prey and predator, host and parasite.

Moving from the tropics towards higher latitudes, particularly the polar regions, and from coastlines into the drier continental interiors, conditions become more unstable because of seasonal and diurnal temperature fluctuations, floods, droughts, etc. These latitudinal gradients are also paralleled to some extent by altitudinal gradients, so that alpine environments show many of the features of arctic ones. At various latitudes there are also particular environments where diversity is low because of extreme physical conditions of heat, cold, aridity or high salinity. Where there are strong environmental fluctuations, the strategy which organisms are most likely to adopt will be a generalized one, for example feeding from a variety of sources rather than becoming specialized to tackle only one particular source. They are then described as having broad niches. A further characteristic is that in stressful environments, such as deserts (hot or cold), high mountains or arctic tundra, most organisms develop special morphological or physiological adaptations which enable them to survive (Chapter 8, Section 8.5.3). In these rather extreme environments, individuals are seldom abundant and productivity is low. Another adaptation to such conditions is behavioural: many animals migrate seasonally to avoid the harshest conditions.

To summarize, high diversity in terrestrial ecosystems does seem to be associated with environmentally stable, i.e. relatively stress-free, ecosystems. It is also associated with ecosystems in which niches are narrow rather than broad.

In the oceans, conditions are more complex. The upper 200 m of the oceans in temperate zones are also relatively unstable, undergoing considerable annual temperature fluctuations; this layer warms up in summer and becomes isolated from the colder deeper water by the formation of a seasonal thermocline (an abrupt zone of temperature change). In autumn, the thermocline gradually breaks down, and upper and lower water masses mix. In tropical oceans, there is usually a permanent thermocline, and deep waters in most parts of the oceans are generally held to be cold, unvarying environments, with a few significant exceptions.

▶ The abyssal plains in the deep oceans are generally very stable environments. Would you expect organisms there to occupy narrow niches?

No, these are also very stressful environments. Resources in the form of biomass and nutrients are generally very scarce in the deep oceans because productivity is low, with a rare exception in the vicinity of submarine volcanic vents. In addition to the problem of low resources, organisms inhabiting the ocean depths must adapt to the physiological problems of survival in an inhospitable, cold, dark, high-pressure environment. They have to adopt strategies for obtaining nourishment from any source they can. These are, however, environments which are still poorly explored and understood, because of their inaccessibility.

In other stressful environments, such as estuaries, where organisms must be able to tolerate rapid changes of salinity, there is a tendency for a relatively small number of species to occur in great abundance, so that productivity is high but diversity low.

In general, within the oceans, availability of nutrients is a more important factor than stability and, as you will learn, the two factors are not necessarily correlated.

(b) Productivity Productivity is strongly linked to the availability, or, often indeed, abundance, of a range of essential resources. Primary production, which determines the amount of biomass resulting from photosynthesis, is the basis for the food-webs of all terrestrial and almost all marine ecosystems. On the continents, moving from the poles to the tropics, the major terrestrial ecosystems exhibit a general increase in primary production, except where this is limited by moisture deficit, as in subtropical desert belts like the Sahara. This latitudinal increase in primary production can be ascribed to the availability of light at suitable temperatures throughout the year. The biomass produced is then available as a resource for primary consumers and ultimately for organisms belonging to other trophic levels in the ecosystem. However, these observations do not by themselves explain why high primary production should be associated with a high diversity of plant species rather than with large populations of individuals of just a few species which might produce as much biomass.

Productivity may also be limited in some terrestrial ecosystems by scarcity of essential inorganic nutrients. There is no particular geographical distribution of nutrient availability, and a latitudinal gradient is anyway improbable, except that in tundra ecosystems, under very cold climatic conditions, nutrient cycling is very slow, and nutrients locked up in permafrost soils remain unavailable. In temperate latitudes, there is a general tendency for forests growing on nutrient-rich brown-earth soils to be richer in species than those growing on nutrient-poor acid podsols. However, in some highly productive ecosystems, particularly outside the tropics, one or a small number of species may achieve a dominant or subdominant status which actually appears to discourage diversity. An example which may be familiar, even though the environment concerned is only semi-natural and ultimately owes its origin to human intervention, is that of the species diversity of grasslands in north-west Europe. On rich, fertile soils, these may support a limited number of very productive grass species which tend to out-compete most other herbs; whereas in chalk and limestone areas, where soils are thin and certain important nutrients are in short supply, a much wider diversity of herbs may occur, as well as a range of grass species.

In the oceans, productivity appears, in general, to be limited not by light or temperature, so much as by the availability of nutrients. Nitrogen, phosphorus, sulphur and iron are all critical and their scarcity may limit the abundance of marine organisms. Tropical waters are often very low in essential nutrients, particularly nitrogen and phosphorus, but despite this, they do contain 'oases' of abundant and diverse life in the form of coral reefs, which will be discussed shortly. Polar waters, such as those surrounding Antarctica, might at first seem unlikely to be productive, but in fact they are characterized by very active mixing of water masses, as the surface freezing of water to form pack-ice results in the production of dense, highly saline water. This sinks and so causes an inflow of surface and intermediate-depth water from lower latitudes. The mixing of these water masses makes available a rich supply of nutrients, which supports an abundance of plankton and small crustaceans (krill) as well as a quite diverse benthic fauna.

A further instance of high productivity in the oceans is found where upwelling of nutrient-rich deep ocean waters takes place, for example off the west coasts of South America and Namibia (south-west Africa). These upwelling zones support huge quantities of phytoplankton, zooplankton and fish, and are an important economic resource. Great hardship is caused in Peru and Chile when this upwelling is disrupted by current reversals occurring every 2–10 years— the so-called 'El Niño' effect—and the supply of nutrients fails. Another unexpected area of productivity and also high diversity is around volcanic vents on the deep ocean floor, found particularly in the vicinity of mid-ocean ridges; these may, however, also be unstable environments, with the vents switching on and off unpredictably.

In terrestrial ecosystems, then, there seems to be some link between latitudinal diversity and productivity, but it is not straightforward because monocultures, such as crops, may also be very productive. In the oceans, some environments, not necessarily tropical ones, are highly productive, depending on their nutrient supply, but they are often unstable and tend to be dominated by large numbers of particular species.

(c) Complexity of habitats Some environments, such as grassy plains or sandy shores, may be extremely uniform, whilst others, like mountain slopes or rocky coastlines, support many different microhabitats and consequently a much greater number of plant and animal species. The physical structure of living communities is also important. In some kinds of temperate woodland, the vegetation is stratified into canopy (tree), shrub, herb and ground layer communities, each accompanied by its dependent fauna. This stratification of course greatly increases species diversity, and the structure of some tropical forests with a three-layered canopy, climbing plants (lianes) and epiphytes is even more complex. Nevertheless, some forest types, such as conifer and beech forest, are by contrast species-poor because deep shade largely excludes any sub-canopy vegetation layers.

The results of a recent detailed analysis of the distribution of North American trees can be presented as a contour map of tree species richness between latitudes 70 °N and 25 °N (Figure 13.5). Only in eastern North America is there any clear

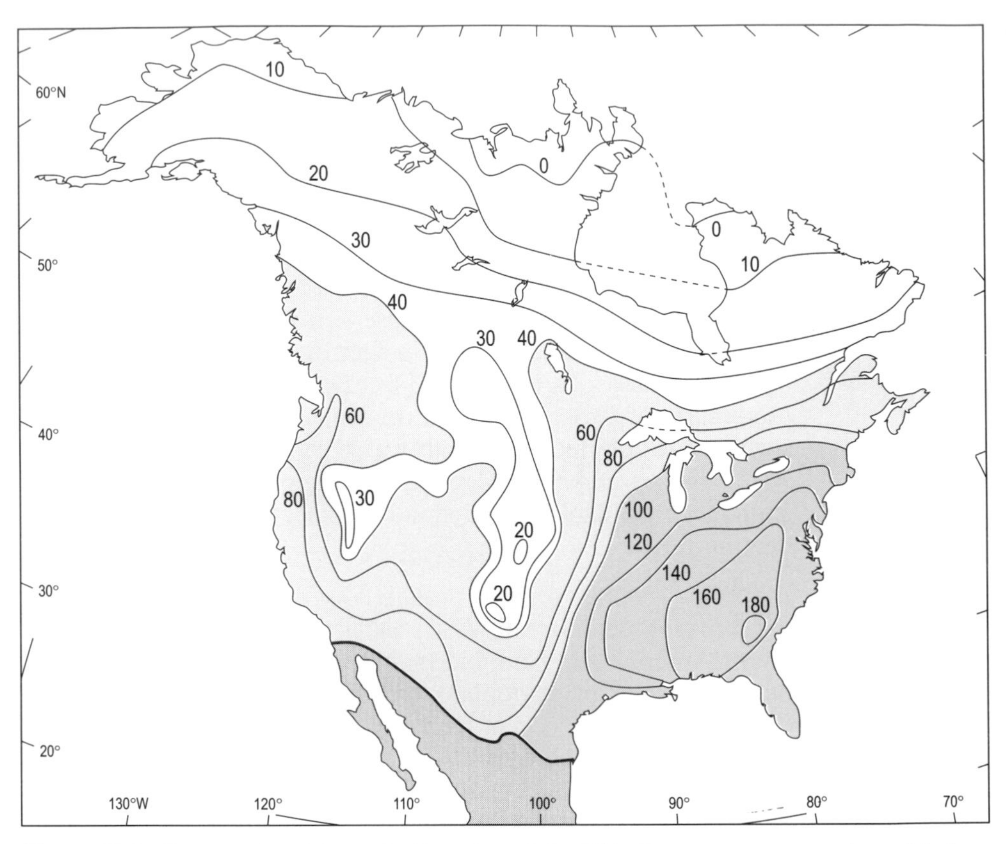

Figure 13.5 *Tree species richness in Canada and the USA. Contours connect points with the same approximate number of species per quadrat based on a regular grid.*

latitudinal gradient. Maximum richness occurs in the south-eastern United States and minima occur in the rain-shadow areas immediately to the east of the Rocky and Sierra Nevada mountain ranges. Statistical analysis of these data showed that, although species richness could be correlated positively with a number of geographical and climatic factors, by far the closest correlation was with 'realized annual evapotranspiration'—all water evaporated from the soil or transpired from vegetation, a factor in turn strongly correlated with primary production. The forests of the south-eastern United States are particularly rich in species, which reflects the species occupying narrow niches. In fact, some of the forests in the area, such as the Everglades swamp forest of Florida, are often described as subtropical, though technically they lie outside the tropics. Such complex forests are likely, and indeed do have further species-rich shrub and herb layers. Furthermore, each of the many species of tree is likely to support its own range of closely adapted invertebrate fauna.

Even though tropical latitudes of the oceans are richer in species than temperate latitudes, their productivity and with it their species diversity, appear to be limited over wide areas. There is, however, one kind of ecosystem in tropical oceans which has a conspicuously higher species diversity: *coral reefs* are characterized by a variety of species of colonial corals, whose tissues contain symbiotic dinoflagellates, unicellular algae belonging to the Pyrrophyta. The growth forms of reef-forming corals, like those of trees on land, provide a very varied range of habitats for other organisms including calcareous algae that also contribute to the reef structure, and the symbiotic association of the corals with photosynthesizing algae provides a resource base and nutrient concentration that supports extensive food-webs. Ocean waters around coral reefs are exceptionally low in nitrogen and phosphorus, but the organisms concerned are very efficient at scavenging these nutrients. Coral reef ecosystems are obviously very well and closely adapted to such nutrient-poor environments, because if nutrient levels are raised, even this mild form of pollution is likely to kill off the reefs and their denizens.

(d) Age of the environment Hypotheses and observations grouped under (a) to (c) have chiefly been concerned with biotic and environmental factors. Hypotheses in this category relate to long-term stability of environments on a geological or at least historical time-scale. Some ecologists have suggested that one of the reasons why the tropics show a greater diversity might be that they have remained relatively free from disturbance for very long periods of time whilst higher latitudes were being affected by long periods of cooling during the Quaternary Ice Age (geological ages are given in Appendix 1). Such arguments were largely the result of 'armchair' reasoning, and evidence from actual geological and palaeoecological studies now suggests that even the tropics were affected by repeated fluctuations of climate, particularly rainfall patterns. During the past 2 Ma there were also significant changes in the distribution of tropical forest, savannah and desert. Nevertheless, some tropical environments must be of great antiquity, and it is suggested that, even though particular habitats have locally been disrupted, their types have survived somewhere in the vicinity and been tracked by the organisms adapted to them, as described in Chapter 10, Section 10.3.3. This view is reinforced by the presence of certain organisms, such as cycad trees and coelacanth fish, which are regarded as 'living fossils', and which in some cases tend to be confined to relatively restricted habitats. These organisms are surviving, somewhat modified representatives of taxonomic groups that were important in Mesozoic or even earlier times, but which have now been almost entirely replaced by more derived groups.

Because of the unstable climate during the Quaternary (discussed in Section 13.5), there has been much debate as to whether, setting aside human disturbance, modern plant and animal communities (and their diversity patterns) at any latitude are really in equilibrium with present-day conditions, or whether they are still reflecting past events. An extreme example can show how present-day forest composition has been influenced by past climatic fluctuations. Consider the behaviour of the common pink rhododendron *Rhododendron ponticum* in woodland and heathland in Ireland and other parts of western Britain. It is an introduced species, which is native to a few scattered areas in southern Europe, but which has become an aggressive colonizer, out-competing native tree and shrub species and therefore regarded as a pretty but dangerous weed species by most conservationists. The irony of the matter is that the fossil record from ancient peat deposits in Ireland shows that this same species of *Rhododendron* was a very common shrub there about 400 000 years ago before it was exterminated by one of the Quaternary glacial periods, when most of Ireland was ice-covered. The extinction of *Rhododendron* (as demonstrated by its reintroduction) must have profoundly altered the composition and character of some natural ecosystems in Ireland.

Summary of Section 13.2

On the continents, there is a gradient of increasing species diversity from high latitudes to the tropics. This is true both for ecosystems, provided you compare like with like (i.e. forests at different latitudes or grasslands at different latitudes), and also for many taxonomic groups. Some groups, though, may not follow this pattern. High species diversity tends to be associated with stable environments, narrow niches, structural complexity of the ecosystems, and sometimes, but not always, an abundance of nutrients, which may be very tightly recycled. Hypotheses attempt to explain predominantly tropical high diversity patterns, or the contrary, by reference to environmental stability on either short or long time-scales, or to conditions of high or low productivity, but often these hypotheses rest on correlation rather than explanation.

Diversity in the oceans is much less obviously latitudinal, but is associated more with nutrient availability. However, the diversity of predominantly tropical coral reefs, which occur in nutrient-depleted waters, bears some comparison with tropical forests, both being structurally complex ecosystems.

The problem with virtually all these hypotheses about species diversity is that, by careful selection, convincing observations can be brought forward both to support or to reject them. They are very difficult to test.

13.3 Analysis of biogeographical patterns

It has been observed that distinctive suites of plants and animals characterize different regions of the world. On the basis of the distribution patterns of mammals and birds, the land surface of the world has been divided into six **biogeographical regions** (see Figure 13.6), a scheme first proposed by the 19th century ornithologist Philip Sclater and later modified by Alfred Russel Wallace.

As an example, compared with other regions and particularly with the adjacent Oriental region the Australian region is highly distinctive in its terrestrial vertebrate fauna, with a high proportion of unique forms, particularly of marsupial mammals—kangaroos, wallabies, koala, opossums—and also the two monotremes, the duck-billed platypus and the echidna. The only native placental mammals are rodents and bats. Whereas these placental mammals show affinities with the fauna of the Oriental region, the closest affinities of the marsupials are with marsupials in the Neotropical region.

Although this subdivision into regions was originally based on the distributions of mammals and birds, it has subsequently been shown to be applicable to many other groups of animals and also plants, so that the distinctiveness of biogeographical regions is a genuine and fundamental feature requiring explanation. The flora of the Australian region again has a very large number of genera and families that occur nowhere else. It also shows interesting links with the Neotropical region as well as with the rich and distinctive flora of the Cape floral province of Southern Africa.

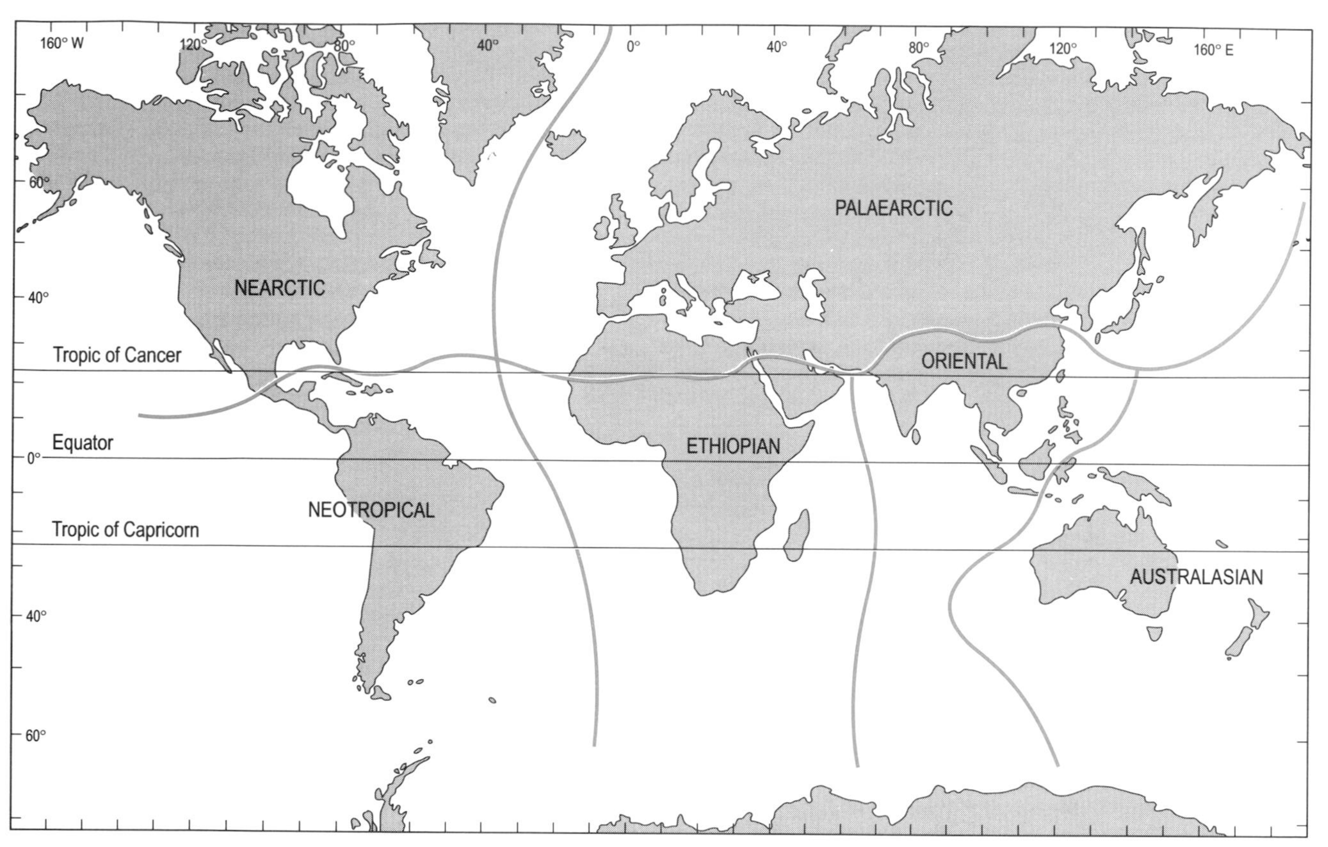

Figure 13.6 *The world divided into six biogeographical regions.*

The boundaries between biogeographical regions are by no means precise, so that different lines have been drawn by specialists working on different groups of organisms. As an example, the boundary between the Australian and Oriental regions lies within the archipelago of the East Indies. Wallace defined a line which separated islands having bird faunas of Asian affinities to the west and Australian affinities to the east (Figure 13.7). However, Weber, who studied the distribution of molluscs and mammals, placed the boundary much farther east because he found that islands such as Sulawesi and Timor possessed mammalian faunas with mostly Asian species or their close relatives (rhinoceros, monkeys, deer, pigs, tarsiers and porcupines), although a few marsupials such as phalangers also occur. The area between **Wallace's line** and **Weber's line** is often referred to as 'Wallacea'.

Examine closely the boundaries of the biogeographical regions (Figure 13.6), particularly on land. They do not, of course, necessarily coincide with the conventional boundaries of the continents.

▶ How would you describe the position of the northern boundary of the Ethiopian region?

The boundary is set near the northern edge of the Saharan and Arabian deserts, but note that the Atlas mountains of Morocco are excluded and form part of the Palaearctic region.

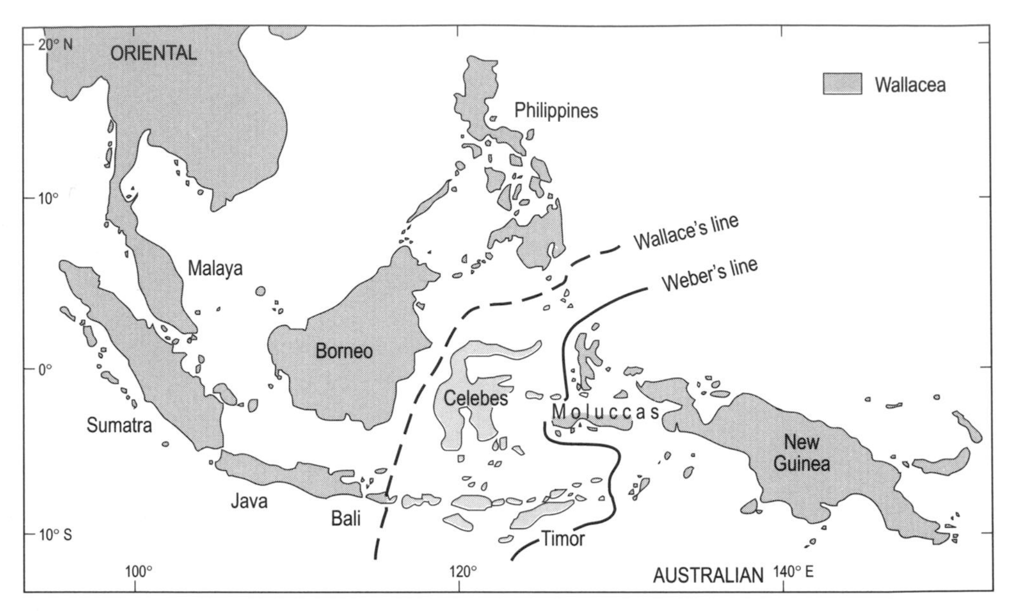

Figure 13.7 *South-East Asia and the East Indies, showing Wallace's and Weber's lines and 'Wallacea'.*

Those deserts are chiefly occupied by plants and animals adapted to otherwise very stressful environmental conditions—low rainfall, rapid and extreme diurnal temperature changes and exposure to strong winds. Such a topographic and climatic belt forms an almost insuperable barrier to dispersal or contact between the more equable ecosystems of the Palaearctic and sub-Saharan Africa.

Other barriers between biogeographical regions (in addition to the wide expanses of ocean) are the huge mountain mass of the Himalayas which effectively limits north–south faunal movements between the Palaearctic and Oriental regions; to some extent, the narrow mountainous isthmus between North and South America; and narrow seaways such as the Bering Strait between Siberia and Alaska, and the straits between the islands of the East Indies.

Such barriers are neither absolute nor unchanging. Land-bridges (or alternatively connecting seaways) may form or disappear during the course of geological time, as will be discussed later in this Chapter. When these bridges permit free migration of fauna and flora in both directions, biogeographers refer to them as **corridors**, but if they act selectively then they are referred to as **filters**. The Bering Strait is an example of a filter. Because of falls in sea-level, a land-bridge has formed between Siberia and Alaska at several periods during the past 3 Ma, but because of the high latitude of this land-bridge, only cold-tolerant species of plants and animals have been able to migrate from Asia to North America or vice versa. Sometimes, individuals succeed in crossing formidable barriers, primarily because of chance events, such as land birds or insects being rafted or blown across the oceans to isolated islands such as the Galápagos (Chapter 1, Section 1.3.1). This is often called **sweepstake dispersal**. Good dispersal mechanisms favour such events.

On a smaller scale, for faunas as well as floras, biogeographical regions may themselves be subdivided into smaller areas—**provinces**—which possess a large proportion of species not found in other regions. For example, the marine molluscan faunas on the coastal shelves of Scandinavia, the British Isles, the Atlantic coast of Spain and Portugal, and the Mediterranean each contain their own distinctive taxa, although there are also many very widespread species. The boundaries between provinces may be sharp or gradual and, indeed, sometimes vary for different faunal groups. Where fossil faunas, particularly of marine invertebrates, have been very well studied over a wide area, it is possible to recognize faunal provinces in the fossil record, which may have interesting implications for both palaeogeography and evolution.

13.3.1 Distribution patterns

The distribution not just of species, but also of genera and families as a whole, can yield important information to evolutionary biologists. Very few species of plants and animals have truly world-wide or **cosmopolitan** distributions through all the biogeographical regions. One that does, however, is the human species, together with certain companions such as dogs, rats, mice and lice! A few plants, like the bracken fern *Pteridium aquilinum* and the grass, common reed *Phragmites australis*, which possess very light, windborne spores or seeds, also have very wide distributions. More interesting to the evolutionary biologist are those taxa which have **disjunct distributions** and occur in widely separated areas (e.g. the tapirs, of which three species occur in South America and one in Malaysia; or the camellia-like shrubs *Stewartia* (Figure 13.8) with two species in eastern North America, three in Japan, two in China and another in Korea). **Endemic species** (or genera or families) are those which are restricted to a single (large or small) area (i.e. a species may be endemic to an island or mountain range, a family to a continent). Although most parts of the world possess some endemic species—a British example being the Scottish primrose *Primula scotica* which occurs only in north-east Scotland and the Orkney Islands—they are particularly characteristic of

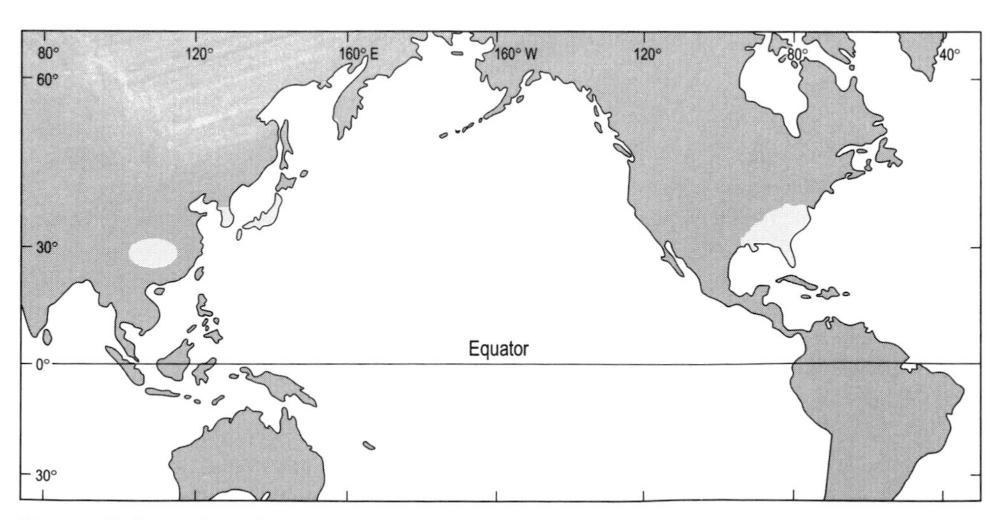

Figure 13.8 *Modern distribution of the genus* Stewartia.

the biota of oceanic islands. The diversity of 'picture-wing' species of *Drosophila* from the Hawaiian archipelago and cichlid fishes in East African lakes have already been discussed in Chapter 9, Section 9.4.4.

Examples of terrestrial molluscs found on islands in the Atlantic can be used to illustrate very different kinds of endemism. On the extremely isolated and geologically comparatively young Tristan da Cunha archipelago in the South Atlantic occur six species of snails assigned to the genus *Tristania*. This genus is defined by slight but significant character differences that distinguish it from a more widespread genus *Balea* which occurs on other Atlantic islands. It is generally believed that chance dispersal of an individual or individuals of *Balea* to the Tristan islands, possibly by adhesion to birds' feet or plumage, was followed by divergence and speciation in the recent geological past. So *Tristania* species are essentially **young endemics**, as are the Hawaiian picture-wing *Drosophila* species. The Azores archipelago in the North Atlantic also has many endemic species in its fauna and flora, including several species of the endemic land snail genus *Lyrodiscus*. However, shells that clearly belonged to very similar species of *Lyrodiscus* have recently been found as fossils in late Tertiary and Quaternary deposits at several sites in Western Europe. This indicates that *Lyrodiscus* is a **relict endemic** genus, that is to say an 'old' taxon which once had a much wider distribution but has now become extinct except in the isolation of the Azores. In very isolated islands or island groups, the proportion of endemic species may be very high indeed. In Hawaii, for instance, 95% of the native plant species are endemic and 98% of the native insects.

13.3.2 Dispersal biogeography

The study of distribution patterns has led to the development of two schools of thought about how they should be interpreted: *dispersal biogeography* and *vicariance biogeography*. The differences between these schools concern not only which processes are most important in explaining distributions, but, particularly with the increasing use of cladistic analysis (Chapter 11, Section 11.2.4), the philosophy of how the interpretation should be approached.

The majority of plants and animals, even where adult organisms are sessile, have a phase within their life cycles when they are able to disperse and have the possibility of colonizing fresh territory. Some species have quite elaborate adaptations (e.g. winged and plumed seeds and fruits in plants) for passive transport by such agencies as wind, water or attachment to other more mobile organisms, while others rely on producing vast quantities of small disseminules (e.g. orchid seeds, fern spores, the planktonic larvae of many marine molluscs and echinoderms) to achieve dispersal. Normally, this does not mean that a species extends its range, but merely that it can occupy all suitable habitats within its range and avoid intraspecific competition between siblings or parents and offspring (e.g. the dispersal of fruits and seeds of trees so that some may germinate beyond the shade of the parent and other trees). Nevertheless, these dispersal mechanisms may permit a species to extend its range by chance occurrences of long-distance transport and they are certainly important when major climatic or environmental changes disrupt existing habitats or when geographical barriers temporarily break down. Examples of such situations will be discussed in Sections 13.4 and 13.5.

▶ In what geographical environments would long-distance dispersal be likely to play a very important role?

Oceanic islands possess faunas and floras whose ancestors are mostly believed to have arrived by long-distance dispersal mechanisms, sometimes passively when carried by wind, water currents, rafting on floating tree trunks, or transport by birds, either internally or externally, or more rarely actively by flying or swimming. Once organisms arrived in these isolated localities, they were able, in the absence of other than intraspecific competition, to diversify and speciate as they occupied specialized niches. There is no other reasonable way by which the origins of taxa such as Darwin's finches on the Galápagos Islands (see Chapter 1, Figure 1.4), the finch-like honeycreepers and the plant group known as silverswords in Hawaii, or *Tristania*—the endemic snails of the Tristan da Cunha archipelago—can be explained, as these isolated volcanic land-masses have never formed part of any major continent.

Most geologists and biologists in the 19th century regarded the distribution of the continents and oceans as fixed. They recognized that many species had adaptations for at least short-distance dispersal and they discovered that certain birds and even insects actually travelled enormous distances during migration flights. Since the study of island faunas and floras by Darwin and others provided convincing evidence that long-distance dispersal was taking place, the concept of dispersal has been used to account for virtually all widespread distribution patterns of organisms.

The central hypothesis of **dispersal biogeography** is that species originate in a particular area or centre and, if successful, spread out from that area to colonize new habitats. However, on a world-wide scale there are many taxa which have obvious relatives, sometimes close, sometimes more distant, in regions that are geographically widely separated and with intervening natural barriers such as oceans, deserts or mountain chains. Taxa with strongly disjunct distributions, such as *Stewartia* (Figure 13.8), are only the more extreme examples. Most of the common tree genera of Europe (including the British Isles)—oak, elm, ash, beech, lime, hornbeam, pine—are represented by closely related species in both North America and Eastern Asia, including Japan. Likewise many mammals, such as brown bear, elk (moose), reindeer (caribou), wolf, fox, beaver, are represented by the same or very closely related species in Northern Europe, Siberia and North America.

Under the dispersal hypothesis, such distributions are the result of successful dispersal, involving extension of range from a centre of origin, the successful crossing of barriers and, in many cases, diversification as a result of adaptation to new habitats.

▶ If a species successfully colonized territory beyond a natural barrier, several processes described in Chapter 9 might occur, such as the founder effect or genetic drift. What would be the net result of such processes?

A population might develop that was reproductively isolated from its parent population. This would then be an opportunity for allopatric speciation to take place.

Given this conceptual framework, many dispersal biogeographers have attempted to determine **centres of origin** or dispersion for different genera and families and to trace their subsequent dispersal pathways over geological time. The evidence selected to identify such centres has been varied and sometimes contentious. Amongst the different criteria that have been considered important are:

1 The area has yielded the oldest known fossils of the group.

2 The area contains a predominance of geologically young taxa.

3 The area has yielded fossils that possess characters regarded as being primitive.

4 The area contains the greatest diversity of living species.

5 The area contains those living species that appear to possess the most primitive characters.

6 The area contains those living species that appear to possess the most advanced characters.

The last two criteria are obviously contradictory, but some biogeographers hold the view that you would expect to find the most highly evolved species at the margins of the dispersal area, because there they have adapted to new and probably different habitats; whilst others believe that it is the centre of origin itself which fuels the dispersal process by the evolution of ever more-advanced forms, so that less competitive, more primitive forms are then forced to move away. You will not be surprised, therefore, that the general approach of dispersal geography is frequently criticized for simply proposing pragmatic or *ad hoc* explanations for dispersal patterns of different individual taxa without any underlying consistency.

13.3.3 VICARIANCE BIOGEOGRAPHY

Vicariant species are closely related species which occupy different geographical or ecological areas. This concept overlaps with that of a disjunct distribution, so that the two *Stewartia* species are vicariants and the genus shows a disjunct distribution pattern. However, the term **vicariance biogeography** has developed a more specialized meaning. There are two main ways in which a disjunct distribution pattern might have arisen. The first, already noted, is by long-range dispersal. The second possibility is that the populations were once continuous, and that disjunction is the result of the disappearance of the taxon from the intervening areas and/or the subsequent creation of natural barriers, which now prevent any possibility of the two populations mixing.

▶ What might be the long-term evolutionary effect of the subdivision of a population in this way?

Once again, this would provide an opportunity for allopatric speciation.

Vicariance biogeographers, whilst accepting that dispersal can take place, believe that the successful crossing of barriers is a rare event and that most species have evolved *in situ* rather than as a consequence of dispersal. Thus, disjunct distributions are the result of reduction rather than extension of range. An example, where it is difficult to decide which is in fact the most likely explanation, will serve to illustrate the differences between these two interpretations.

Gars or garfishes (Family Lepisosteidae) are a group of distinctive long-snouted, carnivorous fishes that occur in fresh and brackish waters in Central and North America. Fossil gars have a wider distribution, being found in Cretaceous rocks not only in North America, but also in Europe, Africa and India, and in the Tertiary of Europe and North and South America. The concern here is that an endemic species occurs on the island of Cuba in the Caribbean. For the 'dispersalists', the origin of this endemic was explained when occasional gars were found in brackish and salt waters along the coast of the Gulf of Mexico. Clearly, they argued, the ancestors of the Cuban species had, as a chance event, swum across the Gulf of Mexico, colonized Cuban rivers and there become differentiated from the parent species. Vicariance biogeographers on the other hand would trace the presence of gars in Cuba back to before the Early Tertiary at which time a land-mass that eventually developed into the Greater Antilles island chain (which includes Cuba) became separated from the rest of Central America. Arguments of this latter kind have been greatly strengthened by the general acceptance of the theory of plate tectonics and continental drift in recent years (Section 13.4).

Vicariance biogeographers, however, also criticize the dispersal hypothesis on other grounds. They argue that critical dispersal events across a barrier would be the result of chance and can be neither predicted nor verified.

▶ How clearly would you expect a dispersal event across a barrier to be recognizable in the fossil record?

Such an event is likely to have been virtually instantaneous in terms of geological time and very difficult to detect in the sedimentary record, particularly since most geological deposits have slow net accumulation rates, as discussed in Chapter 10, Section 10.1.2. Furthermore, bioturbation of sediments blurs the record. More important, the fossil record for most organisms is very poor or non-existent and, even where a particular species has a fossil record, it is very difficult to be sure that it was definitely *absent* from an area. For example, because fossiliferous freshwater deposits are not preserved continuously, but survive only patchily in the sedimentary record, it might be difficult to show that gars had a fossil record in Cuba reaching back to the Lower Tertiary and it would certainly be impossible to demonstrate unambiguously that there was a dispersal event at a particular time when the ancestors of the present species suddenly arrived.

On the other hand, vicariance biogeographers point out that whenever, during the course of geological history, external events have led to the subdivision of a formerly continuously distributed population of a plant or animal, not just single species but also whole communities, ecosystems and groups of ecosystems are likely to have been split up. Leon Croizat (1958), who pioneered the concept of vicariance biogeography, noted that whereas a dispersal 'event' would affect only a single species, vicariant species should form part of a group of different taxa which all had similar, though not necessarily identical, disjunct distribution patterns. He used the method of joining points on a map to connect areas where vicariant taxa occurred (i.e. closely related taxa), and called these lines **tracks**. It is also possible to include information about any fossil occurrences of organisms in the same way.

▶ Figure 13.9 shows the tracks of two closely related tree genera, commonly cultivated in gardens, *Magnolia* and the tulip-tree *Liriodendron*, using both their modern natural distributions and their fossil occurrences. The modern species that occur in North America are, of course, different from those of East Asia. Compare these distribution patterns with that of *Stewartia* (Figure 13.8). What do these maps suggest about the origins of these disjunct distributions?

The similarity of the tracks, which is further emphasized if the information on Tertiary fossil occurrences is taken into account, suggests that the modern disjunct distribution patterns are very unlikely to be the result of coincidental episodes of long-range dispersal. Even if there were no European fossil records, it would still be probable (i.e. the simplest or most parsimonious explanation) that the present distribution of these trees was part of an originally wider distribution area that has subsequently been fragmented by extinction and barrier formation.

The interpretations of vicariance biogeographers have been enormously strengthened by the widespread acceptance of continental drift and the theory of plate tectonics, which are discussed in Section 13.4. With the assumption that the continents had always occupied their present positions, concepts such as land-bridges, over which species could migrate, were essential to explain the similarities of faunas and floras in different regions, for example North America and Eurasia. The idea that the outer crust of the Earth is mobile and that continental land-masses could shift their position, split apart into new continents, as well as collide and become united, has provided a much more compelling mechanism for the isolation of vicariant populations than those previously envisaged. However, it also provides potentially new and complex routes and opportunities for dispersal when previously existing barriers break down. In Section 13.4.5, well-documented examples of dispersal following continental collision will be discussed.

13.3.4 Cladistic biogeography

Attempts to use distribution pattern evidence to investigate the evolutionary history of genera and higher taxa are based not simply on the geographical distribution of living and fossil species, but also on the comparative morphology of those species and on geological history. A further, more recent development has been application of the techniques of cladistic analysis (discussed in Chapter 11) to biogeographical data, namely **cladistic biogeography**.

In this Section, a short case study is presented concerning the southern beeches *Nothofagus*, a genus of forest trees belonging to the same family as the European beech *Fagus*. Unlike the rest of its family, the Fagaceae (oaks, beeches and sweet chestnuts), *Nothofagus* is restricted to the Southern Hemisphere, though certain species have been introduced into western Europe, where they are sometimes grown in forestry plantations. The various species of *Nothofagus*, 36 in all, occur as tall, often dominant, trees in cool–temperate lowland and mountain forests in South America, New Zealand, Australia, Tasmania and also in the cooler upland forests of the tropical islands of New Guinea, New Britain and New Caledonia (Figure 13.10). There are significant morphological differences between the species—some are evergreen, some are deciduous—and they differ particularly in the structure of their flowers.

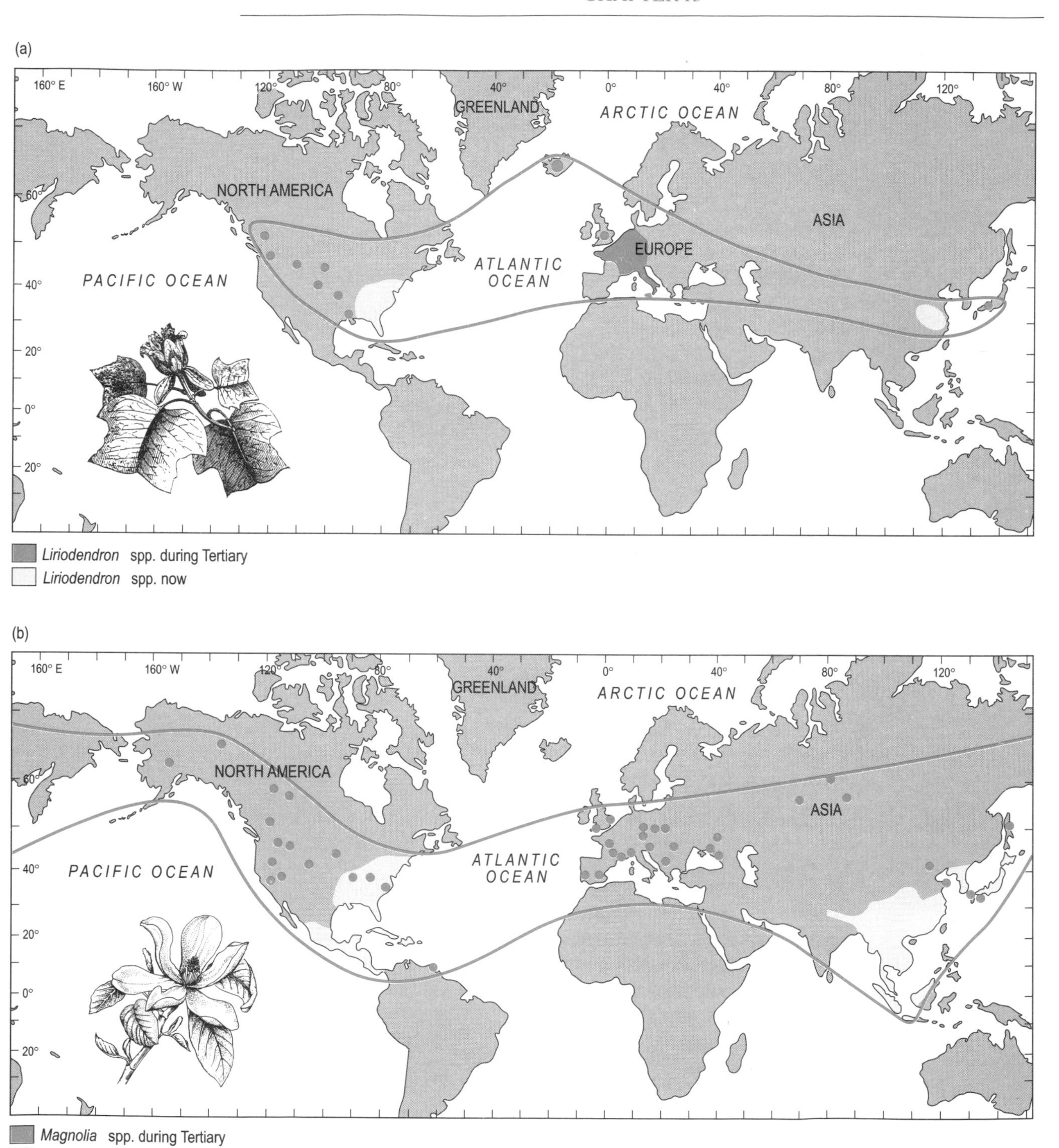

Figure 13.9 *Modern and fossil distribution areas of the tree genera (a)* Liriodendron *and (b)* Magnolia.

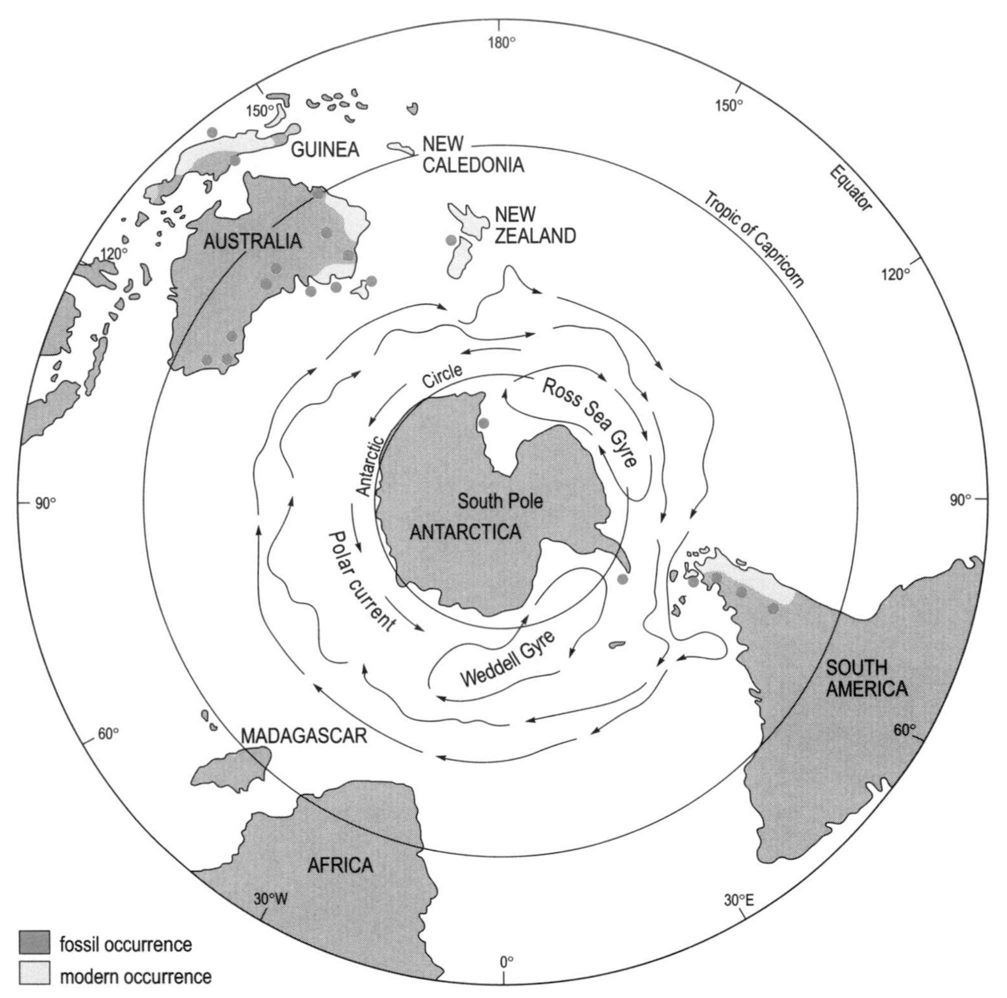

Figure 13.10 *The distribution of modern* Nothofagus *species and of fossil records. Note that the continents on which this genus is now found are separated from Antarctica by oceans dominated by the powerful circum-Antarctic current system.*

Various elaborate hypotheses have been proposed to account for the geographical distribution and evolutionary history of *Nothofagus*, taking note that all species have small seeds, which appear to be poorly adapted for long-range dispersal and are also intolerant of immersion in salt water.

▶ Suggest three possible lines of argument which could explain the occurrence of *Nothofagus* species in both South America and Australasia. Which hypothesis is the most probable and why?

The three possibilities are (i) direct dispersal from South America to Australasia or vice versa; (ii) dispersal via Antarctica; (iii) a vicariance explanation where the areas in which *Nothofagus* now occurs all formed part of a continuous land-mass.

Given the nature of its seeds, *Nothofagus* is unlikely to have been dispersed directly from South America to Australasia. This implies that *Nothofagus* forests once occurred on the intervening continent of Antarctica, an assumption confirmed by the discovery there of fossil *Nothofagus* pollen. Older dispersalist hypotheses proposed that Antarctica was linked to South America and Australia by land-bridges and that the ancestors of modern species migrated either from Australasia via Antarctica to South America or vice versa. A complication to this hypothesis is that it would also be necessary to envisage a land-bridge link to New Zealand.

In the light of continental drift reconstructions, which recognize that the southern continents were once united as a single massive continental land-mass—**Gondwanaland** (Figure 13.11)—that progressively fragmented into smaller continental blocks between the Jurassic and the Early Tertiary, it is no longer necessary to postulate land-bridges. This reconstruction would lend weight to a vicariance hypothesis that *Nothofagus* was widespread on Gondwanaland, and that the present distribution reflects the fragmentation and isolation of existing populations on those land-masses, as a result of continental drift and the development of the Antarctic ice-sheet. The existence of Gondwanaland would not, however, in itself rule out dispersal hypotheses.

A further important piece of evidence comes from the fossil record. Remains of the distinctive pollen grains of *Nothofagus* have been recovered from Cretaceous and Tertiary sedimentary rocks at numerous sites in the Southern Hemisphere, including regions outside the present range of any living species, namely in Antarctica, Western Australia and Patagonia (Figure 13.10). These records confirm that the climate of Antarctica in Cretaceous times was very different from that of today.

Many of the earlier dispersal hypotheses are much concerned with possible centres of origin, both for the Fagaceae in general and for *Nothofagus* in particular. Beeches, oaks and sweet chestnuts have a predominantly Northern Hemisphere distribution, but the most primitive member of the family is a tree called *Trigonobalanus*, a genus with just two species endemic to mountain forests in South-East Asia. The distinctive fruits that characterize this genus have, however, been found fossilized in Tertiary deposits in both Europe and South America.

▶ How would you describe the status of living *Trigonobalanus*?

It is now a relict endemic, since the genus was once widely distributed with species occurring far outside its present distribution area in the highlands of Malaysia, Thailand and Indonesia.

Dispersal biogeographers have tended to suggest South-East Asia as the centre of origin for the Fagaceae, but actual supporting data are very sparse. As for the centre of origin of *Nothofagus* itself, suggestions have included North America, Europe and South-East Asia, but these are all purely speculative because there is no reliable fossil evidence that *Nothofagus* ever occurred in any of these areas. These hypotheses also vary in assessing which characters might suggest that some modern species have retained more ancestral (primitive) features than others, so again contrary assumptions underpin their conclusions.

Humphries (1981) dismissed these hypotheses, in a statement that summarized objections to much of dispersal biogeography:

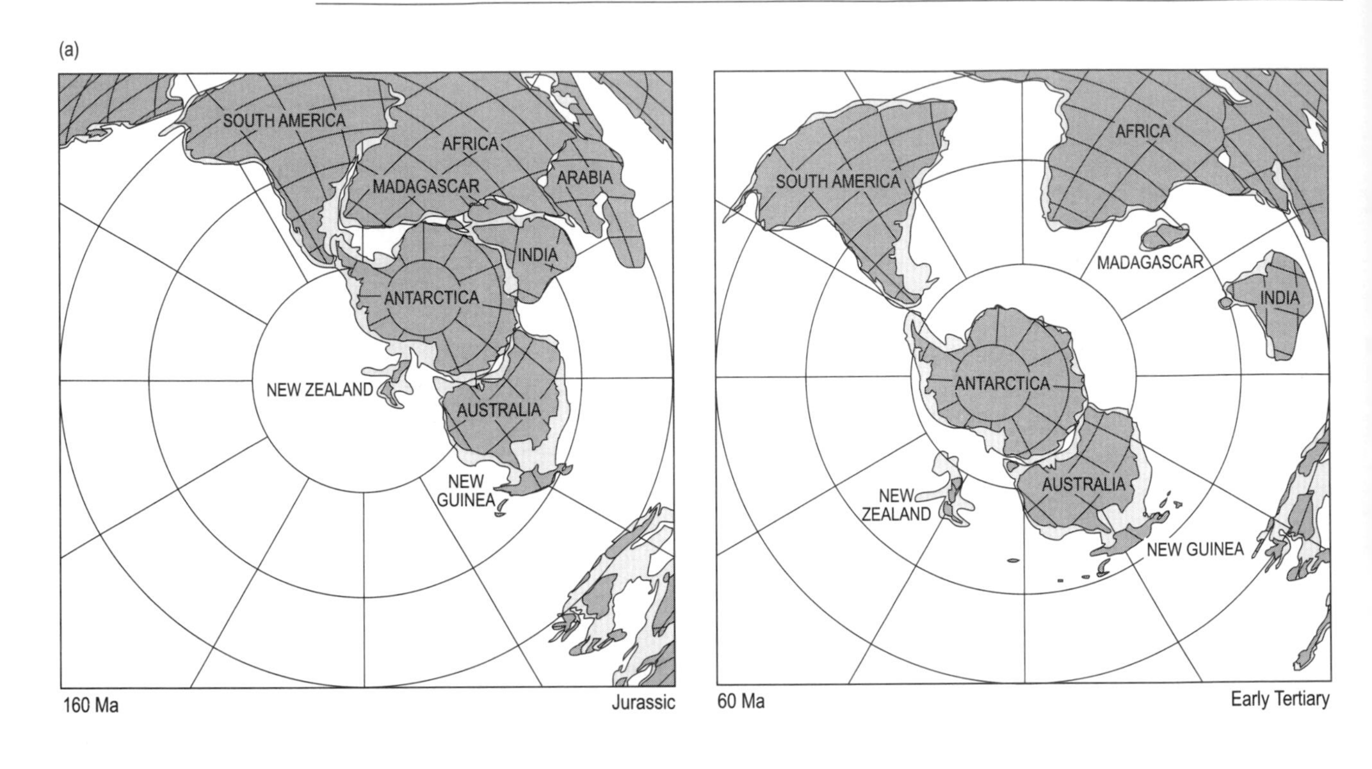

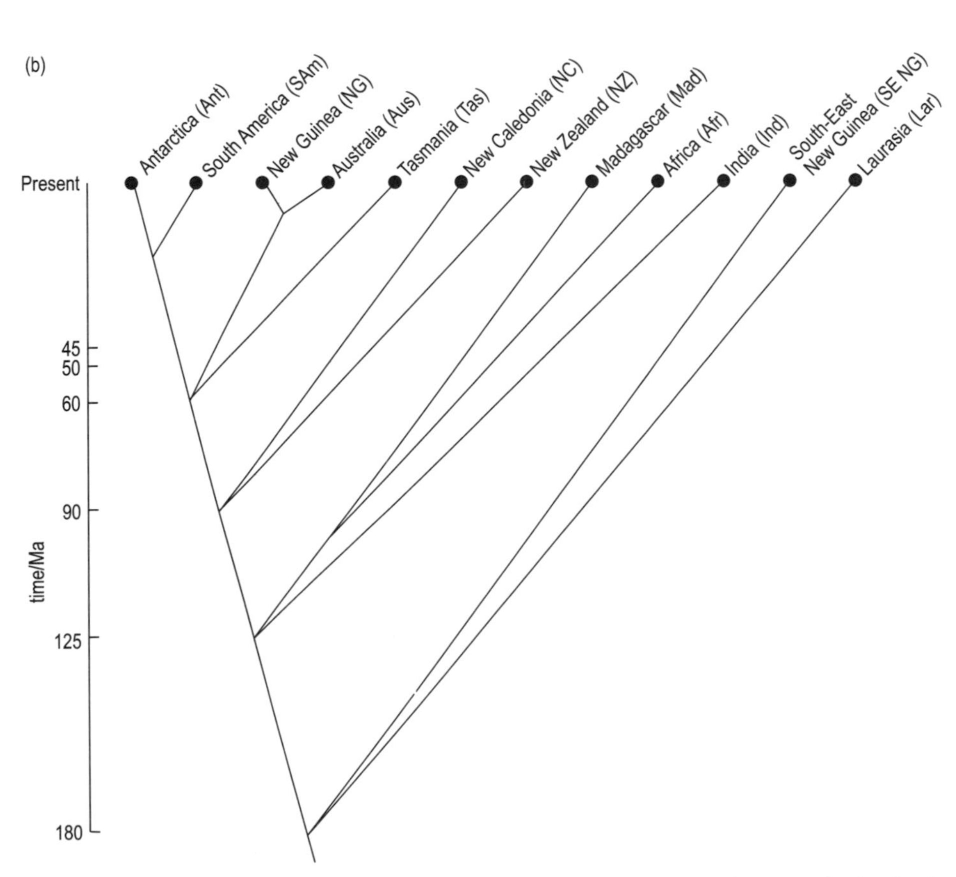

Figure 13.11 *(a) Gondwanaland in Jurassic and Early Tertiary times. (b) Sequence and time-scale for the break-up of Gondwanaland.*

'In all cases the present-day distribution of *Nothofagus* is explained by dispersal in a variety of transoceanic or overland dispersal routes depending on the acceptance or rejection of continental drift. In each case, there is no attempt to use a refined taxonomy showing relationship at the species level and only vague notions are given as to what are "centres of origin". Furthermore, there is little appreciation of the idea that biogeography is more about areas of the globe and their total faunas and floras than about trying to explain how, when and by what route a *particular* group of organisms arrived at its present position.'

The comparative morphology and anatomy of *Nothofagus* species have now been studied in greater detail, so that Humphries was able to create a cladogram to suggest the phylogenetic relationships of the living species of this genus (Figure 13.12) on the basis of the characters listed in Table 13.1 (you do not need to understand some of the more technical botanical terms in this Table).

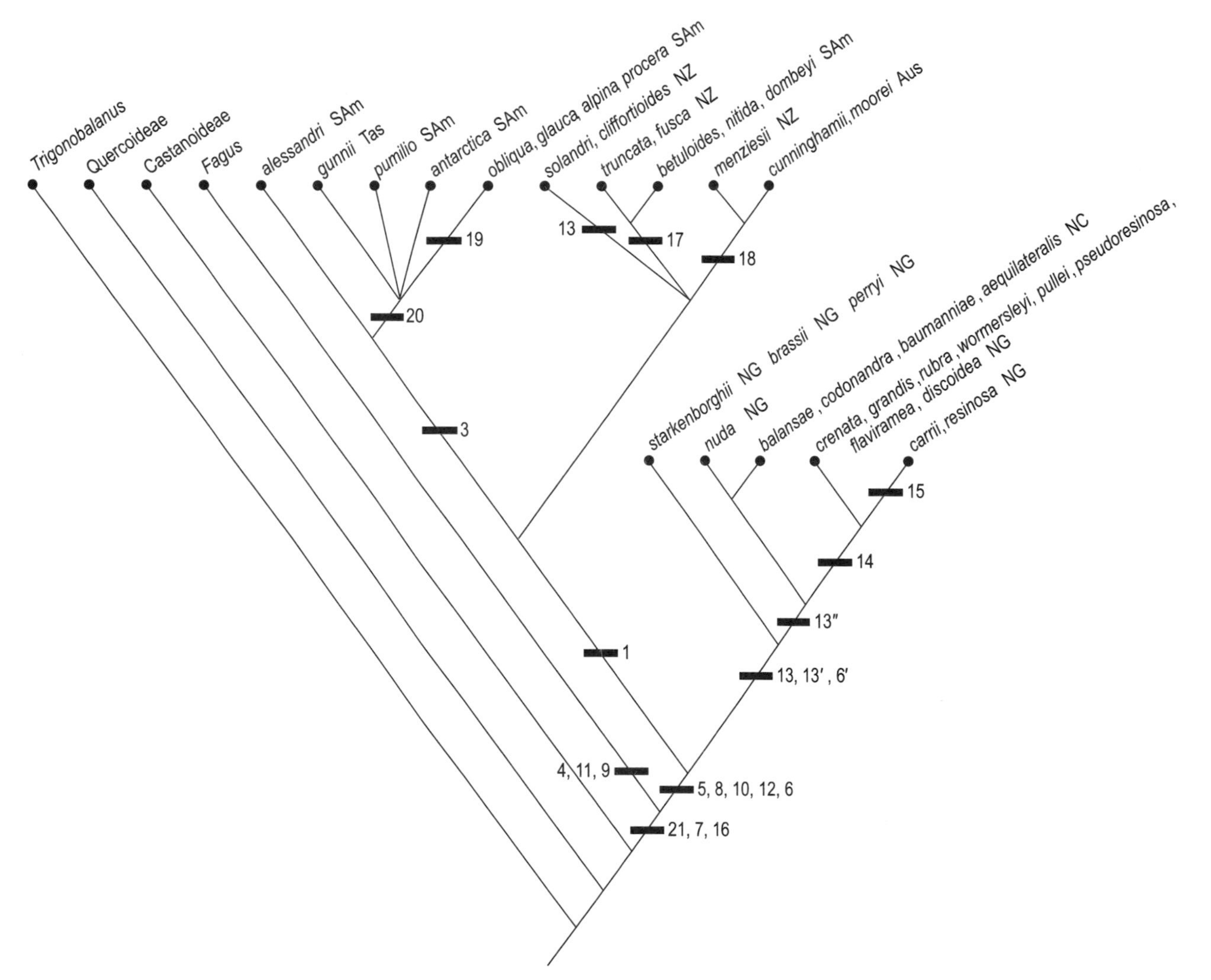

Figure 13.12 Cladogram showing the interrelationships of Nothofagus *species and also other genera and sub-families within the Fagaceae. Black bars indicate derived character states, as listed in Table 13.1.*

Table 13.1 *Characters used to create the cladogram for* Nothofagus *(Figure 13.12). Like oaks and beeches,* Nothofagus *has separate male and female flowers, but these are borne on the same tree.*

Characters	Primitive (–)	Derived (+)
Wood anatomy:		
1 Tracheids	Present	Absent
2 Wood fibre	1.48 mm	0.7–1.16 mm
3 Leaves	Evergreen	Deciduous
Characters of male inflorescences:		
4 Terminal flowers	Present	Aborted
5 Male dichasia	Many flowered	Few flowered
6 Male perianth	6-lobed	Campanulate or fused tubular (a small group of species are more derived in this feature—6′)
7 Male inflorescence	Long spikes	Hanging clusters
8 Pollen aperture no.	3	5–8
9 Pollen apertures	Colpate	Colporate
10 Pollen apertures	Long	Medium or short (10′)
Characters of female inflorescences:		
11 Styles	Flattened with long stigmatic surface	Cylindrical with short stigmatic surface
12 Ovule integments	2	1
13 Cupules	4-lobed	3-lobed (13′)
		2-lobed (13″)
14 Female flowers per cupule	2–7	1
15 Cupules	Reaching nutlet tip	Shorter than nutlet
16 Cupules	Along rhachis	In leaf axils
17 Cupules	Not branched	Branched
18 Cupule lamellae	Not modified	Gland tipped, recurved
19 Cupule lamellae	Entire	Simple branches, elaborate branches
Characters of fruiting structures:		
20 Infructescence	7 fruits	3 fruits
21 Central fruits	Trimerous	Dimerous (or absent)
22 Lateral fruits	Trimerous	Dimerous

▶ Can you recall from Chapter 11 the basis on which clades are separated from their sister groups?

Each clade is regarded as a monophyletic grouping on the basis of shared *derived* character states.

▶ (a) Which important character separates the cool temperate species from the tropical species, found in New Guinea and New Caledonia? Which group possess the primitive character state?

(b) Which character is assumed by this cladogram (on grounds of parsimony) to have evolved independently in both the tropical and cool–temperate lineages?

(a) The cool–temperate species all possess a more 'specialized' wood anatomy, where tracheids are absent (i.e. have been 'lost') (derived character state 1 in Figure 13.12 and Table 13.1). These species are all assumed to be more closely related to one another than to the tropical species which retain a more primitive wood anatomy with tracheids present.

(b) In some species of *Nothofagus*, the cupule lobes are reduced from four to three or two (derived character states 13′ and 13″). This reduction is believed to have taken place independently in the *N. solandri* clade found in New Zealand, as well as in the tropical clade now found in New Guinea and New Caledonia. Cupules are woody structures that surround and protect the fruits in the Fagaceae. In the European beech *Fagus silvatica*, the cupule is spiny and four-lobed, whilst in the oaks, it is reduced to form the acorn 'cup'.

At this point, however, Humphries proceeded to analyse the biogeographical implications of this cladogram, using the same principles and treating the present-day occurrence of a taxon in a particular area as the equivalent of an heritable character. In other words, he accepted the vicariance rather than a dispersal model as an initial assumption.

Examine Figure 13.13 carefully. You will see that it is basically the same cladogram as Figure 13.12, to which areal information (areas of occurrence) has been added.

▶ The ancestral distribution of the three clades represented by *Nothofagus solandri*, *N. truncata* and *N. betuloides* (Figure 13.13) is shown to include Antarctica. Why should this be so? (Note incidentally that the cladogram shows three rather than two branches here, simply because there is insufficient evidence to decide on the most probable sequence of dichotomies.)

It is assumed that, where a common ancestor is postulated for clades, occurring today on the one hand in Australasia and on the other in South America, then that ancestor must have also have occurred in Antarctica to give a continuous geographic distribution, but that subsequently the Antarctic populations were wiped out by climatic change.

seen in other groups with similar distributions in the Southern Hemisphere, so its presence in Malaysia cannot be interpreted as the result of a vicariance event. The most logical interpretation here is that a dispersal event has occurred and that the taxon has succeeded in crossing Wallace's line. Cladistic biogeographical techniques, whilst primarily based on the assumption of vicariance events, can nevertheless sometimes provide well-supported indications of sweepstake dispersal events.

SUMMARY OF SECTION 13.3

Present-day and fossil distribution patterns of organisms are an important source of information about the evolutionary history of both individual taxa and of whole communities. There is a long tradition of interpreting distribution patterns in terms of taxa spreading outwards from centres of origin, using assumptions about diversity patterns and about the presence of species that possess complexes of either primitive or derived characters. This concept—dispersal biogeography—developed at a time when it was believed that the continents had always occupied their present positions, so that former land-bridges or chance dispersal events were necessary to explain the disjunct distribution of obviously related organisms on different continents.

A more recent school of thought—vicariance biogeography—assumes instead that ancestral species were widely dispersed and that fragmentation of populations has taken place with the subsequent development of barriers such as oceans, deserts or mountain ranges. Under those circumstances, the isolated populations have diverged to form separate species. It is argued that whereas dispersal affects individual taxa, vicariance events will influence whole communities, whose members should then show similar disjunct distribution patterns. The vicariance hypothesis has been greatly strengthened by widespread acceptance of the theories of continental drift and plate tectonics. The geological record, however, also provides good evidence for dispersal, as does the distribution of organisms on oceanic islands. Undoubtedly, both dispersal and vicariance events have contributed to the present patterns of species distribution and to the evolution of those species. The application of cladistic techniques of analysis can be extended to distributional data. Combined with good geological data, cladistic analyses may permit actual vicariance events to be recognized and dated, and may distinguish events where a dispersal explanation is more probable.

13.4 CONTINENTAL DRIFT

The similarity of the shape of the continental margins on either side of the Atlantic was noted by a number of early scientists, including Francis Bacon, and in the 19th century some 'catastrophist' geologists formulated hypotheses in which the Old and New Worlds had been pulled or drifted apart. However, it was the German geophysicist and meteorologist Alfred Wegener (1880–1930) who in 1915 first put together the scientific evidence for the theory of **continental drift**.

Although this theory was proposed so long ago, it was not widely accepted until the mid-1960s when it was reformulated as part of the broader concept of plate tectonics. Palaeontologists and biogeographers were amongst the first to rally to this new theory, as it offered a more convincing explanation of their observations on the distribution of plants and animals than did previous hypotheses which assumed that the continents had not moved.

The arguments that Wegener used related not only to continental fit but also to crustal structure, the distribution of palaeoclimatic belts and the apparently disjunct distribution of fossil terrestrial and freshwater faunas. In fact, these arguments have stood the test of time remarkably well and have been reinforced by more recent studies.

Continental fit Wegener himself did not place heavy emphasis on the fit of the shapes of the continental coastlines. Comparison of shapes can easily be subjective, but more recently 'best fit' computer models have produced maps that showed, for example, that the continents bordering the Atlantic could be fitted along the 500 fathom (1 000 m) contour with overlaps and gaps of less than 90 km width (Figure 13.15).

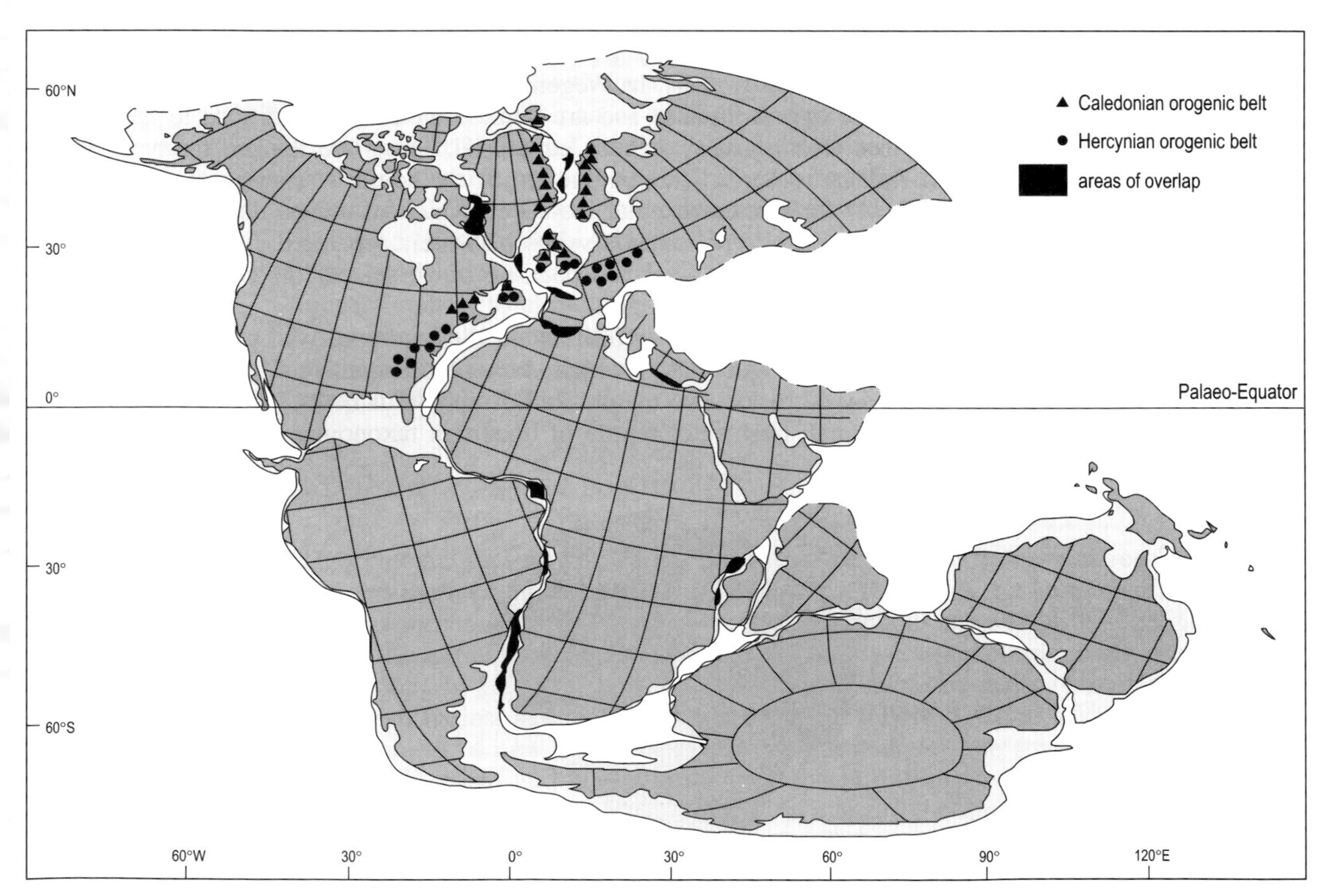

Figure 13.15 Computer reconstruction of continental relationships before the formation of the Atlantic Ocean in the Late Triassic, also showing distribution of the Caledonian and Hercynian orogenic belts.

lithospheric plates which on a geological time-scale move around the surface of the Earth. The present distribution and boundaries of the major plates are shown in Figure 13.18. There is some correspondence with the boundaries of the biogeographical regions (see Figure 13.6), but obvious discrepancies occur.

There is a fundamental distinction between the crust underlying the ocean basins, which has a basaltic composition, with rocks characterized by abundant iron and magnesium silicates and calcium-rich feldspars (density 2 800–2 900 kg m^{-3}), and the lighter continental crust (density 2 600–2 800 kg m^{-3}) which is primarily composed of granitic, quartz-rich rocks or their derivatives. Although it is possible to talk about oceanic and continental plates, particularly where two plates of different composition abut one another, most plates consist both of areas of oceanic crust and of an area of continental crust fused together to form a single integral plate. Others such as the Pacific and Philippine Plates are predominantly oceanic.

Oceanic crust is formed at **constructive plate margins** along the mid-ocean ridge by massive injection of basaltic magmas which melt off from the underlying **mantle**, the much denser layer below the crust. The two plates being formed move laterally away from the ridge (see Figure 13.19(a)). Oceanic crust is destroyed along **destructive plate margins**, where two plates collide and one is **subducted**, i.e. carried down beneath the other into the mantle, and melted. Where continental and oceanic plates collide, it is generally the denser oceanic plate that is subducted

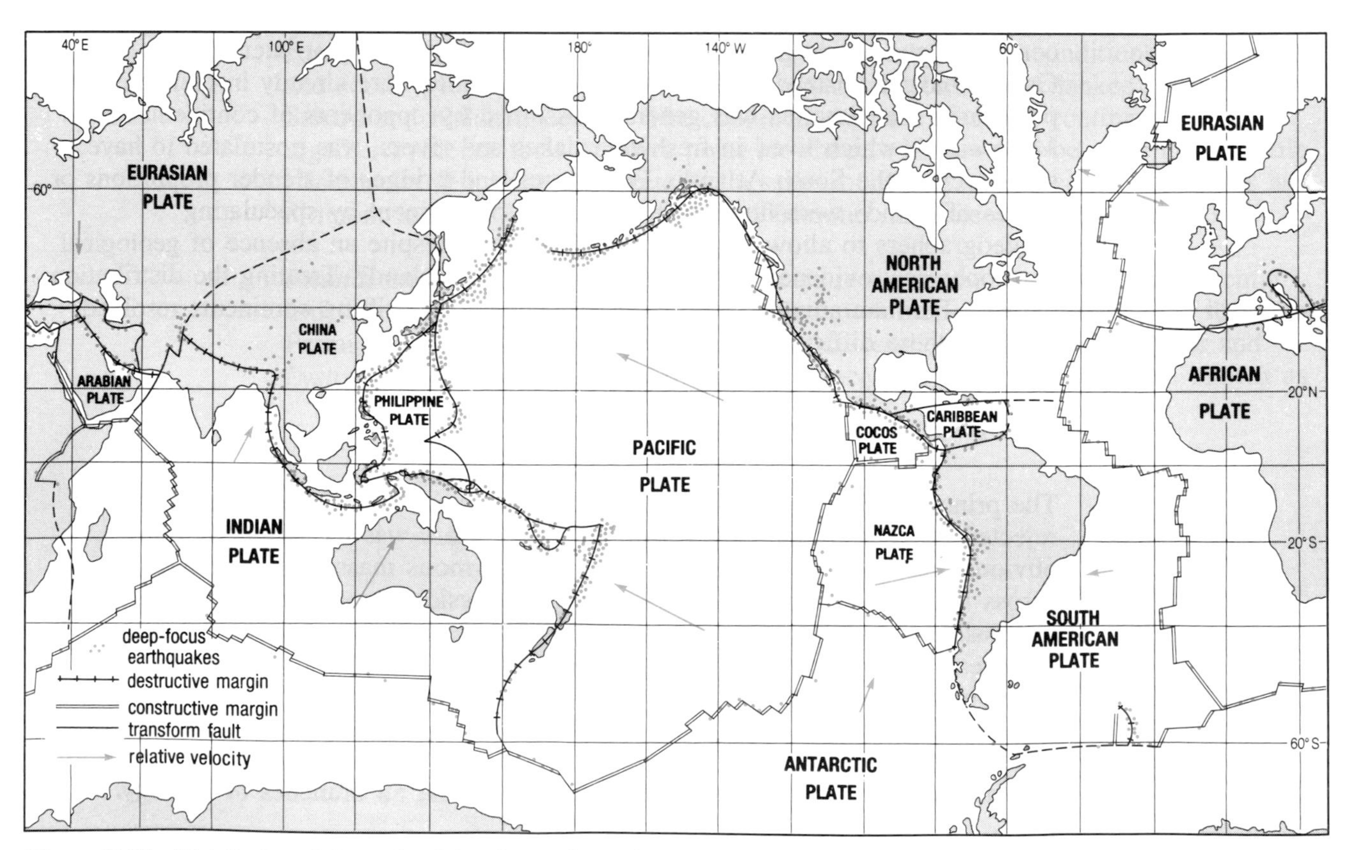

Figure 13.18 *Distribution of the major lithospheric plates showing constructive and destructive plate margins and present-day directions of plate movement. Note the correlation between the occurrence of destructive plate margins and earthquakes.*

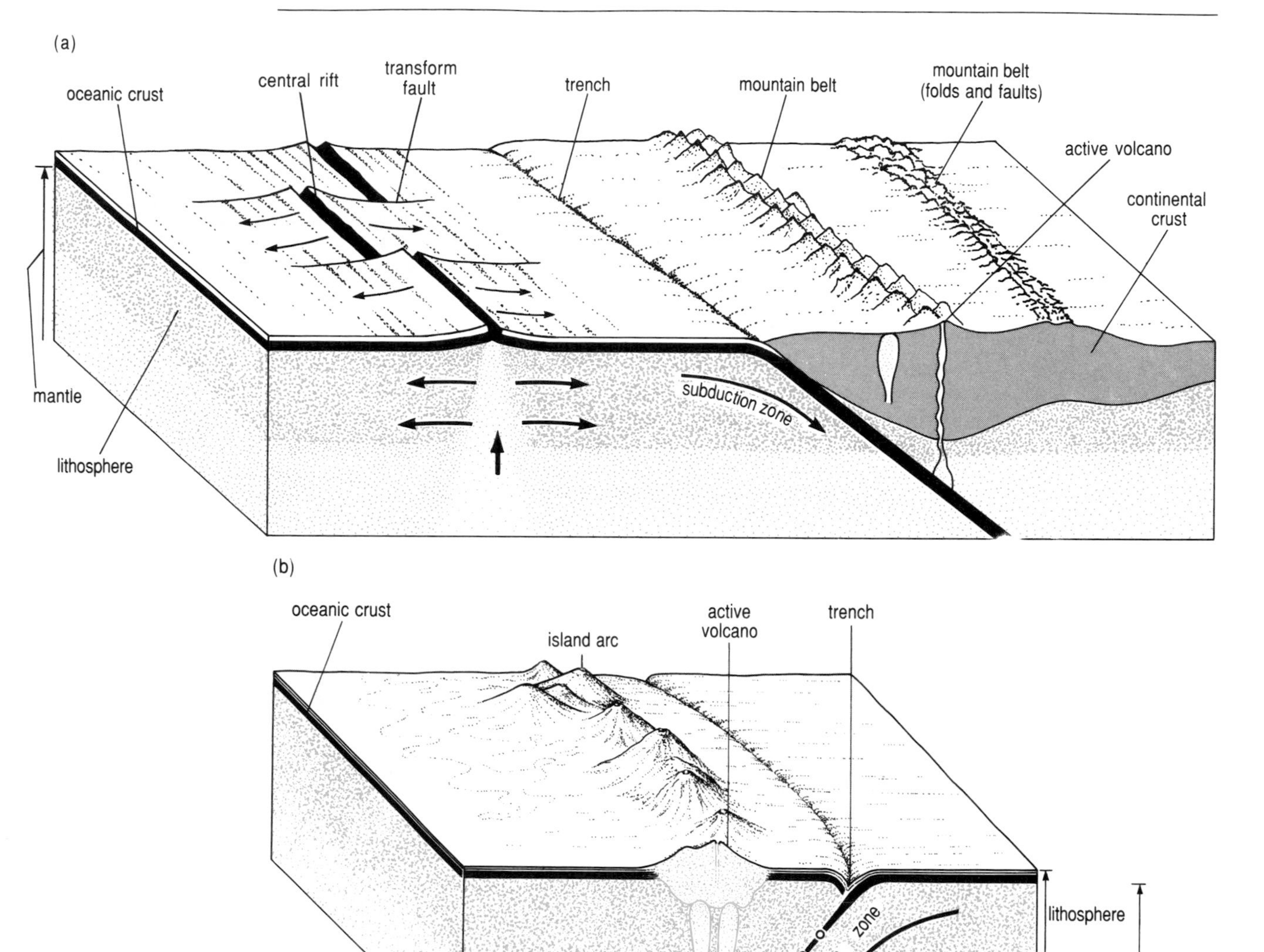

Figure 13.19 *Block diagram showing the main features of (a) a constructive plate margin and an ocean/continent destructive plate margin and (b) an ocean/ocean (island arc) destructive plate margin.*

and then destroyed by melting. Sites of subduction are marked by the occurrence of deep ocean trenches.

▶ Major destructive plate margins occur adjacent to much of the western coasts of both North and South America. What geological features characterize the continental plate margin there?

Chains of very high mountains, from the Andes through Central America to the Rockies and, also in the western USA and Canada, the Cascades (which include Mount St. Helens), occur close to the coast, with many active volcanoes and very intense earthquake activity. The process of mountain building is constantly in

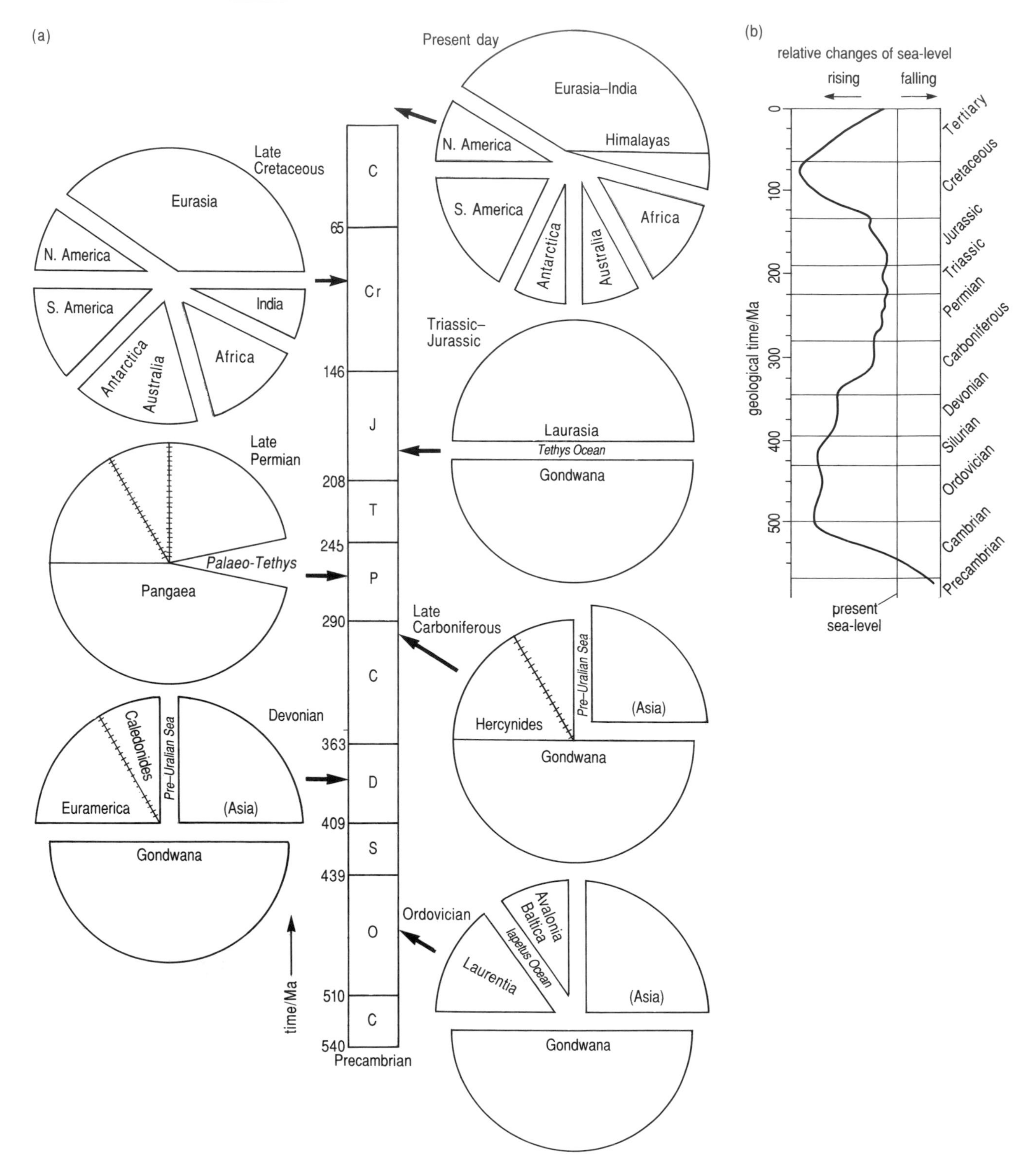

Figure 13.21 *(a) Diagrammatic continental reconstructions for the Phanerozoic; (b) Relative changes in sea-level during the Phanerozoic. (Note that present sea-level is the average for the Pleistocene, a period affected by major glaciations; this is why the curve does not meet the line at 0 Ma.) Note the differences in the scales used for time between (a) and (b).*

Consider the various reasons why these continental movements are likely to have affected the evolution of organisms on land, on fringing continental shelves and in the oceans.

Land-masses will undergo major changes of climate, as they move across different latitudes. For example, during the Precambrian, according to palaeomagnetic measurements, Britain's latitudinal position was between 60° and 90° south, and, indeed, glacial sediments occur in Precambrian rocks in Scotland. By the Late Palaeozoic, it had moved northwards to the Equator and thence to its present latitude of about 55° north (see Figure 13.20). This is also supported by the faunal and sedimentary records. Coral reefs, largely tropical or subtropical ecosystems today, are found in shallow water sediments of various ages from the Silurian to the Jurassic. Coal-swamps are widespread in Carboniferous deposits (note Britain's palaeolatitude at the time), and red-beds, indicating desert conditions, characterize the Permian and Triassic, when much of what is now western Europe was passing through subtropical latitudes (cf. the latitude of modern arid climatic belts).

▶ On the basis of modern observations, how might these changing climatic conditions affect evolutionary processes?

Although the mechanisms are not properly understood, it is clear from Section 13.2 that greater species diversity accumulates under tropical climatic conditions.

From the luxuriant fauna and flora recorded in Carboniferous limestone and coal deposits, this appears to have been true in earlier geological times, but far more important than changes due to latitudinal shift were the evolutionary effects of major changes in the configuration of continents and oceans in the Late Palaeozoic and Mesozoic. Unlike the continents, the size and shape of the ocean basins have changed dramatically with time. Though they may have a long lifetime, they may also disappear virtually completely.

▶ What appears to have been the fate of the Tethys Ocean that existed from the Triassic right through Mesozoic times?

This ocean, which developed from an arm of the great ocean surrounding Pangaea in Triassic times into a critically positioned seaway separating Laurasia from Gondwanaland (Figures 13.20 and 13.21), was finally closed when the African/Arabian Plate moved northwards to collide with the Eurasian Plate. This left a residual basin, the Mediterranean. However, during Miocene (mid-Tertiary) times, the Mediterranean Sea itself was isolated from the Atlantic for a critical period of about a million years, when it dried out and formed an enormous salt lake, so that there was a complete faunal change. Also earlier in the Tertiary, the Indian Plate moved northwards, colliding with Eurasia and progressively eliminating the eastern end of the Tethys.

▶ Try to list the major effects that collision of two continents might have on the diversity of marine and terrestrial organisms.

---

1 As already seen in the case of the Tethys Ocean, the marine faunas, both deep-water and coastal, are likely to suffer destruction of habitat and hence extinction, so that diversity will be reduced. Marine faunas appear to have suffered a mass extinction event during the Permian, which is widely interpreted as the result of the fusion of the smaller continents to form the supercontinent Pangaea—a process that must have eliminated a number of oceans and seas.

2 When land-masses that have had no direct land connections for many millions of years become joined, plant species and also terrestrial animal species that may have become adapted to almost identical niches will be able to disperse onto the other continent. For a short time, this may actually increase diversity on each side (although not total diversity), but then competition and extinction are likely to reduce the total number of species. There is a possibility of vicariant species meeting and being able to hybridize.

3 Continental collision is likely to be accompanied by the formation of new mountain chains such as the Himalayas. If you look carefully at Figures 13.20 and 13.21, you should be able to work out how and when such mountain ranges as the Alps, the Urals, the Appalachians in North America and the mountains of Scotland and Scandinavia formed. These mountains are likely to provide new and diverse habitats and so encourage speciation, and also serve as barriers, promoting vicariance (in contrast to the effect in 2).

4 There will also be far-reaching climatic changes when continents collide, because the circulation systems of both the oceans and the atmosphere are likely to be affected. Before the Tethys was eliminated, the southern parts of Laurasia (see Figure 13.20) supported temperate and subtropical forest communities. However, the closing of Tethys and the rise of the Himalayas cut off the area that is now Central Asia both from moisture-bearing winds and from the moderating influence of an ocean. As a result, like many continental interiors, this immense region developed a typically arid climate with severe winters and hot dry summers; it now supports predominantly desert and semi-desert vegetation. There has thus been a tremendous decrease in species diversity as a result of these changes.

When continents split, however, both geological and evolutionary processes are likely to favour an increase in species diversity. You have already read in Chapter 9 how the African rift valley system, with its series of isolated lakes, contains closely related but extremely diverse cichlid fish faunas. If rifting is succeeded by marine incursion (see Figure 13.23) and the development of a mid-ocean ridge system, the populations of marine fauna that have dispersed into the newly formed coastal habitats will become separated on either side of the ocean, and will be unable to interbreed. Thus, a vicariance event will have taken place, and eventually distinct, albeit related, species may evolve on the new continents. This will equally be possible for the land fauna and for the flora, and such differences are likely to increase if the continents move apart through different climatic zones.

Three other effects of plate tectonic processes may have important consequences for evolution: the development of oceanic islands; changes in sea-level; and changes in oceanic currents.

13.4.2 ISLAND FORMATION

You have seen that plate tectonics can, by two rather different processes, lead to the formation of mountain chains on land. In the oceans, chains of volcanic islands, rising to great heights above the ocean floor, can form as island arcs along destructive plate margins, where two oceanic plates collide and one is subducted beneath the other (see Figure 13.19(b)). Alternatively, as in the Atlantic Ocean, isolated groups of volcanic islands such as the Canaries, Madeira and the Tristan da Cunha archipelago, may form and persist as a result of volcanic activity at constructive plate margins; an even more striking example of this is Iceland. A geologically rather different example of island formation is the Hawaiian chain, which owes its origin to a **hot spot** in the mantle below the crust, so that new volcanic islands are formed intermittently as the crust drifts over the hot spot (Figure 13.22) but within plates, rather than at their boundaries. The floor of the Pacific Ocean is, in fact dotted with seamounts which were formed as intraplate volcanoes that have now mostly subsided or been eroded, though some persist as coral atolls or isolated island groups. The importance of these islands to evolutionary biologists is that many of them certainly seem to act as natural laboratories where speciation processes are extremely active (Chapter 9, Section 9.4.4). They also supply examples of cases where the distribution of organisms is unambiguously the result of dispersal processes rather than vicariance. Nevertheless, as you have seen in the case of the snail *Lyrodiscus* from the Azores (Section 13.3.1), they may provide refuges for relict endemic taxa, as well as for newly evolved ones.

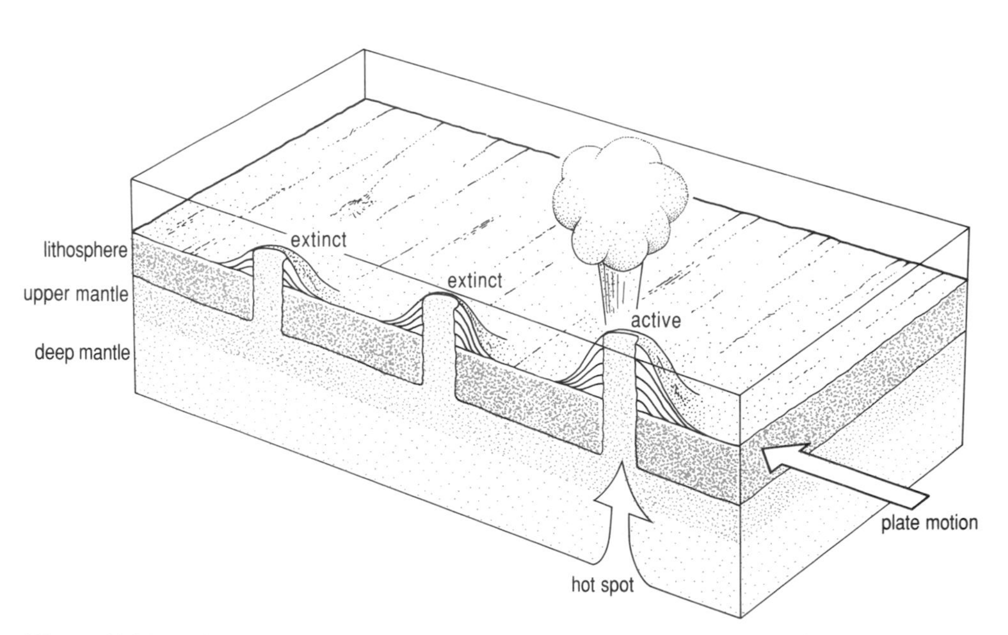

Figure 13.22 *Schematic diagram (not to scale) illustrating how a volcanic chain of islands could be formed by an oceanic plate moving over a stationary hot spot, where a magma plume rises up from the mantle. The age of the islands increases towards the left. New islands will appear on the right as the motion continues, whereas old islands will sink and become smaller and lower in elevation.*

13.4.3 Sea-level changes and ocean circulation

Over the past 600 Ma, sedimentary records show that world sea-levels have been far from constant. For example, they were particularly high for much of the Palaeozoic and also briefly during the Cretaceous (Figure 13.21(b)). These sea-level changes can also be linked to the processes of plate tectonics. A spreading mid-ocean ridge is formed when molten basalt is injected into the oceanic crust from below. As the ridge rises and changes the shape of the sea-floor, ocean water is displaced, causing a rise in sea-level, so that the oceans spill over onto adjacent low-lying land. This is known as a **eustatic** (or global) **marine transgression**; it increases the area available to shallow water marine organisms.

▶ What plate tectonic event is likely to be reflected by a eustatic rise in sea-level?

The splitting up of a continent and the initiation of a new constructive plate margin and mid-ocean ridge system is likely to be accompanied by a marked episode of eustatic marine transgression. Both features, of course, leave a strong imprint on the geological record. Once initiated, mid-ocean ridge systems tend to extend their length, so that this process may be reinforced over a long period of time. In Figures 13.20 and 13.21, you can see how the formation of the Atlantic began in different areas, with the ridge system developing gradually from Middle Jurassic times onwards until, by the Early Tertiary, a total separation of the eastern and western continents had been achieved. Note the changes in global sea-level that occurred over this period (Figure 13.21(b)). Also important is the *rate* of sea-floor spreading. As shown in Figure 13.23, the presence of hot magma at a constructive plate margin causes the ocean floor there to dome up, so that mid-ocean ridges are a very considerable topographic feature. Once the newly formed crust moves laterally away from the axis where magma injection takes place, it both gradually cools and subsides. With a faster rate of lateral spreading, a broader ridge develops because its outer flanks have not had sufficient time to cool and subside. An increase in the spreading rate will, therefore, increase the volume of a mid-ocean ridge and the amount of water it displaces, again leading to eustatic marine transgression.

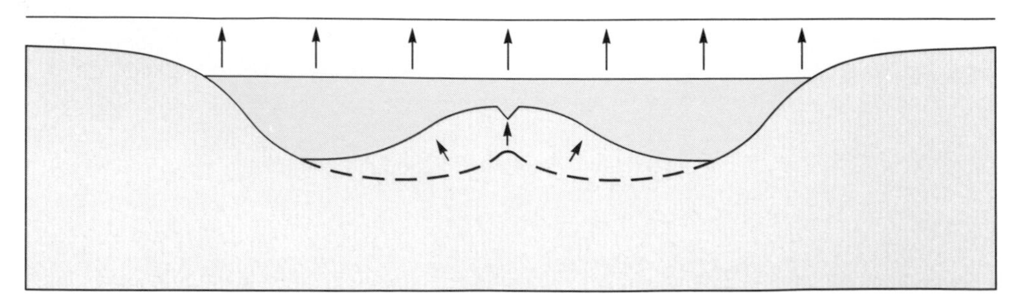

Figure 13.23 *Block diagram showing a constructive plate margin with an increased rate of spreading resulting in a broader ridge and displacement of water, leading to a eustatic marine transgression.*

▶ Under what circumstances would you expect the opposite effect — a **eustatic marine regression** — to occur, i.e. a fall in sea-level and the withdrawal of seawater from continental margins?

Whilst eustatic marine transgressions result from increased sea-floor spreading and continental fragmentation, the cessation of continental motion, especially during collision, will result in the subsidence of spreading ridges. Such an event will severely affect shallow water faunas and may also form land-bridges between formerly separated land-masses, even on plates unaffected by continental collision.

Another factor that has a strong effect on eustatic sea-level changes, namely the growth and melting of continental ice-sheets during ice ages, will be discussed in Section 13.5.

It should be obvious that, when oceans such as Tethys were closing up, a major reorganization of the patterns of ocean circulation systems would be likely to result. These circulation systems have a tremendous effect not only on life in the oceans but also on the climate and biotic diversity of adjacent land-masses and their continental shelves. A comparison of the climate and the low faunal and floral diversity of cold, bleak Labrador in eastern Canada with that of the British Isles, at the same latitude but warmed by the Gulf Stream–North Atlantic Drift current system, illustrates this point. Today, both the Arctic Ocean and the Antarctic continent are cut off from the warm waters which flow polewards from lower latitudes. In the Arctic, the barrier is the ring of land formed by Eurasia, Greenland and North America; whereas in the case of Antarctica it is the cold circum-Antarctic Current which insulates the southern polar region from warming influences (Figure 13.10). There is fossil evidence that until the Early Tertiary Antarctica supported both temperate forest vegetation and a diverse fauna. Studies of fossil skeletons of Foraminifera (microscopic animals important in ocean plankton), collected from deep-sea cores around the edge of the continent, suggest that the freezing up of Antarctica began about 38 Ma ago, soon after rifting had separated the Australian land-mass from Antarctica. At about this time, the strong eastward-flowing ocean current system, that virtually cuts off Antarctica from any southward-flowing warm-water currents, seems to have been initiated.

13.4.4 IAPETUS: AN EARLY PALAEOZOIC OCEAN

Let us examine in more detail an example of the relationship between the movement of continents and the evolution of organisms. The relationship may be seen in the change in biotas, principally marine invertebrates, inhabiting the shallow waters on continental shelves, as continents move together or apart.

Before the theory of plate tectonics became widely accepted, it was realized that there was something decidedly odd about the distribution of certain groups of fossils, especially trilobites and brachiopods, in the Lower Palaeozoic rocks of Britain and eastern Canada. This is best illustrated by collections of fossils from Ordovician rocks in Scotland, which on close examination and identification were shown to have a far greater similarity to the fossil faunas of much of eastern Canada and the USA, e.g. western Newfoundland and Virginia, than to the adjacent fossil faunas of the English Lake District and Wales. Similarly, fossils collected in south-east Newfoundland are comparable to those of England and have little in common with the fossils of equivalent age from the rest of eastern Canada. A further complication is that Ordovician rocks in Scandinavia and Russia contain yet another distinct suite of fossils of this age.

There is hardly any absolute dating evidence for these Ordovician sedimentary rocks, so how can palaeontologists be certain that the different faunas are really of the same age? You should have noticed that the evidence is based on particular fossil groups. Brachiopods and most, but not all, trilobites lived in shallow water environments. Other groups of fossils, particularly pelagic species such as extinct, floating colonial animals called graptolites, and some other genera of trilobites that seem to have had a pelagic life-style, were much more widely distributed and form good zone fossils that can be used at widely separated sites for detailed correlation (Chapter 10, Section 10.3.2).

Various suggestions have been put forward to account for this anomalous distribution of Ordovician faunas. The one most generally accepted today is that in the Early Palaeozoic, an ocean known as Iapetus separated the land-masses which now form most of North America on the one hand and most of Europe on the other (Figure 13.24). The position of the **Iapetus Ocean** was somewhat oblique to that of the present-day Atlantic. A wide branch of this ocean, known as Tornquist's Sea, is thought to have split the 'European' side into two separate continents— Avalonia and Baltica. This would provide an explanation for the three distinct faunal provinces recognized by palaeontologists.

Reconstructions of Early Ordovician palaeogeography have been based on various lines of evidence. The details have been a source of argument for over 25 years, as geologists proposed and rejected different solutions to a very complex jigsaw puzzle. Many Early Palaeozoic rock sequences have been strongly folded and sometimes metamorphosed, but they can yield basic information on geographical patterns of deep and shallow water sediments and volcanic activity. Faunal evidence, as already mentioned, suggests a physical separation of three faunal provinces, but the palaeoecological interpretation of the faunas also allows recognition of shallow water and deep sea facies (sum total of faunal and sedimentary features). Palaeomagnetic measurements can be difficult to interpret in strongly folded rocks but they can yield data on the palaeolatitudes of the continental blocks; they cannot, however, define palaeolongitudes, so that the relative position of the continents must be partly conjectural, as they may have moved laterally. However, recent palaeomagnetic results are consistent with the three-fold division recognized from the faunas.

In the reconstruction shown in Figure 13.24, proposed by Cocks and Fortey (1982), the 'American' continent Laurentia is tropical and lies on the Ordovician Equator. Note that it also includes Scotland, much of Ireland, a small part of Newfoundland and even a fragment of western Norway. Baltica lies at about 40° S, and Avalonia, which forms a strip of continent including small parts of eastern North America as well as the southern British Isles, is attached to the rest of western Europe in high latitudes at about 60° S.

Like all oceans, Iapetus was originally formed by the rifting and splitting of an older plate, perhaps about 1 000 Ma ago. At the new constructive plate margin, upwelling magma would then have created the oceanic crust that floored the growing ocean. A

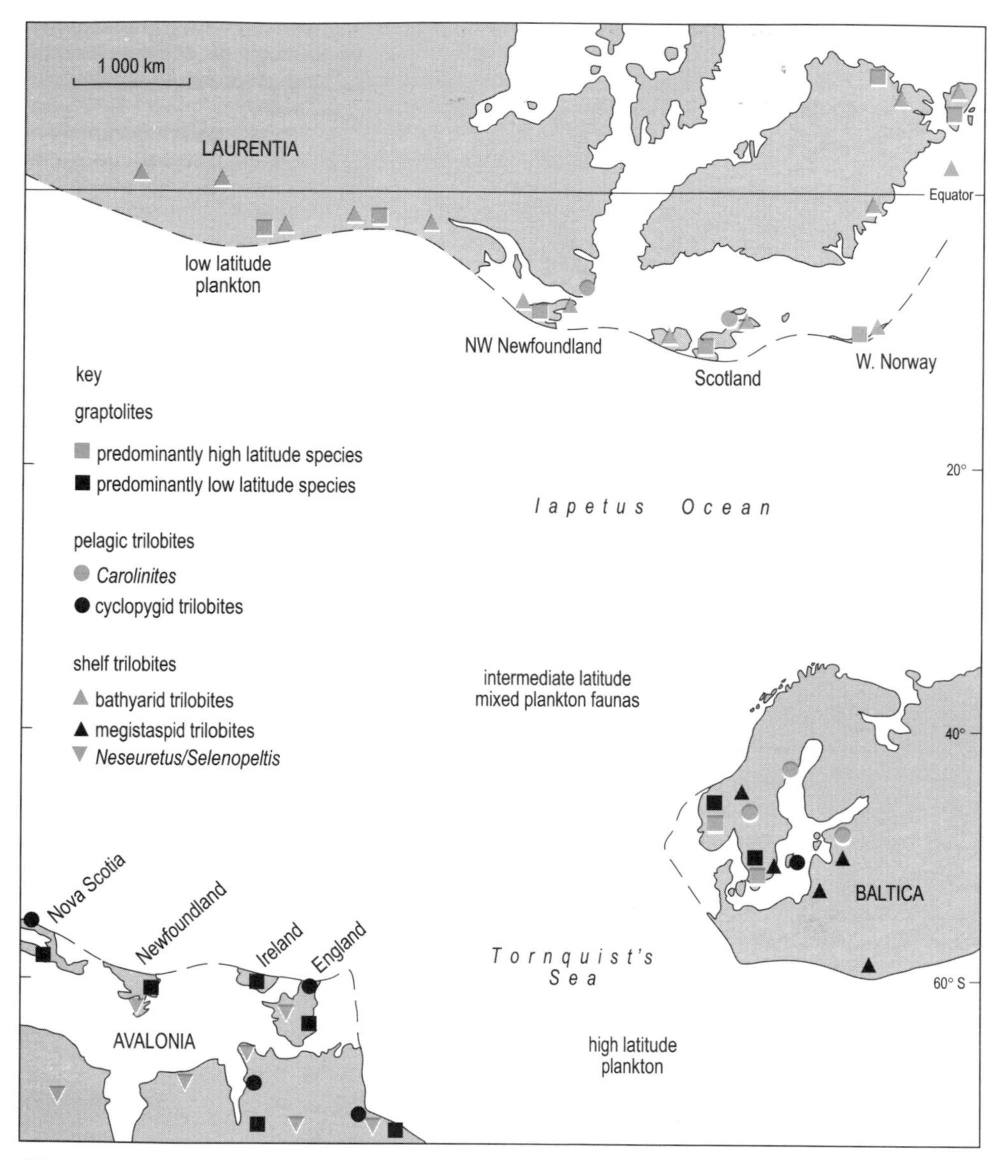

Figure 13.24 *Iapetus Ocean and the surrounding continents. Green symbols refer to the northern biotic province and black symbols to the southern province. The broken lines indicate the probable margins of land at that time. Present-day land areas are shown in grey.*

central theme of plate tectonics is that, over geological time, oceans appear and disappear. Iapetus apparently reached its greatest extent and width in the Early Ordovician. Evidence from a variety of sources suggests that Avalonia and Baltica had moved closer together and coalesced by the Late Ordovician, with the disappearance and subduction of Tornquist's Sea. Iapetus itself was already closing during the Early Palaeozoic, until Laurentia too gradually collided with Avalonia and Baltica during the Silurian and Devonian, when Iapetus ceased to exist. This collision resulted in an extended phase of mountain building, passing through what is now the Appalachians, Newfoundland, Scotland and Scandinavia, and occasioned the widespread development of terrestrial conditions and the destruction of marine habitats. This phase of mountain building is known as the Caledonian orogeny.

▶ What major processes would have been involved in the destruction of the Iapetus Ocean?

The oceanic crust flooring the Iapetus Ocean must have been destroyed by subduction at a destructive plate margin. This should have involved extensive volcanic activity and mountain building along the continental margin, below which subduction was taking place. Adjacent to an ocean floor trench associated with the subduction zone (Figure 13.19(a)), the formation of a characteristic suite of sedimentary deposits would have taken place. Regional analysis of Ordovician and Silurian volcanic and sedimentary rock sequences suggests that destructive plate margins formed at various times on one or both sides of the Iapetus Ocean.

During the opening of the Atlantic, which started many millions of years later (Figure 13.20), the line of opening did not follow exactly the old closure suture where Iapetus was extinguished. Continental segments which formerly had been part of North America, such as Scotland, remained attached to northern Europe, while other areas, e.g. south-east Newfoundland, remained fused to North America. This exchange of marginal areas from each continent explains the anomalous distribution of the Early Palaeozoic faunas. If an exchange took place in this way, then the suture of the closed Iapetus should be traceable through Newfoundland and between southern Scotland and the English Lake District. Evidence for the position of such a suture is in fact available from several lines of geological evidence.

The marine invertebrate faunas of Laurentia, Avalonia and Baltica, principally shallow water species, show little similarity in the Cambrian and much of the Ordovician (Figure 13.24), because both the Iapetus Ocean and Tornquist's Sea appear to have been wide and deep enough to have discouraged migrations. In a few cases, however, rare occurrences of pelagic species allow correlations of faunal assemblages between the different provinces (Figure 13.24). As Tornquist's Sea was subducted and Iapetus itself narrowed, so migration between each continental shelf occurred with increasing frequency, until ultimately the northern and southern provinces merged and the faunas became almost identical. Before studying the history of the migration, two factors must be considered:

1 The period of closure spanned many millions of years, during which time there were many major evolutionary changes in the organisms concerned.

2 In the fossil sequence, the marine fossil faunas on either side of Iapetus become more similar with time. The more species there are in common, the easier it is to correlate rocks on either side. However, for earlier times when Iapetus was still a wide ocean with distinctive faunas on its margins, it is difficult to correlate events across the ocean with much precision.

While Iapetus was gradually closing, there appears to have been a definite sequence in which different groups managed to traverse the ocean to colonize the other side. Figure 13.25 summarizes some data on successive faunal migrations across Iapetus. The earliest group to cross was a genus of floating graptolites (an extinct group of colonial animals) called *Rhabdinopora* (Figure 13.25(g)). This migration was followed relatively soon afterwards by other pelagic graptolites (e.g. didymograptid species) belonging to a different group from *Rhabdinopora* (Figure 13.25(f)). That pelagic animals were the first to migrate is not surprising, as they can drift or swim for enormous distances in favourable ocean currents. Benthic invertebrates, such as trilobites and brachiopods, might appear to have a limited

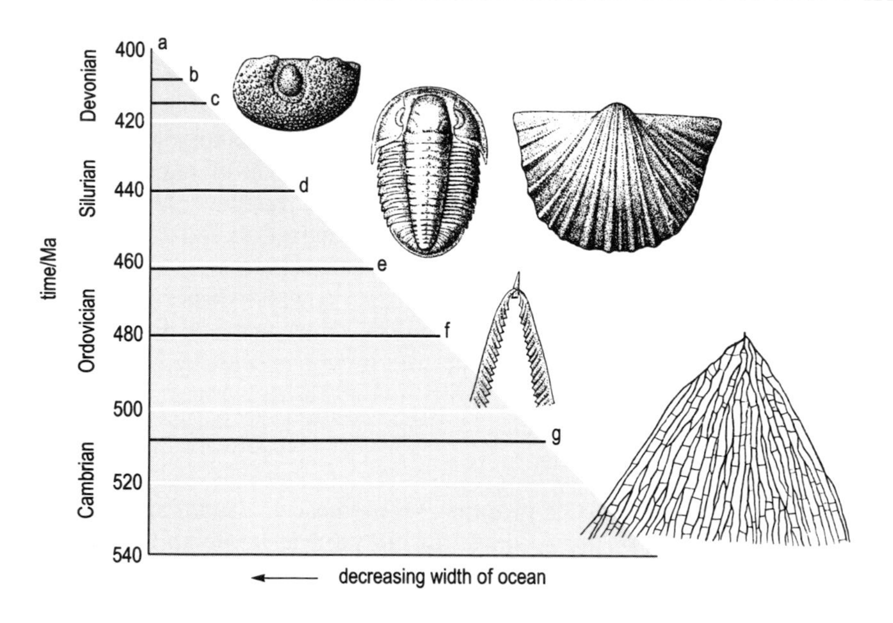

Figure 13.25 *Closure of Iapetus (a); (b–g) times when different organisms became common to both sides of Iapetus: (b) freshwater fish; (c) benthic ostracods; (d) trilobite and brachiopod species; (e) trilobite and brachiopod genera; (f) didymograptid graptolites; (g)* Rhabdinopora.

Note that the various organisms are drawn to different scales.

potential to cross deep oceans from one shelf area to another. However, many invertebrates have swimming larval forms, which, among other functions, allow dispersal. Judging from the larvae of modern brachiopods and certain trilobites (which rather unusually had a mineralized exoskeleton and therefore were preserved as fossils), members of these groups could migrate substantial distances through drifting of larvae in currents.

The distances to which larvae may be transported depend principally on the duration of the larval stage and the strength and direction of the oceanic currents. The first brachiopods and trilobites to migrate across Iapetus appear to have done so in the mid-Ordovician (Figure 13.25(e)). Independent estimates of the width of Iapetus at that time suggest that these first migrations may have been achieved by 'island hopping', for the distances were too great to have been traversed by single larvae. Nevertheless, by the end of the Ordovician, wholesale migration had led to very similar faunas of brachiopods and trilobites on either side of Iapetus.

▶ What kind of processes might have led to the formation of islands in the Iapetus Ocean?

Islands might have been formed either by intraplate volcanism, as in the Pacific today, or by the development of island arcs along subduction zones (see Section 13.4.2).

▶ In Figure 13.25, the benthic ostracods, a group of minute crustaceans with a bivalved shell, are shown to have migrated only just prior to the closure of Iapetus. Can you suggest what sort of life cycle the ostracods probably had?

They probably lacked a swimming larval stage that would have permitted migration across a wider Iapetus earlier in the Palaeozoic. Indeed, modern benthic ostracods lack such a pelagic larval stage.

Shortly before closure, some freshwater fish were able to migrate across the remnants of Iapetus (Figure 13.25(b)). With the final obliteration of Iapetus, marine faunas vanished as terrestrial sediments were deposited. It is important to realize that this scheme of progressive migration of faunas is only a very rough approximation to the actual course of events. This is partly because of uncertainty about the changing width of Iapetus and the pattern of oceanic currents. A further problem is that it is uncertain whether any oceanic islands existed in Iapetus to have acted as staging posts during migration. Island arcs and separate microcontinents (small plates) can move independently and at much faster rates than major plates and so act as 'Noah's Arks' for shallow water species. Even reconstructions like Figure 13.24, though they look convincing and are constantly being refined, cannot take into account the possibility of lateral movement of continental plates within a single climatic zone, as there is much evidence that subduction itself was taking place obliquely.

If the closure of Iapetus marked a period of progressive increase in faunal similarity, you might expect that the widening of the Atlantic during the Mesozoic and Cainozoic would have lead to the reverse situation, with the marine invertebrate faunas on either side of the Atlantic becoming progressively more dissimilar.

The similarity between two related faunas can be measured in terms of the **Simpson coefficient of similarity**, *S*, devised by the American vertebrate palaeontologist G. G. Simpson. This is calculated from the equation

$$S = \frac{C}{N_1} \times 100\% \qquad (13.1)$$

where C is the number of taxa in common and N_1 is the total number of taxa in the smaller sample. For example, if two small islands possess 157 and 105 species of native flowering plants respectively, including 84 species in common, then their coefficient of similarity will be $(84 \times 100) \div 105 = 80\%$.

The Simpson coefficient is only one of several used routinely in such biogeographical studies, but it has the advantage of being conceptually simple; in the case of the widening Atlantic (Figure 13.26), where genera rather than species

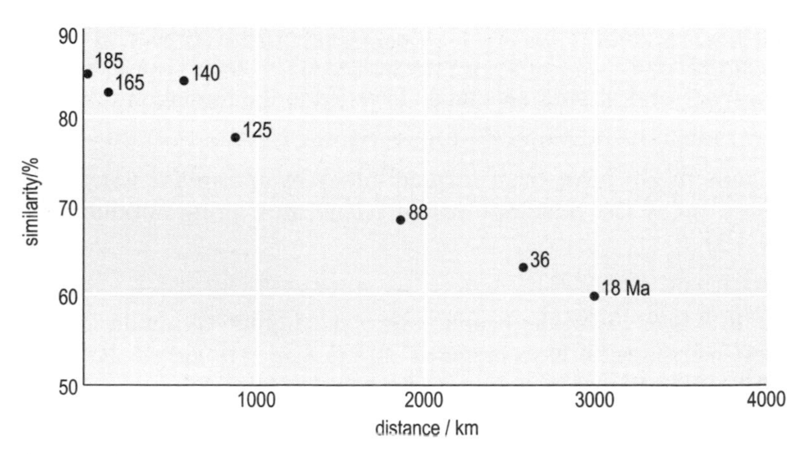

Figure 13.26 *Decreasing similarity of marine invertebrate genera, as measured by the Simpson coefficient of similarity, across the Atlantic Ocean as it widened during the Mesozoic and Cainozoic. The numbers by the points represent Ma before the present.*

are used for the database, it shows the trend of progressive dissimilarity more clearly than do other coefficients.

Note that in Figure 13.26, even with the Atlantic 3 000 km wide, overall similarity is quite high. A more detailed analysis of different groups would show that the trend towards increased dissimilarity was not smooth, but was interrupted by periods of greater cosmopolitanism, i.e. when faunas again became temporarily more uniform. These periods of increased similarity may be linked with eustatic marine transgressions which permitted freer migration, while eustatic marine regressions encouraged the development of endemic faunas with a lower trans-Atlantic similarity.

13.4.5 American vertebrate faunas: isolation and reunion

One of the most thoroughly studied examples of the effects of linking continents, with the consequent mixing of faunas and the rise and fall of different vertebrate groups, is seen in the faunal history of South America and its connection with North America. During the Late Mesozoic and earliest Cainozoic (Paleocene), the Americas were close, and certain turtles and small primitive mammals (marsupials and placentals) were able to colonize South America from the north. The link between the two continents was broken in the early Cainozoic and was not reformed until about 50 Ma later in the Pliocene.

In this period of isolation, fossil evidence shows that several orders of placental mammals evolved which were unique to South America. Some were herbivorous, including the toxodonts, that contained species as large as a rhinoceros and quite similar in general appearance. There were also the litopterns, animals unrelated to but closely resembling horses, both in the development of their dentition and in adaptations for fast running. Detailed studies show, however, that the anatomical parallels were never exact. Other unique placental mammals were ground sloths and armadillos, both of which included giant species. There were no placental carnivores; it was the marsupial stock that gave rise to predators, amongst them the borhyaenids (Chapter 11, Section 11.3.2), some of which bore a remarkable resemblance to the placental sabre-toothed tigers living on other continents. Despite the isolation of South America, three placental groups—monkeys, bats and rodents—managed to migrate from North America; the bats, of course by flying, the others perhaps by rafting. All three groups diversified into a number of endemic families, but their arrival apparently made no detectable impact on the existing endemic orders of mammals.

At the end of the Pliocene, a land-bridge or string of islands linked North and South America. Linkage provided an opportunity for both northward and southward migrations. South America was invaded by a wide range of placental herbivores, including deer, peccaries, camels (even-toed ungulates), horses and tapirs (odd-toed ungulates) and carnivores (cats, bears and raccoons). In the opposite direction, giant ground sloths, armadillos, opossums, some birds (notably hummingbirds) and several families of rodents migrated into North America. Many of the large South American herbivores, such as the giant armadillos and ground sloths, quickly became extinct in both North and South America. Other families indigenous to South America (e.g. monkeys and large rodents such as the capybara) continued to flourish. The ungulates and carnivores from the north spread throughout South America. Few, if any, of these northern placentals became extinct following contact with the southern forms.

The change in total numbers of mammalian families before, during and after the joining of the Americas is summarized in Figure 13.27. In each continent, diversity increased briefly on contact, followed by a return to approximately the same number of families as existed prior to the linkage of the continents. Note, however, that the composition of the assemblages had changed.

In general, when the two faunas mixed, the North American species fared much better than the South American forms, in terms of numbers of families that survived. However, it is important to realize that there are some exceptions to this statement and they cannot easily be explained away. The marsupial common opossum *Didelphis virginiana* has colonized North America as far north as Canada. It has readily adapted to the presence of humans and seems, if anything, to be increasing its range at the present time. Armadillos have spread northwards; various species are common throughout Central America and one has become a state symbol of Texas! Of course, not just mammals but plants and other groups of animals show similar examples. Hummingbirds, belonging to an essentially South American family of birds, the Trochilidae, are now widely established in North America as residents or as migrants. It is still uncertain how rapid was the initial spread of these particular taxa, probably over 2 Ma ago. Their present detailed distribution and abundance are clearly also related to more recent climatic events.

As noted earlier, however, many South American mammals did become extinct. How far is it reasonable to regard their extinction as a direct consequence of invasion by North American mammals? No clear-cut answer to this question can be given. The invasion from the north certainly seems to have been a factor, but at the same time the climate in South America was becoming colder and drier. Next followed the Quaternary Period, when there were major climatic upheavals in both tropical and temperate latitudes as you will read in the next Section. These climatic changes may also account for some of the changes in the indigenous South American fauna.

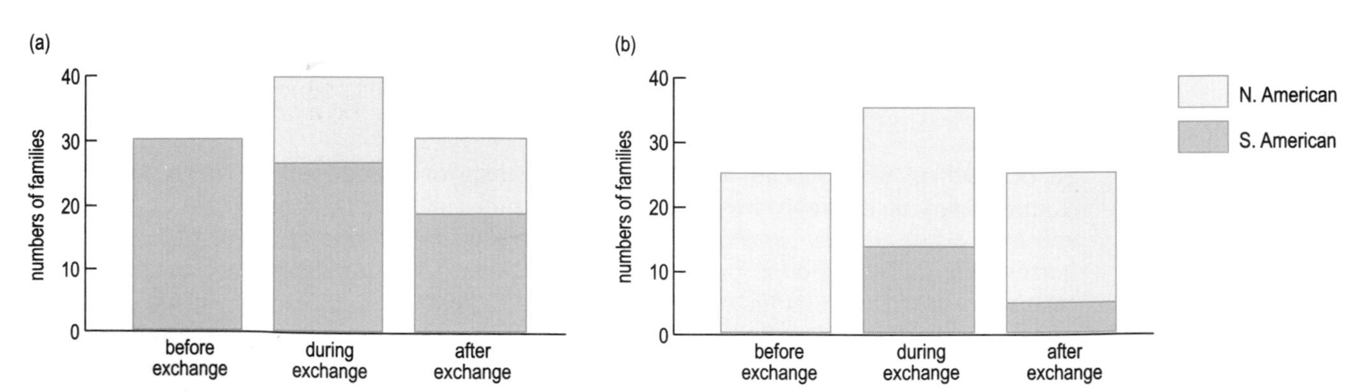

Figure 13.27 *Diversity of mammalian families in (a) South America and (b) North America, before, during and after contact of the two continents.*

13.5 THE QUATERNARY ICE AGE

Whereas continental drift has operated slowly but continuously over a time-scale of hundreds of millions of years, the past 2.5 Ma has been a period of rapid and world-wide climatic change. Ice ages have been characterized by the build-up of ice in mountain and polar areas, which then spread out as massive ice-sheets and covered lowland areas in middle latitudes, such as at present have an essentially temperate climate. Ice ages have occurred at several periods during geological time, including the Precambrian, Ordovician, Carboniferous–Permian and currently during the Quaternary (see Appendix 1 for dates). They are complex phenomena and their causes are still not fully understood, but they seem to have occurred at times when continental land-masses were predominantly clustered in one hemisphere.

The **Quaternary Ice Age** (and probably earlier ones) is not a uniformly cold period but has been characterized by cyclical oscillations of climate whose frequency is regulated by periodic changes in the Earth's orbit around the Sun. During the past 1.7 Ma the major climatic cycles have had a periodicity of about 100 000 years; they are, however, irregular with occasional shorter episodes of rapid change, as more than one orbital parameter is involved. The major climatic and environmental effects can be summarized:

1 In temperate latitudes, particularly northern Europe and much of North America (but also in areas of the Southern Hemisphere, such as New Zealand), relatively short (10–20 000 years) **interglacial periods** with temperate climates like that of the present interglacial have alternated with much longer intervals (70–90 000 years) of intensely cold climate. During these **glacial periods**, vast areas of North America, Scandinavia and intermittently much of the rest of northern Europe, including the British Isles, were covered by thick ice-sheets or, if not glaciated,

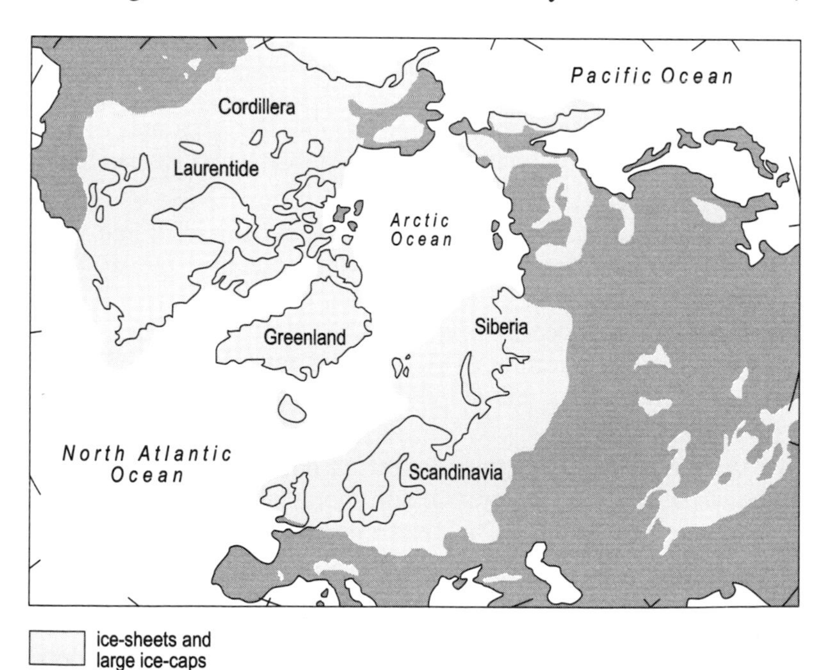

Figure 13.28 *Maximum extent of Quaternary ice-sheets in the Northern Hemisphere.*

were subject to deep permafrost (permanently frozen ground), as is northern Siberia today (Figure 13.28). The present interglacial period, the so-called Post-glacial or Holocene, began about 10 000 years ago, though a brief temporary warming also occurred around 13 000–11 000 years ago.

The Quaternary fossil record is, as mentioned in Chapter 2, Section 2.4, both well preserved and can be interpreted in detail, as it predominantly concerns species still living. It indicates that, in regions such as Europe and North America, temperate faunas and floras were very largely replaced by cold-tolerant ones and vice versa, as the climate changed. Indeed, the critical periods of climatic change were probably very short—a very few thousand years or even just hundreds.

2 Climatic change was equally marked in subtropical and tropical regions, where changes in precipitation were more important than those in temperature, so that the major lake basins of Africa, like Lakes Victoria and Chad, expanded and contracted. At times, the Sahara desert expanded its frontiers; at others, it was considerably less arid than today and supported a much more diverse mammalian and invertebrate fauna. Likewise, the boundaries of the tropical rainforests have undergone considerable change, so that these ecosystems cannot necessarily be regarded as having persisted undisturbed for millions of years, as has often been assumed (see Section 13.2).

3 The accumulation of the Laurentide Ice-Sheet in North America and the Scandinavian Ice-Sheet covering northern Europe and western Siberia had important effects not only on local environments but also on atmospheric circulation patterns and particularly on global sea-levels, which were depressed by over 100 m during extensive periods of glaciation.

▶ How would these falls in sea-level have affected shallow marine faunas?

As with eustatic marine regressions resulting from subsidence of ocean floor ridges, a falling sea-level would have led to subaerial exposure of large areas of continental shelf and a diminution of both the area and diversity of habitat available for shallow water marine vertebrate and invertebrate faunas.

Another important effect of sea-level fall during the glacial periods was that many islands, not least the British Isles, became reunited with adjacent continental mainland, so that terrestrial faunas were able to disperse freely (Chapter 2, Section 2.4.1). Even more significant from this point of view was the closure of the Bering Strait between Siberia and Alaska at these times. The formation of a land-bridge allowed migration of Asian biota into America (Figure 13.28), eventually including the earliest human invaders of the New World, and also American species into Siberia.

4 Marine environments were affected not only by a fall in sea-level, but also by changes in surface temperature and more fundamentally by significant diversions of ocean current systems, which had immense effects on the climate of both the oceans and the adjacent land-masses. In the North Atlantic, for example, the Gulf Stream/North Atlantic Drift current system was severely curtailed so that in winter 18 000 years ago the polar front extended down to the latitude of Portugal, bringing with it pack-ice and effectively blocking warm air from reaching north-western Europe.

In considering the Quaternary, you are looking at a very much shorter time-scale than that studied by most geologists. Rapid climatic changes are clearly registered as distinct horizons, both in the palaeontological and the sedimentary record. Dating methods, such as uranium series and radiocarbon dating, provide a much more detailed time-scale for the Quaternary geological record than is normally available for older periods. Thus, for example, by studying the fossil record of pollen deposition in lake sediments in northern and central Europe, it has been possible to trace the rates of dispersal of forest trees as they recolonized the area after the retreat of the latest ice-sheet. Normally, such dispersal events would be regarded as almost instantaneous in measurable geological time!

In the light of this comparative richness of information about both Quaternary biota and environmental changes, it is worthwhile trying to assess how these changes may have affected the evolution of organisms, particularly in terms of speciation, extinction and the development of diversity. The best evidence comes from the European record, as this has received most detailed study. At the same time there is the opportunity to re-examine the arguments between dispersal and vicariance biogeographers.

In Chapter 9, it was noted that among the most widely accepted models of speciation are those related to allopatric speciation. Two potential situations come to mind at once. The first is the isolation of populations on offshore islands, formed when sea-levels rose during interglacial periods; the second is the restriction of temperate plants and animals to isolated mild, moist refuges in the mountains of southern Europe during the periods of maximum glaciation, when northern Europe was frozen and the Mediterranean region was an arid steppe. The fossil record, at least in Europe, does not support these models very well. There is certainly evidence for widespread population movements in response to climate, but for most groups of organisms there is comparatively little definite evidence from the fossil record of episodes of speciation. If one ignores extinct species, the evidence from plants, molluscs, beetles and many kinds of vertebrates is that modern taxa are very similar morphologically to their presumed ancestors found in deposits of Early Quaternary age, prior to the main episodes of glaciation in Europe. There are exceptions to this statement, notably hominids (*Homo*) and certain rodents—e.g. water voles (*Mimomys* and *Arvicola*), though not bank voles (*Clethrionomys*). However, the opinion amongst specialists is that these are species that have migrated into Europe, the water voles from Asia and hominids from Asia and Africa, so speciation or alternatively the development of new chronospecies has taken place elsewhere. Other examples of apparently recent speciation are plants such as brambles (*Rubus*), roses (*Rosa*), dandelions (*Taraxacum*) and hawkweeds (*Hieracium*), which reproduce apomictically (Chapter 5). In these cases, species may be of very recent origin indeed.

The record of extinctions during the Quaternary is a very different matter. Many geologists place the onset of the Quaternary Ice Age at about 2.4 Ma, when the first really severe episode of cooling took place, instead of *c.* 1.7 Ma as given in Appendix 1. There is no clear geological record of glaciation itself from temperate latitudes of Europe (perhaps that was confined to Scandinavia), but the effect of a drastic climatic deterioration on the tree flora was extremely marked. During the Tertiary, European forests contained many species of trees belonging to genera which are now found mostly in the Far East (China and Japan), and/or in eastern or more rarely western North America. You should recall *Magnolia* and the tulip-tree *Liriodendron* (Figure 13.9), to which can be added black gum *Nyssa*, walnut

Juglans, conifers such as swamp cypress *Taxodium*, redwood *Sequoia*, the Japanese umbrella pine *Sciadopitys*, and many others. These trees vanished abruptly from Europe with the earliest episodes of Quaternary climatic deterioration, but after the onset of the really severe glacial cycles at about 1.7 Ma there were relatively few extinctions, and the interglacial forests were composed of trees that still flourish in Europe today. The pattern with the mammalian faunas is somewhat similar, many exotic species disappearing with the onset of Quaternary cooling. However, other important groups of large mammals characteristic of the interglacial periods, such as straight-tusked elephant, two species of rhinoceros, hippopotamus, and of the glacial periods, namely woolly mammoth, woolly rhinoceros and muskox, that were characteristic of the more recent part of the Quaternary, have become extinct within the past 100 000 years. It has been much debated whether the Late Quaternary extinction of large mammals both in Europe and North America can in any way be related to the spread of flourishing hunting cultures of Upper Palaeolithic humans over this same period (see Chapter 17).

13.5.1 Species diversity and tree floras of the Palaearctic and Nearctic

From the foregoing account of extinctions, it is clear that there has been a general decline in species diversity in both the fauna and the flora of Europe, particularly northern and central, during the Quaternary. The vegetation of northern Europe shows very great floristic similarities with the vegetation of North America and Eastern Asia. Many species of arctic plants, a number of which are also found in the European mountains, have in fact a circumpolar distribution and are widely distributed in Siberia and the American Arctic. Again, many herbaceous species of the ground layer of coniferous forests in northern Europe and some of the dwarf shrubs occur also in the boreal forest zone of North America. In no cases are the trees in the two continents conspecific (i.e. the same species), but vicariant or sister species of pine, spruce, fir, larch and juniper occur, though the actual number of species in the American forests is greater. In the temperate forest zone, the differences are more pronounced and the relationships between European and American species less close, but virtually all the tree genera found in Europe are represented in the American forest flora. Also, as described earlier, in North America there occur many genera that once existed in Europe during the Tertiary but are now extinct there.

What is clear, however, is that, compared to those of North America, the present-day European forests are strikingly impoverished both in terms of numbers of genera and numbers of species. This impoverishment includes not only tree species, but also shrubs and woody climbers. To illustrate this, in the area of Europe north of the Alps and west of European Russia, 45 broad-leaved tree species occur and 5 conifers. The equivalent figure for north-eastern America is 148 broad-leaved and 23 coniferous species! If you next consider Eastern Asia, the discrepancy is even greater—Europe holds 2 species of hornbeam *Carpinus* and 14 species of maple *Acer*, of which 11 occur principally south of the Alps, whereas in one restricted area of China 25 species of maple and a staggering 111 species of hornbeam have been reported! Most surviving European tree genera also have their representatives in eastern Asia.

Four questions arise from this discussion:

1 How did the disjunct distribution patterns for temperate forest tree genera arise?

2 How could one explain the fact that the floras of the boreal forest zones and the arctic zones in Europe are less different from North America than are those of the temperate forest zones?

3 What accounts for the differences in diversity between European and North American forests?

4 What speciation mechanisms might contribute to the extraordinary high diversity of some forest floras in eastern Asia?

It has already been suggested in Section 13.3.3 that, because of the many similar tracks that can be constructed for distribution patterns like those of magnolia and tulip-tree, a vicariance explanation must be preferred to one based solely on dispersal hypotheses. Today, it is the aridity of Central Asia that effectively forms a barrier between, on the one hand, the temperate forests of Europe, including the rather species-rich forests of the Caucasus mountains that lie between the Black and Caspian Seas, and, on the other, the temperate forests of southern China.

▶ After studying Figures 13.20 and 13.21, what possible vicariance events can you recognize that might explain the separation of an ancestral floral province into three distinct areas?

North America and Europe were separated by the opening of the North Atlantic Ocean between about 110 and 50 Ma. It is relevant to note that fossil evidence from Early Tertiary rocks in Greenland, one of the last areas of contact, shows that climatic conditions there were mild enough to support broad-leaved temperate forest. Again if you look at Figure 13.21, you will see that for a very long period Europe and China were linked by the northern shore of the Tethys Ocean. Both the closure of this ocean in mid-Tertiary times and the collision of the Indian and Eurasian Plates, commencing in the Eocene (Early Tertiary) and eventually creating the vast mass of the Himalayas, would, as noted in Section 13.4.1, have cut off moisture-bearing winds to Central Asia and so destroyed any intervening temperate forest belt there. Large land-masses develop constant high pressure areas from which winds blow outwards, leading to desert conditions.

The much greater similarity of floras in the boreal forest and arctic zones must have a different explanation, but these ecosystems have a much shorter and less well-studied geological record than temperate forest communities. It is not yet clear to what extent dispersal and vicariance events may have been involved. The closure of the Bering Strait at intervals both during the Late Tertiary and the Quaternary certainly provided opportunities for dispersal between the Old and New Worlds. What is certain is that during the glacial periods, both in Europe and North America, arctic and boreal forest plant communities enormously expanded their areas of distribution.

One explanation for the low diversity of the European tree flora was put forward by the palaeobotanists Clement and Eleanor Reid as long ago as 1915. They suggested:

'At the beginning of Pliocene times all over the temperate regions of the Northern Hemisphere existed a thermophilous [warmth-loving] flora, consisting mainly of species closely related to the present East Asian and North American vegetation. The progressive cooling of the climate towards the Pleistocene drove this flora southward. In its migration the flora was checked by the barrier of the European East–West mountain ranges and the Mediterranean, causing the less cold-resistant species to die out, whereas in East Asia and North America with mountain ranges running North–South they could migrate far enough southward to survive the extremest cold and to return to their present habitats when the climate improved.'

This hypothesis remains widely accepted, particularly with regard to the wave of plant extinctions at the beginning of the Quaternary. It is now also clear that for many thermophilous tree species, adapted primarily to a mild, moist European interglacial climate, the aridity of the Mediterranean basin during the glacial periods was also a formidable obstacle to survival.

By contrast, there are areas of China where temperate forests still show a very high species diversity today. As already discussed, many biogeographers relate high species diversity to long-term environmental stability (Section 13.3), and indeed these forests occur in regions that have always been well away from the influence of glaciation, but at present very little is known about their detailed Quaternary climatic history or how stable vegetation patterns there have really been.

Obviously, climatic and environmental disturbance during the Quaternary resulted in the extinction of many taxa of plants and animals, particularly in Europe, and indeed throughout the world. On the other hand, it might be thought that such disturbances, which often involved fragmentation of populations, would actually encourage speciation. In general in northern Europe, where Quaternary faunas and floras are well known, this does not seem to have been the case. A broad range of living animals and plants, if they have a fossil record, appear to have persisted without any obvious morphological change, often throughout the Quaternary. It might, therefore, be argued that there are processes which have actually discouraged speciation during the Quaternary. Examination of the history of European tree species during the glacial periods may help to explain this.

Whereas very large populations of temperate forest trees built up both north of the Alps and in southern Europe during interglacial periods (such as the present), they must have crashed dramatically with the onset of each glacial period. It is well established that there was no widespread belt of temperate forest anywhere in southern Europe at the times of the glacial maxima, and that north of the Alps it was generally far too cold for such trees to survive. Their refuges were probably situated in special habitats in mountain areas, such as the steep cliffs of gorges, where precipitation or humidity was greater than in the arid lowlands and where there was some protection from grazing animals, since wild goat or ibex and chamois were abundant. Logically, you might think that these isolated refuges produced ideal conditions for different populations in the mountains of Spain, Italy and Greece to undergo allopatric speciation, but this does not seem to have happened for most species. Instead, when the climate ameliorated again, the trees spread out from their refuges, slowly migrated northwards again and the temperate broad-leafed forests of northern and central Europe were recreated after millennia of frozen steppe tundra.

Perhaps partly because trees are long-lived organisms, particularly under adverse climatic conditions, it seems likely that, whilst isolated in their refuges, there was insufficient time, i.e. insufficient generations, for them to evolve into new species, and the gene pool was regularly remixed at least every 100 000 years when warmer conditions allowed them to re-expand. It is likely that any new variations would then have been swamped or selected against under the rapidly changing conditions. Where isolated endemic taxa have survived, they are found in the mountains of southern Europe. These species too, presumably, have survived the glacial periods in refuges close to their present sites. Some may be young endemics that have undergone recent evolution; others like the five-needled pine *Pinus peuce* from the Balkans, whose closest relatives (with which it is apparently interfertile) live in the Himalayas and North America, are certainly old endemics. Fossil needles and cones of another Balkan endemic tree, the spruce *Picea omorika*, have been found in cool climate Quaternary deposits in northern Europe, emphasizing that it once had a wider range. Apparently, these species either do not possess or alternatively no longer retain the capacity to expand their ranges or migrate rapidly northwards during interglacial periods, and so now remain as small isolated populations.

Much the same explanation could be applied to account for the remarkable uniformity of arctic beetles and plant species which have very extensive circumpolar distributions today, as well as small colonies in the mountains of the temperate zone, but which spread far to the south during periods of glacial cooling. Such a mechanism might explain why many species show little evolutionary change during the last million years, although they have much shorter life cycles than trees. Nevertheless, other taxa, such as the Apollo butterflies *Parnassius* which inhabit high mountain areas in Europe, that you would expect to have been greatly affected by glaciation, have, in fact, evolved many local subspecies and races. Evolutionary patterns are in practice almost never subject to generalizations!

SUMMARY OF SECTIONS 13.4 AND 13.5

These Sections discuss the impact of geological events that affect geography and climate on evolutionary processes. The concept of continental drift was revived and subsumed into the theory of plate tectonics, which attempts to explain a whole range of geological processes within a single cohesive framework. Amongst these processes the following are likely to have had important evolutionary effects.

1 Rifting and the splitting apart of ancient land-masses mean that new continents develop, separated by a constructive plate margin and a widening ocean. Populations of plant and animal species become fragmented and reproductively isolated, and allopatric speciation can take place. New shallow water coastal areas develop on either side of the ocean, but also become progressively more isolated, encouraging the development of provinciality. New deep-ocean habitats are also created.

2 The drifting of continental land-masses through different climatic zones, e.g. passing from tropical humid equatorial to subtropical arid conditions, is likely to have pronounced effects on species diversity and composition.

3 The collision of continents results in the intermixing and migration of faunas and floras between previously separated land-masses. Continental collisions also result in the elimination of intervening oceans with both their deep and shallow water faunas, causing mixing of populations and eventually widespread extinctions. Another feature is the development of mountains, such as the Himalayas, when continents collide. These form substantial physical and climatic barriers to migration and can become boundaries between biogeographical regions, as well as containing specialized environments with many, often isolated, new niches where speciation can occur. The rise of new mountain barriers also affects global atmospheric circulation patterns and therefore the distribution of areas of desert and of high precipitation.

4 The drifting of continents and particularly their collision will also themselves have major effects on oceanic and atmospheric circulation patterns.

5 Islands may form, either as island arcs at a destructive plate boundary where two oceanic plates meet, locally above constructive margins, or as a result of intraplate volcanism above 'hot spots'. The islands are colonized by relatively few species, represented by small numbers of pioneer individuals, so that founder effects and genetic drift may then become important, leading to allopatric speciation. Such islands can also act as stepping stones in the dispersal of species over long distances between otherwise widely separated continents or land-masses.

6 Changes in sea-level, in part related to rates of sea-floor spreading, also lead to the separation and joining of land-masses, and to changes in size and distribution of islands and of shallow water habitats on continental shelves.

The Quaternary Ice Age has produced environmental changes on a much more recent and more rapid time-scale. The repeated, cyclic climatic fluctuations during the Quaternary have affected not only temperate latitudes, where glaciation and permafrost occurred, but also tropical and subtropical regions where there have been repeated major changes in precipitation, and thus in the boundaries between tropical forest and more arid regions. These climatic changes have resulted in almost constant displacements, not just of individual species but of whole ecosystems. There is, therefore, good evidence that large-scale migrations (i.e. dispersal events) have occurred at critical periods of climatic change. In many ecosystems, the resulting stresses have undoubtedly depressed species diversity and many extinctions have taken place, especially in western Eurasia. Periodic species migrations may tend to ensure regular mixing of gene pools and thus discourage speciation, but in other cases the fragmentation and isolation of populations that also occurred during the upheavals of the Quaternary has had the opposite effect.

Frequent sea-level changes during the Quaternary, as a result of the growth and dissolution of huge continental ice-sheets, have both formed and removed isolating barriers between continents and offshore islands, such as Britain, and, most notably, between Siberia and Alaska, thus permitting substantial faunal and floral migrations. The oceans, both along the continental shelves and in the deep-sea, have been almost as strongly affected as the land, with big changes in surface temperature and to major current systems such as the Gulf Stream.

13.6 CONCLUSION

After reading this Chapter, you should understand that, from the time of the earliest appearance of life, geographical factors have influenced the course of evolution. Climatic changes brought about by long-term changes in the Earth's surface temperature and atmosphere, by continental drift and by the recent, or indeed, present Quaternary Ice Age have all had a strong geographical effect on the patterns of evolution of living organisms, and particularly on patterns of diversity. Similarly, the effects of continental drift, mountain and island building, and sea-level change — all an integral part of plate tectonic activity — have led to the isolation and speciation of groups of organisms or brought them together to compete and even interbreed.

At this point, before moving on to the next Chapter, you should ponder briefly on how civilizations during the past 5000 years (particularly the past 100 years) have been affecting the patterns of species diversity and natural geographical distributions discussed in this Chapter. Consider, in this light, the changes that have been and are being wrought so rapidly by deforestation, monoculture and the wholesale world-wide transport of plant and animal species, and by the, until recently almost undetected, progressive effects of industrialization on the atmosphere and oceans and ultimately on global climate. You will return to these themes in the final Chapter of this Book.

REFERENCES FOR CHAPTER 13

Cocks, L. R. M. and Fortey, R. A. (1982) Faunal evidence for oceanic separation in the Palaeozoic of Britain, *J. Geol. Soc. Lond.*, **139**, pp. 465–478.

Croizat, L. (1958) *Panbiogeography*, vols 1, 2a, 2b, Caracas, published by the author.

Humphries, C. J. (1981) Biogeographical methods and the southern beeches, in P. L. Forey (ed.) *The Evolving Biosphere*, pp. 283–297, British Museum (Nat. Hist.), Cambridge University Press.

Reid, C. and E. M. (1915) The Pliocene floras of the Dutch–Prussian border, *Med. Rijksopsporing v. Delfstoffen*, No. 6.

FURTHER READING FOR CHAPTER 13

Forey, P. L. (ed.) (1981) *The Evolving Biosphere*, British Museum (Nat. Hist.), Cambridge University Press.

Good, R. (1974) *The Geography of the Flowering Plants*, 4th edn, Longman.

Humphries, C. J. and Parenti, L. R. (1986) *Cladistic Biogeography*, Clarendon Press, Oxford.

Myers, A. A. and Gillez, P. S. (eds) (1988) *Analytical Biogeography: An Integrated Approach to the Study of Animal and Plant Distributions*, Chapman & Hall, London and New York.

Natural History Museum Publication (1972) *The Story of the Earth*, HMSO.

Nelson, G. and Platnick, N. (1981) *Systematics and Biogeography*, Columbia University Press, New York.

Nelson, G. and Rosen, D. E. (eds) (1980) *Vicariance Biogeography: A Critique*, Columbia University Press, New York.

Valentine, J. W. (1973) *Evolutionary Paleoecology of the Marine Biosphere*, Prentice-Hall Inc., New Jersey.

SELF-ASSESSMENT QUESTIONS

SAQ 13.1 (*Objective 13.1*) Suggest two pieces of evidence which could be used to challenge the hypothesis that high species diversity is strongly correlated with abundant availability of nutrients.

SAQ 13.2 (*Objective 13.1*) Give up to four possible explanations or contributory factors that have been put forward to explain why tropical forests generally support ecosystems with a higher species diversity than temperate forests.

SAQ 13.3 (*Objective 13.2*) How are plate movements likely to have affected the differences and similarities between the biota of the African and Palaearctic regions since the Permian? List the major geological events/processes, and comment on their likely effects and the present position of the boundary between these two biogeographical regions.

SAQ 13.4 (*Objective 13.3*) (a) In what critical respect does cladistic biogeography differ from both dispersal and vicariance biogeography?

(b) Consider a track for a particular group of organisms that show a disjunct distribution pattern. What additional information would be required (two or three ideas should come to mind) for such distributional data to be of use to a cladistic biogeographer in trying to interpret such a pattern?

SAQ 13.5 (*Objective 13.4*) In Figure 13.29, (a) represents the hypothetical distribution of two groups of organisms in the Southern Hemisphere. Suppose A–D represent closely related genera of freshwater fishes and W–Z represent flowering trees. (b) and (c) are cladograms for the fish and plant groups, respectively. Construct area cladograms for each group. What *common* pattern could be resolved from these area cladograms, and what would represent the sister group to the New Zealand and/or South African group of genera in that case? Considering this common pattern and the geological history of the southern continents, what could you conclude about the likely evolutionary history of these groups?

SAQ 13.6 (*Objectives 13.1, 13.5 and 13.6*) Study Figure 13.30, which shows the movement of three imaginary continents A, B and C over a period of 50 Ma (arrows show the direction of movement). Which of the following are probable outcomes?

(i) The biota on continent A would increase in diversity.

(ii) The biota on continent B would increase in diversity.

(iii) The biota on continent C would show little change in diversity.

(iv) The marine invertebrates of B and C would show a progressive decrease in similarity.

(v) The marine invertebrates of A′ and A″ would show a progressive decrease in similarity.

(vi) The biota on B would become less endemic.

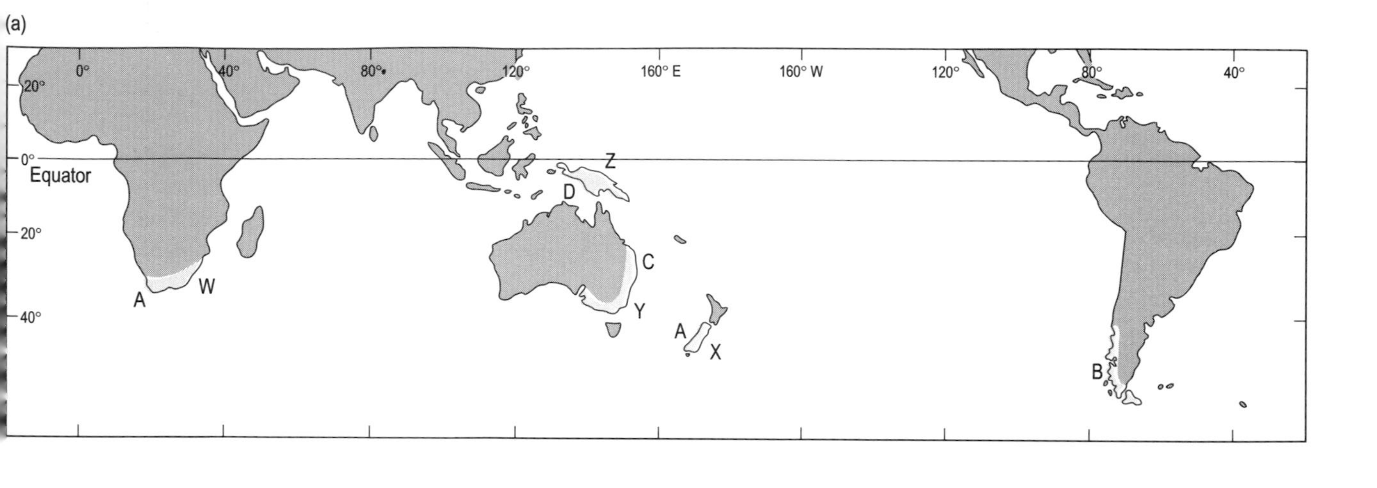

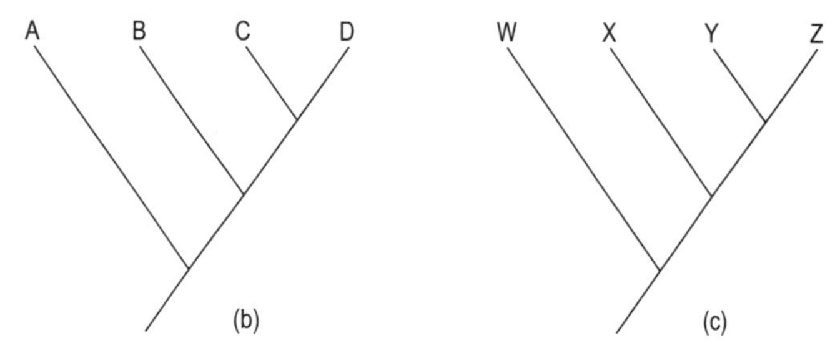

Figure 13.29 For use with SAQ 13.5.

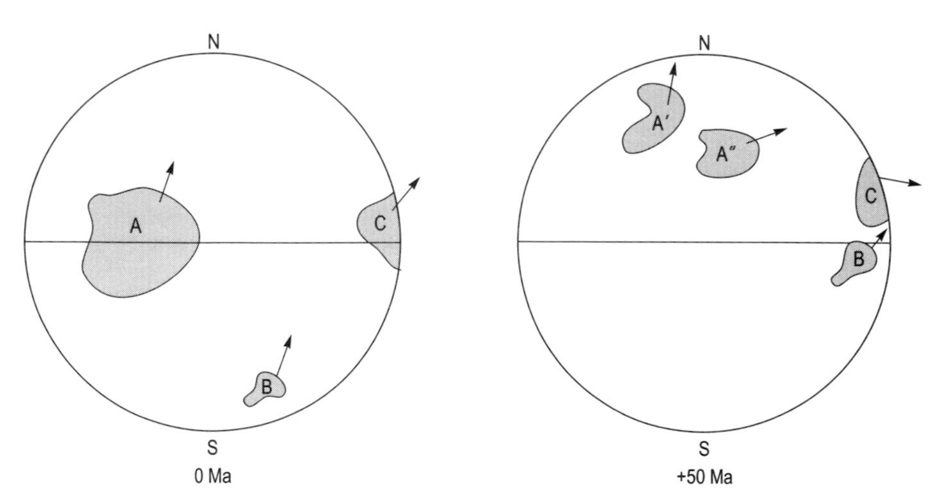

Figure 13.30 For use with SAQ 13.6.

SAQ 13.7 (*Objectives 13.5 and 13.8*) (a) Which geological processes might cause a eustatic rise in sea-level?

(b) List the various ways in which a eustatic rise in sea-level might affect the evolutionary history of marine and terrestrial organisms.

SAQ 13.8 (*Objective 13.7*) Studies have been made of a series of fossil faunas of marine invertebrates of successively younger age which occurred on the coasts of the two (hypothetical) continents of Raesgip and Nosnibor, that were separated by the Evian Ocean. Table 13.2 gives data about the number of genera that were recognized at each time-horizon. From these data, calculate the Simpson coefficients of similarity for the four horizons, and comment on their significance.

Table 13.2 *Numbers of fossil marine invertebrates on the two continents. For use with SAQ 13.8.*

Time/Ma	Total genera on Raesgip	Total genera on Nosnibor	Genera common to Raesgip and Nosnibor	Simpson coefficient /%
74	620	310	260	
62	424	531	310	
54	507	512	345	
44	542	829	320	

SAQ 13.9 (*Objectives 13.3 and 13.8*) The flora of the west of Ireland contains a small number of species ('Lusitanian species') which have interesting disjunct distribution patterns along the Atlantic coast of Europe. These include four species of heather *Erica* spp. and the closely related St. Dabeoc's heath *Daboecia*. Their overall distributions are as follows (Table 13.3):

Table 13.3 *For use with SAQ 13.9.*

Species	Distribution
Erica mackaiana	W Ireland, NW Spain
Erica ciliaris	W Ireland, SW England, W France, Spain, Portugal, Morocco
Erica erigena	W Ireland, SW France (extinct?), Spain, Portugal
Erica vagans	W Ireland, SW England, W France, N Spain
Daboecia cantabrica	W Ireland, SW France, N Spain, Portugal

These species all occur in oceanic lowland heathland, which tends to have mild winters and high annual rainfall. Particularly taking into account information from Sections 13.3 and 13.5, make brief answers to the following questions.

(a) What broad hypotheses might be put forward to explain these distribution patterns?

(b) What facts or observations available to you would support or conflict with these hypotheses?

(c) What further information would (perhaps ideally) be needed to strengthen or test these hypotheses?

(Note that there is still no agreed solution to this problem!)

CHAPTER 14
THE EVOLUTION OF FORM

Prepared for the Course Team by Peter Sheldon

Study Comment *The definition of evolution most often used by Darwin was 'descent with modification'. But what, in essence, is being modified? In fact, it is the form of organisms. Form, as explained in the Introduction, encompasses the entire physical constitution of an organism, from the atomic level up through every observational scale to the complete individual. However, evolution does not have a free hand to modify descendants. There are important constraints that limit the pathways which evolution can follow. The nature of these constraints is a major theme of this Chapter.*

Evolution is characterized here as a tinkerer, fashioning new forms from existing bits and pieces rather than designing from scratch. An analytical approach called theoretical morphology can reveal where there are discrepancies between the range of theoretically possible forms and those actually adopted in nature; this in turn can reveal which constraints are operating. The various ways in which lineages can be released from constraints are also examined. The final Section considers one of these in more detail: heterochrony, which is emerging as an important mechanism that produces large alterations in form with relatively little genetic change.

Objectives When you have finished reading this Chapter, you should be able to do the following:

14.1 Write a short account of: the nature of animal phyla, with special reference to molluscs; the appearance of animal phyla in the late Precambrian and early Cambrian; and the evolutionary significance of the Burgess Shale animals.

14.2 Reproduce 'Seilacher's triangle' of constraints on the generation of form and discuss the phylogenetic, functional and fabricational factors, giving examples of each.

14.3 Given information about the form of an organism, suggest how that form may have been influenced by the factors of 'Seilacher's triangle'.

14.4 Explain in what sense evolution can be said to work like a tinkerer, noting the limitations of this metaphor.

14.5 Discuss the meaning and evolutionary significance of vestigial structures, atavisms, and 'fabricational noise', giving examples.

14.6 Explain how the action of various constraints may limit organisms from achieving optimal form for particular functions.

14.7 Describe and evaluate other factors that may contribute to form (in addition to the main ones of 'Seilacher's triangle'), such as chance and ecophenotypic effects.

14.8 Employ given information on theoretical morphology (including that of spirally coiled shells) to analyse form and the constraints that limit the range of forms adopted in nature.

14.9 Discuss the ways in which lineages can be released from constraints on the generation of form.

14.10 Describe the concept of preadaptation, giving examples.

14.11 Given information on the morphology of ancestor and descendant, determine the nature of any heterochrony, distinguishing between examples of paedomorphosis, peramorphosis, neoteny, postdisplacement, progenesis, acceleration, predisplacement and hypermorphosis.

14.12 Discuss the evolutionary significance of heterochrony, giving examples from the fossil record and from living organisms.

14.1 INTRODUCTION

This Chapter explores how studying the form of organisms can provide insights across the whole spectrum of evolutionary patterns and processes. Using fossils as well as living organisms, a rigorous analysis of form will yield information on aspects as diverse as phylogeny, selection pressures, changes of function, and evolutionary rates. But what is meant by the 'form' of an organism? It is probably best to think of the form of any thing as encompassing its entire physical constitution. As a simple inorganic analogy, take the book you are holding. Its form concerns all its external properties of shape, size, and appearance (e.g. colour and texture) as well as every aspect of its internal structure: its pages, print, binding materials, cover and so on. A full description of form would cover the shape, size, composition and characteristics at every scale from the thickness of a page down through the cellulose fibres to the atoms themselves. And its form is altering with time. This book is not quite as static as it may seem. If the binding smells of glue then the weight of the book is fractionally changing as some of the glue vaporizes; this weight loss may be compensated by the addition of oils and salt from a sweaty palm, notes in the margin, or a fly squashed in the index. Colours may fade, a page get slightly torn, and so on.

Exactly the same kind of exercise could be done with any organism, but, because of the immense degree of internal organization (organ systems, organs, tissues, cells, organelles, genes, etc.), the descriptive task would be far harder. In the same way that a map of a piece of land cannot be more detailed than the land itself, so a complete description of any organism would approach the complexity of the entire organism itself.

Much more than a full description of physical form is, of course, necessary for a full *understanding* of any organism. We would need to know exactly how form related to function and growth at every scale and how that relationship had evolved.

14.1.1 Phyla — a set of bodyplans

A child investigating rock pools at the seaside soon learns to recognize that the different animals there do not make up a continuous spectrum but cluster into limited categories of form (bodyplans), some of which are more closely similar to each other than to others. An observant child might lump a a crab with a shrimp, or even a starfish with a sea-urchin. But appearances can be deceptive. What child would lump a barnacle with a crab and a shrimp? Yet the larvae and internal anatomy of a barnacle reveal a close relationship with them (see Chapter 1, Figure 1.7, for a cirripede larva). What controls such similarities and differences?

All living animals can be grouped within about 32 phyla, the largest subdivisions of kingdoms. The precise number of extant phyla is debatable, depending on how particular (mostly rare) groups are classified. Representative examples of some living animal phyla are shown in Figure 14.1. The general complexity (grade) of their body architecture increases from Porifera through Cnidaria and Ctenophora, to Platyhelminthes, Nemertea, and the remainder— depending on the number of tissue layers, the presence or absence of a body cavity, and other characteristics. Bodyplans of many phyla are mostly so distinct that their evolutionary relationships are uncertain; most sister groupings (i.e. pairs of phyla that share a more recent common ancestor than with any third phylum, as explained in Chapter 11, Section 11.2.4) are controversial.

Only about 10 of these living phyla have ever had parts hard enough to constitute a major component of the fossil record. They are the following, with typical examples from a modern coastline given in brackets: Porifera (sponge), Cnidaria (coral), Bryozoa ('sea-mat' colony), Brachiopoda (two-shelled animal not commonly seen around British coasts), Mollusca (cockle shell), Arthropoda (crab), Echinodermata (starfish), Hemichordata (acorn worm—a distant relative of the extinct colonial hemichordates called graptolites that had a proteinaceous skeleton), and Chordata (herring).

All the subsidiary taxa (e.g. classes, orders, families, genera and species) within each phylum share the same basic bodyplan, albeit modified to a greater or lesser extent. Consider the Mollusca. There are seven living classes of molluscs, and at least one extinct class. Animals as apparently diverse as a snail (class Gastropoda), a cockle shell (class Bivalvia), and a squid (class Cephalopoda) are united together in this phylum. Yet each class is a modification of the same basic bodyplan.

Figure 14.2 illustrates some important features of four living classes of molluscs. In the centre (a) is what was once termed the molluscan archetype, i.e. a hypothetical ancestral mollusc ('HAM') that shows the basic features of the bodyplan shared by the classes. This archetype is believed to resemble the ancestral species of at least the classes shown here, if not all those in the phylum. Two mollusc classes not discussed here *may* be derived from yet more primitive forms with only some of the molluscan features of 'HAM'. Forms similar to this archetype dominate the early fossil record of molluscs, and living molluscs can be easily compared with it; indeed, one living class, the Monoplacophora (b, c), is very similar to it.

The body of the archetype comprises a visceral mass (guts, reproductive organs, etc.) indicated by dark shading on Figure 14.2. Underlying the visceral mass is a

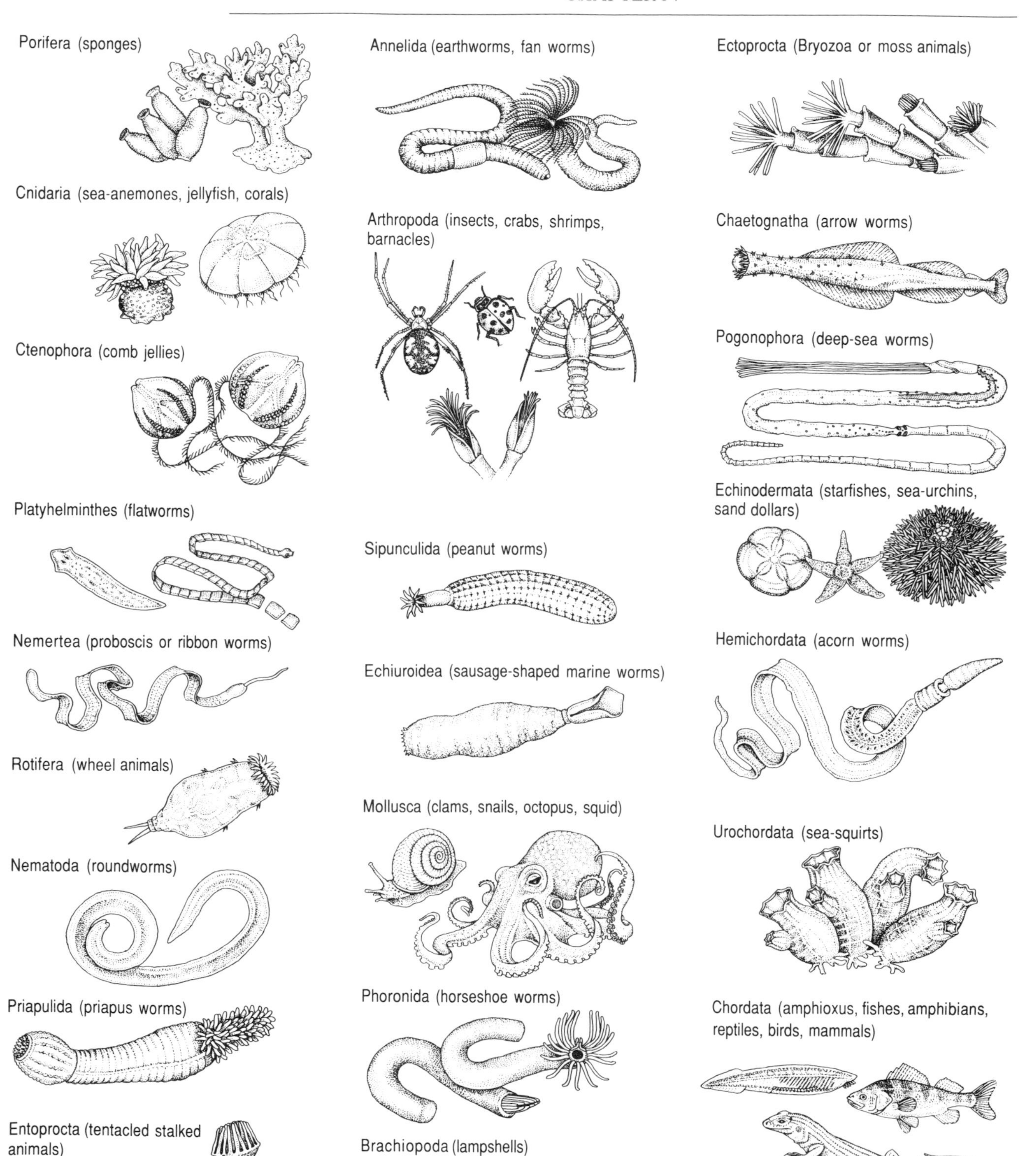

Figure 14.1 *Representative examples of some living animal phyla (drawn to different scales).*

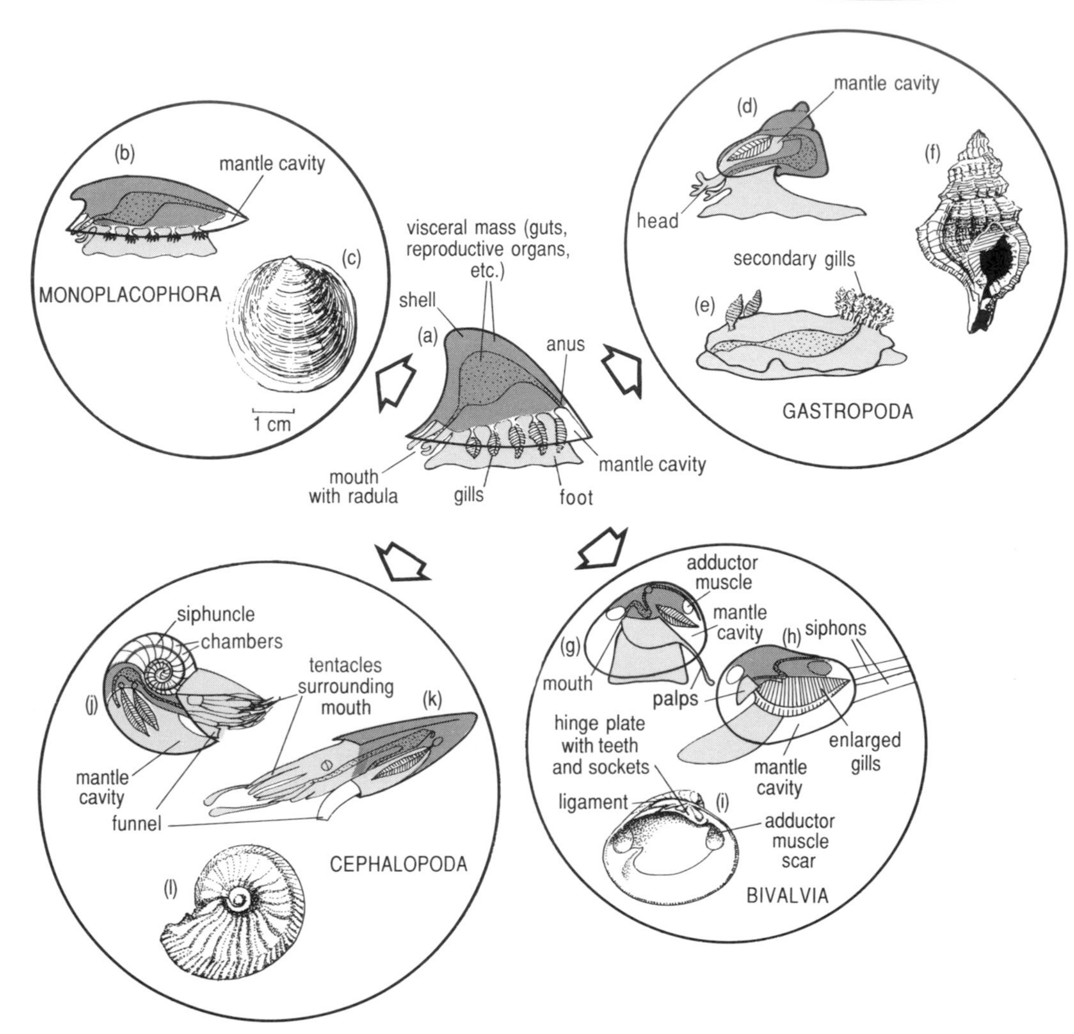

*Figure 14.2 Features of four living classes of the phylum Mollusca. The bodyplan modifications in each class are shown diagrammatically as 'transparent' side views. Examples of their shells are also shown. See text for further explanation. (a) Molluscan 'archetype', a combination of features common to the classes shown here. (b) Monoplacophoran. (c) Monoplacophoran shell (viewed from above). (d) Gastropod. (e) Shell-less gastropod (sea-slug). (f) Gastropod shell. (g) Protobranch bivalve. (h) Suspension-feeding bivalve. (i) Bivalve shell. (j) Cephalopod (*Nautilus*). (k) Cephalopod with reduced shell (squid). (l) Cephalopod shell (ammonite).*

broad creeping foot. Protruding from the front is a snout-like mouth containing the radula (a horny ribbon studied with tiny rasping teeth; see Section 14.3.3 and Figure 14.16), while at the back the gut empties via an anus. Dorsally, the entire archetypal body is covered by a forwardly inclined conical shell. The shell is produced by a sheet of dorsal body-wall tissue that lines its inner surface, called the mantle. The shell keeps pace with body growth by the addition of increments around its margin, which leave growth lines running parallel with the rim on its exterior. The shell and mantle edges extend out beyond the visceral mass, thus enclosing a 'mantle cavity'. This cavity is open ventrally, around the sides and back of the visceral mass and foot. It is somewhat expanded posteriorly, and contains one or several pairs of leaf-like gills which project out sideways from the body wall. The molluscan classes discussed here share all or most of the basic features of this archetype but in modified form, and certain features such as the shell may be reduced or absent in some cases.

▶ Name a group of shell-less land molluscs that are very common in gardens.

Slugs, which are basically snails without a shell. A slug or snail uses its muscular foot to track down lettuces, etc. before ravaging them with its radula.

Each class is characterized by a complex of modified archetypal features peculiar to itself. For instance, in the class Gastropoda (Figure 14.2(d–f)), which includes snails and slugs, the visceral mass is generally connected via a narrow waist to the head-foot. This structure consists of a broad fleshy foot and a head, which is equipped with a mouth, eyes and sensory tentacles. During development as a larva, gastropods undergo torsion: the visceral mass and the mantle are twisted around at the waist, so as to become back-to-front with respect to the head-foot beneath. If a shell is present, torsion allows protective withdrawal of the head, followed by the foot, into the now anteriorly directed mantle cavity inside the shell, as you can confirm for yourself by picking up any garden snail. (Torsion of the visceral mass is quite distinct from the coiling of the shell.) Some gastropods, especially adult shell-less forms, have secondarily straightened out—a process called detorsion.

Members of the class Bivalvia lack a head (see Figure 14.2(g–i)) and the radula has been lost. The shell is divided into two usually symmetrical valves (these 'valves' are not to be confused with devices to control the flow of a fluid), one on each side, which are articulated by a ligament along the top. The ligament is resilient, i.e. capable of regaining its original shape after stretching or compression. In most forms, the valves are kept in mutual alignment by interlocking teeth and sockets situated on hinge plates running around the dorsal margins of the valves. The valves can be drawn together by adductor muscles acting against the opening force of the ligament. The whole body may thus be entirely enclosed by the shell. Adductor muscles are usually indicated by scars on the internal surface of the shell (Figure 14.2(i)). The muscular foot can be protruded between the valves and used for crawling or burrowing. In one small, primitive group of bivalves called protobranchs (Figure 14.2(g)), food particles are collected with a pair of tentacles extended from the lips (palps). In most other bivalves (e.g. Figure 14.2(h)), much expanded gills are adapted for suspension-feeding. Figure 14.6 shows details of one bivalve species, the common cockle *Cerastoderma edule*.

The class Cephalopoda includes octopuses, squids, nautili, and the extinct ammonites (see Figure 14.2(j–l)). The body is generally bilaterally symmetrical and the head is greatly developed, with eyes and a ring of tentacles around the mouth. Cephalopods can move by jet propulsion, water being drawn in through the lower part of the mantle cavity opening (which then closes) and pumped out through the funnel, an extension of the foot beneath the head. In most living forms, the shell is reduced or absent (as in the squid, Figure 14.2(k)), though primitively, as in the nautilus (Figure 14.2(j) and Plate 14.1), it is an extended coiled cone. Most of the shell, when present, is usually divided into chambers by septa (sing. septum) of shelly material, so the body only occupies the part of the shell nearest the aperture, i.e. the body chamber, which is open externally. The chambers are connected to the body by a thin-walled tube called the siphuncle. Cephalopod biology is discussed further in Section 14.4.

So, despite differences between these various classes of molluscs, they all share the same basic bodyplan because they are descended from a single species that has undergone descent with modification; i.e the diversity of molluscan classes is encompassed within the unity of the phylum, and a similar relationship holds for subsidiary taxa within each other phylum.

At this point in a consideration of form, it will help if we gain some perspective on the early evolution of animals. The brief account that follows does not discuss the changing selection pressures that may have influenced the course of evolution (not least because we know so little about them), but concentrates on the changes in form documented by the fossil record. As you will see, fossil-Lagerstätten (Chapter 12, Section 12.4) play a particularly important role in establishing this history.

14.1.2 Precambrian perspective

The first eukaryotic cells known from the fossil record are acritarchs, probably the resting stages of planktonic single-celled algae. Some may occur in sediments as old as 1 600 Ma, but certainly they were abundant by 800 Ma. Acritarchs have thick walls of a substance similar to that of vascular plant spores and thus a far higher preservation potential than the amoeba-like protistans that may well have originated soon after (Chapter 16, Section 16.2.2). The earliest known organisms that can certainly be called animals, i.e. the earliest unequivocal metazoans (multicellular animals other than sponges) in the fossil record, occur in a remarkably widespread fauna from rocks dated at about 620–550 Ma. These rocks form part of the youngest subdivision of the Precambrian called the Vendian Period. Discovered in 1947 in sandstones of the Ediacara Hills in the Flinders Ranges of South Australia, this 'Ediacaran fauna' has also been found in Africa, North America, Europe and Asia. The dominant fossils are large (up to one metre long), cnidarian-like forms resembling jellyfish (Figure 14.3(f, h)) or sea-pens (Figure 14.3(a, b, j)). Some of the forms have also been interpreted as large worms (e.g. Figure 14.3(e)). None of the organisms possessed hard skeletal material (see Chapter 12, Section 12.4.1 for a discussion of their preservation).

Only 20 years or so ago, all the Ediacaran fossils were regarded as belonging to extant invertebrate phyla, but their affinities are now vigorously disputed. Some of the bodyplans cannot be related easily to any living phylum. The enigmatic three-armed *Tribrachidium* (Figure 14.3(c)), the shield-shaped *Parvancorina* (Figure 14.3(d)), and the oval disc form *Praecambridium* (Figure 14.3(k)), for instance, have no very plausible affinities. Preserved as they are in sandstones, the Ediacaran animals seem to have had much tougher surfaces than the jellyfish we are familiar with, and virtually no similar forms have been found in more recent rocks. The German palaeobiologist Adolf Seilacher has argued that they represent a grade of organization that flourished briefly and became extinct before the beginning of the Cambrian Period. He maintains that despite their apparent diversity, nearly all the Ediacaran forms possessed a thin, flattened, round or leaf-like body with a quilted upper surface. There is no indication of a mouth or gut, suggesting they fed by passive absorption of dissolved food over their entire

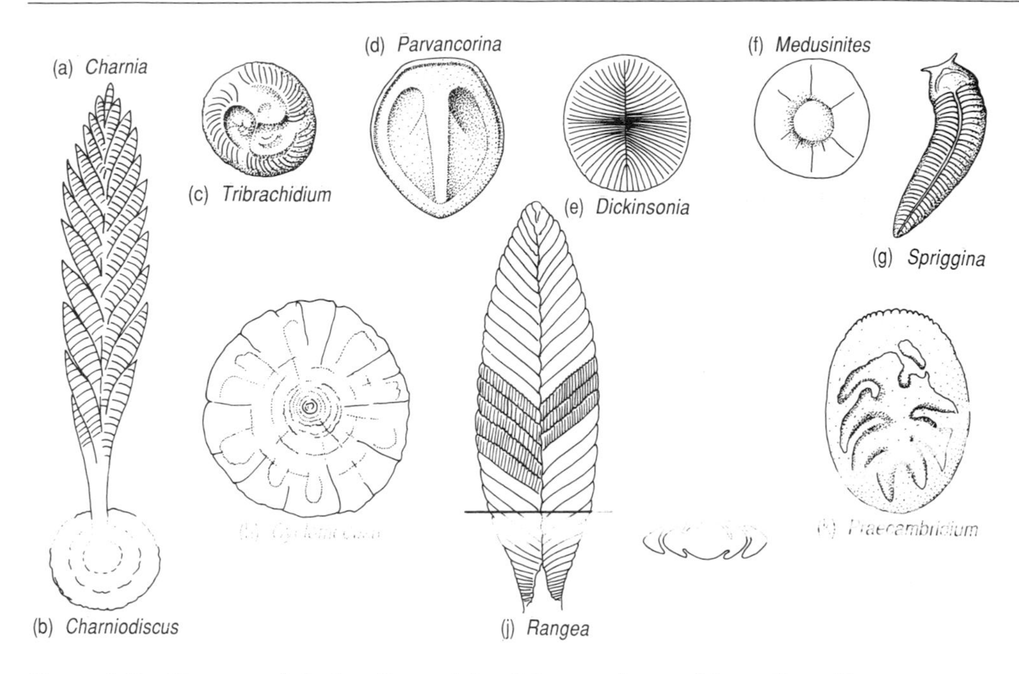

Figure 14.3 *Elements of the late Precambrian Ediacaran fauna of Australia, with a cross-section of* Rangea *(drawn to different scales).* Charnia *was probably anchored to the sea-floor by a basal disc similar to that of* Charniodiscus.

surface. He coined a new term 'Vendozoa' for this grade, possibly a distinct kingdom, although this interpretation is controversial: other authorities recognize a diverse range of bodyplans, some of which are classified as Cnidaria.

Whether they left any direct descendants or not, it seems certain that most of the Ediacaran forms as such did not persist into the Cambrian. A significant gap separates the disappearance of the Ediacaran fauna from the Cambrian fauna, despite the presence in that interval of sediments which otherwise seem suitable for fossil preservation. Close to the Vendian–Cambrian boundary, there is a marked increase in the variety of trace fossils, reflecting novel hunting strategies, ways of penetrating substrates and methods of locomotion. However, in the absence of body fossils at the end of such traces the culprits are seldom clear, and problems of identification are compounded because we know from Recent sediments that similar traces can be made by very different animals. It is possible that the new trace-makers obliterated the taphonomic record of late Ediacaran animals, or perhaps even caused their extinction; the Ediacaran fauna looks poorly equipped either to resist predation or to undertake it.

14.1.3 THE CAMBRIAN 'EXPLOSION'

One of the most important events in the history of life began about 540 Ma — for the first time, organisms with skeletons are found. In the Tommotian Stage, the earliest part of the Cambrian Period, a whole suite of very small shelly fossils appears in the rocks along with larger sponges and some sponge-like forms called archaeocyathids. This widespread 'Tommotian fauna' is dominated by cones and tubes of calcium phosphate typically 1–2 mm long, the affinities of which are often unclear. It is frequently difficult to tell whether a fossil is the complete skeleton of a single organism or a disarticulated plate of some larger creature.

Many phyla make their entry in the Tommotian and the succeeding Atdabanian Stage (also part of the early Cambrian), with a staggering of first appearances. The order and the rate at which the phyla appear is controversial and the subject of intense research. Not only are the affinities of the fossils hard to ascertain but detailed resolution is hampered by difficulties with intercontinental stratigraphic correlation. What seems clear is that most animal phyla had originated by the end of the Atdabanian Stage (*c.* 525 Ma), exceptions with a mineralized skeleton being the Chordata, which are first known from the Middle Cambrian, and the Bryozoa, which appear in the Ordovician Period. The trilobites, which dominate the majority of Cambrian shelly faunas, are absent from the earliest assemblages, and appear within the Atdabanian. A few entirely soft-bodied phyla living today have no known fossil record.

So, over a period that spans no more than about 20 million years, and which may have been significantly less, the stage was set with most of the major themes in the history of animal life—the phyla. The dominant skeletal materials, both in the Cambrian and at the present day, are calcium carbonate, calcium phosphate and silica, all of which generally have a high preservation potential, and the fossil record of skeletonized phyla is generally good (see Chapter 12).

14.1.4 THE BURGESS SHALE ANIMALS

Have any phyla become extinct since the beginning of the Cambrian Period? The answer to this question used to be no, probably not—until a reappraisal of a remarkable fauna from British Columbia discovered in 1909 by Charles D. Walcott of the Smithsonian Institution, USA.

High in the Canadian Rockies are exposures of a fine-grained deposit of Middle Cambrian age called the Burgess Shale. It contains the remains of animals which lived on a muddy sea-floor under a high reef wall built by algae. Occasionally, submarine landslides suddenly transported the animals into deeper, anoxic water, where they died and immediately got covered with mud (Chapter 12, Section 12.4.1). These catastrophic burial events, and the poorly understood processes that led to such exquisite preservation, have given us an exceptional window on Cambrian life. Not only have animals with hard shelly parts been preserved but entirely soft-bodied forms are represented too, as thin shiny films on the sediment surface. About 120 genera have been recognized, mostly represented only by one species each. About 15% of these genera are shelly organisms that dominate typical Cambrian fossil assemblages elsewhere, and include groups such as trilobites and brachiopods that normally have well-mineralized skeletons. But this shelly component was only a tiny part of the living community, probably less than 5% of its individuals.

▶ What is the importance, as far as other fossil assemblages are concerned, of finding that shelly animals probably made up less than 5% of living individuals in the Burgess Shale community?

It offers a cautionary tale to anyone trying to reconstruct entire communities from normal Cambrian assemblages lacking soft-part preservation. Take away the soft-bodied fossils of the Burgess Shale and we are left with a typical Cambrian fossil assemblage. We must therefore assume that any normal Cambrian hard-part fossil assemblage was probably dominated by soft-bodied animals too. There is every reason to believe that the Burgess Shale fauna is exceptional only in terms of its preservation, and that its composition was typical of Cambrian life.

The most important revelation of the Burgess Shale lies in the amazing diversity of bodyplans that were around in the mid-Cambrian *c.* 520 Ma ago. There are representatives of about 11 of the phyla that persist to the present day, including *Pikaia*, the earliest known chordate. However, about 18 types of Burgess Shale fossils are like nothing alive today and are not known from post-Cambrian rocks. They are so bizarre that had they been living now each would certainly have been placed in a separate phylum; their bodyplans are as different from each other as any of the remaining phyla in the Burgess Shale. Three of these forms are shown in Figure 14.4.

Opabinia (Figure 14.4(a)) had five eyes perched on the top of its head and a proboscis-like organ that terminated in a structure suitable for grasping prey. Another particularly interesting animal is *Anomalocaris* (Figure 14.4(b)), fragmentary specimens of which suggest the animal may have reached a metre in length. When found in isolation the anterior jointed appendages (presumably used in

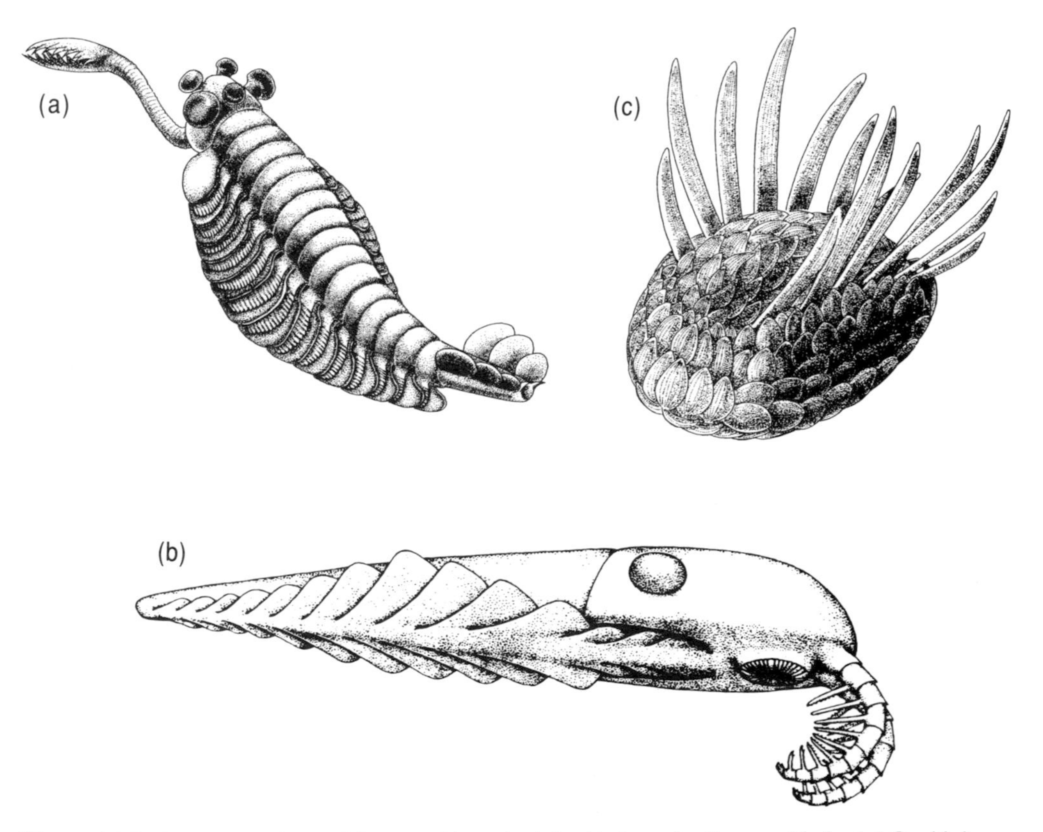

Figure 14.4 *Reconstructions of three problematical fossils from the Burgess Shale. (a)* Opabinia, *length 8 cm. (b)* Anomalocaris, *length 50 cm. (c)* Wiwaxia, *length 6 cm.*

prey capture and manipulation) were identified as belonging to the Arthropoda, but the rest of the anatomy is not comparable with this phylum. Its mouthparts appear to be unique among animals, constructed as they are by a series of contractible spinose plates encircling the mouth. This apparatus, looking a bit like a pineapple 'ring', had been identified as a kind of jellyfish when found in isolation, until painstaking dissection of *Anomalocaris* specimens by Harry Whittington of Cambridge University revealed it for what it was in 1981. This extraordinary jaw probably constricted down on its prey in much the same way that an iris diaphragm cuts down the light in a camera. Wounds on some Cambrian trilobites may have been inflicted by *Anomalocaris*. The curiously plated *Wiwaxia* (Figure 14.4(c)), with its vertical defensive spines, resembles a mollusc in some respects. However, unlike molluscs, each plate seems to have been secreted at a fixed size, and then moulted during growth along with the rest of the plates and spines.

New discoveries have revealed that the Burgess Shale-type faunas not only had a wide geographic distribution (being found from rocks exposed today in North America, Greenland, China and Australia) but were probably present over a time interval encompassing much of the early and middle Cambrian. It is not yet certain whether Burgess Shale-type faunas gradually ebbed away in the succeeding geological periods or came to a relatively abrupt demise before the end of the Cambrian.

The term Cambrian 'explosion' can justifiably be used for this radiation of bodyplans in a very brief interval relative to the immensely long period of simple life that preceded it. As Simon Conway Morris of Cambridge University puts it, 'the term "phylum" is simply a shorthand way of saying that the bodyplan is so distinctive that its nearest relatives cannot be identified with confidence' (Conway Morris, 1989). Were a team of systematists able to travel back to the Cambrian, '. . . what appear to our eyes, in the privilege of hindsight as "extinct phyla", would be shown to be more or less closely related to other species which happened to be successful in ultimately giving rise to myriads of descendent species that together would define the phyla we recognise today'. Taxonomic hierarchies are a product of hindsight, and for the Cambrian radiations orthodox classification schemes are something of a hindrance. Walcott had shoehorned the Burgess Shale fossils into today's distinct phyla, thereby losing the sense of disparity that characterized early Phanerozoic life. The problematic forms of the Burgess Shale are now widely seen to reflect unpruned diversity after the first major metazoan radiation.

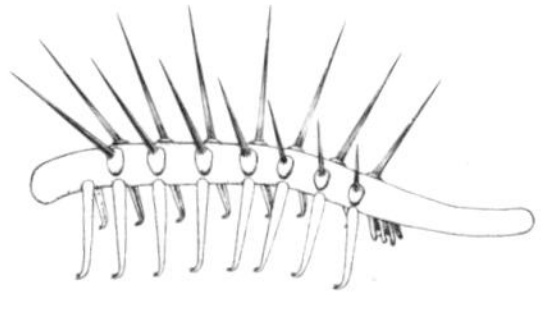

Cladistic analyses should, in theory, eventually produce a better classification based on the serial acquisition of derived characters (Chapter 11). And as collecting proceeds, more complete specimens are revealing hitherto unsuspected relationships between previously disarticulated soft parts and skeletal elements; this information in turn may disclose affinities with living organisms. Such a sequence of revelations has recently occurred with the wonderfully named *Hallucigenia*, previously believed to be an animal that walked on seven pairs of stiff stilts and that had a single row of seven tentacles along its back. Such a bodyplan could not be related to any known phylum. Now, on the basis of new finds from China, *Hallucigenia* is reinterpreted as an 'armoured lobopod' related to living velvet worms in the phylum Onychophora, such as *Peripatus*. The reconstruction has been turned upside down: the 'dorsal tentacles' are now believed to be a *paired* row of eight legs, and the 'stilts' are spines projecting up from paired plates along its flanks (Ramskold and Xianguang, 1991).

As Conway Morris has argued, not only would a Cambrian systematist trying to classify the rapidly diverging 'riot of types' be unable to distinguish what we regard as major groups today, but he or she could not predict which few species would be destined for cladogenetic success, and which (the majority) would be doomed to extinction. Re-run the Cambrian explosion and there is little guarantee that the same phyla would emerge. As has also been argued for some later extinctions (see Chapter 15), the persistence of particular lineages may have been more or less a matter of chance—good luck rather than good genes. In other words, lineage survival may have been more a case of individuals simply being in the right place at a time of drastic environmental changes, rather than survival by rapid *in situ* adaptation to those changing environments.

The reasons for the Cambrian explosion are unclear. Some hold that the genetic control on development of the earliest metazoans was unusually flexible (free of phylogenetic constraints, see below), others that the earliest metazoans simply radiated into an ecological 'vacuum', devoid of competitors. It is not known to what extent (if any) changes in the external physical and chemical environment, or crossing of biochemical or structural thresholds, may have opened the evolutionary floodgates (see also Sections 14.5.2 and 14.5.3). These possibilities are not mutually exclusive; all may have been involved.

SUMMARY OF SECTION 14.1

The form of an organism encompasses its entire physical constitution, from the atomic level up through every scale to the complete individual. Animals living today do not exist in an unlimited range of forms but can be grouped within about 32 phyla, the largest subdivision of life below the level of kingdom. Animals placed within the same phylum share the same distinctive bodyplan, albeit modified to varying extents within the subsidiary taxa.

The earliest known animals occur in the widespread late Precambrian Ediacaran fauna (*c*. 620–550 Ma). Many of the Ediacaran forms are enigmatic and may represent a grade of organization that did not persist into the Cambrian. The Cambrian 'explosion', one of the most important events in the history of life, began about 540 Ma ago at the start of the Cambrian Period. After an initial stage dominated by small phosphatic cones and tubes, nearly all animal phyla made their entry into the fossil record during a period that spans no more than *c*. 20 Ma, and which may have been significantly less.

The Middle Cambrian soft-bodied fauna of the Burgess Shale of British Columbia includes an assortment of bodyplans, some of which are quite unlike those that persist to later geological periods. About 18 forms are so distinct that had they been living now, each would have been placed in a separate phylum (assuming we have identified their characteristics correctly). Soon after the Cambrian 'explosion', there was probably a greater disparity of animal form than at any other time. New finds have shown that Burgess Shale-type faunas had a wide geographic distribution and that they may have existed over a large part of the Cambrian Period. To some extent, it may have been more or less a matter of chance which bodyplans persisted to later geological periods and which became extinct.

14.2 A FRAMEWORK FOR CONSIDERING FORM — SEILACHER'S TRIANGLE

The perspective on early animal life outlined above shows how the major themes of form were established very early on in the Phanerozoic. By the Ordovician, if not sooner, bodyplans had become *faits accomplis*, setting certain limits on form within which later evolution has been constrained. But all organisms, whatever their bodyplan, must both grow and 'work', i.e. continue to function long enough for successful reproduction of the next generation. As well as the legacy of ancestry, the demands of growing and functioning are therefore major controls on the evolution of form.

Using these ideas, in 1970 Adolf Seilacher proposed a new framework for considering the form of organisms. He suggested that any feature is best understood as the interaction between three main factors, each of which impose limitations on its form. Elucidating the relative strengths of these factors, he argued, should be a primary goal in any understanding of form.

The three principal influences are called the phylogenetic, the functional and the fabricational factors. As we shall see, although these factors are often referred to as 'limitations' or 'constraints', changes in any of them, for whatever reason, may present 'opportunities' or 'licences' by which the range of form can be extended.

The three principal limitations or constraints are placed at the corners of a triangle (Figure 14.5). This triangle then provides a conceptual framework within which any particular form can be considered. It is important to realize that the triangle is therefore the arena for placing both the constraints *and* the products of selection within those constraints. Some skeletal features are determined more immediately by one factor than another, but for many individual features the three factors contribute together in an interactive way to determine form. However, it must be stressed that the triangle cannot be used like some triangular diagrams to depict accurate proportions of three variables in percentages; it is simply a framework to aid the understanding of form. The triangle is best viewed as dialectical in nature, i.e. a basis for debate to reveal differences between opposing points of view or forces, rather than to establish one of them as true.

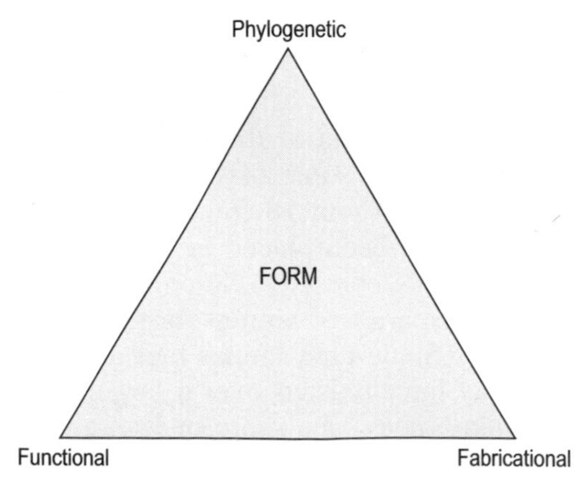

Figure 14.5 *Seilacher's triangle, a framework for the consideration of form. The principal factors are sited at each corner.*

14.2.1 The phylogenetic factor

All organisms have an evolutionary history: any individual is the net result of all the ancestral contingencies that led to it. It has a genetic heritage, present as a hierarchy of historical baggage. Take the common cockle *Cerastoderma edule* (Figure 14.6). As a member of the phylum Mollusca it has inherited a spirally coiled, dorsal calcareous shell, gills, and a muscular, hydrostatic foot, suitable for crawling or burrowing.

▶ What features has the common cockle inherited as a member of the class Bivalvia?

As a member of that class, it has inherited division of the shell into two bilaterally symmetrical valves, loss of the radula, an elastic ligament linking the two valves, and anterior and posterior adductor muscles which pull the valves together.

Continuing on with its genetic heritage, as a member of the subclass Heteroconchia the common cockle has inherited the genetic recipe for a general arrangement of hinge teeth radiating away from the umbo, and gills modified for suspension-feeding. As a member of the superfamily Cardiacea it has inherited features that include a particular arrangement of hinge teeth and ligament insertion, and a characteristic shell shape (globose) and micro-architecture; and so on, through successively finer subdivisions to species-level characters. Assuming that the classification reflects true phylogeny, shared derived characters (i.e. synapomorphies, see Chapter 11) were progressively added at each node in the evolutionary tree.

The basic bodyplans that emerge during development are particularly significant phylogenetic constraints. They have the effect of limiting adaptations that are likely to occur during the evolution of a given clade. Once metazoans had become confined to a limited range of distinct bodyplans in the Cambrian, the possibilities for evolution *within* each of those distinct bodyplans (i.e. within a particular phylum) were also limited; for example, the environmental distribution of many phyla has remained restricted. Examples of the way that phylogenetic inheritance influences the pathways evolution can follow are illustrated in the next Section.

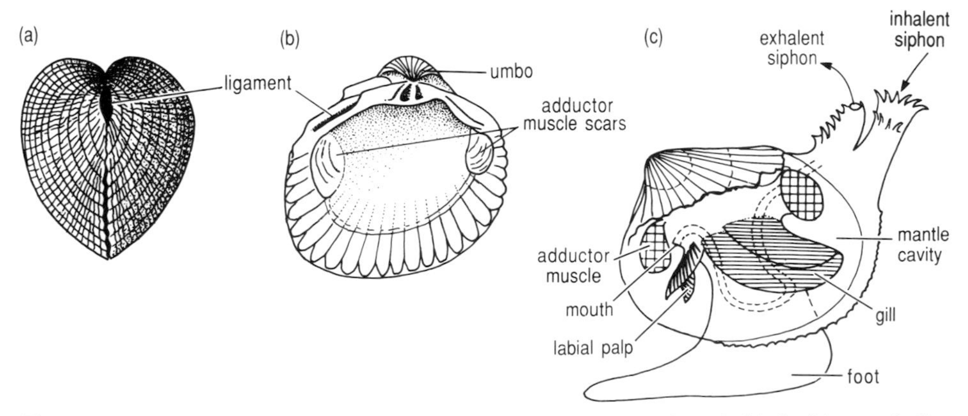

Figure 14.6 *The common cockle* Cerastoderma edule. *(a) View from behind of intact shell. (b) Interior of left valve. (c) View of living animal in life position with shell partially removed to show soft parts.*

14.2.2 Evolution as a tinkerer

In a fruitful metaphor, the French geneticist Francois Jacob described evolution as a tinkerer (Jacob, 1977). A tinkerer works with odds and ends, taking whatever materials are available and fashioning them into new, functioning objects. From a refrigerator shelf he or she might make a plate rack, from a broken table a bookcase, from a cast iron garden gate a fireguard. Similarly, evolution makes a wing from a leg, feathers from scales, or part of an ear from a piece of jaw; but unlike a tinkerer, who is free to obtain extra materials from wherever he or she chooses, evolution is constrained to work on the existing form of the ancestor. Evolution cannot freely combine bits and pieces of different lineages; that is one aspect of the phylogenetic constraint. The analogy is improved by insisting that the tinkerer makes a new object using only the parts of the old one, and keeps the old one working all the time. The time-scales are also different, of course, but the evidence of this process of piecemeal modification, of adding a bit here, taking off a bit there, can be seen by looking at the morphology of organisms.

▶ What is another important difference between the operation of the tinkerer and of the evolutionary process?

A tinkerer has a goal in mind; he/she needs to make, say, a fireguard, from whatever materials are available and *directs* his/her efforts towards that end. Evolution, by contrast, has no purpose to strive for, no ultimate goal, no predetermined direction. Yet it often seems to be that way. This concept is the theme of Richard Dawkins' book *The Blind Watchmaker* (Dawkins, 1986). The title relates to an analogy used by the 18th century theologian William Paley who attempted to show that design in nature was evidence of the intervening hand of a Creator. Paley argued that if he walked across a heath and found a watch lying on the ground, he would observe its intricate cogs and springs and deduce it must have had a maker, someone who had designed its complex interconnections to function as a timepiece. In the same way, Paley insisted that anyone looking at nature would find that 'every indication of contrivance, every manifestation of design, which existed in the watch, exists in the works of nature; with the difference, on the side of nature, of being greater or more, and that in a degree which exceeds all computation'. Dawkins explains, as Darwin himself did, how natural selection has the power to produce forms that have the appearance of having been designed. If natural selection can be said to play the role of watchmaker in evolution, it is the *blind* watchmaker in the sense of lacking vision and foresight. Natural selection is, however, the opposite of a process of blind chance. Variation, the raw material on which selection can act, may be thrown up at random with respect to the needs of present or future circumstances, but the process of selection itself is anything but pure chance.

Natural selection has the *effect* of tending to build up complexity gradually, step-by-step in a cumulative way, using existing materials. In a memorable misunderstanding of natural selection, an astronomer likened the chances of a single cell being evolved to the probability of a tornado blowing through a junkyard and chancing to assemble a Boeing 747. But evolution does not produce novelties from scratch: like the tinkerer, it works on what already exists, slowly transforming one form into another, as long as the all-important thread of genetic material persists from one generation to the next. Extinction—not just loss of life but loss of genes—is forever; species cannot later be reassembled in the way that a company can order a factory to make more aircraft. That, of course is one reason why the extinction of any species is so significant; the contingencies of nearly 4 billion years of accumulated change are irretrievably lost.

Contrary to popular conception, evolution by natural selection does not entail inevitable 'progress' (however we imagine progress might be defined), nor does it *necessarily* lead to increasing morphological complexity. Darwin was fond of pointing out that many parasites are morphologically less complex than their free-living ancestors. Another related misconception is that all evolution is irreversible; on the contrary, the fossil record of individual lineages shows that temporary reversals in long-term character trends are very often seen when the resolution is high, i.e. whenever there are many large, successive fossil samples closely spaced in time (Chapter 10). The step-by-step process by which natural selection adapts different features at different times may produce patterns of change in *individual* characters that are statistically indistinguishable from random patterns.

Because of heredity, each new step in an evolving lineage is partially constrained by preceding ancestral steps. If this were not so, we might expect a snail to be equally likely to evolve into, say, a sea-urchin as to another type of snail. This, of course, illustrates the phylogenetic constraint of Seilacher's triangle. The process of evolution thus has a 'memory'. Organisms have what constitutes that memory in their DNA, which links back in a continuous, unbroken chain of generations to the origin of life and which, looking forward, is potentially immortal. Because of the need to maintain functional integrity at all times in an evolving lineage, selection at any one time is likely to favour relatively minor adjustments, i.e. small steps whose outcome is overwhelmingly influenced by phylogenetic inheritance. In general, dissociated changes in form, where changes in different features occur at different times or different rates in the history of an evolving lineage, are very common, as in our own species; the phenomenon is known as **mosaic evolution**.

As an example of evolutionary tinkering, consider pandas, which spend much of their lives munching bamboo. They hold the stalks of bamboo in their paws and strip off the leaves by passing the stalk between an apparently flexible thumb and the remaining fingers. Yet count the remaining fingers and there are five. Is this a unique example of a modern tetrapod (four-legged animal) having a more than five digit (pentadactyl) arrangement of distal limb bones fixed early in tetrapod evolution? No—as shown in Figure 14.7, the panda's 'thumb' is not, anatomically, a finger at all. It is formed from a bone called the radial sesamoid, normally only a small bone of the wrist. The muscles used in the 'thumb' are also present in the wrist and hand of other Carnivora (Chapter 2, Section 2.3.3). These muscles have also been modified by selection to enable the panda to eat bamboo shoots. It may be somewhat clumsy—not an engineer's optimal design—but it works.

Tinkerers, unlike engineers designing a piece of machinery from first principles, are likely to end up with different solutions to the same problem. So with evolution too, for which 'the problem', viewed in retrospect, is somehow to have continued to function from one generation to the next. For example, the selective advantage of having a visual system is obvious: eyes have evolved independently in many lineages. There is a great variety of eyes in the living world, and even more when

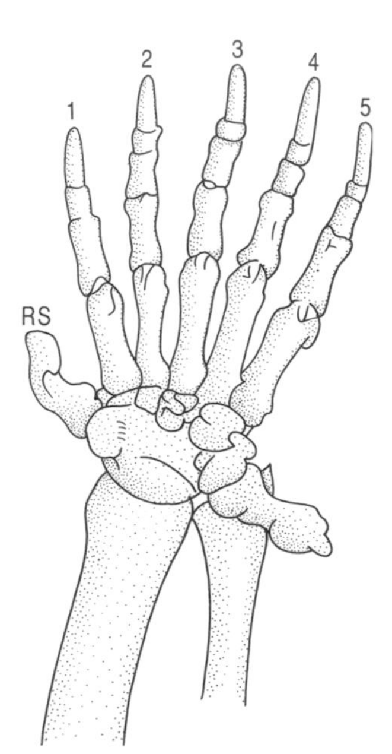

Figure 14.7 The bones in the wrist and hand of the giant panda. The true digits are numbered. The panda's 'thumb' is constructed from the radial sesamoid bone (RS), normally a small component of the mammalian wrist.

extinct species are also considered (see Section 14.2.3). Not surprisingly, visual systems may show a fair degree of convergence, as many are based on the same principles of pinhole and lens used in basic cameras. The human eye is remarkably like that of the octopus and both work on similar lines, yet details reveal that they did not evolve in the same way. In vertebrates the photoreceptor cells of the retina point away from the light, whereas in cephalopods (e.g. an octopus) they point towards it. An engineer starting from scratch would surely have arranged the vertebrate photocells to point *towards* the light. Instead, light has to pass first through a mass of nerves before reaching the retina. And the nerves themselves have to pass over the photocells and plunge down through a hole in the retina to bundle together into the optic nerve—the reason for our 'blind spot'. Evolution does not produce perfection, but selection tends to approach it with the material available. Occasionally, the human eye is cited as an example of perfection. It is, of course, a remarkably efficient organ, but most people need corrective lenses at some time in their lives (though presumably a higher proportion than in the adults of our stone-age ancestors, who probably struggled in a harsh world where poor vision often meant an early death). And the most sharply focusing human eye cannot change magnification and work like a telescope or a microscope, nor see in infrared, all of which might contribute a selective edge.

There are many striking examples of convergence in living and extinct organisms: similar solutions to particular problems that have arisen independently in different lineages. In each case, the convergence is not total. In the eyes of molluscs and vertebrates the pathways and starting materials were different, and the consequences persist to the present day. There are always differences in detail that betray an independent origin and at the same time reveal the power of selection to generate similar adaptations time and time again. One of the best known examples of this phenomenon is the remarkably close resemblance of form between eutherian (placental) and metatherian (marsupial) mammals produced by evolution in similar niches, which is discussed further in Chapter 15.

Evidence of previous stages of phylogeny are particularly clearly seen in anatomical features for which there is no known function, and which are best understood as reduced or degenerate remnants of a functional feature in ancestors.

▶ What is the name given to such features?

Vestigial structures, which were introduced in Chapter 2, Section 2.2.4.

▶ Figure 14.8 shows the skeleton and body outline of a whale. What does the presence of rudimentary hindlimbs suggest to you?

Whales are descended from four-legged ancestors and retain pelvic and leg bones as vestigial structures.

Modern whales live in water and lack hindlimbs entirely, retaining only rod-like vestiges of pelvic bones, femora and rarely tibiae embedded in the musculature of the ventral body wall. Whales evolved from Paleocene land animals that used hindlimbs in locomotion. Recently, specimens of Eocene whales with very small hindlimbs have been found in Egypt. The hindlimbs seem too small to have assisted in swimming and certainly could not have supported the body on land; they may have functioned as guides during copulation. Thereafter, for the past 40 Ma, all known whales have lacked functional hindlimbs.

Snakes are the main group of legless reptiles. But the more primitive snakes—the boa constrictors and pythons, which squeeze their prey to death—still show evidence of hindlimbs. The males have tiny claws used to caress the female during courtship. Some snakes retain externally invisible vestigial hindlimbs within the body wall, like those of whales, and some lizard groups, like the skinks, have tiny, almost useless limbs or no limbs at all. Lizards evolved in the Jurassic from reptiles a bit like the modern tuatara of New Zealand (Chapter 15, Section 15.2.1), whilst snakes probably evolved in the early Cretaceous from a burrowing lizard.

Other well-known vestigial structures include non-functional eyes in cave-dwelling fish, amphibians and arthropods.

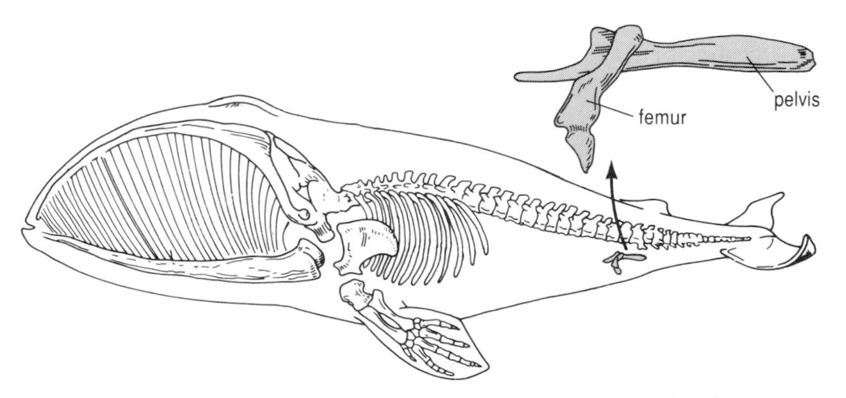

Figure 14.8 *Skeleton and body outline of a whale, with internal rudimentary hindlimbs.*

▶ Suggest a reason for the degeneration of organs into vestigial structures.

Presumably, such organs have been progressively less used or less useful, and selection has favoured their reduction. Individuals that waste metabolic energy maintaining unused organs (perhaps nourished by blood, and invested with associated muscles and nerves) are likely to be at a selective disadvantage. Conversely, variants having such organs that are slightly malfunctioning or less developed may possess other traits that increase their overall fitness. At any time though, given the appropriate conditions, a vestigial organ in the process of elimination might acquire new functions—a case of preadaptation (see Section 14.5.2). A probable case of this is the ostrich wing, which has become vestigial in relation to flight but now plays essential roles in feeding and behaviour (Chapter 2, Section 2.3.4).

In general, we can expect any lineage, at any one time, to have features with a spectrum of degrees of reduced functionality. We are, however, unlikely to be able to assess accurately the relative benefits or costs of such features. Some features vestigial in one individual might benefit another. In humans, for example, wisdom teeth may replace lost early teeth, and the appendix, although itself prone to infection, is lymphoid tissue and may occasionally play a minor role in defence against infection.

Atavisms are the reappearance in an individual of certain characters typical of remote ancestors that have been absent in more recent generations. 'Reversion to a previous evolutionary state' or 'throwback' also express, albeit loosely, the meaning of atavism. For example, a humpback whale was caught in 1919 off Vancouver with externally projecting rudimentary hindlimbs. The femur was nearly complete, although normally it is an internal and diminutive cartilaginous element. Similar atavisms are known in sperm whales, which also usually show only internal vestigial hindlimbs. It seems that the genes to specify such structures have not been lost but that the switch mechanism that controls their expression is normally suppressed until a mutation resurrects it.

Notable examples of atavism occur with horses. Julius Caesar reared and rode a remarkable horse with feet 'almost human, the hoofs being cleft like toes'. The earliest known horses, from the early Tertiary of North America and Europe, had four toes on the forelegs and three on the hindlegs. During the past 55 Ma, the number of functional toes has progressively decreased in some lineages until modern horses retain but a single, greatly elongated third digit on both fore and hindlegs (Figure 14.9(a)). Vestiges of the old second and fourth toes are present as short splints of bone carried high above the hoof on digit three. Occasionally, as with Caesar's horse, extra toes appear which may even be functional. Othniel Marsh, one of the founders of vertebrate palaeontology in America, made a study of these 'polydactyl' horses, and illustrations taken from his 1892 paper are shown in Figure 14.9. About two-thirds of horses with extra toes are not actually reversions to an ancestral form, because the functional third digit has itself simply subdivided, as in Figure 14.9(b). But about one-third are truly atavistic in that it is the vestigial splints of the second toe (Figure 14.9(c)), or even both second and fourth toes (Figure 14.9(d)), that develop into complete hoofed digits. Interestingly, in horses with elongated second and fourth digits there is often a compensatory

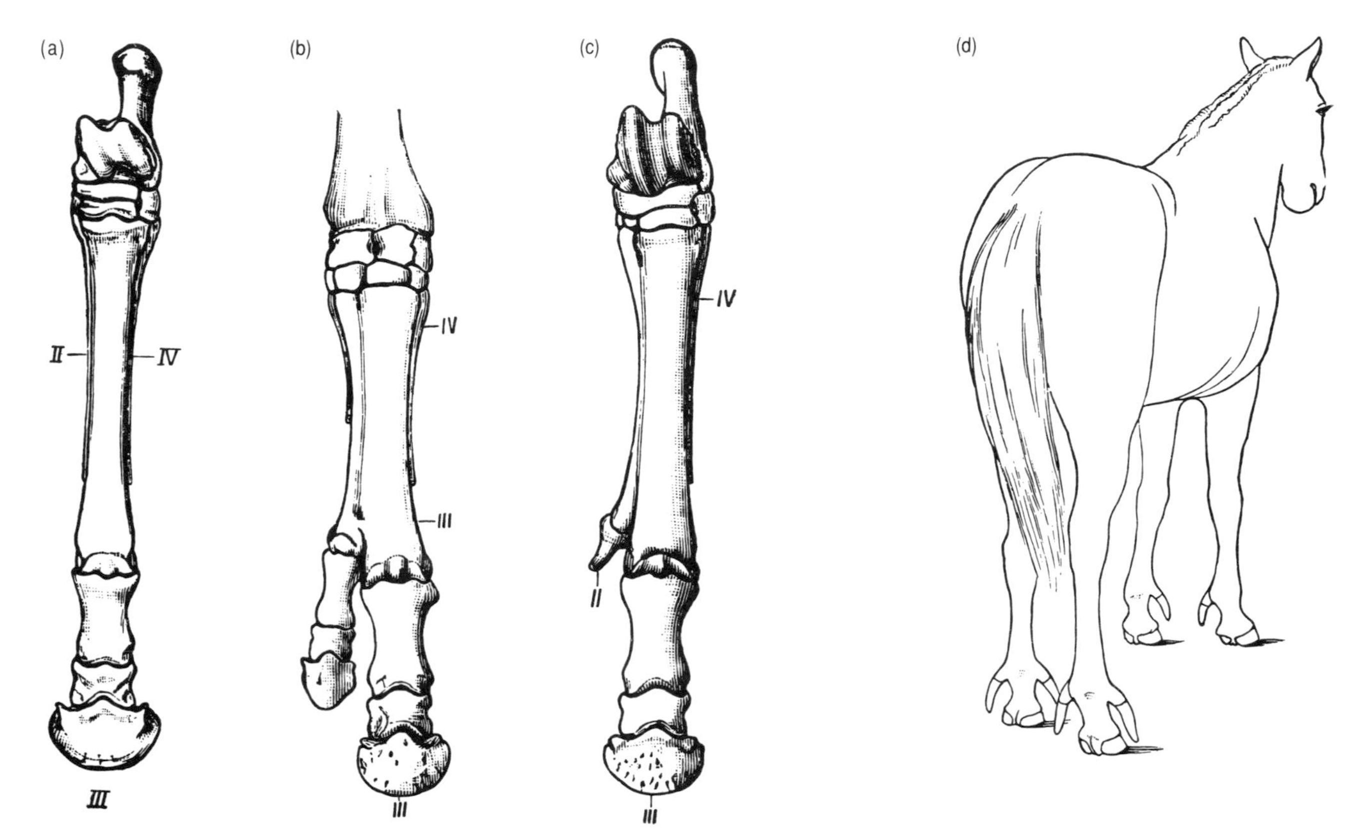

Figure 14.9 *Normal and polydactyl horses, from illustrations in a paper by Marsh, 1892. (a) A normal horse; note the splint remnants of side toes labelled II and IV. (b) Polydactyly by subdivision of the third digit. (c) Polydactyly by atavism; the extra toe is an enlarged side splint. (d) A polydactyl horse from Texas.*

reduced growth of the third digit, enhancing the likelihood of the extra digits making contact with the ground and being functional. This highlights the integrative nature of the development of the limbs in such animals, with associated changes in bones, ligaments, tendons and muscles. Many such atavistic limbs closely resemble the limbs of Tertiary horses such as *Mesohippus* and *Merychippus*, which had three toes on each foot.

Many other spontaneous atavisms occur in natural populations. Examples range from winged individuals in otherwise wingless insect species (such as some earwigs), flatfish pigmented on both sides, and the presence of a tail in humans. Atavisms are also known from the fossil record; for example, first molar tooth cones reappeared in Pleistocene *Lynx*, their ancestors having lost them in the Miocene.

Some atavisms that reveal insight into evolutionary history can be induced artificially. *Archaeopteryx* (Plate 14.2), the earliest known bird, and several other early birds possessed teeth, a feature shared with their reptilian ancestors. But for the past 60 Ma no bird has been known naturally to produce teeth. We use the notion of 'hen's teeth' to denote extreme improbability or rarity: no reports of atavistic teeth in birds are known. Yet in 1980 it was reported that chicken embryos had not lost the ability to produce teeth if their epithelial (outer) tissue was allowed to come into contact with molar mesenchyme (inner embryonic tissue) of a mouse. During normal tooth formation in mammals, a complex series of reactions between epithelium and mesenchyme is required before components of teeth such as dentin and enamel can be deposited. It seems that chick epithelium has lost the ability to induce the formation of dentin in its own mesenchyme, which, in turn would normally induce epithelium to form enamel. The grafted mouse mesenchyme produced the dentin that chick mesenchyme cannot, and well-formed teeth developed. This experiment again suggests that genes can be conserved unaltered but inactivated for millions of years.

▶ How do atavisms differ from vestigial structures?

Vestigial structures, unlike atavisms, are part of the normal range of variation present in most if not all members of a population, whilst atavisms are much rarer and never occur in many individuals of a population. Vestigial structures and atavisms both reflect the persistence of 'phylogenetic memory' after function has been lost.

Vestigial structures and atavisms are important here within the context of phylogenetic constraints. Vestigial structures are one of the clearest witnesses to phylogenetic ancestry and are clear evidence of changed selection pressures. Atavisms, which are often exaggerated vestigial structures, are not simply signs of constraint, imperfections lurking up from the distant past that reflect the inertia to progress suggested by the word 'throwback'. Rather, they illustrate the hidden potential for morphological change that all organisms possess. Atavisms give us glimpses that genomes may maintain a reservoir of latent integrated features that may remain unexpressed for millions of years. If activated by what, in some cases, may amount only to small genetic alterations, such features could provide new sources of variation for rapid morphological change, especially in small populations. Whilst in many cases atavistic individuals may be at a selective disadvantage compared with normal variants, there may be rare but, over geological time-scales, not infrequent occasions when they are at a significant advantage. Indeed, Darwin, writing of a fertilized animal cell, wrote 'we must believe it is crowded with invisible characters ... separated by hundreds or even thousands of generations from the present time: and these characters, like those written on paper with invisible ink, lie ready to be evolved whenever the organisation is disturbed by certain known or unknown conditions'.

We should not, however, get such possibilities out of perspective. Many complex atavistic structures will have taken millions of years and many intermediate stages to evolve before undergoing suppression. Whilst their resurrection may yield novel combinations, the atavistic structure *itself* is what evolution has, by and large, already created. We cannot expect entirely new integrated, complex structures to originate by atavism, although atavistic individuals may indeed be entirely novel *combinations* of form.

It is no coincidence that Seilacher placed phylogenetic inheritance at the top of his triangle of constraints. He refers to this corner as 'the attic' of evolutionary history. Not only are there features of high developmental necessity that cannot be discarded from the phenotype and now constitute the bodyplan, there are also elements of inherited baggage that can lie for a long time as 'forgotten' structures, occasionally to be brought down from the attic and expressed visibly as atavisms.

14.2.3 THE FUNCTIONAL FACTOR

Any new, well-established form must have been derived by viable intermediate states of the whole organism. There are always immediate demands that require an organism to keep functioning, at least until it has produced progeny. In terms of the analogy above, any evolutionary tinkering must be carried out while the machine is running. This constraint is clearly very important, and is ubiquitous when considering the level of an entire organism. In the functional corner there are therefore the influences that relate to the machinery to survive and reproduce in prevailing environmental circumstances. Sometimes this is called the adaptational corner, because the demands of survival and reproduction produce (by natural selection) adaptive solutions to specific functional problems. Of course, most of the *solutions* will have a genetic basis, but many will be relatively recent additions to the inherited ancestral baggage.

For a lineage to evolve, its individuals must remain viable organisms, maintaining fitness. Particular traits may experience strong selection pressures. For example in the event of a new predator, it might be of great survival value for, say, a bivalve to have a less penetrable protective shell. The thicker the shell, the better, might at first seem a plausible solution.

▶ How would functional constraints limit the response of a bivalve to selection for a thicker shell?

There are many extra costs associated with possession of a thicker shell. The biochemical secretion of a shell requires energy itself, but there are many additional energy costs. The heavier shell would need stronger muscles to close it and a stronger foot to move it around, so the bivalve would have to eat more in order to provide that metabolic energy. To gain more food it might have to be more active, requiring a more efficient feeding apparatus, and so on. In other words, changing one feature often requires others to change as well (see also Chapter 2, Section 2.2.3). Clearly, the need to maintain a viable integration of the whole organism will provide checks and balances at all times during the evolution of a lineage. The term given to the action of selection in producing new adaptive *combinations* of traits in a lineage is **coadaptation**.

The shell of the dog whelk *Nucella lapillus*, which is a common gastropod in the intertidal zone of British coasts, displays simple coadapted traits. The species includes a range of shell forms that correlate with the local environment. Two forms are shown in Figure 14.10: the most obvious principal differences are in the thickness of the shell and the size of the aperture.

▶ Considering the difference in shell thickness, which of these two forms do you think lives in the more turbulent environment with heavier wave action?

Figure 14.10 *Shells of the dog whelk* Nucella lapillus. *Note the differences in shell thickness and size of aperture.*

The left form comes from an exposed shore and the right form from a sheltered one. Surprised? Other things being equal, one might have expected a thicker shell to be selected in the more turbulent environment, where resistance to wave battering would be advantageous. But other things are *not* equal. Crabs prey on the whelks, and attempt to crush their shells or extract soft tissue through the aperture. Few crabs can withstand high levels of water turbulence, whereas many thrive on sheltered shores, inflicting severe selection pressures on the whelks. A thicker shell and a narrower aperture (inhibiting the entry of a crab's claws) are therefore favoured on sheltered shores. A large aperture is associated with a bigger and more powerful 'foot', the muscular organ on which the whelk creeps around and with which it holds itself onto rocks.

▶ Is the size of the aperture consistent with expectations, now that we know that the left form lives in a more exposed shoreline, where there are fewer crabs?

Yes. Animals with larger-apertured shells can cling more tightly to the rocks with their more powerful foot, and smaller-apertured, thicker shells are less likely to be probed successfully by hungry crabs.

The differences in shell thickness and aperture size in the dog whelk are examples of form that are strongly influenced by the functional corner of the triangle. The attributes of *having* a shell and an aperture are themselves products of the phylogenetic corner, being major inherited constraints on possible evolutionary forms.

Sometimes the fossil record allows us particularly clear glimpses of how adaptive solutions to specific functional problems were produced in extinct organisms. A close match between observed form and the expectations of theoretical design, constrained only by materials and the laws of physics and chemistry, can be used to support interpretations of functional morphology. Consider the eyes of trilobites, which had the earliest known visual system. Most trilobites, from their first appearance near the start of the Cambrian, had eyes, although in some lineages they are rudimentary or secondarily absent. The eyes were usually carried on crescent-shaped lobes raised up relative to the surrounding exoskeleton (Figure 14.11(a)). Like the compound lateral eyes of modern crustaceans and insects, trilobite eyes had a radial arrangement of visual units often encompassing a wide-angled field of view. The number of lenses in each adult eye generally ranged from several hundred to several thousand. Like the rest of the exoskeleton, the lenses were constructed mainly of the mineral calcite, probably with an admixture of organic material. The lenses therefore had a high preservation potential and were easily fossilized, though less durable parts of the eye, including receptor cells and nerves, were not.

The lenses of most trilobite eyes each consist of a single crystal orientated in such a way as to minimize the problems of refraction in calcite, which otherwise tends to split up the path of light, giving double images. One suborder of trilobites, the Phacopina (Ordovician to Devonian), had a particularly sophisticated visual system unique in the animal kingdom. This group had eyes in which the relatively few lenses were large, isolated, and each covered by a separate cornea (Figure 14.11(b)).

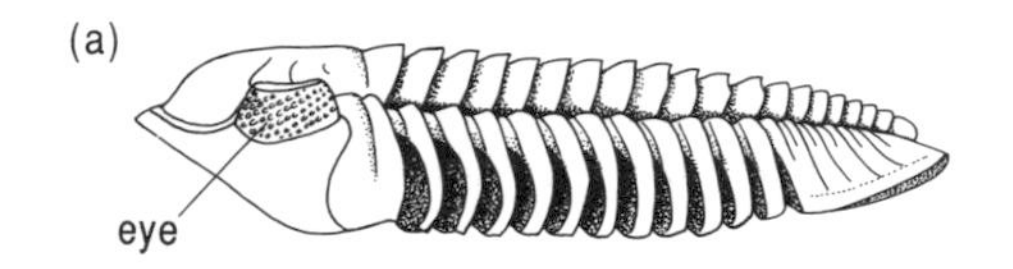

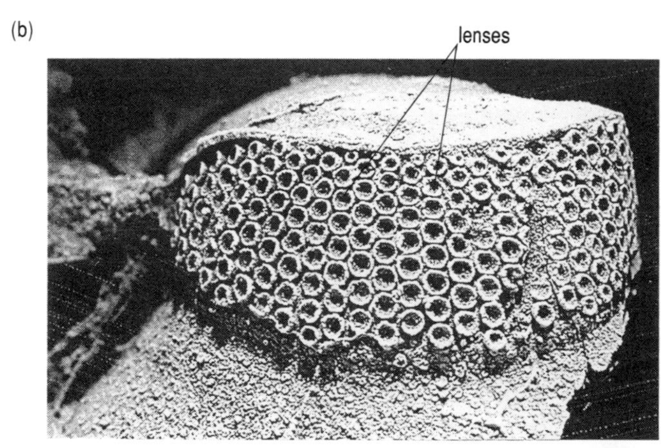

Figure 14.11 *Aspects of the eyes of trilobites. (a) Lateral view of a Silurian trilobite showing the eyes carried on raised, crescent-shaped lobes. (b) Close up of an eye showing individual lenses. (c) Cross-section through a single eye lens showing the lower bowl-like unit separated from the upper unit by a wavy surface.*

Euan Clarkson of Edinburgh University and Riccardo Levi-Setti of Chicago University (1975) discovered a remarkable structure within these lenses that eliminated spherical aberration (the distortion caused by the focus of light rays at the edge of the lens being different to that at the centre). Each biconvex lens had a lower bowl-like unit separated from the upper part of the lens by a wavy contact which acted as a correcting surface that eliminated spherical aberration (Figure 14.11(c)). Humans did not devise a lens to overcome this problem until the 17th century, when Descartes (in 1637) and Huygens (in 1690) published designs for 'aplanatic lenses' that avoid the aberration (Figure 14.12). The chief difference between the trilobite lenses (Figure 14.12(b,d)) and the designs of Descartes (Figure 14.12(a)) and Huygens (Figure 14.12(c)) is that the trilobite lenses have a second (lower) unit. Calculations and model-making have shown that this lower unit (with a slightly lower refractive index) compensates for the relatively high refractive index of seawater compared with air, the medium for which the 17th century philosophers designed their lenses. Trilobites with these eyes probably had excellent, sharp vision free from astigmatism (a defect caused by the curvature of a lens being different in different planes). Natural selection thus came up with a visual system that would require an engineer, designing such a system from first principles, to know such things as the laws of refraction, the optics of crystals, and much else besides.

Given that the alternative to organisms maintaining overall functionality (i.e. sustaining reproductive fitness) is extinction, it is remarkable that earlier this

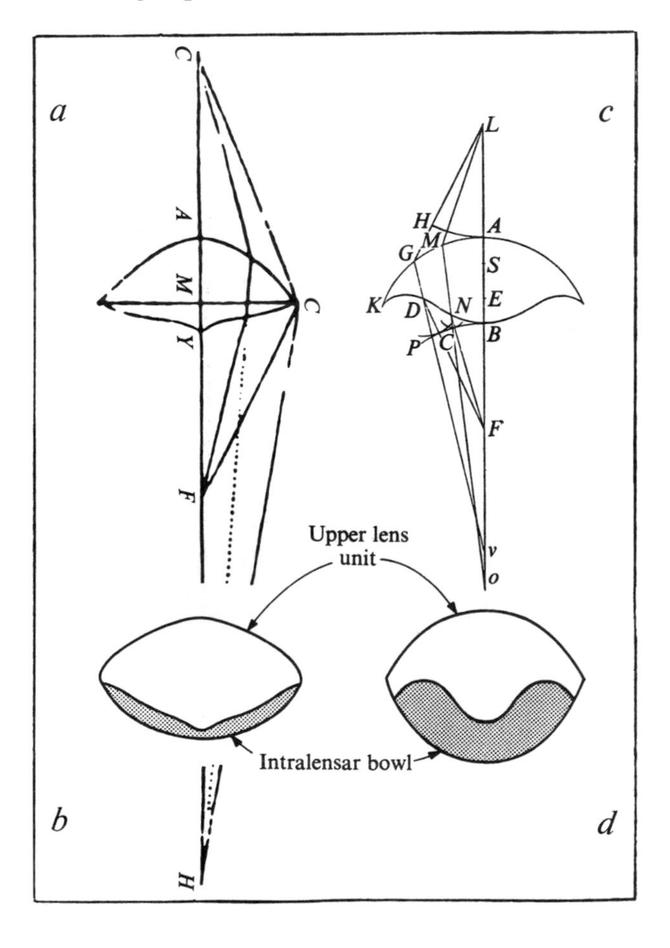

Figure 14.12 *Original constructions of lenses (including lettering) designed to avoid spherical aberration made (a) by Descartes in 1637 and (c) Huygens in 1690, compared with (b) and (d) cross-sections through the lenses of two species of trilobite.*

century a school of thought believed that certain long-sustained trends in extinct lineages were non-adaptive or even maladaptive. The observed trends were supposed to reflect an intrinsic tendency within the organisms themselves, an evolutionary phenomenon called **orthogenesis**. Evolution proceeded, it was supposed, in straight lines that selection could not regulate. Some traits, it was argued, evolved directionally to damaging extremes, reducing the functional adequacy of the entire organism, and eventually leading to the lineage's extinction.

One of the most quoted examples of orthogenetic trends was the increase in size of antlers of the Irish elk *Megaloceros giganteus* (Figure 14.13). This kind of giant deer flourished in the Pleistocene before becoming extinct in Ireland about 11 000 years ago. The 'Irish elk' was not in fact exclusively Irish; it ranged through Europe, Asia and even North Africa. Its enormous antlers, up to 3.7 m across and 40 kg in weight, were believed to have been non-adaptive and eventually maladaptive. The Irish elk's demise supposedly occurred when the antlers eventually became too energy-expensive during growth (like all deer with antlers it shed and regrew them every year), or weighed them down in bogs, or caught them in trees. Stephen Gould of Harvard, however, showed that, compared with other species, the *Megaloceros* antlers were *not* disproportionately large in relation to body size. Antlers of deer in general show interspecific positive allometry (Chapter 10, Section 10.2.3) so that larger and heavier species of deer have relatively longer and heavier antlers. The Irish elk also shows *intra*specific positive allometry. Gould believed that the antlers were too heavy to be used in fighting but argued that there was a simple adaptive explanation: they functioned as organs of display to intimidate rivals and attract females during the rut. Such a strategy of ritualistic combat would eliminate the need for actual battle (with consequent serious injury or death) and establish dominance hierarchies among males.

However, a detailed mechanical analysis by Andrew Kitchener, combined with a knowledge of the ecology and fighting behaviour of living deer, has shown that there are no features of the Irish elk's antlers that are inconsistent with a fighting function (Kitchener, 1987). The antler bone's microstructural fabric has orientations that match the tensile and compressive forces expected to be generated during combat. The Irish elk therefore probably used its antlers primarily for fighting, in common with all living deer with antlers. Field studies in living deer have shown

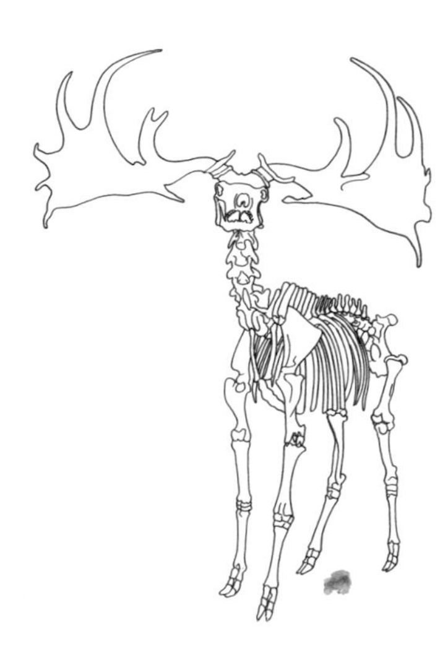

Figure 14.13 *Skeleton of the extinct Irish elk* Megaloceros *(male). The shoulder height was about 1.8 m.*

that the primary function of antlers is as weapons in intraspecific combat between males to compete for access to females in oestrus (see also Chapter 4, Section 4.4.1). The positive allometry of antlers in living deer has been shown to relate to breeding group size: deer within the largest breeding groups fight most frequently and have the heaviest and most complex antlers. It is hard to assess the relative selective advantages of large body size (irrespective of antler size) as opposed to large antlers (irrespective of body size). Selection for large antlers in Irish elks probably occurred because of the greater reproductive success of stags possessing large body size *and* large antlers. Thus, there is no evidence for a maladaptive trend, but rather the opposite. As Gould put it, 'there is no evolutionary advantage more potent than a guarantee of successful reproduction'. As to why the Irish elk died out, that is uncertain but the reason is likely to be more mundane than a trend carried to harmful lengths. Many other large mammals died out in various parts of the world about 11 000 years ago, and the general cause of this is a matter of much debate.

Similar cases of supposed orthogenetic trends have since been refuted. For example, the evolution of enormous canine teeth in sabre-tooth cats did not prevent biting or eating, as once thought. They probably used their huge canines to stab the throats and slash open the soft underbellies of their prey. Mammoths were not disadvantaged by unwieldy and cumbersome tusks but probably used them for clearing snow off forage in winter and for harvesting tree bark.

14.2.4 THE FABRICATIONAL FACTOR

This corner of the triangle refers to aspects of the growth programme and of biological materials that impose their own particular constraints on the evolution of form. Important fabricational constraints result directly from inherited bodyplans. For example, the external shell of molluscs grows by marginal accretion as explained in Section 14.1.1 (see the mollusc shells in Figures 14.2 and 14.17(a–f)). Any new skeletal structure must be compatible with the basic accretionary mode of growth. The juvenile mollusc shell must be suitable to continue on as part of the larger shell of the adult. It is extremely unlikely that a mollusc could ever switch to skeletal growth by periodic moulting as in arthropods.

For swimming cephalopods with a coiled outer shell, such as *Nautilus* (Figure 14.2(j); Plate 14.1), many of the functional requirements are much the same as those faced by a naval architect designing submarine hulls. There are similar problems of maintaining buoyancy, resistance to hydrostatic pressure, stability, manoeuvrability and hydrodynamic efficiency. Yet the shells of swimming cephalopods and submarine hulls do not show strong convergence of form. Most aspects of these structures are not even superficially similar because the external cephalopod shell is constrained to grow by marginal accretion, whereas the submarine hull has virtually no growth requirements, and can therefore be constructed according to functional requirements.

There may be inherent limitations on form because of the properties of the biological materials available for tissue construction. Take the property of hardness. In theory, diamond might be an ideal material for incorporating into vertebrate teeth used for grinding, or for the rasping teeth of a mollusc's radula (e.g. see Figure 14.16), but as far as we know it is biochemically impossible to secrete diamond. (The radulae of chitons—'coat-of-mail' shells of the molluscan class Polyplacophora—are, however, tipped with a hard iron oxide.) Similarly, those skeletal materials that *can* be secreted have their strength limitations, as anyone who has broken a leg while skiing or playing football can testify. Fortunately, for most human activities our bones are usually subjected to forces well below the critical point of failure. (Bone is substantially stronger in compression than tension, hence our proneness to disaster when engaging in tension-inducing activities; in fact, the direction of force in relation to the internal structure of a bone will often determine whether or not it fractures.) As well as limiting the range of functions to which the biological material may be suited, the property of strength is a major constraint limiting size and affecting shape. In vertebrates, for example, one of the most important reasons for allometric change in bone shape is the need to maintain a skeleton of sufficient strength to support the body weight throughout its life. The strength of a bone is approximately proportional to its cross-sectional area. If, as in purely isometric growth, the linear dimensions of an entire organism were to double, then the volume (or weight, assuming uniform density) would be increased eight-fold, because volume increases as the cube of a linear dimension. If the linear dimensions of a bone supporting the weight were also to double, the strength of the bone would only increase by a factor of four (area increases as the square of the linear dimension). Thus, the weight of the body would have increased more rapidly than the effective strength of the bones that support it. The solution, therefore, is change in shape of the bone during growth. It must become relatively thicker in order to carry the increased weight; generally, the weight-supporting bones of adults are relatively short and thick compared with those of juveniles.

One of the more important instances of major fabricational constraints concerns the apparent lack of ability to secrete hard skeletons prior to the Cambrian 'explosion' (Section 14.1.3). Prior to the early Cambrian, the range of metazoan forms was extremely limited. The bodyplans of many phyla depend on being able to achieve a certain amount of rigidity, without which many functions concerned with support, protection and movement are not possible.

In addition to adaptive features of morphology that are constrained by—and arise from—aspects of mode of growth and the biological materials involved, there may be features that Seilacher described as 'non-adaptational elements of low taxonomic significance'. **Fabricational noise** is a term given to morphological features that arise incidentally as a by-product of some particular programme of growth. Seilacher gives the analogy of South American water jars that are now made of plastic instead of pottery. Many of the brightly coloured plastic jars have a seam down their middle that would tell a future archaeologist they were made by a completely different technique. Similarly, a pot turned by hand on a potter's wheel usually has a series of tiny ridges and grooves around its circumference. Such features as the seam and the grooves are examples of fabricational noise, features for which there is not necessarily any functional significance. They are unintended constructional artefacts (though each can be elaborated or emphasized to perform the function of artistic ornamentation).

Several animal groups exhibit polygonal patterns. Such patterns often consist of approximately hexagonal mosaics, and are common, for example, in corals and echinoderms with rigid skeletons. Plate 14.3 shows part of the skeleton of a coral colony with polygonal patterns, some hexagonal, formed by the boundaries between individuals; and Figure 14.14 shows plates of a sea-urchin skeleton. What causes the similarities of pattern? It is not the phylogenetic factor, because the Cnidaria and Echinodermata are about as far apart phylogenetically as any two phyla. Analysing the polygons in three dimensions rather than two and studying them during development would reveal major differences in the mode of growth between the coral and the sea-urchin. It is not the functional factor alone, because the approximately hexagonal pattern does not have a specific function itself in the sense that having a compact skeleton does; there are many other shapes that can and do perform as well in other species or even within the same individual. The reason lies closer to the fabricational corner: the mosaic patterns are often simply the result of close packing of units during growth. The units in both the coral and echinoderm example are of similar size in each animal, approximately equally spaced and grow in a confined area. An individual unit would tend to be circular if not constrained by its neighbours, an argument that could also be applied to the hexagonal pattern of a bee's honeycomb. Clearly, the attribute of 'having an approximately hexagonal pattern' is likely to be of low taxonomic significance, as it will have arisen independently in many different groups; indeed, such a pattern may occur not only in animals but in certain plants too, and can also be seen in the spaces between closely packed familiar objects of similar size, such as oranges, billiard balls or cigarettes!

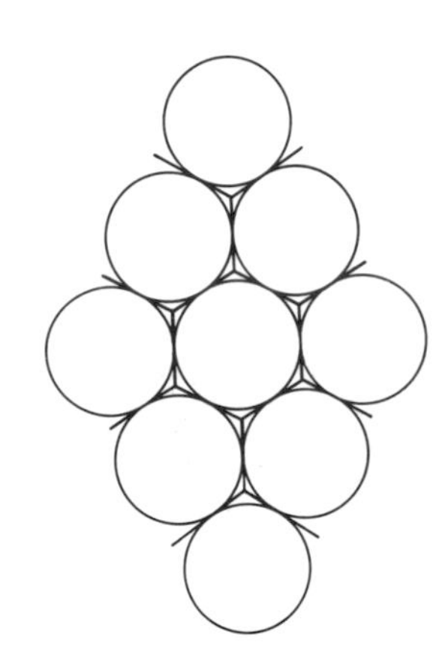

One must be careful not to oversimplify the influences that contribute to forms such as these polygonal patterns. The mosaic pattern in the coral, for example, is not independent of genetic control. But features such as the angles between polygon sides, the lengths of sides and even the number of sides are not specifically encoded in DNA. These details arise during the working through of the growth programme and depend on local conditions during development. Also, the polygons are functional in the sense that they are compatible with the functioning of the organism, but their origin differs from more explicitly adaptive structures such as the spines or tooth plates of a sea-urchin, which would lie closer to the functional corner of Seilacher's triangle.

Other examples of fabricational noise include some developmental asymmetries within individuals—usually small differences in complex patterns that emerge

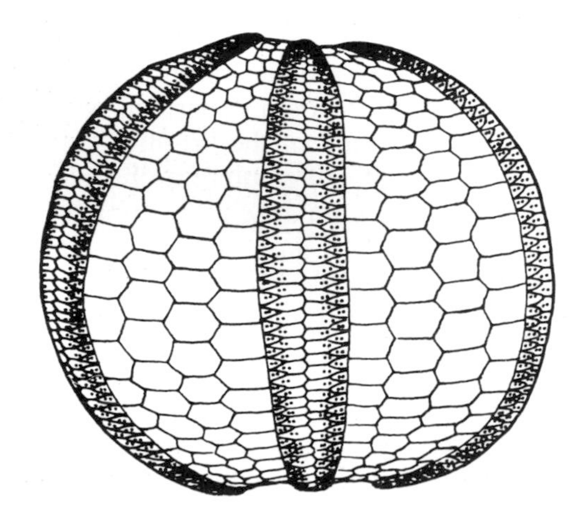

Figure 14.14 *Side view of the skeleton of a Palaeozoic echinoid showing the boundaries between calcite plates.*

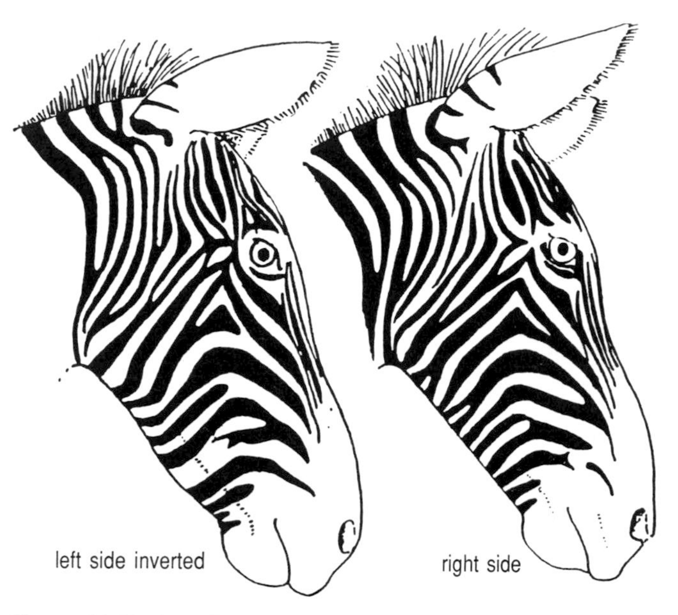

Figure 14.15 *Developmental asymmetry in the striped patterns on either side of the head of a zebra.*

during development but are not determined *directly* by genes (although the general predisposition to show increased incidence of developmental asymmetry may be related to the degree of homozygosity in populations, and therefore be under loose genetic control). Such examples include fingerprints, which not only differ from person to person (including 'identical' twins) but from one finger to another, and the striped patterns of a zebra, which differ from one side of its head to the other (Figure 14.15).

14.2.5 RESTRICTED OPTIMALITY AND FORM

We have already seen that phylogenetic, functional and fabricational constraints may mean that the form of a particular adaptation may not be the optimal one that an engineer would design in isolation, starting from scratch. In general, we can expect the form of adaptations to show more or less **restricted optimality**, depending on the influences of constraints and the intensity of selection. For instance, we owe much of our versatility and athleticism to prehensile hands and nimble limbs. But this also makes us prone to sprains, torn ligaments, and dislocations because structural reinforcement has been compromised for agility. The back problems that many humans have to endure are partly the result of having a skeleton and musculature that are modified from the anatomy of four-legged ancestors; the legacy we inherited is not fully compatible with upright posture. Sometimes, as in the case of the trilobite eyes, there seems to be little difference between an artificial design and the product of natural selection. In many cases, of course, people have tried to copy designs from nature but cannot match them—compare the manoeuvrability of most birds or insects in flight with that of any aircraft. (Actually, most insects are inherently unstable but have very efficient neural control.) The best form for one function may be counter-productive to another function and a trade-off becomes necessary. Consider an organism anchored in strongly flowing water and capturing plankton passing by. A stiff, sieve-like, fan-shaped structure perpendicular to the current might be optimal for feeding but entails maximum danger of breakage; a compromise therefore has to be reached. Also, there may be no single optimal solution but a range of solutions, each equally achievable and suited to perform a particular function. For example, a wide variety of toxic chemicals might be sufficient to harm a predator; a variety of warning colorations or patterns might be equally effective in deterring a potential predator from a real or mimicked poisonous form.

The existence of fabricational noise, vestigial structures and atavisms, as well as mutants and recombinational variants, also raises the possibility that certain morphological features may be non-adaptive in the sense that they have not been selected by virtue of performing some particular function. Aspects of this have already been discussed in Chapter 2, Section 2.2.3.

SUMMARY OF SECTION 14.2

Evolution does not have a free hand to modify descendants. There are constraints that limit the range of forms that can appear in evolution. There are three principal constraints on the generation of form: the phylogenetic, functional, and fabricational factors. These principal influences can be placed at the corners of a triangle, which provides a framework within which any particular form can be considered. In the phylogenetic corner are the constraints imposed by the evolutionary history of the organism. All organisms carry with them a genetic heritage, and genetically coded bodyplans are particularly influential in limiting changes in form. In the functional corner are the immediate demands of survival and reproduction in the prevailing environmental circumstances, and the machinery to cope with them. In the fabricational corner are aspects of the growth programme and of the biological materials used by organisms that impose their own particular constraints and opportunities on the generation of form. Fabricational noise is a term given to features that arise as a by-product of some particular programme of growth, and which may have no functional significance.

Like a tinkerer, evolution always involves modification of an existing form, rather than design from scratch. Evidence for this can be seen in features which reveal a piecemeal employment of existing structures that have been fashioned for new functions. However, unlike a tinkerer, evolution has no goal in mind, no predetermined direction. Any impression of intentional design is an illusion. Natural selection has the effect of building up complexity gradually, step by step in a cumulative way, using existing materials. Selection itself is the very opposite of pure chance, but the variation on which selection can act is thrown up at random with respect to the needs of present or future circumstances.

Vestigial structures—reduced or degenerate structures derived from once-functional features in ancestors—are some of the clearest witnesses to phylogenetic ancestry. Atavisms—the reappearance of certain characters typical of remote ancestors—indicate that some complex ancestral traits may not be lost but merely remain hidden in the genotype. Contrary to popular belief, evolution is not completely irreversible, a misunderstanding linked to the false notion of inevitable progress in evolution. The step-by-step process by which natural selection adapts different features may produce patterns of change in *individual* characters that are statistically indistinguishable from random patterns.

Sometimes, the fossil record allows us clear glimpses of how extinct organisms produced adaptive solutions to specific functional problems. A close match between observed form and the expectations of theoretical design can be used to support interpretations of functional morphology. However, various constraints may limit organisms from achieving optimal form for particular functions. Adaptations are often compromises caused by the interaction of many constraints.

14.3 Other factors influencing form

Seilacher's triangle provides a very useful basic framework for considering the principal factors that constrain or permit the generation of form. In this Section we will see that there may be other factors that deserve separate consideration, whilst at the same time those factors are interacting with one or more of the phylogenetic, functional or fabricational factors.

14.3.1 Ecophenotypic effects

Ecophenotypic effects are non-genetic (non-heritable) modifications of a phenotype that are produced in response to a particular habitat or environmental factor (Chapter 10, Section 10.2.2). In any assessment of form, we need to bear in mind that ecophenotypic effects may be involved. A difference in average size of individuals between populations is one of the most commonly encountered ecophenotypic phenomena, but a wide range of other features can be affected. Ecophenotypic effects can occur at any stage in ontogeny, including after the attainment of maturity, as witnessed by the shape transformation of some people who take up bodybuilding late in life.

Species susceptible to ecophenotypic effects are often displaying a special case of adaptation. There may be strong selection for genotypes that allow plasticity of form in different environments. The factors contributing to the form of a particular ecophenotype would then lie close to the functional corner of Seilacher's triangle. For example, the form of some colonial organisms is highly correlated with the strength of the currents in which they live. Some coral species grow into branched colonies in turbulent water and more platey ones in quiet waters. One of the principal controls seems to be that branched forms offer less resistance to water flow. When transported to deeper water, individuals of some West Indian coral species will 'plate-out', i.e. develop flatter profiles, allowing more sunlight to reach the symbiotic algae that reside in the coral cells and recycle their metabolic products as food. Other ecophenotypes (of any organism) may, however, have lowered fitness as a direct result of adverse local conditions; low oxygen concentrations or introduced pollutants may produce stunted or misshapen forms that are in no way better adapted to their local environment. In a sense, the forms of these ecophenotypes have influences that also lie close to the functional corner by virtue of their *inability* to function correctly; they will probably be selected against and, in effect, pass out of the triangle.

Responses to injury and disease are further examples of ecophenotypic influences on form. An injury itself, by definition, is a harmful physical alteration of phenotype, but the healing response, which is largely under genetic control, will produce ecophenotypic effects that are firmly in the functional corner. An example from human experience would be scar tissue; a more extreme example would be the ready regeneration of entire limbs in many starfish and crustaceans. Some species deliberately shed parts of their body when attacked seriously, a defence mechanism known as autotomy. Yet other ecophenotypic differences in form may be of no functional significance and selectively neutral. For example, many marine invertebrates show differences in the ratios of oxygen isotopes and trace elements in their shells depending on the local environment. Most such differences are not known to have any functional significance, nor to influence fitness. In general, however, many kinds of ecophenotypic variations are known to be under direct genetic control in related species. This suggests that the degree to which a feature is under strict genetic control or is susceptible to ecophenotypic modification (i.e. the degree of plasticity) may vary over evolutionary time-scales, depending on net selective advantage.

Richard Dawkins has proposed the analogy that genes are equivalent to recipes and not to blueprints. A blueprint is a scaled-down diagram of the real thing. A recipe in a cookery book is not a blueprint or a scale model of the cake that will finally emerge from the oven, but a set of instructions, which if correctly obeyed in the right order will result in a cake. The recipe does not specify the coordinates of each sultana and cherry, whereas a detailed blueprint would. Building something from a blueprint is usually reversible; given a building, you could recreate a detailed blueprint from it. But given a cake, it would be impossible to work backwards to the original recipe for it, especially if you've never made a cake before. The development of an organism is, by its very nature, irreversible, like scrambling an egg. The point here is that genes are like recipes, and that lots can happen between reading the recipe (the genes), embarking on making a cake (development) and the final cake itself (the adult organism). Without wishing to push the analogy too far, the recipe can be considered the genotype, and the final cake the phenotype. Many details of the phenotype are not specified within the genotype. In this analogy, a cake can be burnt, comparable perhaps to some unhealable injury, or someone may sneak up and take off the cherries (partial predation!).

14.3.2 Chance

Chance factors of various kinds play a part in all the corners of Seilacher's triangle, and in evolution generally. We saw in Section 14.2.2 that natural selection is the antithesis of a process of pure chance. What people often mean by chance in the context of explaining detailed form is 'not originating nor maintained by natural selection'. Thus, pure fabricational noise may be one manifestation of chance effects, although it usually arises as a by-product of selection for some other feature. It is often appropriate to say that features are random *with respect to* something. For example, an injury to a mouse caused by predation may be random with respect to the colour of its eyes but not the colour of its fur, which may have made it more conspicuous than average. Whether a mouse slips and injures itself falling from a cliff may be random with respect to predation, the colour of its eyes or fur, but not random with respect to the strength of its claws. If strong claws are linked genetically (e.g. by pleiotropy, Chapter 3) to eye colour, then the two variables will not be randomly related. In such instances it is useful to make a distinction between selection *for* strong claws and selection *of* eye colour, the latter being a feature which is hitch-hiked along with strong claws. Normally, though, this distinction is not made, and the phrase 'selection of' is used for any feature that has been directly responsible for maintaining or increasing fitness. At the genetic level, there may be chance changes in allele frequencies (genetic drift). The word 'random' is often used in this context. The neutral theory of molecular evolution, for instance, proposes that many alleles in large populations are maintained by random factors, i.e. that many alleles have no effect on fitness and their frequencies can be explained in terms of mutation and genetic drift (see Chapters 4 and 11, Section 11.3.1).

Chance has always played some part in the phylogenetic corner. As we saw in Section 14.1.4, it may have been more or less a matter of chance (with respect to their adaptations) which of the bodyplans present among the Burgess Shale organisms persisted to become the dominant phyla of post-Cambrian times. The same degree of unpredictability of long-term persistence has been argued for groups that survived certain mass extinctions. That is not to say that the death of each individual of a species becoming extinct did not have a first-hand, proximal reason; it must have had. Rather, on a larger scale, at the taxon level there may have been no inherent susceptibility to extinction. If, for instance, an asteroid was responsible for many of the end-Cretaceous extinctions (Chapter 15), conditions for a while may have been both drastically *and* randomly changed from conditions generally prevailing for millions of years before. Similarly, if it had not been for the extinction of the dinosaurs perhaps mammals might never have become the dominant tetrapods. Chance, therefore, rather than determinism, may well have played a large part in the ascendancy of mammals as opposed to reptiles.

Chance factors may be partly involved in the explanation of why certain forms have *not* evolved, as well as the forms that have. Surprisingly, there don't appear to be any animals that regularly roll themselves up and use gravity to move downhill, certainly none that use any form of wheel (though perhaps rugged terrains virtually preclude wheels), and there are very few species (mainly cnidarians, like the Portuguese man-of-war) that use sails to travel over the surface of water. There is a tendency to accept forms that *have* evolved as the ones most *likely* to evolve, and what has not evolved as unlikely if not impossible. It is doubtful, however, if forms such as the jet-propelled cephalopods, seahorses or the bizarre animals of the Burgess Shale would have been thought likely to evolve were it not for the fact that they did.

14.3.3 ADDITIONAL FACTORS

There is no reason why Seilacher's scheme should not be extended or modified to suit the needs of a particular system under analysis. In a study of the form and function of gastropod radulae, Carole Hickman from the University of California, Berkeley, used a combination of Seilacher's headings and additional ones within which to consider the dominant factors contributing to radular form.

The complexity of gastropod radulae can be gauged from Figure 14.16. Tiny chitinous teeth, sometimes numbering thousands, are supplied in a conveyor belt-like delivery system to the working portion of the apparatus. After use in the working bend of the radula, the worn or broken teeth move on and detach from the disintegrating membrane that carries them. (The conveyor belt analogy therefore falls down because there is no return loop of the membrane and its teeth.) Hickman found that certain aspects of radula form were best considered within a framework that included factors such as 'maturational', 'degenerative' and 'programmatic'. She studied the fine structure of teeth at various positions on radulae and found that the shape of a tooth was 'modified by its contact with the substrate into the appropriate functional form. The tooth is like a new unsharpened pencil; and the form that comes from the "factory" is not the form that tells us the most about how it is used'. Her conclusions had important general implications for the study of form, namely that (i) before a secreted morphological structure is fully formed and before it is used by an organism it can possess fine morphological detail which is related more closely to the secretory process than to ultimate function; (ii) the worn shape of a structure (rather than its pristine form) may be its most efficient working shape; and (iii) wear patterns offer more direct data about function and behaviour than deductive inferences from the unworn shape. 'Programming errors' in radulae were seen as perpetuations of specific flaws in teeth coming off the production line and onto the membrane conveying the teeth to the working bend and beyond. Most of them are believed to be somatic in origin, i.e. developmental abnormalities that are not under genetic control. In our gene/recipe analogy above, such a programming error would be a bit like baking a series of cakes in a battered, misshapen tray. In the gastropod case, the programming errors could be induced experimentally by cold shock, and probably arose naturally in severe winters. Fabricational noise was present as tiny thin ridges of chitin formed on teeth during the pulling apart of one soft, newly secreted tooth from another, or as occasional fused or incompletely separated teeth. These minor constructional errors were inferred to have been selectively neutral.

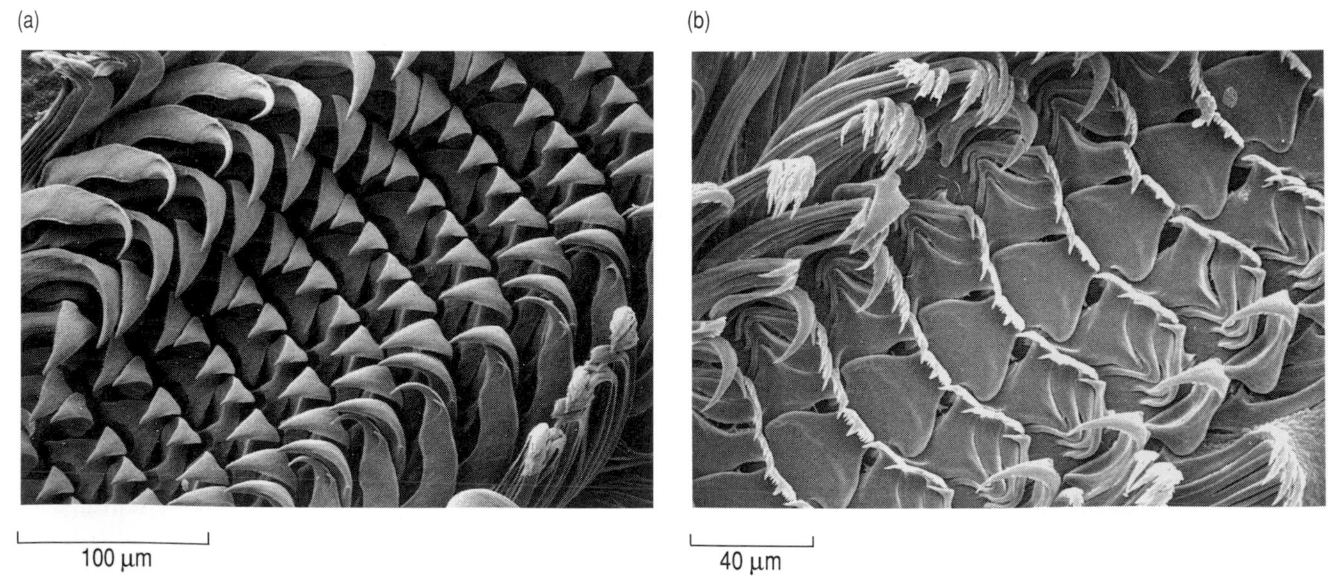

Figure 14.16 *Scanning electron micrographs of gastropod radulae (toothed tongues used for rasping food off the substrate). (a) Radula of a deep-sea limpet* Neomphalus fretterae. *(b) Radula of a pleurotomariid gastropod* Scissurella crispata.

SUMMARY OF SECTION 14.3

There may be factors that influence form which deserve separate consideration from the phylogenetic, functional and fabricational factors of Seilacher's triangle, although they may at the same time interact with them. Seilacher's scheme can be extended or modified to suit the needs of a particular investigation of form.

Ecophenotypic effects are non-genetic (non-heritable) modifications of a phenotype that are produced in response to a particular habitat or environmental factor. There may be strong selection for genotypes that allow plasticity of form in different environments. Ecophenotypic effects within an individual may increase or decrease its fitness, or be selectively neutral. Responses to injury and disease can be considered as ecophenotypic influences on form. So many events can influence the form of an individual during its development that, in many respects, genes can be considered as more like a recipe than a blueprint.

Chance factors play an important part in many aspects of the evolution of form. Although the word 'random' must be used with care, it is often valid in evolutionary biology to speak of features being random *with respect to* something.

The detailed form of a skeletal element may be more related to the secretory process than its ultimate function, and the worn form of a structure may be its most efficient working shape.

14.4 AN EXAMPLE OF FORM ANALYSIS — SPIRALLY COILED SHELLS

The analysis of form and the constraints on its generation can be helped by considering the spectrum of theoretically possible forms that organisms could adopt given their basic mode of growth. The forms that have *not* evolved can then be considered, as well as the forms that have. This approach is known as **theoretical morphology**. To consider extinct forms as well as living ones, it is necessary to study features with a high preservation potential in the fossil record. We will take the form of coiled shells as an example.

Many invertebrate groups secrete a shell which has the form of a spirally coiled cone, for example gastropods, coiled cephalopods, and each valve of bivalves and brachiopods, representatives of which are shown in Figure 14.17. (Brachiopods are not molluscs but members of a separate phylum—their shells have two unlike valves; see also Figure 14.1.) The geometry of these spiral shells is often such that growth occurs without a change in shape. It is obvious at a glance that the shell of a gastropod, or of a cephalopod such as *Nautilus* (Figure 14.2(j); Plate 14.1) or an ammonite (Figure 14.17(e, f)), is coiled, whereas it may be harder to see that the individual valves of bivalves and brachiopods are also spirally coiled. The coiling of, say, a cockle shell is best appreciated by looking at it end on, along the hinge between the valves (Figure 14.6(a)). Similarly, the coiling of brachiopod shells can be seen from the side views in Figure 14.17 (g, h).

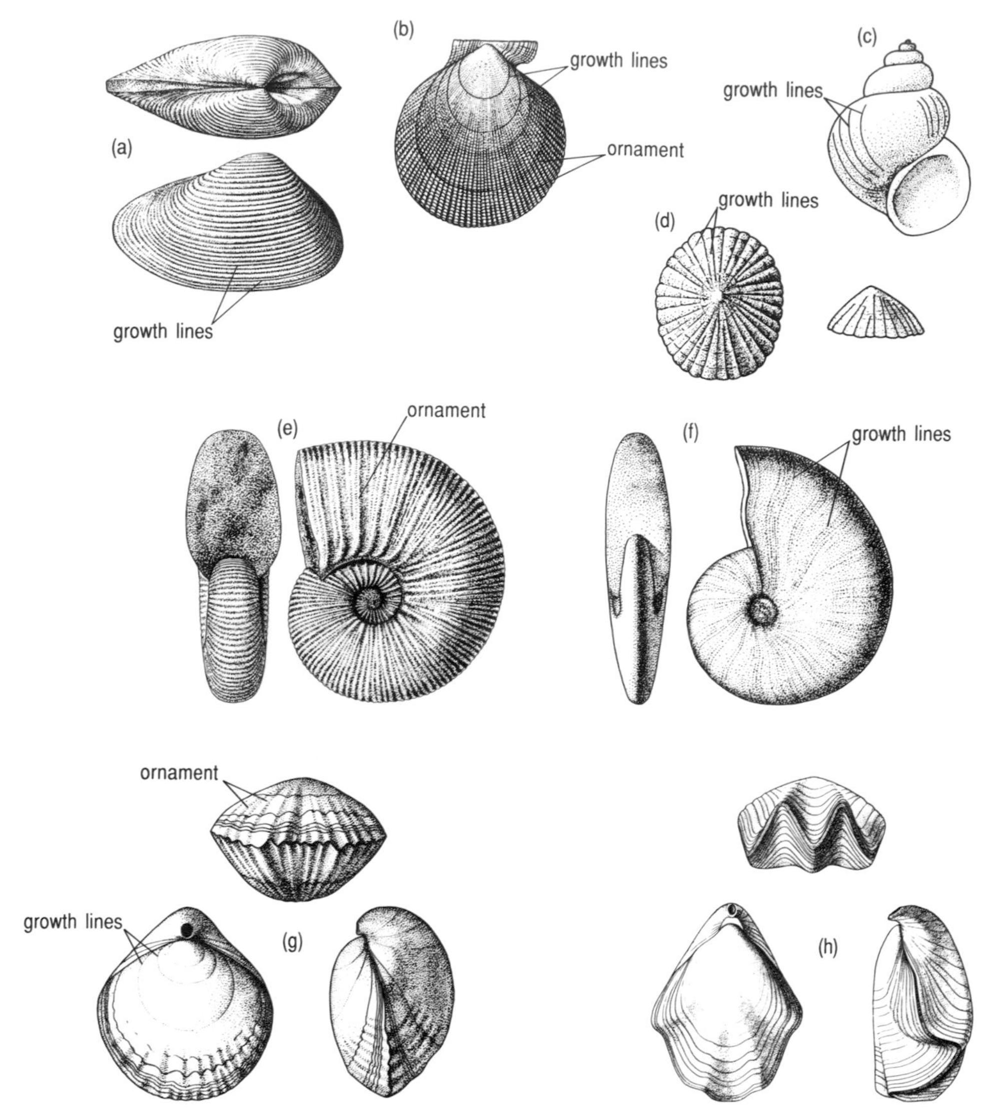

Figure 14.17 *Representative shells for consideration of coiling geometries: bivalves (a, b); gastropods (c, d); coiled cephalopods (ammonites) (e, f); and brachiopods (g, h).*
(a) Pleuromya *(Jurassic). (b)* Mimachlamys *(Cretaceous). (c)* Littorina *(Recent). (d)* Patella *(Recent). (e)* Parahoplites *(Cretaceous). (f)* Tragophylloceras *(Jurassic). (g)* Plectothyris *(Jurassic). (h)* Morrisithyris *(Jurassic).*

The shell, secreted by soft mantle tissue, grows by increments, the most recent increment forming the shell margin. Growth lines record former edges of the shell, and can be clearly seen in most of the shells in Figure 14.17. Growth lines usually lie closely parallel to the edge of the shell, whereas ornament (such as ridges, grooves and spines) often forms a more radial pattern. Growth lines and radial ornament are both particularly clear in the scallop shell in Figure 14.17(b). The patterns on the ammonite shell in Figure 14.17(e), although being roughly parallel to the edge of the shell, are in fact mainly ornament rather than growth lines, which are much finer (see also Chapter 2, Section 2.2.2).

Usually, the growth increments of shells are greater along one part of the margin so that the shell coils about an axis. If coiling is in one plane, the shell is described as **planispiral** (e.g. Figure 14.17(e, f)); if it is in a helical spiral, like the thread of a screw, the coiling is **conispiral** (e.g. Figure 14.17(c). Planispiral shells have a plane of bilateral symmetry; conispiral shells do not. Each complete coil of a shell is called a whorl.

▶ From your general knowledge and Figure 14.2, what type of coiling (planispiral or conispiral) is shown by the majority of (a) gastropods, and (b) coiled cephalopods, such as *Nautilus* and the ammonites?

Most gastropods have conispiral shells, and most coiled cephalopods (with important exceptions discussed later) have planispiral shells.

David Raup of the University of Chicago used a computer to simulate the shapes of planispiral and conispiral shells (Raup, 1966, 1967). He used four parameters to describe the overall form of coiled shells (see Figure 14.18):

1 The shape of the generating curve. This is the shape of the tube in cross-section, more or less equivalent to the shape of the aperture. Moving a circular generating curve would therefore be a bit like moving a luminous hoop through space and photographing the shape it traced out in the dark. In the models illustrated in Figures 14.18 and 14.19, the generating curve is circular, whereas in nature apertures often depart from this shape.

2 The rate of whorl expansion (W) after one complete revolution. If, as in the example of Figure 14.18, the diameter of the tube after a single revolution is twice what it was in the previous whorl, $W = 2$. W rarely exceeds 5 in gastropods, whereas in bivalves and brachiopods W often reaches as high as one million. For shells with up to one whorl (as found in many bivalves and brachiopods), the greater the whorl expansion rate, the less the convexity of the shell.

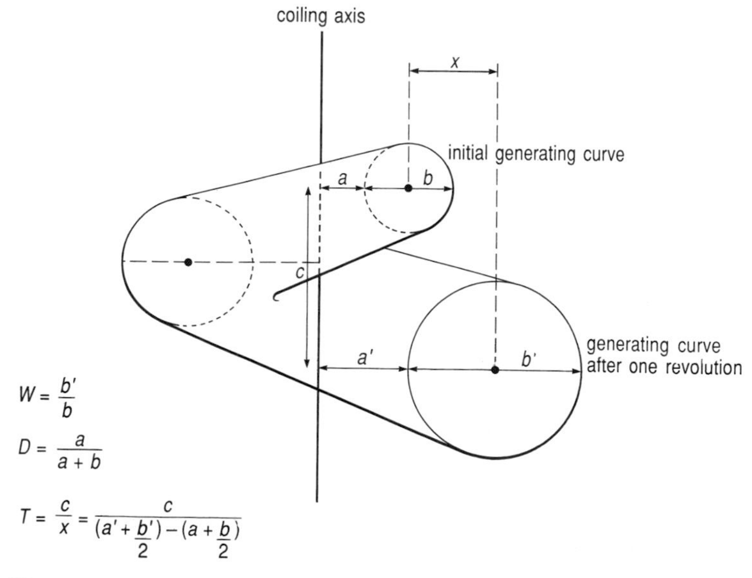

Figure 14.18 *Schematic diagram used in discussion of the theoretical morphology of a coiled shell. See text for discussion.*

3 The relative distance between the generating curve and the coiling axis (*D*). For example, the tube might be separated by half its own diameter from the coiling axis, in which case $D=0.33$ (one-third), as in Figure 14.18. If the generating curve were non-circular, we would also need to specify its orientation with respect to the coiling axis.

4 The movement of the generating curve along the coiling axis, which is known as whorl translation (*T*). Translation is most conveniently expressed by the ratio of movement along the axis to movement away from the axis during any interval of revolution about the axis. The reference point for determining this ratio is the centre of the generating curve (see Figure 14.18).

▶ What is the value of *T* in the example shown in Figure 14.18?

$T=2$. Note that although both *T and* $W=2$ in this example, *T* and *W* are not usually the same. That *T* and *W* have the same value here is a consequence of setting up the parameters for this example so that $a=b/2$ and $c=(2 \times b)=b'$, as can be verified from applying the general formulae given in Figure 14.18.

▶ In planispiral shells, what is the value of *T*?

$T=0$. There is no translation of whorls along the axis. In conispiral shells *T* is greater than zero ($T>0$).

These four variables are often (but not always) approximately constant during the growth of individual shells, but they differ from species to species. Raup set reasonable maximum working values for each parameter *W*, *D* and *T*, and then generated all the possible outcomes of changes in these parameters using a circular generating curve. By doing this he created a region of **morphospace** encompassing the set of all possible results from the given set of instructions (Figure 14.19). He was then able to compare this theoretically available morphospace with the range of forms actually realized during evolution. As explained below, this approach offers insights into the various constraints and opportunities determining the form of coiled shells adopted in nature.

Study Figure 14.19. Whorl expansion rate (*W*) has been set to vary between the theoretical minimum of 1 at the top, to one million at the bottom. Translation rate (*T*) increases from zero on the right to 4 on the left (although in nature the value of *T* may be much higher in long, thin, high-spired gastropods). *D*, the distance between the generating curve and the axis (relative to its own diameter), increases from 0 in the front to 1 at the back.

Examples of various theoretical forms around the block are indicated by photographs taken directly from the screen image generated by the simulation. Labelled regions within the block show the approximate range of forms adopted in nature by *most* bivalves, gastropods, coiled cephalopods, and brachiopods. Note that the complete shells of bivalves and brachiopods each have two valves, whereas in Figure 14.19 only one valve is represented. Remember that the 'open' side of a single valve of a bivalve or brachiopod is analogous to the aperture of gastropods and cephalopods.

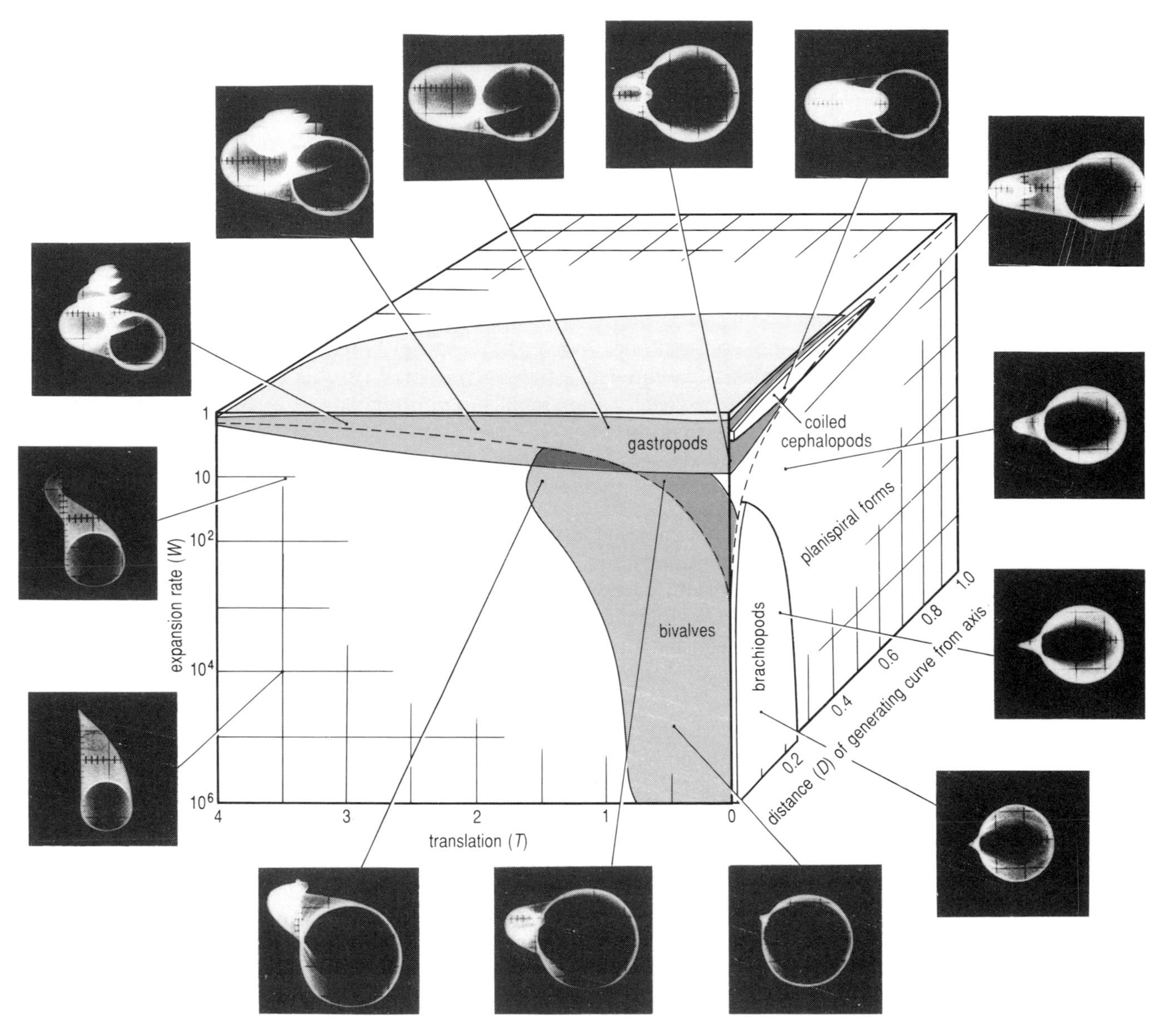

Figure 14.19 *Approximate distribution of form in nature of four major animal groups having a coiled shell: gastropods, coiled cephalopods, bivalves, and brachiopods. The whole block shows the region of available morphospace generated within the restrictions of Raup's model, as discussed in Section 14.4.*

► From Figure 14.19, in general is there much overlap between the shell shapes of bivalves, gastropods, coiled cephalopods, and brachiopods?

No, very little; the shell shapes of each of these groups are not easily confused. We will return to the reasons for this shortly.

▶ In terms of the variables W and T, what are the main similarities and differences between the shell shapes of bivalves and brachiopods?

Both bivalves and brachiopods generally have high values of the whorl expansion rate, W, often as high as one million, and in almost all cases W is higher than 10. Bivalves may have low values of T ($T>0$), but in brachiopods $T=0$.

The finely dashed lines below which the bivalves and brachiopods reside within the block define a surface which is of major functional significance. The surface begins at the bottom right-hand corner of the nearest face and then expands upwards and outwards towards all four corners on the top face of the block. Below that surface the whorls do not overlap (i.e. earlier whorls are not encompassed by later ones), which is essential if a two-valved shell (in this case, a bivalve mollusc or a brachiopod) is to have a functional hinge unimpeded by earlier whorls. Above the surface the whorls do overlap, and the great majority of the univalved shells (gastropods and coiled cephalopods) have overlapping whorls, one consequence of which is increased strength.

Both bivalves and brachiopods have very expanded apertures compared with the vast majority of gastropods and coiled cephalopods, and each valve is generally shaped like a bowl with a curved base. Protection of the soft parts within the 'bowl' is conferred by the presence of another valve with a closely matching margin. It is the need for a relatively unexpanded aperture that confines most univalves to the upper part of the block (which, as we have seen, is also the region of whorl overlap). However, some gastropods, such as limpets (Figure 14.17(d)), also have relatively high expansion rates. In this case, the relatively poor protection that such a wide aperture confers on a single valve is offset by clamping it against a rock surface.

The fact that brachiopods have $T=0$ and bivalves have $T>0$ means that a single brachiopod valve shows a plane of bilateral symmetry whereas the single valve of a bivalve does not; the apex of the latter is turned away out of the plane of symmetry it would have had if $T=0$. The whorl translation rate in bivalves may be minimal, as in scallops (Figure 14.17(b)). Note that most *complete* bivalve mollusc shells do display a plane of bilateral symmetry, but unlike the one in brachiopods, it runs between the valves rather than through them.

Correlations commonly exist between the parameters. For example, in gastropods increasing whorl translation rate (T) is usually correlated with rather low whorl expansion rate (W), and decreasing whorl translation rate with rather high expansion rate.

▶ Study Figure 14.19. In terms of the differences in the parameters of W, D and T, what are the main differences and similarities between the shell shapes of coiled cephalopods and brachiopods?

With significant exceptions discussed below, the vast majority of coiled cephalopods and almost all the brachiopods have $T=0$. The range of D is more in coiled cephalopods than in brachiopods. W in the coiled cephalopods is never normally high, whereas it is usually very high in brachiopods. Because of this difference in *W*, they differ in general shape.

One of the most striking features of Figure 14.19 is the fact that large regions of the block are empty. Why have some conceivable morphologies not been exploited by real organisms? In other words, what may have constrained nature to occupy only a few parts of the block?

There are several possible reasons. The empty spaces may represent forms of low fitness, e.g. the shell might be structurally weak, or too asymmetric for certain soft-part functions. For example, high values of *W* and *D* in planispiral forms create whorl offlap (i.e. non-overlap), tending to give a weaker shell.

▶ If empty spaces in the potential morphospace mainly represent forms of low fitness, in which corner of Seilacher's triangle would the dominant constraints lie?

The functional corner.

▶ Alternatively, there might be properties of the mode of growth, the secretory process, or of the available shell materials, that preclude the construction of certain forms, whether or not they are of low fitness. If this were so, in which corner of Seilacher's triangle would such constraints lie?

The fabricational corner.

In general the basic method of shell secretion is independent of *W*, *D* and *T*, so that fabricational constraints are unlikely to be the dominant cause of the unfilled morphospace. However, an important fabricational constraint in bivalves is illustrated in Figure 14.20. Here, the constraint is related to the fact that the shells have two valves (rather than the one represented in the morphospace block in

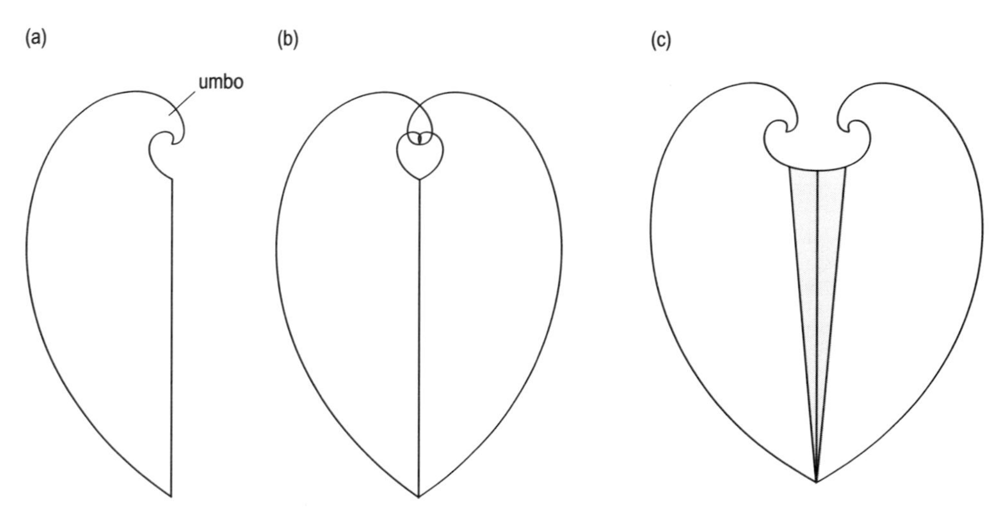

Figure 14.20 *Fabricational constraint in bivalves having a low value of whorl expansion* (**W**). *(a) A single valve with* W = *10. (b) If two such valves are placed together, their umbones overlap—an impossible situation. (c) A zone of additional growth on the side of the aperture nearest the umbones (green shaded area) permits a form with non-overlapping umbones.*

Figure 14.19). As their valves are usually symmetrical, they are normally debarred from having a low value of *W*, because otherwise their valves would overlap (i.e. be attempting to occupy the same space), or at least grind against each other, in the region of the earliest-formed part of the shell, the 'umbo' (Figure 14.20(b)). Bivalves with low values of *W* have escaped from this constraint in a variety of ways: some, like oysters and rudists (see Section 14.5.3), have lost the symmetry of their valves; in others, additional growth takes place in a zone on the side of the aperture nearest the umbones (Figure 14.20(c)).

Alternatively, the inherited bodyplans of the three classes of molluscs and of the phylum Brachiopoda may have kept the range of evolved forms channelled within relatively narrow limits. In general, for example, the vast majority of coiled cephalopods and brachiopods show no whorl translation ($T=0$), though the range of forms within bivalves, gastropods and coiled cephalopods show that variations in the values of all three variables (*W*, *D* and *T*) have been possible. The differences in whorl translation between the groups are largely due to differences in symmetry of the soft parts: cephalopod and brachiopod shells are coiled in the plane of anatomical symmetry; bivalve valves show translation because they are lateral to the body, and gastropods are asymmetrical anyway because of torsion (Section 14.1.1).

▶ If inherited bodyplans have limited the range of evolved forms, in which corner of Seilacher's triangle would the dominant constraints lie?

The phylogenetic corner.

Yet another possibility is that there has so far been insufficient time for evolution to produce the full range of viable geometric forms, and that future evolution may fill progressively more of the block. To some extent, this interpretation might also imply the action of some phylogenetic constraint, because it suggests that the inherited genetic potential has been insufficiently flexible to realize all the possibilities, despite the great antiquity of each bodyplan.

No doubt all the above factors have some part to play in accounting for the relatively small use of potential morphospace. It seems however that, in general, the strongest constraint is probably the functional one; i.e. many of the missing forms may represent forms of low fitness. Supporting evidence for the influence of functional factors comes from taking a small part of the morphospace block and plotting the distribution of individual species of given groups. For example, let us take the ammonoids, i.e. extinct coiled cephalopods that included the ammonites. Ammonoids grew shells partitioned into chambers, like their relatives the nautiloids (see Figure 14.2(j)). A tube called the siphuncle maintained a connection from the body chamber right back to the first formed chambers at the centre of the spiral shell. In modern *Nautilus* (Plate 14.1 and Figure 14.2(j); see also Chapter 2, Section 2.2.2), the siphuncle allows liquid to be removed from individual chambers, to be replaced by gas at a pressure of less than one atmosphere, allowing the shell to remain neutrally buoyant. Given that ammonoids, like other cephalopods, could move by jet propulsion, squirting water out of their funnel, features such as buoyancy, balance and hydrodynamics will have been particularly important to ammonoids.

Consider Figure 14.21 (from the work of Chamberlain, 1980), which shows the effect of whorl expansion rate (W) and whorl position (D) on the drag coefficient of planispiral ammonoid shells with circular cross-sections. (T can be ignored, of course, in the case of planispiral shells.) The drag coefficient, shown by contours, is an indication of the energy required to propel the animal through the water (i.e. the frictional energy lost in water turbulence; see also Chapter 2, Section 2.2.2). Lower values indicate lower drag. The dashed curved line separates forms with overlapping whorls from forms having whorls that do not touch. Other considerations aside, the regions of lowest drag, labelled X, should represent adaptive peaks, because here the ammonoids would have been able to swim through the water with minimum effort. The light-green area, which represents 90% of the total domain occupied by 78 species of smooth-shelled ammonoids with roughly circular cross-sections, approximately coincides with the regions of lowest drag. The dark-green area represents the range of morphologies actually adopted by most species. Note how closely this matches a region of drag

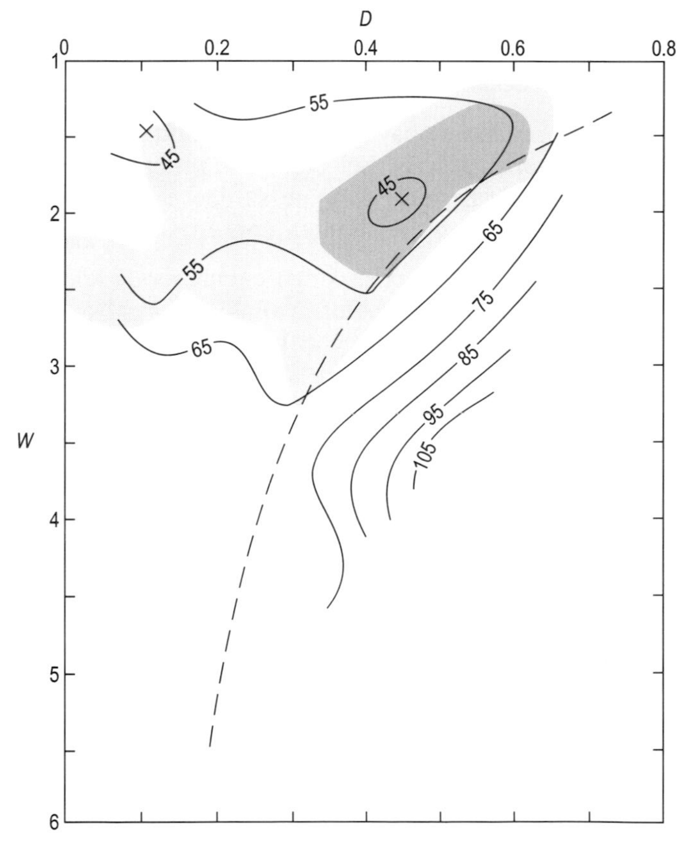

Figure 14.21 *The effect of whorl expansion rate (*W*) and whorl position (*D*) on the drag coefficient of ammonoids with circular cross-sections, indicated by contours. Regions of lowest drag are indicated by X. The dashed curved line separates forms with overlapping whorls (on the left) from forms with non-overlapping whorls (on the right). The light-green area represents the region within which lies 90% of the sample of ammonoids with roughly circular cross-sections, and most of these species cluster in the dark-green area.*

minimum. But why is this peak (i.e. a region of low drag) occupied and not the other peak near the top left corner of available morphospace? Apparently because, according to separate calculations, shapes near the occupied peak are more stable. Thus, for the two variables of W and D, the precise form adopted by the ammonoids at least seems to have been constrained by factors lying in the functional corner of Seilacher's triangle.

▶ Consider the dashed line that separates forms with overlapping whorls from forms with open spiral shells. How does the position of this line relate to values of drag, and why should there be this general relationship?

The drag coefficient increases markedly below the dashed line, i.e. in forms with open spirals. Such shells with non-overlapping whorls would produce more drag (create more eddies, etc.) than shells with overlapping whorls.

▶ How does the position of the dashed line relate to the domain actually occupied by ammonoids? Suggest a reason for this relationship.

Very few of the ammonoids have open spirals and those that do are in the regions of lowest drag for open spirals. Selection has probably not often favoured shells with more open spirals because they are subject to increased drag and are structurally weaker, lacking the strength conferred by overlap between whorls.

In summary, bivalves, gastropods, coiled cephalopods and brachiopods occupy almost mutually exclusive regions of available morphospace. Presumably, this is partly because the groups have very different functional and environmental requirements and partly because there are phylogenetic constraints that generally keep the bodyplans of each group within certain limits. We have seen, for instance, that there are significant differences between the functional requirements of univalved and bivalved shells, largely confining univalved shells to the upper part of the block, and bivalved shells to the bottom. Whether or not a shell has one valve or two is itself primarily determined by the inherited bodyplan, a phylogenetic constraint.

Of course, such studies of theoretical morphology have their limitations. For example, the morphospace in Figure 14.19 is for a circular generating curve. Some of the most striking differences in the form of many planispiral and conispiral shells are in the shape of their apertures (in the case of gastropods and cephalopods) or in the 'open' side of the valves (in bivalves and brachiopods). In addition, most apertures in real gastropods and bivalves are not parallel to the plane that contains the coiling axis.

▶ Look at the brachiopod shell in Figure 14.17(h). What can you say in general about the shape of the aperture of each valve?

Each valve has a strongly non-planar margin (i.e. the opening of each valve does not lie in one plane). The growth rates must be non-uniform along each margin. Therefore, the generating curve will be a complex shape, changing during ontogeny, and is different for each valve. Note that for both the brachiopod shells in Figure 14.17, as with many others, the margin becomes increasingly non-planar during ontogeny. Brachiopods may also show significant departures from the simple logarithmic spiral growth; their valves often show a successively decreasing whorl expansion rate, which can be incorporated into shell simulations.

Even allowing for ontogenetic changes in the shape of generating curves and in the whorl expansion rate, the four parameters listed above encompass by no means all of coiled shell morphology. The model is concerned only with general form. Shell features such as surface sculpture (e.g. spines), shell thickness, internal muscle scars, colour patterns, etc. cannot easily be generated.

Another feature that the above model does not take into account is the coiling direction of gastropods. The conventional way to view a gastropod is to look at it with the aperture facing you and the apex of the shell pointing upwards. Most species (as in Figure 14.17(c) and the models in Figure 14.19) are coiled in a clockwise direction (considered from the apex in the general direction of growth), which means that the aperture is facing you on your right-hand side. This condition is known as dextral coiling. A few species, however, are sinistrally coiled, with their aperture on the left. Some species may even have a mixture of dextrally and sinistrally coiled individuals (Plate 9.1).

The coiled cephalopods included in Raup's analysis were all simple planispiral forms. However, one major group of Mesozoic coiled cephalopods, the ammonites, evolved irregular shells several times in their history. Some examples of these unusual ammonites, which are known as heteromorphs, are shown in Figure 14.22. A few ammonite lineages became very open-whorled, remaining planispiral (Figure 14.22(c)); others, also retaining bilateral symmetry, had early whorls of normal planispiral form, but then had a straight shell which finally recurved at the end so that the aperture faced the first-formed whorls (Figure 14.22(d)). Some of these aberrant forms adopted bizarre, asymmetrical shapes, particularly in the Cretaceous. Several genera became conispiral, looking at first like gastropods (Figure 14.22(e)); using Raup's terminology above, their generating curve underwent translation along the coiling axis ($T>0$). This would have reduced hydrodynamic streamlining and stability during movement.

▶ What does the existence of conispiral ammonites suggest about the relative strengths of the phylogenetic and functional constraints limiting the cephalopod clade to planispiral growth?

The phylogenetic constraint limiting cephalopods to planispiral growth was not inviolate, so functional constraints may have exerted a strong influence to maintain a planispiral shell.

Presumably, the heteromorph ammonites were well adapted to niches somewhat different to those of normal ammonites. A relatively passive lifestyle seems likely, because both hydrodynamic streamlining and stability during motion cannot have been important attributes for many heteromorphs, whereas they were for the majority of ammonites. Interestingly, some of the heteromorph lineages reverted to normal or near-normal coiling later in their history.

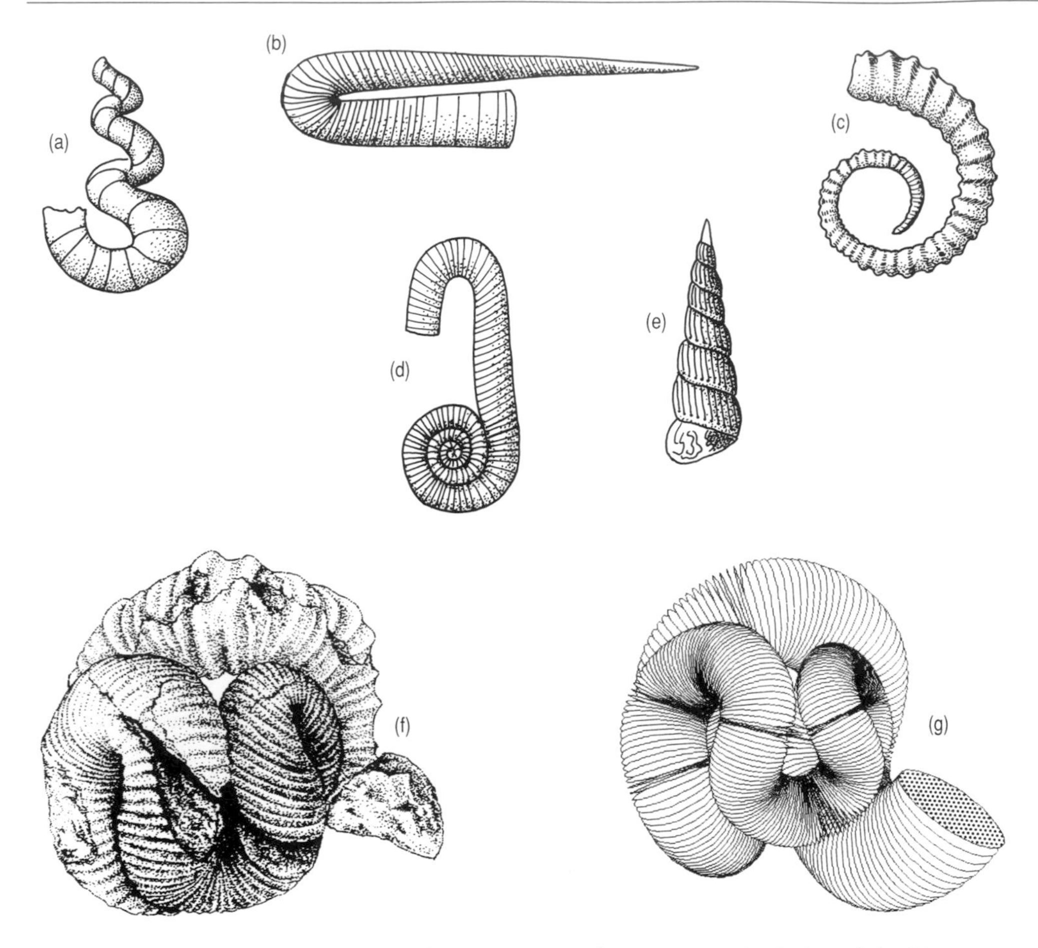

Figure 14.22 *Examples of heteromorph ammonites, and a computer simulation of the bizarre form* Nipponites. *(a)* Hyphantoceras *(Cretaceous). (b)* Hamulina *(Cretaceous). (c)* Spiroceras *(Jurassic). (d)* Macroscaphites *(Cretaceous). (e)* Ostlingoceras *(Cretaceous). (f) Drawing of* Nipponites mirabilis *(Cretaceous). (g) Computer simulation of the form shown in (f).*

The most extreme heteromorphs, such as *Nipponites* from the Cretaceous of Japan, had long tubular shells coiled in an unlikely tangle of 'U'-bends (Figure 14.22(f)). Despite its complexity, the shell form of *Nipponites* has been simulated with remarkable accuracy by Takashi Okamoto of Japan using only a slightly more sophisticated model than Raup's to generate a growing tube. Compare the specimen of *Nipponites* with its simulation (Figure 14.22 (f,g)).

In this Section, we have seen how the analysis of form and the constraints on its generation can be helped by theoretical morphology. Models may be used to simulate the spectrum of theoretically possible forms that organisms could adopt given their basic mode of growth. Within the limits of complexity imposed by the parameters of such models, forms that have *not* evolved can then be more clearly identified and considered, as well as the forms that have. Knowledge of functional morphology, mode of growth and phylogeny can be used to estimate the relative contribution of functional, fabricational or phylogenetic constraints. Conversely, awareness that some constraints are operating (i.e. that there are unfilled regions of potential morphospace) can lead to specific inferences about functional morphology, mode of growth or phylogeny. Pinpointing 'escape routes' from constraints (as in the heteromorph ammonites) can yield insights into the factors influencing the evolution of form.

14.5 Releasing a lineage from constraints

The phylogenetic, functional and fabricational factors each constrain the morphological pathways that a lineage can take. For any organism there are almost countless aspects of form, on different scales and in different dimensions. Details of form are often influenced more by one factor than another, as we have seen.

The existence of variation is essential for evolution. Any system will evolve if there are entities within it with the properties of multiplication, variation and heredity, as long as some of the variation affects the success of those entities in multiplying. It is the free play, the slack between constraints, that supplies much of the variational basis of evolution, in terms both of variation at any one time and of change with time. The range of variation at each corner of Seilacher's triangle provides some of the scope for change by natural selection. Changes in one corner of the triangle may influence what *can* happen in another corner, i.e. there are likely to be knock-on effects.

We will now examine the processes by which a lineage can be released from a particular set of constraints.

14.5.1 A change of environment

Environmental change is widely regarded as the principal driving force behind changes in selection pressures. Here we are not so much concerned with the response of organisms to a change in environment *per se*, but in assessing how such changes may *release* the evolution of form from constraints. The term 'environment' in this context embraces not only physical (i.e. abiotic) factors such as climate, but biotic interactions as well.

A change of biological environment may be particularly important in releasing a lineage from a functional constraint. For example, species cut off on islands might undergo niche expansion under reduced interspecific competition, a phenomenon called *ecological release*. Extinction commonly causes removal of a functional constraint. The removal of a single species may have knock-on effects for all the species that it interacted with; and mass extinctions may open up many niches that were previously occupied. The spectacular early Tertiary radiation of mammals may have been due primarily to the clearing of ecological space by the demise of the dinosaurs. From their origin in the Triassic until the end of the Cretaceous, mammals had remained small, possibly mainly nocturnal and relatively insignificant compared to the immensely successful dinosaurs. The extinction of the dinosaurs may itself have been precipitated by a change in the physical environment, such as by an asteroid impact, volcanic activity and/or major climatic change (Chapter 15). A complex chain of physical and biological interactions, such as the destruction of habitats and disruption of food webs, eventually released mammals from the functional constraint of having to live in a world dominated by large reptiles.

Just as predator–prey interactions provide both important functional constraints *and* a stimulus for change, so the removal of a predator, either by extinction or by the diversion of its attention elsewhere, could release a prey species from the constraint of selection for a particular defence strategy.

▶ Suggest how the release of a bivalve lineage from predation pressure might be detected in the morphology of fossil shells of that lineage.

The rock sequence might first show a decrease in the proportion of shells showing predation damage, and then a reduction in features such as shell thickness, ribbing or spinosity. If the stratigraphic acuity was good enough (see Chapter 10), the abundance of shells might possibly be seen to rise for a while before other factors intervened to limit population increase.

It is easy to envisage how changes in the abiotic (physical) environment may also release a lineage from a functional constraint. For example, a plant species living in an arid desert may no longer require adaptations for drought if the climate becomes much wetter; indeed, it may be counterproductive to keep them, in which case the lineage either evolves *in situ*, or moves its geographic range or becomes at least locally extinct. The release from one functional constraint—adaptation to arid environments in this case—is very likely to switch selection pressures to other morphological features, rather than reduce constraints in general. For instance, the plant may find itself in increased competition with other species for light and nutrients in the new, wetter environment. The point here is that the lineage was released from a *particular* constraint on the evolution of form (adaptation to drought) by a change in physical environment, even though it may have been subjected to new constraints.

14.5.2 Preadaptation

The important concept of preadaptation was introduced in Chapter 2. **Preadaptation** is the phenomenon whereby some structures in ancestors, whether already selected for a certain function or present as selectively neutral features, are *fortuitously* suited for transformation to a new adaptation in descendants. Some, perhaps many, adaptations may have started out as nonadaptations with no particular function—adaptively neutral features that become established by genetic drift or, like fabricational noise, produced as by-products of selection for other features. By good fortune, in retrospect, such nonadaptive features may unexpectedly have served a significant new function and so have started to contribute positively to the organism's fitness, i.e. become adaptations. Preadaptations have the *potential* for playing an important adaptive role in a new context; if they do so, it is only fortuitous because selection cannot plan for the future. We can only tell after the event, with the benefit of hindsight, that a feature was preadapted to perform a radically new function with relatively little change of form.

Without preadaptation it would be hard to see how many fundamentally new 'designs' originated. In 1871, St. George Mivart suggested that Darwin and Wallace's mechanism of natural selection was flawed because it could not explain the incipient stages of useful structures. How could evolutionary novelties such as the feathers on the wings of a bird, or the jaws of vertebrates, get started? What good would be 5% of a wing for flight? The general answer to such questions, which was much invoked by Darwin, is that intermediate stages of a complex, integrated feature were used in ways that were different to their ultimate main function. By this means, novel features can originate through a series of intermediate stages, each of which has some function in the organism's current context. Any failure to conceive the functional operation or selective advantage of intermediate forms is more likely to reflect a failure of our imagination or biological understanding than to raise a genuine objection to the principle of preadaptation.

▶ What is the connection between the idea of evolution as a tinkerer and the principle of preadaptation?

The idea of evolution as a tinkerer (Section 14.2.2) stresses the opportunistic but constrained nature of selection, which has to act on whatever variations are available, and which cannot produce complex adaptations from scratch. Some aspects of an organism's existing form may be fortuitously preadapted for other functions. In the transition from preadaptation to adaptation, selection tinkers with existing form and instils it with new functions.

Feathers, for example, are modified reptile scales (Chapter 11, Section 11.2.4), with which they are developmentally homologous. The earliest known birds, which belong to the genus *Archaeopteryx* (Plate 14.2), very probably evolved from a group of agile, bipedal dinosaurs in the late Jurassic. The primary feathers of *Archaeopteryx* are remarkably similar in structure to the primary feathers of modern flying birds, and—taken with certain features of the skeleton—*Archaeopteryx* could probably fly, albeit fairly weakly. Whether *Archaeopteryx* was a direct ancestor of modern birds rather than an offshoot lineage, or whether flight evolved in fast, ground-running animals or from parachuting or gliding tree-dwellers, are highly contentious issues that need not be considered further here.

It is generally thought unlikely that the earliest birds (i.e. the ancestors of *Archaeopteryx*) first used their feathers for flight. More plausibly, feathers were originally evolved in connection with some other function, perhaps one that is still employed by birds today. They may, for instance, have been an adaptation for insulation (i.e. controlling heat loss) in the presumed endothermic (warm-blooded) reptilian ancestors of birds. One interesting point, however, concerns the feathers of the emperor penguin chick, which has to survive in the dark on the ice-cap of Antarctica in fierce winds without a nest and in temperatures averaging about −25 °C. It has down feathers, which are extremely efficient insulators, like fur (see also Chapter 2, Section 2.4.2 and Figure 2.15). Their structure is, however, very different from flight feathers, display feathers, or the feathers of *Archaeopteryx*. It might be expected that if early feathers evolved initially for insulation alone, they would have come to resemble those of emperor penguin chicks, though, of course, feathers with other structures might be equally effective for thermoregulation.

Alternatively, or additionally, to insulation, early feathers may have been used for camouflage or for display. Another, although perhaps less likely suggestion is that feathers may have been used to capture prey such as insects: a feathered 'insect net' carried on the forelimbs, along with accompanying adaptations such as enlarged pectoral muscles, could have served as an incipient wing, preadapted for powered flight. It is conceivable that birds evolved feathers which served a combination of some or all of these various functions before being co-opted for flight. Unfortunately, the fossil record of early birds is extremely poor and we may never know the sequence by which feathers came to perform their role in flight. Nevertheless, it is easy to see how the principle of preadaptation can be invoked in such cases.

Recent work on the origin of insect flight, involving complex aerodynamic calculations and detailed models based on the sizes and shapes of early insect fossils, has produced clear evidence of a change of wing function as average wing length increased. With small 'proto-wings', aerodynamic advantages are absent or insignificant but thermoregulatory benefits are present with the tiniest of body outgrowths. However, thermoregulatory benefits cease at larger wing dimensions, just at the size aerodynamic advantages rapidly increase. It seems very probable that insect wings thus underwent a functional shift, passing from the domain of waning thermoregulatory gain to increasing aerodynamic advantage. Once across this threshold size, larger wings become increasingly effective for flight because the ratio of lift to drag continues to increase steadily. Thus, initial selection for thermoregulation produced a form fortuitously preadapted for use in flight.

In the Mesozoic, during the well documented and relatively gradual transition from reptiles to mammals, some bones of the jaw were involved in preadaptation. Two bones of the jaw joint of reptiles were incorporated into the middle ear of mammals (Figure 14.23). Reptiles have a single ear bone, the stapes. Mammals, in contrast, have three ear bones — the malleus, incus and stapes. The malleus and incus were derived from two small bones (called the articular and quadrate) that formed the articulation (joint) between the lower and upper jaws of reptiles. The liberation of these bones for use in the mammalian middle ear was intimately linked with modification of the reptilian jaw. The dentary and squamosal bones (see Figure 14.23) that remained in the jaw joint of mammals were being modified, and together with associated musculature the whole joint was under selection for greater efficiency in biting, chewing and manipulating food. The articular and quadrate, while still functioning in the jaw joint, were also in close juxtaposition with the stapes, the single reptilian ear bone. They probably began to function in the transmission of sound while they were being reduced in their role as jaw articulation bones. These bones were thus eventually released from a functional constraint (they were no longer required to perform the function of articulating the jaw with the skull), and also perhaps to some extent from a fabricational constraint (no longer did the growth programme constrain them to be geometrically contiguous with the other jawbones).

It is easy to see that improved hearing to capture prey and to escape predators which were not in the direct line of vision would have been under strong selection, particularly for early mammals and their immediate ancestors who may have become mainly nocturnal, with increased emphasis on auditory perception. Although it is fairly clear from the excellent fossil record of the reptile/mammal

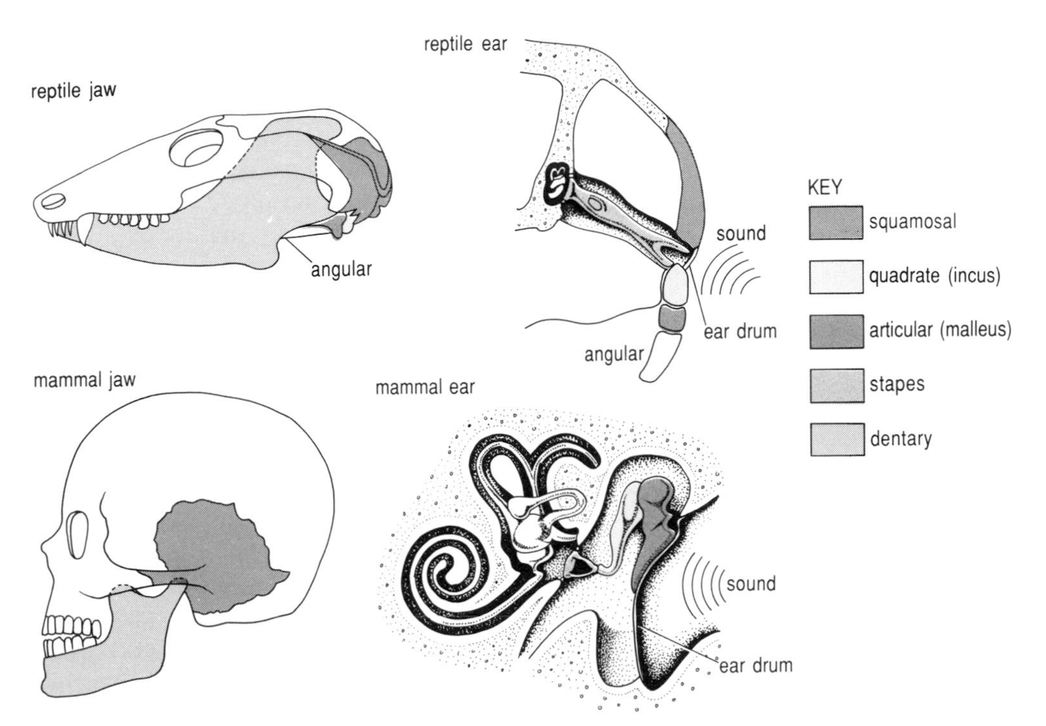

Figure 14.23 *Bones of the jaw and ear of a reptile, above, and of a mammal, below. The articular and quadrate bones that form the articulation of the jaw in reptiles became incorporated into the middle ear of mammals as the malleus and incus.*

transition when the articular and quadrate ceased being part of the jaw joint, it is not certain exactly when these bones became incorporated into the middle ear as the malleus and incus. The latter event may have occurred in the Jurassic, some time after the first mammals appeared in the Triassic. This remarkable innovation seems to have happened independently in several different lineages. The evolutionary transformation is today still reflected by the functional changes of the middle ear bones during the ontogeny of modern marsupials. During post-natal development in the pouch, the malleus and incus retain the reptilian role of the articular and quadrate bones, although at this stage the young are suckling and the jaw is not used for chewing. Only when the young leave the pouch do these bones separate from the lower jaw and enter the middle ear. Incidentally, the jaw bones that remain in humans still (potentially) serve well as sound transmitters; carefully try putting the handle of a ringing (tuning) fork between your teeth!

Darwin gave an invertebrate example of preadaptation from barnacles, which he studied for many years. He suggested that the cementing mechanism by which modern barnacles attach themselves to their substrate was already present as the cementing mechanism by which the barnacle oviduct (egg-carrying tube) coats its eggs so as to attach them to solid objects.

Preadaptation may itself play a part in releasing organisms from a wide range of constraints. A change in construction can predispose organisms to function in new ways, allowing selection to carry them into a new adaptive zone. We have seen how the radiation of birds may have been sparked by a preadaptive cue: the ability to fabricate feathers. Similarly, the appearance and radiation of some phyla in the

Cambrian 'explosion' (Section 14.1.3) was probably connected with acquisition of the ability to secrete hard shells. In these cases, a change of fabricational constraint led to a change in functional possibility.

It is important to appreciate that many complex and interacting factors will be involved in any such evolutionary events. The first rudiments of feathers, or of shells, must have owed their origin to the processes that increase the range of genetic variation in populations, particularly mutations and recombination. A change in the chemistry of seawater in the late Precambrian may have allowed individuals that already possessed the appropriate metabolic pathway to secrete hard parts without their dissolution. Hard parts were probably first secreted as spicules (small rods surrounded by soft tissue) that may have been excretory products, and thereafter selection fashioned spicule-rich tissues into shells. In general, any processes that can release a lineage from constraints may play a part at any stage in the evolution of form, and a release from one constraint may lead to the release *or* imposition of other constraints.

14.5.3 Passing a threshold

Threshold effects in general have played an extremely important part in evolution. There must have been many instances when it was borderline whether some event of evolutionary significance would happen or not. This applies across the entire spectrum of life and its evolution, and at every time-scale. Examples of threshold effects are easy to imagine, such as whether a metabolic reaction has the necessary energy to proceed to the next step in a metabolic pathway, or whether a sense organ will respond to a stimulus of given intensity. Countless individual selection events must depend on such threshold effects. For instance, whether or not a mouse leaves any offspring might depend precisely on having a threshold of hearing low enough to evade an approaching predator at a particular moment; a tiny extra patch of colour might make all the difference to whether or not a stick insect looks sufficiently like a twig on one or many occasions.

The concept of preadaptation is strongly associated with the idea of threshold effects. For example, early in the evolution of birds there must have come a time when the possession of a modified scale coating started to increase fitness, however slightly, whereas before that any such variant scales were selectively neutral. It is easy to imagine that, once under the influence of directional selection, there followed an escalating intensity of selection as the proto-feathers contributed more and more to various functions such as insulation or camouflage, before their eventual use in flight.

A hypothesis involving preadaptation can sometimes be tested when the fossil record is good enough. To do this, it is necessary to predict the outcome, in terms of taxonomic diversification, of the preadaptive model and to show that the predictions of alternative models do not match the evidence so well. Closer inspection of the fossil record can then point to one or other model being the more probable. Peter Skelton of the Open University adopted this approach in testing a preadaptive hypothesis about the evolution of a group of molluscs called rudists (Skelton, 1985).

Rudists are an extinct group of bivalves that became so highly modified that they are hardly recognizable as bivalves at all (Figure 14.24). Some uncoiled, tubular forms even look superficially like certain corals. Rudists thrived during the Mesozoic, especially in the warm clear shallow seas of Tethys, the ancient ocean in which the marine rocks of the Alpine–Himalayan chain were deposited. They became extinct late in the Cretaceous Period.

Skelton had studied the growth geometry of the early rudists and concluded that their particular style of coiled shell growth had imposed a strong fabricational constraint which limited them to only a few adaptive zones. Early rudists had the two valves connected by an external ligament. This arrangement of the ligament constrained the shell to grow such that the earliest-formed parts of each valve, the umbones, grew in a spiral fashion away from each other. These 'spirogyrate' forms were confined to encrusting the sediment surface, as opposed to the later 'elevators' that raised a tall conical shell well above the sediment surface, and 'recumbents' that lay freely but very stably on current-swept shoals (see Figure 14.24). The ability to grow elongate tubular valves, a characteristic of the elevators and recumbents, was not possible with an ordinary external ligament. The shortening and eventual invagination (infolding) of the ligament in one of the spirogyrate lineages was identified as the preadaptive step that allowed the constructional changeover to uncoiled growth, and thereby entry to a new range of possible habitats. Invagination of the ligament to a point where it became internal was postulated as a threshold beyond which an adaptive radiation was triggered. Ligament shortening itself appears to have been selected as an adaptation to

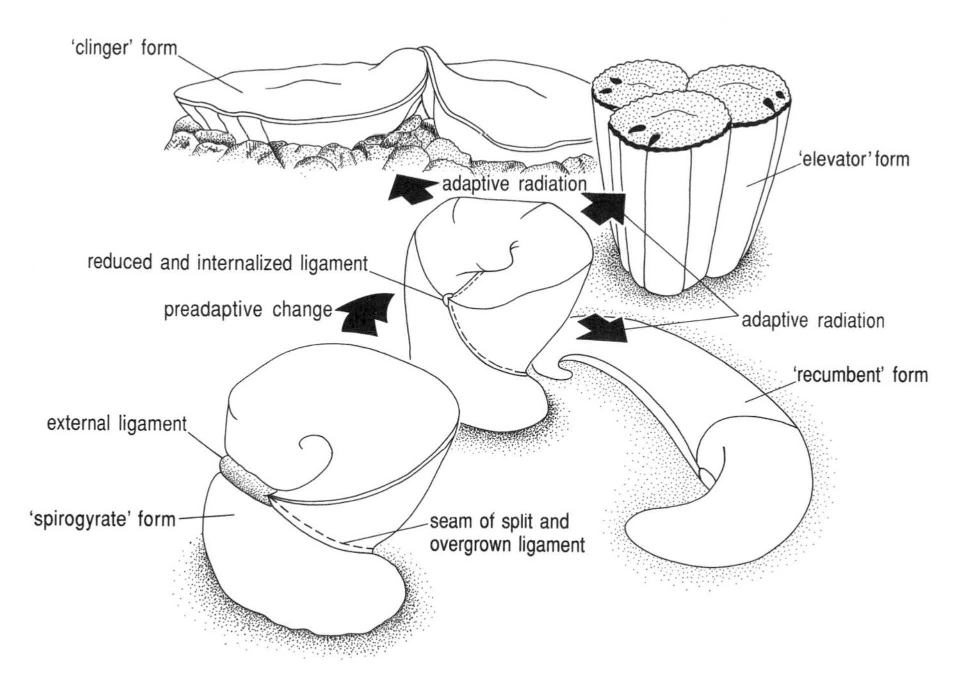

Figure 14.24 *Summary diagram showing the evolutionary history of uncoiling and its consequences in rudist bivalves. In primitive 'spirogyrate' forms (lower left), the external ligament constrained the shell to grow in a spiral fashion. Shortening and invagination (internalizing) of the ligament as shown in the central form allowed uncoiled growth and rapid radiation into new adaptive zones.*

relieve overcrowding in forms that encrusted hard surfaces in dense clusters. Ligament shortening would have promoted a more tightly ascending spiral growth of the valve attached to the sea-floor.

It was possible for Skelton to test his hypothesis, i.e. that radiation into new adaptive zones was constructionally inaccessible prior to the preadaptive cue of ligament invagination. The prediction of the model was that uncoiled taxa should have undergone a rapid exponential diversification on the inception of ligament invagination. Contemporary spirogyrate rudists, in which the ligament remained external, should not have shown any matching change in diversity. Other models, including a null hypothesis of random speciation and extinction, gave different predictions.

▶ Study Figure 14.25, which shows the numbers of genera of uncoiled rudists and spirogyrate rudists over an interval spanning the late Jurassic and early Cretaceous. The first rudists showing ligament invagination appeared in the Tithonian Stage (labelled T on the horizontal axis). Does this evidence support the hypothesis that the shortening and invagination of the ligament was the preadaptive cue for the adaptive radiation of uncoiled rudists?

Yes. Soon after the first appearance of rudists showing ligament invagination there was a striking and persistent increase in the number of genera from 1 to 22. The spirogyrate forms, by contrast, showed very little change in diversity over this interval, remaining at three or four genera.

In general, preadaptation must have played an important role in crossing thresholds, i.e. points just beyond which there are new circumstances and relatively sudden increases in fitness, thereby releasing lineages from a range of constraints and providing new opportunities for evolutionary changes of form.

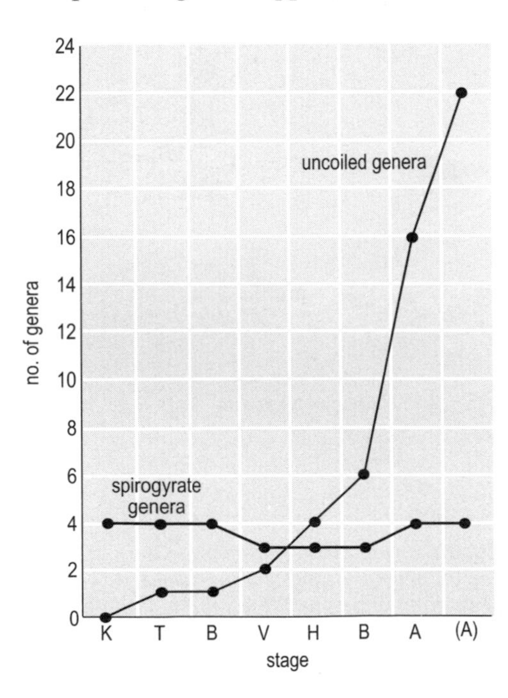

Figure 14.25 *Graph showing the numbers of genera of uncoiled rudists and spirogyrate rudists over an interval spanning the late Jurassic and early Cretaceous. Letters indicate stratigraphic stages. The first rudists showing ligament invagination appeared in the Tithonian Stage, labelled T.*

SUMMARY OF SECTION 14.5

The free play between constraints provides some of the variational basis of evolution. Processes that release a lineage from constraints are therefore very important in the evolution of form. The release of one type of constraint may lead to the release or imposition of other types of constraint. Changes in the physical and biological environment are particularly important in releasing a lineage from functional constraints. But the release from one functional constraint is more likely to switch selection pressures to other features than to reduce constraints in general. Sometimes, however, mass extinctions may release functional constraints on a wide scale, permitting survivors to radiate rapidly into vacant ecological space. A mass extinction may itself be initiated by a change in the physical environment, following which a complex chain of abiotic and biotic interactions culminates in individual extinctions.

Preadaptation is the phenomenon whereby some structures in ancestors, whether already selected for a particular function or present as selectively neutral characters, are fortuitously suited for transformation to a new adaptation in descendants. In retrospect, some features may be seen to have been preadapted to perform a radically new function with relatively little change of form. Without preadaptation it is hard to see how many fundamentally new complex structures originated. Some, perhaps many, adaptations may have started out as nonadaptations with no particular function.

Release from a fabricational constraint may lead to a change in functional possibility. The radiation of many groups may have been sparked by a preadaptive cue involving a relatively slight change in construction that predisposed organisms to function in new ways, allowing selection to carry them into a new adaptive zone. The concept of preadaptation is strongly linked to the idea of a threshold effect. Once past a threshold where an incipient feature starts to increase fitness, directional selection may escalate in intensity, resulting in rapid morphological change. Hypotheses involving preadaptation can sometimes be tested when the fossil record is good enough.

In the next Section we will consider a means by which lineages can be released from phylogenetic constraints.

14.6 HETEROCHRONY

Heterochrony is a word which literally means ‘different time’. It is the name given to a very important evolutionary phenomenon: a change in the timing or rate of developmental events, relative to the same events in the ancestor. The relevance of heterochrony is clear because virtually all evolution involves changes somewhere in the chain of developmental events. Heterochrony thus emphasizes the notion of evolution as a change in patterns of individual development rather than simply change in adult stages. This Section examines different types of heterochrony, discusses the origin and maintenance of heterochronic variation, illustrates evidence for heterochrony in the fossil record, and reviews its evolutionary significance.

14.6.1 Brief historical background

For a long time, biologists have been aware that there is a close relationship between an individual's development (ontogeny) and its evolutionary history (phylogeny). In the late 19th century, the German biologist Ernst Haeckel attempted to formalize this relationship by asserting that ancestral adult forms were encapsulated in the juvenile stages of their descendants. The phrase 'ontogeny recapitulates phylogeny' was supposed to sum this up; i.e. that new evolutionary stages were added on to the end of previous ones so that during ontogeny an individual quickly went through all its earlier evolutionary stages, ending up with the most recently evolved structures. During development an individual was supposed to 'climb its own evolutionary tree', passing through the adult forms of all its ancestors. The existence of a fish-like stage in the embryo of mammals was among the evidence invoked for this general process, which became known as recapitulation (a term now little used).

Haeckel himself coined the term heterochrony in 1875 for exceptions to his rule of recapitulation (interestingly, he also gave us the terms ontogeny, phylogeny and ecology). However, it soon became clear that exceptions to recapitulation were at least as common as the examples used to support it. Furthermore, it was realized that such a process would be inconsistent with natural selection.

▶ Why is it unlikely to be a general rule that, during development, individual organisms pass rapidly in sequence through all the adult stages of their ancestors?

It would mean that natural selection could never modify or remove features formed during the early stages of development; organisms would be restricted to a permanent fixed pattern of development, with successive embryonic and juvenile stages requiring, over a very short time, the same environments that were occupied by successive adult ancestors.

Intensive study of the development of particular organisms failed to reveal successive stages that matched the adult forms of their supposed ancestors. The generalization that emerged relatively unscathed was based on a 'rule' enunciated earlier in the 19th century by the German comparative anatomist Von Baer (Chapter 1, Section 1.3.2), namely that in a large group of animals, general features appear earlier in development than do special features. Thus, the embryo of a 'higher' animal was never seen to resemble the adult of a 'lower' animal, only its embryo. Another expression of this is that two types of organism sharing a common ancestor are usually more similar during early developmental stages than they are as adults. (A reason for this is discussed in Section 14.6.5.) For instance, the embryos of mammals and fish, which are believed to share an early chordate common ancestor, both possess gill slits but, while adult fish retain them, adult mammals do not.

14.6.2 Categories of heterochrony

Heterochrony usually involves an evolutionary change in developmental timing so that the shape and/or size of the descendant no longer matches the ancestor at an equivalent age. (Features other than shape or size, such as behavioural traits, may also be involved, although any with little or no morphological expression will be harder to detect and analyse.)

For example, the rate of shape change can be increased, or its period of operation extended, so that the descendant adult passes beyond the morphological condition of the ancestor. Such a shift leading to *overdevelopment* is called **peramorphosis** (pera- means 'beyond' or 'over'). Conversely, the rate of shape change can be reduced, or its period of operation contracted, so that the descendant adult passes through fewer growth stages and resembles a juvenile stage of the ancestor. Such a shift leading to *underdevelopment* is called **paedomorphosis** (literally meaning 'child formation').

Consider an everyday example to illustrate the difference between peramorphosis and paedomorphosis. There is a wide range of variation in the rate and extent to which human males become bald during their lifetime. Some men keep almost all their hair until very old age, whilst a few boys lose much of their hair in their teens; and most men fall somewhere in the middle of this spectrum.

▶ If, in a future generation, men lived to the same average age but tended to lose their hair earlier, would the heterochronic shift be one of peramorphosis or of paedomorphosis?

Peramorphosis: hair loss was associated with later stages of ontogeny in the ancestor.

Stephen Gould used the evolution of Mickey Mouse as drawn by cartoonists over 50 years to illustrate the concept of heterochrony. Measurements of several stages in the 'lineage' of Mickey (treating new versions as successive descendants, each depicting him at the same age) revealed that later versions had a larger relative head size, larger eyes, an enlarged cranium, wider legs and shorter arms.

▶ Study Figure 14.26. From the human perspective, do the later Mickey Mouse versions described above have more youthful features or more adult features?

Later Mickey Mouse versions appear increasingly more youthful (from the human perspective, at least), i.e. the 'lineage' could be said to show paedomorphosis.

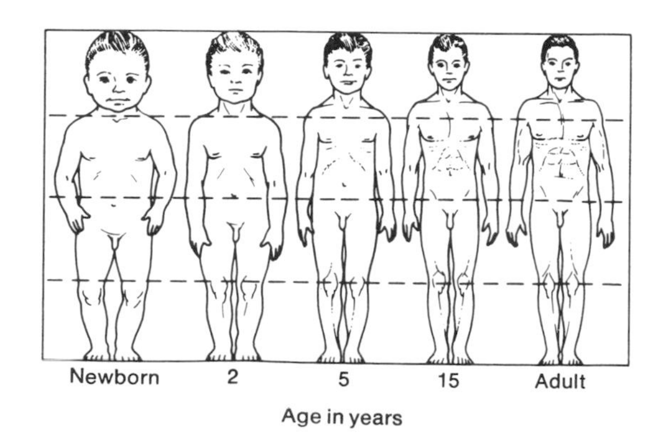

Figure 14.26 *Stages in the ontogeny of a human male, drawn at equal heights.*

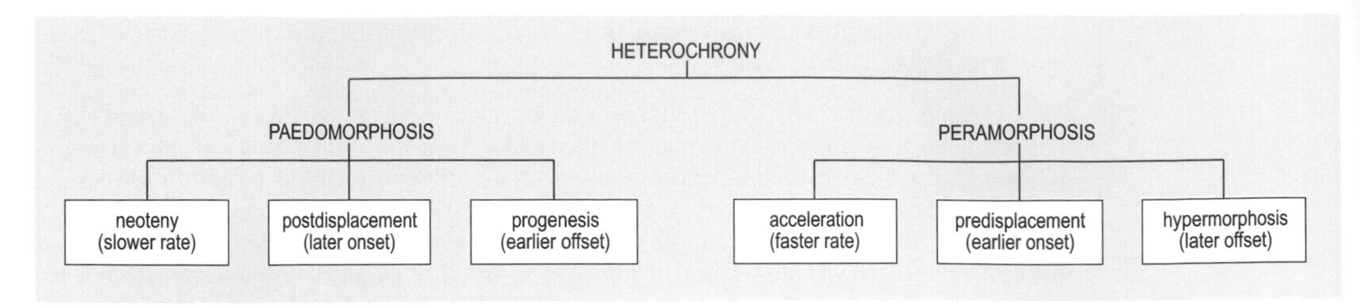

Figure 14.27 The hierarchical classification of heterochrony.

The classification of heterochrony depends on the comparison of ontogenies from ancestor to descendant. Such comparison reveals only three main kinds of possible changes in an organ, trait or whatever is being compared: changes in rate of growth, beginning of growth, and end of growth; i.e. changes in *rate*, changes in *onset* time, and changes in *offset* time. As each change can either be an increase or decrease, a total of only six kinds of 'true' heterochrony can be discerned: **neoteny** (slower rate), **acceleration** (faster rate), **predisplacement** (earlier onset), **postdisplacement** (later onset), **progenesis** (earlier offset), and **hypermorphosis** (later offset). Paedomorphosis (underdevelopment) is caused by slower growth (neoteny—literally 'holding on to youth'), later onset of growth (postdisplacement), or earlier offset of growth (progenesis). Peramorphosis (overdevelopment) is caused by faster growth (acceleration), earlier onset of growth (predisplacement), or later offset of growth (hypermorphosis). Figure 14.27 summarizes these relationships.

Figure 14.28 shows how changes in growth rate, onset time and offset time can be visualized in two dimensions. Growth curves of traits, which can be measured in terms of size, weight or shape, are plotted as a function of time (age). In apes, for example, such traits might include the length of a leg, the weight of the body, or the shape of a foot. Growth curves in nature often follow patterns roughly similar to the curves depicted, with a levelling off (i.e. a stopping of growth) after the phase of most rapid growth (as in Figure 14.30). In each case, the ancestral trajectory is located *between* X and Y.

Paedomorphs ('child shapes') or peramorphs ('shapes beyond') are the terms for individuals displaying heterochronic effects. Paedomorphs and peramorphs may be the same size as their ancestors, or they may be larger or smaller, depending on

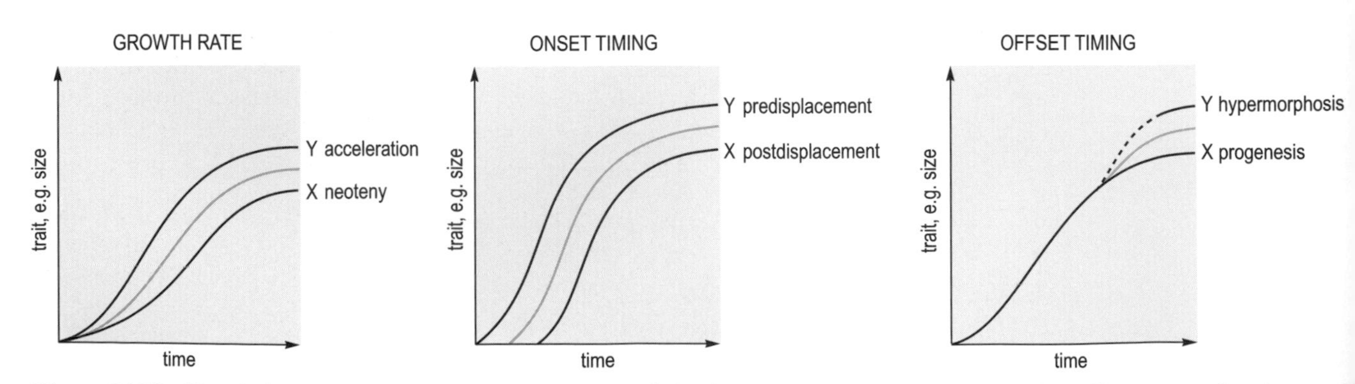

Figure 14.28 *The six basic types of heterochrony in terms of size (or any other trait) versus age plots. The ancestral trajectory is the green line located between X and Y.*

the heterochronic process that has operated. Offset time is often associated with sexual maturation because frequently in groups such as insects, mammals and birds growth essentially terminates then. Some features, however, such as nerve cells, may end growth at times unrelated to sexual maturation.

Global heterochrony refers to heterochronic changes that more or less affect the entire individual, though rarely if ever is every trait affected in exactly the same way. In general, any large-scale changes (e.g. allometric extrapolations from body size increase) that affect many traits in a roughly similar way, may be termed 'global'. However, it is more usual for heterochronic changes to affect only parts of an organism. For example, the length of an organ such as a deer antler might grow at a faster rate in the descendant (i.e. be accelerated) relative to the growth rate of the legs, which remain the same. **Dissociated heterochrony** (as opposed to global heterochrony) refers to such local heterochronic changes that affect some traits but not others, which may remain the same or be affected by another heterochronic process. **Mosaic heterochrony** refers to a complex pattern produced by the operation of opposite heterochronic processes: some characters in the descendant are paedomorphic, whilst others are peramorphic. And as a further complication (in either dissociated or mosaic heterochrony), the descendant may show more than one type of paedomorphosis or peramorphosis: for example, one paedomorphic character may have evolved by neoteny, and another by progenesis. Both dissociated and mosaic heterochrony produce mosaic evolution (Section 14.2.2).

It is important not to confuse *rate* of growth with *duration* of growth. For example, delay in offset time (hypermorphosis) is not the same as a slowing of growth; it is a prolongation of duration of growth so that growth stops later in the descendant. Retardation (slowing) and acceleration are rate concepts, whilst onset and offset of life-history events are timing events. Confusion over such concepts has been responsible for one of the most widely held misconceptions about human evolution, i.e. that humans evolved from apes by neoteny. Neoteny is the process of growing *slower*—a paedomorphic process. Yet humans do not grow slowly overall relative to either the chimpanzee or our ancestors. What we do is *delay the offset* of many developmental events, so that we spend a longer time growing—a peramorphic process. Hypermorphosis seems to be responsible for many traits that make us human: large brain, long learning stage, delayed sexual maturation, large body size, long life-span. Some traits like relative hairlessness are, however, neotenic.

Study Figure 14.29, which illustrates the six general ways that an ancestral form can be modified by heterochrony. The ancestral ontogeny of a hypothetical organism (not very unlike a sea-urchin) is placed at the top of each of the three pairs of processes affecting rate, offset and onset. The ancestor passes through four morphological stages, A to D. The length of each bar labelled A, B, C or D is proportional to the amount of time spent during that growth phase.

The morphological variables that change during the ontogeny of this organism are the size of the main body, indicated by the circle; the size of the mouth, indicated by the central spot; and the number of peripheral spines, which increase from none to six.

▶ Does the mouth of the ancestor show positive or negative allometry?

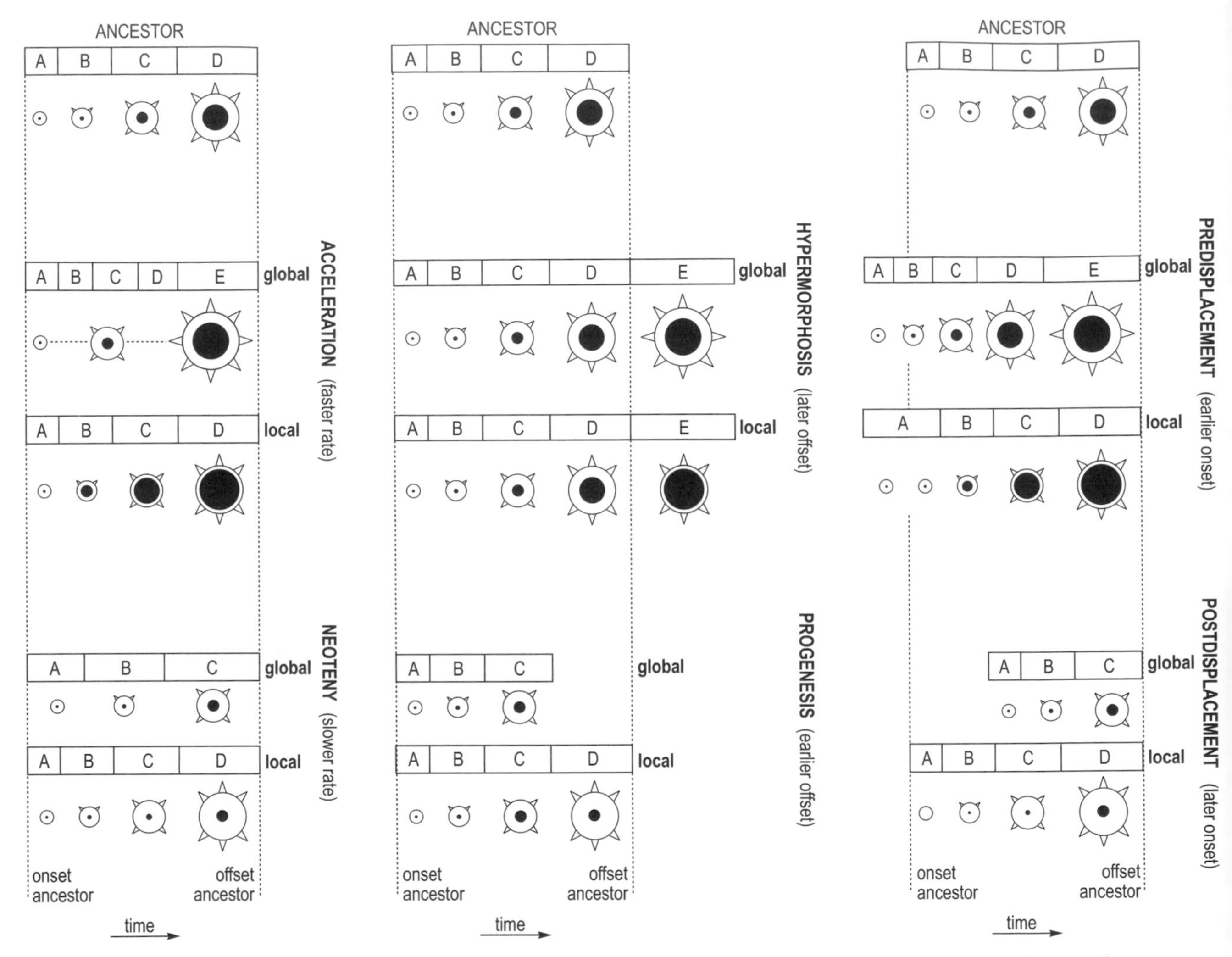

Figure 14.29 Summary of the six basic ways in which an ancestral form can be modified by heterochrony. See text for discussion.

Positive allometry, because during growth it increases in diameter relative to the whole organism (see Chapter 10, Section 10.2.3).

In general, heterochronic change can be either global (affecting all features) or local (affecting only some features).

▶ In global change, the body size, mouth and spines of the organism in Figure 14.29 are affected. What is affected in local change?

Only the mouth.

Each column depicts a pair of heterochronic processes affecting, from left to right, changes in rate, offset and onset. Each process produces a characteristic pattern.

▶ Do the top set of patterns (both global and local) show paedomorphosis or peramorphosis?

Peramorphosis. This is easy to tell at a glance because in every case the descendant has a bigger mouth. Also, in each global case a fifth stage 'E' is added. Conversely, the patterns below reflect paedomorphic processes because the mouth is smaller in every descendant, and in each global case stage D is omitted.

In *differentiative* heterochrony, changes occur before the final differentiation of tissues or organs, i.e. novel features may appear or disappear, as opposed to changes in size or shape of existing features (as in growth heterochrony). In *growth* heterochrony, there is a change in rate or timing of development after cells/tissues have undergone their final differentiation, i.e. it is only the size or shape of a particular feature that changes.

▶ Study the part of Figure 14.29 that illustrates global neoteny. Which feature shows differentiative heterochrony—the mouth or the spines?

The spines. Differentiation is not complete until all spines have been formed. The mouth shows growth heterochrony: it is present in stage A and persists, growing at a slower rate (along with the rest of the organism) until growth is finished.

Note that, in Figure 14.29, the descendants of each of the processes of peramorphosis or paedomorphosis look the same, despite the fact that the processes are different with respect to timing. For example, compare the three adult descendants produced by global neoteny, global progenesis and global postdisplacement.

▶ Which growth stage of the ancestor does each adult descendant resemble?

All look like stage C of the ancestor.

In neoteny, there is a slower rate of growth but growth stops at the same time as in the ancestor, so that the descendant never develops into stage D (though it reaches maturity nonetheless). In progenesis, onset and rate of growth are the same but growth stops earlier. In postdisplacement, the rate of growth is the same but the onset of growth stage A is delayed and growth stops at the same time.

To determine which of the six heterochronic processes was responsible for an observed ontogenetic shift requires knowing information on the *age* of compared individuals. For living organisms, this is in theory at least possible, although accurate age data are not yet available for most species. For most extinct species, however, it is even more difficult to obtain indications of absolute age. Nevertheless, recent work on growth increments added on a regular basis, such as growth lines in the shells of some molluscs, seem to offer a good prospect, at least in some groups (see Chapter 10, Section 10.2.3).

Size has often been used as a proxy for age, in both fossils and living organisms. However, when related species are compared there may be no close relationship between size and age. Some similar-sized Recent bivalves of related species can differ in age by 80 years or more! And larger species may not indicate a longer

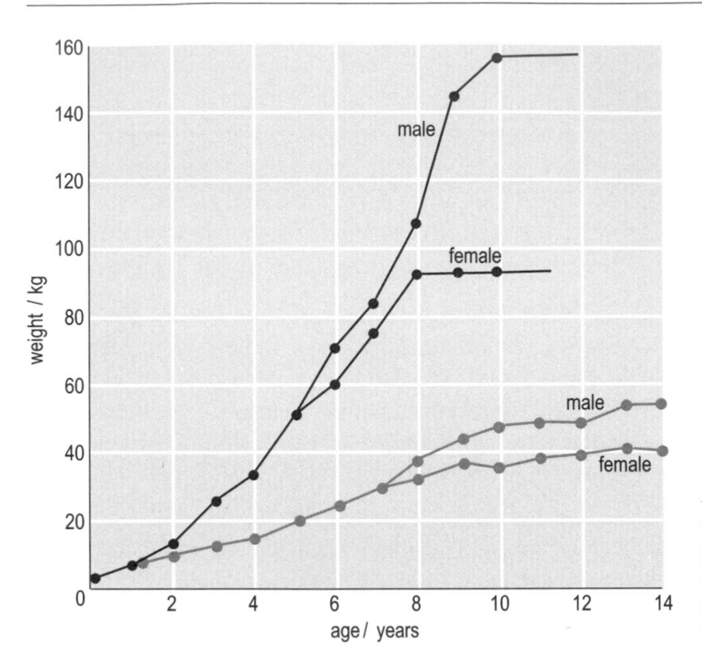

Figure 14.30 *The growth of male and female gorillas* Gorilla gorilla *(black curves) and chimpanzees* Pan troglodytes *(green curves) as measured by cumulative gain in weight.*

growth period. For instance, the gorilla develops in many ways as if it were an accelerated chimpanzee (see Figure 14.30). It grows faster, including body size, but not for longer time periods; onset and offset times in the 'spurt' and 'lag' phases are the same.

We may be able to compare sizes at comparable events in life history such as sexual maturation. For instance, in sea-urchins the size at which genital pores (through which gonads shed their products) appear in a particular skeletal plate indicates size at maturation. Such information may be useful for a variety of reasons (such as indicating a change in life-history strategy, see Section 14.6.5), but without knowing whether maturity occurred earlier or later in absolute time, we will not be able to diagnose the heterochronic process involved. For example, a dwarf adult pterosaur (Mesozoic flying reptile) may have attained its small paedomorphic size either by growing slower (neoteny) and maturing at the ancestral age, or by growing at the ancestral rate and maturing earlier (progenesis). When we can discern the size at maturity, a 'loose' practical definition of paedomorphosis would be: 'the descendent adult resembles a preadult stage of the ancestor (in size or shape), regardless of age'.

Although true age information is frequently missing, information about size is often available against which morphological variables can be plotted. For some purposes, trait change as a function of body size may be as informative as trait change with time. In some animal groups (though not most mammals and birds), many ontogenetic events are size-specific rather than age-specific; i.e. they are programmed to occur at a certain size in development rather than at a certain age. Body size may thus sometimes be regarded as a better measure of 'intrinsic' or 'internal' age than 'external' time. **Allometric heterochrony** is change in a trait as a function of body size, as opposed to a function of age, as in 'true' heterochrony. The same categories apply in this 'size-based heterochrony' as in 'age-based heterochrony', but the adjective 'allometric' is prefixed, e.g. in 'allometric progenesis' the descendant stops growing (i.e. earlier offset) at a smaller size (as opposed to age) than the ancestor. A simple alteration of the axes in Figure 14.28 will make the diagrams applicable to allometric heterochrony: 'time' along the *x*-axis is replaced by 'body size', and the trait along the *y*-axis could be another trait including the size of some feature other than total body size. Bivariate plots to assess allometric heterochrony are very useful for common descriptive purposes, especially in palaeontology.

There is an obvious link between allometry—the relationship between size and shape, particularly how one trait changes relative to other traits during growth (discussed in Chapter 10, Section 10.2.3)—and heterochrony, which addresses trait change relative to time. Allometric differences between ancestors and descendants are differences in growth pattern caused by heterochronic processes.

The relevance of allometry in heterochrony can be illustrated by an example from familiar animals. The range of head morphology exhibited by domesticated dogs is enormous compared with the range exhibited by cats. The reason lies in the difference in degree of skull allometry between dogs and cats. Figure 14.31 shows new-born and adult skulls of a typical cat and a typical domestic dog such as a hound.

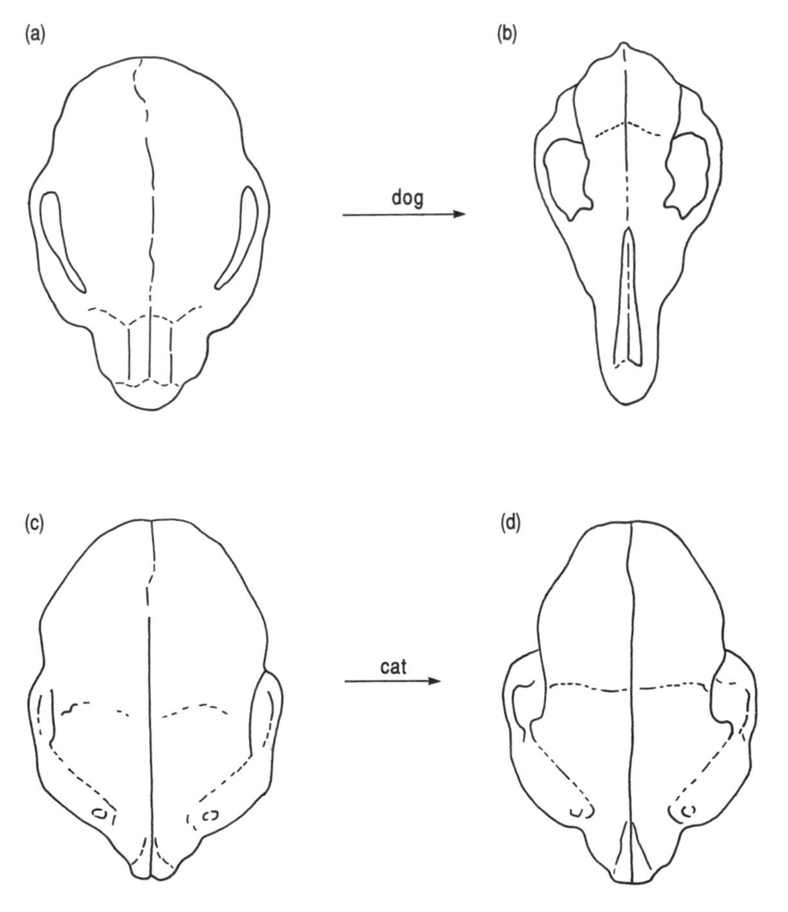

Figure 14.31 *Dorsal view of the skulls of (a) a domestic new-born dog and (b) an adult dog, contrasted with that of (c) a new-born cat and (d) an adult cat. Not drawn to the same scale.*

▶ Which shows more allometric skull growth—the cat or dog?

Whereas the cat skull grows almost isometrically, the dog skull shows highly allometric growth. Not surprisingly, in dogs slight extensions or contractions of the juvenile growth period, or slight changes in the rate of local fields of growth, will produce marked differences in skull shape. Other dog breeds (produced by artificial selection) will have ontogenetic trajectories with different allometries. In cats, however, extension or contraction of the growth period will have little effect on skull shape; compared with dogs, cat breeds (big or small) tend to have rather similar-shaped skulls.

14.6.3 THE ORIGIN AND MAINTENANCE OF HETEROCHRONIC VARIATION

Understanding the origin of variation is essential for a full understanding of evolution. Since Darwin, the emphasis has been on selection, but selection can only act on variation created by genes and developmental processes. Much intraspecific morphological variation is, in fact, generated by heterochronic processes, i.e. much phenotypic variation is caused by slight differences in inherited developmental trajectories (ontogenies). Consider variation in human populations. In any school classroom there are children exhibiting different degrees of development towards adulthood; they differ in features such as limb proportions or body shape from others of the same age, and differ from their parent of the same sex at the same age. Some of this variation is non-genetic and due to differences in nutrition, but part of it can be viewed as reflecting growth heterochronic processes at fine spatial and temporal scales. Recombination can commonly produce gene combinations that will change the rate or timing of ontogenetic processes in the next generation. Slight intraspecific differences in allometries produced by heterochrony account for a large amount of the normal range of phenotypic variation seen in vertebrates and invertebrates. Individuals can be compared to assess whether a feature is relatively more paedomorphic or peramorphic in one individual than another. For instance, an individual that has a bone with a slightly higher positive allometry than the same bone in another member of the population is relatively peramorphic in that feature. Alternatively, the *mean* value of the allometric coefficient of the bone can be compared between populations to assess relative degrees of interpopulation heterochrony.

Intraspecific variation expressed as polymorphism and as sexual dimorphism can often be interpreted as having a heterochronic origin. For instance, slight changes in the timing of release of growth hormones are known to have profound effects on anatomical proportions and on the timing of reproductive maturity. Dimorphic size and shape differences caused by different timing of maturation in males and females are common in many groups.

It used to be believed that there were simple 'rate' genes or 'timing' genes that were normally responsible for heterochronic changes. However, the complexities of development and gene/cell interactions (Chapter 3) are such that a wide range of genetic changes can affect the cascade of development. Most cases of heterochrony probably result from a set of genetic interactions that are rarely initiated by a simple 'rate' or 'time' gene. However, genetic analysis of heterochronic mutants shows that mutations that cause heterochrony are not necessarily complex changes in gene structure and function. For instance, in the case of the roundworm *Caenorhabditis elegans*, single-gene mutations have been identified that change the time of onset of developmental events (rather than change growth rates). It seems likely that many different kinds of genes can produce heterochronic changes; for example, many genes are responsible for the production of numerous growth hormones that stimulate, inhibit or enhance cellular activity, and many genes determine the properties of cell membranes across which signals are transmitted. What is clear is that relatively small genetic changes affecting the early stages of development can translate into major heterochronic effects in the adult phenotype.

Ecophenotypic effects during development can sometimes mimic those of genetic-based heterochrony. For instance, living cichlid fish may undergo pronounced neoteny or acceleration of shape, depending on their diet. In general, organisms developing in adverse environments often grow at a slower pace and stop growing at a smaller size. In some groups, if conditions improve, the adult forms may ultimately converge on the normal 'programmed' size and shape by later acceleration of growth; in others, the morphological results of stressed environments are permanent, irrespective of improvement.

The capacity to show marked ecophenotypic plasticity (by which a single genotype may exhibit wide morphological variation in different environments) may often increase fitness and will itself have been selected (Section 14.3.1). Surface water temperature is known to be a major determinant of the onset of maturation in molluscs, with an increase in temperature often inducing early maturity at small size. Pheromones released by certain members of social insect communities may retard the onset of maturity, as in adult honeybees, or advance it, as in desert locusts. The important point is that such developmental flexibility is frequently adaptive and ensures that there is usually a range of heterochronic phenotypes in the population on which selection can act. External controls often simply alter rates and onset/offset times of the same, one developmental pathway; sometimes, however, as in the case of the axolotl discussed below, cues in the external environment may trigger *alternative* developmental pathways that are already genetically programmed, and thereby cause heterochrony.

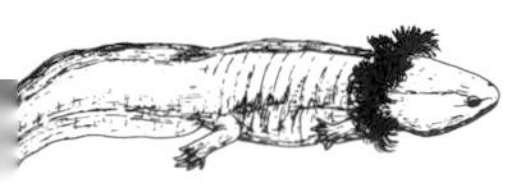

Some of the most striking examples of heterochrony are from organisms that undergo metamorphosis, such as arthropods, amphibians, and many marine invertebrates. Perhaps the most familiar cases are those of paedomorphic salamanders, especially the Mexican axolotl, sometimes kept as a pet. These strange animals, up to 30 cm long with small weak limbs, reach sexual maturity and reproduce when morphologically in the larval stage with external gills, and do not normally metamorphose into the adult. Failure of the thyroid gland to produce the hormone thyroxine causes neoteny, and the onset of metamorphosis is usually permanently delayed. The genetic basis for this in the axolotl is simple: a single gene with two alleles. Some individuals in captivity, however, can be induced to metamorphose into the adult by the addition of thyroxine. In the wild, with much increased population densities and/or polluted waters, some paedomorphic salamanders, including the axolotl, will metamorphose into adults capable of a (largely) terrestrial existence. Presumably, individuals that retain the ability to metamorphose and escape from a deteriorating aquatic environment, such as a drying up pond, will be at a selective advantage and such lineages with facultative paedomorphs (i.e. with paedomorphs as an optional alternative) will resist extinction. The adaptive significance of paedomorphosis is clear from field observations: reproduction at the larval stage is most common where the surrounding terrestrial environment is harsh (e.g. severe fluctuations in temperature and low humidity), and where water is permanent and predatory fish rare or absent. Some species of salamanders, however, have apparently committed themselves to eternal youth and their individuals will not metamorphose (i.e. they are obligate paedomorphs, always reproducing at the larval stage). Note that if the genetic changes causing paedomorphosis were not extensive, and if there has been little other evolutionary change, it is conceivable that future mutations might resurrect the potential to metamorphose.

14.6.4 AN EXAMPLE OF HETEROCHRONY FROM THE FOSSIL RECORD

Determination of whether features of a species in the fossil record have arisen by peramorphosis or paedomorphosis is dependent on being able to assess ancestor/descendant relationships (see Chapter 11), *and* on knowledge of ontogeny, particularly of the ancestor. Fortunately, for groups that grow a skeleton by accretion, evidence of successive stages of ontogeny is preserved within the shell.

Figure 14.32 shows an example of heterochrony in Cainozoic brachiopods from the waters around Australia and New Zealand. *Tegulorhynchia boongeroodaensis* was ancestral to a series of descendent species, two of which persist to the present day—*T. doederleini* and *Notosaria nigricans*. *T. doederleini* is so similar to *T. squamosa* that it may represent a case of stasis in shell morphology over *c*. 45 million years (see also Chapter 15, Section 15.2.1).

▶ Compare *N. nigricans* with the adult form of *T. boongeroodaensis*. What are the main differences between the two species? (The black area at the 'top' of the shell is called the pedicle foramen; it is a hole in one valve through which projected the fleshy pedicle, an organ of attachment.)

N. nigricans has fewer ribs, a relatively narrow shell, a smoother anterior margin (i.e. at the 'bottom' of the shell), and a larger pedicle foramen.

▶ Do these features look more like those of the juvenile of *T. boongeroodaensis*, or do they appear to be an extension of its growth, beyond the adult form?

For each feature, the adult of *N. nigricans* resembles the juvenile of *T. boongeroodaensis*.

▶ Which heterochronic process has been dominant—paedomorphosis or peramorphosis?

Paedomorphosis. (A study of growth lines yielding age information should, in future, reveal which of the three paedomorphic processes was mainly responsible.)

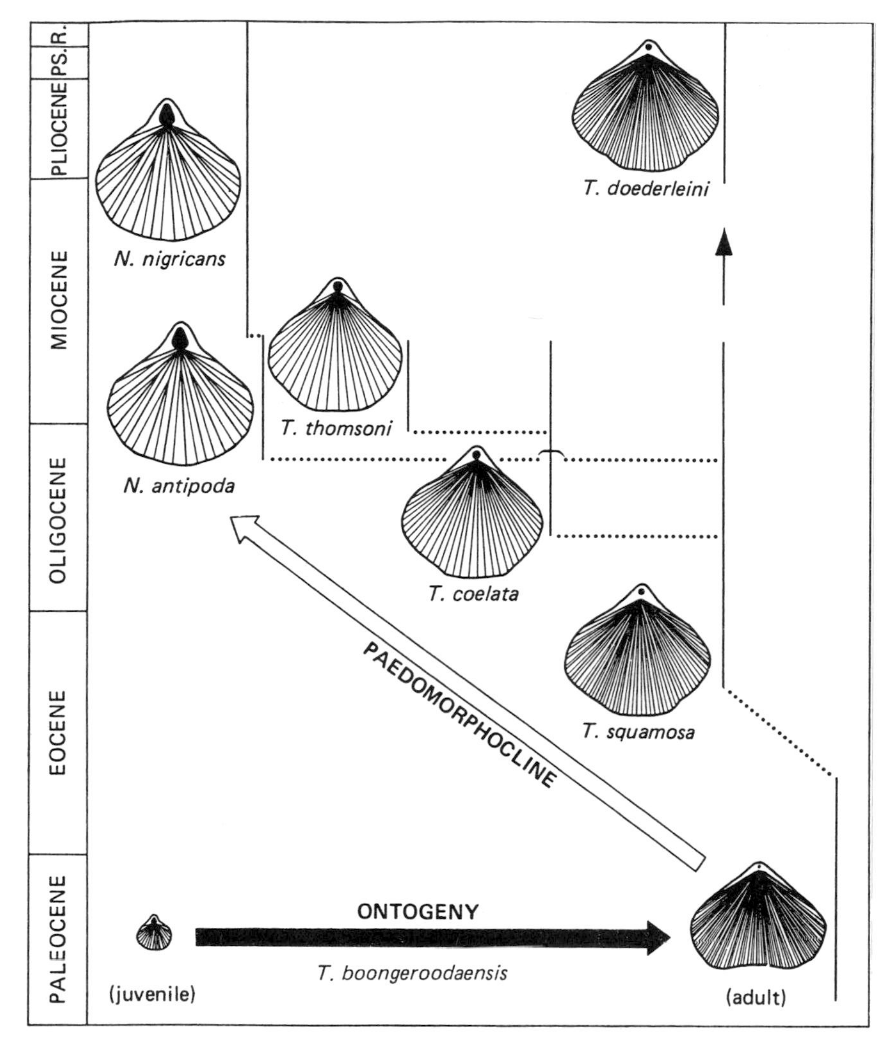

Figure 14.32 *Evolution of species of the brachiopods* Tegulorhynchia *and* Notosaria *from the waters around Australia and New Zealand.*

The successive species *T. boongeroodaensis*, *T. squamosa*, *T. coelata*, *T. thomsoni*, and the related species of *Notosaria* form what is termed a paedomorphocline, i.e. an evolutionary sequence wherein adult morphologies become progressively more paedomorphic. Often, such a sequence (or its converse, a peramorphocline) appears to reflect adaptation to an environmental gradient. In the case of these brachiopods, the heterochronic changes reflect successive adaptations from deep to shallow water. For example, a wider pedicle foramen reflects a thicker pedicle for firmer attachment to the substrate, as required in the stronger currents of shallower waters. Being relatively unstable, the ancestral juvenile had a relatively wide foramen compared with the adult, which on gaining size had little need for the stability conferred by attachment in the deep, rather quiet waters. Heterochronoclines, the collective term for paedomorphoclines or peramorphoclines, have been reported from a number of groups, principally marine invertebrates. In general, however, patterns in the fossil record are often more complex than the global paedomorphocline described above, reflecting dissociated and mosaic heterochrony.

14.6.5 The evolutionary significance of heterochrony

Given that various heterochronic mechanisms generate much of the variation within populations, it is clear that heterochrony may play an important part in events that lead to speciation (see Chapter 9). Furthermore, because major morphological change can be generated by minor genetic change in a few generations, it seems probable that heterochrony has been responsible for some macroevolutionary changes above the species level, i.e. the founding of new clades over short geological time-scales. Many cases have been made for the role of heterochrony in the Cambrian 'explosion' and in the later evolution of major groups, ranging from fish and birds to parasitic worms and conifers, to name but a few. Lineages can be released from phylogenetic constraints by heterochrony, particularly by paedomorphosis, whereby the descendant is no longer genetically programmed to pass through as many stages as the ancestor. Paedomorphosis, particularly allometric progenesis, has been seen as the dominant process, and paedomorphosis was argued to have more evolutionary potential than peramorphosis. But paedomorphosis can only use existing structures, which themselves had to have an origin. Recent re-evaluation has led to the belief that peramorphic forms have the same general evolutionary potential as paedomorphs, and that the origin of many major groups, including phyla, is currently so unclear and controversial that the relative influence of different heterochronic processes is unknown.

As an example of a recent reinterpretation, consider ratites, a group of flightless birds (Chapter 2, Section 2.2.2). Attention has usually focused on the reduced wings and fluffy juvenile feathers clearly brought about by paedomorphosis. However, in view of the susceptibility to predation, flightlessness hardly seems adaptive in itself; more likely, two *peramorphic* features—the development of larger, more powerful legs and large body size—were also involved in selection, a case of mosaic heterochrony.

Ken McNamara of the Western Australian Museum, Perth, found that of 155 of the best documented cases of heterochrony in the fossil record, 90 were of paedomorphosis, and 65 of peramorphosis. However, the survey did not address the issue of which processes are responsible for the origin of novel clades, for which ancestors are often unknown or uncertain. A general result of McNamara's survey is that the relative importance of paedomorphosis and peramorphosis varies between groups, and over time within groups. Mosaic heterochrony is common and widespread.

Several reasons support the idea that small organisms form the ancestors of most clades (see Chapter 15). Small-bodied species are much more common in most clades (probably because small organisms divide the environment and its resources more finely); therefore, other things being equal, a small species is more likely to start a clade than a large one. Also, during major extinctions, large-bodied species suffer disproportionate losses, so any surviving species radiating into empty ecospace is likely to be relatively small. But even if a descendant is smaller than its ancestor, small size does not necessarily indicate a fully global paedomorph, and a size-paedomorphic organism may be peramorphic for many other traits of high evolutionary potential.

Natural selection appears to modify the adult stages of development more easily than the early embryonic stages, which tend to have a closer resemblance to those of ancestral forms. Changes in genes affecting early stages of development have a greater effect on the phenotype than changes in later-acting genes, and, *on average*, the greater the effect that a mutation has on the phenotype, the less likely the mutation will be selectively advantageous (Chapter 3, Section 3.3.2). This is one major reason why the embryos of a wide range of groups are more similar than the corresponding adults. Nevertheless, genetic changes affecting early stages of development without any loss of fitness are likely to translate into relatively large and rapid changes in the adult phenotype.

The fact that successful developmental mutations tend to modify only the least disruptive, later parts of ontogeny has had an important consequence: since the early Palaeozoic, it has become less and less easy for new bodyplans to originate. This 'hardening' of ontogenies, with a build up of contingencies in particular developmental programmes (i.e. increasing interdependence of steps), is possibly one major reason why no new phyla have appeared since the Ordovician. The dominant form of evolutionary creativity has been permutations on pre-existing themes, with origination of lower level clades. This process of making change more difficult has been called the 'evolutionary ratchet'.

Heterochrony can play a role in the evolution of behaviour: adult and juvenile behaviours can change in the timing of appearance. Some adults in bird populations will sing in ways characteristic of juveniles of their species, such as singing continuously, with an undefined repertoire, as opposed to discontinuous, well defined songs. Similar *inter*specific differences in degrees of song specialization suggest that heterochrony may have played a part in bird speciation. As well as differences in predominantly innate behaviour, differences in learned behaviour may also be influenced by heterochrony, e.g. a peramorphic increase in the number of brain cells used in information processing.

A heterochronic change in morphology may precede a change in behaviour. For example, it was long believed that the feeding behaviour of certain finch species had caused sustained selection pressures that modified beak sizes and shapes for optimum intake. However, it was later shown that the finches learn through trial-and-error experience which seeds are most suited for their particular beak! Thus, young finches born with marked heterochronic variation in beak size or shape may adopt a different feeding behaviour from their ancestors.

It is likely that many traits produced by heterochrony may not themselves have been individual targets of selection, but are *incidental* morphological by-products of heterochronic changes induced by selection for particular reproductive or life-history strategies. Some tentatively proposed relationships have been proposed that link life-history strategy, body size, water depth, and heterochrony in certain aquatic animals.

▶ Looking at Figure 14.33, what heterochronic processes appear to be associated with animals in more unstable, unpredictable onshore environments?

Progenesis and acceleration.

The concept of *r*-*K* life-history strategies was discussed in Chapter 8. The predicted characteristics of *r*-selected species are small size, early maturity, single reproduction, and high numbers of small offspring (and the converse in each case

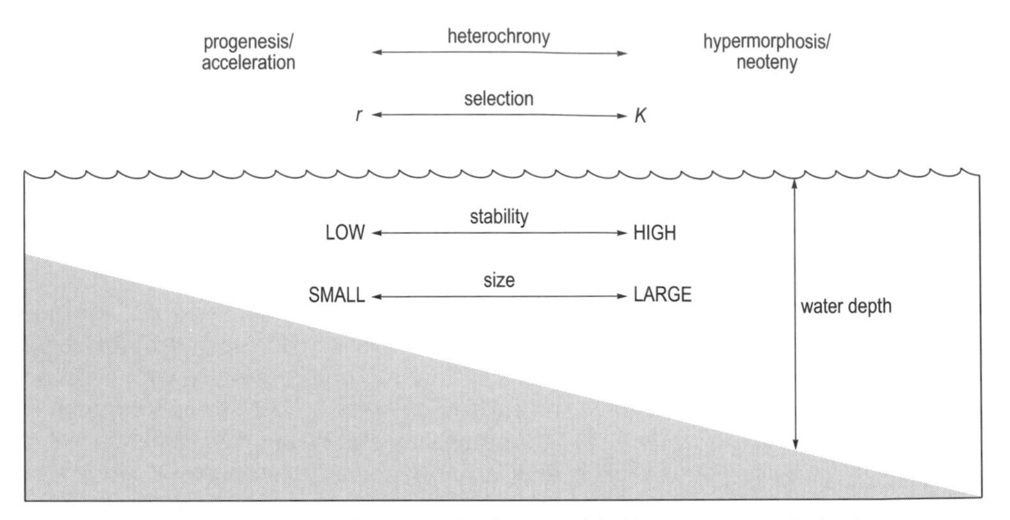

Figure 14.33 *Tentative proposed relationship between life-history strategy, body size, water depth, and heterochrony.*

for *K*-selected species). Evidence from a wide range of fossil groups, including trilobites, sea-urchins, bryozoans, bivalves, ammonites and brachiopods, indicates fair agreement with the scheme. Palaeontological observations suggest that many major clades originated nearshore (see Chapter 15), possibly from small, *r*-selected, allometrically progenetic forms. Thereafter, subsequent diversification into more stable, offshore environments was often accompanied by larger body size, increased longevity and other *K*-selected traits produced by hypermorphosis (delayed offset of growth).

Predation can be a powerful selective agent that alters life-history strategies by heterochrony. For example, a few species of mayfly spend less than two hours as adults out of a total lifespan of about two years. Since originating in the Devonian, mayflies have extended the juvenile phase at the expense of the adult phase. Postdisplacement, i.e. late onset of a number of features including the onset of maturity, seems to have been the mechanism responsible. As juveniles, most species of mayfly are aquatic and predators. However, as adults, during mass mating in the dawn hours, mayflies often suffer heavy predation. Conversely, for other organisms with more intense predation during the juvenile stages, selection has favoured earlier attainment of the safer, adult stage.

Clearly, heterochrony is of immense importance in generating evolutionary changes in form; and, as the main agent of change in releasing lineages from the (dominant) phylogenetic constraint, it is a key aspect of macroevolution.

Summary of Section 14.6

Heterochrony is a genetically determined change in the timing or rate of developmental events, relative to the same events in the ancestor. Such changes in timing or rate can occur at any scale (molecular, cellular, tissue, organism), and heterochronic effects can be compared between individuals, between populations, and between species. Only six kinds of 'true' heterochrony can be discerned:

neoteny (slower rate), acceleration (faster rate), predisplacement (earlier onset), postdisplacement (later onset), progenesis (earlier offset), and hypermorphosis (later offset). Paedomorphosis ('underdevelopment') results from neoteny, postdisplacement, or progenesis. Peramorphosis ('overdevelopment') results from acceleration, predisplacement, or hypermorphosis. To distinguish between the categories of 'true' heterochrony requires a knowledge of age in compared ontogenies. However, studies of 'size-based' allometric heterochrony can be useful, especially in the fossil record where age data are often lacking.

Much of the normal morphological variation within species populations and between closely related species has a heterochronic basis. Changes in many different kinds of genes and gene combinations can produce heterochronic effects. Furthermore, ecophenotypic influences during development can mimic, amplify or reduce heterochronic variation otherwise under genetic control.

As development is a highly orchestrated cascade, small timing or rate changes at the lower levels will often produce complex results at the level of the whole organism. Heterochrony has the most creative evolutionary potential if changes in developmental timing or rate occur early in ontogeny. But natural selection generally appears to modify the least disruptive, later stages of development more easily than early embryonic stages, which tend to have a closer resemblance to those of ancestral forms. This 'evolutionary ratchet' is probably one reason why no new phyla have appeared since the Ordovician. Heterochrony has been implicated in the origin of major new clades, especially during the Cambrian 'explosion'. Lineages can be released from phylogenetic constraints by heterochrony. Peramorphs have as much potential for originating clades as paedomorphs.

Evidence from the fossil record of established clades indicates that the relative importance of paedomorphosis and peramorphosis varies between different clades, and over time within clades. Often, the target of selection may not have been individual morphological traits so much as reproductive and life-history strategies. Mosaic heterochrony, a complex pattern produced by the operation of different heterochronic processes, is widespread.

14.7 Conclusion

Studying the form of organisms, whether living or fossil, can provide insights across the whole spectrum of evolutionary patterns and processes. Evolution does not have a free hand to modify descendants, but works like a tinkerer, fashioning new form from existing features. Phylogenetic, functional and fabricational constraints impose limitations on the generation of form, and may prevent organisms from achieving optimal form for particular functions. Theoretical morphology can reveal the operation of particular constraints in nature. Lineages can be released from constraints in various ways, of which environmental change, preadaptation and heterochrony are particularly important. The next Chapter looks at patterns of change in form and in taxonomic diversity within clades.

References for Chapter 14

Chamberlain, J. A. (1980) Hydromechanical design of fossil cephalopods, in House, M. R. and Senior, J. R., *The Ammonoidea*, Systematics Association Special Volume No. 18, Academic Press.

Clarkson, E. N. K. and Levi-Setti, R. (1975) Trilobite eyes and the optics of Descartes and Huygens, *Nature,* **254**, pp. 663–67.

Conway Morris, S. (1989) Burgess Shale faunas and the Cambrian explosion, *Science*, **246**, 339–46.

Dawkins, R. (1986) *The Blind Watchmaker*, Penguin.

Jacob, F. (1977) Evolution and tinkering, *Science*, **196**, 1161–66.

Kitchener, A. (1987) Fighting behaviour of the extinct Irish elk, *Modern Geology*, **11**, 1–28.

Ramskold, L. and Xianguang, H. (1991) New early Cambrian animal and onychophoran affinities of enigmatic metazoans, *Nature,* **351**, 225–28.

Raup, D. M. (1966) Geometric analysis of shell coiling: general problems, *Journal of Paleontology*, **40**, 1178–90.

Raup, D. M. (1967) Geometric analysis of shell coiling: coiling in ammonoids, *Journal of Paleontology*, **41**, 43–65.

Skelton, P. W. (1985) Preadaptation and evolutionary innovation in rudist bivalves, in Cope, J. C. W. and Skelton, P. W. (eds) Evolutionary case histories from the fossil record, *Special Papers in Palaeontology*, **33**, 159–73.

Further Reading for Chapter 14

Clarkson, E. N. K. (1992) *Invertebrate Palaeontology and Evolution*, 3rd edn, Chapman and Hall.

Gould, S. J. (1980) *The Panda's Thumb*, Penguin.

Gould, S. J. (1984) *Hen's Teeth and Horse's Toes*, Penguin.

Gould, S. J. (1989) *Wonderful Life*, Penguin.

Gould, S. J. (1991) *Bully for Brontosaurus*, Penguin.

Hickman, C. S. (1988) Analysis of form and function in fossils, *American Zoologist*, **28**, 775–93.

McKinney, M. L. and McNamara, K. J. (1991) *Heterochrony— The Evolution of Ontogeny*, Plenum Press.

Self-Assessment Questions

SAQ 14.1 (*Objective 14.1*) For each of statements (a) to (c) below, indicate whether it is true or false. If false, give reasons for your answer.

(a) Descent with modification has ensured that subsidiary taxa within each phylum do not share the same basic bodyplan.

(b) The late Precambrian Burgess Shale fauna, which preceded the Cambrian 'explosion', contains about 18 types of organisms that are hard to place within extant phyla.

(c) The Ediacaran fauna includes tough, quilt-like forms that may represent a grade of organization that flourished briefly and became extinct before the Cambrian Period.

SAQ 14.2 (*Objectives 14.2 and 14.3*) State which of the factors of 'Seilacher's triangle' is the dominant constraint in each of the following cases, giving your reasons.

(a) Bivalves cannot obtain food with a radula.

(b) No gastropods have a radula made of diamond.

(c) The apertures of dog whelk shells in sheltered shorelines tend to be small compared with apertures of individuals from exposed shorelines.

SAQ 14.3 (*Objective 14.4*) Evolution is often characterized as a tinkerer, but there are important limitations to this metaphor. Suggest three key differences between the work of a tinkerer and the action of evolution.

SAQ 14.4 (*Objectives 14.5 and 14.6*) Of what evolutionary phenomenon is each of the following an example?

(a) Small, more or less random constructional differences in morphological features that arise as a by-product of some particular programme of growth.

(b) A rare modern horse that, compared with other individuals of the species, has elongated second and fourth digits, and slightly reduced growth of the third digit.

(c) A species of bryozoan ('sea-mat') having a stiff fan-shaped colony that filters organic particles from turbulent seawater but which is characteristically not orientated perpendicular to the direction of maximum current flow.

(d) Small, non-functional pelvic and leg bones embedded within the ventral body wall of whales.

SAQ 14.5 (*Objective 14.7*) For each statement (a) to (c) below, indicate whether it is true or false. If false, give reasons for your answer.

(a) Ecophenotypic effects are heritable modifications of a phenotype that are produced in response to a particular environmental influence.

(b) A recipe in a cookery book is a better analogy for genes and their operation than is a detailed blueprint of an organism.

(c) Natural selection is purely a chance process because it operates on variations thrown up at random with respect to the present and future needs of organisms.

SAQ 14.6 (*Objective 14.8*) (a) Why are bivalves usually debarred from having a low value of W, the whorl expansion rate?

(b) Why is it of interest to know what regions of morphospace have *not* been filled in nature, as well as the regions that have?

SAQ 14.7 (*Objective 14.9*) (a) From which general type of constraint are survivors of a mass extinction most commonly released?

(b) From which general type of constraint are lineages most often released by heterochrony?

SAQ 14.8 (*Objective 14.10*) For what particular aspect of evolution is it frequently necessary to invoke the principle of preadaptation?

SAQ 14.9 (*Objective 14.11*) Correct any erroneous statement in the following paragraph about heterochrony.

Heterochrony refers to different sexes having different rates of development. The classification of heterochrony depends on the comparison of ontogenies from ancestor to descendant. To distinguish between the various categories of 'true' heterochrony, we need information on the size of compared individuals. Peramorphosis (overdevelopment) is caused by faster growth (neoteny), earlier onset of growth (predisplacement), or later offset of growth (hypermorphosis). A complex pattern of changes produced by the operation of different heterochronic processes is called global heterochrony.

SAQ 14.10 (*Objective 14.12*) (a) Do changes in genes affecting early stages of development generally have a greater or lesser effect on the adult phenotype than changes in later-acting genes?

(b) Explain why it follows from the correct answer to (a) that embryos of a wide range of groups are more likely to be similar than the corresponding adults.

CHAPTER 15
PHYLOGENETIC PATTERNS

Prepared for the Course Team by Peter Sheldon and Peter Skelton

Study Comment *Following consideration of the evolution of form in Chapter 14, this Chapter looks at how patterns of change in form and in taxonomic diversity can be analysed, both in theory and using data obtained from the fossil record. The emphasis is on macroevolution, i.e. evolution above the species level, and on the history of entire clades. The extent to which random events may generate observed evolutionary patterns is considered, along with a simple test of randomness in patterns. Several major patterns in the fossil record that demand deterministic explanations are discussed, such as mass extinctions, radiations, and a tendency for the evolution of large size. Evidence for changes in global diversity throughout the Phanerozoic is reviewed, as are the factors that may promote the radiation of particular groups. The Chapter ends by asking whether macroevolution can be accounted for simply by an extrapolation of microevolutionary principles, or whether other selective processes operate on properties that emerge at the species level and above.*

Objectives *When you have finished reading this Chapter, you should be able to do the following:*

15.1 Illustrate with diagrams, or interpret from diagrams or other sources, various aspects of phylogenetic history, identifying cladogenesis, anagenesis, stasis, and extinction.

15.2 Test whether or not a purported evolutionary trend can be attributed to a deterministic cause on its pattern alone, qualifying the result as appropriate.

15.3 Distinguish between, give examples of, and suggest possible explanations for, the phylogenetic patterns of divergence, convergence, parallel evolution, iterative evolution and radiation, and discuss the nature of evolutionary trends.

15.4 Discuss what is meant by the informal term 'living fossil', giving examples, and explain why it is often difficult to assess the extent to which an organism qualifies as a living fossil.

15.5 Discuss the evolution of body size, and give possible reasons why the attainment of large body size is a common evolutionary trend, both within lineages and over clades as a whole.

15.6 Measure rates of origination and extinction and plot survivorship curves for groups of taxa, given information about their durations, and discuss explanatory models for the resulting patterns.

15.7 Discuss the nature and possible causes of the principal Phanerozoic mass extinctions, reviewing evidence concerning their intensity, selectivity, and timespan, and describe the various sources of bias that can distort evidence of extinctions in the fossil record.

15.8 Diagnose an evolutionary radiation and identify its likely cue, from suitable data, and distinguish between adaptive and non-adaptive radiations.

15.9 Describe and assess evidence for changes in global taxonomic diversity in the Phanerozoic, including Sepkoski's 'evolutionary faunas', suggesting possible explanations for such changes in diversity.

15.10 Discuss the evolutionary hierarchy and the part that species sorting and species selection may play in macroevolution.

15.1 INTRODUCTION— THE EVOLUTION OF CLADES

The evolutionary history of any clade (Chapter 11, Section 11.2.1) and its component lineages can be described or analysed in terms of the basic features illustrated in Figure 15.1. Five events or features of particular significance in clade evolution are: *cladogenesis* — the splitting of one lineage into two or more lineages; *extinction* — the termination of a lineage or an entire clade; *diversity* — the number of independent lineages present within a clade at a particular time; *anagenesis* — evolutionary changes within a single unbranching lineage; and *stasis* — a period when a lineage shows no significant net evolutionary change in one or more aspects.

It is important to realize that diagrams such as Figure 15.1 which depict generalized morphology on one axis and time on the other are highly schematic; the position of a lineage along the morphology axis cannot be anything other than generalized. Imagine, for a moment, trying to plot the whole of human morphology along one axis! An accurate representation of morphology would, of course, be multidimensional.

The boxes on the right-hand side of Figure 15.1 indicate the number of lineages within the depicted clade at different times in its history.

▶ Insert the appropriate numbers in the three blank boxes at the top right of Figure 15.1. How did the diversity of the clade change during this interval?

The diversity of the clade decreased from a maximum of 4 to 3, and then to 2. 'Turnover' is a useful term, often loosely applied, for the rate at which taxa are gained and lost within a clade. Thus, a clade with a high turnover rate would have more gains and losses per given time interval than a clade with a lower

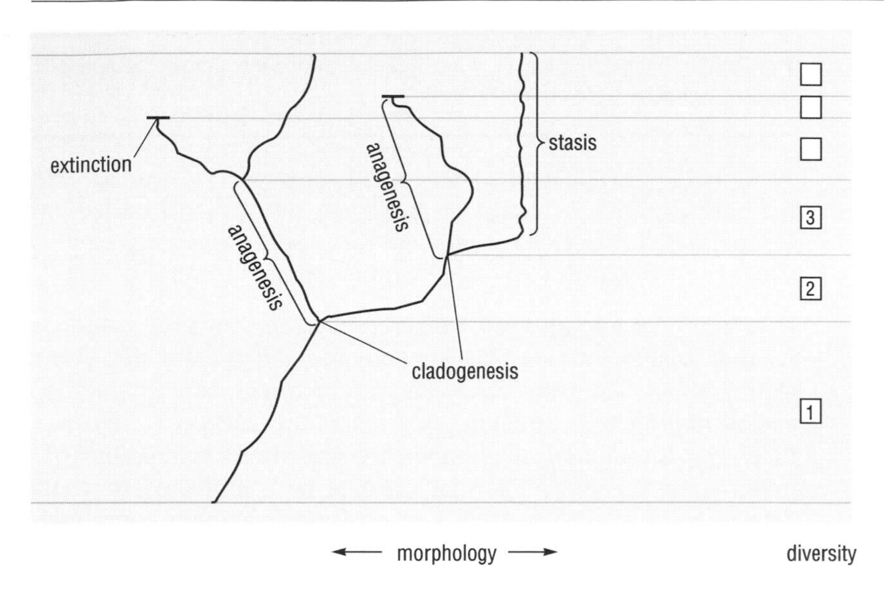

Figure 15.1 *Events or features of particular significance in clade evolution, including changes in diversity. See text for discussion.*

turnover rate. High turnover rates need not be associated with any net change in diversity; the number of gains in a given time interval could equal the number of losses.

Armed with an appreciation of some of the basic features of clades, we can now discuss various aspects of macroevolutionary patterns, some of which have already been touched on in earlier Chapters.

15.1.1 MODELS OF EVOLUTION BASED ON CHANCE

Patterns of macroevolution, as discussed above and in earlier Chapters, are often attributed to particular, pervasive causes, such as environmental change, competitive or other interactions between taxa, or the appearance of evolutionary novelties which open up new adaptive opportunities for their possessors. Hypotheses of this kind are said to be **deterministic**. As noted in Chapter 11 (Section 11.1), the problem for investigating such hypotheses is that they concern past events, the causes of which can be inferred only from the pattern of evidence that remains today — that is, the direct but incomplete testimony of the fossil record and, when available, the richer, though indirect, evidence of the comparative morphology, genetics and distributions of any extant relatives of the organisms in question. Unless tested somehow, these explanations cannot amount to more than informed speculative rationalizations of history — sometimes characterized by critics as '*Just So Stories*', after Rudyard Kipling's fanciful accounts of how various animals acquired their distinctive features. A scientific approach to macroevolutionary hypotheses requires that they be tested in order to qualify as acceptable theories. Experimental tests are limited to exceedingly short-term simulations employing comparable living organisms or other analogues (e.g. the study of defensive adaptations in sessile bivalves discussed in Section 12.3.1 of Chapter 12); but the long-term effects of evolutionary processes, operating over geological time-scales, are beyond the reach of experiment.

▶ Recall from Chapter 11 what general approach can be adopted for testing macroevolutionary hypotheses.

Section 11.1 of Chapter 11 discussed retrospective testing, which involves seeking the corroborative evidence of predicted, but hitherto undocumented consequences of postulated historical events. As this approach forms an important aspect of this Chapter, it is worth exploring it rather more thoroughly here.

Deterministic hypotheses of macroevolution aim to provide simple explanations for what in reality are complex phenomena, and therein lies the main difficulty for testing them: the same observed pattern might be explained by many different causal hypotheses. In nature, organisms are subject to myriads of influences. How, then, can the relative likelihood of competing explanations of their evolutionary patterns be assessed? The first point to be established is whether the search for a deterministic explanation of a given pattern is even warranted. The pattern itself may be merely a chance effect — the 'luck of the draw'.

▶ Cast your mind back to earlier Chapters. What general procedure is followed to test the possibility of a given outcome being due to chance alone?

The first step is to erect an appropriate null hypothesis (Chapter 3, Section 3.5.1), with which the observed data may be compared. In the present context, the null hypothesis needs to simulate the basic patterns of evolution, but in such a way that each successive event is random in character (i.e. not subject to some persistent determining influence). For example, in a sequence of coin tosses, the outcome of each next toss, 'heads' or 'tails', is independent of the preceding one (assuming a fair coin and coin-tosser!). A simple model such as this could readily serve as a null hypothesis for many evolutionary processes, with the 'heads' and 'tails' being equated respectively with, for example, successive increases or decreases in mean size of individuals or taxonomic diversity of a clade, as discussed later. If a pattern observed in the record can be simulated by the coin-tossing model more frequently than a pre-selected level of significance (e.g. $P = 0.05$), then the null hypothesis cannot be rejected, and a deterministic explanation is deemed not to be warranted. Only where the observed pattern differs significantly from the null hypothesis (i.e. $P < 0.05$) is the search for a deterministic explanation justified.

It is important to realize, however, that a failure to reject the null hypothesis in such cases does not mean that the small-scale evolutionary events making up the pattern are themselves considered to be without cause. Each speciation and extinction, or evolutionary change in size or shape, for example, presumably had its own particular causes. However, the latter are assumed to have been so many, and so complex, that their combined effects are *unpredictable* from one instance to another. Nevertheless, the evolutionary events themselves may be found empirically to occur with certain probabilities. The null hypothesis model can thus be set up to generate simulations of evolution based on sequences of events occurring according to these probabilities (i.e. *as if* random). Such a model is said to be **probabilistic**, in contrast to the alternative hypothesis of a deterministic model.

The probability of a given pattern emerging from a null hypothesis model can either be deduced analytically, from equations, or found empirically by generating large numbers of simulations. The latter approach — known as the 'Monte Carlo method' — is now widely used, as computers can generate huge numbers of simulations very rapidly, whereas the maths involved in the analysis of even relatively simple models is often ferociously complicated.

A crucial aspect of any model that is intended to simulate patterns of evolution is that each successive event, no matter how unpredictable in itself, must start from the outcome of the previous event, because evolution is itself a continuous process. Chapter 14 stressed this aspect of historical contingency in its discussion of the phylogenetic factor in 'Seilacher's triangle' (Section 14.2.1). Null hypothesis models of this kind are said to be *path-dependent*.

A simple example of such a process is a **random walk** through time, which can be illustrated using the coin-tossing exercise mentioned earlier. Figure 15.2 shows a grid with a central horizontal axis representing time, divided into ten successive steps by the vertical lines. Now start tossing a coin, and for each toss, move one step to the right (starting from the point indicated at the left), and one step up if the coin shows heads, or down if it shows tails. Mark the resulting point for each toss at the appropriate position on the grid. After ten tosses, stop and connect up the points. You will then have produced a random walk which fluctuates vertically along a determined horizontal axis (time).

Most of you will find that the final (tenth) point has shifted from the centre line; indeed, only about a quarter will have arrived back on it at the tenth step. That may not surprise you. However, some one in every 500 should produce a walk which leads straight out for the tenth line either above or below the central line (as shown in Figure 15.2): order can indeed emerge by chance, given enough trials. Note here that the probability of such an outcome (P = 0.002) is less than the conventional significance level of P = 0.05 (or even that of P = 0.01). If confronted by such an instance without knowing that it was a chance effect, we would conventionally attribute it to some deterministic cause. This example

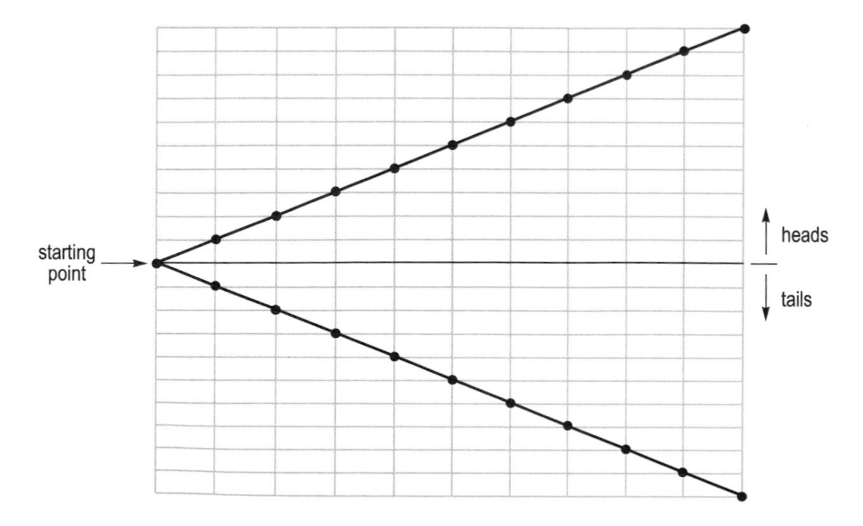

Figure 15.2 *Grid for drawing a random walk. See text for explanation.*

illustrates the arbitrary nature of the chosen significance level. We merely set up the test so that, at an arbitrary level of significance, if a pattern has a sufficiently small probability of being due to chance alone, then it is deemed worthwhile searching for a deterministic explanation. It is also worth noting, however, that path-dependent processes can quite frequently yield what *appear* to be moderately distinct trends. It is therefore important to test any given pattern observed from the fossil record for significant deviation from the null hypothesis, in order at least to be consistent about what is to be attributed to a deterministic cause.

As noted earlier, the precise analytical calculation of probabilities can be quite complex, even for random walks. For convenience, Figure 15.3 shows a grid for all possible random walks of up to 30 steps, with the critical areas for significance levels of $P = 0.1$ and $P = 0.05$ marked on it. These are the critical areas for a two-tailed test (see Chapter 10, Section 10.2.3), i.e. where a significant deviation may be *either* above *or* below the central line, with $P = 0.05$, for example, comprising 0.025 in each tail. If you wished to test only for significance in one tail (e.g. for a trend of increase only), then the critical areas shown in that tail would be for half of the P values indicated. Thus, the area for $P = 0.1$ in the two-tailed test becomes that for $P = 0.05$ in the one-tailed test.

To illustrate the use of such a test, refer back to Figure 14.25 in Chapter 14. The pattern of increasing generic diversity of the clade of uncoiled rudist bivalves shown there was held to confirm the preadaptive significance of ligamentary invagination in opening up new adaptive opportunities. But what is the probability that the pattern shown is due to chance? The random walk model can provide a simple form of test. The null hypothesis would be that the origination and extinction of genera were equally probable events (see Section 15.3.1). If this were so, generic diversity could either increase or decrease at each step with equal probability, as in a random walk. In this case, we are interested only in the *direction* of change in diversity at each step, not the *amount*. The five successive steps of diversity increase (from the first stage 'B') shown in Figure 14.25 can now be plotted onto Figure 15.3 (as you did for the coin-tossing exercise in Figure 15.2).

▶ Is the trend of increasing diversity significant according to this test (at $P = 0.05$)?

Yes. Remember that this is a one-tailed test (for a significant trend of *increasing* diversity only), so that the critical area for $P = 0.05$ is that shown as $P = 0.10$ for two-tailed tests. The five successive steps of increase just enter that critical area, and so the null hypothesis can be rejected.

In this instance, the test is somewhat crude, but it has the merit of simplicity and hence intelligibility. More sophisticated tests are, of course, possible. It is also in fact rather conservative, for two reasons. The first is that the critical areas have been shown on the assumption that the probability of each increase is exactly 0.5 (as with coin-tossing). However, there is, in reality, some probability of diversity remaining the same at each step, which has been ignored here; so, if the probability of an increase is to be the same as that of a decrease, that probability must be < 0.5. Secondly, the amount of diversity change has been ignored. It would be possible for a large decrease to wipe out all the remaining genera; such

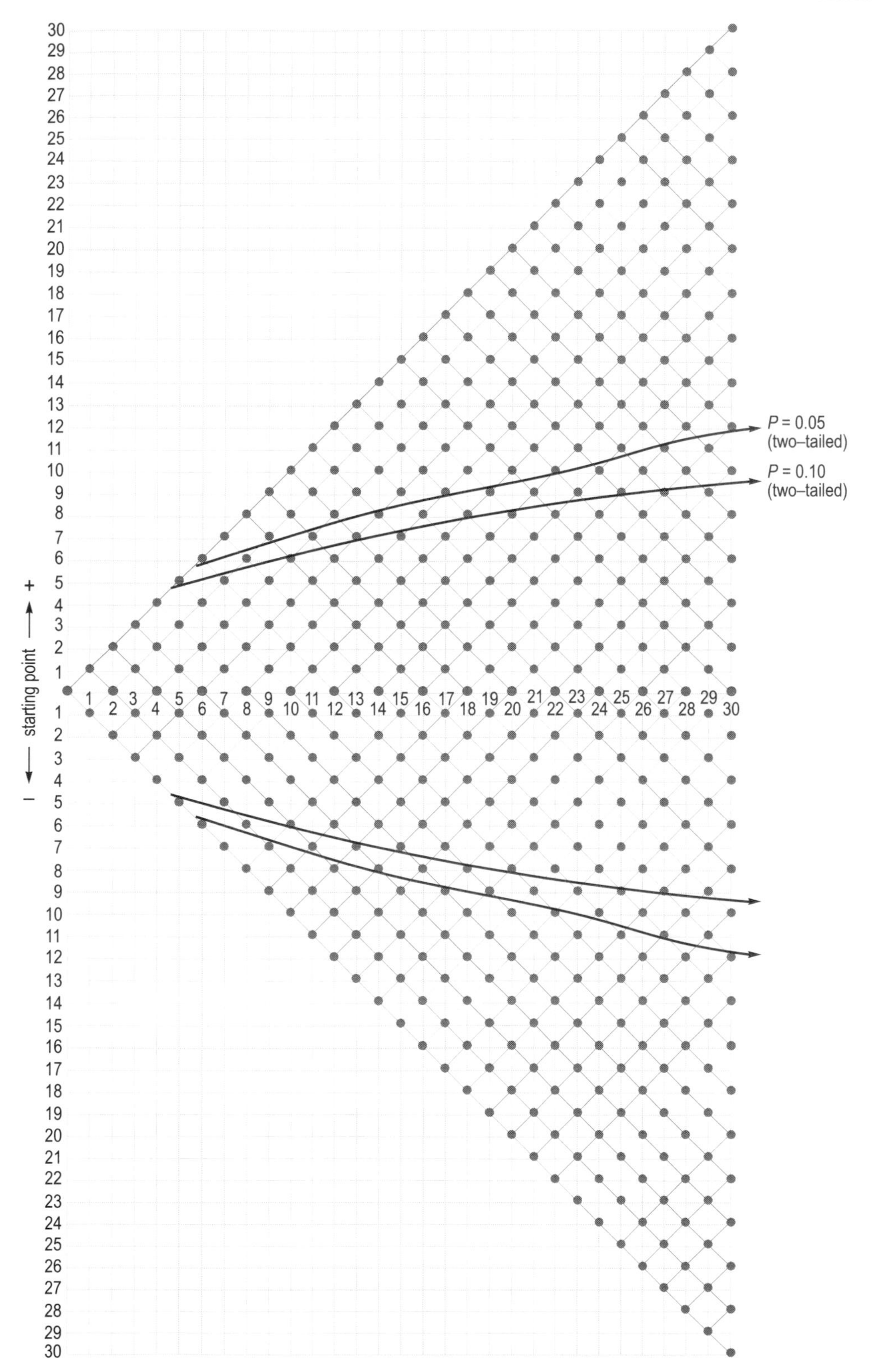

Figure 15.3 *Grid showing the probability distribution for random walks of up to 30 steps. The critical areas shown are for two-tailed tests (P = 0.10 = 0.05 in each tail; and P = 0.05 = 0.025 in each tail).*

a path would be shown as terminating before achieving even five steps, and would not be included in the probability distribution of walks that achieved the appropriate number of steps. Both these qualifications mean that the critical areas shown for diversity increase are overestimates for random walks (which actually strengthens the conclusion, in this case, that the pattern shown by the uncoiled rudists is significant).

So far, only small numbers of steps have been discussed. For patterns of change involving > 30 steps, the probability distribution of random walks approximates to a normal distribution with a standard deviation (measured in vertical units away from the central line in Figure 15.3) of $\sqrt{N}$, where N is the total number of steps. Critical areas for given significance levels can therefore readily be estimated (beyond $1.96\sqrt{N}$ for $P = 0.05$ and $1.645\sqrt{N}$ for $P = 0.1$, in two-tailed tests).

So far, we have only considered tests against a null hypothesis. In the rudist example, all that can be concluded from the test is that it is legitimate to seek a deterministic explanation for the pattern of change in generic diversity of the uncoiled genera. However, several alternative possibilities spring to mind, of which the preadaptive hypothesis is only one. The pattern might otherwise reflect either the expansion of favourable environments for rudists in general, or the increasing accessibility of strata from which specimens may be recovered (and hence genera described). This is where the graph for the spirogyrate genera comes in, serving, so to speak, as a control. The fact that no matching pattern of increase is shown there allows one to reject the two last possibilities cited above: the significant pattern is a unique feature of the clade of uncoiled rudists. Finally, the coincidence of the start of the trend for increase with the inception of uncoiling itself (around the Jurassic–Cretaceous boundary) renders another hypothesis, i.e. that the cue was some feature of the uncoiled rudists other than that of ligamentary invagination—unparsimonious, at the very least. Thus, the preadaptive hypothesis can be accepted as the *most probable* one. This example illustrates a point that applies to all scientific tests, and which can never be overstated: competing hypotheses can be rejected or accepted only on the basis of an assessment of their relative probabilities, because absolute proof or disproof can never truly be achieved. This is the spirit in which the macroevolutionary hypotheses discussed later in this Chapter must be approached.

SUMMARY OF SECTION 15.1

Five events or features of particular significance in clade evolution are cladogenesis, extinction, diversity, anagenesis, and stasis. Deterministic hypotheses explain patterns of evolution in terms of particular, pervasive causes, but there are usually several possible explanations. Retrospective tests are therefore required to assess the relative probabilities of the competing explanatory hypotheses. Where possible, one should attempt to test for deviation from a null hypothesis of randomly generated patterns of evolution to justify any appeal to a deterministic hypothesis. The probability distribution of such random patterns can be determined, either analytically or empirically, from some appropriate probabilistic model of evolution. Path-dependent processes, in which each step starts from the outcome of the previous step, yet is itself random, are particularly suitable for probabilistic simulations of evolution. An example is the model of random walks through time, which can be applied to a variety of evolutionary variables, such as mean body size or shape, or taxonomic diversity within a higher taxon. On rejection of the null hypothesis, it may be possible to choose between competing alternative hypotheses on the basis of further corroborative evidence, such as predicted similarities or differences in the patterns of evolution of other, control taxa.

15.2 Patterns of change in form

Patterns of change in form within *individual* lineages have been discussed earlier (e.g. Chapter 10, Section 10.3). This Section considers some commonly encountered general patterns of change that concern form in more than one lineage. Figure 15.4 illustrates some of these patterns: divergent evolution, convergent evolution, parallel evolution, iterative evolution and radiation. Such patterns normally concern only a few aspects of morphology, such as brain volume, shell ribbing, and so on, rather than total morphology, which would be impossible to quantify in any case, as explained in Section 15.1.1. The patterns can be applied to evolutionary changes in form at any scale, including that of gene sequences.

Divergent evolution results in increasing morphological difference between initially more similar lineages. Divergence is, of course, extremely common in evolution; were it not for divergence, all lineages would remain similar.

Convergent evolution is the evolution of two or more lineages towards similar morphology so that morphological differences between them decrease. As

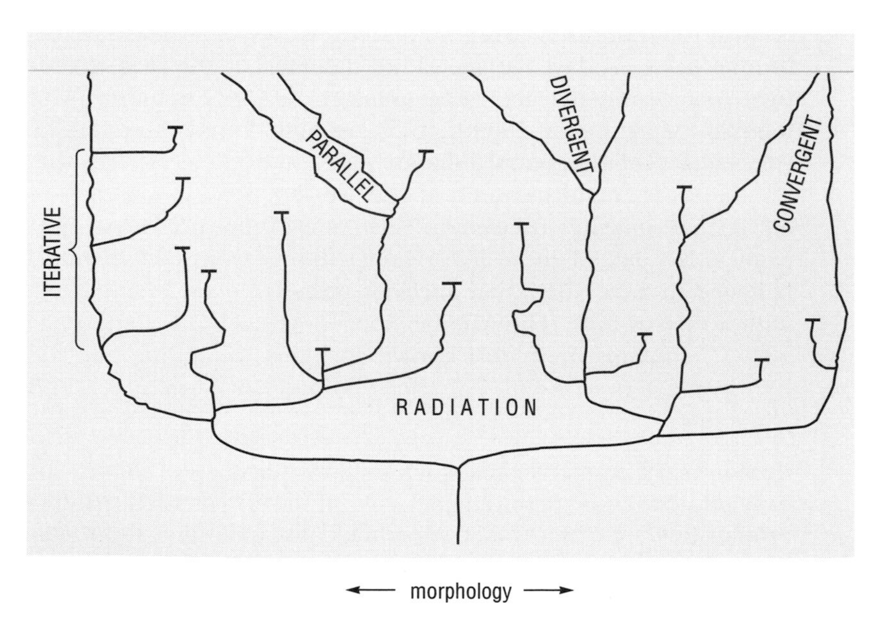

Figure 15.4 General patterns of change in form concerning more than one lineage. Horizontal bars represent extinctions.

discussed in Chapter 2, Section 2.3.3, convergence may sometimes produce very similar adaptations in organisms of quite different ancestry, subject to the various constraints on evolution discussed in Chapter 14.

▶ What is the term applied to features that show similarities due to convergence rather than to inheritance from a common ancestor?

Such features are analogous, as opposed to homologous (Chapter 2, Section 2.2.2 and Chapter 11, Section 11.2.1).

One of the best known general cases of convergent evolution is that of the striking similarity between certain marsupial mammals of Australia and placental mammals of other continents. Each of the marsupials on the right-hand side of Figure 15.5 is more closely related (by recency of common ancestry) to a kangaroo than to its placental counterpart in the other column. The phenomenon is essentially one of convergence rather than parallelism (see below), as was once maintained, because a greater degree of morphological difference once separated the ancestors of the particular marsupial and placental lineages being compared. The explanation for convergence in these cases, as in others, seems to be that strong selection pressures in repeated, specific environments can sometimes produce very similar adaptations. The convergences apparent in Figure 15.5 presumably reflect the continued existence, not only of similar constraints, but also of similar evolutionary *opportunities* afforded to the clade of marsupials + placentals by virtue of their sharing certain features of the mammalian bodyplan.

The capacity of evolution by natural selection to generate similar adaptations independently in different lineages is remarkable. For example, in the case of mammals, at least five orders evolved convergently the specialized habit of eating ants; all show morphological features linked with their peculiar diet—long snouts, reduced or little-used teeth, and strong claws on forelimbs for digging into ant or termite nests. Today, the ant-eating members of these five orders are distributed over four continents: the spiny anteater, or spiny echidna (Monotremata) and the numbat (Marsupialia; Figure 15.5) are both found in Australia; the aardvark (placental: Tubulidentata) inhabits parts of tropical Africa; the pangolin (placental: Pholidota) is found in northern and western Africa, and in south-east Asia; and the ant bear or anteater (placental: Edentata; Figure 15.5) lives in the South American tropics. No pair of these five distinct living groups of mammalian anteaters is believed to have shared an ancestor possessing the specialized features associated with a diet of ants. The Middle Eocene Messel Oil Shale (Chapter 12, Section 12.3.3), contains the oldest known anteaters, including one form related to the living anteaters of South America and another form very similar to modern pangolins.

Evolutionary convergences, such as those discussed above, should not be exaggerated; convergent adaptations are never identical in every respect. Similarities are often rather superficial; for example, the marsupial mice in Figure 15.5 are carnivorous and much fiercer than the placental mice, which are mainly seed-eaters. No doubt some parallelism has occurred *within* the marsupial and placental clades themselves, although without a good fossil record and detailed knowledge of phylogeny it is often difficult to decide whether similarities are due

Figure 15.5 *Convergent evolution between placental mammals and Australian marsupial mammals.*

to convergence or to parallelism. The use of molecular data to assess the closest marsupial relatives of the recently extinct marsupial wolf *Thylacinus cynocephalus* was discussed in Chapter 11, Section 11.3.2.

Parallel evolution is the phenomenon whereby the morphologies of two or more closely related lineages change together in roughly similar fashion, rather than diverge or converge. The term is used loosely and is not meant to imply precise coincidence of pattern, nor, of course, that the paths traced are straight lines. Parallelism is fairly common among closely related species because they have similar developmental programmes that are subject to similar evolutionary constraints and flexibilities. Closely related species thus often respond to similar selection pressures with similar heterochronic shifts. The distinction between parallelism and convergence is, in practice, somewhat arbitrary as there are no rules that limit how recent a common ancestor must be for a pattern to be parallel evolution, and all convergent lineages have common ancestors, albeit distant ones. The basic rule of thumb is that if lineages start off rather similar in many respects, and their degree of resemblance remains about the same during and after evolutionary change, then that is parallel evolution. Although the lineages representing parallel evolution in Figure 15.4 are separated along the morphology axis, the lineages could easily overlap *on the diagram* as far as individual features are concerned; this would be the case, for example, if 'number of vertebrae' was being plotted along the *x*-axis, and the number of vertebrae in two lineages started the same and increased together at the same rate.

An example of parallel evolution in trilobites was discussed in Chapter 10, Section 10.3.2 and Figure 10.26. There was a general increase in the mean number of ribs on the trilobite tails in eight lineages, although in the short term there were numerous reversals and the changes were not closely parallel in different lineages.

▶ Given that, in any lineage, countless aspects of variation are potential candidates for selection at any one time, would you expect that a feature showing a great degree of short-term parallelism among many related fossil lineages is more likely to be reflecting true evolutionary change or ecophenotypic (i.e. non-genetic) change?

Ecophenotypic change. Given sufficiently intense selection pressure for a particular trait, short-term parallel genetic changes *can* be expected to occur, at least for a while, as, for example, has been the case with industrial melanism in moths, spiders, and ladybirds. However, as the vast majority of heritable traits that organisms possess are not preserved in the fossil record, only in very rare cases are we likely to find ourselves plotting traits subject to intense directional selection over that period. In general, therefore, the greater the degree of short-term parallelism in a temporarily reversible trend, the more likely that the short-term changes are due to ecophenotypic effects.

Numerous examples of parallel evolution are known, over varying time-scales and in various groups. For instance, the inner walls that subdivided the shells of ammonites into chambers became increasingly complex in very many lineages; and almost identical patterns of wing pigmentation have evolved independently in different families of living butterflies, as is also true of moths.

Iterative evolution is the repeated origination of lineages with generally similar morphology at different times in the history of a clade. Such lineages often evolve from a single persistent lineage — or group of lineages — a 'basic stock' that changes relatively little, and the time between the origin of each iterative lineage often amounts to millions of years. Iterative evolution can only be established with an exceptionally good fossil record, with wide geographical coverage and high-resolution sampling, as is the case with Cainozoic planktonic foraminifera. These organisms lived, as their modern relatives do, mostly in the upper 200 m of the oceans, and have chambered calcareous shells which are generally less than 1 mm

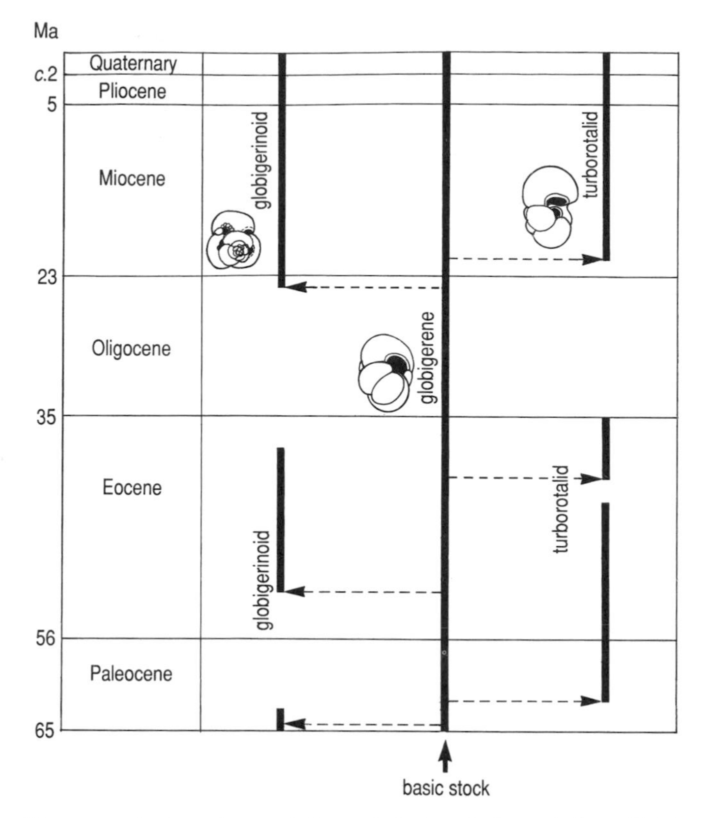

Figure 15.6 *Iterative evolution of planktonic foraminifera. Thick vertical lines indicate approximate ranges; dashed lines indicate probable lines of descent; and black areas in the shells represent apertures.*

in diameter. Figure 15.6 depicts a simplified history of the repeated evolution of certain forms of planktonic Foraminifera from a basic stock that persisted throughout the Cainozoic. The basic globigerine form has a somewhat snail-like morphology with a single aperture (indicated in black); the form is very common in the extensive deep-sea deposits known as *Globigerina* ooze. Two other forms arose on a number of occasions: a globigerinoid form with multiple apertures, and a turborotalid form, like a compressed globigerine, with a flattened periphery. Other morphologies such as planispirally coiled and keeled forms also evolved iteratively (not shown).

▶ How many times did the globigerinoid and turborotalid forms evolve from the basic globigerine stock, as depicted in Figure 15.6?

Three times each (recent research in fact suggests more).

Iterative evolution has been reported in a variety of groups, including the Palaeozoic trilobites and graptolites. In many cases, the likely explanation for iterative evolution relates to the fact that the basic stock is a relatively unspecialized form from which more specialized forms evolved. The more specialized features of the latter (e.g. more complex life cycles, higher temperature requirements) apparently rendered these lineages more susceptible to extinction. However, conditions favourable to the evolution of such forms arose again on several occasions, but each time the new lineages were more prone to extinction than the ancestral stock.

Radiation is a pattern of relatively rapid increase in diversity within a clade. Radiations necessarily involve a sustained net excess of cladogenesis over extinction. When radiations are accompanied by wide divergences in morphology, and the morphological innovations are interpreted as indicating adaptations in new circumstances, they are often called *adaptive* radiations. The 'Cambrian explosion' discussed in Chapter 14 was arguably the largest radiation in the history of the animal kingdom. Radiations are discussed further in Section 15.5.

▶ How would a time of radiation be reflected in the general shape of a spindle diagram (Chapter 12, Section 12.4.2)?

The spindle would become much wider, indicating a rapid increase in diversity.

Evolutionary trends The term **evolutionary trend** is used loosely for any sustained tendency for evolutionary change in a particular direction. Depending on the context, trends can span as little as a few generations (as in increased tolerance of a plant lineage to heavy metal pollution on mine spoil heaps) or as much as tens of millions of years (as in the increase in number of types of body cell in early metazoan evolution). Whether a pattern of change has been sufficiently sustained in a particular direction to deviate significantly from a random walk can be tested formally — albeit crudely — using the procedures discussed in Section 15.1.1.

Evolutionary trends can concern all kinds of morphological features, e.g. the number of chromosomes in a wheat plant lineage, an increase in the average brain volume of primates, a reduction in the number of toes on the front feet of horses. Trends may also relate to more abstract concepts, such as a trend of decreasing rate of origination of taxa.

It is important to distinguish trends within single lineages, i.e. *anagenetic trends*, from trends that relate to an entire clade containing many lineages, i.e. *whole-clade trends*. In an anagenetic trend (also called a phyletic trend), there is a net direction of change in whatever variate, or combination of variates, is being considered in a particular lineage. By contrast, a whole-clade trend concerns a net shift in some variate when all lineages in the clade are taken into account. Note that anagenetic trends may be uncoupled from whole-clade trends; there is no need for every constituent lineage to undergo the trend of the clade as a whole; indeed, that is rare.

▶ Is a trend of decreasing origination rate an anagenetic trend or a whole-clade trend?

A whole-clade trend, because origination must, by definition, involve new lineages, and so the trend concerns more than one lineage.

Trends within clades as a whole can be generated in a variety of ways, especially by biased speciation and biased extinction, as illustrated in Figure 15.7. All the

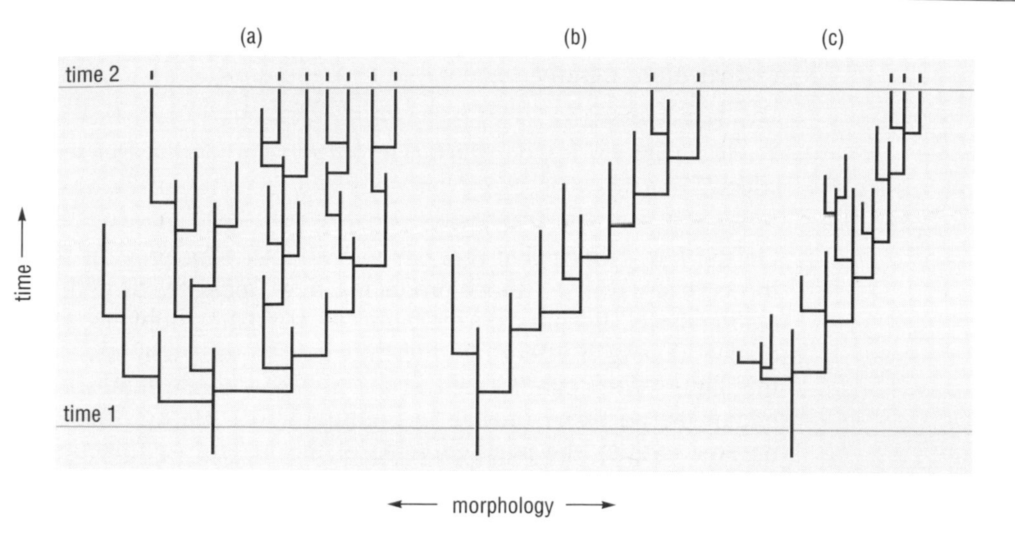

Figure 15.7 *Evolutionary trends within clades produced by biased speciation, (a) and (b), and by biased extinction, (c).*

patterns in Figure 15.7 show speciation (i.e. cladogenesis or branching) as being punctuational, with anagenesis focused in a brief interval, followed by stasis in each lineage. This has been done just to make it easier to visualize what is meant by biased speciation and biased extinction, but each can produce similar, if less obvious, effects in clades having more gradualistic patterns.

Biased speciation can yield whole-clade trends in two ways: lineages with particular morphological characteristics may speciate more often (Figure 15.7(a)), or lineages may tend to speciate (i.e. branch off) more often in one direction than another (Figure 15.7(b)).

▶ In Figure 15.7(a), do lineages on the left or the right of the diagram tend to speciate more often?

Lineages on the right. As a result, the average value of the morphological variate for the whole clade moves to the right between time 1 and time 2.

▶ Do lineages in Figure 15.7(a) also tend to branch off more often in one direction than the other?

No; there are 13 speciation events to the right, and 13 to the left.

▶ In Figure 15.7(b), how many speciation events move morphology to the right, and how many move morphology to the left?

Seven speciation events move morphology to the right, and three move morphology to the left.

In biased extinction, lineages with particular morphological characteristics become extinct more rapidly.

▶ In Figure 15.7(c), on which side, left or right, do lineages tend to become more quickly extinct?

On the left. Notice that this has two effects. Lineages on the left are more likely to become extinct without producing new lineages. In addition, even without cladogenesis, lineages on the right, by persisting longer, 'carry' that morphology on for longer, in effect moving the clade to the right with time. An example of this latter effect can be seen in the first new clade that arises after time 1, on the left of the ancestral lineage in Figure 15.7(c). Before that small, short-lived clade became extinct, the average value of the variate (within that small clade) had shifted to the right.

Note that a whole-clade trend would also occur if there tended to be a greater amount of anagenetic change in lineages that branched off to one side compared with those that branched off to the other. This is, however, not the case with any of the three patterns in Figure 15.7; with each branching event, morphology shifts, on average, approximately equal distances to left or right.

All sorts of combinations of anagenesis in individual lineages, and of biased speciation and biased extinction, can contribute towards a whole-clade trend; untangling the relative contribution and dynamics of such effects may be a difficult enough task even in hypothetical clades, let alone real ones. In reality too, morphology is, of course, multidimensional, rather than one dimensional, as in Figure 15.7.

Stephen Gould (1988) has emphasized that some — perhaps many — trends in the fossil record actually reflect change in variance. In other words, rather than a simple shifting of means, with the plotted distribution of the variate maintaining a similar shape or dispersion, the shape of the distribution also changes, such as by flattening out or becoming more peaked. The mean may, indeed, remain the same whilst variance changes. Study Figure 15.8, which shows generalized frequency distributions for encephalization quotients (EQ) for ungulates and carnivores during the Cainozoic. (An encephalization quotient is the ratio of actual brain size to the brain size expected for an *average* mammal at that body size.) The time sequence of successive groupings is 'archaic', then Palaeogene (Paleocene–Oligocene), Neogene (Miocene–Pliocene), and Quaternary.

▶ Does the mean EQ of ungulates and of carnivores increase, decrease, or stay the same during the Cainozoic? What happens to the shape of the generalized frequency distributions?

The data show that the means for both ungulates and carnivores all increased with time. But, as Gould commented: 'the major phenomenon is not linear increase, but expansion and flattening of the distribution as numbers of taxa grow: in other words an increase in variance. The left ends of the frequency distributions remain anchored at the small size of ancestral lineages, while the curve of EQ spreads out, as numbers of species increase, into the previously unoccupied right end of the range'.

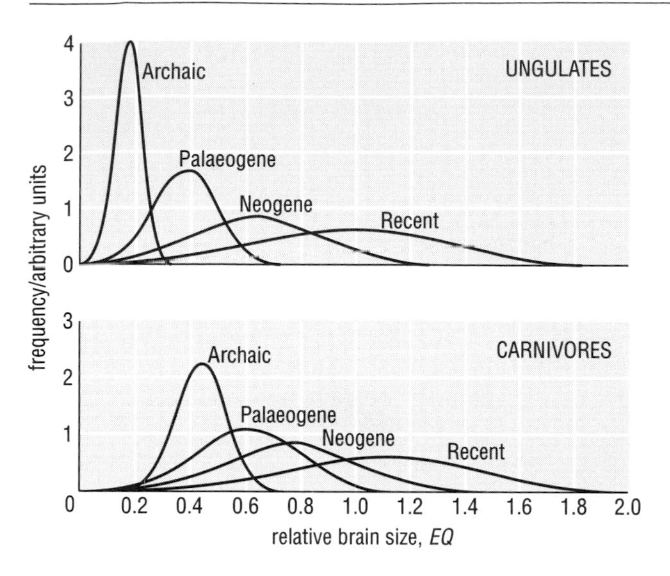

Figure 15.8 *Generalized frequency distributions for encephalization quotients (EQ) for two mammal groups (ungulates and carnivores) during the Cainozoic.*

Care should be taken to specify the precise nature of a trend. Consider a statement such as 'one evolutionary trend in horses was an increase in the length of the femur'. The statement is rather ambiguous. Is it the mean length of the femur, the maximum length, the modal length within a range of size classes, or some other measure? Is it absolute length, or proportional length, such as the length of the femur in relation to the tibia? Does the trend relate to the clade as a whole, to a succession of chronospecies in a single lineage, or what? Over what time-scale was it? And so on. Ideally, enough information, if known, should be given to avoid such confusion.

15.2.1 Living fossils

'**Living fossil**' is an informal term, originally introduced by Darwin, that generally denotes an extant species which is morphologically very similar to a species from the ancient past. In popular imagination, living fossils have probably attracted more attention than groups displaying exceptionally high rates of evolution, perhaps because evolutionary history is usually expected to involve change and it is surprising when it does not. Interest is reinforced because living fossils somehow seem to have escaped extinction, despite apparent lack of change.

There is, however, no agreed definition of living fossils. The central concept is survival to the present day with minimum morphological change over long periods — several tens or even hundreds of millions of years. Such time intervals would be far longer than the usual duration of species, including most of those that fit the punctuated equilibrium model (Chapter 10, Section 10.3.1). However, few if any convincing cases of species-level identity spanning such intervals have been documented, so the term living fossil tends to relate to supraspecific taxa (especially genera, but sometimes families and higher categories) that have shown unusual morphological conservatism *once a particular form has evolved.*

To put such genera in context, Figure 15.9 shows the distribution of the lifespans of 17 505 fossil marine genera. Note that some genera span more than 100 Ma, and a few even as much as 150 Ma. The average duration of the genera plotted is

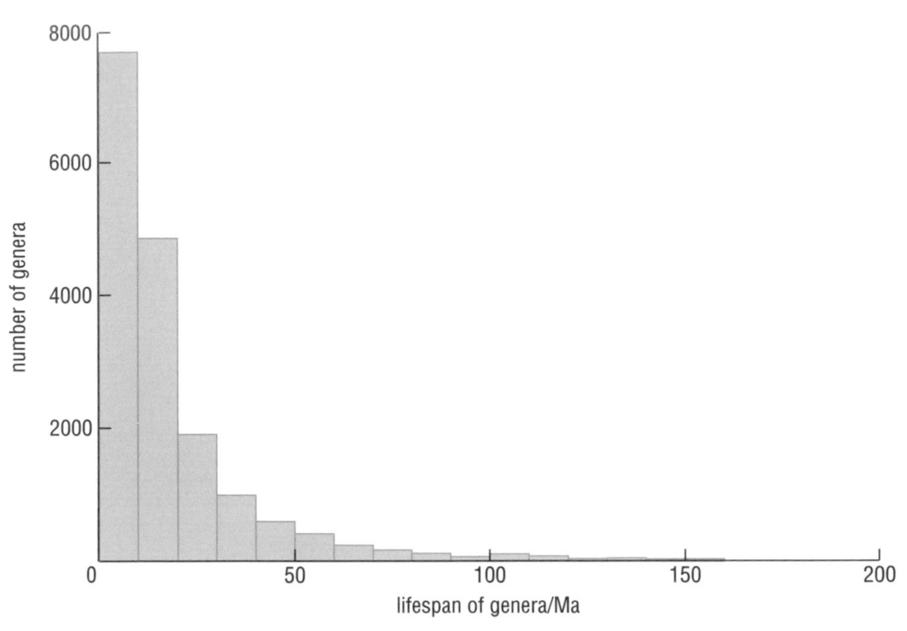

Figure 15.9 *Histogram showing the distribution of lifespans for 17505 fossil marine genera.*

about 20 Ma. One of the key aspects of Figure 15.9 is that the histogram is highly skewed, with many more genera having lifespans less than the mean than those having lifespans more than the mean. As a very approximate figure, the average duration of *species* described from the fossil record is about 4 or 5 Ma, although there is much variation about this mean, and the distribution of lifespans is probably skewed, as for genera, with more species having durations less than this average figure.

Depending on the author, a living fossil might fit one or more of the following descriptions: (i) a living species or group that has persisted over very long intervals of geological time; e.g., genera on the right of the distribution in Figure 15.9 would justly be regarded as living fossils; (ii) a living species, or group within a clade, that has a maximum of primitive character states (plesiomorphies), and a minimum of shared derived ones (synapomorphies), in contrast to other clade members, which may or may not be extinct; (iii) a living species or group that was once thought to be extinct.

Neopilina galatheae, specimens of which were first dredged up from the deep Pacific Ocean in 1952, is a mollusc of the class Monoplacophora. Figure 14.2(b,c) illustrates *Neopilina*.

▶ What is the general significance of monoplacophorans to the evolution of the Mollusca? (Recall Chapter 14, Section 14.1.1.)

The form of monoplacophorans is very similar to that of the molluscan archetype, i.e. the hypothetical mollusc that may have been ultimately ancestral to most if not all of the molluscan classes.

Before the discovery of the living representatives, monoplacophorans were known only from Cambrian to Devonian shallow water sediments, which yield over 80 described genera. Since 1952, two genera and eight species of living neopilinids have been described, all found in relatively deep water, from the shelf edge (*c.* 200 m) down to *c.* 6 500 m. The shells of the Palaeozoic species are very much thicker than those of modern forms.

▶ From the above information, suggest two reasons, in addition to reduced diversity and abundance, why post-Devonian monoplacophorans may have had a low fossil preservation potential.

Sedimentary rocks from deep water environments, to which post-Devonian monoplacophorans may have become confined, are much rarer than those from shallow water, and thinner shells have a lower preservation potential than thick ones.

▶ With which of the above categories (i to iii) of living fossils does *Neopilina* conform?

All three categories.

Living fossils are often the only living remnants of groups that have otherwise become extinct. An example is *Sphenodon*, the living tuatara (Figure 15.10(a)), a lizard-like, mainly nocturnal reptile that is now confined to small islands off New Zealand. It lacks, however, the derived skull features of lizards and snakes, to which it forms the sister group today, and it is the sole survivor of the sphenodontids, a group first known from the Triassic. The modern species shows no significant differences in the post-cranial skeleton from Upper Jurassic genera, but *Sphenodon* itself has no known fossil record. Sphenodontids were diverse in the Mesozoic, but have no fossil record in the Cainozoic, reflecting the general decline of the group, in contrast to lizards, which became progressively more common and diverse since the late Jurassic. *Sphenodon* spends most of the daytime in a burrow; it can operate at unusually low temperatures for a reptile, remaining active at only 12 °C, and can attain an age of over 100 years, qualities that conceivably may have influenced its survival to the present day.

In December 1938 a large steel-blue and white fish was landed off the coast of South Africa and identified as a coelacanth, one of a group of fishes that were believed to have died out in the late Cretaceous, about 80 Ma ago. Subsequently, other coelacanths have occasionally been caught in waters of the Comores Archipelago north-west of Madagascar, normally at depths between 60–400 m. Their size, up to 1.8 m long, is far greater than that of fossil forms. Named *Latimeria chalumnae* (Figure 15.10(b)), it was not only hailed as a living fossil but also as a 'missing link', for it had long been held that coelacanths were close to the ancestry of tetrapods. Subsequent research, however, has failed to support the theory that coelacanths and tetrapods are more closely related to each other than either is to any other living group, and *Latimeria* has shed little light on the mysteries of the fish–amphibian transition. Although its precise phylogenetic position is controversial, there are remarkably few features that link *Latimeria* with

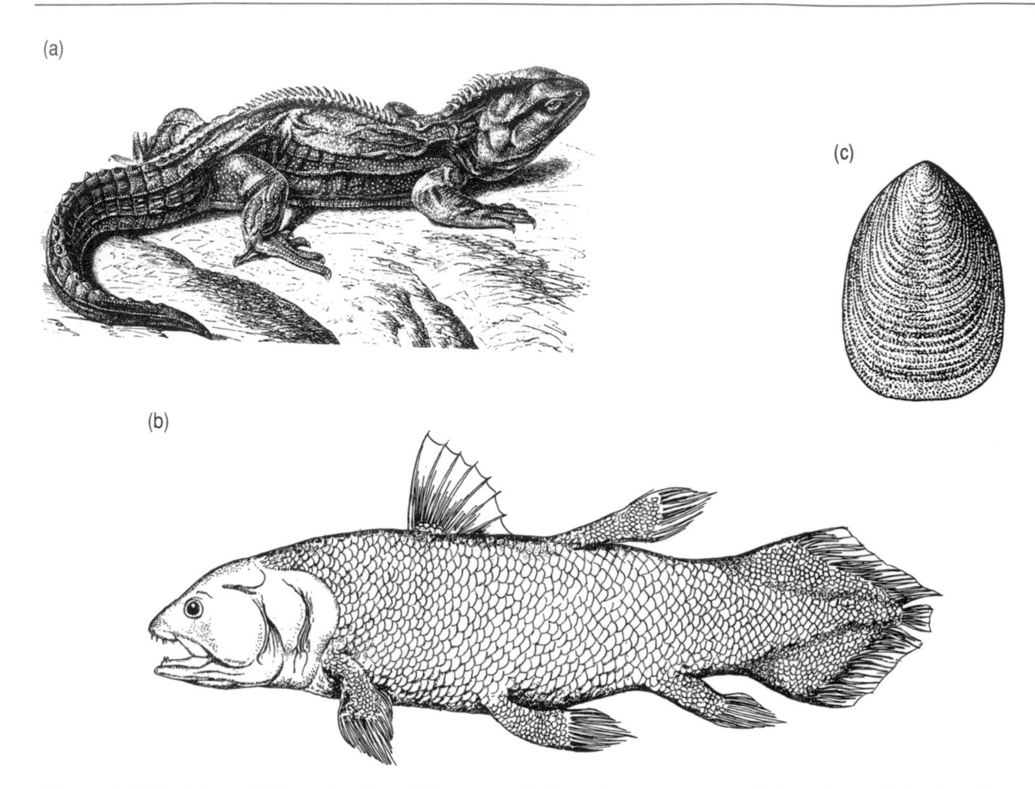

Figure 15.10 *Three 'living fossils': (a) tuatara* Sphenodon punctatus*; (b) coelacanth* Latimeria chalumnae*; (c) a brachiopod,* Lingula.

tetrapods. According to Peter Forey of the Natural History Museum in London, '*Latimeria* is best considered as a bony fish which has some features resembling those in chondrichthyans [sharks, rays, and skates] (presumably through convergence) as well as retaining many primitive vertebrate features. But in other respects it seems uniquely specialised...' (Forey, 1988). Some features seem to be the result of adaptation to life in deep water at relatively low temperatures, whilst others, perhaps adaptive in other ways, appear to result from paedomorphosis. The presence of many unique specializations (autapomorphies) alongside many retained primitive characters (plesiomorphies) is typical of many so-called living fossils: evolution may in many respects have been slow but was far from absent. It is possible that some coelacanths have lived in deep waters from early in the history of the group, as may also have been the case with monoplacophorans, and that the absence of relatively recent fossil records in both groups is largely due to a combination of the loss of species from shallow-water habitats, and the low long-term preservation potential of deep water sediments and their fossils.

Nautilus (Chapter 2, Section 2.2.2; Figure 14.2j, Plate 14.1), of which there are four recognized extant species, is often called a living fossil because it is the only remaining cephalopod genus with an external shell. Externally shelled cephalopods, which also include the ammonites (Plate 2.1) — a separate group from the nautiloids — were, by contrast, enormously diverse and abundant in the Palaeozoic and, especially, in the Mesozoic. In addition, the shell of coiled nautiloids has in general changed little since the early Mesozoic. Precisely when the genus *Nautilus* itself appeared is controversial, but forms very similar if not identical to the genus arose in the Oligocene. The estimated mean duration of 26 genera of post-Triassic nautiloids is 45 Ma. Species ranges appear to be far shorter. Peter Ward concluded (in Eldredge and Stanley, 1984) that 'rather than being a prime example of a living fossil, the nautiloids may be examples of rapidly speciating organisms that change only slightly during each event, and return to the same form over and over. The result would be apparent stasis, but the actual history would be similar to that of any other rapidly speciating group — except that the net morphological change over time would be small, rather than large'.

Nautiloids were the only externally shelled cephalopod group to survive the Late Cretaceous extinctions. Each female *Nautilus* produces only about a dozen large eggs per year, while other living cephalopods produce thousands, some even tens of thousands. *Nautilus* eggs are laid in relatively deep water (100–300 m), and after a year the young probably develop directly, bypassing a free-living larval stage in the plankton. This reproductive pattern of deep-hatching young may have permitted the nautiloids to survive, while the ammonites, which probably spent a long juvenile phase as shallow-water plankton, were seemingly more susceptible to ecological collapse of the plankton at the end of the Cretaceous (see Section 15.4.1).

Assessing the validity of an organism's status as a living fossil is beset with pitfalls. For example, living fossils are sometimes recognized on the basis of only a few features of the skeleton or shell, whereas soft-part anatomy and physiology may be as advanced as in any other members of the clade. External appearance may be maintained whilst internal features undergo extensive modification, a phenomenon sometimes referred to as the 'Volkswagen syndrome', because although the general appearance of the Volkswagen 'Beetle' car hardly altered for decades, this belied major mechanical changes that took place inside. Similar dissociated changes in organisms are likely to be missed in the fossil record if there is no indication of underlying (or overlying) soft tissues in the parts that are fossilized.

▶ Recalling Chapter 14, what is the term given to dissociated changes in form within an evolving lineage?

Mosaic evolution.

The monotremes, duck-billed platypus and echidna (Chapter 11, Section 11.1.1), demonstrate another complication, somewhat the reverse of the 'Volkswagen syndrome'. They are said to be the most primitive living mammals on the basis of their reproductive pattern (evidence for which would not easily fossilize), but their skulls are extremely specialized, with many autapomorphies.

Another bias in assessing the status of living fossils is that evolutionary change in groups with complex morphology is more likely to be recognized than change in taxa with simple morphology, giving apparently shorter ranges per taxon. For example, the brachiopod genus *Lingula* (Figure 15.10(c)) is known from Ordovician rocks to the present day, and, having a range of *c*. 450 Ma, is not plotted on Figure 15.9. *Lingula* is usually called a living fossil, but its shell has very few describable features. *Lingula* may conceivably be an example of the

'Volkswagen syndrome', its internal tissues possibly having changed much more than its shell; but without detailed soft-part fossil preservation this must remain speculation.

Steven Stanley, of Johns Hopkins University, invoked the punctuated equilibrium model (Chapter 10, Section 10.3.1) to explain many living fossils. Under punctuated equilibrium, morphological evolution is associated with cladogenesis, and species, once evolved, remain in stasis; therefore any persistent lineage, or long-lived clade that experiences few speciation events, can be expected to show little morphological change. In Stanley's framework, therefore, most living fossils are single lineages, or small, persistent clades of low diversity, that are 'simply champions at warding off extinction'.

Plethodon is a genus of salamander that has persisted for more than 60 Ma, apparently accommodating environmental perturbation without responding by structural change. This capacity is attributed to behavioural, physiological and developmental plasticity. The limited morphological change is, however, not correlated with a low rate of speciation; on the contrary, there is evidence in this and some other salamander genera of a high speciation rate accompanied by trans-specific morphological stasis (see Section 15.5).

▶ Does evidence of trans-specific morphological stasis in these salamanders support or contradict Stanley's general hypothesis to explain living fossils, and why?

It contradicts it, because Stanley's hypothesis assumes that speciation follows the punctuated equilibrium model, with morphological change mainly coincident with cladogenesis, and that living fossils are generally associated with very persistent individual lineages or small, long-lived clades with low speciation rates.

The persistence of primitive members of a clade long after more advanced members have become extinct is a fairly common phenomenon, of which *Sphenodon* (see above) is a good example. One hypothesis to explain this suggests that lineages which first invade a new adaptive zone (Section 15.5) are generalists compared with the many descendent clades that specialize in utilizing finer subdivisions of the environment. Such generalists, albeit few in species diversity, are postulated to be often more resistant to extinction than the many specialists. Living fossils may thus often have relatively few derived characters compared with the extinct specialists. Alternatively, some living fossils may be species that are highly specialized in their use of resources, but the appropriate environment has persisted for tens or even hundreds of millions of years, and for some reason, no other lineages have succeeded in displacing them.

Some living fossils may, in contrast to most of the above examples, simply be species that exhibit an abundance of primitive characters within a clade which has otherwise undergone major changes and is still flourishing. The tree shrews (placental mammals) and the opossum (a marsupial) retain skeletal anatomies similar to forms of the late Cretaceous, whereas the remaining placental and marsupial lineages underwent dramatic changes in the early Tertiary radiation (Section 15.5). The tree shrews and the opossum, therefore, remain living fossils within high-diversity clades.

Other commonly cited examples of living fossils include horseshoe crabs, the sturgeon, tapirs, the common tree squirrel; and, among plants, the gingko and horsetails. As with the examples given earlier, the status of these organisms as living fossils, and the biological attributes associated with each one, vary considerably.

In summary, there is no evidence to support the idea that living fossils form a category of evolutionary rates quite distinct and separate from those of other organisms; they merely represent the tail end of a wide spectrum of taxon durations, as is clear for genera in Figure 15.9. A compendium of case histories of living fossils (Eldredge and Stanley, 1984) revealed no one prevailing reason for their existence; a lineage or small clade may persist over immense intervals because of great niche breadth, broad geographical distribution, specialization in a persistent, sometimes cloistered habitat, plasticity, and/or various other factors that are not mutually exclusive. Claims that certain fossils are indistinguishable from living species often reflect inadequate documentation based only on hard-part morphology; the necessary comparisons of softer tissues are simply not possible. There may, in some cases, be an association between living fossils and a low speciation rate, which, in the punctuated equilibrium model would allow few opportunities for morphological change, restricting a clade to a relatively low rate of change averaged over the long term. There are, however, exceptions to this pattern, too.

15.2.2 THE EVOLUTION OF BODY SIZE

Palaeontologists have obtained from the fossil record more information on size than on any other trait. Reasons for this are numerous. Body size, or some proxy for it, such as molar tooth area, is readily preserved in fossils, easy to measure, conspicuous and ecologically important, comparable across taxa (unlike many other traits), and very changeable (palaeontologists naturally tend to focus on traits that change the most). Body size is, however, a variable concept, and can pertain to maximum linear dimensions, total weight, total volume, or some other measure.

Body size has a heritability of over 0.5 in many animals, but is typically a polygenic trait (see Chapter 3, Sections 3.2.1 and 3.5.1): mouse size, for instance, is determined by at least 100 loci. Single gene mutations can, however, have dramatic size effects, usually by interfering with the hormone production-response system: the small size of the African human pygmy apparently results from neoteny due to such a mutation.

▶ Heterochrony (Chapter 14, Section 14.6) is a major source of heritable size variation. Which heterochronic processes are likely to be associated with an evolutionary increase in size?

The three peramorphic processes: acceleration, i.e. growing at a faster rate; hypermorphosis, i.e. growing for a longer period, with later offset; and predisplacement, i.e. beginning growth sooner, with earlier onset. Rarely, however, is an evolutionary increase in size likely to be due purely to one peramorphic process; heterochronic shifts often involve combinations — for example, growing both faster *and* for a longer period.

Living organisms span an enormous range in size: the blue whale is 10^{21} times heavier, and about 10^{8} longer, than the smallest microbe. In general, size is positively, albeit loosely, correlated with organic complexity: the more complex the organism (as estimated, for example, by the number of cell types), the larger it tends to be. The diversification of life has certainly produced a general increase in *maximum* body size, from bacteria in the early Precambrian to sauropod dinosaurs, blue whales, and giant sequoia trees within the last 150 Ma. As far as the whole biosphere is concerned, therefore, there has tended to be a whole-clade trend of size increase, though microbes are still by far the most numerous organisms. The modal organisms in a size–frequency distribution of the biosphere would be prokaryotes: you probably contain more of the bacteria *Escherichia coli* in your gut than the globe contains people!

One of the most commonly reported evolutionary trends in animals is an increase in body size, i.e. a general tendency for descendants to attain a larger average size than their ancestors. This became known as Cope's Rule, after the American palaeontologist Edward Drinker Cope who, in the 1890s, pointed out the widespread tendency for vertebrate lineages to evolve towards larger size. Cope used this finding as evidence for orthogenesis, an idea now discredited (Chapter 14, Section 14.2.3).

Remember that there is no necessary coupling of anagenetic trends and whole-clade trends. For example, a clade might show a tendency for later species to be larger, by biased speciation or biased extinction, but individual lineages would not necessarily have to undergo anagenetic size increase (see Section 15.2, and Figure 15.7).

Study Figure 15.11, which shows the estimated body mass for 40 species of fossil horses from the New World (principally North America), following the origin of the horse clade at about 58 Ma ago. Body mass, the measurement of size used here, was estimated from a variety of dental and skeletal characters.

▶ Treating the forms in Figure 15.11 as a clade, was there a whole-clade trend of size increase or size decrease over this period? If so, did this happen at a continuous, steady rate?

The data indicate an overall increase in average size, for these New World horses at least, but not at a steady, uniform rate. Rather, there was a long period of relative stasis in body size, followed by diversification of species and an increase in mean and maximum size.

▶ At what time did this diversification and size increase begin?

After about 25 Ma ago. In other words, size was relatively static for about 33 Ma, until the onset of a major taxonomic diversification of both browsing and grazing horses (see Section 15.5).

▶ By how much had the maximum size of horse species increased between 58 Ma and 10 Ma?

---

10 Ma ago, the species labelled 17 in Figure 15.11 was over eight times larger than the species labelled 1 (58 Ma); body mass increased from *c.* 50 kg to 400 kg.

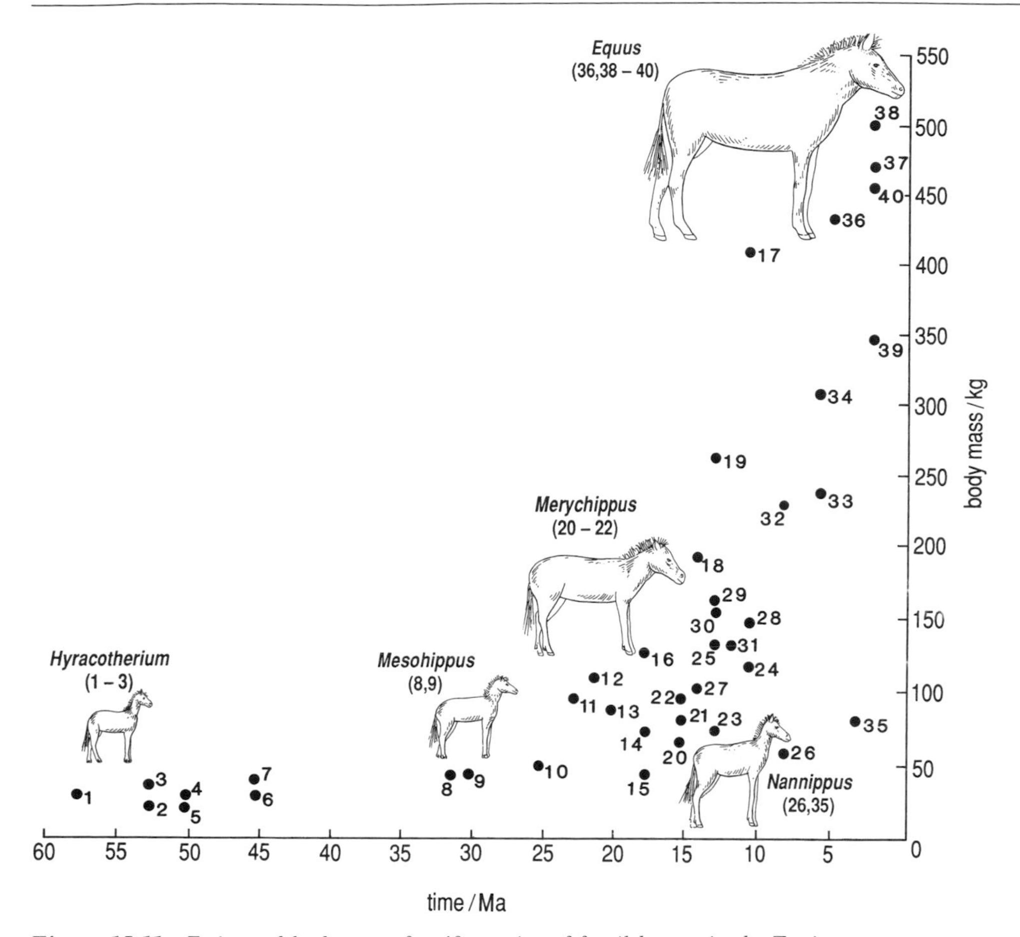

Figure 15.11 *Estimated body mass for 40 species of fossil horses in the Tertiary.*

Note that small species continued to be common while the maximum size of species increased.

▶ Is there any evidence within Figure 15.11 for anagenetic size changes?

No. There is no information concerning the pattern of size changes within individual lineages, and ancestor–descendant relationships are not indicated on Figure 15.11. Bruce MacFadden of the University of Florida, who compiled the diagram, concluded that of 24 ancestral-descendent species pairs that could be identified, 19 showed a size increase, and 5 showed a size decrease, suggesting that for these horses Cope's Rule was valid only as a general tendency and that there were significant exceptions.

In general, body size is subject to a very wide and complex variety of selection pressures. In a major review, Michael McKinney (in McNamara, 1990) emphasized the following points:

(a) There are no intrinsic, automatic advantages to larger body size, as is commonly supposed. The 'bigger is better' assumption ignores the fact that small organisms are much more abundant than larger ones.

(b) There is no one overriding cause of anagenetic size increase, and large size may confer advantages for many reasons, such as improved ability to capture prey and escape from predators, competitive benefits, increased intelligence, or increased physiological efficiency. In the past, researchers have tended to emphasize one particular advantage of large size, whereas there are always multiple selection pressures acting on the size of individuals; influences and trade-offs must be teased apart before their relative importance can be established. For example, body size in certain zooplankton increases significantly in the absence of predators, because the larger-bodied plankton are more efficient feeders; predators, however, often prefer larger prey, which suppresses size increase. In the case of increased intelligence, a larger brain is not necessarily associated with a larger body; evolutionary changes in brain size can be decoupled from changes in body size by allometric heterochrony (Chapter 14, Section 14.6.2).

(c) Change in average adult size may, to a large extent, be an incidental by-product of selection for something else, such as for developmental rates, for shape, and for features such as life history traits which can affect final adult size. Selection for size will often produce incidental shape changes, and *vice versa*; this makes it difficult to determine if size is the prime target of selection.

(d) Selection acts not only on adults, but on all ontogenetic stages, and this will often affect average adult size.

Some of the most important biotic and abiotic influences that favour larger-sized individuals, i.e. promote anagenetic size increase, are as follows (the influences are, however, only general tendencies revealed by correlations, and different generalizations will be valid for different groups):

1 *Regularly abundant food*. When food is regularly abundant and varied, larger individuals often tend to: (a) eat live animals that smaller individuals cannot obtain; (b) have foraging advantages such as lower locomotor costs per unit mass; and (c) actively impede the feeding of smaller forms. However, when isolated in resource-limited environments, such as small islands, forms that are normally larger do less well, having greater absolute needs.

2 *Low-nutrient food*. The length and capacity of the digestive tract correlates quite well with body size, and larger animals can therefore eat low-nutrient foods that require more processing but which are generally much more abundant. For example, large primates (e.g. gorillas) eat lower-nutrient foods such as leaves, while progressively smaller ones eat progressively more high-nutrient foods such as fruits, nuts and insects.

3 *Cool ambient temperatures*, favouring large warm-blooded animals, and

4 *Warm, stable ambient temperatures*, favouring large cold-blooded animals.

Ambient climatic temperatures generally have opposite effects on warm-blooded (endothermic) and cold-blooded (ectothermic) animals. Cooler climates tend to favour increased body size of mammals (endotherms) because the surface area/body volume ratio decreases as body volume increases, enabling a large body to lose proportionally less heat than a small one. Hotter, stable, climates favour increased size of cold-blooded ectotherms (e.g. reptiles) because large ectothermic individuals can retain heat for longer periods, therefore staying active longer, putting them at an advantage over smaller ectotherms. Climates with pronounced seasonality tend to select against large ectotherms, because the ectotherms can only survive by finding shelters in microhabitats during seasonal extremes (especially winters).

5 *Predation.* There is usually selection on prey to increase in size when predators prefer, or can more easily obtain, smaller prey. Commonly, the optimal foraging strategy is not to seek the largest prey, because the greater energy costs in obtaining it are not compensated by the greater energy gained. For prey species, large size can thus be a form of refuge from predation, with immediate fitness benefits.

6 *Prey size.* There is usually reciprocal selection on predators to increase in size as their prey does.

7 *Seasonality.* Major seasonality of food availability (irrespective of ambient temperatures) promotes size increase because larger individuals can survive longer on body stores.

8 *Sexual selection.* Selection of relatively large mates, especially by females, is common and can influence average size.

9 *Female fecundity.* In environments where reproductive output (e.g. number of eggs) is under strong selection, females with larger body size will tend to be favoured because reproductive output is often correlated with individual size.

Note that although the above factors can contribute to evolutionary size changes in lineages, many can also promote *ecophenotypic* size changes. For example, for some organisms, abundant food, or the lack of it, can dramatically change the mean size of individuals from one generation to the next, without any change in genes which influence size. This can sometimes make it difficult to ascertain the extent to which a change in mean size is genetic. However, as in the case of many of the horse species in Figure 15.11, when the mean and maximum size of earlier *time-averaged* fossil samples are so different from later ones, there is no doubt that, assuming equal chances of preservation, such changes to a very large extent reflect genetic changes.

The fossil record of a variety of marine invertebrates and terrestrial vertebrates shows that for periods of over 1 Ma, anagenetic increase in size is more common than anagenetic decrease in size. Despite the numerous situations favouring larger average size given above, size decrease on time-scales of less than 1 Ma is however, very common, and reversals of long-term anagenetic trends of size increase are seen in the fossil record of many lineages. Presumably, size decrease sometimes occurs because of direct selection for smaller size, but often size reduction is an indirect consequence of selection for some other feature(s). Rates of evolutionary size change, including size decrease, can be very fast: Adrian Lister of University College, London documented the evolutionary dwarfing of a form of red deer on the island of Jersey (off northern France) to one-sixth of its ancestral body weight in less than 6000 years.

Data from the fossil record suggest that small-bodied species tend to be less susceptible to global (as opposed to local) extinction than larger-bodied species. Large-bodied species may be more prone to extinction for several related reasons: they tend to be more specialized, i.e. more restricted in their niches, more demanding of resources, and therefore more susceptible to major environmental perturbation; they tend to have smaller population sizes, and hence smaller gene pools, and may be slower to adapt; they tend to have a lower intrinsic rate of population increase, and to be *K*- rather than *r*-strategists. The relationship between population extinction and species (i.e. global) extinction is, however, little understood.

The species which originate clades often seem to have been small. This can be expected because (a) many clades originate after major extinctions, and survivors of extinctions tend to be small-bodied species; and (b) small-bodied species are usually more common than large ones at any given point in clade evolution. Data on 46 clades of vertebrates and invertebrates showed that constituent members within clades tend to be small or medium-sized rather than large (although the most common body size in each clade is not usually the smallest category). Therefore, even if extinction were random with respect to size, small species would be more likely to initiate a clade.

Figure 15.12 shows the maximum diameter for 133 species of planktonic foraminifera ('forams') at the time of each species' first appearance in the Cretaceous. (The Figure does not indicate changes in diversity, just the size at which each new species appears.) Notice how, with time, the maximum size of new species increases, as does the mean, but one end of the distribution remains anchored at the smallest size. In the late Cretaceous, all but a few of the smallest planktonic forams became extinct (Section 15.4), after which, on recovery, there was a similar slow expansion of variance in size in the Paleocene and Eocene, followed by another series of extinctions leaving mostly small species. Then, yet again, the maximum size increased whilst the domain of small size remained continually occupied. The planktonic foraminifera thus provide a good example both of trends produced by an increase in variance and of the selective extinction of large-bodied species.

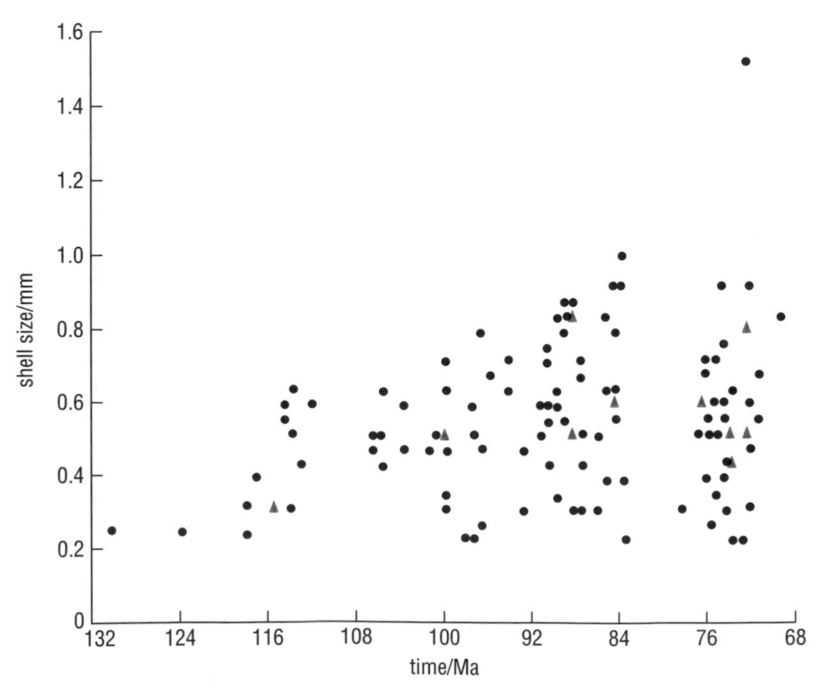

Figure 15.12 *Size at first appearance for Cretaceous species of planktonic foraminifera. Triangles indicate that two or more species occupy the same point.*

The various phylogenetic, functional and fabricational factors discussed in Chapter 14 not only *affect* the average adult size of any species, but they also *constrain* the maximum and minimum size any species may achieve.

▶ In the case of terrestrial vertebrates, two major constraints limiting maximum size are the strength of bone and the power of muscles. With which of the factors of Seilacher's triangle are these particular constraints most closely associated?

The fabricational factor (Chapter 14, Section 14.2.4).

Structural, physiological, energetic, and ecological considerations are particularly important in limiting size. For all species, there may in effect be reflecting barriers beyond which individuals of smaller or larger sizes either never appear (for phylogenetic or fabricational reasons), or are of very low fitness if they do (for functional reasons). The position of such 'barriers' can be expected to vary at different times in the history of the lineage or clade. Although it is impossible for us to judge where such reflecting barriers might be, there is little doubt they exist. As McKinney (in McNamara, 1990) puts it: 'A horse for example, can only become so small before its digestive system, metabolism, bone structure or some other aspect prevents it from running, eating or performing some other activity necessary to its existence. Even if radical "down-sizing" could occur, the anatomical modifications needed to meet the scaling demands would eventually lead to deviations so novel that the organism would cease to be a horse'. As small organisms are much more common than large ones, sustained selection towards ever smaller sizes is likely to lead to size worlds where there is more intense competition for resources. This is probably one reason for the apparent existence of a 'reflecting barrier' at small size in many clades. Such a barrier would promote trends of size increase within whole clades because, as far as size is concerned, there is nowhere else to go but up. One consequence of such a size increase, however, is a tendency for smaller population sizes.

The evolution of body size is thus an exceedingly complicated phenomenon, and can be recognized both within individual lineages (anagenetic size change), and within clades as a whole, after intervals of speciation and extinction. Anagenetic size increase, especially over periods more than 1 Ma, tends to be more common than size decrease, but there is a stronger tendency for whole-clade trends of size increase, a process that can be viewed as diffusion to larger sizes away from the small size at which many clades originate.

SUMMARY OF SECTION 15.2

Patterns of change in form concerning more than one lineage include divergent evolution, convergent evolution, parallel evolution, iterative evolution, and radiation. The term 'evolutionary trend' is used loosely for any sustained tendency for evolutionary change in a particular direction. It is important to distinguish anagenetic trends, which occur within single lineages, from whole-clade trends, which concern a net shift in some variate when all constituent lineages in a clade are taken into account. Whole-clade trends, which may be uncoupled from anagenetic trends, can be generated by biased speciation (in which lineages with particular morphological characteristics split more often, or lineages tend to branch off more in one direction than another), and by biased extinction (in which lineages with particular morphological characteristics become extinct more rapidly). Trends in the fossil record may be associated with a change in variance, rather than simply a change in mean.

'Living fossil' is an informal term with various meanings, the central concept of which is survival to the present day with minimum morphological change over very long periods. Some genera span well over 100 Ma, but there is little evidence of species spanning tens of millions of years. Many pitfalls beset attempts to assess the validity of an organism's status as a living fossil, including: (i) the difficulty of ascertaining whether or not soft parts have undergone significant change; (ii) organisms may qualify as living fossils by virtue of having a maximum number of primitive character states (plesiomorphies), but at the same time have many highly specialized, unique derived characters (autapomorphies); (iii) the more distinct measurable and preservable characters that a group possesses, the more likely that relatively minor change will be documented, giving an exaggerated impression of change. Living fossils represent the tail end of a wide and continuous spectrum of evolutionary rates, and there appears to be no one prevailing reason for their existence.

The evolution of body size is well documented from the fossil record, but is a highly complex phenomenon, as body size is subject to a very wide range of selection pressures, and may change as a consequence of selection for some other feature. For periods of over 1 Ma, *net* anagenetic increase in mean size is more common than size decrease. There is a stronger tendency for whole-clade trends of size increase. One reason for this may be that, because clade originators often seem to have been small, and because smaller organisms are far more common than large ones, over geological time-scales, there is, in effect less competition for resources at larger sizes, as long as populations are small. Large-bodied species are, however, more susceptible to extinction.

15.3 Clade dynamics: patterns of change in taxonomic diversity

The idea of using changes in taxonomic diversity as measures of the relative evolutionary fortunes of major groups of organisms was introduced in Chapter 11 (Section 11.1). Many macroevolutionary studies have been devoted to examining such patterns, as the postulated effects of, say, the appearance of evolutionary innovations or of changes in the environment, including the evolution of other taxa. The focus of interest here is thus upon the evolutionary histories of entire clades, rather than of specific lineages.

15.3.1 Modelling clade dynamics

The theoretical basis for analysing clade dynamics is remarkably simple (although, as you will see, the practical application of the theory encounters many problems). Suppose that the taxonomic diversity of a clade at one time is D (where D = the number of species, genera or whichever taxonomic category is to be analysed), and that, over a given interval of time (Δt), S additional new taxa arise, by cladogenesis (= speciation, hence the symbol S), and E taxa become extinct. Then:

the *average* rate of origination of new taxa = $S/\Delta t$; and

the *average* rate of extinction of taxa = $E/\Delta t$.

Clearly, however, these overall values will depend upon the total number of taxa (D) in the clade at the start of the interval: the more taxa there are at the outset, the greater the total numbers of taxa that can be expected to arise, and to become extinct. Thus, in order to compare clades on a standardized basis, it is necessary to calculate the rates of origination and of extinction of taxa *per pre-existing taxon*, as follows:

$$\text{per taxon rate of origination } (r_s) = \frac{S/\Delta t}{D} \qquad (15.1)$$

$$\text{per taxon rate of extinction } (r_e) = \frac{E/\Delta t}{D} \qquad (15.2)$$

Any change in the taxonomic diversity of a clade is simply a budgetary consequence of the originations and extinctions of taxa within it. Hence a net rate of change in diversity per pre-existing taxon (r_d) can be calculated:

$$r_d = r_s - r_e \qquad (15.3)$$

The rate of change in total diversity in the clade, per interval of time, is thus r_dD.

We will explore the possible influences on values of r_s and r_e later, but before doing so it is useful to consider a null hypothesis for the dynamics of clades which is free of any such determining influences. As noted in Section 15.1.1, such a model can provide a consistent means for testing whether or not there is any justification for seeking a deterministic explanation in a given case. If, in the null hypothesis, the origination and extinction of taxa are considered to be random events, then there is no reason to expect r_s to be necessarily greater or less than r_e. In other words, the mean values of r_s and r_e over an extended period of time can be expected to be approximately equal. Thus, if r_s and r_e are treated as *probabilistically* equal rates, each showing random independent variation around the same mean value, then randomly evolving clades (referred to as *stochastic phylogenies*) can be generated, for comparison with real clades.

▶ What is the *mean* value of r_d in the null hypothesis?

Since the mean values of r_s and r_e are taken to be equal, the mean value of r_d, according to Equation 15.3, should be 0. In terms of the randomly generated clades, this means that the per taxon rate of diversification is expected to fluctuate randomly between positive and negative values, over successive time intervals, around a mean of zero.

Exactly such a model was created by Raup *et al.* (1973). They described the way it worked as follows:

'An arbitrary time scale is established and a single lineage is entered at the beginning of the time sequence. The initial lineage is made to advance through time; as it advances, its fate is determined by chance, controlled by automatically generated random numbers. At each time unit, the lineage can (1) become extinct, (2) persist to the next time unit and branch to produce a second lineage, or (3) persist to the next time unit without branching. The probabilities of these three fates are supplied as input constants. When and if a second lineage is produced by the second alternative, it also becomes subject to randomly controlled development in time.'

In order to establish clades with suitable early values of D (to avoid too many becoming extinct early in the simulation), the initial probability of branching was set higher than that of extinction.

▶ What effect would this have upon (a) r_d, and (b) the pattern of change in total diversity?

According to Equation 15.3, r_d would be positive if r_s exceeded r_e. Total diversity (D) would thus tend to rise in each successive interval of a simulation (though remember that the probabilistic nature of the simulations means that occasional drops of D could be expected when, by chance, r_s fell below r_e). The pattern of increase, moreover, would be approximately exponential.

When the number of co-existing lineages reached a predetermined level, the probability of branching was then automatically lowered to equal that of extinction, and from then on the programme generated stochastic phylogenies as discussed above (Figure 15.13(a)). Phylogenies were arbitrarily divided into higher taxa (each containing more than a minimum number of lineages), and the changing diversity of each higher taxon shown as a spindle (Figure 15.13(b); see also Chapter 12, Section 12.4.2).

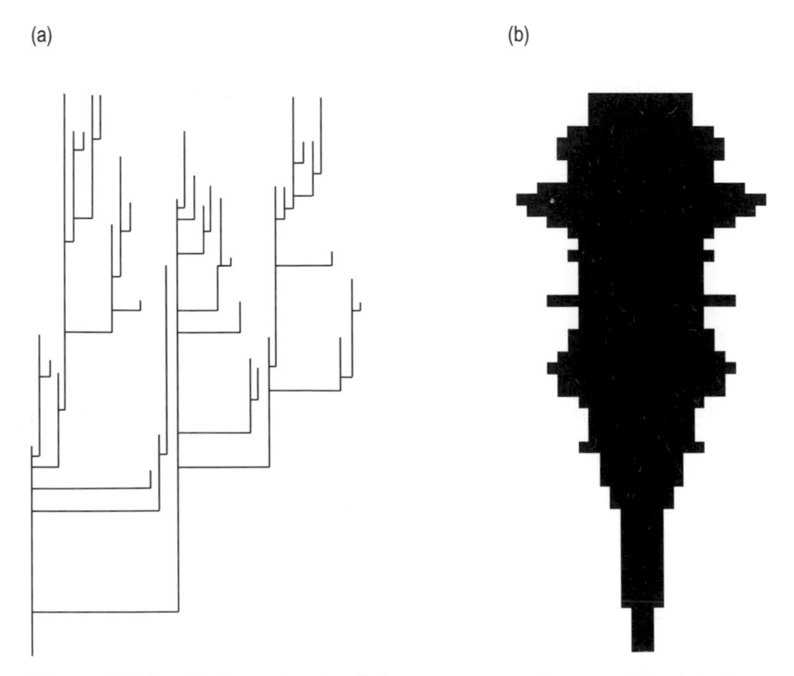

Figure 15.13 *(a) A stochastic phylogeny generated as explained in the text. (b) A spindle showing the changing diversity of the phylogeny shown in (a).*

By generating large numbers of such stochastic phylogenies, the likelihood of certain patterns of change in diversity being produced by chance alone could be assessed (i.e. by adopting the 'Monte Carlo' approach mentioned in Section 15.1.1). Figure 15.14(a), for example, shows just four such stochastic phylogenies produced by Raup *et al.* Compare these with Figure 15.14(b), which shows spindles (without phylogenetic connections) for Carboniferous to Recent reptiles. The reptile data have been somewhat unconventionally organized to allow them to be plotted in as similar a fashion to the computer simulations as possible. Only two taxonomic levels have been employed, corresponding to lineages and higher taxa in the simulations, although the actual taxonomic categories chosen varied from group to group so as to yield spindles of broadly comparable sizes. For example, some are plotted as families within orders, and others as orders within subclasses. Moreover, stratigraphical intervals have been treated as representing equal units of time, which, strictly, they do not.

Notwithstanding the various qualifications given above, comparison of the real higher taxa with the simulations is instructive. The similarities of many aspects are striking: for most of the real examples, counterparts of similar shape can been seen in Figure 15.14(a). There are certain features of the real data, however, which stand out as being almost consistently absent (i.e. of very low probability) in the stochastic phylogenies. In particular, notice how none of the simulations in Figure 15.14(a) shows such a coincidence of extinctions as that seen at the Cretaceous–Tertiary boundary for the reptile taxa in Figure 15.14(b). As noted earlier (Chapter, 11, Section 11.1), this is an example of a mass extinction, for which the search for a deterministic cause is thus evidently justified (mass extinctions are discussed further in Section 15.4).

However, comparisons between real and random taxa beg certain questions. To serve as suitable test data, the null hypothesis simulations need values of D, r_s and r_e which are themselves similar to those exhibited *on average* by the real taxa. The early simulations of Raup *et al.* have been criticized as having inappropriately low values of D and high values of r_s and r_e for comparison with the real taxa investigated: the fundamental units of the branching and extinction process should really be species, and species diversity within the higher taxa considered was much higher than the taxonomic diversities employed in the simulations.

▶ What would be the expected effect of higher values of D and of lower values of r_s and r_e on the patterns of change shown by stochastic phylogenies?

The simulated taxa would tend to show greater stability through time. Higher taxonomic diversity gives increased scope for gains and losses to cancel each other out, so that proportional changes in taxonomic diversity are less, while lowered values of r_s and r_e mean relatively fewer gains and losses in any case. Thus, some of the rapid radiations to which the fossil record bears witness, as with the mass extinctions, are also deserving of deterministic explanation — a point that will be followed up in Section 15.5.

By definition, every monophyletic or paraphyletic taxon commences with a single species, so the expansion to higher values of species diversity requires an episode of net excess of speciations over extinctions. Conversely, the demise of any taxon

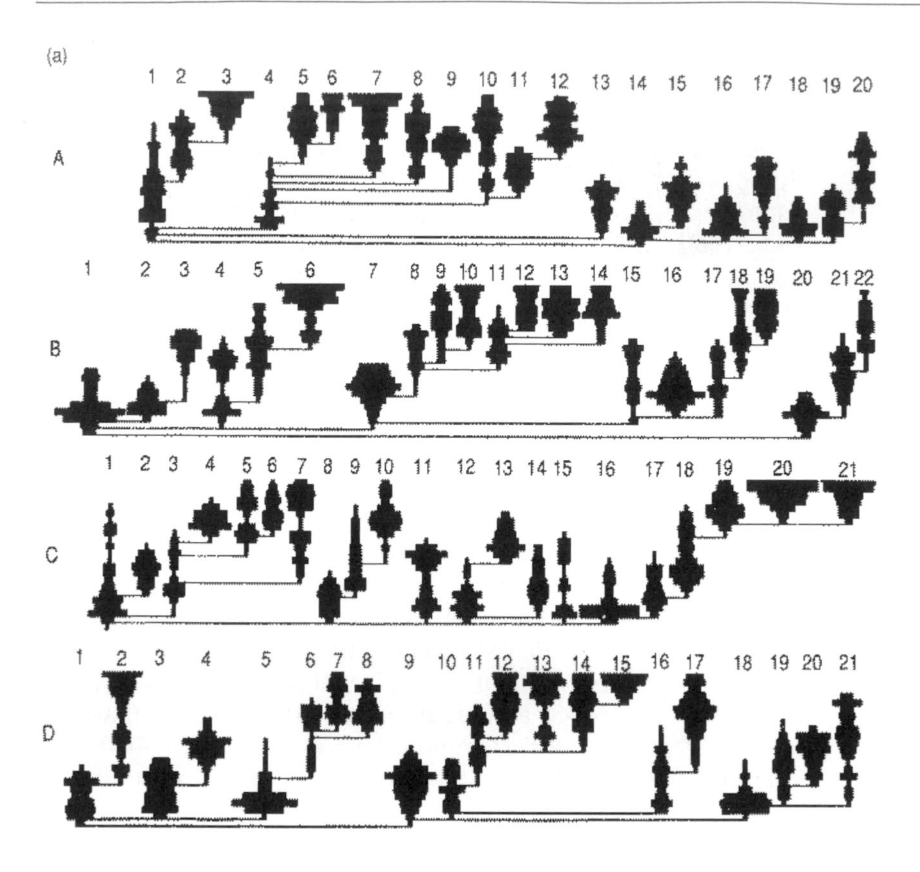

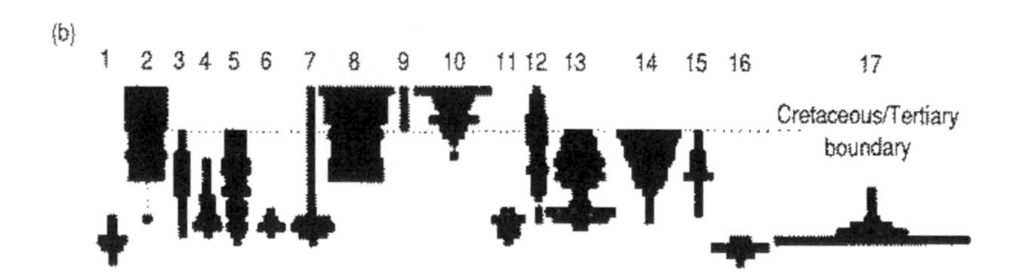

Figure 15.14 *(a) Four examples of simulated phylogenies (shown as evolutionary trees of spindles, simulating higher taxa) produced by the computer program of Raup* et al. *(1973). (b) Spindles showing diversity changes in 17 higher taxa of reptiles from the Carboniferous to the Recent.*

to total extinction requires an episode of net excess of extinctions over speciations. The stochastic phylogenies of Figure 15.14(a) show that both expansions and declines may readily arise by chance, even with equal probabilities of speciation and extinction (the basis of the null hypothesis). But how, *on average*, do the episodes of expansion compare with those of contraction, towards extinction? The relationship can be analysed by finding the average value, for a given set of higher taxa, of a measure of temporal asymmetry of each taxon (between its episodes of growth and decline), termed its *centre of gravity* (*CG*). The *CG* is calculated as the

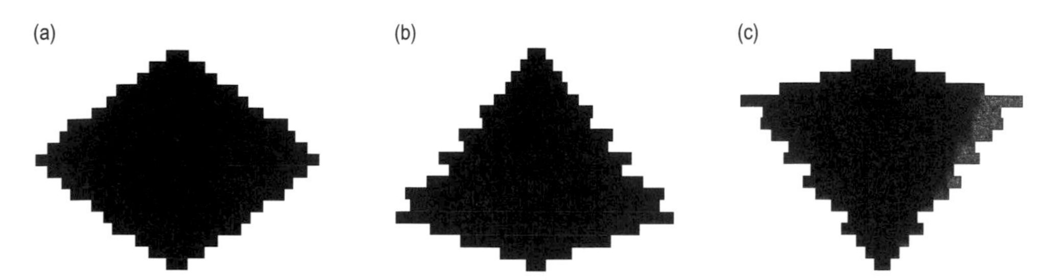

Figure 15.15 *(a) Symmetrical spindle for a higher taxon, with a* CG *of 0.5 (see text for explanation). (b) Bottom-heavy spindle with* CG *<0.5. (c) Top-heavy spindle with* CG *>0.5.*

mean position in time of taxonomic diversity, measured at successive equal intervals, expressed as a proportion of the temporal duration of the higher taxon:

$$CG = \frac{\Sigma N_i t_i}{\Sigma N_i} \tag{15.4}$$

where Σ means 'sum of', N_i is the taxonomic diversity in each time interval, t_i, with i running from the first (earliest) interval to the last for the higher taxon in question. The term t_i is thus a number ranging from 1 to the total number of intervals. Symmetrical taxa therefore have a CG of 0.5 (Figure 15.15(a)), bottom-heavy taxa a CG of <0.5 (Figure 15.15(b)), and top-heavy taxa a CG of >0.5 (Figure 15.15(c)). (Although the interval showing the maximum diversity value in a taxon may frequently coincide with the CG, it is less reliable as a guide to the taxon's asymmetry, because of the possibility of chance fluctuations.)

▶ What *mean CG* value would you expect for a large sample of randomly generated higher taxa such as those shown in Figure 15.14(a)?

A mean value of about 0.5 would be expected, because episodes of expansion and decline should, on average, be symmetrical, as a result of r_s and r_e being probabilistically equal.

Measuring the mean CG value of real taxa can therefore provide a useful means of testing whether these probabilities were indeed equal for real clades variously distributed over given periods of time.

▶ If r_s (probabilistically) exceeded r_e for the entire set of clades throughout their history, would you expect the *average* spindle shape to be 'bottom-heavy' or 'top-heavy', as in Figure 15.15(b) or (c), respectively?

The expected outcome is an average spindle shape that is bottom-heavy (Figure 15.15(b)). The bias for positive values of r_d (Equation 15.3) would promote the rapid expansion of higher taxa through a chance succession of diversity increases, but retard their decline by decreasing the probability of diversity values showing a succession of falls.

Investigation of Palaeozoic marine metazoan taxa shows that the average spindle pattern (for generic diversity of families) is bottom-heavy for taxa which evolved in Cambrian–Ordovician times, but is symmetrical for those which evolved thereafter. It may thus be inferred that rates of origination of new genera significantly outstripped rates of extinction in the early Palaeozoic (see also Chapter 14, Section 14.1.3), but that they later became more similar. This inference is only valid, however, as long as the Cambrian–Ordovician taxa are not dominantly paraphyletic (see below).

It is unlikely, however, that this equilibration in rates of diversification occurred as in the somewhat crude manner employed for launching the stochastic phylogenies described earlier (where r_s and r_e were simply equalized when D reached a predetermined level). An elegant model for the underlying pattern of change has been proposed by Jack Sepkoski (1978) of the University of Chicago. He suggested that the per taxon rates of origination and extinction might themselves have been diversity-dependent (Figure 15.16(a)). At low diversities, competition could have been at a minimum, such that the survivorship of incipient species was enhanced and the chances of their extinction, relatively low. With rising levels of diversity, and thus perhaps of competition, it was argued, the probability of successful speciation would have started to decline, and that of extinction, to increase.

▶ Study Figure 15.16(a). What eventually happens to r_s and r_e as D rises, and what is the implication for r_d, according to this model?

As D rises, r_s is seen to decline while r_e increases until, at the point labelled $\hat{D}$ (the '*equilibrium diversity*'), they become equal. As a consequence, r_d declines to 0 at $\hat{D}$ and so total diversity becomes stabilized at that level.

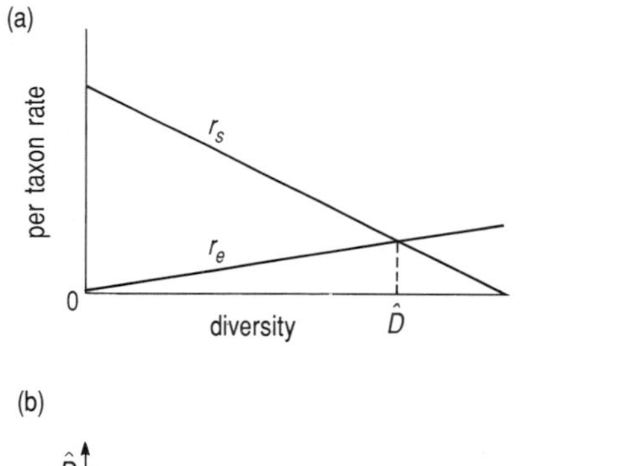

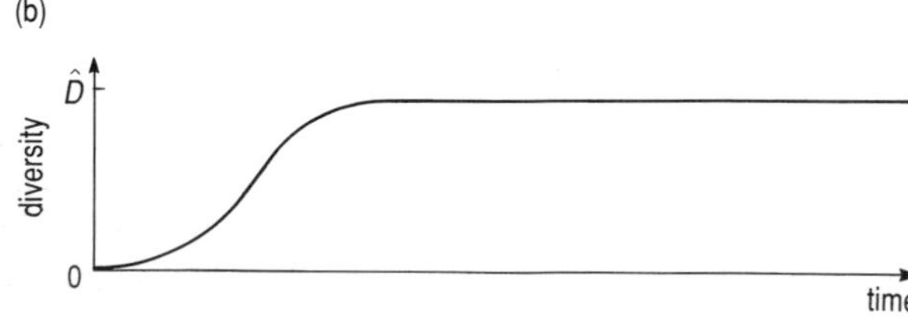

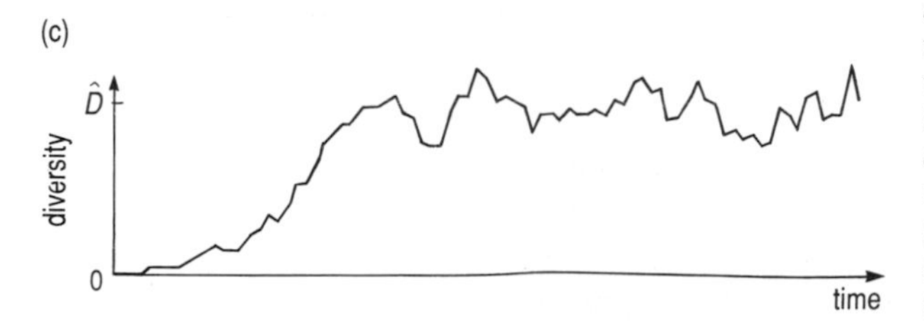

Figure 15.16 *(a) Diversity-dependent model of taxonomic diversification where* r_s *declines, and* r_e *increases, with increasing total diversity* (D)*, becoming equal at the equilibrium diversity,* $\hat{D}$*. (b) Logistic diversification resulting from the model in (a). (c) Logistic diversification with random fluctuations introduced into the values of* r_s *and* r_e*.*

The pattern of change in diversity through time predicted by this model is shown in Figure 15.16(b). Such a pattern of diversification, with an exponential early phase followed by a phase of decelerating increase in the numbers of taxa towards an eventual equilibrium value, is termed **logistic diversification**. Figure 15.16(b) illustrates a smooth, deterministic form of the model. One outcome of a probabilistic version of the model (with r_s and r_e showing random fluctuations around the diversity-dependent values indicated in Figure 15.16(a)) is shown in Figure 15.16(c). We will return to consider attempts to apply logistic models to the fossil record of taxa in Section 15.6.1.

Before leaving this general discussion of clade dynamics, a systematic question which is still a source of considerable contention needs to be raised: what kinds of taxa (in terms of the differing methods for erecting them explained in Chapter 11) are suitable for quantitative analyses of this sort? Cladists assert that only monophyletic taxa (Chapter 11, Section 11.2.1) should be employed in analyses of clade dynamics, because paraphyletic taxa are regarded as artificial (Chapter 11, Section 11.2.4); any pattern of taxonomic turnover which is based upon paraphyletic taxa, it is argued, is likely to say more about the working practices of taxonomists than about the history of life.

The cladists' criticisms of paraphyletic taxa may at first seem quite reasonable, but such a purist approach cannot be sustained in practice, because paraphyletic taxa cannot be avoided without artificially excluding at least some species from consideration. If monophyletic taxa alone were to be compared, then any fossil species which happened in fact to be a common ancestor to two or more such clades could not be assigned to any one of them because it would lack the relevant diagnostic synapomorphies (Chapter 11, Section 11.2.4). Nor would it possess any autapomorphies to justify placing it in its own clade, distinct from the others. If the other monophyletic taxa were to be retained in the analysis, the ancestral species (and indeed any morphologically unchanged lineage persisting from it after the other clades had branched off) would simply have to be ignored, because it could not be accommodated in a comparable monophyletic taxon. In practice, however, such ancestral species are rarely, if ever, identified. A similar (though weaker) argument might be applied to subsequent species that largely retained the primitive character states of the ancestral species (e.g. the Soluta, discussed in Chapter 12, Section 12.1). The alternative to ignoring such entirely plesiomorphic fossil taxa would be to recognize just one monophyletic taxon embracing both the ancestral (and similar) species and its descendent clades. Logically, however, this process of agglomeration would ultimately lead to the recognition of only one monophyletic taxon — that comprising the entire phylogeny of life — which would defeat the original objective of comparing the dynamics of different clades! To illustrate this point, consider again the stochastic phylogenies in Figure 15.14(a)).

▶ Are the higher taxa, shown as spindles in the Figure, monophyletic, paraphyletic, or a mixture of the two?

They are a mixture of the two. In the top row of the Figure, for example, spindles 3, 6, 7, 8, 9, 12, 13, 15, 17 and 20 are monophyletic (giving issue to no further spindles). All the remaining spindles, in contrast, have had the descendent clades listed above removed from them and are thus paraphyletic.

▶ What would be the result if you were to reclassify the phylogenies cladistically, so that *all* the paraphyletic taxa (spindles) were combined with their monophyletic offshoots?

Each entire phylogeny (i.e. each of the four rows) would end up as a single clade spindle, because in these simulations all the ancestral taxa, down to the original species, are present and thus have to be incorporated in the classification.

In addition to the theoretical problem outlined above, it is also frequently difficult in practice to establish whether a given higher taxon of real organisms is monophyletic or paraphyletic, especially when morphologically simple fossils are considered. So the cladists' insistence on the use only of monophyletic taxa for the analysis of clade dynamics cannot be accomplished unless certain taxa are selectively ignored, thereby introducing no less an element of artificiality than that introduced by the acceptance of paraphyletic taxa.

Finally, notwithstanding their arbitrary phylogenetic limits, can paraphyletic taxa be considered to have any other kind of natural identity? Pheneticists would say 'yes', in that they may comprise relatively distinct morphological clusters of species, and thus indeed natural ecological groupings. Building upon the idea of a population occupying a single adaptive peak in an adaptive landscape (Chapter 4, Section 4.5.2), G. G. Simpson suggested that higher taxa might be conceived of as occupying constellations of associated adaptive peaks, separated from other such constellations by major depressions, rather like distinct mountain chains. He termed these clusters of adaptive peaks, **adaptive zones**: higher taxa, whether monophyletic or paraphyletic might thus have some natural identity by virtue of occupying distinct adaptive zones. An example mentioned in Chapter 11, Section 11.1.1, is the mammalian Order Carnivora, all members of which, as the name suggests, are to a greater or lesser extent carnivorous. But if mere ecological similarity were to be the criterion for recognizing higher taxa, why then should not polyphyletic taxa, comprising convergent but unrelated forms (Chapter 11, Section 11.2.1), also be considered? The reason they are not depends upon an important distinction that should be made between the origins and the ecological deployment of taxa. Organisms are the products of evolution and so phylogenetic divergence is the natural source of higher taxa. Their adaptive deployment in nature is, by contrast, governed by their ecological relationships, although these are greatly influenced by phylogenetic constraints, as Chapter 14, Section 14.2.1 made clear. Higher taxa as traditionally conceived, then, combine elements both of the phylogenetic and the ecological hierarchies in nature, and it is in this, albeit only vaguely defined, context that paraphyletic, though not polyphyletic, taxa might be considered to comprise natural entities. By restricting attention to monophyletic taxa, and hence only to the phylogenetic hierarchy, the cladist risks losing touch with the ecological contingencies in the history of life.

If the discussion above seems inconclusive, that is because there is still little agreement about what kinds of higher taxa should be used in the analysis of clade dynamics. For example, although cladistics has certainly provided a rigorous method of analysing the phylogenetic hierarchy, a correspondingly rigorous and consistent (i.e. non-anecdotal) method for recognizing adaptive zones (and hence, paraphyletic taxa which can be considered natural) is still wanting. Debates on the issue look set to continue for some time yet. In this Chapter, however, paraphyletic taxa are considered in analyses, where suitable, for the various reasons given above. Following these theoretical considerations, we turn next to how rates of turnover among taxa of real organisms are calculated in practice.

15.3.2 CALCULATION OF TAXONOMIC RATES OF EVOLUTION

For calculations of taxonomic turnover in clades, interest focuses upon the originations and terminations of lineages: only cladogenesis and true extinction are relevant. The appearance of new chronospecies (Chapter 10, Section 10.2.1), by anagenetic evolution, and the simultaneous pseudoextinction of their ancestors are theoretically irrelevant.

▶ Why?

As was explained in Chapter 10, chronospecies are arbitrary subdivisions of single lineages. Their comings and goings testify to microevolutionary change but do not alter the numbers of co-existing lineages. Unfortunately, however, it is often difficult to ignore them in practice. Because of the incompleteness of the fossil record, coupled with the problems of correlating widely separated sedimentary sequences, different segments of a single lineage (i.e. different chronospecies) may sometimes be treated erroneously as distinct lineages. Such errors can lead to over-estimates of diversity and of taxonomic turnover — a problem to which we will return later in this Section.

The rates of taxonomic turnover can be derived from range charts, the construction of which was explained in Section 12.4.2 of Chapter 12. Ideally one might wish to conduct all such analyses using numbers of species lineages, as these are the basic units of phylogenies. For the practical reasons already discussed in Chapter 11 (Section 11.1.1), however, higher taxa are frequently employed as proxies for relative species numbers. In this Section, for example, the methods of calculation will be illustrated by reference to families of the ammonoid Suborders Ammonitina and Ancyloceratina, from the Cretaceous (Figure 15.17).

If changes in the rates of turnover through time are to be properly identified, it is important that the range chart be plotted on an absolute time-scale (in Ma), as in Figure 15.17. Most fossil data have been recorded by reference to the traditional stratigraphical scale of relative time (as illustrated by the succession of geological periods shown in Appendix 1). That scale pre-dated the advent of the absolute time-scale (and is still the most practical system for dating fossils), and, for purely historical reasons, the various periods and their constituent stages are not of equivalent length in absolute time. Thus, to treat the conventional stratigraphical intervals as if they were of equal duration could lead to biases in the calculated rates of taxonomic turnover.

Although the Ammonitina and Ancyloceratina probably represent a single clade (commencing in the latest Triassic), phylogenetic relationships have not been shown on Figure 15.17. For our present purposes, we simply want to know what the *average* rates of taxonomic turnover were within the clade, over different time intervals, and so the families have been arranged in order of first appearance. If a

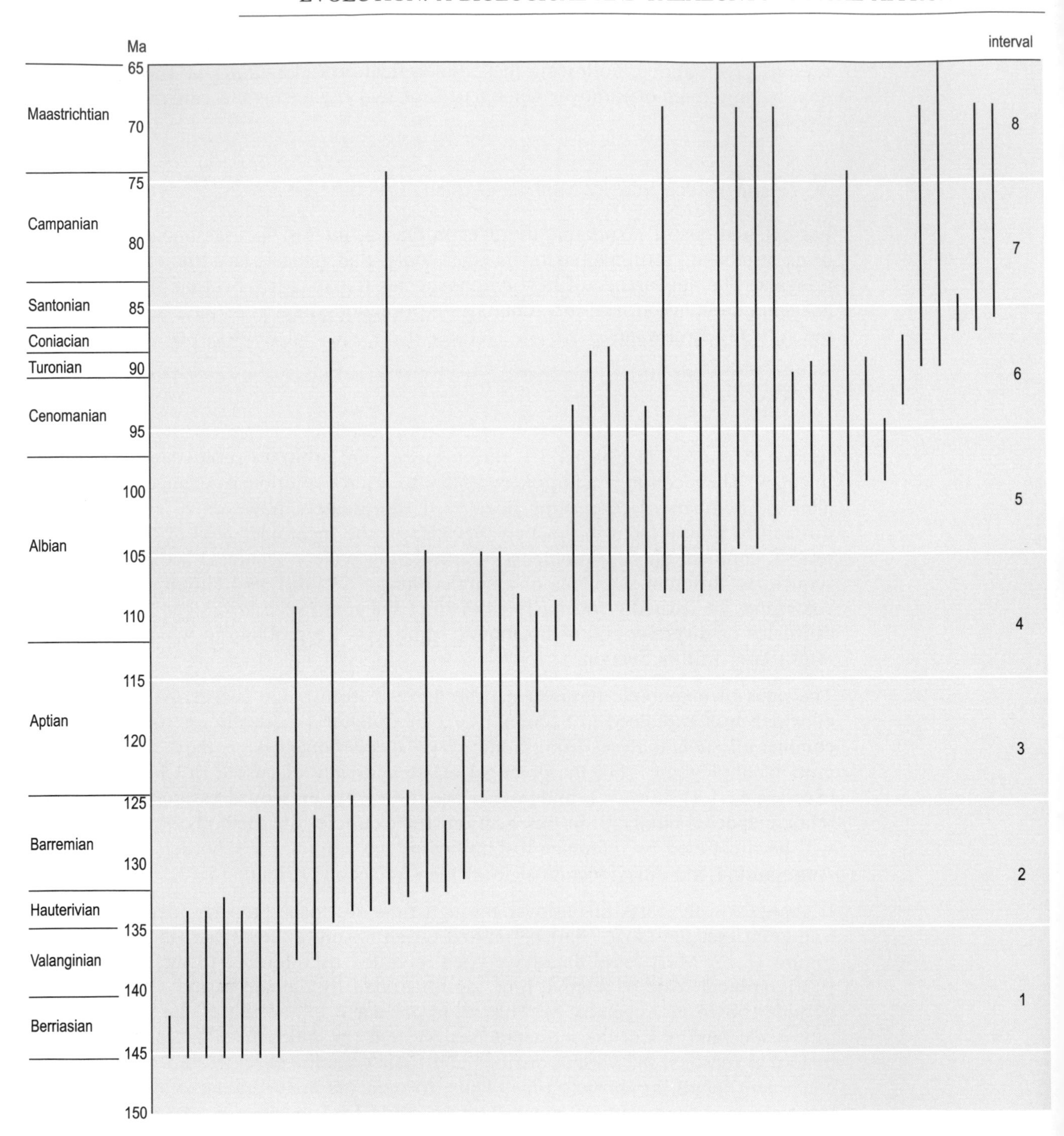

Figure 15.17 *Range chart for Cretaceous families of the ammonoid Suborders Ammonitina and Ancyloceratina. Absolute time is divided into eight 10Ma intervals. The seven initial families all range down variously into the Jurassic.*

more detailed analysis were required, showing, for example, differences in the rates for different subordinate clades, then further phylogenetic information (and re-organization of the range chart) would be necessary.

As noted in the previous Section, the two rates that need to be measured are that of the appearance of new taxa (r_s) and that of the extinction of taxa (r_e), both per pre-existing taxon. For each of the 10 Ma intervals shown in Figure 15.17, these rates can be simply calculated by adding up the total of appearances or of extinctions, respectively, shown in any given interval and dividing by the number of families shown to enter that interval from the previous one, finally also dividing by 10 Ma, to yield an average rate per Ma. Thus, for example, eight families are shown to enter interval 2 from interval 1. During interval 2, eight new families appear, but seven families become extinct. So the per taxon rates for interval 2 are as follows:

$$r_s = \frac{8}{8} \times \frac{1}{10\,\text{Ma}} = 0.1 \text{ families per Ma}$$

$$r_e = \frac{7}{8} \times \frac{1}{10\,\text{Ma}} = 0.0875 \text{ families per Ma}$$

Of course, these are relatively crude *average* measures with a minimum limit of resolution of 10 Ma. It is clear from the range chart itself that the extinctions in interval 2 are concentrated in two episodes, one near the beginning of the interval and the other near its end, while the originations are concentrated in an episode following the first of the extinction episodes. In order to pick out these patterns, a more refined subdivision of the range chart would be necessary (though the same basic method of calculation would be used).

▶ Calculate the values of r_s and r_e for interval 8.

Eleven families are shown entering the interval, no new ones appear within it, and all eleven are shown to have become extinct by its end. Thus:

$$r_s = \frac{0}{11} \times \frac{1}{10\,\text{Ma}} = 0 \text{ families per Ma}$$

$$r_e = \frac{11}{11} \times \frac{1}{10\,\text{Ma}} = 0.1 \text{ families per Ma}$$

Notice that the extinction rate is little greater than that calculated for interval 2. The main reason for the complete extinction of the clade is the lack of compensating originations (in contrast to interval 2). This example thus also illustrates the important point that both originations and extinctions need to be analysed when considering major changes in diversity.

Once rates have been calculated for successive intervals, they can themselves be plotted against time, alongside total diversity, so as to show the relative contributions of cladogenesis and extinction to change in diversity (Figure 15.18).

For much of the time, the per taxon rates of origination and extinction shown in Figure 15.18 fluctuate up and down in tandem, but there are some notable exceptions. In particular, the steep rise in origination rate in interval 4 is not matched by a rise in extinction rate, while the reverse holds for interval 8: the

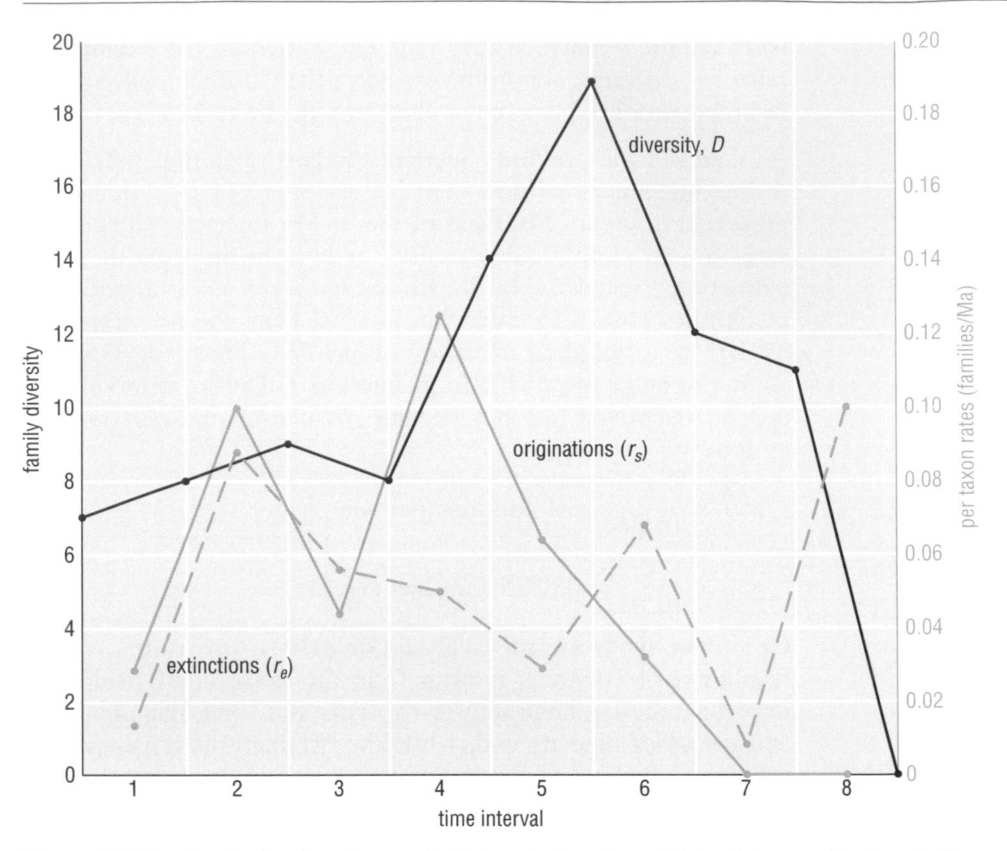

Figure 15.18 *Graph showing changes in* D *(vertical scale to left) and in* r_s *and* r_e *(vertical scale to right) for the ammonite families shown in Figure 15.17, over the eight successive time intervals shown there.*

former relationship is associated with a sharp rise in total diversity, while the latter is linked with the final extinction of the clade. The possible reasons for such changes in rates are another consideration altogether, and need not concern us here: the purpose of the present illustration is simply to demonstrate how the rates may be derived and plotted, and their contributions to change in diversity identified.

Another approach to the analysis of extinction, borrowed from population ecology, is that of **survivorship analysis**, which involves plotting the relative survival of taxa as a function of time. There are several methods of survivorship analysis, of which two will be discussed here — *cohort survivorship* and *dynamic survivorship*.

Cohort survivorship monitors the progressive decline of a sample of taxa, drawn from a single time interval (the cohort), over successive time intervals. Consider, for example, the 14 families that are shown entering interval 5 on Figure 15.17. Their survivorship in the subsequent intervals can be recorded in a *life table* (Table 15.1), which shows, for each interval (x), the number (or proportion) of surviving (i.e. 'living') taxa at the start of the interval (l_x) and that of those taxa becoming extinct ('dead') during that interval (d_x).

Table 15.1 Life table for the cohort of ammonite families entering interval 5 on Figure 15.17.

Interval (x)	Survivors at start (l_x)	(%)	Taxa becoming extinct (d_x)	(%)
5	14	100	4	29
6	10	71	7	70
7	3	21	0	0
8	3	21	3	100
	(0)	(0)		

The d_x numbers provide a useful check on the successive l_x numbers, as each next l_x value on the life table should be equal to the previous one minus d_x from the previous interval. The l_x percentages may now be plotted against x, to yield a **survivorship curve** (Figure 15.19(a)).

A more useful way of showing survivorship than the arithmetic plot shown in Figure 15.19(a), however, is to plot the l_x values on a logarithmic scale (Figure 15.19(b)). As x remains on an arithmetic scale, such a plot is termed semilogarithmic (or 'semilog' for short). The advantage of this form of plot is that equal intervals on the l_x axis represent equal *proportionate* (rather than arithmetic) increases (e.g. 1, 10, 100, etc.), so any changes in the relative proportions of taxa becoming extinct show up as different slopes on the curve. Note how the steeper slopes for intervals 6 and 8 on Figure 15.19(b), for example, correspond with the relatively higher per taxon rates of extinction for the whole clade already shown on Figure 15.18.

Cohort survivorship curves are thus a useful means for detecting changes in extinction rates over time within a single sample of taxa, and have been much exploited in the study of mass extinctions (Section 15.4).

Dynamic survivorship monitors the relative decay of taxa drawn from many different time intervals as a function of taxon duration. The time-scale used is therefore time elapsed since the origin of each taxon — 'survivorship time' — rather than the real time-scale used in cohort survivorship. This is essentially the

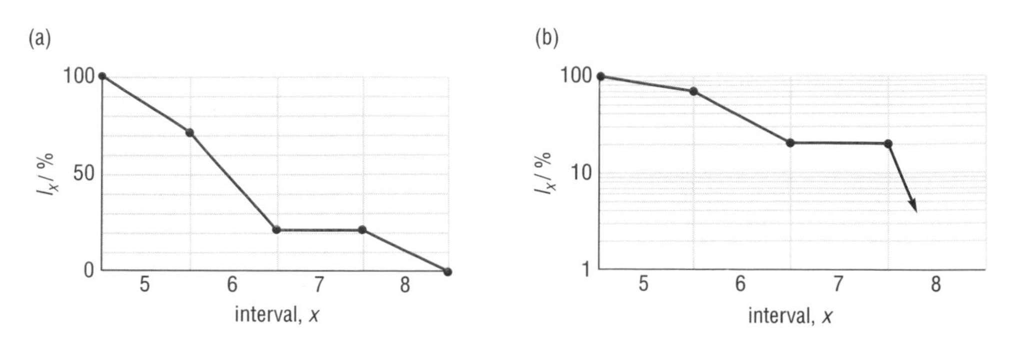

Figure 15.19 *(a) Cohort survivorship curve (arithmetic plot), over successive 10Ma intervals, for the 14 ammonite families entering interval 5 on Figure 15.17. (b) Semilog plot of the same data as in (a).*

method used by life insurance companies to assess the average risk of death for people of given ages. Imagine, for example, that the ages at death of 100 people, born and dying at different times, were recorded. Again, a life table (Table 15.2) can be drawn up showing, for successive individual age intervals (e.g. 0–19.9 years, 20–39.9 years, 40–59.9 years etc.), the numbers who died within each interval (d_x) and the numbers of those surviving into each interval (l_x).

Table 15.2 *Life table for a hypothetical sample of 100 people, born and dying at different times, for successive 20-year age intervals.*

Age intervals (x)	Survivors into interval (l_x)	Deaths (d_x)
0 –19.9 years	100	3
20–39.9 years	97	2
40–59.9 years	95	35
60–79.9 years	60	50
80–99.9 years	10	10
	(0)	(100)

Plotted as a semilog survivorship curve (Figure 15.20), these data show the expected pattern of old age being accompanied by an increasing probability of death (reflected by the steepening slope of the curve to the right, giving it a convex shape).

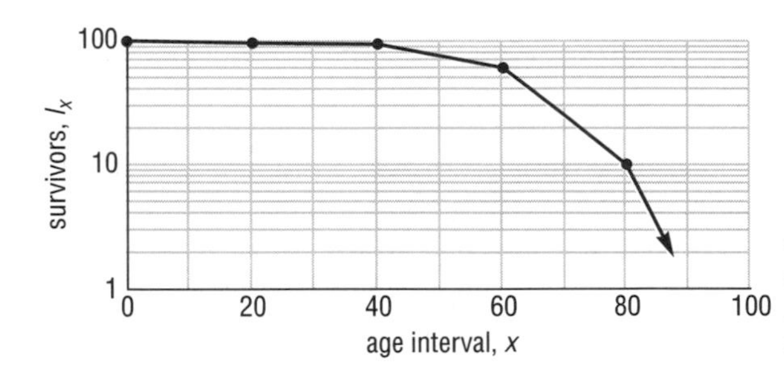

Figure 15.20 *Dynamic survivorship curve (semilog plot) for an hypothetical sample of 100 people born and dying at different times. Data from Table 15.2.*

When applied to the stratigraphical ranges of fossil taxa, dynamic survivorship analysis shows some intriguing patterns, which have been a source of much debate in recent years.

▶ Measure the durations (in Ma) of all the ammonite families shown in Figure 15.17, except the first seven at the bottom left of the chart (the ranges of which are incomplete, extending back into the Jurassic): you should thus measure 39 families in all. Then prepare a life table for them, like that shown in Table 15.2, distributing the family durations into 'age' (stratigraphical range) intervals of 10 Ma each (0–9.9 Ma, 10–19.9 Ma, 20–29.9 Ma, etc.).

Your life table for the ammonite families should have looked something like Table 15.3. A corresponding survivorship curve is shown in Figure 15.21(a).

Table 15.3 Dynamic survivorship of Cretaceous ammonite families, based on Figure 15.17.

Age interval (x)/Ma	Survivors into interval (l_x)	Extinctions (d_x)
0–9.9	39	10*
10–19.9	29	14
20–29.9	15	8
30–39.9	7	4
40–49.9	3	2
50–59.9	1	1
	(0)	(39)

*i.e. ten families have ranges between 0 and 9.9 Ma, and so on.

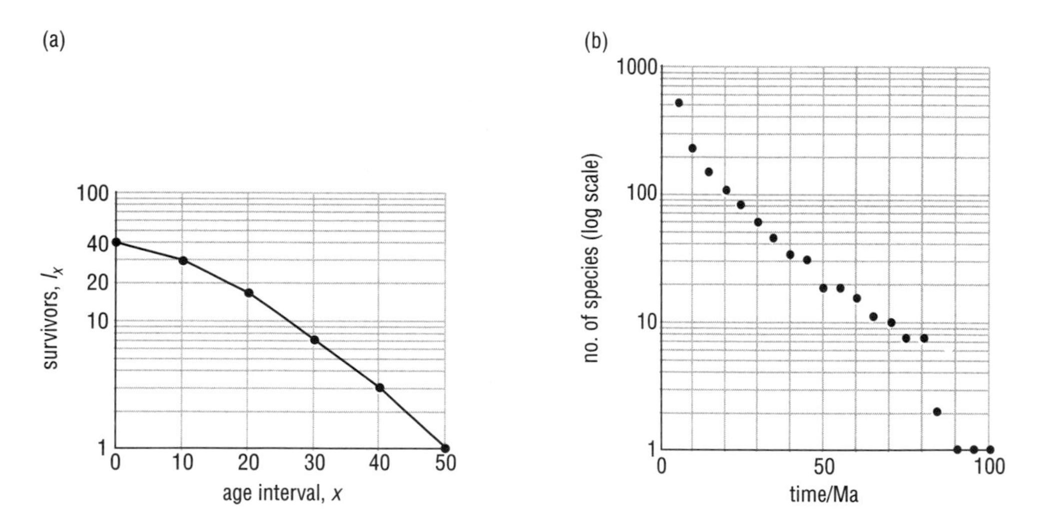

Figure 15.21 *(a) Dynamic survivorship curve (semilog plot) for the ammonite families, of which the complete ranges are shown in Figure 15.17. (Data from Table 15.3). (b) Dynamic survivorship curve (semilog plot) for extinct species of dinoflagellate cysts, assembled by Van Valen (1973).*

The form of the survivorship curve contrasts rather markedly with that shown in Figure 15.20. Though still slightly curved, it is more nearly straight than that for the hypothetical sample of human individuals.

▶ What would a completely straight descending line imply about the relationship between the probability of extinction and taxon duration?

It would imply that, *on average*, the probability of extinction did not change with taxon duration — i.e. that no matter how long a taxon had already persisted, it had the same average probability of becoming extinct as any other taxon. Hence, the line descends with a constant slope, denoting the same proportional decay for each age interval.

Van Valen (1973) plotted dynamic survivorship curves for taxa of several different groups of organisms, and concluded that most showed an approximately linear (semilog) pattern, though with some deviation at each end of the line which he attributed to various sampling problems. One of his examples, for extinct species of dinoflagellate cysts (the hard outer coverings of a group of protists), is illustrated in Figure 15.21(b). Different slopes appeared to characterize different major groups of organisms, however, and plots for higher taxonomic categories (e.g. families) showed gentler slopes than those for lower taxa (e.g. species). The latter pattern is readily explained by the more inclusive and hence longer-lasting nature of higher taxa (a genus does not become extinct until its last species goes, and a family persists until the demise of the last species of its last genus, and so on). The differences in slope between groups at similar taxonomic levels, on the other hand, were taken to reflect differences in their natural proneness to extinction. Van Valen thus enunciated a 'law of constant extinction rates': for any ecologically homogeneous group of organisms (i.e. members of a single adaptive zone), the probability of extinction is approximately constant, varying only randomly around some characteristic value. The intended implication was that, as noted above, the chances of a taxon becoming extinct at any given time were unrelated to its prior duration. Hence species could not be considered analogous in this sense to individuals: they do not degenerate towards extinction with age like the latter, nor indeed do they show the opposite effect of increased resistance to extinction (through, say, improved adaptation).

To explain this 'law', Van Valen further proposed what he termed '**the Red Queen's hypothesis**': no matter how well members of a species are adapted at any one time, their environment can be said to be constantly deteriorating with respect to their adaptations. Adaptive responses to one selective pressure may be deleterious to adaptations to other, less immediate pressures, so that mean fitness declines whenever the balance of pressures changes. Indeed, the balance of selective pressures is constantly varying, not least because of the continuing adaptive responses of different species to each other: any temporary advantage gained by one species may correspond to a worsening of the environment experienced by the others. Thus, mean fitness in a species can be expected to fluctuate randomly with time (Chapter 4, Section 4.5.2). The risk of extinction is therefore ever-present (and is similar for all species in the adaptive zone). As noted in Chapter 4, the hypothesis derives its name from the Red Queen's remark to Alice in *Through the Looking Glass*: 'Now here, you see, it takes all the running you can do, to keep in the same place'.

Van Valen's conclusion provoked considerable discussion, and much criticism. At an early stage it was pointed out that even if the law of constant extinction did obtain for species, it would be unlikely to apply as well to higher taxa. The longer any given higher taxon survives, the more species it can be expected to bud off, so that it should acquire greater resistance to extinction simply through having more species as a buffer against total extinction, so to speak. Therefore, higher taxa should show decreased probabilities of extinction with respect to duration.

▶ What shape of survivorship curve (on a semilog plot) would then be expected for higher taxa?

As the probability of extinction decreased with taxon age, so the curve should decrease in slope, yielding a concave curve. There is some evidence that this is so, but exceptions are common (e.g. Figure 15.21(a)): bias introduced by mass extinctions (which seem prematurely to terminate many longer-ranging higher taxa) is one explanation for such exceptions.

A much more damaging criticism, however, has recently been marshalled by Paul Pearson (1992). He drew attention to earlier complaints that the law of constant extinction rates was a fallacy based upon the method of plotting. Although many of the published survivorship plots are indeed acceptably straight (by various statistical criteria), it is important to remember that the time-scale used is that of 'survivorship time', i.e. time from origination of the taxa drawn from different stratigraphical intervals, not real time. Hence, the law of constant extinction rates strictly applies only to survivorship time. However, because the latter amalgamates extinctions from many different intervals, it artificially averages out the variations of extinction rate in real time. It is therefore not true to say that rates of extinction have varied only randomly around some constant value in real time. Indeed, cohort survivorship analysis (as discussed earlier) suggests that extinction rates have fluctuated in a decidedly non-random way through time (see also Section 15.4). So Pearson went on to argue that, because the taxa in such a plot had been drawn from different times with different rates of extinction (and thus, probably, environmental conditions), they could not be said to constitute an 'ecologically homogeneous group': the comparisons are not of like with like. It is necessary to apply a correction factor to each taxon duration, to allow for the real-time fluctuations in extinction rate, before dynamic survivorship analysis can be applied. Pearson proposed a relatively simple correction factor, to yield a *corrected survivorship score (CSS)*, which is derived as follows:

$$\mathrm{CSS} = L \times r_e\,(\mathrm{extant})/r_e\,(\mathrm{total}) \qquad (15.5)$$

where L is the measured duration (stratigraphical range) of a taxon; r_e (extant) is the average per taxon rate of extinction in the group being investigated, *during the duration of the taxon in question*; and r_e (total) is the average per taxon rate of extinction for the group during the entire time interval under consideration. The efficacy of this method of correction was confirmed by analysing artificial data sets with known patterns of extinction.

Applying the correction factor to the ranges of species of planktonic foraminifera of early Tertiary age, which had previously been cited as exemplifying the law of constant extinction rates, Pearson found that the larger part of the survivorship curve for their CSS values took on a markedly convex form (Figure 15.22).

▶ What is the implication of this finding for the relationship between species duration and the probability of extinction?

A convex survivorship curve, with the slope steepening downwards to the right, suggests that the probability of extinction increased as species aged. However, Pearson was also at pains to point out that this finding may itself be due to an artefact of taxonomic practice. At the beginning of this Section, it was noted that

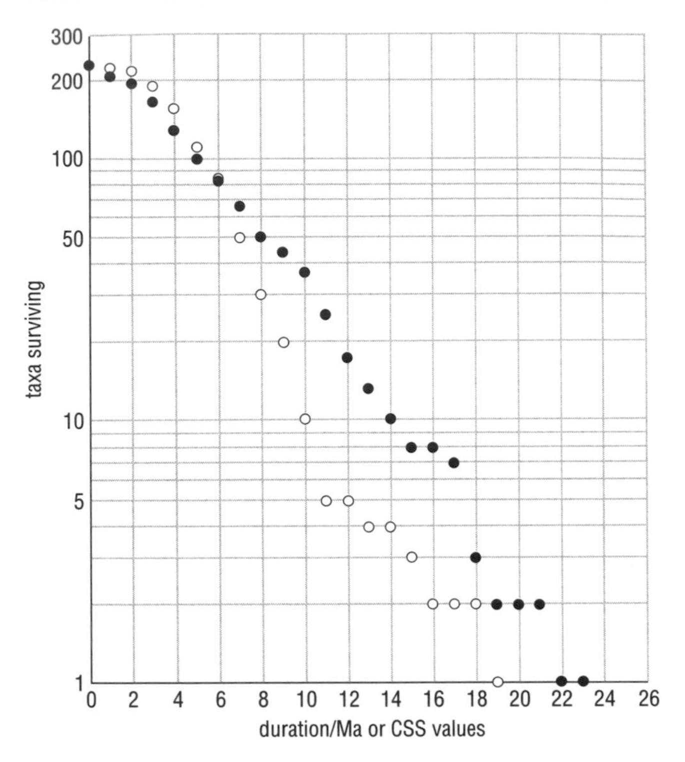

Figure 15.22 *Dynamic survivorship curves (semilog plot) for Palaeogene species of planktonic foraminifera. Black points show uncorrected data and white points show the same data corrected for variations in extinction rate according to the method of Pearson (1992), yielding CSS values.*

chronospecies are theoretically irrelevant to the analysis of turnover within clades, but that it is not always possible to exclude them from consideration for practical geological reasons.

▶ If chronospecies were mistakenly treated as discrete species lineages, what bias might thereby be imposed upon the *apparent* relation between species duration and probability of 'extinction'?

Because of sustained microevolutionary change within lineages, the evolution of new chronospecies (with pseudoextinction of the ancestors, by definition) becomes increasingly probable as lineages age. If pseudoextinctions are not distinguished from true lineage extinctions, then the *apparent* probability of extinction will tend to increase over survivorship time. How widespread this problem is remains to be seen. At the time of writing (1992), it is not possible to give any clear cut conclusions from dynamic survivorship analysis of the fossil record: the results of all earlier studies need to be revised using CSS values. Nevertheless, the method holds considerable promise for revealing the natural patterns of taxonomic turnover in clades, and is likely to be widely employed (in its revised form) in future years.

SUMMARY OF SECTION 15.3

Clade dynamics are determined by the interaction of per taxon rates of origination and extinction of taxa. A null hypothesis can be erected, with probabilistically equal rates of origination and extinction, which allows random or stochastic phylogenies to be generated. The record of higher taxa can then be compared with the stochastic phylogenies to establish which aspect of the real patterns are worthy of deterministic explanation. Mass extinctions and radiations are two such patterns encountered in the real record.

The relative values of mean rates of origination and extinction can be determined from the average shapes of diversity spindles for higher taxa: symmetrical spindles imply an equality of mean rates, while bottom-heavy spindles characterize episodes when originations exceeded extinctions, as with the Cambrian to Ordovician marine fauna. One model to explain the apparent equilibration of rates invokes diversity-dependent decline in the rates of origination, and increase in the rates of extinction, yielding a logistic pattern of diversification.

There is much debate about what sort of higher taxa should be used in analyses of clade dynamics. Cladists argue that only monophyletic taxa can be considered natural, and so only these should be used. However, this is frequently impracticable (because of taxonomic uncertainties) and is also debatable from theoretical considerations. First, some ancestral, or stem group taxa would have to be arbitrarily excluded from analysis if monophyletic taxa alone were compared. Secondly, a case can be made that at least some paraphyletic taxa can be recognized as natural entities from an ecological perspective, in so far as their members all belong within some distinct adaptive zone.

For any given higher taxon, per taxon rates of origination and extinction can be derived from a range chart. It is important to consider the contribution of both rates to patterns of change in taxonomic diversity. Another approach to measuring extinction rates is to plot survivorship — the relative survival of a given sample of taxa over successive intervals. Cohort survivorship shows the decay of a sample drawn from a single starting time and decaying over real time. It identifies fluctuations in extinction rates and can be useful in the study of mass extinctions. Dynamic survivorship shows the pattern of decay of taxa drawn from different times, with respect to their own durations ('survivorship time'). Early studies suggested that the probability of extinction was independent of taxon age, a pattern for which 'the Red Queen's hypothesis' was proposed as an explanation. However, such plots are invalid without first correcting for variations in extinction rates in real time. Preliminary analyses of corrected data give results that conflict with the Red Queen's hypothesis, though an apparent pattern of increasing risk of extinction with taxon age amongst some organisms may be an artefact of confusing chronospecies with distinct lineages.

15.4 MASS EXTINCTIONS

Early in the 19th century it was recognized that, from time to time, a large number of groups disappeared from the fossil record more or less together, never to be found again in younger strata. It was partly for this reason that many of the boundaries between one geological period and another were erected. The most conspicuous and severe extinctions of all in the Phanerozoic — those at the end of the Permian and at the end of the Cretaceous — were used to mark the end of the Palaeozoic Era and the Mesozoic Era, respectively. Most of the extinctions were real, not merely pseudoextinctions. Ever since their recognition, 'mass extinctions' have exerted a grip on the imaginations of many, and challenged palaeontologists to document the evidence and account for the disappearances. Recently, one development has generated so much public interest and research activity that some now talk of the 'extinctions industry'. Research is proceeding at such a pace that barely a week goes by without publication of a paper discussing mass extinctions.

Mass extinctions can be defined loosely as geologically rapid, major reductions in organism diversity on a global scale. They are characterized by having substantial magnitude and global extent, broad taxonomic effect, and relatively short duration — although magnitude, extent, etc. are not specified. There is general agreement that *major* extinctions occurred in the late Precambrian, the late Cambrian, the late Ordovician, the late Devonian, the late Permian, the late Triassic, the late Cretaceous, the Eocene–Oligocene, and the late Pleistocene (see Appendix 1). The most severe in the Phanerozoic — the 'Big Five' — are indisputably mass extinctions: i.e. those at, or near, the end of the Ordovician, Devonian, Permian, Triassic and Cretaceous. The late Devonian mass extinction is often known as the Frasnian–Famennian extinction, coming as it does mainly between the last two stages of the Devonian: the Frasnian (pronounced 'Frarnian') and the Famennian Stages.

The 'Big Five' Phanerozoic extinctions, and their principal casualties, are summarized below:

Late Ordovician. Major groups of the following: trilobites, brachiopods, graptolites, echinoderms, conodonts (fish-like chordates represented by tiny teeth), corals, and bryozoans.

Late Devonian (Frasnian–Famennian). Extensive communities of low-latitude reef-dwelling organisms collapsed. Loss of 21% of all marine families, especially from the following groups: corals, brachiopods, bivalves, sponges, fish, phytoplankton.

Late Permian. Loss of 57% of marine families, especially from low latitudes. Reefs were eliminated. Higher taxa that became extinct include rugose corals and tabulate corals, eurypterids (water-scorpions), trilobites. Major losses within the following groups: foraminifera, crinoids, bryozoans, brachiopods, bivalves, gastropods, ammonoids, ostracods, fish. Loss of 27 tetrapod families, especially of amphibians and mammal-like reptiles (though these extinctions probably spanned a considerable time in the Permian).

Late Triassic. Major losses within the following groups: cephalopods (58 families became extinct, especially ammonoids, with only one family of ammonoids surviving into the Jurassic), gastropods, brachiopods, bivalves, sponges, various marine reptiles. Conodonts became extinct. On land, most genera of mammal-like reptiles (therapsids), thecodont reptiles (see Figure 11.1), and large amphibians, were lost, as were many insect families. The late Triassic extinctions seem to have been stepped, possibly in three closely successive events.

Late Cretaceous. Groups becoming extinct include: ammonites, belemnites (cephalopods), rudist and inoceramid bivalves, plesiosaurs, ichthyosaurs, mosasaurs, pterosaurs, dinosaurs. Groups suffering major losses include: planktonic forams, calcareous nanoplankton, especially coccoliths and radiolarians, brachiopods, gastropods, echinoids, bryozoans, sponges, gymnosperms (a group of plants including cycads). Groups apparently little affected include: corals, amphibians, crocodiles, snakes, turtles, birds, placental mammals (marsupial mammals suffered more losses). Flowering plants, including hardwood trees, suffered substantial losses, especially in the Northern Hemisphere, although changes to terrestrial floras in the Southern Hemisphere were more gradual. The abbreviation 'K–T' is often used when referring to this extinction ('K' is the international symbol for the Cretaceous; 'T' stands for Tertiary).

Since the late Cretaceous, there have been several less severe extinctions, similar in intensity to many other minor extinction episodes of the Palaeozoic and Mesozoic. The Eocene–Oligocene extinctions were the last large-scale events to have affected organic diversity in both terrestrial and marine environments. The extinctions were apparently stepped, with diversity declining in a series of about five pulsed events spanning *c.* 10 Ma. Relatively few higher taxa disappeared but many genera and species were lost, especially gastropods and bivalves, calcareous microplankton, and mammals. The late Pleistocene extinctions have already been discussed in Chapter 13, and Recent extinctions are described in Chapter 17. The best known of the late Pleistocene extinctions (mainly between 15 000 and 10 000 years before present) were of large mammals and large flightless birds, and there is considerable debate as to the relative extent to which human hunting activities and/or climate changes were responsible. The late Pleistocene was thus *not* a time of mass extinction, although some would argue that if people continue to deplete the globe of species at the current rate, this general time in Earth history will appear as a mass extinction in the geological record.

Such a compilation as the one above can give a misleading impression of abrupt loss of a flourishing group, whereas that was not always the case. For example, the decline of trilobites and tabulate corals was very much underway before their eventual demise in the late Permian.

Many of the extinction statistics referred to in this Section derive from a remarkable database of the stratigraphical ranges of over 32 000 marine animal genera (and their families), assembled by Sepkoski. This database is of great use in analyses of both extinction and origination.

Estimates of the percentage extinction of marine genera and species for the 'Big Five' extinctions are as follows, with genera given first, then species: late Ordovician (61;85); late Devonian (55;82); late Permian (84;96); late Triassic (47;76); late Cretaceous (47;76). The estimates for species loss are upper estimates, and are likely to be highly inaccurate; they are based on the taxonomic structure of living organisms and assume, for instance, that the species-to-family ratios have not changed significantly over the Phanerozoic, and that species became extinct at random, regardless of which higher taxon they belonged to. The estimates for genera are also subject to considerable error, but are somewhat more reliable, being based on empirical counting, as opposed to extrapolation. Despite the associated errors, such estimates probably give a reasonable idea of the intensity of mass extinctions.

▶ Why is much of our understanding of extinction patterns based on animals rather than plants, and, among animals, those that lived in the sea?

Simply because the vast majority of fossils are of marine animals (Chapter 12, Section 12.2).

▶ Are estimates of the proportions of taxa that become extinct within a higher taxon likely to be more accurate for lower taxa or for higher taxa?

Higher taxa (see Chapter 11, Section 11.1.1). For example, only one species is necessary to demonstrate the continued existence of a higher taxon, whereas every species must be found to confirm that no species became extinct.

▶ Why, on average, is it more likely that a genus would become extinct than a family?

Again, this is because of the hierarchical nature of taxa, where the higher taxonomic units usually accommodate more than one lower unit (Section 15.3.2). Thus, a family normally encompasses several genera, and all these genera would have to become extinct before the family itself would. Patterns of extinction often differ for different taxonomic levels, depending on the severity of the crisis: minor extinction events are less likely to eliminate families than are mass extinctions.

One way of revealing times of elevated extinction rates is to plot cohort survivorship patterns, as was done in Section 15.3.2 for ammonites. Figure 15.23 shows 77 successive cohort survivorship curves for a total of 2316 *extinct* marine families during the Phanerozoic. Each curve traces the survivorship of families that were extant during a particular geological stage (i.e. the cohort). There is a new cohort for each geological stage. The same family often belongs to several cohorts because it is included in the cohort for any stage in which it is present. Thus, the extinction of a single family is often reflected in more than one survivorship curve, which tends to exaggerate the impression of intensity of mass extinction episodes. Steep portions of the curve represent intervals of elevated extinction rates. Despite the limitations of this somewhat crude method, it is clear that, as far as marine families are concerned, particularly severe extinctions occurred at the end of the Permian, the end of the Cretaceous, and the Eocene–Oligocene, with others such as the late Ordovician, late Devonian and late Triassic also showing up. The apparent mass extinction at the end of the Tertiary is an artefact. In order to avoid distortions due to 'the pull of the Recent' (Chapter 12, Section 12.4.2), only marine families that had become extinct by the end of the Tertiary have been plotted.

Are mass extinctions discrete phenomena? Two very important questions concerning mass extinctions are: (1) Are mass extinctions the end of a smooth spectrum of extinction intensity, or are they quantitatively quite different, i.e. discrete phenomena? (2) If mass extinctions are discrete phenomena separated in intensity from other 'background' extinctions, do they demand qualitatively different explanations? Again these questions cannot yet be answered with certainty, but strong clues have recently emerged, especially concerning whether mass extinctions are discrete phenomena.

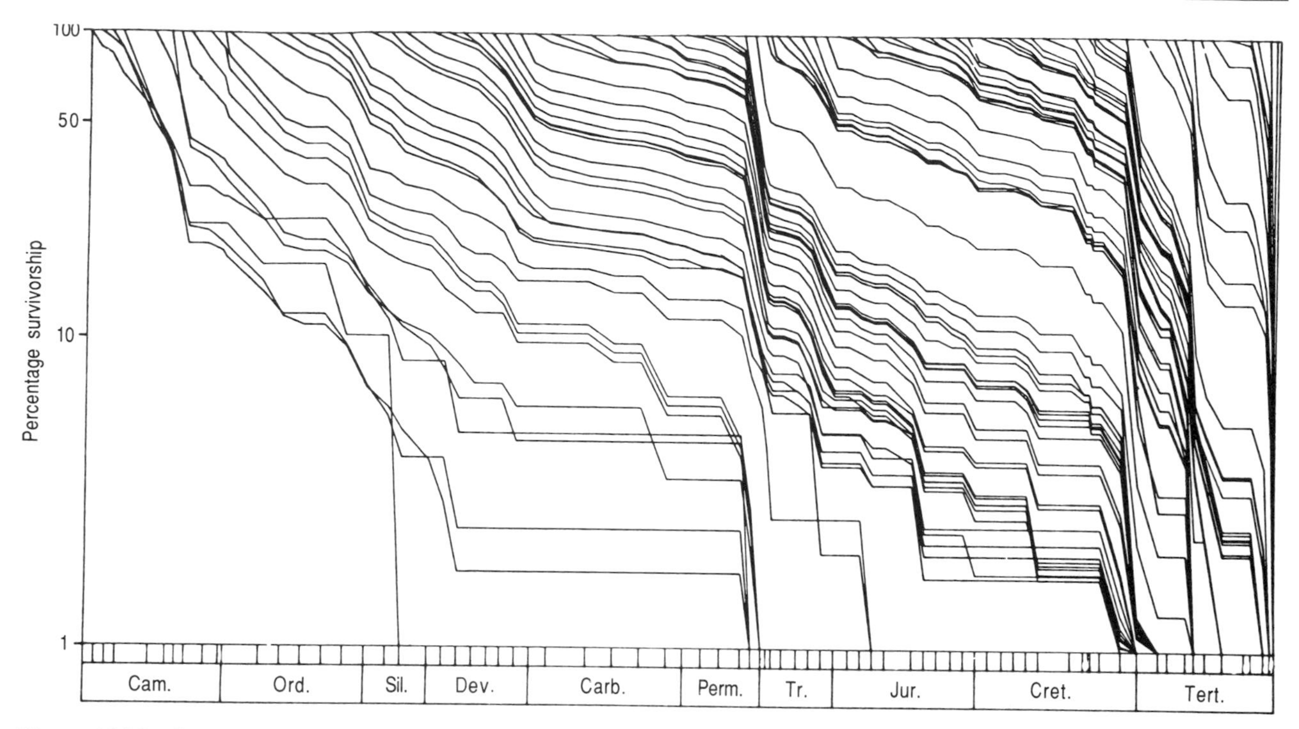

Figure 15.23 *Successive cohort survivorship curves for 2316 extinct families during the Phanerozoic.*

Figure 15.24 (from Raup, 1991a) is a histogram showing variations in intensity of extinction during the Phanerozoic. It plots the percentage extinction for fossil marine genera in each of 106 time intervals (geological stages, or parts of stages, of varying duration). Data for the five largest extinctions fit, as expected, towards the right. (Some mass extinctions span more than one of the time intervals used in the analysis, with the result that there appear to be some differences with the percentage extinctions of genera given earlier.)

▶ On the evidence of Figure 15.24, does it seem likely that mass extinctions are set discretely apart from low-intensity, background extinctions or do they grade smoothly into them?

The histogram suggests that mass extinctions are located towards one end of a fairly smooth distribution of extinction intensity.

Raup (1991b) analysed the stratigraphical ranges of 17 621 marine genera from Sepkoski's database and, after making a number of interpolations to derive data for species, found that the *average* extinction rate for a Phanerozoic species was 0.25/Ma. The average species duration (the reciprocal of species extinction rate) was therefore 4 Ma, a result that falls close to, or well within, other estimates of the mean duration of marine animal species. These findings mean that if there were a million species in existence at a given time, the extinction rate would be equivalent, on average, to one species dying out every four years.

Raup found further evidence to support a continuum of extinction intensities from 'background' rates through to mass extinctions. He developed the idea of a 'kill

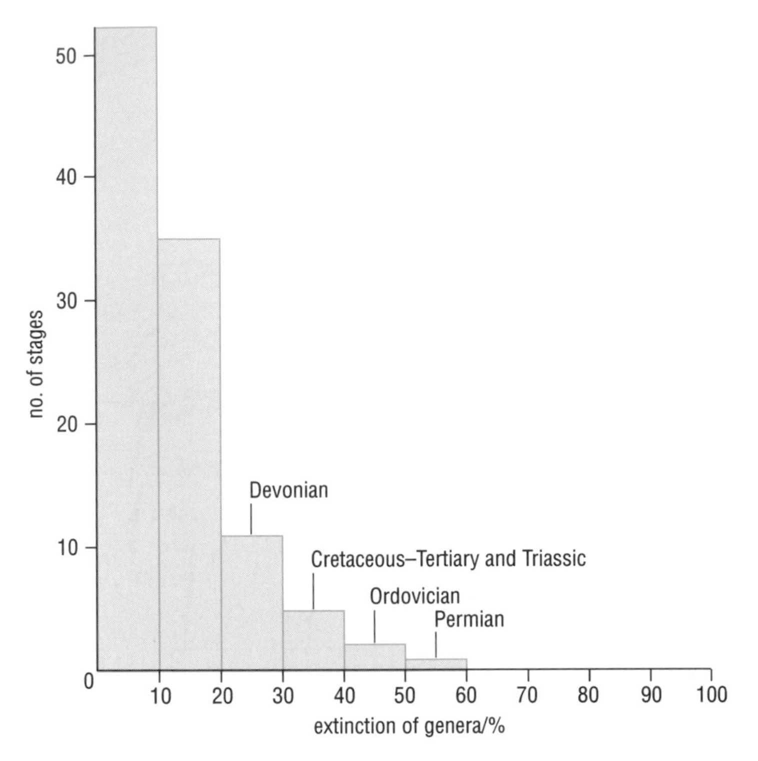

Figure 15.24 *Histogram of percentage extinction for fossil marine genera in 106 geological time intervals (stages or parts of stages). The 'Big-Five' mass extinctions are labelled.*

curve' which depicts the average level of species extinction for a series of waiting times. This method is like the approach hydrologists adopt to deal with flood prediction, and indeed it can be used for any other phenomenon where events become more rare as they become larger, such as earthquakes and storms. Hydrologists, for example, assemble all available records for the flow rate for a river so that intensity can be expressed as that flood level which is equalled or exceeded (on average) every so many years. Thus, the waiting (or return) time is the time interval one can expect, on average, to wait for the return of a given flood level. For example, the 300-year flood is the flood level that is equalled or exceeded every 300 years, on average. Methods to estimate the frequency of high-magnitude events having waiting times longer than the available record have been developed, but these predictions are associated with much greater errors.

Raup's 'kill curve', the best-fit curve based on available evidence to date, is given in Figure 15.25. It shows the average waiting time — the spacing — for events of varying extinction intensity. For example, it suggests that a 30% extinction of species occurs, on average, every 10 Ma. It must be emphasized that although the 'kill curve' was modelled to match the Phanerozoic record as well as possible, it necessarily involves many assumptions and approximations, which, although regarded as reasonable, may or may not be valid. The analysis did not address the *duration* of extinction events. For the analysis, the time interval within which the extinctions occur was made arbitrarily short (10 000 years), but the use of this sampling interval does not imply that mass extinctions occur over such a short interval, but neither is this precluded by the results. Similarly, the kill curve says nothing directly about the mechanisms of extinction, nor how many mechanisms were operating at any one time, nor whether extinctions are regularly periodic. The curve can, however, be used to constrain choices of extinction mechanisms in that the action of any cause or causes of extinction must be distributed in time in a manner compatible with the curve.

▶ From Figure 15.25, approximately how often can an event be expected that would 'kill': (a) 65% of species; (b) 50% of species; (c) 5% of species?

(a) 100 Ma; (b) about 40 Ma, i.e just over half-way between 10^7 and 10^8 years (note the log scale); (c) about 1 Ma.

Raup's findings help to explain why palaeontologists have such difficulty in defining the term 'mass extinction'. There were no major discontinuities within the curve. He writes: 'One should not expect to find a clear demarcation between high- and low-intensity events, any more than a seismologist expects to recognise a clear boundary between large and small earthquakes, or a meteorologist expects an obvious boundary between a hurricane and a severe storm'. It seems that we will not be able to define mass extinction except by agreeing on an arbitrary cut-off, such as when at least 65% of species become extinct within a period of less than five million years. But why is it that palaeontologists gain a strong intuitive impression of two discrete kinds of extinction: mass extinction and background extinction? Raup (1991a) argues that 'to the extent that this perception is real, it is probably due to the steepness of the middle region of the kill curve. In a typical 100 Ma interval of geological time, there may be one or two large events, a few

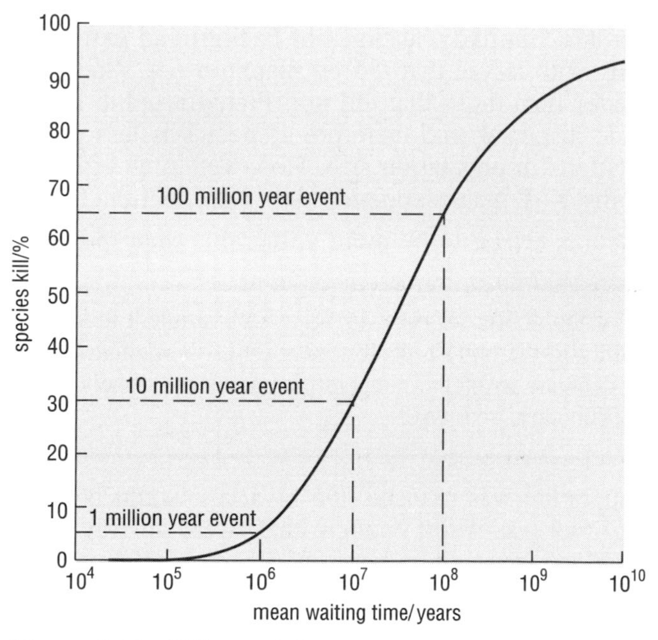

Figure 15.25 *'Kill curve' showing the average waiting time for extinction events of varying intensity, based on evidence from Phanerozoic marine organisms.*

small but identifiable events, and hundreds or thousands of extinctions too small to distinguish stratigraphically. Thus with the relatively small [and patchy] sample available (the Phanerozoic record), one gets an impression of a choppier distribution of the risk of extinction than actually exists'.

The selectivity of mass extinctions In general, mass extinctions cut across many ecological lines, pervading many habitats and their ecosystems. However, all mass extinctions exhibit some selectivity, in that certain groups remain unaffected, or virtually unaffected, whilst others may be decimated or completely eliminated.

Several important, though tentative generalizations seem to apply to a number of mass extinctions:

1 Among marine invertebrates at least, widespread genera preferentially tend to survive mass extinctions, whereas geographically restricted genera are particularly vulnerable. In other words, wide geographical distribution of the different species within a genus may increase the chances of survival of that genus during mass extinctions *regardless* of the geographical range of individuals within constituent species. This may be an example of a feature that emerges only at the level of taxa, rather than populations and individuals. Selection during mass extinction may be expressed at higher hierarchical levels than selection during times of 'background' extinction. Such concepts — involving the emergent properties of species and higher taxa — will be discussed further in Section 15.7.

Another attribute emerging at the level of entire taxa is species diversity, i.e. the number of species within the higher taxon. However, although high species diversity seems to have contributed to genus survival at 'background' times, it did not necessarily ensure survival during mass extinctions: for example, for late Cretaceous bivalves, gastropods, and echinoderms, and for late Cambrian trilobites, high species diversity was not enough to get some genera over those mass extinction hurdles. Similarly, at times of background extinction, species of bivalves and gastropods with larvae that fed on plankton (i.e. planktotrophic larvae) seem to have fared better than those that did not. Planktotrophic larvae have a higher chance of wide dispersal, and therefore a species is less prone to extinction from chance fluctuations in population size. However, such effects made no difference to species at the K–T mass extinction (see also Section 15.7).

2 Tropical biotas appear to be more vulnerable than those of other latitudes.

▶ Imagine a group of Recent bivalves for which ambient temperature is a limiting factor; they can live only in the hottest shallow seas found immediately around the Equator. Would an increase or decrease in global temperature seem more likely to threaten these bivalve lineages with extinction, and why?

If global temperature was reduced, the bivalves would be more vulnerable, having nowhere to go that was warm enough; they were already in the hottest environment. If, however, there was an increase in temperature, the bivalves might well, over generations, be able to migrate north and south along the continental shelves away from the Equator to where the water was cooler, or perhaps migrate into deeper water on the Equator. In general, as climate changes, so does the distribution of organisms, with shifts towards the Equator during phases of cooling, and shifts towards the poles when the climate warms up. Extinctions can be expected when global climatic belts shrink or disappear, and regional extinctions are likely to occur when organisms cannot migrate with their favoured environment because of geographical barriers. Climatic cooling is, in fact, strongly implicated as a major factor in a number of regional extinctions, such as the Pliocene loss of more than 50% of north-eastern Atlantic and of 75% of north-western Atlantic bivalve species.

It is for this reason — the 'nowhere else to go' hypothesis — that tropical biotas may be particularly vulnerable to extinction at times of global cooling. The data on tropical forms, however, tend to be dominated by marine species of reef communities, which seem particularly susceptible at times of mass extinctions; because of this, and the lack of fossil data on tropical terrestrial organisms, it is not certain whether tropical forms in general are especially vulnerable compared with forms at higher latitudes. The inverse effect should be expected near the poles during times of global warming: marine species adapted to the coldest temperatures would have nowhere else to go except perhaps into deeper water. Species diversities at very high latitudes are, however, generally very much lower than at the Equator, and are poorly known (Chapter 13, Section 13.2).

3 Large-bodied species seem to be more vulnerable than small-bodied forms, as discussed in Section 15.2.2.

As far as the K–T extinctions are concerned, there is also some evidence that detritus-feeders are less prone to extinction than other trophic groups, as are smaller, simpler planktonic forams compared with larger, more complex forms (Section 15.4.1). Terrestrial plants have sometimes been described as exempt from mass extinctions but this is true only at the highest taxonomic levels; in the K–T extinctions, as many as 30% of plant species may have been lost. Mechanisms of long dormancy may conceivably have been important if there was a sudden, very short-lived crisis; phytoplankton, for example, can produce and occupy a protective cyst to ride out hostile circumstances.

Establishing that extinction is selective as far as particular features are concerned is very difficult, not least because our knowledge of extinct organisms will always be limited. Organisms have a virtually unlimited number of characteristics that might confer resistance — or increase susceptibility — to extinction: anatomical, behavioural, physiological, geographical, ecological, etc. Even if we eventually find a feature that appears to unite either the survivors or the victims, but not both, the correlation may be totally spurious as far as cause and effect are concerned. (However, corroboration by multiple correlations is the only way we can test such historical hypotheses; see Section 15.1.1.)

The time-span of mass extinctions Much controversy surrounds the time-span over which the various mass extinctions occur; detailed information on this would help to constrain theories concerning causes. With the possible exception of the K–T event (see below), it seems likely that mass extinctions were not geologically instantaneous but were composite or stepped events that encompassed several million years. However, for the pre-Pleistocene geological record, even events lasting, say, a day cannot usually be distinguished from those spanning 10^5 years or even 10^6 years, and establishing that a very short-lived event is synchronous on a global scale is notoriously difficult (Chapter 10, Section 10.3.2).

Estimates of the duration of the Frasnian–Famennian (late Devonian) extinction have varied widely in the last few decades, from about 15 Ma to less than 0.5 Ma. Current evidence suggests an extended period of high extinction rates in the latter half of the Devonian Period over at least 3 Ma, with a peak at the end of the Frasnian. The Permian extinctions have long been thought to span about 5–10 Ma, but recent evidence suggests that they may have been much more abrupt. A major problem with unravelling events at the end of the Permian is that it was a time of major regression: waters retreated from large areas of continental shelves, with the result that very few shallow water sequences with fossils are preserved. Reef environments are unknown from the latest Permian and earliest Triassic, and few marine benthic and pelagic species are found at this time. Many taxa which are known from the late Permian and middle Triassic have no fossil record in the early Triassic. It therefore seems very possible that some taxa suspected of extinction at the end of the Permian actually persisted for a while into the Triassic, but in the absence of exposed strata from suitable environments, an accurate chronology of disappearances cannot be made.

15.4.1 THE CRETACEOUS–TERTIARY (K–T) MASS EXTINCTION

Although it was not the most severe, no extinction has attracted more attention than that which ended the Cretaceous Period 65 million years ago, and since 1980 no other subject in evolutionary palaeontology has stimulated so much interdisciplinary research. The trigger for renewed interest in events at the end of the Cretaceous was the discovery in Italy of a layer at precisely the K–T boundary containing an anomalously high concentration of the element iridium. In 1980, an American team proposed that this iridium-rich clay layer was the fallout from material ejected into the atmosphere by the impact of an asteroid roughly 10 km in diameter travelling at more than $10\,\mathrm{km\,s^{-1}}$ (Alvarez *et al.*, 1980). The Alvarez team suggested that the impact ejected materials into the atmosphere, darkened the skies, and led to suppression of photosynthesis, the break up of food chains and a rapid series of extinctions.

Subsequent searches have found unusually high levels of iridium at over 100 K–T sites throughout the world, especially from cores taken during ocean drilling programmes, but also from exposures on land. Iridium is very rare in the Earth's crust (averaging about 0.03 parts per billion, i.e. much rarer than gold), but is relatively abundant in extraterrestrial bodies. Small amounts of iridium from a continual rain of micrometeorites ('cosmic dust') were expected to be incorporated into deep sea sediments — such as those in Italy where the anomaly was first identified — but never to reach the concentrations present. The ratio of iridium to elements with similar chemical properties, such as platinum, osmium, and gold, is the same in the boundary layer as it is in some meteorites. To account for such a large concentration of iridium, the Alvarez team invoked the impact of a body large enough to be called an asteroid — one of numerous celestial bodies that move around the Sun mainly between the orbits of Mars and Jupiter.

The iridium is enriched (often to concentrations several hundred times its usual abundance) in a thin distinctive clay layer that often coincides with the extinction events, within the limits of stratigraphical resolution. Numerous factors can blur the picture, though, such as the vertical mixing of successive fossil populations caused by burrowing organisms. No technique can yet resolve the time taken for the boundary clay and its associated features to form; that far back in the geological record, a resolution of less than about 10 000–100 000 years is not yet possible. In other words, some events could, in theory, have occurred in less than a year or over perhaps as much as 100 000 years, and independent corroboration of synchroneity across the globe is, as yet, impossible.

Other evidence of impact includes small spheres and splash-shaped droplets of glassy material found in the boundary layer at some sites. The droplets are believed to have originated by rapid cooling of shock-melted rock ejected into the atmosphere by an impact, and the spherules may have condensed from vaporized rock. Strong support for an impact also comes from grains of quartz in the boundary layer that show symptoms of very high velocity shock. Such grains are otherwise found only in known impact craters, at nuclear test sites and in quartz subjected to extreme shock in laboratories. A form of quartz called stishovite, which is characteristic of pressures far higher than those that arise in volcanic eruptions, has been identified in the boundary layer. In 1991, tiny diamonds (*c.* 5×10^{-9}m) were found in the boundary clay at several sites and interpreted as the result of the shock metamorphism of carbon-containing material in the target rocks or in the asteroid itself. In addition, soot particles are abundant in some of the boundary layers, and have been interpreted as the result of extensive wildfires triggered by the heat from material re-entering the atmosphere after condensing from a vapour plume that was shot up through the atmosphere on impact. In many non-marine sections in North America, a sudden increase in abundance of a single type of fern spore immediately above the K–T boundary has been interpreted as being analogous to the early successional vegetation dominated by colonizing ferns that often follows mass kill of plants by volcanic eruptions (though in this case mass kill supposedly resulted from some physical consequence of an impact). Many species of hardwood trees and other flowering plants appear to have become extinct close to thc boundary, but the record of plant turnover in the Southern Hemisphere appears to have been less sudden, with a more gradual replacement of Cretaceous floras by those characteristic of the early Tertiary.

Impact events of greatest intensity are expected to be the least frequent, as with other natural phenomena such as earthquakes and floods. There is no doubt that the Earth receives occasional bombardment by large meteorites, as witnessed by numerous craters on the Earth's surface; there are at least 28 craters with diameters of over 10 km in rocks under 250 million years old, and five exceed a diameter of 50 km. The well-known Meteor Crater in Arizona is actually one of the smallest craters (about 1.2 km in diameter) and one of the youngest (*c.* 50 000 years old). Many more craters are sure to exist under the sea but are not yet identified. It is estimated that about two asteroids 10 km or more in diameter can be expected to strike the Earth every 100 Ma. By cosmic standards, asteroids often pass near to the Earth: on March 23, 1989, a fair-sized asteroid, perhaps as large as 0.8 km across, missed the Earth by only 800 000 km (*c.* two Earth–Moon distances). Several promising candidate sites for the K–T impact have been suggested, but so far all have had their Achilles' heels. Currently under intense investigation is a structure known as the Chicxulub Crater, 180 km in diameter, at the coastline of northern Yucatan, Mexico. The crater is now filled with Tertiary and Quaternary sediments and is not conspicuous. The location of the

Chicxulub Crater appears to be consistent with the size, distribution, and nature of shocked and glassy materials, and with a deposit of huge boulders at the K–T boundary in the Caribbean area interpreted by some as materials which were ejected during an impact or deposited by a tidal wave.

The originally simple hypothesis that ejected material darkened the skies and suppressed photosynthesis has been replaced by a variety of more complex schemes. The shift from description to causal dynamics reflects fast-accumulating data, but many scenarios remain entirely conjectural, with some more plausible than others. Shock heating of the atmosphere by the impact of an extraterrestrial object may have led to production of nitrogen oxides, and hence to nitric acid and acid rain, with the possibility of reduction in alkalinity of the oceans and dissolution of the shells of calcareous organisms in the resulting weak acid. Carbon and oxygen isotope data from rocks in southern Spain suggest a rise in sea surface temperatures of at least 8 °C. A greenhouse atmosphere might have been promoted by the release of trapped natural gas (mainly methane), following disturbance of submarine sediments, by the direct release of carbon dioxide gas through impact with limestones, and by the release of carbon dioxide into the atmosphere during worldwide weathering of limestones with acid rain. A short-term, opposite, 'antigreenhouse effect' could have occurred from emission of sulphur dioxide from sulphur-rich target rocks such as evaporites (sediments deposited from evaporated seawater). Some isotopic data suggest a relatively short-term *atmospheric* cooling prior to more prolonged warming, a so-called 'impact winter' similar to the 'nuclear winter' scenarios for the aftermath of all-out nuclear war. In addition, some fossil aquatic plants in a K–T boundary section in Wyoming have certain structural features that have been attributed to freezing, yet, by comparison with their close living relatives, the preserved reproductive stages of the plants suggest that freezing (if correctly recognized) probably took place around June. It must be stressed that, although this approach is a valid line of enquiry, current evidence for a mid-year impact is thin.

Some researchers argue that the physical evidence used to support an impact hypothesis is equally consistent with an episode of intense volcanism. They maintain, for instance, that certain types of eruption can produce large amounts of iridium and that Earth-driven iridium 'anomalies' may be much more common in the stratigraphic record than currently appreciated. More examples of sections from different parts of the geological record *lacking* iridium are certainly needed to establish the true rarity of iridium anomalies. However, the case for an impact at the K–T boundary is very strong. Variations on the theme of asteroid impact, each of which is supported by some evidence, albeit of variable strength, include the possibility of: multiple impacts; impact by a comet, causing destruction of the protective stratospheric ozone layer and poisoning by cyanide; impact somehow triggering massive eruptions (such as the enormously thick basalt lava flows that erupted in the Deccan plateau in India at this time); and an impact into the oceans, injecting large amounts of water vapour into the atmosphere, and causing a 'humid greenhouse'.

A few other well-defined iridium anomalies have been identified in the geological record, including one in the late Devonian, one in the late Triassic, and one in the late Eocene, but, as yet, attempts to find a consistent correlation between major extinctions and iridium anomalies have failed. Those who wish to support a widespread role of impacts in mass extinctions argue that some types of impacts, especially those of icy comets, need not be associated with iridium.

It is crucial to separate the evidence of an impact from the evidence for extinctions: correlation is not the same as cause and effect. It is quite feasible that an impact did indeed occur at the end of the Cretaceous Period, but that it had nothing to do with extinctions. Alternatively, an impact may have occurred, but it was only a *coup de grâce* a final blow, the biosphere being already perturbed and the majority of extinctions already over. At the time of writing, there is no consensus view.

Apparent vs. real extinction horizons Methods of sampling and the nature of the stratigraphical and fossil records can conspire to distort the history of extinction. All manner of spurious impressions of synchroneity or gradualness of pattern can be created by various combinations of biases. If ever there was a need for brilliant detective work, it is in establishing details about the nature and causes of mass extinctions!

A hiatus, i.e. a break in a stratigraphical succession caused by non-deposition or erosion, can create an impression of simultaneous extinction and origination where none existed. Figure 15.26 shows, on the left, the ranges of 11 numbered taxa that in reality crossed a time-plane without any two extinctions or originations coinciding; i.e. the pattern is one of relatively gradual extinction and origination. The stratigraphic record has a large hiatus in section A and a smaller hiatus in section B. Write in the appropriate number beside each taxon range in sections A and B; taxa 1 and 2 are labelled to start you off.

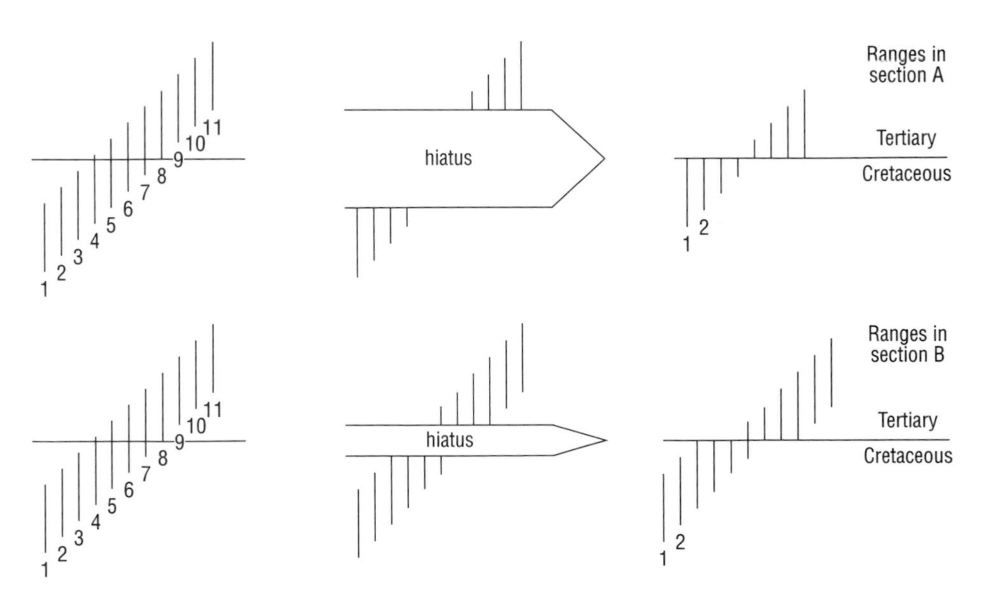

Figure 15.26 *Diagram showing how a hiatus in the stratigraphic record can affect the apparent pattern of extinction and origination. The ranges of 11 numbered taxa on the left cross a time-plane without any extinctions or originations coinciding. On the right are the apparent ranges in two sections, one with a larger hiatus (A), and another with a smaller hiatus (B).*

▶ How many more taxa appear to have become extinct simultaneously in A than in B, and which are the taxa concerned in each section (give their numbers)? How many more taxa appear to have originated simultaneously in A than in B, and, again, which are the taxa concerned?

One more taxon appears to have become extinct simultaneously in A than in B, i.e. four taxa in A (Nos. 1, 2, 3, 4) versus three taxa in B (Nos. 3, 4, 5). One more taxon appears to have originated simultaneously in A than in B, i.e. four taxa in A (Nos. 8, 9, 10, 11) versus three taxa in B (Nos. 7, 8, 9).

▶ How many taxa have no record in section A? Which are they?

Three taxa are missing in section A: numbers 5, 6 and 7.

In theory, *any* possible pattern of origination and extinction can be totally lost if there is no stratigraphic record, and therefore no fossil record. Within the time-span of a long hiatus, a large number of species could originate rapidly and become extinct simultaneously. Given the hiatus, we would have no indication of these events.

Many factors tend to create the impression that an extinction occurred earlier than it did, and 'backward smearing' of extinction records due to imperfect sampling is very common. The observed stratigraphical ranges in measured sections are always underestimates of true ranges. The principal problems are failure to collect specimens, misidentification, local periods of non-deposition or erosion, change of local conditions causing shifts in a species' geographical range, and failure of fossil preservation. In addition, as a species declines in population size approaching extinction, the chances of finding evidence of that species decreases.

▶ What effect does the conventional method of plotting range charts tend to have on the apparent ranges of taxa? (Recall Chapter 12, Section 12.4.2.)

Range charts tend artificially to lengthen the true ranges of taxa because ranges are systematically extended to stratigraphical boundaries (e.g. between zones, stages, etc.). Several taxa which became extinct at different times within a unit would, when plotted up, appear to show simultaneous extinction at the end of that unit. As the Maastrichtian stage lasted about 9 Ma, species plotted only to stage level might, in theory, show apparent synchronous extinction but have died out nearly 9 Ma apart! A spurious impression of synchronous extinction of individual species across large areas can arise from circular reasoning when correlating strata: stratigraphical levels are sometimes matched up using the extinction levels of particular species, in the absence of independent evidence. The importance of having many measured stratigraphical sections spanning extinction periods, with every taxon occurrence indicated (rather than simply extrapolated as a solid line through known occurrences), is obvious.

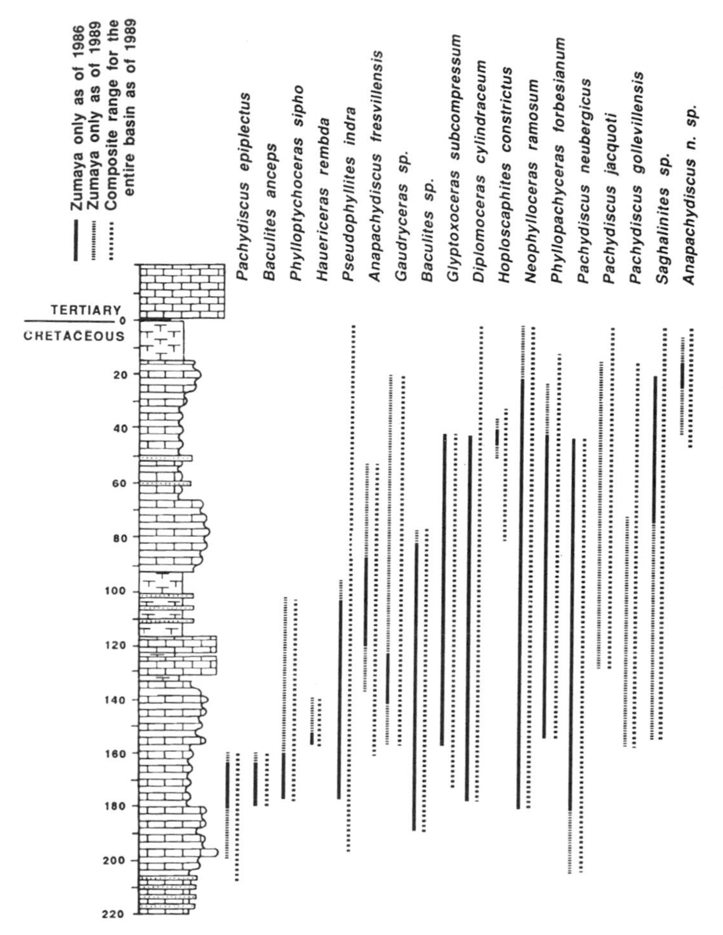

Figure 15.27 *Ranges of ammonite species near the K–T boundary in Zumaya, Spain, and composite ranges for the entire sedimentary basin. Stratigraphic thickness in metres.*

Data from K–T boundary sections Figure 15.27 shows data for one of the best known sections recording the extinction of the ammonites, which is at Zumaya in Spain. Peter Ward and colleagues at the University of Washington reported in 1986 that the ammonites died out gradually over a 5 Ma interval. After two field seasons of collecting, no ammonites had been found in the top 12 metres, and the ranges in the Zumaya section were as indicated by the solid lines in Figure 15.27. After three further field seasons, in 1989, the ranges were plotted again, this time as a closely dashed line. Composite ammonite ranges from the entire sedimentary basin of France and Spain, including the section at Zumaya, were then plotted with less closely dashed lines, slightly to the right of the ranges for Zumaya alone.

▶ What generally happened to the perception of the ammonite ranges between 1986 and 1989?

The ranges increased in length, with many species being found closer to the K–T boundary.

▶ Which species were identified in the Zumaya section for the first time between 1986 and 1989?

Pachydiscus jacquoti and *Pachydiscus gollevillensis.*

▶ What general effect does combining data from other sections in the sedimentary basin have on the ammonite ranges?

It extends many of the ranges still further, to the extent that six species range right up to the K–T boundary. Clearly, the more careful the searching for fossils in individual sections, and the wider the geographical area providing those sections, the closer the true (i.e. global) extinction level is approached. In this case, the new data strengthen the hypothesis of a relatively abrupt termination of ammonite lineages at the K–T boundary.

Study Figure 15.28, which plots the ranges of 50 brachiopod species in a K–T boundary sequence in Denmark.

▶ How many species become extinct within ± 1 m (see caption for scale) of the K–T boundary?

20 species (numbers 1 to 20).

▶ Which of these 20 species have ranges that terminate exactly at the boundary?

Species 4, 5, 6, 9, 16, 18, 19, 20.

▶ What is special about the ranges of species numbered 21 to 26?

All these species seem at first glance to have become extinct within ± 1 m of the boundary (as did species 1 to 20), yet they reappear much later in the sequence. Such species, which disappear for a while close to an extinction horizon and then reappear somewhat surprisingly, are known as *Lazarus taxa* (after the Biblical character, Lazarus, who was supposedly raised from the dead).

▶ Which 'Lazarus species' has the longest stratigraphical absence between apparent extinction and reappearance?

Species 21.

▶ Which of the 'Lazarus species' would most probably have been regarded as part of the K–T extinctions with less adequate sampling, i.e. which 'Lazarus species' seems to have reappeared for the shortest period?

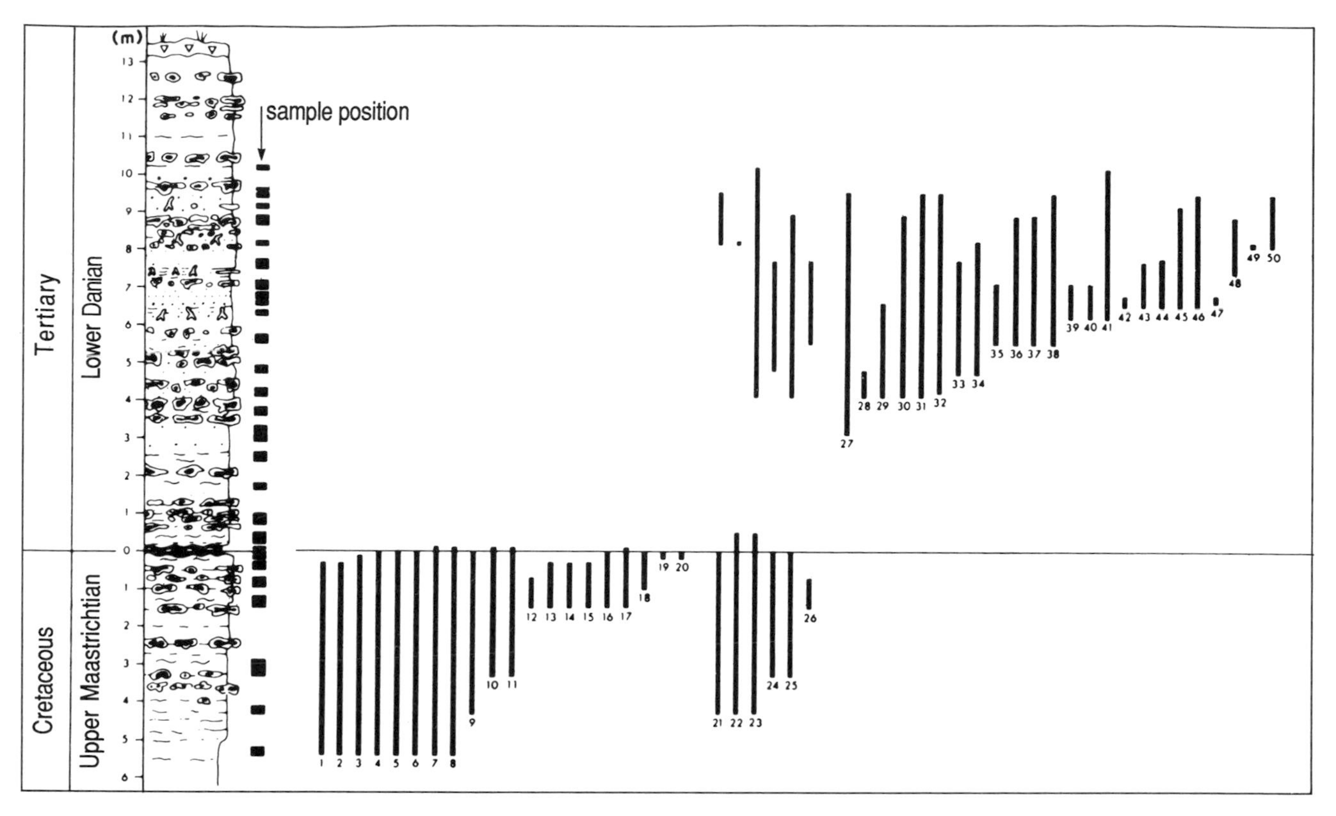

Figure 15.28 *Ranges of 50 brachiopod species over about 15 metres of a K–T boundary section in Denmark.*

Species 22.

In this Danish section, the boundary itself is a 3 cm thick clay, representing a cessation of chalk production as a result of sudden decline of the calcareous microplankton. Many brachiopod species specialized in various ways to the Chalk substrate suffered rapid extinction. The six 'Lazarus' brachiopod species in Figure 15.28 were, however, morphologically unspecialized. They are interpreted as environmentally tolerant and generalized forms that were able to survive in undiscovered 'refuges' somewhere on the sea-floor during whatever environmental disturbances were associated with the K–T extinctions. When chalk sedimentation was eventually resumed in the early Danian, albeit by the accumulation of a whole new plankton fauna and flora, radiation of surviving brachiopod groups led to the appearance of 23 new species.

▶ Imagine that a search for fossils is made in a 10 cm diameter core taken through oceanic sediments spanning the K–T extinction horizon. Would you generally expect very small organisms (e.g. forams or radiolarians) or larger organisms (such as fish) more clearly to reveal a true extinction horizon, and why?

In general, smaller organisms are far more numerous in ecosystems than large ones and are therefore much more likely to be present in sediment samples right up to a true extinction horizon. In addition, any large fossils are likely to be broken in a random core sample, making identification more difficult. As is usual for the stratigraphic record of the past 200 Ma, the pelagic record is more complete than the shallow marine record, and both of these are more complete than the terrestrial record. Excluding soft-bodied organisms for which the record is always the poorest, the sampling problem is most serious for large terrestrial vertebrates such as dinosaurs. The extent to which dinosaurs were already in decline before the K–T boundary is, as a consequence, particularly difficult to assess.

It must be emphasized that several groups were in certain decline long before the end of the Maastrichtian, such as rudists and inoceramids (bivalves), and belemnites (cephalopods). The evidence so far for dinosaurs strongly suggests a reduction in diversity several million years before the end of the Maastrichtian. The ichthyosaurs (marine reptiles) are not even known from the Maastrichtian. Such long-term declines mean that a simplistic all-embracing catastrophic scenario for late Cretaceous extinctions cannot be justified.

Planktonic microorganisms suffered the most intense casualties at or very close to the K–T boundary, with coccoliths (minute calcite shells of phytoplankton that accumulated on the late Cretaceous sea to form the Chalk ooze), radiolarians (minute zooplankton with a siliceous skeleton), and planktonic forams all reduced by at least 90% at the species level.

Some of the best records of the history of microorganisms across the K–T boundary come from relatively expanded marine sequences (with high sediment accumulation rates) exposed on land in Tunisia and Texas. Here the planktonic forams apparently underwent a 30%–45% loss of species during about 350 000 years *before* the K–T boundary. Further extinctions take place at the boundary itself (at least in Tunisia), with yet others spanning at least another 50 000 years. Thus, the general picture, here at least, seems to be a stepwise pattern of extinction staggered over perhaps 400 000 years. Complex, large, ornate forams disappear first, and smaller, less ornate forms survive longer, and indeed, only very small species are recorded in the earliest Tertiary. This sequential decline contrasts with the apparently much more abrupt record of planktonic foram extinctions from sections deposited in the deeper sea. There, however, an impression of abruptness may derive from the fact that the record is more condensed, with less scope for resolving the fine-scale pattern. There is no doubt, however, that the K–T boundary event initiated a dramatic and sustained environmental change, with more stable conditions resuming only several hundred thousand years later.

15.4.2 The causes of mass extinctions

In Chapter 9, Section 9.8, it was emphasized that extinction was a normal part of evolution, but that the relationship between the extinction of local populations and the global extinction of entire species was poorly understood. A similar lack of understanding surrounds the causes of most if not all mass extinctions. Given that, in the face of environmental changes, organisms normally respond (by natural selection) with new adaptations, one way of looking at the extinction problem is to try to answer the question: what kind of environmental changes (in the broadest sense) are so severe that large numbers of species prove unable to migrate away from or 'evolve out of' extinction?

Until recently, there was little hard empirical evidence either to corroborate or to contradict most hypotheses for the causes of mass extinctions, and few constraints to limit the imagination. Constipation, senility, stupidity, and less seriously even AIDS, have at times been invoked as causal agents in the extinction of dinosaurs. A variety of extraterrestrial mechanisms had been proposed from time to time long before the discovery of an iridium anomaly at the K–T boundary. They include: impacts of asteroids or comets causing a huge range of environmental disturbances, especially in the atmosphere and oceans; intense cosmic radiation from exploding supernovae; and increase in cosmic radiation reaching Earth due to a sudden weakening of the Earth's magnetic field, which normally acts as a protective shield.

The Earth-bound causes most often proposed for mass extinctions include: climatic change, especially cooling and drying; sea-level rise or fall; intense volcanic activity; changing continental and oceanic plate configurations; changes in oceanic circulation and chemistry (particularly reduced levels of oxygen and salinity); changes in atmospheric chemistry; predation; disease; interspecific competition. Most of these causes are not mutually exclusive; for example, volcanic activity (itself usually associated with new plate configurations) could promote a climate change and a shift in atmospheric chemistry, leading to unfavourable conditions and increased competition for resources.

In almost all cases, we are unlikely to know the full sequence of events leading to the extinction of a species. Establishing the chain of cause and effect is an exceptionally difficult or impossible task, even for extinctions taking place in the 20th century. If, say, a particular species of parrot becomes extinct when part of a rainforest is cut down, the proximate cause may be clear: bulldozing by humans. But precisely which biological attributes — or lack of them — led most to the demise of that species may be unclear or even unfathomable, depending ultimately on a chain of historical contingencies passing back not just to the origin of that species but to many species before it. The positive and negative feedback mechanisms in the Earth's ocean–atmosphere system, and in its ecosystems, are immensely complex. The eruption of the island of Krakatoa in 1883, a small event on geological time-scales, threw up so much debris and aerosols into the atmosphere that the effects, such as spectacular sunsets, were noticeable for many months around the world, and for a while global temperature may have dropped by nearly 1 °C. The physical and chemical effects of huge impacts are difficult to calculate; biological consequences far harder still. It is quite possible that a major impact, albeit virtually instantaneous in itself, might have environmental consequences lasting millions of years.

Any explanatory hypotheses for mass extinctions must embrace as many as possible of the groups concerned, not just one group such as the dinosaurs, less still a single species. Each particular mechanism has its advocates. The most persistent hypotheses are those that relate to environmental variables of great survival value to the largest number of organisms, such as rapid change in mean global temperature or oxygen levels, as opposed to, say, disease or predation, which seem most unlikely to affect great numbers of species simultaneously. In general, changes in the physical environment such as temperature, sea-level or volcanic activity are relatively easy to study in the geological record, compared with biological factors such as disease and predation. However, establishing a clear *causal link* between changes in the physical environment and global species extinction is often likely to remain problematic. If a physical mechanism is also known to have a dramatic and direct effect in living relatives of extinct organisms, then it receives particular attention. Consider temperature change and dinosaurs. As explained in Chapter 5, Section 5.4, many *living* reptiles have environmental sex determination, in which the sex of an individual is not determined genetically but by the temperatures experienced by the embryo at a critical stage in development. *If* the same were true of dinosaurs, then a widespread change in ambient incubation temperature could have led to a generation that was entirely male or entirely female — a sure path to rapid extermination!

Current thinking about the probable causes of major extinctions can be gained from Table 15.4, which is based on a compilation by Donovan (1989). Note that, in addition to the extinctions discussed above, the Table includes one in the late Precambrian, when most if not all the Ediacaran fauna was lost (see Chapter 14, Section 14.1.2), and one in the late Cambrian, when many trilobite families became extinct, especially in North America.

Table 15.4 *Principal extinctions and probable causes, based on a review of mass extinctions edited by Donovan (1989).*

Late Pleistocene	Post-glacial warming plus predation by humans
Eocene–Oligocene	Stepwise extinction associated with severe cooling, glaciation and changes of oceanographic circulation, driven by the development of the circum-Antarctic current
Late Cretaceous	Asteroid impact producing catastrophic environmental disturbance
Late Triassic	Climatic changes possibly linked to marine regression
Late Permian	Gradual reduction in diversity produced by sustained period of refrigeration, associated with widespread marine regression and reduction in area of warm, shallow seas
Late Devonian	Global cooling associated with (or causing?) widespread oxygen deficiency of shallow seas
Late Ordovician	Controlled by the growth and decay of the Gondwanan ice-sheet following a sustained period of environmental stability associated with high sea-level
Late Cambrian	Habitat reduction, probably in response to a rise in sea-level, producing a reduction in the number of component communities
Late Precambrian	Complex, including widespread marine regression, physical stress (restricted circulation and oxygen deficiency) and biological stress (increased predation, scavenging and bioturbation)

▶ Which two general mechanisms are most frequently invoked as a cause of extinctions in Table 15.4?

Changes in sea-level, i.e. marine transgressions or regressions, and climatic changes.

▶ Recalling Chapter 13, what are the most important mechanisms for eustatic, i.e. global, changes in sea-level?

Changes in the rates of sea-floor spreading at mid-ocean ridges and/or in the extent of continental fragmentation, both of which, when increased, raise sea-level (causing transgressions), and when reduced, lower sea-level (causing regressions); and the melting and freezing of ice-sheets in a glacial phase. In the case of the Permian regression, for example, the most probable explanation is a temporary reduction in sea-floor spreading as Pangaea formed (Chapter 13, Section 13.4.1), with expansion of the volume of the ocean basins.

Perhaps the most important effect of changes in sea-level is to change the area of habitat suitable for many species. For example, a marked regression removes water from a large area of continental shelf, and regressions are implicated in many extinctions, especially of shallow marine organisms. The idea that the number of species that can occupy an area correlates with the area of suitable habitat is called *the species area-effect*. The effect is not, however, linear: normally the rate of increase in number of species with increase in habitat area tails off exponentially so that, for example, doubling an area might add 35% to the number of species, whilst another doubling adds only an extra 20%. In general, though, the larger the area of continental shelves covered by shallow seas, the greater the diversity that can be accommodated on the shelves.

A major regression occurred near the end of the Cretaceous and, together with increased tectonic activity, may well have been responsible for the more gradual extinctions of benthic groups such as rudists and inoceramid bivalves. Marine regression is, however, clearly flawed as an explanation for all extinctions in the marine ecosystem. Mass extinctions, many of which do occur at times of regression, often affect not only shallow marine benthos but also pelagic life, i.e. drifting plankton and swimming nekton that live far out at sea or in deep water. The relatively sudden late Cretaceous decline of plankton, for example, is most unlikely to have been caused by regression. Some mass extinctions of shallow water faunas, such as those of the Frasnian–Famennian, were not associated with regression. Nor has every major regression been accompanied by mass extinctions of shallow-water faunas, as witnessed, for example, by the very few such extinctions associated with Pleistocene sea-level fluctuations. The Pleistocene regressions may, however, have been too brief to have had an effect, and the proportion of the total area of shallow sea habitats lost was much less than at times such as the late Ordovician, when a vast area of shallow sea would have shrunk away with a small fall in sea-level.

Oceanic islands are probably important as refuges during regression. Many oceanic islands, being volcanic, are conical in shape, and so the amount of habitable shallow marine area around them actually increases during a regression: the coastline moves outward and remains narrow but its circumference increases. Of 276 living families of marine molluscs, echinoderms and corals analysed by David Jablonski of the University of Chicago, 87% have at least one species around the 22 steep-sided oceanic islands studied; were there now to be a major regression, most of these marine families would be protected from its effects by their presence on such islands.

Sea-level changes also appear to influence the oxygen concentrations in seawater. For reasons that are not yet clear, rapid rises in sea-level are often associated with anoxia. Anoxia is linked with several extinctions, both major (e.g. late Devonian) and minor (e.g. at the Cenomanian–Turonian boundary in the late Cretaceous). Habitat areas can be as severely reduced by oxygen starvation during transgression as by loss of shallow marine shelf habitats during regression. In addition, there is some evidence that rapid transgressions may, in effect, 'drown' major reef ecosystems.

Evidence from oxygen and carbon isotopes within sedimentary rocks can sometimes provide important information to constrain or suggest theories of mass extinction mechanisms. For example, a reduced concentration of oxygen in the atmosphere, and a build up of carbon dioxide, seems to have occurred in the late Permian. In this case, the most plausible mechanism for atmospheric change is related to regression. Vast amounts of carbonaceous matter had accumulated as coal in the preceding Carboniferous Period. With the Permian regression, however, these coal-bearing beds were exposed to aerial erosion. Oxidation of this organic matter could have increased the carbon dioxide in the atmosphere at the expense of oxygen, the consequences of which might eventually have been fatal for many animals. Again, such is the complexity of feedback loops in the Earth's atmospheric chemistry that details remain conjectural at best.

Even in the unlikely event that the trigger (or triggers) for extinction have been the same in each case, each mass extinction must have been associated with a unique set of physical, chemical and biological conditions. As Antoni Hoffman said (in Donovan, 1989) when reviewing work on the influence of physical factors: 'Perhaps then mass extinctions are not the biotic consequences of any single phenomenon of one or another sort, but rather rare incidences of more than one major change in the physical environment accidentally clumped together within relatively short intervals of time, say, 2-4 million years in duration'. As the word 'accidentally' in the sentence implies, mass extinctions could be the summed outcome of a range of chance events. For small populations of individuals, chance events such as volcanic eruptions, floods and hurricanes are often sufficient for local species extinction. For a species already confined to a small population, and having a short individual lifetime and a low intrinsic rate of increase, such chance events may be the last straw. But such local chance events are likely to have been experienced many times during a species' history. It may be that, on average, a species which an ecologist reports as very susceptible to local extinction is highly resistant to *global* extinction; the species may well be extinction-resistant on a global scale in order for us to observe it frequently becoming extinct on a local scale.

There is an increasing realization among geologists, palaeontologists and biologists that if something is 'an exceedingly unlikely event' it becomes highly likely to happen given enough time. The long battle in the early 19th century between catastrophists — those who thought that sudden upheavals moulded the planet and its life — and uniformitarians — who explained Earth's features by extrapolation from present-day gradual change — has been resolved with general acceptance of a wide spectrum of rates. If an impact did trigger a mass extinction at the end of the Cretaceous, then such catastrophes must play an extremely important role in evolution. Selection can never hone species to an environment containing such rare, haphazard events. Species cannot have any opportunity to evolve defences to stresses sufficiently rare as to be beyond the reach of natural selection. It may be very much a matter of chance whether a particular lineage survives, as might be the case for any others further along the food chain. Survival may be more a matter of good luck rather than good genes, as Raup has suggested.

Discussion of the causes of mass extinctions tends to be dominated by mechanisms that might eliminate existing species rather than suppress origination rates. This is no trivial distinction: in theory, major 'extinction' could occur simply by a massive suppression of species origination rates (i.e. reduced rates of cladogenesis), whilst extinction rates remained roughly as usual (see Section 15.3.1). This is at least testable: the prediction would be that species ranges, on average, show no abrupt truncation at times supposed to be 'mass extinction', whereas origination rates dropped dramatically. Existing lineages would have persisted with the same average duration as usual but, in the case of extinct clades, no new species appeared to take their place. Clearly, once increased extinctions were underway, there would be fewer and fewer clades from which to originate, and the effect would be exponential. Although deemed very unlikely as the principal mechanism for any mass extinction, it is feasible that 'mass decline in origination' could have been a significant *component* in some cases.

Periodicity of extinctions In 1984, Raup and Sepkoski subjected data on the extinction of marine animal families to a series of statistical analyses. They created great interest by concluding that the temporal spacing of extinction peaks since the end of the Permian was best explained by an approximately 26 million year periodic signal. Further work using generic data seemed to support the 26 Ma periodicity. Figure 15.29 shows the percentage extinction for marine animal genera from the middle Permian to the Recent, a total of 44 stratigraphical intervals. Of the 10 383 genera represented, 6 350 are extinct. A 26 Ma periodicity, represented by the vertical lines, has been superimposed in best-fit position to illustrate conformity of the data to the hypothesis of periodic extinction.

▶ Which are the three most severe extinctions, according to Figure 15.29?

The extinction of the late Permian was the most severe, followed by the late Cretaceous, and the late Triassic. Note that these mass extinctions have peaks that span several stages, which means that the total percentage extinction of genera exceeds that of the single central value.

An extraterrestrial control on the inferred periodicity was postulated because no convincing terrestrial driving mechanism was (or is) known that could produce such a pattern. Periodic impact of comets derived from a cloud of comets at the outer fringes of the Solar System was suggested as the most likely possibility.

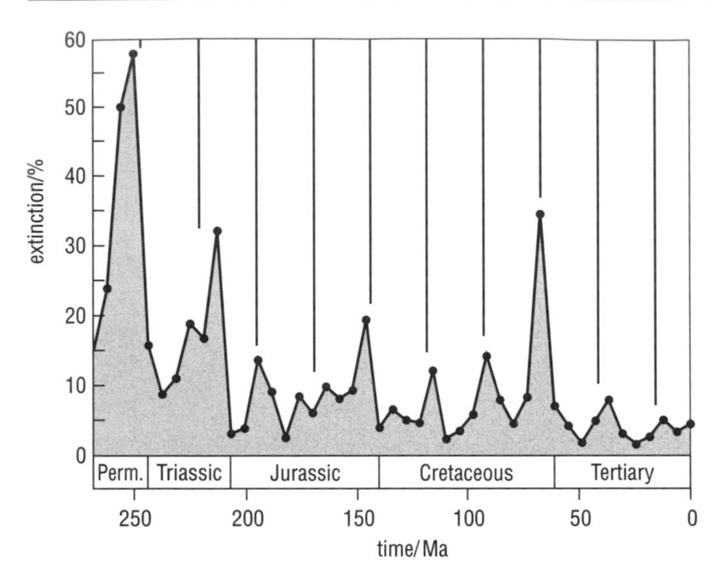

Figure 15.29 *Percentage extinction of marine animal genera from the middle Permian to the Recent. The total number of genera represented is 10383, of which 6350 are extinct. A 26 Ma periodicity, represented by the vertical lines, has been superimposed.*

Various hypothetical ways of inducing a disturbance in a comet or asteroid cloud have been proposed, including a postulated dim and distant companion star to the Sun with a highly eccentric orbit, dubbed 'Nemesis' — but neither this nor the other possibilities are regarded as very plausible. Vigorous debates over whether or not the data show significant periodicity followed the initial claim of Raup and Sepkoski, and seem set to continue, with protagonists on both sides using sophisticated statistical techniques to buttress their view. A major problem remains the extent to which the radiometric dates of extinctions are accurate. Another problem was pointed out by Colin Patterson and Andrew Smith of the Natural History Museum, London. They examined the data for fish and echinoderm families, which together made up about 20% of Raup and Sepkoski's data, and concluded that only 25% of the fish and echinoderm extinctions were valid. The remainder included paraphyletic families that underwent pseudoextinction, polyphyletic families, monophyletic families attributed the wrong extinction time, families with only a single species, and other sources of error. However, as argued in Section 15.3.1, a case can at least be made for including paraphyletic taxa in analyses of evolutionary patterns if the taxa represent ecological entities (e.g. they are all members of the same adaptive zone). If, on the other hand, the taxa said to have become extinct are mere artefacts of stratigraphical or taxonomic practice, then clearly they should be removed from the analysis.

Recent work by Sepkoski on extinction patterns *within* individual groups such as corals, brachiopods and gastropods offers continued support that periodic extinctions are real phenomena (Sepkoski in Sharpton and Ward, 1990). At the moment, as with many other issues concerning extinction, the jury is still out, with lack of cyclicity seeming slightly the more likely verdict. But if Raup and Sepkoski *are* right about the cyclicity, the next extinction peak is not due (without humans inducing one earlier) for about another 13 Ma!

SUMMARY OF SECTION 15.4

Mass extinctions are geologically rapid, major reductions in global diversity. The five most severe mass extinctions of the Phanerozoic were in the late Ordovician, late Devonian, late Permian, late Triassic, and late Cretaceous. There appears to have been a continuum of extinction intensities from 'background' rates through to mass extinctions. The time-spans of the 'Big Five' mass extinctions are uncertain, but in most cases probably ranged from *c.* 0.5 to 5 Ma.

Evidence of a probable asteroid impact at the K–T boundary includes a layer rich in iridium, shocked quartz grains and tiny diamonds. An impact (and/or intense volcanic activity) may have ejected materials into the atmosphere, suppressing photosynthesis for a while, causing a wide range of physical and chemical effects leading to a series of extinctions, of which perhaps the most significant was a reduction of oceanic plankton apparently spanning about 400 000 years. Various terrestrial and marine groups that became extinct close to the K–T boundary had, however, been in decline for millions of years before, and for these any K–T 'event' was just the final blow.

Various sources of bias can distort evidence of extinctions in the fossil record, especially hiatuses in the stratigraphic record, correlation difficulties, sampling deficiencies, imprecise plotting of stratigraphic ranges, and failure to recognize 'Lazarus taxa', which seem to disappear at times of mass extinction, only to reappear later.

Mass extinctions cut across many ecological lines, and although there is always some selectivity, it is usually impossible to ascertain why some groups survived and others did not. Tropical biotas appear to be particularly vulnerable, as do large-bodied species. Wide geographical distribution may assist survival of species at times of background extinction, and survival of genera at times of mass extinction.

There are many possible causes of mass extinction, and different mass extinctions may have had different primary causes. Changes in sea-level and climate are implicated in most cases; however, these and other mechanisms, such as changes in plate configurations, volcanic activity, and oceanic and atmospheric chemistry, are not mutually exclusive. Suppression of origination rates may have played a part in some mass extinctions. Establishing the chain of cause and effect leading to the extinction of any species is extremely difficult. If mass extinctions are linked to rare extraterrestrial events, survival may be largely a matter of luck, because adaptations to the stresses of such rare, haphazard events would be beyond the reach of natural selection. Evidence that major extinctions have a periodicity of about 26 Ma is inconclusive at present.

15.5 RADIATIONS

The Darwinian concept of the proliferation of species being driven by novel opportunities for the adaptive divergence of populations — now termed *adaptive radiation* (Chapter 9, Section 9.4.2) — is frequently invoked to explain the expansion of higher taxa. Episodes of marked diversification in certain groups of organisms, as revealed by the fossil record (e.g. Figure 11.1, which shows part of the phylogeny of vertebrates), have been attributed to the appearance of various kinds of ecological opportunities. For example, the 20 Ma interval from 65 to 45 Ma ago (i.e. from the end of the Cretaceous to the mid-Eocene) saw the origination of about 78% (*c.* 14 out of *c.* 18) of the living orders of placental mammals. This early Tertiary radiation of mammals (along with certain other taxa) has been widely interpreted as reflecting the opening up of new possibilities for adaptive divergence following the demise of the dinosaurs and other previously dominant taxa at the end of the Cretaceous (Chapter 11, Section 11.1.1). By contrast, the initial radiation of the reptiles in the late Palaeozoic has been interpreted as ensuing from the evolution of the shelled egg, which allowed the penetration of terrestrial habitats inaccessible to the ancestral amphibians. Thus, the postulated cues for radiation include both extrinsic and intrinsic factors.

Such hypotheses, which relate phylogenetic patterns to changes either in the environments of organisms, or in the organisms themselves, or possibly both, are plausible and consistent with the theory of evolution by natural selection. They have long been paraded through the literature as major features of evolution. Yet the very plausibility of the principle has fostered a tendency for uncritical application of it as a general explanation for the diversification of higher taxa: evolutionary '*Just-So Stories*' (Section 15.1.1) abound. For example, was it really the demise of the dinosaurs that launched the mammals on their early Tertiary radiations, or had the latter group coincidentally reached some threshold (e.g. of size, physiological capability or behavioural complexity) that permitted their exploitation of new habitats? Or, might the pattern of radiation of the early reptiles be more directly linked with, say, climatic change rather than the evolution of the shelled egg? Indeed, is it even reasonable to seek such simplistic explanations for the evolutionary histories of entire higher taxa, encompassing organisms of quite disparate ecology and distribution?

Moreover, even simulated stochastic phylogenies (Section 15.3.1), in which no allowance is made for adaptive differences between taxa, may exhibit phases of expansion, as was illustrated in Figure 15.13.

These considerations prompt two questions: (1) what should constitute an evolutionary radiation, as opposed to a mere 'luck-of-the-draw' increase in taxonomic diversity? And (2) are all radiations necessarily adaptive (in which case, the word 'adaptive' would be redundant), or are there also *non-adaptive radiations* (i.e. radiations resulting from factors other than the adaptive differences between individuals)? These questions in turn demand retrospective tests for hypotheses concerning evolutionary radiations.

An approach to answering the first question, concerning the definition of radiation, was outlined in Section 15.1.1. There, the probability distribution of random walks was used as a null hypothesis to test for significantly sustained trends of increase in taxonomic diversity. On the basis of this test, a **radiation** can be defined as 'an episode of significantly sustained excess of cladogenesis over extinction' (Skelton, in press). A clade expansion which fails this test cannot be distinguished from a random pattern of increase, and is thus deemed not to merit description as a radiation, ensuing from some distinct cue.

A qualification that should be emphasized is that the time of inception of a postulated radiation must be specified before the test is applied, and must therefore be based upon some independent line of evidence (such as an important morphological change or geological evidence for a radical environmental change). Otherwise, there is no independent means of deciding from which point to apply the test. The test grid in Figure 15.3 shows the probability distribution of random walks that start immediately from the 0 position: only 1 in 20 (on average) is then expected to reach the critical area for $P = 0.05$. However, even a random walk is likely to contain some segments which, taken by themselves, would reach the critical area for significance if the test grid were arbitrarily zeroed upon them (i.e. if the grid were shifted along the random walk until it encountered such segments). But this would be an improper use of the test, because the segments in question would not have been independently identified.

▶ What was the independently identified cue for the radiation of uncoiled rudists discussed in Chapter 14, Section 14.5.3, and in Section 15.1.1?

Invagination of the ligament, which permitted uncoiled shell growth, was itself specified as the cue for the postulated radiation of uncoiled rudists: a morphological feature was thus detected independently of the assessment of taxonomic diversity.

Turning to the second question, if a radiation is to be recognized as an adaptive radiation, simple logic demands that there must have been an intimate association between adaptive divergence and the episode of taxonomic diversification identified as the radiation. It would not be sufficient merely to demonstrate that extant members of a given clade are adaptively diverse (a commonly used criterion for recognizing adaptive radiations in an over-generalized sense). Adaptive divergence in the latter case might have been but a secondary effect long after an earlier phase(s) of prolific speciation that involved negligible differentiation of niches (and therefore negligible morphological change).

▶ Cast your mind back to Chapters 9 and 13, and suggest how reproductive isolation between populations (i.e. speciation) might arise without necessarily involving adaptation to different niches.

In plants, especially, polyploidy is an important route to speciation that does not involve prior adaptive divergence, although ecological differentiation usually soon ensues (Chapter 9, Section 9.6.3). For animals, the most likely first step is vicariance — the splitting-up of populations through the appearance of barriers to dispersal — or long-distance dispersal (Chapter 13, Section 13.3.3). Allopatric speciation might then follow without necessarily involving adaptation to different niches, perhaps as a result of changes in mating behaviour, for example (e.g. Kaneshiro's model for speciation in the Hawaiian drosophilids, discussed in Section 9.4.5 of Chapter 9).

Plethodon, a genus of salamander, was discussed in the context of living fossils in Section 15.2.1. In a study of the numerous sibling and other closely similar species of plethodontid salamanders in North and Central America, Larson (1989) proposed a model that combined the limited tendency for dispersal of these animals with episodes of vicariance. In favourable conditions, dense expanding populations probably spread gradually into suitable habitats in contiguous regions. Periods of climatic deterioration could then have imposed geographical fragmentation of the populations, with allopatric speciation resulting from divergence of their mate recognition systems. This example introduces another factor, that of species characteristics which may themselves promote high rates of cladogenesis, a topic that will be discussed further in Section 15.7.

The possibility of non-adaptive radiations means that it is therefore necessary to establish a direct link between taxonomic and adaptive diversification in order to verify a hypothesis of **adaptive radiation**. The latter can now be defined as 'an episode of significantly sustained excess of cladogenesis over extinction, with adaptive divergence cued by the appearance of some form of ecological stimulus' (Skelton, in press).

In fact, there is likely to be a gradation in nature between adaptive and non-adaptive radiations. The precise nature of the link between speciation and adaptive divergence may be difficult, if not impossible, to pin down in most cases. Sometimes, adaptive divergence may have led to speciation, as in the classic Darwinian model, but in other cases speciation may have been closely followed by adaptive divergence (through resource-partitioning), when the newly separated species became secondarily sympatric (as with the Hawaiian drosophilids). In both cases, there is an intimate association between speciation and adaptive divergence, and so both can legitimately be termed adaptive radiations. But a case in which adaptive divergence is essentially post-speciational can be thought of as grading into non-adaptive radiations, in which there is no close linkage between speciation and any adaptive divergence. Detection of non-adaptive radiations may be difficult in the fossil record because in the absence of new adaptations there may be few morphological changes associated with speciation.

One form of diagnostic test for adaptive radiations in extant organisms that has been proposed involves multiple cladistic comparison (Figure 15.30). The starting point is the proposal of a **key evolutionary innovation (KEI)**, a novel feature which is hypothesized to have been preadaptive for new ecological opportunities. Radiation is then predicted to have characterized each clade possessing the KEI, but not its sister clade (lacking the KEI). Jensen (1990) has proposed a two-step test. First, the taxonomic diversity of a clade characterized by possession of the KEI (i.e. as a synapomorphy, see Chapter 11, Section 11.2.4) is compared with that of its sister group. If (as predicted) the former is markedly greater than that of the latter, the second step of the test is implemented whereby corroboration is sought in comparisons of other clades in which the KEI has evolved independently, with their sister groups, to see if the disparity is maintained.

Jensen illustrated the test by reference to some meticulous studies of the jaws of cichlid fish that had been conducted by the Harvard biologist Karel Liem (some cichlids are shown in Figure 9.12). Liem had shown that, in cichlids, a set of bones around the back of the throat had become operationally decoupled from the main jaw mechanism around the front of the mouth, so as to form an independent *pharyngeal jaw apparatus*. This change, Liem hypothesized, had allowed a much wider repertoire of specialized feeding mechanisms (e.g. Figure 9.14 in Chapter 9) than was possible with the primitive coupled jaw system found in most other percoid fishes (the major group from which the cichlids are derived). In fact, the

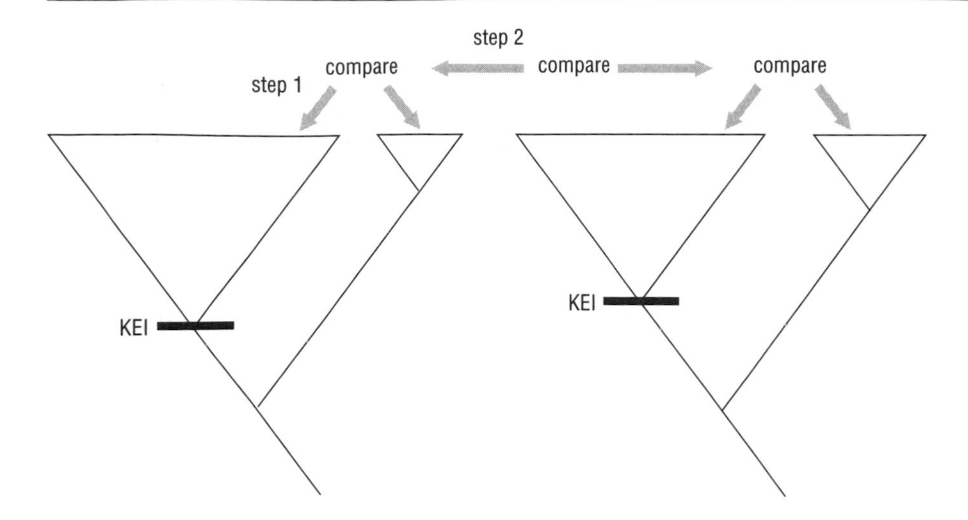

Figure 15.30 *Diagrammatic summary of Jensen's test for the phylogenetic consequences of a postulated KEI. Clade diversities are indicated by triangle widths on the cladograms.*

cichlid arrangement was later found to characterize a somewhat broader clade, incorporating the cichlids, called the Labroidei. Noting that no other percoid taxon that might be a sister group to the Labroidei appeared to contain anything like the 1 800 or so species of the latter, Jensen concluded that the first step of the test had been satisfied.

A distinct group of fishes, the Beloniformes, proved suitable for the second step. One clade within these, the Exocoetoidea, is characterized by a synapomorphic pharyngeal jaw apparatus analogous to that in the Labroidei. The Exocoetoidea today contains at least 130 species, while its sister group (the Scomberesocoidea) contains only 36 species.

▶ Does this finding corroborate the result from the first step of the test?

Yes it does. The clade with the KEI (the Exocoetoidea) again turns out to have a higher taxonomic diversity than that without it. Liem's original hypothesis thus seems vindicated.

There are, however, some problems with this approach. Examples of analogous KEIs in ecologically comparable but unrelated taxa are likely to be limited; the greater the differences in other respects between the clades being compared, the more difficult it is to be sure that the KEI in question is the main reason for any differences in taxonomic diversity.

▶ What other problem can you detect in comparing numbers only of extant species in the clades in question?

From such data, we know nothing of the historical patterns of diversification. Notwithstanding the differences in extant species numbers between the clades, we cannot tell if the clades with the KEI did indeed show significantly sustained trends of increasing diversity, nor even if the difference between each clade and its sister group can be considered significant. The lack (so far) of any statistical test for the comparisons is thus a major weakness of the method of cladistic comparison of extant taxa. At present, only where an adequate fossil record exists to furnish stratigraphical range data (as with the rudist example discussed earlier) can radiations be positively diagnosed as such, and their inceptions specifically linked in time with a cue such as the evolution of a KEI. In the absence of such data, it may yet be possible partially to reconstruct the history of diversification using molecular distance data (Chapter 11, Section 11.3.1), but this approach has two shortcomings. First, it depends upon a reliable molecular clock, and secondly, it can take little or no account of any extinct taxa involved in a radiation.

The question of the time of inception of a radiation is important because even if possession of a KEI permitted the eventual radiation of clades possessing it, it might not necessarily have been the immediate cue that sparked off the pattern of radiation. Adaptive radiations can only occur when the progenitors possess suitable preadaptations (Chapter 14, Section 14.5.2) *and* when environmental conditions are receptive — i.e. when the right species is in the right place at the right time! The prior existence of one factor thus serves only as an enabling circumstance, so to speak, while the subsequent emergence of the other then acts as the immediate cue to radiation. Thus, the effective cue, as noted at the beginning of this Section, can be either intrinsic or extrinsic.

▶ Apart from the ecological opportunities opened up by mass extinction, mentioned earlier, can you think of other environmental changes which might serve as extrinsic cues for adaptive radiation?

Many of the classic examples of adaptive radiation, such as that of Darwin's finches (see Figure 1.4 in Chapter 1) were based upon the chance colonization of unoccupied habitats created by the birth of new oceanic islands, for example (Chapter 13, Section 13.3.2). However, probably a much more important extrinsic factor on the macroevolutionary scale has been the evolution of other organisms.

The spectacular and well documented radiation of grazing horses in North America, for example (Figure 15.31), was initiated in mid-Miocene times, with the appearance of species of *Merychippus*, some 17 to 18 Ma ago (see also Section 15.2.2). Their incipiently high-crowned teeth, elongated lower limbs, and increasing body size are all consistent with a switch to grazing in more open grassland habitats (Chapter 12, Section 12.1). Independent evidence of fossil grasses points to a substantial expansion of true grasslands (i.e. savannas, rather than parkland-like mosaics of woodland and grass patches) around that time. The coincidence of the evidence for the latter environmental change and the inception of the radiation of grazing horses (along with those of other grazing mammals) strongly hints at the former having been the cue for the latter. When, as in this instance, several unrelated clades of organisms with analogous adaptations show synchronous radiations, it is more parsimonious to interpret them as having responded to a single extrinsic cue than as having coincidentally radiated because of the simultaneous independent evolution in each of some KEI. This argument, incidentally, can also be used to tackle the problem cited at the beginning of this Section concerning the interpretation of the early Tertiary history of the mammals.

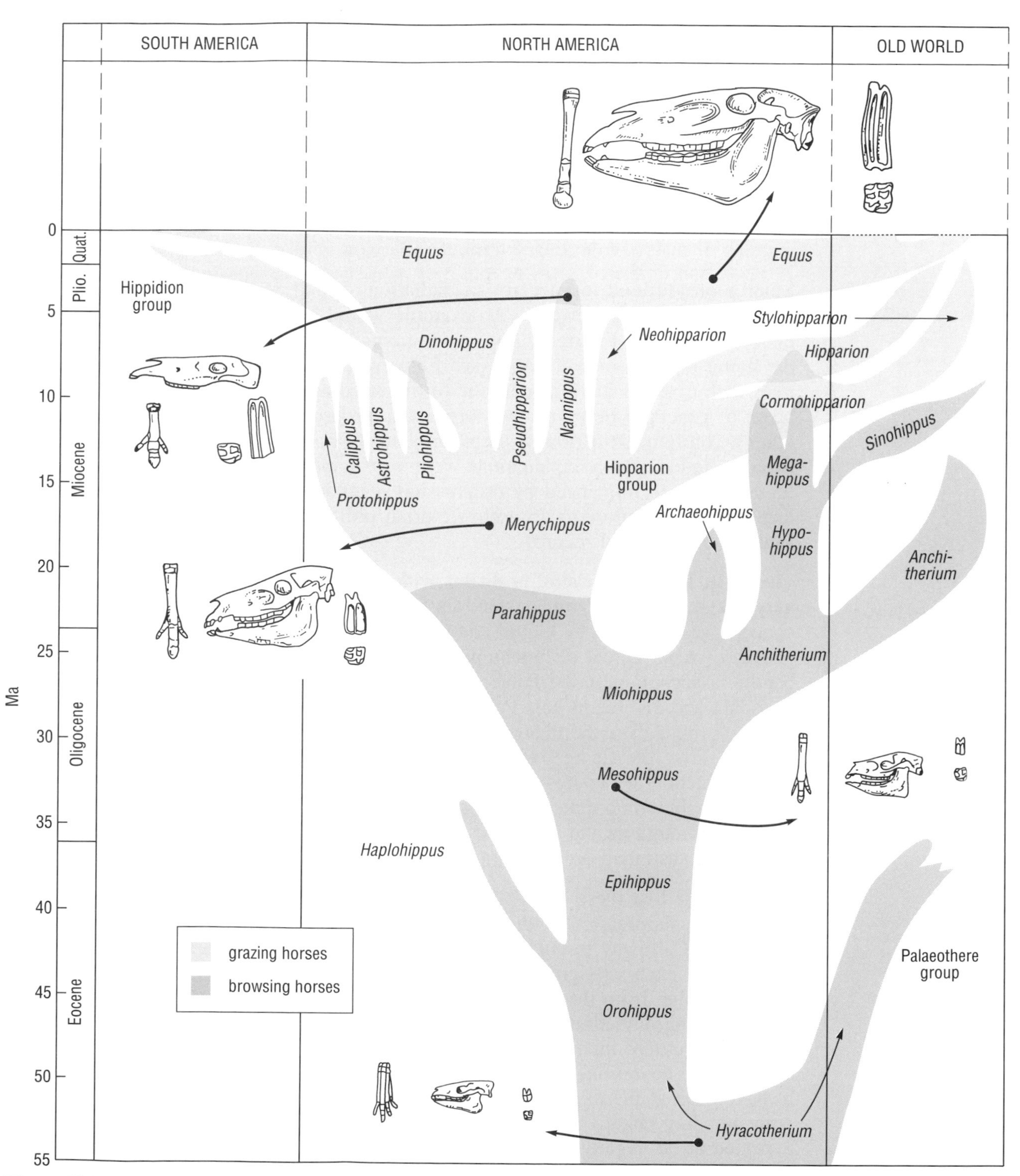

Figure 15.31 *Synoptic history of horse evolution. The relative sizes of the skulls, of the lower forelimb bones and of the cheek teeth are shown for selected forms. Note that the dating of some epochs differs slightly from Appendix 1.*

The fact that several already distinct clades of mammals (which had evolved in Cretaceous or even earlier times) radiated together at this time, again strongly implicates a common, extrinsic cue — for which the immediately preceding mass extinction is the most plausible candidate.

So far, we have restricted attention to the radiations of clades following some cue, but in the case of mass extinctions this can involve the wholesale replacement of some higher taxa by others. Here, there is obviously no role for competition in the pattern of replacement, since the radiations of the later taxa came after the demise of the earlier taxa. Such a pattern seems to have been quite common in the fossil record, and it highlights the important influence of mass extinctions on gross patterns of macroevolutionary turnover (Section 15.4).

Much more difficult to interpret are radiations which coincide with the decline of other, adaptively similar taxa. An example would be the early radiation of the birds (Chapter 12, Figure 12.16), which coincided with the progressive decline of the flying reptiles, or pterosaurs (see Figure 11.1 in Chapter 11). It is tempting to treat such cases as having been due to direct competitive replacement, with one taxon of generally better-adapted organisms progressively driving the species of the other taxon into extinction as their members competed over shared resources and habitats. Indeed, such explanations were very popular in the older evolutionary literature, much influenced by progressionist models of macroevolutionary turnover that perhaps had more to do with Victorian political fashions than with objective analysis of the fossil record.

There are, however, several problems with competitive replacement models, and very few purported examples stand up to closer scrutiny. An important assault on the concept was mounted by Gould and Calloway (1980), who studied the long-term pattern of replacement of brachiopods (which tended to dominate shallow marine benthic associations in the Palaeozoic) by bivalves (which became enormously abundant and diverse in post-Palaeozoic seas). From a plot of generic diversity in the two groups over the Phanerozoic (Figure 15.32), Gould and Calloway noted two patterns. First, the changeover in relative dominance was largely attributable to the mass extinction at the close of the Permian: the brachiopods suffered a greater relative decline then than did the bivalves. Secondly, while the bivalves continued to diversify throughout the Mesozoic and Cainozoic, the brachiopods, in contrast, showed no sustained diversification from their post-extinction levels.

Is it possible that these post-Palaeozoic patterns might have resulted from competitive suppression of the brachiopods by the burgeoning stocks of bivalves? Gould and Calloway tested that idea by plotting a graph (Figure 15.33) that showed, for each stratigraphical interval considered, the deviation of the bivalve diversity value from the regression line for all post-Palaeozoic bivalves versus the corresponding deviation for the brachiopods in that interval. Had there been a direct competitive interaction between bivalves and brachiopods, one would expect any positive deviations (i.e. above-average increases in diversity) among bivalves to have been at the expense of brachiopods, which should thus show a corresponding negative deviation: i.e. the deviations for the two groups would be expected to be negatively correlated.

▶ Does the graph in Figure 15.33 support the competitive model?

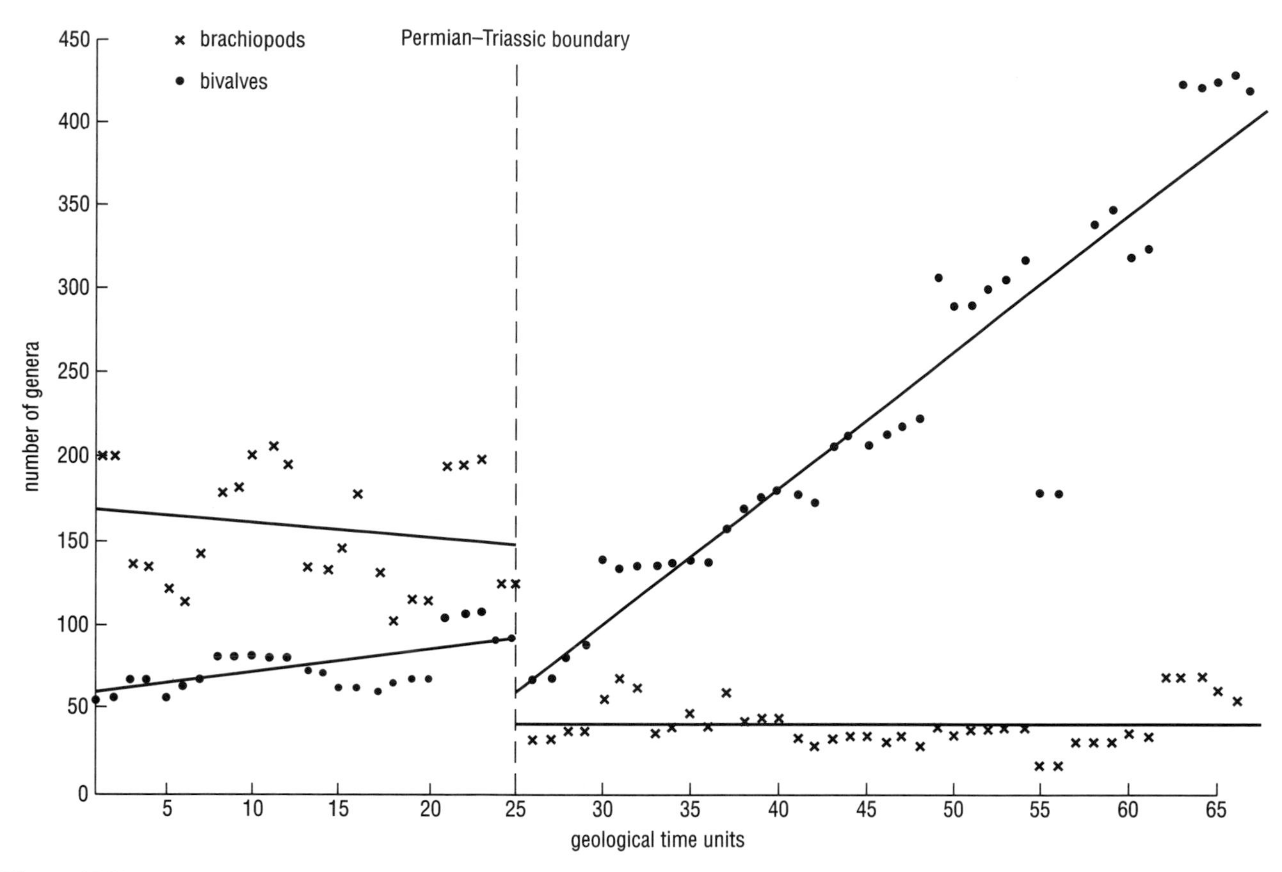

Figure 15.32 *Numbers of genera of brachiopods and bivalves for successive stratigraphical intervals in the Phanerozoic treated as (approximately equal) units of time — numbered along the* x-*axis. Separate regression lines are shown for the two groups in the Palaeozoic and in the post-Palaeozoic.*

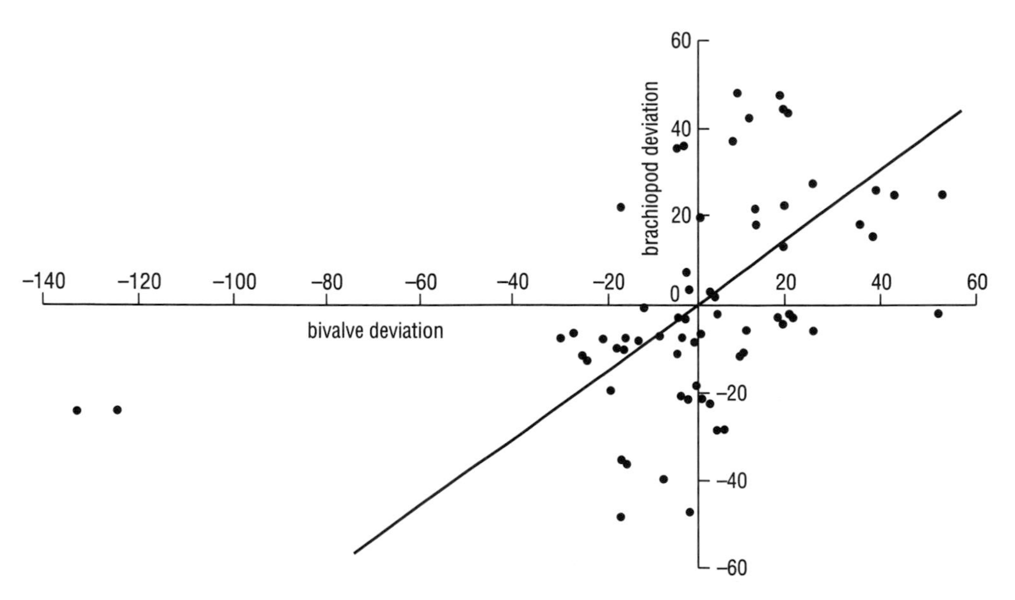

Figure 15.33 *Plot of the deviations from their respective regression lines of the species diversity values of bivalves* versus *brachiopods, derived from Figure 15.32. Note that most data fall in the upper right and lower left quadrants, yielding a positive correlation (for which a reduced major axis is shown).*

No, the results are the opposite of those predicted by the competitive model. The scatter shows a positive, not negative, correlation of the deviations. Thus, abnormal flourishings of bivalves were matched by those of brachiopods, and likewise declines in one group by declines in the other (presumably reflecting environmental changes). We must look elsewhere for an explanation of the relatively poor showing of post-Palaeozoic brachiopods *vis-à-vis* the successes of the bivalves (see Section 15.6 for a discussion of extinction patterns in sessile suspension-feeding organisms). The title that Gould and Calloway gave to their paper, 'Clams and brachiopods: ships that pass in the night', provides an apt and graphic metaphor for the alternative model of independent clade histories. A theoretical consideration also makes the above result hardly surprising. Recall from Chapter 8, Section 8.3.3, that interspecific competition results when two or more species occur in the same habitat and exploit the same resources such that individuals of at least one of the species suffer a reduction of fitness. Thus, to assert that an entire higher taxon can competitively displace another requires the somewhat unlikely circumstance of comprehensive overlap of the totality of niches embraced by each of the higher taxa, with no remaining uncontested niche dimensions along which species of the 'victim' taxon might take evolutionary refuge. It also requires an improbable universal adaptive superiority of members of the invading taxon over those of the victim taxon. The concept seems really to be an example of the inappropriate extrapolation of a principle that is of undoubted importance in the microevolution of species to the scale of macroevolutionary patterns (a point taken up again in Section 15.7).

A much more plausible, compromise model that could explain patterns of higher taxonomic replacement (involving coincident expansion of one group and decline of another), without invoking competitive displacement, has been proposed by Rosenzweig and McCord (1991): they postulated a process of **incumbent replacement**. An already established, 'incumbent' species, they noted, would be finely adapted to its existing niche. Its complete displacement by an invading species population, which has not had the benefit of adaptation to precisely that niche, would be most unlikely. If, however, the incumbent species happened to become extinct due to a chance combination of adverse factors (as discussed in Section 15.3.2), its vacated niche is open to invasion by members of either group (or indeed from any other group in that adaptive zone). Let us suppose that members of one taxon possess some KEI that has the effect of lessening the fitness costs of adaptive compromises resulting from contrary selective pressures experienced in common by all the taxa in the adaptive zone (as discussed in Sections 4.3.2 and 4.4.5 of Chapter 4). In this circumstance, members of that taxon are likely to have the competitive edge on invaders belonging to the other higher taxon (or taxa), and so eventually to take over the resources of the old niche. The consequent higher rate of successful speciation of the taxon with the KEI will, given enough time, eventually lead to a progressive replacement of the other formerly dominant taxon (or taxa) as the species of the latter are gradually whittled away by random extinction (completely unconnected with competitive inferiority). Such a model could also account for switches in dominance of higher taxa associated with mass extinctions, as with the example of the brachiopods and the bivalves cited above.

The incumbent replacement model provides a satisfactory explanation for the comprehensive replacement of one taxon by another in different areas and at different times, without requiring the theoretically improbable phenomenon of direct competitive displacement. As an example of such a replacement, Rosenzweig and McCord cited the evolutionary history of the turtles. Primitive turtles could not retract their heads and necks into their shells. At different times in different parts of the world, they have been progressively replaced by members of two clades of turtle which (in different ways) are capable of retracting their heads and necks. The independent repetition of the pattern of replacement strongly suggests some form of overall advantage in the filling of niches possessed by the turtles capable of retraction, in keeping with the incumbent replacement model.

Not all radiations involve replacement of niches by the radiating group. In radiations that *do* involve replacement, however, there is a spectrum of possible degrees of interaction, ranging from no interaction at all, through the pattern of incumbent replacement where competition may exist, with the replacing species having some adaptively advantageous KEI, but which has nothing to do with the extinction of the replaced species, to the unlikely case in which a species directly drives another to extinction by competitive displacement.

In conclusion, you will no doubt have been struck by a general problem that often afflicts the interpretation of patterns of clade growth: there are so many alternative ways of explaining the same historical phenomenon — too many hypotheses chasing too few facts. The only way to avoid mere speculative arm-waving, as this Section has attempted to illustrate, is to set up retrospective tests of the competing hypotheses wherever possible, so that their relative probabilities can be assessed. By this means we can progress from '*Just-So Stories*' to science.

SUMMARY OF SECTION 15.5

Adaptive radiations, triggered by extrinsic (environmental) or intrinsic (pre-adaptive) cues, have often been invoked to explain the expansion of higher taxa. Many causes for radiations can be postulated, however, so retrospective testing of hypotheses is required. The persistence of taxonomic increase (tested against a null hypothesis of random walks) can be used as a criterion for confirming a radiation, though the time of its inception must be postulated before the test is applied. Some radiations may be non-adaptive (i.e. due to factors other than adaptive divergence, such as vicariant speciation without a differentiation of niches), so an intimate linkage between taxonomic and adaptive diversification must be demonstrated for an adaptive radiation to be diagnosed.

The method of multiple cladistic comparison seeks to find an association in a given clade between synapomorphic possession of a key evolutionary innovation (KEI), and increased numbers of extant species relative to a sister clade. Repetition of that pattern in other taxa in which the KEI has independently evolved provides corroboration. As yet, however, the method lacks a statistical foundation, in the absence of stratigraphical range data.

A KEI need not have been the effective cue for a radiation. Synchronous polyphyletic radiations indicate a likely environmental cue, such as the opening up of new ecological opportunities following a mass extinction, the appearance of new habitats, or evolutionary changes in other organisms.

Radiation of one higher taxon at the apparent expense of another has often been attributed to competitive displacement. Supporting evidence is lacking, however, and the idea is theoretically improbable. More plausible is a model of incumbent replacement, whereby members of an invading taxon have a competitive advantage in taking over the resources of niches as they become vacated through chance extinctions.

15.6 Changes in global diversity

What does the fossil record have to tell us about changes in diversity throughout the Phanerozoic? For instance, has the total number of families increased, decreased, or remained much the same since early in the Phanerozoic? How do changes in families compare with changes in species? In practice, such questions can best be approached by examination of the marine fossil record of readily preserved organisms.

15.6.1 Evolutionary faunas: the history of family diversity

In 1981, Sepkoski analysed historical patterns within his large database of marine families. The statistical analysis he employed ('factor analysis') grouped together classes that attained their maximum diversity (in terms of constituent families) around the same time. As a result, he identified what he called 'three great evolutionary faunas' in the Phanerozoic marine record. **Evolutionary faunas** are sets of higher taxa (especially classes) that have similar histories of diversification and that together dominate the biota for a long interval of geological time. Each successive evolutionary fauna displays a slower rate of diversification — but a higher level of maximum diversity — than those preceding it. As each new fauna expands, so the previously dominant one declines. The declines are much slower than the initial diversifications, giving each evolutionary fauna an asymmetrical history. The evolutionary faunas overlap in time, and their names refer to only a subset of all the taxa present at a particular time.

The three evolutionary faunas (EFs) that Sepkoski identified are:

I The Cambrian EF, important during the Cambrian.

II The Palaeozoic EF, dominant from Ordovician to Permian.

III The Modern EF (or Mesozoic–Cainozoic EF), dominant in the post-Palaeozoic.

Figure 15.34(a) shows changes in the total diversity of marine animal families from the Vendian (at the end of the Precambrian, marked 'V') to the end of the Tertiary. The top line, or 'curve', represents the total number of marine families known from the fossil record. The grey field represents the known diversity in the fossil record of families with poor preservation potential, i.e. those with members lacking well-mineralized skeletons.

▶ What is the middle Cambrian spike in the grey field of Figure 15.34(a) most likely to represent (recall Chapter 14)?

The spike represents the contribution from the Burgess Shale soft-bodied fauna (Chapter 14, Section 14.1.4). Below this field there are two curves that divide the diversity of heavily skeletonized families into the three constituent evolutionary faunas. Note that the plot in Figure 15.34(a) stops at the end of the Tertiary. Today, there are about 1 900 marine families, many of which are entirely soft-bodied and have little or no fossil record.

Figure 15.34(b, c and d) show separately the history of diversity for each evolutionary fauna.

► Approximately how many families were present (a) in the Cambrian EF at its peak diversity, and (b) in the Palaeozoic EF at its peak diversity? (c) To what diversity has the Palaeozoic EF dropped today?

(a) About 90 families; (b) about 350 families; (c) about 90 families.

► Which EF was most responsible for the massive increase in family diversity during the Ordovician?

The Palaeozoic Fauna.

From Figure 15.34 it can be seen that, in general, as the second EF expanded, so the first one slowly declined, and as the third EF expanded, so the second one declined. The Modern Fauna underwent the slowest diversification, but its diversity is much the highest yet.

The Cambrian EF was dominated by trilobites (77% of families present in the EF), inarticulate brachiopods, and worms, with various other elements, mostly small-bodied groups. Note that the Cambrian Evolutionary Fauna is only a subset of total Cambrian taxa. Some elements of the Burgess Shale fauna, for example, are not classified as part of the Cambrian Fauna because they belong to classes that achieved their peak later.

The Palaeozoic EF was much more diverse. It was dominated by articulate brachiopods, and other important components included crinoids, corals, ostracods, cephalopods, bryozoans and graptolites. Maximum diversity was reached in the late Ordovician to Devonian. A slow decline was followed by a very severe fall at the end of the Permian, particularly of epifaunal, fixed suspension-feeders. A partial recovery in the Triassic, and again in the Jurassic, was insignificant relative to the diversification of the Modern Fauna.

The Modern Fauna is even more diverse ecologically, and is dominated by gastropods and bivalves (together 58% of families), fish, bryozoans (a different group to that of the Palaeozoic Fauna), malacostracans (including crustaceans), and echinoids (sea-urchins). Most of the classes concerned appeared during the Cambrian and Ordovician but diversified only slowly during the Palaeozoic. They suffered relatively little in the late Permian extinction, and became the dominant EF in the Triassic. Thereafter, the Modern Fauna continued a rather slow but steady diversification. On the assumption that the pattern is that of logistic diversification (explained in Section 15.3.1), equilibrium diversity, $\hat{D}$, has apparently not yet been reached.

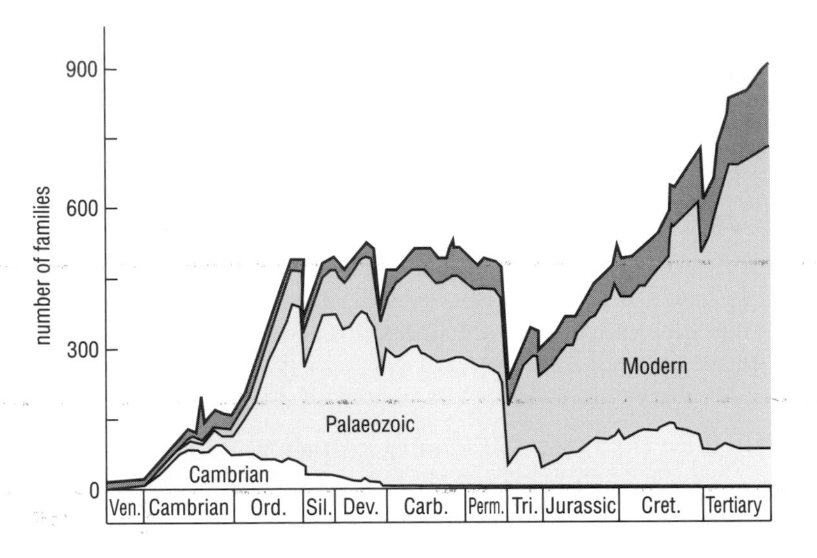

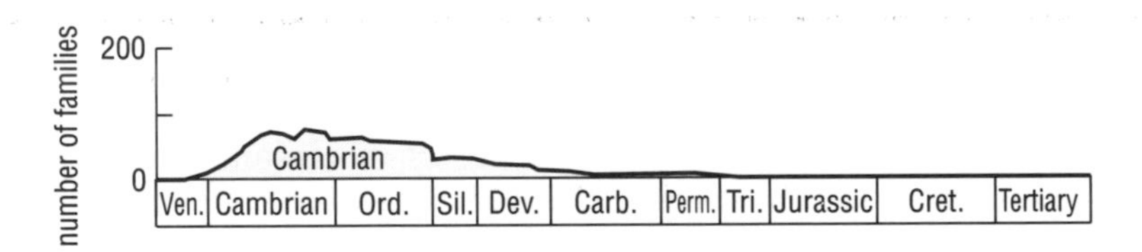

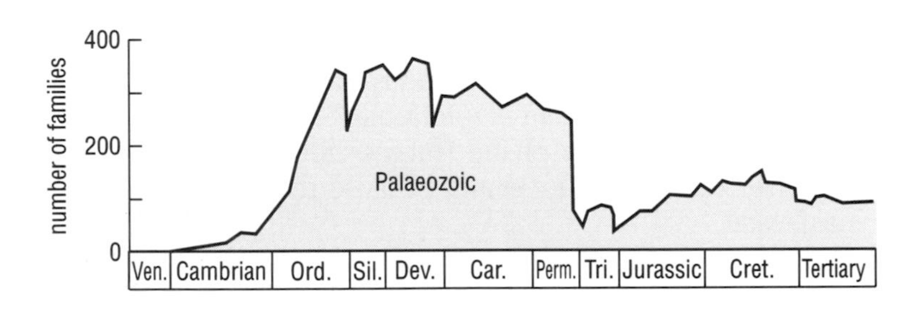

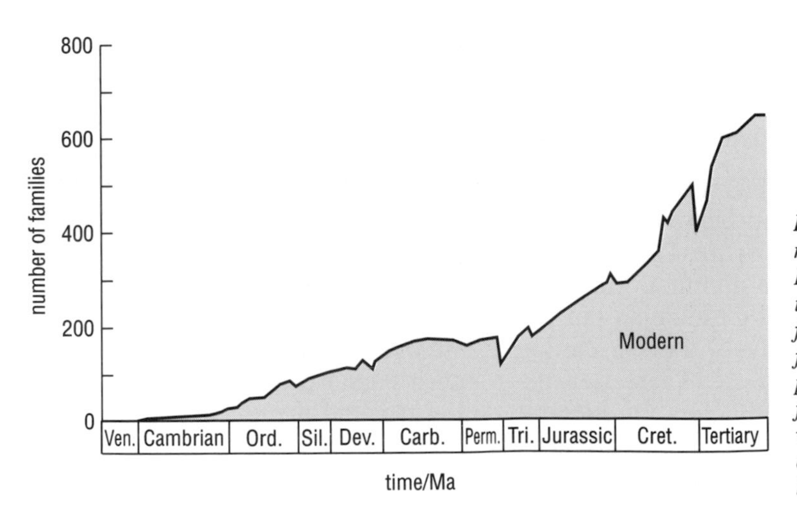

Figure 15.34 *(a) The diversity of marine animal families during the Phanerozoic. The top curve shows the total number of marine families known from the fossil record, and the grey field below represents families with poor preservation potential. Below that field, two curves divide families with well-mineralized skeletons into three evolutionary faunas, shown separately below in (b), (c), and (d).*

To assist in the description and analysis of evolutionary faunas, Sepkoski developed a three-phase model, based on overlapping patterns of logistic diversification (Sepkoski, 1984). Three coupled logistic equations were proposed, one for each fauna. Each EF was assumed to have diversified logistically as a consequence of early exponential growth and of later slowing as the ecosystems became filled. In the model, each EF can diversify and replace the preceding fauna only if its initial diversification rate (r_d) is lower and its equilibrium diversity ($\hat{D}$) is higher than the preceding fauna's. The model assumes interaction among the EF's such that expansion of the combined diversities of all three faunas above any single EF's equilibrium caused that EF's diversity to begin to decline. Mass extinctions can be incorporated in the model as time-specific perturbations in which extinction rates are increased and equilibrium diversities decreased for a short interval.

Figure 15.35 shows a solution of Sepkoski's model, with two perturbations (marked by arrows) that simulate the late Permian and late Triassic mass extinctions. The dashed lines in Figure 15.35 represent the trajectory of the unperturbed three-phase system. Notice how well the model fits the general features of the empirical pattern of Figure 15.34(a).

Sepkoski's three-phase model of constituent evolutionary faunas, each undergoing logistic diversification, and each linked to the fortunes of the others, thus simulates closely the history of marine family diversity. Explaining the pattern is, however, another matter. Some of the more plausible hypotheses relate to differences in the ecological characteristics of the EF's.

The Cambrian Fauna was dominated by generalized deposit-feeders and grazers (although predators and scavengers were also present), and benthic suspension-feeders lay close to the sea-floor. The spatial development of communities both above and below the sediment surface is called *tiering* (see Figure 12.12 for an illustration concerning tiering on hardgrounds in the Palaeozoic and Mesozoic). The Cambrian Fauna thus had low epifaunal tiering, and also low infaunal tiering, i.e. only shallow burrowing. The succeeding Palaeozoic EF communities were dominated by epifaunal fixed suspension-feeders such as articulate brachiopods, crinoids and bryozoans that were much reduced in the late Permian. Tiering became more complex in the Palaeozoic Fauna, with more animals feeding at different levels; crinoids formed the highest tier, with stems often over 1 metre. The Palaeozoic Fauna was more varied ecologically than the Cambrian Fauna, and the Modern Fauna was yet more diverse, with very complex feeding interactions. Swimming predators such as cephalopods and fish, many of highly specialized habits, increased enormously in the Modern Fauna, as did burrowing animals, such as burrowing bivalves, burrowing echinoids, and burrowing crustaceans. However,

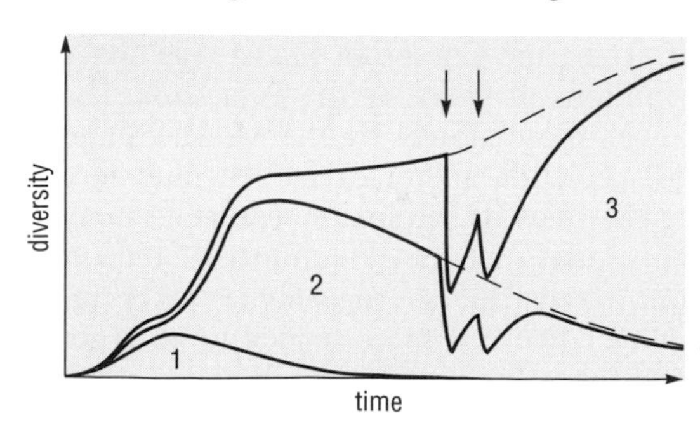

Figure 15.35 *A solution of Sepkoski's three-phase model for the history of evolutionary faunas. Arrows mark perturbations simulating mass extinctions. The three phases are explained in the text.*

epifaunal tiering was reduced in height and complexity in most communities by the Cretaceous, with far fewer fixed suspension-feeders than in the Palaeozoic Fauna.

One very important factor in the diversification of the Modern Fauna was the rapid rise in the Mesozoic of marine predators specializing in hard-shelled prey (which they crush, smash, drill, force open, or swallow whole). At the same time, there were important changes in the shell structure and life habits of other bottom-dwelling animals. Marine gastropods, for example, evolved many modifications of shell structure that conferred some resistance to shell-destroying predators, such as thick shells with prominent external sculpture, and narrow apertures with reinforced rims (e.g. dog whelks in Figure 14.10). The diversification of shell-destroying predators has certainly been a source of considerable increase in species diversity especially since the Cretaceous. The increases in specialized predators, in burrowers, and in predator avoidance and defence mechanisms, were so significant that together they have been called the **Mesozoic Marine Revolution (MMR)** by Geerat Vermeij of the University of California, Davis. The MMR was introduced in Chapter 12, Section 12.3.1, in which a predatory influence on the evolutionary proliferation of cemented bivalves was discussed.

Another idea, the 'biological bulldozer hypothesis', relates the decline of fixed suspension-feeders to the massive increase in grazing pressure, particularly by certain fish and echinoids, and in churning over of the sediment by burrowers (some of which probably evolved the burrowing habit under increased pressure from predators). Deep-burrowing bivalves, for example, although appearing in the Palaeozoic, only became very diverse in the Cretaceous. With an increase in bioturbation (disturbance of sediment by organisms), fixed suspension-feeders would find it difficult to attach as larvae, or stay anchored as juveniles and adults, and frequent clouds of disturbed sediment would tend to clog their filters. Today, fixed suspension-feeders are mainly restricted to hard substrates, whilst suspension-feeders on soft substrates tend to be mobile forms. Thus, a combination of increased predation and 'bulldozing' by elements of the Modern Fauna has probably been at least partly responsible for the failure of the Palaeozoic Fauna (especially the fixed epifauna) to recover since its demise in the late Permian.

It is not clear why shallow water environments should have favoured more kinds of ecological communities, as was the case with successive evolutionary faunas. The reasons for this are currently a source of much debate: the high productivity of shallow water communities, linked to patterns of nutrient supply, is one possibility being considered.

Onshore–offshore patterns. In a general way, each EF seems to have replaced its predecessor in shallow-water communities, gradually displacing it towards the edge of the continental shelf. Thus, the Cambrian Fauna was apparently 'pushed offshore' over tens of millions of years by the Palaeozoic Fauna, which in its turn was 'pushed offshore' even more slowly by the Modern Fauna. The pattern could be linked to differential speciation and/or extinction rates in different clades and different settings, along the lines of the incumbent replacement model discussed in Section 15.5. In general, clades of later evolutionary faunas tended to have lower average rates of turnover, so that the average higher taxon in successive faunas became progressively longer lived. If there tended to be a general increase in extinction intensity toward the shore, as seems the case, at least in the Palaeozoic, the more extinction-resistant clades of later evolutionary faunas may have simply persisted there at times of crisis and then expanded in the only direction they could — offshore.

One of the most interesting findings is that, in the Mesozoic, higher marine taxa, especially orders, tended to appear first in onshore environments. This onshore origination of higher taxa has been documented in a variety of groups, including crinoids, echinoids, bivalves and bryozoa. Novelties at lower taxonomic levels (families and genera) show no such bias. Why major evolutionary novelties should tend to appear in shallow water environments is not clear. In some Mesozoic cases at least, the general pattern seems to be: (1) *onshore origination* of higher taxa, followed by (2) *offshore expansion* into progressively deeper environments, followed by (3) *retreat offshore* of higher taxa, which progressively disappear from onshore environments until they remain only in offshore environments. Onshore–offshore patterns can thus be recognized among individual higher taxa (orders), as well as ensembles of taxa (evolutionary faunas).

The significance of Sepkoski's evolutionary faunas has been queried on several grounds, one of which is that they are largely abstract statistical constructs. The history of the Cambrian Fauna, for example, is largely that of trilobites, which form 77% of its total family diversity; such a dominant group, purely by being diverse, swamps other clades, which are pulled along during the statistical factor analysis and grouped within that fauna. The logistical models used in simulating the history of marine family diversity have also been criticized. The feature central to the concept of logistical diversification may be invalid, i.e. that taxonomic diversity will tend to an equilibrium state, supposedly due to the diversity dependence of rates of extinction and origination of taxa. The idea that the Earth contains a limited number of niches is an old one, and this 'principle of plenitude' was accepted by Darwin. There is, however, little evidence for this notion, and there may be no limits to diversification, which, in the past and the future, may *tend* to continue *ad infinitum*. It can also be argued that, although Sepkoski's three-phase logistical model fits the empirical data, certain assumptions may not be valid, and other models with different assumptions could be made to fit the data too. Chance may have played a major role throughout; there may, in reality, have been no shared biological features that specially united the main groups within each evolutionary fauna. Another argument is that, because of mass extinctions, equilibrium diversity was actually never reached in either of the first two evolutionary faunas, even if the balancing forces of origination and extinction were continually at work to achieve it. Again, these are difficult issues, and ones that are actively being researched.

One of the main features of Figure 15.34(a) is that the total number of marine families has increased to a maximum as the present day is approached. It is important to bear in mind that the changes in total family diversity are the result of the dynamics of both origination and extinction. There is evidence of an overall Phanerozoic trend toward declining turnover rates at the family level. Rates of extinction decrease, as do rates of origination, although total diversity has increased. In other words, clades have become less volatile. A number of reasons have been proposed for this. Jim Valentine of the University of California, Santa Barbara, suggested that there is 'a fundamental asymmetry in clade dynamics: the upper limit of diversity permitted by resources may be tested repeatedly, but the lower limit may be tested only once; extinction is forever. Therefore in the long run, the clades with superior extinction resistance are more likely to endure than those with high extinction rates, no matter that the origination rates of the latter are correspondingly high. So, in fact, clades with low turnover rates are more likely to persist, other things being equal, than clades with high turnover rates' (Valentine, in Ross and Allmon, 1990). Genera, as well as families, also tend to show a decrease in the rate of background extinction, at least since the Permian, as can be seen from Figure 15.29.

The pattern of change in the diversity of families in Figure 15.34(a) does *not* reflect a proportional change in the total number of classes or orders. Most marine classes originated early in the Phanerozoic and the total number has remained much the same since. The number of orders exceeded 100 in the Ordovician, and since then has only fluctuated between limits of about 90 and 130, averaging about 115, with times of change roughly coinciding with those of families.

15.6.2 Global species diversity

Does the history of species diversity correspond to that of family diversity? In other words, would the shape of the curve for the total number of marine species match that for the total number of marine families in Figure 15.34(a)? Has the number of species remained more or less constant since, say, the Ordovician (except for the influence of mass extinctions), in accordance with an equilibrium model? This is a difficult question, principally because the data for species are so much less complete than those for higher taxa. Nevertheless, estimates can be made that attempt to take into account some of the biases concerned in extrapolating from the available data. The most reliable data, as usual, concern fossilizable marine invertebrates.

Figure 15.36(a) shows apparent species diversity of fossil marine invertebrates through the Phanerozoic, plotted as the estimated average number of described species per million years of each period. Figure 15.36(b) shows the area of sedimentary rock mapped for each period, excluding the Quaternary, in km^2 per year. Figure 15.36(c) shows a measure of the stratigraphical distribution of palaeontologists' interests: the fossils of some periods have attracted more attention than others. The histograms in Figure 15.36(a) to (c) show amounts per unit time because the periods lasted for different intervals, and the longer periods would otherwise be disproportionately represented.

▶ Is there much similarity between the three histograms for apparent species diversity, the mapped area of sedimentary rock, and 'palaeontological interest units'? If so, what are the main implications for changes in species diversity?

Yes, there is a strong correspondence between the three histograms in Figure 15.36(a) to (c). At first glance, it would seem that sampling bias could be entirely responsible for an impression of species diversity increasing to a maximum in the Cainozoic, whereas in reality diversity might have been in equilibrium for most of the Phanerozoic. The relatively low sampled diversity of earlier periods could merely be due to the general decline of exposed or mapped strata with increasing

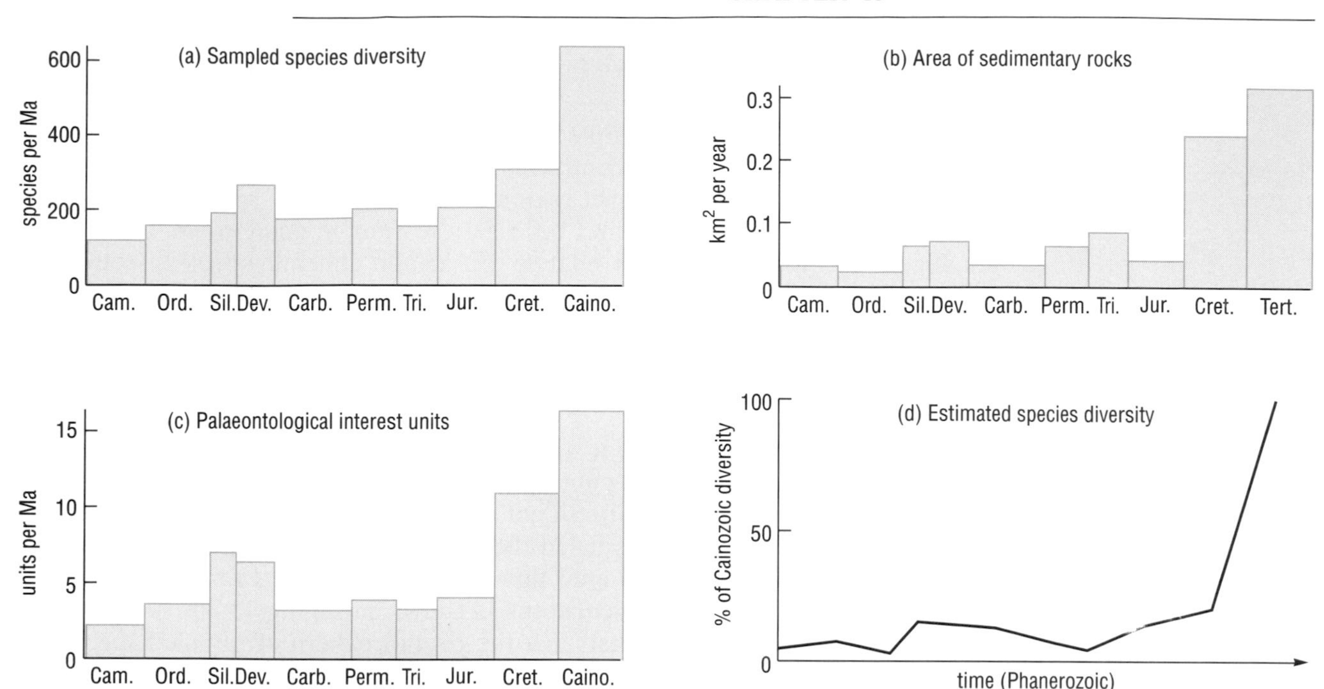

Figure 15.36 *Aspects of global species diversity. (a) Sampled (i.e. apparent) species diversity for well-mineralized marine invertebrates. (b) Mapped area of sedimentary rock (Quaternary excluded). (c) Distribution of palaeontologists' interests (units based on publications). (d) Estimated diversity as a percentage of Cainozoic diversity, taking sampling bias into account.*

age, as shown in Figure 15.36(b). This decline arises because the older a sedimentary rock is, the more likely it will have been eroded, or buried under younger sediments. The quality of fossil preservation also tends to deteriorate with increasing age. In addition, if palaeontologists are not so interested in the rocks of a particular period, for whatever reason, then species of that period are less likely to be found and described. Untangling the biases is complex, however. Palaeontologists are naturally interested in describing fossils, and are attracted to strata with rich fossil pickings. Also, the dominant kinds of sedimentary environments, both actual and preserved, have varied greatly, and this can affect apparent diversity. For example, a single lump of reef limestone may contain more species than a desert sandstone hundreds of metres thick. Times of low sea-level are associated with erosion of earlier deposits, and so on.

Several lines of evidence, although only partially independent, suggest that the major feature of Figure 15.36(a), i.e. a marked increase in species diversity in the Cretaceous and Cainozoic, is not simply an artefact:

1 It is clear that the number of marine families has risen greatly, as shown in Figure 15.34(a). There is also empirical evidence that the number of genera per family has increased markedly since the Palaeozoic, at least for easily fossilized marine invertebrates. If the number of species per genus also went up, the increase in families would translate into an immense increase in species from the end of the Mesozoic.

2 Studies of species diversity within a large number of individual assemblages, originally from nearshore and open marine environments, have suggested that within-community diversity has almost doubled from about the end of the Mesozoic.

3 Trace fossils show the highest diversity within individual habitats in the Cainozoic.

4 Provinciality probably rose to unprecedented levels over the Mesozoic and Cainozoic as a result of extensive continental break-up (Chapter 13, Section 13.4.1), resulting in numerous distinct regions containing many species not found elsewhere. This means that it becomes necessary to sample many more geographical areas to discover even relatively common endemic species, so true Cainozoic diversity is likely to be under-represented unless fossil collections are correspondingly more widespread.

5 Other factors, such as the Marine Mesozoic Revolution, increased behavioural complexity, etc., are expected to have promoted high species diversity.

An attempt to estimate the percentage variation in diversity of skeletonized marine invertebrate species, taking into account the sampling bias evident in Figure 15.36(b) and (c), was made by Philip Signor of the University of California, Davis. Making certain assumptions, he estimated the total number of fossilizable species that had lived in the Cainozoic and used this to calibrate estimates for earlier periods. A typical result of such calculations is shown in Figure 15.36(d). These findings lend support to the hypothesis that the overall pattern of sampled species diversity in Figure 15.36(a) is a true signal of the underlying trend. Note that the late Cretaceous mass extinction is not evident here because the plotted curve is generalized for whole geological intervals such as the Cretaceous and Cainozoic.

The species diversity for terrestrial vertebrates and for plants seems also to have increased substantially in the Cainozoic. The pattern at species level is, however, even less clear than for marine invertebrates. The number of terrestrial vertebrate orders, in contrast to marine invertebrate orders, has almost doubled in the Cainozoic compared with the Mesozoic, largely because of the radiation of birds.

The numbers of higher taxa such as classes, and to a lesser extent orders, are well damped against variations in species diversity, as expected. In general, for marine invertebrates in the Phanerozoic, there was an increasing discrepancy between the numbers of higher taxa and the species included within them. Although neither the details of the pattern of change in species diversity, nor the controls on it, are clear, it seems almost certain that by Recent times there were more species on Earth than ever before.

SUMMARY OF SECTION 15.6

The fossil record shows a general but uneven increase in the number of marine families throughout the Phanerozoic, with reductions at mass extinctions, especially in the late Permian. Three evolutionary faunas — sets of higher taxa, especially classes, that have similar histories of diversification and that together dominate the biota for a long interval — have been identified: the Cambrian Fauna, the Palaeozoic Fauna, and the Modern Fauna. Each one displays a slower rate of diversification, but a higher level of maximum diversity, than any preceding it, which declines as each new fauna expands. A three-phase model of constituent evolutionary faunas, each undergoing logistical diversification, and each linked to the history of the others, simulates quite closely the history of marine family diversity.

A marked increase in marine family diversity accompanied the proliferation of the Modern Fauna in the post-Palaeozoic. The Modern Fauna is the most diverse, taxonomically and ecologically, and equilibrium diversity has apparently not yet been reached. The diversification of shell-destroying predators in the Mesozoic Marine Revolution, and an associated increase in deep burrowing ('biological bulldozing'), were probably instrumental in the failure of the Palaeozoic Fauna (especially fixed suspension-feeding brachiopods and crinoids) to recover after its massive demise in the late Permian.

A pattern of movement from onshore to offshore environments can be recognized among whole evolutionary faunas and some individual higher taxa. Some higher taxa seem to have originated in shallow water, expanded into deeper water, and subsequently retreated from shallow water, remaining only offshore.

Most marine phyla and classes originated early in the Phanerozoic and their diversity has remained very similar since. Family turnover rates have declined: rates of extinction and origination have both decreased, although total family diversity has increased. Estimates of changes in species diversity are much less reliable than those for higher taxa. Nevertheless, there seems to have been a general but uneven increase in species diversity throughout the Phanerozoic, and higher taxa tend to contain increasingly more species as the Phanerozoic proceeds. After attempting to correct for various sampling biases, such as the progressive loss of exposed sedimentary rocks with increasing age, and the uneven distribution of palaeontologists' interests in fossils of different periods, a spectacular increase in species diversity seems to have occurred in the Cretaceous and Cainozoic. This conclusion is supported by various data, including an increase in diversity within communities.

15.7 THE EVOLUTIONARY HIERARCHY

So far in this Book, microevolution and macroevolution have been distinguished merely as successive scales of evolution, with macroevolution referring to evolution above the species level. It has been convenient to discuss them separately, because each scale has its own topics of interest, calling for somewhat different forms of evidence and methods of study. In this Section, however, we stand back to consider the nature of the relationship between the two scales, and thus the workings of the evolutionary hierarchy as a whole.

At issue is the question: is macroevolution merely an extrapolation of microevolution, or are there additional principles that come into play at the macroevolutionary scale, which could not have been predicted from an understanding of microevolution alone? In other words, is neo-Darwinian theory, based upon the genetic consequences of individual selection and associated processes, *sufficient* to explain patterns of macroevolution? The orthodox view from around the middle of the 20th century has been that it is. The whole thrust of G. G. Simpson's monumental contribution to the 'Modern Synthesis' of neo-Darwinian theory (see Chapter 1, Section 1.1.3 and Chapter 10, Section 10.3.1) was to show that patterns of evolution at all scales were consistent with the theory, and that no additional principles were called for. As Section 10.3.3 in Chapter 10 concluded, his view can still be upheld at the level of evolving species populations: today, notions of saltation by macromutation, or of orthogenetic change, seem as improbable, and unsupported by any evidence, as they did to Simpson. The dominant biological models of speciation being discussed also still fall within the scope of neo-Darwinian theory, albeit enriched with modern findings in such areas as genetic mechanisms and behavioural studies (Chapter 9).

It is true that the recognition of major perturbations affecting rates of extinction and origination, with consequential mass extinctions and radiations, requires much historical knowledge over geological time-scales in addition to an understanding of microevolution. However, it could be argued that such evolutionary events merely qualify the context within which neo-Darwinian principles can be supposed to have operated, rather than undermine them.

In these respects, then, the Modern Synthesis has provided a satisfactory perspective for insights into many evolutionary patterns. Here and there in this Book, however, some warnings have been sounded about the dangers of simplistic extrapolation from the microevolutionary to the macroevolutionary scale. In Section 15.5, for example, the principle of competition, which is clearly an important influence on the microevolution of species populations (as illustrated in Chapter 8), was found to be inappropriate as an explanation for the extinctions of species in the turnover of higher taxa.

Likewise, a caution was given in Chapter 9 and Section 15.4 concerning extrapolation from the patterns of extinction among local populations to the level of entire species populations. Populations of *r*-selected organisms, for example, may be typically ephemeral, springing up as conditions allow, but frequently being snuffed out by environmental disturbances. (As Section 8.5.3 of Chapter 8 showed, their adaptations are characteristically such as to promote rapid repopulation.) Yet there is no evidence at the macroevolutionary scale that such species are any more prone to extinction than, say those of *K*-selected forms. Their opportunistic patterns of repopulation evidently compensate for the ephemeral existence of their local populations. Thus, patterns evident at the microevolutionary scale cannot necessarily be transposed in a simplistic fashion to the record of macroevolutionary change.

The nub of the problem of extrapolation is the issue of the focus of selection, which was discussed in Chapter 6. The only necessary beneficiaries of the operation of natural selection on individuals are the fittest genotypes (or more properly, those with the greatest inclusive fitness). Adaptations may influence rates of speciation, or of extinction, but only by way of accidental *effect*: they cannot evolve the *function* (Chapter 2, Section 2.2.4) of promoting speciation, or of resisting extinction, say, because adaptations do not operate 'for the good of the species' (Chapter 6, Section 6.5.6). Nevertheless, if the adaptive differences between the members of different higher taxa happen to influence the relative rates of speciation and/or extinction of the taxa in a consistent fashion, then such differences can produce a *sorting* effect (Chapter 6, Section 6.1), whereby some taxa simply diversify more than others. The South African vertebrate palaeontologist Elizabeth Vrba has termed this the **effect hypothesis**.

▶ Can you think of an example, mentioned earlier in this Chapter, of a model for the replacement of one higher taxon by another that was based upon the effects of the adaptive differences between the individuals of the two taxa?

The incumbent replacement model, described in Section 15.5. According to this model, taxonomic replacement results from the relatively greater rate of speciation of an invading taxon arising from some competitive advantage over remaining members of the incumbent taxon in the taking over of the resources of vacated niches.

The effect hypothesis poses no fundamental problem for neo-Darwinian theory; it simply recognizes that the macroevolutionary expression of microevolutionary processes may sometimes be counter-intuitive. A feature may both promote the fitness of its individual possessors and, in the longer term, prove to be disadvantageous for the survival of species comprising such individuals.

▶ Cast your mind back to Chapter 5 (Reproductive Patterns). Can you recall a feature discussed there that, it was argued, had contrasting implications for individuals and for species?

In Chapter 5, Section 5.1.3, it was argued that asexual reproduction offers several advantages over sexual reproduction at the individual level, but that species of asexually reproducing organisms are more prone to extinction than their sexually reproducing relatives, because of their lack of genotypic variety with which to confront environmental change.

In the example above, the increased probability of extinction of the asexual species is due to the aggregate effect of lowered individual fitnesses in the circumstance of a changed environment. In so far, then, as the *sorting* of the species (asexual versus sexual) is an effect of *selection* acting on individuals, the phenomenon is still consistent with the neo-Darwinian axiom of individual selection.

In contrast, it has been cogently argued by Vrba and Gould (1986), among others, that some sorting of species may also arise as a result of direct selection of the species themselves. In this model, the features upon which selection acts are those of entire species populations, which are irreducible to the level of the individual. An example might be the size of the geographical range of a species: that is a property of the entire population, not of its individuals. Of course, such properties are necessarily emergent from those of the individuals making up the population — the dispersal of the individuals determines the geographical range of the population — but the emergent property in question is nevertheless still a characteristic of the species as an entity. In so far as any such property is heritable by daughter species, the theoretical possibility exists for evolution through selection acting on that property of the species: the process is analogous to evolution by the natural selection of individuals, though operating at a higher level in the evolutionary hierarchy. **Species selection**, as this postulated process is termed, does contradict the neo-Darwinian axiom of individual selection, and thereby introduces the possibility of a higher-level process that may shape patterns of macroevolution.

A possible example of species selection, operating indeed upon the relative geographical ranges of species, has been put forward by Thor Hansen. In a study of some Tertiary gastropods from the American Gulf Coast region, he noted a correlation between the geographical ranges of species and their stratigraphical ranges — more widely distributed species tended to last longer. One way to explain this is that populations which become extinct in one region, through a chance combination of factors, may be re-established if other populations exist elsewhere to permit restocking. The geographical ranges were themselves found to be strongly linked with the mode of larval development and the environmental tolerances of the individuals comprising the species (larval type can be recognized from the form of the initial part, or *protoconch*, of the shell). Hansen found that species with planktonic larvae, combined with adult tolerance of a broad range of habitat types, tended to achieve wider geographical ranges, and also stratigraphical ranges, than those showing development in fixed egg masses (Figure 15.37). In so far as these individual traits tended to be passed on to daughter species, the emergent property of size of realized geographical range could be considered heritable. Here, the individual traits would have acted for species analogously to genes in organisms, causing them to develop certain (emergent) properties.

Later work, by David Jablonski, has shown that this kind of link between the distribution and duration of benthic molluscan species broke down during the mass extinction at the end of the Cretaceous (Section 15.4). This is intelligible if species had indeed been the units of selection: the catastrophic events associated with the mass extinction may well have been of virtually global extent, such that the differing geographical ranges of individual species would have had little influence on their susceptibility to extinction. However, as mentioned in Section 15.4, wide geographical distribution of the different species within a genus may increase generic survival during mass extinctions.

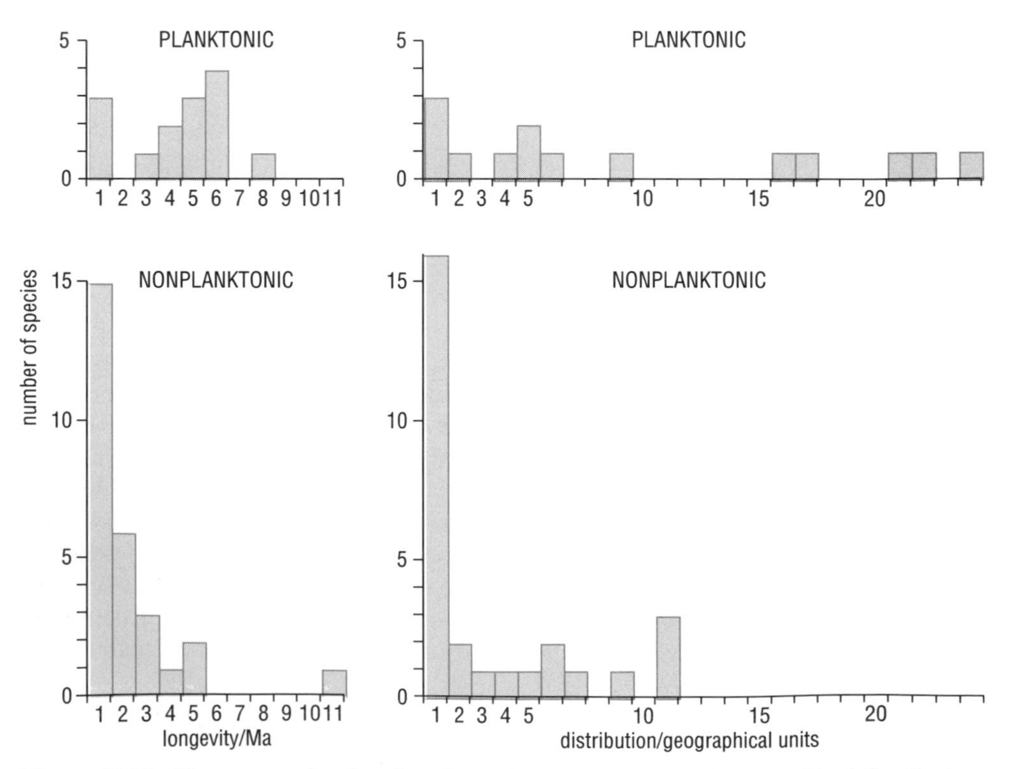

Figure 15.37 *Histograms showing the relative longevities (left) and geographical distributions (right) of species of volutid gastropods, with planktonic larvae (above) and non-planktonic larvae (below) in the Lower Tertiary of the Gulf Coast of the USA. Size of geographical distribution is measured in arbitrary units of area.*

There are several possible criticisms of this example, which call for much further testing. For example, sampling biases alone may account for at least some part of the correlation between geographical and stratigraphical range, given the incomplete nature of the fossil record. Another possibility, that the difference in extinction rates is directly attributable to the aggregate effect of individual fitness differences — i.e. the effect hypothesis — needs also to be rejected before the hypothesis of species selection can be accepted in this case.

On the other hand, there are various theoretically plausible candidates for other emergent species properties which might cause species selection, not only through differential extinction but also through differential rates of speciation. For example, the tendency for limited dispersal of the plethodontid salamanders mentioned in Section 15.5 means that gene flow between neighbouring populations is highly susceptible to disruption during episodes of climatic deterioration, promoting the probability of allopatric speciation, as discussed earlier. Highly specialized fertilization systems (Chapter 9, Section 9.2.3) may make populations susceptible to the disruption of gene flow as a consequence of temporary separation of populations (as with Kaneshiro's model for Hawaiian drosophilids: see Section 15.5), thus favouring high rates of speciation. In both examples, a property of populations (restricted patterns of gene flow), which is emergent from individual properties, may promote speciation.

As yet, no example of species selection has been unambiguously established, with all rival hypotheses rejected. Yet the idea is theoretically valid, and the testing of it, relying as it does on the detection of differential rates of speciation and/or extinction, linked with certain species properties, poses an exciting challenge to evolutionary palaeontologists. If confirmed, it would refute the neo-Darwinian assumption that all patterns of macroevolution are merely the scaled-up consequences of microevolutionary processes. That prospect may yet be some way off, but it is salutary to remember that Darwin's own theory of natural selection took more than half a century to be adequately tested and generally accepted by students of evolution.

Summary of Section 15.7

An important question is: can the extrapolation of neo-Darwinian principles of microevolution adequately explain patterns of macroevolution? In many respects the answer still seems to be 'yes', as G. G. Simpson argued in his major contributions to the Modern Synthesis half a century ago, though allowances must be made for the disruptive effects of mass extinctions. However, simplistic extrapolations of pattern from microevolution to macroevolution are rarely appropriate, because the adaptations fashioned by natural selection necessarily benefit only the fittest genotypes in a population, not the entire population. Probabilities of speciation, or of extinction, in different clades may nevertheless be influenced in a consistent fashion by the adaptive differences between their members, but only by way of effect, perhaps with counter-intuitive consequences for the sorting of species (the 'effect hypothesis').

One theoretically possible exception to the sufficiency of neo-Darwinian theory is species selection. In this hypothesis, the sorting of species may also arise from direct selection for irreducible (though emergent) features of species populations with some degree of heritability in daughter species. Size of geographical range is one such feature, and there is some evidence consistent with it being a target of selection. Another possibility is a characteristic tendency for a restricted pattern of gene flow, which promotes high rates of speciation.

The sorting of species, i.e. the differential representation of species with time, may thus occur either as an indirect effect (the effect hypothesis) or by direct selection (i.e. species selection). The idea of species selection has been little tested, but if confirmed will require a hierarchical expansion of our understanding of evolutionary *processes*.

REFERENCES FOR CHAPTER 15

Alvarez, L. W., Alvarez, W., Asaro, F., and Michel, H. V. (1980) Extraterrestrial cause for the Cretaceous–Tertiary extinction, *Science*, **208**, 1095–1098.

Donovan, S. K. (1989) *Mass Extinctions — Processes and Evidence*, Belhaven Press, London.

Eldredge, N. and Stanley, S. M. (eds) (1984) *Living Fossils*, Springer Verlag, New York.

Forey, P. L. (1988) Golden jubilee for the coelacanth *Latimeria chalumnae*, *Nature*, **336**, 727–32.

Gould, S. J. (1988) Trends as changes in variance: a new slant on progress and directionality in evolution, *Journal of Paleontology*, **62**, 319–29.

Gould, S. J. and Calloway, C. B. (1980) Clams and brachiopods — ships that pass in the night, *Paleobiology*, **6**, 383–96.

Jensen, J. S. (1990) Plausibility and testability: assessing the consequences of evolutionary innovation, in Nitecki, M. H. (ed.) *Evolutionary Innovations*, 171–190, University of Chicago Press.

Larson, A. (1989) The relationship between speciation and morphological evolution, in Otte, D. and Endler, J. (eds) *Speciation and its Consequences*, 579–598, Sinauer Associates.

McNamara, K. J. (ed.) (1990) *Evolutionary Trends*, Belhaven Press, London.

Pearson, P. N. (1992) Survivorship analysis of fossil taxa when real-time extinction rates vary: the Paleogene planktonic foraminifera, *Paleobiology*, **18**, 115–31.

Raup, D. M. (1991a) A kill curve for Phanerozoic marine species, *Paleobiology*, **17**, 37–48.

Raup, D. M. (1991b) *Extinction: Bad Genes or Bad Luck?*, Norton and Co., London.

Raup, D. M., Gould, S. J., Schopf, T. J. M. and Simberloff, D. S. (1973) Stochastic models of phylogeny and the evolution of diversity, *Journal of Geology*, **81**, 525–42.

Raup, D. M. and Sepkoski, J. J. (1984) Periodicity of extinctions in the geologic past, *Proc. Nat Acad. Sci. USA*, **81**, 801–5.

Ross, R. M. and Allmon, W. D. (eds) (1990) *Causes of Evolution*, University of Chicago Press.

Rozenzweig, M. L. and McCord, R. D. (1991) Incumbent replacement: evidence for long-term evolutionary progress, *Paleobiology*, **17**, 202–13.

Sepkoski, J. J. (1978) A kinetic model of Phanerozoic taxonomic diversity. I. Analysis of marine orders, *Paleobiology*, **4**, 223–51.

Sepkoski, J. J. (1981) A factor analytic description of the Phanerozoic marine fossil record, *Paleobiology*, **7**, 36–53.

Sepkoski, J. J. (1984) A kinetic model of Phanerozoic taxonomic diversity. III. Post-Paleozoic families and mass extinctions, *Paleobiology*, **10**, 246–67.

Sharpton, V. L. and Ward, P. D. (eds) (1990) *Global Catastrophes in Earth History*, Geological Society of America Special Paper 247.

Skelton, P. W. (in press) Adaptive radiation: definition and diagnostic tests, in Edwards, D. and Lees, D. R. (eds), *Evolutionary Patterns and Processes*, Symposium Volume of the Linnean Society, London.

Van Valen, L. (1973) A new evolutionary law, *Evolutionary Theory*, **1**, 1–30.

Vrba, E. S. and Gould, S. J. (1986) The hierarchical expansion of sorting and selection: sorting and selection cannot be equated, *Paleobiology*, **12**, 217–28.

FURTHER READING FOR CHAPTER 15

Briggs, D. E. G. and Crowther, P. R. (eds) (1990) *Palaeobiology: A Synthesis*, Blackwell Scientific Publications, Oxford.

Gould, S. J., Gilinsky, N. L. and German, R. Z. (1987) Asymmetry of lineages and the direction of evolutionary time, *Science*, **236**, 1437–41.

Levinton, J. (1988) *Genetics, Paleontology, and Macroevolution*, Cambridge University Press, Cambridge, UK.

Taylor, P. D. and Larwood, G. P. (eds) (1990) *Major Evolutionary Radiations*, Systematics Association Special Volume No. 42, Clarendon Press, Oxford, UK.

SELF-ASSESSMENT QUESTIONS

SAQ 15.1 (*Objectives 15.1 and 15.3*) For each of the cases given in (a)–(f) below, choose the phenomenon listed in 1–7 that most closely relates to it.

(a) A lineage of gastropods in which the mean size of the population increased markedly within ten generations as a direct result of a steady increase in food availability, rather than directional selection or genetic drift.

(b) A group of rudist bivalves that, during evolution, came to look increasingly like a group of cone-shaped corals.

(c) A group of primate lineages that underwent an increase in mean brain size with the shape of the variate's distribution neither flattening out nor becoming more peaked.

(d) A bivalve lineage that underwent sufficient morphological change for earlier and later parts of the lineage to have been called different chronospecies.

(e) A long-lasting trilobite lineage that, over a period of about 10 Ma, budded off four successive, short-lived species, each with very similar morphology.

(f) A mammal clade in which small-bodied species tended to have longer durations than larger-bodied species.

1 Iterative evolution.

2 Anagenesis.

3 Biased extinction.

4 An evolutionary trend concerning a change in variance.

5 Convergent evolution.

6 A trend that is not evolutionary in nature.

7 None of the above.

SAQ 15.2 (*Objectives 15.2, 15.6 and 15.8*) Many ammonite lineages became extinct towards the end of the Jurassic. Can the pattern of rediversification in the Cretaceous be termed a radiation, on the strength of the data on changes in family diversity over successive 10 Ma intervals given in Figure 15.18? What further investigations might be appropriate to investigate this question?

SAQ 15.3 (*Objectives 15.4 and 15.5*) For each statement (a)–(d) below, indicate whether it is true or false. If false, give reasons for your answer.

(a) Data on the durations of fossil marine genera show that many more genera have lifespans in excess of the mean than have lifespans less than the mean.

(b) Living fossils do not represent a separate category of evolutionary rates quite distinct from rates in other organisms.

(c) The dominant cause of anagenetic size increase is greater physiological efficiency.

(d) Small-bodied species tend to be more susceptible to extinction than larger-bodied species.

SAQ 15.4 (*Objective 15.7*) Correct any errors, if present, in the following statements about mass extinctions.

(a) The 'Big Five' mass extinctions of the Phanerozoic include those near the end of the Ordovician, the Silurian, the Permian, the Jurassic, and the Cretaceous.

(b) Some Cretaceous–Tertiary boundary sections show 'Lazarus taxa', i.e. taxa whose fossils are continuously present through strata that record extinctions in other species.

(c) Although the late Permian mass extinction probably saw the loss of proportionally more species than the late Cretaceous mass extinction, the latter was more remarkable in that until the very end of the Cretaceous there was no sign of decline in any of the groups that became extinct.

(d) Most of the causal mechanisms proposed for mass extinctions are not mutually exclusive, which is one reason among many why it is difficult to establish the chain of cause and effect leading to the extinction of species.

(e) If a large (10 km) asteroid did strike the Earth at the end of the Cretaceous, and the evidence of the rubidium-rich layer strongly suggests that it did, this would undoubtedly have caused a vast number of extinctions.

SAQ 15.5 (*Objective 15.9*) For each statement (a)–(d) below, indicate whether it is true or false. If false (or partially false), give reasons for your answer.

(a) The Palaeozoic Evolutionary Fauna had become extinct by the end of the Permian, at which time epifaunal, fixed suspension-feeders suffered huge losses.

(b) Evolutionary faunas are sets of higher taxa, especially classes, that have similar histories of diversification and that together dominate the biota for a long interval.

(c) Equilibrium diversity in families of the Modern Evolutionary Fauna appears to have been reached by the early Tertiary, soon after the Mesozoic Marine Revolution.

(d) It is very likely that sampling bias is primarily responsible for the apparent massive increase in species diversity during the Cainozoic.

SAQ 15.6 (*Objective 15.10*) There are many more species of placental mammals today than there are species of marsupial mammals. Can we therefore conclude that the placentals are somehow adaptively superior to the marsupials?

Part IV
Case Studies

CHAPTER 16
ORIGINS

Prepared for the Course Team by Irene Ridge and Iain Gilmour

Study Comment *This Chapter is about evolution during the first three-quarters of the Earth's history. During this period, life—the first cells—originated from non-living molecular systems; the prokaryotes evolved into three major groups; and eukaryote cells originated by combining different prokaryotes. Much of what happened is informed guesswork or based on very indirect evidence but, to follow the arguments, you need some basic knowledge of chemistry and biochemistry.*

Objectives When you have finished reading this Chapter, you should be able to do the following:

16.1 Distinguish between the living and the non-living, and recognize the main characteristics of prokaryotic cells and eukaryotic cells.

16.2 Recognize and describe the main features of the Earth's early environment during the Hadean and be able to point out the implications for the origin of life by chemical evolution.

16.3 Give an account of chemical evolution describing and evaluating some of the solutions suggested for the following problems: the sources of chemical building blocks; protection and concentration of molecules; sources of energy for early and later systems; and replication.

16.4 Describe and assess evidence that photo-autotrophs were present 3 800–3 500 Ma ago and list the implications of this for the origin of life and for early heterotrophs.

16.5 List the two types of evidence which suggest that prokaryotes separated into three main lines early in the Archaean and describe the main characteristics, including energy metabolism, of modern representatives of these lines.

16.6 Discuss and give evidence for and against the ideas that the first prokaryotes were (a) fermenters; (b) chemo-autotrophs; (c) photo-autotrophs.

16.7 Give an account of the role of endosymbiosis in the origin of eukaryotes, citing or evaluating evidence for: (a) the autogenous origin of the nucleus; (b) the endosymbiotic origin of mitochondria, chloroplasts and flagella (or cilia).

16.8 Compare and contrast the endogenous hypothesis and Margulis' hypothesis about the origin of mitosis.

16.9 Describe ways in which diploidy, meiosis and the sexual cycle could have arisen and assess the advantages and evolutionary significance of each.

16.1 INTRODUCTION

In the study of evolution, the question of when and how life originated on this planet is one of the most fundamental and most difficult to answer. Earlier Chapters have been concerned mainly with evolution over the past 500 million years, the best documented and arguably the most interesting period, but only the tail end of the story. Our concern now is with the beginning and middle of the story, a period of some 3 500 million years, during which living cells appeared, evolved and profoundly altered the geology and climate of the planet.

Life was probably present on Earth 4 000 million years ago (4 giga years (GA) BP, or 4 billion years in popular speech) and, if you look at Appendix 1, you can see that this was a relatively short time after the origin of the Earth itself. What the Earth was like then is discussed in Section 16.2. It is a kind of chemical and geological detective story which describes the kinds of evidence used to chart the early history of the Earth. How the basic chemical building blocks of cells might have originated in this environment is discussed in Section 16.3. Section 16.4 considers in more detail how and in what order chemical substances might have come together to form simple cells—the first grade of cell organization which occurs now in prokaryotes. Finally, in Sections 16.5 and 16.6, we discuss the diversification of these prokaryotes, their effects on climate, and the origin from them of more complex eukaryotic cells.

16.1.1 WHAT IS LIFE?

Any discussion about the origin of life must face the problem of distinguishing the living from the non-living. Life as we know it is based on the unit of the cell and it is useful to start by examining a very simple unicellular organism to try to tease out the essential features of being alive. Figure 16.1 shows a generalized bacterial cell, a prokaryote.

Living bacteria are commonly 0.5–5 μm long and, in addition to water, each cell contains millions of macromolecules such as lipids (in membranes, for example), proteins and polysaccharides, some thousands of RNA molecules and one large molecule of DNA. Unlike the eukaryotic cell, the prokaryote cell has no nucleus, mitochondria or chloroplasts (see Chapter 3, Section 3.2.2). It can do two basic things:

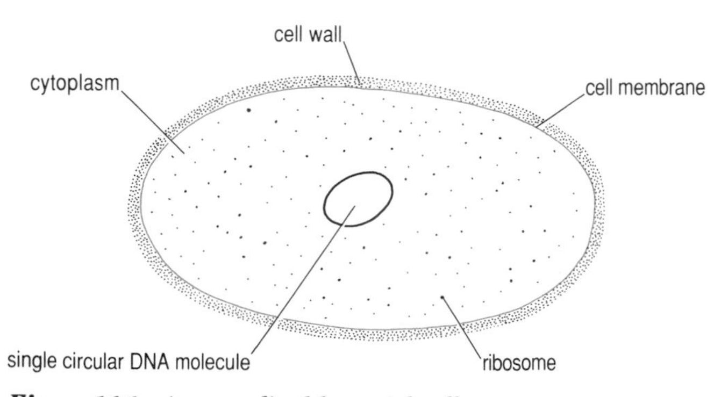

Figure 16.1 ***A generalized bacterial cell.***

1 Replicate to produce two smaller copies (daughter cells) which then grow to the size of the original (parent) cell. Central to this process is the single and usually circular DNA molecule. As you know from Chapter 3, DNA is both a *replicating molecule* and an *information store*—the sequence of its four components (nucleotides) encodes all the information needed to control and manufacture other cell components. Replication, whether of cells or DNA molecules, is generally faithful but an essential feature of living systems is that *there must be occasional mistakes in replication*. Change, i.e. evolution, can occur only if variants arise which can either occupy a new environment or out-perform and displace the original type in the original environment. Rare mistakes in replication or mutations are the source of variation and life could not have evolved without them.

2 Maintain itself, which involves obtaining energy and chemical raw materials and using them to manufacture everything needed for replication, growth and repair. The cell must also maintain an internal environment which is suitable for all this metabolism and for replication—appropriate conditions of pH and levels of ions, for example, which are often very different from those outside the cell. This last process is largely a function of the cell membrane, which controls molecular traffic into and out of the cell. Energy in the chemical form of **ATP** (**adenosine triphosphate**) can be used to pump materials across the membrane through ion- or molecule-specific channels.

The 'simple' bacterial cell is, therefore, an extremely complicated system where the fundamental living processes, replication (plus occasional mistakes) and metabolism, are inextricably intertwined. DNA replication requires energy (ATP); nucleotides manufactured by the cell; enzymes to catalyze the process; and a suitable microenvironment—purified DNA in a test tube cannot replicate unless provided with all these factors. In the same way, complex multicellular organisms like ourselves normally cannot reproduce or metabolize unless they are whole, because individual cells have become specialized. Some cells, such as mammalian red blood cells or phloem transport cells in plants, lose their nucleus and so clearly cannot divide, whilst all cells depend for essential supplies of energy and nutrients on the functioning of the entire organism. This is why the basic unit of life now is the whole system—the cell or organism—and not the replicating molecule alone. Here lies a real problem for theories about the origin of life because how could such a complex system as a cell, with such a high degree of interdependence between its parts, ever have arisen? How *did* life originate on Earth? The main types of theories are reviewed briefly in the next Section.

16.1.2 THEORIES ON THE ORIGIN OF LIFE

Most cultures have their own views concerning the origin of life. One such view is familiar to many of us in the teachings of the first Chapter of the Book of Genesis. Significantly, nearly all religious creeds embody the idea of some kind of omnipotent being capable of creating life from inanimate matter. This is *divine creation*, but because the existence of a creator is untestable the hypothesis is not scientific and therefore lies outside the scope of this Book.

Some early Greek philosophers believed that they could account for creation just as well by leaving 'the gods' out of it. They, and others before them, were convinced that lifeless decaying matter could turn into living things by a process called *spontaneous generation*—frogs formed from May dew, birds and insects from twigs, and fruits of trees and maggots from putrefying flesh. However, simple observations and experiments, such as those made by the Italian physician Francesco Redi in 1668, had disproved most of these ideas by the 17th century. Redi showed that the maggots in rotting meat developed from the eggs of flies. But the theory of spontaneous generation lingered on until the 19th century following the discovery of micro-organisms by the microscopist van Leeuwenhoek in 1683. Only when Louis Pasteur in 1861 showed that the so-called spontaneous generation of micro-organisms was due to the air and fluids not being sterile was it finally abandoned. Pasteur concluded that all living things have their origin in other living things with a total and sharp division between living and non-living. There was simply no scientific answer to the question of how the first living things arose.

Extraterrestrial theories Science entered the 20th century without any convincing ideas as to how life originated on Earth. At that time geology and astronomy provided some evidence that the Earth was originally molten and from this it followed that it would have been impossible for life to have existed at the time of its origin. The difficulty was side-stepped in 1907 by the Swedish chemist Arrhenius who proposed the theory of *panspermia* (germs everywhere). As its name implies, this theory suggests that life on Earth came from outer space as micro-organisms drifted from planet to planet propelled by the pressure of starlight. Although the hypothesis had, and continues to have, a few eminent supporters, the dangers of space travel (e.g. the lethal effects of radiation to organisms) are considered by most scientists to preclude such cosmic wanderings. Furthermore, the assumption that life is ubiquitous seems to be inconsistent with the current view that stars and galaxies originate as clouds of hydrogen or helium. Only as stars age are the heavier elements necessary for the origin of life formed. For these reasons, panspermia was discarded in favour of the theory that life originated in those parts of the Universe where conditions are or were favourable. Earth is the only place where we know that life arose, but the possibility that it evolved independently in other parts of the Universe or even in other planets of our solar system cannot be ruled out.

The chemical theory If an extraterrestrial origin for life is ruled out, just one possibility remains—that life arose on Earth from organic molecules produced abiogenically, i.e. not manufactured by organisms. This is the essence of the **chemical theory**, which states that the appearance of life followed a period of *chemical evolution* at an early stage in the Earth's history when the atmosphere contained no oxygen. During this period, simple inorganic molecules reacted to give organic molecules such as amino acids and sugars from which more complex organic compounds and eventually macromolecules were formed. Organized replicating systems, then protocells, and finally true cells evolved.

These basic ideas remain unchallenged to the present day, but up to the 1980s thinking was dominated by the first version of the chemical theory, which included more detail about where chemical evolution occurred and the nature of early metabolism. This is the **Oparin–Haldane theory**, so named because the idea was conceived quite independently by the Russian and British biologists, A. I. Oparin and J. B. S. Haldane, in the 1920s. They were much influenced by a casual remark made by Charles Darwin in a letter written in 1871, suggesting that '... in some warm little pond, with all sorts of ammonia and phosphoric salts, light, heat, electricity etc. present ... ' the chemical changes which preceded the origin of life could have occurred. Along similar lines, the Oparin–Haldane scenario envisaged that:

1 Organic molecules formed on the early Earth accumulated in the oceans, which came to resemble a warm, dilute soup.

2 Protocells and the earliest cells obtained energy by consuming this soup, i.e. they were *heterotrophic* (other-feeding), and fermented substances in the absence of oxygen.

You need to remember that this theory was suggested at a time when little was known about cell metabolism and nothing at all about the nature of the genetic material. As biochemical knowledge increased, the Oparin–Haldane version of chemical evolution was elaborated and modified. Some of these ideas together with more radical alternatives to the Oparin–Haldane theory will be discussed in Section 16.3. In the next Section, we consider evidence about the nature of the environment on Earth during the period of chemical evolution.

Summary of Section 16.1

The essential properties of life are replication (with occasional mistakes) and metabolism. Modern organisms are based on cellular units and the properties of both replication and information storage reside in nucleic acid molecules. There is a high degree of interdependence between metabolism and nucleic acids. Scientific theories about the origin of life include spontaneous generation (discarded), panspermia or extraterrestrial origin (discarded by most people), and the chemical theory. The chemical theory suggests that life arose on Earth from non-living molecules after a period of chemical evolution. Oparin and Haldane put forward the first detailed version of this chemical theory. They argued that life evolved in the oceans (the primordial soup), the first cells being **anaerobic heterotrophs** that fermented organic molecules within the oceanic soup.

16.2 Prebiotic environments

The Precambrian rock record (see Appendix 1) is divided into the **Hadean, Archaean** and Proterozoic. The Archaean begins with the oldest known rocks on Earth, 3 900 Ma ago, and ends at about 2 500 Ma ago, shortly after a major period of continental growth and stabilization. Above it lies the Proterozoic which spans the interval from 2 500 Ma to the base of the Cambrian at about 540 Ma. There is good evidence, described in Section 16.4.4, that prokaryotic life was already present at the start of the Archaean, which means that chemical evolution occurred during the Hadean and that life had probably originated by around 4 000 Ma ago. Since there are no Hadean rocks, our knowledge of the Earth's history during this period is necessarily based on indirect evidence, such as that derived from meteorites and lunar samples, together with plausible assumptions based on our understanding of more recent geological processes. With little direct evidence, many of the ideas concerning this period in Earth history are speculative and often depend more on the persuasiveness of their proponents than on the weight of any relevant data. Some of these ideas are described next.

16.2.1 EVIDENCE FROM THE ROCKS

Isotopic dating of the Moon and meteorite samples seems to indicate that the formation of the solar system started some 4600 Ma ago, when gravitational collapse of interstellar matter produced a dust cloud. This dust cloud or *accretion disc* was probably composed almost entirely of gaseous hydrogen and helium together with small concentrations of water, ammonia and methane ices, and solid iron and silicate particles. The greatest mass of this proto-solar system (from more than half to as much as 99%) became concentrated towards the centre of the disc, creating conditions in which thermonuclear reactions started and the Sun formed. As the collapse continued, condensation of these components and gravitational instabilities within the resulting cloud of gas and dust led to the formation of *planetessimals* (rocky bodies 1–10 km in size) that within 1–100 Ma accreted to form protoplanets, including the Earth. This primordial phase of formation of the solar system was probably completed by about 4500 Ma ago.

Heat played a major role in the next stage of Earth's history. The young Earth became hot, partly because *accretion* generates heat, and partly because of two other processes—radioactive decay and meteorite impacts. Radioisotopes of elements such as uranium, potassium and thorium were more abundant in the newly formed Earth than they are now, and a significant amount of heating due to *radioactive decay* would have occurred directly following the accretionary stage. Another source of heating was *meteorite impacts*, which are very rare events today, but were probably common on the early Earth. All traces of these early impacts have been obliterated on Earth (by erosion, for example), but traces *are* preserved on other planets and in the rocks of the lunar highlands on the Moon. Some of these lunar rocks are as old as 4200–4400 Ma, and they are heavily cratered—evidence of continued, intense meteorite bombardment (see Figure 16.2). Similar cratering is observed on Mars and on the satellites of the outer planets, so it seems reasonable to suppose that the Earth, together with Mercury and Venus, also experienced this bombardment, which must have generated considerable amounts of heat and may have persisted for much of the Hadean.

Since most planetary materials, except for metals, are poor thermal conductors, the combination of accretion, radioactive and impact heating would have resulted in the partial melting of the more massive planetary bodies such as the Earth. Because iron melts at a lower temperature than do silicates under high pressure, a dense iron-rich melt would have formed and would have percolated downward, displacing silicates upwards. Eventually, a molten metal **core** became separated from a silicate **mantle** (see Figure 16.3). The slight cooling that has occurred since has resulted in the formation of the Earth's solid inner core; differential flow in the outer liquid portion of the core is regarded as the source of the dynamo responsible for the generation of the Earth's magnetic field.

Figure 16.2 *The far side of the Moon from Apollo 8, showing the lunar highlands and impact craters. Some craters have been partially flooded by later outpourings of lunar lavas.*

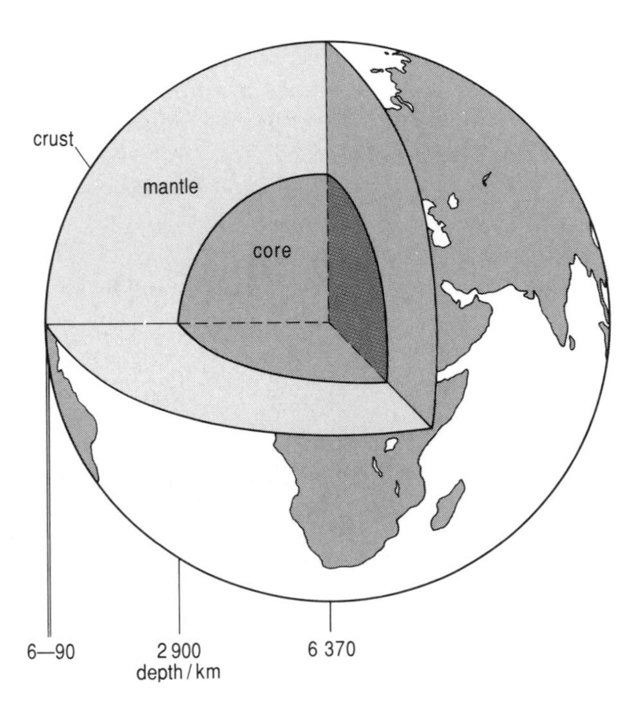

Figure 16.3 *Diagram showing the Earth's internal structure.*

Following the core–mantle separation, it is thought that an oceanic-type crust formed and, by the end of the Hadean, there is good evidence that light granitic continental crust had formed from the mantle. This evidence is preserved in the ancient rocks that marked the Hadean–Archaean boundary. The oldest, dated at 3960 Ma, were found in Canada in 1989, but better known are those dated at 3800 Ma old from the Isua region of western Greenland, which include granites and associated volcanic rocks.

16.2.2 THE OCEANS AND THE EARLY ATMOSPHERE

Conditions at the Earth's surface and in the atmosphere during the Hadean must have had a powerful influence on chemical evolution. The separation of the Earth's core and mantle resulted in the release of a considerable amount of heat over a period of only a few hundred million years with an associated release of gases that had been trapped in the Earth's interior (outgassing). The composition of the resulting primitive atmosphere depended on the precise timing and nature of heat-generating events on the Earth. If outgassing of the mantle occurred before the formation of the core (i.e. before the removal of metallic iron and as a result of energy released via impact events), the atmosphere may have contained substantial concentrations of hydrogen (H_2); with carbon occurring as carbon monoxide (CO) and methane (CH_4); nitrogen as N_2 and ammonia (NH_3); and sulphur as hydrogen sulphide (H_2S). If outgassing occurred later, however, the mantle would have been more oxidized, with metallic iron having been removed to the core, and the principal outgassed constituents would then have been carbon dioxide (CO_2), N_2 and sulphur dioxide (SO_2).

▶ Which major constituent of the present-day atmosphere is absent from both of the primitive atmospheres just described?

Oxygen. With an outgassing model for the origin of the Earth's atmosphere, we must accept that the primitive atmosphere was oxygen-free (anoxic) because modern volcanic gases are devoid of oxygen. An important side effect of this lack of oxygen is that there would have been *no ozone* layer in the primitive atmosphere. Ozone (O_3) is produced from oxygen (O_2), by the action of solar ultraviolet radiation (u.v.), and it has accumulated in the lower stratosphere 10–15 km above the Earth's surface. The most significant property of the ozone layer is that it is very effective in absorbing most of the short-wave solar u.v. which is particularly damaging to living organisms.

This account still leaves unresolved the question of which of the two types of primitive atmosphere actually existed on the early Earth. The atmosphere formed by early outgassing is much more reducing than that formed by later outgassing, i.e. it would promote more strongly reactions in which electrons or hydrogen atoms were added to reactants. The consensus view today is that late outgassing occurred, producing an atmosphere dominated by carbon dioxide and nitrogen. However, when we discuss the synthesis of pre-biotic materials (Section 16.3) you will see that there are real problems because the molecular precursors of life would not have formed readily in a CO_2/N_2 atmosphere! A suggested compromise is that the primitive atmosphere was slightly more reducing than are present-day volcanic gases having greater amounts of hydrogen, ammonia and methane.

What about the early oceans, the *hydrosphere*? There is good evidence that some of the ancient Isua rocks (see Section 16.2.1) were water-deposited sediments called conglomerates. These were changed (metamorphosed) over time and are now metamorphic 'metasediments'. Furthermore, the associated lava flows in the Isua rocks have what is termed a 'pillow' structure, which indicates that they cooled under water. So there is little doubt that the Earth possessed a hydrosphere by about 3 800 Ma ago.

The chemical composition of this early ocean would have reflected that of the atmosphere, modified by the inclusion of materials derived from the weathering of igneous rocks. These materials include clay particles and iron-rich complexes. Like the atmosphere, the ocean would have been anoxic and so iron would have been present in the reduced (ferrous) form, Fe(II) or Fe^{2+}. Because reduced iron is much more soluble than the oxidized (ferric) iron, Fe(III) or Fe^{3+}, which rapidly produces insoluble precipitates in oxygenated water, the ancient ocean probably contained far more iron in solution—perhaps 1 000 times more—than the present-day oceans. As you will see later, this feature is significant in discussions about the origin of life and early biological evolution.

A more difficult problem arises in assessing when the hydrosphere formed during the Hadean. Woese (1979) argues that the Earth's surface was too hot to support large bodies of water during the early stages of atmosphere formation and, instead, water would have been present mainly as steam. In addition, however, many of the large meteorites that bombarded Earth during the Hadean would have generated enough heat to boil the surface of the ocean as well as to throw large clouds of dust and molten rock into the atmosphere. The general picture is that, whenever the ocean formed, it was for a long time a much less stable environment than are the present-day oceans.

16.2.3 Earth's early climate

The average surface temperature of a planet like the Earth depends on three factors, all of which have varied with time. The first is the radiant energy emitted by the Sun, the *solar luminosity*. The second is the fraction of solar energy incident on the planet that is reflected back into space, the *albedo*. The third factor is the *greenhouse effect* of the atmosphere—a proportion of the heat emitted by the Earth is absorbed and re-radiated by the atmosphere depending on atmospheric composition. Theories of stellar evolution suggest that there has been an increase of about 25–30% in solar luminosity over geological time. A cool Sun in the Hadean would have resulted in surface temperatures being well below the freezing point of water, which may have been further exaggerated by the increased albedo of ice reflecting more heat back into space. The resulting feedback loop would have led to a completely ice-covered planet! For the oceans to have been liquid, therefore, the atmospheric greenhouse effect must have been larger than it is today. Water vapour by itself is not a sufficiently good absorber of energy to have provided the necessary enhanced greenhouse effect so an additional greenhouse gas was required.

▶ From general knowledge and information given in Section 16.2.2, which gas might this have been?

Carbon dioxide. This is a well-known greenhouse gas and the primitive atmosphere contained a lot of CO_2—rising CO_2 levels today are causing much concern about possible global warming. Work by Tobias Owen in the late 1970s showed that the combined greenhouse effect of water and CO_2 is great enough to 'solve' the cool Sun problem.

The Earth in Hadean times was not freezing cold, therefore, and, by the beginning of the Archaean, 3 800 Ma ago, there was enough liquid water at the surface to sustain such normal geological processes as weathering, erosion and sedimentation.

▶ What is the evidence for this?

The nature of the Isua metasediments (Section 16.2.2) clearly shows that these processes were occurring. Since 3 800 Ma ago there have been areas of the Earth where average temperatures have been below the boiling point of water and above the freezing point. Because of the albedo feedback-loop, it is unlikely that the average temperature for the whole Earth has ever dropped below freezing; even during later 'Ice Ages' the average temperature was above 0 °C, and the Earth was only partially glaciated.

It is more difficult to define an upper limit for the average temperature of the early Earth. The fact that life existed 3 800–3 500 Ma ago and must have originated during the Hadean does not mean that temperatures then were necessarily the same as today. This is because **thermophilic bacteria** exist now that can survive temperatures of 100 °C or more, and contain enzymes that are not denatured by such high temperatures. If the first living organisms were thermophilic, the Earth could have been a very hot place. D. J. Chapman has argued against this idea because, if true, you would expect thermophilic properties such as heat-resistant enzymes to be widespread today, but they are not. However, others argue that thermophily is a shared primitive character state (see Chapter 11) among the most primitive living bacteria and that life could indeed have originated in hot environments, either near the surface or associated with hydrothermal vents in the oceans. Such arguments depend, of course, on whether modern micro-organisms are representative of their Archaean ancestors or whether subsequent evolution and adaptation have completely changed them.

The general consensus is that Archaean and probably late Hadean surface temperatures were not greatly different from those of today. Nevertheless, frequent severe storms probably occurred and together with continued bombardment by meteorites and a high influx of solar u.v., the Earth's surface was certainly not a hospitable environment for early life.

SUMMARY OF SECTION 16.2

The environmental conditions under which life arose on the primitive Earth were very different from those of today. The atmosphere resulted from outgassing of the mantle shortly after the accretion of the planet. Its composition was linked to the evolution of the planet and the point at which the core formed. Early outgassing would have produced a strongly reducing mixture of gases, but with later outgassing, which seems more likely, a less reducing mixture of nitrogen, CO_2 and SO_2 would be present, possibly with small amounts of hydrogen, ammonia and methane. There was no free oxygen and, therefore, no ozone layer in the primitive atmosphere.

Evidence from the ancient rocks at Isua in Greenland indicates that the ocean was present by 3 800 Ma ago and that weathering, erosion and sedimentation were occurring. It probably contained high levels of dissolved, reduced iron, Fe(II), but there must have been frequent disturbance, including surface boiling, due to meteorite bombardment throughout the Hadean. Once the Earth had cooled, after its formation, average temperatures were probably similar to those of today. Low solar luminosity would have been counteracted by a strong greenhouse effect.

16.3 Chemical evolution

The assumption that the organic compounds necessary for life would have been formed naturally by non-biological processes is the foundation of the chemical evolution hypothesis. The questions addressed now are:

1 'Were environmental conditions in the Hadean suitable for this pre-biotic chemistry?'

2 'What were the main steps in chemical evolution, and how did the transition from non-living to living occur?'

Two lines of evidence suggest that the answer to Question 1 is probably yes. The first comes from the study of extraterrestrial organic matter (see Section 16.3.1) which is widespread in the solar system and beyond, and may even have reached the early Earth in significant quantities. The second comes from experiments which simulated conditions on the primitive Earth on a micro-scale, and examined the chemical reactions which occurred (see Sections 16.3.2 and 16.3.3). Possible answers to Question 2, all of which should be regarded as informed guesswork, are discussed in Sections 16.3.3 and 16.3.4.

16.3.1 Extraterrestrial evidence

Since the 19th century, certain meteorites have been known to contain organic compounds which were generally considered to be extraterrestrial in origin. However, there was always a risk that the meteorites had been contaminated with terrestrial material and not until the early 1970s was unambiguous evidence obtained through a series of serendipitous events. As the United States was preparing for the return of the first lunar samples in 1969 two meteorites fell, one at Pueblo de Allende in Mexico, just over the border from the lunar receiving laboratory in Houston, the other at Murchison in Australia. By a stroke of luck, both meteorites were rich in carbon and organic molecules, so the whole battery of new techniques that had been developed to study the lunar samples was brought to bear on these two meteorites. The **Murchison meteorite**, in particular, was found to contain hydrocarbons and, significantly, amino acids which were mixtures of both D and L isomers. Isomers are different forms of a molecule, the same atoms being arranged in alternative ways, and on Earth only L isomers of amino acids are produced by living organisms. The presence in the Murchison meteorite of D isomers together with the later discovery of over 70 different amino acids, many of which are not found associated with living organisms on Earth, is strong evidence for an extraterrestrial origin. The presence of a large variety of other organic compounds in the Murchison meteorite (e.g. hydroxy, mono-, and dicarboxylic acids; urea and amides; ketones and aldehydes; hydrocarbons and alcohols; amines) leaves little doubt that chemical evolution in the extraterrestrial environment passed well beyond the synthesis of simple molecules. However, the chemical processes involved in these syntheses and where they occurred remain poorly understood.

As long ago as the 1930s, astronomers deduced from absorption lines found in the visible spectra of distant hot stars that organic molecules existed in interstellar space. More recently, techniques involving radio spectral lines have revealed the existence of some 70 different **interstellar molecules** in space (see Figure 16.4.), many of which contain only carbon, nitrogen, hydrogen and oxygen. They range in complexity from having two atoms (e.g. OH, CO, CN, etc.) to some having 13 atoms such as cyano-penta-acetylene [$H(C_2)_5CN$]. The environment of a dense interstellar cloud with very low temperatures (<–170 °C) and fewer than 106 particles per cubic metre, would seem an unlikely place for chemical reactions, yet it exhibits a rich chemistry manifest in the production of numerous types of organic compounds.

Organic compounds are also ubiquitous in the solar system. On Jupiter, the presence of simple hydrocarbons in the atmosphere, and the colours of its clouds, are strong indicators of active organic chemistry. The hydrocarbons are readily explained as the products of ultraviolet photochemistry and lightning activity. Harold Urey was the first to suggest that the colour of Jupiter's atmosphere was due to organic molecules. One of Saturn's moons, Titan, contains methane and small amounts of other hydrocarbons in its atmosphere; it has a reddish-brown colour due to the presence of organic matter. Although the surface of Titan is very

Hydrogen	H_2		
Molecules containing only C and H	CH_4 $H_3CC{\equiv}CH$	$C{\equiv}C$ $C{\equiv}CH$	$(C{\equiv}C)_2H$ $H_3C(C{\equiv}C)_2H$
Molecules containing O	OH CO	H_2O H_2CO	HCO_2H CH_3OH
Molecules containing N	CN CH_3NH_2	HCN $HC{\equiv}CCN$	NH_3 CH_3CH_2CN
Molecules containing O and N	NO	HNCO	NH_2CHO
Molecules containing S and Si	SO SiO	H_2S SiC_2	SO_2 SiH_4

Figure 16.4 *Examples of some molecules detected in interstellar clouds and in the shells around stars.*

cold (*c.*–180 °C) and it is the only moon in the solar system with such an atmosphere, some analogies may be drawn between the processes occurring in Titan's atmosphere and those that may have prevailed on a pre-biotic Earth.

All such investigations seem to indicate that the building blocks of proteins (amino acids) and some components of nucleic acid bases (purines and some pyrimidines) are being formed by chemical evolution on a cosmic scale. The chemistry thought to have given rise to life, therefore, is not some exotic, improbable set of reactions but the normal chemistry of carbon, hydrogen, oxygen, nitrogen and sulphur. However, the potential pathways for the chemical evolution of organic matter are governed by the conditions of a particular environment, be it an interstellar cloud, a meteorite, or a planet; progress towards the evolution of life will be diverted at different stages depending on the physical and chemical constraints imposed by each environment as we can see from the planets Venus, Earth and Mars. All of these experienced a similar early history but only on the Earth did life evolve and persist: we cannot rule out the possibility that life arose on Venus and Mars also, but did not persist because of later changes in their environments.

Could extraterrestrial organic molecules have been a significant or even a major source of building blocks for prebiotic synthesis on Earth? One problem with this hypothesis arises from the manner by which extraterrestrial material was delivered to the early Earth. Large meteorite impacts would largely destroy the organic matter in the resulting fireball; smaller meteorites would be destroyed by burning up in the atmosphere. However, there would have been a considerable amount of material with a size of only a few tens of microns (10^{-6} m) floating around the early solar system. Recent work suggests that much of this material could have reached the surface of the Earth unaltered and contributed to the stock of organic molecules, but by how much is not yet known.

16.3.2 Sources of energy

The second approach to unravelling chemical evolution was to create, in the laboratory, conditions similar to those on the primitive Earth or in its atmosphere and see what chemical reactions occurred. The likely composition of the atmosphere and oceans was described in Section 16.2.2, but another factor that must be considered is the possible energy sources available for pre-biotic chemistry on the early Earth. These sources are summarized in Figure 16.5.

By far the largest input of energy came from the Sun. However, it is not the amount of energy but rather the *type of radiation* that is important, because only radiation with a wavelength of less than 200 nm can dissociate water and methane into reactive components (radicals) and initiate the synthesis of larger organic molecules. Just 1.2% of ultraviolet radiation was of this type, but it was still the major source of energy for the synthesis of organic molecules.

▶ Recall why virtually none of this energetic, short-wave u.v. radiation reaches the Earth's surface today?

It is now absorbed by the ozone layer (Section 16.2.2) which formed later as oxygen accumulated in the atmosphere. Next in importance to u.v. were *electric*

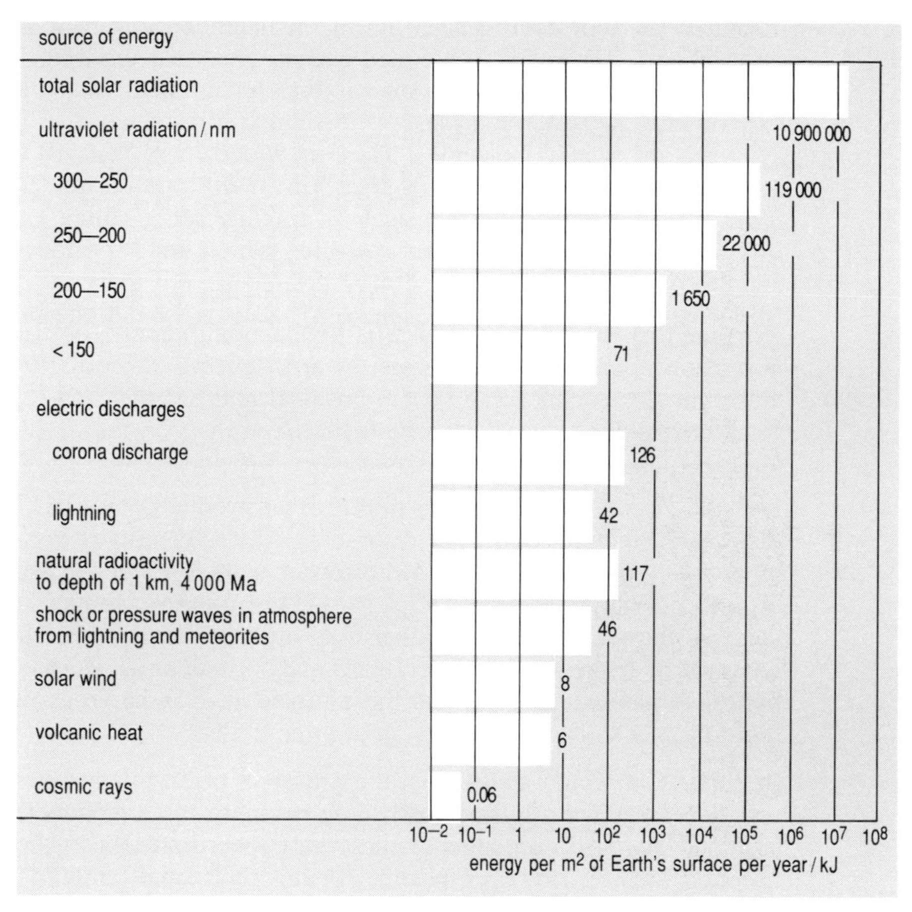

Figure 16.5 *Sources of energy for chemical synthesis on the primitive Earth.*

discharges which were probably just as frequent in the Earth's early atmosphere as they are today. Much *heat* was also generated by the decay of radioactive elements, meteorite impacts (Section 16.2.1) and outpourings of volcanic lava at temperatures in excess of 1 000 °C—hot enough to dissociate methane. The alpha and beta particles released by the radioactive decay of elements such as uranium and potassium may also have played a role by knocking off electrons from surrounding atoms, thus rendering them chemically active.

16.3.3 Experiments in pre-biotic synthesis

The scenario for chemical evolution according to the Oparin–Haldane model (and its subsequent versions) is now one where an anoxic, slightly reducing atmosphere is acted on mainly by ultraviolet radiation and electrical discharges. Simple organic molecules are deposited in the oceans, where further reactions occur. We consider next how this model stands up to experimental testing.

Synthesis of small molecules In 1953, Stanley Miller working at the University of Chicago recreated what was thought to represent the atmosphere of the primitive Earth in a sealed glass apparatus. He filled the vessel with 200 cm^3 of water (the ocean), then, after evacuating the air from the apparatus, he added

10 cm pressure of hydrogen, 20 cm of methane and 20 cm of ammonia (the atmosphere). The water was boiled and an electrical discharge, simulating lightning, was passed through the mixture for a week. At the end of the experiment, the water and glass were stained a deep red as a result of the synthesis of organic molecules. These included more than 10 of the 20 *amino acids* that occur in proteins, together with many that do not; *fatty acids*, *hydroxy aldehydes* and *sugars*; *purines* such as adenine and guanine, as well as the *pyrimidines* uracil, thymine and cytosine, which are important constituents of RNA and DNA. These results looked very promising, but there is a major snag with the Miller experiments—the atmosphere he used is not the one now thought to have existed on the primitive Earth.

▶ What is wrong with Miller's atmosphere?

It is too strongly reducing and mimics that produced by early outgassing whereas the weakly reducing atmosphere produced by later outgassing is thought to be more likely (Section 16.2.2). More recent work by Stanley Miller showed that electric discharges through CO_2—N_2—H_2O mixtures produced nitrous and nitric acids as the major products rather than organic compounds. However, small amounts of hydrogen cyanide (HCN) and formaldehyde formed, especially if some hydrogen gas was present, and these molecules can serve as precursors for building blocks such as sugars and amino acids.

In general, it would appear that the presence of a reducing gas (H_2, CH_4, or CO) was necessary for organic synthesis to occur in the atmosphere, and so it has been assumed that small amounts of these gases were present in the early atmosphere (Section 16.2.2). However, whether or not such small amounts would have been sufficient to permit atmospheric organic synthesis remains to be evaluated. If not, and if other highly reducing species were unavailable, how might the basic chemical building blocks of life have been produced?

Extraterrestrial inputs represent one possibility but the interaction of inorganic minerals is another. In solutions containing reduced iron, (Fe(II)), CO_2 can be reduced to formaldehyde (HCHO) by irradiation with ultraviolet light, suggesting that dissolved iron might have provided an important source of reducing power on the pre-biotic Earth. Other experiments have shown that ultraviolet irradiation of water containing suspended clay particles and dissolved CO_2 results in the production of methanol (CH_3OH) and other simple organic compounds.

▶ Were clay particles and Fe(II) available on the primitive Earth?

Both could have been made available by the weathering of rocks and then have been washed into the ocean (Section 16.2.2); hydrothermal vents release considerable amounts of Fe(II) and dust and water thrown up into the atmosphere by storms and meteorite impacts would allow their presence in the atmosphere as well. But, even if formed in the atmosphere, the first organic molecules must have ended up in the oceans according to the Oparin–Haldane model—the highly reactive conditions in the atmosphere would simply have destroyed them too rapidly to allow any further organic syntheses.

Synthesis of larger molecules Supposing that molecules such as HCN (hydrogen cyanide) and HCHO (formaldehyde) were the major products of early pre-biotic synthesis and supposing that they were protected from u.v. irradiation in the oceans, the next question to answer is: 'How did they reach sufficiently high concentrations to combine into more complex, stable compounds?' It is difficult to accept that the whole ocean became a concentrated broth of these first reactants. Bernal (1967) suggested that *clay particles* provided surfaces where HCN, HCHO and other molecules could be adsorbed, concentrated and react further. He argued that these reactant-laden clays were concentrated in surface foam and driven into estuaries where even more concentrated muds were deposited—the 'warm, shallow pond scenario'.

Using u.v. irradiation or spark discharges, a remarkable array of more complex molecules can be produced from HCN and HCHO. One example is the **formose reaction** which could have led to the pre-biotic synthesis of sugars from formaldehyde; a simplified version is shown in Figure 16.6.

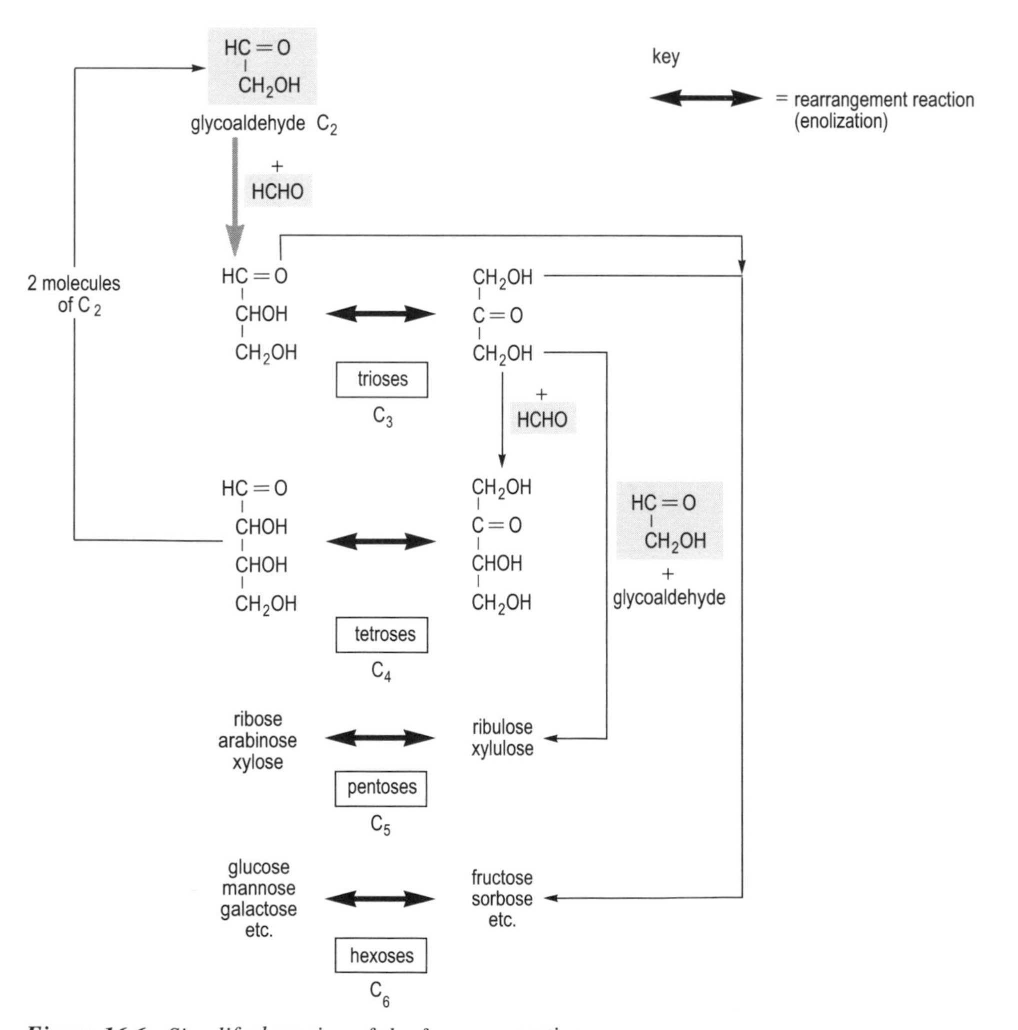

Figure 16.6 *Simplified version of the formose reaction.*

The starting point, *glycoaldehyde*, may form spontaneously or by a condensation reaction of HCHO. It initiates a cascade of reactions that converts most of the available formaldehyde into trioses (3-carbon sugars), tetroses (4-carbon sugars) and higher sugars in both D and L isomeric forms. Furthermore, the cascade proceeds autocatalytically because the early-formed tetroses can cleave to give two molecules of glycoaldehyde.

Starting off with HCN and mildly reducing conditions, spark discharges produce an array of amino acids, although in low yield. In the 1950s, Sidney Fox of the University of Florida showed that heating mixtures of amino acids can result in *thermal polymerization*—when the melt is cooled, protein-like polymers called **proteinoids** are formed with many of the properties of polypeptides. For example, proteinoids and polypeptides break down under identical acid conditions into their constituent amino acids, and proteinoids may also have catalytic properties similar to those of enzymes. Another interesting feature was that the amino acids in thermal proteinoids showed a marked tendency to become 'self-ordered'. In one experiment a mixture of glutamic acid, glycine and tyrosine (including isomers) was heated for 10 hours. If these amino acids were to combine on a random basis, some 36 tyrosine-containing tripeptides should form. In fact, only two such tripeptides were formed, and they were calculated to be 19.2 times more abundant than would be expected in a random synthesis. The probability that this result was due to chance is less than 1 in 106.

If placed in warm water and then cooled, Fox found that his proteinoids formed spherical structures with a remarkably uniform diameter of about 2 μm (Figure 16.7). These **proteinoid microspheres** were insoluble, very stable in water and were suggested as a possible model for protocells. Fox further suggested that all these syntheses occurred in the hot water associated with submarine vents and volcanoes—the 'hot spring scenario'.

Figure 16.7 *Scanning electron micrograph of microspheres developed from proteinoids.*

Compared with proteins, the pre-biotic route to *nucleic acids* is much more difficult. Purines, including adenine and guanine, can be derived quite readily from HCN, although it is altogether more difficult to form pyrimidines (thymine, cytosine and uracil) and there is no obvious pre-biotic route. The formation of nucleosides (purine or pyrimidine base plus the 5-carbon sugar ribose), phosphorylation of nucleosides to give nucleotides (e.g. ATP) and polymerization of these units to give nucleic acids capable of self-replication is even more difficult to envisage. Reviewing the problems, Joyce (1989) concluded that they were insuperable and that life did not start with nucleic acids—they came later after a simpler replicating system had introduced some degree of order. Two possible models for this system are described next, both of them (and especially the second) differing considerably from the original ideas of Oparin and Haldane.

16.3.4 Other routes to life

Replicating clays Based on the assumption that some replicating template system was essential for the origin of life, and faced with the intractable problems of nucleic acid synthesis, Cairns-Smith (1985) came up with a daring new idea. He suggested that the electrically charged surfaces of clay particles could form replicating templates. These **replicating clays** adsorbed and thereby concentrated and organized other organic molecules, acting rather like catalysts. Genetic information was stored as a distribution of electrical charges on the clay surface, and replicated by ionic interactions with newly formed surface layers.

This suggestion fitted in rather well with another idea that gained strength during the 1980s, namely that the first type of metabolism that arose was **autotrophic** (self-feeding). In autotrophic metabolism, energy and building blocks for synthesis are not obtained from other organic molecules—energy may be obtained from light (*photo-autotrophy*), or from simple, easily oxidized molecules (*chemo-autotrophy*), and building blocks are synthesized from molecules such as carbon dioxide and nitrogen gas or nitrates.

▶ How does this idea differ from the Oparin–Haldane theory?

This theory suggested that heterotrophy was the first type of metabolism (see Section 16.1.2). One reason for favouring autotrophy is that the first signs of life, 3 500 or even 3 800 Ma ago, clearly indicate that autotrophic organisms existed (discussed in Section 16.5.4). It seems unlikely that autotrophs could have evolved from heterotrophs in the relatively 'short' time between about 4 400 Ma ago when the Earth's surface cooled and 3 800 Ma.

One possibility supporting early autotrophy is that energy was provided by the oxidation of ferrous iron in iron-rich clays. Another suggestion was that a primitive kind of photosynthesis arose when iron-rich clays were exposed to u.v. radiation, with CO_2 being converted to oxalic and formic acids and, subsequently, nitrogen gas being reduced to ammonia (NH_3) and thence to a range of organic compounds. Quite how nucleic acids were synthesized on clay surfaces, how they eventually took over the replicating template role (a stage described by Cairns-

Smith as the 'genetic takeover'), and how cells evolved from such clay proto-life are unresolved questions.

Autotrophy in the atmosphere All of the models for chemical evolution so far described share certain characteristics; they assume that:

1 The first reactions occurred in the atmosphere or in surface waters and the products accumulated in the ocean.

2 Further reactions occurred in some concentrated micro-environment (e.g. adsorbed to clay particles, linked to replicating clays or inside proteinoid microspheres); alternatively, everything happened close to submarine hot springs (hydrothermal vents).

All, except the replicating clay model just described, further assume that essential substrates for later reactions came from the surrounding medium and that the systems were basically heterotrophic.

A model which does not share these assumptions and departs from the Oparin–Haldane theory even more sharply than the replicating clay model is that developed by Woese and Wächterhäuser (1990). They dismiss totally the idea of an oceanic soup of reactants, arguing that the ocean was always too dilute. They suggest, not only that life had an autotrophic origin, but also that chemical evolution proceeded for some considerable time in the *atmosphere*.

The scenario proposed is that at a very early stage in Earth history, before the surface had cooled sufficiently for oceans to form and when it was surrounded by vast cloud banks, dust particles were swept into the atmosphere and water vapour condensed on them. Early reactions occurred inside these droplets, or on hydrated dust particles, the most significant reactions being on the surface of minerals such as pyrite, which contains iron disulphide, FeS_2. Here an autotrophic system evolved, with CO_2 being fixed into simple organic molecules powered by energy from the oxidation of hydrogen sulphide by ferrous iron:

$$Fe^{2+} + H_2S \rightarrow Fe^{3+} + S + 2H^+$$

The ferric iron and sulphur react further to form FeS_2 which adds to the pyrite surface. On this surface a negatively charged organic monolayer is bound and develops autocatalytically, growing and spreading onto vacant surfaces. Because this system is based on surface chemistry, Woese and Wächterhäuser argue that evolution towards greater complexity would be favoured, whereas the equilibrium in solution reactions tends to favour degradation. Beyond the initial monolayer state, they suggest that a second stage, semicellular system, arose in which the pyrite surface was surrounded by a lipid layer. Inside the 'cytosol', detached products of surface metabolism could be trapped; proteins could form from amino acids (if temperatures were high), and bind and organize other molecules. Eventually,

'In the third stage the pyrite support is abandoned and true cellular organisms arise.'

As yet there has been virtually no experimental testing of this breathtakingly comprehensive hypothesis. No doubt there will be, and no doubt the ideas will be modified, just as the Oparin–Haldane theory has been.

16.3.5 The RNA world

In all living organisms today, the genetic machinery consists of self-replicating DNA molecules. The DNA is copied (transcribed) into RNA, some of which contains coded information that is translated into proteins; the translation machinery involves both protein and RNA molecules. It is an elegant, highly complex system. A crucial question relevant to the origin of life is: 'How did it arise?'

An early idea, that life arose essentially as a naked gene, foundered because there is no feasible way of synthesizing nucleic acids at an early stage of chemical evolution (Section 16.3.3). Nevertheless, whether we start with replicating clays, autocatalytic pyrite systems or something completely different, at some protocellular or very early cellular stage, nucleic acids must have evolved and taken on their role in replication and translation. There are now good reasons for believing that RNA and not DNA was the first nucleic acid to perform this role.

The most cogent reason for this belief was provided by the discovery that RNA can have *catalytic activity*, similar to that of protein enzymes. The discovery was made independently by the American biochemists Sidney Altman and Thomas Cech; it led to their being jointly awarded the Nobel prize for chemistry in 1989. Even though no modern organisms use RNA as a genetic template, some viruses do, so RNA can combine both a genetic (informational) role and a functional (enzymic) role — genotype and phenotype are combined in one molecule. There are other reasons for believing that RNA not only preceded DNA as the genetic material but also once played a more prominent role in cell metabolism. These reasons are clearly summarized in a review by Joyce (1989) and include the following facts:

1 DNA monomers (deoxyribonucleotides) are synthesized *from* RNA monomers (ribonucleotides), rather than by an independent pathway; this suggests the primacy of RNA over DNA.

2 RNA plays a role in several key cellular processes, including DNA synthesis, translation and protein synthesis.

The general idea, therefore, is that RNA was once the genetic material of early cells or protocells, and may have been involved in many more cellular activities than it is now. How the RNA world arose is, as usual in origin of life studies, uncertain. Two possibilities suggested by Joyce are illustrated in Figure 16.8.

In Figure 16.8(a), RNA nucleotides first play essential roles as metabolic co-factors. These polymerize to small polymers that eventually become self-replicating. In Figure 16.8(b), the polymerization of RNA nucleotides, again functioning as co-factors, is directed by an early, pre-nucleic acid template. The template directs both its own replication and the 'transcription' of RNA, which eventually becomes autonomously self-replicating and displaces the original template. Reverse transcription, whereby DNA is synthesized from RNA, is an obvious route by which DNA could have arisen, eventually to displace RNA as the genetic material. Certainly no living organisms today use RNA as their genetic material, although RNA viruses do occur; it has been suggested that these viruses are molecular fossils of the RNA world.

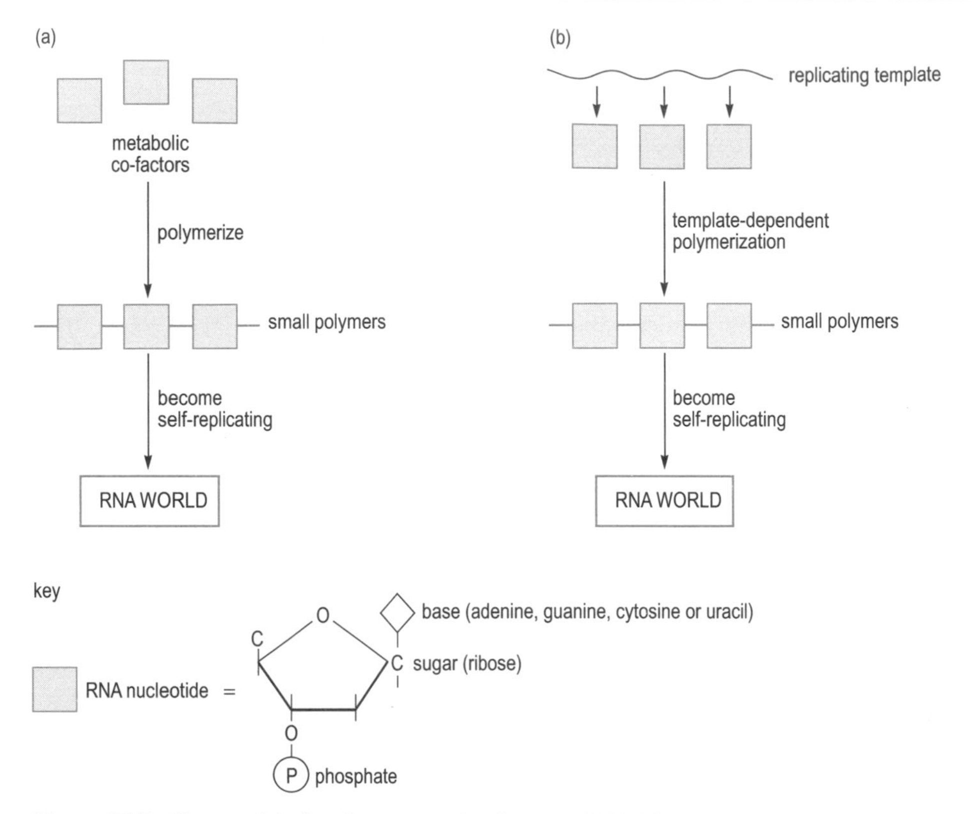

Figure 16.8 *Two models for the progression from an RNA-like to an RNA world.*

However, before the transition from the RNA to the DNA world, the complex machinery for translating nucleic acid into protein sequences must have evolved. Translation requires transfer RNAs (tRNAs), ribosomal RNAs (rRNAs) and messenger RNA; the tRNAs and rRNAs associate in highly specific ways with particular proteins (tRNA synthetases and ribosomal proteins). Detailed biochemical arguments about how the translation system might have evolved are outside the scope of this Book but, if you are interested to learn more, there is a clear account in Watson *et al.* (1987).

The end result of all this biochemical evolution was that by the time the first organisms appeared they had a genetic code based on DNA, and translation machinery based on RNA plus proteins. The majority of modern organisms investigated share the same genetic code (i.e. the same codon of three DNA bases codes for the same amino acid) but there are small variations (e.g. in ciliate protistans and the mitochondrial DNA of eukaryote cells), so the code is not completely universal. Recent evidence suggests that the code itself was subject to selection which acted to minimize the damage caused by point mutations or errors in translation. Whether this selection occurred in proto-living systems or in the first prokaryotes is not known, but we consider next how evolution proceeded in these early prokaryotes.

SUMMARY OF SECTION 16.3

Organic molecules are widespread in interstellar space, throughout the solar system and on meteorites. Minute dust particles carrying such molecules could have reached the surface of the primitive Earth and this extraterrestrial material could have provided substrates for early chemical evolution.

The initial sources of energy for chemical evolution were mainly short-wave ultraviolet radiation and electrical discharges, with heat and the products of radioactive decay playing a smaller role. Spark discharge experiments in a mildly reducing (CO_2—N_2) atmosphere indicate that hydrogen cyanide and formaldehyde were the most likely substrates for pre-biotic chemistry. Yields would have been much higher if traces of reducing gases were present in the atmosphere, and ferrous iron in solution could also have served as a reducing agent.

If HCN and HCHO were deposited in the oceans, they must have been concentrated in some way for further reactions such as the formose reaction (producing sugars) to occur. Concentration might have been by adsorption to clay particles which were deposited in estuarine muds—the 'warm pond scenario'.

For the formation of macromolecules, Fox provided a model for proteins. Heating up mixtures of amino acids can give peptide-like proteinoids and, if warmed and cooled, these form proteinoid microspheres. Such reactions might have occurred close to submarine vents—the 'hot spring scenario'.

Pre-biotic synthesis of nucleotides and nucleic acids in an unstructured, soup-like and heterotrophic system is much more difficult, probably impossible. One alternative might be that they evolved after a simple replicating system, with organization at the protocell level, came into existence. Two models for early replicating systems have been provided by the replicating clays of Cairns-Smith and the atmospheric pyrite-based system of Woese and Wächterhäuscr. Both are linked to autotrophic metabolism and, for the pyrite system, it is the surface-bound metabolic arrays that actually grow.

When, at the protocell or early cellular stage of evolution, nucleic acids arrived on the scene, RNA probably served as the first replicating (genetic) system. It was later displaced by DNA but RNA remained as the basis of the translation system. The genetic code itself appears to have undergone adaptive evolution in such a way that the consequences of base changes or mis-translation are minimized.

16.4 THE AGE OF PROKARYOTES

The first true organisms, i.e. those which have left living descendants, were simple, unicellular *prokaryotes* which were probably present on Earth by about 4000 Ma ago. For the next 2000 Ma (more than 40% of the Earth's history), prokaryotes evolved and diversified and this period, which spans the whole of the Archaean, is commonly referred to as the *Age of Prokaryotes*. In this Section we consider the major changes that occurred then, ideas about phylogeny—what gave rise to what—and the kinds of evidence on which these theories are based. Broadly, this evidence is of four types:

1 Fossils, which are few and often difficult to interpret (Section 16.4.3).

2 Comparative studies of structure and biochemistry among living prokaryotes (Sections 16.4.1 and 16.4.2).

3 Molecular phylogeny studies (Chapter 11), based on the relative amounts of divergence in proteins and nucleic acids (Section 16.4.1).

4 Geological data about climate, conditions in the atmosphere and oceans and biogeochemical cycles during the Archaean (Section 16.4.3).

For prokaryotes, molecular phylogeny studies have been by far the most important tool in revealing evolutionary relationships and led, in the late 1970s, to a complete revolution in ideas about the subject. We discuss this topic in the next Section followed by a brief review of energy metabolism in prokaryotes (Section 16.4.2) and a review of modern prokaryotes and their inter-relationships (Section 16.4.3), so that you have a clear idea of the end products of prokaryote evolution. In Section 16.4.4 we consider geological evidence about prokaryote evolution, not only from fossils but also from isotope ratios and sediments which indicate, for example, when photosynthesis first occurred and when oxygen was being released by prokaryotes. Only then, in Section 16.4.5, can we return to one of the most tantalizing questions: 'What were the *first* prokaryotes, the first living cells, like?', before we go on to look at the origin of eukaryotes in Section 16.5.

16.4.1 MOLECULAR PHYLOGENY AND THE EARLY EVOLUTION OF PROKARYOTES

Until the late 1970s, all prokaryotes were classified into a single kingdom which was divided into two phyla by some taxonomists and into as many as 16 by others. The two-phyla split had on the one hand 'bacteria', with a wide range of morphology and energy metabolism and, on the other, the **cyanobacteria** or blue–greens (once called blue–green algae), all of which are photo-autotrophs (discussed in Section 16.4.2). Clearly, there was much disagreement about the detailed subdivision of prokaryotes, but the kingdom as a whole was generally regarded as cohesive. In about 1980, however, prokaryote classification underwent a revolution. Three groups of bacteria were shown to differ from all other prokaryotes in such fundamental ways that they are now placed in a separate kingdom, the **Archaebacteria**, all other prokaryotes being grouped in the kingdom **Eubacteria**. The origin of these two kingdoms cannot be dated precisely but must be very ancient and probably — the current best guess — shortly after the origin of life. Perhaps even more surprising is evidence that the *eukaryote ancestor* diverged as a separate line at the same time. This statement needs some explanation because, as described in Section 16.5, the great majority of modern eukaryotic cells arose through the association of two or more different prokaryotes — mitochondria and chloroplasts, for example, were once free-living organisms that were taken into the cytoplasm of other prokaryotes and there became established in stable associations. It is the 'host prokaryote', which became the cytoplasm and nucleus of the eukaryote cell, that diverged from the Archaebacteria and Eubacteria at such an early date and is shown in Figure 16.9 as the nuclear line. We consider next how molecular evidence was obtained for the early divergence of three lines.

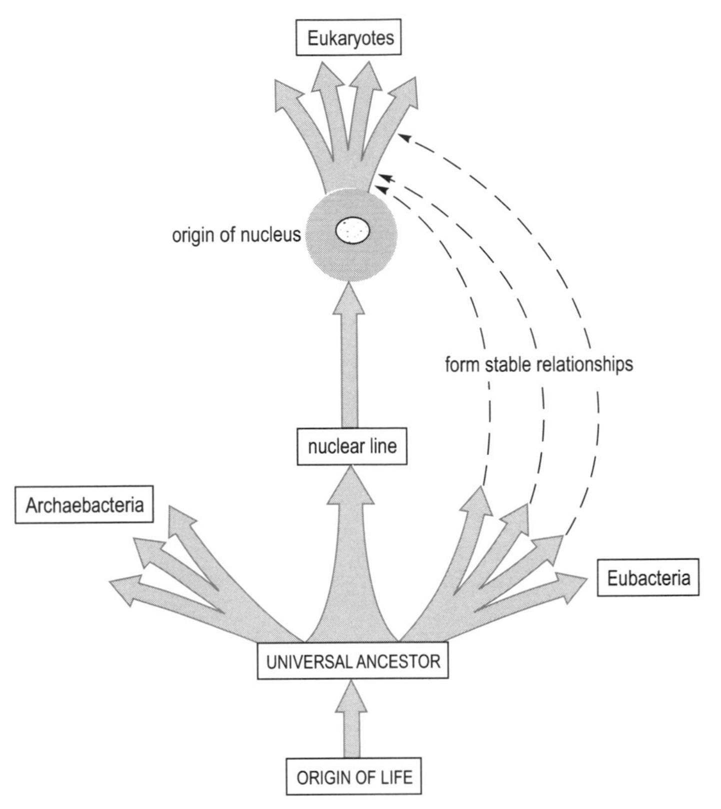

Figure 16.9 *Early divergence of three prokaryotic lines and the endosymbiotic origin of modern eukaryotes.*

Molecular sequencing and phylogeny During the 1970s, Carl Woese in the USA used molecular sequence data for ribosomal RNA (as described in Chapter 11) to unravel the ancient phylogeny of prokaryotes.

▶ Recall from Section 11.3.2 why rRNA was a good choice for these studies.

The molecules compared in different species must perform the same function and, when constructing ancient phylogenies, must have very slow rates of change. Both of these conditions are fulfilled by rRNA.

The universal evolutionary tree based on rRNA sequences and supported by various types of analysis is shown in Figure 16.10. This is an *unrooted* tree (Chapter 11, Section 11.2.4) because, from RNA data alone, there is no way of knowing where the root should be.

▶ If the arrow indicating the possible position of the root in Figure 16.10 is correct (and there is growing evidence that it is), which organisms have diverged *least* from the common ancestor?

Using the method described in Chapter 11 (summing branch lengths from the arrow), the answer is the Sulphobacteria, an archaebacterial group. The

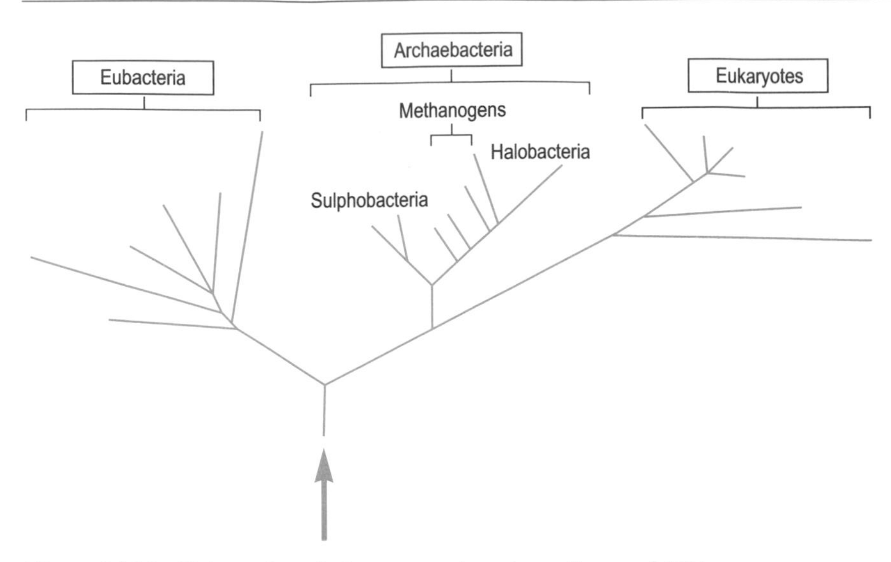

Figure 16.10 *Universal evolutionary tree based on ribosomal RNA sequence comparisons. The arrow indicates the possible position of the root.*

Archaebacteria as a whole have diverged less than either Eubacteria or Eukaryotes from the universal common ancestor so, although they are now highly specialized organisms, they are closer to the earliest cells than any other group.

Other evidence In addition to rRNA sequencing, there are several other kinds of evidence which support the early branching of prokaryotes into three lines. These are all based on biochemical or molecular characteristics and a few are summarized in Table 16.1 and discussed briefly below.

Table 16.1 *Some of the features that characterize living Eubacteria, Archaebacteria and the nucleo-cytoplasm of eukaryotes.*

Characteristic	Eubacteria	Archaebacteria	Eukaryote nucleo-cytoplasm
Nucleus	Absent	Absent	Present
Membrane lipids	Ester linked	Ether linked	Ester linked
Cell wall contains muramic acid	Yes	No	No (where walls present)
Ribosome sedimentation coefficient*	70S*	70S	80S
Types of RNA polymerase	One (4 subunits)	Several (8–12 subunits)	Three (12–14 subunits)
Initiator transfer-RNA	Formylmethionine	Methionine	Methionine
Ribosomes sensitive to diphtheria toxin	No	Yes	Yes
Protein synthesis inhibited by chloramphenicol	Yes	No	No

* The sedimentation coefficient is measured in Svedberg units (S) and relates to the speed of 'settling' in a gravitational field, i.e. during centrifugation. This gives a measure of the size and shape of particles: light 'spread out' objects (e.g. pancakes) settle slowly; heavy, compact particles (e.g. ball bearings) settle rapidly.

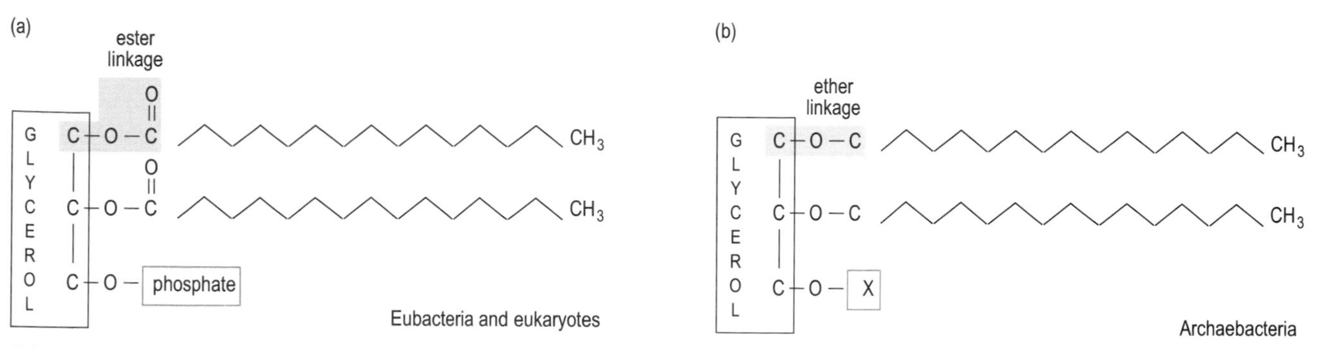

Figure 16.11 *Membrane lipid units in: (a) Eubacteria and eukaryotes; and (b) Archaebacteria, where X may be a phosphate, sulphate or carbohydrate group. The zig-zag lines represent fatty acid chains.*

Perhaps the most striking feature which distinguishes the Archaebacteria from both Eubacteria and the nucleo-cytoplasmic part of eukaryotes (excluding the organelles) is their *membrane lipids*. In all organisms the basic building block of such lipids consists of the 3-carbon alcohol, glycerol, to which two fatty acids (hydrocarbon chains with a carboxyl, COOH, groups on the ends) are attached (Figure 16.11). In Eubacteria and eukaryotes, the linkage between fatty acid and glycerol is of the *ester* type but in Archaebacteria it is an *ether* linkage (Figure 16.11). In addition, the third carbon of glycerol is always attached to a phosphate group in Eubacteria and eukaryotes but in Archaebacteria there may be a phosphate, sulphate or carbohydrate attached.

Cell walls differ in all three groups and so does *RNA polymerase*, the enzyme which catalyses the synthesis of RNA from a DNA template. In one characteristic—the *size of ribosomes*—Archaebacteria resemble Eubacteria, but in the remaining characteristics in Table 16.1, all of which relate to protein synthesis, Archaebacteria and eukaryotes resemble each other and differ from Eubacteria. It is now generally agreed that none of the evidence from comparative biochemistry or rRNA sequencing supports the evolution of Eubacteria or the eukaryote line *from* Archaebacteria. The only safe conclusion is that three separate lines diverged from an ancient prokaryotic stock, probably a relatively short time after the origin of life.

16.4.2 Energy metabolism

Following the origin of the three prokaryote lines, the Eubacteria and Archaebacteria diversified greatly. What happened in the ancestral eukaryote (nuclear-cytoplasmic) line is a complete mystery because it left no prokaryotic descendants, only eukaroytes. In the next Section, therefore, we look at the different groups of modern prokaryotes and try to trace their origins in the Archaean, concentrating particularly on the different ways in which they obtain energy. The main reason why the energy metabolism of early prokaryotes is so important is that, first, it provides clues about the mode of life and possible evolution of early prokaryotes and, secondly, it influenced profoundly the Earth's environment. The shift from an anoxic to oxygen-containing atmosphere, for example, resulted entirely from the oxygen released by photosynthetic cyanobacteria (Eubacteria). This Section is a brief review of prokaryotic energy metabolism.

Basic principles Heterotrophy and autotrophy are the two broad types of energy metabolism with which you are probably familiar. *Heterotrophs* utilize complex organic molecules ('food') as a source of both energy and carbon for biosynthesis and, today, although not necessarily in the early Archaean, the food molecules are always synthesized by other organisms. *Autotrophs* use a source of energy which is not synthesized by other organisms (light or inorganic molecules, for example) and CO_2 is usually their source of carbon. Aerobic respiration and photosynthesis, as carried out by eukaryotic animals and green plants, are classic examples of heterotrophy and autotrophy, respectively. Among prokaryotes however, there is much more variation in energy metabolism and, to appreciate the differences without going into too much detail, you need to understand some basic principles.

Figure 16.12 illustrates the fundamentals of energy metabolism. Electrons at a high energy level—which means that they are chemically reactive and part of a molecule, atom or ion that readily donates electrons—are transferred in a series of small steps in an energetically downhill direction. Energy released on the way is coupled to the synthesis of ATP, the energy currency that is essential to sustain life.

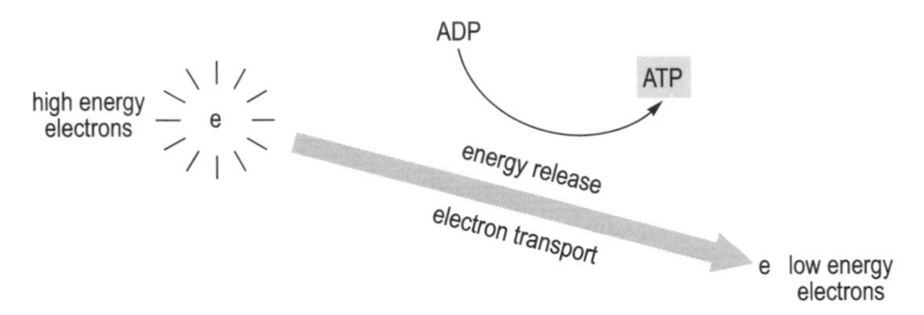

Figure 16.12 *Basic principles of energy metabolism.*

The key questions from which all the variations in energy metabolism arise, are:

1 'Where do the high energy electrons come from?'

2 'What happens to the terminal, lower energy electrons?'

There is remarkably little variation in the mechanism of energy conservation, i.e. the way in which energy released during electron transport is linked to ATP synthesis. With one exception, the fermenters which are discussed later in this Section, energy is first used to pump protons (H^+ ions) across a membrane which is represented by the green, upward-pointing arrows in Figure 16.13. This pumping creates a gradient of both electrical charge and proton concentration, i.e. an electro-chemical gradient or *proton motive force* (PMF). The PMF is discharged constantly by allowing protons to flow back across the membrane through special channels or *coupling factors*, which act like turbines, coupling the energy stored in the PMF to the synthesis of ATP (Figure 16.13).

Autotrophy Returning to Questions 1 and 2 raised in the previous paragraph, and starting with the autotrophs, Figure 16.14 shows two versions of *photo-autotrophy* which occur in photosynthetic bacteria where light is the source of energy.

▶ For the cyclic and non-cyclic schemes shown in Figure 16.14, what are the answers to Questions 1 and 2?

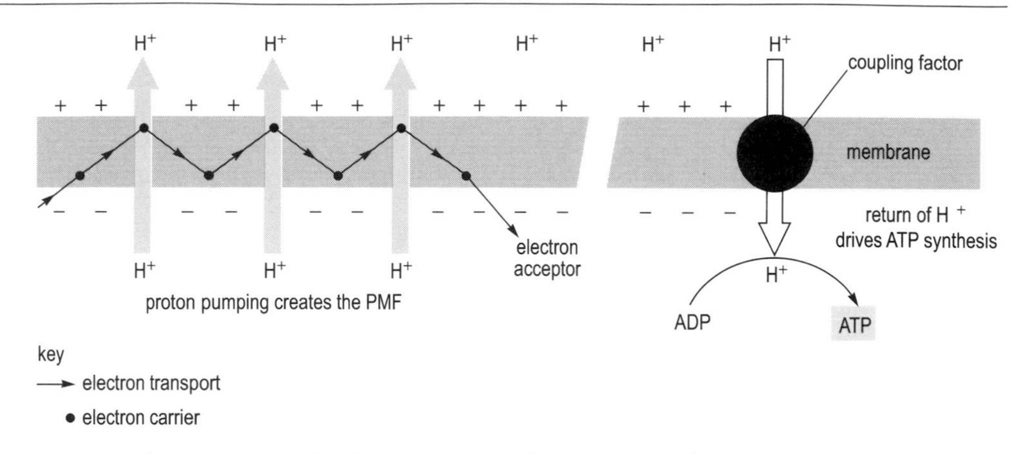

Figure 16.13 *Formation of a PMF and its coupling to ATP synthesis.*

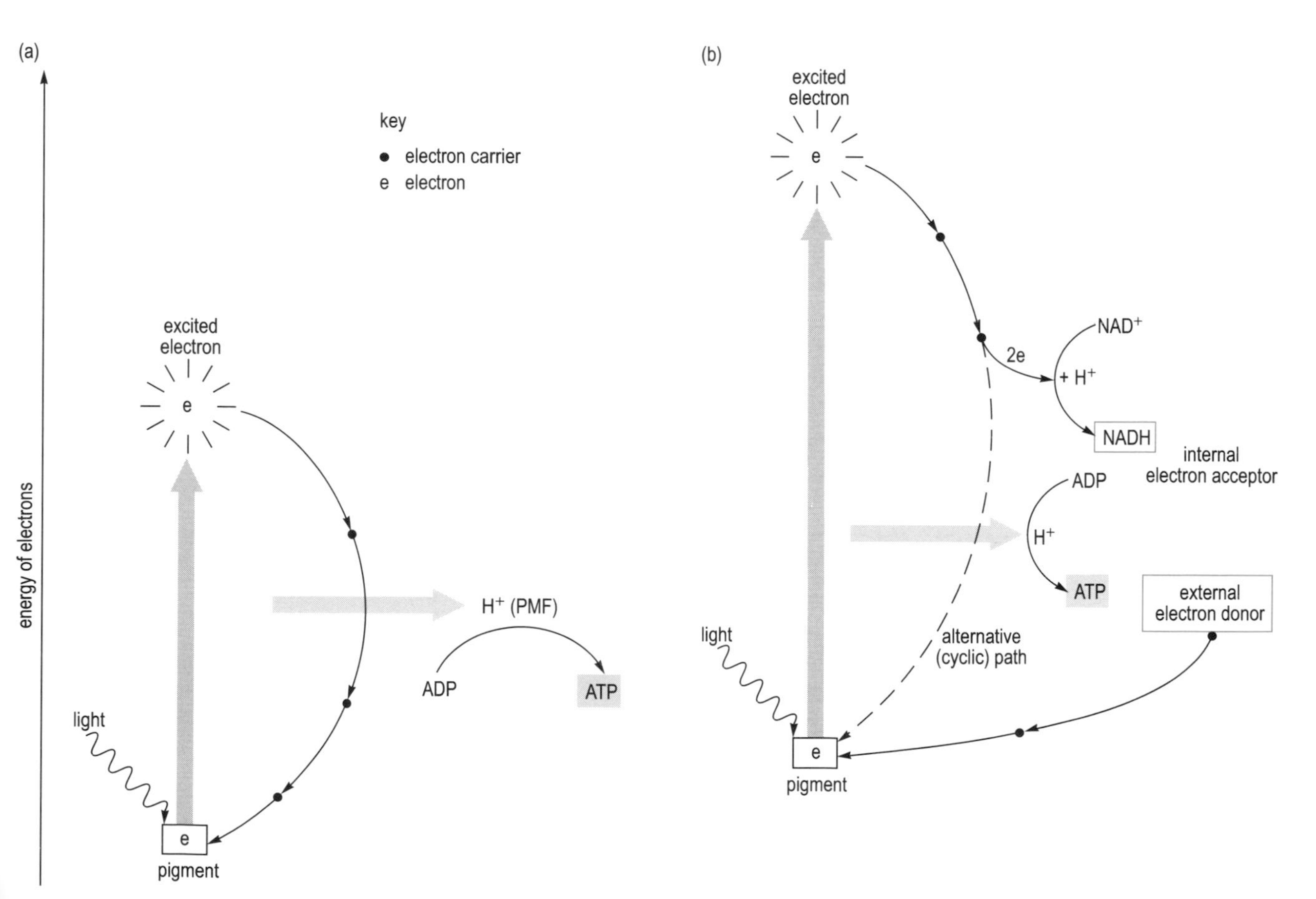

Figure 16.14 *Photo-autotrophy: versions of (a) cyclic, and (b) non-cyclic electron transport in photosynthetic bacteria. Both photosynthetic pigments and the electron carriers (shown as solid black circles) are associated with the cell membrane. Note that in (b), cyclic electron flow may occur as an alternative to non-cyclic electron flow.*

For both schemes, electrons are energized through the absorption of light by a special photosynthetic pigment and then transferred to other electron carriers. In the cyclic scheme, electrons return to the pigment molecule; so they do not come 'from' (Question 1) or go 'to' (Question 2) anywhere. However, in the non-cyclic scheme, electrons lost by the pigment molecule are replaced from an external electron donor and they are transferred finally to an internal electron acceptor, the co-factor NAD^+. The reaction is:

$NAD^+ + 2$ electrons $\rightarrow NAD^-$,

$NAD^- + H^+ \rightarrow NADH$

NADH is a useful source of reducing power for the cell, i.e. readily donating electrons or a hydrogen atom (the two are equivalent). Many biosynthetic reactions, including the fixation of CO_2, require reducing power.

For *chemo-autotrophy*, which occurs in chemosynthetic bacteria, high energy electrons are obtained from an external, inorganic substance. The basic scheme resembles Figure 16.12, or the upper right 'arm' of Figure 16.14(b). A typical example of electron transport is shown in Figure 16.15 (for simplicity, the comings and goings of protons, H^+, as in the formation of NADH, are not shown).

► What is the source and the fate of high energy electrons in the energy-releasing example of Figure 16.15?

The energy source is hydrogen sulphide gas, H_2S, or a sulphide which, having donated electrons to a membrane-bound electron transport chain, is oxidized to sulphur. The fate of electrons at the end of this chain is transfer to oxygen (the terminal electron acceptor), which combines with two protons and is reduced to water. Note that this process, unlike non-cyclic photosynthetic electron transport, does not generate reducing power. The dashed black line in Figure 16.15 shows what happens when reducing power is required, to fix CO_2, for example. In a

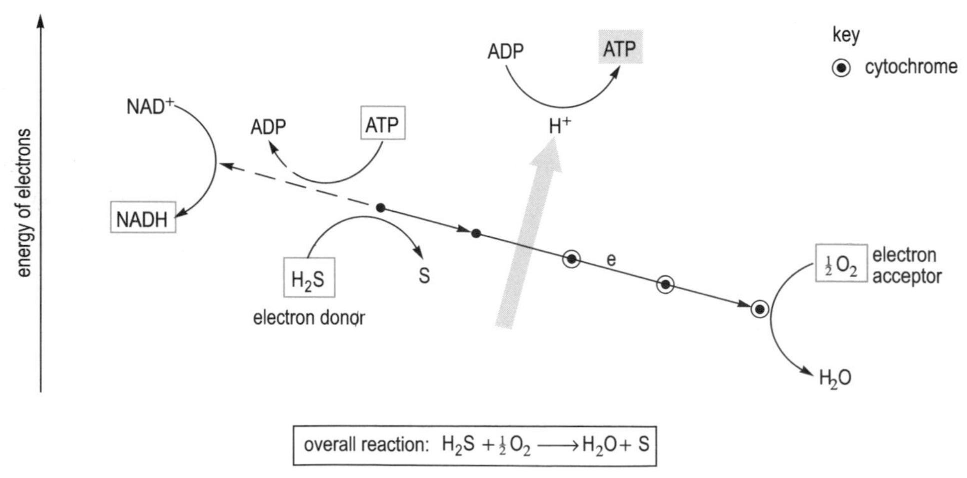

Figure 16.15 *An example of chemo-autotrophy. All electron carriers are bound to the prokaryotic cell membrane. Solid black lines show the energy-releasing process and dashed black lines show the energy-consuming process.*

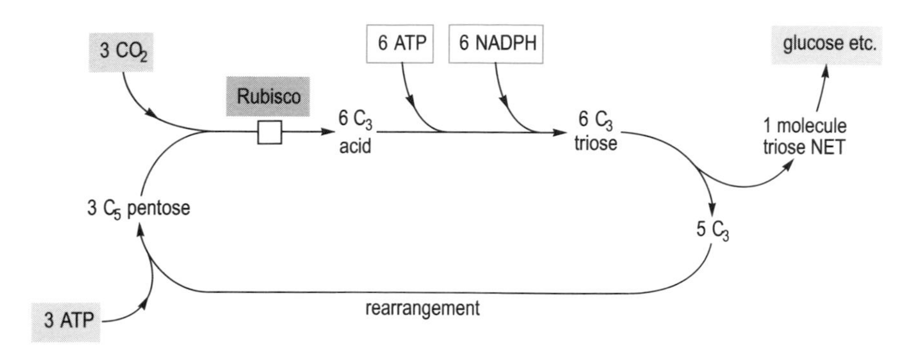

Figure 16.16 *An outline of the Calvin cycle for CO_2 fixation. The number of carbon atoms is shown by the subscript numbers.*

process called *reverse electron transport*—electrons from H_2S are driven in an energetically uphill direction to NAD^+ or its phosphorylated form $NADP^+$, reducing them to NADH or NADPH. A wide range of electron donors are used in such chemosynthetic bacteria, including hydrogen gas (H_2), sulphur, ferrous ions (Fe^{2+}) and ammonium ions (NH_4^+), most of which donate electrons of lower energy than sulphides, i.e. they feed into the electron transport chain lower down the slope and there is then less energy available for ATP synthesis. The range of terminal electron acceptors, however, is more restricted, with oxygen by far the most widely used; in its absence, some species can use substances such as nitrate (NO_3^-), reducing it to nitrite (NO_2^-).

Apart from energy metabolism, the other important characteristic of autotrophs is that they can *fix* CO_2, i.e. assimilate carbon from a simple inorganic molecule into complex organic molecules. There are various pathways of carbon fixation but by far the most widespread is the **Calvin cycle** or C_3 pathway (so named because the first-formed product has three carbon atoms). It is illustrated schematically in Figure 16.16; note that both ATP and reducing power (NADPH, the phosphorylated form of NADH) are required. The enzyme which catalyses the first step is the most abundant protein on Earth, and the name Rubisco is short for ribulose biphosphate carboxylase oxygenase! The overall reaction for the reduction of CO_2 to a sugar such as glucose is:

$$6CO_2 + 12NADPH + 12H^+ + 18ATP \rightarrow C_6H_{12}O_6 + 12NADP^+ + 18ADP + 6H_2O$$

Heterotrophy Variation in the terminal electron acceptor is a hallmark of *respiration* in prokaryotes, the process by which the majority of heterotrophs obtain energy. The final stage, *respiratory electron transport*, is basically the same as electron transport in chemotrophs (Figure 16.15), and they even use some of the same membrane-bound carrier proteins, which include cytochromes. Figure 16.17(a)–(c) shows three versions where the donor of high energy electrons is NADH or lactate (whose formation is discussed below), but different electron acceptors are used. If you have studied any biochemistry previously, you should recognize that Figure 16.17(a) shows the familiar aerobic respiration which occurs in eukaryote mitochondria as well as in many prokaryotes. The anaerobic respiration pathways shown in Figure 16.17(b) and (c), however, occur only in prokaryotes, and illustrate their much greater metabolic flexibility, particularly their

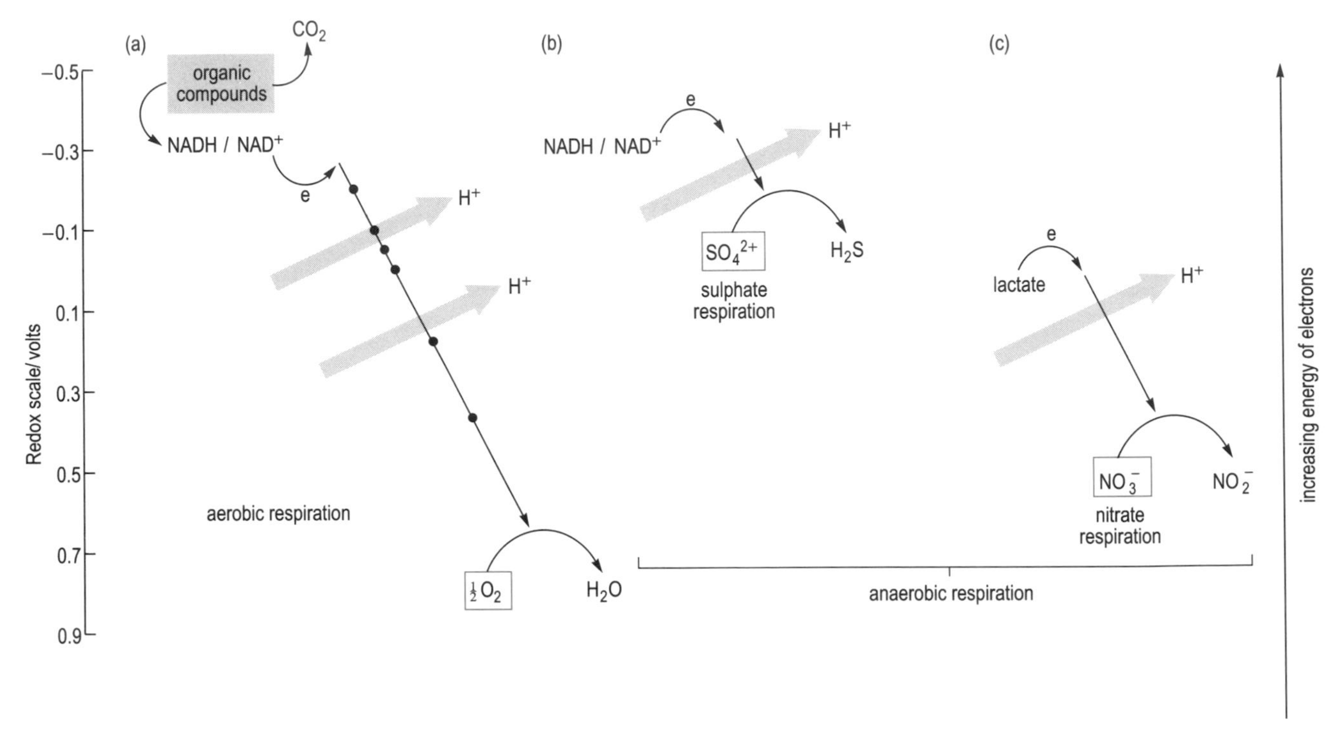

Figure 16.17 *Three pathways of respiratory electron transport using as the terminal electron acceptor (a) oxygen, (b) sulphate, and (c) nitrate. Green arrows denote proton pumping, generating a PMF which is coupled to ATP synthesis.*

capacity to live in the absence of oxygen. Other electron acceptors include organic molecules such as acetate, ferric iron (Fe^{3+}) and even CO_2, which is reduced to methane (CH_4).

The electron donor NADH is produced in prokaryotic heterotrophs in the same way as in eukaryotes. Complex organic food molecules are, if necessary, broken down to simple units such as sugars or amino acids which are then reduced in a series of small steps equivalent to a controlled combustion process. For sugars, these steps start off with the glycolytic pathway with combustion to CO_2 occurring in the tricarboxylic acid (TCA) or Krebs cycle (Figure 16.18). The electrons removed during this last process are transferred mainly to NAD^+, providing the NADH substrate for respiratory electron transport. If the TCA cycle is not operating, the end product of glycolysis (a 3-carbon acid, pyruvate) is converted instead to substances such as lactate, another 3-carbon acid which may also act as an electron donor, as shown in Figure 16.17(c).

As you have seen, this complex, elegant process of respiration can occur in both the presence or absence of oxygen but there must always be a suitable terminal electron acceptor. Sometimes, however, no acceptor is available and this is when the anaerobic process of **fermentation** occurs, essentially shown by the right side of Figure 16.18. Figure 16.19 shows in more detail the fermentation of glucose to lactate, via the glycolytic pathway.

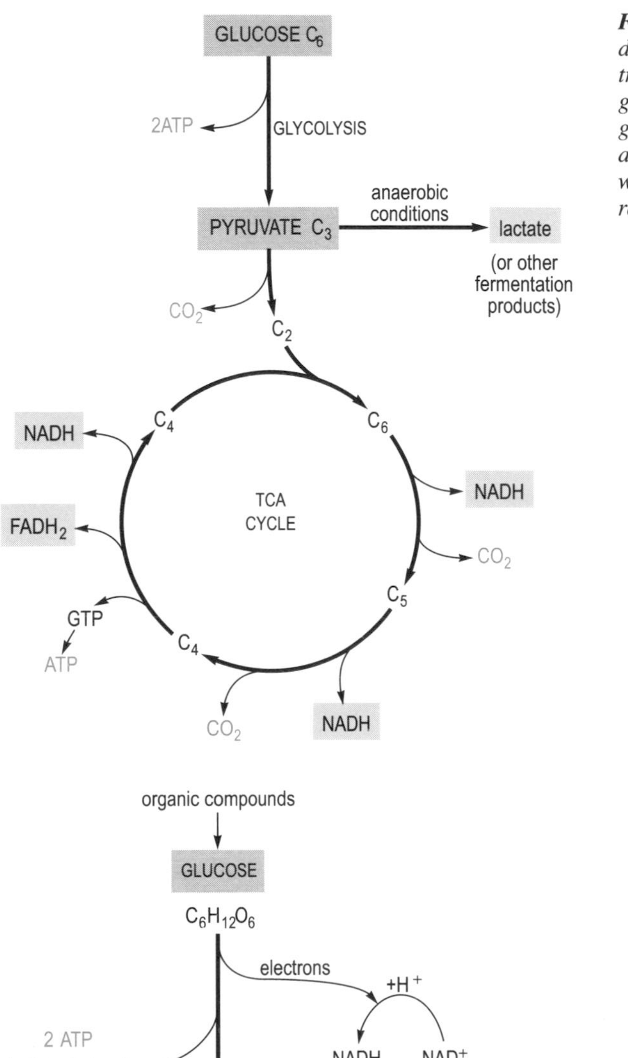

Figure 16.18 Formation of electron donors for respiratory electron transport. Lactate is produced via the glycolytic pathway and NADH via glycolysis plus the TCA cycle. $FADH_2$ is an electron donor, similar to NADH, which can donate electrons to the respiratory electron transport chain.

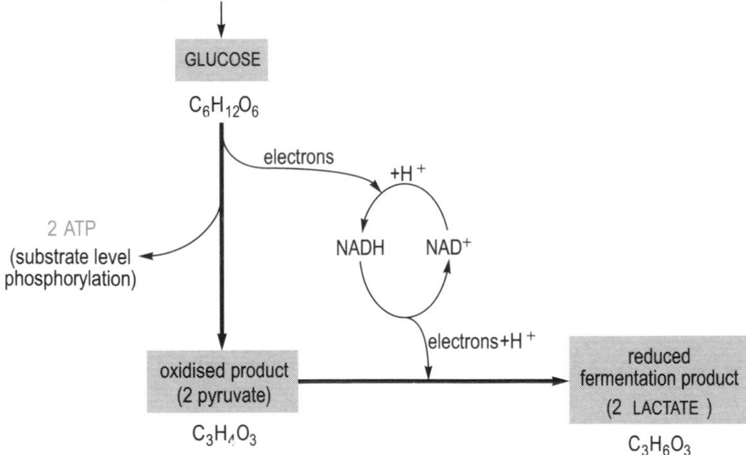

Figure 16.19 *An example of fermentation: the formation of lactate from glucose.*

▶ In fermentation, what happens to electrons removed during glucose reduction to pyruvate?

They are transferred to NAD^+, which is reduced to NADH, which in turn reduces pyruvate generating the fermentation product, lactate—so no external electron acceptor is involved. The ATP produced during fermentation does not utilize a proton motive force but instead is synthesized by *substrate level phosphorylation* (transfer of a phosphate group from an organic compound to ADP). Compared with the complete oxidation of glucose in aerobic respiration (Figure 16.17(a)), the ATP yield of fermentation is pathetically small, two instead of 38 ATP molecules, so it is often regarded as a primitive and inefficient system of energy metabolism. In some prokaryotes, fermentation is simply an emergency system which is used when oxygen is unavailable; but in others it is the only system they have. The interesting question is whether these obligate fermenters are the most primitive prokaryotes, resembling the first living cells most closely.

16.4.3 Diversification of prokaryotes

Archaebacteria The three groups of Archaebacteria (Figure 16.10) are a remarkably heterogeneous lot, which perhaps reflects the extinction of many intermediate groups. Their main common feature is that they all inhabit very *extreme environments*. Figure 16.20 shows the phylogeny of Archaebacteria based on 16S rRNA sequence comparisons (Section 16.5.1) and you can see that although there are three clear groups, genera such as *Archaeoglobus* and *Methanococcus* are intermediates which do not fit tidily within them.

The **Sulphobacteria,** or sulphur-dependent extreme thermophiles (heat-lovers), occur in the hot springs around volcanoes, both on land and under the sea. Their temperature optima are commonly in the range 85–100 °C, and sulphur is always involved in some way in energy metabolism. The *Thermococcus* lineage and the three side branches nearest the base of the main branch, *Thermoproteus*, *Pyrodictium* and *Desulphurococcus*, are strict anaerobes, and all use sulphur as an electron acceptor (similar to the process shown in Figure 16.17(b)). But, whereas *Pyrodictium* is a chemo-autotroph, using hydrogen gas as an energy source, the

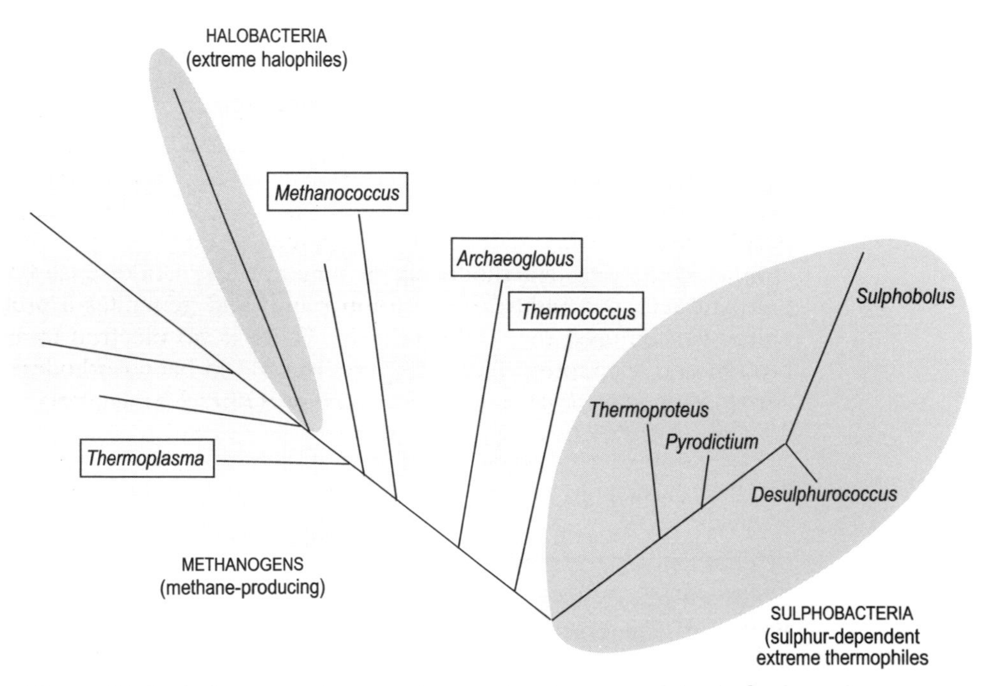

Figure 16.20 *A distance-matrix tree showing the phylogeny of the Archaebacteria.*

other three are sulphur respiring heterotrophs (although *Thermoproteus* can also function chemotrophically). *Sulphobolus*, the most highly evolved genus of the group, is, by contrast, an obligate aerobe which functions chemotrophically as shown in Figure 16.15. It inhabits acidic, near-boiling volcanic springs and fixes CO_2, although *not* by the Calvin cycle—the standard pathway in most photosynthetic organisms. Bearing in mind that evolutionary distance is measured as total branch length from the starting point of a tree (Chapter 11), Figure 16.20 suggests that the Sulphobacteria as a whole have evolved very slowly. For this reason it has been suggested that they resemble the earliest prokaryotes most closely and that life itself may have originated in conditions similar to the present habitats of Sulphobacteria, i.e. the hot spring scenario mentioned in Section 16.3.3.

The methanogenic (methane (CH_4)-forming) and extremely halophilic (salt-loving) Archaebacteria are related more closely to each other than to the Sulphobacteria (Figure 16.20). **Methanogens** are all obligate anaerobes and grow in swamps, the guts of ruminants and anaerobic parts of the soil. The basic type of metabolism, although there are many variants, can be summarized as:

$$CO_2 + 4H_2 \rightarrow CH_4 + 2H_2O$$

▶ Compare this with Figure 16.15 and then work out the energy source (electron donor), electron acceptor and general description of energy metabolism in methanogens.

The energy source is hydrogen gas, which is oxidized to water. CO_2 is the electron acceptor, being reduced to methane, and overall these Archaebacteria are chemo-autotrophs. CO_2 also serves as a carbon source although, as for Sulphobacteria, fixation is not by the Calvin cycle. The need for hydrogen gas and the toxicity of oxygen for these bacteria explains why they are restricted to such strange, anaerobic environments.

As their name suggests, the extremely halophilic **Halobacteria** grow in *very* salty environments, places such as the Dead Sea, Great Salt Lake (Utah), salt pans and even salted foods. Usually, they are heterotrophic and obligate aerobes but, again, there are many exceptions and some of these are particularly interesting from an evolutionary point of view. *Halobacterium halobium*, for example, has so-called purple patches in the cell membrane, which contain the protein **bacteriorhodopsin**, similar to the pigment rhodopsin in human eyes. Under anaerobic conditions, this pigment acts as a light-driven proton pump and generates a proton motive force which can be used for ATP synthesis. There is no electron transport chain, just one protein and a coupling factor (Figure 16.13), so bacteriorhodopsin represents the simplest known photo-autotrophic system. Other Halobacteria can function as anaerobic fermenters and some respire anaerobically (Figure 16.17) but, until 1988, the one thing it was thought that they could not do was fix CO_2. The Indian workers Kelkar and Altekar showed, however, that one species of Halobacteria contains Rubisco, the first enzyme of the Calvin cycle (Figure 16.16), and can indeed fix carbon by this pathway. This sudden appearance of a very complex enzyme in a relatively advanced archaebacterial lineage is not the only puzzling feature of this group. One species contains an enzyme nitrate reductase which, unlike all other halobacterial enzymes studied in detail, is not salt-dependent, i.e. it does not *require* high salt concentration for stability. The archaebacterial nitrate reductase appears to be identical to that found in Eubacteria and a strong possible explanation for the presence of this enzyme, and perhaps for the presence of Rubisco in a halobacterium, is that they were acquired by *lateral gene transfer* from Eubacteria. The same process goes on today among prokaryotes (discussed in Chapter 3, Section 3.2.4); it has been responsible for the spread of antibiotic resistance among bacteria. This means that novel or anomalous characteristics may not have evolved within a lineage but may have been grafted on from the outside.

At present, we have to assume that *most* archaebacterial features reflect evolutionary relationships and, as we said earlier, Archaebacteria are a very heterogeneous group. Every type of energy metabolism occurs, with the emphasis being on heterotrophy and chemo-autotrophy of types basically similar to those found in Eubacteria. Only the photo-autotrophs have a unique archaebacterial system lacking both the pigments and the electron transport chain typical of photosynthetic bacteria. The general absence of the Calvin cycle (the CO_2 fixation system found in many photosynthetic Eubacteria and all green eukaryotes) similarly distinguishes the Archaebacteria and suggests that their ancestral stock did not possess it.

Before leaving the Archaebacteria, two 'misfit' lineages deserve special mention. *Archaeoglobus* (see Figure 16.20), both from rRNA sequence analysis and from other features, seems to form a link between the Sulphobacteria and the methanogens. It grows optimally at 83 °C and uses sulphate, rather than sulphur, as an electron acceptor. However, in addition to these Sulphobacteria-like properties, *Archaeoglobus* can produce methane and carries out certain reactions which are otherwise confined to methanogens. It looks like a classic 'missing link' organism.

Thermoplasma (see Figure 16.20), the second misfit lineage, evolved from the methanogen/extreme halophile stock according to rRNA sequence analysis, but shares no other common characteristics. It grows as an aerobic heterotroph in hot coal refuse heaps, typically at pH 2 and 55 °C, and, unlike all other Archaebacteria, has no cell wall. Another unique feature of *Thermoplasma* is that its DNA is associated with proteins which are remarkably similar to the DNA-binding histones in *eukaryotic* nuclei and which occur in no other prokaryotes. Do these histone-like proteins represent parallel evolution with the ancestral eukaryotes (the nuclear 'host' lineage), or is it possible that the eukaryote lineage evolved from *Thermoplasma* stock? The answer is not known at present and *Thermoplasma* remains something of an enigma.

Eubacteria Modern Eubacteria are much more diverse than Archaebacteria and they have evolved further from the universal prokaryote ancestor. Figure 16.21 shows a phylogenetic tree for the main groups which is based on 16S ribosomal RNA; different shading indicates the types of energy metabolism found in these groups.

Notice that anaerobic fermentation (dark-grey shading in Figure 16.21), once regarded as the most primitive type of metabolism, occurs in five unrelated groups. Perhaps significantly, it is the only type of metabolism that occurs in the least evolved group, which contains the single genus *Thermotoga*. Like many Archaebacteria, *Thermotoga* species are thermophiles, living in geothermally heated marine sediments. In general, however, several types of energy metabolism

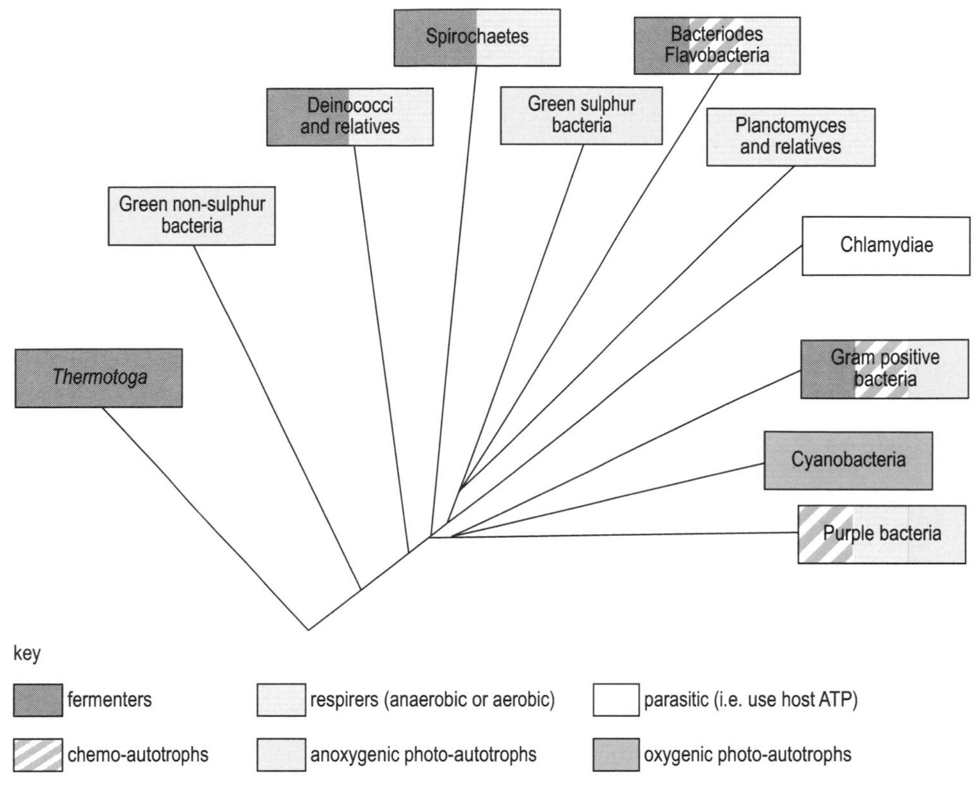

Figure 16.21 *Eubacterial phylogenetic tree based on 16S ribosomal RNA.*

have evolved in most groups. One group, the purple bacteria, has been studied in some detail by ribosomal RNA sequencing and the results suggest that aerobic heterotrophs and chemo-autotrophs evolved *from* early photo-autotrophs. So, using light as an energy source may be just as ancient as fermenting organic compounds.

▶ Are the photosynthetic Eubacteria closely related to each other?

No. Photosynthesis occurs in five scattered groups. Four of these (light-green shading in Figure 16.21) have light reactions of the type shown in Figure 16.14, i.e. they can generate ATP by cyclic electron flow and reducing power by non-cyclic flow coupled to the oxidation of an external electron or hydrogen donor. Commonly used donors are hydrogen sulphide (H_2S) and hydrogen gas (H_2), both of which occur mainly in anaerobic conditions. Indeed, photosynthetic bacteria in these groups are strict anaerobes which are poisoned by oxygen. Their abundance has probably always been limited to some extent by the availability of these electron donors, which is why cyanobacteria (dark-green shading in Figure 16.21) represent a major evolutionary advance. For non-cyclic electron flow, these Eubacteria use water (H_2O), the most plentiful electron donor on Earth, oxidizing it to oxygen (O_2):

cyanobacteria: $2H_2O \rightarrow O_2 + 4H^+ + 4e^-$

Most of the oxygen in the Earth's atmosphere derives from this reaction, so the evolution of oxygen-producing, or **oxygenic**, **photosynthesis** not only allowed a

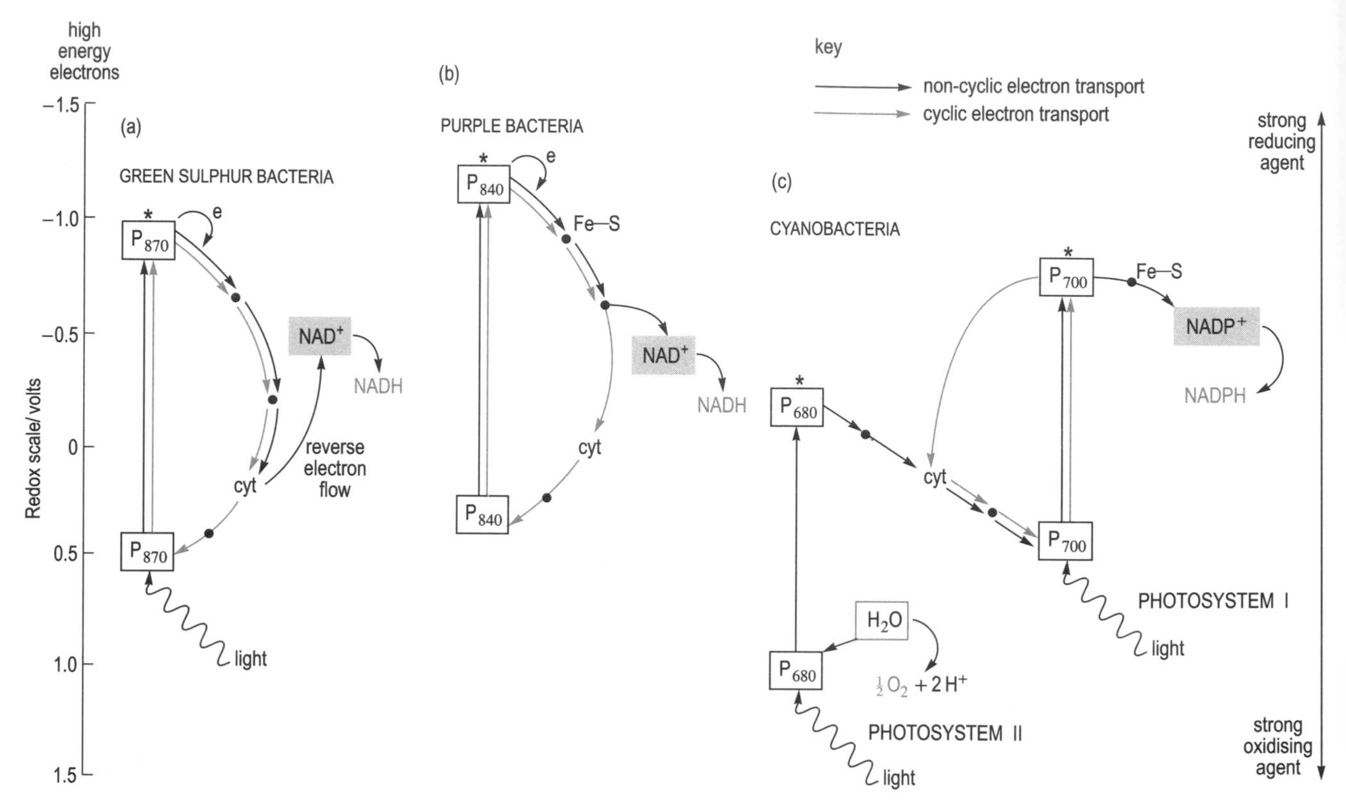

Figure 16.22 *Photosynthetic electron flow in: (a) green sulphur bacteria; (b) purple bacteria; and (c) cyanobacteria. The green pathways indicate cyclic flow; cyt is a cytochrome, and Fe–S is an iron-sulphur protein. P is a pigment complex that emits an energized electron after absorbing light—an 'excited' state indicated by an asterisk; the numbers indicate the wavelength of light absorbed.*

huge increase in bacterial production but also led to an environmental change from which followed the evolution of all oxygen respiring organisms.

▶ Study Figure 16.22 and work out the special features of oxygenic light reactions compared with anoxygenic reactions in purple and green photosynthetic bacteria.

The most striking feature apparent from Figure 16.22 is that cyanobacteria have *two linked light reactions* which are named Photosystems I and II, instead of the single light reaction found in other bacteria. To appreciate the significance of the two light reactions, look at the scale on the left in Figure 16.22—oxidation of water requires a very powerful *oxidizer*, low down the scale, and this is what Photosystem II has. Neither of the other bacterial pigment complexes (Figure 16.22) can oxidize water. However, to reduce $NADP^+$ and generate the reducing power for CO_2 fixation, a powerful *reducing agent* high up the scale on the left is required; this is what Photosystem I has—there is some resemblance to the system shown in Figure 16.14(b). Overall, transferring electrons from water to $NADP^+$ is rather like trying to kick a football to the top of the Eiffel Tower: impossible with one kicker but possible with a second (powerful) kicker half-way up to give the ball a booster kick!

In addition to similarities between Photosystem I and green sulphur bacteria (e.g. both use an iron-sulphur protein as the primary electron acceptor), there are some similarities between Photosystem II and purple bacteria (e.g. both use a quinone as the primary electron acceptor). These similarities have led to suggestions that cyanobacterial photosynthesis arose by combining and modifying the two existing bacterial systems at an early stage in their evolution. Another link between cyanobacteria and purple bacteria is their mechanism for carbon fixation, because both use the Calvin cycle to reduce CO_2 to carbohydrates. Chemosynthetic Eubacteria and all autotrophic eukaryotes also use this pathway, but the archaebacterial methanogens and some photo-autotrophs, in particular the green sulphur bacteria, do not. This distinction is important because organic matter produced by the Calvin cycle contains a particular ratio of carbon isotopes by which it can be recognized in sediments. Such data (described in the next Section) indicate that the Calvin cycle appeared very early in prokaryote evolution.

The remaining Eubacteria are all heterotrophic with either anaerobic or aerobic respiration (light-grey shading in Figure 16.21). The former could have evolved before, or in parallel with, cyanobacteria when the atmosphere was anoxic, but the latter clearly evolved after cyanobacterial photosynthesis produced an oxygen-rich atmosphere. The Chlamydiae stand out as an anomaly because they have no energy metabolism at all, being greatly reduced parasites of eukaryotic cells and dependent on their hosts for ATP. The other heterotrophic Eubacteria now depend on dead or living organisms, mostly eukaryotes, for their source of energy (food), so we can pinpoint three important factors that influenced the diversification of Eubacteria:

1 The supply of organic molecules for fermenters or respiring heterotrophs. Initially, these molecules might have been provided in the primordial 'soup', but if this was as dilute as many people now think (Section 16.3.4), then autotrophs and later the eukaryotes must have become the main sources of supply.

2 The supply of electron donors for anoxygenic photosynthetic bacteria and suitable substrates for chemosynthetic bacteria (Figure 16.15).

3 Adaptation to rising levels of atmospheric oxygen. For strict anaerobes, oxygen is highly toxic, but once bacteria evolved oxygen tolerance they were able to oxidize substrates more efficiently (thereby obtaining more energy) and occupy an increasing range of habitats.

Another effect of rising atmospheric oxygen was that it led to the formation of the ozone layer in the upper atmosphere.

▶ Recall from Section 16.2.2 the significance of the ozone layer for living organisms.

It absorbs most of the short wavelength ultraviolet radiation from the Sun, which damages nucleic acids and causes considerable other cellular damage. Before the ozone layer appeared, some prokaryotes were probably confined to 'protected' environments below water or in sediments, although others, like the modern Deinococci, which has remarkable powers of DNA repair and extreme resistance to radiation damage, probably lived at the surface.

With this background, we examine next some of the geological evidence for the timing and sequence of events in prokaryote evolution.

16.4.4 Geological evidence of prokaryote evolution

Although phylogenies based on nucleic acid sequences can tell you what evolved from what, they provide no information about *when* things happened. For that we need geological evidence from rocks of known age and, for the Archaean, such evidence is not only sparse — only about a dozen good sites are known — but it also needs very careful interpretation. It does exist, however, and has been added to so significantly during the late 1970s–1980s that ideas about when life originated and the metabolism of early cells have required radical re-thinking.

Until the 1970s, most textbooks suggested a date between 3 000 and 3 500 Ma ago for the origin of life. What the newer data show is that *autotrophic* life was undoubtedly flourishing 3 500 Ma ago and probably existed as far back as 3 800 Ma, which means that life must have originated by at least 4 000 Ma, a comparatively short time after the formation of the Earth (*c.* 4 600 Ma).

The two most reliable types of evidence for early autotrophic life are fossil **stromatolites** (Chapter 12) and carbon isotope ratios indicating photosynthesis. We describe this evidence next and then consider the implications for the early evolution of heterotrophic prokaryotes.

Stromatolites Certain modern cyanobacteria living in shallow coastal waters form characteristic bun-shaped structures called stromatolites (Figure 16.23(a)). These build up as sediments accumulate around gelatinous mats which are produced by cyanobacteria. Remarkably, identical structures occur as fossils in Precambrian rocks. Modern stromatolites grow only in conditions where grazing and burrowing metazoan predators are excluded (e.g. very saline areas) but Precambrian stromatolites were widespread and abundant by 3 500 Ma ago (Figure 16.23(c)), becoming highly diversified in the late Precambrian. In section, they have a layered structure like a stack of pancakes.

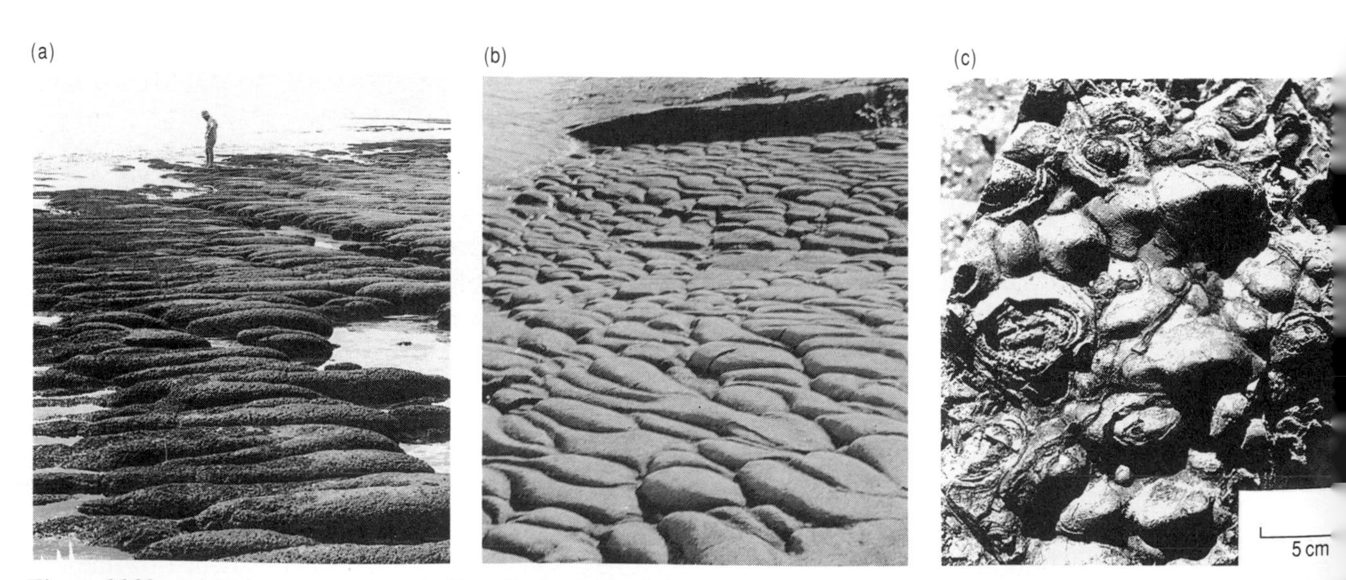

Figure 16.23 *(a) Living stromatolites in Hamelin Pool, Shark Bay, eastern Australia. (b) 2000 Ma old stromatolites in the Pathe[r] Group, Great Slave Lake, Canada. (c) 3500 Ma old stromatolites from Western Australia.*

It is likely that the organisms which formed these stromatolites were cyanobacteria, but this is not absolutely certain because some modern photosynthetic bacteria form rather similar structures. This is an important point with far-reaching implications for the nature of the environment then.

▶ Why? What effect do cyanobacteria, but not photosynthetic bacteria, have?

Cyanobacteria (but not photosynthetic bacteria) carry out oxygenic photosynthesis and release oxygen into the environment (Section 16.4.3). No-one disputes, however, that highly organized communities of autotrophs were present at this early date and other evidence supports this conclusion. For example, it is now reasonably certain that microfossils associated with ancient stromatolites were *cells* and not artefacts or proteinoid microspheres (Section 16.3.3). In some ancient Australian stromatolites there are filaments and spherical structures which look remarkably like modern cyanobacteria but this interpretation is still controversial. No firm conclusion can be drawn from fossil evidence as to whether oxygenic cyanobacteria or anoxygenic photosynthetic bacteria were the dominant autotrophs 3 400 to 3 500 Ma ago.

Evidence of early photosynthesis from carbon isotope ratios Although biochemical remains are exceedingly scarce in Archaean sediments, a portion of the organic matter is often preserved as an insoluble (i.e. not recoverable from the rocks using organic solvents) macromolecular material called *kerogen* (from the Greek *keros* meaning wax, and used to describe the organic matter present in sediments). Most Archaean sediments have generally been subjected to heating in the more than 3 000 Ma since they were deposited and the kerogen occurring in such sequences is often greatly altered, rendering undecipherable much of the structural chemical information it once contained. However, various workers have used the abundance of the carbon isotopes in Archaean kerogen as a tool for unravelling the biochemistry of the ancient Earth.

Carbon consists principally of a mixture of two stable isotopes, ^{12}C and ^{13}C and, in photosynthesis, fixation of the lighter $^{12}CO_2$ is favoured over that of the heavier $^{13}CO_2$. The reasons are that $^{12}CO_2$ is taken up and diffuses into cells more rapidly and also reacts more readily in the first carboxylation reaction. As a result of this isotope discrimination, organic matter produced by photosynthesis is enriched in ^{12}C and depleted in ^{13}C relative to the inorganic pool of carbon (mainly CO_2, carbonate and bicarbonate). Enrichment or depletion of ^{13}C is expressed in terms of a **delta (δ)^{13}C value**, which is calculated as follows. The ratio of $^{13}C/^{12}C$ for the sample being investigated is compared with that of a carbonate standard with a ratio of 1/88.99, i.e. (ratio of sample)/(ratio of standard). One is subtracted from this value and the whole is multiplied by 1 000 to give a $\delta^{13}C$ value in terms of parts per thousand (‰) ^{13}C relative to the standard:

$$\delta^{13}C = \left[\frac{(^{13}C/^{12}C)\text{sample}}{(^{13}C/^{12}C)\text{standard}} - 1\right] \times 1\,000 \text{ (‰, standard)}$$

▶ With this formula, will the value of $\delta^{13}C$ be negative, positive or zero when (a) the ratio $^{13}C/^{12}C$ is equal for standard and sample; (b) greater in the sample; (c) lower in the sample?

Because one is subtracted from the ratio of ratios, for (a), the value will be zero; for (b) it will be positive, and for (c) it will be negative. So when the sample is organic matter enriched in ^{12}C and depleted in ^{13}C relative to the standard, *the value of $\delta^{13}C$ will be negative*. Now look at the left side of Figure 16.24 which shows the range of $\delta^{13}C$ values for living autotrophs (dark-green shading), recent marine (organic) sediments (pale-green shading) and inorganic, oxidized carbon compounds (grey shading). There is quite a range of values among autotrophs, reflecting their different types of carboxylation reaction, but the $\delta^{13}C$ value is consistently negative and mostly between −20 and −30‰.

▶ What are the $\delta^{13}C$ values for the carbon *sources* for photosynthesis (carbon dioxide and bicarbonate)?

Values of −7 and −1‰ respectively, while the value for recent marine carbonate, which is formed from marine bicarbonate, is approximately zero, like the standard.

It was noted as long ago as the 1960s that the mass balance of the carbon isotopes indicated that an approximately constant partitioning occurred between organic carbon and carbonate carbon in the Earth's sediments since the start of the sedimentary record in the Isua sediments (Section 16.2.1). In addition, the average organic carbon content of Phanerozoic sediments (~0.5%) is observed in most Precambrian sediments, including those from Isua, thereby establishing the continuity of the organic record from the oldest known sediments to the present

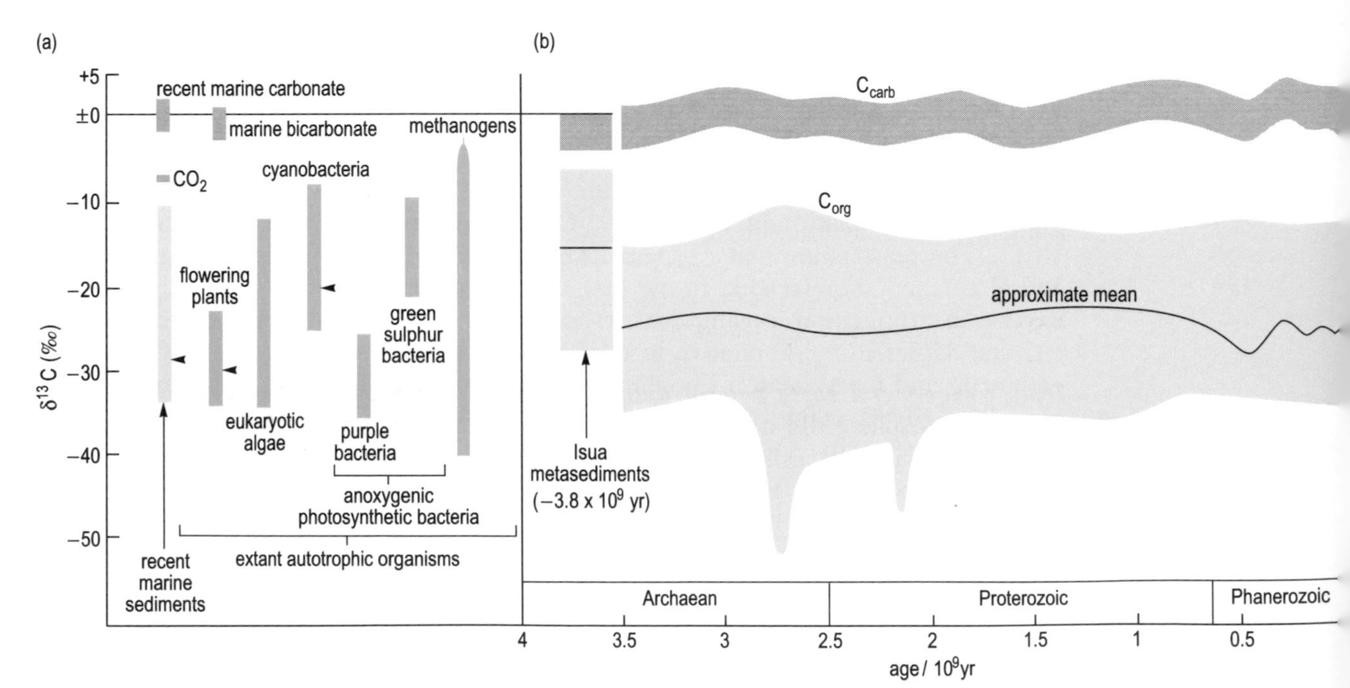

Figure 16.24 *Carbon isotope composition as $\delta^{13}C$ values for (a) extant autotrophs, recent marine (organic) sediments, and oxidized inorganic carbon. All organisms except green sulphur bacteria and methanogens use the Calvin cycle for carbon fixation. Triangles indicate the approximate means. (b) Sedimentary carbonate (C_{carb}) and organic carbon (C_{org}) over 3800 Ma of the Earth's history*

time. Figure 16.24(b) summarizes more than 10 000 measurements of carbon isotope ratios in sediments through time (Schidlowski, 1988). For 3.5 Ga, the mean $\delta^{13}C$ value for organic sediments lies between −24 and −28‰, which you can see from Figure 16.24(a) is the same degree of fractionation presently observed in living autotrophs. The majority of these autotrophs fix carbon by the Calvin cycle (Section 16.4.2).

▶ What does this imply about organisms present in the early Archaean?

It implies that autotrophs fixing carbon via the Calvin cycle must have existed for 3.5 billion years. Schidlowski argues that there is evidence for carbon fixation by the Calvin cycle even in the oldest material examined, the Isua metasediments (i.e. altered sediments) from Greenland, dated at 3.8 Ga. Figure 16.24(b) shows that these metasediments have both more negative carbonate and less negative organic $\delta^{13}C$ values than later sediments but their extensive metamorphism, i.e. subjection to high temperature and pressure, is known to cause release of CO_2 and a shift in $\delta^{13}C$ values. It seems highly probable that the *original* $\delta^{13}C$ values for these rocks, before metamorphism, were the same as for the later rocks. If so, autotrophs appear to have been thriving 3 800 Ma ago! This conclusion would only be in error if there was some other global process capable of mimicking the isotope fractionations in photosynthesis.

These attempts to extend the record of biological evolution back in time have met with some scepticism. Some scientists believe that the carbon isotope signature in the Isua sediments is simply too faint to interpret. Assuming that Schidlowski is correct, however, then these earliest autotrophs must have used the Calvin cycle and must have had a system of non-cyclic photosynthetic electron transport, requiring an external electron donor (Section 16.4.2), in order to generate the necessary reducing power. We cannot rule out completely the possibility that substances such as hydrogen sulphide were used as the external donor, but geological evidence does not support this view. Large deposits of oxidized H_2S (i.e. sulphur, S, or more highly oxidized sulphate) do not accompany all deposits of organic carbon and it is more likely that water was used as the external electron donor, i.e. oxygenic photosynthesis as practised by cyanobacteria was occurring 3 500 and perhaps 3 800 Ma ago.

Iron, oxygen and early heterotrophs As discussed earlier in Section 16.4.3, the Earth's atmosphere did not become rich in oxygen until about 2 000 Ma ago. If, however, oxygen-releasing cyanobacteria were abundant by 3 500 Ma ago, as suggested above, what became of this oxygen? The question is still being debated in 1992, but, whatever the answer, there are some important implications for the early evolution of heterotrophs.

The first explanation to be widely accepted for the low oxygen or oxygen-free atmosphere during the Archaean was put forward by Preston Cloud in 1976. He suggested that O_2 was mopped up by soluble iron present in the oceans in its reduced or ferrous form (Fe(II), Fe^{2+}), with the formation of oxidized ferric iron (Fe(III), Fe^{3+}). The ferric iron formed insoluble ferric oxides or hydroxides which were precipitated and contributed to the laying down of **banded ironstone**

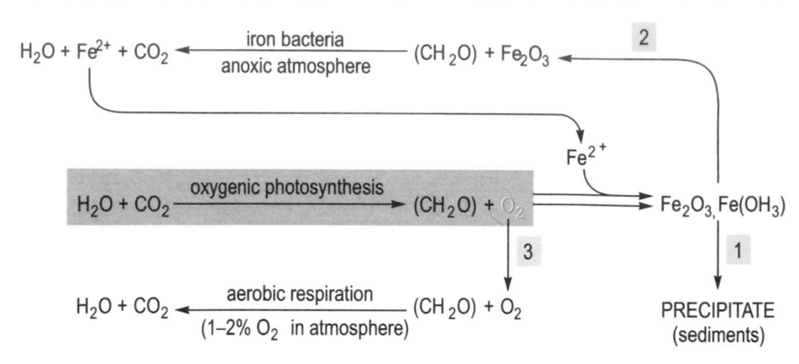

Figure 16.25 *Three possible mechanisms for removing oxygen generated by oxygenic photosynthesis in the Archaean.*

formations (BIFs for short). Thus, iron was viewed as an oxygen depressor or **oxygen sink** at this period, and thin BIFs do indeed occur in the earliest rocks. However, they are most abundant in the period 2500 to 2000 Ma ago when atmospheric oxygen levels began to rise. Presumably, and logically, this rise occurred because the increasing amounts of oxygen released by the cyanobacteria exceeded the capacity of the iron 'sink' to absorb it and excess oxygen was released into the atmosphere.

The basic tenet of Cloud's theory—that the Archaean atmosphere was anoxic—is still widely accepted, but there are certain problems associated with it. For example, even if global productivity by cyanobacteria is assumed to have been very low, removal of all oxygen released so as to maintain an anoxic atmosphere still requires the sedimentation of far more oxidized iron than is actually found. A possible solution, suggested by Walker (1987), is that oxidized iron was reduced and recycled by anaerobic 'iron-breathing' heterotrophs (following the principles shown in Figure 16.17); this is illustrated as Pathway 2 in Figure 16.25.

▶ Describe the solution shown as Pathway 3 in Figure 16.25 and identify the ways in which it differs from Pathway 2.

In Pathway 3, suggested by Towe (1990), oxygen is removed by aerobic, i.e. oxygen-breathing, heterotrophs as happens today. According to Towe, this system could have worked provided that the atmosphere contained a low level of oxygen (1–2%) instead of being totally anoxic; oxygen removal by oceanic iron (Pathway 1) would still have acted as an important subsidiary brake on oxygen build-up in the atmosphere.

Debate about the significance of Pathways 2 and 3, particularly the feasibility of Pathway 3, still continues but, clearly, both have implications for the evolution of heterotrophic prokaryotes. Anaerobic respirers could have evolved early in the Archaean, and aerobic respirers long before 2000 Ma ago when atmospheric oxygen levels began to rise. If aerobic respiration was possible, then oxygen-requiring chemo-autotrophy (Figure 16.15) would also have been possible. We are left with a situation therefore where every type of energy metabolism described in Section 16.4.2 could have been found in the period 3500 to 3000 Ma ago.

16.4.5 THE ANCESTRAL PROKARYOTES

Having covered some background information about the metabolism and types of prokaryotes present both now and in the Archaean, we can return to Figure 16.10 and speculate about the universal prokaryote stock — the earliest organisms from which the Archaebacteria, Eubacteria and eukaryotic nuclear lines evolved. Speculate is the operative word here because nobody actually *knows* and there are many differing views. Until the 1970s–1980s it was almost universally believed that the first prokaryotes were anaerobic fermenters, obtaining energy by rather inefficient breakdown of organic molecules in the primordial 'soup'. This view is still widely held, but with a better understanding of energy metabolism and, in particular, the ways in which ion pumping can be coupled in relatively simple ways to energy-requiring processes, there has been growing support for the view that the earliest prokaryotes were autotrophs.

The geological evidence (Section 16.4.4) and study of modern Eubacteria (Section 16.4.3) suggest that photo-autotrophy is very ancient and it has, indeed, been suggested that the first cells obtained energy in this way (Section 16.3.4). But there is little to support this view from the Archaebacteria (Section 16.4.3); only in the relatively advanced Halobacteria does a primitive kind of photo-autotrophy occur, and less evolved members of the kingdom such as the Sulphobacteria are mainly anaerobic respirers or chemo-autotrophs.

▶ Recall two characteristics shared by anaerobic Sulphobacteria.

All tolerate very high temperatures and sulphur plays a central role in their energy metabolism (Section 16.4.3). The 'hot spring scenario' discussed in Section 16.3.3, suggested that life originated in hot, submarine vents, an environment which is also rich in sulphur. If true, then the Sulphobacteria may bear the closest resemblance to the ancestral prokaryotes. Reduced sulphur also plays a role as an electron donor in non-cyclic photophosphorylation in the photosynthetic Eubacteria (Section 16.5.3), which must have evolved before water-splitting cyanobacteria. Figure 16.26 attempts to draw these threads together without making any firm statements that *this is how it was.*

▶ Look at Figure 16.26 with a critical eye. In what way does the proposed origin of fermenters in the ancestral stock seem unlikely?

With both scenarios, fermenters arose by the *loss* of an autotrophic system for obtaining energy, which involved light or chemically driven membrane pumps. This loss is an odd thing to have happened and yet, unless fermenters were the first forms of life, it must have occurred, either in the ancestral stock or later, and independently in several eubacterial lines. Answers to puzzles like this are only likely to come from sequencing more nucleic acids, combined with more studies of sediment chemistry which give data about the environment and the biological processes affecting it.

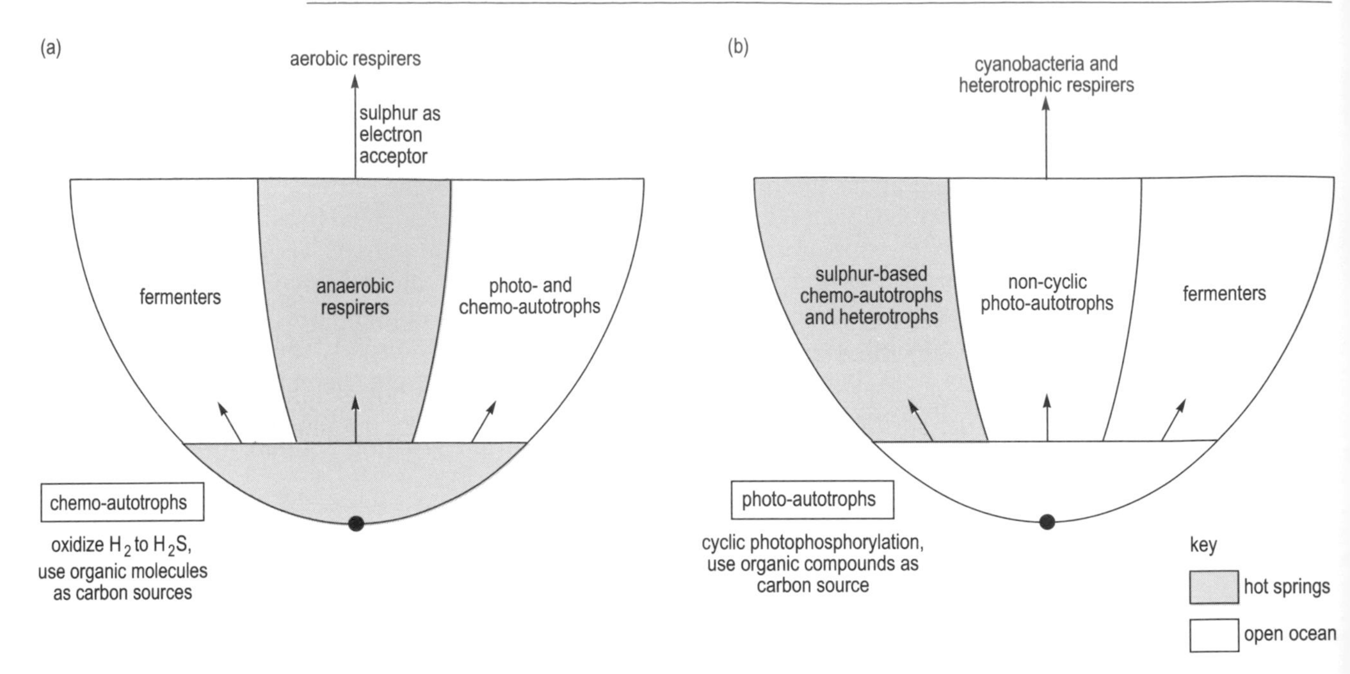

Figure 16.26 *Two scenarios for the possible nature of the ancestral prokaryote stock.*

SUMMARY OF SECTION 16.4

Molecular sequence data and comparative biochemistry suggest that prokaryotes divided soon after the origin of life into three lines: the Archaebacteria, the Eubacteria and the ancestral (nucleo-cytoplasmic) eukaryotes. There is great variation in the ways that prokaryotes obtain energy (energy metabolism), but the principle common to all except fermenters is that of electron transport linked to proton pumping. High energy electrons are transferred along a chain of one or more membrane-bound electron carriers; energy released during transport is used to pump protons across the membrane, and the electro-chemical gradient (proton motive force, PMF) formed drives protons back across the membrane through coupling factors (the turbines), which synthesize ATP.

In autotrophs, high energy electrons are acquired by the absorption of light (photo-autotrophs) or by the oxidation of simple, usually inorganic molecules (chemo-autotrophs). Electron transport may be cyclic (only ATP formed) or non-cyclic (both ATP and reducing power, e.g. NADH produced). Carbon fixation, e.g. by the Calvin cycle, requires both ATP and reducing power.

In photosynthetic bacteria, molecules such as hydrogen sulphide provide a source of electrons for non-cyclic electron transport (anoxygenic photosynthesis). In cyanobacteria, the electron donor is water, which is oxidized to oxygen (oxygenic photosynthesis). Nearly all the oxygen in the atmosphere derives from this reaction.

In heterotrophs, organic molecules (food) broken down by the central metabolic pathways (glycolysis and the TCA cycle) provide the source of high energy electrons for respiratory electron transport. This source is mainly NADH from the TCA cycle. Energy release is maximal (i.e. most ATP synthesized) if the terminal electron acceptor is oxygen (aerobic respiration); but many other acceptors (e.g. iron, sulphate, sulphur or nitrate) can be used in anaerobic respiration. Anaerobic fermenters release only a small fraction of the energy available in food, using substrate level phosphorylation for ATP synthesis in glycolysis.

Living Archaebacteria all tend to inhabit extreme environments and can be classified into three groups:

(i) Sulphobacteria are the least evolved and occur in hot, sulphur-rich places such as volcanic springs; they are either chemo-autotrophs or sulphur-respiring (anaerobic) heterotrophs.

(ii) Methanogens are anaerobic, methane-producing chemo-autotrophs which use hydrogen gas as an energy source.

(iii) Halobacteria, the most 'advanced' group, live in very salty places and, although mostly aerobic heterotrophs, show a wide range of energy metabolism. One species displays a unique kind of photo-autotrophy.

The Eubacteria are more diverse and highly evolved than the Archaebacteria. All types of energy metabolism occur and photo-autotrophy is found in five separate groups. Oxygenic photosynthesis occurs only in cyanobacteria. Diversification was probably influenced strongly by adaptation to rising levels of atmospheric oxygen and by the availability of essential molecules.

Evidence from fossils and carbon isotope ratios suggests that photo-autotrophs were abundant by 3 500 Ma ago and possibly by 3 800 Ma. These early photosynthesizers apparently fixed carbon by the Calvin cycle and used water as an electron donor, so they were probably cyanobacteria. The atmosphere contained little or no oxygen in the early Archaean and oxygen released by cyanobacteria may have been disposed of by three routes: aerobic respiration by heterotrophs, or by the formation of ferric oxides or hydroxides which were either deposited in sediments or recycled via iron-breathing heterotrophs.

What the first (ancestral) prokaryotes were like is still a mystery but the available evidence suggests that they were probably autotrophs. They might have been chemo-autotrophs confined to sulphur-rich habitats, such as hot submarine vents, and resembling modern Sulphobacteria (Archaebacteria). Or they could have been photo-autotrophs with a simple kind of cyclic electron transport system.

16.5 THE ORIGIN AND EARLY EVOLUTION OF EUKARYOTES

The move from the prokaryote to the eukaryote grade of organization was certainly of immense significance in biological evolution. Virtually all organisms that you see are eukaryotes and, in this sense, they dominate the Earth—although microscopic prokaryotes are still abundant and ecologically important as agents of decomposition, disease and primary production. After looking in more detail at the differences between eukaryotes and prokaryotes, the main questions tackled in this Section are:

1 *When* did eukaryotes first appear in the fossil record?

2 *How* did the eukaryote cell evolve?

3 *How many times* did this happen?

16.5.1 CRITERIA FOR DISTINGUISHING PROKARYOTES FROM EUKARYOTES

Basic differences between prokaryotes and eukaryotes were described in Chapter 3, Section 3.2.2, and Figure 16.27 should remind you what these are. The difference in cell size is striking, and so is the presence of organelles—nucleus, mitochondria and chloroplasts—in eukaryote cells. These and other key differences are summarized in Table 16.2.

Modern eukaryote cells are much more complex than those of prokaryotes (characteristics 2, 5, 7 and 8 in Table 16.2 are especially relevant here). In eukaryotes there are usually clear divisions of labour within the cell (characteristic 2)—aerobic respiration and photosynthesis occur only in mitochondria and chloroplasts, respectively, with the electron transport proteins bound to the inner, highly convoluted membranes of these organelles. By contrast, in prokaryotes the electron carriers are bound to the cell membrane or infoldings from it. However, certain unicellular eukaryotes, that have nuclei and flagella, *lack* mitochondria—the protistan parasite *Giardia lamblia*, for example, which lives in the anaerobic environment of animal guts.

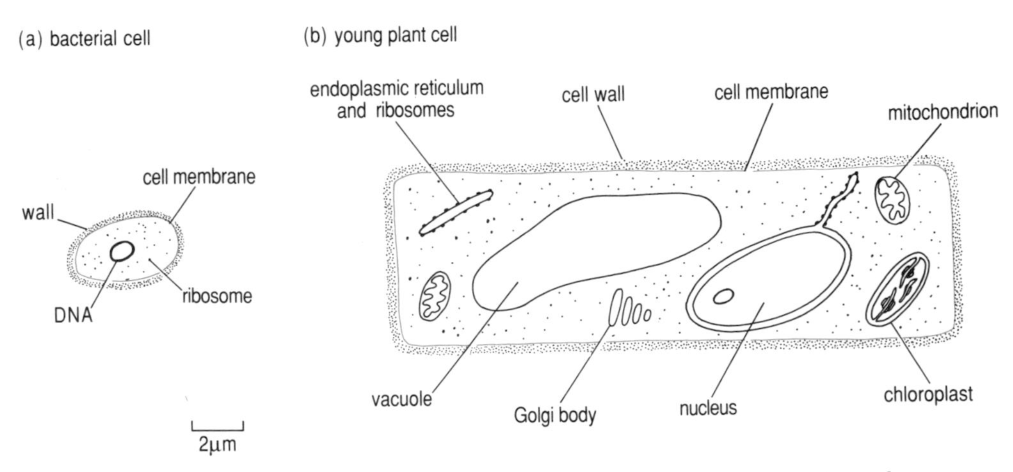

Figure 16.27 *Typical prokaryote and eukaryote cells. Green lines indicate cell membranes. Note the double membrane around mitochondria and chloroplasts.*

Table 16.2 A comparison between prokaryotes and eukaryotes.

Characteristic	Prokaryotes (Archaebacteria and Eubacteria)	Eukaryotes
1 Cell size	Mostly < 10 μm in diameter; bacteria generally < 1 μm	Mostly 10–100 μm; some unicellular algae are only 4–5 μm
2 Organelles	No nucleus or organelles bounded by a double membrane	Nucleus present and usually organelles bounded by a double membrane (chloroplasts and/or mitochondria)
3 Metabolism	Anaerobic or aerobic; heterotrophic, chemo-autotrophic or photo-autotrophic	Mostly aerobic; heterotrophic or photo-autotrophic
4 Motility	Non-motile, or with flagella composed of flagellin protein	Usually motile at some stage (e.g. sperm); cilia or flagella composed of tubulin protein
5 Genetic material	Single loop of DNA in cytoplasm	Several linear DNA molecules associated with histone proteins to form chromosomes and contained in the nucleus
6 Cellular organization	Mostly unicellular; some cyanobacteria form multicellular filaments	Some unicells, but mostly multicellular with differentiation of cells
7 Cell reproduction	Usually by simple fission	By mitosis; diploid cells may undergo meiosis and form haploid gametes
8 Cytoskeleton*	Absent	Present

* The cytoskeleton is a three-dimensional network of fibrous proteins that gives order and structure to the cytosol. The proteins form either delicate microfilaments or more robust microtubules, which are composed of the protein tubulin.

▶ What are the two alternative explanations for the absence of mitochondria in *Giardia*?

First, *Giardia* may once have possessed mitochondria but may have lost them because they were selectively disadvantageous in the anaerobic gut environment, i.e. the species is *secondarily amitochondrial*. Secondly, *Giardia* may represent an early stage of eukaryote evolution when nuclei but not mitochondria were present, i.e. the species is *primitively amitochondrial*. Recent work (Sogin *et al*., 1989) using sequence analysis of rRNA suggests that the second alternative is true, which has the important implication that possession of mitochondria can no longer be regarded as definitive for eukaryotes. We shall return to this point later.

The evolution of *mitosis* in eukaryotes (characteristic 7 in Table 16.2) is linked to their greater quantity of DNA, organized into chromosomes (characteristic 5)—mitosis ensures absolutely equal distribution of the replicated chromosomal DNA to the two daughter cells and is illustrated in Chapter 3, Figure 3.14. In prokaryotes, cell division is achieved by the simpler process of *binary fission*—DNA replicates, the two loops move apart and the cell splits in two. Of equal evolutionary significance for eukaryotes is the evolution of *meiosis* which, as we discuss later, is essential for sexual reproduction. Finally, there is the presence in eukaryote cells of a *cytoskeleton* (characteristic 8 in Table 16.2). This complex protein scaffold is remarkably dynamic, able to break down or grow locally, and it plays vital roles in controlling cell shape, the movement of organelles, mitosis, cell division and differentiation.

16.5.2 Eukaryote origins: when?

The first eukaryotes were unicellular and, like the prokaryotes, occur only rarely as well-preserved microfossils. These have been found mainly in shales and cherts (Chapter 12). Distinguishing between prokaryote and early eukaryote microfossils, however, is not at all easy and there have been some heated disagreements about this, and, consequently, about when eukaryotes first evolved.

▶ From Table 16.2, identify: (a) one or two cell characteristics which could be preserved in Precambrian microfossils and would identify any eukaryote cell; (b) one cell characteristic which would be indicative but not definitive for eukaryote cells.

The obvious answer to (a) is, first, the *presence of cell organelles* and, secondly, any evidence of mitosis or meiosis—the end product of meiosis, for example, is commonly a group of four cells arranged as a tetrahedron, i.e. like four marbles, with three on a flat surface and one placed on top. The problem with such criteria is that cell preservation is usually imperfect and it is difficult to be certain that dark spots inside cells are really cell organelles or that the absence of dark spots indicates a prokaryote. Similarly, tetrahedral cells can arise because prokaryote cells failed to separate after undergoing fission (see Figure 16.28(c)). Therefore these cell characteristics are not easy to apply. Figure 16.28 shows examples of microfossils ranging from a generally accepted eukaryote (a) about 1 400 Ma old, to a doubtful eukaryote (b), and a tetrahedral tetrad (c) once thought to be a eukaryote but now regarded as a cluster of cyanobacteria.

The answer to (b) is simply *cell size*. 90% of living prokaryotes have cell diameters of less than 15 μm and although there is clearly size overlap between the two grades from Table 16.2, larger size appears to be a useful indicator of eukaryotes. In the 1970s, a marked shift was noted in the cell diameters of microfossils after about 1 600 Ma ago, approximately the middle of the Proterozoic (Appendix 1). Up to this time, average sizes were mostly in the range 1–30 μm but subsequently, in the period 1 200–1 400 Ma ago, were in the range 40–60 μm. This shift was attributed to the growing numbers of eukaryotes. It correlates also with the early period of increasing atmospheric oxygen levels, which reached about 15% of present levels by 1 900 Ma ago.

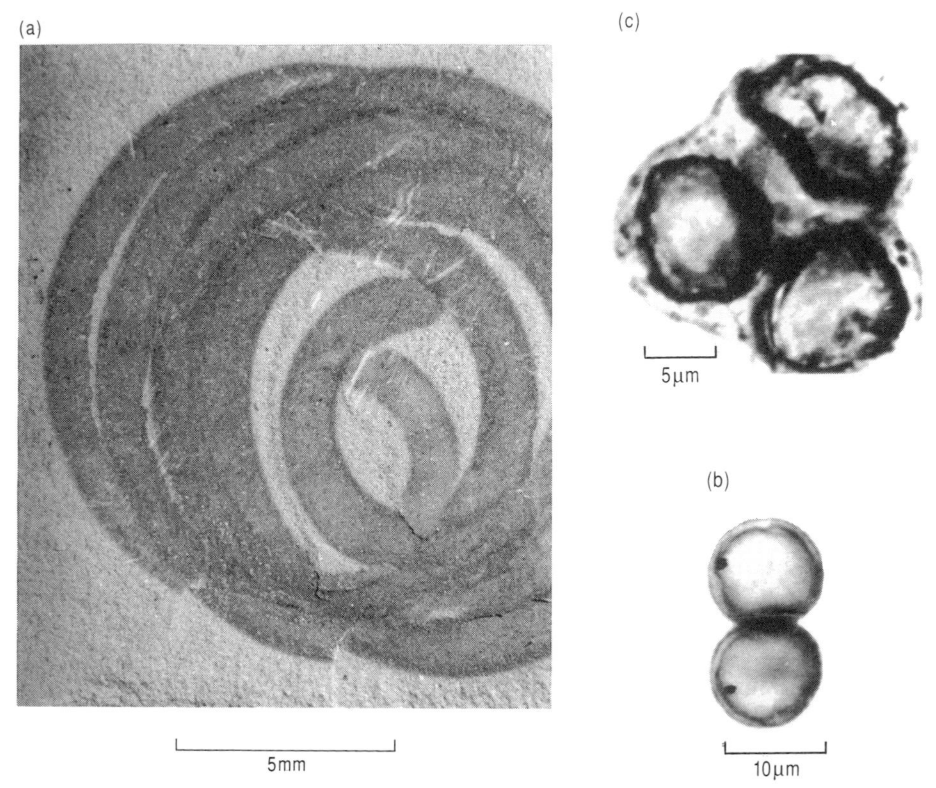

Figure 16.28 *Precambrian microfossils. (a) Spiral, filamentous eukaryote,* Grypania spiralis, *from* c. *1400 Ma old rocks in China. (b)* Glenobotrydion, *a doubtful eukaryote. (c) A tetrahedral tetrad now regarded as prokaryotic (cyanobacterial) and not eukaryotic. (Both (b) and (c) are from the Bitter Springs Chert in Australia, 800–900 Ma old.)*

▶ Why would you expect eukaryotes to increase at a time of increasing atmospheric oxygen?

Because nearly all eukaryotes are *aerobic* (Table 16.2, characteristic 3). However, we noted earlier that some primitive eukaryotes are anaerobes and, furthermore, the fact that eukaryotes were plentiful after 1 600 Ma does not mean that they *originated* at this time. The oldest microfossils which have been interpreted as eukaryotes are from rocks 2 100 Ma old in Michigan, USA. They are spiral filaments which have been interpreted as a species of *Grypania* (Figure 16.28a), which is thought to be an alga.

A different kind of evidence, derived from geochemistry rather than palaeontology, suggests that eukaryotic cells were present around 1 700 Ma ago. This evidence relates to the presence of substances (termed *steranes*) in bitumen, associated with Australian rocks of this age. Steranes are formed from membrane lipids called *sterols* that occur only in eukaryotic cells. The present consensus, therefore, is that eukaryotes probably existed in significant numbers 1 900–2 100 Ma ago; but they may have *originated* much earlier than this without leaving any trace in the rock record.

16.5.3 Eukaryote origins: how?

By the time the eukaryotes appeared, two of the prokaryote lines (the Archaebacteria and the Eubacteria) were highly diversified. Very little is known about the third line, which was mentioned in Section 16.4.1 as the eukaryote ancestor, but the line probably included large, anaerobic cells with a flexible surface (i.e. lacking any rigid wall or capsule). Thanks to some brilliant detective work, however, we have a clear and well supported hypothesis as to how more complex eukaryote cells originated from this line. The key to the problem lies in the origin of the membrane-bound organelles (mitochondria and chloroplasts). As hinted at in Section 16.5.1, it is now believed that mitochondria and chloroplasts were once free-living prokaryotes which entered and took up residence in the cytoplasm of a larger cell from the nucleo-cytoplasmic line. The relationship between guests, which were Eubacteria, and the host, which was a descendant of the third prokaryotic line, became stable and permanent and can be described as an **endosymbiosis** (from *endo*, within, and *symbiosis*, living together). There was a progressive loss of independence by the guests until they became fully integrated components of the host cell—the organelles. This is a bare outline of the **endosymbiotic hypothesis** for the origin of eukaryotic organelles and you may find it an altogether fanciful idea. To try to persuade you otherwise, we describe next some modern parallels of organelle symbiosis, further details of the hypothesis and some of the evidence which supports it.

Modern equivalents of organelle symbiosis Symbiosis, the living together of two or more organisms, sometimes—but not necessarily—to their mutual advantage, is very widespread in nature (Chapter 8, Section 8.4.1). For example, lichens which encrust rocks and trees are symbiotic associations between certain species of algae and fungi. An example of endosymbiosis, where one partner lives inside the cells or body of another, is shown in Figure 16.29. It involves two unicellular eukaryotes (i.e. protistans), the heterotrophic ciliate *Paramecium bursaria*, inside which are numerous green algae from the genus *Chlorella*. Photosynthetic products from the algae provide the host *Paramecium* with carbon compounds which supplement or, in hard times, replace the ciliate's external sources of food. A *Paramecium* which

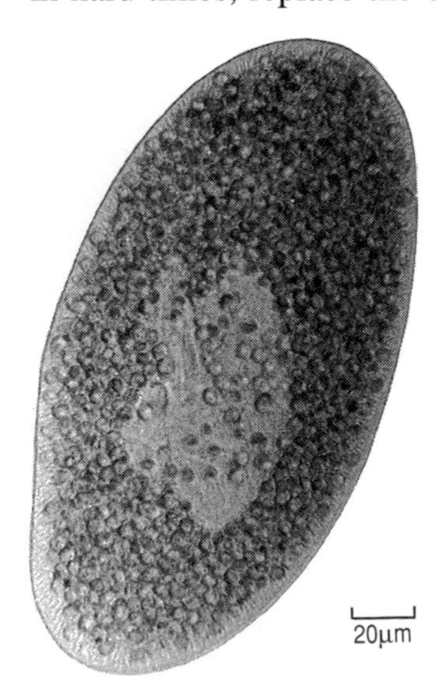

Figure 16.29 *The protistan ciliate* Paramecium bursaria *containing hundreds of green* Chlorella *symbionts.*

lacks algae will ingest them and they multiply within the host cell, but only up to a certain level. If additional, free-living *Chlorella* are encountered, the *Paramecium* ingests them but then promptly digests or ejects them, although the original symbionts remain. This situation provides a rough parallel with the early stages of chloroplast evolution.

An even more striking parallel, but this time with the endosymbiotic origin of mitochondria, is provided by the giant amoeba *Pelomyxa palustris*. On its own, this protistan is an anaerobe which lacks mitochondria and obtains energy by fermenting organic substances to lactic acid. However, *Pelomyxa* readily forms a symbiosis with certain aerobic bacteria and, when it does, requires oxygen (presumably for its partners) to survive. Under these conditions, the bacteria must be supplying energy to their host in much the same way that mitochondria do in other eukaryote cells.

A third type of modern symbiosis has been suggested as a model for the origin of eukaryotic cilia and flagella. These are hair- or whip-like surface structures and they either cause cells to move (as in sperm and many protistans) or, on fixed cells such as those lining the human nose and wind-pipe (trachea), cause materials to move past. The symbiosis involves protistans that live symbiotically in the gut of wood-eating termites, where they are essential for digestion of wood. These protistans possess flagella but they glide about in a smooth way and at a speed which their own flagella could not possibly achieve. The reason is that they form a second association with flexible, spiral Eubacteria called **spirochaetes**, (Figure 16.30)

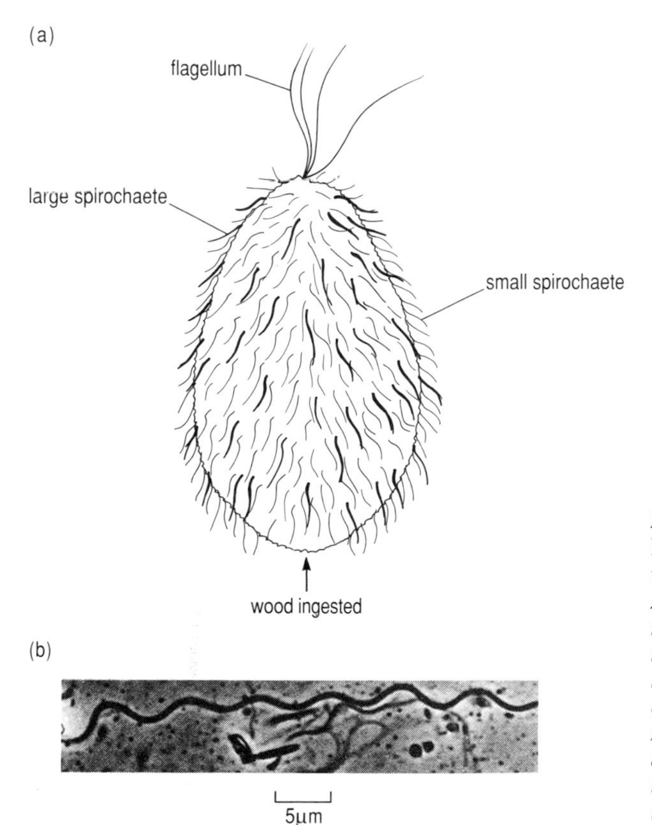

Figure 16.30 *(a) A symbiotic protistan* Myxotricha paradoxa *which facilitates wood digestion in the gut of certain Australian termites and forms a motility symbiosis with two kinds of spirochaetes. (b) Unattached spirochaetes,* Pillotina *sp., from the gut of an American desert termite. Neighbouring spirochaetes, whether attached or unattached to a protistan, beat synchronously simply because of their physical proximity.*

which undulate rather like isolated flagella. Spirochaetes may become attached by special structures to 'docking' sites on the protistan's surface and their coordinated beating helps to propel the host along, an association described as a **motility symbiosis** .

These examples, three of many, illustrate that symbiosis is common and may involve intracellular or surface partners which become integrated with their host to varying degrees. Once you appreciate this, the endosymbiotic hypothesis for the origin of eukaryotes does not seem such a strange idea. The resemblance of chloroplasts to free-living algae was recognized over a century ago and various endosymbiotic hypotheses for chloroplast origin were suggested around the turn of the century. Little attention was paid to these ideas however, until about 1970, when Lynn Margulis produced fresh evidence and revived interest.

16.5.4 THE ENDOSYMBIOTIC HYPOTHESIS

The left side of Figure 16.31 summarizes the main points of Margulis' original hypothesis and the right side shows a modified, more recent scheme.

▶ What is the main difference between the left and right sides of Figure 16.31, and what evidence cited earlier in the text supports the scheme represented by the right side?

According to the original Margulis hypothesis, eukaryotes originated when a prokaryote host, A, acquired mitochondria (stage B); the nucleus and flagellum appeared at a later stage. By contrast, the modified hypothesis suggests that eukaryotes originated when the nucleus and flagellum appeared (B'), with acquisition of mitochondria at a later stage, C. The fact that some eukaryotes which have both nucleus and flagella are primitively amitochondrial (Section 16.5.2) supports this modified scheme.

Most scientists agree with Margulis' ideas about the origins of mitochondria and chloroplasts, but there is considerable disagreement with her more speculative proposals about the origins of eukaryotic flagella and the nucleus—hence the question marks in Figure 16.31. Scientists (including Margulis) agree that the nucleus did *not* originate endosymbiotically but opinions differ about when the nucleus arose. Nevertheless, the integration of completely different kinds of organisms through endosymbiosis undoubtedly played a major role in eukaryote evolution. We shall examine the different steps one at a time.

Origin of mitochondria The original host was probably an anaerobic fermenter with a flexible surface, i.e. it was surrounded only by a membrane and lacked a cell wall. This host, which was either a prokaryote or an early eukaryote (Figure 16.31), acquired as symbiotic partners a number of smaller, oxygen-respiring Eubacteria, which ribosomal RNA sequence analysis (Chapter 11) suggests were related most closely to the purple bacteria (Figure 16.21). At first these symbionts may have utilized some of the end products of fermentation, such as lactate (Figure 16.32(a)), as appears to happen in the *Pelomyxa*–bacterial symbiosis described in the previous Section. Eventually, symbionts and host became totally dependent on each other—

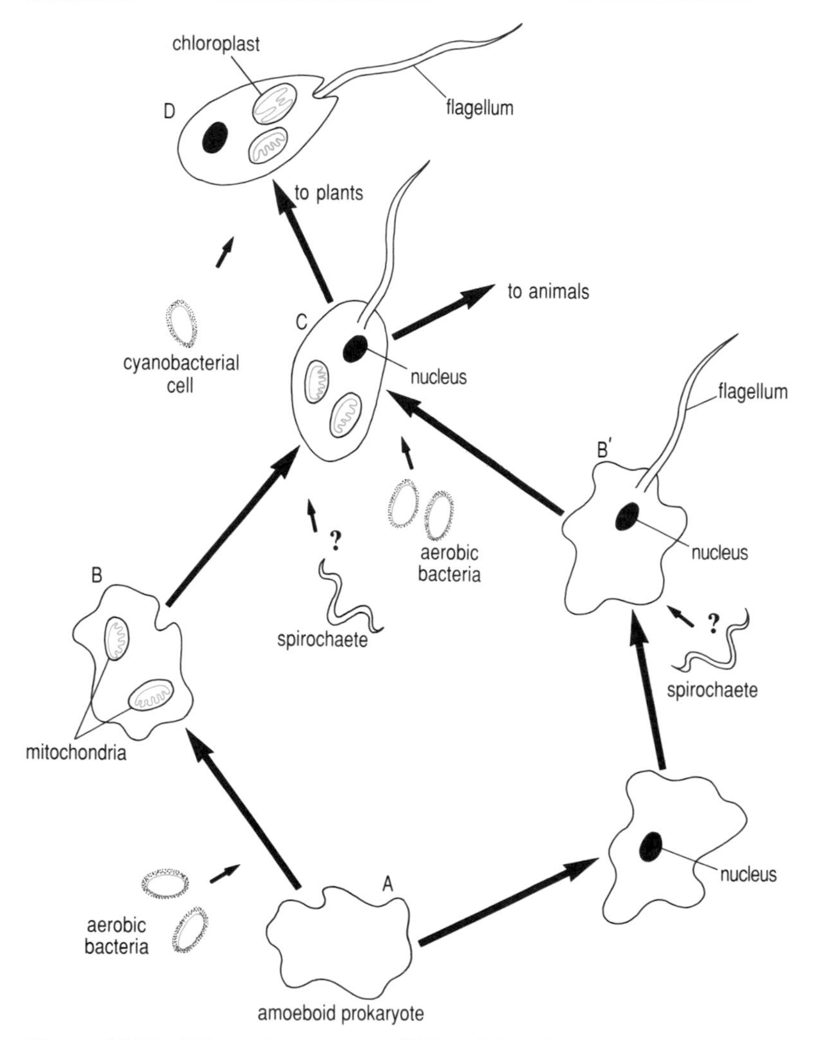

Figure 16.31 *Schematic summary of Margulis' endosymbiotic hypothesis for the origin of eukaryotic organelles and the origin of the nucleus (left side), and a modified, more recent hypothesis (right side).*

symbionts, which can now be called mitochondria (Figure 16.32(b)), lost their cell walls and supplied the host with ATP, fatty acids and certain by-products of the TCA cycle. Some of their genetic material was transferred to the host, which then passed the protein products back to the mitochondria, the end result being an integrated 'eukaryotic co-operative', stage B or B′ in Figure 16.31. We shall see later that a considerable amount of evidence supports this model.

▶ Using Figures 16.31 and 16.32 as guides, what are the origins of the two membranes that surround mitochondria?

The inner mitochondrial membrane, where respiratory electron transport occurs, is the original *outer* cell membrane of the symbiont. The outer mitochondrial membrane represents a host membrane that initially enclosed the bacterial symbiont.

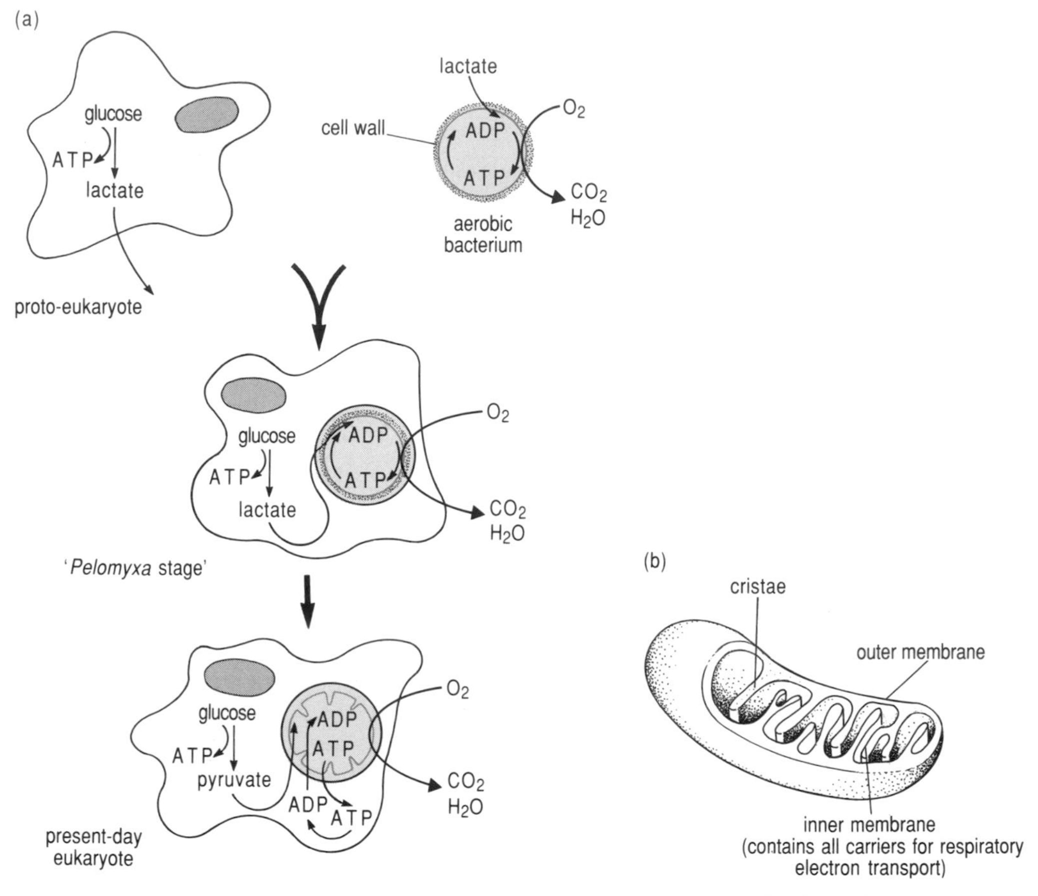

Figure 16.32 *(a) A possible evolutionary transition from a free-living aerobic bacterium to a mitochondrion. (b) Diagrammatic representation of a mitochondrion.*

▶ Bearing in mind what environmental conditions were like when eukaryotes first appeared (Section 16.5.2), why do you think it might have been advantageous for the anaerobic host to form a partnership with aerobic bacteria?

Remember that atmospheric oxygen levels were rising in the early stages of eukaryote evolution. Aerobic respiration is a much more 'efficient' type of energy metabolism than fermentation (which requires anaerobic conditions), and the symbiotic partners allowed the host to take a short cut to aerobic life, providing a tremendous advantage over other anaerobes. At present there is no evidence to suggest that mitochondria originated more than once.

Origin of motility When early eukaryotes were faced with the problem of finding food, there was probably strong selective pressure for more efficient mechanisms of locomotion or for creating water currents from which food particles could be collected. Margulis suggests that these problems were solved by a second symbiotic union being made with a type of eubacterium similar to modern spirochaetes (stage C or B' in Figure 16.31).

▶ Recall an example of this kind of motility symbiosis discussed earlier.

The association between protistans and spirochaetes in the termite gut is an exact parallel and Margulis' idea is that eukaryotic flagella and cilia were originally symbiotic spirochaetes. She goes even further than this and has suggested that the cytoskeleton (see Table 16.2) arose from 'internalized' flagella. Since the cytoskeleton has a major influence on cell shape and was probably essential for engulfing the cells that gave rise to mitochondria, this provides more support for the modified hypothesis depicted by the right side of Figure 16.31. We must emphasize, however, that the evidence (discussed later) for the endosymbiotic origin of flagella and the cytoskeleton is extremely tenuous. These structures may well have arisen *autogenously* (from the Greek *autos*, self), but at what stage, or in what way, is still a mystery.

Origin of the nucleus The nature of the selective forces that led to the origin of the eukaryotic nucleus are not at all clear but, as mentioned earlier, there is widespread agreement that the nucleus arose autogenously and not by endosymbiosis. A possible mechanism, by infolding of the outer membrane of the 'host' cell, is illustrated in Figure 16.33.

The evidence supporting this autogenous origin is, first, the chemical similarity and continuity between the nuclear membrane and the major, internal membrane system of eukaryotic cells, i.e. the endoplasmic reticulum. Secondly, the ribosomal RNA sequence data mentioned in Section 16.4.1 showed that the eukaryotic nucleus is quite different from the Eubacteria and the Archaebacteria and that it has descended from the third, nucleo-cytoplasmic line of prokaryotes. If, as seems likely, the nucleus was the first eukaryotic organelle, then it can be argued that the origin of the nucleus marks the origin of eukaryotes from prokaryotes. Perhaps the DNA in this first eukaryote cell was already complexed with histone proteins, as occurs in all eukaryotes (Chapter 3, Section 3.2.2) but in only one group of Archaebacteria among the prokaryotes—we do not know.

Origin of chloroplasts Because they are usually much larger than mitochondria and have a more independent existence inside the eukaryote cell, chloroplasts are generally assumed to have been acquired *after* mitochondria, as both models in Figure 16.31 show. Unlike mitochondria, there is much evidence which suggests that chloroplasts were acquired independently by more than one organism, i.e. that their origin is *polyphyletic* (Chapter 11, Section 11.2.1) and we shall consider the

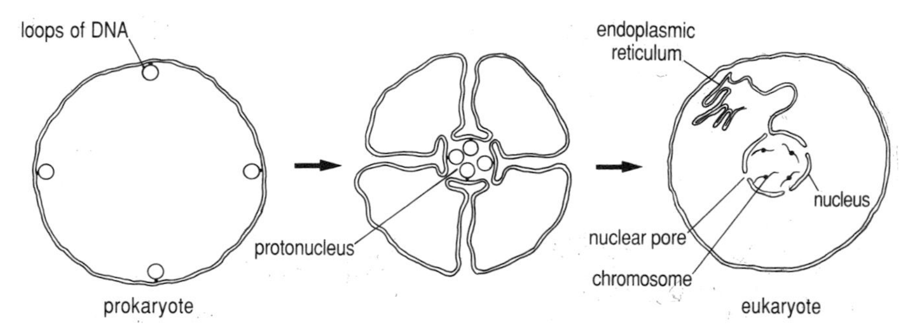

Figure 16.33 *Hypothetical scheme for the autogenous origin of the nucleus.*

multiple origins of the algae in more detail later in this Section. However, multiple origins mean that we have to ask the question: 'Have all chloroplasts originated from the *same kind* of photosynthetic prokaryote?' It seems quite possible that they have not, although there is much debate and disagreement on this point. Some scientists maintain that all the primary endosymbiotic events involved various kinds of cyanobacteria, as illustrated in Figure 16.31. But others suggest that all or at least one of these events involved a different and quite recently discovered group of oxygenic photosynthetic bacteria, the **prochlorophytes** or **Prochlorobacteria**, which possibly evolved from cyanobacteria. The first known member of this group, *Prochloron*, is a green bacterium, gigantic for a prokaryote (9–30 μm), which lives symbiotically in the cells of certain colonial sea-squirts. Other symbiotic and, in 1988, one free-living prochlorophyte have since been discovered, and in terms of their photosynthetic pigments and membrane arrangements, they look much more like a chloroplast than do any cyanobacteria. However, these characteristics may have evolved in parallel with the chloroplast and more biochemical detective work is needed to resolve the question.

Whatever the prokaryote(s), the sequence of events from pro-chloroplast to chloroplast seems to have been much the same as for mitochondria.

▶ Reconstruct this sequence.

Following the mitochondrial model, the first step would be some kind of nutritional exchange (e.g. carbon compounds supplied by the pro-chloroplast and mineral nutrients such as nitrogen or phosphorus compounds supplied by the host); secondly, increasing interdependence and coordination of division cycles with loss of cell walls by the symbiont; thirdly, total interdependence and transfer of genetic material from symbiont to host. This is the story—we look next at some of the evidence which supports the endosymbiotic hypothesis for the origin of mitochondria, chloroplasts and (possibly) flagella.

16.5.5 Evidence for the endosymbiotic origins of organelles

If all eukaryotic organelles had arisen autogenously like the nucleus, as many people once claimed, then you might expect fundamental processes such as protein synthesis and the structures of key macromolecules in cytoplasm and organelles to be basically similar. The first two parts of this Section provide evidence which clearly refutes this idea. The third part discusses more equivocal evidence about the origins of flagella from spirochaetes.

Mitochondria and chloroplasts are biochemically similar to prokaryotes
Ribosomes, the RNA–protein structures on which messenger RNA is 'read' during protein synthesis (see Figure 3.6 in Chapter 3), differ in size in prokaryotes compared with eukaryote cytoplasm. There are also differences in sensitivity to inhibitors of protein synthesis. These differences are summarized in Table 16.3, which also provides data about the characteristics of ribosomes and protein synthesis in chloroplasts and mitochondria.

Table 16.3 *Ribosomal characteristics and sensitivity of protein synthesis to various inhibitors in eukaryote cytoplasm, organelles (chloroplasts and mitochondria) and prokaryotes (Eubacteria). Sedimentation coefficient, S, is defined in Table 16.1.*

	Eukaryote cytoplasm	Eukaryote organelles	Prokaryotes (Eubacteria)
Sedimentation coefficient of:			
ribosomes	80S	70S	70S
ribosomal sub-units	60S and 40S	50S and 30S	50S and 30S
Inhibition of protein synthesis by:			
cycloheximide	Yes	No	No
anisomycin	Yes	No	No
tetracycline antibiotics	No	Yes	Yes
chloramphenicol	No	Yes	Yes

▶ Do the data in Table 16.3 suggest that organelles are more closely related to eukaryote cytoplasm or to eubacterial prokaryotes?

Quite clearly the data suggest the latter, and this is an important piece of evidence supporting the endosymbiotic origin of organelles from Eubacteria.

Other evidence relates to the base sequence in ribosomal RNAs. There are remarkable similarities between some prokaryote sequences and those in organelles, particularly when comparing cyanobacteria and chloroplasts. Molecular phylogenies, constructed as described in Chapter 11, show that mitochondria and chloroplasts are derived from eubacterial ancestors whereas the eukaryote cytoplasm is not.

Comparative studies of enzymes and of membrane components all point in the same direction—organelles are much more like Eubacteria than like the eukaryote cytoplasm, which is entirely consistent with the endosymbiotic hypothesis.

Organelle DNA It has been known since the 1960s that chloroplasts and mitochondria contain DNA and that they can multiply by binary fission. But mere possession of DNA is not evidence for an endosymbiotic origin because DNA obviously occurs in autogenous organelles such as nuclei. However, the structure and functioning of organelle DNA resembles much more closely that of Eubacteria than of eukaryotic nuclear DNA—organelle and eubacterial DNA exist as closed circles and are not complexed to histone proteins for example, whereas nuclear DNA is in the form of linear chains and is complexed to histones.

Organelle DNA usually codes for only about 10% of organelle proteins, the remaining proteins being encoded in nuclear DNA. The endosymbiotic hypothesis explains this *partial genetic autonomy* of organelles in terms of gene transfer to the host nucleus and there is now evidence that gene loss can occur quite readily

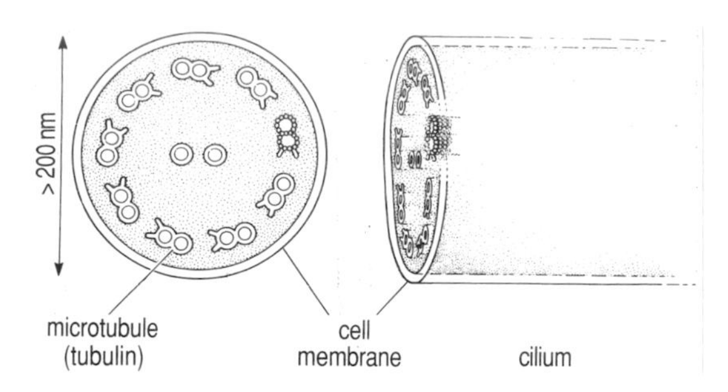

Figure 16.34 *Structure of the eukaryotic flagellum.*

from organelles. For example, the chloroplast genome of *Epifagus*, a non-photosynthetic higher plant that is parasitic on roots, is only about half the size of typical chloroplast genomes and lacks most of the genes concerned with photosynthesis. Gene transfer to the nucleus has also occurred long after eukaryotes originated—in the green algae, a gene involved in chloroplast protein synthesis is found on the chloroplast genome but in land plants, which evolved from green algae, this gene is found in the nucleus.

Similarities between spirochaetes and eukaryotic flagella Many of the Eubacteria have flagella. They are about 13 nm in diameter and composed of a protein called *flagellin*—completely different from eukaryotic flagella (and cilia, which are just smaller versions). These eukaryotic organelles (Figure 16.34) are much larger and composed of the protein *tubulin*, which forms rod-like structures called **microtubules** that are arranged in a distinctive 9+2 pattern—nine pairs of microtubules around the outside and two in the middle.

So bacterial flagella were never serious candidates as the ancestors of eukaryotic flagella and the reason why spirochaetes are a candidate is that some, those from termite guts, alone among prokaryotes, contain microtubules or, at least, axial fibrils that *look* like microtubules.

▶ What evidence is needed to confirm that axial fibrils really are a kind of microtubule?

They must be shown to consist of tubulin. Proteins from spirochaete axial filaments cross-react in immunological tests with antibodies for tubulin and, during purification, separate into the same fractions as tubulin; but these data do not demonstrate conclusively that axial filaments *consist* of tubulin. Unfortunately, spirochaetes from the termite gut cannot be cultured outside the gut, which makes studying them a remarkably difficult business.

Another important feature of eukaryotic flagella is their anchoring region or *basal body*. This has a 9+0 arrangement of microtubules, i.e. the central pair are missing, and directs the synthesis of new flagella. Basal bodies divide when the parent cell does (Figure 16.35) and they contain nucleic acid (RNA), which Margulis suggests represents the remnants of a spirochaete genome.

If this suggestion were true, however, one might expect to find base sequence homologies between basal body RNA and spirochaete DNA, a prediction that has still not been tested. Nor is it known whether nuclear tubulin genes are

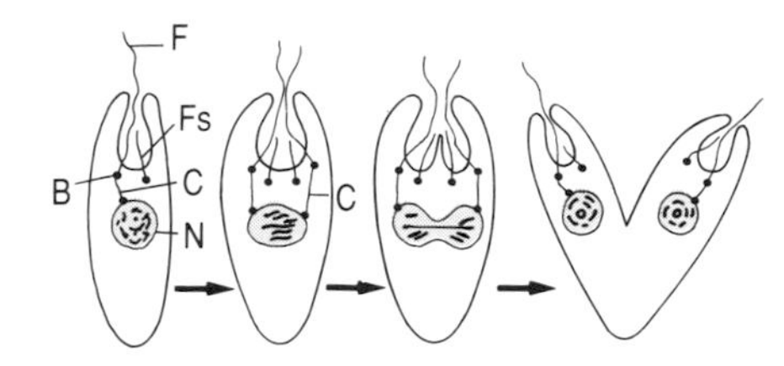

Figure 16.35 *Mitotic cell division in a euglenoid flagellate (a protistan) illustrating the parallel but independent replication of flagella and basal bodies with the nucleus. B = basal body; C = connection between nucleus and basal body; F = large locomotory flagellum; Fs = small flagellum; N = nucleus.*

homologous with genes that code for similar proteins in spirochaetes. The endosymbiotic origin of flagella, therefore, is still a relatively untested hypothesis and an autogenous origin for this organelle cannot be ruled out. The matter has a bearing on another important aspect of eukaryote evolution, the origin of mitosis and the cytoskeleton. Microtubules and a microtubule organizing system similar to basal bodies play important roles in mitosis, and the microtubules are an important component of the cytoskeleton. According to Margulis, these intracellular microtubules all originated from an internalized flagellar system descended from a spirochaete. We argued earlier (Section 16.5.4) that the capacity to engulf other cells, which is the necessary first step in endosymbiosis, could not have been achieved without a cytoskeleton.

▶ Accepting this argument and Margulis' idea that the cytoskeleton evolved from flagella, which side of Figure 16.31 (right or left) is more likely to be correct?

The modified hypothesis on the right—this shows acquisition of flagella *before* that of mitochondria.

16.5.6 How did mitosis, meiosis and sex evolve?

Mitosis was identified earlier as an essential process for the regular distribution of DNA between daughter cells, especially when there is a lot of DNA, and it was certainly essential for the evolution of multicellular eukaryotes. The question is how did mitosis evolve and following on from that, how or why did meiosis, the haploid–diploid cell cycle and sex evolve?

Mitosis You are probably familiar with mitosis as it occurs in animals and plants. Figure 3.14 in Chapter 3 illustrates the process in outline. Before delving into its origins in more detail, however, a number of basic points must be made. First, it must be stressed that mitosis is concerned with the separation of replicated chromosomes and is quite distinct from replication of DNA. DNA replication occurs during **interphase** (the period between one mitosis and the next) and, usually, cells cannot enter mitosis until DNA replication is complete, which cannot occur until mitosis has ended. These different phases of the cell cycle are illustrated in Figure 16.36 and only recently, in 1991, has the elaborate biochemical switch which controls this system been discovered.

A second point to be made about mitosis is the central role played by microtubules and certain structures which may resemble, sometimes to a remarkable degree, the basal bodies of flagella. The *mitotic spindle*, on which chromosomes line up and then separate, is composed largely of microtubules. In

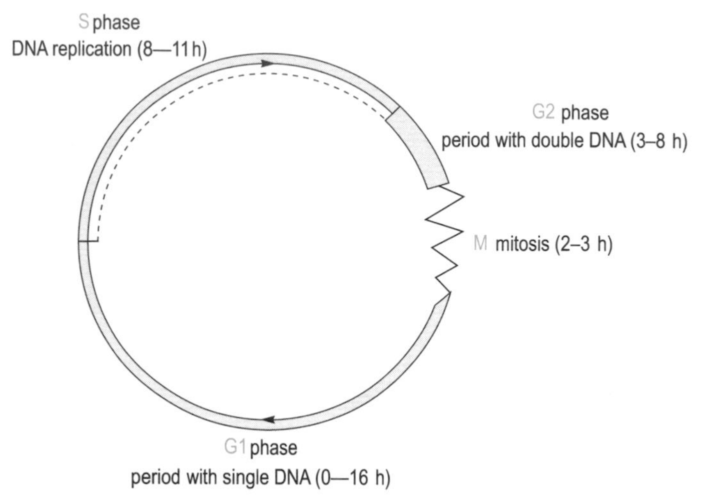

Figure 16.36 *The eukaryotic cell cycle showing the four phases and their approximate lengths in dividing cells of plant roots.*

animal cells and certain protists, algae and fungi, the spindle is organized by and microtubules grow out from *centrioles* (Figure 16.37). These contain RNA and have the 9+0 structure of basal bodies; indeed, flagella sometimes develop from centrioles. However, other organisms lack centrioles although they may have other, quite complex structures, and centrioles *per se* are clearly not essential for mitosis. What is essential is some system for organizing microtubules and in the 1970s Pickett-Heaps gave this the logical name **microtubule organizing centre**, or MC—centrioles and basal bodies are then just one kind of MC. Finally, there are the *centromeres* which are a special region on chromosomes that attach them to the spindle and guide their movement during separation on the spindle. The whole system is clearly complex but the essential features are that (1) a special relationship exists between flagella and the mitotic apparatus, and (2) microtubule organizing centres play a key role.

Lynn Margulis again has developed an interesting hypothesis about the origin of mitosis that is based on the spirochaete origin of flagella and their subsequent internalization. Her basic thesis is that the spirochaete genome, which evolved into

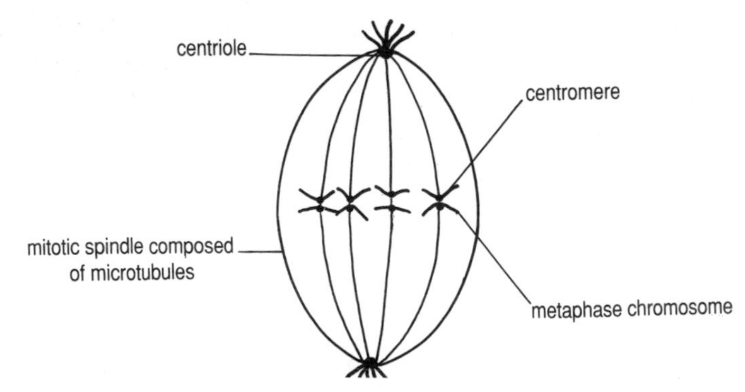

Figure 16.37 *Diagrammatic representation of the mitotic spindle, centrioles and centromeres in an animal cell.*

the basal body of the eukaryotic flagellum, was taken over by and replicated inside the host cell. Many essential functions, including the genes for tubulin synthesis, were integrated fully into the host genome but relic spirochaete genomes persist as microtubule organizing centres, which play such an important role in mitosis. According to this theory, mitosis evolved only *after* eukaryotes acquired flagella and Margulis suggests that it evolved independently in several lines of protists.

▶ Following this line of argument, suggest an explanation for the observation that, compared with animals and plants, many protist and algal groups have unusual and sometimes bizarre types of mitosis.

Mitosis could have evolved separately in each of these groups, i.e. convergent evolution, different types of mitosis representing various attempts at 'getting the process right' and solving certain problems. One problem, for example, is that microtubule organizing centres seem to be incapable of functioning simultaneously in mitosis and as basal bodies. Some protists solved this problem by becoming amoeboid, i.e. losing cilia and flagella but retaining mitosis. Others are motile at one stage in the life cycle and divide at another. The ciliates, which are covered with small cilia, evolved a highly unusual kind of mitosis which allowed division to occur whilst retaining an independent control system for the cilia. Multicellular eukaryotes evolved the most elegant solution, however, because they split the functions between different cells: some cells, which lack flagella or cilia, can divide whilst others, which are ciliated or have flagella, never divide once they are mature. If this interpretation is correct, then the origin of multicellularity in several protist lines may have been an adaptation which allowed simultaneous motility and cell division.

It has to be said, however, that there is little firm evidence to support Margulis' hypothesis about the origin of mitosis and many biologists do not believe it. There is no single, well-articulated alternative hypothesis but the main suggestions are that genes for tubulin and other components of the mitotic apparatus were present in the nucleo-cytoplasmic prokaryote host and that mitosis evolved *before* motility. In this scheme, which can be described as the *endogenous hypothesis*, flagella would have evolved from the mitotic apparatus and not the other way round.

It is interesting to compare the explanations offered by the endosymbiotic and endogenous hypotheses for some experimental results obtained by L. R. Cleveland in the 1950s. Cleveland worked with large protistan flagellates from termite and cockroach guts and found that treatment with high concentrations of oxygen destroyed the chromosomes but not centrioles, centromeres or basal bodies. Cells without chromosomes formed spindles to which centromeres attached and separated, and basal bodies subsequently replicated. The cells divided to form two daughters without chromosomes.

▶ How would the endosymbiotic hypothesis explain these results?

Margulis argues that the results demonstrate the semi-autonomous nature of all systems controlled by microtubule organizing centres—MCs have high oxygen tolerance and are not controlled by chromosomal genes. The endogenous hypothesis, by contrast, would argue that genes controlling mitosis, which might be on chromosomes or exist separately in the cell, simply have an unusually high resistance to oxygen. As yet, no-one can say with certainty which (if either) interpretation is correct.

Ploidy, meiosis and sex The first eukaryotes presumably had only one copy of each chromosome, i.e. they were haploid. The majority of modern eukaryotes, however, spend some part of their life cycle as diploids; in most metazoans only the gametes are haploid and in flowering plants haploid nuclei are present only in pollen grains and egg cells. So we need to ask how and, for that matter, why diploidy evolved?

Diploidy Cells may double the number of chromosomes or amount of DNA so that they have two copies of each gene in two basic ways. The first way is simply by juggling with the lengths or timing of phases in the cell cycle (Figure 16.36).

▶ How could the cycle be altered so that cells spent most of it in a diploid state and were diploid once they had stopped dividing?

The obvious answer is to shorten G1 so that DNA replication occurs immediately after mitosis and then, until the next mitosis or for the rest of that cell's life, it is diploid (Figure 16.38(a)). Bearing in mind that centromeres are part of the mitotic apparatus and replicate during mitosis, this mechanism would not alter the number of chromosomes but each chromosome, apart from its centromere, would be double. A cell that simply inserted an extra S phase (DNA replication) during G2 would produce the same result and mutant yeasts which do this if given a brief exposure to high temperature during G2 are known (Figure 16.38(b)). Now that the nature of the molecular switch controlling the cell cycle is better understood, it

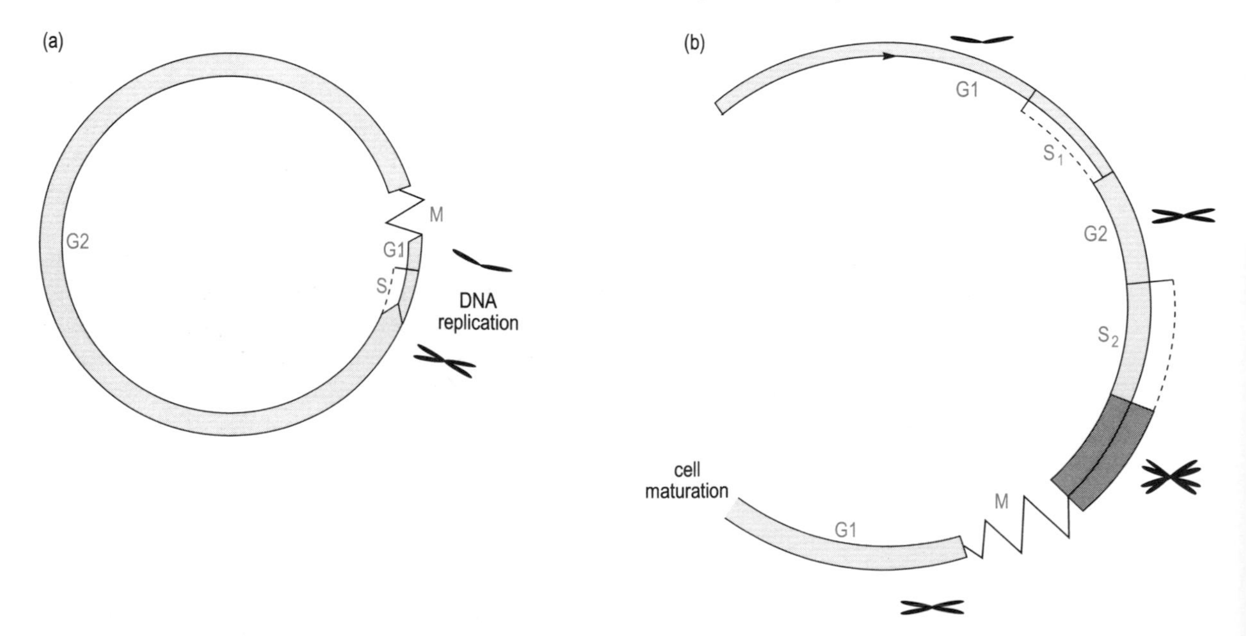

Figure 16.38 *Two ways in which diploidy can be achieved by changes in the cell cycle: (a) shortening G1; (b) inserting an extra S phase (DNA replication) during G2 (a once-off mechanism).*

has been suggested that early eukaryotes may have evolved mechanisms for altering ploidy either upwards (extra S phases), or downwards (omitting S phases).

▶ Why is (b) in Figure 16.38 described as a 'once-off mechanism'?

Diploidy in G1 is achieved by an extra S phase on one occasion only. If the sequence shown in Figure 16.38(b) were repeated a second time, cells in G1 would be tetraploid, with four times the original amount of DNA—the DNA content in G1 doubles at each cell cycle.

For unicellular organisms, a second mechanism for acquiring another set of chromosomes is to fuse with another cell of the same species, with or without nuclear fusion; it is equivalent to the first stage of sexual reproduction (fusion+meiosis) but, on its own, it is a 'once-only process'. This mechanism, unlike the first, may introduce new genes, which could increase the chances of survival in difficult conditions. However, there is another advantage of being diploid which applies to both mechanisms: deleterious mutations are usually recessive, which means that a diploid unlike a haploid is shielded from their effects and the higher the mutation rate per genome, the greater the advantage of diploidy. Mathematical models which calculate the relative fitness of haploids and diploids given reasonable assumptions about rates of mutation and the proportion that are recessive, generally support this view. So, based on these genetical considerations, diploidy is good for you. However, there are other considerations which relate to, for example, the larger cell size of diploids—this usually means that they have slower growth rates than haploids and may be at a disadvantage in highly competitive situations.

Another feature of the second route to diploidy is that it introduces a second set of centromeres, which means that the cell cannot revert to the haploid state simply by adjusting the cell cycle. Here may lie the origin of the selective pressures that led to the evolution of reduction division, i.e. meiosis, and the full sexual cycle.

Meiosis and the sexual cycle Meiosis occurs only in diploid cells where nuclei have fused. It involves two cell divisions which produce four haploid daughters from one diploid cell and thoroughly mix up genetic material on homologous chromosomes (see Figure 3.15, Chapter 3). It must have evolved after mitosis and, until it did, cell fusion would have been a once-only event or, alternatively, would have caused chromosome doubling every time it occurred. The first stage in the evolution of meiosis must have been a mutation which allowed cells to carry out mitosis *without* replication of centromeres. When coupled with an association of homologous chromosomes on the equator at division, this leads to the formation of two cells that are haploid with respect to chromosome number but still diploid with respect to the amount of DNA, i.e. like a haploid cell in G2 phase. Because of random assortment of chromosomes, however, these cells are unlike either of the original haploid cells that fused (Figure 3.15, Chapter 3). Chromosome association was probably quite loose at first and only later did close association and crossing over, which is essential for genetic recombination, evolve. Margulis and Sagan (1986) argue that the enzymes which mediate crossing over originated in prokaryotes as DNA repair enzymes which snipped out lengths of damaged DNA and inserted new bits. When this first meiotic division was coupled with a normal mitosis, the end result was four cells with a haploid DNA content (phase G1) and a completely new mixture of genes. Meiosis may have evolved independently in several protist lines and there are some groups (e.g. the euglenoids) which apparently lack the process.

A cycle in which fusion of haploid cells alternates with meiosis is, by definition, a *sexual cycle* (Chapter 5). However, whether and to what extent mitosis occurs in the diploid, after fusion and before meiosis, or in the haploid, after meiosis and before fusion, determines whether the 'normal' state of a protist is haploid or diploid (Figure 16.39). In multicellular eukaryotes the bulk of the life cycle may be haploid (e.g. some algae), or diploid (flowering plants and animals), or there may be an alternation of multicellular haploids and diploids (some algae and 'lower' land plants). In general, increasing complexity of organisms on land has been accompanied by increasing dominance of the diploid state. Margulis and Sagan (1986) offer the interesting suggestion that alternation of haploid and diploid phases in the life cycle was originally an adaptation to cyclical changes in the environment. They argue that harsh conditions favour diploids (e.g. because deleterious mutations are shielded and alleles may 'complement' each other in advantageous ways), while haploids are favoured in 'good' conditions suitable for rapid (asexual) multiplication. Similar arguments were advanced in Chapter 5 about the relative advantages of sexual and asexual reproduction.

The increase in genetic variability, mixing up genes and generating new genotypes, is an obvious advantage of sex (fusion + meiosis), but *only* if the haploid cells that fuse are genetically different. A vast array of mechanisms which ensure that this is exactly what happens has evolved in the eukaryotes and these were discussed in

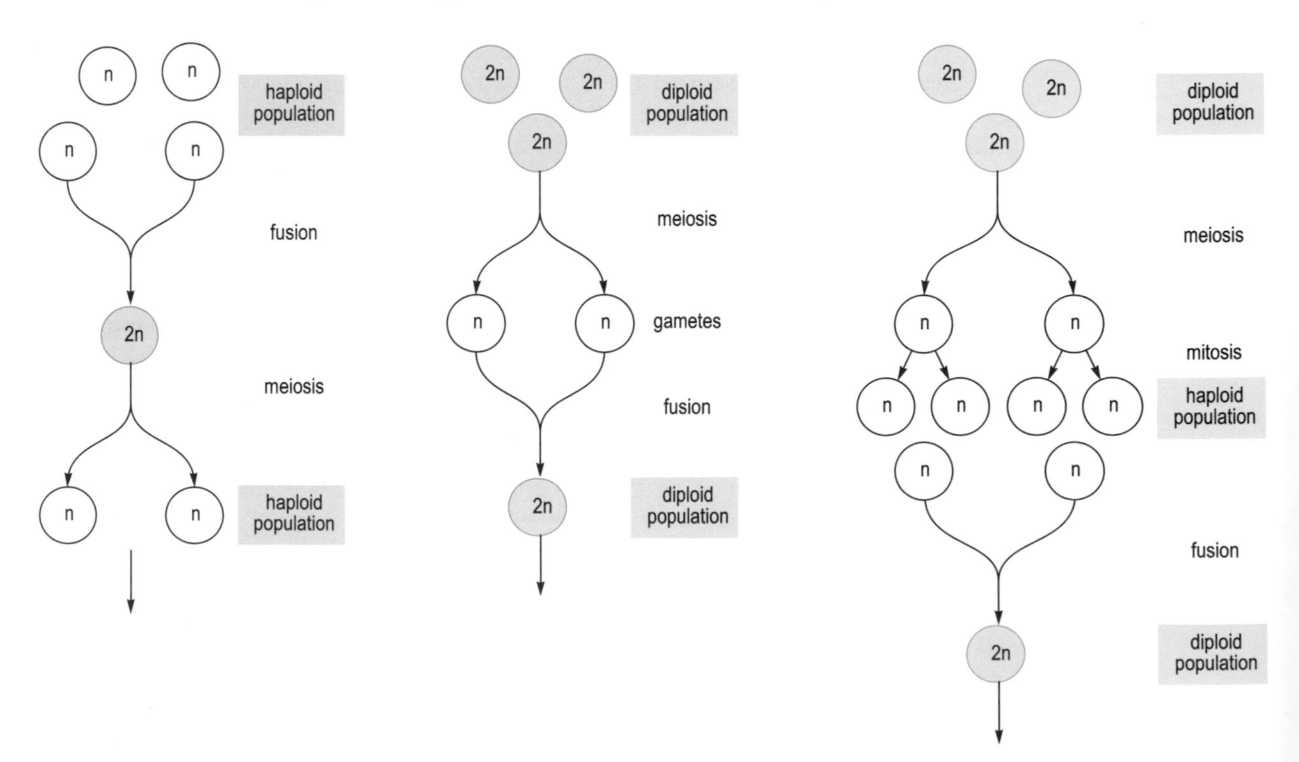

Figure 16.39 *Variations in the timing of meiosis, fusion and mitosis in unicellular organisms.*

Chapter 5. This, coupled with multicellularity, certainly contributed to the rapid diversification which occurred among eukaryotes. We consider next some of the early stages in this diversification.

16.5.7 DIVERSIFICATION IN EARLY EUKARYOTES

By 580 Ma ago, there was a diverse array of multicellular animals, as revealed by the Ediacaran fauna (Chapter 14). However, the oldest multicellular eukaryotes in the fossil record were plants—red algae found in Canadian rocks with an age of 1 260–950 Ma. During the period 1 000–540 Ma, evidence from fossils and geochemical evidence from biomarkers suggests that single-celled eukaryotes diversified greatly and several lines of multicellular algae evolved (Knoll, 1992).

At least 27 distinct lines, which have left modern descendants, evolved in early eukaryotes, and at least 17 of these independently evolved the multicellular grade of organization. Figure 16.40 is a phylogenetic tree of early eukaryote evolution which shows some (but not all) of the 27 lines and is based mainly on evidence from ribosomal RNA sequencing, amino acid sequences in enzymes and comparisons of mitosis. The whole tree is speculative and many scientists would disagree with parts of it. Nevertheless, we have not covered it with question marks and you can regard it as something which gives the flavour of what was going on in the late Precambrian.

A striking feature of Figure 16.40 is that the branches which form the crown of the tree all separate from the main trunk at about the same time, approximately 1 200–1 000 Ma ago. The tree is based mainly on rRNA sequence analysis, but geological evidence also indicates a sudden increase in eukaryote diversity with the appearance of new kinds of scales, protective coats and cell walls. The explanation for this 'big bang' of eukaryote evolution is unknown but one suggestion is that it relates to the establishment of outbreeding, sexual populations. Certainly, living protists such as Diplomonads and euglenoids that originated before the big bang are all effectively asexual.

The emphasis in Figure 16.40 (which you do not need to memorize) is on the origins of autotrophic protistans, the lines which gave rise to the modern groups of algae, and clearly these have multiple origins—several lines independently acquired chloroplasts by endosymbiosis.

▶ Was the same symbiont involved in all lines?

No it was not. Figure 16.40 suggests that there were at least four different symbionts. One line, red algae, acquired cyanobacteria (C); other lines not shown in the Figure may have done the same. The green algae acquired a cyanobacterium or a *Prochloron*-like symbiont, P (Section 16.5.4), and possibly the euglenoids (flagellated protistans belonging to a group which includes colourless heterotrophs) did the same. Surprisingly, however, the remaining five algal lines acquired as a symbiont not a prokaryote but a eukaryote, something that was long suspected because of the peculiar morphology of the chloroplasts in these lines. In the aptly named *Cryptophytes*, for example, the 'chloroplast' has an outer double membrane

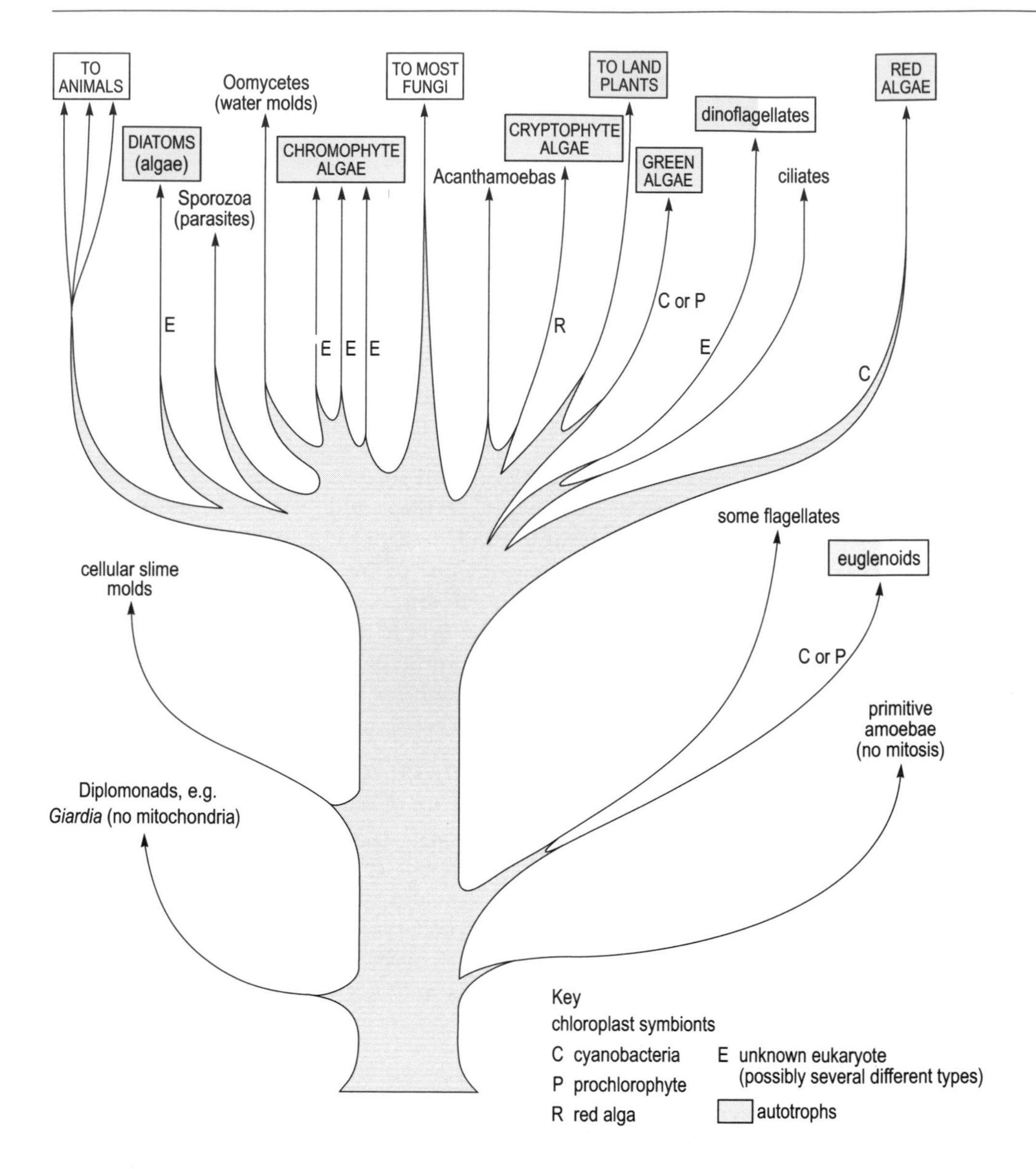

Figure 16.40 *Possible phylogeny of eukaryotes. Shading indicates wholly or partially autotrophic groups.*

within which lies first a small body like a nucleus (the nucleomorph), some cytoplasm and then a standard chloroplast with the usual double membrane. In effect the 'chloroplast' has four membranes with cytoplasm and a nucleomorph interposed between the outer and inner pair. The pigments and membrane arrangements of the 'inner' chloroplast are like those of red algae and recently, in 1991, rRNA gene sequencing of the nucleomorph and nucleus from a cryptophyte showed conclusively that the nucleomorph is indeed closely related to red algae (Figure 16.41), whereas the nucleus is not. Something similar happened in the Chromophyte group of algae, which includes the brown algae commonly seen as seaweeds on the shore. Their chloroplasts usually have four surrounding

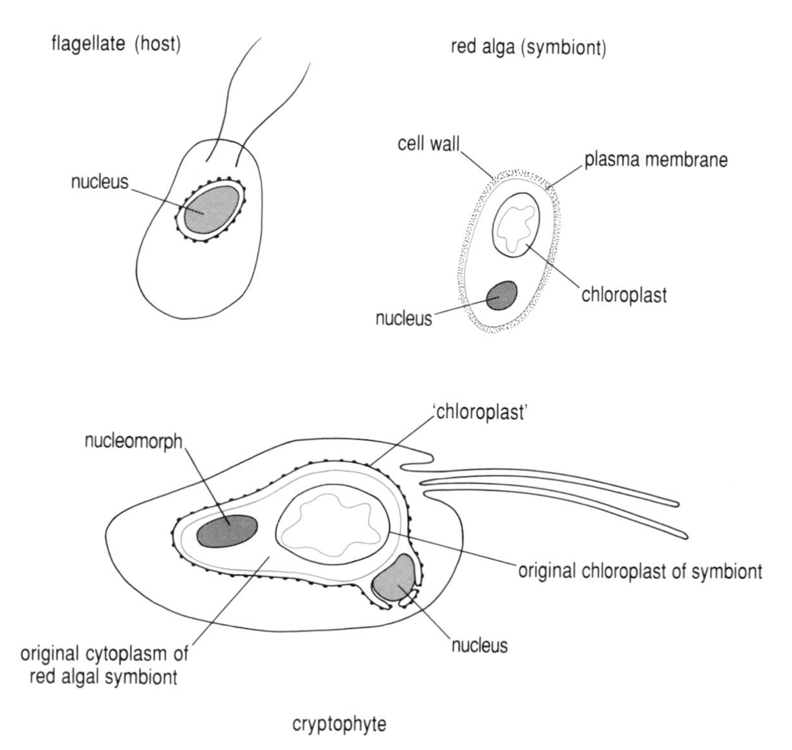

Figure 16.41 *Origin of the cryptophyte algae from a flagellated host and a red algal (eukaryotic) symbiont.*

membranes and all traces of the symbiont nucleus and cytoplasm, again possibly derived from red algae, have disappeared. Chloroplasts, therefore, were acquired independently by several eukaryotic lines, each of which is now classified as a separate algal group.

It is useful to note here that all land plants evolved from the green algae, or Chlorophyta. Notice also that organisms such as the water molds, which are now classified as fungi, evolved separately from the main line of fungi. Multicellular animals, however, all seem to have originated from the same line and the larger metazoans diversified about 540–580 Ma ago when atmospheric oxygen levels seem to have undergone a further increase. Clearly there was enormous diversification among early protists and multicellular eukaryotes and this is on the basis of those lines which have survived to the present day—one wonders how many other lines evolved and then became extinct without leaving any trace of their existence.

Summary of Section 16.5

Geological evidence suggests that eukaryotes were numerous by 1900–2100 Ma ago, when atmospheric oxygen reached 15% of current levels. They originated before this time from the eukaryotic (nucleo-cytoplasmic) line of prokaryotes, possibly as anaerobes with a nucleus, flagella and cytoskeleton.

Eubacteria formed stable associations within the cytoplasm of early eukaryotes (the endosymbiotic hypothesis). Eukaryotic organelles evolved from the symbionts and many parallel examples of such endosymbiosis exist among modern organisms. Mitochondria were acquired when aerobic Eubacteria were engulfed by anaerobic host cells. The Eubacteria probably utilized fermentation products as a respiratory substrate and supplied ATP to the host. The partners became increasingly interdependent and gene transfer from symbiont to host completed the transformation to mitochondrial organelle.

According to Lynn Margulis, mitochondria were acquired before the nucleus and the next stage was the evolution of eukaryotic cilia and flagella from endosymbiotic spirochaetes (a motility symbiosis). Later still, several eukaryotic lines independently acquired photosynthetic endosymbionts (either cyanobacteria, Prochlorobacteria or other eukaryotic cells) which evolved into chloroplasts.

The origin of the eukaryotic nucleus is almost certainly autogenous and not as a result of endosymbiosis. Many scientists believe that eukaryotic flagella also originated endogenously and were present before the acquisition of mitochondria. Evidence for the endosymbiotic origins of eukaryote organelles comes mainly from comparative biochemistry, coupled (for mitochondria and chloroplasts) with base sequence similarities between organelle and eubacterial rRNA and ultrastructural studies. Evidence for the endosymbiotic origin of flagella is much weaker. The microtubules and basal bodies of flagella are likened to spirochaete axial filaments and genome respectively. Margulis speculates that internalized flagellar systems became highly integrated with the host cell and provided the essential machinery for mitosis, i.e. microtubules and microtubule organizing centres (MCs). The alternative hypothesis is that first mitosis evolved autogenously in eukaryotes and then flagella evolved from the mitotic apparatus.

Diploidy in eukaryotes may have arisen by changes to the cell cycle or by cell fusion. It shields the cell from the effects of deleterious, recessive mutations, and cell fusion introduces new and possibly useful genes. In general, diploidy is regarded as advantageous in 'difficult' conditions, and haploidy as advantageous in 'good' conditions that allow rapid multiplication.

The evolution of meiosis allowed cells to switch between diploid and haploid states. The first step was coupling an aberrant mitosis (i.e. one lacking centromere replication) with a normal mitosis, producing four haploid cells from one diploid. The second step linked closer and closer chromosome pairing during division 1 with genetic recombination, utilizing enzymes originally used for DNA repair in prokaryotes.

Coupling cell fusion with meiosis is how the eukaryotic sexual cycle originated. It is possible that the 'big bang' of eukaryotic evolution 1 200–1 000 Ma ago was triggered by the establishment of many sexually outbreeding populations. During the Proterozoic, eukaryotes diversified greatly, mainly in the sea. A variety of protective exoskeletons and cell walls evolved; some lines showed increasingly complex grades of multicellular organization, and 'algae' evolved in several lines as different photosynthetic endosymbionts were incorporated in cells.

16.6 Conclusion

In a very condensed way, this Chapter covers biological evolution in the first three-quarters of the Earth's history. It is far more speculative than any of the other Chapters because there is so little evidence of what went on and you need to look at our story as just that—a story and not a factual account. With this important proviso, early evolution can be divided into three stages: the origin of life; the diversification of prokaryotes; and the origin and diversification of eukaryotes.

Life originated by a process of chemical evolution remarkably soon after the formation of the Earth, probably about 4 000 Ma ago. Organic matter from space may have provided some of the chemical building blocks, others being synthesized in Earth's early atmosphere, which was anoxic but not strongly reducing. Autotrophic metabolic systems protected in some way from the extreme harshness of the early environment, and replicating by a mechanism that was not based on nucleic acids, were probably the first forms of proto-life. Nucleic acids, initially RNA and later DNA, took over the role of replicator and information store in the first cells. So far as anyone can tell, all modern organisms originate from a common ancestor, so either life originated only once or other forms of life became extinct and left no traces.

From the ancestral prokaryote stock, the Archaebacteria, Eubacteria and nucleo-cytoplasmic line of the eukaryotes evolved. The Archaebacteria, which today live in various extreme environments, have diverged least from the ancestral stock and are mostly chemo-autotrophs or heterotrophs. The Eubacteria diversified much more, and by 3 500 Ma ago oxygen-evolving cyanobacteria were fixing CO_2. Their activities gradually raised oxygen levels in the atmosphere, paving the way for the evolution of aerobic prokaryotes and eukaryotes.

Eukaryotes may have started as simple anaerobes with a nucleus but the origin of the modern eukaryote cell represents a quantum leap in evolution because it originated by combining whole genomes through endosymbiosis. Mitochondria and chloroplasts represent intracellular remnants of once independent Eubacteria. Margulis suggests that flagella and perhaps even key parts of the mitotic apparatus were also acquired by endosymbiosis, although it is equally possible that they, like the nucleus, arose endogenously. Evolution of mitosis, diploidy, meiosis and the sexual cycle facilitated great diversification and an increasing complexity of multicellular organisms. In all this flurry of biological evolution, the mixing and matching of genes in the sexual cycle and the mixing and matching of organisms through endosymbiosis played vital roles.

REFERENCES FOR CHAPTER 16

Bernal, J. D. (1967) *The Origin of Life*, World Publishing Co., Cleveland, Ohio.

Cairns-Smith, A. G. (1985) *Seven Clues to the Origin of Life*, Cambridge University Press, Cambridge.

Joyce, G. F. (1989) RNA evolution and the origins of life, *Nature*, **338**, 217–223.

Knoll, A. H. (1992) The early evolution of eukaryotes: a geological perspective, *Science*, **256**, 622–627.

Margulis, L. and Sagan, D. (1986) *Origins of Sex: Three Billion Years of Genetic Recombination*, Yale University Press.

Schidlowski, M. (1988) A 3 800-million-year isotopic record of life from carbon in sedimentary rocks, *Nature*, **333**, 313–318.

Sogin, M. L., Gunderson, J. H., Elwood, H. J., Alonso, R. A. and Preattie, D. A. (1989) Phylogenetic meaning of the kingdom concept: an unusual ribosomal RNA from *Giardia lamblia*, *Science*, **243**, 75–77.

Towe, K. M. (1990) Aerobic respiration in the Archaean?, *Nature*, **348**, 54–56.

Walker, J. C. G. (1987) Was the Archaean biosphere upside down?, *Nature*, **329**, 710–712.

Watson, J. D., Hopkins, N. H., Roberts, J. W., Steitz, J. A. and Weiner, A. M. (1987) *Molecular Biology of the Gene* (4th edn), Ch. 28, The Benjamin/Cummings Publishing Co. Inc.

Woese and Wächterhäuser (1990) Origin of life, in: D. E. G. Briggs and P. R. Crowther (eds) *Palaeobiology, A Synthesis*, 3–9, Blackwell Scientific Publications.

Woese, C. R. (1979) A proposal concerning the origin of life on the planet Earth, *Journal of Molecular Evolution*, **13**, 95–101.

FURTHER READING FOR CHAPTER 16

Margulis, L. (1981) *Symbiosis in Cell Evolution*, W. H. Freeman and Co.

Briggs, D. E. G. and Crowther, P. R. (eds) (1990) *Palaeobiology, A Synthesis*, Blackwell Scientific Publications.

Horgan, J. (1991) In the beginning ..., *Scientific American*, February, 101–109.

Schopf, J. W. and Packer, B. M. (1986) Early Archean (3.3–3.5 billion-year-old) microfossils from Warrawoona Group, Australia, *Science*, **237**, 70–73.

SELF-ASSESSMENT QUESTIONS

SAQ 16.1 *(Objective 16.1)* Which of (i) to (vi) are present in, or characteristic of, (a) prokaryotic cells, (b) eukaryotic cells, (c) inorganic crystals?

(i) DNA.

(ii) A capacity to grow or replicate.

(iii) A nucleus.

(iv) A cell membrane.

(v) Organelles bounded by a double membrane.

(vi) A capacity to transform external sources of energy into chemical forms which can be used for repair, replication or growth.

SAQ 16.2 *(Objective 16.2)* (a) Choose four items from (i) to (vii) which correctly describe conditions during the Hadean.

(b) In what, if any, ways would these four items have influenced chemical evolution?

(i) The atmosphere was anoxic.

(ii) The atmosphere contained about 1% oxygen.

(iii) There was no ozone layer.

(iv) Intense bombardment by meteors occurred.

(v) The atmosphere was strongly reducing.

(vi) The atmosphere was mildly reducing.

(vii) Low solar luminosity resulted in average temperatures below 0 °C.

SAQ 16.3 *(Objective 16.3)* Assuming that the primitive atmosphere was mildly reducing, describe two ways in which the basic building blocks for chemical evolution could have originated.

SAQ 16.4 *(Objective 16.3)* Compare and contrast the classical Oparin–Haldane theory of chemical evolution for the origin of life with the model proposed by Woese and Wächterhäuser by completing Table 16.4

Table 16.4 *For use with SAQ 16.4.*

	Oparin–Haldane	Woese & Wächterhäuser
Site of early reactions		
Site of later reactions		
Sources of reactants for later synthesis		
Type of metabolism in proto-life		

SAQ 16.5 *(Objective 16.4)* What evidence is there for the existence of cyanobacteria 3 500–3 800 Ma ago? If they were present, how might this have affected heterotrophic prokaryotes?

SAQ 16.6 *(Objective 16.5)* Decide which of (a)–(e) are true, partially true, or false, and give reasons for your answers.

(a) Molecular sequence data suggest that the Archaebacteria evolved first from the ancestral prokaryote stock.

(b) Ether-linked membrane lipids are a characteristic unique to Archaebacteria.

(c) Sulphobacteria are the least evolved group of Archaebacteria and are characteristically thermophilic, sulphur-respiring heterotrophs.

(d) Both aerobic respiration and photosynthesis appear to have evolved independently in several groups of Eubacteria.

(e) Cyanobacteria probably became the dominant photo-autotrophs in the Archaean because their mode of photosynthesis is the most efficient.

SAQ 16.7 *(Objective 16.6)* (a) Present one argument against the view that anaerobic fermenters were the first living organisms.

(b) Present an argument in support of the view that the first living organisms may have been chemo-autotrophs.

SAQ 16.8 *(Objective 16.7)* For each of (a) to (c), decide whether or not it serves as a parallel for some stage in the endosymbiotic origin of modern eukaryote cells. If it does, describe the nature of the parallelism.

(a) Certain marine slugs suck chloroplasts out of the (very large) cells of the algae on which they live. The chloroplasts enter cells on the outer surface of the animal where they persist for several weeks, carrying out photosynthesis and supplying carbohydrates to the slug.

(b) Members of the protistan phylum Archamoebae live in anaerobic environments and contain a nucleus but no mitochondria. They obtain energy by substrate level phosphorylation in glycolysis and oxidize the end product, pyruvate, by an enzyme that is regarded as very ancient. This enzyme occurs otherwise only in anaerobic prokaryotes, including the methanogenic Archaebacteria.

(c) In certain algae and fungi which have a filamentous structure, there are cell-like compartments that contain many nuclei.

SAQ 16.9 *(Objective 16.8)* Assume that the following, hypothetical situation is true. 'An amoeboid protistan was discovered which, on biochemical grounds, appeared to be a very primitive eukaryote. This amoeba had only two chromosomes, which lacked centromeres, and an unusual type of mitosis. A spindle-like structure formed which contained some tubulin protein but no recognizable microtubules; the mitotic chromosomes attached themselves, apparently in random positions, to the 'spindle' before chromatid separation occurred.'

How would these observations be explained by supporters of:

(a) The endogenous hypothesis.

(b) The endosymbiotic hypothesis for the origin of mitosis.

In your view, to which, if either, of these hypotheses do the observations lend the strongest support?

SAQ 16.10 *(Objective 16.9)* Which of (a) to (d) are true, and which are false?

(a) A diploid cell is one in which there are two copies of each chromosome and/or gene.

(b) In the absence of crossing over (recombination), meiosis has no evolutionary significance.

(c) Nuclear fusion is necessary before either diploidy or meiosis can occur.

(d) The sequential evolution of mitosis, nuclear fusion and meiosis led to the origin of the sexual cycle.

CHAPTER 17
HUMAN FACTORS IN EVOLUTION

Prepared for the Course Team by Caroline Pond

Study Comment *This Chapter is about events that happened in the recent past and that may occur in the future. Although drawing heavily on historical records, this Chapter is not a history text and there is no attempt to identify human causes or to apportion blame: we are concerned only with the facts and their biological consequences. However, although we have tried to be objective and concentrate on mechanisms and other scientific aspects of the topics discussed, there are inevitably some political and sociological dimensions. The text includes descriptions of events and aspects of human behaviour that many people would now regard as unreasonable, even barbaric, but it is essential to understand how past and present attitudes and habits have shaped the evolution of our own species and many of those with which we have come into contact. Here, we view humans as a biological species, and use the same scientific and unemotive language as in the rest of the Book. You may not agree with our conclusions, but we hope that you will follow the scientific reasoning that led us to them. Discussions of extinction usually concentrate on well-publicized species such as elephants, whales and rainforest trees, but as evolutionists we are interested in the whole biosphere, and so have included some less familiar examples. Similarly, examples from different human populations and parts of the world are discussed, although inevitably there is more information about the 'western' world, i.e. Europe, North America and Australasia. Nothing dates so rapidly as our perception of the future: new facts will be discovered and new theories will be advanced, so this Chapter will probably become outdated more quickly than the rest of this Book.*

The major concepts are illustrated by particular species and case histories but you should not attempt to memorize all the details. Most of the examples are vertebrates and higher plants, but as always in biology, other examples would have served just as well: many invertebrates and micro-organisms have also been affected by human activities. To help you place the biological processes in the context of human history, certain historical figures and events are mentioned briefly, but you should not be concerned if you are unfamiliar with them. Unless your geography is very strong, you will probably find it useful to have an atlas handy. Modern place names are used throughout, even if the localities were known by a different name when the events described took place.

Objectives When you have finished reading this Chapter, you should be able to do the following:

17.1 Describe how cultural attitudes and technological advances can influence the course of biological evolution in humans.

17.2 Outline the prehistoric and historic spread of humans and the impact of such migrations on human evolution.

17.3 Outline the role of humans as hunters, and describe the circumstances under which such predation is likely to lead to extinction of the prey.

17.4 Describe some ways in which humans and other large animals alter the environment for themselves and other species.

17.5 Describe the consequences of intentional or accidental introduction of exotic species into the wild and human attempts to control the resulting situations.

17.6 Outline the main causes of pollution and its impact on the course of evolution.

17.7 Explain how selective breeding and gene manipulation can produce non-adaptive evolution in domesticated species.

17.8 Describe the main steps in the domestication of livestock and crops.

17.9 Outline the main biological features of keeping pets and cultivating gardens and describe how these habits have influenced the course of evolution of wild and captive species.

17.10 Describe the aims and limitations of the preservation of wild species and domesticated breeds, and the conservation of natural communities.

17.1 INTRODUCTION

In the late 18th century, the Scottish physician, meteorologist and geologist, James Hutton, concluded from his study of the formation and erosion of rocks in Scotland that mountains, rivers, sand-banks, mud-flats and other topographic features were formed in the past by the same physical and biological processes that shape the landscape now. This concept, known as the Principle of Uniformitarianism, was a major step forward in both theory and methodology: by establishing that elucidating processes happening now would explain events that took place long ago, it was a major impetus to research into geophysics, climatology and sedimentology. By extension of the same principle, the study of the mechanism of evolution and its course in the recent past should help us to understand evolutionary changes that are taking place now, and so predict the future state of our planet and its organisms. One of the major factors influencing contemporary evolution is humans: their habits, needs and tastes are affecting the future course of evolution of many other species as well as themselves. The first topic of this Chapter is evolution in modern human populations.

Culture is the body of knowledge, beliefs, attitudes and learned skills that is accumulated and modified as it is passed on from generation to generation. Many of the most ancient and strictly followed cultural practices concern the acquisition or, more recently, the production of food, clothing and shelter, and marriage, birth and death. As described in Section 17.2, attitudes to birth, marriage and death are important to biologists too, because they, as well as natural selection, determine which individuals breed successfully, and how many offspring they contribute to the next generation and hence the course of evolution in human populations.

Archaeology, the study of human artefacts and remains, is the main source of information about prehistoric events and the development of culture and technology. Archaeological terms like Palaeolithic (Old Stone Age) and Neolithic (New Stone Age) refer to stages in human cultural development, often characterized by the manufacture of particular kinds of tools and pottery. People in different parts of the world did not reach the same stage of cultural development simultaneously, so these terms do not indicate exact times in the same way as geological Periods like 'Pleistocene' or dates like 6000 y BP (years Before Present). Thus, people who live entirely by hunting and gathering, and make only simple tools and clothes, are said to have a Palaeolithic culture. Among the most important cultural advances were the invention of writing and of comprehensive and accurate methods of recording time. Events for which there are written records are, by definition, 'history', and the standard system of historical chronology, BC and AD, is used to describe them.

Like many relatively large animals, humans have influenced the course of evolution of many of the other organisms with which they have come into contact. Humans have domesticated certain fungi (for baking and brewing) as well as many species of plants and animals, wheat, apples, cattle and dogs, to name but a few. Rats, cockroaches, lice and many other species have evolved adaptations that enable them to exploit such human habits as storing food and wearing clothes. Such organisms were never deliberately domesticated, but human habits and migrations have permanently changed the course of their evolution, and that of many other species with which they came into contact. Because such verminous species deplete or contaminate food supplies and shelters, people have developed ways of expelling or exterminating them, but in many cases the vermin have evolved counter-adaptations.

Deliberate elimination or cultivation of particular individuals is usually called 'artificial' selection, while accidental or unanticipated effects of human actions on other species are regarded as elements of natural selection; but, as explained in Section 17.4, this distinction is blurred. There are many similarities between the way organisms have responded to artificial selection and evolutionary change under natural selection. The parallel between natural and artificial mechanisms of evolution was clear to Darwin. He devoted a long section of *The Origin of Species* to discussion of the changes brought about by artificial selection in cultivated plants and domesticated animals, and later wrote a book called *The Variation of Animals and Plants under Domestication* (1868). It is fashionable to regard modern industrial society as the root of all deleterious effects on 'the environment'. So it is useful to examine carefully the palaeontological, archaeological and historical evidence to assess what the impact of humans has been, and how it could change in the future.

17.2 Evolution in modern human populations

Humans are still evolving in a biological as well as a cultural sense. Selection, adaptation and the migration and interbreeding of formerly separated populations determine the course of evolution in humans just as they do in other organisms. Cultural attitudes and technological advances, as well as biological factors, may affect the course and rate of evolution.

17.2.1 Selection

Compared to other animals and plants, negative natural selection, i.e premature mortality and reproductive failure, seems to be very weak in modern humans. Even among the most underprivileged peoples, about half of all infants born become adults, and most adults breed; in the western world, a couple who raise three children can expect to have at least four grandchildren. Nonetheless, natural selection is clearly still operating in human populations: the reproductive rate is very low among people with severe genetic diseases, such as cystic fibrosis, muscular dystrophy and Down's syndrome (until the last 50 years, most such people did not survive childhood). In humans, the action of natural selection is partially obscured, even reversed, by other factors that determine who breeds and how many children they raise, as the following examples show.

Twins The number of offspring per litter, and their size and state of maturity at birth, are an integral part of the reproductive strategy of a species (see Chapter 8). Most small mammals, such as rodents (e.g. rats, mice, squirrels) and primitive orders such as insectivores (e.g. hedgehogs and moles) have multiple births. Some species, such as marmosets and polar bears, normally have twins, though singletons and triplets are quite common. Many species that normally have a single offspring occasionally have multiple births: horses, cattle, deer, hippopotamuses and even elephants sometimes give birth to twins, although they are very rare indeed among rhinoceroses and almost all species of bats, seals and whales.

There are two kinds of multiple births: multizygotic pregnancies, including **dizygotic** twins, trizygotic triplets, etc., which involve two or more different zygotes, each of which forms its own placenta and develops in its own set of embryonic membranes. Such twins are normally no more similar than ordinary siblings and may be of the same or different sex. In some cases, the two placentas may join, permitting blood to flow between the genetically dissimilar foetuses from early in development, causing the twins to share more characters, particularly endocrinological and immunological characters, than siblings would. In cattle, such twins are normal if they are the same sex, but if the sexes are mixed, the females of such pregnancies, called freemartins, are nearly always sterile, due to the adverse effects of the male hormones on the development of the female reproductive organs. **Monozygotic** multiple births arise from anomalous division of a single zygote. Instead of remaining together and eventually differentiating to form a single integrated embryo, each daughter cell forms a separate, but genetically identical, embryo. The fetuses normally share a placenta and embryonic membranes. Because they develop from a single zygote, monozygotic multiple births are always the same sex and, unless one develops abnormally, are very similar in both appearance and behaviour.

▶ How would you determine whether multiple births in wild mammals are monozygotic or multizygotic?

It is not easy. Particularly in species that breed in burrows or dens, it is very rarely possible to observe the arrangement of the embryonic membranes at birth. Unless there is some obvious genetically based difference, such as fur colour, between the offspring, it is difficult to determine whether they are monozygotic or multizygotic (except by expensive new techniques such as measuring the similarities in the DNA, see Chapter 3), but the limited evidence suggests that in most wild mammals each fetus develops from a different zygote.

▶ Why would this situation be adaptive?

Multizygotic litters increase the genetic diversity of the offspring in each litter, and would therefore normally be favoured by natural selection (see Chapter 5). Armadillos are among the few mammals known to have monozygotic multiple births, but this only came to light following observations on specimens breeding in captivity. The nine-banded armadillo of Central America, *Dasypus novemcinctus*, routinely has identical quadruplets (see Section 17.4.3).

All the living species of great apes normally have singletons, but chimpanzees, gorillas, orang-utans and humans occasionally produce twins, and even more rarely triplets or quadruplets. Human twins may be monozygotic or dizygotic, and triplets may be a pair of monozygotic twins with the third child developing from a separate zygote, or may arise from three different zygotes. Identical triplets or quads that develop from a single zygote, like those of armadillos, are very rare. The registration of births, that has been compulsory in most countries for the past 100–150 years, better observation of infants at birth, and the development of modern methods of quantifying genetic relationships, have enabled demographers to build up a very accurate picture of the frequency of multiple births in different populations, and how it is affected by cultural attitudes, economic factors and health care policies.

Natural selection is one of the factors that determine the frequency of twins in humans. Giving birth to twins increases the mother's chances of dying in childbirth, so women who bear twins are also less likely to survive to produce more children. There are also risks for the infants: twins are almost always smaller at birth and are very often at least three weeks premature. There is no detailed information about the outcome of twin pregnancies in prehistoric times but more recently the Master of a Dublin maternity hospital kept records of perinatal and maternal mortality in his institution between 1757 and 1784. His data, and similar information from Britain in 1950, are shown in Table 17.1.

Table 17.1 Perinatal and maternal mortality of single and multiple births in 18th century Dublin and 20th century Britain. The average numbers of survivors per pregnancy calculated from these data are also shown.

	Mortality (% births)			Surviving infants per pregnancy			Maternal mortality (%)	
	Singletons	Twins	Triplets	Singletons	Twins	Triplets	Singletons	Twins
18th C Dublin	20	39	No data	0.8	1.22	No data	1	5
20th C Britain	5	17	38	0.95	1.66	1.86	< 0.01	

▶ Is there any selective advantage (or disadvantage) in twinning?

From the point of view of fecundity per pregnancy only, significantly more babies are produced from twin pregnancies than from singleton pregnancies. However, until the 20th century, maternal mortality following twin births was also higher, which would directly affect the life expectancy of the infants, and, of course, eliminate any chance of the woman producing further offspring. Taken together, these figures suggest that until modern medical methods were developed, the net selection for twins was about the same as that for singletons. Until recently, successful triplet pregnancies were probably very rare.

▶ Will multiple births in developed countries be increasingly favoured by natural selection?

Yes. In the 20th century, significantly more surviving infants are produced from twin and triplet pregnancies than from those involving singletons, and maternal mortality is so low that it would have little effect on the fitness of women who bear twins.

▶ What other conditions are necessary for the evolution of a higher frequency of multiple births?

The tendency to produce twins must be heritable. At the beginning of the 20th century, German sociologists analysed the family history of more than 67 000 births in relation to twinning. Their data are shown in Table 17.2.

Table 17.2 *The frequency of twins (twin births per 1000 population) among first degree relatives of twins and triplets compared to that of the population as a whole.*

	Numbers of pairs of twins born	Total number of maternities	Twinning rate (relatives of twins)	Twinning rate in the whole population
Female relatives of mothers of twins and triplets	537	28 318	19.0	12.1
Female relatives of fathers of twins and triplets	95	8 896	10.7	11.2
Female relatives of mothers of unlike-sexed twins	174	7 892	22.0	11.9
Female relatives of mothers of like-sexed twins	255	15 201	16.8	11.9
Female relatives of mothers of monozygotic twins	81	7 309	11.1	11.9

▶ Is twinning inherited through either parent? Is monozygotic and dizygotic twinning equally heritable?

Twinning is inherited only through the female line. Mothers, sisters and daughters of all types of twins are more likely to produce twins than the general population as a whole, but the family history of the male has no detectable effect on his chances of fathering twins. Heritability is particularly strong for unlike-sex twins, which of course are certain to be dizygotic. The data suggest that heritability of monozygotic twinning is slight, perhaps non-existent.

▶ What do these data suggest about the mechanism of twinning?

There must be a heritable component in the control of the formation of dizygotic twins. Conception of dizygotic twins depends upon the release of two ova in the same menstrual cycle, and the tendency for this to happen seems to be at least partly determined by inheritance, probably by means of minor differences in the hormonal control of ovulation.

▶ Why would such factors not affect the frequency of monozygotic twinning?

Monozygotic twinning arises from anomalous division of a single zygote and so is not determined by the ovulation rate. Very little is known about the physiological control of monozygotic twinning, but the fact that it is non-heritable is consistent with the observation that the frequency of such twins is almost constant in all modern human populations for which there are data (see Table 17.3).

▶ Is the tendency to have monozygotic twins always non-heritable in mammals?

No. Bats, seals and whales have evolved means of avoiding twins entirely, probably because such pregnancies are always maladaptive, so in these groups, monozygotic twinning must be heritable to some degree. Conversely, most armadillos have monozygotic multiple births. Certain other features of the biology of twins in humans suggest that the trait has arisen fairly recently in evolution and the mechanisms involved do not yet function perfectly. Of 963 pairs of dizygotic twins born in and around Aberdeen, Scotland, between 1951–1985, 356 (37%) were both boys, 308 (32%) were both girls and 299 (31%) were a boy and a girl.

▶ Would you expect these frequencies of the like-sex and unlike-sex pairs of twins?

No. The sex distribution of singleton births is about 50% of each sex. If the same principle applies to dizygotic twins, there should be similar numbers of like-sex and unlike-sex twins, i.e. M/M, M/F, F/M, F/F equally frequent. These data suggest that conditions favouring simultaneous conception of male and female zygotes are unusual, or that such pregnancies are less likely to be successful, but the reasons for such anomalies are poorly understood. This situation, combined with the fact that monozygotic twins are always of the same sex, combine to create the general impression that boy/girl twins are rare.

▶ Do other mammals have difficulty sustaining mixed-sex pregnancies?

No. Most litters of kittens, puppies, piglets, etc., include both sexes. However, in domestic sheep, like-sex and unlike-sex dizygotic twins are also born with about equal frequency (see Section 17.4.1).

The occurrence of twins in different populations provides further insight into how the trait has evolved in humans. Table 17.3 shows the frequencies of twin births in various countries.

▶ How do the incidences of the two kinds of twins differ between groups?

The frequency of monozygotic twins is about the same in all peoples for which there are data, but that of dizygotic twins varies up to 20-fold. Among the Yoruba of West Africa, as many as one in ten people is a twin, 90% of them dizygotic, but among some Oriental peoples, twins are rare (about one in a hundred adults) and the majority are monozygotic. The limited data indicate that these racial differences persist even among migrants, suggesting that they arise from evolutionary changes within isolated populations. Where racial groups interbreed (e.g. in modern America), the twinning rate is intermediate between those of the ancestral populations.

Table 17.3 *The twinning rate (pairs of twins born per thousand full-term pregnancies) in various countries during the first half of the 20th century, standardized as far as possible for maternal age.*

Country or region	Tribe	Dizygotic	Monozygotic
Europe:			
Spain		5.9	3.2
France		7.1	3.7
Western Germany		8.2	3.3
Norway		8.3	3.8
Italy		8.6	3.7
England and Wales		8.9	3.6
Asia:			
India and Pakistan		7.8	4.3
Japan		2.5	4.5
China		3.4	5.0
Korea		6.3	5.4
Malaysia		3.4	6.1
Hawaii		3.9	3.7
Africa:			
West Nigeria	Yoruba	49	5
North Nigeria	Hausa	10	5
East Nigeria	Ibo	22	5
Cameroun		13	3
South Africa	Bantu	16	4
Zimbabwe	Mashona	27	2
Zaire		19	3

▶ Do these data suggest that the evolution of differences in the frequency of twins arises from cultural rather than physiological factors (e.g. adaptation to different climates or diets)?

Yes. In Nigeria, differences between the twinning rates of separate tribes living in adjacent areas are as large as those between widely separated peoples, e.g. western Europe and Japan.

In many groups, such as Eskimos, North American Indians and Africans, the mother carries the infant on her back until it can walk efficiently, at two or three years old. It is almost impossible to secure two infants satisfactorily using a traditional sling, so twins would be very impractical, particularly for nomadic tribes that walked long distances carrying their possessions. Conversely, twins may pose fewer problems for sedentary peoples who do not carry their infants over long distances, and indeed may be an efficient way of increasing fecundity when a woman's life expectancy, and hence her lifetime reproductive output, is likely to be short. However, such factors cannot account entirely for racial differences in the frequency of twins. In humans, cultural beliefs and practices have also influenced the frequency of twins. In some cultures, particularly Oriental people such as the Japanese, the birth of twins has long been regarded as an evil omen, particularly if the infants are female. Most of the populations in which twins are now rare have long practised discrimination against multiple births, which in many cases takes the form of infanticide of one or all the infants. Cultural discrimination against twins tends to reinforce itself as they become rare: most rare traits are regarded with suspicion, simply because they are different.

▶ Would these selective mechanisms operate equally against monozygotic and dizygotic twins?

Yes, but they would only result in evolutionary change (i.e. twins becoming rarer) in the latter case, because monozygotic twinning is not heritable (Table 17.2). In many cultures, a female infant is more likely to be disposed of than a male infant.

▶ How would such a bias affect long-term changes in the frequency of dizygotic twins?

The trait would become rare faster than if males and females were eliminated equally, but selection would still be slow, because some women, such as the sisters or aunts of twins, may carry the alleles for the tendency for twinning, without experiencing such births themselves (and thereby identifying themselves as carriers of alleles that promote twinning).

Evolution in contemporary human populations is slow, both compared to that of many other organisms and to the rate of change of cultural attitudes and technological competence. The frequency of people who are twins in Europe and other developed countries has increased over the last 100 years, but this change is due almost entirely to the higher proportion of premature and otherwise disadvantaged infants who survive to adulthood (Table 17.1). There is as yet no evidence that the frequency of multiple pregnancies has changed, except as a result of cultural factors such as later marriage and technological advances in the artificial control of conception (see Section 17.4). Although more such babies survive, other factors still operating lower the inclusive fitness of twins. In all populations for which there are data, genetic defects, birth injuries and perinatal mortality are more common among twins. Women who were born as twins, particularly monozygotic twins, experience more complications of conception and pregnancy than singletons. Twins achieve lower scores on most intelligence tests and tend to have shy, introverted personalities. They marry somewhat less frequently than the population average and are rare in socially prominent occupations (except hereditary positions). However, these effects are slight and are believed to be due mainly, if not entirely, to the facts that twins are more often born to older women and into large families, and their parents are busier with routine care, so the children receive less individual attention. Overall, the net contribution of twins to future generations is only slightly greater than if the same pregnancies had produced only one infant.

This example illustrates some important points about modern human populations: human populations are both genetically and culturally very heterogeneous, and there are many traits that, although rare, are still present in some populations. Habits and cultural attitudes, as well as 'natural' agents of selection (such as predation and susceptibility to disease), sometimes play a major part in the evolution of differences between races.

The effect of medical technology on selection in human populations Many recent developments in medicine and public health, if applied consistently, have some impact on the course of human evolution. Infectious diseases such as measles, diphtheria and pneumonia, and illness such as diarrhoea caused by dirty water or improperly prepared food, were, until very recently (and in many parts of the world still are), major causes of mortality, particularly among children. Any heritable resistance to such diseases would confer a selective advantage on a family. But at least in western countries, standards of hygiene are now much improved, children can be protected from many diseases by immunization, and more effective medicines (particularly antibiotics) and improved nursing mean that those who become infected are much more likely to recover and grow up to breed.

▶ What effects would mass immunization have on the frequency of alleles that confer innate resistance to disease?

Selection for such alleles would be relaxed and genotypes that produce little or no resistance to disease would become more common. In the long term, more people will become dependent upon medical procedures rather than genetic adaptation for resistance to many diseases.

The technology is now available for eliminating a few inherited diseases by embryo selection. For example, cystic fibrosis is caused by a single gene defect: severe symptoms only appear in homozygous people, although the defective allele can be identified by biochemical methods in healthy heterozygotes. It is now possible to identify the allele for cystic fibrosis in early embryos as well. If the gametes from parents who carry the allele are removed and fertilized *in vitro*, and the zygotes cultured until they become 2 or 4 cell embryos, those that lack the allele for cystic fibrosis can be selected and replaced in the uterus to continue development. Such technology ensures that the resulting child, and all its descendants, are free from the disease.

▶ What would be the evolutionary consequences of such procedures?

The allele would gradually disappear. Homozygous children very rarely survive long enough to breed, and so in evolutionary terms would be eliminated by natural selection anyway, but the new procedures eliminate potentially healthy heterozygotes as well, thereby removing the allele entirely. The genes involved in other heritable diseases, notably muscular dystrophy, seem to be unstable and to mutate frequently, so embryo selection would have to be practised continuously to eliminate the disease entirely. Such advanced technology is currently available to very few people, so it will be many generations before it has a significant impact on gene frequencies in the human population as a whole.

17.2.2 Adaptation

Biologists often speak of organisms being 'adapted to their environment' (see Chapter 2). Humans clearly live in many different environments, and some migrant people complain of their adopted land that 'the climate doesn't suit us'.

Fuegans Tierra del Fuego is a group of mountainous, partially glaciated, islands off the southern tip of South America, first reported to the western world by Magellan in 1520, but not explored in detail by Europeans until the early 19th century. One of the most thorough surveys was conducted by Robert FitzRoy, captain of HMS *Beagle*. The ship's naturalist on FitzRoy's second voyage around South America was Charles Darwin, and both men made numerous drawings and took detailed notes about the terrain, wildlife and native peoples. When they landed there on 16 December 1832, the islands were inhabited by three distinct tribes descended from people who had probably migrated from the north, perhaps 8 000 years ago. All three groups, then numbering around 20 000, lived in small huts made from tree trunks and hides, and dressed only in capes of animal skins. They had no agriculture but ate seabirds' eggs and other wild animals and plants, and two tribes also constructed canoes, from which they caught fish and small marine mammals. Even in the summer, when the *Beagle* was there, the weather on Tierra del Fuego is cold, with frequent gales, heavy rain and snow, particularly on the western side. The European visitors marvelled at the Fuegans' tolerance of cold and damp, evident even in newly born infants. In spite of the harsh climate, the people were tall, strong and apparently healthy in their scanty clothing and crude shelters, though, as Darwin emphasized, lean and dirty. Following FitzRoy's work, other European explorers and missionaries established bases on the islands, bringing with them the clothes, boots and housing necessary to survive the severe climate, and in the 1880s, Argentine miners arrived in search of alluvial gold. These exotic people brought measles, tuberculosis, smallpox and other infectious diseases from which the native Fuegans died in large numbers. By the end of the 19th century, all three tribes were decimated, and two have since become extinct.

▶ What can you conclude from these facts about evolutionary changes in the native Fuegans?

The Fuegans were clearly much better adapted to withstanding prolonged, severe cold than the Europeans, but had little resistance to introduced diseases, presumably because they were not regularly exposed to them. Many similar sad stories point to the same conclusion: humans, like other organisms, evolve adaptations to the environment in which they live. Natural selection can only operate effectively if the population is continuously, or at least regularly, exposed to the selective agent. Measles, tuberculosis and smallpox are ancient diseases (evidence of them can be detected in some Stone Age skeletons), so the Fuegans' distant ancestors were probably exposed to them, and hence acquired some resistance (see Section 17.4.2). Such adaptations seem to have largely disappeared during their period of isolation and inbreeding. With the introduction of these infectious diseases, selection suddenly became very severe, and variation, in the form of genetic resistance to the diseases, was insufficient in a small population with a low reproductive rate.

It is important to emphasize that not all populations suffer exceptionally severely from the introduction of exotic diseases. Humans are highly social animals and factors such as the loss of political leaders or breakdown in social order following an unexpected epidemic can contribute as much to civil chaos as the disease itself. For example, up to 40 000 people died from various causes during an outbreak of measles in Fiji in 1875, but later medical research, and the monitoring of subsequent epidemics, suggest that the Fijians' resistance to the disease is not greatly different from that of western people. Panic and civil disorder triggered by the epidemic seem to have augmented the impact of the disease itself.

Adaptation to diet The adoption of an omnivorous diet is one of the main factors that enabled *Homo* to diversify the habitats in which it lived and thereby to extend its range. So far as can be deduced from remains around human settlements and patterns of tooth wear, until about 9 000 years ago, people ate nuts, roots, leaves, fruit, large insects, shellfish and many different kinds of fish, birds and mammals in various proportions depending upon the locality and the season. The diet of many modern humans is similar, but some groups have much more specialized eating habits. Until the 1950s, most Eskimos ate mainly seals, reindeer and fish, the Masai live largely on cattle blood and milk products, while other peoples in west Africa and south-east Asia live mainly or entirely on plant food. With a few notable exceptions relating mainly to mineral deficiencies, most such foods seem to be adequate to maintain health, but people raised on one kind of diet rarely adapt successfully to a very different diet. The food people eat is determined mainly by what wild species are available and can be hunted or collected and by which crops or livestock can be cultivated efficiently. However, cultural attitudes and traditions also play an important part in food selection and food preparation. Indeed, dietary habits and taboos are often an integral part of racial or religious identity. There is evidence that some people in the Amazon rainforest forego certain foods *because* they are an important part of the diet of neighbouring tribes, thereby leading to resource partitioning of different groups (see Chapter 8, Section 8.3.2).

Like any other animal in a rapidly changing habitat, major changes in the human environment, whether or not caused by people's own actions, may affect the availability of traditional foods. Therefore, understanding people's capacity to adapt to dietary change must be an essential part of formulating policies for producing alternative foods. There are very few clear-cut examples in this biochemically complicated topic but the incidence of lactose intolerance is one of the best studied.

All female mammals secrete milk after giving birth, and all infants can digest this food so completely that very little faecal material is generated. The digestive system matures at weaning, and, while the capacity to break down solid foods increases, the synthesis of lactase, the enzyme that digests lactose (the principal carbohydrate in milk), wanes. Thereafter, although some animals, e.g. cats, hedgehogs and certain birds, sometimes drink small quantities of milk as adults, they cannot digest it well enough to live on it. Some adults, including the Chinese, aboriginal Australians and many west and central African people, find whole milk indigestible (as well as distasteful), but to pastoralist peoples, notably the Masai, the Bedouin, many Indian groups and most Europeans (and their descendants in America and Australia), it is wholesome and nourishing as well as appetizing.

▶ Would you expect that ancestral adult humans could digest milk?

No. Like all other adult mammals, pre-agricultural humans probably could not digest lactose properly after weaning. When people started harvesting milk as well as meat and skins from livestock (see Section 17.4.1), their digestive system must have evolved adaptations to the new food. The change is physiologically quite simple: the infantile capacity to synthesize lactase is retained into adult life, enabling pastoralists to digest milk as efficiently as babies can. Such people are able to consume large quantities of milk without ill-effect, and some groups, such as the Masai, live mainly on this 'unnatural' food.

Soon after people began to collect and store starchy plants (grain, rice, potatoes) and fruit in significant quantities, the process of fermentation, and with it, the pleasures of alcoholic intoxication, were discovered. Fermentation in the gut produces small quantities of alcohol in most herbivorous mammals, including humans, and it is detoxified in the liver by an enzyme called alcohol dehydrogenase. This enzyme is more abundant and active in peoples who are exposed regularly to additional sources of alcohol, and increases further in individuals who drink heavily. Because the liver breaks down the alcohol efficiently as soon as its concentration in the blood rises, such people can drink substantial quantities without becoming seriously intoxicated.

▶ Would all humans have this physiological capacity?

No. It would be absent in people who do not grow or collect starchy or sugary plants. Eskimos (who lived almost entirely by hunting), Australian aboriginals (who have never grown crops) and certain American Indian groups have a very low tolerance for alcoholic drinks. Grown men may become severely intoxicated on half a pint of beer.

Similarly, people such as southern Europeans, Indians and many Africans who traditionally eat mainly grains, vegetables, fruit and fish, find the high proportion of animal fat and protein of Eskimo diets very indigestible. The Pima Indians in southwestern USA and some Polynesians readily become obese and a high proportion (up to 40% in some groups) develop diabetes in middle age if they eat a western diet containing large quantities of refined sugar and animal fat. Although cultural attitudes and upbringing play a role, most racial differences in the capacity to digest and metabolize these foods are at least partially genetic, but the genes involved have only been identified in the case of lactose intolerance.

▶ Would comparable physiological differences have evolved in subspecies of animals and plants that occur in different habitats?

Very probably, but such differences would rarely be detected without detailed study. Human cultural attitudes now encourage people to eat 'modern' and 'fashionable' foods even if they are harmful to them. Animals would not be subjected to similar pressure to eat foods that made them feel unwell. The evolution of such minor physiological differences often proves to be a major, and for the most part, unforeseen hazard when previously isolated groups of people suddenly adopt each other's habits and cultural values, as inevitably happens following migrations.

17.2.3 Migration

With the possible exception of certain micro-organisms, humans are now the most widespread of all species. Their migrations have had major and irreversible effects on their own evolution and that of many other species with which they have come into contact. There are several rival theories about the exact course and dates of colonization, and the evidence is sparse and conflicting. Hominids originated in eastern Africa and remains such as skeletons and tools suggest that *Homo erectus* then spread through the rest of Africa and to Asia and Europe, reaching Java about 1 Ma ago and central China about 0.5 Ma ago. *H. erectus* probably entered southern Europe via Greece during glaciated periods when the Aegean Sea was much shallower than it is now, and reached France, Italy and Spain about 0.4–0.7 Ma ago.

Several hundred thousand years later, *H. sapiens* invaded Europe and Asia probably via similar routes and has been widespread there for at least 40 000 years. At least 11 000 years and perhaps as much as 30 000 years ago, people from eastern Siberia crossed the land-bridge over the Bering Strait to Alaska when the sea-level was low near the end of the last glaciation. They quickly spread throughout North and South America and western Greenland. There were probably several waves of such immigrants, the most recent groups travelling short distances by sea. Long-distance seafaring first appeared in south-east Asia, and was the means by which humans reached New Guinea and Australia from Indonesia about 35 000 to 40 000 years ago, the West Indies and the larger Mediterranean islands about 10 000 years ago, and the Polynesian Islands and Hawaii about 6 000 to 3 000 years ago. Madagascar and New Zealand were the last large islands outside the arctic regions to be colonized, about 1 500 years ago. The range of *H. sapiens* is still extending: continuous habitation of Antarctica and some islands, notably in the southern oceans and in the Svalbard archipelago in the Arctic, began only in the 20th century.

Geographical isolation has restricted gene flow (see Chapter 9) between these populations long enough for the evolution of racial differences in stature, the distribution of fat (adipose tissue), various features of the skeleton, particularly of the face and jaw, and the colour and texture of the skin and hair. Cultural differences in language, religious beliefs and social organization also emerged and further promoted inbreeding within social groups, even where migration and hence intermarriage were possible. Such evolutionary processes led to the formation of genetically distinct lineages that differ in both biological and cultural characters but have not been operating long enough for speciation (see Chapter 9) to have occurred.

There are historical records of numerous voluntary and forcible migrations, mostly over relatively short distances, within Europe and southwestern Asia (e.g. those reported in the Bible). Beginning in the 15th century, and reaching a maximum rate in the 19th and early 20th centuries, large numbers of people from Europe, and more recently from eastern Asia, colonized America, Australia, New Zealand and southern and eastern Africa. For at least 600 years during the last millennium, there was also forcible migration of people from east Africa to the Persian Gulf region, and for about 350 years, from west Africa to the Americas and the West Indies. Within the past 50 years, people have moved into Europe from Asia, Africa and the Caribbean in significant numbers. Migration to areas already populated has radically changed the ecology and evolution of both the immigrants and the indigenous peoples. In most areas, migration led to interbreeding, causing intermingling of both genes and culture; for example, Arabs (of Eurasian stock) and Negroes have interbred extensively on coastal areas of east Africa, where Swahili, a mixture of Arabic and Bantu, is now the most widely spoken language.

In other cases, whole groups of people and their culture failed to survive the impact of immigration. For example, when Columbus reached the West Indies in 1492, there were more than five million American Indians, organized into scores of distinct tribes, the larger of which lived by hunting buffalo and other mammals and birds, and cultivating crops such as maize, beans and squash. All the major Caribbean islands were also densely populated, with at least a million people on Hispaniola (now Haiti and the Dominican Republic), where one of the first European settlements was established. Within 20 years of Columbus's voyage, Spanish merchants began bringing slaves from west Africa, initially to work in the gold mines and then mainly on plantations of sugar, tobacco and cotton grown for export to Europe. Military action, enslavement and the introduction of exotic diseases and unfamiliar food and drink (see Section 17.2.2) eliminated the Indians in an amazingly short time, and the modern Caribbean population is almost entirely of African and European descent.

Many other native Indian tribes suffered similar fates, as European settlement, cultivation of cash crops and importation of African slaves followed quickly in Jamaica, Cuba and other islands of the Caribbean and on the mainland of North, South and Central America. In much of South and Central America, intermarriage between Indians, Europeans and Negroes has retained some of the genetic stock, if not the cultural integrity, of the original inhabitants. But only a tiny minority of those now living in eastern USA, California or the Caribbean Islands can claim any direct descent from the peoples who lived there for at least 10000 years before they arrived. Such recent invasions leading to extinction of human groups may be better documented, and probably also occurred much faster, aided by modern ships and firearms, but they are certainly not unique. Over several million years, species and subspecies of *Homo* replaced each other in Africa, and *H. sapiens sapiens* either wiped out, or perhaps interbred with, *H. sapiens neanderthalensis* in Europe about 40000 years ago.

Although migration sometimes led to local extinction of certain groups, overall human numbers have increased throughout the last one million years. The rate of increase was relatively slow at first, from an estimated world population of less than 5 million (about the same as Greater London or New York City in 1980) when people first reached America, to about 500 million by the time the Pilgrim Fathers settled on the north-eastern coast of America about 10 000 years later (in AD 1620). During the past 400 years, world population has increased tenfold to about 5 billion, and is still rising very rapidly, having tripled within the past century.

SUMMARY OF SECTION 17.2

Cultural attitudes and practices, as well as natural and sexual selection, influence the course and rate of evolutionary change among modern people. Humans are physiologically as well as culturally very diverse with some groups evolving adaptations to different diets and climates. Technological advances, such as the manufacture and use of ships and other vehicles and more efficient shelters and clothing, have accelerated the rate of migration and permitted colonization of islands and other remote and inhospitable areas. The present range of the species is among the largest of all organisms and is still extending. The total human population has increased more than a thousand fold in the past 10 000 years.

17.3 THE IMPACT OF HUMANS ON OTHER ORGANISMS

Throughout the fossil record, species have become extinct, and new species have appeared. There have also been several mass extinctions (Chapter 15). During the last 100 000 years, many species have become extinct on continental land-masses. The rate of extinction has not been uniform, being most rapid about 10 000 years ago and at the present time, nor have all taxa or all geographical areas been equally affected. Several theories have been proposed to explain this situation. The climate was changing rapidly, and with it changes in sea-level and vegetation (see Chapter 2, Section 2.4 and Chapter 13, Section 13.5), which might have contributed to such extinctions. Other theories blame the emergence and spread of humans. Although there are many dissenters, some authorities believe that prehistoric people were a major factor in most recent extinctions. Several different aspects of human activities were involved, and should be considered separately.

17.3.1 PREDATION

Humans, like their hominid ancestors, are omnivores, eating a variety of plant food and preying on animals. *Homo* has long been a predator, often hunting in groups and using elaborate tools. Their ability to kill animals much larger than themselves is impressive, though not more so than that of wolves, which are only half the size of humans. The role of communication, planning and social organization in human hunting has been much discussed. But humans are not unique in this respect: careful observation of hyenas also suggests that they 'decide' which species they are seeking before the hunt begins, and alter group size and tactics accordingly. Social organization of hunting groups of chimpanzees is also elaborate, with the most difficult tasks being performed by the strongest and most experienced males.

Neither of these species is known to have hunted any species to extinction, so have humans done so? The role of human hunting in the extinction of prey species is very controversial.

The most persuasive evidence comes from the minority of cases that have occurred relatively recently and in small, isolated areas. Islands are particularly vulnerable to human intervention because by definition they are isolated and many are small, supporting small populations of an impoverished fauna and flora that often includes very few predators. On remote volcanic islands such as the Galápagos and Hawaii, evolution proceeds in isolation for long periods, producing species that are adapted to their unique habitat. For example, terrestrial gastropods on continents have many predators, including birds and tortoises, and most species produce dozens or often hundreds of eggs with all but the largest forms becoming adult within a year or two of hatching. But the lack of predators and limited resources on islands promotes the evolution of slow juvenile growth and low fecundity (see Chapter 8, Section 8.5). The tree snail *Partulina proxima* occurs only at high elevations on Molokai, Hawaii. Although only about 2 cm long as adults, *P. proxima* does not begin to breed until it is 5–7 years old, and produces only about six offspring per year. Since juvenile mortality is about 98%, natural longevity must be at least 18–19 years to maintain the population. The demographic structure of many other island species seems to be similar.

► In the presence of efficient predators, which traits would be most adaptive: high fecundity, rapid growth to maturity, or longevity?

High fecundity and rapid growth enable populations to recover rapidly from heavy predation and to reoccupy the niches temporarily left vacant when its numbers were reduced. Unless older animals can evade predation better or produce more offspring, the selective advantage of longevity is very small.

Dodos In many cases, the disappearance of large species on islands has been closely correlated with the arrival of humans. The dodo was a large flightless relative of the pigeon found only on Mauritius and Réunion, two adjacent volcanic islands in the western Indian Ocean. In 1598, when the dodo was first described by Portuguese sailors, the islands were uninhabited, and there is no trace of prehistoric human settlement there. Being conveniently situated about half-way between Europe and the Far East round the Cape, merchant ships began calling there to restock with water and fresh food. Dodo meat, like that of most herbivorous birds, proved to be delicious. The sailors had firearms but they probably did not need to waste ammunition: having no natural predators, dodos were slow and reacted to strangers with more curiosity than fear. The total land area of Mauritius and Réunion is about 4400 km^2, much of it steep mountains, so the total population of this endemic species was probably never more than a few tens of thousands. Human predation quickly removed them, and none was seen alive after 1670. A century later, a similar fate befell the solitaire, another large flightless bird endemic to the nearby island of Rodriguez, only about 100 km^2 and also then without human inhabitants. Sailors also plundered giant tortoises from all three islands, and from Aldabra, north of Madagascar, and the Galápagos archipelago in the eastern Pacific; all have recently been greatly reduced in numbers, and some of the subspecies on the Galápagos have become extinct.

Like pigeons, dodos ate seeds, some of which passed through the intestine undigested and were deposited elsewhere with the faeces. At least one species of tree, the tambalacoque, is believed to have been dispersed mainly by dodos. Some large, old specimens are still extant, but the absence of their natural seed-dispersal agent and the introduction of exotic seed predators, such as rats and pigs (see Section 17.3.3), are preventing young trees from getting established, so the species will probably become extinct in the next few hundred years. The elimination of one species from an integrated ecosystem often affects the viability or reproductive success of many others. If the species are inconspicuous or, as in the case of trees, very long-lived, such effects may not be detected easily.

Flightless birds in New Zealand Although the depredations of modern western people may be the best documented, they are not the only, or even the most damaging, cases of human predation on other species. New Zealand was colonized from Polynesia in several waves of immigration, starting about 1500 years ago. There is clear evidence that the settlers killed moas (see Figure 2.4), particularly the larger species, in great numbers. Moa bones (see Chapter 2, Section 2.2.2), some fashioned into artefacts such as fish hooks, spear points and beads, have been found in over a hundred archaeological sites and the birds are represented in rock paintings. The flesh of these herbivorous, flightless birds was probably very tasty and there is every reason to believe that moas were a major, possibly a staple, item in the Maoris' diet. Many eggs were also collected and used as containers for water and other liquids.

Although other explanations, such as climate change, have been advanced, most scholars now agree that predation, combined with habitat destruction due to burning of the forests (see Section 17.3.2), were the main causes of the extinction of most of the 34 species of birds that disappeared in the first 800 years of human occupation, including a flightless goose, a giant rail and a swan. The process probably went much more slowly than in the case of the dodo, because the land area was larger and the terrain more diverse. As the prey becomes rarer, it becomes increasingly hard work to locate and catch them, and most predators switch to alternative, more accessible species, thereby avoiding extermination of the last few specimens. Humans did turn to alternative sources of food: fishing increased in importance and, from the 14th century onwards, the Maoris cultivated crops such as taro and yams, introduced from Polynesia. However, by the time the birds had become rare, many species had assumed cultural importance to the Maori people. The bones were the raw material for tools and ornaments and the feathers featured prominently in ceremonial clothing (e.g. moas, the huia, see Chapter 7 Figure 7.17). Tribal customs as much as hunger then became the motive for pursuing them into remote areas.

A few of the recently extinct species, such as the giant eagle, probably disappeared as a result of extermination of their prey: their remains are rare in archaeological sites, but in the absence of land mammals, they probably preyed mainly on large birds, including moas. At least five more species of bird have become extinct since the Europeans arrived, and several others, notably the kiwi (a small relative of the moas) and the kakapo (a large flightless parrot), have become very rare.

Prehistoric extinctions The role of human predation in the extinction of wild animals on continents in the late Pleistocene and in historic time is more difficult to establish clearly. Other factors, such as changes in temperature and rainfall, and the spread or decline of other predators, obscure the picture. Climatic change would alter the habitat of plants and animals of all sizes about equally, while predatory humans would probably have concentrated their efforts on large species. However, although large species are usually less abundant than small animals, their remains are more likely to survive, so evidence incriminating human activities is selectively preserved. One way round this problem is to look for a close temporal correlation between human presence and mammalian extinctions.

Such a link has been sought in the recent evolutionary history of the mammalian order Proboscidea. Proboscideans are very large and have exceptionally massive, hard teeth and long, curved tusks, so their fossil record is unusually complete. There are only two living species, the African elephant *Loxodonta africana* and the Indian elephant *Elephas maximus*, but as recently as 100 000 years ago there were ten species that occurred over Europe, North America and almost the whole of Asia, as well as Africa. Several arctic and subarctic proboscideans, known as mammoths, were moderately abundant all over the Northern Hemisphere for most of the Pleistocene. As the ice retreated after the last glacial maximum about 20 000 years ago, many animals and plants including mammoths recolonized areas that they had inhabited during the previous interglacial period. Modern humans were the main newcomer to this group. In Europe, several cave paintings, some as far south as northern Spain, clearly represent mammoths, with their distinctive domed head, long curved tusks and shaggy hair. In Europe and northern Asia, several sites contain hundreds, even thousands, of mammoth skeletons, almost all of them less than 50 000 years old and many of more recent date. Most are disarticulated and a few, from the Palaeolithic of Poland, Ukraine and Siberia, were arranged in such way as to suggest that the bones were used to construct shelters. At about the same time, there is clear evidence for use of mammoth tusks for making tools and other artefacts, suggesting that humans were killing (or scavenging) this species for raw materials as well as, or possibly instead of, for food.

▶ How does the ecological impact of using animals and plants for making artefacts differ from that of eating them?

Raw materials, such as ivory, can only be obtained from particular species but a wide range of animals are edible. Food consumption is proportional to population size but demand for materials is not necessarily related directly to the numbers of 'consumers'. Humans have used animal skins for clothing and shelter, and leather and bones for tools, for at least 50 000 years, but it is not clear when these habits led to increased predation, i.e. to killing animals primarily for materials rather than for food.

There are no cave paintings from the same period in America, but other evidence shows that humans certainly did kill mammoths there: a few sites with mammoth fossils also contain spearheads and other human artefacts, but many others have yielded no such evidence. In Eurasia, several other large predators, including hyenas, sabre-toothed tigers, lions and wolves that were in the same areas at the same time, may have contributed more to mammoth mortality. Nonetheless, mammoths (and the straight-tusked elephants) disappeared quite rapidly shortly after *Homo* spread into these regions. In general, extinction in Eurasia seems to have proceeded from south-west to north and north-east, with a few mammoths surviving to less than 10 000 years ago in Estonia and Siberia, and even later in Canada and northern USA. The human population of these arctic and subarctic areas was probably very small until well into historic time so it is unlikely that their actions could have been solely responsible for extinction of the mammoths.

▶ Is there any other evidence that large proboscideans can withstand predation by humans?

The two surviving species of elephants occur in areas with a long history of human habitation. Indeed, *Homo* probably originated, and its hunting techniques developed, among large concentrations of *Loxodonta*. As shown in Table 17.4, large mammals generally seem to have survived better in Africa, where *Homo sapiens* evolved as part of the fauna, than in continents to which humans are relative newcomers. It seems more likely that the same climatic and faunal changes which facilitated extension of the primate's range promoted the extinction of the proboscideans, but that human predation may have accelerated their decline.

Table 17.4 *The total number and proportion (%) of species of large mammals that have become extinct during the past 100 000 years on four continents.*

Continent	Extinct	Living	Total	% extinct
Africa	7	42	49	14
North America	33	12	45	73
South America	46	12	58	80
Australia	19	3	22	86

Modern hunting on and around continents Compared to islands, there are relatively few instances of entire species being exterminated on continents solely by hunting. Herding or flocking species seem to be most at risk, particularly if they are diurnal in habits and spend time exposed in open country. The North American bison (*Bison bison*) is one of the most spectacular examples. Although hunted by pre-Columbian people for at least 10 000 years, as recently as 300 years ago there were about 50 million bison in North America, in mountains and open forest as well as on the prairies. Shooting for meat and skins, and to protect wheat production and other agricultural interests, reduced the population to fewer than 1 000 individuals by 1890, most of them in western Canada. From this group, small populations have been reintroduced into national parks and other protected areas. Human hunting has also exterminated several other ungulates, such as muskoxen *Ovibos moschatus* (Plate 2.4), from large parts of their former range, including the whole of the European Arctic, during the past millennium. They too are being reintroduced and are managed for production of wool and meat.

The grey wolf *Canis lupus* (Figure 17.1) and the common red fox *Vulpes vulpes* used to be among the most widespread of all mammals. At the end of the Pleistocene, wolves occurred, with some regional differences in size and colour, throughout the Northern Hemisphere except in tropical forests and arid deserts.

Figure 17.1 *The grey wolf* Canis lupus, *showing adults with cubs.*

They normally live and hunt in large packs, maintained by elaborate social interactions, and are scavengers as well as predators on a wide variety of species. Prehistoric people seem to have had little effect on the wolf population, and may indeed have aided them by providing carrion (see Section 17.4.1). Since agriculture became established, the actual and imagined threat that wolves pose to people and their livestock has led to their persecution by hunting, trapping and poisoning. Nonetheless, they remained widespread until firearms came into general use. Solitary animals and depleted packs are much less efficient as predators, and very small populations are probably not viable. Wolves disappeared from England and Wales before the 16th century and from Scotland and Ireland in the 18th century; they are now exterminated throughout their former range except in a few remote areas of central Europe, the Balkans, northern USA, Canada and Russia.

Similar processes have led to the near extinction of another social carnivore, the lion (see Chapter 2, Section 2.3.3). About 10 000 years ago, its range was similar to that of the wolf, but it disappeared from America, northern Europe and eastern Asia when humans became well established there, from southern Europe in Roman times and from southwestern and central Asia during the past few hundred years. A few hundred lions survive in the Gir forest of India and several larger populations in Africa south of the Sahara.

Solitary species have fared much better, particularly if they are adapted to a varied diet and habitat. For 1 Ma, from the middle Pleistocene to this millennium, the brown (grizzly) bear *Ursus arctos* occurred over most of Europe, Asia and North America and on many offshore islands, including parts of Japan. Although long hunted as vermin and for its fur, and taken alive for bear-baiting from Roman times until the 19th century, small populations have survived in mountains, the far north, on islands and a few other pockets of inaccessible territory in Russia, Scandinavia, Spain, the Alps, the Balkans, USA, Canada and Mexico. Other species hunted in modern times, such as beaver, badger, otter, puma, lynx, tiger and wolverine, although now rare, have also escaped complete extinction; small, scattered populations persist, from which these species could recolonize parts of their former range if hunting and habitat destruction were stopped. Red foxes are among the few large mammals that, in spite of intensive hunting, are still abundant as wild animals throughout most of their natural range. As omnivores and scavengers, foxes have adapted successfully to the presence of humans, and are common in many cities.

Examination of the circumstances surrounding recent extinctions of birds points to similar conclusions. Until 150 years ago, dense flocks of passenger pigeons, often consisting of millions of birds, were common all over the eastern USA. Shooting and trapping by European settlers decimated the flocks, and the remaining birds seemed to be unable to breed successfully in pairs or small groups. Its numbers declined quickly and the last one died in captivity in 1914. The Carolina parakeet became extinct at about the same time for a similar combination of reasons. But, although the ranges and population densities of almost all of the scores of species of hawks, eagles, buzzards and ospreys have declined recently due to hunting, poisoning and habitat destruction, very few have disappeared completely since AD 1600.

Aquatic organisms Marine mammals and birds were almost unaffected by the wave of extinctions that destroyed so many large terrestrial mammals around 10 000 to 11 000 years ago. Most living groups of people eat some aquatic plants and animals, but such foraging was restricted to river banks, shores and reefs until the technology for building boats from wood and skins was developed. Nonetheless, when their habits made predation feasible, prehistoric people tackled some remarkably large marine species. For example, the largest known sirenian, Steller's sea cow *Hydrodamalis gigas*, weighed around 4 000 kg when adult, and fed on kelp and other marine algae in shallow, sheltered waters. Fossils and evidence from archaeological sites suggest that in the late Pleistocene, the species occurred on many coasts and islands around the northern Pacific from Japan to California. It swam slowly and probably had few predators until humans invaded its habitat. Hunting by prehistoric people appears to have exterminated it except around a few small islands off the west coast of Alaska, where a population of 1 000 to 2 000 specimens was first reported to western scientists by Steller, a German biologist with Captain Bering's Russian-financed expedition of 1741. Bering's ship was stranded there and the crew survived the winter by eating *Hydrodamalis*. Other hunters followed, and the last specimen was seen in 1768.

During the past 400 years, similar hunting for materials such as skins, ivory and 'whale bone' as well as for meat exterminated a few marine mammals, such as the Atlantic grey whale and the Caribbean monk seal; while others, notably the sea otter, were reduced to a few dozen specimens. However, with the exception of the grey whale, nearly all total exterminations have taken place at breeding grounds on land or in shallow waters or estuaries. Many fish, particularly littoral and shallow water species, e.g. herring and haddock in the North Sea and invertebrates such as abalone (a gastropod) and oysters (bivalves), have also declined drastically in the past 50 years. But until the rapid improvements in ships and fishing technology in the 20th century, humans were not efficient hunters at sea. Even now, as on continental areas, although species may be decimated, it is simply too difficult to eliminate all of them by predation alone. Drift nets and environmental modification from chemical pollution are now believed to pose a much greater threat to more marine species than direct predation.

Freshwater species are much more vulnerable to disturbance from human activities. Lakes and, to a lesser extent, rivers are like islands in reverse, being isolated and often small areas of habitat. As in the case of islands, many species are endemic to only one or a few adjacent lakes, and may have evolved with few predators and are thus vulnerable to extinction from overhunting. Large, isolated, ancient lakes contain many endemic species: for example, Lake Baikal in southern Siberia had, until recently, about 2 500 endemic species, mostly fish and invertebrates, but also a unique species of seal. In this lake, as in Lake Victoria (see Chapter 9, Section 9.4.4), dozens of species have disappeared within the last 100 years, at least in part from overfishing.

17.3.2 Habitat modification

Everyone is aware that the presence of humans has had a major impact on the ecology of many other organisms with which they have come into contact. It is important to emphasize that in this respect, humans are not unique: the presence of most large animals affects many other species and aspects of their physical environment, and even small organisms such as ants, locusts and termites can cause spectacular changes in the landscape and vegetation if they are sufficiently numerous.

The impact on the environment of very large herbivorous mammals such as elephants, rhinoceroses and hippopotamuses has been studied in detail. Adult elephants eat about 200 kg of foliage and grass each day, often uprooting small trees and pulling branches and bark off larger ones. Bark stripping weakens trees, making them susceptible to wood-boring insects and fungi. Elephants are particularly fond of baobab trees, which have nutritious bark and soft, pithy wood. In some areas, baobabs are almost extinct, due to damaging of mature trees and uprooting of young ones. Elephants defecate every 1–2 hours, producing about 150 kg of faeces per day, which support several species of dung-beetles and many smaller invertebrates and micro-organisms. They also wallow in pools and rivers, often trampling nearby vegetation, damaging banks and stirring up the bottom as they do so. The presence of large numbers of elephants transforms woodlands into open grassland, and clear pools into mud holes. These changes drastically affect the abundance of many other species, from other grazers such as antelopes to tree-nesting birds, and fish to mud-feeding worms, and are so conspicuous and distinctive that experienced observers can tell if there have been elephants about within the past year or so just by walking around. None of these effects does irreversible harm as long as elephants remain at low density and roam over a large area. But if they are confined, as, for example, within a park or reserve, their impact can be as drastic as that of humans, and often results in less food and an impoverished habitat for the elephants themselves.

Humans also cut down trees, disturb the soil and contaminate water. Like elephants, early humans and probably most other hominids wandered over a large area, and most of the effects of their presence on the environment had disappeared by the time they revisited an area. However, beginning perhaps as much as 1.7 Ma ago, the impact of human activities on their environment differed from those of other large mammals in one very important respect: the use of fire. Fires started naturally by lightning and volcanic eruptions have always been an integral part of ecological cycles. By breaking down dead wood and grass, burning releases nutrients and thereby stimulates new growth; it also promotes the dispersal of the seeds of certain conifers and destroys parasites, pathogens and pests that flourish in bark and undergrowth. Humans are the only animals that can start fires deliberately or control their spread. The use of fire meant that vegetation, mainly wood, was used for fuel as well as for food and as a material for making tools and other artefacts. Accidental or deliberate burning also enabled humans to achieve rapid, widespread and lasting ecological changes to the flora and fauna. The method is still in use: farmers burn stubble to destroy weeds, and gamekeepers burn old, woody heather to promote the growth of young, nutritious shoots, which are preferred by grouse and partridge. Young hardwood trees, such as oak and ash, are also destroyed by such fires, so this practice, together with sheep grazing, effectively prevents the replacement of moorlands by forest.

It is important to emphasize that such long-lasting effects of human activities are not new, nor are they always connected with agriculture or urbanization. Fires can flush out animals, thereby facilitating hunting, and promote regeneration of certain edible plants. Australian aboriginals have never kept herds nor grown crops, and their technology is only a little more advanced than that of Stone Age Europe, but deliberate burning of forests and grasslands during the last 40 000 years is thought to have drastically changed the ecology of much of the continent, and may have contributed to the extinction of several species of plants and animals. The Maori people from Polynesia settled in New Zealand only about a thousand years ago. By the time Europeans arrived some 800 years later, almost all the forest on the eastern side of the South Island, much of that on the wetter, western side and large areas of the North Island, had been burned, and were maintained as scrub and grassland by repeated burning. Large areas of both forest and grassland in North America, and the area around the Mediterranean, were also burned many times both before and after agriculture and animal husbandry replaced hunting as the basis for human ecology.

Although such fires had a major impact on the fauna and landscape, their contribution to the concentration of carbon dioxide in the atmosphere was probably very small compared to that from natural sources. Fires of human origin did not have a significant effect on the composition of the atmosphere until people started to burn fossil fuels (coal, oil and natural gas) in substantial quantities (see Section 17.3.5).

Manipulation of fresh waters is a longstanding form of habitat modification that has increased greatly during the past 500 years. Since agriculture and commerce became established (see Section 17.4.1), people have tried to control natural sources of freshwater by damming rivers, building canals and reservoirs, draining lakes and marshes and digging wells. Such habitats often have unique and diverse fauna and flora, and many are important temporary feeding grounds for migratory birds. Demands for water for agriculture, industry and domestic use have drastically altered many freshwater habitats: the Aral Sea in southern Kazakhstan, which has shrunk to half its former area in less than a century, is a particularly spectacular example of very many instances of major and rapid changes to inland waters. Although some species have successfully colonized artificial canals, reservoirs and sewers, many wetland faunas have been severely damaged as a result of human exploitation of the water. The construction of canals has also promoted mixing of naturally isolated fauna and flora, e.g. completion of the Suez Canal in 1869 linked the Red Sea to the Mediterranean Sea after millions of years of separation, and the Welland Canal now permits organisms (e.g. fish) that could not ascend Niagara Falls to enter the Great Lakes of North America.

Human activities have extended or improved the habitat of many other organisms. Apart from domesticated animals and cultivated plants (Section 17.4), a wide variety of organisms have evolved habits that depend directly on human artefacts or practices. Swifts and swallows probably originally nested in banks and sheltered cliffs, but now almost always breed in or on buildings, as do many owls, storks and bats, which previously nested in trees. Cockroaches first appeared in the Carboniferous, and have always been mainly tropical, omnivorous insects. They readily colonize the warm, damp environments created by humans, and eat scraps of many different human foods. Several species have colonized the sewers, central heating ducts and kitchens of many large cities, where they flourish, in spite of strenuous efforts to exterminate them. Human activities have also opened up new habitats for rats (see Section 17.3.3), mice, houseflies, dust mites, carpet beetles, clothes moths (formerly restricted to nests of small mammals and birds), booklice, deathwatch beetles, woodworm, bakers' and brewers' yeast, dry rot (a wood-consuming fungus) and many others. Most of these species have been introduced wherever humans and their possessions have travelled, greatly extending their natural range. The range and abundance of many plants, particularly those that evolved as colonizers of disturbed habitats, have also increased. Naturally disturbed habitats are quite rare, arising mainly from forest fires, river bank erosion, landslides, volcanic eruptions and excavations of burrowing animals such as moles and termites. Humans burn forests, fell trees and plough and dig the ground, creating more habitat suitable for such rapidly colonizing species that we call 'weeds'.

Finally, we should mention the efforts to eradicate pathogenic organisms that cause disease in humans, their livestock and crops. Habitat destruction, in the form of isolation of infected persons and effective immunization of exposed populations, was started by Edward Jenner in the 1790s. Such practices have reduced the range and abundance of many pathogens, notably polio, tuberculosis and yellow fever, and have caused the extinction of the smallpox virus in the mid-20th century. Although few people would mourn the disappearance of this dangerous and disfiguring disease, it was a component of the natural biosphere that has been deliberately eradicated by humans. Unfortunately, there are very few other examples of extinction of pathogens, mainly because, like prey species on continents, there are too many places where they can 'hide' from human intervention. Efforts to eliminate them with antibiotics and antiseptics are usually also ineffective in the long term, because the micro-organisms readily evolve resistance to such agents (see Chapter 8, Section 8.4.3).

17.3.3 Artificial introduction of exotic species

Migratory animals have also been important agents in spreading sedentary species to remote or isolated areas (e.g. Hawaii): they carry their parasites and deposit seeds and other organisms in their faeces. *Homo sapiens* has long been nomadic, and since humans turned to agriculture and animal husbandry (see Sections 17.4.1 and 17.4.2), they have transported animals and plants wherever they went, for purposes ranging from food and transport to sport and ornamentation. Numerous parasites, pathogens, weeds and vermin were transported with their crop plants and livestock. The domestic livestock were usually intended to be kept in captivity, but they often escaped and, in a surprisingly large number of cases, have bred in the wild. Such populations are called **feral** in the case of animals, and **naturalized** in the case of plants.

We do not begin to have a comprehensive, much less a predictive, theory of evolution following artificial introduction of exotic species. Such events have, by definition, occurred in the very recent past, many within the past few hundred years. The evolution of adaptations in both immigrant and native species to the new situations is far from complete: only fragments of the story are available for study. This Section includes several case histories describing the outcome (so far) of artificial introductions; the examples are chosen because they illustrate evolutionary principles that might be important to developing a synthetic theory that accounts for all aspects of the process.

Rats The two species associated with humans in Europe and America, the brown rat *Rattus norvegicus* and the black rat *R. rattus,* are among the biggest of the 78 species in the genus. Both species breed rapidly, producing up to 20 offspring as often as once a month on unlimited food, and the young quickly disperse to new habitats. Their diet is very varied, and includes almost all forms of human food. Rat populations build up wherever large quantities of food are stored, e.g. in barns, warehouses and docks, hence their association with ships. *R. rattus* in particular is an expert climber, boarding and going ashore along the docking ropes whenever the opportunity presents itself. By this means, rats have spread to almost every island that has been visited by large ships. These species are effective colonizers; with very few exceptions, a healthy breeding population has developed from a few founder individuals, with *R. rattus* doing best in tropical areas, and *R. norvegicus* in the temperate zone, especially in towns.

Their presence has nearly always proved destructive to wildlife, as well as to the food stores of the human population. The best documented cases are those that occurred recently and in small areas. In 1918, a ship was wrecked on the coast of Lord Howe Island in the South Pacific and a few black rats reached land. Within a few years, five of the eight native songbirds became extinct, probably because the climbing ability of *R. rattus* enabled it to raid the birds' nests up to a height of 3 m above ground. Black rats also reached the major Hawaiian islands successively between 1873 and 1932, and their arrival correlated strongly with the steep decline in the numbers of several native birds and the endemic land snails (see Chapter 9, Section 9.4.3, and Section 17.3.1). These species had long coexisted with the native *R. exulans* but were unable to withstand predation from the larger, more aggressive introduced species. Although they are obviously in competition with *R. rattus* over much of their former range, the indigenous rats survive, albeit in smaller numbers.

Some time before the 15th century, the Maoris brought rats and dogs to New Zealand from Polynesia, the former probably by accident, the latter probably as hunting and companion animals and perhaps for food. The Polynesian rat *R. exulans* has never been as closely associated with humans as the species that are common in Europe and is only about one-tenth the size of *R. rattus*, but it seems to have done as much or more damage there than the dogs. Except for a few bats and marine forms, New Zealand had no native mammals (see Chapter 13). The absence of rodents and mammalian carnivores is believed to be the main reason why large birds, many of them flightless and ground-nesting, and a variety of reptiles evolved there. Rats and dogs probably contributed to the extinction of many of these endemic species by destroying eggs and young. Bones of the tuatara *Sphenodon,* the only surviving genus of the largely Mesozoic reptilian order Rhynchocephalia, have been found in archaeological sites on both major islands of New Zealand. But by the time Europeans reached Australasia, tuatara occurred on only about a dozen small islands, and are common only where rats and other mammalian predators are absent. Rats were probably also responsible for the species extinction of a large frog and a gecko and for the rarity of other reptiles and large invertebrates. Similar forms persist on a few small islands, where rats, at least for the time being, are absent.

There are 30 endemic species of rodent, including several species of *Rattus*, in Australia: their ancestors must have reached that continent long before humans appeared. Australian native rats were almost certainly among the prey of the marsupial native 'cats' *Dasyurus*, the Tasmanian devil *Sarcophilus* and the thylacine *Thylacinus cynocephalus* (Chapter 11, Section 11.3.2) also called the Tasmanian wolf. (Both of the latter two species occurred throughout most of Australia as well as in Tasmania.) When humans introduced other eutherian predators, such as dogs, cats and foxes, the native rodents as well as the marsupials declined in numbers and a few species became extinct, along with ten species of kangaroo, and more than 30 other marsupials. Even genera such as *Rattus*, that are abundant and highly successful in many environments, became rare in the presence of efficient predators after long periods of evolution in isolation from them. The capacity to adapt rapidly to the presence of efficient competitors and predators is a quality that dodos, moas and many marsupials apparently lacked, but that rats, houseflies and humans seem to have in abundance. The genetic and physiological bases for such differences are discussed in Section 17.5.

Although European travellers have done so much to spread *Rattus* almost throughout the world, its association with people on that continent is relatively recent. The genus evolved in south-east Asia and modern scholarship suggests that *R. rattus* may have originated in Malaysia or Indonesia, and *R. norvegicus* probably in northern China. The former probably became associated with humans and spread to other continents in prehistoric times, but there is no evidence for *R. norvegicus* in Europe until 1553, and it did not reach North America until 1775. The relationship between humans and mice is much more ancient; *Mus musculus*, the common house mouse in Europe, and, since its introduction by humans, in most other parts of the world, is native to southwestern Asia and the

Mediterranean area and has probably frequented human settlements ever since these areas were colonized. It is one of only three (of more than 30) *Mus* species to associate closely with humans.

Feral domestic animals Many kinds of domestic livestock, such as goats, sheep, cattle and rabbits (see Section 17.4.1), have been introduced to all continents except Antarctica and to numerous islands. Many such introductions happened so long ago, and the animals are so familiar and seem to be so well integrated into the habitat, that we tend to forget that their presence is artificial. The 'wild' horses of the Camargue and the New Forest are probably descended from animals released by the Romans, who may also have introduced macaque monkeys to Gibraltar and fallow deer to much of central and northern Europe. Nonetheless, feral domestic livestock have had a major impact on the natural flora and fauna. Starting in the early 16th century, the Spanish imported a whole farmyard of animals, including cattle, horses, poultry, pigs, sheep and goats, into the West Indies, Mexico and their other American colonies. Military action, epidemics and simple carelessness resulted in the livestock being released, and most species established feral populations and they and their diseases spread. The introduced species radically altered the native fauna and flora as well as the human economy: much of the native flora was eaten or trampled and native mammals suffered severely from certain introduced diseases.

One of the best documented cases of evolution in a feral domestic animal is the common rabbit *Oryctolagus cuniculus* (Chapter 2, Figure 2.11(a)). The species is native to Spain and the area around the western Mediterranean, where wolves, lynx, foxes and predatory birds such as eagles, hawks and buzzards prey on it. Even though many of these predators are now rare due to human activities, the rabbit is not intrusively abundant in its native habitat. Rabbits were introduced into Britain only about a thousand years ago, probably by the Normans, who raised them for food. For several centuries, rabbits rarely bred outside specially constructed warrens, where they were fed and tended. They gradually adapted to conditions in Britain and, by the 18th century, were widespread and in many areas common on moors and meadows. In spite of regular hunting for meat and fur, sometimes intensified, as during and just after the Second World War, rabbits have spread almost everywhere, including the Scottish highlands and on all the larger offshore islands. We know very little about the physiological and genetic basis for the evolution of its adaptations to the cooler, wetter conditions in Britain.

Rabbits were introduced into Australia from Britain for use as game animals (see Chapter 2, Section 2.3.1). They thrived in the hot, arid climate that resembles their native Spain, and quickly became very abundant, their spread being facilitated by intensive shooting and trapping of native and introduced predators undertaken to protect livestock sheep and game birds. Within 30 years, they were a serious pest over most of the continent from the Snowy Mountains to the central desert. They tunnelled through the subsoil and overgrazed the land, causing soil erosion and destroying sheep pastures, as well as competing with indigenous wild herbivores such as kangaroos. Vast sums of money were spent on many different projects aimed at exterminating them, or at least limiting their spread: thousands of people worked full-time, shooting, trapping or poisoning millions of rabbits. In 1907, a fence over 1 800 km long was built to cordon off much of western Australia from rabbits. It was effective for a few years but was eventually breached. The problem was finally contained, if not completely solved, in the 1950s by the deliberate breeding and dissemination of the lethal virus, myxomatosis, and the fleas that transmit it from one rabbit to another. This disease, and introduced predators (dingos, foxes and feral cats), keep the population at a level acceptable to sheep ranchers, although wild herbivores are still adversely affected by competition from rabbits.

Goats eat almost any vegetation including tough, bitter plants (see Section 17.4.1). They also tolerate dry conditions and close confinement, making them convenient as a source of milk and meat on long sea voyages. They were accidentally, or in many cases deliberately, released in exotic places, where they frequently established feral populations and have severely damaged the native flora on many tropical islands, particularly those lacking large predators. Captain James Cook released a few goats on Hawaii (then called the Sandwich Islands) in 1778. Their descendants have done more damage to the natural vegetation, bringing some endemic species almost to extinction, than a thousand years of human habitation there. Pigs also readily become feral and have an even more varied diet. They have adversely affected many indigenous species of plants and animals, especially in Hawaii and several Caribbean islands.

Cats are another human introduction that has devastated many wild animals, particularly non-venomous reptiles and amphibians, flightless birds and the eggs and young of ground-nesting species. Little changed from the ancestral wild cat (see Section 17.4.2), domestic cats harass wild birds and rodents even if fed regularly. They adapt readily to changes in climate and diet and their hunting strategy has proved effective on a wide variety of prey. Feral cats are very abundant in parts of the Australian outback, and along with other small carnivores such as foxes and ferrets, have greatly reduced the numbers of many indigenous species there and on many islands in the Pacific (e.g. New Zealand), South Atlantic and Caribbean.

► Are there any places in Britain where cats and foxes have excluded certain birds?

Yes. Many seabirds, e.g. puffins, guillemots, kittiwakes and most water fowl, usually breed on rocks or small islands offshore or in lakes, where foxes and cats cannot reach them. Heavy predation on the eggs and young often prevents successful breeding on the mainland. In Britain, such predation does not seriously affect numbers or species diversity, because there are plenty of suitable predator-free islands within easy flying distance of the mainland and other feeding grounds, but in very remote islands, or when flightless species are involved, the effects can be disastrous. The flightless wren *Xenicus lyalli*, endemic to Stephen Island off the north coast of New Zealand's South Island, was exterminated (in 1894) by a single domestic cat imported by the keeper of the lighthouse that had just been built there. Cats, rats and pigs reached Mauritius in the 17th century, and while the sailors hunted the adult dodos (see Section 17.3.1), the introduced mammals ate their eggs and juveniles. Most of the endemic reptiles also disappeared, probably destroyed by these introduced predators. The native Australian fauna included few predators, and the introduction of the dingo dog (by Aboriginal settlers) more than 10 000 years ago, and cats and foxes (by Europeans) about 200 years ago, has depleted populations of birds, small marsupials and reptiles.

Most of the best known examples of feral livestock harming indigenous fauna and flora relate to tropical islands, but the process is by no means confined to such areas. Figure 17.2 shows the distribution of artificially introduced mammals on islands in the southern ocean around Antarctica. Food animals such as rabbits and goats were often released by sailors who harvested them for food on subsequent visits. Only four out of twenty islands, all of them very small, are still free of exotic species. In many cases, there had been little scientific study of the native biota before the introductions, so it is impossible to quantify what their effects have been. Nonetheless, comparison between the vegetation in heavily grazed areas and that in inaccessible places such as cliffs, shows that herbivorous mammals have damaged the native flora. There are very few trees on these exposed subarctic islands and most of the birds, including many long-lived, slow-breeding seabirds, nest on the ground or on cliffs. Like Mauritius and New Zealand, most

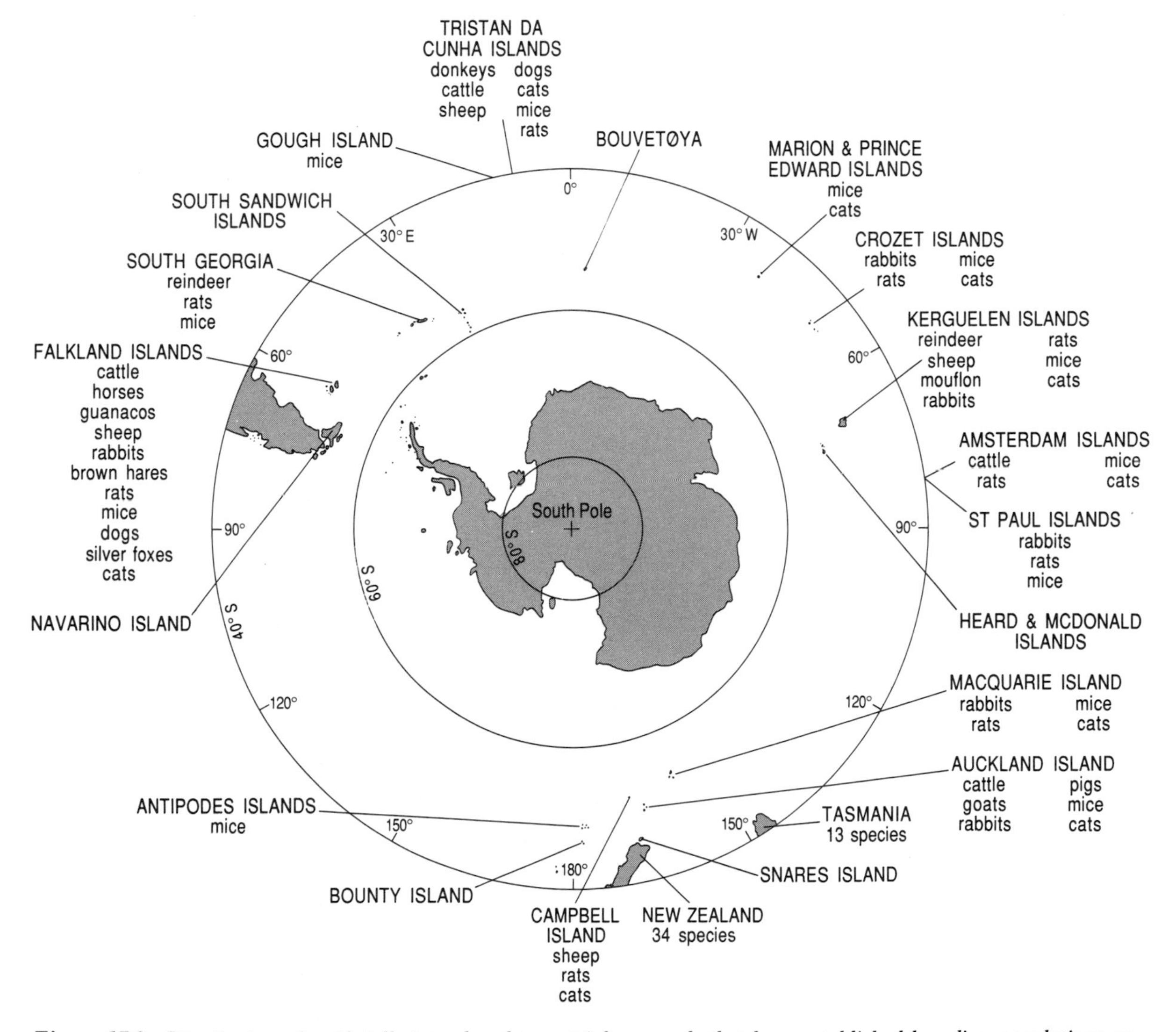

Figure 17.2 *Distribution of artificially introduced terrestrial mammals that have established breeding populations on islands in the southern ocean. Where no species are indicated, the islands are free from artificially introduced mammals.*

such islands were never colonized by large predators (or the tiny populations that such small areas could sustain quickly became extinct), so there was little selection against ground-nesting, which made the eggs and nestlings very vulnerable to predation from introduced cats and rats.

Once established, feral animals are very difficult to remove. One of the few successful operations was the elimination from East Anglian wetlands of the coypu, a large South American rodent that became feral in the 1920s after escaping from fur farms. Intensive trapping, aided by high mortality during hard winters in the 1960s, exterminated the species from Britain by the early 1980s.

Squirrels *Sciurus* is primarily a New World genus; of the 28 species, only one, *S. vulgaris*, the 'common' or red squirrel of Europe and northern Asia, occurs naturally outside this area. Small numbers of specimens of *S. carolinensis*, the grey squirrel of North America, were released in various parts of Britain and Ireland between 1876 and 1929. They spread rapidly and by the 1950s they were much more abundant than the native species over most of southern Europe and Britain. But although *S. vulgaris* is now very rare over much of its former range, red squirrels are still abundant, and greys rare or absent, in the Scottish highlands, the Lake District, the northern Pennines, northern and western Ireland and some forested areas of central and northern Europe. These facts suggest that red squirrels are better adapted than grey squirrels to boreal forests, similar to the vegetation that covered much of Europe at the end of the Quaternary Ice Ages, but that the American species seems to do best in more temperate areas. These squirrels illustrate the general principle that, unless the invading species is competitively superior (see Chapter 8) to the indigenous species in all habitats, it cannot by itself eliminate them entirely. The invader merely restricts the range of the native species.

Cane toads The introduction of exotic lower vertebrates and invertebrates, particularly insects, has had just as far-reaching effects as that of importing non-native mammals and birds, as the case of the cane toad illustrates clearly. In the second half of the 19th century, European settlers in Queensland, northern Australia, began cultivating sugar cane in the moist, tropical climate. Two native species of herbivorous beetles that originally fed on indigenous trees began damaging the introduced crops. By the early 1930s, losses due to the adult beetles eating the leaves, and the soil-dwelling grubs destroying the roots, were enormous. The extermination methods tried—fumigation, picking them off the cane—had proved to be very expensive and largely ineffective.

The American giant toad *Bufo marinus* (Plate 17.1) was suggested as a remedy. This large toad is common in warm lowland areas throughout its native Central and South America, where it feeds on large arthropods and carrion. It has a tough, warty skin from which foul-tasting toxins are secreted, and seems to have little fear of large animals, including people, and is often seen near dwellings at night, catching insects attracted to the lights. Like all amphibians, the egg and tadpole stages are aquatic, but development is very rapid compared to that of other toads, and *B. marinus* can breed in small, temporary pools and puddles. Females lay up to 20 000 eggs three times a year. The species had already been introduced into Jamaica, Puerto Rico and several smaller Caribbean islands, and from these areas into agricultural land on Hawaii, the Philippines, Fiji and some other Polynesian islands. In nearly all these places, it quickly became established and seemed to be curtailing certain insect pests. So in June 1935, biologists from the Bureau of Sugar Experiment Station caught 102 toads in Hawaii and shipped them to Queensland, where they were fed and tended in a specially prepared pond. One died, but the rest soon started to breed, and in the following months, thousands of tadpoles and toadlets were distributed to farmers in beer bottles and jam jars, to be released in the sugar cane fields. Without further assistance from humans, the toads proliferated and spread at the amazingly fast rate of more than $9\,\mathrm{km\,y^{-1}}$. Soon they were common on streets, in gardens and even inside houses over much of north-eastern Australia, where people started calling them 'cane toads'. The numbers of toads increased alarmingly, but those of the beetle pests were undiminished. It soon became clear that *B. marinus* was eating other insects, small vertebrates and almost any kind of carrion, but few, if any, of the beetles that damaged the sugar cane. Within six months of their arrival, harmful effects of the toads were reported: honey-bees were eaten and dogs became sick after biting them. Dingos, introduced and native cats, indigenous birds such as the ubiquitous kookaburra, large lizards such as the goanna (*Varanus* spp.), and many snakes, including several highly venomous species, attacked them but were repelled, and in many cases fatally poisoned, by the toxins in the toads' skin. Even crocodiles ate them only when very hungry.

By 1940, the failure of the introduction was clear, and the toads themselves were declared to be vermin. Millions have been clubbed with baseball bats, squashed by vehicles, shot or drowned: some are cooked and eaten, others are collected and used as dissection specimens in biology classes. Fifty years later, cane toads are still a major pest in warm, wet areas of Australia. The mammalian and avian predators seem to have learnt to avoid them, but several indigenous species of amphibians and reptiles have become rare wherever cane toads are common, probably from competition for food, as well as from direct encounters with them.

▶ Is it likely that genetic diversity has been important in the successful invasion of cane toads?

No. The entire population is descended from only 101 specimens, themselves taken from a population derived from a small number of founders. In this species, adaptability does not seem to depend upon high genetic diversity. Although lack of genetic diversity in small, inbred populations such as dodos, and humans on Tierra del Fuego, may have contributed to their inability to adapt to sudden changes in their ecological circumstances, the converse is not necessarily true: a surprising number of spectacularly successful invading populations are derived from very few specimens.

A few deliberate introductions, usually insects imported to control the spread of exotic plants, have been effective and have had few side-effects. For example, the introduction of the South American *Cactoblastis* moth into Australia in the 1920s curtailed the spread of prickly pear, a cactus that had been imported from the same continent 50 years earlier. But many other introductions have had wide-ranging and long-lasting effects on the indigenous ecosystem, permanently altering the course of evolution of many other species. For example, the introduction of the Nile perch to Lake Victoria and Lake Malawi in eastern Africa in 1960 has caused a drastic decline in the populations of many indigenous fish, including several commercially important species (see Chapter 9, Section 9.4.4).

Plants Hundreds of species of plants have been introduced as crops or ornamentals all over the world. Although many such imports are artificial varieties and cannot survive in the wild, some have become naturalized, sometimes in habitats that seem to be very different from those to which they are native. For example, *Senecio squalidus*, the 'Oxford' ragwort, is an annual herb native to Sicily and southern Italy that was introduced into Britain early in the 18th century and cultivated in the Oxford Botanic Gardens. It was first reported growing without human assistance in 1794 but for most of the 19th century, it was confined to the Oxford area. During the last hundred years, it has spread rapidly and is now a major weed in lowlands over much of the British Isles.

▶ What do these factors suggest about the time-course of evolutionary changes in this species?

Ragwort, like rabbits, seems to have adapted gradually to its new habitat over the scores of generations during which it was cultivated in Britain. However, the history of the species *appears* to have two major phases of rapid evolution: when the plant first escaped from cultivation, and when the feral population extended its range. As in the case of rabbits, we know almost nothing about what the new adaptations consisted of (e.g. better defences against herbivory, biochemical adaptation to the climate, etc.), or how they evolved (whether from mutation, hybridization, etc.). Buddleia (from China), rhododendrons (from southern Europe and the Himalayas), Michaelmas daisy (*Aster* sp. from USA), giant hogweed (*Heracleum* spp. from southwestern Asia) and many more were introduced to Europe as garden plants, but have established themselves and are spreading, often to the detriment of the native flora. Many herbivorous animals cannot eat the exotic plants, so the fauna is also adversely affected. Such importations and naturalizations are clearly as important to the long-term evolution of the invasive and native species as any natural colonization, but we know very little about the genetic and physiological mechanisms involved.

Weeds are often introduced as seeds with crops and livestock (as undigested material in their guts) and frequently become naturalized even faster than the species that were imported intentionally.

▶ Why would weeds be particularly likely to become naturalized?

Weeds are plants that evolved as rapid colonizers of newly disturbed habitats. Most weeds in Europe are native, having evolved into the 'weedy' habit as more land was ploughed and crop plants were cultivated, but the majority of weeds in North America are introduced, most arriving from Europe or Asia with human immigrants and their crops. Over 700 exotic species are now established there, growing with crops or in pastures. Some, such as the aptly named 'White Man's Foot' (*Plantago lanceolata*), and several species of thistles (*Carduus* spp.), cause serious economic losses as well as radically altering the native vegetation.

As in the case of animals, such exotic species often exclude native forms. Our deciduous woodlands originally consisted mainly of oak, ash, elm, willow, beech, birch, alder, hawthorn and hornbeam. Many other familiar trees such as the horse chestnut and common sycamore, all native to highlands of southeastern Europe, were introduced to Britain and northern Europe during the last thousand years, as were several kinds of fruit trees, including mulberries, apricots and some kinds of cherries. Although many species are still propagated mainly by human action, some, such as sycamore and horse chestnut, are becoming naturalized and their presence is changing the composition of 'wild' woodland. Similar introductions and naturalization have happened almost every place that people have colonized. Species native to continental areas of Europe and Asia are usually more successful in Australia and islands than the converse, but there are exceptions. For example, the tree of heaven (*Ailanthus altissima*) from China is now common over much of the USA, especially in cities, and has invaded some Mediterranean islands to the exclusion of native species such as oaks. Gorse (*Ulex europaeus*) has been widely introduced over the past 200 years and is spreading rapidly in New Zealand, Tasmania, Chile and other places with a climate similar to its native western Europe.

Introduced aquatic plants also drastically alter the habitat for the native flora and fauna, and for the people who use the water for agriculture, fishing and transport. For example, the fern *Salvinia molesta* and water hyacinth *Eichhornia crassipes* (both native to tropical South America) have been introduced (usually by accident) and have become naturalized in Africa, Australia and southern USA where they are spreading rapidly, choking lakes, rivers and artificial waterways such as reservoirs and canals. Canadian pondweed *Elodea canadensis* is having similar effects on ponds and lakes in northern Europe and Asia.

Diseases Diseases spread most efficiently when the host organisms are sedentary and live at high density. Humans are no exception, and infectious diseases such as cholera, typhoid, tuberculosis, measles and smallpox were, and in some parts of the world still are, rife in large cities, and in military and naval installations. Until the mid-19th century, the death rate in cities was notoriously high (the population of most cities was sustained only by continual immigration from the countryside) but most of those who survived to adulthood had some immunity to the common infectious diseases. Until the 17th century, most densely populated areas, and hence the most effective immunity of the population, were in Europe, India and China. Military conquest, economic exploitation, missionary activity and other forms of colonization spread diseases such as smallpox, syphilis, measles and typhoid to native peoples in North and South America, Australia, New Zealand and Polynesia, where their effects were sometimes devastating (see Section 17.2.3).

Exotic diseases have also caused epidemics among animals and plants. During the past 150 years, several outbreaks of rinderpest, a viral disease of cattle, caused extensive mortality to many species of antelope and other artiodactyls, including buffalo, warthog and giraffe in Africa. In some areas, kob, buffalo and giraffe disappeared completely and because these species breed quite slowly, it was many years before they recolonized their former range (see Chapter 9, Section 9.8). About 1900, the chestnut blight fungus was introduced to North America from Europe, where it is indigenous and generally harmless to its host, the European chestnut *Castanea sativa*. It spread rapidly and killed so many sweet chestnut trees (*Castanea dentata*) that this once abundant woodland species is now quite rare. Dutch elm disease (also a fungus) decimated elms in America and Britain in the 1960s and 1970s. Such epidemics probably happened from time to time in the past, but human activities have increased their frequency, and have made their effects more likely to lead to permanent ecological changes. Even in undisturbed habitats, it would take several hundred years for large, slow-growing trees such as elms and chestnuts to recover their numbers. Logging, urbanization, the spread of agriculture and the elimination of animals that disperse the seeds retard the re-establishment of such species.

Overview There have been so many cases of introduced species having established breeding populations that biologists have looked for common patterns in the characters of successful colonizers. Fast breeding is clearly an advantage in the short term, i.e. the period most likely to be recorded by scientists, but may be of little significance in the long run. Thus feral populations of camels and horses, both slow-breeding species, have become established in northern Australia, where they are damaging the native flora and competing with kangaroo populations. In general, animals and plants native to islands, and areas long isolated from the rest of the world such as Australia, seem to be less successful as colonists, and more adversely affected by invasions, than those native to large continents. But there are many exceptions to this generalization: dozens of species of Australian eucalyptus trees have been planted in many areas, including California, India and east Africa, and in most cases have flourished and become naturalized. A few Australian animals have also become established outside the continent: wallabies (escaped from zoos) have been breeding in the Peak District of Britain since the 1950s, and are a pest in parts of New Zealand. Budgerigars, the western world's most popular cage bird, are becoming feral (see Section 17.4.2) and the canary, a species of finch native to small islands in the eastern Atlantic, the Azores, Canaries and Madeira, has not only been very successful as a cage bird since the 16th century, but has been observed to breed in the wild in a few mainland areas. Taxa that have recently (on a geological time-scale) undergone rapid speciation, such as rodents and songbirds, seem to be more successful as colonizers than long established, slower evolving groups such as marsupials, reptiles and amphibians, but the data do not provide unqualified support for such theories. Cane toads proved to be effective colonizers, even though many (some biologists believe most) other species of amphibians are in decline all over the world. Most species of pigeon-like birds are doing well: collared doves are a pest on farmland and pigeons flourish in cities all over the temperate zone, but the dodo and the passenger pigeon are among the most spectacular examples of recent extinctions.

The course of evolution over almost the entire world has been permanently affected by artificial introductions of exotic organisms. The importation to Antarctica of all exotic species, whether as pets, food or draught animals, is now forbidden under the Antarctic Treaty of 1991. But unfortunately, no amount of legislation can completely prevent the accidental introduction of seeds, spores or small organisms with people and their food. One of the most alarming conclusions to arise from the study of the long-term consequences of artificial introductions is that we do not understand the processes involved sufficiently well to predict the outcomes, much less to control or reverse them. The response to human impact seems to be one of the very few topics in biology for which phylogeny is irrelevant: ancestry is a poor guide to whether a species proves to be adaptable to new environments.

17.3.4 Other activities: pollution

Pollution is a vast and complex topic and it is not feasible or appropriate to discuss here the details of sources of pollutants or the mechanisms by which they affect humans and other organisms. This Section is a brief overview of the biological perspective of pollution from human activities, and its impact on the course of evolution.

All organisms produce waste products, and many large animals damage their environment (see Section 17.3.2), but in the wild, there are nearly always other organisms which can utilize the materials and return them to the ecological cycle. For example, 150 kg of faeces per elephant per day might seem a lot, but normally it is quickly broken down and dispersed by bacteria, fungi, dung-beetles and other organisms adapted to live off this abundant source of food. Where such organisms are absent, the dung becomes a pollutant. For example, Australian dung-feeding organisms, adapted to living on the faeces of indigenous mammals, proved unable to adapt rapidly to breaking down cattle dung. Nonetheless, people bred large numbers of cattle there, and their dung accumulated, causing deterioration of the habitat, not least for the cattle themselves. The problem was only solved when species of beetle that utilize cattle dung were introduced from Africa.

Pollution from human activities is more harmful than that caused by other large animals in two main ways: metals and other minerals from ores are extracted and concentrated from their natural highly diluted state, often less than 0.1% of the ore, into an almost pure form, and people synthesize substances that are absent or at least very rare, in natural systems. Pure elements are sometimes as harmful to biological systems as chemically elaborate synthetic molecules. For example, chlorine, one of the 12 most abundant elements in the Universe, is normally found as an ion or as a compound, in which state it is usually harmless (e.g. salt, sodium chloride) but as an element, it is toxic to most forms of life. Arsenic, mercury and to a lesser extent, copper and lead are toxic to many plants and animals at high concentrations. Mining and refining ores containing these and most other minerals almost inevitably leads to unnaturally high concentrations of them, as well as creating extensive mechanical disturbance of the soil and landscape. All the ingredients of artificial fertilizers are, by definition, nutritious to plants in small quantities; but at high concentrations, and in the wrong places such as in rivers and shallow seas, some of the components are highly poisonous to many organisms.

Many toxic substances are synthesized naturally by biological metabolism (usually by bacteria or insects) and in geological and atmospheric processes. Cyanide, ammonia, hydrogen sulphide, sulphuric acid, ozone and many other substances normally occur at high concentrations only in a few localized parts of the biosphere (e.g. in plants that synthesize cyanogenic glycosides and around volcanoes). Human activities have created much larger quantities of such substances, and, equally important, have distributed them much more widely so that they are now concentrated in ecosystems in which they are naturally rare.

▶ Suggest some different ways in which such pollution would affect species composition.

Indigenous species would become rarer as they were killed or rendered less efficient by the toxins. Organisms that were formerly rare may became more abundant, and new species may become established, e.g. bacteria and other organisms that can utilize, or at least tolerate, the pollutants. The effect of altering the distribution and abundance of such naturally occurring substances is not to extinguish life entirely, but to change the direction of natural selection. Organisms can adapt to moderate amounts of certain forms of pollution: the spread of industrial melanism in peppered moths and other insects, and of heavy metal tolerance (see Chapter 4, Section 4.3.1) in certain plants, have provided some of the most convincing evidence for evolution by natural selection. The frequency of different genotypes in the affected species, and often the species composition of communities as well, are altered by the pollution. Such changes often lead to ecological instability, as well as eliminating many of the species that are economically and aesthetically valuable to the humans themselves: most people prefer eagles and condors to rats and houseflies, trees and crops to weeds, and salmon or polar bears to bacterial slime. In the long term, these effects alter the course of evolution of most, if not all, the species in the polluted area.

Humans also bring to the surface and release into the biosphere large quantities of materials including coal, oil, and water from underground deposits that are not normally accessible to living organisms. Crude oil occurs naturally in both terrestrial and marine ecosystems: it seeps from oleiferous rocks near the surface through faults. So, as you might expect, several kinds of bacteria, notably species of *Pseudomonas*, long ago evolved the capacity to break it down, just as other such organisms utilize the hydrogen sulphide and other gases emitted from volcanic vents.

▶ Why are oil spills so harmful?

Because the concentration of oil is far greater than that which normally occurs naturally, and spills do not necessarily happen in places where oil-utilizing bacteria occur. The bacteria cannot break down the oil fast enough to prevent it accumulating. Many of the harmful effects of oil are indirect: the floating oil impedes the exchange of gases between the atmosphere and the sea, reducing oxygen levels and inhibiting photosynthesis, and it alters the surface tension between water and fur or feathers, reducing their insulative capacity, and prompting birds and mammals to groom themselves, thereby ingesting some of the oil. Nonetheless, oil-utilizing micro-organisms are bred in captivity (see Section 17.4.4) and have been deployed successfully in cleaning up some oil spills, particularly those at sea.

Another uniquely human activity is the synthesis of entirely new substances, including artificial elements such as plutonium, and a great many artificial molecules, particularly organic compounds, e.g. chlorofluorocarbons (CFCs), polyethylene and many other organic polymers and artificial isotopes. Only a few such substances have proved to be toxic in the sense that they actively impede biochemical processes.

▶ Why would such artificial substances be even more dangerous than increased concentrations of natural ones?

Because no organisms would have evolved the capacity to break them down. Fortunately, many such substances, including 'nerve gas' and other chemical weapons, oxidize, hydrolyse or otherwise degrade to harmless breakdown products after a few hours or days of exposure to the atmosphere, and some artificial materials are sufficiently similar to natural ones for bacteria to break them down slowly. But others, notably PCBs (polychlorinated biphenyls) and some pesticides such as DDT, decompose very slowly, so even small quantities remain a hazard for a long time.

One of the most widely discussed forms of pollution is that caused by radioactivity. There are many natural radioactive isotopes, and low levels of radioactivity can be measured almost everywhere. Cosmic rays are similar to radioactivity and occur continuously all over the surface of the Earth. So all biological systems have, since their inception, lived with radioactivity. Most molecules are susceptible to damage from radiation but induced changes in the structure of long-lived, chemically stable molecules such as DNA are by far the most important effect of radiation on evolution. Almost all organisms are adapted to cope with such damage: they have enzymes that repair many of the breaks and substitutions in the genetic material, although, like most biological processes, repair is not 100 per cent efficient. Most natural and artificial isotopes decay in fractions of a second, and many others produce radiation that is only slightly harmful to living systems, except in large doses; only a few, such as ^{238}uranium and ^{239}plutonium, are long-lived and produce the more harmful forms of radiation. As with most chemically toxic substances, only prolonged exposure to unnaturally high concentrations of radioactivity poses an exceptional hazard to most organisms. In such cases, the DNA repair enzymes are 'swamped' with damaged genetic material and more of the alterations in it become permanent, i.e. there are more genetic mutations and breakage of chromosomes.

Many other synthetic substances and materials, e.g. plastics, polystyrene and polythenes, are a pollution problem not because they are toxic—they are too inert chemically to do much harm unless burnt to more volatile products—but, because no organisms break them down, they accumulate in rubbish dumps. Structures formed from such materials sometimes provide breeding sites for a variety of organisms such as rats, cane toads (Section 17.3.3) and mosquitoes.

▶ Would paper pose a similar problem?

No. Being composed mainly of wood pulp, micro-organisms (and insects such as termites) adapted to utilizing rotting wood break it down eventually, particularly if it is damp. Dyes, particularly those containing metals, and synthetic coatings applied to glossy magazines and photographs, may greatly delay the breakdown of paper and the recycling of the nutrients in it, because such artificial materials may be toxic to the micro-organisms that digest wood pulp.

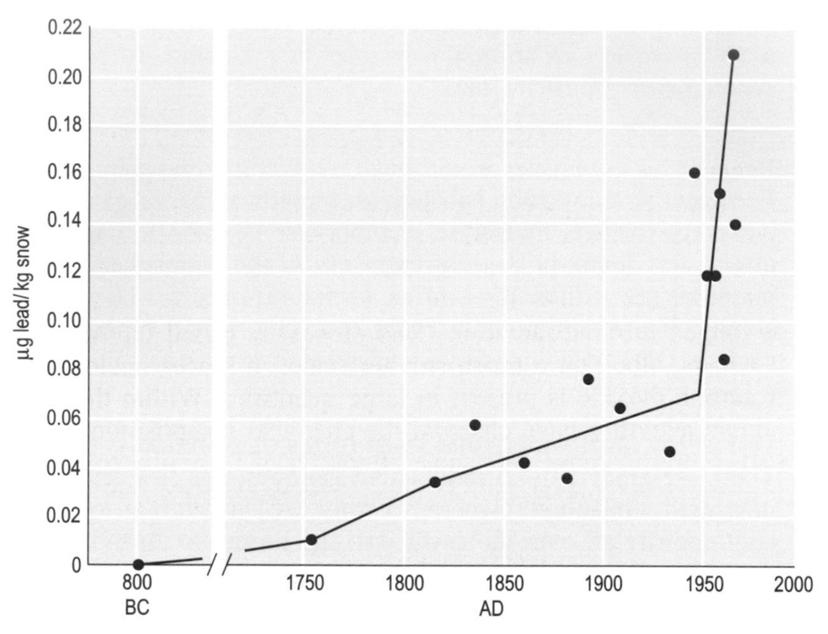

Figure 17.3 Lead concentration in the snow accumulated during the last 2800 years in an inland area of north-west Greenland.

Many pollutants spread widely, as well as accumulating near their source. In most environments, air and water currents as well as biological activity make it almost impossible to determine how long a substance has been present, and hence to calculate its rate of accumulation. Snow and ice trap small pockets of air and airborne particles, so very cold areas with a moderately high annual snowfall provide a unique record of the past composition of the atmosphere. The ice-caps of central Greenland and Antarctica have proved to be useful sources of data such as those in Figure 17.3. Lead levels began to increase about 1750, about the time that the industrial revolution in northern Europe got underway. There was no industry in or near Greenland at the time, so the lead must have originated elsewhere. The quantities of lead deposited in the arctic snow have increased even more rapidly during the past 40 years, almost certainly due to the widespread use of leaded petrol. Similar data suggest that minute particles of mercury, copper, cadmium and other heavy metals enter the atmosphere as a result of mining, metal working and the burning of coal and oil, and are carried over large areas by winds and currents before being deposited far from their site of origin. Pollution from such sources is particularly widespread in the Northern Hemisphere, where more than 90% of all industrial activity takes place.

The capacity to tolerate or even utilize a wide variety of chemicals has appeared in many different organisms, especially micro-organisms, and there is no reason to suppose that similar adaptations to artificial substances would not evolve eventually in many other species. The problem is mainly one of time-scale: during the past few hundred years, people have produced a great many 'new' substances, and abnormally high concentrations of natural ones; in most organisms, adaptations to deal with the new conditions do not evolve fast enough, particularly in slow-breeding species with low genetic variation.

17.3.5 THE 'GREENHOUSE' EFFECT

The surface of the Earth is about 33 °C warmer than would be expected from the solar radiation that falls on it and the heat generated by geothermal processes in its interior. The surface of the planet Venus is even hotter, about 450 °C warmer than calculated from its distance from the Sun. Certain gases in the atmosphere of these (and other) planets, notably water vapour, carbon dioxide, methane and the oxides of nitrogen, are almost transparent to the shorter wavelengths of incoming solar radiation, but absorb much of its energy when it is reflected back and re-emitted as longer, infrared radiation. This process is called the 'greenhouse' effect because it occurs, albeit on a much smaller scale, in spaces enclosed by glass, particularly if carbon dioxide is present in large quantities. Within the last few hundred years, human activities have changed the chemical composition of the atmosphere sufficiently to cause detectable alteration of its ability to retain heat from the Sun, thereby promoting global warming. The phenomenon might therefore be described as habitat modification on a global scale arising from chemical pollution. The two main processes involved, the change in the chemical composition of the atmosphere and the resulting increase in the surface temperature of the Earth, directly affect all living systems. Both its causes and its biological and geological consequences are very complex and highly controversial, so this Section is a brief summary of the main facts and the principal mechanisms.

Table 17.5 *Natural and artificial sources and sinks (i.e. natural means of removal) of the principal gases (other than water vapour) that cause the 'greenhouse' effect.*

Gas	Natural sources	Natural abundance (ppm)	Natural sinks	Artificial sources	Increase per year (%)
Carbon dioxide	Respiration, decomposition	280	Photosynthesis, biomineralization	Burning fossil fuel, deforestation	0.4
Methane	Anaerobic decomposition and digestion	0.65	Chemical breakdown	Use of natural gas, rice production, cattle and sheep	0.9
Nitrous oxide	Bacterial metabolism	0.28	Chemical breakdown	Use of fertilizers, burning fossil fuel	0.25
Chlorofluoro-carbons (CFCs)	None	0	Chemical breakdown	Refrigerants, aerosols, foam	4

Some features of the principal atmospheric gases involved in the greenhouse effect are listed in Table 17.5. CFCs are completely artificial but the other gases are natural components of the atmosphere. Changes in their concentration arise mainly from activities such as agriculture, animal husbandry and burning, rather than from the presence of the people themselves. Methane ('marsh gas') is both produced and utilized by certain micro-organisms, but the anaerobic conditions in stagnant rice paddies and inside the rumen (a greatly enlarged part of the stomach) of cattle promote its production in much greater quantities than can be broken down naturally. As an essential substrate for photosynthesis, carbon dioxide is one of the longest established and biologically most important gases. Its concentration naturally undergoes annual fluctuations, being higher in winter when fewer plants are photosynthesizing, but during the past 200 years, its average concentration has been rising because fossil fuels (coal and oil), which were produced by photosynthesis long ago, are being burnt. Fossil fuel is now burnt at 30 times the rate of a century ago. At the same time, trees and other green plants that would reabsorb the carbon dioxide by photosynthesis are being destroyed in increasing numbers (see Section 17.3.2).

▶ Can this increase be due only to the growth in the human population?

No. It is ten times faster than the rate of population increase (see Section 17.2.3). Each person is burning much more carbonaceous fuel than ever before: the worldwide average is about 2 kW per person, but in industrialized countries such as the USA, it is five times as much.

The contributions of the different gases to the greenhouse effect are not equal: one molecule of CFC is 14 000 times as effective as carbon dioxide in retaining solar heat in the atmosphere, nitrous oxide 200 times and methane 30 times. Although only very small quantities of CFCs (about 0.4 parts per billion) are present, they account for about 17% of the extra energy retained, and thus make a significant impact on global warming. As with other synthetic substances, CFCs are not utilized by any living organism. They are also chemically stable, and degrade only slowly in the atmosphere.

Past climates can be inferred from fossil fauna and flora, particularly the distribution and abundance of marine algae and protistans such as foraminifera, from changes in sea-level and from the presence of glacial sediments and structures. Such information was used to compile Figure 17.4.

▶ Are higher world temperatures unnatural or unprecedented?

No. For most of the history of life, the Earth has been warmer than it is now. The difference is that the rate of change of temperature seems to have been faster in the Quaternary than in most previous ages. During the past one million years, calculations from changes in sea-level and the distribution of fossil pollen indicate that the maximum rate of change was about 0.3 °C per century. During this period, many species became extinct and others evolved a variety of adaptations to the cooler climate (see Chapter 2, Section 2.4 and Chapter 13, Section 13.5). The course of evolution during the Quaternary Ice Age suggests that the evolution of most organisms only just 'kept up' with the relatively rapid changes in climate and sea-level. During the past 200 years, the rate of change in climate has accelerated further. Estimates of the current rate of temperature change vary, but some authorities believe that it could be as high as ten times the maximum rate during

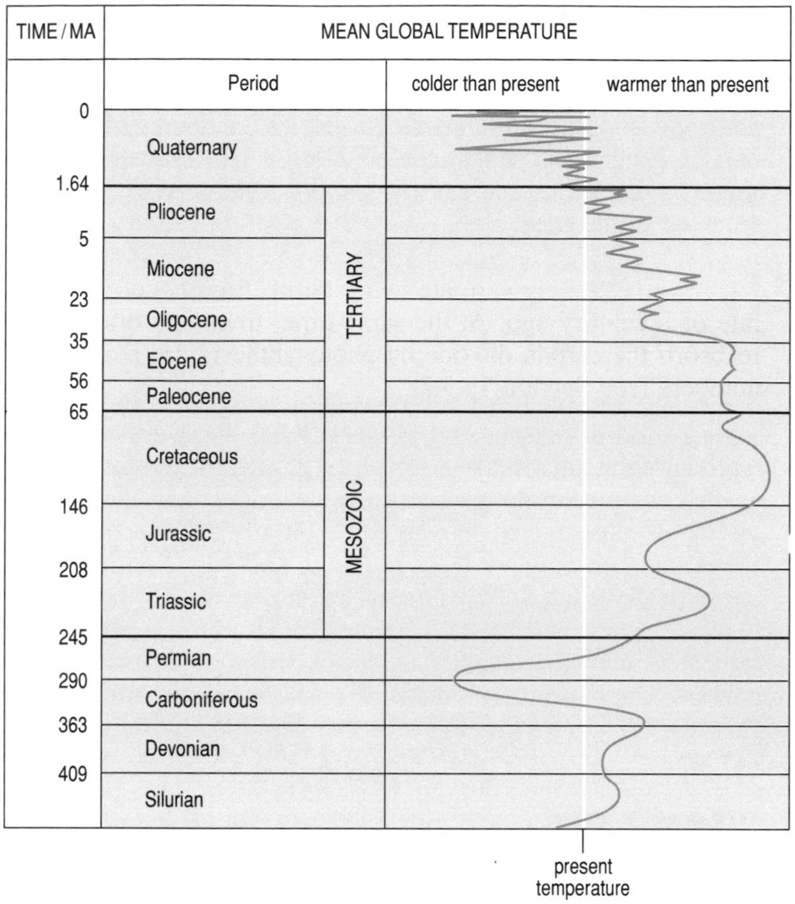

TIME / MA	MEAN GLOBAL TEMPERATURE		
	Period		colder than present / warmer than present
0	Quaternary		
1.64	Pliocene	TERTIARY	
5	Miocene	TERTIARY	
23	Oligocene	TERTIARY	
35	Eocene	TERTIARY	
56	Paleocene	TERTIARY	
65	Cretaceous	MESOZOIC	
146	Jurassic	MESOZOIC	
208	Triassic	MESOZOIC	
245	Permian		
290	Carboniferous		
363	Devonian		
409	Silurian		

present temperature

Figure 17.4 *Relative global temperatures during the past 440Ma.*

the Quaternary Ice Age. The mean sea-level is now rising at about 1–2 mm y^{-1}, partly from melting of polar ice-caps, but mainly from thermal expansion of the oceans. Some biologists believe that many organisms, particularly slow-breeding species such as trees and large mammals, would not evolve fast enough to adapt to the rapid rise in temperature and all the changes in flora and fauna that it causes. The ranges of such species would be curtailed, and many would become extinct.

SUMMARY OF SECTION 17.3

Many large animals modify their habitat, but the impact of humans is more widespread, penetrating and long-lasting, mainly because of technological capabilities such as the control of fire, tool manufacture, husbandry of other species and the ability to synthesize and concentrate new substances. Humans eat a wide variety of food and, like many other social animals, can kill prey much larger than themselves. Although tool use also increased the efficiency of hunting, killing for food has only rarely led to extermination of the prey species, usually in confined areas such as islands and lakes. However, people often hunt animals and collect plants for materials long after harvesting them for food has become inefficient. Accidental or intentional introduction of exotic predators, herbivores and plants has often been more damaging than the activities of the people themselves. Many organisms evolve adaptations that enable them to tolerate, or in a few cases destroy, pollutants created or concentrated by people. Human activities are changing the chemical composition of the atmosphere, thereby altering its heat retention properties and causing a more rapid rise in global temperatures than has occurred in the past.

17.4 Evolution in captivity

A few species, such as the fungus-cultivating ants, nurture other organisms and harvest food or other useful products from them. By keeping such organisms 'in captivity', the ants influence the course of their evolution, creating 'domesticated' species. Very recently, probably for no longer than the past 9 000 years, humans have kept other species in captivity, initially for food, then for clothing and recreation, and now for these purposes and for scientific enquiry. Domestication has radically altered the course of evolution of the species involved. Manipulation of the structure and habits of domesticated animals and plants was among the earliest of human technologies, and is still a sophisticated and rapidly developing science. The capacity to control the course of evolution of other species is undoubtedly one of the major factors that have promoted the dominance of humans.

Artificial breeding Both the genotype and phenotype of many captive species have been modified by artificial control of breeding. By far the longest established means of human control over the evolution of domesticated organisms is **artificial selection**. The basic mechanism of artificial selection is similar to that of natural selection (see Chapter 4), but people, not some combination of natural agencies, determine which individuals become parents and which of their offspring survive to breed. Humans select for the characters that increase the organisms' usefulness to them, so the process generates adaptations (see Chapter 2) to human needs and desires, and to the conditions of husbandry or cultivation, not to the evolutionary fitness of the organisms concerned. By promoting migration, people may also increase the fecundity of individuals with desired traits far beyond their natural limits: for example, prize bulls or rams may be put with one herd of cows or ewes, then moved to other herds a few days later. With only a few individuals chosen as parents, and their offspring nurtured, even if they would not normally survive in the wild, evolution by artificial selection may be unnaturally rapid. Darwin, and many biologists since, exploited this property, studying changes under artificial selection as a speeded-up model for evolution by natural selection. Artificial selection often, though not invariably, reduces the organisms' viability in the wild and the adaptations that it creates are often not as well integrated with other aspects of its structure and habits as those that evolved under natural selection (Chapter 2, Section 2.2.3).

As in natural evolution, mutation is an important source of variation upon which artificial selection can operate. Breeders of domestic animals and plants have also tried to increase variation by artificial hybridization.

► Why would such artificial hybridization often be difficult to achieve?

Sometimes, hybridization can be achieved by simply putting members of naturally separated subspecies or races in close proximity, but many plants and animals have behavioural, mechanical or physiological mechanisms that prevent hybridization between species or subspecies (see Chapter 9, Section 9.4.3). Artificial plant hybrids were (and still are) often created by manual transfer of pollen, but for animals, particularly those with internal fertilization, such hybridization is more difficult. Excluding normal mates and manipulating natural mechanisms of sexual recognition increase the chances of unnatural hybridization.

Mules have been bred in western Asia since at least the 7th century BC; they are larger and stronger than donkeys, and hardier and more sure-footed than horses. The Romans used mules for a wide range of civilian and military purposes, and refined the technology for producing them. A selected male donkey is removed from its natural mother as soon as it is born and put with a lactating mare. If raised among horses, a sexually mature jackass responds to mares coming into oestrus and may attempt to mount them. Mares often reject such advances from a male of another species (see Chapter 9, Section 9.4.3), so to avoid injury to valued sires, dispensable teaser donkeys are sometimes used to test the receptivity of the larger and stronger mares. Almost all mules (and hinnies, the offspring of a stallion and a female donkey) are sterile, so unlike many hybrid plants, breeding mules cannot affect the course of evolution of the parental species. All the living species of *Equus* seem to be able to interbreed: hybrids between zebras and donkeys or horses are sometimes produced accidentally in zoos, but nearly all such offspring are sterile.

Table 17.6 *The diploid chromosome complement of the living species of* Equus.

Species	No. of chromosomes
Equus caballus (domestic horse)	64
E. caballus przewalskii (Przewalski's wild horse)	66
E. asinus (domestic donkey)	62
E. africanus (African wild ass)	62
E. hemionus (onager)	56
E. burchelli (common zebra)	46

► From the data in Table 17.6, can you suggest why hybrid equids are infertile?

The species have different numbers of chromosomes. Hybrid cells cannot divide satisfactorily by meiosis and so very rarely produce viable gametes (see Chapter 3).

Until very recently, artificial selection and hybridization were the only means of modifying the genetic composition of the population, but within the past 30 years, advances in cell biology and genetics have permitted the development of even faster methods of altering the genes of captive populations. The technology for artificial insemination was developed in the 1940s and is now used routinely for cattle and several other domesticated species of birds and mammals: a few specially selected males are raised to maturity and sperm is collected from them artificially. Each ejaculate is subdivided into scores of small samples, which are separately packaged, stored and transported frozen, to be inserted into the females as required.

▶ Which important evolutionary mechanism is eliminated by this procedure?

Sexual selection (Chapter 7); the cows never meet the fathers of their calves, and bulls never test their strength against each other. Artificial insemination also enables sperm from the same male to fertilize far more females than would be possible by any natural means: one bull might father up to 80 000 calves during ten years as a favoured sire. In the wild, a male bovid would remain dominant over a harem comprising a maximum of 100 females, for at the most five years.

▶ What effect would such practices have on the genetic diversity of the population?

Genetic diversity would be greatly reduced. In many modern dairy herds, most of the cows are half-sisters, and often their mothers were related as well. Within the past 20 years, technology for artificially increasing the fecundity of females has been developed. Mature ova can be removed surgically from cows (or even from freshly dead carcasses) and fertilized *in vitro* before being reimplanted into the same or another female. As well as sperm, ova and, in some species, early embryos can be frozen to very low temperatures and stored for weeks, even years. Although viability is always impaired, a few such embryos develop normally after being thawed and implanted into a female.

In some cases, ova can be cloned in the early stages of development, in a process similar to the natural mechanism of formation of monozygotic twins (see Section 17.2.1). By reimplanting fertilized ova into surrogate mothers, several dozen offspring of a valued cow can be bred simultaneously. An extension of similar technology is **gene transfer**. Small pieces of DNA, taken from the same or another species and replicated in micro-organisms, are injected into an excised zygote or early embryo before it is reimplanted to continue its development. In the case of plants, foreign genes are injected into single cells or clusters of embryonic cells maintained in tissue culture and entire plants are raised from them. Genes for such traits as disease or pest resistance that were identified and selected in one species are inserted directly into the genome of other species.

▶ How does this process differ from natural mechanisms of causing genetic change in populations?

It is radically different. Selection is not involved at all, and instead of being mutant forms of existing genes, the artificially inserted genes may be completely new to the species. In fact, similar processes occur naturally, at least in bacteria and other prokaryotes: certain viruses incorporate bits of DNA from one host cell and replicate it along with its own genetic material. When the virus particles released from the first host infect another cell, the foreign DNA is incorporated into the genome of the new host. Such virus-mediated transfer of genes, called transduction, is probably widespread among prokaryotes, where it may be an important source of genetic variation (see Chapters 3 and 16). In multicellular eukaryotic organisms, the process has been best described in somatic cells, where the 'new' gene sometimes brings about striking phenotypic changes. But nothing is known of how widespread the process is in gametes, and what contribution, if any, the mechanism has made to natural evolution in higher organisms.

A major problem with genetic engineering of multicellular eukaryote organisms is ensuring that the foreign DNA is successfully incorporated into a chromosome. Proper alignment of chromosomes is essential for successful meiosis, without which eukaryote organisms would be infertile (see Chapter 3, Section 3.3); isolated fragments of chromosomes also often get lost in mitosis. So-called 'jumping genes' that move from one chromosome to another, and perhaps also between cells, were described in maize in the 1940s and more recently in *Drosophila*. They may prove as useful as viruses in acting as vectors for the artificial modification of eukaryote genomes.

Research into gene transfer and its commercial development are moving very fast: by 1992, transgenic strains of goats, sheep, cattle, pigs, poultry, catfish and numerous plants, including food crops such as tomatoes, beans and maize, and ornamental species are being cultivated on a commercial scale.

17.4.1 Livestock and crops

Animal husbandry and agriculture have had far-reaching and permanent effects on the course of human evolution as well as that of the species that were domesticated. Farming greatly increased the food supply, thereby enabling people to live at much higher densities, leading to the appearance of cities, and to a rapid rise in population: there was a 100-fold increase in human numbers during the first 8000 years of agriculture (see Section 17.2.3). One family could produce enough food for several others, so more people were emancipated from the demands of food acquisition and were able to devote themselves to arts and crafts, religion, trade, politics and, more recently, science. These activities, and agriculture itself, greatly increased the demand for materials such as stone, wood and later metals and coal, leading to more mining and smelting, and for animal products such as horn, ivory, pearl (from oysters) and coral and thereby affected the course of evolution of the species involved. Animal husbandry and agriculture began more or less simultaneously and separately in different parts of the world, and although the circumstances and the species involved differ widely, it is possible to see some consistent patterns in the kinds of organisms selected for domestication and the consequences for their evolution and that of their wild relations.

Sheep and goats The first evidence for animal husbandry in Eurasia is from southwestern Asia about 9000 years ago. This area was probably cooler and wetter then than it is now and was rich in wild ungulates: there were fallow deer, boar, camels, asses and several species of Bovidae (order Artiodactyla), including the

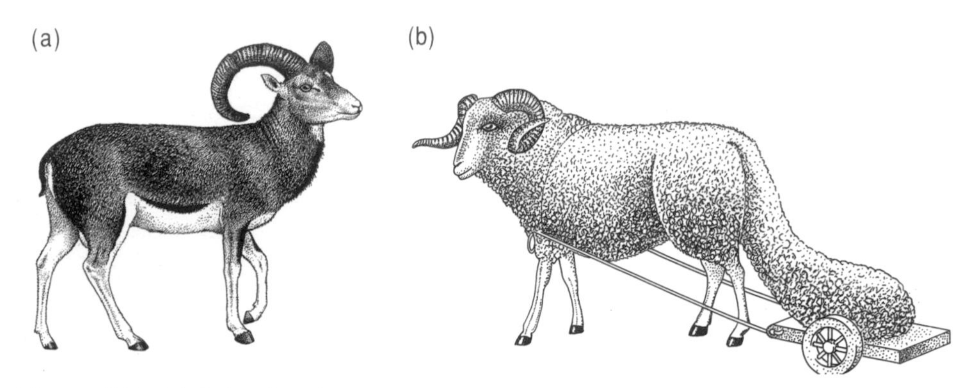

Figure 17.5 *(a)* Ovis orientalis, *the Asiatic mouflon; (b) a fat-tailed sheep fitted with a trailer to carry its heavy tail.*

Arabian oryx, oxen, goats and sheep. People probably hunted all these animals from time to time, but the first to be domesticated were the two smallest species of bovids, sheep and goats. Most (some authorities believe all) modern domestic sheep are derived from the Asiatic mouflon *Ovis orientalis* (Figure 17.5(a)) and goats from *Capra aegragus,* both native to the highlands of eastern Turkey and northern Iraq and Iran.

▶ Why would it be easier and more efficient to domesticate small-bodied species?

Small animals are physically easier to handle and they run more slowly. They can also be harvested one at a time, each animal providing food for a small group of people, thereby minimizing wastage. As well as their small size, their unusual social structure made these species particularly suitable for domestication: while many ungulates, including most deer and antelopes, live in herds and, particularly during the breeding season, the males defend territories, sheep and goats are nomadic, although they may remain within a defined home range, and herds are led by a single dominant male. Shepherds exploit this behaviour by establishing themselves, or sometimes a tamed, tethered 'lead' animal, as the flock leader, and the rest 'follow like sheep'. Hundreds of animals can easily be guarded and shepherded to new pastures by a few people. Although rams sometimes fight when sexually receptive females are present, flocks of sheep may be merged or separated, or led into pastures recently grazed by other flocks, without suffering any distress. Another advantage was their adaptable and complementary feeding habits: sheep graze grasses and herbaceous plants, but goats are mainly browsers, eating from bushes and the lower branches of trees. They thrive on tough and toxic plants that most deer and antelopes would not be able to digest efficiently, and they tolerate arid conditions.

Convenient though these little bovids were for Neolithic shepherds, they set about 'improving' the sheep by selective breeding. Limb bones found near ancient human settlements were shorter, and more of the skulls (probably those of ewes) were hornless, than those of the same species found elsewhere, indicating that domestication was producing distinct 'breeds'. A **breed** is a distinct, self-perpetuating strain of organisms produced artificially, whether by selection, hybridization or gene manipulation. The term **variety** is used for the same concept in plants, and many such forms are given Latinized names, similar to those of natural subspecies, but breeds are normally known by common names. Most of the distinctive features of breeds and varieties (Figure 17.5(b) and Plate 17.2) are those desired by people and are not necessarily adaptive (see Chapter 2). As in the case of natural subspecies (Chapter 9), most artificial breeds hybridize successfully with each other and with the ancestral wild forms. Sheep are no exception, so according to the biological definition of species, they should be classified as *Ovis orientalis*. However, by convention, all domestic sheep are classified as a separate species, *Ovis aries*.

The coat of most mammals, including wild sheep and goats, consists of long, coarse guard hairs over short, fine underfur that grows mainly in the winter and is shed each spring. This feature is retained in primitive breeds such as the Soay sheep (Plate 17.2(a)), which have long been feral on the island of St. Kilda (Outer Hebrides) and are probably descended from stock imported to Britain by Neolithic settlers about 3 000 years ago. Most modern sheep retain their fleece until it is sheared, and in wool-producing breeds, the underfur is long as well as dense, making the animal appear very massive (Plate 17.2(b)). A statuette found in Iran shows a woolly sheep, suggesting that selection for thicker, longer, unsheddable underfur was practised 8 000 years ago, and that the fleece was already being used for spinning and weaving.

Differences in texture and colour (Plate 17.2(c) and (e)) of the hair, comparable to those that have evolved in different human races (Section 17.2.3), have appeared under artificial selection in sheep and other livestock. Wild sheep (Figure 17.5(a)) and primitive breeds (Plate 17.2(a)) are brown, black or beige, but, with the exception of the occasional 'black sheep', most of the fleece of modern breeds is nearly white, which makes the wool much more suitable for dyeing, although some breeds have dark faces (Plate 17.2(d) and (e)) or legs, like the ancestral forms. Breeds also have different nutritional and husbandry requirements: the sheep kept by the Bedouin Arabs are small and their fleece is coarse but they flourish on the sparse grazing and in the harsh climate of southern Jordan (Plate 17.2(e)).

As well as wool, sheep yield meat, milk, horn and bone, the latter being useful raw materials for making tools and ornaments until metals became widely available. Fatty meat was valued for its flavour and nutrient content, and tallow, made by rendering adipose tissue, was essential for cooking and for lamps. Sheep (and other livestock, notably pigs) were selected for greater abundance of fat, and, as in the case of certain human races (see Section 17.2.3), particular distributions of adipose tissue. The fat-tailed sheep (Figure 17.5(b)) was well established as a breed by 500 BC: in some cases, the tails become so heavy and susceptible to mechanical damage that shepherds fit individuals with small trailers to carry the tails.

Most wild bovids normally have singletons, although in the smaller species, twins are fairly common and triplets have been recorded very occasionally. Intensive selection for higher fecundity has increased the frequency of twins in both sheep and goats, so that in some modern breeds, twins are the norm. The Finnsheep, an all-purpose breed developed in Scandinavia, averages 2.6 lambs per ewe per year, and sometimes produces quadruplets or quintuplets. Although fertility is modulated by the quantity and quality of food, inheritance is the major determinant, and the breed is often crossed with others to increase the fecundity of the progeny.

▶ Why does the frequency of multiple births increase under artificial selection but not under natural selection?

Mothers of twins may be at greater risk from predators while pregnant, and in animals that feed the young, larger litters may suffer higher mortality (see Chapter 2, Section 2.2.1). But in captivity, animals are protected from predators, they are not required to run fast and they often have a more nutritious diet. Unless there is selection against higher fecundity, the most fertile animals have the largest number of descendants, i.e. they are favoured by selection. However, the anomalous sex ratio (see Section 17.2.1) suggests that, as in humans, the physiological mechanisms involved in twinning are not as fully integrated in these bovids as in mammals such as rodents and carnivores in which multiple births are long established. In spite of similar intense selection for high fecundity, twins are still rare in cattle, and such pregnancies are even more prone to complications (see Section 17.2.1). The family Suidae (pigs) are the only artiodactyls to have multiple births routinely, with wild boar normally having litters of four to eight, and warthogs two or three. Their higher fecundity, combined with relatively small size and omnivorous diet, may have been important reasons for domesticating pigs. All modern breeds average at least 10 piglets per litter (and several litters per year), and more than 20 is not rare.

The value to humans of features such as high fecundity, pale-coloured wool and more fat are obvious, but rapid evolution under artificial selection has also perpetuated some non-adaptive characters: some Jacob's sheep from western Asia and certain Hebridean breeds from western Scotland have up to six horns (usually four). There has also been evolutionary divergence in head shape: the oval head of the black-faced sheep (Plate 17.2(d)) contrasts with the slender snout of the curly-fleeced breed (Plate 17.2(c)). Such features were probably not the main focus of selection, either artificial or natural, but appeared in association with desired traits such as the quality of wool, milk and meat, and were not actively selected against.

People have taken sheep to all continents except Antarctica and to many islands (see Figure 17.2), and they have become feral in numerous places. For example, the Soay sheep (Plate 17.2(a)) have probably been feral on St. Kilda, a small, remote island in the north-eastern Atlantic, for several thousand years, and they breed with minimal management on mountains and moorland. Their presence permanently alters the flora (see Section 17.3.2).

Dogs In the western world, dogs are kept mainly as pets, but almost everywhere they contribute to the management of other livestock. Eskimos use them as draught animals and in some places, notably China, parts of Polynesia and in pre-Columbian Central America, they are also raised for meat. Dogs were probably the first animals to be domesticated, possibly in several areas at more or less the same time. Bones of African hunting dogs *Lycaon pictus* and in America of coyotes *Canis latrans* are occasionally found in association with human remains, but most authorities believe that the wolf is the main progenitor of modern dogs.

It is not difficult to imagine how wolves and humans came to be associated: they may have taken carrion from each others' kills. However, several other species, notably vultures and hyenas, have similar habits, but were not domesticated. The scavenging habits and omnivorous diet of wolves contribute to their suitability as domestic livestock, but as in the case of sheep, behavioural as well as anatomical and physiological features were important factors predisposing the species to domestication. Except for a few elderly males (the proverbial 'lone wolf'), wolves normally live in packs of a dozen or more animals of both sexes and all ages. The hierarchical social structure of the pack is coordinated by elaborate visual and vocal communication: tail wagging, snarling, rolling over and many other sounds and actions indicate mood, intention and social status which can be understood by humans as well as by other wolves. The dominant male wolf leads the pack, has priority access to food, and fathers most of the cubs. Continual social interaction, in the form of snarling and mock fighting, maintains his position in relation to other members of the pack, which form a dominance hierarchy, with juveniles having the lowest rank. Dogs readily accept a person as higher in the dominance hierarchy than themselves, particularly if that status is established before the animals reach sexual maturity. Such dogs assume lifelong subordinate status, following the 'pack leader' and obeying his or her orders, often indicating their submission by rolling over or crouching. It is much more difficult to train adult animals, particularly males, to obey orders, especially if they have been allowed to become dominant over humans or other dogs. Some of the communication and social behaviour is innate but much of it is learnt from other members of the pack. This situation can be exploited to train dogs to respond to orders, guard property, herd livestock and communicate with blind or deaf people.

Dogs have a much more acute and discriminating sense of smell and better hearing than humans (though their eyesight is weaker), and can run faster, and often further. People have long exploited these capacities, training dogs to track and hunt prey, to herd sheep and cattle, and more recently to detect and locate outlawed drugs and explosives. Juvenile male wolves may leave the natal pack and join other packs, but once the dominance hierarchy is established, they do not tolerate the presence of rivals in their territory, and attack adult intruders. This behaviour is exploited by humans: a dog threatens strangers who enter its (i.e. its owner's) territory or seem to be molesting its pack (i.e. its owner's family). As expected of a social animal, they also alert other members of the pack to the presence of danger, intruders or hunting opportunities. The main vocalizations in wolves and wild dogs are whining, snarling, growling and howling, but they rarely bark. Barking seems to be a behaviour that has increased greatly under artificial selection, probably to improve the animals' efficiency as guard dogs. The character was probably established in prehistoric times, and may have been one of the features distinguishing domesticated dogs from wild wolves.

Domestic dogs are now given species status (*Canis familiaris*) although the larger breeds still sometimes interbreed with wild wolves. Since such offspring are viable and fertile, *C. familiaris* and *C. lupus* are not true species in the strict sense (see Chapter 9).

► Would dogs have hybridized with wolves after they were domesticated?

Yes. For the first 10 000 years of dog domestication, wolves and humans were sympatric over much of their range (see Section 17.3.1), and some interbreeding almost certainly took place. As well as protecting livestock, extermination of wolves also increased the genetic isolation of domesticated dogs.

The feral dingo in Australia is regarded as a sub-species, *C. f. dingo*. It is one of the most primitive of the living dog breeds, having a long, straight snout, prick ears and a weak, squeaky bark.

▶ Would dingos have hybridized with wolves after they became feral?

Almost certainly not. *Canis lupus* never occurred in Australia so the two forms would only meet under artificial circumstances, e.g. in zoos.

Anatomically, *C. familiaris* must be one of the most diverse species ever to have evolved. There are over 400 modern breeds of dogs (Figure 17.6), ranging in adult body mass from less than 1 kg (similar to a guinea-pig) to more than 75 kg (as big as a man). Selective breeding has produced a huge range of behavioural and anatomical modifications. Some traits, such as the short legs and long back of dachshunds, were selected to adapt the animals for special roles, in this case entering the dens of badgers and foxes. But the body form makes the breed particularly susceptible to back injuries. Many pug-nosed breeds (e.g. Pekinese) have difficulty in breathing, which would obviously be maladaptive for an animal living wild (Figure 17.6(a)). Breeds such as boxers (Figure 17.6(b)) and bulldogs in which the head is wide and round and the hindquarters are straight and slim often cannot give birth naturally (the enlarged foetal head cannot pass easily through the narrow maternal pelvis), and often have to be delivered surgically.

▶ Could such a combination of traits have evolved naturally?

No. Natural selection would eliminate traits that preclude giving birth easily and safely.

Many of the features introduced by artificial selection arise from the retention into adulthood of juvenile characters (see Chapter 14, Section 14.6). For example, all canids are born with floppy external ears (pinnae), but in wild forms they become erect at about the time that the juveniles become independent (Figure 17.1). In this posture, the pinna reflects and focuses sound onto the middle ear, thereby improving hearing, and is also important in visual communication: when frightened or submissive, adult canids lay their ears back, concealing the pricked form. Some modern breeds, particularly sociable, non-aggressive forms such as retrievers, spaniels and hounds (Figure 17.6(c)), retain the juvenile form of the pinna throughout life. The character is clearly not adaptive: floppy-eared breeds tend to harbour more ear mites than prick-eared dogs, and to get more ear infections. Some breeds have other characters such as large eyes, short noses and fluffy hair that produce exaggerated infantile appearance, and juvenile behaviour such as docility and playfulness. Breeds with such characters are sometimes inept as mothers, and require human assistance in raising their pups.

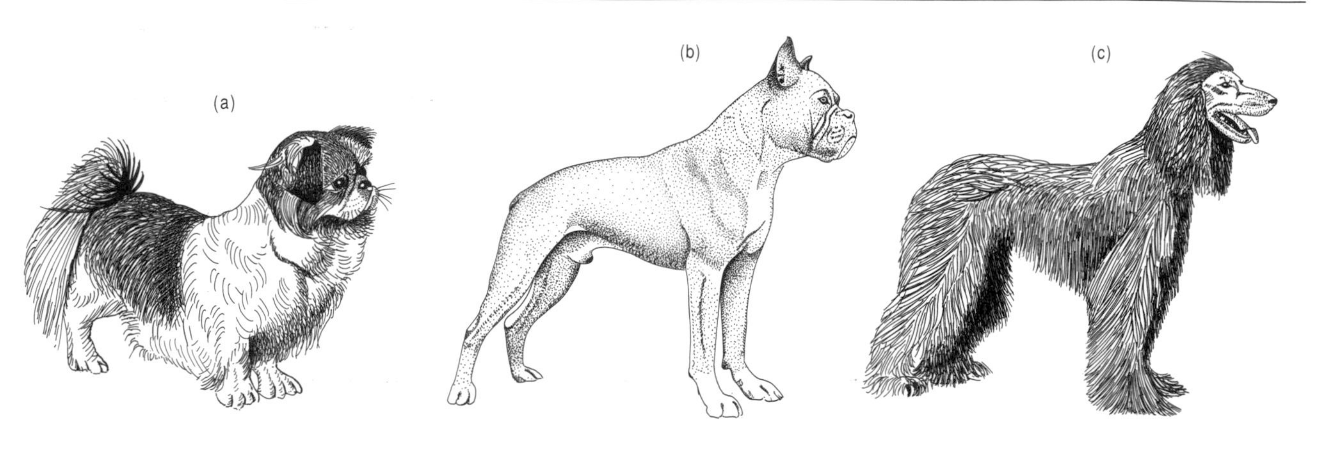

igure 17.6 Some artificial breeds of dogs (not to scale!). (a) Pekinese, which originated in the Chinese Imperial Court about D 1000. (b) Boxer, with a large, round head, narrow hips, prick ears and short hair. (c) Afghan hound, an ancient breed with a ıt head, long floppy ears, and long hair. (d) Relief carving from Nineveh (Iraq) about 645 BC, showing dogs resembling odern mastiffs.

▶ Are differences between breeds as firmly established as those between natural species?

No. All breeds can hybridize, and the resulting mongrels usually lack the more extreme characters, and are often hardier and longer lived. Although most modern breeds were developed during this millennium, some within the past 200 years, at least two kinds of canine skeletal remains can be identified in archaeological sites from 10 000 years ago and decorations in Ancient Egyptian tombs from about 1900 BC show several distinct breeds of dog. Bronze Age civilizations in southwestern Asia had dogs similar to modern mastiffs (Figure 17.6(d)).

▶ Comparing Figures 17.1 and 17.6(d), is there any evidence for selective breeding for juvenile characters?

The Assyrians' dogs, although large, and from their skeletal proportions, clearly adult, had floppy ears like wolf cubs, but adult wolves have prick ears. Although many breeds of dog are anatomically very different from the wild forms, the species still readily becomes feral (see Section 17.3.3). Packs of stray dogs are a serious nuisance in some cities, and there are feral populations in rural areas of Australia (i.e. the dingo) and on many oceanic islands, where they are major predators of the native vertebrates.

The effects of domestication of livestock on wild animals When practised on a small scale, agriculture can be beneficial to certain wild species. Figure 17.7 shows the abundance of bones of various mammals in two archaeological levels near Jericho in the valley of the River Jordan.

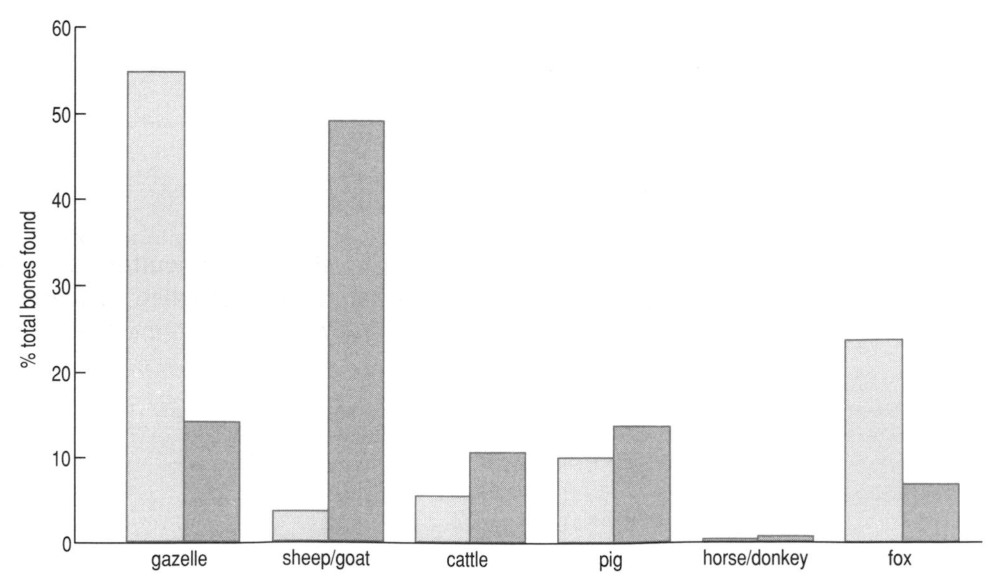

Figure 17.7 *The proportion of bones of various mammals in two archaeological levels near Jericho: (light green)* c. *8000 BC, before the development of animal husbandry and agriculture; (dark green)* c. *7000 BC, after the domestication of sheep and goats.*

▶ What effect has husbandry of sheep and goats had on the wildlife?

Far fewer wild antelopes were killed after these species were domesticated. As well as radically altering the human diet and economy, animal husbandry and its attendant changes in attitude towards wild species led to major changes in species composition. Predation pressure on many wild mammals, particularly herbivores, was reduced after people began to keep livestock and grow crops. The evidence is clearest in America, where many large mammals and birds became extinct around the time that humans first colonized the continents, but none is known to have disappeared between 8 000 and 300 years ago. Agriculture, possibly combined with a greater understanding of what we would now call ecology, supported fairly large numbers of Incas and American Indians without major depletion of wild populations of mammals, birds or fish. However, in most areas, predation pressure on wildlife was only briefly relaxed because agriculture led to rapid increases in human numbers, and to the active exclusion of wild animals from pastures and ploughed land. Competition between the wild and domesticated forms often led to the decline of the former: the aurochs, the wild ancestor of domestic cattle, had been rare since Roman times and the last ones were killed in Poland in 1627. Both the mouflon and wild goats are extinct over much of their former range, and now occur only in a few remote mountainous areas.

Following the development of agriculture, a few wild species became more abundant and extended their range, at least until the use of herbicides and insecticides became widespread. As mentioned in Section 17.3.2, plants and animals adapted to rapid colonization of disturbed habitats readily became weeds and pests in ploughed or heavily grazed land, sometimes undergoing evolutionary changes such as becoming polyploid or self-fertilizing. Some grain-eating animals such as harvest mice *Micromys minutus*, and birds such as corncrakes *Crex pratensis* that nest in long grass, also became more abundant following the spread of cereal cultivation. However, these species have declined drastically during the 19th and 20th centuries: modern strains of cereals grow faster, have shorter stems and mature more rapidly, and mechanical harvesting is quicker and leaves much shorter stubble than traditional methods, so there is less time and less suitable habitat for such animals to breed. The plants have evolved faster under artificial selection than the animals, leaving the latter poorly adapted to the rapidly changing habitat.

In many other ways, animal husbandry and agriculture have been harmful to wildlife: large areas of natural habitat are ploughed or fenced, trees are felled and wild predators that interfere with livestock are hunted as vermin (see Section 17.3.1). The natural vegetation of most of southern and central Europe, Asia and north Africa is forest; during the past 10 000 years, more than 95% has been cut down for building materials and firewood, and to produce pasture and agricultural land. Once felled, forests do not regenerate on heavily grazed land, even if there is a seed source nearby. Almost as soon as animal husbandry became of major economic importance, the unnaturally high densities of livestock often led to overgrazing, which, if prolonged and severe, causes desertification (see Section 17.3.3).

Mixed flocks of sheep and goats are particularly likely to cause desertification because the sheep graze the grass and small plants, often almost down to the roots, and the goats eat all but the most inaccessible shrubs and young trees. Once mature trees become so rare that pollination is inefficient and viable seeds scarce, forests regenerate only very slowly, often not at all. Overgrazed land supports fewer herbivores, both wild and domesticated, and numbers and diversity of all species, including humans, decreases drastically. Mosaics, paintings and other artefacts show that as recently as AD 500, bears, lions, ostriches and several species of deer were widespread in the mountainous areas of what is now Jordan, Israel and Lebanon. Now, the landscape is bare rock interspersed with small patches of herbaceous plants and bushes that support little more than sheep and goats (Plate 17.2(e)).

Non-sustainable agriculture and pastoralism were probably major determinants of the prosperity of many ancient settlements, as well as being an impetus to population migration and colonization of virgin lands (see Section 17.2.3). Overgrazing by domestic livestock is still a major cause of recurring famine (as well as soil erosion and dust-storms) around the Sahara and Kalahari deserts in Africa, and in southwestern Asia. So, while a few species such as weeds, certain insects and grain-feeding birds have increased as a result of farming and animal husbandry, many more, particularly large predators and forest-dwelling species, have become rare or extinct, and the landscape has been radically altered. In general, species diversity on heavily grazed and cultivated land is much less than that found in undisturbed habitats such as rainforests.

Interbreeding between wild and domestic forms, whether from deliberate human action or as a result of animals becoming feral, also affects the genetic composition of wild species. Some species, although long domesticated in the sense that they are bred in captivity and are widely and frequently used in the human economy, are phenotypically and presumably genotypically, almost unchanged from their wild ancestors. An example is the Indian buffalo *Bos bubalus* which closely resembles the local wild species and still regularly interbreeds with it.

Humans have also taken their flocks and herds with them wherever they went and such migrations also cause mixing of genotypes and the breaking down of differences between subspecies. For example, modern Scandinavian reindeer are the product of cross-breeding between several naturally separated subspecies (see Chapter 2, Section 2.4.3). Such accidental or intentional hybridization between subspecies, and between domesticated and wild forms, has made it almost impossible to sort out the phylogeny of some livestock, such as the guinea-pig in South America.

Most livestock were domesticated so long ago that the impact of artificial breeding and hybridization on the genetic stock of their wild relations and on the course of their evolution are now impossible to determine exactly. Commercial exploitation and 'improvement' of ostriches in the last two centuries provides a model of the likely course of events. *Struthio camelus* (see Chapter 2, Figure 2.4(b)) is the largest living species of bird, although the recently extinct giant moas (Section 17.3.1) and the Madagascan elephant birds (a relative of the African ostrich) were considerably larger. The demand in Europe and America for feathers for hats, fans and other fancy goods led to widespread and, in places intensive, hunting in the 18th and early 19th centuries. To ensure supplies in the face of increasing rarity, commercial ostrich farming began in South Africa in 1860, producing feathers, meat and leather. The captive stock was bred from a small number of birds obtained from the local wild population already much depleted by hunting, so at the end of the century, it was cross-bred with birds imported from north Africa (where it has recently become extinct in the wild). These artificial hybrids of the northern and southern subspecies have since become feral both in southern Africa and in Australia, thereby permanently altering the genetic composition of the natural population as well as introducing the species to areas where it is not native.

Animal husbandry has almost certainly saved some lineages from extinction. For example, the perissodactyl family Equidae (horses), although widespread and in places numerous for much of the Tertiary, is now almost extinct in the wild: of the eight Recent species, four have become extinct in historic time, and three others (Grevy's zebra, the ass-like onagar and the kiang) are now rare. Only the common zebra *Equus burchelli* is still abundant in its natural range. Horses *E. caballus* and donkeys *E. asinus* have declined greatly since the last glaciation and are probably now extinct in the wild, although a few Przewalski's wild horses have been recently returned from zoos to their native habitat (see Section 17.5). Nonetheless, there are millions of horses and donkeys in captivity and feral populations of the domesticated breeds have become established in several areas of Europe and America, where the wild forms once occurred, and in Australia.

Fewer than 20 species of animals have been domesticated for food or clothing, all of them mammals or birds except the silkworm moth *Bombyx mori* which has been cultivated in China by hybridization and selective breeding since before 1000 BC. For nearly all species, the processes of genetic isolation from the wild ancestors and evolutionary changes arising from artificial selection were essentially completed more than 2000 years ago (although in most cases, the rates of change under artificial selection have greatly accelerated during the past 250 years).

Several other animals, among them honey-bees, certain deer and various freshwater fish, have been semi-domesticated for a long time: they are fed and maintained, sometimes at unnaturally high densities, and they or their products (e.g. honey) are used for human purposes, but there has been no significant artificial selection. Nonetheless, the value of such species to the human economy affects people's attitudes to them. For example, elephants (see Section 17.3.1) have been used for transport, heavy lifting and in warfare for at least 2500 years. Like dogs, horses and other 'working' animals, elephants are social species and communicate with each other using a wide repertoire of elaborate visual signals and calls (Figure 17.8), so they can be tamed and trained more easily than most wild mammals. *Elephas* is still valued in India and south-east Asia and, until exterminated from north Africa, the African elephant was used in southwestern Asia and the Mediterranean area (elephants were deployed in battle both by Alexander the Great against the Persians in 331 BC and by Hannibal against the Romans in 218 BC). However, domestication of the elephant is still at the stage that the domestication of sheep was 10000 years ago: most specimens are born in the wild, and caught

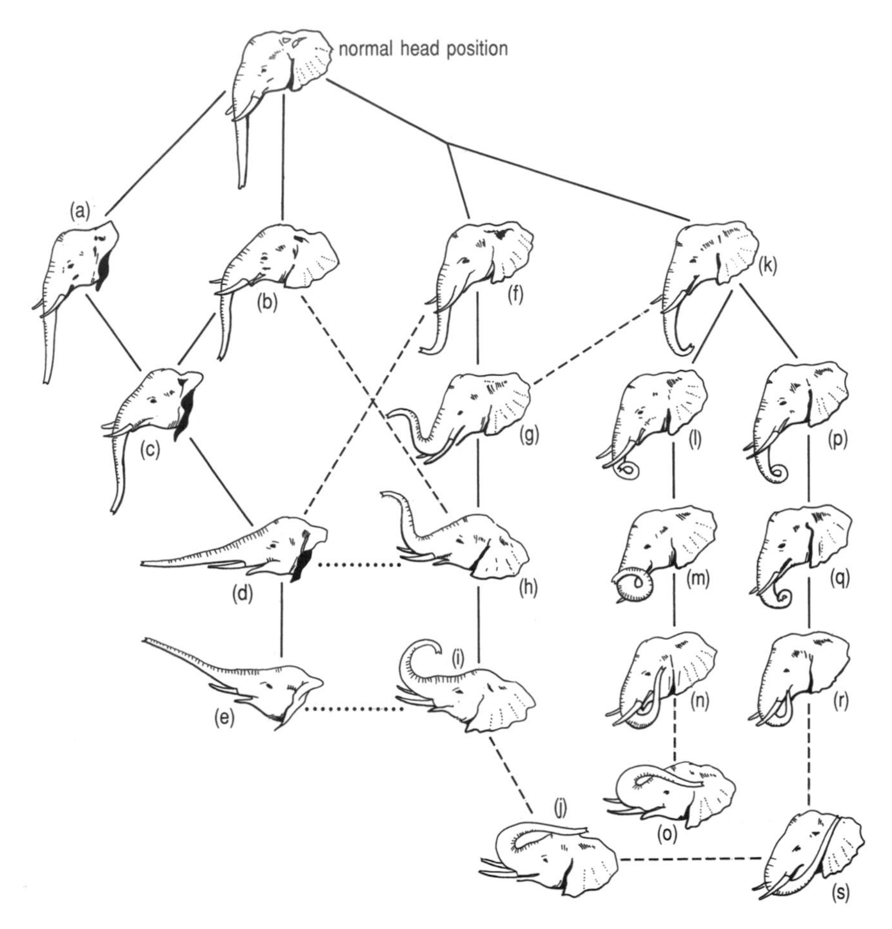

Figure 17.8 *Postures used in social communication in elephants. As aggressiveness increases, the head is raised (as in f–i), the trunk is directed forwards (as in a–e) and the ears are directed forwards (as in a). Thus (e) indicates a very high level of aggression, and (s) a very low level of aggression.*

and tamed as juveniles aged 5–15 years. Females are preferred: they are more readily trained, are capable of almost as much work as males and do not become intransigent in the mating season.

▶ How would this preference affect the wild population of the species?

Adversely. Removal of young females would reduce the net reproductive rate of the population, but elephant damage to plants (see Section 17.3.2) would be only slightly reduced, as the males would remain. The use of elephants as draught animals, as well as managing them for their valuable ivory, may forestall the extinction of the whole order Proboscidea (see Section 17.3.1).

Cultivated plants Selective breeding and cultivation of plants began at about the same time, and in many of the same areas, as the domestication of livestock, and often involved similar processes. Among the first plants to be structurally altered by human actions, and still the world's most important crops, are the grasses:

wheat, maize, rice, barley, rye, oats. Sugar cane and sorghum also belong to this plant family, as do bamboo and reeds, which are valuable as building materials. All wild grasses are wind-pollinated, and many are also dispersed by wind. Some grasses (bamboo is a major exception) are annuals, remaining dormant as seeds during cold or dry periods and maturing and setting seed during a few months of rapid growth. Wild forms of wheat are still found in western Asia and the presence of flint sickles and grinding stones near ancient stands of the plant suggest that it was harvested on a fairly large scale long before cultivation began. Cultivated wheat is derived from the hybridization of two species, einkorn wheat and the emmer which are sympatric for a small part of their range. The hybrid subsequently became polyploid (see Chapter 9, Section 9.6.3) and hundreds of varieties have been developed. The two principal modern strains are hard, or durum, wheat used for pasta, which is a tetraploid, and the hexaploid bread wheat.

▶ How would polyploidy affect the progress of domestication?

Polyploids are new species which cannot interbreed with the ancestral forms (see Chapter 9, Section 9.6.3). Domestication would be hastened by stopping such cross-breeding. We shall probably never know exactly whether hybridization was natural or caused by deliberate cross-pollination. Once established, cultivation proceeded rapidly and has produced evolutionary changes in the phenotype as well as the genotype. Wheat (and rice and many other grains) have lost the capacity to disperse naturally: instead of being shed, the greatly enlarged seeds remain attached to the stem (ear) and must be shaken (threshed) to separate them from the inedible chaff. This feature makes harvesting the grain much more efficient, and, instead of dispersing themselves, the plants grow only where they are deliberately sown. Many other changes, such as adaptations to different climates, resistance to fungal disease and shorter stems, have been developed under artificial selection.

Maize has been cultivated in Mexico for at least 8 000 years, originally as human food, but now mainly as fodder for domestic animals. The cultivated plant is so highly modified that its ancestry is obscure: several contrasting theories have been put forward, all involving hybridization and polyploidy. Wheat and many other cereals are naturally self-pollinating, so outbreeding is unusual, but in maize the male and female flowers are separate and do not mature simultaneously, so cross-pollination is the rule. A huge range of different varieties of the plant have been developed, producing products such as maize oil as well as grain.

Mutation, as well as outbreeding and polyploidy, has played an important role in the evolution of maize and many other cultivated plants. Many mutant strains are very ancient, but some are surprisingly modern. A species of Cruciferae, *Brassica oleracea*, has long been cultivated in Europe in the forms of kale, kohlrabi, cabbage, broccoli and cauliflower. Another form, the Brussels sprout, was unknown until 1750, when it appeared suddenly, apparently as a mutant, and is now one of the most widely cultivated brassicas. Several other closely related species of *Brassica* have been cultivated as turnip, swede and oil-seed rape.

Similar evolutionary mechanisms underly the domestication of scores of other plants. Many different parts of the plant have been modified. Cereals are seeds but many other foods, including pears, peaches, plums, berries, tomatoes, marrow, aubergine and pepper, are fruits. The fruit may have greatly enlarged under artificial selection, as in apples, peaches and avocado pears, or seedless forms may have evolved, as in bananas.

▶ How would seedlessness affect the evolutionary potential of the species?

All reproduction would be asexual, probably by transplantation of roots or cuttings, and hence completely controllable by humans. Without seeds, sexual reproduction and hence outbreeding would be impossible and evolutionary changes (other than those arising from mutation or artificial genetic change, see Section 17.4.4) would be impossible.

Foods such as potatoes, yams, manioc, taro, carrot, parsnip and many others are modified roots or stems. Many such crops, e.g. potatoes, are normally propagated vegetatively, but for the development of new varieties, the plant may be allowed to flower and set seed. Thus, the commercially available 'seed' potatoes are not seeds in the botanical sense: they are tubers in which one or more 'eyes' develop shoots under appropriate conditions. Plant breeders seeking to improve the genetic stock of this important crop cross-pollinate the flowers and raise the plant from seed.

Many morphological and chemical properties of seeds, fruits, roots, etc. that humans regarded as desirable have been perpetuated by artificial selection. Much more recently, characters such as cold tolerance, or resistance to disease and herbivory, have been generated in crop plants by direct modification of the genome (see Section 17.4). Such changes make the plant more vigorous, more nutritious or more palatable as human food, but they often also make it more vulnerable to herbivory. In many cases, the structures or chemicals that protect the plant from herbivory are removed or depleted by artificial selection: for example, the caterpillar of cabbage white butterflies, like humans, avoids tough, strong-tasting brassica leaves and relishes mild-tasting artificial varieties. Cooking, one of the earliest applications of the control of fire, enables people to eat potatoes, rice, cassava and other plants that are indigestible and distasteful when raw. Such plants are also unpalatable to other animals, so many crops that require cooking are very resistant to destruction from herbivorous insects and mammals, and so are easily stored.

Newly evolved artificial changes in the structure or life cycle of the plant may interfere with the natural regulation of its herbivores. For example, the codling moth lays its eggs in the core of ripening apples and the caterpillar feeds on the fruit. Its abundance is controlled naturally by a parasitic wasp that lays its egg in the growing caterpillar. But modern apples are so big that codling moth caterpillars can burrow out of reach of the wasp's long ovipositor. The wasp cannot easily parasitize its host and is therefore no longer as effective in regulating moth numbers. Artificial insecticides have to be used to protect large modern apples from codling moth caterpillars.

Cultivated varieties of hundreds of plants have been developed as crops or as ornamentals (Section 17.4.2) and attempts are still being made to 'improve' other wild species. A few peoples in a few areas of the world (see Table 17.7) have successfully domesticated many of the livestock and crops that are now important worldwide.

Table 17.7 Some plants and animals domesticated in the three primary zones of agricultural development before 5000 BC.

Zone	Plants	Animals
Southern Europe &Western Asia	Almonds Apricots Barley Dates Figs Grapes Lentils Olives Peas Wheat Flax	Cattle Goats Horses Donkeys Camels Pigs Sheep
China & India	Bananas Coconuts Millet Rice Soybeans Sugar cane	Poultry Cattle Pigs
South & Central America	Avocados Beans Chilli Cocoa Corn (maize) Potatoes Tomatoes Squashes	Llamas Guinea-pigs

The chance occurrence of potential cultivars as well as human foresight and ingenuity must have played a part in determining which species were domesticated. Although crops and livestock clearly contributed to the cultural development of the people who cultivated them, the most abundant and politically powerful groups have not necessarily made the greatest contribution to domestication. Many crop plants were first cultivated in Central America and northern South America; sunflowers are the only important crop to originate in what is now the USA, although the numbers of American Indians, and the diversity and sophistication of their culture, matched that of tribes further south. In spite of the successes in domesticating plants, native people in America had very few domestic animals except guinea-pigs and llamas and their close relatives, the alpaca (see Section 17.3.3). Horses, donkeys, cattle, sheep, goats, pigs and poultry were domesticated in Asia or Europe and were introduced into America only a few hundred years ago.

Migrations of human populations have affected the evolution and distribution of cultivated plants at least as much as that of domestic animals. One result of such introductions is that many crops are now much more important outside their place of origin. For example, peanuts were first cultivated in Mexico but are now grown mainly in China, India and the USA. Coffee is native to Ethiopia, was cultivated in Arabia but is now grown mainly in Brazil and in Africa south of the Equator. Cacao, the principal raw material for chocolate, is native to rainforests of Central and South America where people have long collected the seeds for making drinks, but 75% of world supplies are now grown in west Africa. New crop plants often have radical and lasting effects on the human ecology. The introduction of maize into Africa (from the dry plains of Central America), potatoes into Ireland (from the cool, moist highlands of South America) and bananas into the Caribbean area (from tropical Asia) have fundamentally changed the local economy, social structure and ultimately the landscape and the history of the human inhabitants.

Cultivation, of course, is always at the expense of the native vegetation, which is often disposed of as 'weeds'. On large continents, only the more productive areas are farmed intensively enough to exclude large numbers of native species entirely. But on many small tropical islands, much of the often very diverse native flora has been completely destroyed to grow introduced crops, such as pineapple, sugar cane and coconut. As with animals, the domesticated forms of most cultivars are now far more widespread and abundant than their wild ancestors. Wild strawberries and blackberries are still common in England, and the wild forms of oats, wheat, coffee and many other crops still exist, albeit in a few small areas, but the wild ancestors of cassava, peas and probably also tea are believed to be extinct. These species, like domestic horses and cattle, now exist only as crop plants, and their future evolution is determined entirely by cultivation and artificial selection.

17.4.2 Pets and gardens

People keep animals as pets and cultivate plants in gardens generally because they take pleasure in their beauty or companionship. Many pets are chosen and often selectively bred for their infantile appearance and habits (Figure 17.6(a) and (c)), and it is undoubtedly true that childless adults and the elderly value pets more highly than do the people occupied with young children. Intensive and prolonged infant care has been a major element in the evolution of human technical and intellectual ability, and parental instincts are firmly established. Until a few hundred years ago, only about half of all infants born survived to adulthood, and average life expectancy was not much more than 40 years, so most people had dependent children for most of their adult lives. Modern medicine has greatly reduced juvenile mortality and prolonged average lifespan (see Chapter 8, Figure 8.30), so fewer children are necessary to maintain stable population sizes. In future, more people are likely to spend a smaller proportion of their longer lives with dependent children, and there will be more childless adults. If we are not going to deny these people an outlet for caring, parental behaviour, a realistic and humane attitude towards pets must be fostered. Familiar though such activities may be, it would be unwise to dismiss them as frivolous or irrelevant to the course of evolution in the recent past and the future. Changes in the structure and habits of pet and garden species not only illustrate how natural selection may work; in some cases, they have a direct impact on the evolution of the wild species.

Pet-keeping may now be most conspicuous in affluent westernized city dwellers, but the habit is by no means restricted to such people: 16th and 17th century explorers reported that several North American Indian tribes tamed dogs, birds and even bear cubs and moose or bison calves. The pets were fed and fondled, and followed their owners on journeys. Many of the peoples of the Amazon rainforest tamed monkeys, agoutis, sloths and several kinds of wild birds, particularly parrots, upon which they lavished hours of attention. In some tribes, valued pets were buried following rituals similar to those used for dead infants. Polynesians were (and still are) enthusiastic pet-keepers: when first visited by European travellers, Fijians were keeping parrots, bats and lizards, as well as dogs, and Samoans had pet pigeons and even eels. One widespread habit particularly impressed (male) western visitors to New Guinea, Hawaii, Malaysia and North America: women who breast fed puppies, piglets, baby monkeys or bear cubs alongside, or sometimes instead of, their own children. In such societies, keeping animals as pets does not necessarily preclude harvesting the same species for food. Thus, in New Guinea, pigs are kept both as pets and as the main source of meat, and several Oriental peoples raise dogs for both purposes.

In most polytheistic religions, certain animals and plants are believed to have mystical influences over human affairs. Such attitudes were echoed in the views about the role of cats, bats, toads and other common animals in witchcraft that were widespread in Europe until the late 17th century. Such beliefs have led to some species, e.g. wolves and certain reptiles and amphibians, being exterminated from human settlements, and, in England, to the active persecution of pet-keepers. Thousands of people, mostly elderly, impoverished women, were condemned as witches, often partly on account of their 'unnatural' association with cats, rabbits and various wild animals. In other cultures, these and other species were revered as sacred, and harming them, even accidentally, was regarded as shameful and inauspicious, e.g. cats in ancient Egypt. Some sacred animals were accorded a status similar to that of pets, captive populations being fed and maintained by special attendants at public expense.

▶ How would these beliefs and habits have affected the evolution of the wild species?

If large numbers were collected from the wild, population size could be affected. But if most pets were orphaned or abandoned by their parents, the impact would be minimal. The genetic composition of the population would be altered only if the animals were bred in captivity or certain specimens (e.g. the most colourful) were selectively collected from the wild.

In Europe and China during the past 4000 years, and more recently in modern America and Australia, the breeding of dogs and other companion and sporting animals in captivity has been carefully controlled and they have often been subject to intensive selection for 'improved' features. Probably because of the expensive facilities and labour required to regulate the animals' breeding, until recently, the habit was almost confined to royalty and the wealthy classes. As well as many different breeds of dogs, pet-breeders produced fancy varieties of goldfish, pigeons, poultry, rabbits and many more, charging very high prices for specimens with traits that fashion deemed desirable. The breeds differ in size, colour, the form of the body (e.g. bulging eyes, elongated fins, flop-ears) and in the colour and texture of

(a)

(b)

Figure 17.9 *Artificial evolution in pigeons: (a) the rock dove* Columba livia, *the wild ancestor of fancy breeds of pigeons; (b) a print published in 1864 showing various fancy breeds of pigeons that would have been familiar to Darwin.*

the fur or feathers. Captive breeding revealed a great deal of variation in most species, much more than would be expected from the range of colour and form observed in adult wild specimens.

► How does evolutionary theory help to explain these facts?

In most species, there is much genetic variation (see Chapter 3), but it is not easily detected in wild populations because deviant specimens are normally eliminated by natural selection. In captivity, the animals are protected from predators and provided with ample food, so unusually coloured or maladaptively shaped specimens, and timid or docile individuals, can prosper and breed. In the wild, sexual selection would also prevent the perpetuation of many such traits, but pet-breeders, like those who breed livestock, often strive to eliminate sexual selection by segregating breeding pairs. From the European rock dove *Columba livia* (Figure 17.9(a)), sometimes hybridized artificially with other species of *Columba* (e.g. ring doves *C. palumbus*), pigeon fanciers have bred racing pigeons, pouters (Figure 17.9(b), upper centre), fantails (Figure 17.9(b) left) and birds with greatly elongated feathers on their legs (Figure 17.9(b) foreground) or expanded skin about the nose (Figure 17.9(b) left centre). While formulating his theory of evolution by natural selection, Darwin befriended several pigeon fanciers and took a great interest in how they bred specimens with unnatural body form, colour and behaviour.

When artificial selection of locally available species does not produce the characters desired in companion animals, people have imported exotic species. Some social birds such as parrots and starlings readily adapt to human company for many of the reasons that wolves were suited to domestication (see Section 17.4.1): they respond to human movements and postures as well as to verbal commands, and attempt to communicate with people by gestures and imitating speech. However, although certain parrots imitate human speech and other artificial sounds with remarkable accuracy, they are not known to imitate other species in the wild (although the habit has been described in several other bird species).

► Could the habit have arisen following association with humans, like barking in dogs?

Only if its breeding had been subject to artificial selection. Most parrots and parrot-like birds, including the species that talk readily, have only rarely been bred successfully in captivity, and no deliberate selection has been applied. The best talking birds are the African grey parrot, the yellow-naped Amazonian parrot and the Australian budgerigar.

► What can you conclude from these facts about the evolutionary origin of talking ability?

It seems to have arisen independently in only a few species that, being native to different continents, cannot be very closely related. All such species are social, and live in large flocks. However, many other birds with equally elaborate social behaviour imitate human sounds very little or not at all, even with intensive training. The origin and evolutionary significance of this trait remain a mystery (see Chapter 2, Section 2.3.4). Difficulties in breeding the birds in captivity has meant that the principal effect of their association with humans has been depletion of wild populations, mainly by capture of nestlings. Budgerigars *Melopsittacus undulatus* are the exception: unknown in Europe or America until 1840, and not popular until the mid-20th century, millions are now been bred in captivity for use as pets. Wild budgerigars live in huge flocks in arid areas of central Australia, eating fruit and small seeds, and are green with a yellow head and neck and barred wings. Selective breeding in captivity has produced blue, violet, mauve, white and yellow forms.

Some artificial breeds have proved unable to survive in the wild even in their ancestors' natural habitat, but others, such as cats, have been little altered from the wild form after thousands of years of domestication, and readily become feral in their natural range and elsewhere (see Section 17.3.3). Since the 1960s, feral populations of budgerigars, presumably escaped pets, have become established in Miami, Florida. Clearly, a new chapter in the evolutionary history of budgerigars is just beginning.

Disease Disease is an aspect of human contact with animals that is rarely emphasized. Many infectious diseases in humans (some authorities believe as many as 80%) have arisen from contact between people and domestic livestock or pets: for example, rabies is much more common in many wild mammals than in humans, who usually acquire the disease from contact with infected dogs; cowpox, a viral disease similar to smallpox, is transmitted from cattle to humans (and the converse) and psittacosis, a pneumonia-like disease, is present in many wild birds including fulmar petrels and pigeons, but sometimes infects humans who come into contact with poultry and pets such as budgerigars and other parrot-like birds. The virus associated with AIDS occurs in several species of primates and some authorities believe that it entered the human population through eating and handling wild monkeys. In itself, there is nothing unusual about acquiring parasites and other pathogens from contact with other species: most parasites are transmitted through carnivory (i.e. when one species eats another), and ectoparasites such as fleas and lice act as vectors for disease. But most non-human animals avoid contact with other species except that arising from predation or symbiosis, thereby limiting opportunities for transmission of pathogens. Paradoxically, transmission of diseases via food is much less efficient among people than in nearly all other carnivores: the uniquely human practice of cooking kills nearly all the parasites in meat and fish.

Gardens For at least 5000 years, people have cultivated ornamental plants and fruits in sheltered places, protected from grazing animals (e.g. the Hanging Gardens of Babylon, Garden of Eden), but initially, only native species were planted. Deliberate importation of exotic decorative plants probably began in Egypt at about 2000 BC. Greek, Roman and Arab armies, and mediaeval friars and pilgrims, spread medicinal, culinary and decorative plants throughout Europe, southwestern Asia and north Africa. Beginning in the 16th century, overseas exploration and the desire for the unusual led to a fashion for collecting and cultivating exotic plants from the Americas, India, China, eastern Asia, Australia and the Pacific islands. The search for novel garden plants was one of the main reasons why biologists such as Sir Joseph Banks (1743–1820) travelled first to Newfoundland and Labrador, and then with Captain Cook to Australia and the South Seas, naming the inlet where they landed, south of the modern city of Sydney, 'Botany Bay'. At first, such activities had little impact on wild populations, but as a result of more intensive and efficient collecting, combined with habitat destruction, some species, particularly orchids, are now rare in their natural habitat.

Most modern garden plants are derived, at least partly, from imported species. Rhododendrons, azaleas, lilac, tulips, aubrieta, wallflowers, begonias, nasturtium, asters, 'Virginia' creeper and many more owe their presence in northern Europe and North America and their modern form to human intervention. Many have hybridized with native species, or with other subspecies with which they are not naturally sympatric. Nearly all modern garden plants have been subject to artificial selection for characters such as longer or earlier flowering, larger flowers, different colours of flowers or foliage, improved perfume and different life forms (climbers, creepers, etc.).

▶ Would garden plants be an adequate substitute for native species in supporting wild animals?

Generally not. Most herbivorous insects and many seed-eating birds thrive only on the plants that they are adapted to eat. So, although imported plants may do better separated from their natural herbivores, the native animals often do worse, because there is less food for them. Many insects, such as butterflies and bugs, that feed on only one or a few species of plant decline rapidly when natural vegetation is replaced by exotic garden plants, and with them, insectivorous birds and mammals. Exotic insects, mites, worms, fungi and micro-organisms were often imported with garden plants, some of which have proliferated and proved harmful to native species (see Section 17.3.3), further damaging the natural vegetation and the animals that feed on it.

A few wild species have adapted successfully to breeding in or near gardens: following the draining of most natural pools and marshes, garden ponds are now major breeding places for many amphibians in western Europe. Most other horticultural practices are harmful to wildlife. Growing plants in pure stands and selective breeding for edibility often favours the proliferation of herbivores such as insects, gastropods and birds, leading to the use of pesticides and other drastic measures to control them (see Section 17.3.4). Extermination of the passenger pigeon and the Carolina parakeet, the only parrot native to North America, was motivated partly by the desire to protect crops and gardens (Section 17.3.1). In fact, these species were probably not much more destructive than many other seed-eating birds, but both formed large flocks, so the damage that they did was more obvious and they were easy to shoot in large numbers.

As described in Section 17.3.3, many garden plants have 'escaped' and become naturalized, sometimes hybridizing with indigenous species and thereby radically altering the course of their evolution and the composition of the vegetation. In fairness to the horticulturalists who imported and propagated such plants, we should point out that few, if any, of these long term consequences were foreseen.

Sir Joseph Banks was one of the most experienced and influential biologists of his time: he was President of The Royal Society for more than 40 years and advised the Government on many scientific issues, among them the expansion of the Royal Botanic Gardens at Kew, an institution devoted mainly to the cultivation of exotic plants. But ecology and evolutionary biology barely existed at the time, and such naturalists were unaware of the major and permanent effects of their actions on the course of evolution of both imported and native species.

17.4.3 Research organisms

Only a tiny fraction of all the animals and plants kept in captivity are used for research. Nonetheless, because much of our understanding of genetics, physiology, biochemistry and psychology is derived from laboratory experiments, it is important, particularly for students and practitioners of science, to understand something of evolutionary processes in these species, both before and after they become 'model' organisms. Such information can help us to assess whether common laboratory organisms are really typical of wild species, that for practical and ethical reasons cannot be investigated directly.

▶ Most laboratory experiments (in Britain, over 85%) involve rats (mostly mutant strains of *R. norvegicus*), house mice, guinea-pigs or rabbits. Is there anything unusual about the natural evolution of these species?

Apart from guinea-pigs, these species have all proved to be effective colonizers when artificially introduced (see Section 17.3.3). Adaptability in diet and habits, fast breeding, and the capacity to live in close association with humans have made the Muridae (rats, mice and hamsters) and rabbits some of the most widespread and rapidly increasing of all mammals. These same qualities have made these species convenient for scientific research. They breed very rapidly, tolerate living at high densities in small spaces, and can be fed on chow made cheaply from grain or grass.

▶ Could evolutionary changes take place in laboratory species?

Yes. Adaptive modifications have occurred in rabbits introduced to Australia (Chapter 2, Section 2.3.1) and heritable changes have evolved in livestock (Section 17.4.1) and pets (Section 17.4.2). Similar changes could easily happen in the laboratory. Indeed, circumstances favour such evolution: most laboratory colonies constitute small, isolated breeding populations, usually consisting of scores, at the most thousands, of animals. As with farm animals, the 'predator-free, unlimited resource' environment of the laboratory means that those that breed soonest and have the largest litters contribute most offspring to the next generation, leading to 'selection' for early maturity and greater fecundity. Docility is also selected for: excitable animals are more likely to escape and when excess males are culled, aggressive specimens are more likely to be eliminated. Laboratory managers are aware of the effects of inbreeding, and take steps to avoid it by introducing new stock from time to time. Nonetheless, the litter size of rats has increased from an average of eight in wild forms to as many as 20 in captive populations, and the average age of attaining sexual maturity has decreased.

Sometimes, scientists deliberately inbreed laboratory animals to produce homozygous strains, in which all specimens are genetically identical.

▶ Why should scientists want laboratory animals to be homozygous?

The standard paradigm for much experimental research is to take animals that are identical in as many ways as possible, and to manipulate only one or a very few variables between the 'control' and 'experimental' groups.

▶ Is homozygosity usually adaptive in the evolutionary sense?

No. Few wild vertebrates are completely homozygous, and species that lose genetic variability probably become extinct quite quickly (Chapter 5). The sexual behaviour of many species promotes outbreeding, and thereby increases heterozygosity (Chapter 7). In the laboratory, such natural mechanisms are thwarted and maladaptive characters are not eliminated by natural selection. To overcome these problems, naturally homozygous organisms are sometimes studied.

▶ Which mammalian species routinely produces genetically identical offspring?

Armadillos nearly always produce identical quadruplets (see Section 17.2.1). They are being used for research, but they are larger, breed less frequently and require a more specialized diet than rats and mice, so are less convenient as laboratory animals.

Mutations that would be selected against in the wild are retained, and individuals carrying them are carefully tended and bred. Thousands of different mutant strains of rats and mice, with traits such as hairlessness, obesity, blindness, deafness, spasticity, immune deficiency, congenital muscular dystrophy or a very high chance of developing malignant cancers, have been widely, and very effectively, used for medical and biological research. More recently, scientists have produced transgenic rats and mice (see Section 17.4).

▶ Most mutants and transgenic strains are difficult to breed and require expert handling. How can you explain those special properties?

The mutants may have changes in many genes other than those that control the character under scrutiny, some producing harmful effects. Transgenic animals have completely new genes obtained from other strains, or even other species. Mutant strains would also be highly inbred, again promoting the expression of deleterious genes. Nearly all of the characters thus evolved would probably prove to be maladaptive in the wild, but the possibility that humans may inadvertently breed a laboratory animal that would be more successful in the wild than the ancestral form cannot be excluded.

Summary of Section 17.4

Evolution under artificial selection produces organisms that are adapted to human exploitation. Domestication involves controlling the organism's breeding: natural evolutionary processes such as sexual selection may be eliminated and species and subspecies may be hybridized. Features of social and territorial behaviour as well as body size and dietary habits make certain animals particularly suitable for domestication. Far more plants than animals have been successfully domesticated. Polyploidy and other rapidly acting species isolating mechanisms are important for evolution under domestication among plants, but rare in animals. Although much altered in anatomical structure and behaviour from the ancestral form, many domesticated animals and plants may become feral or naturalized, particularly on islands and predator-free habitats.

Pet-keeping and the cultivation of ornamental plants are very ancient and widespread human habits, but formerly involved mainly native species. Until the last 500 years, only indigenous plants and animals were used, and the practice had little impact on the course of evolution of the species involved, and on the rest of the ecosystem. However, importation and subsequent selective breeding and sometimes release of exotic species depletes natural populations of the pet or garden species and may adversely affect the fauna and flora of areas to which they are introduced. Accidental or intentional selective breeding of laboratory species often results in rapid and substantial evolutionary divergence of the captive from the wild form.

17.5 Human intervention in the future

As illustrated in Section 17.2, human technology and cultural attitudes are changing the way that natural selection operates in human populations. Immunization against many pathogenic bacteria and a few viruses is more efficient and more widespread so fewer people die young from infectious diseases. However, natural immunity which has developed following exposure to the pathogen is still the most important form of protection from such diseases. Therefore, reducing its contribution to evolutionary fitness may also, in the long term, increase the incidence of ill-health and promote maladaptation in the evolutionary sense. Easier world-wide travel means that fewer people can remain totally isolated from contact with the rest of the world for very long, so devastating epidemics of newly imported diseases are less likely to happen. On the other hand, when infectious diseases such as AIDS do appear, they can spread rapidly over long distances, often far outstripping the technological capacity to develop and administer medical protection or treatment to those affected. Many sufferers from some formerly lethal disorders can now be treated sufficiently effectively for them to be able to live a normal life, including breeding. The curtailing of selection against carriers of genes for heritable defects, while raising overall heterozygosity, may also increase the prevalence of such diseases in the population as a whole. Indeed, social trends such as later marriage, small families and intensive care of the defective and chronically sick mean that all forms of natural selection are becoming slower and less effective.

It is clear from Sections 17.3 and 17.4 that in both the recent and the remote past, and in many different ways, people have had major impacts on the course of evolution of many other species. Many of the most destructive influences are now so well-established that they are very difficult to curtail. For example, modern communications such as air travel, and the desire for more crops, pets and garden plants, seem certain to result in more accidental introductions of exotic species (and their diseases and parasites) in spite of increasingly elaborate and stringently enforced restrictions on the importation of animals and plants. Scientific study of natural and artificial evolution has at least elucidated some of the mechanisms involved, if not shown the way to preventing or reversing its progress, and has made people aware that the environment is changing much faster than most organisms evolve adaptations to it.

Deforestation and other forms of ecological destruction continue almost unabated, made necessary primarily because of the rapidly expanding human population, now almost universally dependent on agriculture and animal husbandry. Draining, irrigation and levelling are converting more and more areas formerly thought unsuitable for agriculture into arable land and pasture and in the process excluding many of the indigenous species. Large deserts, mountains, and arctic regions are least affected by human activities mainly because they are not amenable to such exploitation. In many areas, the problem of vermin seems to be getting worse, as denser cities produce higher concentrations of rubbish. The extermination of natural predators, whether of urban vermin or of crop pests, and the introduction of exotic pests, makes hunting, pesticides and other artificial means of destruction even more necessary. In other words, as people meddle with the course of evolution, they cause themselves more and more work trying to control the relative abundance of species that were formerly regulated by the natural relationships between herbivores and plants and predators and prey.

17.5.1 Preservation of endangered species

The aim of wildlife preservation is to protect certain species from extinction. If a species appears to be endangered in its natural habitat, some specimens may be kept in captivity, as individuals and breeding groups in zoos, or even as frozen embryos (Section 17.4) or seeds. Sometimes, it is necessary to keep the species indefinitely under such artificial conditions, but often it is intended that they or their progeny be reintroduced into the wild.

One of the first, and most successful, preservation projects concerns Père David's deer *Elaphurus davidianus* (Plate 17.3). This monospecific genus of deer formerly occurred throughout the lowlands of north-eastern China, but for reasons that are not entirely clear, it disappeared from the wild, and for about 2000 years the only population was in the Imperial hunting park near Peking. This herd was largely destroyed by a flood in 1894, and the last native-born specimen died in Peking Zoo in 1922. However, shortly before the disaster, a few dozen animals had been exported to zoos and private menageries in Europe, among them Woburn Abbey in Bedfordshire, at the instigation of the Jesuit missionary Père David, after whom the species is named. In the 1950s, descendants of these specimens were exported from Woburn Abbey to Peking, where they are once again breeding. In spite of such efforts, the total world population of the species is still only about 1 000 individuals, in nearly 100 separate institutions.

Zoos and other artificial refuges have preserved several other species, but usually action was taken only when a handful of specimens remained. The range of the European bison *Bison bonasus* has been reduced progressively for thousands of years, mainly from destruction of its forest habitat (Section 17.3.3). The last two wild, though protected, populations in Poland and western Russia were destroyed (for food) during the First World War. A few scores of animals remained in zoos, and from them more than a thousand have now been bred, and a few returned to part of their former range in Poland. Przewalski's wild horse (Section 17.4) also became extinct in the wild during the 19th century; the current captive population of about 1 000 specimens is descended from 13 that were preserved in zoos. The Arabian oryx *Oryx leucoryx* occurred in historic time over much of southwestern Asia (Section 17.4.2) but became extinct there in 1972. About 150 specimens were then in captivity, and a few of their progeny have recently been re-introduced into the wild, but it remains to be seen whether they will establish a viable population in the long term. Similar attempts are being made to preserve other species, notably the Californian condor, the né-né goose in Hawaii, the giant panda of China and chimpanzees, but so far not all such efforts have been successful, and the re-introduced specimens have failed to establish breeding populations.

▶ What biological factors are likely to limit the success of such policies?

The founder populations are nearly always very small, sometimes fewer than a dozen specimens, and there is no opportunity for outbreeding unless new stock are imported. Populations descended from a few animals are unlikely to have retained the natural genetic diversity of the species. To quantify the loss of genetic diversity in captive populations, we need to know something about the natural situation. Unfortunately, it is often difficult to measure genetic diversity in wild populations, particularly in large mammals and birds, so there are extensive data for only a few species. About a third of the world population of about 30 000 polar bears (Plate 2.3) occur around North America (most of the rest are in the Russian Arctic), where they have been studied intensively for more than 15 years. The proteins in blood samples from 460 polar bears from Canada, Alaska, Svalbard (see Chapter 2, Figure 2.17) and Greenland were compared. Only a handful of the 75 gene loci studied had more than one allele, indicating that genetic diversity is low, even between widely separated populations. Similar analysis of 24 genetic loci in 159 northern elephant seals *Mirounga angustirostris* in four widely separated breeding populations in Mexico and California also revealed no heterozygosity.

The very sparse data suggest that genetic diversity of vertebrates in general, and of mammals and birds in particular, is lower than that of plants and invertebrate taxa such as insects, but the reasons for this situation are not entirely clear. Several other rare mammals, such as black bears, walruses and cheetahs, have also been found to be genetically homogeneous. Such situations could arise from bottlenecks in the population—periods in which numbers fell drastically so that the existing population is descended from a few survivors— or from intense selection pressure in a very specialized environment, or from other, unknown causes. Thus, intensive hunting for oil and skins reduced *M. angustirostris* to only about 100 specimens in 1892, from which the present population of about 60000 seals is descended. The range and density of black bears and cheetahs has been restricted by human invasion, but there is no historical record of a drastic decline in the numbers of polar bears, so the origin of their genetic homogeneity is unclear.

As well as reducing the genetic variation necessary for long-term evolutionary change, deleterious recessive alleles are more likely to be expressed in highly inbred species and such populations are also more susceptible to infectious diseases. Natural selection does not operate as efficiently in captivity and specimens with deleterious genes that would have been eliminated naturally may be able to breed. It is also very difficult to avoid the evolution of incipient domestication because zoos prefer animals that seem content with the conditions in captivity and cooperate with their keepers. However, intentional or inadvertent selection for such traits must be avoided if wild populations are to be successfully re-established.

Modern techniques (Section 17.4) of *in vitro* fertilization and embryo storage and transfer are used to maximize fertility and promote outbreeding while minimizing stress on the animals, but without information about the extent of genetic exchange in natural populations such procedures are mainly inspired guesses.

▶ Under what circumstances is outbreeding unsuccessful, or, if successful, produces maladapted offspring?

Members of different races and subspecies (Chapter 9) do not naturally interbreed and even within a single population, there is an optimum level of outbreeding and inbreeding (Chapter 5, Section 5.1.6). The qualities required of ideal mates may be almost undetectable to human eyes (particularly if wild populations are poorly known because they are almost extinct), but behavioural or physiological barriers may have evolved that preclude successful breeding. Alternatively, such artificial hybrids may be viable but sterile, and therefore evolutionarily useless, e.g. mules (Section 17.4.1).

There is often neither the time nor the space to maintain the natural age structure of the population, which can be important to its prosperity in rather subtle ways. The social structure of elephants, for example, seems to require several 'middle-aged' or elderly females to lead each herd: groups of young animals, although physiologically adult, do not breed so successfully and many cows select only mature bulls as mates. Such animals always have the largest, and hence the most sought-after, tusks. Killing them not only eliminates those that have survived to an advanced age, but renders the rest of the herd non-viable. Lack of appropriate social environment is also believed to be the main reason why small groups of elephants, great apes and other long-lived, highly social animals rarely breed well in captivity. Even if mating is successful, female mammals often reject or maltreat their offspring under artificial conditions. Breaking up established pairs or larger groups to promote outbreeding often exacerbates these problems.

▶ Would hand-rearing or cross-fostering to another species be a satisfactory alternative?

No. Mammals and birds reared by other species often fail to form normal sexual relationships as adults (e.g. male donkeys raised to father mules) and so do not breed themselves. For example, much effort has been put into breeding whooping cranes *Grus americana*, once widespread in North America but reduced to only 15 pairs by 1941. Although hand-rearing and cross-fostering has now produced about 150 birds, there are only 25 pairs breeding in the wild, all in a single migratory population. Their migratory habits may also make them less amenable to such preservation efforts: the young learn the migration routes from their parents and return to breed where they themselves were raised. Without such experience, they are unlikely to breed successfully. As with designing breeding programmes, recreating suitable conditions for long-term preservation in captivity is almost impossible unless the species has been thoroughly studied in the wild. It is very difficult to provide some birds with all the right foods to feed to their nestlings.

▶ In what other ways are very small captive populations vulnerable to extinction?

Infectious diseases or parasites (Sections 17.2.2 and 17.3.2) would spread easily through small, genetically homogeneous populations, especially if they were kept at high density. The fossil record provides little information about the role of disease in evolution, but the distribution of physiological specialization of parasites such as tapeworms and schistosomes indicate that they have been around for a very long time. Some fossil human bones show clear evidence of syphilis. Epidemics of these and other diseases probably happen naturally from time to time and may cause local extinctions, but affected areas are normally recolonized from other populations that escaped infection, either through being physiologically different, or because the disease did not reach them.

▶ Is it possible to exterminate the pathogens completely?

Generally not. Many pathogens infect several hosts (Section 17.3.2). Rare species such as Père David's deer and Arabian oryx could be at risk from diseases such as rinderpest and brucellosis that occur in cattle and in other deer. Many primates, including chimpanzees and gorillas, catch infectious human diseases (e.g. yellow fever, chickenpox and measles), and carnivores such as wolves, tigers and cheetahs suffer from diseases (e.g. canine distemper) that are endemic among pet and feral dogs and cats. Keeping such animals isolated from people (and their pets and livestock) is not only impractical, it also defeats a major objective of preserving species in captivity, extending and disseminating knowledge about them. Valuable captive birds such as parrots sometimes catch psittacosis and avian malaria from indigenous species such as pigeons, among which such diseases are widespread, although not prevalent enough to be a serious threat to their numbers. But for very rare species, the loss of even a few specimens to disease has serious consequences for their long-term evolution. Keeping animals at unnaturally high densities, in stressful social situations or on diets lacking essential nutrients further increases the risk of disease. For example, a disease introduced with domestic turkeys made a major contribution to the extinction (in 1932) of the last remaining colony of heath hens *Tympanuchus cupido*, which were once common in eastern USA.

Attempts are also being made to preserve artificial breeds of livestock and plants which have become rare because they are no longer economically efficient under modern husbandry and market conditions. Although not as productive as more widely cultivated varieties, such breeds often have potentially useful features such as resistance to disease or the capacity to thrive in severe climates.

▶ Why is preservation more likely to be successful for livestock and crops than for wild species?

Domesticated animals and cultivated plants are artificial species: their evolution was already under human control. Because the species are economically important, and in many ways already adapted to breeding successfully in captivity, the prospects for preserving domesticated breeds seem fairly good.

Even its most optimistic sponsors do not believe that species preservation provides a long-term, universally applicable solution to the rapid increase in the rate of extinction of wild animals and plants. Successful preservation depends upon genetic diversity, so the ancestors of the preserved populations should be as numerous and diverse as possible: in other words, extinction must be foreseen, and action taken, well before it happens. Declines in numbers of all but the most conspicuous species or economically important species are likely to go unnoticed unless they are very rapid. Such preservation is extremely expensive: while captive breeding of giant pandas and chimpanzees readily wins public support, it is difficult to imagine as much enthusiasm for similar facilities for bats, toads or snakes. Nonetheless, with nearly a thousand species, many of them essential as pollinators or dispersers of tropical plants, bats are far more important to the ecosystem as a whole, and scores of species are known to be declining rapidly.

Although there are more captive breeding programmes for mammals and birds than for any other taxa, only a small minority are likely to be preserved by such policies. By the 1970s, only about 10% of more than 250 mammals identified as being close to extinction were being bred in captivity in sufficient numbers for there to be any realistic hope of maintaining a viable population, or of returning any specimens to the wild. The Hawaiian tree snails (Section 17.3.1) are among the very few invertebrates that are being bred in captivity with a view to increasing their numbers in the wild. Captive breeding of small populations is unlikely to preserve rare wild species indefinitely and therefore the policy only delays, not averts, their extinction.

17.5.2 Conservation of natural ecosystems

Conservation means protecting whole communities of organisms in ways that retain their potential for natural evolution. The first problem is to verify that at least the majority of species in an area to be conserved are indeed indigenous.

▶ Which areas of Britain would qualify as truly wild?

Almost none, except a few very small remote islands. There are rabbits (Section 17.3.3) and sheep almost everywhere that is not ploughed or fenced, large predators have been exterminated and almost all areas have been logged, if not completely deforested.

▶ Could any conservation areas ever become as they were before the arrival of humans?

No. Even if the reintroduction of wolves and lions was socially acceptable, the former prey species have adapted to the absence of these large predators. Introduced animals and plants cannot be eliminated except by measures so drastic that many indigenous species would be seriously harmed. Nonetheless, some areas are clearly worth conserving for their natural beauty, if not for the authenticity of their wildlife.

Such wildlife reserves maintain the genetic diversity of the species more effectively than breeding in captivity, mainly because they are larger and ecologically more diverse. The minimum population size that avoids immediate loss of genetic variability has been estimated at 500 specimens. Smaller populations, and those of this size and larger with demographic anomalies such as unbalanced sex ratio, a high proportion of elderly or closely related individuals, quickly suffer from reduced vigour and fecundity in the same way as small groups breeding in captivity. Small areas would be sufficient to maintain such populations of insects or small fish, but for larger, less abundant species, much more extensive reserves are required. For example, the American mountain lion or puma, *Felis concolor*, occurs naturally at a density of about one per 26 km^2, so a reserve of at least 13 000 km^2 is necessary to support a population of 500 animals, believed to be the minimum sustainable size. The minimum area for conservation of a breeding population of wolves is estimated at between 39 000 and 78 000 km^2.

▶ What other habits indicate that conservation can only be achieved in large reserves?

Migratory species require very large areas, and 'safe passage' between their major habitats. Some birds migrate over thousands of kilometres and, as well as appropriate habitats at both ends of their journey, they require suitable places, often wetlands or woodland, where they can rest and feed *en route*. Many other vertebrates, including many newts and toads, and mammals such as wildebeest, and certain insects, also migrate regularly: delays and alterations of route or destination severely disrupt breeding. Elephants wander over very large areas, and curtailing their migration often results in lasting modifications to the habitat (Section 17.3.2). Unfortunately, less than 4% of existing wildlife reserves (outside arctic areas) are more than 10 000 km^2 in area, and the great majority are less than 1 000 km^2. Furthermore, most are designated as nature reserves *because* they are mountainous, windy, drought-prone or for some other reason are unsuitable for agriculture, and they support fewer organisms than fertile lowlands. Such reserves may be sufficient for some species but not for others. For example, the Serengeti National Park in Tanzania, one of the largest and most famous in Africa, was estimated in 1972 to contain about 2 000 lions but only 30 hunting dogs (*Lycaon pictus*).

▶ What will be the long-term outcome of this situation?

Hunting dogs will become extinct in the park. Their disappearance will alter the ecology of their potential prey, their predators and that of other any other species with which they formerly interacted. Ecological diversity will decrease, and the long-term course of evolution of the entire ecosystem will change. Some common species may increase in numbers, extending their habits to occupy the niches vacated by the extinct species, but many others will decline to extinction. Conservationists estimate that, even in the largest wildlife reserves now in existence, inbreeding, unnatural social and demographic structure of populations and factors such as epidemics and famine, will lead to the extinction of at least 30% of all the large terrestrial mammals within 500 years. Smaller reserves will be depleted much more rapidly. Similar calculations suggest that still larger areas are necessary for there to be any chance of natural sympatric speciation: at least half a million square kilometres (about the size of Madagascar) for animals the size of small monkeys.

Unfortunately, biological diversity is highest in the tropics, particularly in tropical forests, where human populations are expanding fastest and the demands of agriculture are most intense, so species will be lost from these areas at an even greater rate.

▶ Have mass extinctions on this scale occurred in the past?

Yes, but the difference is that mass extinctions in the past did not involve the simultaneous increase in the numbers of a few dominant species (i.e. humans and their livestock).

SUMMARY OF SECTION 17.5

Satisfactory preservation of a species involves maintaining its natural genetic diversity and avoiding evolutionary adaptation to the artificial conditions, i.e. accidental domestication. These objectives are very difficult to achieve and keeping the few remaining specimens of nearly extinct species in captivity and later re-introducing them into their former range are rarely successful. Only a few species thus managed have re-established self-sustaining breeding populations in the wild. Conservation is protecting whole ecosystems from the effects of human activities. Such areas must be large enough to support all the indigenous species in sufficient numbers to retain the natural genetic diversity of the population.

17.6 CONCLUSION

Humans have intervened extensively with the evolution of many species largely without an adequate understanding of the long-term outcome of such meddling. They have exterminated many species, produced some animals and plants that have become totally dependent on humans, and others that run wild, causing unimagined changes to the course of evolution of themselves and other species with which they come into contact. The situation is becoming more and more complicated and many of the remedies tried have proved more harmful than the original situation. Active conservation can delay the decline in numbers of wild populations, and enable a few more people to enjoy something of the natural diversity of plants and animals, but such policies are unlikely to prevent many species becoming extinct eventually, unless much larger areas are set aside as reserves. Limiting human numbers and applying policies based upon thorough understanding of the mechanism of evolution offer the best prospects for controlling the situation and maintaining the diversity of organisms.

REFERENCES FOR CHAPTER 17

Ashmore, M. (1990) The greenhouse gases, *Trends in Ecology and Evolution*, **5**, 296–297.

Bulmer, M. G. (1970) *The Biology of Twinning in Man*, Oxford University Press, Oxford.

Clutton-Brock, J. (1987) *A Natural History of Domesticated Mammals*, Cambridge University Press, Cambridge.

Leader-Williams, N. (1988) *Reindeer on South Georgia*, Cambridge University Press, Cambridge.

Martin, P. S. and Klein, R. G. (eds) (1984) *Quaternary Extinctions: A Prehistoric Revolution*, University of Arizona Press, Tucson, Arizona.

Murozumi, M., Chow, T. J. and Patterson, C. C. (1969) Chemical concentrations of pollutant lead aerosols, terrestrial dusts and sea salts in Greenland and Antarctic snow strata, *Geochimica et Cosmochimica Acta*, **33**, 1247–1294.

Spicer, R. A. and Chapman, J. L. (1990) Climate changes and the evolution of high-latitude terrestrial vegetation and floras, *Trends in Ecology and Evolution*, **5**, 279–284.

FURTHER READING FOR CHAPTER 17

Bolin, B., Döös, B., Jäger, J. and Warrick, R (1990) *The Greenhouse Effect, Climatic Change and Ecosystems*, Wiley Press, Chichester.

Frankel, O. H. and Soulé, M. E. (1981) *Conservation and Evolution*, Cambridge University Press, Cambridge.

MacGillivary, I., Campbell, D. M. and Thompson, B. (1988) *Twinning and Twins*, Wiley Press, Chichester.

Trends in Ecology and Evolution, **5**, September 1990.

SELF-ASSESSMENT QUESTIONS

SAQ 17.1 (*Objective 17.1*) (a) From a strictly biological point of view, what are the advantages and disadvantages of twinning?

(b) In the context of human culture, what are the advantages and disadvantages of twinning?

SAQ 17.2 (*Objective 17.2*) Arrange the following places in the order in which they were probably inhabited by breeding populations of *Homo sapiens:* eastern Africa; western Africa; Antarctica; North America, South America; south-east Asia; western Asia; Australia; China; southern Europe; northern Europe; Greenland; India; Indonesia; Madagascar; Mediterranean islands; New Guinea; New Zealand; Polynesia; Siberia, South Atlantic islands; Svalbard; West Indies.

SAQ 17.3 (*Objective 17.3*) Describe two ways in which the manufacture of tools and clothing has had a direct impact on the ecology and evolution of other animal species.

SAQ 17.4 (*Objective 17.4*) Which of statements (a)–(g) are valid reasons why the presence of humans has a major impact on the environment?

(a) By the time they made a detectable impact on the environment, the human population was increasing very rapidly and was approaching its contemporary size.

(b) By the time they made a detectable impact on the environment, humans ate more large animals than other mammals of similar size.

(c) Humans always have a more specialized diet than other mammals of similar size, so their presence has a major impact on particular species that they eat.

(d) Human foraging involves destruction of large trees.

(e) Humans can start and control fires.

(f) Humans use large quantities of materials for purposes other than food and shelter.

(g) Humans can concentrate and synthesize materials.

SAQ 17.5 (*Objective 17.5*) List five features of the biology of brown rats and black rats that may have enabled them to colonize many different exotic habitats.

SAQ 17.6 (*Objective 17.6*) Which of the statements (a)–(g) are generally true of pollution?

(a) All artificially synthesized substances are toxic, at least to some organisms.

(b) Very few natural substances are toxic to most organisms.

(c) Organisms are unable to adapt to the presence of toxic artificial substances.

(d) Only higher organisms such as humans can tolerate high concentrations of artificially synthesized substances.

(e) Species differ greatly in their response to the presence of toxic substances.

(f) Severe pollution usually exterminates all living organisms.

(g) Even a very little pollution can alter the species composition of the ecosystem.

SAQ 17.7 (*Objective 17.7*) In which of the ways (a)–(g) does artificial selection differ from natural selection?

(a) Genetic variation is essential for natural selection but not for artificial selection.

(b) Artificial selection requires much more genetic variation to produce significant evolution than natural selection.

(c) Artificial selection always produces faster evolutionary changes than natural selection.

(d) Organisms that have undergone significant evolution under artificial selection cannot survive in the wild.

(e) Artificial selection acts only on genes artificially added to the organism's genome.

(f) Artificial selection does not involve sexual selection.

(g) Most of the organisms subjected to artificial selection become infertile.

SAQ 17.8 (*Objective 17.8*) Describe three reasons why far more plants than animals have been domesticated.

SAQ 17.9 (*Objective 17.9*) Describe some features of the biology of wolves that may have predisposed them to domestication as (a) hunting and herding dogs (b) pets and guard dogs.

SAQ 17.10 (*Objective 17.10*) Which of the conditions (a)–(g) are essential for the establishment and maintenance of a genuinely natural conservation area?

(a) There must be comprehensive information about which species occurred in the conservation area before contact with humans.

(b) There must be comprehensive information about the relative abundance of each species in the conservation area before contact with humans.

(c) All introduced species must be exterminated and further introductions prevented.

(d) Endemic diseases must remain but new or introduced diseases eliminated.

(e) The area must be large enough to maintain genetically diverse breeding populations of all species.

(f) Alternative and transitory habitats of migratory species must be sustained.

(g) The area must be monitored regularly, and species composition checked against (a) and (b), to ensure that conditions (c), (d) and (f) are maintained.

APPENDIX 1 Geological time-scale, with variable scales for different eras

	ERA	PERIOD	EPOCH	AGE/MA		MAJOR FEATURES DISCUSSED
PHANEROZOIC	CAINOZOIC	QUATERNARY	Pleist. + Holocene			Major glaciations
				1.64		
		TERTIARY	Pliocene			Earliest hominid fossils
				5		
			Miocene			Major spread of grasslands in N. America
					20	Earliest hominoid fossils
				23		
			Oligocene			
				35		
			Eocene		40	
						Messel biota, Germany
				56		
			Paleocene		60	Major radiation of mammals
					65	Mass extinction (incl. dinosaurs and ammonites)
	MESOZOIC	CRETACEOUS			85	
					105	Santana Formation biota, Brazil
					125	Earliest angiosperms
				146	145	Solnhofen biota (with *Archaeopteryx*), Germany
		JURASSIC			165	
					185	Major radiation of ammonites
				208	205	Mass extinction
		TRIASSIC			225	Earliest mammals and dinosaurs
					245	Mass extinction
	PALAEOZOIC	PERMIAN				
				290	300	Major glaciations
		CARBONIFEROUS			340	Earliest reptiles
				363		
		DEVONIAN			380	Mass extinction
						First evidence for land vertebrates
				409	420	Rhynie Chert plants, Scotland
		SILURIAN				Earliest evidence for land plants
				439		
		ORDOVICIAN			460	Mass extinction
						Major glaciations
					500	
				510		
		CAMBRIAN				Burgess Shale fauna, Canada
					540	Early Cambrian radiation of metazoans
PRECAMBRIAN	PROTEROZOIC	VENDIAN				Ediacaran fauna
				610	1 000	Bitter Springs Chert (with eukaryotes), Australia
					2 000	Gunflint Chert (prokaryotes only), N. America
	ARCHAEAN				3 000	
						Earliest evidence for life, Australia
	HADEAN				4 000	Isotopic evidence for autotrophs
					4 600	Origin of the Earth

ANSWERS TO SELF-ASSESSMENT QUESTIONS

CHAPTER 1

SAQ 1.1 Statement (c) is correct. Neither statement (a) nor (b) adequately defines 'biological evolution'. The changes mentioned are not explained: they could be changes due to age, changes associated with injury or changes which are part of the normal development of the organism or changes in populations arising from changes in nutrition. Very importantly, neither (a) nor (b) refer to changes which are heritable and passed on from one generation to the next.

SAQ 1.2 (a) This is a correct statement. Biological variability of the kind involved in evolution arises independently of environmental changes in the sense that the variations that occur are not adaptively directed. But those that happen to be adaptive are favoured by natural selection.

(b) Some inorganic changes occur rapidly, e.g. many chemical or physical interactions, but others, e.g. geological changes, may be very slow. Equally, not all evolutionary changes occur slowly, e.g. evolutionary changes in bacteria or viruses may occur in as little as one generation— in these two cases about 20 minutes or so.

(c) Evolutionary changes are certainly not always progressive or directional in the sense of always occurring in the same characteristics in the same direction. Also, there are no grounds for supposing that recently evolved organisms are necessarily better adapted than their ancestors.

SAQ 1.3 (a) Fitness is not an absolute quality in the sense implied in the statement; rather it is a relative measure and should refer not just to the relative number of offspring produced but to the number of offspring which themselves reproduce—see Section 1.1.2.

(b) and (c) As is made clear in Section 1.1.2, fitness is the outcome of the process of natural selection and carries with it consideration of the reproductive performance of the survivors. Note also that fitness is not a property of a population of organisms but of the individual organisms of which the population is made up.

(d) Correct.

(e) It is incorrect to think that natural selection always favours diversity within a species: instead, it may promote constancy of characteristics.

SAQ 1.4 (a) Darwin believed that very large variations were unlikely to be advantageous (see Section 1.1.3), and so would not play an important role in evolution.

(b) This statement does represent Darwin's views (Section 1.1.3).

(c) Both Lamarck and Darwin believed in the now discredited notion of the inheritance of acquired characters, although Darwin considered natural selection to be paramount.

(d) and (e) The orthodox view was that the terms 'variety' and 'species' differed in meaning, a view that Darwin successfully challenged.

(f) Species were not defined with reference to their reproductive isolation by Darwin, but this is a modern view—see Section 1.1.4.

(g) Darwin's theory was attacked on this point in his lifetime, but opposition has died away as estimates of the age of the Earth have changed—see Section 1.1.4.

Chapter 2

SAQ 2.1 (a) Collecting provided evidence of the large number of different species that are and have lived, and illustrated the immense range of different forms. It also provided information about characteristics of organisms from different kinds of habitat, e.g. tropical versus arctic, grassland versus forest forms, which suggested ways in which organisms might be adapted to different environments.

(b) Observation of organisms in their natural habitat demonstrated the relationship between structures and their natural functions.

(c) Features that were shown to be adaptive in living species could be recognized in extinct forms, thereby improving reconstruction of the habits and habitats of fossil species.

SAQ 2.2 (a) If food suitable for the nestlings is available only for part of the year, clutch size (and whether the birds breed at all) is adapted to the season.

(b) In variable climates, the weather at any one date and place is not necessarily the same each year. If food abundance or foraging ability is significantly affected by the weather, the adaptiveness of clutch size will depend upon the local climate.

(c) The abundance and availability of the prey of species such as owls fluctuate in time and in space. Egg laying is staggered over several weeks and the largest chicks are fed first, thereby adjusting effective clutch size to the availability of food at the time.

(d) Different pairs of birds lay clutches of widely different sizes, depending upon the circumstances mentioned in (a)–(c) above and probably many other unknown factors.

SAQ 2.3 (a) The possible adaptive significance of fossilized organs and body parts can be assessed by comparing them with similar structures in living species, the adaptive significance of which can be determined by direct observation and/or experimentation.

(b) The physical properties or mechanical performance of extinct species can be measured from actual fossils or from reconstructions or models of fossils, providing insight into what their habits could have been.

SAQ 2.4 (a) The heart has to generate higher blood pressure, so it must be more powerful (and hence probably larger).

(b) The blood vessels in the neck and head must be able to adjust to very large, rapid changes in blood pressure, caused by abrupt and extensive changes in the position of the head in relation to that of the heart.

(c) Giraffes do not perform threats or attacks which would involve the head being down. They generally avoid confrontation by running away from intruders.

(d) The long legs complement the long neck adaptation, so they are as long as possible, but they cannot be longer than the neck or the animal would be unable to drink at ground level. The long neck increases the animal's body mass, so the legs are stronger (hence thicker) than in similar short-necked species.

SAQ 2.5 (a) Features contribute to the adaptation of the organism to its environment, whether they are genetic or arise in response to the local conditions under which the organism is living.

(b) The adaptation makes a lasting contribution to the long-term evolution of the species only if it has a genetic, heritable basis.

SAQ 2.6 Only (a) and (e) are correct. Coevolution describes reciprocal, adaptive evolution between two or more species that interact closely, but not necessarily exclusively, with each other.

SAQ 2.7 It is often difficult to see how a novel structure can increase its bearer's fitness, and hence be favoured by natural selection, in the early stages of its evolution.

SAQ 2.8 Non-shivering thermogenesis involves producing *more* internal heat, and using *more* energy (derived from food or from fat reserves). Torpor and hibernation involve producing *less* internal heat, leading to cooling of the body, and thereby minimizing demands on body energy reserves and enabling the animal to fast for long periods.

SAQ 2.9 Wolves are primarily terrestrial animals that live and hunt in packs, remaining active throughout the winter. These features make them poorly adapted to colonizing offshore islands: they cannot swim far and a fairly high density of prey animals is necessary to sustain a pack. In the absence of large mammalian predators, reindeer on Svalbard have become smaller and slower than those on the mainland, and they do not form large herds, even when breeding.

CHAPTER 3

SAQ 3.1 (a) Lamarckian inheritance is the inheritance of acquired phenotypic characters, i.e. characters that are acquired during an individual's lifetime are then passed on to progeny. In sexually reproducing individuals, changes in the DNA of gametes (i.e. mutations) are inherited by the progeny. The organism carrying the gamete with the mutation will not have acquired any phenotypic character as a result of bearing that gamete. So this is *not* Lamarckian inheritance.

(b) The central dogma is that information (by which is meant the base sequence of DNA, that is replicated, passed on to progeny and is ultimately translated into protein) can pass from DNA to protein but that any change in a protein cannot be translated back into a new sequence of DNA bases. Mutations usually occur at random with respect to their potential usefulness and are not directed by proteins. So the central dogma is not challenged.

SAQ 3.2 A gene can be identified as a section of DNA that can be transcribed to a complementary strand of mRNA. The mRNA can be translated to specify a sequence of amino acids which form a protein. But it is rare to find a protein that acts alone to control a particular morphological character. The example of the control of eye colour in *Drosophila* given in this Chapter involved at least two proteins. Variations in a phenotypic attribute are usually the result of the expression of several genes.

It is also very rare to find a protein that is produced as a result of the activity of a single gene because:

(i) genes at several loci may act together;

(ii) a gene's expression may be affected by genes at other loci (epistasis), even though these genes do not affect the character directly;

(iii) there may be interaction with the cellular environment that affects the processes that intervene between the DNA being transcribed and the protein being produced.

SAQ 3.3 Mutation, because this introduces novel genetic material. Recombination merely creates and destroys gene combinations, although this does allow allele frequencies to change independently of one another. A particularly well adapted phenotype that has resulted from favourable recombinations will not be preserved intact in succeeding generations except in asexually reproducing organisms.

SAQ 3.4 (a) The frequency p of the E allele in the population is $(3\,969 \times 2) + 3\,174 = 11\,112$ divided by the total number of alleles in the population, i.e. $2 \times 8\,070 = 16\,140$.

Therefore $p = 11\,112/16\,140 = 0.688$

From Equation 3.2, $q = 1 - p$

Thus $q = 0.312$

The Hardy–Weinberg ratio is $p^2{:}2pq{:}q^2$. Hence the expected genotype frequencies are:

0.474 *EE*:0.429 *Ee*:0.097 *ee*

(b) For 8 070 individuals the expected numbers of each genotype are:

(0.474×8070) *EE*:$(0.429 \times 8\,070)$ *Ee*:$(0.097 \times 8\,070)$ *ee*

= 3 825 *EE*:3462 *Ee*:783 *ee*

So the observed values of 3 969:3174:927 are rather far from the expected values.

Possible reasons for this discrepancy are listed below.

(i) Mating may be positive assortative.

(ii) There are fewer heterozygotes than expected so natural selection could be operating.

(iii) The flies might not have been from a single population—there might have been migration of homozygous flies from a neighbouring population(s).

(iv) Allele frequencies may be subject to large random fluctuations (genetic drift) though the large sample size makes this unlikely.

SAQ 3.5 (a) Proteins consist of sequences of amino acids that are specified by codons (sequences of three bases in DNA). Thus a change in the base sequence of DNA may produce a change in the amino acid sequence of the protein. If this involves an alteration in net charge, this will change its electrophoretic mobility.

(b) Electrophoretic variation provides an underestimate of variation, for the following reasons:

(i) Only amino acid changes involving an alteration in net charge are likely to be identified (about one-third of the amino acid substitutions).

(ii) Synonymous codon changes in the DNA, i.e. those which do not result in an altered amino acid, will not be recognized.

(iii) Sequences of DNA without a protein as a gene product (e.g. genes for ribosomal or transfer RNAs) cannot be studied electrophoretically.

SAQ 3.6 (a) The elephant seal is known to have had its population reduced to a very low level at the end of the 19th century. A similar population reduction is believed to have occurred about 100 000 years ago for the cheetah. So the low genetic diversity could be because:

(i) The original founding/remaining population would have reduced genetic variability due to sampling error (founder effect).

(ii) Thereafter, variability may have been further reduced by genetic drift and fixation of the alleles which survived.

(iii) Certain genetic variants may not be distinguishable by the electrophoretic methods employed.

(b) The most recent cheetah population crash resulted from an infectious disease. The nature of the immune response (Section 3.2.4) is such as to render populations with low genetic diversity particularly susceptible to devastation from infectious diseases. It will be a matter of chance when epidemics occur and the extent of the damage they wreak. Perhaps the elephant seals have not been exposed to such a disease since the end of the 19th century.

CHAPTER 4

SAQ 4.1 The modern definition of fitness emphasizes reproductive success as a major component of fitness, whereas Darwin stressed survival.

SAQ 4.2 (i) – (b); (ii) – (a); (iii) – (a).

SAQ 4.3 (i) Yes, condition (b). There is a relationship between a phenotypic character (territory quality) and a component of fitness (survival of the young).

(ii) Yes, condition (a). There is variation in a phenotypic character (singing).

(iii) Yes, condition (b). There is a relationship between a phenotypic character (coloration) and a component of fitness (mating success).

(iv) No. The statement suggests that a change has occurred over time. It thus does not refer to the conditions for natural selection.

SAQ 4.4 1 The development of defensive armour in sticklebacks varies from one locality to another and is correlated with the abundance of predators at those localities.

2 The grass *Agrostis capillaris* varies in its tolerance to copper in relation to the local level of copper pollution.

SAQ 4.5 1 Male guppies exposed to more intense predation became less colourful over time; those exposed to less intense predation became more colourful.

2 In peppered moth populations in woods exposed to industrial pollution, the frequency of the dark morph increased.

SAQ 4.6 (a) 0; (b) 0.18. The proportions of frogs surviving from one year to the next are: males—0.85, females—0.70. Males have a higher survival rate and, by definition, have fitness = 1 and the coefficient of selection on males = 0. The relative fitness of females is 0.70/0.85 = 0.82. Therefore the coefficient of selection on females = 1.0 – 0.82 = 0.18.

SAQ 4.7 (i) Toads are long-lived animals and breed over several years. Hence, estimates of mating success based on a single season are unlikely to be good estimates of lifetime reproductive success. (You would also be correct in thinking that the sample size for mating males in 1977 in Table 4.1 is rather too small to be the basis of a firm conclusion.)

(ii) Clutch size is only one component of a female's reproductive success. The number of a female's eggs that actually survive to fledging is a much better estimate. Furthermore, some female blue tits breed in successive years, so that reproductive success in one year may not be a true reflection of lifetime reproductive success.

SAQ 4.8 It is necessary, first, to calculate the frequencies of the two alleles, B and b, as follows:

Phenotype	No. of individuals	Genotype	No. of B alleles	No. of b alleles
Black	44	BB	88	—
Intermediate	70	Bb	70	70
White	16	bb	—	32
Total	130		158	102

The relative frequency of the rarer allele (b), $q = 102/(158 + 102) = 102/260 = 0.392$. Therefore, the relative frequency of the commoner allele (B), $p = 1 - 0.392 = 0.608$.

From these values, the Hardy–Weinberg ratio can be determined as:

$p^2 = 0.370 : 2pq = 0.477 : q^2 = 0.154$.

For a population of 130 individuals, this gives an expected ratio of:

$130 \times 0.370 : 130 \times 0.477 : 130 \times 0.154 = 48 : 62 : 20$.

This does not differ greatly from the proportions of the three forms found in the sample, so the conclusion is that the population is not subject to natural selection.

SAQ 4.9 (i) Yes: dominant alleles spread faster than recessive alleles.

(ii) No: the rate of change of frequency of an allele is slow if that allele is very rare.

SAQ 4.10 (a) Variation will tend to increase if there is negative frequency-dependent selection because, as an allele becomes rarer, its relative fitness increases. Rarer alleles are thus maintained within the population.

(b) Variation will tend to decrease if there is positive frequency-dependent selection because, as an allele becomes more common, its relative fitness increases, leading to the elimination of rarer alleles.

SAQ 4.11 The greater the variation in a character, the stronger is the potential for selection on that character; but strong selection on the character will tend to reduce the heritability of that character.

SAQ 4.12 The selection differential (S). This is obtained by comparing the mean value of the character under investigation before and after selection has taken place.

CHAPTER 5

SAQ 5.1 A—3. Recombination during sexual reproduction produces genetically diverse progeny.

B—3. Recombination during sexual reproduction produces some progeny that do not carry deleterious mutations.

C—2. Asexual reproduction, because there is no recombination, cannot eliminate deleterious mutations and so they tend to accumulate over many generations.

D—4. Recombination during sexual reproduction tends to break up favourable combinations of genes.

E—1. Asexual species can reproduce faster than comparable sexual species because mitosis is faster than meiosis. Furthermore, all their progeny are female and so the overall reproductive rate will be higher in an asexual than a sexual population.

F—4. Producing sons imposes a two-fold cost on sexually reproducing females that produce sons and daughters in equal numbers.

SAQ 5.2 Because a larger number of progeny usually leads to more intense competition between those progeny. Only if there is competition, such that only some progeny can survive, will there be differential survival among progeny with different genotypes.

SAQ 5.3 Sexual reproduction produces genetically more diverse progeny. A sexual species is thus more likely to produce novel genotypes that have a competitive advantage over competing species than is an asexual species.

SAQ 5.4 (a) Because it leads to an increase in the frequency with which deleterious recessive genes occur in homozygous form and are thus expressed.

(b) If the difference between the dispersal distances of the two sexes is very large, there is a danger that individuals will mate with unrelated individuals, leading to outbreeding depression.

SAQ 5.5 Because two extreme types of gamete are favoured at the same time, whereas intermediate types are selected against.

SAQ 5.6 Selection will favour individuals that produce more sons than daughters. In other words, the population is open to invasion by individuals that produce young in a ratio different from that prevailing in the population as a whole.

CHAPTER 6

SAQ 6.1 Figure 6.7 shows the relationship between two cousins (solid green circles). From this diagram, it can be seen that there are four generation links (L = 4) and two alternative routes (n = 2) between cousins. So from Equation 6.1, $r = 2 \times (0.50)^4 = 0.125$ (see Figure 6.7).

SAQ 6.2 (a) False. Parents can typically rear young on their own but do so less successfully than when they have helpers.

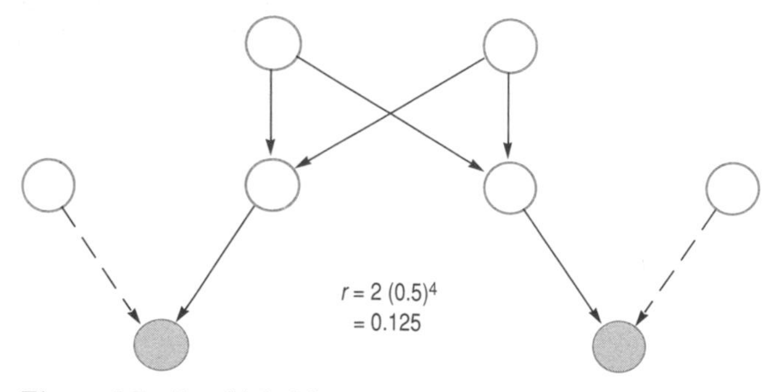

Figure 6.7 *For SAQ 6.1 answer.*

(b) False. Because helpers derive non-genetic benefits from belonging to a cooperatively breeding group (e.g. experience of breeding, mutual defence against predators), it may be adaptive for them to help, even if they are not related to the young.

SAQ 6.3 (a) Helping will evolve.

(b) Helping will not evolve.

This question requires you to use Formula 6.4:

$$B/C > r_{\text{donor to own offspring}} / r_{\text{donor to recipient's offspring}}$$

For the potential helper, the benefit of helping (B) = the *additional* young that the mother rears as a result of receiving help. This is 5 − 2 = 3. The cost (C) = the number of chicks the helper fails to have by not breeding herself. This = 2. Therefore, B/C = 3/2 = 1.5.

In condition (a) helper and recipient are full siblings, so that r between helper and recipient = 0.5. In condition (b) they are half-sibs (same mother, different fathers), so that r = 0.25.

Thus, in condition (a) the ratio of coefficients of relatedness between donor and her own offspring and between donor and recipient = 0.5/0.5 = 1. In condition (b) the ratio is 0.5/0.25 = 2. Thus, in (a) B/C is greater than the ratio of r values, in (b) it is less.

SAQ 6.4 (a) Yes, individual selection and group selection are incompatible. If the behaviour is explicable in terms of individual selection, group selection need not be invoked. Group selection is unacceptable because, if natural selection is operating, behaviour based on group advantage is not evolutionarily stable.

(b) No, individual and kin selection may be compatible. It is possible that, by calling, an individual benefits not only itself, through the confusion effect, but also its relatives, who are warned about the predator.

CHAPTER 7

SAQ 7.1 (a) Anisogamy is the existence of two kinds of gamete that differ in size and behaviour (eggs and sperms).

(b) Polygyny is a mating system in which an individual male mates with more than one female during a breeding episode but each female mates with only one male.

(c) Somatic effort refers to all the physiological and behavioural activities that promote the survival and growth of an individual as opposed to its reproduction.

(d) A satellite male is one that shows an alternative mating strategy. In the case of frogs he is to be found close to a calling male, and he attempts to intercept and mate with females that approach the calling male.

SAQ 7.2 (a) Figure 7.6(a) Some female turtles lay over a thousand eggs in a breeding season (Section 7.2.2). Clearly more reproductive effort goes into gamete formation than into parental care so the answer cannot be either Figure 7.5(a) or Figure 7.6(b). The text does not tell us anything about the mating behaviour of female turtles but, in general, females put more reproductive effort into gamete production than into mating, so Figure 7.6(a) is more likely to represent the allocation of reproductive resources than Figure 7.5(b).

(b) Figure 7.6(b) There is only one diagram that relates to males (Figure 7.5(b)) but this does not accurately reflect the situation for modern man. The allocation of resources between gametes, mating, and parental care is probably closer to that shown in Figure 7.6(b) than that shown in Figure 7.5(b), though the allocation of the gamete resource between number and size obviously remains as in Figure 7.5(b).

SAQ 7.3 In mammals only the female provides milk so the male cannot provide this form of parental care; he can, however, provide other forms of care. In birds, either parent can incubate the eggs so the males can be as effective a parent as females in this respect. (Moreover, in pigeons, and some other species of bird, both parents produce crop milk to feed their young). In birds, individual males are more predisposed to share parental care than are male mammals which often contribute in more indirect ways such as by defending a territory.

SAQ 7.4 (a) The distance between z and y is less than B, so the difference in quality between y and z is less than the polygyny threshold and the second female should chose territory z. The third female should choose territory y.

(b) Territory quality can depend on resources such as food, nesting sites and protection from predators.

(c) According to the model shown in Figure 7.9, the second female on territory y has a reduced fitness compared to her fitness if she were the only female on territory y. This could be because the male spends all (or most) of his time with the first female, and the second female, being alone, is unable to provide adequate food and protection for her young. If the first female suffers reduced fitness, she may try to drive away the second and any subsequent females. If her fitness is not affected, she may tolerate other females. The third female on such a territory may therefore have about the same fitness as the second female, if the male spends all or most of his time with the first female. But, if the presence of a third female reduces the fitness of the other two females, they may then both try to drive her away.

SAQ 7.5

1. In jacanas, females mate with several males so that individual males have no option but to care for the young, only some of which may be their own.
2. In many species, dragonflies for example, males guard females between mating and egg-laying with the result that they cannot mate with other females during that period.

SAQ 7.6 An individual's fitness is measured by reproductive success. Anisogamy means that, all else being equal, a male can leave more descendants than can a female of the same species.

In species that show parental care, a female can increase her reproductive success if several males rear her young whilst a male can increase his reproductive success if several females rear his young. Thus there is a conflict of interest. In hostile environments, where two parents are required to rear young successfully, males that help rear young will have higher reproductive success than those that do not. If young are very heavily preyed upon, then increased egg production and hence large body size in the female will be selected for (Section 7.3.2). Polyandry can arise from this only if the female can, by her behaviour, control several males and hold a territory which can contain them. In practice, these conditions are met only rarely.

SAQ 7.7 Males with long tails should obtain more matings than those with shorter tails.

Relative tail-length must have a heritable component.

CHAPTER 8

SAQ 8.1 (a) Either. Animals defend territories against both conspecifics and heterospecifics.

(b) Nothing to do with competition; indeed, suggests an absence of competition.

(c) Intraspecific competition. Sexual dimorphism can lead to reduced competition for resources between the sexes within a species.

(d) Interspecific competition.

(e) Intraspecific competition. Feeding differences between age classes may, however, evolve for other reasons, such as differences in body size or in overall life-style, as in the case of animals with larval stages.

SAQ 8.2 (a) When the fitness benefits to be gained by fighting are very high.

(b) When the fitness benefits to be gained by fighting are low, relative to the fitness costs involved in fighting.

SAQ 8.3 (a) Hypothesis (i)—current competition.

(b) Hypothesis (ii)—past competition.

(c) Both hypothesis (ii) and hypothesis (iii) are supported.

SAQ 8.4 (a) + – In parasitism, one partner (the parasite) gains at the expense of the other (its host).

(b) + 0 In commensalism, one partner gains but there is no cost to the other.

(c) + + In mutualism, both partners gain from the interaction.

SAQ 8.5 Predators commonly prey on those prey types that are most common in their environment. A prey type that is rare will thus tend to be at an advantage because it suffers lower predation than commoner prey types. If a prey species exists as several distinct types, i.e. it is polymorphic, frequency-dependent selection will tend to maintain the polymorphism because, when any one prey type becomes relatively rare as a result of heavy predation, selection against it will be relaxed as predators switch to commoner types.

SAQ 8.6 Hosts will tend to evolve counter-measures, such as an improved immune system, that reduce the fitness cost of carrying parasites. Parasites will tend to evolve reduced virulence, because the continued survival of their host will tend to increase their (the parasites') fitness.

SAQ 8.7 (a) This could refer to Batesian or aggressive mimicry, but not to Müllerian mimicry.

(b) Batesian mimicry.

(c) Aggressive mimicry.

(d) Müllerian mimicry.

SAQ 8.8 (a Early breeding is favoured because the longer a female delays breeding, the more likely is it that she will die before she breeds.

(b) Late breeding is favoured because fecundity (number of eggs) in female toads is positively correlated with body size and toads continue to grow throughout life.

Factor (a) applies to male toads as much as it does to females. With respect to factor (b), the situation is more complex for males. Males that delay breeding will tend to be larger and, therefore, more likely to win fights over females. However, they will tend to breed earlier than females because their fecundity is less affected by body size; sperm, being very cheap to produce, can be produced in large quantities, even by very small males.

SAQ 8.9 (a) Semelparity.

(b) Iteroparity.

(c) Semelparity.

SAQ 8.10 (a) K-strategist.

(b) r-strategist.

(c) r-strategist.

(d) K-strategist.

(e) K-strategist.

CHAPTER 9

SAQ 9.1 (a) Speciation is the multiplication of species—the division of one species during evolution into two or more separate species. (b) Extinction generally refers to the loss of a species—when a species becomes extinct, it ceases to exist—but the term can also be applied to populations.

SAQ 9.2 The RSC is more biologically meaningful than the BSC for species separated in space or time—the fertilization system (which is the basis of the RSC) of different species can sometimes be compared on the basis of morphology of genitalia for example, whereas it is impossible to tell whether geographically separated or fossil species could actually interbreed (which is the basis of the BSC). Both concepts apply only to species that reproduce sexually and neither allows for the existence of hybrids between species.

SAQ 9.3 (a) Post-zygotic isolation—this is an example of *hybrid breakdown*. (b) Pre-zygotic isolation—this is an example of *mechanical isolation* preventing mating. (c) Neither—feeding on different things can, in certain circumstances prevent interbreeding, such as when insects mate and lay their eggs on their food plant, but does not *necessarily* do so and there would be no reason to expect diet alone to be an isolating factor in birds. (d) Pre-zygotic isolation—this is an example of *ecological* or *habitat isolation* as each fish species can only live in its own habitat, so the two species can never meet to breed.

SAQ 9.4 A cline is a continuous gradation in some measurable character across the range of a species, or a hybrid zone showing such continuous gradation between two types. A mosaic hybrid zone, on the other hand, does *not* show a gradual change from one type to another. Instead, it consists of a mosaic of patches, with some patches containing one type, or the hybrids most like it, and other patches containing the other type, or the hybrids most like it. There will generally be a gradual change in the *proportion* of the two types of patch across the mosaic hybrid zone, however.

SAQ 9.5 (a) Parapatric—the Scottish and English populations of worm share a common boundary and there is no barrier to dispersal across it. (b) Allopatric—the French colonists are geographically separated from the British population by an extrinsic factor—the

English Channel—and, before they come into contact again, the French population has diverged sufficiently to prevent interbreeding between them. (c) Sympatric—the offspring of the female that lays her eggs in the nest of a new host will themselves choose to mate near a nest of the new host and lay their eggs in that nest. So, they will be reproductively isolated from the parent generation and will effectively be a new species.

SAQ 9.6 The main requirement for speciation is that gene flow between two diverging populations be sufficiently reduced to allow further divergence. This means that interbreeding between individuals from each population must be greatly reduced—unlikely if both populations continue to be sympatric, i.e. occupy the same geographical area, because there should be nothing to prevent them meeting and mating successfully.

SAQ 9.7 (a) Less liable to extinction because species that are not highly specialized should find it easier to adapt to changing circumstances. (b) More liable to extinction because highly specialized species are 'stuck' with their highly specific adaptations—in this case, adaptations that enable the *Drosophila* to feed entirely on the sap of a particular tree—and are less likely to be able to adapt quickly if conditions change. (c) The areas of mountain top higher than 3 000 m are likely to be quite small so the species will be separated into many small populations. Small populations are more likely to become extinct so the species as a whole may be more likely to become extinct. (d) As the mammal is slow to reproduce, this makes it more likely to become extinct, and as the flea lives exclusively on this particular mammal, if the mammal becomes extinct, so may the flea. (e) Species that reproduce asexually may not be able to evolve as quickly as sexual species because they lack the possibility of genetic recombination each generation, so are more likely to become extinct if conditions suddenly change. On the other hand, they do not have the problem of finding a mate and this can be a great advantage when numbers of conspecifics are very low. (f) A rare plant that is wind-pollinated is more likely to become extinct because each plant would have a very small chance of being pollinated by another member of the same species—each plant would have to produce an enormous quantity of pollen to have any chance of fertilizing another. (g) A rare plant that is animal-pollinated would be less likely to become extinct because the animal pollinator tends to seek out plants of the same species and thereby help fertilization to take place.

CHAPTER 10

SAQ 10.1 (a) The moulds could be filled with a suitable substance such as latex rubber to form artificial casts that could then be compared with the museum specimens (see Section 10.1.1).

(b) Further specimens of the ammonites could be sought in the concretions in the shale, for these might have been protected from the compaction as in Figure 10.6(b). Alternatively, the original morphology of the deformed specimens might be reconstructed diagrammatically by applying the inverse of the compaction factor (c), derived from the concretions. (This would be an elaborate procedure, however, even with the aid of computer graphic simulations.)

SAQ 10.2 For the answers to both (a) and (b), values of the compaction factor (c) and short-term rates of sedimentation (s) must first be obtained. The value of c (for both deposits) is estimated from Figure 10.6(b): $c = l_0/l_1 = 2$. The value of s for each deposit, at a limit of resolution of 100 years, can be estimated from the appropriate line on Figure 10.5, by reading off the sedimentation rate which corresponds to a time-span of 100 years (do not forget that all values of sedimentation rate for whatever time-span are standardized to m/10^3 years). For the inland sea deposits, s is approximately 0.8 m/10^3 years, and for the coastal wetland deposits, it is approximately 4 m/10^3 years.

(a) The values of stratigraphical completeness can now be calculated from Equation 10.3.

For the inland sea deposits:

$$S.C. = \frac{10\,\text{m} \times 2}{(0.8\,\text{m}/10^3\ \text{years}) \times 100\,000\ \text{years}} \times 100\% = 25\%$$

For the coastal wetland deposits:

$$S.C. = \frac{10\,\text{m} \times 2}{(4\,\text{m}/10^3\ \text{years}) \times 100\,000\ \text{years}} \times 100\% = 5\%$$

(b) As the temporal scope and thickness of each sequence are the same, both sequences will have the same minimum acuity for a 10 cm interval, according to Equation 10.1:

$$m_{\text{min}} = \frac{0.1\,\text{m} \times 100\,000\ \text{years}}{10\,\text{m}} = 1\,000\ \text{years}$$

However, the estimates of maximum acuity will differ according to s. Hence, from Equation 10.2:

$$m_{\text{max}}\ (\text{inland sea deposits}) = \frac{0.1\,\text{m} \times 2}{(0.8\,\text{m}/10^3\ \text{years})} = 250\ \text{years}$$

$$m_{\text{max}}\ (\text{coastal wetland deposits}) = \frac{0.1\,\text{m} \times 2}{(4\,\text{m}/10^3\ \text{years})} = 50\ \text{years}$$

For palaeontological sampling purposes, the inland sea sequence offers the better, that is, the higher value of stratigraphical completeness: less of the history of change will be missing. In contrast, the microstratigraphical acuity of the coastal wetland sequence is superior, since the estimate of m_{max} (i.e. the *shortest* likely time-span represented by any 10 cm sample) is less than that for the inland sea sequence. Fossils from such an interval would sample the variability of their original populations from over a relatively shorter period of time in the coastal sequence.

The reason for these differences is that the coastal deposits can be expected to have suffered more frequent bouts both of erosion and of rapid deposition than those of the inland sea. Hence, in the long term, the coastal deposits are less complete, and comprised of packets of sediment deposited over shorter periods of time (see Section 10.1.2 for further discussion).

SAQ 10.3 According to the evolutionary species concept, species are limited to 'those individuals that share a common evolutionary history' (Chapter 9, Section 9.2.1). Thus, each lineage between branching points (speciations) should be considered a separate species. Such a definition is impractical for palaeontologists, who must recognize species on morphological grounds, from an incomplete record. Consider a case where there had been speciation in the budding mode (Figure 9.1(b)), with no morphological change in the main population. Without knowing precisely when the smaller population budded off (an event unlikely to be registered in the fossil record), there is no way of knowing where to separate the main population into a *parent species* and a *daughter species*, as these are morphologically indistinguishable. Moreover, the discovery of other related species would require further splitting up of the main population to accommodate the new inferred branching points. Finally, sibling species, despite following their own evolutionary histories are unlikely to be detected among fossils. All these problems are discussed in Section

10.2.1, where it is concluded that overall morphological difference is the most practical criterion for recognizing species in the fossil record, except where there is evidence in favour of including more than one morphological grouping in a single species (Section 10.2.2).

SAQ 10.4 (a) Attachment of the shell to the substrate evidently took place only in the early stages of growth, with later parts of the shell (shown by the growth lines and the ribbing) growing free. The size of the scar was thus independent of the age of any adult oyster, and can be considered an age-independent variate.

(b) The size of the scar is unlikely to be a reliable guide to species differences. It would depend very much upon the size and shape of the hard surface to which the oyster attached, and therefore be prone to much ecophenotypic variation. (Different genera of oysters, however, may differ broadly in the extent of attachment, but the amount of variation within species is still usually considerable.)

SAQ 10.5 (a) Following the procedure explained in Section 10.2.3:

$$F = \frac{(14.564)^2}{(12.312)^2} = 1.40$$

This is less than the critical value of F for the null hypothesis, and so the difference in the sample variances is not significant.

To calculate t the following procedure is followed (Section 10.2.3). From Equation 10.7:

$$s_p = \sqrt{\frac{(62-1)(14.564)^2 + (68-1)(12.312)^2}{62+68-2}} = 13.432$$

Hence, from Equation 10.6:

$$s_e = 13.432\sqrt{\frac{1}{62}+\frac{1}{68}} = 2.359$$

Finally, from Equation 10.5:

$$t = \frac{50.45-51.51}{2.359} = -0.449$$

Again, this is within the critical limits of t for the null hypothesis, and so the difference in the sample means is not significant.

(b) The immediate implication of the results in (a) is that we cannot reject the null hypothesis that the two samples were drawn from populations which were equivalent with respect to standard length. Given the significant differences in the other variates, however, we are faced with two possibilities. If they were indeed different species populations succeeding one another in the sampling area—a possibility raised in Section 10.3.2—then the lack of difference in length is a curious coincidence. Alternatively, the two samples may have come from the same species population, though with an episode of punctuational evolution in the other features between sampling levels 20 and 21. On present evidence, it is not possible to decide between these alternatives, although the latter avoids the similarity of lengths in the two samples being merely coincidental.

SAQ 10.6 If growth is allometric, then the ratio between two variates will itself vary with size (Section 10.2.3). A difference in the mean value of the ratio in two samples may then arise simply because of a difference in mean size of the specimens, not through any difference in the pattern of relative growth. (That is not to say that there has been no evolutionary change, since size itself may be subject to evolutionary change.) Only where growth can be shown to have been isometric can ratios be compared.

SAQ 10.7 (a) The net logarithmic change shown for *E. matuyamai* (i.e. the difference between the first and last values on the *x* axis) is 0.2. The time-span for the change was 1.9 Ma − 0.7 Ma = 1.2 Ma. Thus, the rate of change in darwins (Equation 10.14) was 0.2/ 1.2 Ma = 0.17 darwins. Confirm this by comparing the slope of a line drawn between the first and last mean values for *E. matuyamai* with the rate diagram shown at bottom right.

(b) Similar comparison of the slope of the line just below 300 cm in the core with the rate diagram yields a rate of just under 1 darwin. Assuming a constant rate of sedimentation, the time-span of this change may be estimated at 0.2 Ma (in fact this corresponds to the minimum estimate of microstratigraphical acuity, as explained in Section 10.1.2, but in such oceanic sediments the value of m_{max} is expected to be little different).

(c) If the two sets of results (0.17 darwins over 1.2 Ma in (a) and ~1 darwin over ~0.2 Ma in (b)) are plotted on Figure 10.27, they will be found to fall comfortably within the expected range of data, and closely following the overall trend of inverse relationship between rate and time-span.

(d) The pattern shown corresponds well with the *punctuated gradualism* model shown in Figure 10.20(b). The 'classical' phyletic gradualist model (Figure 10.20(a)) implies uniformly slow rates of phyletic evolution which are contradicted by the inverse relation between measured rate and time-span indicated in (c). Nor is the punctuated equilibrium model (Figure 10.20(c)) applicable here, since morphological divergence continues well after the probable time of speciation at ~1.9 Ma: there is evidently no stasis in this instance.

CHAPTER 11

SAQ 11.1 (a) Outgroup comparison, with more distantly related primates such as monkeys, for example, suggests that the relative hairlessness of humans (i) is derived. Using the embryological criterion (Section 11.2.4), having eight adult carpals appears to be the derived state of character (ii). The latter thus constitutes a synapomorphy for humans and chimps, while relative hairlessness is merely an autapomorphy for humans. On these grounds, the cladogram shown in Figure 11.22(a) may be constructed.

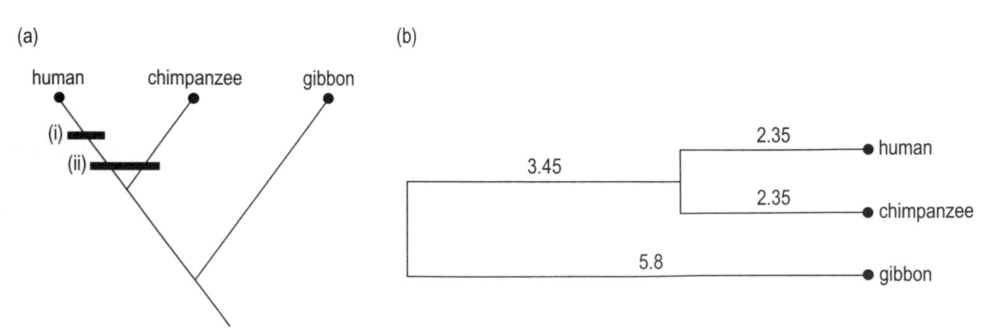

Figure 11.22 *(a) Cladogram for SAQ 11.1. (b) Phenogram for SAQ 11.2.*

(b) An 'ape' grouping, including chimps and gibbons but excluding humans would be a paraphyletic taxon (Section 11.2.1), united only by common possession of a primitive character state (symplesiomorphy). Such a taxon would be misleading as a guide to relationships, implying a relatively closer kinship between chimps and gibbons; the cladogram shows, in contrast, that chimps and humans probably shared the more recent common ancestor.

SAQ 11.2 (a) Your grid should have looked like Table 11.5.

Table 11.5 *Mean immunological distances between albumins of humans, chimpanzees and gibbons, derived from Table 3.7.*

	Human	Chimpanzee	Gibbon
Human	—	4.7	10.9
Chimpanzee		—	12.2
Gibbon			—

The resulting phenogram is shown in Figure 11.22(b). Note that the human–chimpanzee distance is equally split between their two branches, and that the mean of the distances gibbon–human and gibbon–chimpanzee (11.6) is equally split between the gibbon and human/chimpanzee branches.

(b) Stochastically constant rates of molecular evolution (the molecular clock theory: Section 11.3.1) must be assumed if the phenogram is to be equated with a phylogeny. The fact that the distances between the (outgroup) gibbon and the human, on the one hand, and the chimpanzee, on the other, are similar suggests that the assumption is legitimate in this case (the relative rate test: Section 11.3.1). However, the slight difference in their values does suggest some variation in rates.

(c) The phenogram serves as a retrospective test of the phylogenetic implications of the cladogram, which it corroborates. The paraphyletic 'ape' taxon discussed in part (b) of SAQ 11.1 should be rejected for the purposes of phylogenetic analysis.

(d) The conclusion of both the cladogram and the molecular phenogram is that the human and chimp species should be placed together. The gibbon is thus a sister group to these.

CHAPTER 12

SAQ 12.1 (a) C, 1. The hardening of the burrow walls, allowing colonization by encrusters, and the presence of borings imply early cementation, and hence both water supersaturated with respect to $CaCO_3$ and slow sediment accumulation to give plenty of time for the cement to form near to the sediment surface (see hardgrounds, in Section 12.3.1).

(b) A, 3. As dead fish sank to the sea-floor, conditions there would have had to have been such as to have excluded scavengers and potentially disruptive burrowers. Anoxia is the most likely of the factors cited to have had this effect (see Section 12.2.3).

(c) B, 4. The mammoths' teeth would have had to have been eroded from some earlier deposit, and coastal erosion (through B) is the most likely of the factors cited.

(d) B, 1. The frequent current activity would have led both to rapid losses, and gains, of sediment, necessitating adjustments to the changing sediment surface level by burrowing organisms.

(e) B, 2. The intact nature of the skeletons (like that in Figure 12.1(a)) implies a lack of disturbance by scavengers and burrowers, although the presence of the animals themselves suggests that conditions were not anoxic. Rapid introduction of a blanket of sediment, swept offshore during a storm, might readily have swamped the organisms concerned, thus protecting their remains from later disturbance.

SAQ 12.2 (a) A blanket of sediment (as in (e) in SAQ 12.1) would have buried living and dead mussels alike. Yet, for the survey, the palaeontologist needs to know what proportion of the already dead mussels, alone, were originally killed by drilling gastropods. The mussels exceptionally preserved in life position ought therefore to be excluded.

(b) An accumulation, as a residue of the shells of dead mussels, would satisfy the criterion above, but the differential destruction of shells as they lay on the sea-floor (Section 12.2.1) might have distorted the relative frequencies of drilled and undrilled shells in some instances.

In both (a) and (b) a further source of bias would have been the unrecognizable extent of destruction of other shells by crushing and smashing predators (Section 12.2.1).

SAQ 12.3 (a) The completed spindle diagram is shown in Figure 12.19.

(b) The sharpest increase in diversity is shown in the Upper Eocene. Recalling the comments made earlier on the sensitivity of the fossil record of birds to data from conservation-Lagerstätten (Section 12.3.3), it is quite likely that the sharp rise here may

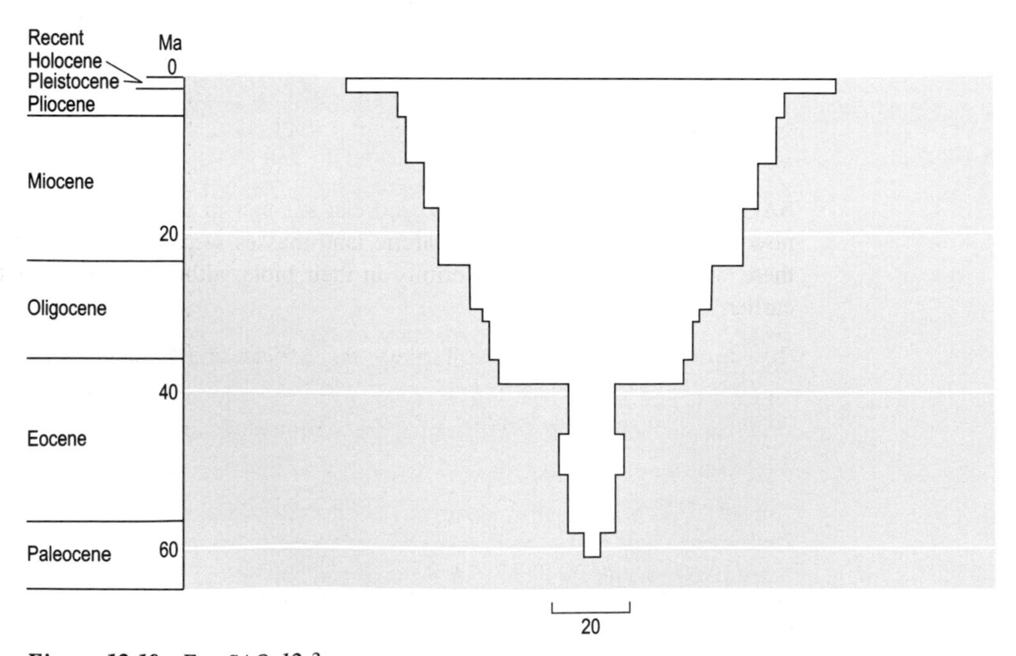

Figure 12.19 *For SAQ 12.3 answer.*

reflect a major deposit of this kind (with the record of diversity maintained thereafter by the pull of the Recent, as explained in Section 12.4.2). Many of the records from the Upper Eocene do indeed come from one deposit, the Phosphorites du Quercy in France. Further work on the taxonomy of the plentiful fossil birds from the Middle Eocene of Messel (Section 12.3.3), for example, may yet require a downward revision of the first records of many of the families concerned.

CHAPTER 13

SAQ 13.1 Coral reefs (the most diverse communities in the oceans) develop in waters that are conspicuously low in some essential nutrients. The constituent organisms are clearly well adapted to 'scavenging' and recycling these nutrients, but nevertheless if nutrients are added artificially (i.e. the waters are fertilized, e.g. with sewage effluent), many of the organisms die, so the organisms are adapted to low nutrient availability. A second example is chalk grassland, where there is a high species diversity, but nevertheless certain important nutrients, especially nitrate and phosphate, are in short supply, so that these have to be constantly added as fertilizer when such areas are ploughed and cultivated.

SAQ 13.2 Suggested hypotheses relate to short- and long-term environmental stability (i) and (ii), high productivity (iii) and complexity of habitats (iv).

(i) High species diversity is favoured in tropical forests because, on an annual and even seasonal time-scale, there is a more constant and dependable supply of resources available.

(ii) Tropical forests occur at latitudes and in areas that have been affected relatively little by Quaternary climatic fluctuations, so that ecosystems have been relatively stable for very long periods. In fact, recent investigations suggest that climatic and consequently environmental changes may have been much greater in the tropics than previously believed.

(iii) High species diversity is (in some undefined way) related to the high primary productivity of tropical forests that is favoured by high annual temperatures and high light intensity over most of the year.

(iv) Tropical forest ecosystems are not only more stratified than temperate ones (i.e. have more layers of vegetation) but are highly partitioned, with organisms occupying quite narrow niches.

SAQ 13.3 (a) In Figure 13.20, you can see that in the Permian/Triassic the plates which now carry the African and Palaearctic land-masses were united as part of Pangaea, so that there was probably some uniformity in their biota, although the plates had been separated earlier.

(b) From the Jurassic to the Tertiary, the African and Eurasian Plates were separated by the Tethys Ocean, so that the biotas would have evolved independently and also in different latitudes (and so in part under different climates).

(c) During the Tertiary, the subduction of Tethys, leading to collision between the African and Eurasian Plates, combined with events such as the desiccation of the Mediterranean, would have permitted migration of faunas and floras between the two land-masses. Indeed, a land connection via Arabia and the Middle East occurs today. Note, however, that the collision between Africa and Eurasia has actually been a very complex event in time and space, because a number of intervening microplates existed, and some still persist, in the Mediterranean basin.

(d) The northward drift of Africa during the Tertiary has resulted in the development of an arid subtropical belt, the Sahara, which effectively separates the African tropics from the sub-temperate northern (Mediterranean) margin of the continent. Many of the plants and animals that now occupy both the northern and southern coastal regions of the Mediterranean tend to be Palaearctic taxa. Consequently, it is now the northern edge of the Sahara rather than plate boundaries or associated features (the Atlas mountains being the exception) that forms the accepted boundary between the two biogeographical regions.

SAQ 13.4 (a) Dispersal and vicariance biogeography are essentially concerned with data on geographical distributions and its interpretation. Cladistic biogeography also takes into account the phylogenetic relationships within the groups of taxa under scrutiny (i.e. through cladograms).

(b) Cladistic biogeographical analysis requires information not on the distribution pattern or track of a single species, but on coincident distribution patterns involving several unrelated groups. Each of these groups needs to be monophyletic. Furthermore, it is necessary to have sufficient comparative morphological, anatomical or biochemical information available for different species within each group to be able to develop a cladogram for each group. Very often it is this lack of basic information which makes it difficult to apply cladistic analysis effectively.

SAQ 13.5 The area cladograms derived from Figure 13.29(b) and (c) are shown as Figure 13.31(d) and (e). The common pattern resolved from these area cladograms is given in Figure 13.31(f). The South American component from the fish cladogram has been deleted, because it is not corroborated by evidence from the flowering trees (though it might have been, if we had cladograms for other groups of organisms with complementary distribution patterns in the Southern Hemisphere). From the point of view of strict logic (which is basically what the cladistic method is trying to apply), New Zealand and South Africa should not *necessarily* be split off as separate branches on the reduced cladogram, as the possibility arises, given the occurrence of the same fish genus A in both areas (a unique feature in this data set), that these two land-masses might initially have split off together as a single area and separated later. For both fishes and flowering trees, the South Africa/New Zealand genera are the sister group to the Australia/New Guinea group.

If you consider the proposed sequence of break-up of Gondwanaland (Figure 13.11(b)), you will see that it closely reflects the area pattern of evolutionary separation suggested by the cladograms. It would also therefore be possible to suggest minimum ages for the divergence of the hypothetical ancestors of the different genera.

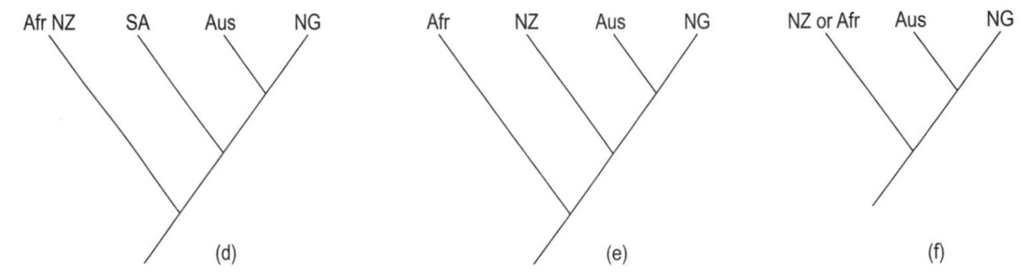

Figure 13.31 *(d,e) Area cladograms for each group and (f) reduced area cladograms for the groups shown in Figure 13.29. For use with answer to SAQ 13.5.*

SAQ 13.6 (i) False. The biota on continent A would probably show a decrease in diversity, as the continent moved to a higher latitude down the latitudinal diversity gradient. This effect would outweigh any increase in diversity due to splitting or to continued isolation.

(ii) True. Movement of continent B to the equatorial regions would probably result in an increase in diversity.

(iii) True. A slight decrease in diversity might be expected as continent C moves slightly northwards into higher latitudes.

(iv) False. As the continental shelves of B and C become progressively closer, so migration of marine invertebrates would be facilitated, resulting in increasing similarity of marine faunas.

(v) True. Rifting of continent A into A′ and A″ would result in a progressive decrease in similarity as the continental shelves are separated by a widening ocean.

(vi) True. Movement of continent B towards C will result in a loss of endemism of marine faunas; moreover, if collision subsequently occurs, there will be an exchange of terrestrial biotas.

SAQ 13.7 (a) The principal causes of a eustatic rise in world sea-levels are (i) an increase in spreading activity at constructive plate margins or (ii) climatic change involving the melting of large ice-sheets. An increase in spreading activity might involve either an increase in the rate of spreading along important ridges (cf. Figure 13.23) or the creation of new ridges by rifting and subdivision of a continent.

(b) A rise in sea-level could lead to the separation of continental land-masses and also to the formation of offshore islands, with consequent splitting and isolation of population groups, which might lead to allopatric speciation. At the same time, oceanic islands might be completely overwhelmed by rising sea-level and endemic taxa driven to extinction.

With a rise of sea-level in marine environments, the opposite effects could occur. Continental shelf marine habitats would be expanded, and previously separated coastal provinces might become accessible to one another, leading to migration and mixing of faunas and floras, competition and extinction. Rising sea-levels might also alter the climate of coastal areas of the continents, encouraging faunal and floral change.

SAQ 13.8 The Simpson coefficients for the four horizons are given in Table 13.4. The decreasing values of the Simpson coefficients of similarity of the marine invertebrates suggest that migration became increasingly difficult between the continental shelves of Raesgip and Nosnibor. One explanation is that migration was hindered by a progressive widening of the Evian Ocean between them, although the introduction of pronounced environmental barriers such as temperature should also be considered.

Table 13.4 *For use with answer to SAQ 13.8.*

Time/Ma	Simpson coefficient /%
74	84
62	73
54	68
44	59

SAQ 13.9 (a) The initial hypotheses must be that these dispersal patterns are the result of dispersal or of a vicariance event.

(b) The fact that no less than five species show similar disjunct distribution patterns, which could be combined into a generalized track, would quite strongly support a vicariance hypothesis. However, during the Quaternary, patterns of vegetation in Europe have repeatedly been disrupted by climatic change. Figure 13.28 suggests that Ireland was almost completely covered by glaciation at times during the Quaternary. Certainly oceanic heathland would have been unlikely to persist, so that, at least in Ireland, these plant species are unlikely to have survived in their present area of distribution.

(c) Three questions that you might want to ask are:

(i) What dispersal mechanisms do these plants have, and do they disperse freely?

(ii) Are there any other taxa, perhaps animals, which also show a similar distribution pattern?

(iii) Is there any fossil evidence for earlier distribution of any of these taxa?

The answers to these questions are:

(i) The seeds are small, probably wind-dispersed but without any special adaptations and travel only over short distances. None of the taxa seems to disperse freely.

(ii) The Kerry slug *Geomalacus maculosus* occurs in SW Ireland, Brittany, NW Spain and Portugal.

(iii) Interglacial deposits, dating from before the last glacial period (but younger than the most extensive Quaternary glaciation), have yielded fossil leaves, seeds and flowers not only of *Erica mackiana* but also of *Rhododendron ponticum*, now extinct in Ireland but persisting in Portugal, and so having had a comparable distribution in the past.

During the glacial periods, sea-level probably fell by about 100 m and would have exposed considerable areas of the continental shelf. This is sufficient for land-bridges to have existed between Ireland and Britain and between Britain and continental Europe across the present floor of the North Sea, but Ireland and France would still have been separated by a marine barrier. It seems unlikely that all these taxa, given their ecological preferences, migrated from the Atlantic coasts of Europe via Britain into Ireland in the early (post-glacial) Holocene. At present, the recent history of these taxa remains a mystery, with the most frequent speculation being a combination of the vicariance and dispersal hypotheses, i.e. that during the glacial periods these plants may have occupied sheltered refuges on what is now the flooded continental shelf to the south or west of Ireland, away from the influence of glaciation.

CHAPTER 14

SAQ 14.1 (a) False. Subsidiary taxa within each phylum *do* share the same basic bodyplan, albeit in modified form (Section 14.1.1).

(b) False. The Burgess Shale fauna is of middle Cambrian age, and thus illustrates an effect of the Cambrian 'explosion' (Sections 14.1.3 and 14.1.4).

(c) True (Section 14.1.2).

SAQ 14.2

(a) Phylogenetic constraint. One of the diagnostic features of the Class Bivalvia is absence (loss) of radula, in contrast to all the other mollusc classes, which possess one (Sections 14.1.1 and 14.2.1).

(b) Fabricational constraint. As far as we know, it is biochemically impossible to secrete diamond (Section 14.2.4).

(c) Functional constraint. In sheltered shorelines where crabs are more numerous, shells with smaller apertures are less likely to be probed successfully by hungry crabs (Section 14.2.3).

SAQ 14.3 Evolution has no ultimate goal, no predetermined direction, unlike a tinkerer directing efforts towards a particular end. Unlike a tinkerer, evolution cannot freely combine bits and pieces from anywhere, but is constrained to modify the existing form of the ancestor, step by step. During evolution, lineages have to keep functioning continuously from one generation to the next, whereas a tinkerer does not always have to work on a machine that is running (Section 14.2.2).

SAQ 14.4 (a) Fabricational noise (Section 14.2.4).

(b) Atavism (Section 14.2.2).

(c) Restricted optimality. A stiff fan-shaped colony held perpendicular to the maximum current flow is probably optimal for gathering food but this orientation entails maximum danger of breakage. The form selected is thus a compromise (Section 14.2.5).

(d) Vestigial structure(s) (Section 14.2.2).

SAQ 14.5 (a) False. Ecophenotypic effects are *non*-heritable modifications of a phenotype that are produced in response to a particular environmental influence. However, the capacity to show ecophenotypic effects, i.e. for plasticity of form in different environments, may itself be under genetic control (Section 14.3.1).

(b) True (Section 14.3.1).

(c) False. Natural selection *does* operate on variations thrown up at random with respect to the present and future needs of organisms, but is itself the very opposite of a process of pure chance (Sections 14.2.2 and 14.3.2).

SAQ 14.6 (a) Most bivalves have symmetrical shells, an attribute that imposes a fabricational constraint. With low values of W, each valve would overlap, i.e. be attempting to occupy the same space, in the region of the umbo (the first formed part of the shell) (Section 14.4).

(b) Analysis of unfilled regions of morphospace may indicate the operation of phylogenetic, functional or fabricational constraints, which in turn may lead to specific inferences about phylogeny, functional morphology and mode of growth (Section 14.4).

SAQ 14.7 (a) Functional constraint (Section 14.5.1).

(b) Phylogenetic constraint (Sections 14.5 and 14.6.5).

SAQ 14.8 The incipient stages of complex structures (adaptations). Such stages are seen in retrospect to have been fortuitously suited to perform a radically new function with relatively little change of form (Sections 14.5.2 and 14.5.3).

SAQ 14.9 Heterochrony refers to *a genetically determined change in the timing or rate of developmental events, relative to the same events in the ancestor*. The classification of heterochrony depends on the comparison of ontogenies from ancestor to descendant. To distinguish between the various categories of 'true' heterochrony, we need information on the *age* of compared individuals. Peramorphosis (overdevelopment) is caused by faster growth (*acceleration*), earlier onset of growth (predisplacement), or later offset of growth (hypermorphosis). A complex pattern of changes produced by the operation of different heterochronic processes is called *mosaic* heterochrony (Section 14.6.2).

SAQ 14.10 (a) Genes affecting early stages of development generally have a *greater* effect on the adult phenotype than changes in later-acting genes (Sections 14.6.1 and 14.6.5).

(b) On average, the greater effect that a mutation (or other cause of heritable variation) has on the phenotype, the less likely it will be selectively advantageous. In general, it seems easier for natural selection to modify the later stages of development than early embryonic stages, a 'hardening' of ontogenies that has been called the 'evolutionary ratchet'. The embryos of a wide range of groups that once shared a common ancestor are therefore more likely to be similar than the corresponding adults (Sections 14.6.1 and 14.6.5).

CHAPTER 15

SAQ 15.1 (a): 6. The apparent trend is ecophenotypic, rather than evolutionary in origin, because there is no evidence of genetic change (Section 15.2).

(b): 5; convergent evolution (Section 15.2).

(c): 7; there has been no change in variance (Section 15.2), and none of the other alternatives fits.

(d): 2; anagenesis (Section 15.1).

(e): 1; iterative evolution (Section 15.2).

(f): 3; biased extinction (Section 15.2).

SAQ 15.2 According to Section 15.5, a radiation may be defined as an episode of significantly sustained excess of cladogenesis over extinction. The pattern of successive changes in diversity shown in Figure 15.18 must therefore be tested for deviation from the null hypothesis of a random walk, as was done for the diversity history of the uncoiled rudists in Section 15.1.1. Figure 15.18 shows the following sequence of changes in diversity over its eight successive intervals: + + − + + − − −. When plotted on Figure 15.3 this sequence does not enter the critical area for $P = 0.05$ (one-tailed test), and so cannot be considered significant. Hence, the data on Figure 15.18 do not justify use of the term radiation in this case. However, the compilation of data in Figure 15.18 is very crude, comprising only family diversities, with a minimum resolution of 10 Ma. Thus, many subsidiary episodes have been subsumed in the pattern shown (as discussed in Section 15.3.2), and any patterns that may exist at lower taxonomic levels have been masked. If data on species ranges were recorded and analysed over much shorter time intervals, a different conclusion might be justified, with distinct minor episodes of radiation and extinction becoming apparent.

SAQ 15.3 (a) False. Many more fossil marine genera have lifespans *less than* the mean than have lifespans more than the mean (Section 15.2.1; Figure 15.9).

(b) True (Section 15.2.1).

(c) False. There is *no* one dominant cause of anagenetic size increase, and increased physiological efficiency is one among many (Section 15.2.2).

(d) False. Small-bodied species tend to be *less* susceptible to extinction than larger-bodied species (Section 15.2.2).

SAQ 15.4 (a) The 'Big Five' mass extinctions of the Phanerozoic include those near the end of the Ordovician, the *Devonian*, the Permian, the *Triassic*, and the Cretaceous (Section 15.4).

(b) The fossils of 'Lazarus taxa' are *not* continuously present through strata that record extinctions in other species; rather, 'Lazarus taxa' disappear for a while close to an extinction horizon, only to reappear later (Section 15.4.1).

(c) Many groups *do* show signs of decline long before their final extinction in the late Cretaceous (Section 15.4.1).

(d) The statement is true (Section 15.4.2).

(e) It is the evidence of a layer rich in *iridium* that suggests a large asteroid struck the Earth at the end of the Cretaceous. However, even if such an impact did occur, it would not necessarily have caused a vast number of extinctions (Section 15.4.1).

SAQ 15.5 (a) False. The Palaeozoic Evolutionary Fauna had *not* become extinct by the end of the Permian, although it did suffer severe losses in epifaunal, fixed suspension-feeders at that time (Section 15.6).

(b) True (Section 15.6).

(c) False. Equilibrium diversity in families of the Modern Evolutionary Fauna does *not* yet appear to have been reached (Section 15.6.1).

(d) False. Despite the undoubted influence of various sampling biases, there is independent evidence for a massive increase in species diversity during the Cainozoic (Section 15.6.1).

SAQ 15.6 No, we cannot draw such a conclusion. As Section 15.7 makes clear, adaptations directly relate only to individual fitness, not to the relative fortunes of species and higher taxa. The number of species in a higher taxon is the result of many factors, including the effects of countless accidents of geography and history and, perhaps, species selection, in addition to any *effects* of individual adaptations (which may influence clade patterns in a counter-intuitive way). If placental species were introduced to habitats where marsupials are currently found, there is no reason to suppose that, in general, they should outcompete the marsupials.

CHAPTER 16

SAQ 16.1 (a) (i), (ii), (iv) and (vi) apply to prokaryotic cells.

(b) All of (i)–(vi) apply to eukaryotic cells.

(c) Crystals can grow (i) but no other characteristics apply. The information to answer this question is mainly in Sections 16.1.1 and 16.5.1.

SAQ 16.2 (a) Items (i), (iii), (iv) and (vi) all apply to the Hadean (see Section 16.2).

(b) The reactions necessary for chemical evolution occur best under anoxic, strongly reducing conditions. Hence (i) can be said to 'favour' chemical evolution whereas (vi) does not. In the absence of an ozone layer (iii), there would have been a high flux of u.v. radiation which, initially, would be a major energy source for chemical evolution but, subsequently, would tend to destroy organic molecules (hence the need for protection). Intense meteor bombardment (iv) would create highly unstable conditions, especially in the oceans, and was probably inimical to chemical evolution.

SAQ 16.3 The two ways are from extraterrestrial sources (Section 16.3.1), or from chemical reactions in the Earth's atmosphere. Energy for these last reactions would have been mainly solar u.v. radiation or electrical discharges (Section 16.3.2), and the first products were most likely to be formaldehyde and hydrogen cyanide. Traces of reducing gases in the atmosphere (e.g. hydrogen), or the presence of ferrous iron (FeII) in water droplets, on atmospheric dust or at the ocean surface would have facilitated the reduction of CO_2 and N_2 (Section 16.3.3) to these first products. Building blocks such as sugars and amino acids may be formed from HCHO and HCN (e.g. the formose reaction for sugar synthesis, Section 16.3.3).

SAQ 16.4 See Table 16.5. The necessary information to answer this question is in Sections 16.1.3, 16.3.3, 16.3.4 and 16.4.2.

Table 16.5 *Completed Table 16.4.*

	Oparin–Haldane	Woese & Wächterhäuser
Site of early reactions	The atmosphere	The atmosphere, in association with pyrite dust
Site of later reactions	The open oceans (regarded as a concentrated soup of reactants)	The atmosphere, on pyrite surfaces protected by membrane-like structures
Sources of reactants for later synthesis	The external medium (the ocean 'soup')	Synthesized autotrophically from simple precursors from the atmosphere (e.g. CO_2)
Type of metabolism in proto-life	Anaerobic heterotrophy (fermenters)	Anaerobic autotrophy (chemo- or photo-autotrophs)

SAQ 16.5 The two kinds of evidence are carbon isotope ratios and fossils, chiefly fossil stromatolites (Section 16.4.4). The carbon isotope data suggest that autotrophic life that fixed carbon by the Calvin cycle was present 3 500 Ma ago and probably (if you accept the arguments about shifts in the isotope ratios in the Isua metasediments) 3 800 Ma ago. However, these data provide no evidence that the autotrophs were cyanobacteria—they could have been other types of photo- or chemotrophic Eubacteria which used the Calvin cycle. Fossil stromatolites and microfossils dated at around 3 500 Ma ago, however, suggest that cyanobacteria probably were present at this time because the fossils look very similar to modern cyanobacteria. This has two major implications for heterotrophs. First, large

populations of autotrophs (and it does not matter whether these were cyanobacteria or other types) provide a source of food for heterotrophs and thus support further evolution and diversification of heterotrophs. Secondly, cyanobacteria (but not other photo-autotrophs) carry out oxygenic photosynthesis which inevitably changes the environment in which heterotrophs live. It would cause removal of ferrous iron (end of Section 16.4.4), and there may even have been enough oxygen to support aerobic respiration by some heterotrophs, thus facilitating the evolution of aerobic heterotrophs.

SAQ 16.6

(a) False. Molecular sequence data suggest that Archaebacteria have diverged least from the ancestral prokaryotes but not that they evolved first (Section 16.4.1).

(b) True (see Table 16.1).

(c) This statement is true except for the last bit — Sulphobacteria have a wider range of energy metabolism (including chemo-autotrophy) and are not just sulphur-respiring heterotrophs (Section 16.4.3).

(d) True (see Section 16.4.3 and Figure 16.21).

(e) False. Cyanobacterial photosynthesis is not more 'efficient' but it uses a much more widely available hydrogen (or electron) donor (i.e. water) than other types of prokaryotic photosynthesis. Shortage of suitable electron donors is the main reason why other photo-autotrophs did not become dominant.

SAQ 16.7 (a) If one accepts the view that the ancient oceans were not a rich primordial 'soup' (Section 16.3.3), then the main argument against fermenters as the first forms of life is that they would have had no source of energy (food) (Section 16.4.5).

(b) If the primordial soup was too dilute to support heterotrophs, then the first organisms must have been autotrophs and a number of scenarios for chemical evolution lead logically to chemo-autotrophy as their mode of energy metabolism. In the 'hot springs scenario' (Section 16.3.3), life originated in deep submarine vents where there was no light but abundant sulphur compounds were available as oxidizable substrates for chemo-autotrophy. In the Cairns-Smith replicating clays model, either photo- or chemo-autotrophy is equally feasible as the first type of energy metabolism but chemo-autotrophy is suggested as being the more likely in the atmospheric model of Woese and Wächterhäuser (Section 16.3.4).

SAQ 16.8 (a) This example appears to provide a parallel with the endosymbiotic origin of chloroplasts. It certainly demonstrates that chloroplasts can exist as semi-autonomous entities outside the parent (algal) cell and within another organism for a considerable time. However, because the endosymbionts here (i.e. chloroplasts) are not truly independent and cannot, for example, multiply in the slug, they could also be regarded as long-lived 'food' rather than endosymbiotic partners. The situation, as so often occurs, is not clear cut.

(b) The example here provides a parallel with a very early stage in the evolution of eukaryotes — before the acquisition of mitochondria but after the origin of the nucleus. This follows if one accepts that these amoebae are primitively amitochondrial and did not once possess mitochondria but 'lost' them at some stage in their evolution. Possession of an ancient pyruvate oxidizing enzyme supports this view.

(c) This example has no parallel with or bearing on the endosymbiotic origin of eukaryote cells.

SAQ 16.9 Information relevant to this question can be found in Section 16.5.6.

(a) Supporters of the endogenous hypothesis would argue that this primitive amoeba represents an early stage in the evolution of mitosis. The gene for tubulin proteins is present in the nucleus but assembly into microtubules occurs at a later evolutionary stage. Nevertheless, chromosomes can still be crudely organized on a spindle without microtubules and in the absence of centromeres (or presumably any microtubule organizing centres, MCs).

(b) Supporters of the endosymbiotic hypothesis would have difficulty explaining these observations. There is no evidence that this amoeba ever possessed flagella or MCs, and they would have to concede that the tubulin gene did not originate in endosymbiotic spirochaetes. However, they could well argue that later acquisition of these endosymbionts led to great improvements in mitosis and perhaps the origin of microtubule structures and MCs. Thus although, on balance, the observations lend stronger support to the endogenous hypothesis of mitosis, they are not inconsistent with an endosymbiotic contribution to the process.

SAQ 16.10 The information to answer this question can be found in Section 16.5.6.

(a) True.

(b) False. Meiosis has evolutionary significance on two counts—it changes ploidy from diploid to haploid and, if these have differences in relative fitness (e.g. in different environmental conditions), the change has evolutionary significance. Meiosis also generates new genetic combinations but not solely through recombination as the statement implies—random assortment of (non-identical) homologous chromosomes at meiosis means that, even in the absence of recombination, the four haploid products of meiosis will be genetically unlike the original diploid cell.

(c) False. Diploidy can arise without nuclear fusion although meiosis cannot occur without it.

(d) True.

CHAPTER 17

SAQ 17.1 (a) Twinning yields more births per pregnancy, but infant mortality is higher among twins than among singletons. The mother is also more likely to die in childbirth, so the lifetime reproductive output of women who give birth to twins may be smaller than those who produce only singletons. The mother may be able to provide less milk and other food for each twin than for singletons born consecutively, thereby increasing the risk of slower growth and delayed sexual maturity.

(b) Among some groups, twins are regarded as inauspicious and may be actively eliminated. Women who give birth to twins may have less access to resources for themselves and their offspring than those who produce only singletons. Even without such factors, each twin may receive less parental attention and other resources than singletons born consecutively, which might impair or delay the cultural development of twins.

SAQ 17.2 *Homo sapiens* arose in eastern Africa and probably spread to western Africa, western Asia and southern Europe at about the same time. From there to India, China, south-east Asia and northern Europe, and from there to Polynesia and Indonesia, from which colonists reached New Guinea and on to Australia. Probably much later, humans reached Siberia and from there to North America, South America, the West Indies and Greenland. The Mediterranean islands were colonized from Europe, North Africa and western Asia during the past 10 000 years. People reached New Zealand from Polynesia and Madagascar from eastern Africa and southern Asia during the last millennium, and many South Atlantic islands, Antarctica and Svalbard during the 20th century.

SAQ 17.3 Tools such as spears and axes increase the efficiency of hunting and butchering, particularly of large animals. Many tools and clothes are made from bone, horn, skins and other animal materials, leading to animals being slaughtered mainly or solely for these products.

SAQ 17.4 Only (e), (f) and (g) are true. Humans affected the environment long before the species became numerous, so (a) is false. Their diet is very varied and includes much plant food, so (b) and (c) are false. They eat fruit, leaves and seeds and do not destroy trees, so (d) is false.

SAQ 17.5 Rats eat almost anything but can survive on a monotonous diet of grain or other dried food; their rate of breeding is variable but can be very rapid when food supply and other circumstances allow; they can climb ropes and trees; their genetic make-up is such that large populations descended from a few founders are not adversely affected by inbreeding; they tolerate close association with humans.

SAQ 17.6 Only (e) and (g) are correct. Many synthetic materials (e.g. most plastics) are not chemically toxic but natural substances such as cyanide and carbon monoxide are poisonous, so (a) and (b) are wrong. Many animals (e.g. peppered moths), plants and micro-organisms have evolved adaptations that minimize the effects of pollutants and synthetic toxins, so (c) and (d) are wrong and (e) is correct. However, only a few species, usually micro-organisms, can adapt quickly to high concentrations of pollutants, so (f) is wrong but (g) correct.

SAQ 17.7 Only (f) is always true. The role of genetic variation in the response to selection is similar in natural and artificial selection so neither (a) nor (b) is true. (c) is often, but not invariably, true. Natural selection also sometimes produces very rapid evolutionary changes (see Chapter 4). If (d) were true, domestic livestock and pets would not become feral, and cultivated plants would not become naturalized, but many forms have done so. Gene transfer speeds up evolution under artificial selection but is not essential to it, so (e) is false. Artificial selection does not necessarily affect fertility, so (f) is false.

SAQ 17.8 Plants readily form polyploids, which instantly precludes breeding with the ancestral wild forms, while such instant species isolating mechanisms are rare among animals. Most domestic animals (e.g. dogs) can interbreed with their wild ancestors even after tens of thousands of generations in captivity. It is easier to control where a plant grows and how it is pollinated and its seed dispersed, than it is to confine an animal and regulate its breeding. The natural behaviour and social organization of only a few species of animals predisposes them to domestication. Except for very small species, strongly territorial or aggressive animals are too much trouble for people to handle. Similar problems do not occur in plants.

SAQ 17.9 (a) Wolves have a keen sense of smell and hearing which can be used to identify and track prey. They normally hunt in packs maintained by a strong dominance hierarchy and produce a variety of calls. These habits can be manipulated so that dogs communicate with each other and with human handlers to locate and chase prey, and handlers can establish themselves as the 'top dog' who leads the pack and has priority access to the food. Dogs can thus be trained to stand off kills and retrieve prey for their handler without eating it themselves. They can also be trained to chase livestock without doing them any harm, i.e. to herd them.

(b) Wolves have a hierarchical social organization which can be manipulated so that animals of both sexes assume a life-long subordinate position in relation to human handlers, whom they follow everywhere and accept orders from. Such animals always respond affectionately to their handlers, although they can be trained to be aggressive towards strangers. They readily learn to use sounds and gestures to communicate with humans. Wolves are a genetically diverse species: there were several distinct geographical subspecies that differed in body size, coat colour and probably many other characters. Most of these subspecies have been exterminated during the past 3000 years (Section 17.3.1), but the genetic variation formed the basis for artificial selection for many different anatomical and behavioural characters, such as differences in structure and colour of the fur, body size and shape of the skull and limbs. Small body size and the retention of juvenile characters into adulthood are among the most important features that make them suitable as pets.

SAQ 17.10 All these features are essential for the establishment and maintenance of a genuinely natural conservation area, but such conditions are almost unattainable in practice.

ACKNOWLEDGEMENTS

Grateful acknowledgement is made to the following for permission to reproduce material:

Figure 1.1 President & Council of Royal College of Surgeons; *Figure 1.2* H. Lewis McKinney, *Wallace and Natural Selection*, Yale University Press; *Figure 2.2a* D. Lack (1968) *Ecological Adaptations for Breeding in Birds*, Methuen; *Figure 2.2b* D. W. Snow (1958) *A Study of Blackbirds*, British Museum of Natural History; *Figure 2.2c* R. A. Pettifor *et al.* (1988) *Nature*, **336**, copyright © 1988 Macmillan; *Figure 2.6* J. A. Chamberlain and G. E. G. Westermann (1976) *Paleobiology*, **2**, Paleontological Society, University of California; *Figure 2.10a,b* C. M. Pond, Open University; *Figures 2.13 & 2.14* G. Heldmaier *et al.* in J. Bligh and K. Voigt (1990) *Thermoreception and Temperature Regulation*, Springer-Verlag; *Figure 2.15* I. Moote (1955) *Textile Research Journal*, **25**, 832–37, Textile Research Institute; *Figure 2.16* C. M. Pond and M. A. Ramsay in *Canadian Journal of Zoology*, National Research Council, Canada; *Figure 3.7* R. A. Raff and T. C. Kaufman (1983) *Embryos, Genes and Evolution*, Macmillan; *Figure 3.13* B. H. Yoo (1980) *Genetical Research*, **35**, 1–31, Cambridge University Press; *Figure 3.22* J. Maynard Smith (1989) *Evolutionary Genetics*, Oxford University Press; *Figure 3.23* T. H. Dobzhansky and O. Pavlovsky (1957) *Evolution*, **11**, 311–19; *Figures 4.1 & 4.18* J. A. Endler (1986) *Natural Selection in the Wild*, Princeton University Press; *Figures 4.2 & 7.13* N. B. Davies and T. R. Halliday (1979) *Animal Behaviour*, **27**, 1253–67; *Figures 4.3 & 7.14* N. R. Davies and T. R. Halliday (1977) *Nature*, **269**, 56–8; *Figure 4.4a* D. M. McFarland (1985) in *Animal Behaviour*, Pitman; *Figures 4.4b,c & 4.5a,b* H. P. Gross (1978) *Journal of Zoology*, **56**, 398–413; *Figure 4.6* F. A. Huntingford (1982) *Animal Behaviour*, **30**, 909–16; *Figure 4.7* courtesy Prof. S. J. Gould, Harvard University; *Figure 4.8* M. McNair in J. A. Bishop and L. M. Cook (1981) *Genetic Consequences of Man-Made Change*, Academic Press; *Figures 4.9–4.13* J. A. Endler (1980) *Evolution*, **34**, 76–91; *Figures 4.14 & 4.15* L. M. Cook *et al.* (1986) *Science*, **231**, 611–13, AAAS; *Figure 4.16* T. H. Clutton-Brock *et al.* (1982) *Red Deer: Behavior and Ecology of Two Sexes*, Chicago University Press; *Figures 4.17, 5.4 & Table 3.2* D. J. Futuyma (1986) *Evolutionary Biology*, Sinauer Associates; *Figure 4.19* L. Partridge and T. Halliday in J. Krebs and N. B. Davies (1984) *Behavioural Ecology: An Evolutionary Approach*, Blackwell; *Figure 4.20* D. G. Reid (1987) *Biological Journal of the Linnean Society*, **30**, 1–24, Academic Press © 1987 Linnean Society of London; *Figure 4.22* S. J. Arnold and M. J. Wade (1984) *Evolution*, **38**, 720–34; *Figure 5.6* M. Woods (1984) in B. Meeuse and S. Morris (1984) *The Sex Life of Flowers*, Faber & Faber/Rainbird by permission of Penguin Books Ltd; *Figure 5.7* M. V. Price and N. M. Waser (1979) *Nature*, **277**, 294–97; *Figure 5.12* E. L. Charnov *et al.* (1981) *Nature*, **289**, 27–33; *Figure 5.13* D. Crews (1988) in *Psychobiology*, **16**, 321–34, Psychonomic Society Inc; *Figure 5.14* R. R. Warner (1984) *American Scientist*, **72**, 128–36, Scientific Research Society; *Figure 6.2* J. L. Brown *et al.* (1978) *Behavioural Ecology & Sociobiology*, **4**, 43–59, Springer-Verlag; *Figure 6.3* J. L. Brown and E. R. Brown (1982) *Science*, **215**, 421–2, AAAS; *Figure 6.4* Ardea; *Figure 6.5* W. H. Thorpe in R. A. Hinde (ed.) (1972) *Non-Verbal Communication*, Cambridge University Press; *Figure 6.6* J. L. Hoogland (1983) *Animal Behaviour*, **31**, 472–9 © J. L.Hoogland; *Figures 7.1, 7.10, 7.19 & Table 4.1* Prof. T. R. Halliday, Open University; *Figure 7.2* A. J. Bateman (1948) *Heredity*, **2**, 349–68, Blackwell; *Figure 7.3* O. R. Floody and A. P. Arnold (1975) *Z. Tierp.*, **37**, 192–212; *Figure 7.7* P. H. Harvey *et al.* (1978) *Journal of Zoology*, **186**, 475–86; *Figure 7.8* D. W. Kitchen (1974) *Wildlife Monographs*, **38**, 1–96; *Figures 7.9 & 8.36* J. R. Krebs and N. B. Davies (1978) *Behavioural Ecology*, Blackwell; *Figure 7.12* M. Daly and M. Wilson (1978) *Sex Evolution and Behaviour*, Wadsworth Publishing; *Figure 7.15* A. P. Møller (1988) *Nature*, **332**, 640–42; *Figure 7.16* A. P. Møller (1988) *Nature*, **339**, 132–35; *Figure 7.17* T. Halliday (1980) *Sexual Strategy*, Oxford University Press; *Figure 7.18* C. R. Cox and B. J. Le Boeuf (1977) *American Naturalist*, **111**, 317–35; *Figure 8.1* J. B. C. Jackson

(1983) in M. J. S. Tevesz and P. L. McCall, *Biotic Interactions in Recent Fossil Benthic Communities*, Plenum Press; *Figures 8.2 & 8.21* M. Begon *et al.* (1986) *Ecology*, Blackwell; *Figure 8.3* M. P. Hassell (1976) *The Dynamics of Competition and Predation*, Edward Arnold; *Figure 8.4a* C. W. Fowler (1981) *Ecology*, **62**, 602–10, Ecological Society of America; *Figure 8.4b* A. R. Watkinson and A. J. Davy (1985) *Vegetatio*, **62**, 487–97, Kluwer Academic Publishers; *Figure 8.4c* G. L. Mackie *et al.* (1978) *Ecology*, **59**, 1069–74, Ecological Society of America; *Figure 8.6* T. D. Price (1984) *American Naturalist*, **123**, 500–18, University of Chicago Press; *Figure 8.7* H. B. Cott (1961) *Trans. Zool. Soc. Lond.*, **29**, Zoological Society; *Figure 8.9* J. R. Krebs and N. B. Davies (1987) *An Introduction to Behavioural Ecology*, Blackwell; *Figure 8.10* M. Petrie (1988) *Animal Behaviour*, **36**, 1174–79, Academic Press; *Figures 8.11 & 15.5* T. Dobzhansky *et al.* (1977) *Evolution*, W. H. Freeman & Co; *Figure 8.12a,b* D. Otte and K. Stayman in M. S and N. A. Blum (eds) (1979) *Sexual Selection and Reproductive Competition in Insects*, Academic Press; *Figure 8.13a* J. R. Krebs (1971) *Ecology*, **52**, 2–22, Ecological Society of America; *Figure 8.13b* A. R. E. Sinclair (1977) *The African Buffalo: A Study of Resource Limitation of Populations*, University of Chicago Press; *Figure 8.14* G. M. Spooner (1947) *Journal of the Marine Biological Association*, **27**, 1–52, Cambridge University Press; *Figure 8.15b* P. A. Haefner (1970) *Physiological Zoology*, **43**, 30–37, University of Chicago Press; *Figure 8.17* R. E. Ricklefs (1979) *Ecology*, Thomas Nelson & Sons; *Figure 8.18* E. Pianka in R. M. May (ed.) (1977) *Theoretical Ecology*, Blackwell; *Figure 8.19* R. A. Griffiths (1986) *Journal of Animal Ecology*, **55**, 201–14, Blackwell; *Figure 8.20* T. W. Schoener in D. R Strong Jr *et al.* (eds) (1984) *Ecological Communities: Conceptual Issues and the Evidence*, Princeton University Press; *Figure 8.23* T. Fenchel (1975) *Oecologia*, **20**, 19–32; *Figure 8.24* J. C. Welty (1975) *The Life of Birds*, W. B. Saunders & Co; *Figure 8.25* J. H. Lawton *et al.* in M. B. Usher and M. H. Williamson (eds) (1974) *Ecological Stability*, Chapman & Hall; *Figures 8.26 & 8.27* D. McFarland (1981) *The Oxford Companion to Animal Behaviour*, Oxford University Press; *Figure 8.29* C. A. Lanciani (1975) *Ecology*, **56**, 689–95, Ecological Society of America; *Figure 8.31* J. A. Yorke and W. P. London (1973) *American Journal of Epidemiology*, **98**, 469–82; *Figure 8.32* R. M. Anderson and R. M. May (1980) *Science*, **210**, 658–61, AAAS; *Figure 8.33* W. Wickler (1968) *Mimicry in Plants and Animals*, Weidenfeld & Nicolson; *Figures 8.34 & 8.35* D. Owen (1980) *Camouflage and Mimicry*, Oxford University Press; *Figure 8.37a* D. Levin and W. J. Leverich (1979) *American Naturalist*, **113**, 881–903, University of Chicago Press; *Figure 8.37b* F. S. Barkalow (1970) *Journal of Wildlife Management*, **34**, 489–500; *Figure 8.38* T. H. Clutton-Brock (1984) *American Naturalist*, **123**, 212–29; *Figure 9.2* A. C. Wilson *et al.* (1974) *Proc. Nat. Acad. Sci*, **71**, 3028–30, US National Academy of Sciences; *Figure 9.3* J. A. Endler (1977) *Geographic Variation, Speciation and Clines*, Princeton University Press; *Figures 9.4 & 9.9* R. G. Harrison and D. M. Rand in D. Otte and J. A. Endler (eds) (1989) *Speciation and its Consequences*, Sinauer Associates; *Figure 9.6* G. M. Hewitt (1975) *Heredity*, **35**, 375–87, Genetical Society of Great Britain; *Figure 9.10* E. Mayr (1963) *Population, Species and Evolution: An Abridgement of Animal Species and Evolution*, Harvard University Press © 1963, 1970 President & Fellows of Harvard College, adapted from E. Mayr in J. Huxley (ed.) (1954) *Evolution as a Process*, Allen & Unwin; *Figures 9.12, 9.13 & 9.14* P. H. Greenwood (1974) *Cichlid Fishes of Lake Victoria, East Africa*, Natural History Museum; *Figure 9.15* H. L. Carson (1987) *Trends in Ecology and Evolution*, **2**, No 7, Elsevier; *Figure 9.16* H. L. Carson and A. R. Templeton (1984) *Annual Review of Ecology and Systematics*, **15**, 97–131 © 1984 Annual Reviews Inc.; *Figure 9.19* M. Ridley (1985) *The Problems of Evolution*, Oxford University Press; *Figure 9.20* W. R. Rice and G. W. Salt (1988) *American Naturalist*, **131**, 911–17, University of Chicago; *Figure 9.21* T. Ehlinger and D. S. Wilson (1988) *Proc. Nat. Acad. Sci.*, **85**, 1878–82, US National Academy of Sciences; *Figure 9.22* courtesy T. Ehlinger; *Figures 9.25 & 9.26* C. Patterson (1978) *Evolution*, Routledge, by permission of Natural History Museum; *Figure 10. 2* A. L. A. Johnson and C. D. Lennon (1990) *Palaeontology*, **33**, 453–85, Palaeontological Association; *Figure 10.5* D. E. Schindel (1982) *Paleobiology*, **8**, 340–53,

Paleontological Society; *Figure 10.9* A. B. Smith and C. R. C. Paul (1985) *Special Papers in Palaeontology*, **33**, 29–37, Palaeontological Association; *Figure 10.10* W. J. Kennedy and W. A. Cobban (1976) *Special Papers in Palaeontology*, **17**, Palaeontological Association; *Figure 10.12* B. Kurtén (1988) *On Evolution and Fossil Mammals*, Columbia University Press; *Figure 10.14* A. W. Owen, University of Glasgow; *Figure 10.16* S. Crasquin-Soleau (1989) *Geobios*, **22**, 537–41; *Figures 10.17 & 10.18* J. C. Brower (1973) *Palaeontographica Americana*, Paleontological Research Institution, Ithaca, NY; *Figure 10.21* P. G. Williamson (1981) *Nature*, **293**, 437–43; *Figure 10.23* P. D. Gingerich (1985) *Paleobiology*, **8**, 340–53, Paleontological Society; *Figure 10.24* courtesy C. F. Helmke, State University of New York; *Figure 10.25* M. A. Bell *et al.* (1985) *Paleobiology*, **11**, 258–71, Paleontological Society; *Figure 10.26* P. R. Sheldon (1987) *Nature*, **330**, 561–63; *Figure 10.27* P. D. Gingerich (1983) *Science*, **222**, 159–61, AAAS; *Figures 11.4 & 11.6* M. Ridley (1986) *Evolution and Classification—The Reformation of Cladism*, Longman; *Figure 11.21* R. H. Thomas *et al.* (1989) *Nature*, **340**, 465–67; *Figure 12.1a* R. P. S. Jefferies (1990) *Palaeontology*, **33**, 631–79, Palaeontological Association; *Figure 12.1b* David Norman, Sedgwick Museum, Cambridge; *Figure 12.3* R. G. Bromley, Institute for Geology and Palaeontology, Copenhagen University; *Figure 12.4* P. R. Sheldon, Open University; *Figure 12.5* H. P. Banks (1972) *Evolution and Plants of the Past*, Wadsworth Publishing © H. P. Banks; *Figures 12.6, 12.9 & 12.11* P. W. Skelton, Open University; *Figure 12.7a* Dianne Edwards, University of Wales College of Cardiff; *Figures 12.7b & 12.14* Dave Martill, Open University; *Figure 12.8* R. Fortey (1991) *Fossils: The Key to the Past*, Natural History Museum; *Figure 12.10* R. Goldring (1991) *Fossils in the Field*, Longman; *Figure 12.12* T. J. Palmer (1982) *Lethaia*, **15**, 309–23, Norwegian University Press; *Figure 12.13a* C. Heyden (1862) *Palaeontographica*, **10**, 62–82; *Figure 12.13b* A. Menge (1866) *Schriften der Naturforchenden Gesellschaft in Danzig*, **8**, 12–15; *Figure 12.13c* R. Forster (1969) *Miteilungen der Bayerischen Staatssammlung für Palaeontologie und Ilistongche Geologie*, **9**, 45–49; *Figure 12.13d* A. Radwanski (1972) *Acta Geologica Polonica*, **22**, 499–506; *Figure 12.15* A. Seilacher (1990) in D. E. G. Briggs and P. R. Crowther (eds) *Palaeobiology — A Synthesis*, Blackwell, for the Palaeontological Association; *Figure 12.17* J. Franzen (1985) *Phil. Trans. Roy. Soc.*, **311**, 181–6; *Figure 13.2* J. Iversen (1954) *Geologiska Föreningens, Stockholm Förhandlingar*, **66**, Swedish Research Councils; *Figures 13.3 & 13.4* S. Stehli in E. T. Drake (ed.) (1966) *Evolution and Environment*, Yale University Press; *Figure 13.5* D. J. Currie and V. Paquin (1987) *Nature*, **329**, 326–27; *Figure 13.8* E. C. Pielou (1979) *Biogeography*, Wiley; *Figures 13.10, 13.11b, 13.12–14* C. J. Humphries in P. L. Forey (ed.) (1981) *The Evolving Biosphere*, Natural History Museum; *Figure 13.17* © Edwin H. Colbert, Museum of Northern Arizona; *Figure 13.20* R. S. Dietz and J. C. Holden (1970) *J. Geophys. Res.*, **75**, American Geophysical Union; *Figure 13.21a* Ilil Arbel in J. W. Valentine and E. M. Moores (1974) *Scientific American*, **230**, W. H. Freeman & Co; *Figure 13.21b* C. E. Payton (1977) *Seismic Stratigraphy—Applications to Hydrocarbon Exploration*, AAPG; *Figure 13.24* L. R. M. Cocks and R. A. Fortey (1982) *J. Geol. Soc.*, **139**, Geological Society, London; *Figure 13.25* W. S. McKerrow and L. R. M. Cocks (1976) *Nature*, **263**, 304–6; *Figure 13.26* K. W. Flessa (1980) *Bioscience*, **30**, 518–24, American Institute of Biological Sciences; *Figure 13.27* G. Kiss in R. M. May (1978) *Scientific American*, **239**, W. H. Freeman & Co; *Figure 13.28* B. S. John (1977) *The Ice Age Past and Present*, Harper Collins; *Figure 13.29* C. J. Humphries and L. R. Parenti (1986) *Cladistic Biogeography*, Oxford University Press; *Figure 14.1* Patricia Wynne in J. W. Valentine (1978) *Scientific American*, **239**, W. H. Freeman & Co; *Figures 14.3, 14.6, 14.11a & 14.22a–e* E. N. K. Clarkson (1986) *Invertebrate Paleontology and Evolution*, Chapman & Hall; *Figure 14.4a* S. J. Gould (1991) *Wonderful Life, The Burgess Shale and the Nature of History*, W. W. Norton & Co; *Figure 14.4b* M. A. S. McMenamin in K. Allen and D. E. G. Briggs (eds) (1989) *Evolution and the Fossil Record*, Belhaven Press; *Figure 14.8* S. E. Luria *et al.* (1981) *A View of Life*, Benjamin/Cummings Publishing; *Figure 14.10* courtesy John Crothers, Field Studies Council; *Figure 14.12* E. N. K. Clarkson and R. Levi-Setti (1975) *Nature*, **254**, 663–67;

Figure 14.14 A. Smith (1984) *Echinoid Palaeobiology*, Chapman & Hall; *Figure 14.15* A. Seilacher in N. Schmidt-Kittler and K. Vogel (eds) (1991) *Constructional Morphology and Evolution*, Springer-Verlag; *Figure 14.16a* C. S. Hickman (1983) *The Veliger*, **26**, 73–92, California Malacozoological Soc.; *Figure 14.16b* C. S. Hickman (1984) *The Veliger*, **27**, 29–36, California Malacozoological Soc.; *Figure 14.17a,b,e–h* Natural History Museum; *Figure 14.17c,d* R. Black (1988) *The Elements of Palaeontology*, Cambridge University Press; *Figure 14.19* courtesy Prof. David M. Raup, University of Chicago, Dept of Geophysical Studies; *Figure 14.21* J. A. Chamberlain in M. R. House and J. R. Senior (eds)(1980) *The Ammonoidea*, Special Vol. 18, Systematics Association; *Figure 14.22f,g* T. Okamoto (1988) *Palaeontology*, **31**, 35–51, The Palaeontological Society; *Figure 14.23* L. B. Halstead (1978) *The Evolution of Mammals*, Peter Lowe © Eurobook, Wallingford; *Figure 14.25* P. W. Skelton in J. C. W. Cope and P. W. Skelton (eds) (1985) *Evolutionary Case Histories From The Fossil Record, Special Papers in Palaeontology*, **33**, The Palaeontological Association; *Figure 14.26* D. Sinclair (1969) *Human Growth after Birth*, Oxford University Press; *Figures 14.27–29, 14.32 & 14.33* M. L. McKinney and K. J. McNamara (1991) *Heterochrony: The Evolution of Ontogeny*, Plenum Publishing; *Figures 15.9, 15.24 & 15.25* D. M. Raup (1991) *Extinction: Bad Genes or Good Luck?*, Academic Press; *Figure 15.11* B. J. McFadden (1986) *Paleobiology*, **12**, 355–69, Paleontological Society; *Figure 15.12* S. J. Gould (1988) *Journal of Paleobiology*, **62**, 319–29, Paleontological Society; *Figure 15.14* D. M. Raup *et al.* (1973) *Journal of Geology*, **81**, 525–42, University of Chicago Press; *Figure 15.16* J. J. Sepkoski (1978) *Paleobiology*, **4**, 3, Paleontological Society; *Figure 15.22* P. N. Pearson (1992) *Paleobiology*, **18**, 115–31, Paleontological Society; *Figure 15.23* D. M. Raup and D. Jablonski (1986) *Patterns and Processes in the History of Life*, Springer-Verlag; *Figures 15.26, 15.27 and 15.29* V. L. Sharpton and P. D. Ward (1990) *Special Paper 247*, Geological Society of America; *Figure 15.28* F. Surlyk (1984) *Science*, **223**, AAAS; *Figures 15.32 and 15.33* S. J. Gould and C. B. Galloway (1980) *Paleobiology*, **6**, 383–96, Paleontological Society; *Figures 15.34 and 15.35* J. J. Sepkoski (1984) *Paleobiology*, **10**, 246–47, Paleontological Society; *Figure 15.36a,b* D. M. Raup (1976) *Paleobiology*, 279–97, Paleontological Society; *Figure 15.36c* P. M. Sheehan (1977) *Paleobiology*, **3**, 325–28, Paleontological Society; *Figure 15.36d* P. W. Signor (1985) in J. W. Valentine, *Phanerozoic Diversity Patterns: Profiles in Macroevolution*, Princeton University Press; *Figure 15.37* H. Ansen (1980) *Paleobiology*, **6**, 193–207, Paleontological Society; *Figure 16.5* R. E. Dickerson (1978) *Scientific American*, **239**, W. H. Freeman & Co; *Figure 16.7* courtesy Dr. Steven Brooke, University of Miami; *Figure 16.10* C. Woese *et al.* (1990) *Proc. Nat. Acad. Sci.*, **87**, 4576–79, National Academy of Sciences; *Figures 16.20, 16.21 & 16.22* T. D. Brock and M. T. Madigan (1988) *Biology of Microorganisms*, Prentice Hall; *Figure 16.23a* courtesy Dr Phillip Playford, Geological Survey Dept. of Mines, Western Australia; *Figures 16.23b & 16.28a* courtesy M. R. Walter, MacQuarie University, NSW, Australia; *Figure 16.23c* D. R. Lowe (1980) *Nature*, **284**, 441–13; *Figure 16.24* M. Schidlowski (1988) *Nature*, **333**, 313–18; *Figure 16.28b* P. Cloud (1976) *Paleobiology*, **2**, Paleontological Society; *Figure 16.30b* L. Margulis *et al.* (1978) *Science*, **200**, 1118–24, AAAS; *Figure 16.32a* P. John and F. R. Whatley (1975) *Nature*, **254**, 495–8; *Figure 17.1* illustration by P. Barrett in N. Duplaix and N. Simon (1976) *World Guide to Mammals*, Crown Publishers Inc; *Figure 17.2* N. Leader-Williams (1988) *Reindeer on South Georgia*, Cambridge University Press; *Figure 17.3* M. Murozomi *et al.* (1969) *Geochimica et Cosmochimica Acta*, **33**, 1247–94, Pergamon; *Figure 17.4* R. A. Spicer and J. L. Chapman (1990) *Trends in Ecology and Evolution*, **5**, 279–84, Elsevier; *Figure 17.6d* British Museum; *Figure 17.7* J. Clutton-Brock (1987) *A Natural History of Domesticated Animals*, Cambridge University Press; *Figure 17.8* R. A. Hinde (1970) *Animal Behaviour*, McGraw-Hill; *Table 4.3* H. B. D. Kettlewell (1955) *Heredity*, **9**, Blackwell; *Table 17.2* M. Bulmer (1970) *The Biology of Twinning in Man*, Oxford University Press.

GLOSSARY

Acceleration A category of heterochrony (peramorphosis) in which the descendant shows a *faster rate* of developmental events.

Acclimatization Physiological adaptation to natural changes in the environment, e.g. temperature, altitude.

Adaptive radiation An episode of rapid and sustained increase in the numbers of species in a higher taxon, linked with marked adaptive diversification. *See also* **radiation**.

Adaptive zones Clusters or constellations of adaptive peaks in an adaptive landscape, separated from each other by major adaptive troughs, yielding natural ecological groupings of species.

Adenosine triphosphate (ATP) The molecule used as a source of chemical energy in almost all living cells. It is synthesized by phosphorylation of adenosine diphosphate (ADP), and converted back to ADP when used for energy-requiring processes or reactions.

Age-dependent variates Variates which show increase in size or number during the development of an organism.

Aggression Behaviour which results in one individual acquiring a contested resource at the expense of another.

Allele One of the different versions of a gene that can occur at the locus occupied by that gene. Often there are just two versions, one frequently dominant over the other. Sometimes, however, there are several versions, only two of which can be present in any given diploid organism.

Allochthonous Describes a fossil of an organism that may have been introduced to a site from elsewhere by one of a number of means.

Allometric exponent (*a*) The exponent of relative growth of one age-dependent variate, x, with respect to another, y, in the **allometry equation**. Allometry is said to be *positive* when $a>1$, and negative when $a<1$.

Allometric heterochrony A heterochronic change in a trait as a function of body size, as opposed to a function of age, as in 'true' heterochrony.

Allometry equation The growth relationship of two age-dependent variates, x and y, usually conforms to the allometry equation $y=bx^a$, where b and a (the allometric exponent) are constants.

Allometry The differential relative growth of two age-dependent variates. *See also* **allometry equation**, **allometric exponent** and **isometry**.

Allopatric speciation Speciation that occurs as a result of two or more populations diverging while allopatric, or allopatric for at least part of the speciation process.

Allopatry(ic) Occupying completely separate non-adjacent areas.

Allopolyploidy Formation of polyploids after hybridization between two species.

Allozyme One of two or more forms of an enzyme arising from different alleles at a locus.

Alternation of generations The alternation within a life cycle of two distinct forms, one haploid, the other diploid.

Alternative hypothesis The prediction that there is a difference between two populations in respect of some measurement. *See also* **null hypothesis**.

Alternative mating strategy A pattern of behaviour displayed by certain individuals of a given sex (commonly males) by which access to mates is gained, that differs from the mate acquisition behaviour of the majority of members of that sex.

Altruism The performing of behaviour that benefits another individual at a cost to the performer's own individual fitness.

Anaerobic heterotrophs Organisms that obtain energy from organic molecules in the absence of oxygen. They may be fermenters or respire anaerobically, and are mostly prokaryotes.

Anagenesis Descent with modification within a single evolutionary lineage. *See also* **cladogenesis**.

Analogous Similarities between organisms which are due to convergent evolution are said to be analogous. *See also* **homologous**.

Anisogamy The existence of gametes in two morphologically distinct forms (i.e. eggs and sperm).

Annual Plant that completes its life cycle, from seed germination to seed production to death, within a single year.

Anoxic Lacking oxygen.

Antibody A protein that specifically recognizes and binds to a foreign substance.

Antigen A foreign substance that triggers the production of a unique antibody.

Apomixis A form of reproduction in plants in which a progeny develops from an unfertilized cell.

Apomorphy A **derived** character state. *See also* **plesiomorphy**.

Aposematic coloration Conspicuous colour patterns associated with noxious or harmful qualities of an animal.

Archaean The geological era spanning the period from 3800 Ma to 2500 Ma ago. Often referred to as the Age of Prokaryotes.

Archaebacteria A prokaryote kingdom that has diverged least from the presumed ancestral prokaryote stock. Modern Archaebacteria live in extreme environments such as hot sulphur springs.

Archaeology The study of prehistoric human artefacts and remains.

Area cladogram A cladogram in which areas where taxa occur have been substituted for the taxa themselves.

Arms race A pattern of evolution involving two interacting organisms in which adaptation in one organism leads to counter-adaptation in the other, e.g. predator and prey.

Artificial selection Selective breeding of organisms by humans.

Asexual reproduction Reproduction that does not involve meiosis or the production or fusion of gametes.

Atavism The reappearance in an individual of certain characters, typical of remote ancestors, that have been absent in more recent generations.

Autapomorphy A derived character state unique to one taxon. *See also* **synapomorphy**.

Autochthonous Describes a fossilized organism preserved in life position.

Autopolyploidy Formation of polyploids by a straightforward self-doubling in chromosome number without hybridization taking place. An organism whose normal diploid genome is *AA* would thus produce the autopolyploid *AAAA*.

Autotrophic A description applied to organisms that obtain energy from light, or by oxidation of inorganic molecules, and synthesize organic molecules from simple, inorganic precursors.

Bacteriorhodopsin A purple pigment found in the outer cell membrane of a halobacterium, which can act as a light-driven proton pump. This is the simplest known type of photo-autotrophy.

Banded ironstone formations (BIFs) A geological formation in which oxidized iron has been deposited in sedimentary layers.

Bedding plane The interface between two layers of sediment.

Beds Layers of deposited sediment.

Benthic Describes bottom-dwelling (marine or freshwater) organisms.

Biennial Plant that completes its life cycle, from seed germination to seed production to death, in two years.

Biogeographical regions Major subdivisions of the Earth's land surface characterized by distinctive suites of plants and animals.

Biogeography The study of the geographical distribution of animals and plants.

Biological Species Concept (or Isolation Species Concept) A definition of a species as a group of actually or potentially interbreeding natural populations which are reproductively isolated from other such groups.

Biostratigraphical zones Ranges in the stratigraphical column defined by the presence of one or more species, which can be used for correlating sedimentary sequences on the basis of their fossils.

Bioturbation The churning of sediment by burrowing organisms.

Bivariate scatter A graph on which are plotted measurements of two variates from a sample of specimens, with one variate on each axis.

Body fossils The original or altered remains of organisms which have been preserved, usually by burial in sediment or more rarely by other means.

Breed An anatomically or physiologically distinct lineage of animals created and maintained in captivity by artificial selection.

***c*-value** The amount of DNA in the unreplicated haploid genome of a species.

Calvin cycle A cycle of reactions which is the most common mechanism by which carbon dioxide is fixed (reduced) to give organic molecules in autotrophs.

Carrying capacity (*K*) The number of adult individuals of a given species that a habitat can support.

Casts Fossil replicas formed by the filling of a mould, either naturally by mineral cement, or artificially with some suitable substance such as latex rubber.

Categorical level measurement Alternatively called *nominal level measurement*. The object measured can be placed in a particular category for the characteristic being scored, e.g. male or female, smooth or wrinkled, red or yellow, but the categories cannot be ranked.

Central dogma The concept of the one-way flow of genetic information, a term first used by Francis Crick. The central dogma is that information (by which is meant the base sequence of DNA, that is replicated, passed on to progeny and is ultimately translated into protein) can pass from DNA to protein but that any change in a protein cannot be translated back into a new sequence of DNA bases.

Centres of origin Areas where the ancestors of particular plant or animal taxa evolved and from where they subsequently spread out to colonize other regions, according to dispersal biogeography hypotheses.

Character displacement The phenomenon whereby competition causes two closely related species to become more different in regions where their ranges overlap than in regions where they do not. *See also* **character release**.

Character release The phenomenon whereby a reduction in competition causes two closely related species to become more alike in regions where their ranges do not overlap than in regions where they do. *See also* **character displacement**.

Chemical fossils Organic molecules, or their breakdown products, that are specific to certain taxa.

Chemical lag Term used to describe a fossil accumulation which originally contained skeletal material of a range of compositions, but which has had more soluble elements removed by dissolution.

Chemical theory The theory that life originated on Earth by elaboration of simple organic molecules through a process of chemical evolution.

Chronospecies Arbitrary divisions of a single evolutionary lineage, defined on the basis of morphological change.

Clade A single whole branch of a phylogeny; i.e. a grouping of *all* the descendants of any given species. *See also* **grade**.

Cladistic biogeography The application of cladistic analysis to geographical distribution data for taxa as well as to phylogenetic characters.

Cladistics A systematic method whereby taxa are hierarchically clustered according to inferred relative recency of common ancestry, based upon their shared derived character states.

Cladogenesis The division of evolutionary lineages, by speciation, causing a proliferation of species. *See also* **anagenesis**.

Cladogram A diagram depicting the cladistic relationships of any three or more taxa, which are placed at the terminations of a tree comprising successive dichotomous branches with each branching node recognized from one or more shared derived character states.

Cleistogamy The production of flowers that do not open, and in which self-fertilization occurs.

Cline A continuous gradation in some measurable character, whether morphological, physiological, genetic, chromosomal or behavioural, across the range of a species, or any hybrid zone between species or sub-specific types where there is such continuous gradation from one type to another.

Coadaptation (genetic) *See* **fitness epistasis**.

Coadaptation (morphological) The action of selection in producing new adaptive *combinations* of traits in a lineage.

Codon Each codon is a specific DNA sequence of three bases coding for a particular amino acid.

Coefficient of relatedness (***r***) The probability that a particular allele in one individual will be inherited from a common ancestor by another individual, or the proportion of the parental and other ancestral genotypes inherited by an individual. The value of r for any pair of related individuals can be calculated using the equation $r = n(0.5)^L$ where n is the number of genetic pathways by which a particular allele can be acquired and L is the number of meioses or generation links between the individuals concerned.

Coevolution The evolution of reciprocal adaptations of two or more species that have prolonged, close interactions.

Cohort survivorship This shows the progressive decline of a sample of individuals or taxa drawn from a single time interval (the 'cohort'), over successive time intervals. *See* **survivorship curve**.

Commensalism An interaction in which one organism lives in or on another organism (its host) at no detriment to the host.

Competition An interaction between individuals, brought about by a shared requirement for a resource in limited supply, and leading to a reduction in the survivorship, growth and/or reproduction of the competing individuals concerned.

Competitive exclusion Prevention of one species from maintaining a viable population as a result of competition with another species.

Competitive exclusion principle A principle which states that ecologically identical species cannot coexist in the same habitat.

Competitive speciation Speciation in sympatry as a result of divergence in resource utilization brought about by intraspecific competition.

Component of fitness A measure of one aspect of fitness, i.e. one aspect of either survival or reproductive success.

Concentration Lagerstätten Large accumulations of fossils, such as shell beds, in which it is possible to obtain large numbers of individual specimens.

Conispiral Coiling in a helical spiral (i.e. with whorl translation), giving a shell that lacks a plane of bilateral symmetry.

Conservation Lagerstätten Deposits in which fossils, although not necessarily common, are exceptionally well preserved.

Constructive plate margin A boundary between two **lithospheric plates** at a mid-ocean ridge, where crust is generated by upwelling of magma from the mantle.

Continental drift The theory that the continental land-masses have not remained in fixed positions but have moved around over the Earth's surface apparently independently.

Convergent evolution Evolution of two or more lineages towards similar morphology or adaptations so that differences between the lineages decrease.

Core The centre of the Earth, composed largely of molten metal (mainly iron).

Correlation coefficient (*r*) A measure of the strength of linear correlation between two variates. r varies between +1 (perfect positive correlation) and −1 (perfect inverse correlation). When $r=0$, there is an absence of linear correlation.

Corridors Land-bridges (or narrow connecting seaways) which link previously unconnected biogeographical provinces and thus allow free migration of fauna and flora in both directions.

Cosmopolitan species Species having a world-wide distribution.

Culture The customs, knowledge and beliefs that are distinctive of particular groups of people and are transmitted from one generation to the next by teaching and social imitation.

Cyanobacteria A group of Eubacteria that carry out oxygen-producing (oxygenic) photosynthesis, using water as an electron donor for photosynthesis.

Darwin A standardized unit of proportional increase in any linear measurement in an evolving lineage. One darwin represents an increase by a factor of 2.718 (e), expressed as a natural logarithm, per million years. Hence any change in a variate, x, can be expressed in darwins as $(\ln\bar{x}_2 - \ln\bar{x}_1)/(t_2 - t_1)$, where $\bar{x}_1$ and $\bar{x}_2$ are the mean values, respectively, of samples from populations at times t_1 and t_2 (in millions of years).

Delta (δ) ^{13}C value A measure of the enrichment or depletion of the isotope ^{13}C with respect to the more common isotope ^{12}C in a sample (relative to a standard).

Density-dependent competition Competition in which the more competitors there are, the greater the deleterious effects.

Derived fossils Fossils that originally formed in rocks older than those in which they are found.

Derived A character state interpreted as being evolutionarily modified with respect to that found in other compared taxa is said to be derived—an **apomorphy**. *See also* **primitive** character state.

Destructive plate margin A boundary where two **lithospheric plates** meet and where one is subducted below the other down into the mantle to be melted and destroyed.

Deterministic hypotheses Hypotheses that seek to explain evolutionary patterns as the consequences of pervasive causes. *See also* **probabilistic hypotheses**.

Detritivory The consumption by a variety of organisms, including animals, plants, fungi and bacteria, of organic debris derived from dead organisms or from the waste products of living organisms.

Diagenesis Physical and chemical changes that take place within sediment. Especially important are geochemical reactions, many bacterially mediated, which produce minerals that cement sediment to form rock.

Dioecious A condition, in plants, in which male and female flowers are carried on different individuals.

Diploid A cell or organism possessing two sets of chromosomes, which are paired. *See also* **haploid**.

Directional selection Selection in which individuals at one end of a phenotype distribution are at an advantage over other individuals.

Disjunct distribution The discontinuous occurrence of species or higher taxa in two or more widely separated geographical areas.

Dispersal biogeography The interpretation of distribution patterns of organisms in terms of outward dispersal from centres of origin.

Disruptive selection Selection in which individuals at both extremes of a phenotype distribution are at an advantage over intermediate individuals.

Dissociated heterochrony Local heterochronic changes that affect some characters of the descendant but not others, which may remain the same or be affected by another heterochronic process. *See also* **global heterochrony**.

Divergent evolution Evolution that results in increasing morphological difference between initially more similar lineages.

Dizygotic Twins (or triplets, quadruplets, etc.) that develop from different zygotes.

DNA hybridization The formation of double-stranded DNA between single strands derived from different sources.

Dynamic survivorship This shows the relative decay of individuals or taxa drawn from many different time intervals as a function of individual or taxon duration. Taxon durations should be corrected for fluctuations in extinction rate over real time before being employed in dynamic survivorship. *See* **survivorship curve**.

Ecological (or habitat) isolation Existence of a reproductive barrier between two species or sub-specific types due to them occupying different habitats in the same geographical area.

Ecological biogeography The study of the distribution patterns of species and higher taxa in relation to environmental factors, including climatic factors, soils and interactions with other species.

Ecophenotypic effect Non-genetic (non-heritable) modification of a phenotype in response to a particular habitat or environmental factor.

Ecophenotypy Phenotypic differentiation of specimens due to local circumstances of growth, but not to genetic differences between them.

Ecosystem An ecological unit usually comprising several habitats and their many communities of organisms, together with their physico-chemical environment.

Ecotone A boundary where ecological conditions change abruptly.

Ecotype A distinct form, within a species, adapted to a different set of environmental conditions to the rest of that species.

Effect hypothesis Individual adaptations may have the (accidental) effect of bringing about differential rates of speciation and/or extinction, and hence sorting between clades; this is termed the effect hypothesis.

Electrophoresis A technique used to detect differences in proteins and polypeptides based on differences in electrical charge of the molecules.

Embryology The study of the embryo and its development in animals and plants.

Endemic species Species confined to one particular country or region.

Endosymbiosis Symbiosis (living together) where one organism lives inside the cells of another (the host).

Endosymbiotic hypothesis The hypothesis that modern eukaryote cells originated by a series of endosymbiotic associations which gave rise to eukaryotic organelles (excluding the nucleus).

Environmental sex determination Phenomenon whereby the sex of an individual is determined by environmental factors, most commonly temperature, during the course of its development.

Epigamic selection Synonym of **intersexual selection**.

Epistasis A synergistic effect on the phenotype of two or more gene loci, whereby their joint effect differs from the sum of the effects of the loci taken separately.

Ethological (or sexual) isolation Reproductive isolation between two species or sub-specific types as a result of the sexual attraction between males and females of the different species or types being reduced or absent due to a mismatch in behaviour or physiology.

Ethology The scientific study of animal behaviour.

Eubacteria A prokaryote kingdom that includes most of the modern prokaryotes and is highly diverse.

Eukaryote An organism with cell(s) having a well-defined nucleus with the chromosomes enclosed by a nuclear membrane, and having mitochondria and in some cases chloroplasts. *See also* **prokaryote**.

Eustatic marine regression Seaward shifting of coastlines as a result of a global fall in sea-level.

Eustatic marine transgression Landward shifting of coastlines as a result of a global rise in sea-level.

Evolutionarily stable strategy (ESS) A pattern of behaviour or some other phenotypic character which, when adopted by all other members of a population, cannot be bettered by an alternative strategy.

Evolutionary equilibrium A population is said to be in evolutionary equilibrium with respect to a given gene at the point where there is no effect of natural selection on that gene.

Evolutionary fauna A set of higher taxa (especially classes) that have similar histories of diversification and that together dominate the biota for a long interval of geological time. The three Phanerozoic evolutionary faunas are the Cambrian, the Palaeozoic, and the Modern.

Evolutionary novelties Structures, biochemical properties or habits that appear to evolve without obvious precedent.

Evolutionary Species Concept A definition of a species as all those individuals that share a common evolutionary history; i.e. they are of the same lineage which is distinct from other lineages.

Evolutionary systematics A systematic method whereby taxa are hierarchically clustered according to all available forms of evidence for evolutionary relationships; the standard approach until the appearance of **phenetics** and **cladistics** in the 1950s and 1960s.

Evolutionary trend A term used loosely for any sustained tendency for evolutionary change in a particular direction. Anagenetic trends occur within single lineages, whilst whole-clade trends, which may be uncoupled from anagenetic trends, concern a net shift in some variate when all the constituent lineages in a clade are taken into account.

Exotic fossils Allochthonous elements from a fossil assemblage, which are more or less contemporaneous with any autochthonous or parautochthonous elements in the same assemblage, but representative of other habitats elsewhere.

Exploitation competition Competition that occurs between individuals, in which there is no direct contact between those individuals.

Extinction The loss of species or higher taxa. A taxon becomes extinct when all the populations of that taxon have died out (global extinction). The term may also apply to a single population (local extinction).

***F* test** A test of the equivalence of variance of different populations based on the statistic $F = s_1^2 / s_2^2$, where s_1^2 and s_2^2 are the greater and smaller sample variances, respectively. Critical values of F are tabulated in statistical texts.

Fabricational noise Morphological features that arise incidentally as a non-adaptive by-product of some particular programme of growth.

Fecundity The number of eggs or offspring produced by an organism.

Feral A domesticated animal that is released or escapes and breeds in the wild without assistance or regulation by humans.

Fermentation The process by which organic molecules are broken down by glycolysis in anaerobic conditions yielding **adenosine triphosphate (ATP)**, through substrate level phosphorylation, and a range of organic end products.

Filters Land-bridges (or narrow connecting seaways) which link previously unconnected biogeographical provinces but only allow restricted or selective migration of fauna and flora across them.

Fisher's fundamental theorem The rate of increase in fitness is equal to the additive genetic variance in fitness.

Fitness The relative ability of an individual organism or genotype to survive and leave offspring that themselves can survive and leave offspring.

Fitness epistasis When the fitness conferred on an organism by a particular allele depends on what alleles are present at other loci. *See also* **coadaptation**.

Fixation The situation in which an allele has attained a frequency of 1 (i.e. 100%); in other words, it has spread through a gene pool, so that the only other alleles present at the same locus in the gene pool arise through mutation.

Formose reaction A cascade of reactions by which a range of sugars may be formed from formaldehyde (HCHO). It may have been an important reaction in pre-biotic chemistry.

Fossil-Lagerstätten German term (sing. fossil-Lagerstätte) meaning 'fossil bonanza', introduced by the German palaeontologist Adolf Seilacher to describe sedimentary deposits that are an especially rich source of fossil material. *See* **concentration Lagerstätten** and **conservation Lagerstätten**.

Founder effect speciation *See* **peripatric speciation**.

Founder effect The effect on genetic variability of the small size of a colonizing, or 'founder', population (which may consist of only one or a few individuals). Such a population can never contain more than a fraction of the total genetic variability of the parent population.

Frequency-dependent selection Selection in which the fitness of a genotype or phenotype is not constant but varies according to the frequency of that genotype or phenotype relative to others.

Fundamental niche The niche which can be occupied by a species in the absence of competitors, predators and other species that adversely affect it.

Gametes Haploid reproductive cells produced by organisms that reproduce sexually. Male gametes are called sperms or spermatozoa in animals, and sperm cells (formed within pollen grains) in plants. Female gametes are called eggs or ova in animals, and egg nuclei (formed within ovules) in plants.

Gametic isolation Reproductive isolation between two species or sub-specific types as a result of fertilization not taking place, even though gamete transfer occurs, either because the male or female gametes of different types fail to attract or unite with each other, or because male gametes are inviable in the reproductive organ of the female of the other type.

Gametophyte The haploid phase of a plant with a life cycle showing alternation of generations. *See also* **sporophyte**.

Gender The phenotypic sex of an individual, referring to all the morphological, physiological and other differences between males and females.

Gene flow The spread of alleles from one breeding population to others, by the dispersal of gametes or zygotes or by the emigration of adults.

Gene pool The total set of genes present in a population.

Gene transfer The injection of DNA from the same or another species into a zygote or early embryo.

Genetic bottleneck Loss of genetic variability in a population following a reduction in population size.

Genetic distance A measure of the degree of genetic difference between populations, sub-species or species, based on differences in allele frequencies.

Genetic drift Random changes in the frequencies of alleles within a population.

Genetic load A decrease in mean fitness due to the existence of alleles that confer a reduction in fitness in the homozygous form.

Genetic revolution A major reorganization of the genome to a new coadapted combination of alleles—effectively a shift from one adaptive peak to another.

Genome The totality of all the genes (or nuclear DNA) of a cell.

Glacial period Intervals of climatic cooling during an ice age, characterized by the expansion of ice-sheets or intense permafrost into areas that today have a temperate climate.

Global heterochrony Heterochronic changes that affect many characters of the descendant in a roughly similar way. *See also* **dissociated heterochrony**.

Gondwanaland A very extensive Southern-Hemisphere supercontinent that eventually broke up progressively through the Mesozoic and Cainozoic to give rise to Africa, India, South America, Australasia and Antarctica.

Gonochoristic A condition, in animals, in which individuals exist as males or females.

Grade A grouping of taxa which show similar modifications with respect to their ancestors. Grade taxa may be **polyphyletic**, **paraphyletic** or **monophyletic**. *See also* **clade**.

Group selection The theory put forward to explain certain types of behaviour, such as altruism, in terms of advantages that accrue to a group of organisms. It is a necessary condition for group selection to operate that the species in question lives in isolated groups and that there is minimal or no gene flow between groups.

Growth lines Periodic lines found on the outer surfaces of many shells, running parallel with the shell margin. They result from the progressive addition of increments of shell material around the margin, as, for example, in the shells of molluscs, brachiopods and corals.

Guild A group of species that exploit a particular environmental resource in a similar way, e.g. seed-eating or insect-eating birds.

Habitat isolation *See* **ecological isolation**.

Habitat selection Choosing to live or breed in a habitat to which an individual is adapted.

Hadean The geological era spanning the period from the origin of the Earth (4 600 Ma ago) to the deposition of the first known sedimentary rocks (3 800 Ma ago).

Halobacteria A group of Archaebacteria that live in very salty environments. They are the most highly evolved Archaebacteria.

Hamilton's rule A formula that defines whether or not altruistic behaviour will evolve in a given situation. Altruism will evolve if the ratio of benefits (B) to costs (C) incurred by being altruistic is greater than the ratio of the **coefficient of relatedness** r for the donor and its own offspring to r for the donor and the offspring of the recipient of the altruism. Thus, altruism will evolve if $B/C > r_{\text{donor to own offspring}} / r_{\text{donor to recipient's offspring}}$.

Haploid The presence of one set of chromosomes in a cell, shown typically by gametes. *See also* **diploid**.

Hardy–Weinberg ratio Ratio of genotypes expected in a population of effectively infinite size containing a genetic locus with two alleles (frequencies p and q) and where there is no selection or immigration, and random mating occurs. It can be shown that $p^2 + 2pq + q^2 = 1$, where p^2 and q^2 are the frequencies of the two homozygotes and $2pq$ that of the heterozygote. This is known as the Hardy–Weinberg equation.

Heritability The proportion of the total phenotypic variance among individuals in a trait, that is attributable to differences in genotype, $h^2 = \dfrac{\text{genetic variance}}{\text{total phenotypic variance}}$

Heterochrony A (genetically determined) change in the timing or rate of developmental events, relative to the same events in the ancestor. The two major categories are **paedomorphosis** and **peramorphosis**.

Heterozygosity When the members of a homologous pair of chromosomes carry different alleles at a particular locus, the organism is said to be heterozygous for that locus. Heterozygosity, H, is the average proportion of heterozygous loci per individual within a population.

Heterozygous advantage Situation in which individuals that are heterozygous at a given locus have higher fitness than either of the homozygotes.

Hexaploid A polyploid with six sets of the haploid number of chromosomes ($6n$).

Hibernation Active physiological processes that result in cooling of the body below normal body temperature but above freezing. It may last for days or months and typically it takes place during winter.

Historical biogeography Study of the distribution patterns of species and higher taxa in relation to their fossil record and historical and geological events.

Homologous Similarities of organization between organisms which are inferred to be due to inheritance from a common ancestor are said to be homologous. *See also* **analogous**.

Hot spot A fixed point on Earth which is anomalously hot because of heat flow from the deep mantle, over which a chain of volcanoes forms as a **lithospheric plate** moves across it.

Hybrid breakdown A reproductive barrier between two species or sub-specific types due to the decreased viability and/or fertility of the offspring of any hybrids.

Hybrid inviability A reproductive barrier between two species or sub-specific types due to the death of any hybrids. The egg is fertilized but it does not develop, or the development of the embryo becomes arrested at some stage, or the hybrid dies before it reaches sexual maturity.

Hybrid maintenance A mechanism such as allopolyploidy which maintains the hybrid state, i.e. prevents the hybrid breakdown that would normally occur after hybridization between two species.

Hybrid sterility A reproductive barrier between two species or sub-specific types due to the sterility of any hybrids. Hybrids survive but fail to produce functioning sex cells or gametes.

Hybrid zone An area of overlap between two species or sub-specific types where hybridization takes place.

Hypermorphosis A category of **heterochrony** (**peramorphosis**) in which the descendant shows *later offset* of developmental events.

Iapetus Ocean An early Palaeozoic ocean that separated Laurentia (much of present North America, as well as Scotland) from Avalonia and Baltica.

Inbreeding Mating between close relatives.

Inbreeding depression Reduction in offspring fitness resulting from mating between close relatives. *See* **outbreeding depression**.

Incumbent replacement A model to explain a pattern of overlapping taxonomic replacement. As the incumbent species of a pre-existing higher taxon suffer random extinction, their vacated niches may be taken over preferentially by species of an invading higher taxon, the members of which possess some key evolutionary innovation which gives them a fitness advantage over competitors from other higher taxa.

Individual selection The theory put forward to explain apparently altruistic behaviour, such as the hawk alarm call. This theory does *not* interpret that alarm call as an altruistic act. Instead, it argues that individuals manipulate the behaviour of their fellow flock members and so decrease their own risk, relative to that of the rest of the flock.

Insulation Any material, living or non-living, that limits heat flow.

Interference competition Competition that occurs between individuals, in which there is some form of direct interaction, such as aggression, between those individuals.

Interglacial period Intervals of climatic warming during an ice age, when climate was as warm or warmer than the present. Most climatologists regard the present 'post-glacial' interval as simply an uncompleted interglacial period.

Intermediate host An organism that carries a parasite, or its infective stage, between host individuals. Also called a *vector*.

Interphase Period in the cell cycle between one mitosis and the next during which DNA replication occurs.

Intersexual selection Selection arising from variation between individuals of the same sex (most commonly males) in terms of competing to attract and sexually stimulate members of the other sex. Synonym: *epigamic selection*.

Interspecific competition Competition that occurs between individuals of different species.

Interstellar molecules Molecules that occur in the space between stars, and which include quite complex organic compounds.

Interval level measurement The exact measurement of a phenotypic character on a continuous scale of equal units.

Intrasexual selection Selection arising from variation between individuals of the same sex (most commonly males) in terms of competing directly (e.g. by fighting) for access to females.

Intraspecific competition Competition that occurs between individuals of the same species.

Intrinsic rate of natural increase (*r*) The maximum rate of growth of a population (i.e. when mortality is zero).

Island arc An arc of volcanic islands created when two oceanic **lithospheric plates** collide and one is subducted and melted beneath the other.

Isogamy Condition in which reproduction involves the production of only one type of gamete.

Isolating mechanism A reproductive barrier between two species or sub-specific types that prevents gene flow between them.

Isolation Species Concept *See* **Biological Species Concept**.

Isometry Proportionate growth of two age-dependent variates according to the equation $y = bx + c$, where y and x are the two variates, and b and c are constants. This equation represents a special case of the **allometry equation**, where the allometric exponent $a = 1$.

Iterative evolution The repeated origination of lineages with generally similar morphology at different times in the history of a clade. Such lineages often evolve from a single persistent lineage that changes relatively little.

Iteroparity A form of life history in which individuals engage in more than one breeding episode during their life.

***K*-strategist** An organism, the life history of which is adapted to maximize its competitiveness and adult survival.

Karyotype The number and constitution of the chromosomes in each cell of an individual eukaryote.

Key evolutionary innovation (KEI) A novel feature in an organism that is hypothesized to have been preadaptive for new ecological opportunities (by introducing some major fitness benefit).

Kin selection A form of natural selection operating indirectly on an allele by favouring the relatives of the individual having the allele, as these are likely also to be carrying the allele concerned by descent from a common ancestor.

Laurasia A Northern-Hemisphere supercontinent comprising most of the present continents of North America and Eurasia, which split up in the late Cretaceous to early Tertiary.

Lek A mating system in which males gather in dense groups which are visited by females for the sole purpose of mating.

Life position A fossilized organism is is said to be in life position when it is preserved at the site, and in the original orientation of life.

Line of relative growth A line on a bivariate scatter showing the relative growth of two age-dependent variates, based either on measurements from a single individual at different stages of growth, or from many individuals in a single population, provided that a wide range of individual ages is represented.

Linkage disequilibrium Occurrence, in a population, of two or more alleles that occur together within individuals more often than is to be expected from random reassortment of alleles.

Linkage group The cluster of genes carried on the same chromosome. During meiosis they remain together except when separated by crossing over.

Lithospheric plate The outer part of the Earth is essentially rigid and formed of plates. These plates include the Earth's crust, both oceanic and continental, and the part of the upper mantle down to about 100 km depth.

Living fossil An informal term with various meanings, the central concept of which is survival to the present day with minimum morphological change over long periods—several tens or even hundreds of millions of years.

Locus The location of a gene on a chromosome.

Logistic diversification A pattern of taxonomic diversification involving an initial phase of exponential increase followed by a phase of decelerating increase, towards an eventual equilibrium level of diversity.

Macroevolution Evolution above the species level; that is, the patterns of change in the forms of organisms and in their diversity arising from the appearance and disappearance of species, and the processes involved in such changes.

Mantle The part of the Earth's interior lying below the crust, at a depth of between 6 km and 90 km, and above the **core** at a depth of 2900 km. It is largely composed of iron and magnesium rich silicates.

Mass extinction An episode of large-scale extinction affecting many different groups of organisms within a short interval of geological time.

Mass mortality The death of large numbers of individual organisms in a single event.

Maternal effect Situation in which an aspect of the phenotype is inherited from the mother only, either genetically, or in the non-genetic component of the egg.

Mating system Term that describes the pattern of mating relationships within a population, such as whether individuals have one mate or several.

Mean The arithmetic mean or average: the sum of n values divided by n.

Mean fitness ($\bar{W}$) The sum of the fitnesses of all the genotypes in a population, each multiplied by the frequency of that genotype.

Mechanical isolation Reproductive isolation between two species or sub-specific types due to differences in the mechanical structure of the reproductive organs or genitalia which impede or prevent the transfer of gametes between them.

Mesozoic Marine Revolution (MMR) An episode of considerable re-organization in benthic marine communities with, especially, an increase in levels of predation on shelly prey.

Methanogens A group of Archaebacteria that live in anaerobic environments and obtain energy by the reduction of carbon dioxide with hydrogen, producing methane gas.

Microstratigraphical acuity The estimated time that elapsed during the deposition of a given interval in a sedimentary sequence. Minimum acuity (m_{min}) is calculated as iT/X, where i is the interval thickness, and T and X, the temporal scope and total thickness, respectively, of the entire sedimentary sequence investigated. Maximum acuity (m_{max}) is calculated as ic/s, where c is the compaction factor for the sediment, and s is the shorter-term rate of deposition, at a given limit of resolution, estimated from sedimentation rates observed in equivalent environments today.

Microtubule organizing centre (MC) Any system for organizing microtubules, e.g. centrioles in the mitotic spindle and basal bodies in flagella.

Microtubules Eukaryotic, rod-like structures composed of the protein tubulin. They have many functions, e.g. in flagella and the mitotic spindle.

Mimicry The resemblance of one organism (the *mimic*) to another organism (the *model*), such that the two organisms are confused by a third organism.

Molecular clock theory The theory of stochastically constant rates of change in given amino acid and nucleotide sequences based upon the prediction that new alleles will become fixed in an equilibrium population at an average rate equal to that of mutation.

Monoecious A condition, in plants, in which male and female flowers are carried on the same individual.

Monogamy Condition in which individuals have one mating partner over a certain time period such as one breeding season or for a lifetime.

Monomorphic Term used to describe a population in which only one form (morph) of a particular character is present.

Monophyletic A taxon which comprises an entire single clade of organisms. *See also* **polyphyletic** and **paraphyletic**.

Monozygotic Twins (or triplets, quadruplets, etc.) that develop from atypical division of one zygote.

Morphology The (study of the) structure and form of organisms and their parts.

Morphometrics The quantitative description of morphology in organisms.

Morphospace In **theoretical morphology**, morphospace encompasses the set of all possible results from a given set of instructions to generate form. Parts of this morphospace may or may not be occupied in nature.

Mosaic evolution Evolution within a lineage such that different features change at different times or different rates.

Mosaic heterochrony A complex pattern produced by operation of opposite heterochronic processes: some characters in the descendant are paedomorphic, whilst others are peramorphic.

Mosaic hybrid zone A hybrid zone that consists of a mosaic of patches of different hybrid types rather than a gradual change from one 'pure' type to another through a series of hybrids.

Motility symbiosis A symbiotic association in which one partner increases the mobility of the other.

Moulds Hollow spaces left in rock resulting from the dissolution of organic remains. Secondary filling of moulds forms **casts**.

Murchison meteorite A meteorite that fell at Murchison, Australia, in 1969. It was rich in organic molecules, including amino acids, and demonstrated that such molecules can be of extraterrestrial origin.

Mutation Any heritable change brought about by an alteration in the genetic material of an organism. In its narrowest sense it is the alteration, deletion or insertion of a single base (or sequence of bases) in a DNA molecule.

Mutualism An interaction in which two organisms live in a close association, to the benefit of both.

Natural logarithm (ln) A logarithm of any number to the base e (≈ 2.718). Hence $e^{\ln x} = x$.

Naturalized A domesticated or artificially introduced plant that propagates in the wild without assistance or regulation by humans.

Neo-Darwinism The version of Darwin's theory of evolution by natural selection refined and developed in the light of modern biological knowledge (especially genetics) in the mid-20th century.

Neoteny A category of heterochrony (paedomorphosis) in which the descendant shows a *slower rate* of developmental events.

Neutral mutation A mutation that neither increases nor decreases an organism's fitness relative to another allele at the same locus.

Niche The *n*-dimensional hypervolume within which a species can maintain a viable population, the dimensions being all those environmental factors that affect survival and reproduction.

Nominal level measurement *See* **categorical level measurement**.

Non-shivering thermogenesis (NST) The production of additional body heat without muscular movement, normally involving a specialized tissue called brown adipose tissue.

Normal distribution The spread of measurements of a character often falls symmetrically around the mean value. When the shape of the distribution of measurements is a bell-shaped curve, it is known as a normal distribution.

Null hypothesis The prediction that there is no difference between two populations in respect of some measurement. *See also* **alternative hypothesis**.

Oparin–Haldane theory The first detailed scientific theory about the origin of life.

Ordinal level measurement Measurements where it is possible to say that one is, for example, bigger than another, and so to rank them with respect to each other, but where it is impossible to assess the size of the differences.

Orthogenesis The now discredited idea that certain evolutionary trends reflect an intrinsic tendency within organisms to evolve in certain directions that selection cannot regulate, reducing the functional adequacy of individuals, and eventually leading to the lineage's extinction.

Orthology Homology between corresponding nucleotide or amino acid sequences in different organisms. *See also* **paralogy**.

Outbreeding depression Reduction in offspring fitness resulting from mating between unrelated individuals. *See* **inbreeding depression**.

Oxygen sink A substance such as ferrous iron (FeII, Fe^{2+}) which reacts with and reduces the level of oxygen in the environment.

Oxygenic photosynthesis Photosynthesis which yields oxygen by the splitting of water. It occurs in cyanobacteria and in all photosynthetic eukaryotes, and is the source of oxygen in the Earth's atmosphere.

Paedomorphosis A category of heterochrony in which the descendant shows *under*development relative to the ancestor. Paedomorphosis results from **neoteny**, **postdisplacement** or **progenesis**.

Palaeoecology Interpretation of the ecology of past organisms from fossils and their enclosing sediments.

Palaeolatitude The latitude of a specific point on the Earth's surface in the geological past, with respect to the position of the Equator and poles at that time. It is usually determined by palaeomagnetic measurements of rock samples.

Palaeomagnetism The magnetization recorded in ancient rocks at the time of their formation.

Palaeontology The study of life in the past as recorded by fossil remains.

Pangaea A massive supercontinent formed by the coming together in the Permian of land-masses that now form all of the present five continents. This split up during the Mesozoic into **Laurasia** and **Gondwanaland**.

Panmictic population A population in which individuals are mating randomly with respect to phenotype and/or genotype.

Parallel evolution Evolution whereby the morphologies of two or more closely related lineages change together in roughly similar fashion, rather than diverge or converge.

Paralogy Homology between corresponding nucleotide or amino acid sequences in the same organism, arising from duplication of a single original sequence. *See also* **orthology**.

Parapatric Occupying adjacent areas, i.e. occupying separate areas that share a common boundary.

Parapatric speciation Speciation that occurs as a result of two or more populations diverging while occupying adjacent (parapatric) areas.

Paraphyletic A taxon which comprises only a part of a single clade, from which one or more subordinate clades have been removed because they are deemed to represent distinct grades of organization. *See also* **monophyletic** and **polyphyletic**.

Parasitism An interaction in which one organism lives in (endoparasite) or on (ectoparasite) another organism, its host, obtaining nourishment at the latter's expense. Typically, the host is not killed by the parasite.

Parasitoid An insect that lays its eggs inside the eggs, larvae or pupae of another insect species, where the parasitoid larvae develop, finally killing the host.

Parautochthonous Describes a fossil whose orientation may have been disturbed post-mortem, but which remains at or near the original site of life.

Parsimony The scientific convention whereby the simplest (i.e. that involving the fewest hypothesized steps or components) from among a number of explanations, e.g. alternative phylogenies, that are consistent with the available evidence, is accepted in preference to the others.

Parthenogenesis A form of reproduction in which an egg develops without being fertilized.

Partial fitness A measure of fitness that does not take account of all components of fitness.

Pathogen A disease-causing parasite, usually a micro-organism (virus or bacterium).

Pelagic Describes free-swimming or floating organisms.

Peramorphosis A category of heterochrony in which the descendant shows *over*development relative to the ancestor. Peramorphosis results from **acceleration**, **predisplacement** or **hypermorphosis**.

Perennial A plant with a life cycle extending over several years that typically continues to grow and reproduce each year.

Peripatric Occupying a completely separate, non-adjacent local area on the periphery of the species' range.

Peripatric speciation (or founder effect speciation) Speciation that occurs as a result of a small, peripheral, allopatric population diverging from the main population, while the main population remains relatively unchanged. A mode of speciation in which chance effects such as the founder effect and genetic drift play an important part.

Phenetics A systematic method whereby taxa are hierarchically clustered according to their relative overall morphometric similarities.

Phenogram A diagram depicting the relative phenetic distances between taxa, in the form of a hierarchy constructed according to some chosen cluster statistic.

Philopatry A tendency among certain animals to return to breed at or close to the location where they were born.

Photic zone Zone in which light penetrates water with sufficient intensity for photosynthesis to occur.

Phyletic evolution Evolution within species lineages.

Phyletic gradualism A model in which gradual evolutionary change is expected within species lineages, and not exclusively in association with speciation. Some earlier authors portrayed an exaggeratedly gradual pattern of change extending over millions of years, but this was rejected with the rise of neo-Darwinian theory, when anything from abrupt to gradual change was admitted (punctuated gradualism).

Phylogeny An evolutionary tree showing the inferred relationships of descent and common ancestry of any given taxa.

Planispiral Coiling in one plane (i.e. with no whorl translation), giving a shell with a plane of bilateral symmetry.

Plate tectonic theory A theory which states that the outer 100 km of the Earth consists of rigid slabs, called **lithospheric plates**, which are in motion relative to each other and to the interior of the Earth.

Pleiotropy The effect of a gene on more than one phenotypic character.

Plesiomorphy A **primitive** character state. *See also* **apomorphy**.

Polyandry Condition in which an individual female has more than one mate within a breeding season whilst males have a maximum of one mate.

Polygamy Condition in which individuals have more than one mating partner within a single breeding season.

Polygenic trait A character, the variation in which is based wholly or in part on allelic variation at several loci.

Polygyny Condition in which an individual male has more than one mate within a breeding season whilst females have a maximum of one mate.

Polymorphic A population in which discretely different forms (morphs) of a character coexist.

Polyphyletic A taxon which assembles species with independently evolved similarities (derived, therefore, from separate ancestors). Such taxa are rejected from modern systems of classification. *See also* **monophyletic** and **paraphyletic**.

Polyploidy The possession of more than two sets of the haploid number of chromosomes (n). The 'normal' diploid has $2n$ chromosomes, while a polyploid has $3n$ or more.

Positive assortative mating Non-random mating that involves like pairing with like.

Postdisplacement A category of heterochrony (paedomorphosis) in which the descendant shows *later onset* of developmental events.

Post-mating isolating mechanism An isolating mechanism that operates after mating has taken place.

Post-zygotic isolating mechanism An isolating mechanism that operates after fertilization has taken place.

Preadaptation The phenomenon whereby some structures or processes in ancestors, whether already selected for a certain function or present as selectively neutral features, are *fortuitously* suited for transformation to a new adaptation in descendants. Such structures or processes are seen in retrospect to have been suited to perform a radically new function with relatively little change of form.

Predisplacement A category of heterochrony (peramorphosis) in which the descendant shows *earlier onset* of developmental events.

Pre-mating isolating mechanism An isolating mechanism that operates before mating can take place.

Pre-zygotic isolating mechanism An isolating mechanism that operates before fertilization can take place.

Preservation potential The likelihood of the remains of an organism becoming incorporated into the fossil record.

Primary contact Contact, either sympatric or parapatric, between two populations that has been continuous during any period of differentiation between them.

Primitive Describes a character state interpreted as having been present in the latest common ancestor to any given set of taxa being compared — a **plesiomorphy**. *See also* **derived** character state.

Probabilistic hypotheses Hypotheses that assign probability values to evolutionary events such as speciation and extinction, without invoking any pervasive cause(s), thus allowing random simulations of evolution, which can serve as null hypotheses for the investigation of real evolutionary patterns. *See also* **deterministic hypotheses**.

Prochlorophytes (Prochlorobacteria) A recently discovered group of **Eubacteria** with **oxygenic photosynthesis**. They may have given rise to chloroplasts in at least one group of eukaryotes.

Progenesis A category of heterochrony (paedomorphosis) in which the descendant shows *earlier offset* of developmental events.

Prokaryote A unicellular organism in which the hereditary material is not separated from the cytoplasm by a nuclear membrane, nor is it differentiated into chromosomes, and there are no membrane-bound organelles. *See also* **eukaryote**.

Protandrous Describes those animals, capable of changing sex during their lives, that begin life as males. *See also* **protogynous**.

Proteinoid microspheres Spherical structures formed when **proteinoids** are placed in warm water and then cooled; they are regarded by Sidney Fox as models for protocells in the origin of life.

Proteinoids Protein-like polymers which form when mixtures of amino acids are heated and cooled. Sidney Fox believed that a similar process occurred during chemical evolution.

Protogynous Describes those animals, capable of changing sex during their lives, that begin life as females. *See also* **protandrous**.

Provinces Different areas, each containing distinctive assemblages of species, some of which may not be found elsewhere.

Pseudoextinction The arbitrary termination of a **chronospecies** involving the disappearance from the record only of a morphologically defined species, though with the (changed) descendants continuing as a new chronospecies.

Punctuated equilibrium A model in which species are expected to arise by rapid (punctuational) evolution of small, isolated populations, but thereafter to show negligible morphological change (**stasis**).

Quaternary Ice Age The period of global climatic instability, particularly characterized by repeated growth and disintegration of huge ice-sheets over parts of North America, Europe and Asia, which began in the latest part of the Tertiary and continued throughout the Quaternary to the present day.

r-strategist An organism, the life history of which is adapted to maximize its reproductive rate.

Radiation An episode of significantly sustained excess of cladogenesis over extinction in a clade of organisms.

Random sample An unbiased collection of measurements of a quantity taken to be representative of the total variation of the same quantity within the population.

Random walk Any path traced out by moving a point *either* up *or* down by some amount, with equal probability, while moving it step by step along a horizontal axis. A random walk may serve as a null hypothesis for many evolutionary patterns, in which case the horizontal axis represents time.

Range charts Compilations of the stratigraphical ranges of several taxa, shown alongside a geological time-scale.

Realized niche The limited set of conditions and resources (a subset of the **fundamental niche**) within which a species maintains a viable population.

Recognition Species Concept A definition of a species as all those individuals of a biparental organism that share a common fertilization system.

Recombination The production of new combinations of alleles during gamete production by meiosis. Recombination usually involves both crossing over and independent assortment of chromosomes.

Red Queen's hypothesis A hypothesis to explain the constant average survivorship of species with respect to prior duration. For a given group of species (within an adaptive zone), the environment is constantly deteriorating, in terms of the organisms' adaptations, because of changes both in the physical environment and in the adaptations of other organisms. Thus, mean fitness is postulated to fluctuate randomly, such that any species is equally likely to become a candidate for extinction.

Reduced Major Axis (RMA) A straight line fitted to mutually dependent bivariate data by drawing a line of slope s_y/s_x (where s_y and s_x are the standard deviations for the two variates) through the point $\bar{y}$, $\bar{x}$ corresponding to the mean values for the two variates.

Reinforcement The evolution of isolating mechanisms through natural selection when two incipient species come into secondary contact.

Relict endemics Endemic taxa which were once geographically much more widespread but have now become extinct elsewhere.

Replicating clays Clay particles which were postulated by Cairns-Smith to play a central role in the origin of life. The electrically charged surface layer was able to replicate and to adsorb and organize organic molecules.

Reproductive character displacement Increased differentiation between the reproductive systems (i.e. any morphological, physiological or behavioural aspect of reproduction) of two species living in sympatry, which comes about as a result of natural selection because any hybrids between the two species are totally inviable or totally sterile.

Reproductive effort That part of the total available energy of an organism which is expended on reproductive activities as opposed to growth and survival (**somatic effort**).

Reproductive isolation The situation in which members of a group of organisms breed with each other but not with members of other groups.

Reproductive success The number of an individual's progeny that survive to adulthood relative to other individuals' progeny.

Reproductive value (V_i) A value for the reproductive potential of an individual at time i that takes into account both its current reproduction and its projected reproductive success over the rest of its life.

Resolution analysis The estimation of how time is represented in a sedimentary sequence by the determination of three factors: **temporal scope**, **microstratigraphical acuity** and **stratigraphical completeness**.

Resource partitioning The possession of adaptations that enable two or more species to coexist by exploiting a common resource in different ways.

Restricted optimality The interaction of constraints that limits organisms from achieving optimal form for particular functions.

Retrospective testing A procedure for testing a hypothesis concerning historical events, involving the prediction of additional consequences, which differ from those of competing hypotheses, and the search for evidence relating to the contrasting predictions.

Ring species A species distributed continuously in a circle, with overlap between the extreme ends of the distribution, such that there is clinal variation around the circle and all adjacent populations can interbreed, except for those at the two overlapping extremes.

Sampling error The random deviations of the statistics of samples of measurements from the parameters of the total population.

Seasonal (or temporal) isolation Reproductive isolation between two species or sub-specific types due to them breeding at different seasons or at different times of day, or becoming sexually mature at different times of the year.

Secondary contact Contact, either sympatric or parapatric, between two populations that have previously been separated, i.e. allopatric, for at least part of a period of differentiation between them.

Secondary sexual characters Characters in which the two sexes of a species differ, excluding the gonads, their ducts and associated glands.

Sedimentary basins Depressions produced by relative subsidence of the Earth's crust on a local or regional scale, within which thick sedimentary sequences have accumulated.

Sedimentary sequence A coherent succession of beds of sediment deposited in a specific environment.

Sedimentology The study of sedimentation and the interpretation of sedimentary sequences.

Selection coefficient (s) A measure of the relative **fitness** penalty incurred by a particular genotype subject to natural selection, relative to the fittest genotype. Calculated as $s = 1 - W$, where W is the fitness of the genotype in question.

Selection differential (S) The difference between the mean value of a phenotypic trait before and after selection.

Semelparity A form of life history in which individuals engage in only one breeding episode during their life.

Serial monogamy Condition in which individuals have one mating partner over the course of a single breeding season but change partners from one breeding season to the next.

Sessile Habitually fixed to a particular spot in life. *See also* **vagile**.

Sexual dimorphism The existence, within a species, of differences in morphology between the sexes.

Sexual isolation *See* **ethological isolation**.

Sibling species Species which are morphologically identical but behaviourally, genetically and/or karyotypically different and which do not hybridize.

Significance level An arbitrary minimum limit of probability for tests of a **null hypothesis**; if the probability of the null hypothesis being correct is below this level, then it is rejected in favour of the **alternative hypothesis**.

Simpson coefficient of similarity (S) A simple statistical method for comparing communities in two provinces in terms of the similarity of their species compositions.

Somatic effort That part of the total available energy of an organism which is expended on non-reproductive activities, principally growth and survival. *See also* **reproductive effort**.

Speciation The multiplication of species, i.e. the division of one species during evolution into two or more separate species.

Species flock (or species swarm) A group of closely related species confined to a small geographical area, often with no close relatives elsewhere.

Species selection A postulated process of species sorting based upon the direct selection of those emergent properties of entire species populations which show some degree of heritability in daughter species.

Spindle diagram A diagram showing changes in the numerical abundance of fossil taxa within discrete time slots, based on data from a range chart.

Spirochaetes A group of **Eubacteria** postulated by Lyn Margulis to have given rise to eukaryotic flagella and cilia by endosymbiosis.

Sporophyte The diploid phase of a plant with a life cycle showing alternation of generations. *See also* **gametophyte**.

Stabilizing selection Selection in which individuals that are intermediate with respect to a phenotype distribution are at an advantage over individuals at either extreme.

Standard deviation (s) The square root of **variance**.

Stasis Negligible morphological change in an evolutionary lineage. A major component of the **punctuated equilibrium** model of evolution, in which established species are expected to have shown evolutionary inertia and hence stasis.

Step cline A cline where the change from one type to another is particularly rapid in the centre of the hybrid zone.

Stratigraphical completeness The proportion of the **temporal scope** of a sedimentary sequence physically represented by beds. It is estimated at a given limit of time resolution, and is expressed as a percentage.

Stromatolites Banded columnar stony structures built up as layers of fine-grained carbonate sediment are trapped by mats of filamentous microbial organisms such as blue–green algae. Stromatolites may be several metres across, and over one metre high. They are among the first easily recognized fossils of life on Earth.

Subduction The drawing down of one **lithospheric plate** below another at a **destructive plate margin**.

Sulphobacteria A group of **Archaebacteria** that are thermophilic (living in hot environments), and whose energy metabolism involves sulphur. They are the least evolved Archaebacteria.

Survivorship analysis This involves plotting **survivorship curves**, in order to reveal patterns of death or extinction.

Survivorship curve A curve showing the relative survival of individuals or taxa as a function of time. It is frequently plotted on semilogarithmic graph paper (which shows equal slopes for equal rates of death or extinction). *See also* **cohort** and **dynamic survivorship**.

Sweepstake dispersal The long-range dispersal of organisms across natural barriers by chance events.

Symbiosis The living together in permanent or prolonged association of members (symbionts) of two different species with beneficial or deleterious consequences for at least one of the parties.

Sympatric speciation Speciation that occurs as a result of two or more populations diverging while occupying the same geographical area.

Sympatry(ic) Occupying the same geographical area.

Symplesiomorphy A **primitive** character state shared by two or more taxa.

Synapomorphy A **derived** character state shared by two or more taxa and held to reflect their common ancestry. *See also* **autapomorphy**.

Systematics The study of the diversity amongst animals and plants, their classification, and the relationships amongst them.

***t* test** A test of the equivalence of the mean values of different populations, based on the statistic $t=(\bar{x}_1-\bar{x}_2)/s_e$, where $\bar{x}_1$ and $\bar{x}_2$ are the sample means and s_e is the standard error calculated from the statistics of the two samples. Critical values of t are tabulated in statistical texts.

Taphonomy The study of those processes that act on the remains or traces of organisms.

Taxonomic diversity The number of constituent taxa (e.g. species, genera, families, etc.) in a higher taxon at any given time in geological history.

Taxonomy The study aimed at describing, and producing systems of classification of, animals and plants.

Temporal isolation *See* **seasonal isolation**.

Temporal scope The total time elapsed during deposition of a given sedimentary sequence.

Tension zone A hybrid zone maintained by selection against hybrids.

Territorial behaviour Behaviour that leads to individual animals or groups of animals being spaced out more than would be expected from a random occupation of suitable habitats.

Tethys Ocean A wide east–west trending ocean that separated **Laurasia** from **Gondwanaland** from Triassic to early Tertiary times.

Tetraploid A polyploid with four sets of the haploid number of chromosomes ($4n$).

Theoretical morphology An analytical approach that can reveal discrepancies between the range of theoretically possible forms and that actually adopted in nature, i.e. reveal unfilled regions of **morphospace**; this may indicate the operation of constraints, leading in turn to inferences about phylogeny, functional morphology and mode of growth.

Thermophilic bacteria Bacteria (prokaryotes) that live in hot environments, typically more than 80 °C.

Time-averaging The pooling together of fossilized remains (or data from them) from an extended period of time.

Torpor Active physiological processes that result in cooling of the body below normal body temperature but above freezing for periods of hours or a few days.

Trace fossils Traces of the activity of organisms, such as burrows, footprints, borings, bite marks and droppings, preserved in the sedimentary record.

Tracks Lines drawn on a map which connect the known distributions of related taxa in different areas.

Trade-off Situation in which allocation of effort to one component of **fitness** (e.g. survival) yields a benefit in terms of that component, but incurs a cost in terms of another component of fitness (e.g. reproductive success).

Transposable elements Sequences of DNA that can replicate and then move to new positions within the genome.

Triploid A polyploid with three sets of the haploid number of chromosomes ($3n$).

Unrooted tree A branching diagram linking three or more taxa, with boundaries between their differing character states indicated on the appropriate branch lines, and lacking any free stem to other, outgroup taxa.

Vagile Mobile. *See also* **sessile**.

Variance (s^2) A measure of the amount of variability in a trait within a population or sample; the average squared deviation of measurements from the arithmetical mean, $s^2 = \frac{\sum_{i=1}^{i=n} (x_i - \bar{x})^2}{n}$, where x_i is the value of the ith measurement of x, $\bar{x}$ is the mean value of x, and n is the number of measurements. In calculating the variance of a *sample* from the population, the denominator $(n-1)$ is used, instead of n.

Variates Homologous features or dimensions in organisms which can be consistently measured for morphometric analysis.

Variety An anatomically or physiologically distinct lineage of plants created and maintained under cultivation by artificial selection.

Vector *See* **intermediate host**.

Vestigial Applied to a structure or process that is believed to have been functional in an ancestor, but which has no known function in the living species.

Viability An organism's ability to survive from being a fertilized egg to reproductive maturity.

Vicariance biogeography The interpretation of the distribution patterns of organisms in terms of a widespread ancestral population, which has been split up by the development of barriers, isolating descendent populations which then evolved to form new taxa.

Vicariant species Closely related species which occupy different geographical or ecological areas.

Wallace's line The proposed geographical boundary line between the Oriental and Australian biogeographical regions, based on the distribution of bird faunas.

Weber's line The proposed geographical boundary line between the Oriental and Australian biogeographical regions, based on the distribution of molluscan and mammalian faunas.

Young endemics Endemic taxa which have evolved usually as a result of geologically recent isolation or radiation.

***Z* statistic** A normally distributed statistic with a **mean** of 0 and a **standard deviation** of ±1. It can be used for certain inferential statistical tests, such as that of isometry in a line of relative growth on a double logarithmic plot. Here, $Z=(a-1)/s_e$, where a is the allometric exponent and s_e the standard error of a.

INDEX

Notes

Notes